P9-AOT-954

KIRK-OTHMER ENCYCLOPEDIA OF

CHEMICAL TECHNOLOGY

Fifth Edition

VOLUME 1

KIRK-OTHMER ENCYCLOPEDIA OF CHEMICAL TECHNOLOGY, FIFTH EDITION

EDITORIAL STAFF

KIRK-OTHMER ENCYCLOPEDIA OF

CHEMICAL TECHNOLOGY

Fifth Edition

VOLUME 1

Kirk-Othmer Encyclopedia of Chemical Technology
is available Online in full color and with additional content at
http://www3.interscience.wiley.com/cgi-bin/mrwhome/104554789/HOME.

A John Wiley & Sons, Inc., Publication

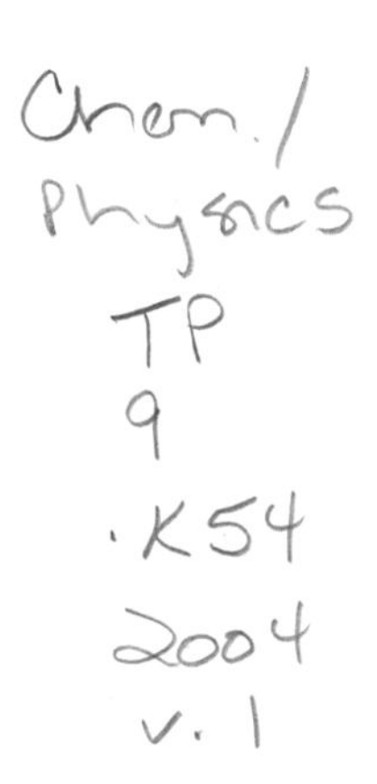

Published by John Wiley & Sons, Inc., Hoboken, New Jersey.
Published simultaneously in Canada.

Library of Congress Cataloging-in-Publication Data:

Kirk-Othmer encyclopedia of chemical technology. – 5th ed.
p. cm.
Editor-in-chief, Arza Seidel.
"A Wiley-Interscience publication."
Includes index.
ISBN 0-471-48494-6 (set) – ISBN 0-471-48522-5 (v. 1)
1. Chemistry, Technical–Encyclopedias. I. Title: Encyclopedia of chemical technology. II. Kroschwitz, Jacqueline I.
TP9.K54 2004
660′.03–dc22 2003021960

Printed in the United States of America

10 9 8 7 6 5 4 3 2 1

CONTENTS

PREFACE

The chemical industry, a dominant and diverse field of manufacturing and scientific research, includes a wide range of rather different industries, from the established petrochemicals to the emerging advanced materials industry. The chemical industry manufactures organic and inorganic chemicals, plastics, agrochemicals, dyes and colorants, paints and coatings, pharmaceuticals, cosmetics, and much more. Moreover, the chemical industry plays a key role in the supply chain of many other sectors, and provides services such as testing and control of materials, to other industries. However, regardless of product, production scale, or the employed manufacturing process, basic scientific principals and common practices prevail throughout.

The industry as a whole employs a large number of professionals who depend on chemical and engineering knowledge: chemist and chemical engineers, researchers in academia and government institutions, educators and students of related disciplines, as well as patent attorneys and other consultants. Accurate, up-to-date information on chemical technology is, therefore, in great demand.

The ***Kirk-Othmer Encyclopedia of Chemical Technology*** is designed to present the versatile field of chemical technology to professionals who wish to learn about technologically important materials, established as well as cutting edge methods, and relevant phenomena.

Although a large amount of information regarding chemistry and chemical technology is now available via various communication channels, the ***Kirk-Othmer Encyclopedia of Chemical Technology*** has kept its unique role as "the best reference of chemical technology" for over 50 years, by providing the necessary perspective and insight into pertinent aspects, rather than merely presenting information that could be found elsewhere. Written by prominent scholars from industry, academia, and research institutions, the Encyclopedia brings together and treats systematically facts on the properties, manufacturing, and uses of chemicals and materials, processes, and engineering principals, coupled with insights into current research, emerging technologies, economic aspects, as well as addresses environmental and health concerns.

The fifth edition of the Encyclopedia is based on the content and format of previous editions with additions, adjustments, and modernization of the content reflecting changes and developments in the field. In particular, emphasis is put on sustainable and environmentally conscious chemical technology, and on advanced characterization and analytical techniques, synthesis and fabrication methods, and new materials made available with modern technology. The 27 volumes of the fifth edition of the Encyclopedia (including an Index and a Supplement volume) contain over one thousand articles, organized alphabetically. The ***Kirk-Othmer Encyclopedia of Chemical Technology*** is available also online as part of the electronic Reference Works content of **Wiley InterScience**.

Acknowledgement

The ***Kirk-Othmer Encyclopedia of Chemical Technology***, 5th edition, was made possible by the hard work of many hundreds of individuals, most notably contributors, and many reviewers and other advisors, who dedicated their precious time and expertise to this publication.

We wish to acknowledge and thank all those–new to the task and veterans alike–who took part in the preparation of the 5th edition of this Encyclopedia.

The Editors

CONTRIBUTORS

David T. Allen, *University of Texas, Austin, TX,* Air Pollution

Donna Bange, *3M Company, St. Paul, MN,* Abrasives

William Bauer, Jr., *Rohm and Haas Company, Spring House, PA* Acrylic Acid and its Derivatives

John C. Bost, *The Dow Chemical Company, Midland, MI,* Acetic Acid, Halogenated Derivatives

James Brazdil, *BP Amoco Chemicals, Naperville, IL,* Acrylonitrile

Gary L. Dickerson, *Solutia Inc, Cantonment, FL,* Adipic Acid

C. M. Dietz, *Consultant,* Acetylene

William Etzkorn, *Union Carbide Corporation, South Charleston, WV,* Acrolein and Derivatives

David S. Farinato, *Cytec Technology Corporation, Stamford, CT,* Acrylamide Polymers

S. K. Gaggar, *GE Plastics Technology Center, Pittsfield, MA,* Acrylonitrile-Butadiene-Styrene (ABS) Polymers

Richard E. Gannon, *Textron Defense Systems Wilmington, MA,* Acetylene

Roger Gary, *Milacron Inc. Cincinnati OH,* Abrasives

Stanley A. Gembicki, *UOP LLC, Des Plaines, IL,* Adsorption, Liquid Separation

C. E. Habermann, *Dow Chemical Midland, MI,* Acrylamide

H. J. Hagameyer, *Texas Eastman Company, Longview, TX,* Acetaldehyde

Darleane Hoffman, *University of California, Berkeley, Oakland, CA,* Actinides and Transactinides

Eugene V. Hort, *GAF Corporation, Wayne, NJ,* Acetylene-derived Chemicals

William L. Howard, *The Dow Chemical Company, Freeport, TX,* Acetone

Sun-Yi Huang, *Cytec Technology Corporation, Stamford, CT,* Acrylamide Polymers

James Johnson, *UOP LLC, Des Plaines, IL,* Adsorption, Liquid Separation

Edmond I. Ko, *Carnegie Mellon University, Pittsburg, PA,* Aerogels

Donald Kulich, *GE Plastics Technology Center, Pittsfield, MA,* Acrylonitrile-Butadiene-Styrene (ABS) Polymers

Manuel Laso, *Universidad Politecnica, ETSII, Madrid, Spain,* Absorption

Diana M. Lee, *Lawrence Berkeley National Laboratory, Berkeley, CA,* Actinides and Transactinides

David W. Lipp, *Cytec Technology Corporation, Stamford, CT,* Acrylamide Polymers

V. Lowry, *GE Plastics Technology Center, Pittsfield, MA,* Acrylonitrile-Butadiene-Styrene (ABS) Polymers

Robert M. Manyik, *Union Carbide Corporation, South Charleston, WV,* Acetylene

Earl D. Morris, *The Dow Chemical Company, Midland, MI,* Acetic Acid, Halogenated Derivatives

Sanford Moskowitz, *Villanova University, Roslyn, PA,* Advanced Materials, Economic Evaluation

Alvin W. Nienow, *University of Birmingham, Edgbaston, Birmingham, United Kingdom,* Aeration, Biotechnology

Judith P. Oppenheim, *Solutia Inc, Cantonment, FL,* Adipic Acid

Anil Oroskar, *UOP LLC, Des Plaines, IL,* Adsorption, Liquid Separation

S. E. Pederson, *Union Carbide Corporation, South Charleston, WV,* Acrolein and Derivatives

Alphonsus V. Pocius, *3M Adhesive Technologies Center, St. Paul, MN,* Adhesion

James Rekoske, *UOP LLC, Des Plaines, IL,* Adsorption, Liquid Separation

Douglas M. Ruthven, *University of Maine, Orono, ME,* Adsorption

H. B. Sargent, *Consultant,* Acetylene

R. P. Schaffer, *Consultant,* Acetylene

Christopher J. Sciarra, *Sciarra Laboratories, Inc., Hicksville, NY,* Aerosols

John J. Sciarra, *Sciarra Laboratories, Inc., Hicksville, NY,* Aerosols

Glenn T. Seaborg, *University of California, Berkeley, Oakland, CA,* Actinides and Transactinides

Rob Slone, *Rohm and Haas Company, Spring House, PA,* Acrylic Ester Polymers

Thomas E. Snead, *Union Carbide Corporation, South Charleston, WV,* Acrolein and Derivatives

R. Stepien, *GE Plastics Technology Center, Pittsfield, MA,* Acrylonitrile-Butadiene-Styrene (ABS) Polymers

Urs von Stockar, *Laboratoire de Genie Chimique et Biologique, Lausanne, Switzerland,* Absorption

Paul Taylor, *GAF Corporation, Wayne, NJ,* Acetylene-derived Chemicals

R. O. Thribolet, *Consultant,* Acetylene

W. Gene Tucker, *James Madison University, Harrisonburg, VA,* Air Pollution and Control, Indoor

Frank S. Wagner, *Nandina Corporation, Corpus Christi, TX,* Acetic Acid and Derivatives

Mike Wu, *BP Amoco Chemicals, Naperville, IL,* Acrylonitrile Polymers, Survey and SAN

Carmen M. Yon, *UOP, Fitzwilliam, NH,* Adsorption, Gas Separation

Elaine Yorkgitis, *3M Company, Automotive Division, Mendota Heights, MN,* Adhesives

CONVERSION FACTORS, ABBREVIATIONS, AND UNIT SYMBOLS

SI Units (Adopted 1960)

The International System of Units (abbreviated SI), is implemented throughout the world. This measurement system is a modernized version of the MKSA (meter, kilogram, second, ampere) system, and its details are published and controlled by an international treaty organization (The International Bureau of Weights and Measures) (1).

SI units are divided into three classes:

BASE UNITS

length	meter[†] (m)
mass	kilogram (kg)
time	second (s)
electric current	ampere (A)
thermodynamic temperature[‡]	kelvin (K)
amount of substance	mole (mol)
luminous intensity	candela (cd)

SUPPLEMENTARY UNITS

plane angle	radian (rad)
solid angle	steradian (sr)

DERIVED UNITS AND OTHER ACCEPTABLE UNITS

These units are formed by combining base units, suplementary units, and other derived units (2–4). Those derived units having special names and symbols are marked with an asterisk in the list below.

[†] The spellings "metre" and "litre" are preferred by ASTM; however, "-er" is used in the *Encyclopedia*.

[‡] Wide use is made of Celsius temperature (t) defined by

$$t = T - T_0$$

where T is the thermodynamic temperature, expressed in kelvin, and $T_0 = 273.15$ K by definition. A temperature interval may be expressed in degrees Celsius as well as in kelvin.

Quantity	Unit	Symbol	Acceptable equivalent
*absorbed dose	gray	Gy	J/Kg
acceleration	meter per second squared	m/s^2	
*activity (of a radionuclide)	becquerel	Bq	1/s
area	square kilometer	km^2	
	square hectometer	hm^2	ha (hectare)
	square meter	m^2	
concentration (of amount of substance)	mole per cubic meter	mol/m^3	
current density	ampere per square meter	A/m^2	
density, mass density	kilogram per cubic meter	kg/m^3	g/L; mg/cm^3
dipole moment (quantity)	coulomb meter	C · m	
*dose equivalent	sievert	Sv	J/kg
*electric capacitance	farad	F	C/V
*electric charge, quantity of electricity	coulomb	C	A · s
electric charge density	coulomb per cubic meter	C/m^3	
*electric conductance	siemens	S	A/V
electric field strength	volt per meter	V/m	
electric flux density	coulomb per square meter	C/m^2	
*electric potential, potential difference, electromotive force	volt	V	W/A
*electric resistance	ohm	Ω	V/A
*energy, work, quantity of heat	megajoule	MJ	
	kilojoule	kJ	
	joule	J	N · m
	electronvolt†	eV†	
	kilowatt-hour†	kW · h†	
energy density	joule per cubic meter	J/m^3	
*force	kilonewton	kN	
	newton	N	$kg \cdot m/s^2$

†This non-SI unit is recognized by the CIPM as having to be retained because of practical importance or use in specialized fields (1).

Quantity	Unit	Symbol	Acceptable equivalent
*frequency	megahertz	MHz	
	hertz	Hz	1/s
heat capacity, entropy	joule per kelvin	J/K	
heat capacity (specific), specific entropy	joule per kilogram kelvin	J/(kg · K)	
heat-transfer coefficient	watt per square meter kelvin	$W/(m^2 \cdot K)$	
*illuminance	lux	lx	lm/m^2
*inductance	henry	H	Wb/A
linear density	kilogram per meter	kg/m	
luminance	candela per square meter	cd/m^2	
*luminous flux	lumen	lm	cd · sr
magnetic field strength	ampere per meter	A/m	
*magnetic flux	weber	Wb	V · s
*magnetic flux density	tesla	T	Wb/m^2
molar energy	joule per mole	J/mol	
molar entropy, molar heat capacity	joule per mole kelvin	J/(mol · K)	
moment of force, torque	newton meter	N · m	
momentum	kilogram meter per second	kg · m/s	
permeability	henry per meter	H/m	
permittivity	farad per meter	F/m	
*power, heat flow rate, radiant flux	kilowatt	kW	
	watt	W	J/s
power density, heat flux density, irradiance	watt per square meter	W/m^2	
*pressure, stress	megapascal	MPa	
	kilopascal	kPa	
	pascal	Pa	N/m^2
sound level	decibel	dB	
specific energy	joule per kilogram	J/kg	
specific volume	cubic meter per kilogram	m^3/kg	
surface tension	newton per meter	N/m	
thermal conductivity	watt per meter kelvin	W/(m · K)	
velocity	meter per second	m/s	
	kilometer per hour	km/h	
viscosity, dynamic	pascal second	Pa · s	
	millipascal second	mPa · s	
viscosity, kinematic	square meter per second	m^2/s	
	square millimeter per second	mm^2/s	

Quantity	Unit	Symbol	Acceptable equivalent
volume	cubic meter	m^3	
	cubic diameter	dm^3	L (liter) (5)
	cubic centimeter	cm^3	mL
wave number	1 per meter	m^{-1}	
	1 per centimeter	cm^{-1}	

In addition, there are 16 prefixes used to indicate order of magnitude, as follows

Multiplication factor	Prefix	symbol	Note
10^{18}	exa	E	
10^{15}	peta	P	
10^{12}	tera	T	
10^{9}	giga	G	
10^{6}	mega	M	
10^{3}	kilo	k	
10^{2}	hecto	h[a]	[a]Although hecto, deka, deci, and centi are SI prefixes, their use should be avoided except for SI unit-multiples for area and volume and nontechnical use of centimeter, as for body and clothing measurement.
10	deka	da[a]	
10^{-1}	deci	d[a]	
10^{-2}	centi	c[a]	
10^{-3}	milli	m	
10^{-6}	micro	μ	
10^{-9}	nano	n	
10^{-12}	pico	p	
10^{-15}	femto	f	
10^{-18}	atto	a	

For a complete description of SI and its use the reader is referred to ASTM E380 (4) and the article UNITS AND CONVERSION FACTORS which appears in Vol. 24.

A representative list of conversion factors from non-SI to SI units is presented herewith. Factors are given to four significant figures. Exact relationships are followed by a dagger. A more complete list is given in the latest editions of ASTM E380 (4) and ANSI Z210.1 (6).

Conversion Factors to SI Units

To convert from	To	Multiply by
acre	square meter (m^2)	4.047×10^3
angstrom	meter (m)	$1.0 \times 10^{-10\dagger}$
are	square meter (m^2)	$1.0 \times 10^{2\dagger}$
astronomical unit	meter (m)	1.496×10^{11}

†Exact.

To convert from	To	Multiply by
atmosphere, standard	pascal (Pa)	1.013×10^5
bar	pascal (Pa)	$1.0 \times 10^{5\dagger}$
barn	square meter (m^2)	$1.0 \times 10^{-28\dagger}$
barrel (42 U.S. liquid gallons)	cubic meter (m^3)	0.1590
Bohr magneton (μ_B)	J/T	9.274×10^{-24}
Btu (International Table)	joule (J)	1.055×10^3
Btu (mean)	joule (J)	1.056×10^3
Btu (thermochemical)	joule (J)	1.054×10^3
bushel	cubic meter(m^3)	3.524×10^{-2}
calorie (International Table)	joule (J)	4.187
calorie (mean)	joule (J)	4.190
calorie (thermochemical)	joule (J)	$4.184^\dagger$
centipoise	pascal second (Pa · s)	$1.0 \times 10^{-3\dagger}$
centistokes	square millimeter per second (mm^2/s)	$1.0^\dagger$
cfm (cubic foot per minute)	cubic meter per second (m^3s)	4.72×10^{-4}
cubic inch	cubic meter (m^3)	1.639×10^{-5}
cubic foot	cubic meter (m^3)	2.832×10^{-2}
cubic yard	cubic meter (m^3)	0.7646
curie	becquerel (Bq)	$3.70 \times 10^{10\dagger}$
debye	coulomb meter (C · m)	3.336×10^{-30}
degree (angle)	radian (rad)	1.745×10^{-2}
denier (international)	kilogram per meter (kg/m)	1.111×10^{-7}
	$tex^\ddagger$	0.1111
dram (apothecaries')	kilogram (kg)	3.888×10^{-3}
dram (avoirdupois)	kilogram (kg)	1.772×10^{-3}
dram (U.S. fluid)	cubic meter (m^3)	3.697×10^{-6}
dyne	newton (N)	$1.0 \times 10^{-5\dagger}$
dyne/cm	newton per meter (N/m)	$1.0 \times 10^{-3\dagger}$
electronvolt	joule (J)	1.602×10^{-19}
erg	joule (J)	$1.0 \times 10^{-7\dagger}$
fathom	meter (m)	1.829
fluid ounce (U.S.)	cubic meter (m^3)	2.957×10^{-5}
foot	meter (m)	$0.3048^\dagger$
footcandle	lux (lx)	10.76
furlong	meter (m)	2.012×10^{-2}
gal	meter per second squared (m/s^2)	$1.0 \times 10^{-2\dagger}$
gallon (U.S. dry)	cubic meter (m^3)	4.405×10^{-3}
gallon (U.S. liquid)	cubic meter (m^3)	3.785×10^{-3}
gallon per minute (gpm)	cubic meter per second (m^3/s)	6.309×10^{-5}
	cubic meter per hour (m^3/h)	0.2271

†Exact.

‡See footnote on p. xii.

To convert from	To	Multiply by
gauss	tesla (T)	1.0×10^{-4}
gilbert	ampere (A)	0.7958
gill (U.S.)	cubic meter (m^3)	1.183×10^{-4}
grade	radian	1.571×10^{-2}
grain	kilogram (kg)	6.480×10^{-5}
gram force per denier	newton per tex (N/tex)	8.826×10^{-2}
hectare	square meter (m^2)	$1.0 \times 10^{4\dagger}$
horsepower (550 ft · lbf/s)	watt (W)	7.457×10^2
horsepower (boiler)	watt (W)	9.810×10^3
horsepower (electric)	watt (W)	$7.46 \times 10^{2\dagger}$
hundredweight (long)	kilogram (kg)	50.80
hundredweight (short)	kilogram (kg)	45.36
inch	meter (m)	$2.54 \times 10^{-2\dagger}$
inch of mercury (32°F)	pascal (Pa)	3.386×10^3
inch of water (39.2°F)	pascal (Pa)	2.491×10^2
kilogram-force	newton (N)	9.807
kilowatt hour	megajoule (MJ)	$3.6^\dagger$
kip	newton (N)	4.448×10^3
knot (international)	meter per second (m/S)	0.5144
lambert	candela per square meter (cd/m^3)	3.183×10^3
league (British nautical)	meter (m)	5.559×10^3
league (statute)	meter (m)	4.828×10^3
light year	meter (m)	9.461×10^{15}
liter (for fluids only)	cubic meter (m^3)	$1.0 \times 10^{-3\dagger}$
maxwell	weber (Wb)	$1.0 \times 10^{-8\dagger}$
micron	meter (m)	$1.0 \times 10^{-6\dagger}$
mil	meter (m)	$2.54 \times 10^{-5\dagger}$
mile (statue)	meter (m)	1.609×10^3
mile (U.S. nautical)	meter (m)	$1.852 \times 10^{3\dagger}$
mile per hour	meter per second (m/s)	0.4470
millibar	pascal (Pa)	1.0×10^2
millimeter of mercury (0°C)	pascal (Pa)	$1.333 \times 10^{2\dagger}$
minute (angular)	radian	2.909×10^{-4}
myriagram	kilogram (Kg)	10
myriameter	kilometer (Km)	10
oersted	ampere per meter (A/m)	79.58
ounce (avoirdupois)	kilogram (kg)	2.835×10^{-2}
ounce (troy)	kilogram (kg)	3.110×10^{-2}
ounce (U.S. fluid)	cubic meter (m^3)	2.957×10^{-5}
ounce-force	newton (N)	0.2780
peck (U.S.)	cubic meter (m^3)	8.810×10^{-3}
pennyweight	kilogram (kg)	1.555×10^{-3}
pint (U.S. dry)	cubic meter (m^3)	5.506×10^{-4}

†Exact.

To convert from	To	Multiply by
pint (U.S. liquid)	cubic meter (m^3)	4.732×10^{-4}
poise (absolute viscosity)	pascal second (Pa · s)	0.10†
pound (avoirdupois)	kilogram (kg)	0.4536
pound (troy)	kilogram (kg)	0.3732
poundal	newton (N)	0.1383
pound-force	newton (N)	4.448
pound force per square inch (psi)	pascal (Pa)	6.895×10^{3}
quart (U.S. dry)	cubic meter (m^3)	1.101×10^{-3}
quart (U.S. liquid)	cubic meter (m^3)	9.464×10^{-4}
quintal	kilogram (kg)	1.0×10^{-2}†
rad	gray (Gy)	1.0×10^{-2}†
rod	meter (m)	5.029
roentgen	coulomb per kilogram (C/kg)	2.58×10^{-4}
second (angle)	radian (rad)	4.848×10^{-6}†
section	square meter (m^2)	2.590×10^{6}
slug	kilogram (kg)	14.59
spherical candle power	lumen (lm)	12.57
square inch	square meter (m^2)	6.452×10^{-4}
square foot	square meter (m^2)	9.290×10^{-2}
square mile	square meter (m^2)	2.590×10^{6}
square yard	square meter (m^2)	0.8361
stere	cubic meter (m^3)	1.0†
stokes (kinematic viscosity)	square meter per second (m^2/s)	1.0×10^{-4}†
tex	kilogram per meter (kg/m)	1.0×10^{-6}†
ton (long, 2240 pounds)	kilogram (kg)	1.016×10^{3}
ton (metric) (tonne)	kilogram (kg)	1.0×10^{3}†
ton (short, 2000 pounds)	kilogram (kg)	9.072×10^{2}
torr	pascal (Pa)	1.333×10^{2}
unit pole	weber (Wb)	1.257×10^{-7}
yard	meter (m)	0.9144†

†Exact.

Abbreviations and Unit Symbols

Following is a list of common abbreviations and unit symbnols used in the Encyclopedia. In general they agree with those listed in *American National Standard Abbreviations for Use on Drawings and in Text* (*ANSI Y1.1*) (6) and *American National Standard Letter Symbols for Units in Science and Technology* (*ANSI Y10*) (6). Also included is a list of acronyms for a number of private and government organizations as well as common industrial solvents, polymers, and other chemicals.

Rules for Writing Unit Symbols (4):

1. Unit symbols are printed in upright letters (roman) regardless of the type style used in the surrounding text.
2. Unit symbols are unaltered in the plural.
3. Unit symbols are not followed by a period except when used at the end of a sentence.
4. Letter unit symbols are generally printed lower-case (for example, cd for candela) unless the unit name has been derived from a proper name, in which case the first letter of the symbol is capitalized (W, Pa). Prefixes and unit symbols retain their prescribed form regardless of the surrounding typography.
5. In the complete expression for a quantity, a space should be left between the numerical value and the unit symbol. For example, write 2.37 lm, *not* 2.37lm, and 35 mm, *not* 35mm. When the quantity is used in an adjectival sense, a hyphen is often used, for example, 35-mm film. *Exception:* No space is left between the numerical value and the symbols of degree, minute, and second of plane angle, degree Celsius, and the percent sign.
6. No space is used between the prefix and unit symbol (for example, kg).
7. Symbols, not abbreviations, should be used for units. For example, use "A," not "amp," for ampere.
8. When multiplying unit symbols, use a raised dot:

   $$\mathrm{N \cdot m} \text{ for newton meter}$$

   In the case of $\mathrm{W \cdot h}$, the dot may be omitted, thus:

   $$\mathrm{Wh}$$

   An exception to this practice is made for computer printouts, automatic typewriter work, etc, where the raised dot is not possible, and a dot on the line may be used.
9. When dividing unit symbols, use one of the following forms:

   $$\mathrm{m/s} \quad \text{or} \quad \mathrm{m \cdot s^{-1}} \quad \textit{or} \quad \frac{\mathrm{m}}{\mathrm{s}}$$

   In no case should more than one slash be used in the same expression unless parentheses are inserted to avoid ambiguity. For example, write:

   $$\mathrm{J/(mol \cdot K)} \quad \textit{or} \quad \mathrm{J \cdot mol^{-1} \cdot K^{-1}} \quad \textit{or} \quad \mathrm{(J/mol)/K}$$

   but *not*

   $$\mathrm{J/mol/K}$$

10. Do not mix symbols and unit names in the same expression. Write:

joules per kilogram *or* J/kg *or* $J \cdot kg^{-1}$

but *not*

joules/kilogram *nor* Joules/kg *nor* $Joules \cdot kg^{-1}$

ABBREVIATIONS AND UNITS

A	ampere
A	anion (eg, HA)
A	mass number
a	atto (prefix for 10^{-18})
AATCC	American Association of Textile Chemists and Colorists
ABS	acrylonitrile–butadiene–styrene
abs	absolute
ac	alternating current, *n.*
a-c	alternating current, *adj.*
ac-	alicyclic
acac	acetylacetonate
ACGIH	American Conference of Governmental Industrial Hygienists
ACS	American Chemical Society
AGA	American Gas Association
Ah	ampere hour
AIChE	American Institute of Chemical Engineers
AIME	American Institute of Mining, metallurgical, and Petroleum Engineers
AIP	American Institute of Physics
AISI	American Iron and Steel Institute
alc	alcohol(ic)
Alk	alkyl
alk	alkaline (not alkali)
amt	amount
amu	atomic mass unit
ANSI	American National Standards Institute
AO	atomic orbital

AOAC	Association of Official Analytical Chemists
AOCS	American Oil Chemists' Society
APHA	American Public Health Association
API	American Petroleum Institute
aq	aqueous
Ar	aryl
ar-	aromatic
as-	Asymmetric(al)
ASHRAE	American Society of Heating, Refrigerating, and Air Conditioning Engineers
ASM	American Society for Metals
ASME	American Society of Mechanical Engineers
ASTM	American Society for Testing and Materials
at no.	atomic number
at wt	atomic weight
av(g)	average
AWS	American Welding Society
b	bonding orbital
bbl	barrel
bcc	body-centered cubic
BCT	body-centered tetragonal
Bé	Baumé
BET	Brunauer-Emmett-Teller (adsorption equation)
bid	twice daily
Boc	*t*-butyloxycarbonyl
BOD	biochemical (biological) oxygen demand
bp	boiling point
Bq	becquerel

C	coulomb
°C	degree Celsius
C-	denoting attachment to carbon
c	centi (prefix for 10^{-2})
c	critical
ca	circa (Approximately)
cd	candela; current density; circular dichroism
CFR	Code of Federal Regulations
cgs	centimeter-gram-second
CI	Color Index
cis-	isomer in which substituted groups are on some side of double bond between C atoms
cl	carload
cm	centimeter
cmil	circular mil
cmpd	compound
CNS	central nervous system
CoA	coenzyme A
COD	chemical oxygen demand
coml	commerical(ly)
cp	chemically pure
cph	close-packed hexagonal
CPSC	Consumer Product Safety Commission
cryst	crystalline
cub	cubic
D	debye
D-	denoting configurational relationship
d	differential operator
d	day; deci (prefix for 10^{-1})
d	density
d-	*dextro*-, dextrorotatory
da	deka (prefix for 10^{-1})
dB	decibel
dc	direct current, *n.*
d-c	direct current, *adj.*
dec	decompose
detd	determined
detn	determination
Di	didymium, a mixture of all lanthanons
dia	diameter
dil	dilute
DIN	Deutsche Industrie Normen
dl-; DL-	racemic
DMA	dimethylacetamide
DMF	dimethylformamide
DMG	dimethyl glyoxime
DMSO	dimethyl sulfoxide
DOD	Department of Defense
DOE	Department of Energy
DOT	Department of Transportation
DP	degree of polymerization
dp	dew point
DPH	diamond pyramid hardness
dstl(d)	distill(ed)
dta	differential thermal analysis
(*E*)-	entgegen; opposed
ϵ	dielectric constant (unitless number)
e	electron
ECU	electrochemical unit
ed.	edited, edition, editor
ED	effective dose
EDTA	ethylenediaminetetra-acetic acid
emf	electromotive force
emu	electromagnetic unit
en	ethylene diamine
eng	engineering
EPA	Environmental Protection Agency
epr	electron paramagnetic resonance
eq.	equation
esca	electron spectroscopy for chemical analysis
esp	especially
esr	electron-spin resonance
est(d)	estimate(d)
estn	estimation
esu	electrostatic unit
exp	experiment, experimental
ext(d)	extract(ed)
F	farad (capacitance)
F	fraday (96,487 C)

f	femto (prefix for 10^{-15})
FAO	Food and Agriculture Organization (United Nations)
fcc	face-centered cubic
FDA	Food and Drug Administration
FEA	Federal Energy Administration
FHSA	Federal Hazardous Substances Act
fob	free on board
fp	freezing point
FPC	Federal Power Commission
FRB	Federal Reserve Board
frz	freezing
G	giga (prefix for 10^9)
G	gravitational constant $= 6.67 \times 10^{11} \mathrm{N \cdot m^2/kg^2}$
g	gram
(g)	gas, only as in $H_2O(g)$
g	gravitatonal acceleration
gc	gas chromatography
gem-	geminal
glc	gas–liquid chromatography
g-mol wt; gmw	gram-molecular weight
GNP	gross national product
gpc	gel-permeation chromatography
GRAS	Generally Recognized as Safe
grd	ground
Gy	gray
H	henry
h	hour; hecto (prefix for 10^2)
ha	hectare
HB	Brinell hardness number
Hb	hemoglobin
hcp	hexagonal close-packed
hex	hexagonal
HK	Knoop hardness number
hplc	high performance liquid chromatography
HRC	Rockwell hardness (C scale)
HV	Vickers hardness number
hyd	hydrated, hydrous
hyg	hygroscopic
Hz	hertz
^{i}(eg, Pr^i)	iso (eg, isopropyl)
i-	inactive (eg, *i*-methionine)
IACS	international Annealed Copper Standard
ibp	initial boiling point
IC	integrated circuit
ICC	Interstate Commerce Commission
ICT	International Critical Table
ID	inside diameter; infective dose
ip	intraperitoneal
IPS	iron pipe size
ir	infrared
IRLG	Interagency Regulatory Liaison Group
ISO	International Organization Standardization
ITS-90	International Temperature Scale (NIST)
IU	International Unit
IUPAC	International Union of Pure and Applied Chemistry
IV	iodine value
iv	intravenous
J	joule
K	kelvin
k	kilo (prefix for 10^3)
kg	kilogram
L	denoting configurational relationship
L	liter (for fluids only) (5)
l-	*levo*-, levorotatory
(l)	liquid, only as in $NH_3(l)$
LC_{50}	conc lethal to 50% of the animals tested
LCAO	linear combnination of atomic orbitals
lc	liquid chromatography
LCD	liquid crystal display
lcl	less than carload lots

LD_{50}	dose lethal to 50% of the animals tested
LED	light-emitting diode
liq	liquid
lm	lumen
ln	logarithm (natural)
LNG	liquefied natural gas
log	logarithm (common)
LOI	limiting oxygen index
LPG	liquefied petroleum gas
ltl	less than truckload lots
lx	lux
M	mega (prefix for 10^6); metal (as in MA)
M	molar; actual mass
$\overline{M}_w$	weight-average mol wt
$\overline{M}_n$	number-average mol wt
m	meter; milli (prefix for 10^{-3})
m	molal
m-	meta
max	maximum
MCA	Chemical Manufacturers' Association (was Manufacturing Chemists Association)
MEK	methyl ethyl ketone
meq	milliequivalent
mfd	manufactured
mfg	manufacturing
mfr	manufacturer
MIBC	Methyl isobutyl carbinol
MIBK	methyl isobutyl ketone
MIC	minimum inhibiting concentration
min	minute; minimum
mL	milliliter
MLD	minimum lethal dose
MO	molecular orbital
mo	month
mol	mole
mol wt	molecular weight
mp	melting point
MR	molar refraction
ms	mass spectrometry
MSDS	material safety data sheet
mxt	mixture
μ	micro (prefix for 10^{-6})
N	newton (force)
N	normal (concentration); neutron number
N-	denoting attachment to nitrogen
n (as n_D^{20})	index of refraction (for 20°C and sodium light)
^{n}(as Bu^n), *n*-	normal (straight-chain structure)
n	neutron
n	nano (prefix for 10^9)
na	not available
NAS	National Academy of Sciences
NASA	National Aeronautics and Space Administration
nat	natural
ndt	nondestructive testing
neg	negative
NF	*National Formulary*
NIH	National Institutes of Health
NIOSH	National Institute of Occupational Safety and Health
NIST	National Institute of Standards and Technology (formerly National Bureau of Standards)
nmr	nuclear magnetic resonance
NND	New and Nonofficial Drugs (AMA)
no.	number
NOI-(BN)	not otherwise indexed (by name)
NOS	not otherwise specified
nqr	nuclear quadruple resonance
NRC	Nuclear Regulatory Commission; National Research Council
NRI	New Ring Index
NSF	National Science Foundation
NTA	nitrilotriacetic acid

NTP	normal temperature and pressure (25°C and 101.3 kPa or 1 atm)
NTSB	National Transportation Safety Board
O-	denoting attachment to oxygen
o-	ortho
OD	outside diameter
OPEC	Organization of Petroleum Exporting Countries
o-phen	*o*-phenanthridine
OSHA	Occupational Safety and Health Administration
owf	on weight of fiber
Ω	ohm
P	peta (prefix for 10^{15})
p	pico (prefix for 10^{-12}
p-	para
p	proton
p.	page
Pa	Pascal (pressure)
PEL	personal exposure limit based on an 8-h exposure
pd	potential difference
pH	negative logarithm of the effective hydrogen ion concentration
phr	parts per hundred of resin (rubber)
p-i-n	positive-intrinsic-negative
pmr	proton magnetic resonance
p-n	positive-negative
po	per os (oral)
POP	polyoxypropylene
pos	positive
pp.	pages
ppb	parts per billion (10^9)
ppm	parts per milion (10^6)
ppmv	parts per million by volume
ppmwt	parts per million by weight
PPO	poly(phenyl oxide)
ppt(d)	precipitate(d)
pptn	precipitation
Pr (no.)	foreign prototype (number)
pt	point; part
PVC	poly(vinyl chloride)

pwd	powder
py	pyridine
qv	quod vide (which see)
R	univalent hydrocarbon radical
(*R*)-	rectus (clockwise configuration)
r	precision of data
rad	radian; radius
RCRA	Resource Conservation and Recovery Act
rds	rate-determining step
ref.	reference
rf	radio frequency, *n.*
r-f	radio frequency, *adj.*
rh	relative humidity
RI	Ring Index
rms	root-mean square
rpm	rotations per minute
rps	revolutions per second
RT	room temperature
RTECS	Registry of Toxic Effects of Chemical Substances
^{s}(eg, Bu^s); *sec*-	secondary (eg, secondary butyl)
S	siemens
(*S*)-	sinister (counterclockwise configuration)
S-	denoting attachment to sulfur
s-	symmetric(al)
S	second
(s)	solid, only as in $H_2O(s)$
SAE	Society of Automotive Engineers
SAN	styrene-acrylonitrile
sat(d)	saturate(d)
satn	saturation
SBS	styrene–butadiene–styrene
sc	subcutaneous
SCF	self-consistent field; standard cubic feet
Sch	Schultz number
sem	scanning electron microscope(y)
SFs	Saybolt Furol seconds
sl sol	slightly soluble
sol	soluble

soln	solution
soly	solubility
sp	specific; species
sp gr	specific gravity
sr	steradian
std	standard
STP	standard temperature and pressure (0°C and 101.3 kPa)
sub	sublime(s)
SUs	Saybolt Universal seconds
syn	synthetic
^{t} (eg, Bu^t), *t*-, *tert*-	tertiary (eg, tertiary butyl)
T	tera (prefix for 10^{12}); tesla (magnetic flux density)
t	metric to (tonne)
t	temperature
TAPPI	Technical Association of the Pulp and Paper Industry
TCC	Tagliabue closed cup
tex	tex (linear density)
T_g	glass-transition temperature
tga	thermogravimetric analysis
THF	tetrahydrofuran
tlc	thin layer chromatography
TLV	threshold limit value
trans-	isomer in which substituted groups are on opposite sides of double bond between C atoms
TSCA	Toxic Substances Control Act
TWA	time-weighted average
Twad	Twaddell
UL	Underwriters' Laboratory
USDA	United States Department of Agriculture
USP	*United States Pharmacopeia*
uv	ultraviolet
V	volt (emf)
var	variable
vic-	vicinal
vol	volume (not volatile)
vs	versus
v sol	very soluble
W	watt
Wb	weber
Wh	watt hour
WHO	World Health Organization (United Nations)
wk	week
yr	year
(*Z*)-	zusammen; together; atomic number

Non-SI (Unacceptable and Obsolete) Units		Use
Å	angstrom	nm
at	atmosphere, technical	Pa
atm	atmosphere, standard	Pa
b	barn	cm^2
bar†	bar	Pa
bbl	barrel	m^3
bhp	brake horsepower	W
Btu	British thermal unit	J
bu	bushel	m^3; L
cal	calorie	J
cfm	cubic foot per minute	m^3/s
Ci	curie	Bq
cSt	centistokes	mm^2/s
c/s	cycle per second	Hz
cu	cubic	exponential form

†Do not use bar (10^5 Pa) or millibar (10^2 Pa) because they are not SI units, and are accepted internationally only in special fields because of existing usage.

Non-SI (Unacceptable and Obsolete) Units		Use
D	debye	C · m
den	denier	tex
dr	dram	kg
dyn	dyne	N
dyn/cm	dyne per centimeter	mN/m
erg	erg	J
eu	entropy unit	J/K
°F	degree Fahrenheit	°C; K
fc	footcandle	lx
fl	footlambert	lx
fl oz	fluid ounce	m^3; L
ft	foot	m
ft · lbf	foot pound-force	J
gf den	gram-force per denier	N/tex
G	gauss	T
Gal	gal	m/s^2
gal	gallon	m^3; L
Gb	gilbert	A
gpm	gallon per minute	(m^3/s); (m^3/h)
gr	grain	kg
hp	horsepower	W
ihp	indicated horsepower	W
in.	inch	m
in. Hg	inch of mercury	Pa
in. H_2O	inch of water	Pa
in.-lbf	inch pound-force	J
kcal	kilo-calorie	J
kgf	kilogram-force	N
kilo	for kilogram	kg
L	lambert	lx
lb	pound	kg
lbf	pound-force	N
mho	mho	S
mi	mile	m
MM	million	M
mm Hg	millimeter of mercury	Pa
mμ	millimicron	nm
mph	miles per hour	km/h
μ	micron	µm
Oe	oersted	A/m
oz	ounce	kg
ozf	ounce-force	N
η	poise	Pa · s
P	poise	Pa · s
ph	phot	lx
psi	pounds-force per square inch	Pa
psia	pounds-force per square inch absolute	Pa
psig	pounds-force per square inch gage	Pa
qt	quart	m^3; L
°R	degree Rankine	K
rd	rad	Gy
sb	stilb	lx
SCF	standard cubic foot	m^3
sq	square	exponential form
thm	therm	J
yd	yard	m

BIBLIOGRAPHY

1. The International Bureau of Weights and Measures, BIPM (Parc Saint-Cloud, France) is described in Ref. 4. This bureau operates under the exclusive supervision of the International Committee for Weights and Measures (CIPM).
2. *Metric Editorial Guide* (*ANMC-78-1*), latest ed., American National Metric Council, 900 Mix Avenue, Suite 1 Hamden CT 06514-5106, 1981.
3. *SI Units and Recommendations for the Use of Their Multiples and of Certain Other Units* (*ISO 1000-1992*), American National Standards Institute, 25 W 43rd St., New York, 10036, 1992.
4. Based on IEEE/ASTM-SI-10 *Standard for use of the International System of Units* (*SI*): *The Modern Metric System* (Replaces ASTM380 and ANSI/IEEE Std 268-1992), ASTM International, West Conshohocken, PA., 2002. See also www.astm.org
5. *Fed. Reg.*, Dec. 10, 1976 (41 FR 36414).
6. For ANSI address, see Ref. 3. See also www.ansi.org

ABRASIVES

1. Introduction

An abrasive is a substance used to abrade, smooth, or polish an object. If the object is soft, such as wood, then relatively soft abrasive materials may be used. Usually, however, abrasive connotes very hard substances ranging from naturally occuring sands to the hardest material known, diamond.

Abrasives were literally as old or older than the Egyptian pyramids; in ancient times, humans used a variety of materials to refine or polish surfaces. For example, the Chinese used corncob skins for polishing. Shark skin, with its dermal denticles known as placoid scales, naturally provided abrasive properties. Humans also harvested the most common and abundant minerals in the earth's crust, namely, the quartz family of minerals. The ancient Greeks called quartz "crystal" and this quartz took the form of sandstone, loose sand, and flint. This crystal material was used in the abrading of stone, wood, metal and grinding grains, and limestone. Additionally, the ancient Greeks developed the use of corundum (naturally occurring aluminum oxide) and garnet, which were superior to quartz. Other abrasive materials known were hematite, now known as Jeweler's rouge, in 325 BC by Theophratus (1). Diamond as a polishing material was referenced in India in 800 BC (2) and its exceptional hardness was referred to in ancient Hindu proverbs (3) and in the Bible (4).

During the Industrial Revolution of the 1800s, the development of abrasive articles went hand in hand with the metal-working industry. In early years, sandstone rocks were mined out of the earth and carved into grinding wheels. Flint and naturally occurring corundum were bonded to paper to form sandpaper. The abrasive grains used were primarily mined materials such as sandstone, quartz, naturally occurring corundum, and garnet. These minerals had significant amounts of impurities, including iron, silica, and silicates, which lowered their abrasive grain hardness and hindered their performance. During the late 1800s, and early 1900s, synthetically manufactured abrasive grains revolutionized the abrasive industry. The synthetic abrasives tended to be harder, tougher and purer than mined abrasive grains. Edward Acheson is credited with inventing synthesized silicon carbide in 1891 (5). Excluding diamond, silicon carbide was the hardest abrasive grain available for years to come. Even today, silicon carbide is produced under essentially the same process and furnace that Acheson invented. The silicon source (typically from very pure sand) and carbon source (usually graphite) are reacted at temperatures in excess of 2000°C to cause the reduction of silica by carbon.

Around the turn of the twentieth century, synthetically manufactured fused alumina was invented. In this process, alumina-based raw materials are heated above its melting point, typically ~2000°C, and subsequently cooled to form fused alumina. This basic process was originally patented by Werlein (6) in France in 1893 and by Hasslacher (7) in Germany in 1894. This process was further advanced by C. M. Hall through the addition of iron borings into the fusion melt to remove metallic impurities (8). The resulting aluminous abrasive grain was purer. A. C. Higgins then developed an improved furnace design that involved the use of a water-cooled shell container (9). This new furnace design used a solid, thin alumina coating on the furnace walls that prevented the molten alumina from attacking the steel furnace walls. This basic Higgins furnace design is still widely used today in manufacturing many fused alumina grains.

Over these multitude of centuries, abrasive articles were employed, quite simply, to change a surface. These abrasive articles relied on a broad range of technologies including ceramics, inorganic chemistry, paper, textiles, organic chemistry, polymer science, and related process technologies. During the past 150 years, as these technologies grew, so did the advancement of abrasive technologies to create abrasive articles with even higher efficiencies.

There are four major forms of abrasive articles. A bonded abrasive is a three-dimensional (3D) composite of abrasive grains dispersed in a bond system. This bond system may be organic (eg, resinoid wheels) glassy inorganic bond (vitrified wheels), or metallic. Bonded abrasives are commercially available in a wide variety of forms including wheels (most popular), stones, mounted points, saws, segments, and the like. Coated abrasives are generally described as a plurality of abrasive grains bonded to a backing. Nonwoven abrasives comprise a plurality of abrasive grains bonded into and onto a porous non-woven web substrate. Nonwoven and coated abrasives are also available in a wide variety of converted forms of belts, sheets, disks, cones, flap wheels, etc. Loose abrasive slurries comprise a plurality of abrasive grains dispersed in a liquid medium, such as water. Loose abrasive slurries are typically employed in polishing type applications where a very fine surface finish is desired.

These abrasive articles are used in a plethora of different refining processes including metal degating, grinding, shaping, cutting, deburring, finishing, sanding, cleaning, polishing, and planarizing. Today abrasive articles are employed in some aspect in most manufactured goods sold. The abrasive article may be used in the supporting equipment to make the manufactured good and/or the direct process to produce the manufactured good. For examples, resinoid grinding wheels are used to remove metal gates from castings, where cut rate is measured in mm^3/s. Coated abrasive belts grind away parting lines from a forged hand tool and consequently reshape the surface of the hand tool. Metal bonded saws are employed to cut through concrete in road construction. Nonwoven flap wheels debur the surface of metal components. Coated abrasive belts sand wood boards to a finished surface such that these wood boards are ready for staining. Nonwoven abrasives are popular for cleaning the surface of metal components including cooking utensils. Loose abrasive slurries are utilized in precision polishing, eg, in semiconductor planarization or polishing glass to optical clarity.

2. Properties of Abrasive Materials

2.1. Hardness. Table 1 lists the various scales of hardness used for abrasives. The earliest scale was developed by the German mineralogist Friedrich Mohs in 1820. It is based on the relative scratch hardness of one mineral compared to another ranging from talc, assigned a value of 1, to diamond, assigned a value of 10. Mohs' scale has two limitations; it is not linear

Table 1. **Scales of Hardness**

Material	CAS Registry Number	Mohs' scale	Ridgeway's[a] scale	Woodell's[b] scale	Knoop hardness[c] kN/m^2,[d]
talc	[14807-96-6]	1			
gypsum	[13397-24-5]	2			
calcite	[13397-26-7]	2			
fluorite	[7789-75-5]	4			
apatite	[1306-05-4]	5			
orthoclase	[12251-44-4]	6	6		
vitreous silica	[60676-86-0]		7		
quartz	[14808-60-7]	7	8	7	8
topaz	[1302-59-6]	8	9		13
garnet	[12178-41-5]		10		13
corundum	[1302-74-5]	9		9	20
fused ZrO_2	[1314-23-4]		11		11
fused Al_2O_3/ ZrO_2[e]					16
fused Al_2O_3	[1344-28-1]		12		21
SiC	[409-21-2]		13	14	24
boron carbide	[13069-32-8]		14		27
cubic boron nitride	[10043-11-5]				46
diamond	[7782-41-3]	10	15	42.5	78

[a]Ref. 10.
[b]Ref. 11.
[c]At a 100-g load (K-100) average.
[d]To convert kN/m^2 to kgf/mm^2 divide by 0.00981.
[e]39% ZrO_2 (NZ Alundum).

and, because most modern abrasives fall between 9 and 10, there is insufficient delineation. Ridgeway and co-workers (10) modified Mohs' scale by giving garnet a hardness value of 10 ($H = 10$) and making diamond 15. Woodell (11) extended the scale even further by using resistance to abrasion, where diamond equals 42.5. This method is dynamic and less affected by surface hardness variations than the other methods, which involve indentation.

Knoop developed an accepted method of measuring abrasive hardness using a diamond indenter of pyramidal shape and forcing it into the material to be evaluated with a fixed, often 100 g, load. The depth of penetration is then determined from the length and width of the indentation produced. Unlike Woodell's method, Knoop values are static and primarily measure resistance to plastic flow and surface deformation. Variables such as load, temperature, and environment, which affect determination of hardness by the Knoop procedure, have been examined in detail (12).

A linear relationship exists between the cohesive energy density of an abrasive (13) and the Woodell wear resistance values occurring between corundum ($H = 9$) and diamond ($H = 42.5$). The cohesive energy density is a measure of the lattice energy per unit volume.

2.2. Toughness. An abrasive's toughness is often measured and expressed as the degree of friability, the ability of an abrasive grit to withstand impact without cracking, spalling, or shattering. Toughness is often considered a measure of resistance to fracture and given the symbol K_c. This value is directly related to the load on an indenter required to initiate cracking and leads to a brittleness index defined as hardness/K_c (14).

A practical industry friability test (15) for abrasives involves careful sizing of subject grains to pass a given sieve size while being retained on the next finer screen. A unit weight of this grain is then ball-milled using a standard steel ball load for a given time. The percentage of milled grain retained on the original screen is a measure of toughness or lack of friability. Other methods of evaluating this property involve centrifugally impacting sized grits and then evaluating the debris (16).

2.3. Refractoriness (Melting Temperature). Instantaneous grinding temperatures may exceed exceptionally high temperatures at the interface between an abrasive and the workpiece being ground (17). Hence, melting temperature is an important property. Additionally, for alumina, silicon carbide, B_4C, and many other materials, hardness decreases rapidly with increasing temperature (18). Fortunately, ferrous metals also soften with increasing temperatures and do so even more rapidly than abrasives (19).

2.4. Chemical Reactivity. Any chemical interaction between abrasive grains and the material being abraded affects the abrasion. Endurance scratch tests made on polished glass and iron rolls using conical grains of aluminum oxide and silicon carbide (20) showed that silicon carbide produced a long scratch path on the glass roll and a short path on the steel roll. Exactly the opposite was true for aluminum oxide. These effects are explained by the reactivities of the two abrasives toward glass and steel. Silicon carbides resist attack by glass but readily dissolve in steel, whereas aluminum oxide may be attacked by glass and is relatively inert to steel. The advent of boron carbide, harder that either fused aluminum oxide or silicon carbide, brought grand hopes for its use in grinding wheels and belts. Boron carbide's ease of oxidation and its reactivity toward both metals and ceramics prevented wide commercialization.

2.5. Thermal Conductivity. Abrasive materials may transfer heat from the cutting tip of the grain to the bond posts, retaining the heat in a bonded wheel or coated belt. Fused zirconium oxide has a relatively low thermal conductivity compared to other abrasive materials. It also has a lower hardness than aluminum oxide, yet it performs quite well on hard-to-grind materials. This finding is attributed in part to the decreased heat flow from the grinding interface into the grain (whose hardness decreases as its temperature rises) and to the bond (subject to heat degradation).

2.6. Fracture. Fracture characteristics of abrasive materials are important, as well as the resulting grain shapes. Equiaxed grains are generally preferred for bonded abrasive products and sharp, acicular grains are typically preferred for coated abrasives. How the grains fracture in the grinding process determines the wear resistance and self-sharpening characteristics of the wheel or belt.

2.7. Microstructure. Crystal size, porosity, and minor phases play an important role in determining the fracture characteristics and toughness of an abrasive grain. As an example, an alumina sol–gel abrasive grain containing zirconia or a spinel as a minor phase promotes toughness for heavy-duty grinding applications, whereas the same composition without the minor phase is more friable and thus tends to be more suited for medium-duty grinding.

3. Abrasive Materials

3.1. Silicon Carbide. The first artificial abrasive was silicon carbide [409-21-2], SiC, which was produced by a process developed by Edward Acheson in 1891 (5). In this process, silicon carbide is produced from quartz sand and carbon in a large electric furnace in which the charge acts as the refractory container and thermal insulator for the ingot being formed. Reaction temperature range from 1800 to 2200°C, melting the quartz sand, which then reacts with the solid carbon to form crystalline silicon carbide. There are two basic types of silicon carbide: one is gray or black in color and the other is green. The green form is somewhat purer, slightly harder, but is more friable. It is particularly suited for the grinding of tungsten carbides; the black form is commonly used on cast iron, nonferrous metals, and nonmetallic materials. In 2000, the United States and Canada produced 45,000 metric tons of crude silicon carbide valued at $26,300,000.

3.2. Fused Aluminum Oxide. Fused aluminum oxide [1344-28-1] was the next manufactured abrasive to be produced (6,7). By 1900, the first commercially successful fused alumina of controlled friability was produced from bauxite [1318-16-7]. The Norton Company obtained the rights and patent in 1901 and constructed the first plant to produce this material on a commercial basis. In 1904 the Higgins furnace, which used a water-cooled steel shell instead of a refractory lining, was first used (9). A typical Higgins furnace is ~ 2 m in diameter by 2 m in height, producing an ingot of ~ 5.5 ton. The fusion and slow cooling of a mixture of bauxite, coke, and iron turnings gives a coarse crystalline product of about 95% alumina and 0.7% titania [13403-67-7], designated regular aluminum oxide. By adding more coke to the charge, greater reduction is obtained and the percentage of residual titania is reduced, producing a semifriable alumina (~97% alumina). As the name implies, this material is less tough than the regular alumina. After crushing, further heat treatment in rotary furnaces is used to alter the valence state and solubility of titania, producing an exsolved dispersed phase that affects the impact strength of the resulting product. Most aluminum oxide is now fused in tilting furnaces and poured into ingots of sizes suitable for the desired rate of cooling and resulting crystal size. Small ingots cool rapidly and produce microcrystalline alumina, which is tougher and stronger than regular.

Bayer alumina (see ALUMINUM COMPOUNDS), containing ~0.5% soda as its only significant impurity, is also the starting material for the production of fused white aluminum oxide abrasive. During the fusion process much of the soda is volatilized, producing small bubbles and fissures in the final product and giving a slightly less dense, much more friable abrasive than regular or semifriable aluminum oxide. This white abrasive is widely used in tool grinding as well as in other applications requiring cool cutting, self-sharpening, or a damage-free workpiece. Special pink or ruby variations of the white abrasive are produced by adding small amounts of chromium compounds to the melt. The chromium addition improves the suitability of white abrasive for tool and precision grinding.

In 2000, 90,000 tons of regular fused alumina and 10,000 tons of high purity fused alumina were produced in the United States and Canada, valued at U.S. $33 and 5.4 million, respectively (21).

3.3. Fused Alumina–Zirconia. The performance of fused alumina abrasives can be greatly improved for many grinding applications by the formation during the fusion process of a eutectic alloy with zirconium oxide [1314-23-4]. Extremely rapid cooling of the melt, achieved by casting into a bed of steel balls or between steel plates, is important to obtain optimum properties of this grain. Typical applications include the production of large resin-bonded snagging wheels for the heavy-duty conditioning of steel slabs and billets and for weld-bead removal in pipeline construction. This grain is also used extensively in coated abrasive applications, but cannot be used in vitrified bonded wheels because of thermal instability. There are two principal varieties of alumina–zirconia abrasives: a near eutectic combination of 40% ZrO_2 and 60% Al_2O_3, and a less costly 25% ZrO_2 and 75% Al_2O_3 (22, 23).

3.4. Sol-Gel Sintered Aluminum Oxide. Since the early 1980s, sol–gel technology has been used to improve the performance of aluminum oxide abrasives and has had a major impact on both the coated and bonded abrasive business. Sol–gel processing permits the microstructure of the aluminum oxide to be controlled to a much greater extent than is possible by the fusion process. Consequently, the sol–gel abrasives have a crystal size several orders of magnitude smaller than that of the fused abrasives and exhibit a corresponding increase in toughness (24).

In the sol–gel process, a colloidal dispersion or sol is first formed by dispersing synthetic boehmite (aluminum oxide monohydrate) [1344-28-1] in acidified water. There are two types of modifiers that may next be added to the sol that will subsequently permit densification of the abrasive to occur during the final sintering step. The first type of modifier is a water-soluble salt (preferably a nitrate) and during sintering these modifiers form a secondary metal oxide phase within the alpha alumina matrix. In some instances, the metal oxide modifiers react with alumina, eg, $MgAl_2O_4$ or $MgLaAl_{11}O_{19}$. Examples of common modifiers include magnesia, silica, zirconia, yttria, rare earth metal oxides, nickel oxide, and titania. The second type of modifier is a seed particle that is isostructural with alpha alumina such as alpha iron oxide, alpha chromic oxide, or alpha alumina itself (25). These seed particles function as a template to facilitate the transformation from a transitional alumina into alpha alumina or corundum, the hard abrasive form of aluminum oxide (26).

After the addition of modifiers, the sol is then dried to a soft, friable solid, which may then be easily crushed into the desired particle sizes. Alternatively, the sol may be partially dried into a gel state that may then be extruded into rods or other controlled shapes. The final step is to fire the particles at temperature ranging from 1200 to 1500°C (see Fig. 1).

The 3M Company commercialized the first sol–gel abrasive under the trade name Cubitron in 1981 (27). There are several types of sol–gel abrasive grains manufactured by 3M. Cubitron 321 grain (28) contains magnesium and certain rare earth oxides that form a distinct platelet phase. This platelet phase increases the toughness of the grain (29). Cubitron 321 grain is particularly suited for use in grinding wheels and coated abrasives under heavy duty grinding conditions and on stainless steel and exotic alloys. Another Cubitron (30) grain is seeded but does not contain the spinel phase. This grain is used in low to moderate pressure grinding of metal, paint, and wood.

Saint-Gobain abrasives also produces a sol–gel abrasive under the name Cerpass and SG (31). This grain uses alpha alumina seed crystals to produce a microstructure containing a submicron growth of alpha alumina. SG and Cerpass abrasive grains are used in both bonded and coated abrasive applications and is also produced in a rod form for use in specialized grinding wheels (32).

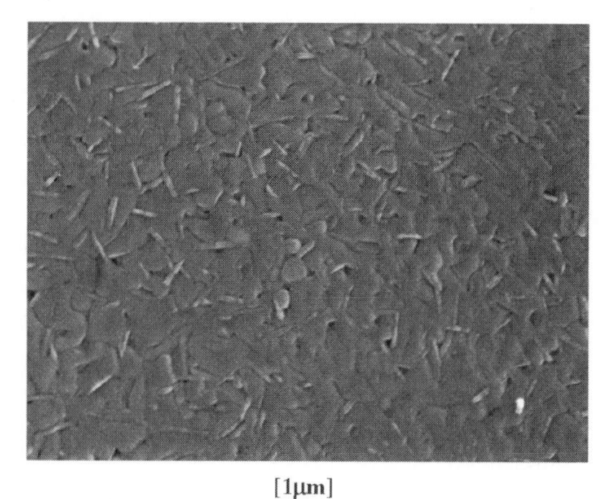

[1μm]

Fig. 1. SEM photomicrograph of polished and thermally etched section of Norton SG sol–gel alumina abrasive grain.

3.5. Diamond. Diamond [7782-40-3] is the hardest substance known with a Knoop hardness value of 78 – 80 kN/m^2 (8000 – 8200 kgf/m^2). The next hardest substance is cubic boron nitride with a Knoop value of 46 kN/m^2. The United States was the world's largest consumer of both natural and synthetic industrial diamond in 2000 with an estimated usage of 484 million carats (1 g = 5 carats). Of this amount, $> 90\%$ was synthetic diamond (21).

Abrasive applications for industrial diamonds include their use in rock drilling, as tools for dressing and trueing abrasive wheels, in polishing and cutting operations (as a loose powder), and as abrasive grits in bonded wheels and coated abrasive products.

3.6. Cubic Boron Nitride. Cubic boron nitride [10043-11-5] or CBN, is a synthetic mineral not found in nature. It was first produced by the General Electric Company using the same equipment used for the production of synthetic diamond and is designated as "Borazon". CBN is nearly as hard as diamond, yet it does not perform as well in the usual diamond grinding applications such as abrading ceramics, rock, or cemented tungsten carbide. However, CBN is an extremely efficient abrasive for grinding steel (33). Although its cost is comparable to that of synthetic diamond, it successfully competes with the inexpensive fused aluminum oxide in grinding steel tools. CBN improves grinding wheel life by as much as 100 times over that of alumina, thus increasing productivity, reducing downtime for the wheel changes and dressing, and improving the quality of parts (34).

3.7. Boron Carbide. Boron carbide [12069-32-8], B_4C, is produced by the reaction of boron oxide and coke in an electric arc furnace (70% B_4C) or by that of carbon and boric anhydride in a carbon resistance furnace (80% B_4C) (see Boron compounds; Refractory boron compounds). It is primarily used as a loose abrasive for grinding and lapping hard metals, gems, and optics (35). Although B_4C is oxidation prone, the slow speed of lapping does not generate enough heat to oxidize the abrasive.

3.8. Metallic Abrasives. Metallic abrasives are most commonly used as a blast medium to clean or to improve the properties of metallic surfaces. Industries that utilize metallic abrasives include foundries, steel manufacturers, machine tool industries, and metal working plants. In 2000, ~269,000 metric tons of steel shot and cut wire shot were produced in the United States with an estimated value of $125 million (21).

3.9. Natural Abrasives. With the introduction of the first manufactured abrasives at the beginning of the twentieth century, the use of natural abrasives has steadily declined.

Diamond. Natural diamonds account for ~10% of the industrial market (21). The two types of natural diamonds used by industry include diamond stone (typically coarser than 60 mesh/800 μ) and diamond bort (smaller, fragmented pieces). Abrasive applications for industrial diamonds include their use in rock drilling, as tools for dressing and trueing abrasive wheels, in polishing and cutting operations (as a loose powder), and as abrasive grits in bonded wheels and coated abrasive products.

Garnet. Garnet [12178-41-5] is the name given to a group of silicate minerals possessing similar physical properties and crystal forms but differing in chemical composition. Seven species exist but the two most important are pyrope, a magnesium aluminum silicate, and almandine, an iron aluminum silicate. Industrial uses for garnet include the lapping, manufacturer of coated abrasives, hydrocutting, and the finishing of wood, leather, hard rubber, felt and plastics. In 2000, the domestic production of crude garnet was 60,200 tons with a value of >$7 million (21).

Silica. In the past, silica has found wide use as an abrasive, particularly as an inexpensive coated abrasive for woodworking. The term sandpaper is still used as a generic term for coated abrasives although the use of sand in coated abrasives has very little commercialize presence.

Tripoli. Tripoli [1317-95-9] is a fine grained, porous, decomposed siliceous rock produced mainly in Arkansas, Illinois, and Oklahoma. It is widely used for polishing and buffing metals, lacquer finishing, and plated products. Since Tripoli particles are rounded, not sharp, it has a mild abrasive action particularly suited for polishing. Tripoli is also used in toothpastes, in jewelry polishing, and as filler in paints, plastics, and rubber. Rottenstone and amorphous silica are similar to Tripoli and find the same uses. In 2000, the abrasive use of Tripoli in the United States totaled 73.3 million tons and was valued at ~ $3.8 million (21).

4. Sizing, Shaping, and Testing of Abrasive Grains

4.1. Sizing.

Manufactured abrasives are produced in a variety of sizes that range from a pea-sized grit of 4 (5.2 mm) to submicron diameters. It is almost impossible to produce an abrasive grit that will just pass through one sieve size yet be 100% retained on the next smaller sieve. Thus a standard range was adopted in the United States that specifies a screen size through which 99.9% of the grit must pass, maximum oversize, minimum on-size, maximum through-size, and fines. The original Bureau of Commerce size standards, although non-metric, have been internationally recognized. These standards have been updated by ANSI Standard B74.12-1982 in the United States (36) and by FEPA Standard 42-GB-1984 in Europe (37). Table 2 shows the average

Table 2. **Bonded Abrasive Grit Sizes**

Grit	Diameter, μ	Grit	Diameter, μ	Grit	Diameter, μ
4	5200	24	775	120	115
5	4500	30	650	150	95
6	3650	36	550	180	80
7	3050	40	460	220	69
8	2550	46	388	240	58
10	2150	54	328	280	48
12	1850	60	275	320	35
14	1550	70	230	400	23
16	1300	80	195	500	16
18	1100	90	165	600	8
20	925	100	138	1200	3

diameter of grit sizes ranging from 4 to 1200. Designations for the finest grit sizes vary in the United States, Europe, and Japan.

The permissible variation in the openings of U.S. standard test sieves varies from 15% for the coarser sizes to 60% for the range of 200–400 grit. To reduce this built-in error, the diamond industry has developed precision, electroformed test screens that are produced by a combination of photoengraving and electroplating. These screens, in which the accuracy and uniformity of aperture size can be tightly controlled, are too expensive for routine testing; they are used instead to calibrate standard wire sieves (38). The sizing of diamond abrasive grains is much tighter that that of other abrasives; details can be found in (39).

4.2. Shaping. Screening is a two-dimensional (2D) process and cannot give information about the shape of the abrasive particle. Desired shapes are obtained by controlling the method of crushing and by impacting or milling. Shape determinations are made optically and by measuring the loose-packed density of the abrasive particles; cubical-shaped particles pack more efficiently than acicular-shaped particles. In general, cubical particles are preferred for grinding wheels, whereas high aspect ratio acicular particles are preferred for coated abrasive belts and disks.

4.3. Testing. Chemical analyses are done on all manufactured abrasives, as well as physical tests such as sieve analysis, specific gravity, impact strength, and loose poured density (a rough measure of particle shape). Special abrasives such as sintered sol–gel aluminas require more sophisticated tests such as electron microscope measurement of alpha alumina crystal size, and indentation microhardness.

5. Coated Abrasives

5.1. Manufacture. In a broad sense, a coated abrasive is defined as a plurality of abrasive particles adhered to a substrate, commonly called a backing. There are several typical constructions in this coated abrasive family, namely, conventional coated abrasives, lapping film, and structured abrasives. A conventional coated abrasive comprises a backing with a first binder called a make coat; the make coat adheres the abrasive particles to the backing. It is generally preferred to orient the abrasive particles, such that the majority of abrasive particles are substantially perpendicular to the backing. Over and in between the abrasive particles is a second binder, called a size coat; the size coat reinforces the abrasive particles. There may optionally be a third coating, called a supersize coating applied over the size coating. This supersize coating typically comprises a grinding aid or antiloading additive. A second coated abrasive construction is commonly referred to as "lapping film", where a plurality of abrasive particles are randomly dispersed in a binder. This abrasive particle/binder composite is applied over the front surface of the backing. Likewise, these products are typically employed in precision polishing applications, where previously loose abrasive slurries were used. Recently, a new coated abrasive construction has emerged, named a structured abrasive, which comprises a plurality of shaped abrasive composites are bonded to a backing.

5.2. Sizes. Coated abrasive articles are commercially in abrasive particle sizes ranging from grit sizes ANSI 16 (particle size of $\sim$1180 μ) to finer than JIS 10,000 (particle size $<$0.5 μ). Likewise, coated abrasives come in a plethora of converted forms including belts, disks, sheets, rolls, flap wheels, and mop wheels. Coated abrasive belts may range in size from 1 to $>$200 cm width and lengths 5 to $>$500 cm. For example, 1 $\times$ 30-cm belts are employed in a portable tool to grind the inside corner of a sink. Larger belts, for example stroke sanding belts, are 15 wide $\times$ 450-cm long are employed in furniture markets. Coated abrasive discs come in sizes from 1 to $>$100 cm in diameter. The 1-cm coated abrasive disk may be employed to remove defects from a painted surface. Coated abrasives made with a vulcanized fiber backing are used in leveling welds on a surface. Lapping film in the form of a daisy shaped disc is used to polish ophthalmic surfaces.

5.3. Backings. There are several types of popular coated abrasive backings: paper, cloth, vulcanized fiber, polymeric film, and foam-like substrates. Paper backings are designated by their basis weight. The lightweight papers tend to be more flexible and are used in sheet and disk products. The heavier papers are stiffer and are typically used in belts, and disk products.

Fourdrinier paper is the most common type. This paper may be treated to seal the front surface, make the product waterproof, or modify some other physical property of the paper. During the 1920s a major advancement in coated abrasives was the invention and development of waterproof coated abrasives (40). This enabled operators to sand automobile painted surfaces wet to reduce the airborne dust. Eighty years later, waterproof paper backings are a mainstay in coated abrasives.

Another large volume backing is cloth. Cloth typically is much stronger than paper, while still being flexible. The majority of cloth is employed in coated abrasive belts, with a small percentage in disks and sheet rolls. The cloth may be woven or stitch bonded (41). Like paper, cloth backings are designated by their weights.

Typical, cloth fibers used are cotton, polyester, rayon, nylon, or combinations thereof. Rayon tends to be used when it is desired to have a flexible coated abrasive belt, eg, in polishing the contours of cast parts such as golf clubs and hand tools. Conversely, polyester and nylon fibers are commonly utilized in heavy-duty applications such as lumber planing and metal de-gating. Cotton fibers are versatile; cotton fibers are employed in a range of applications from general purpose belt grinding to coated abrasive disk sanding. The cloth is typically treated to seal the backing and modify the physical properties of the cloth. These treatments are either a thermoplastic or thermosetting resin, along with optional additives. Typical thermosetting polymers include: phenolics, epoxies, acrylates, aminoplast resins, urethane, latexes, melamine formaldehyde, and mixtures thereof. There are numerous cloth treatments; the cloth may be first saturated to fill in the intersices of the cloth. The cloth may contain one or more coating(s) over its front surface, knows as a presize or subsize coating. The presize coating may provide a smooth coating onto which the abrasive particles are coated. The cloth may contain a backing coat on the backside of the backing. The backsize coating may serve to protect the cloth fibers during abrading. These treatments seal the cloth, enhance the heat resistance of the cloth, and modify the cloth to the desired flexibility or stiffness. For example, in mold sanding of intricate furniture it is necessary to have a flexible cloth backing. Conversely, in lumber planning it is desired to have a stiff backing.

A third common backing is vulcanized fiber. This backing is essentially exclusively used in coated abrasive disks, ranging in size from 5 to 25 cm in diameter. These fiber disks are employed extensively to remove weld lines, deburring, shaping, and other metal fabrication operations. Fiber disks are commonly used in coarse grades, where metal removal and life tend to be more important than surface finish.

A fourth backing is polymeric film, such as polyester, polypropylene, and polyamide films; a popular film backing is a primed polyester film. Film backings have advantages in that they provide a very smooth front surface for coating abrasive particles and have a very uniform surface finish. Polymeric films are commonly employed in precision polishing with fine grade abrasive particles. In polishing applications, this backing uniformity and smooth surface may translate into a more consistent surface finish.

Another abrasive backing is a foam or other compressible backing. The abrasive particles may be coated directly onto the foam or laminated to a foam. The foam may be an open- or closed-cell polyurethane. The foam thickness may range from 0.2 to 20 mm. These foam backings are generally employed in hand sanding applications and some disk applications.

5.4. Resin Binders. In the majority of coated abrasive constructions, the binder is a thermosetting polymer. During manufacturing, a thermosetting resin is coated and exposed to an appropriate energy to initiate the curing of the resin into a cross-linked polymer. Examples of typical resins include resole phenolic, epoxy, acrylate, urea–formaldehyde, melamine formaldehyde, urethane, aminoplast resins having pendant alpha, beta unsaturated groups and mixtures thereof. For heavy duty coated abrasive applications, for many years, thermally cured resins such as phenolic resins, epoxy resins, urea–formaldehyde resins, and urethane resins were popular. Coated abrasive articles can use resins that are cured upon exposure to a radiation energy source such as electron beam or ultraviolet (uv) light, these resins include epoxy, acrylates, acrylated epoxies, acrylated urethanes, and mixtures thereof.

Phenolic resins are considered to be the workhorse of coated abrasives and widely used in heavy duty-to-medium duty grinding applications. Coated abrasive products utilizing phenolic based binders were first commercialized in the 1930s (42). There are two types of phenolic resins: resole and novolac. Resole phenolic have the ratio of formaldehyde to free phenol >1, where in novolac phenolics this ratio is <1. Resole phenolic resins are traditionally employed in coated and non-woven abrasives, whereas novolac phenolic resins are employed in bonded abrasives. Resole phenolic resins polymerize via a condensation mechanism. During phenolic polymerization, water is released. In the coated abrasive manufacturing, the oven temperatures are controlled such that not too much water is released in a short time so as to result in defects in the cross-linked phenolic binder. Resole phenolic resins are typically polymerized on temperatures $<175°C$, usually $<125°C$. Phenolic resins have excellent physical properties in regard to abrasive articles. They are relatively low cost, hard, heat resistant, and tough. One major drawback of phenolic resins is the processing. Phenolic resins, along with other thermally curable resins, use a festoon type oven to cure the resin into a cross-linked polymer. After coating, the abrasive web is hung on racks and transported through a large oven. It is common for this festoon cure to take anywhere from 30 to 300 min to advance the phenolic resin into a "B" stage type cure to the make and size coats. During this festoon cure, the coated web is carefully handled to maintain a good abrasive grain coating. After the festoon cure, the resulting product is wound into a jumbo to either be converted or further thermal cure.

In the medium to finer grades, other resins are commercially used including epoxy, urea–formaldehyde, plasticized phenolic, and urethane resins. These resins can be thermally cured using a festoon type oven. In some instances, the made and size coats have different polymer chemistries to achieve a desired abrasive characteristic.

In some light duty product, hyde glue is used for the make and/or size coat. Hyde glue is not a thermosetting polymer, but rather it is coated out of water and subsequently dried. These products cannot be used wet, only dry sanding conditions. Hyde glue binders are generally softer in comparison to thermosetting binders. The hyde glue-type binders are traditionally used in light duty sanding where a fine surface finish is desired.

Recently, radiation curable resins have emerged, the radiation energy source can be electron beam, uv light on visible light. These binder systems are rapidly cured in comparison to phenolic resins that are usually festoon cured. Likewise it is possible to blend a radiation curable binder with a thermally curable binder (43). There are two common types of uv light curable resins, cationically cured epoxy resins and free radically cured-based acrylate resins. In some instances, these resins may be mixed together or in the same molecule (44). For visible and uv light binders, a photoinitiator is added to the resin. Upon exposure to the light, the photoinitiator cleaves and generates a free radical, which initiates the polymerization process.

5.5. Additives to Binder Systems. The abrasive article binder system may further contain additives that modify the polymer physical properties polymer and/or positively affect the abrading performance of the resulting abrasive article. Examples of such additives including fillers, grinding aids, antiloading additives, coupling agents, dyes, pigments, wetting agents, plasticizers, and combinations thereof.

Fillers are commonly used in bonded and coated abrasives, especially in the coarse-to-medium grades. Fillers are particulate matter, generally softer and smaller in size than the abrasive particles. Fillers offer at least one of the following advantages, lower cost, change binder system friability, increase binder system toughness, hardness or, heat resistance. Fillers that tend to inorganic particulates have a particle size between 5 and 120 μ. Examples of popular fillers include metal carbonates (eg, calcium carbonate), silicates (eg, calcium metasilicate), silica (eg, glass beads, glass bubbles), graphite, carbon black, mica, and the like.

Grinding aids are another class of additives generally preferred in dry metal grinding applications. It is theorized that grinding aids will either (*1*) reduce the grinding temperature, (*2*) act as a lubricant, and/or (*3*) minimize the amount of metal chips from rewelding to the abrasive grains. Examples of popular grinding aids include metal halide salts (eg, potassium tetrafluoroborate, sodium aluminum fluoride), poly(vinyl chloride), iron pyrites, waxes, halogenated waxes, and phosphate salts (45). Typically, the grinding aid is included in the outermost layer of the abrasive article. In the case of bonded abrasives, the grinding aid may be incorporated into the abrasive particle and binder mix. In coated abrasives, the grinding aid is typically dispersed in a binder and incorporated into either the size or supersize coating.

Where grinding aids are generally preferred in dry metal grinding, antiloading materials are sometimes preferred in dry paint, wood sanding. During sanding of soft substrates, the debris removed can become lodged in between adjacent abrasive particles. This phenomena, known as "loading" may reduce the cutting ability and ultimately shorten the coated abrasive life. It is theorized that nonloading additives may either act as a lubricant, inhibit the debris from adhering to the abrasive article outer surface and/or the anti-loading material flakes off during abrading and take the debris along with it. Common examples of anti-loading materials include metal salts of fatty acids, graphite, mica, talc, and fluoro chemicals. The most popular materials are metal stearates such as lithium stearate, calcium stearate, and zinc stearates (46). These metal stearates are typically dispersed in a binder and applied as a supersize coating.

5.6. Coating. Regarding typical coated abrasive manufacturing, the backing is first coated with a make coat resin to the desired weight. Immediately afterward, the abrasive grains are either drop coated or electrostatically coated into the make coat resin. The electrostatic coating process tends to orient the abrasive grains. In some instances, there may two different abrasive grains coated. The first abrasive grain may be drop coated and the second abrasive grain electrostatically coated. Alternatively, both types of abrasive grains may be electrostatically coated simultaneously. Next the make coat resin is either dried or exposed to an energy source to at least partially cross-link the resin. Following this, a size coat resin is coated over the abrasive grains to the desired weight. Likewise, the size coat is either dried or exposed to an energy source to cross-link the resin. Similarly, if desired, a supersize coating is applied and solidified. After the binders are cured, the resulting coated abrasive may be flexed to break the thermosetting polymers in a controlled manner. This flexing process alters the flexibility and abrasive performance of the product.

5.7. Structured Abrasives. In recent years, a new coated abrasive construction has emerged, known as a structured abrasive. The structured abrasive comprises a plurality of shaped abrasive composites adhered to a backing; these shaped abrasive composites may be precisely or irregularly shaped (47). These shapes may be any geometric shape such as pyramidal, ridge-like, hemisphere, cube-like, and block-like (48). Structured abrasives offer flexibility in designing and optimizing the abrasive article in that both the composite shape, size, and density, along with the abrasive composite chemistry may be optimized to achieve superior performance. Structured abrasive products have gained acceptance in medium grinding to very fine polishing applications.

To make a structured abrasive article, a slurry is first prepared comprising a plurality of abrasive grains dispersed in a resin, along with optional additives. A production tool is generated comprising cavities having the desired shape, density, and size of abrasive composites. For precisely shaped abrasive composites, this abrasive slurry is coated onto the backing and the resulting construction is brought into contact with the production tool. The abrasive slurry flows into the cavities of a production tool. Next, an energy source, such as uv light, is transmitted through the production tool and into the resin. The resin is at least partially cross-linked and the resulting abrasive slurry is solidified to form a plurality of abrasive composites bonded to a backing (49). For nonprecisely shaped abrasive composites, the abrasive slurry is coated into the cavities of a production tool, such as a rotogravure roll. The backing is brought into contact with the abrasive slurry and the rotogravure roll imparts a pattern to the abrasive slurry. The abrasive slurry is removed and then exposed to an energy source to at least partially cure the resin to form the structured abrasive (50).

6. Lapping Papers and Films

6.1. Manufacture.

Lapping papers and films differ from other coated abrasives in several regards. The products are made by casting a slurry of abrasive grain and binder on the backing. The slurries are usually not 100% solids, but are either aqueous or solvent based. In addition to the binders and the abrasive grain they also frequently contain modifying agents, such as dispersants, antistats, wetting agents, catalysts, etc.

The common binder classes employed are polyester resins, acrylates, epoxies, and phenolics, as well as blends of two or more resin classes. In most cases, the solvent is dried from the binder systems and cured. Some systems, however, merely dry the slurry leaving an uncured binder film to hold the abrasive grain to the backing. (The binder systems without solvent, of course, always require a cure.)

The abrasive grain is frequently described in microns, designating the mean particle size, rather than the ANSI, FEPA, or JIS grading systems employed with coarser coated products. The abrasive grains are almost invariably $<60\ \mu$, and most commonly $< 10\ \mu$ mean particle size. The abrasive grain types vary from very hard, such as diamond, to the conventional coated abrasive materials (ie, silicon carbide, fused alumina, sol–gel aluminas), to softer abrasive grains like ceria, rouge, etc. As with other forms of coated abrasives, the coated article may contain a single abrasive grain or a blend of abrasive grains.

The backings used for coated lapping products may be a paper or other nonwoven cloth, a woven cloth, or a polymeric film. Of these choices, polymeric films are by far the most common as the film provides a smooth uniform base, in turn allowing precise control of the roughness of the abrasive coating. The most common polymeric film used is poly(ethylene terephthalate). The thickness of the film varies with the end use application, but is most commonly between 0.5 and 3.0 mil ($\sim$12.5–75 μ).

The slurries may be coated onto the backing via a wide variety of coating techniques; common techniques are knife coating, knife over roll coating, gravure roll coating, three/four roll coating, and extrusion coating. The drying is accomplished with conventional forced air ovens. The rate of the drying, as well as the formulation, is frequently controlled in order to affect the presence or absence of Benard Cells, and hence the surface roughness in the dried product.

6.2. Applications. The lapping products are used in a host of fine polishing applications. They are used in tape format for the polishing of large industrial calender and mill rolls, for the "buffing" of both floppy and rigid memory disks, for the contouring of magnetic heads, etc. Disks and sheets are used for the polishing of fiber optic connectors, metallurgical laboratory sample preparations, defect removal/polishing of glass, the fining of ophthalmic lenses, die polishing, etc. The specific application, of course, determines the thickness of the backing, the type and size of the abrasive grain, and the class of the binder system. Many applications, such as fiber optic polishing, are sequential and involve steps with two, three, or four progressively finer and less aggressive lapping products.

7. Nonwoven Abrasives

Nonwoven abrasives are unique forms of abrasives that find use in many aspects of material finishing and surface cleaning. This category of abrasives provides a different interaction with the workpiece than coated abrasives or grinding wheels, and are, therefore, commonly called surface conditioning abrasives. A great deal of study has gone into the cause/effect relationship of these abrasives with a surface. Much evidence exists suggesting that these articles generally provide compressive, rather than tensile stresses, on surfaces. Nonwoven abrasives also create a very clean surface that enhances long-term corrosion prevention and superior wetting properties for subsequent coating.

Nonwoven abrasives are so named because of the random fibrous matrix on which they are based. This matrix provides a sturdy backbone on which can be coated a variety of resins and minerals in such a way as to provide the desired surface finish for many diverse types of articles. These abrasives are available as disks, belts, pads, brushes or wheels, and can be adapted for use on many power tools as well as large industrial cleaning equipment.

As mentioned above, the compressive stresses are imparted by the interaction of the springy nonwoven matrix with the surface being worked upon. This spring-like action works as a tiny peening hammer on the surface. What this does is to relieve the tensile stresses induced by the actions of machining metals or heavy stock removal using grinding wheels or coated abrasives. Also, adsorbed surface contaminants, carbonaceous deposits, or surface segregated components are effectively removed using nonwoven abrasives. The very light surface abrasion removes undesirable surface contaminants or chemistry that may affect coating adhesion or corrosion resistance. This is useful when a very clean and uniformly wettable surface is desired prior to priming or similar processes. Actual stock removal is very minimal, but burrs and sharp corners are effectively removed and rounded by the spring action of the matrix. Effective removal of burrs and sharp edges is also desirable for long-term protection against fatigue failure and corrosion resistance.

7.1. Manufacture. The first step in the manufacture of nonwoven abrasives is the creation of the carrier web. A mechanical carding, melt bonded, or air laid process in which crimped polymeric fibers are laid down on a carrier belt and passed through a variety of resin baths to give desired characteristics forms this web. The coated web is then passed through a drying oven to provide structural integrity. This web is commonly called "prebond", as it does not have any abrasive added at this time. In some cases, an additional scrim layer may be mechanically bonded to provide even greater structural integrity.

In the second step, the "make coat" is applied, which can be a two-step resin/abrasive process or can be applied as a one-step resin/abrasive slurry. In the former, the prebond web is passed through a resin bath and the abrasive can be gravity applied or blown through the web using an air stream. In the latter process, the slurry can be roll coated or applied by spraying. Many of the webs that are produced by either of these two methods can then be converted into the usable forms listed above.

In order to create wheels, another resin coat, called the "size coat", is added. In this case, a "make coated" web is passed through a resin bath and lightly oven cured to produce a tacky web. This web is then treated in one of two ways to make wheels. In one process, winding the size-coated web into a roll that is cured before wheels are sliced from it makes "convolute" wheels. These tightly wound rolls can have a variety of densities depending on the degree of conformity desired. In the second process, "unitized" wheels are cut from slabs that have been compacted and cured under pressure. Density is a function of the number of layers and resins used for a given thickness.

Ganging together nonwoven rings on a suitable core material makes large industrial "brushes", 8–80 in. in width. Alternatively, long strips of nonwoven material can be gathered radially around a core to form "flap brushes". These assemblies are compressed to form a brush. The resulting brush is then surface dressed and balanced before being put into service. They are used primarily for the surface cleaning of coiled metals in the removal of rust, carbon, and other surface contaminants.

7.2. Nonwoven Abrasives Sales. Nonwoven abrasives were commercialized in the 1960s (51). Nonwoven abrasives have rapidly emerged into three major markets: consumer, commercial, and industrial markets.

8. Bonded Abrasives

Grinding wheels are by far the most important bonded abrasive product both in production volume and utility. They are produced in grit sizes ranging from 4, for steel mill snagging wheels, to 1200, for polishing the surface of rotogravure rolls. Wheel sizes vary in diameter from tiny mounted wheels for internal bore grinding to > 1.8-m wheels for cutting-off steel billets. Bonded shapes other than wheels, such as segments, cylinders, blocks, and honing stones, are also widely used.

8.1. Marking System. Grinding wheels and other bonded abrasive products are specified by a standard marking system that is used throughout most of the world. This system allows the user to recognize the type of abrasive, the size and shaping of the abrasive grit, and the relative among and type of bonding material. The individual symbols chosen by each manufacturer may vary, but the relative position for each item in the marking system is standard.

Grain Size. The surface finish, or degree of roughness, produced by a grinding wheel on the workpiece being ground is roughly proportional to the size of abrasive grains in the wheel (see the section Sizing and Table 2).

Grade. The grade, or grinding hardness, of a bonded abrasive product is determined by the bond content. It is represented alphabetically with A being the softest (least bond) and Z being the hardest (most bond).

Structure. The structure designation is a numeric indication of the relative volume of abrasive in a unit volume of wheel. A low number indicates a high volume of abrasive grain, thus the grains are closely packed. As the structure number increases there is less abrasive grain in the wheel, and the abrasive grains tend to be more widely spaced.

8.2. Bond Type. Most bonded abrasive products are produced with either a vitreous (glass or ceramic) or a resinoid (usually phenolic resin) bond. Bonding agents such as rubber, shellac, sodium silicate, magnesium oxychloride, or metal are used for special applications.

Vitrified (Glass or Ceramic) Bond. Vitrified wheels probably account for about one half of all conventional abrasive wheels. Vitrified bonds are formed from mixtures of clay, feldspar, and frit. The clays and feldspars are naturally occurring materials and the amounts of these materials are mainly determined by the nature of the wheel to be built, but also affected by the mineralogy and detailed chemistry of the clays and feldspars. The frit is synthetic glass and its composition is under better control. The bond mixtures soften and melt in the temperature range from 950 to 1400°C, with mixtures richer in clay melting at higher temperatures, those with more frit melting at lower temperatures.

It is possible to prepare bond mixtures with different viscosities, and hence different surface tensions at a given temperature, and so tailor the bond to the required structure of the final wheel. In particular, it becomes possible to help control the porosity in, and provide the strength to, the wheel by careful choice of bonding mix. Vitrified wheels may be compounded to include organic "spacer" materials. These organic materials are lost during the firing step and thus would not show up in the finished wheel. However, these organic materials can have a significant effect on the structure, properties, and grinding performance of the wheels.

Resin (Resinoid). The resinoid bond, originally called Bakelite, was named for its inventor, Leo Bakeland (see PHENOLIC RESINS). Bakeland's original patent was issued in 1909, but it was not until the 1920s that this type of bonding was perfected for use in abrasives. The resin consists basically of phenol and formaldehyde. It is a thermosetting polymer, with a combination of properties that make it particularly suitable for a number of grinding operations. It withstands heat better than rubber or shellac, and resists thermal and mechanical shock better than vitrified bonds. A drawback of resinoid bond is its susceptibility to degradation by water. Phenolic resins may be modified with other resin types to reduce brittleness and extend characteristics of toughness and strength.

Resin bonds commonly cure at temperatures in the range of 150–200°C, so there are no problems with grain reactivity, and fiberglass and metal reinforcements may be molded to make a much safer product for high speed or abusive applications. This low processing temperature also allows the use of inert fillers to strengthen the bonded product, or of "active" fillers to increase the efficiency of grinding. Active fillers, also known as grinding aids, include such materials as cryolite, pyrites, potassium fluoroborate, sodium and potassium chloride, zinc sulfide, antimony sulfide, and tin powder. Such materials, alone or in combination, aid grinding by acting as extreme pressure lubricants or as reactants for the metal being ground; this prevents rewelding of the chips being removed.

Resinoid-bonded wheels find wide use in heavy-duty snagging operations, where large amounts of metal are generally removed quickly; in cutting-off operations; in portable disk grinding (as for weld beads); in grinding of steel mill rolls; and in vertical spindle disk grinding.

Rubber. Both natural and synthetic rubber are used as bonding agents for abrasive wheels. Rubber-bond wheels are ideal for thin cut-off and slicing wheels and centerless grinding feed wheels. They are more flexible and more water- resistant than resinoid wheels. In manufacture, the abrasive grain is mixed with crude rubber, sulfur, and other ingredients for curing, then passed through calender rolls to produce a sheet of desired thickness. The wheels are stamped from this sheet and heated under pressure to vulcanize the rubber.

Shellac. Shellac wheels are limited to a few applications where extreme coolness of cut is required and wheel life is immaterial. They are produced by mixing shellac [9000-59-3] and abrasive grain in a heated mixer, then rolling or shaping to the desired configuration.

Magnesium Oxychloride. A mixture of abrasive grains, MgO, water, and $MgCl_2$, placed in an appropriate mold will cold-set to form a grinding wheel that is then cured for a long period of time in a moist atmosphere. This type of bond finds some use in disk grinding applications and cutlery grinding. These wheels have been largely replaced by soft-acting resinoid bonds.

9. Special Forms of Bonded Abrasives

There are many specialized forms and uses of bonded abrasives, a detailed discussion of which is found in reference (52).

9.1. Honing and Superfinishing. Honing and superfinish stones are produced from large vitrified-bonded abrasive blocks that are diamond sawed to smaller rectangular pieces suitable for mounting in metal or plastic holders. Honing stones, used to true engine and hydraulic cylinders, can vary in grit size from 36 (0.55 mm) to 600 (8 μ); superfinish stones, used to polish the external diameters of machine and automotive parts, vary in grit size from 600 to 1200 (3 μ). Both types of stones are quite soft; steel ball indentation hardness and density are often used as quality control to measure grade.

9.2. Pulpstone Wheels. Grinding wheels play an important role in the production of paper pulp. Massive pulpstone wheels are made from vitrified abrasive segments, bolted and cemented together around a reinforced concrete central body. They may be up to 1.8 m in diameter and have a breadth of 1.7 m. In operation, debarked wood logs are fed into a machine and forced against the rotating pulpstone, which shreds the wood into fibers under a torrent of water. The ground fibers are then screened and passed through subsequent operations to produce various types of paper.

9.3. Crush-Form Grinding. In crush-form grinding, a rotating, contoured crushing wheel is forced into the face of a revolving vitrified wheel, crushing the face to the exact contour needed on the metal object to be ground. The contoured wheel is then placed in production and when wear or dulling occurs, the face is again crushed to regain proper contour. Many parts formerly turned with metal-cutting tolls and then surface ground are now shaped and surface finished in one pass of a crush-formed wheel.

9.4. Creep Feed Wheels. Creep feed grinding is an abrasive machining process, characterized by the use of slow (creep) workpiece velocities and extremely large depths of cut that are hundreds or even thousands of times greater than those in regular grinding applications. With this process, it may be possible to grind complex profiles or deep slots in only a few or even a single pass. A shape is generated in the face of an open-structure vitrified wheel by diamond tooling or a crush form roller. The profiled wheel is fed into the part under a flood of coolant.

Creep feed wheels are manufactured by incorporating in the wheel mixture a substance that is burnt-off during the firing process to leave voids throughout the wheel structure. By this means, the level of porosity can be increased from 35% of wheel volume for a conventional wheel to ~60% for some of creep feed wheels. This additional porosity is ideal for the creep feed process, as the voids can be saturated with coolant prior to entering the grinding zone so that a large volume of coolant can be taken into the grinding zone by the wheel. This is important since the contact arc is much longer than for reciprocating grinding. Also, as the feed rate is comparatively low the heat generated can easily build up in the work piece. This must not be allowed to happen, so the heat must be removed rapidly from the grinding zone and it is essential to have an efficient supply of coolant throughout the zone.

As well as making coolant application difficult, the long arc of contact also has the effect of promoting the formation of wear flats on the grinding grits. These wear flats give rise to rubbing forces between the wheel and work piece, and additional power is required to provide this rubbing energy. The size of the wear flats is dependent on the grade of the wheel. The softer the grade, the sooner the force on a grit will build up to a level that will either fracture the grit or remove it from the bond. Thus, a soft wheel will give a lower wear flat area, and hence lower rubbing energy than a hard wheel. It is thus usual to choose a wheel with a soft grade for creep feed grinding (53).

For significant improvements in grinding performance, soft grade wheels containing thermally conductive solid particles are utilized. In these grinding wheels, the thermally conductive particles are held by the vitreous matrix with a binding force (or strength) weaker than the strength of the bond between the abrasive grain and the vitreous matrix. Thus, the thermally conductive particles are generally more readily lost from the grinding wheel during the grinding of a work piece and act as heat sinks to conduct heat away from the grinding zone (interface between the grinding wheel and work piece during grinding).

When these thermally conductive particles are separated from the grinding wheel, they distribute and dissipate the heat in and from the grinding wheel to thereby assist in reducing and preventing the risk of a burn of the metal work piece and thermally induced breakdown of the grinding wheel (54).

10. Superabrasive Wheels

10.1. Diamond Wheels. Little natural diamond is used in grinding wheels today. Synthetic diamonds are manufactured in many versions that have been tailored for specific grinding applications and can be produced with excellent quality control. Diamonds with more perfect crystal structure are used in metal bonds for sawing stone and concrete and for grinding glass and ceramics. Diamonds for grinding of carbide are designed to be multicrystalline, giving the diamond the property of friability. These friable diamonds are used in resin bonds. To improve the retention of the diamond, the grains can be coated with metal, usually nickel. The metal adheres well to the grain, and the specially controlled rough surface of the coating adheres very well to the resin. This reduces pullout of the crystal from the matrix. Other metals (copper, silver) are used to improve the heat transfer properties of the wheel, keeping the resin cooler when heat of grinding is a concern.

In addition to resin and metal, vitrified bonds are also used. Resinoid wheels are hot pressed to zero porosity to maximize grain retention. Grit sizes range from 16 to 1800. Wheels are constructed with the diamond in a thin rim on the periphery of the wheel body, or core, which may be metal- or fiber-filled plastic, steel, aluminum, ceramic, or bronze. Intricate shapes are produced by electroplating a single layer of superabrasive on a shaped core.

The amount of superabrasive in the wheel rim is expressed as concentration. A wheel with 100 concentration contains 25% by volume of superabrasive crystal. Concentrations usually range from 50 to 200. As concentration is increased wheel cost increases, but tool life will increase and finish will improve.

Diamond wheels are not recommended for grinding ferrous materials. Under the temperature and pressure conditions of grinding, diamond (carbon) is chemically soluble in iron; thus diamond will wear excessively and be uneconomical.

10.2. Cubic Boron Nitride Wheels. Although CBN is only about two-thirds as hard as diamond, it is about twice as hard as aluminum oxide and silicon carbide abrasives. This property, plus its non-carbon chemistry and its high thermal conductivity, make it an excellent abrasive for steel. As with diamond, wheels are made using metal, resin, and vitrified bonds and as plated products. The porous nature of a vitrified bonded wheel makes it particularly suitable for automated, high production grinding; when trued, the vitrified wheel is ready to grind. It is not necessary to dress the wheel, ie, to remove some of the bond and expose the abrasive. Like diamond, most CBN used with resin bond is coated with metal to increase grain retention.

CBNs high thermal conductivity helps prevent heat buildup in the part. This reduces the chance of work-piece metallurgical damage, and may improve the quality of the ground surface by leaving it in a neutral or compressive state. When the surface is heated then cooled, as when grinding with aluminum oxide (a thermal insulator), it may be in tension, leading to cracking and failure.

11. Metal-Working Fluids

Metal-working fluids may be divided into four subclassifications: metal forming, metal removal, metal treating, and metal protecting fluids. In this discussion, the focal point will be on metal removal fluids, which are those products developed for use in applications, such as grinding and machining, where the material, typically metal, is removed to manufacture a part.

11.1. Types of Fluids. Metal removal fluids are generally categorized as one of four product types: (*1*) straight (neat) oils, (*2*) soluble oils, (*3*) semisynthetics, or (*4*) synthetics. The distinctive difference between each type is based mainly on two formulation features: the amount of petroleum oil in the concentrate and whether the concentrate is soluble in water. Straight oil as defined by Childers (55) is petroleum or vegetable oil used without dilution. Straight oils are often compounded with additives to enhance their lubrication and rust inhibition properties. Straight oils are used "neat" as supplied to the end user.

Soluble Oil. Soluble oil (or emulsifiable oil) is a combination of oil, emulsifiers, and other performance additives that are supplied as a concentrate to the end user. A soluble oil concentrate generally contains 60–90% oil. They are diluted with water, typically at a ratio of 1 part concentrate to 20 parts water or 5% (56). When mixed with water they have an opaque, milky appearance. They generally are considered as general purpose fluids, since they often have the capability to be used with both ferrous and nonferrous materials in a variety of applications.

Semi-Synthetic Fluids. These fluids have much lower oil content than soluble oils. The concentrate typically contains 2–30% (56) oil. When mixed with water, characteristically at a ratio of 1 part concentrate to 20 parts water or 5%, the blend will appear opaque to translucent. These fluids usually have lubricity sufficient for applications in the moderate-to-heavy-duty range (ie, centerless, internal, and creep feed grinding). Their wetting and cooling properties are better than soluble oils, which allow for faster speeds and feed rates.

Synthetic Fluids. These fluids contain no mineral oil. Most synthetic fluids have a transparent appearance when mixed with water. There are some synthetic fluids that are categorized as synthetic emulsions that contain no mineral oil, but appear as an opaque, milky emulsion when mixed with water. Synthetic fluids have the capability to work in applications ranging from light (ie, double disk, surface) to heavy duty (ie, creep feed, threading). Synthetic fluids generally are low foaming, clean, and have good cooling properties allowing for high speeds and feeds, high production rates, and good size control.

11.2. Functions of a Grinding Fluid. Grinding fluids provide two primary benefits, namely, cooling and lubrication.

Cooling. A tremendous amount of heat is produced in the metal removal process making it important to extract that heat away from the part and the wheel. Dissipating heat from the work piece eliminates temperature related damage to the part such as finish and part distortion. Removing heat from the grinding wheel extends wheel life while preventing burning and smoking. The metal removal fluid carries away most (96%) of the input energy with its contact to the work piece, chips, and grinding wheel. The input energy ends up in the fluid where it will be transferred to its surroundings by evaporation, convection, or in a forced manner, by a chiller. Methods for cooling metal-working fluid are discussed in detail by Smits (57).

Lubrication. Fluids are formulated to provide lubrication that reduces friction at the interface of the wheel and the part. The modes of lubrication are described as being physical, boundary, or chemical. Physical lubrication in metal removal fluid is provided by a thin film of a lubricating component. Examples of these components may be a mineral oil or a nonionic surfactant. Boundary lubrication occurs when a specially included component of the metal removal fluid attaches itself to the surface of the work piece. Boundary lubricants are polar additives such as animal fats and natural esters. Chemical lubrication occurs when a constituent of the fluid (ie, sulfur, chlorine, phosphorous) reacts with a metallic element of the work piece, resulting in improved tool life, better finishes, or both. These additives are known as extreme pressure (EP) additives.

In addition to the primary functions performed by a fluid as previously described, there are other functions required from a fluid. These include, providing corrosion protection for the work piece and the machine, assisting in the removal of chips or swarf (build up of fine metal and abrasive particles) at the wheel work piece interface (grinding zone), transporting chips and swarf away from the machine tool, and to lubricate the machine tool itself.

11.3. Application of Metal-Working Fluids. Understanding the application of metal removal fluids to the grind zone is an important aspect of the grinding process. Metal removal fluids are held in a fluid reservoir and pumped through the machine to a fluid nozzle, which directs the fluid to the grinding zone. Smits (57) describes the various nozzle types: Dribble, Acceleration Zone, Fire Hose, Jet, and Wrap Around. Smits (57) discusses how fluid flow rate, fluid speed entering the flow gap, fluid nozzle position, and grinding wheel contact with the work piece all influence the results of the grinding process. The fluid leaves the cut zone and flows back to the fluid reservoir where it should be filtered before being circulated back to the cut zone.

12. Summary

Abrasive articles and abrading processes are interwoven into a vast variety of manufactured goods produced today. Long-term sustained sales and sales growth in the abrasive industry will depend on many factors. Over the years, manufacturers through improvements in their production practices have reduced the inherent demand of abrasive articles. Near net manufacturing, eg, has reduced the amount of coarse grade abrasives needed for metal degating. Conversely, increased demand for finer and more consistent surface finishes has contributed to the growth of finer abrasive articles.

What does the future hold for abrasives with its strong dependence on other manufactured goods? Quite simply the abrasive's future may be intertwined with continued technology development in abrasive articles and abrading processes with the goal to lower the overall manufacturing cost. As becomes evident form this article, there are a variety of technologies affiliated with abrasives; these technologies include organic chemistry, polymer science, inorganic chemistry, glass chemistry, and ceramics and related process technologies.

Advances in abrasive technologies will most likely be achieved through advances in these related fundamental technologies. Thus, achieving higher abrasive efficiencies and correspondingly lower abrasive costs through technology development should inherently advance the abrasives industry.

BIBLIOGRAPHY

"Abrasives" in *ECT* 3rd ed., Vol. 1, pp. 26–52 by, W. G. Pinkstone; in *ECT* 4th ed., Vol. 1, pp. 17–37, by Charles V. Rue, Norton Company; "Abrasives" in *ECT* (online), posting date: December 4, 2000, by Charles V. Rue, Norton Company.

CITED PUBLICATIONS

1. A. H. Baumgartner, *Theophrastus von den Steinen aus dem Griechischen*, Nurnberg 210, 1770.
2. L. Coes, Jr., *Abrasives*, Springer-Verlag, New York, 1971, p. 2.
3. Hindu proverbs "Diamond cuts Diamond," and "The heart of a magnate is harder that diamond." For other ancient Hindu references to diamond see also The Hindu Vedas (1100–1200 BCE) and Brhatsanhita (sixth century).
4. "Like a diamond, harder than flint, I have made your forehead..." Ezekeil 3:9 (New World Translation of the Holy Bible, Watchtower Bible and Tract Society, New York).
5. U. S. Pats. 492,767; 527,826; 650,291; 615,648; 718,891; 718,892; 722,792; 722,793; and 723,631, E. G. Acheson.
6. Fr. Pat. 233,996 (1893), I. Werlein.
7. Ger. Pat. 85,021 (1894), F. Hasslacher.
8. U.S. Pat. 677,207, C. M. Hall.
9. U.S. Pat. 775,654 (1904), A. C. Higgins.
10. R. R. Ridgeway, A. H. Ballard, and B. L. Bailey, *Trans. Electrochem. Soc.* **63**, 369 (1933).
11. C. E. Woodell, *Trans. Electrochem. Soc.* **68**, 111 (1935).
12. J. T. Czernuska and T. F. Page, *Proc. Br. Ceram. Soc.* **34**, 145 (1984).
13. J. N. Plendl and P. J. Gielisse, *Phys. Rev.* **125**, 828 (1962).
14. B. Lamy, *Trib. Int.* **17**(1), 36 (1984).
15. "Procedure for Friability of Abrasive Grain," *ANSI Standard B74.8-1987*, American National Standards Institute, New York, 1987.
16. See Ref. 2, pp. 155–156.
17. W. J. Sauer, in M. C. Shaw, ed., *New Developments in Grinding, Proceedings of the International Grinding Conference 1972*, Carnegie Press, Carnegie-Mellon University, Pittsburgh, Pa., 1972, pp. 391–411.
18. See Ref. 2, p. 55.
19. See Ref. 2, pp. 152–153.
20. L. Coes, Jr., *Ind. Eng. Chem.* **47**, 2493 (1955).
21. D. W. Olson, "Manufactured Abrasives", in *Minerals Yearbook 2000*, Vol. 1, *Metals and Minerals*, U.S. Dept. of the Interior, p. 5.2.
22. U.S. Pat. 3,891,408 (June 1975), R. A. Rowse and G. R. Watson (to Norton Company).
23. U.S. Pat. 5,009,676 (Nov. 1976), J. J. Scott (to Norton Company).
24. W. Konig, Th. Ludewig, and D. Stuff, *Produk. Management* **85**, 22 (1995).
25. U.S. Pat. 4,744,802 (May, 1988), M. G. Schwabel (to 3M).
26. D. D. Erickson, T. E. Wood, and W. P. Wood, *Ceramic Trans.* **95**, 73 (1999).

27. U.S. Pat. 4,314,827 (Feb. 1982), M. A. Leitheiser and H. Sowman (to 3M).
28. U.S. Pat. 4,881,951 (Nov. 1989), W. P. Wood, L. D. Monroe, and S. L. Conwell (to 3M).
29. D. D. Erickson and W. P. Wood, *Ceramic Trans.* **46**, 463 (1994).
30. U.S. Pat. 5,611,829 (March 1997), L. D. Monroe and T. E. Wood (to 3M).
31. U.S. Pat. 4,623,364 (Nov. 1986), T. E. Cottringer, R. van de Merwe, R. Bauer (to St. Gobain).
32. U.S. Pat. 5,009,676 (April 1991), C. V. Rue, R. H. Van de Merwe, R. Bauer, S. W. Pellow, T. E. Cottringer, and R. J. Klok (to Norton Company).
33. R. H. Wentorf, Jr., *1986 Proceedings of the 24th Abrasive Engineering Society Conference*, Abrasive Engineering Society, Pittsburgh, Pa., 1986, pp. 27–31.
34. *Norton CBN Wheels, Form 4800 LPBXM 8-83*, Company Bulletin, Norton Abrasives, Saint-Gobain Corporation.
35. P. Harbin, *Ind. Min.* 49 (Nov. 1978).
36. *ANSI Standard B74.12-1982 (macro sizes) and ANSI Standard B74.10-1977* (R1983) (micro sizes), American National Standards Institute, New York, 1982 and 1977.
37. *FEPA Standard 42-GB-1984*, British Abrasive Federation, London, 1984.
38. B.T.G. O'Carroll, *Ind. Diamond Rev.* **4**, 129 (1973).
39. *A Review of Diamond Sizing and Standards*, IDA Bulletin, Industrial Diamond Association of America, Columbia, S.C., 1985.
40. U.S. Pat. 1,565,027 (Dec. 1925), F. G. Okie (to 3M).
41. U.S. Pat. 4,867,760 (Sept. 1989), W. L. Yarborough (to St. Gobain).
42. U.S. Pat. 2,310,935 (Feb. 1943), R. P. Carlton and B. J. Oakes (to 3M).
43. U.S. Pat. 4,903,440 (Feb. 1990), E. G. Larson, and A. Kirk (to 3M).
44. U.S. Pat. 4,751,138 (June 1988), M. L. Tumey, D. W. Bange, and A. F. Robbins (to 3M).
45. U.S. Pat. 5,441,549 (Aug. 1995), H. J. Helmin (to 3M).
46. U.S. Pat. 2,768,886 (Oct. 1956), J. F. Twombly (to Norton).
47. U.S. Pat. 5,152,917 (Oct. 1992), J. R. Peiper, R. M. Olson, M. V. Mucci, G. L. Holmes, and R. V. Heiti, (to 3M).
48. U.S. Pat. 5,454,844 (Oct. 1995), L. D. Hibbard, S. B. Collins, and J. D. Haas (to 3M).
49. U.S. Pat. 5,435,816 (July 1995), K. M. Spurgeon, S. R. Culler, D. H. Hardy, and G. L. Holmes (to 3M).
50. U.S. Pat. 5,833,724 (Nov. 1998), Wei (to St. Gobain).
51. U.S. Pat. 2,958,593 (Nov. 1960), H. L. Hoover, E. J. Dupre, and W. L. Rankin (to 3M Company).
52. W. F. Schleicher, *The Grinding Wheel*, 3rd ed., The Grinding Wheel Institute, Cleveland, Ohio, 1976.
53. K. B. Southwell, *Development of High Porosity Grinding Wheels for Creep Feed*, International Conference on Creep Feed Grinding, Bristol, 1979, pp. 103–119.
54. U.S Pat. 5,536,282 (July 1996), S. C. Yoon and R. A. Gary (to Milacron Corporation).
55. J. C. Childers, in J. Byers, ed., *Metalworking Fluids*, Marcel Dekker Inc., New York, 1994, pp. 170–177.
56. G. Foltz, *Waste Minimazation and Wastewater Treatment of Metalworking Fluids*, R. M. Dick ed., Independent Lubrication Manufacturers Assoc., Alexandria, Va., 1990, pp. 2–3.
57. C. A. Smits, in J. Byers, ed., *Metalworking Fluids*, Marcel Dekker Inc., New York, 1994, pp. 100–132.

Donna Bange
3M Company
Roger Gary
Milacron Inc.

ABSORPTION

1. Introduction

Absorption, or gas absorption, is a unit operation used in the chemical industry and increasingly in environmental applications to separate gases by washing or scrubbing a gas mixture with a suitable liquid. One or more of the constituents of the gas mixture dissolves or is absorbed in the liquid and can thus be removed from the mixture. In some systems, this gaseous constituent forms a physical solution with the liquid or the solvent. In other cases, it undergoes a chemical reaction with one or more components of the liquid.

The purpose of such scrubbing operations may be any of the following: gas purification (eg, removal of air pollutants from exhaust gases or contaminants from gases that will be further processed), product recovery, or production of solutions of gases for various purposes. Several examples of applied absorption processes are shown in Table 1.

Gas absorption is usually carried out in vertical countercurrent columns as shown in Figure 1. The solvent is fed at the top of the absorber, whereas the gas mixture enters from the bottom. The absorbed substance is washed out by the solvent and leaves the absorber at the bottom as a liquid solution. The solvent is often recovered in a subsequent stripping or desorption operation. This second

Table 1. **Typical Commercial Gas Absorption Processes**

Treated gas	Absorbed gas, solute	Solvent	Function
coke oven gas	ammonia	water	by-product recovery
coke oven gas	benzene and toluene	straw oil	by-product recovery
reactor gases in manufacture of formaldehyde from methanol	formaldehyde	water	product recovery
drying gases in cellulose acetate fiber production	acetone	water	solvent recovery
natural and refinery gases	hydrogen sulfide	amine solutions	pollutant removal
flue gases	sulfur dioxide	water	pollutant removal
	carbon dioxide	amine solutions	by-product recovery
wet well gas	propane and butane	kerosene	gas separation
wet well gas	water	triethyleneglycol	gas drying
ammonia synthesis gas	carbon monoxide	ammoniacal cuprous chloride solution	contaminant removal
roast gases	sulfur dioxide	water	production of calcium sulfite solution for pulping

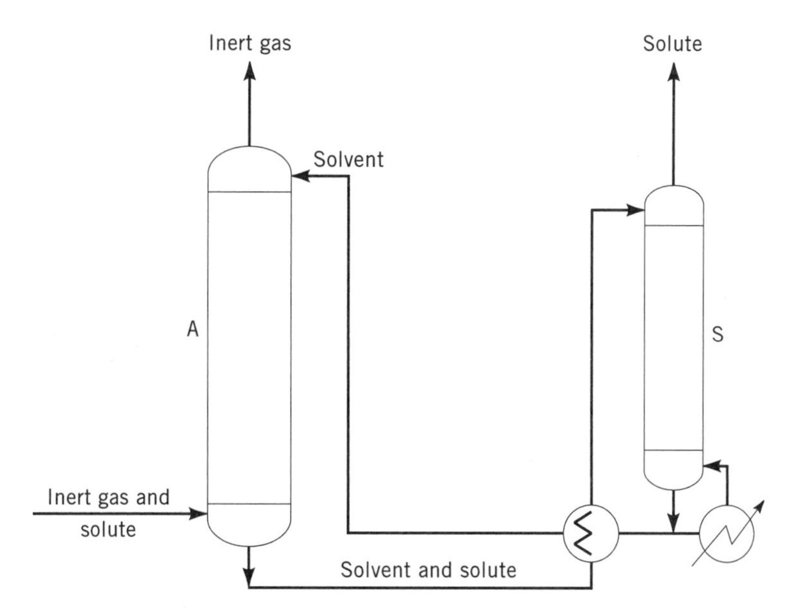

Fig. 1. Absorption column arrangement with a gas absorber A and a stripper S to recover solvent.

step is essentially the reverse of absorption and involves countercurrent contacting of the liquid loaded with solute using an inert gas or water vapor. Desorption is frequently carried out at higher temperatures and/or at lower pressure than the absorption step. The absorber may be a packed column, plate tower, or simple spray column, or a bubble column. The packed column is a shell either filled with randomly packed elements or having a regular solid structure designed to disperse the liquid and bring it and the rising gas into close contact. Dumped-type packing elements come in a great variety of shapes (Fig. 2**a–f**) and construction materials. These elements are intended to create a large internal surface but a small pressure drop. Structured, or arranged packings may be made of corrugated metal or plastic sheets providing a large number of regularly arranged channels (Fig. 2**g**), but a variety of other geometries exists. In plate towers, liquid flows from plate to plate in cascade fashion and gases bubble through the flowing liquid at each plate through a multitude of dispersers (eg, holes in a sieve tray, slits in a bubble-cap tray) or through a cascade of liquid as in a shower deck tray (see DISTILLATION).

The advantages of packed columns include simple and, as long as the tower diameter is not too large, usually relatively cheaper construction. These columns are preferred for corrosive gases because packing, but not plates, can be made from ceramic or plastic materials. Packed columns are also used in vacuum applications because the pressure drop, especially for regularly structured packings, is usually less than through plate columns. Tray absorbers are used in applications where tall columns are required, because tall, random-type packed towers are more prone to channeling and maldistribution of the liquid and gas streams. The sensitivity of packed columns to maldistribution is especially high in applications where the density difference between gas and liquid is small and the

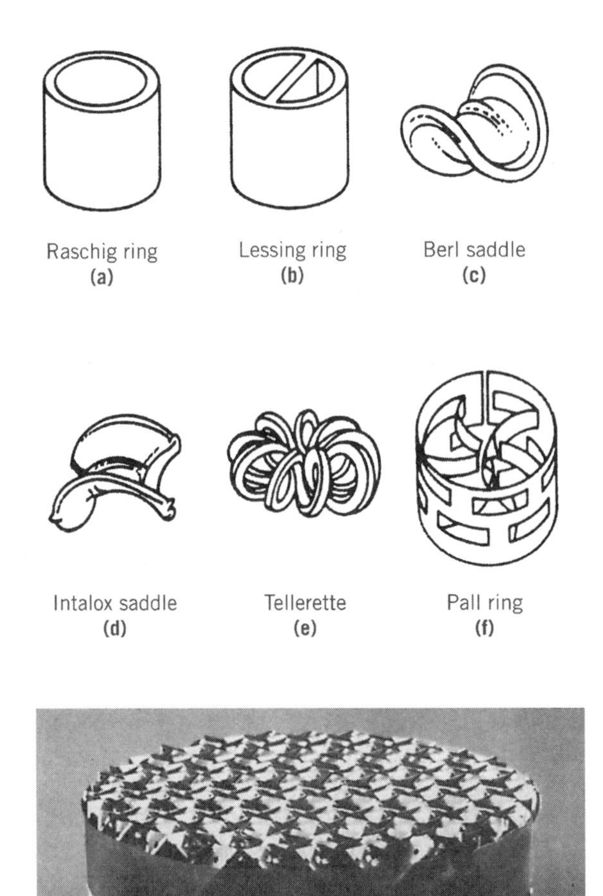

Fig. 2. Packing materials for packed columns. (**a**–**f**) Typical packing elements generally used for random packing; (**g**) example of structured packing. (**g**) Courtesy of Sulzer Bros. S.A. Winterthur, Switzerland.

interfacial tension between the liquid and the gas or vapor is lower than about 0.01 N m^{-1}. However, packings tend to have significantly higher capacity than trays for a given separation duty. This increased capacity makes them useful in revamping applications. On the other hand, plate towers can be more easily cleaned. Plates are also preferred in applications having large heat effects since cooling coils are more easily installed in plate towers and liquid can be withdrawn more easily from plates than from packings for external cooling. Bubble trays can also be designed for large liquid holdup.

The fundamental physical principles underlying the process of gas absorption are the solubility of the absorbed gas and the rate of mass transfer. Information on both must be available when sizing equipment for a given application. Additionally, in the very frequent case of the design of countercurrent columns, it is also necessary to have information on the hydraulic capacity (eg, entrainment, loading, flooding) of the equipment. In addition to the fundamental design concepts based on solubility and mass transfer, many other practical details have to be considered during actual plant design and construction which may affect the performance of the absorber significantly. These details have been described in reviews (1) and in some of the more comprehensive treatments of gas absorption and absorbers (2–5) (see also DISTILLATION; HEAT EXCHANGE TECHNOLOGY).

2. Gas Solubility

At equilibrium, a component of a gas in contact with a liquid has identical fugacities in both the gas and liquid phase. For ideal solutions Raoult's law applies:

$$y_A = \frac{P_s}{P} x_A \tag{1}$$

where y_A is the mole fraction of A in the gas phase, P is the total pressure, P_s is the vapor pressure of pure A, and x_A is the mole fraction of A in the liquid. For moderately soluble gases with relatively little interaction between the gas and liquid molecules Henry's law is often applicable:

$$y_A = \frac{H}{P} x_A \tag{2}$$

where H is Henry's constant. Usually H is dependent upon temperature, but relatively independent of pressure at moderate levels. In solutions containing inorganic salts, H is also a function of the ionic strength. Henry's constants are tabulated for many of the common gases in water (6).

A more general way of expressing solubilities is through the vapor–liquid equilibrium constant m defined by

$$y_A = m x_A \tag{3}$$

The value of m, also known as equilibrium K value, is widely employed to represent hydrocarbon vapor–liquid equilibria in absorption and distillation calculations. When equation 1 or 2 is applicable at constant pressure and temperature (equivalent to constant m in eq. 3) a plot of y vs x for a given solute is linear from the origin. In other cases, the y–x plot may be approximated by a linear relationship over limited regions. Generally, for nonideal solutions or for nonisothermal conditions, y is a curving function of x and must be determined from experimental data or more rigorous theoretical relationships. In a y–x plot, when applied to absorber design this function is commonly called the equilibrium line.

Obtaining reliable data on gas solubility for gas absorber design, in the form of Henry constats, m values or equations for the equilibrium line represents a major issue. Gas solubility has been treated extensively (7). Methods for the prediction of phase equilibria and actual solubility data have been given (8,9) and correlations of the equilibrium K values of hydrocarbons have been developed and compiled (10). Several good sources for experimental information on gas and vapor liquid equilibrium data of nonideal systems are also available (6,11–13). Several of these sources contain not only experimental data but also include models for correlating experimental data and for gas solubility prediction are based on equilibrium phase thermodynamics, involving equation of state and fugacities or Gibbs excess functions.

In addition to these phenomenological approaches, there exists a novel route based on statistical mechanics for the prediction of gas–liquid equilibrium based on a molecular description of gas and liquid. The field of molecular simulation (14–16) has advanced sufficiently in recent years so as to allow the computation of thermophysical data and multiphase equilibrium behavior. The key idea of such methods is to describe two or more phases in thermodynamic equilibrium by an atomistically detailed collection of interacting molecules. These simulation boxes contain a relatively small number of molecules (typically between a hundred and a million), which are endowed with proper geometry and interaction potentials (also known as the force field). A large number of individual configurations of these molecules is generated using methods such as molecular dynamics (MD) or Monte Carlo (MC). Macroscopic properties (eg, PvT) are obtained by averaging over the set of individual configurations. Depending on the constraints under which the MD or MC calculation is performed, different statistical mechanical ensembles can be generated. For equilibrium calculations, sophisticated techniques based on the so-called Gibbs ensemble (17) have been developed.

In a typical gas–liquid or vapor–liquid equilibrium simulation, two computational boxes, one representing the gas or vapor, the other representing the liquid, are "brought into equilibrium" by performing a series of MC moves, one of which is a particle exchange move. Although the compositions of the boxes fluctuate along the simulation due to particle interchanges, they do so about well-defined average values. In many cases, these average compositions closely match the values one would obtain in an equilibrium cell in a laboratory. A series of such calculations allows one to construct the full equilibrium line as required for absorber design. This procedure bypasses the need to use an equation of state for gas nonideality and an activity model for the liquid phase. The entire physicochemical complexity of the system is treated at the molecular level and is entirely represented by the geometry and the interaction potentials.

Such molecular techniques allow the prediction of thermophysical properties and phase equilibria under conditions that cannot be achieved easily in practice (eg, very high pressure, highly toxic or carcinogenic substances), or for compounds that have not been synthesized yet. Although they do not replace either traditional phase equilibrium calculations or experimental work, molecular simulations are a useful complement to both. They are particularly valuable in the screening stages of new solvents for absorption. A molecular simulation can frequently help assess the suitability of a new compound or a new mixture of compounds for a given absorption application. They are also useful for new families of solvents or for mixtures of solvents that are not amenable to a group contribution approach, either because they contain a group that has not been tabulated yet or because there are strong interactions between groups that cannot be properly predicted by standard contribution mixing rules.

As an illustrative example a typical phase equilibrium calculation using molecular modeling techniques will be described in some detail. We will consider a binary system composed of a short-chain, linear hydrocarbon of length N_s as the solute (eg, ethane) and a long-chain linear hydrocarbon of length N_l as the solvent. We will consider two particular cases: $N_s = 2$, $N_l = 7$ and $N_s = 5$, $N_l = 150$. In the first case, we will be dealing with the phase equilibrium behavior of the system ethane–n-heptane and therefore with the solubility of ethane in n-heptane. The second system will demonstrate how the same approach can automatically be extended to the calculation of the solubility of low molecular weight compounds (pentane in this case) in molten polymers or high molecular weight solvents.

In standard molecular modeling, alkane molecules are considered classical mechanical systems in which atoms or atom groups, also called united atoms, are represented by interaction sites (see Fig. 3). Along a molecule, all pairs of sites i and j separated by more than three bonds along the chain and all intermolecular sites interact, in one of the simplest cases, via the Lennard–Jones pair potential:

$$V_{ij}^{\mathrm{LJ}}(r) = 4\epsilon_{ij}\left[\left(\frac{\sigma_{ij}}{r_{ij}}\right)^{12} - \left(\frac{\sigma_{ij}}{r_{ij}}\right)^{6}\right]$$

with r_{ij} being the scalar minimum-image distance between sites i and j and ε_{ij} and σ_{ij} being pair-specific interaction constants. In the case of polar molecules,

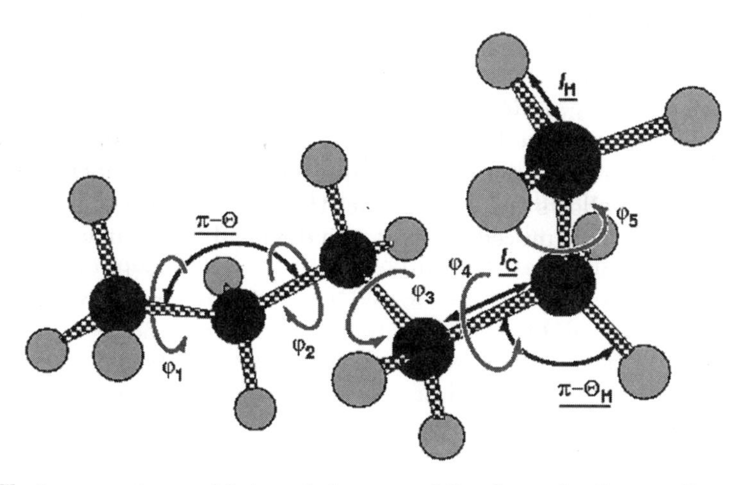

Fig. 3. Chain geometry and internal degrees of freedom of a linear alkane molecule. External degrees of freedom include translation and rotation as a rigid object.

additional electrostatic terms are added, using partial charges associated with interaction sites.

Besides the site–site interaction potential, additional (in the simplest case, harmonic) potentials are associated with bond stretching (18)

$$V_{\text{stretching}}(l) = \frac{1}{2} K_l (l - l_0)^2$$

where l is the bond length and l_0 the equilibrium bond length. Similarly, bond angle bending is endowed with a potential of the form:

$$V_{\text{bending}}(\theta) = \frac{1}{2} K_\theta (\theta - \theta_0)^2$$

where θ is the bond angle and θ is the equilibrium bond angle (19). Lastly, associated with each dihedral angle ϕ is also a torsional potential of the form (20):

$$V_{\text{torsion}}(\phi) = c_0 + c_1 \cos(\phi) + c_2 \cos(\phi)^2 + c_3 \cos(\phi)^3 + c_4 \cos(\phi)^4 + c_5 \cos(\phi)^5$$

where ϕ is a specific torsion angle along the hydrocarbon chain backbone.

Several other terms (contributions) to the energy of a molecular system have been proposed. They represent higher order corrections and cross terms which couple, eg, bond stretching and bending, or account for out-of-plane deviations in planar molecules. In commercial molecular modeling software, the specific functional form of the interaction terms and the set of all parameters appearing in them is known as the "force field" and can contain several thousand parameters. For solubility calculations it is seldom necessary to consider the full set of interaction terms and those presented above are sufficient in many cases.

In order to evaluate each of the previous contributions to the energy of the system, the coordinates of all interaction sites of all molecules present in the system must also be computed. These are determined as a function of the external and internal degrees of freedom. External degrees of freedom are typically the absolute position of one of the sites of each molecule plus the absolute orientation in space of one bond of each molecule, expressed by means of Euler angles or by quaternions, with respect to a fixed laboratory frame of reference. Internal degrees of freedom are usually bond lengths, bond angles and torsional angles.

Once the absolute position and orientation of a reference site and a reference bond of each molecule and the internal degrees of freedom are known, it is possible to determine all site positions for all sites in all molecules using, e.g. the Eyring transfer matrix technique (20). In a last step, in order to eliminate edge effects which would mask the true behavior of the system, periodic boundary conditions, as originally introduced by Born (21) are applied to all site coordinates. Periodic boundary conditions are a necessity, since available computational power limits the characteristic size of systems that can be simulated to a few nanometers at the most.

Once the geometry of all chemical species and the force field describing their interactions are defined, the calculation of phase equilibrium is performed in the Gibbs ensemble, which was succinctly described above. The Gibbs ensemble is a combination of the canonical (*NVT*), the isobaric–isothermal (*NPT*), and the grand-canonical (*NμT*) ensembles. In a molecular modeling Monte Carlo (MC) calculation, moves are performed sequentially in order to generate a Markov chain of system microstates. Macroscopic properties are obtained by simple averaging the corresponding properties for each of the microstates. In a Gibbs ensemble calculation, three types of MC are performed:

- *NVT* moves, in which molecular conformations in each of the simulation boxes are modified by one or more types of moves such as simple displacements, pivots, reptation, Continuous Configurational Bias, End Bridging, etc. (22).
- *NPT* moves, in which the volume of both boxes is changed in such a way that the total volume remains constant, ie, if one of the boxes is expanded by a given amount, the other is shrunk by the same amount.
- *NμT* moves, in which particles are exchanged between boxes, ie, solute and solvent molecules can be transferred from one box to the other.

Each of these different types of moves is accepted or rejected with a certain probability. This probability depends directly (and strongly) on the change in the energy in the system induced by the move but also on the changes in volume and particle number. For example, the probability for a particle exchange (*NμT* move) between boxes, denoted by I and II, to be accepted is given by:

$$min\left[1, \exp\left(\frac{1}{kT}\left[\Delta U^{I} + \Delta U^{II} + kT\ln\left[\frac{V^{II}(N^{I}+1)}{V^{I}N^{II}}\right]\right]\right)\right]$$

where ΔU is the change in the total energy (nonbonded intermolecular and intramolecular, bond stretching, bond bending and torsional) of a box (the superindex labels the box), V is the box volume, and N is the number of particles in the box. Analogous acceptance criteria are used for the other types of move. These criteria are defined so as to fulfill macroscopic equality of temperature, pressure and chemical potential between the boxes, i.e. thermodynamic phase equilibrium.

In a typical phase equilibrium calculation, a large number of MC steps of the three types described above are performed and system microstates (coordinates of atoms) are stored periodically together with properties of each of the simulation boxes, such as a density, potential energy, molar fraction of each component, etc. After a sufficiently large number of such MC steps have been performed, estimates of macroscopic properties, for example density, of each phase is obtained as the average of the densities of all the microstates generated in the MC run. Similarly, the composition of each phase is determined as the average of the compositions of each phase, which are trivially determined from the number of molecules of each species in each box for every microstate. Thus, a Gibbs ensemble calculation produces a pair of thermodynamically equilibrated phases, ie, a pair of points on the phase diagram of the two-component system (extensions to multicomponent systems are straightforward). A series of such calculations yields an equilibrium envelope describing the phase equilibria and therefore the mutual solubility of the two phases. In absorption calculations, the branch of the equilibrium envelope describing the solubility of the light species in the heavy one is the most interesting one, although the other branch is of course useful, eg, in order to determine solvent losses.

A common feature of all molecular modeling techniques is that system non-idealities, high concentration effects, etc, are incorporated automatically and in a natural way in the calculation. In classical, equation of state (EOS)-based methods, saturation pressures, fugacities, activities and high-pressure correction factors enter the calculation separately. In molecular modeling, the full physico-chemical complexity enters via the geometry and forcefield and needs never be split in separate contributions like fugacity, activity coefficients, etc.

The Gibbs ensemble calculation, provided the molecular description and the MC algorithms are correct, directly yields two phases in equilibrium. For this reason it is especially attractive for highly non-ideal systems, high-pressure systems or both. Molecular modeling is however not devoid of adjustable parameters: the values of the constants describing nonbonded interaction and bonded potentials (the force field parameters) are de facto adjustable parameters associated with specific atom types, just like parameters are associated with whole molecules, atom groups or single atoms in activity coefficient models and group contribution techniques. It is, however, generally felt that the force field parameters are better transferable among widely varying chemical families. This is especially justified in cases in which the parameters of the force field stem from a quantum chemical calculation.

Figure 4 shows the vapor–liquid equilibrium envelope (experimental data from (23)) for the system ethane–n-heptane at 58.71 mol% ethane. Symbols

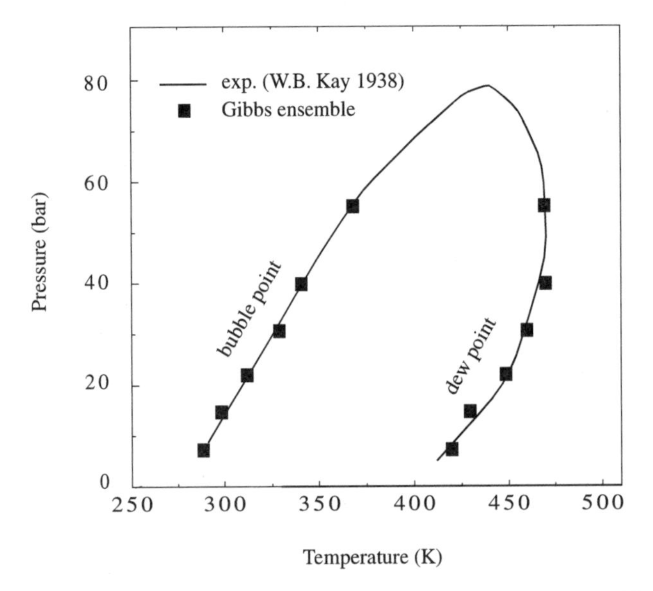

Fig. 4. Comparison of experimental (continuous line) and Gibbs ensemble simulation results for the vapor–liquid equilibrium of the ethane/heptane system. This diagram was constructed from a series of Gibbs ensemble simulations aimed at locating the dew and bubble points of a mixture of given composition. The construction of his diagram requires a great deal more effort than a typical solubility calculation (ie, given pressure and temperature, unknown compositions of the phases) which involves a single Gibbs ensemble simulation.

denote the results of a series of MC calculations in the Gibbs ensemble for the same system ($N_s = 2$, $N_l = 7$).

For this system, the agreement between experimental measurements and Gibbs ensemble calculations is remarkably good. In this relatively simple case, however, an advanced EOS approach (24–26) would also work very well and probably surpass the molecular modeling results in accuracy. The advantages of molecular modeling are more obvious when dealing with highly nonideal, high pressure systems or with new compounds or complex mixtures for which no experimental data are available and group contribution methods are known to be unreliable.

One such situation is the calculation of the solubility of short-chain alkanes (pentane) in molten linear polyethylene, $N_s = 5$, $N_l = 150$. Although hydrocarbon chains with $N_l = 150$ carbon atoms in the backbone are, strictly speaking, not yet polymers but high oligomers, they are however a reasonably good representation of a simple linear polymer such as polyethylene (PE).

Figure 5 shows experimental values of the solubility of pentane in polyethylene at infinite dilution (Henry's constant) and a series of Gibbs ensemble results at increasingly high pressures (27,28). At low pressures the results of Gibbs-ensemble and infinite-dilution data are consistent. At moderate pressures, deviations from Henry's law become significant. Finally, at pressures >20 bar, the polymer reaches the saturation limit and the solubility of pentane does not grow as rapidly as predicted by the linear extrapolation implicit in Henry's law.

These two examples of molecular modeling-based calculation of solubility illustrate the range of applicability of such atomistic techniques to phase equilibrium. However, they are also applicable to the determination of transport

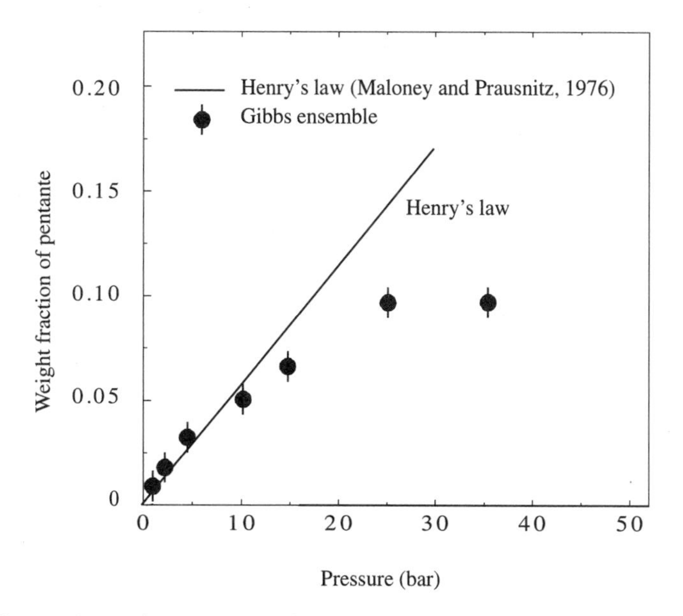

Fig. 5. Comparison of experimental (straight line) and Gibbs ensemble simulation results for the solubility of pentane in polyethylene. As the solvent saturates with pentane at high pressures, solubility deviates from the linearity predicted by Henry's law.

properties. An interesting application of these techniques to absorption technology is in the calculation of molecular diffusivities by means of molecular dynamics (MD). Its detailed treatment is out of the scope of this article but the interested reader can find an extensive description in (14).

It must, however, be kept in mind that the great power and generality of molecular modeling is achieved at the expense of massive computational effort: the calculation of a single pair of data points in the ethane–n-heptane diagram presented above requires several hours of calculation on state-of-the-art hardware, whereas the same calculation can be performed following the standard EOS approach in a few milliseconds. Molecular modeling techniques should therefore be applied judiciously and be reserved for those cases in which conventional methods fail or where their reliability is questionable.

The possibility to test a large number of alternative formulations of mixtures of solvents or pure compounds for their suitability for a particular application has made it possible to actually "design" solvents on a purely computational basis. Needless to say, such "designed" compounds or formulations must be subsequently tested in a laboratory or in a pilot plant in order to verify their effectiveness. Tests have shown molecular calculations to be remarkably successful at predicting solvent performance in many cases (29–31).

User-friendly molecular simulation packages already exist that allow phase equilibria calculations to be set up and performed relatively rapidly and with a modicum of effort on the part of the user. Nevertheless, it must be emphasized that molecular modeling approaches to solubility calculation will remain many orders of magnitude slower than EOS methods for the foreseeable future. It is therefore very doubtful that they will ever be used directly in single column let alone flowsheet simulators, where solubility calculations must be performed a very large number of times. A single MC calculation of gas solubility in a binary system takes nowadays significantly longer than the calculation of an entire plant flowsheet of average complexity. Increases in computer peformance will reduce the time required for MC or MD calculations, but so will they reduce the time required for EOS calculation, so that the ratio of computational effort will very likely remain close to its present value. For this reason, the greatest utility of molecular modeling methods is in generating a database of "computer-experimental" solubility values to which a suitable EOS can be fitted. It is this EOS that finds use in column and flowsheet simulators.

In spite of this computational disadvantage, the molecular computational approach to phase equilibrium and solvent design will probably see increased use in coming years.

3. Mass Transfer Concepts

3.1. Mass Transfer Coefficients and Driving Forces. In order to determine the size of the equipment necessary to absorb a given amount of solvent per unit time, one must know not only the equilibrium solubility of the solute in the solvent, but also the rate at which the equilibrium is established; ie, the rate at which the solute is transferred from the gas to the liquid phase must be determined. One of the first theoretical models describing the process proposed an essentially stable gas-liquid interface (32). Large fluid motions are presumed to exist at a certain distance from this interface distributing all material rapidly and equally in the bulk of the fluid so that no concentration gradients are developed. Closer to this interface, however, the fluid motions are impaired and the slow process of molecular diffusion becomes more important as a mechanism of mass transfer. The rate-governing step in gas absorption is therefore the transfer of solute through two thin gas and liquid films adjacent to the phase interface. Transfer of materials through the interface itself is normally presumed to take place instantaneously so that equilibrium exists between these two films precisely at the interface. Although this assumption has been confirmed in experiments utilizing many systems and different types of phase interface (17,33–36), interfacial resistances can develop in some situations (37–42).

The resulting concentration profile is shown in Figure 6. With the passage of time in a nonflowing closed system, the profiles would become straight horizontal lines as the bulk gas and bulk liquid reached equilibrium. In a flowing system, Figure 6 represents conditions at some countercurrent flow point, eg, at a certain height in an absorption tower where, as gas and liquid pass each other, the bulk materials do not have sufficient contact time to attain equilibrium. Solute is continuously transferred from the gas to the liquid and concentration gradients develop when this transfer proceeds at only a finite rate.

The experimentally observed rates of mass transfer are often proportional to the displacement from equilibrium and the rate equations for the gas and liquid films are

$$N_A = k_G(p_A - p_{Ai}) = k_G P(y_A - y_{Ai}) \tag{4}$$

$$N_A = k_L(c_{Ai} - c_A) = k_L\bar{\rho}(x_{Ai} - x_A) \tag{5}$$

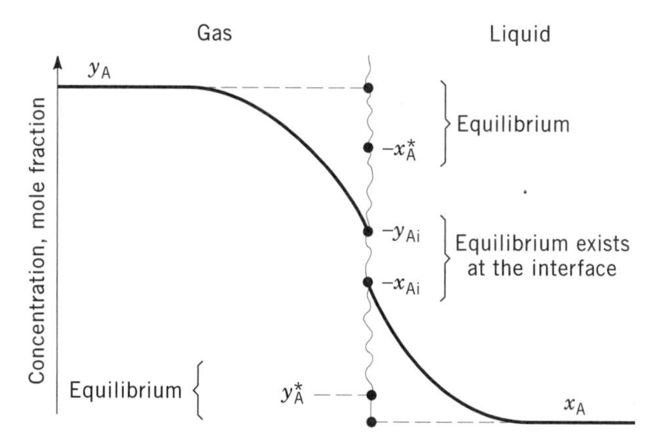

Fig. 6. The two-film concept: y_A and x_A are the concentrations in the bulk of the phases; y_{Ai} and x_{Ai} are the actual interfacial concentrations at equilibrium; y^*_A and x^*_A are the hypothetical equilibrium concentrations which would be in equilibrium with the bulk concentration of the other phase.

where $y_A - y_{Ai}$ and $x_{Ai} - x_A$ are concentration driving forces, k_G is the gas-phase mass transfer coefficient, and k_L is the liquid-phase mass transfer coefficient.

Mass transfer rates may also be expressed in terms of an overall gas-phase driving force by defining a hypothetical equilibrium mole fraction y^*_A as the concentration which would be in equilibrium with the bulk liquid concentration ($y^*_A = mx_A$):

$$N_A = K_{OG} P(y_A - y^*_A) \tag{6}$$

The relationship of the overall gas-phase mass transfer coefficient K_{OG} to the individual film coefficients may be found from equations 4 and 5, assuming a straight equilibrium line:

$$N_A = k_G P(y_A - y_{Ai}) = k_L \bar{\rho} \frac{1}{m} (y_{Ai} - y^*_A)$$

and by comparison with equation 6,

$$\frac{1}{K_{OG}} = \frac{1}{k_G} + \frac{mP}{k_L \bar{\rho}} \tag{7}$$

Expressions similar to equations 6 and 7 may be derived in terms of an overall liquid-phase driving force. Equation 7 represents an addition of the resistances to mass transfer in the gas and liquid films. The analogy of this process to the flow of electrical current through two resistances in series has been analyzed (43).

A representation of the various concentrations and driving forces in a y–x diagram is shown in Figure 7. The point representing the interfacial

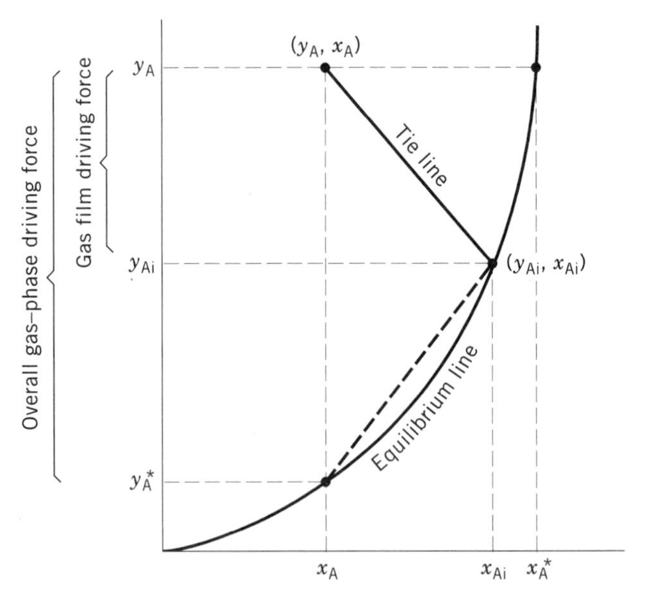

Fig. 7. The driving forces in the y–x diagram.

concentrations (y_{Ai}, x_{Ai}) must lie on the equilibrium curve since these concentrations are at equilibrium. The point representing the bulk concentrations (y_A, x_A) may be anywhere above the equilibrium line for absorption or below it for desorption. The slope of the tie line connecting the two points is given by equations 4 and 5:

$$\frac{y_A - y_{Ai}}{x_A - x_{Ai}} = -\frac{k_L \bar{\rho}}{k_G P} \tag{8}$$

In situations where the gas film resistance is predominant (gas film-controlled situation), $k_G P$ is much smaller than $k_L \rho$ and the tie line is very steep. y_{Ai} approaches y^* so that the overall gas-phase driving force and the gas-film driving force become approximately equal, whereas the liquid-film driving force becomes negligible. From equation 7, it also follows that in such cases $K_{OG} \approx k_G$. The reverse is true if the liquid film resistance is controlling. Since the example depicted in Figure 4 involves a strongly curved equilibrium line, equation 7 is only valid if the slope of the dashed line between x_A and x_{Ai} is substituted for m. Overall mass transfer coefficients may vary considerably over a certain concentration range as a result of variations in m even if the individual film constants stay essentially constant.

3.2. Mass Transfer Coefficients and Molecular Diffusion. Many theories have been put forth to explain experimentally measured mass transfer coefficients and to link them to more fundamental phenomena and parameters. There are two mechanisms that contribute to mass transfer and are reflected in the mass transfer coefficients. As described above (Fig. 6) at some distance from the interface molecules are transported towards or away from the interface mainly by turbulent convection, whereas molecular diffusion dominates mass transfer increasingly as these molecules approach the interface more closely. For analyzing the effects of molecular diffusion it is best to separate it conceptually from turbulent convection by approximating the real situation at the interface by hypothetical stagnant gas and liquid films. This model is proposed by the film theory (see below). The fluid is assumed to be essentially stagnant within these "effective" films making a sharp change to totally turbulent flow where the film is in contact with the bulk of the fluid. As a result, mass is transferred through the effective films only by steady-state molecular diffusion and it is possible to compute the concentration profile through the films by integrating Fick's law:

$$J_A = -D_{AB} \frac{dc_A}{dz} \tag{9}$$

where J_A is the flux of component A relative to the average molar flow of the whole mixture, D_{AB} is the diffusion coefficient of A in B, z is the distance of diffusion, and c_A is the molar concentration of A at a given point in the film. The bulk concentrations are denoted y_{Ab} and x_{Ab} in the following three sections whereas y_A and x_A stand for the concentration at a particular point within the films.

Equimolar Counterdiffusion in Binary Cases. If the flux of A is balanced by an equal flux of B in the opposite direction (frequently encountered in binary distillation columns), there is no net flow through the film and N_A, like J_A, is directly given by Fick's law. In an ideal gas, where the diffusivity can be shown to be independent of concentration, integration of Fick's law leads to a linear concentration profile through the film and to the following expression where $(P/RT)y_A$ is substituted for c_A:

$$N_A = \frac{D_{AB}P}{z_0 RT}(y_{Ab} - y_{Ai}) \tag{10}$$

thus

$$k_G^0 = \frac{D_{AB}}{z_0 RT} \tag{11}$$

where k_G is labeled with a zero to indicate equimolar counterdiffusion and z_0 is the effective film thickness. This same treatment is usually adopted for liquids, although the diffusion coefficients are customarily not completely independent of concentration. Substituting $\bar{\rho} x_A$ for c_A, the result is

$$k_L^0 = \frac{D_A}{z_0} \tag{12}$$

Equations 11 and 12 cannot be used to predict the mass transfer coefficients directly, because z_0 is usually not known. The theory, however, predicts a linear dependence of the mass transfer coefficient on diffusivity.

Unidirectional Diffusion Through a Stagnant Medium in Binary Cases. An ideally simple gas absorption process involves diffusion of only one component through a nondiffusing medium, either the inert gas or the solvent. There exists a net flux of material through the film in this case, and therefore the mixture as a whole is not at rest. The total flux of A is now the sum of the flux with respect to the average flow of the mixture, that is still given by Fick's law of diffusion, plus the flux of A caused by the average bulk flow of the mixture itself:

$$N_A = J_A + y_A \sum_j N_j = -\frac{D_{AB}P}{RT}\frac{dy_A}{dz} + y_A N_A \tag{13}$$

The derivation of this result may be found in various texts (44). Rearranging and integrating equation 13 yields

$$N_A = \frac{D_{AB}P}{z_0 RT}\frac{1}{y_{BM}}(y_{Ab} - y_{Ai}) \tag{14}$$

where y_{BM} is the logarithmic mean of the stagnant gas concentration through the film:

$$y_{BM} = \frac{(1 - y_{Ab}) - (1 - y_{Ai})}{\ln[(1 - y_{Ab})/(1 - y_{Ai})]} \tag{15}$$

Therefore

$$k'_{\mathrm{G}} = \frac{D_{\mathrm{AB}}}{z_0 RT y_{\mathrm{BM}}} \tag{16}$$

For liquids,

$$k'_{\mathrm{L}} = \frac{D_{\mathrm{A}}}{z_0 x_{\mathrm{BM}}} \tag{17}$$

where

$$x_{\mathrm{BM}} = \frac{(1 - x_{\mathrm{Ab}}) - (1 - x_{\mathrm{Ai}})}{\ln[(1 - x_{\mathrm{Ab}})/(1 - x_{\mathrm{Ai}})]} \tag{18}$$

As y_{BM} and x_{BM} are smaller than unity, they predict an increase of mass transfer caused by the average bulk flow of the mixture as a whole. The effect is known as drift flux correction. The values of y_{BM} and x_{BM} are near unity for dilute mixtures.

Any Degree of Counterdiffusion in Binary Cases. In cases where counterdiffusion is not exactly equimolar nor zero, but somewhere in between or even outside these two cases, the influence of bulk flow of material through the films may be corrected for by the film factor concept (45). It is based on a slightly different form of equation 13:

$$N_{\mathrm{A}} = J_{\mathrm{A}} + y_{\mathrm{A}} t_{\mathrm{A}} N_{\mathrm{A}} \tag{19}$$

where

$$t_{\mathrm{A}} \equiv \frac{\sum_j N_j}{N_{\mathrm{A}}} \tag{20}$$

Applying the same derivation as for unidirectional diffusion through a stagnant medium, the results turn out to be

$$k_{\mathrm{G}} = \frac{D_{\mathrm{AB}}}{z_0 RT Y_{\mathrm{f}}} \tag{21}$$

$$k_{\mathrm{L}} = \frac{D_{\mathrm{AB}}}{z_0 X_{\mathrm{f}}} \tag{22}$$

where

$$Y_{\mathrm{f}} \equiv \frac{(1 - t_{\mathrm{A}} y_{\mathrm{Ab}}) - (1 - t_{\mathrm{A}} y_{\mathrm{Ai}})}{\ln(1 - t_{\mathrm{A}} y_{\mathrm{Ab}})/(1 - t_{\mathrm{A}} y_{\mathrm{Ai}})} \tag{23}$$

$$x_{\mathrm{f}} \equiv \frac{(1 - t_{\mathrm{A}} x_{\mathrm{Ab}}) - (1 - t_{\mathrm{A}} x_{\mathrm{Ai}})}{\ln(1 - t_{\mathrm{A}} x_{\mathrm{Ab}})/(1 - t_{\mathrm{A}} x_{\mathrm{Ai}})} \tag{24}$$

The parameters Y_f and X_f are called the film factors. They are generalized y_{BM} and x_{BM} factors, respectively, and are reduced to them in the case of unidirectional diffusion through a stagnant medium because $t_A = 1$ in this case. The film factors Y_f and X_f correct the mass transfer coefficients for the effect of net flux, or the drift flux, through the films. In situations having strong counterdiffusion giving rise to a net flow opposed to the diffusion of A, the film factor becomes larger than one and therefore decreases the mass transfer coefficient and the flux of A. For weak or negative counterdiffusion producing a bulk flux parallel to the diffusion of A, the film factor is smaller than unity and thus increases N_A. In extreme situations, counterdiffusion may become large enough to reverse the direction of transport of a given component and force it to diffuse against its own driving force. These situations are characterized by a negative film factor, and hence a negative k_G. If equimolar counterdiffusion prevails, t_A becomes zero, the film factor is unity, irrespective of the concentration, and $k_G = k_G^0$.

Except for equimolar counterdiffusion, the mass transfer coefficients applicable to the various situations apparently depend on concentration through the y_{BM} and Y_f factors. Instead of the classical rate equations 4 and 5, containing variable mass transfer coefficients, the rate of mass transfer can be expressed in terms of the constant coefficients for equimolar counterdiffusion using the relationships

$$k_G Y_f = k'_G y_{BM} = k_G^0 \tag{25}$$

$$k_L X_f = k'_L x_{BM} = k_L^0 \tag{26}$$

This leads to rate equations with constant mass transfer coefficients, whereas the effect of net transport through the film is reflected separately in the y_{BM} and Y_f factors. For unidirectional mass transfer through a stagnant gas the rate equation becomes

$$N_A = k_G^0 P (y_{Ab} - y_{Ai}) \frac{1}{y_{BM}} \tag{27}$$

For any degree of counterdiffusion,

$$N_A = k_G^0 P (y_{Ab} - y_{Ai}) \frac{1}{Y_f} \tag{28}$$

Equation 28 and its liquid-phase equivalent are quite general. Similarly, the overall mass transfer coefficients may be made independent of the effect of drift flux through the films and thus nearly concentration independent for straight equilibrium lines:

$$N_A = K_{OGP}^0 \left(y_{Ab} - y_A^*\right) \frac{1}{y_{BM}^*} \tag{29}$$

$$= K_{OGP}^* \left(y_{Ab} - y_A^*\right) \frac{1}{Y_f^*} \tag{30}$$

where the logarithmic means in y^*_{BM} and Y^*_f must be taken between y_{Ab} and y_A^*.

3.3. Fundamental Description of Multicomponent Diffusion using Maxwell-Stefan Equations. Fick's law cannot describe multicomponent diffusion because it links the driving force for diffusion of a given species to one single flux. If a driving force in terms of chemical potential gradient for a given species exists in a multicomponent mixture, it will cause this species to move relative to all other types of molecules in the mixture. It is not an absolute flux, as stipulated by Fick's law, but the movements of solute molecules A relative to all the other molecules j that dissipate Gibbs energy (see eq. 31). It is therefore necessary to sum over all these relative movements. Each one of them will need an amount of driving force determined by the importance of the interactions between molecules A and j, expressed by the ratio of RT and the binary diffusion coefficient A-j. The correct sum over all individual relative movements has already been worked out a long time ago and is known as the Maxwell-Stefan equations (44,46):

$$-\frac{d\mu_{\mathrm{A}}}{dz} = \sum_{j=B}^{n} \frac{RT}{D_{Aj}} x_j \left(U_A - U_j\right) \tag{31}$$

where U_j is the moving velocity of the ith species and D_{Aj} the binary diffusion coefficient between A and the jth other species in the mixture. As the Maxwell-Stefan equations are about binary interactions, there are only $n-1$ independent such equations in a n species mixture. In order to calculate all the n diffusion velocities U_{i}, one more constraint must be formulated in terms of a velocity balance such as $\Sigma\ U_i = 0$, $U_{\mathrm{B}} = 0$ or the like. After transforming the gradient of chemical potential into a mole fraction gradient and by linking U_{A} to the fluxes by $N_A = U_A \bar{\rho} x_A$ the Maxwell-Stefan equations may be written as

$$-\Gamma \frac{dx_A}{dz} = \sum_{j=B}^{n} \frac{N_A x_j - N_j x_A}{D_{Aj}\rho} \tag{32}$$

where

$$\Gamma = 1 + x_A \frac{d\ln\gamma_A}{dx_{\mathrm{A}}} \tag{33}$$

Γ is called the thermodynamic correction factor and reduces to unity for ideal mixtures. It is easily shown that for ideal binary situations equation 32 reduces to either Fick's law, equation 13 or 19 if $N_{\mathrm{B}} = -N_{\mathrm{A}}$, $N_{\mathrm{B}} = 0$, or $\Sigma\ N_{\mathrm{i}} = N_{\mathrm{A}} + N_{\mathrm{B}} = t_{\mathrm{A}} N_{\mathrm{A}}$ are substituted, respectively.

Using Maxwell-Stefan equations instead of Fick's law has the advantage of taking all the binary diffusional interactions correctly into account. The Maxwell-Stefan diffusivities can be measured in binary diffusion experiments and for gaseous mixtures are concentration independent. Another important advantage is the fact that thermodynamic nonidealities are explicitely accounted for, whereas Fick's law lumps them into the diffusion coefficient, which therefore depends more on concentration than Maxwell-Stefan diffusivities. Maxwell-Stefan equations may also readily be extended to cases where diffusion is also driven by forces other than concentration gradients, such as pressure gradients, centrifugal force and electrical fields.

On the other hand, Maxwell-Stefan equations are not easy to integrate in order to find the concentration profiles. The fact that they are implicit in the fluxes further complicates all numerical calculations.

Treating Complex Mixtures as Pseudobinaries. The traditional way of dealing with multicomponent mixture has always been to describe the diffusion of any given type of molecule through the mixture by lumping all other species into a single pseudospecies and to compute the diffusion flux based on Fick's law. This procedure requires evaluating an effective diffusivity of A through the rest of the mixture. Eliminating the mole fraction gradient from Fick's law and from the Maxwell-Stefan equation one finds the effective diffusivity to be

$$D_{A_{\mathrm{eff}}} = \frac{N_A}{\sum_{j=B}^{n} \frac{N_A x_j - N_j x_A}{D_{Aj}}} \tag{34}$$

Effective diffusivities calculated in this way do already contain the drift-flux correction. They may be positive or negative. In general cases, they are too difficult to compute to be of practical use.

There are, however, several situations frequently occurring in practice in which mixtures do indeed behave like binaries. The most obvious case is a dilute mixture of solutes in a carrier gas or solvent, as often occurring in gas absorption. If the mole fractions of all solutes are small, the sum in equation 31 will contain only a single interaction term and equation 32 reduces to

$$-\frac{dx_{\mathrm{A}}}{dz} = \frac{N_{\mathrm{A}} x_{\mathrm{B}}}{d_{\mathrm{AB}} \rho} \tag{35}$$

Where the subscript B denotes the solvent or carrier. In this case, the Maxwell-Stefan approach for multicomponent mixtures reduces to the familiar equations treated earlier. In case x_{B} is sufficiently close to unity, equation 35 corresponds to Fick's law for equimolar counterdiffusion and the mass transfer coefficients may be evaluated as shown by equations 11 and 12. If x_{B} deviates sufficiently from unity to induce a drift flux, equation 35 reduces with $x_{\mathrm{B}} = 1 - x_{\mathrm{A}}$ to the liquid equivalent of equation 13 and the drift flux may be corrected as shown by equation 27.

Another frequently occuring case is the diffusion of a solute through a mixture of nondiffusing substances such as air. In this case, the effective diffusivity becomes

$$D_{\mathrm{A_{eff}}} = \frac{1}{\sum_{j=\mathrm{B}}^{n} \frac{x_j}{D_{\mathrm{A}j}}} \tag{36}$$

$D_{\mathrm{A_{eff}}}$ represents the average diffusivity of A through the stagnant mixture and can directly be used in Fick's law. Mixtures in which the binary diffusion coefficients of one of the constituents are markedly lower than the others, also may be described as pseudobinaries because only the interactions with this particular component need to be accounted for and all other terms can be neglected in equation 32. However, the evaluation of $D_{A_{eff}}$ from equation 34 remains intricate.

Other Ways to Use Maxwell-Stefan Equations. Wesselingh and Krishna (47) have proposed to integrate the Maxwell-Stefan equations by assuming linear concentration profiles through the diffusion film (eq. 37).

$$\frac{\Delta x_A}{z_o} = \sum_{j=B}^{n} \frac{N_A \bar{x}_j - N_j \bar{x}_A}{D_{Aj} \rho} \tag{37}$$

Although still not explicit in the fluxes, this technique is highly useful for numerical calculations. Wesselingh and Krishna propose to use these equations, eg, in the following form:

$$\Delta x_A = \sum_{j=B}^{n} \frac{N_A \bar{x}_j - N_j \bar{x}_A}{k_{Aj} \rho} \tag{38}$$

where k_{Aj} are binary mass transfer coefficients defined as

$$k_{Aj} = \frac{D_{Aj}}{z_0} \tag{39}$$

If the molar fraction in the bulk and at the interface are known, the fluxes in a mixture containing n species may be calculated by solving $n-1$ linear equations of the type seen in equation 38 together with one mass balance constraint such as, eg, $N_B = 0$. This procedure is thus useful for obtaining an approximately correct estimation of all fluxes through a diffusion film.

Rigorous computation of multicomponent mass transfer may be achieved resorting to the text by Taylor and Krishna (48 and references cited therein). At its heart lies a matrix formulation of the generalized Maxwell-Stefan equations (49,50). This approach is quite rigorous and consistent with the thermodynamics of irreversible processes and in the case of plate columns does not rely on the use of lumped phenomenological concepts like Murphree tray efficiencies. It also eliminates the assumption of thermal equilibrium between phases at a given stage within the column. Although it has been primarily used in the design of distillation and reactive distillation columns, it is increasingly being used for the design of absorption equipment. Reference 51 is strongly recommended as a very comprehensive and up-to-date treatise on the design and calculation of absorption columns using advanced techniques. Although the complication of some of these numerical techniques precludes its use in hand calculations, these rigorous methods have already found their way into commercial flowsheet simulators.

The use of sophisticated modeling tools is warranted when large multicomponent molar fluxes or large temperature differences between phases or both (52) are expected. In the majority of industrial design cases, however, it is satisfactory to apply simpler methods of sufficient approximation.

3.4. Mass Transfer Coefficients and Convection. As explained at the beginning of the last section, many theories have been developped in attempts to model mass transfer rates under the combined effects of molecular diffusion and turbulent convection. The classical model has been the film theory (53) effectively assuming completely stagnant layers of a given thickness z_0 adjacent to the interface and a sudden change to the completely turbulent conditions prevailing in the bulk of the phase. Mass transfer is thus assumed to occur through these films only by molecular diffusion at steady state. Equations 14–30 may then used directly to correlate mass transfer coefficients and rates.

The film model has often received criticism because it appears to predict that the rate of mass transfer is directly proportional to the molecular diffusivity. This dependency is at odds with experimental data that shows the mass transfer to be proportional to the molecular diffusivity raised to an exponent which varies between 0.5 and 0.7 (54–56), and was one of the reasons why other models were developed. In spite of this criticism, the film theory continues to be used very widely in the design of gas absorption equipment.

In contrast to the film theory, other approaches assume that transfer of material does not occur by steady-state diffusion. Rather there are large fluid motions which constantly bring fresh masses of bulk material into direct contact with the interface. According to the penetration theory (57), transient diffusion proceeds from the interface into the particular element of fluid in contact with the interface. This is an unsteady state, transient process where the rate decreases with time. After a while, the element is replaced by a fresh one brought to the interface by the relative movements of gas and liquid, and the process is repeated. In order to evaluate N_A, a constant average contact time for the individual fluid elements is assumed (57). This leads to relations such as

$$k_L = 2\sqrt{\frac{D}{\pi\tau}} \tag{40}$$

If, on the other hand, it is assumed that contact times for the individual fluid elements vary at random, an exponential surface age distribution characterized by a fractional rate of renewals may be used (58). This approach is called surface renewal theory and results in

$$k_L = \sqrt{Ds} \tag{41}$$

Neither the penetration nor the several variations on the surface renewal theory can be used to predict mass transfer coefficients directly because τ and s are not normally known. Each suggests, however, that mass transfer coefficients should vary as the square root of the molecular diffusivity and thus reflects experimental data better than the film theory.

The penetration model is also superior to the film model in predicting the influence of thermodynamic nonidealities on gas–liquid mass transfer (59). In the case of gas absorption in dilute systems, these differences are irrelevant because both transport models lead to the same expression for the transport rate as a function of the concentration difference. But in the case of gas absorption in concentrated systems an error is introduced when the film method is used instead of the more appropriate penetration model. The mass transfer rates will be influenced by convective contributions and thermodynamic nonidealities. Mass transfer rates computed with the film method can easily differ by >50% from those obtained from the penetration model, which seems to be in better agreement with available experimental data. This criticism, while essentially correct, has been shown not to have severe consequences in practice. The simple ad hoc modification suggested in some cases (59) seems to have wide validity and does show that, for rapid design purposes, the theoretically criticized film theory can still be used very successfully.

Finally, the film-penetration theory (60) combines features of the film, penetration and surface renewal theories and predicts a dependency of the mass transfer coefficient on the diffusivity raised to a power that varies continuously between 0.5 and 1. This theory assumes the entire resistance to mass transfer to reside in a film of fixed thickness. Eddies move to and from the bulk fluid and this film. Age distributions for time spent in the film are of the Higbie (penetration) or Danckwerts (surface renewal) type. For high rates of surface renewal, it reduces to the surface renewal theory. For low rates of renewal it reduces to the film theory. Although the application of the latter theories is difficult because of lack of data on film thickness or on fractional rate of surface renewal or on both, they have found some application in the design of nonisothermal absorption with chemical reaction (61,62).

Another concept sometimes used as a basis for comparison and correlation of mass transfer data in columns is the Chilton-Colburn analogy (63). This semiempirical relationship was developed for correlating mass and heat transfer data in pipes and is based on the turbulent boundary layer model and the close analogy between momentum and mass transfer. It must be considerably modified for gas-absorption columns, but it predicts that the mass transfer coefficient varies with D raised to the two-thirds power (4,64) which is in good agreement with experimental data.

3.5. Absorption and Chemical Reaction. In instances where the solute gas is absorbed into a liquid or a solution where it is able to undergo chemical reaction, the driving forces of absorption become far more complex. The solute not only diffuses through the liquid film at a rate determined by the gradient of the concentration, but at the same time also reacts with the liquid at a rate determined by the concentrations of both the solute and the solvent at the point of interest. Calculating the concentration profiles through the liquid film requires formulating a differential mass balance over an infinitesimal control volume in the film which accounts for both diffusion and reaction of the solute gas and subsequently integrating it. The calculations show that these profiles are steeper and the rate of mass transfer higher than without chemical reaction. Thus the results are often expressed as an enhancement factor ϕ defined as the fractional increase of the liquid film mass transfer coefficient resulting from the chemical reaction (k^{r}_{L}/k^{0}_{L}). The solutions that have been developed in this manner based on the film, penetration, and surface renewal theories are quite similar for a given type of reaction (65,66). Solutions and estimations of enhancement factors may be found in the literature (4,51,65–85).

An illustration of how concentration profiles are calculated in columns for gas absorption with chemical appears at the end of the section on packed column design.

4. Design of Packed Absorption Columns

Discussion of the concepts and procedures involved in designing packed gas absorption systems shall first be confined to simple gas absorption processes without complications: isothermal absorption of a solute from a mixture containing an inert gas into a nonvolatile solvent without chemical reaction. Gas and liquid are assumed to move through the packing in a plug-flow fashion. Deviations such as nonisothermal operation, multicomponent mass transfer effects, and departure from plug flow are treated in later sections.

4.1. Standard Absorber Design Methods

Operating Line. As a gas mixture travels up through a gas absorption tower, as shown in Figure 8, the solute A is transferred to the liquid phase

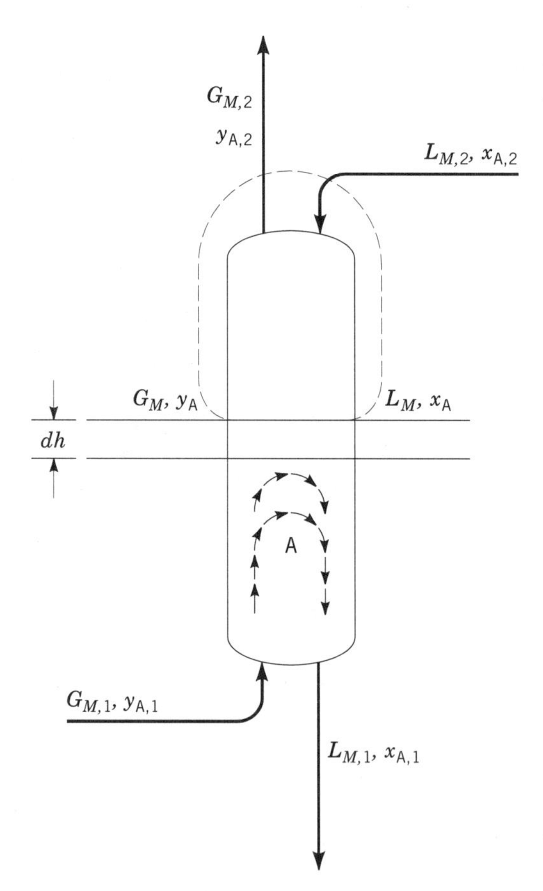

Fig. 8. Mass balance in gas absorption columns. The curved arrows indicate the travel path of the solute A. The upper broken curve delineates the envelope for the material balance of equation 42.

and thus gradually removed from the gas. The liquid accumulates solute on its way down through the column so x increases from the top to the bottom of the column. The steady-state concentrations y and x at any given point in the column are interrelated through a mass balance around either the upper or lower part of the column (eq. 43), whereas the four concentrations in the streams entering and leaving the system are interrelated by the overall material balance.

Since the total gas and liquid flow rates per unit cross-sectional area vary throughout the tower (Fig. 8) rigorous material balances should be based on the constant inert gas and solvent flow rates G'_M and L'_M, respectively, and expressed in terms of mole ratios Y' and X'. A balance around the upper part of the tower yields

$$G'_M Y' + L'_M X'_2 = G'_M Y'_2 + L'_M X' \tag{42}$$

which may be rearranged to give

$$Y' = \frac{L'_M}{G'_M}\left(X' - X'_{A,2}\right) + Y'_{A,2} \tag{43}$$

where G'_M and L'_M are in kg.mol/(h.m^2) [lb.mol/(h.ft^2)] and $Y' = y/(1-y)$ and $X' = x/(1-x)$. The overall material balance is obtained by substituting $Y' = Y'_1$ and $X' = X'_1$. For dilute gases the total molar gas and liquid flows may be assumed constant and a similar mass balance yields

$$y = \frac{L_M}{G_M}(x - x_2) + y_2 \tag{44}$$

A plot of either equation 43 or 44 is called the operating line of the process as shown in Figure 9. As indicated by equation 44, the line for dilute gases is straight, having a slope given by L_M/G_M. (This line is always straight when plotted in $Y' - X'$ coordinates.) Together with the equilibrium line, the operating line permits the evaluation of the driving forces for gas absorption along the column (Fig. 7). The farther apart the equilibrium and operating lines, the larger the driving forces become and the faster absorption occurs, resulting in the need for a shorter column (Fig. 9).

To place the operating line, the flows, composition of the entering gas y_1, entering liquid x_2, and desired degree of absorption y_2, are usually specified. The specification of the actual liquid rate used for a given gas flow (the L_M/G_M ratio) usually depends on an economic optimization because the slope of the operating line may be seen to have a drastic effect on the driving force. For example, use of a very high liquid rate (line A in Fig. 9) results in a short column and a low absorber cost, but at the expense of a high cost for solvent circulation and subsequent recovery of the solute from a relatively dilute solution. On the other hand, a liquid rate near the theoretical minimum, which is the rate at which the operating line just touches the equilibrium line (line B in Fig. 9), requires a very tall tower because the driving force becomes very small at its bottom (the extreme case in which which the operating line just touches the equilibrium line is usually designated as a "pinch point"). Use of a liquid rate on the order of one

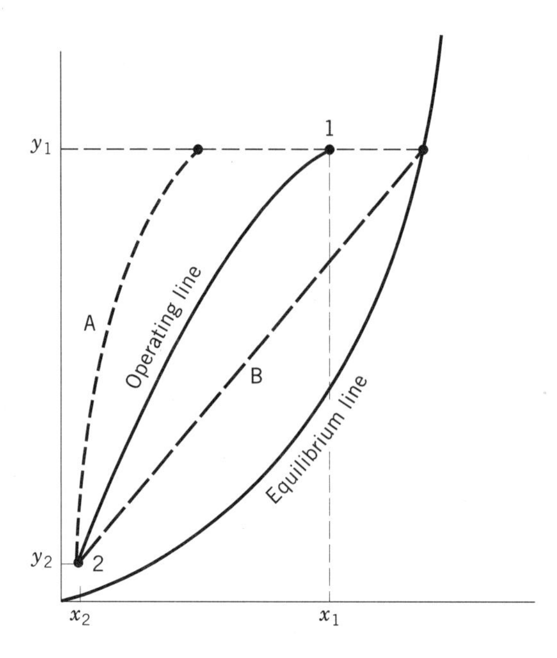

Fig. 9. Operating lines for an absorption system: line A, high L_M/G_M ratio; solid line, medium L_M/G_M ratio; line B, L_M/G_M ratio at theoretical minimum necessary for the removal of the specified quantity of solute. Subscript 1 represents the bottom of tower, 2, the top of tower.

and one-half times the theoretical minimum is not unusual. In the absence of a detailed cost analysis, the L_M/G_M ratio is often specified at 1.4 times the slope of the equilibrium line (86).

Design Procedure. The packed height of the tower required to reduce the concentration of the solute in the gas stream from $y_{A,1}$ to an acceptable residual level of $y_{A,2}$ may be calculated by combining point values of the mass transfer rate and a differential material balance for the absorbed component. Referring to a slice dh of the absorber (Fig. 8),

$$N_A a\, dh = -d(G_M y) = -G_M dy - y\, dG_M \tag{45}$$

and

$$dG_M = -N_A a\, dh \tag{46}$$

where a is the interfacial area present per unit volume of packing. Combining equations 45 and 46,

$$-dh = \frac{G_M\, dy}{N_A a(1-y)} \tag{47}$$

Substituting for N_A from equation 27 and integrating over the tower,

$$h = \int_{y_2}^{y_1} \frac{G_M}{k_G^0 aP} \frac{y_{BM}\, dy}{(1-y)(y-y_i)} \tag{48}$$

Equation 48 may be integrated numerically or graphically and its component terms evaluated at a series of points on the operating line. y_i is found by placing tie lines from each of these points; the slopes are given by equation 8. Thus equation 48 is a general expression determining the column height required to effect a given reduction in y_A.

Equation 48 can often be simplified by adopting the concept of a mass transfer unit. As explained in the film theory discussion earlier, the purpose of selecting equation 27 as a rate equation is that k_G^0 is independent of concentration. This is also true for the $G_M/k_G^0 aP$ term in equation 48. In many practical instances, this expression is fairly independent of both pressure and G_M: as G_M increases through the tower, $k^0{}_G$ increases also, nearly compensating for the variations in G_M. Thus this term is often effectively constant and can be removed from the integral:

$$h = \left(\frac{G_M}{k_G^0 aP}\right) \int_{y_2}^{y_1} \frac{y_{BM}\, dy}{(1-y)(y-y_i)} \tag{49}$$

The parameter $G_M/k^0{}_G aP$ has the dimension of length or height and is thus designated the gas-phase height of one transfer unit, H_G. The integral is dimensionless and indicates how many of these transfer units it takes to make up the whole tower. Consequently, it is called the number of gas-phase transfer units, N_G. Equation 49 may therefore be written as

$$h = (H_G)(N_G) \tag{50}$$

where

$$H_G = \frac{G_M}{k_G^0 aP} \tag{51}$$

and

$$N_G = \int_{y_2}^{y_1} \frac{y_{BM}\, dy}{(1-y)(y-y_i)} \tag{52}$$

The same treatment for the liquid side yields

$$h = (H_L)(N_L) \tag{53}$$

where

$$H_L = \frac{L_M}{k_L^0 a \bar{\rho}} \tag{54}$$

and

$$N_{\mathrm{L}} = \int_{x_2}^{x_1} \frac{x_{\mathrm{BM}}\, dx}{(1-x)(x_{\mathrm{i}}-x)} \tag{55}$$

A similar treatment is possible in terms of an overall gas-phase driving force by substituting equation 29 into equation 47:

$$h = (H_{\mathrm{OG}})(N_{\mathrm{OG}}) \tag{56}$$

where

$$H_{\mathrm{OG}} = \frac{G_M}{K_{\mathrm{OG}}^0 aP} \tag{57}$$

and

$$N_{\mathrm{OG}} = \int_{y_2}^{y_1} \frac{y_{\mathrm{BM}}^* \, dy}{(1-y)(y-y^*)} \tag{58}$$

Both H_{OG} and N_{OG} are called the overall gas-phase height of a transfer unit and the number of overall gas-phase transfer units, respectively. In the case of a straight equilibrium line, K_{OG}^0 is often nearly concentration-independent as explained earlier. In such cases, use of equation 56 is especially convenient because N_{OG}, as opposed to N_{G}, can be evaluated without solving for the interfacial concentrations. In all other cases, H_{OG} must be retained under the integral and its value calculated from H_{G} and H_{L} at different points of the equilibrium line as

$$H_{\mathrm{OG}} = \frac{y_{\mathrm{BM}}}{y_{\mathrm{BM}}^*} H_{\mathrm{G}} + \frac{mG_M}{L_M} \frac{x_{\mathrm{BM}}}{y_{\mathrm{BM}}^*} H_{\mathrm{L}} \tag{59}$$

To use all of these equations, the heights of the transfer units or the mass transfer coefficients $k_{\mathrm{L}}^0 a$ and k_{L}^0 must be known. Transfer data for packed columns are often measured and reported directly in terms of from H_{G} and H_{L} and correlated in this form against from G_M and L_M.

Sometimes the height equivalent to a theoretical plate (HETP) is employed rather than from H_{G} and H_{L} to characterize the performance of packed towers. The number of heights equivalent to one theoretical plate required for a specified absorption job is equal to the number of theoretical plates, N_{TP}. It follows that

$$h = (\mathrm{HETP})(N_{\mathrm{TP}}) \tag{60}$$

which is similar in form to equation 56. The HETP is a less fundamental variable than the heights of the transfer units, and it is more difficult to translate HETPs from one situation to another. Only for linear operating and equilibrium lines can they be related analytically to H_{OG} as shown later by equation 86.

4.2. Simplified Design Procedures for Linear Operating and Equilibrium Lines

Logarithmic-Mean Driving Force. As noted earlier, linear operating lines occur if all concentrations involved stay low. Where it is possible to assume that the equilibrium line is linear, it can be shown that use of the logarithmic mean of the terminal driving forces is theoretically correct. When the overall gas-film coefficient is used to express the rate of absorption, the calculation reduces to solution of the equation

$$L_M(x_1 - y_2) = G_M(y_1 - y_2) = K_{\mathrm{OG}}aPh(y - y^*)_{\mathrm{av}} \tag{61}$$

where

$$(y - y^*)_{\mathrm{av}} = \frac{(y_1 - y_1^*) - (y_2 - y_2^*)}{\ln\left[(y_1 - y_1^*)/(y_2 - y_2^*)\right]} \tag{62}$$

In these cases, a quantitative significance can be given to the concept of a transfer unit. Because $H_{\mathrm{OG}} = G_{\mathrm{M}}/K_{\mathrm{OG}}aP$, it follows from equations 61 and 56 that

$$N_{\mathrm{OG}} = \frac{y_{\mathrm{i}} - y_2}{(y - y^*)_{\mathrm{av}}}$$

Therefore, in this case, one transfer unit corresponds to the height of packing required to effect a composition change just equal to the average driving force.

Number of Transfer Units. For relatively dilute systems the ratios involving y_{BM}, y^*_{BM}, and $1 - y$ approach unity so that the computation of H_{OG} from equation 59 and N_{OG} from equation 58 may be simplified to

$$H_{\mathrm{OG}} = H_{\mathrm{HG}} + \left(\frac{mG_M}{L_M}\right)H_{\mathrm{L}} \tag{63}$$

$$N_{\mathrm{OG}} \approx N_{\mathrm{T}} = \int_{y_2}^{y_1} \frac{dy}{y - y^*} \tag{64}$$

Equation 64 is a rigorous expression for the number of overall transfer units for equimolar counterdiffusion, in distillation columns, for instance.

For cases in which the equilibrium and operating lines may be assumed linear, having slopes L_M/G_M and m, respectively, an algebraic expression for the integral of equation 64 has been developed (86):

$$N_{\mathrm{OG}} \approx N_{\mathrm{T}} = \frac{\ln\left[\left(1 - \dfrac{mG_M}{L_M}\right)\left(\dfrac{y_1 - mx_2}{y_2 - mx_2}\right) + \dfrac{mG_M}{L_M}\right]}{1 - \dfrac{mG_M}{L_M}} \tag{65}$$

The required tower height may thus be easily calculated using equation 56, where H_{OG} is given by equation 63 and N_{OG} by equation 65.

4.3. Rapid Approximate Design Procedure for Curved Operating and Equilibrium Lines. If the operating or the equilibrium line is nonlinear, equation 65 is of little use because mG_M/L_M will assume a range of values over the tower. The substitution of effective average values for m and for L_M/G_M into equations 59 and 65 obviates lengthy graphical or numerical integrations and leads to a quick, approximate solution for the required tower height (4).

The effective average values of m and L_M/G_M were determined in a computational study covering hundreds of hypothetical absorber designs for gas streams containing up to 80 mol% of solute for recoveries from 81 to 99.9%. By numerical integration, precise values were obtained for N_{OG} and N_T. By solving equation 65 numerically for each of the design cases, average values of the slope of the equilibrium line m and average flow ratios $L_M/G_M = R_{av}$ were found which gave the same N_T when substituted into equation 65 as the graphical or numerical integration.

It was found that the effective average L_M/G_M ratio, R_{av}, could be correlated satisfactorily as a function of the terminal values R_1 and R_2, of the change in the mole fraction of the absorbed component over the tower, and of the fractional approach to equilibrium y_1^*/y_1 between the concentrated gas entering the tower and the liquid leaving. Figure 10 shows the resulting correlation for cases with $L_M/G_M > 1$. No correlation was obtained when this ratio was less

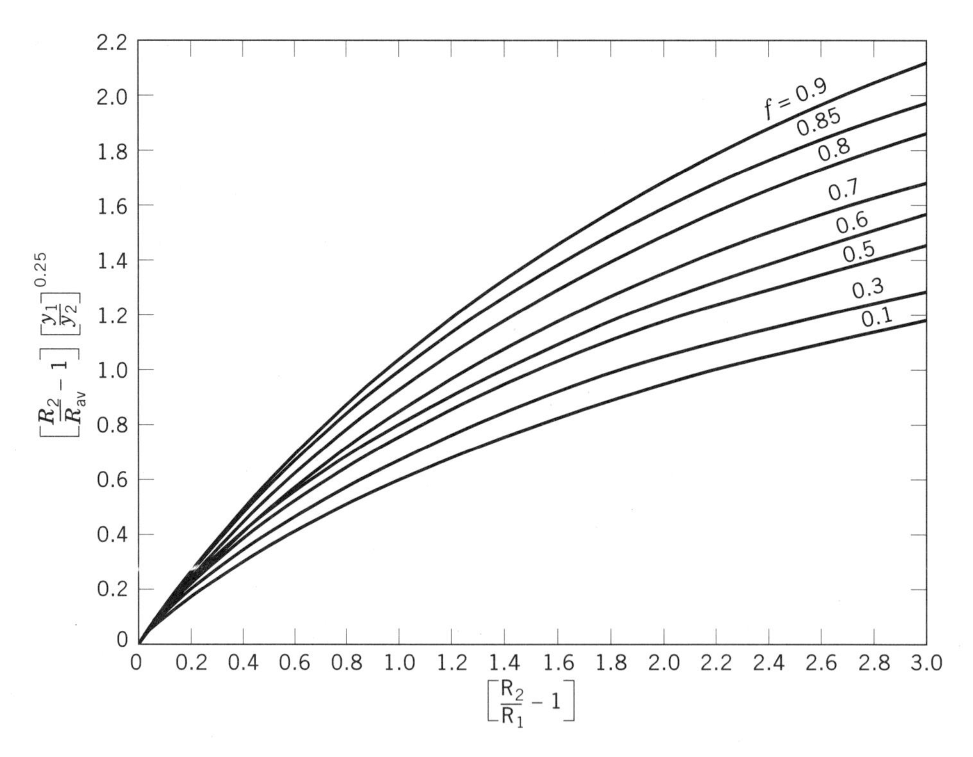

Fig. 10. Design chart for estimation of average-flow ratio in absorption (4). $R_2 = L_M/G_M$ at gas outlet; $R_1 = G_M/L_M$ at gas inlet; y_2 = mole fraction in outlet gas; y_1 = mole fraction in inlet gas; R_{av} = effective average L_M/G_M; $f = y^*{}_1/y_1$ = fractional approach to equilibrium.

than unity. The effective average slope of the equilibrium line, m, was correlated as a function of the initial slope m_2, of the slope m_c of the chord connecting the points on the x-y diagram with the coordinates (x_1, y_1^*) and (x_2, y_2^*), and of various other parameters as shown in Figure 11. Figure 11**a** applies when the equilibrium line is concave upward, ie, m_c/m_2; and Figure 11**b** applies when the curvature is concave downward, m_c/m_2.

The recommended design procedure uses the values of $(L_M/G_M)_{av}$ and m from Figures 10 and 11 in equation 65 and yields a very good estimation of N_T despite the curvature of the operating and the equilibrium lines. This value differs from N_{OG} obtained by equation 58 because of the $y^*_{BM}/(1-y)$ term in the latter equation. A convenient approach for purposes of approximate design is to define a correction term ΔN_{OG} which can be added to equation 64:

$$N_{OG} = N_T + \Delta N_{OG} \tag{66}$$

For cases in which y^*_{BM} may be represented by arithmetic mean (87),

$$\Delta N_{OG} = \frac{1}{2} \ln \frac{1-y_2}{1-y_1} \tag{67}$$

Equation 67 is sufficiently accurate for most situations.

The average slopes R_{av} and m from Figures 10 and 11 may also be used in equation 63 to compute H_{OG} although equation 59, with some suitable averages of y^*_{BM} and x_{BM}, should be preferred. Use of point values at an effective average liquid concentration given by equation 68 is suggested.

$$\bar{x} = \left(\frac{R_2}{R_{av}} - 1\right)/(R_2 - 1) \tag{68}$$

In many situations, however, especially when $m > 1$, the results using the simpler equation 63 are virtually the same. The required tower height is finally calculated by means of equation 56.

4.4. Drift Flux Correction for Pseudobinary Cases

Equimolar Counterdiffusion. Just as unidirectional diffusion through stagnant films represents the situation in an ideally simple gas absorption process, equimolar counterdiffusion prevails as another special case in ideal distillation columns. In this case, the total molar flows L_M and G_M are constant, and the mass balance is given by equation 44. As shown earlier, no y_{BM} factors have to be included in the derivation and the height of the packing is

$$h_T = \int_{y_2}^{y_1} H_G \frac{dy}{y - y_i} \tag{69}$$

N_{OG} is given by N_T:

$$N_{OG} = \int_{y_2}^{y_1} \frac{dy}{y - y^*} \tag{70}$$

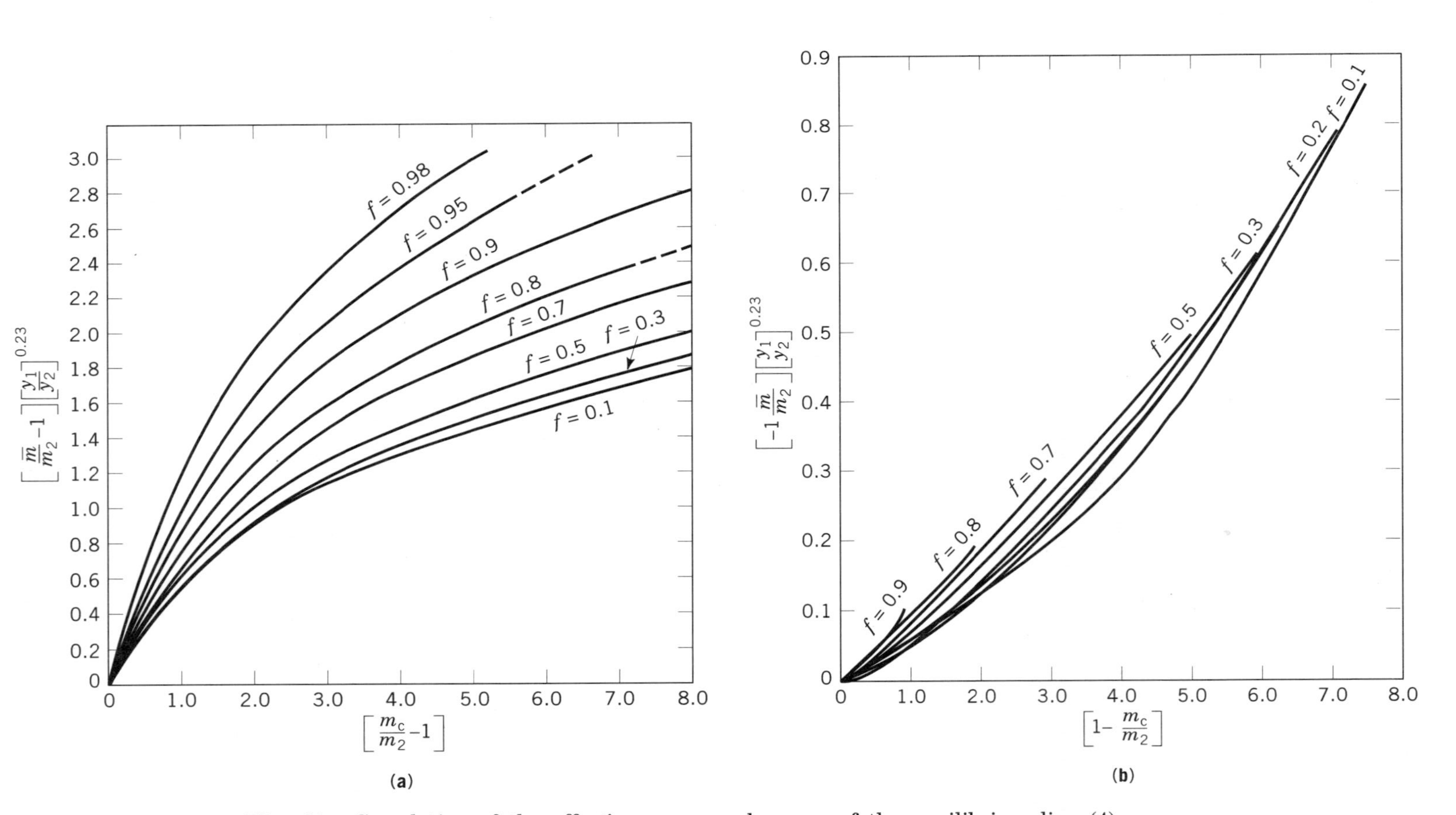

Fig. 11. Correlation of the effective average slopes m of the equilibrium line (4). (**a**) Equilibrium line curved concave upward; (**b**) equilibrium line curved concave downward.

and H_{OG} is rigorously defined by equation 63. It must, however, be retained under the integral because m usually changes over the tower:

$$h = \int_{y_2}^{y_1} H_{OG} \frac{dy}{y - y^*} \tag{71}$$

General Situation. Both unidirectional diffusion through stagnant media and equimolar diffusion are idealizations that are usually violated in real processes. In gas absorption, slight solvent evaporation may provide some counter-diffusion, and in distillation counterdiffusion may not be equimolar for a number of reasons. This is especially true for multicomponent operation.

A simple treatment is still possible if it may be assumed that the flux of the component of interest A through the interface stays in a constant proportion to the total molar transfer through the interface over the entire tower and by treating multicomponent absorption as pseudobinary cases:

$$\frac{\sum N_j}{N_A} = t_A = \text{constant} = \frac{\sum \Delta g_j}{\Delta g_A} \tag{72}$$

where Δg_j = total moles of component j absorbed over the tower. It will generally suffice to compute t_A from preliminary estimates of Δg_j and Δg_A, the total mass transfer of each component over the tower.

The mass balance for A is best represented as a straight line in hypothetical coordinates Y^0 and X^0:

$$Y_A^0 = \frac{L_M^0}{G_M^0}\left(X_A^0 - X_{A,2}^0\right) + Y_{A,2}^0 \tag{73}$$

where $Y^0{}_A = y_A/(1 - t_A y_A)$, $X^0{}_A = x_A/(1 - t_A x_A)$, $G^0{}_M = G_M/(1 - t_A y_A)$, and $L^0{}_M = L_M/(1 - t_A x_A)$, $G^0{}_M$ and $L^0{}_M$ are always constant, whereas G_M and L_M are not. For unimolecular diffusion through stagnant gas ($t_A = 1$), Y^0 and X^0 reduce to Y' and X' $G^0{}_M$ and $L^0{}_M$ reduce to G'_M and L'_M; equation 73 then becomes equation 43. For equimolar counterdiffusion $t_A = 0$, and the variables reduce to y, x, G_M and L_M, respectively, and equation 73 becomes equation 44 . Using the film factor concept and rate equation 28, the tower height may be computed by

$$h_T = \int_{y_2}^{y_1} H_G \frac{Y_f}{(1 - t_A y_A)} \frac{dy_A}{(y_A - y_{Ai})} \tag{74}$$

y_{Ai} is found as usual through tie lines of the slope

$$\frac{y_A - y_{Ai}}{x_A - x_{Ai}} = -\frac{L_M}{G_M}\frac{H_G}{H_L}\frac{Y_f}{X_f} \tag{75}$$

where

$$\frac{L_M}{G_M} = \frac{L_M^0(1 - t_A y_A)}{G_M^0(1 - t_A x_A)} \tag{76}$$

It may be noted that the above system of equations is quite general and encompasses both the usual equations given for gas absorption and distillation as well as situations with any degree of counterdiffusion. The exact derivations may be found elsewhere (4). The system of equation is, however restricted to cases that may be treated as pseudo binary (see Section 3.3).

4.5. Nonisothermal Gas Absorption

Nonvolatile Solvents. In practice, some gases tend to liberate such large amounts of heat when they are absorbed into a solvent that the operation cannot be assumed to be isothermal, as has been done thus far. The resulting temperature variations over the tower will displace the equilibrium line on a $y-x$ diagram considerably because the solubility usually depends strongly on temperature. Thus nonisothermal operation affects column performance drastically.

The principles outlined so far may be used to calculate the tower height as long as it is possible to estimate the temperature as a function of liquid concentration. The classical basis for such an estimate is the assumption that the heat of solution manifests itself entirely in the liquid stream. It is possible to relate the temperature increase experienced by the liquid flowing down through the tower to the concentration increase through a simple enthalpy balance, equation 77, and thus correct the equilibrium line in a $y-x$ diagram for the heat of solution as shown in Figure 12,

$$T_{\mathrm{L}} \approx T_{\mathrm{L2}} + \frac{(x_{\mathrm{A}} - x_{\mathrm{A2}})H_{\mathrm{OS}}}{x_{\mathrm{A}}C_{\mathrm{qA}} + (1 - x_{\mathrm{A}})C_{\mathrm{qB}}} \tag{77}$$

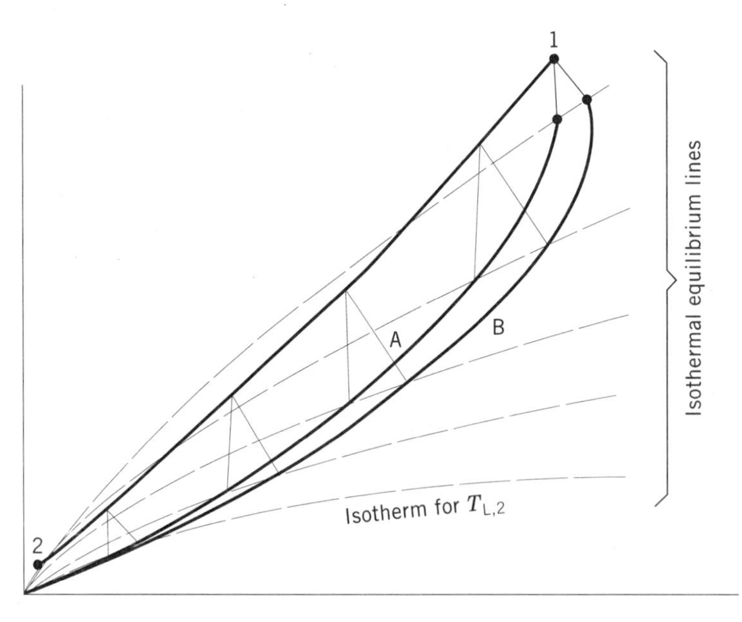

Fig. 12. Simple model of adiabatic gas absorption. (A) nonisothermal equilibrium line for overall gas-phase driving force: $y^* = f(x, T_{\mathrm{L}})$; (B) nonisothermal equilibrium line for individual gas-film driving force: $y_{\mathrm{i}} = f(x_{\mathrm{i}})$.

where T_L is the liquid temperature, °C; T_{L2} is the temperature of the entering liquid, °C; C_{qj} is the liquid molar heat capacity of component j; H_{OS} is the integral mean heat of solution of solute. For each pair of values for x_A and T_L obtained from equation 68 it is possible to evaluate $y^*(x, T_L)$, the concentration in equilibrium with the bulk of the liquid phase, and to place the equilibrium line for the overall driving force (Fig. 12A). The line connecting the actual interfacial concentrations (y_i, x_i), Figure 12B, does not coincide with line A unless there is no liquid mass transfer resistance. However, because the interfacial temperature T_i and the bulk liquid temperature T_L usually are virtually equal, the equilibrium concentration y^* and the actual interfacial concentration y_i are connected by an isotherm. Line B may therefore be constructed as shown on the basis of line A, tie lines, and isothermal equilibrium lines. Line B may be used in conjunction with equation 48 to compute the required depth of packing.

General Case. The simple adiabatic model just discussed often represents an oversimplification, since the real situation implies a multitude of heat effects: (*1*) The heat of solution tends to increase the temperature and thus to reduce the solubility. (*2*) In the case of a volatile solvent, partial solvent evaporation absorbs some of the heat. (This effect is particularly important when using water, the cheapest solvent.) (*3*) Heat is transferred from the liquid to the gas phase and vice versa. (*4*) Heat is transferred from both phase streams to the shell of the column and from the shell to the outside or to cooling coils.

In the general case, the temperature profile is determined simultaneously by all of the four heat effects. The temperature influences the transfer of mass and heat to a large extent by changing the solubilities. This turns the simple gas absorption process into a very complex one and all factors exhibit a high degree of interaction. Computer algorithms for solving the problem rigorously have been developed (51,88,89) and several implementations are available in kinetic or rate-based column modules in flowsheet simulators . Figure 13 depicts typical profiles through an adiabatic packed gas absorber from one of these algorithms (88,89). The calculations were carried out to solve a design example calling for the removal of 90% of the acetone vapors present in an air stream by absorption into water at an L_M/G_M ratio of 2.5. The air stream contained 6 mol % acetone and was saturated with water; the ambient temperature was 15°C.

It is a typical feature of such calculations that the shapes of the liquid temperature profiles are highly irregular and often exhibit maxima within the column. Such internal temperature maxima have been observed experimentally in plate and packed absorbers (88,90,91), and the measured temperature profiles can be shown to agree closely with rigorous computations. The appearance of an internal hot spot (and eventually an internal pinch-point) in the absorber is relatively common in natural gas sweetening with amines and in the absorption of water vapor from air with organic salts of alkali metals. The temperature maximum occurs in part because the heat of solution causes the entering liquid stream to be heated. In the lower part of the tower, however, the heat of absorption is smaller than the opposite heat effects of solvent evaporation and heat transfer to the cold entering gas, so that the net effect is a cooling of the liquid phase. These transfers are reversed in the upper part of the column, as is obvious from Figure 13: The gas gives up heat to the liquid, is cooled, and some of the solvent condenses from the gas stream into the liquid stream, which is heated

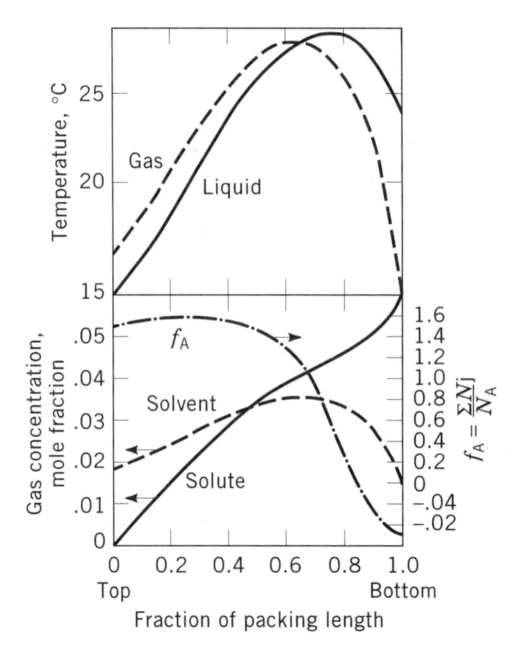

Fig. 13. Computed rigorous profiles through an adiabatic packed absorber during the absorption of acetone into water (89).

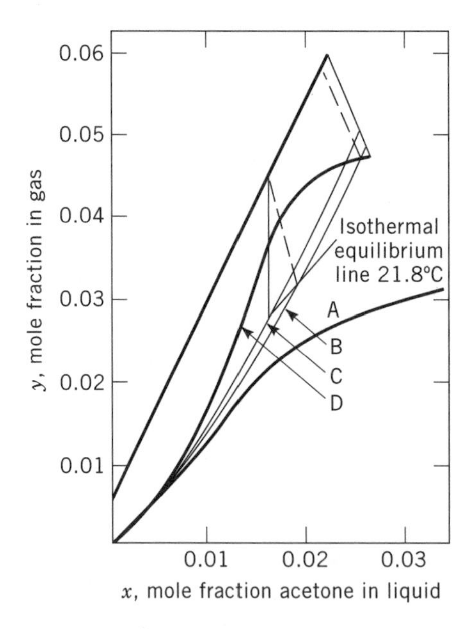

Fig. 14. The $y - x$ diagram for adiabatic absorption of acetone into water. (A) isothermal equilibrium line at T_{L2}; (B) equilibrium line for simple model of adiabatic gas absorption, gas-film driving force; (C) equilibrium line for simple model of adiabatic gas absorption, overall gas-phase driving force; (D) rigorously computed equilibrium line, gas-film driving force (89).

Table 2. **Comparison of Results of Different Design Calculations**

Method used	N_{OG}	Required depth of packing, m
rigorous calculation	5.56	3.63
isothermal approximation	3.30	1.96
simple adiabatic model	4.01	2.38

much faster in this part of the column than would be the case with the absorption alone. Figure 14 shows the rigorously computed $y-x$ diagram for the same example. The temperature maximum within the column produces a region of reduced solubility reflecting itself in the typical bulge in the middle of the rigorous equilibrium line. Since less acetone is absorbed in this part of the equipment, the gas concentration curve exhibits a slight plateau (Fig. 13). This example may also serve as a demonstration of the difficulty in estimating the required depth of packing using simplifying assumptions (Table 2). The isothermal approximation failed completely in this case and yielded 1.95 m of required packing as opposed to the rigorously determined value of 3.63 m. Neglecting the temperature increase completely, this model assumes a solubility which is much too large, reflected by equilibrium line A, and thus underestimates the rigorous result by 90%.

The standard way to correct for the heat of solution approximately is the simple adiabatic model described on the preceding pages, which yields equilibrium line B if the gas-phase driving force is used and line C on the basis of the overall driving force. This model, however, is a poor representation of the conditions prevailing in the absorber, as demonstrated by the deviation of its equilibrium line B from the rigorous line D. The approximation underestimates the true packing depth value of 3.63 m by more than one-third yielding 2.4 m (Table 2).

These calculations emphasize how important it is to account for heat effects even when performing quick, preliminary calculations. Since correct numerical calculations are not quite straightforward due to the highly non-linear character of the problem, a rapid approximate design procedure has been developped permitting to assess the importance of deviations from isothermal conditions without the need to resort to a computer (6,92).

4.6. Axial Dispersion Effects

Effect of Axial Dispersion on Column Performance. Another assumption underlying standard design methods is that the gas and the liquid phases move in plug-flow fashion through the column. In reality, considerable departure from this ideal flow assumption exists (4) and different fluid particles travel through the packing at varying velocities. The impact of this effect, which is usually referred to as axial dispersion, on the concentration profiles is demonstrated in Figure 15. The effect counteracts the countercurrent contacting scheme for which the column is designed and thus lowers the driving forces throughout the packed bed. Neglect of axial dispersion results in an overestimation of the driving forces and in an underestimation of the number of transfer units needed. It may therefore lead to an unsafe design.

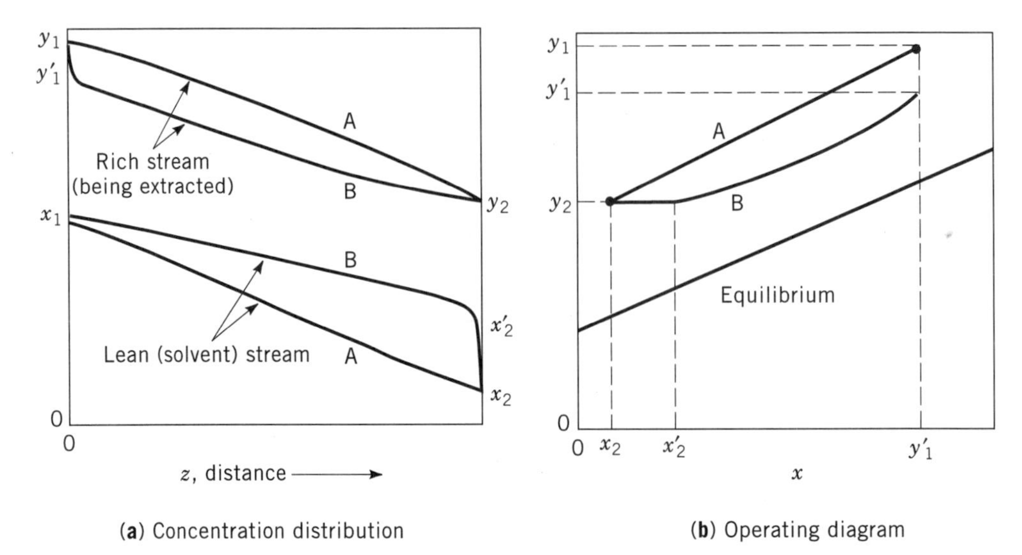

Fig. 15. Effect of axial dispersion in both phases on solute distribution through countercurrent mass transfer equipment. A, piston or plug flow; B, axial dispersion in both streams (diagrammatic). Reprinted with permission (4).

Determination of separation efficiencies from pilot-plant data is also affected by axial dispersion. Neglecting it yields high H_G or H_L values. Literature data for this parameter have usually not been corrected for this effect.

The extent of axial dispersion occurring in a gas absorber can be determined by measuring the residence time distribution of both gas and liquid. Based on classical flow models of axial dispersion, the result is usually expressed in terms of two Peclet numbers (Pe), one for each phase. Peclet numbers tending towards infinity indicate near-ideal plug flow, whereas vanishing values of Pe indicate axial dispersion to such an extent that the phase begins to become well back-mixed. When designing packed columns, the Peclet numbers are usually estimated from literature correlations (93–95). Correlations for predicting Peclet numbers in large scale gas-liquid contactors are quite scarce, but contributions have been made (96–100). Some of the available data have been described (4) and a review of published correlations for liquid-phase Peclet numbers is also available (100). Figure 16 reproduces some data (96). Axial dispersion data for bubble columns and other less common systems have also been published (101–105).

When designing packed towers, axial dispersion can be accounted for by incorporating terms for axial dispersion of the solute into the differential mass balance equation 45. The integration of the resulting differential equations is best effected by computer. Analytical solutions for cases having linear equilibrium and operating lines have been developed (106,107). They are, however, not explicit for the design case and are of such complexity that application for design also requires a computer.

Rapid Approximate Design Procedure. Several simplified approximations to the rigorous solutions have been developed over the years (101–105,107–110),

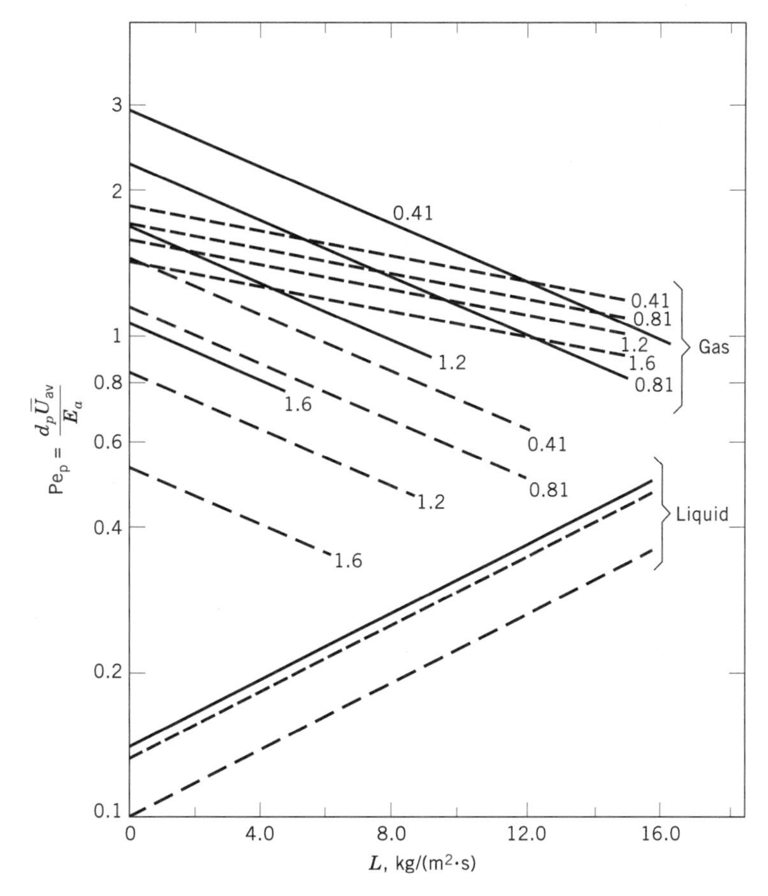

Fig. 16. Peclet numbers in large scale gas–liquid contactors using 2.54-cm Berl saddles (—) or 2.54 cm (- -) or 5.08 cm (– – –) Raschig rings (96). Numbers on lines represent G values = gas flow in kg/(m^2·s); d_p = nominal packing size; U_{av} = superficial velocity. To convert kg/(m^2·s) to lb/(h-ft^2) multiply by 737.5. Reprinted with permission (4).

but they all remain too complicated for practical use. A simple method proposed in 1989 (111,112) uses a correction factor accounting for the effect of axial dispersion, which is defined as (107)

$$\text{correction factor} = \frac{\text{NTU}_{\text{ap}}}{\text{NTU}} \tag{78}$$

The parameter NTU_{ap} is the "exterior apparent" overall gas-phase number of transfer units calculated neglecting axial dispersion simply on the basis of equation 65, whereas NTU stands for the higher real number of transfer units (N_{OG}) which is actually required under the influence of axial dispersion. The correction factor ratio can be represented as a function of those parameters that are actually known at the outset of the calculation

$$\frac{\mathrm{NTU_{ap}}}{\mathrm{NTU}} = f\left(\mathrm{NTU_{ap}}, \left(\frac{mG}{L}\right), \mathrm{Pe}_x, \mathrm{Pe}_y\right) \tag{79}$$

Equation 79 is shown graphically in Figure 17**a** for a given set of conditions. Curves such as these cannot be directly used for design, however, because the Peclet number contains the tower height h as a characteristic dimension. Therefore, new Peclet numbers are defined containing H_{OG} as the characteristic length. These relate to the conventional Pe as

$$\begin{aligned} \mathrm{Pe_{HTU}} &= \frac{uH_{OG}}{D_{ax}} = \frac{uh}{D_{ax}}\frac{H_{OG}}{h} \\ &= \mathrm{Pe}\frac{1}{N_{OG}} \end{aligned} \tag{80}$$

The correction factor $(\mathrm{NTU})_{ap}/\mathrm{NTU}$ as a function of $\mathrm{Pe_{HTU}}$ rather than Pe is shown in Figure 17**b**. The correction factors given in Figures 17**a** and 17**b** can roughly be estimated as

$$\frac{\mathrm{NTU_{ap}}}{\mathrm{NTU}} \approx 1 - \frac{\mathrm{NTU_{ap}}}{\dfrac{\ln S}{S-1} + \dfrac{\mathrm{Pe}_x\mathrm{Pe}_y}{\mathrm{Pe}_y + S\mathrm{Pe}_x}} \tag{81}$$

$$\frac{\mathrm{NTU_{ap}}}{\mathrm{NTU}} \approx \frac{\mathrm{Pe}_{\mathrm{HTU}\,y}\mathrm{Pe}_{\mathrm{HTU}\,x}}{\mathrm{Pe}_{\mathrm{HTU}\,y}\mathrm{Pe}_{\mathrm{HTU}\,x} + \mathrm{Pe}_{\mathrm{HTU}\,y} + S\mathrm{Pe}_{\mathrm{HTU}\,x}} \tag{82}$$

In these equations, S denotes the stripping factor, mG_M/L_M. Equation 82 is only valid for a sufficiently high number of transfer units so that the correction factor becomes independent of $\mathrm{NTU_{ap}}$.

In the original study (111), $\mathrm{NTU_{ap}}/\mathrm{NTU}$ was calculated for thousands of hypothetical design cases as a function of both Pe and $\mathrm{Pe_{HTU}}$. The results were correlated and empirical expressions were given that can be evaluated on a handheld calculator, just as equations 81 and 82, but which approximate the computer calculation much better, to within about +/− 5%.

The recommended rapid design procedure consists of the following steps: (*1*) The apparent N_{OG} is calculated using equation 65. (*2*) The extent of axial dispersion is estimated from literature correlations for each phase in terms of Pe numbers and transformed into $\mathrm{Pe_{HTU}}$ values. (*3*) The correction factor $\mathrm{NTU_{ap}}/\mathrm{NTU}$ is estimated on the basis of the correlation given in the literature (111). A reasonable, conservative estimate may also be obtained using equation 82, provided $\mathrm{NTU_{ap}} > 5$. When the apparent number of transfer units is divided by this correction factor, the value of N_{OG} actually required under the influence of axial dispersion is obtained. (*4*) The packed tower height is found by multiplying N_{OG} by the true $\mathrm{H_{OG}}$. In order to obtain values for the latter, pilot-plant data has to be

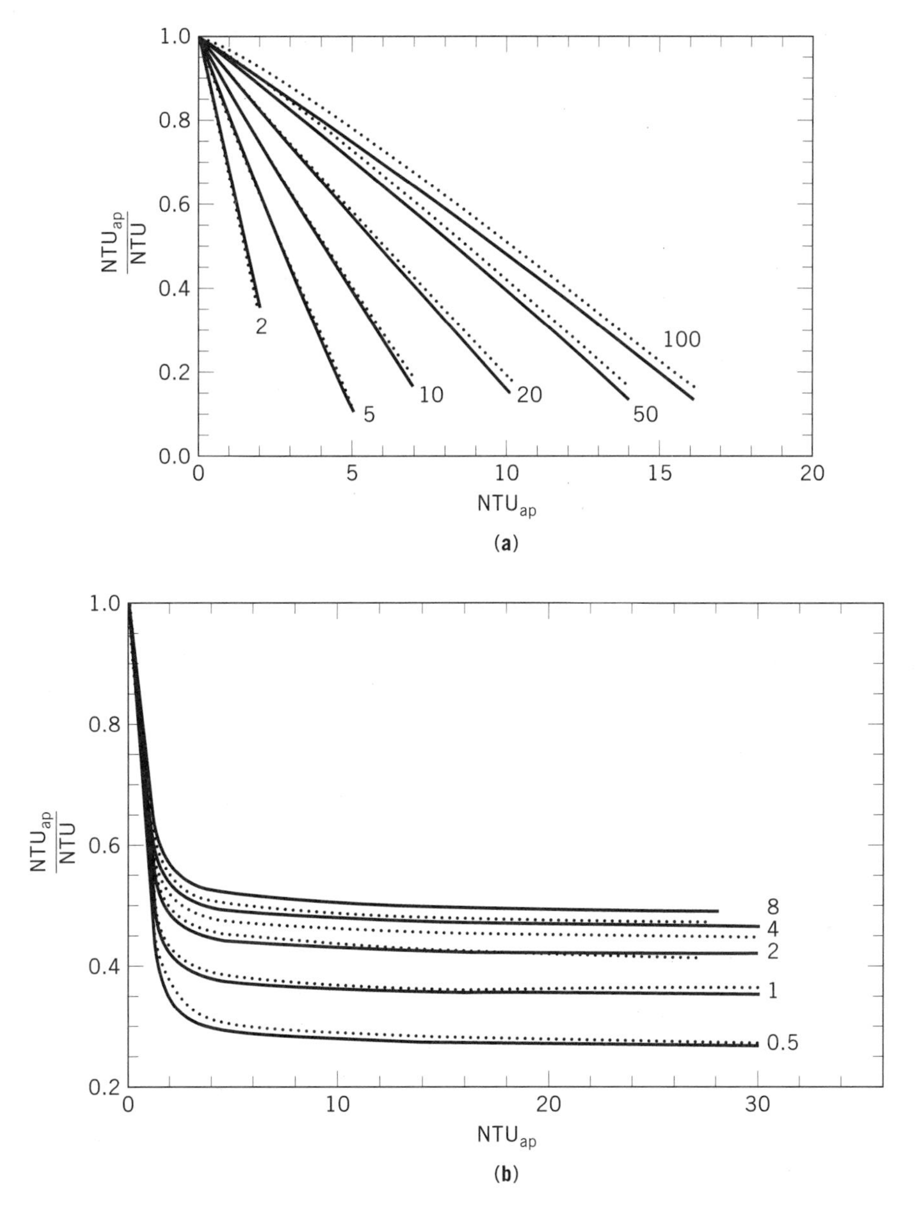

Fig. 17. Correction factor for axial dispersion as a function of NTU_{ap}. Solid lines are rigorous calculations; broken lines, approximate formulas according to literature (111). (**a**) Numbers on lines represent Pe_x values; $Pe_y = 20$; $mG_M/L_M = 0.8$. (**b**) For design calculations. Numbers on lines represent $Pe_{HTU,x}$ values; $Pe_{HTU,y} = 1$; $mG_M/L_M = 0.8$.

corrected for the influence of axial dispersion. This correction may be made in a manner similar to that described above, but equation 81 would be used to estimate the correction factor rather than equation (82).

4.7. Experimental Mass Transfer Coefficients. Hundreds of papers have been published reporting mass transfer coefficients in packed columns.

For some simple systems which have been studied quite extensively, mass transfer data may be obtained directly from the literature (6). The situation with respect to the prediction of mass transfer coefficients for new systems is still poor. Despite the wealth of experimental and theoretical studies, no comprehensive theory has been developed, and most generalizations are based on empirical or semiempirical equations.

Liquid-Phase Transfer. It is difficult to measure transfer coefficients separately from the effective interfacial area; thus data is usually correlated in a lumped form, eg, as $k_{L}a$ or as H_{L}. These parameters are measured for the liquid film by absorption or desorption of sparingly soluble gases such as O_2 or CO_2 in water. The liquid film resistance is completely controlling in such cases, and $k_{L}a$ may be estimated as $K_{OL}a$ since $x_i \approx x^*$ (Fig. 17). This is a prerequisite because the interfacial concentrations would not be known otherwise and hence the driving force through the liquid film could not be evaluated.

The resulting correlations fall into several categories. Some are essentially empirical in nature. Examples include a classical correlation proposed by Sherwood and Holloway (113).

$$H_L = \frac{1}{\alpha}\left(\frac{L}{\mu}\right)^n \left(\frac{\mu_L}{\rho_L D_L}\right)^{0.5} \tag{83}$$

The values of α and n are given in Table 3; typical values for D_L can be found in Table 4. The exponent of 0.5 on the Schmidt number ($\mu_L/\rho_L D_L$) supports the penetration theory. Further examples of empirical correlations provide partial experimental confirmation of equation 83 (3,114–118). The correlation reflecting what is probably the most comprehensive experimental basis, the Monsanto Model, also falls in this category (118,119). It is based on 545 observations from 13 different sources and may be summarized as

$$H_L = \phi C_{fl}\left(\frac{h}{3.05}\right)^{0.15} \mathrm{Sc}^{0.5}{}_L \tag{84}$$

The packing parameter $\phi(m)$ reflects the influence of the liquid flow rate as shown in Figure 18. C_{fl} reflects the influence of the gas flow rate, staying at

Table 3. **Values of Constants for Equation 83**

Packing	Size, cm	α^a	α^b	n
Raschig rings	0.95	3120	550	0.46
	1.3	1390	280	0.35
	2.5	430	100	0.22
	3.8	380	90	0.22
	5.1	340	80	0.22
Berl saddles	1.3	685	150	0.28
	2.5	780	170	0.28
	3.8	730	160	0.28

[a] Valid for units kg, s, m.
[b] Valid for units lb, h, ft.

Table 4. **Diffusion Coefficients for Dilute Solutions of Gases in Liquids at 20°C**

Gas	Liquid	D_L, $m^2/s \times 10^{-9}$
CO_2	water	1.78
Cl_2	water	1.61
H_2	water	5.22
HCl	water	0.61
H_2S	water	1.64
N_2	water	1.92
N_2O	water	1.75
NH_3	water	1.83
O_2	water	2.08
acetone	water	1.61
benzene	kerosene	1.41

unity <50% of the flooding rate but beginning to decrease above this point. At 75% of the flooding velocity, $C_{fl} = 0.6$. Sc_L is the Schmidt number of the liquid.

Other correlations based partially on theoretical considerations but made to fit existing data also exist (121–128). A number of researchers have also attempted to separate k_L from a by measuring the latter, sometimes in terms of the wetted area (56,129,130). Finally, a number of correlations for the mass transfer coefficient k_L itself exist (131,132). These are based on a more fundamental theory of mass transfer in packed columns (133–136). Although certain predictions were verified by experimental evidence, these models often cannot serve as design basis because the equations contain the interfacial area a as an independent variable. Only few correlations for the interfacial area in structured packing are available (122,124). Based on a mechanistic model for mass transfer, a way to estimate HETP values for structured packings in distillation columns has been proposed (137), yet there is a clear need for more experimental data in this area.

Gas-Phase Transfer. The height of a gas-phase mass transfer unit, or $k_G a$, is normally measured either by vaporization experiments of pure liquids, in which no liquid mass transfer resistance exists, or using extremely soluble gases. In the latter case, m is so small that the liquid-film resistance in equation 7 is negligible and the gas-film mass transfer coefficient can be observed as $k_G a \approx K_{OG} a$ and $y_i \approx y^*$. The experiments are difficult because they have to be carried out in very shallow beds. Otherwise, all of the highly soluble gas is absorbed and the driving force cannot be evaluated. The resulting end effects are probably the main reason for the substantial disagreement of the published data, which have been reported to vary some threefold for the same packing and flow rates (137). Furthermore, the effective interfacial areas seem to differ for absorption and vaporization (138). During the absorption experiments, the many stagnant or semistagnant pockets of liquid which exist in a packing tend to become saturated and thus ineffective. This is not the case in vaporization experiments where the total effective interfacial area consists of the surface area of moving liquid plus the semistagnant liquid pockets.

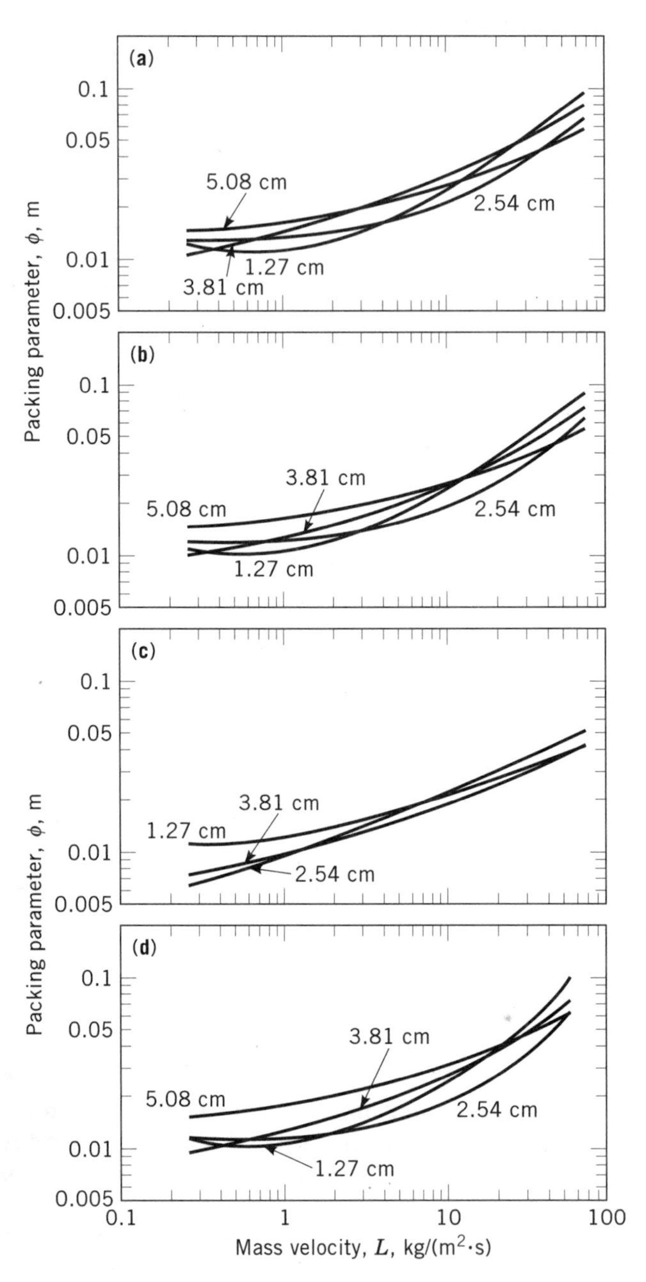

Fig. 18. Improved packing parameters ϕ for liquid mass transfer: (**a**) ceramic Raschig rings; (**b**) metal Raschig rings; (**c**) ceramic Berl saddles; (**d**) metal Pall rings (119). Reprinted with permission (120).

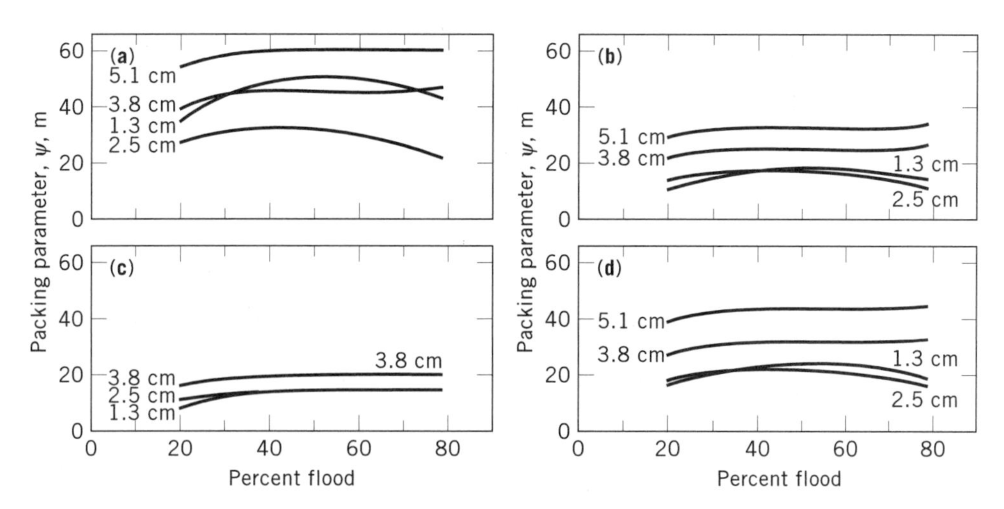

Fig. 19. Improved packing parameters ψ for gas mass transfer: (**a**) ceramic Raschig rings; (**b**) metal Rashig rings; (**c**) ceramic Berl saddles; (**d**) metal Pall rings (119). Reprinted with permission.

The correlation of H_G based on the most extensive experimental basis is again the Monsanto model (119):

$$H_G = \psi \frac{(3.28d'_c)^m (h/3.05)^{1.3}}{(737Lf_\mu f_\rho f_\sigma)^n} Sc_G^{0.5} \tag{85}$$

where d'_c is the lesser of 0.61 or column diameter, m; $f_\mu=(\mu_L/\mu_w)^{0.16}$; $f_\rho=(\rho_L/\rho_w)^{-1.25}$; $f_\sigma=(\sigma_L/\sigma_w)^{-0.8}$; Sc_G = Schmidt number of the gas. The packing parameter ψ in m, depends on the gas flow rate as shown in Figure 19. Values of the diffusivity of various solutes in air and typical Schmidt numbers for use in equation 85 are found in Table 5. The exponents m and n adopt the values of 1.24 and 0.6, respectively, for rings as packing materials, and 1.11 and 0.5, for saddles.

Mass transfer coefficients in the gas phase for traditional (dumped) packings are often computed from the correlation of Onda, either in the original form (56) or in a more recent and improved form (55). Useful correlations for structured packings have been published by Bravo and co-workers (53) and Spiegel and Meier (139). Software packages implementing kinetic design methods typically offer a number of alternatives for mass transfer correlation (123).

Another type of experiment to measure k_G separately from other factors consists of saturating packings made from porous materials using a volatile liquid and subsequently drying it by passing a stream of inert gas through the packing (140–143). Since the surface of the packing is normally known in these experiments, k_G can be computed. Application of these kinds of data to gas absorption design is difficult, however, because of the different, unknown effective interfacial areas when two phases are flowing through the packing. A similar approach was used by evaporating naphthalene from a packing made

Table 5. **Values of the Diffusion Coefficient D_A and of $\mu_G/D_{A\rho G}$ for Various Gases in Air at 0°C and at Atmospheric Pressure**[a]

Gas	CAS Registry Number	D_A, $m^2/s \times 10^{-5}$ Calculated	D_A, $m^2/s \times 10^{-5}$ Experiment	$\frac{\mu_G}{\rho_G D_A}$
acetic acid	[64-19-7]		1.05	1.26
acetone	[67-64-1]	0.83		1.60
ammonia	[7664-41-7]	1.62	2.17	0.61
benzene	[71-43-2]	0.72	0.78	1.71
bromobenzene	[108-86-1]	0.67		1.71
butane	[106-97-8]	0.75		1.77
n-butyl alcohol	[71-36-3]		0.69	1.88
carbon dioxide	[124-38-9]	1.19	1.39	0.96
carbon disulfide	[75-15-0]		278.00	1.48
carbon tetrachloride	[56-23-5]	0.61		2.13
chlorine	[7782-50-5]	0.92		1.42
chlorobenzene	[108-90-7]	0.61		2.13
chloropicrin	[76-06-2]	0.61		2.13
2,2′-dichloroethyl sulfide (mustard gas)	[505-60-2]	0.56		2.44
ethane	[74-84-0]	1.08		1.22
ethyl acetate	[141-78-6]	0.67	0.72	1.84
ethyl alcohol	[64-17-5]	0.94	1.03	1.30
ethyl ether	[60-29-7]	0.69	0.78	1.70
ethylene dibromide	[106-93-4]	0.67		1.97
hydrogen	[1333-74-0]	5.61		0.22
methane	[74-82-8]	1.58		0.84
methyl acetate	[79-20-9]		0.94	1.57
methyl alcohol	[67-56-1]	1.22	1.33	1.00
naphthalene	[91-20-3]		0.50	2.57
nitrogen	[7727-37-9]	1.33		0.98
n-octane	[111-65-9]		0.50	2.57
oxygen	[7782-44-7]	1.64	1.78	0.74
pentane	[109-66-0]	0.67		1.97
phosgene	[75-44-5]	0.81		1.65
propane	[74-98-6]	0.89		1.51
n-propyl acetate	[109-60-4]	0.67	0.67	1.97
n-propyl alcohol	[71-23-8]	0.81	0.86	1.55
sulfur dioxide	[7446-09-5]	1.03		1.28
toluene	[108-88-3]	0.64	0.72	1.86
water	[7732-18-5]	1.89	2.19	0.60

[a] The value of μ_G/ρ_G is that for pure air, 1.33×10^{-5} m^2/s. Diffusion coefficients may be corrected for other conditions by assuming them proportional to $T^{2/3}$ and inversely proportional to P. The Schmidt numbers depend only weakly on temperature (139).

from this material (130,138). The mass transfer coefficient k_G measured in this manner was then combined with k_G a data (144) to determine the effective area, which was found to be fairly independent of gas rate up to the loading point. The data bank underlying the Monsanto Model and literature correlations for k_G and k_L (130,138) have also been used (145) to develop a new correlation for packed distillation columns.

Height Equivalent to a Theoretical Plate. Provided both the equilibrium and operating lines are straight, HETP values may be estimated by combining the H_G and H_L values predicted by the above correlations and by translating the resulting H_{OG} into HETP by combining equations 56, 60, and 65 with equation 95, which is discussed under bubble tray absorption columns:

$$HETP = \frac{\ln(mG_M/L_M)}{(mG_M/L_M) - 1} H_{OG} \tag{86}$$

The HETP values obtained in this way have been compared to measured values in data banks (119) and statistical analysis reveals that the agreement is better when equations 84 and 85 are used to predict H_G and H_L than with the other models tested. Even so, a design at 95% confidence level would require a safety factor of 1.7 to account for scatter. The use of HETP in the special case of absorption with chemical reaction is discouraged, since the equilibrium line for many systems of chemical absorption is virtually horizontal. Therefore the absorption factor defined in the usual way L_M/G_Mm tends to infinity , the number of theoretical plates required apparently goes to zero and the height equivalent to a theoretical plate predicted by equation 86 tends to infinity. The more natural HTU/NTU approach is preferred (even for horizontal equilibrium line, the integral that defines the required NTU is well behaved).

5. Case Studies

In real-world situations one will sometimes be confronted simultaneously with several of the complications treated schematically above. Standard design proceedures may then yield little more than a first approximate idea of packing requirements and column operation. Even a trustwothy preliminary design should be based in such situations on a computer calculation of the concentration profiles. As an illustration of how the principles of mass transfer and packed column design discussed earlier may be used in computer models, we present in what follows two relatively complex real-world cases of gas absorption processes.

5.1. Absorption with Chemical Reactions. The number and variety of industrial applications of absorption with chemical reaction has grown significantly in the last decade, partly driven by environmental considerations and partly by the increased use of natural gas as an energy source. Some typical applications include the absorption of NO_x in aqueous solutions in the synthesis of nitric acid (68) or in caustic in the synthesis of sodium nitrite (69), selective and unselective absorption of H_2S from sour gas streams, possibly containing large amounts of CO_2 using amines, refinery sour water stripping operations, regenerative Wellman-Lord desulfurization of flue gases by means of sodium sulfites (70).

Such applications almost invariably involve a complex set of reactions, the chemical reaction kinetics and equilibrium data of which are not always available or reliably known. These difficulties represent a serious obstacle to the rigorous design of chemical absorbers. Consequently, designs are very often complemented by extensive laboratory and pilot plant tests. In spite of design uncertainties, the benefits to be gained from gas absorption with chemical reaction have not been overlooked by the chemical and petrochemical industries and new applications and processes appear continually (71).

The selective absorption of H_2S from CO_2-rich sour natural or process gases is a good illustration of the significant advances achieved in the field and motivated by the increased use of sour natural gases: up to approx. the early 1980s large amounts of sweet natural gases, containing little if any H_2S and CO_2 were readily available. Such gases could be treated with unselective solvents that absorbed both acid components. As production from sweet gas wells started to dwindle, other sour natural gas sources had to be tapped.

Treating such sour gases with a conventional amine (like monoethanolamine MEA) in order to reduce H_2S concentrations to an acceptable level would lead to the absorption of a very large fraction of the accompanying CO_2. The resulting absorption plant would then be unnecessarily overdesigned, both in the size of the equipment and in the energy requirements for amine regeneration. Through the use of amines such as methyl-diethanolamine (MDEA) or of the sterically hindered 2-amino-2-methyl-1-propanol (AMP) it is nowadays possible to meet H_2S specifications while co-absorbing a moderate amount of CO_2 thus greatly improving the economics of the process. Hindered amines are finding increased application in this area since it is generally believed that steric effects reduce the stability of the carbamates formed by the amine with CO_2. The carbamates readily undergo hydrolysis and release free amine that again reacts with CO_2 thus increasing the loading capacity of the amine. In addition, hindered amines demonstrate very good selectivity towards H_2S in the presence of CO_2. Consequently, hindered amines have found their way into commercial gas treating processes such as Flexsorb SE (72). However, even in well documented cases, like H_2S/CO_2 absorption with AMP (73,74) the chemical complication of the system can be appreciable. As an illustration of the type of calculations involved in gas absorption with subsequent chemical reaction in the liquid phase, we will consider the absorption of moderate amounts of carbon dioxide CO_2 and hydrogen sulphide H_2S (in a molar ratio of 3:1) from a low pressure fuel gas stream into a methyl-diethanolamine (MDEA) solution. This application is a good representative of a typical sweetening application for a semi-sour refinery fuel gas (71). Absorption will be carried out in a column packed with structured packing. At the relatively low specific liquid load required (15 m^3/m^2 h), there is no appreciable hydrodynamic gas-liquid interaction and liquid flow over the corrugated metal packing sheets is laminar to a very good approximation (see Refs. 54, 139 for a detailed description of the packing geometry). For the time being, we will assume that no appreciable heat effects need to be taken into account, ie, we will consider isothermal absorption only.

Mass transfer in the gas phase will be handled by means of mass transfer coefficients valid for structured packings (53,132,139). We will focus on a short section of packing between two bends. Additionally, we will assume the diffusive contribution to the flux in the direction of the liquid flow to be negligible with respect to convection, ie, the mechanism of transfer across the liquid film is diffusion, whereas along the film convection is the dominant mechanism.

The following table lists the chemical species present in the system:

species	H_2O	OH^-	H_3O^+	CO_2	HCO_3^-	CO_3^{2-}	H_2S	HS^-	S^{2-}	MDEA	$MDEA^+$
species label	1	2	3	4	5	6	7	8	9	10	11

The reactions to be considered are:

1. Water dissociation:

$$2\ H_2O \leftrightarrow H_3O^+ + OH^-$$

2. CO_2 first dissociation:

$$CO_2 + 2\ H_2O \leftrightarrow H_3O^+ + HCO_3^-$$

3. CO_2 second dissociation:

$$HCO_3^- + H_2O \leftrightarrow H_3O^+ + CO_3^{2-}$$

4. H_2S first dissociation:

$$H_2S + H_2O \leftrightarrow H_3O^+ + HS^-$$

5. H_2S second dissociation:

$$HS^- + H_2O \leftrightarrow H_3O^+ + S^{2-}$$

6. MDEA protonation:

$$H_3O^+ + MDEA \leftrightarrow MDEAH^+ + H_2O$$

Proper kinetics will be taken into account for all the previous reactions, i.e. we will not assume chemical equilibrium for any of these reactions. We have kept the previous set of chemical reactions as complete as possible for illustration purposes and because for a general system it is not always obvious a priori which, if any, of a set of reactions can be neglected or treated in simplified manner. For H_2S and CO_2 absorption in MDEA, it is usual to reduce the number of chemical reactions (and accompanying kinetic parameters) by neglecting the concentrations of S^{2-} and $CO_3{}^{2-}$. This would be reasonable in the present case because, for both gases, the second dissociation constant is at least a factor of 1000 smaller than the first one. A usual second assumption is to neglect OH^- and H^+ concentrations at all gas loadings (146) because both MDEA and acid gases are a weak base and weak acids in water, respectively.

Extensive transport, kinetic and thermodynamic data for this system are available in Refs. 146–153. The assumptions stated above of isothermal operation, liquid laminar flow, use of mass transfer coefficients for the gas phase and no axial dispersion due to diffusion allow us to write balance equations for all components in the liquid control volume defined in Figure 20. This figure represents the flow between two bends in a packing element. Upon reaching the next bend, the fluid is mixed and concentration gradients across the film

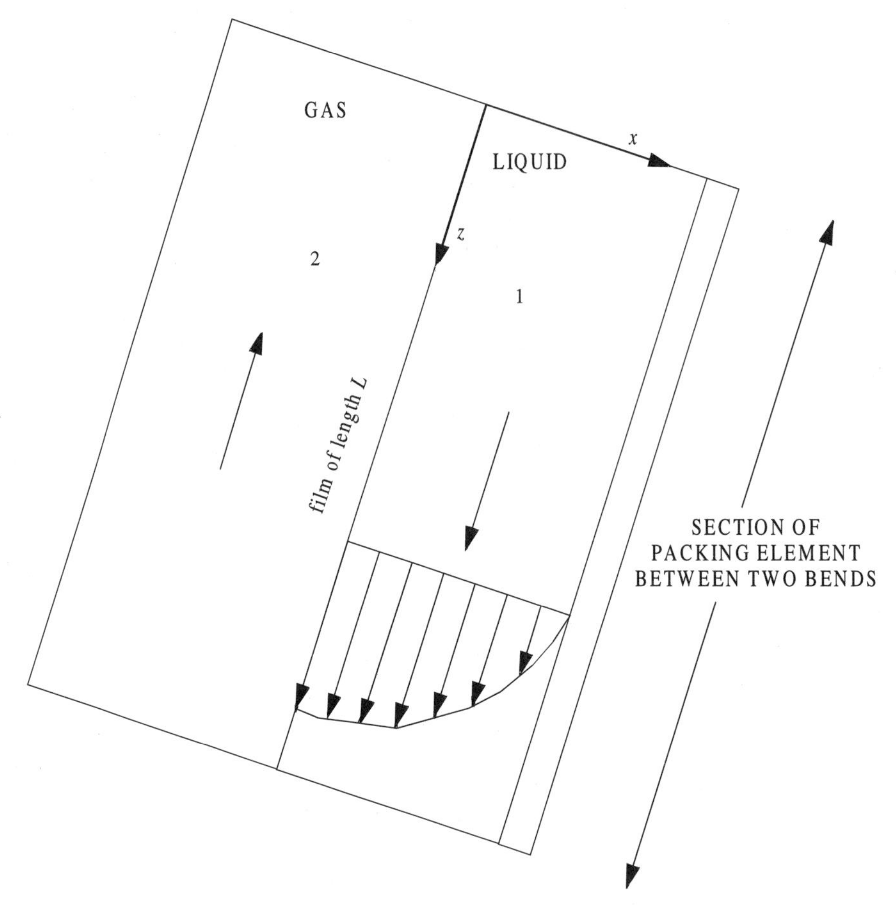

Fig. 20. Schematic of contercurrent absorption from a gas into a laminar film flowing along a structured packing.

equalized. A succession of such laminar flows followed by complete mixing is a widely accepted mechanism for liquid side mass transfer in structured packings (22,145). Mass transfer and complex chemical reaction in each of the laminar flow sections can therefore be treated naturally by means of the following scheme:

For each of the species, the following differential mole balance equation must hold in the liquid domain Ω_1:

$$-D_i \frac{\partial^2 c_1}{\partial x^2} + v_z \frac{\partial c_j}{\partial z} - \sum_j r_{ij} = 0 \qquad i = 1, 11 \quad j = 1, 6 \tag{87}$$

where $v_z = 3/2\, \bar{v}\, (x/\delta)^2$ is the velocity profile across the film, $\bar{v}$ is the average liquid velocity (proportional to the liquid throughput), δ is the film thickness and r_{ij} is the reaction velocity for species i in reaction j (here we follow the nomenclature and sign conventions of Ref. 154).

In the gas domain Ω_2, the following differential mode balance equations must hold for the two species undergoing transfer:

$$\frac{dp_{co_2}}{dz} = -\frac{Pk_{g,CO_2}}{G}\,(p_{CO_2} - H_{CO_2}c_{CO_2}) \tag{88}$$

$$\frac{dp_{H_2S}}{dz} = -\frac{Pk_{g,H_2S}}{G}\,(p_{H_2S} - H_{H_2S}c_{H_2S}) \tag{89}$$

where P is the total pressure in the column and G the gas load (moles per unit area per unit time).

In addition, for Eq. (87) the following boundary conditions must be imposed:

$$c_i(x,0) = 0 \qquad \text{for} \quad i = 4-9, 11$$

$$c_j(x,0) = c_i^0 \qquad \text{for} \quad i = 1,2,3,10$$

which correspond to inlet of unloaded MDEA solution at the top. Boundary conditions for mass transfer across the film must also be imposed:

$$-D_i\,\frac{\partial c_j}{\partial x}\bigg|_{\substack{x=\delta\\ \forall z}} = 0 \qquad \text{for} \qquad i = 1,2,3,5,6,8-11$$

$$-D_{CO_2}\,\frac{\partial c_{CO_2}}{\partial x}\bigg|_{\substack{x=\delta\\ \forall z}} = k_{g,CO_2}\,(p_{CO_2} - H_{CO_2}c_{CO_2}) \tag{90}$$

$$-D_{SH_2}\,\frac{\partial c_{SH_2}}{\partial x}\bigg|_{\substack{x=\delta\\ \forall z}} = k_{g,SH_2}(p_{SH_2} - H_{SH_2}c_{SH_2}) \tag{91}$$

And finally,

$$-D_i\,\frac{\partial c_i}{\partial x}\bigg|_{\substack{x=0\\ \forall z}} = 0 \qquad \text{for} \quad i = 1-11$$

at the surface of the packing (packing impenetrable to mass transfer).

For Eqs. (88) and (89) the following gas inlet boundary conditions must be imposed:

$$p_{CO_2}(\mathrm{L}) = p^0_{CO_2} \qquad \text{and} \quad p_{H_2S}(\mathrm{L}) = p^0_{H_2S}$$

where $p^0_{CO_2}$ and $p^0_{H_2S}$ are the partial pressures of the sour components at the gas inlet ($z = L$).

The solution sought are thus the concentrations of all species throughout the liquid domain and the partial pressures of the sour components in the gas domain.

Although these equations are quite simple to write formally, they contain terms such as Henry's constants, which are reather tedious to compute, since they are a complicated function of composition, temperature and pressure (eg, in the present example, the Clegg-Pitzer equation of state was used to describe sour gas solubility).

It is important to keep in mind that this approach leads to a rather large system of mixed partial (for the liquid phase, where reaction and diffusion take place) and ordinary (for the gas phase where the mass transfer formalism is applied) differential equations. The equations in the liquid domain are strongly coupled through the kinetic terms r_{ij}. The solution of the system is further complicated by numerical difficulties which appear if variables are not scaled properly. Marching or multiple shooting methods (155) are not very successful in this type of calculations. Nevertheless, the geometry of the example is simple enough so that rather straightforward Finite Differences (FD) methods can be used. In a general case with a more complex geometry, however, a Finite Element (FE) approach is mandatory. Figure 21 shows the concentration profiles of the sour components in the film at four different positions along the packing computed using a FE method to solve the set of Eqs. 87–89 with their boundary conditions.

Doubly ionized species are present in very low concentrations, as expected. Hence the usual assumption of neglecting them in practical calculations. Additionally, the selective absorption of CO_2 over H_2S by the amine is also evident (notice the differences in scales in the ordinates of both plots). Although the amount of CO_2 in the gas feed is three times larger than that of H_2S, the selectivity of MDEA results in a enhanced absorption of hydrogen sulfide, as is shown by the concentration of HS^- being approximately ten times higher than that of bicarbonate at all locations within the film. Higher point selectivities can be reached in practice by careful selection of solvent, column internals and operating conditions. For column and internal optimization, the problem has to be solved repeatedly, and especialized software packages (see below under Design of Nonisothermal Scrubbers with Chemical Reactions) are a requirement.

Although computationally expensive, the advantage of this approach is that it is rate-based and yields full information on concentration profiles for all components. In addition, concentration dependent diffusivities can be incorporated in the calculation easily.

5.2. Nonisothermal Scrubbers with Chemical Reactions. Amine scrubbers may serve as an example for gas absorption processes involving both chemical reactions and heat effects. In addition to detailed knowledge of transport, equilibrium and reaction kinetics, the design of amine scrubbers has to take into account large absorption heat effects (see section on Nonisothermal Gas Absorption above). Complex design cases like gas sweetening require an understanding of both phase behavior coupled with chemical equilibrium and mass transfer accompanied by chemical reactions (75,76). The complexity of the design makes it necessary to resort to highly specialized software packages (AMSIM from DBR Software Inc, Hysys from Hyprotech/AEA Technology plc, PROSIM and TSWEET from Bryan Research and Engineering, Inc, etc) or to special packages within general purpose flowsheet simulators. Most flowsheet simulators contain nowadays optional data packages in which full kinetic and equilibrium data are contained (ASPEN PLUS from Aspen Technologies, PRO/II from SimSci, Design II from WinSim, Inc., etc).

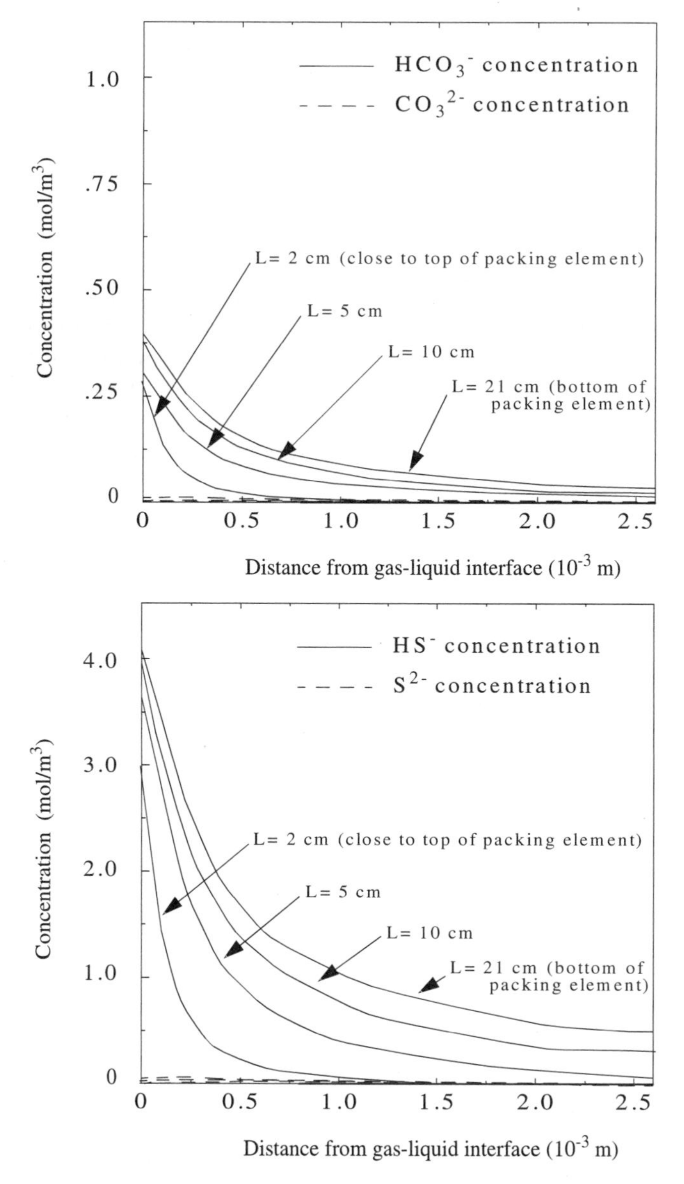

Fig. 21. Concentration profiles across liquid film for sour components. Isothermal case.

These special-purpose programs or packages typically utilize a modified form of the Murphree Stage Efficiency model to calculate the stage efficiency at each stage in the column, rather than using an ideal stage approach for the entire column. This allows the calculation of individual component stage efficiencies as a function of pressure, temperature, phase compositions as well as kinetic and mass transfer parameters. Alternatively, they use more advanced and rigorous approaches, based on nonequilibrium models and even including the possibility of dynamic modeling (51,77). It must be said that these advanced design techniques are very demanding in terms of computation and its industrial application is therefore not very widespread yet. Nevertheless given the rapid growth of performance/cost ratio for computing hardware, they will very probably see increased usage in coming years. The sophistication of numerical design procedures has reached a point where the current limiting factor in practice, from the user's point of view is the availability of reliable thermodynamic, chemical and transport properties.

In order to illustrate a rigorous calculation procedure for chemical reaction with heat effects, we will again consider the sour gas sweetening example already presented but we will now remove the assumption of isothermal operation. This situation arises when the reaction enthalpies and the rate of absorption of sour gases (i.e. the heat release rate) are sufficiently high. Assuming the reaction heat to remain entirely in the liquid, the system of equations (87–89 plus boundary conditions) are still valid if we now consider temperature dependent physical properties (reaction rates, specific heats, diffusivities, etc).

In addition, that system must be augmented by an energy balance equation with takes the form:

$$-k_{\text{sol}} \frac{\partial^2 T}{\partial x^2} + v_z \rho C_p \frac{\partial T}{\partial z} - \sum_j r_{ij}(T) \Delta H_{\text{R}_{ij}}(T) = 0 \qquad j = 1, 6$$

in which k_{sol}, ρ, and C_{p} are the thermal conductivity, the density of the solution and the specific heat of the solution (themselves weak functions of composition and temperature) and $\Delta H_{\text{R}_{ij}}(T)$ is the molar enthalpy of reaction number j referred to component i. Component i is the component to which chemical reaction rates and enthalpy of reaction are referred (154). Unlike the other thermophysical properties the term r_{ij} is very strongly temperature dependent (usually an Arrhenius dependence).

Under the assumption of no heat transfer between gas and liquid, the proper boundary conditions for the liquid temperature field are

$$\left.\frac{\partial T}{\partial x}\right|_{\substack{x=0 \\ \forall z}} = 0$$

at the surface of the packing (by symmetry, since both sides of the packing are covered by liquid),

$$\left.\frac{\partial T}{\partial x}\right|_{\substack{x=\delta \\ \forall z}} = 0$$

at the gas–liquid interphase and

$$T(x, 0) = T_0$$

at the inlet, where T_0 is the known inlet temperature of the liquid. The augmented system of equations necessitates numerical solution in order to obtain the concentration and temperature fields. Figure 22 shows the effect that taking

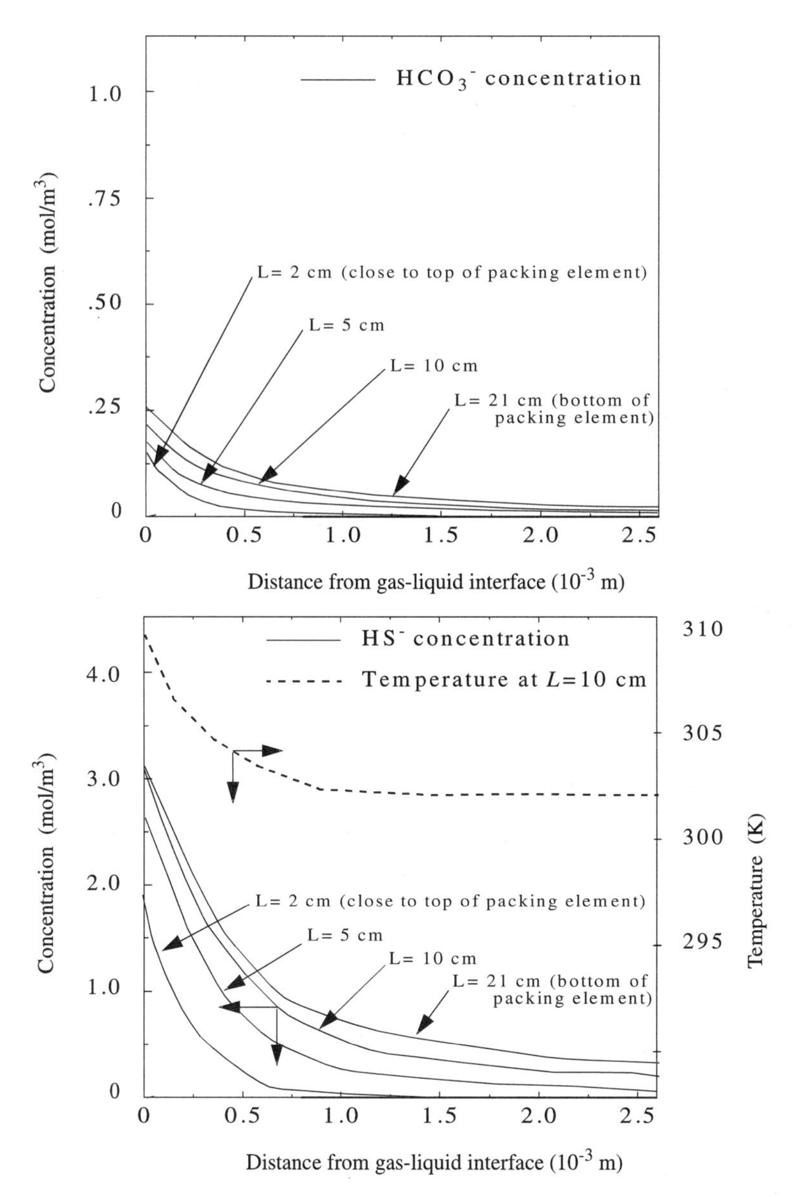

Fig. 22. Concentration profiles a cross liquid film for sour components. Nonisothermal case.

the heats of reaction into account has on the concentration profiles for the species associated with sour components:

Since the formation of both bicarbonate and hydrosulfide are both exothermic, temperature increases throughout the liquid film, but most appreciably close to the interphase. Sour component solubilities in this region drop correspondingly, leading to a reduction in the amounts absorbed. The overall rate of absorption drops by 28%. Selectivity, however, remains approximately the same as in the isothermal case.

The results presented in Figure 21 were obtained under the assumption that no heat transfer takes place by conduction or convection between both phases. This approximation can be removed, if interphase heat transfer coefficients are known, by adding another heat balance (ordinary differential) equation for the gas. In industrial practice it is however quite usual to work under the adiabatic assumption for the liquid phase in the cases where the gas is cooler than the liquid or where solvent evaporation takes place. The assumption of adiabatic liquid phase eliminates the cooling effect of the gas and thus leads to higher liquid temperatures, less effective absorption and to correspondingly more conservative design of contacting equipment.

A similar situation is often encountered for heat effects due to evaporation or condensation of solvent: the latent heat can be a cause of cooling or heating of the liquid phase that can be incorporated in the equations above adding the term for release of latent heat in the energy balance equation. Standard commercial software is generally able to take this solvent effect into account. But preliminary designs are often performed excluding latent heat terms in the cases where this exclusion leads to more conservative and, hence, safer design.

However, in those cases where the gas stream enters the column at a higher temperature than the liquid, taking proper account of heat transfer between phases is mandatory.

In all cases, during the final design stage, where tighter designs are required, the full mass and energy balance equations must be considered.

It is important to emphasize that including heat effects in the formulation increases the number of equations by one (or by one partial and one ordinary differential equation if heat transfer between gas and liquid is to be considered as well), ie, in a multicomponent system, it represents only a moderate increase in computation. Furthermore, experience shows that the heat balance equation itself is well behaved and seldom causes additional numerical problems.

Complex as the situation is in sour gas treating, it is not nearly as involved as in other cases, like sour water strippers (78) or flue gas treating (79) where the product stream to be treated is a very complex reacting mixture of a large number of species of not always exactly known composition. As an example, a widely used flue gas treating modeling package involves a set of 36 chemical reactions in which 26 nonionic species participate. Additional ionic species are produced from these aqueous phase ionic reactions so that the system actually contains 22 additional components (80).

By now a significant body of specialized literature dealing with the general subject of absorption with chemical reaction is available (51,71,81–84). Nevertheless, it is probable that an even larger amount of information, mainly data on thermophysical and equilibrium properties and column internals, is kept as proprietary by the companies which have developed specific processes. As is obvious from the preceding example, the rigorous design of chemical absorption columns, whether isothermal or not, requires considerable computational effort.

6. Capacity Limitations of Packed Absorption Columns

Thus far the discussion has been confined to factors affecting the tower height required to perform a specific absorption job. The necessary tower diameter, on

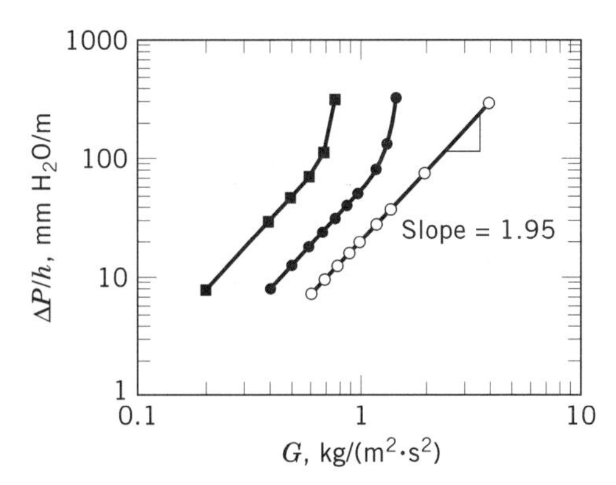

Fig. 23. Schematic representation of typical pressure drop as a function of superficial gas velocity, expressed in terms of $G = \rho_G u_G$, in packed columns. ○, Dry packing; ●, low liquid flow rate; ■, higher liquid flow rate. The points do not correspond to actual experimental data, but represent examples.

the other hand, depends primarily on the total amount of gas and liquid that must be handled. At a given set of flow rates, the diameter of the packing can only be decreased at the expense of a large pressure drop, which in turn generates higher operating costs because more power is needed to blow the gas through the packing. The reason for this is the fact that handling a given total gas flow rate in a smaller tower diameter increases the superficial velocity at which the gas has to be pushed through the packing.

The relationship between the pressure drop per unit of packed height and the superficial gas velocity given in terms of the gas flow rate is shown schematically in Figure 23. In a dry packing, ΔP increases almost as the square of the gas velocity, which is in accord with the turbulent nature of the flow. At low liquid flow rates, the curves are somewhat shifted upwards because the presence of liquid films restricts the free section available for gas flow and thus increases the linear gas velocity somewhat. Because the liquid hold-up remains independent of G, the slope of the curve in this log–log plot remains close to 2. At higher pressure drops, however, the upflowing gas impairs the downflow of liquid and excess liquid starts to accumulate in the packing, thereby increasing the hold-up. In this operating region, called the loading zone, the increasing liquid hold-up restricts the free section available for the gas flow further as G becomes larger. Hence the linear gas velocities increase faster than G and the power dependence of ΔP on G starts to rise >2. The G value at which the curve begins to deviate from the straight line has been defined as the loading point.

If the tower diameter is made too small for a given total gas flow rate, that is, if u_G and G are increased above a certain critical value, ΔP becomes so great that the liquid cannot flow downward anymore over the packing, but is blown out the top of the packing. The vertical asymptotes on Figure 23 indicate the gas rates at which this condition, called flooding, occurs. This gas flow rate at flooding, G_F, determines the theoretical minimum diameter at which the tower is operable, and knowledge of it is therefore very important. Flooding rates have been correlated (156–158).

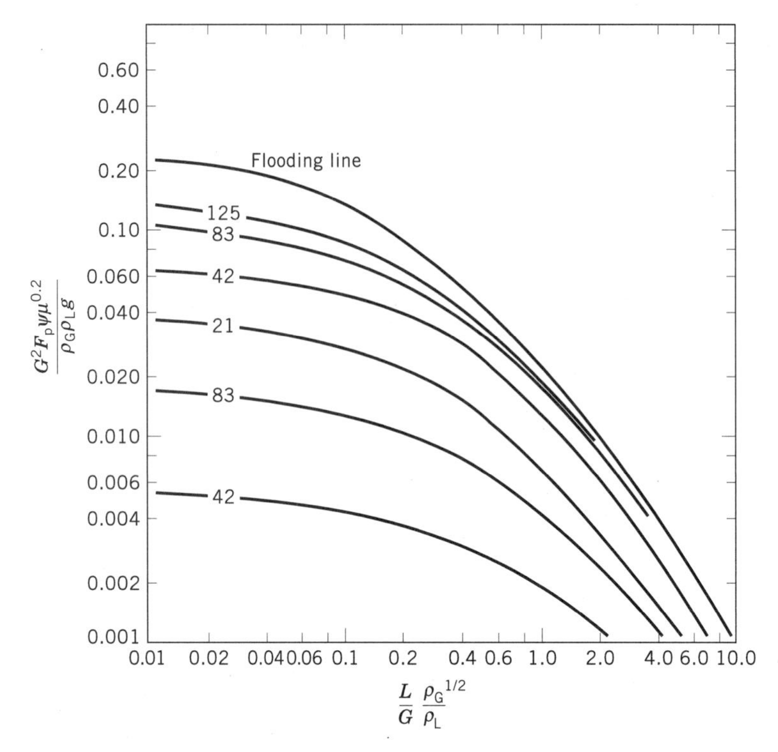

Fig. 24. Pressure drop and flooding correlation for various random packings (159). $\psi = \rho_{H_2O}/\rho_L$, g (standard acceleration of free fall) $= 9.81 m/s^2$, μ = liquid viscosity in mPa · s; s; numbers on lines represent pressure drop, mm H_2O/m of packed height; to convert to in. H_2O/ft multiply by 0.012. Packing factors for various packings have been published (160) and are reproduced in part in Table 6.

Table 6. **Characteristics of Dumped Tower Packings**[a]

Packing type	Nominal size, mm	Surface area, m^2/m^3	Packing factor, F_P, m^{-1}
Raschig rings, ceramic	6	710	5250
	13	370	2000
	25	190	510
	50	92	215
Raschig rings, steel	25	185	450
	50	95	187
Berl saddles, ceramic	6	900	2950
	13	465	790
	25	250	360
	50	105	150
Pall rings, metal	25	205	157
	50	115	66
Pall rings, polypropylene	25	205	170
	50	100	82

[a] Ref. 160.

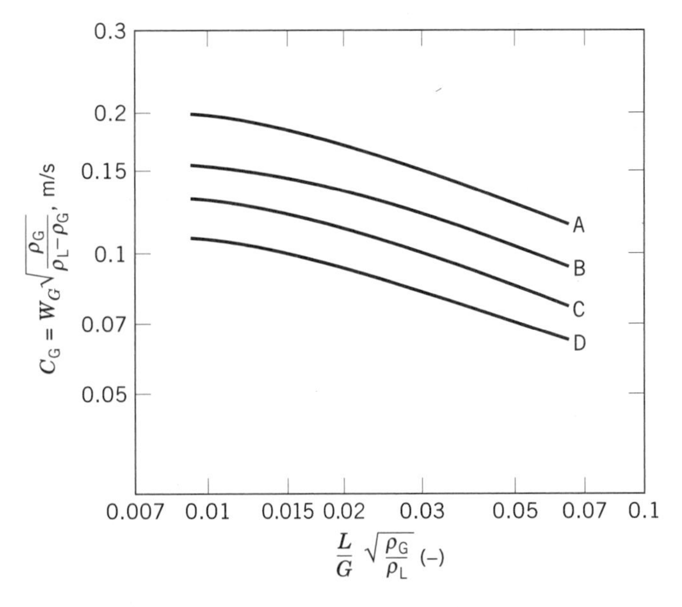

Fig. 25. Souders load diagram for capacity limit determination for four structured packings of the Sulzer-Mellapak type. The solid lines represent the capacity limits of the respective packings as defined by a pressure drop of 1.2 kPa/m: A, 125 *Y*; B, 250 *Y*; C, 350 *Y*; D, 500 *Y*. Flooding rates are ~5% higher. Reprinted with permission (139).

Both the pressure drop per unit length of packed tower and the gas flooding rate have been correlated for random packings as shown in Figure 24 (159). Such correlations enable predicting the gas flow rate G that will flood the packing at a given L/G ratio. In practice, the tower has to be operated at flow rates considerably less than the flooding rates for safety reasons. It is generally accepted that 50–80% of the flooding flow rates can be permitted. The curves plotted on Figure 24 thus also enable prediction of the pressure drop at any chosen operating value of G. The diameter of the column must then be evaluated by comparing this value to the total quantity of gas that the tower is supposed to handle. The correlation shown in Figure 24 can be applied to predict the hydraulic performance of many different packings owing to an adjustable parameter known as the packing factor F_P. Values for F_P have been compiled (160); a few examples are listed in Table 6. Similar flooding rate correlations are available for arranged packings. Figure 25 reports results for four examples of Mellapak types (139). Comparison of Figures 25 and 24 reveals higher capacity limits in structured than in many of the dumped packings. At similar loads, structured packings tend generally to give rise to less pressure drop.

7. Bubble Tray Absorption Columns

7.1. General Design Procedure. Bubble tray absorbers may be designed graphically based on a so-called McCabe-Thiele diagram. An operating line and an equilibrium line are plotted in y-x, Y'-X', or Y^0-X^0 coordinates using

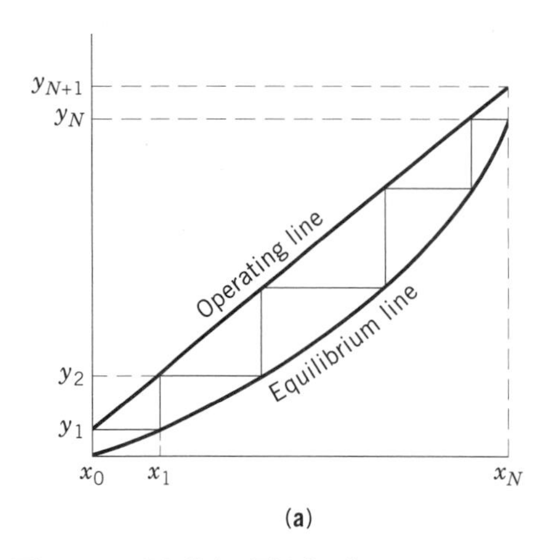

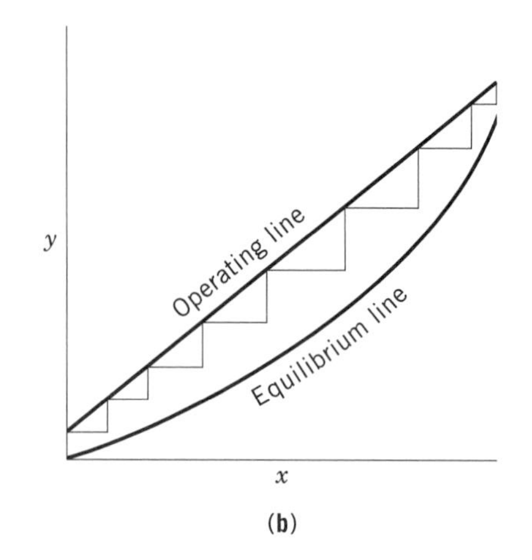

Fig. 26. McCabe-Thiele diagram. (**a**) Number of theoretical plates, 5; (**b**) number of actual plates, 8.

the principles for packed adsorbers outlined above (see Fig. 26). The minimum number of plates required for a specified recovery may be computed by assuming that equilibrium is reached between the two phases on each bubble tray. Thus the gas and the liquid leaving a tray are at equilibrium and a hypothetical tray capable of equilibrating the phase streams is termed a theoretical plate. Starting the calculation at the bottom of the tower, where the concentrations are y_{N+1} and x_N (see Fig. 27), the concentration leaving the lowest theoretical plate y_N may be found on the design diagram (Fig. 26**a**) by moving from the operating line vertically to the equilibrium line, because y_N is at equilibrium with x_N. Since the concentrations between two plates are always related by the operating line, x_{N-1} may be found from y_N by moving horizontally to the operating line. By repeating this sequence of steps until the desired residual gas concentration y_1 is reached, the number of theoretical plates can be counted.

The required number of actual plates, N_P, is larger than the number of theoretical plates, N_{TP}, because it would take an infinite contacting time at each stage to establish equilibrium. The ratio N_{TP}: N_P is called the overall column efficiency. This parameter is difficult to predict from theoretical considerations, however, or to correct for new systems and operating conditions. It is therefore customary to characterize the single plate by the so-called Murphree vapor plate efficiency, E_{MV} (161):

$$E_{MV} \equiv \frac{y_n - y_{n+1}}{y_n^* - y_{n+1}} \tag{92}$$

which indicates the fractional approach to equilibrium achieved by the plate. An efficiency of 80% means that the reduction in solute gas concentration effected by the plate is 80% of the reduction obtained from a theoretical plate. Corresponding actual plates may therefore be stepped off by moving from the operating line vertically only 80% of the distance between operating and equilibrium line (Fig. 26**b**). In some special cases having negligible resistance in the gas phase, E_{MV} values may become unreasonably small. It is then more logical to define a

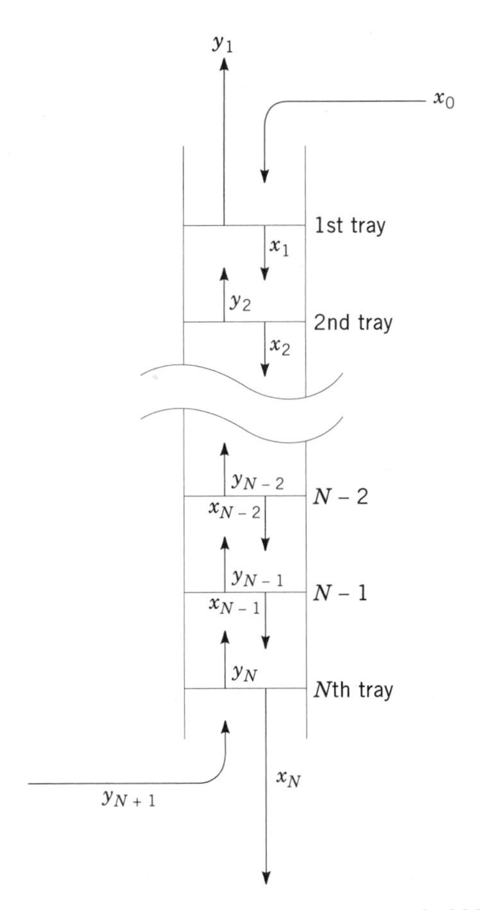

Fig. 27. Numbering plates and concentrations in a bubble tray column.

Murphree liquid plate efficiency, E_{ML}, simply by reversing the role of liquid and gas and by focusing on the change in liquid composition across the plate with respect to an equilibrium given by the leaving vapor.

7.2. Simplified Design Procedure for Linear Equilibrium and Operating Lines. A straight operating line occurs when the concentrations are low such that L_M and G_M remain essentially constant. (The material balance is obtained from eq. 44). In cases where the equilibrium K value does not depend too much on concentration, the use of absorption and stripping factors (162–164) allows rapid calculations for absorption design. One of the simplifying assumptions made in the development of this so-called Kremser-Brown method involves the use of the absorption factor A. The following algebraic expression describes the liquid and vapor flows from the plate (165):

$$A_n = \frac{L_n}{K_n G_n} \quad \text{or} \quad A_{av} = \frac{L_{av}}{K_{av} G_{av}} \tag{93}$$

where A_n is the absorption factor for each plate n and L_n and G_n are the liquid and vapor flows from the plate. The fractional absorption of any component by an absorber of N plates is expressed in a form similar to the Kremser equation (163,164,166),

$$\frac{Y_{N+1} - Y_1}{Y_{N+1} - K_{x_0}} = \frac{A^{N+1} - A}{A^{N+1} - 1} \tag{94}$$

where Y_{N+1} = moles of absorbed component entering the column per mole of entering vapor and Y_1 = mole of absorbed component leaving the column per mol of entering vapor. The calculation of plate efficiency (163) is quite sensitive to the choice of equilibrium constants (4,167).

For linear equilibrium and operating lines, an explicit expression for the number of theoretical plates required for reducing the solute mole fraction from y_{N+1} to $y_1$has been derived (86):

$$N_{TP} = \frac{\ln\left[\left(1 - \frac{mG_M}{L_M}\right)\left(\frac{y_{N+1} - mx_0}{y_1 - mx_0}\right) + \frac{mG_M}{L_M}\right]}{\ln\frac{L_M}{mG_M}} \tag{95}$$

This is the one case where the overall column efficiency can be related analytically to the Murphree plate efficiency, so that the actual number of plates is calculable by dividing the number of theoretical plates through equation 96:

$$\frac{N_{TP}}{N_P} = \frac{\ln\left[1 + E_{MV}\left(\frac{mG_M}{L_M} - 1\right)\right]}{\ln\frac{mG_M}{L_M}} \tag{96}$$

7.3. Nonisothermal Gas Absorption. The computation of nonisothermal gas absorption processes is difficult because of all the interactions involved as described for packed columns. An very large number of plate calculations is normally required to establish the correct concentration and temperature profiles through the tower. Suitable algorithms have been developed (90,168) and nonisothermal gas absorption in plate columns has been studied experimentally and the measured profiles compared to the calculated results (91,169). Figure 28 shows a typical liquid temperature profile observed in an adiabatic bubble plate absorber (170). The close agreement between the calculated and observed profiles was obtained without adjusting parameters. The plate efficiencies required for the calculations were measured independently on a single exact copy of the bubble cap plates installed in the five-tray absorber.

A general, approximate, short-cut design procedure for adiabatic bubble tray absorbers has not been developed, although work has been done in the field of nonisothermal and multicomponent hydrocarbon absorbers. An analytical expression has been developed which will predict the recovery of each component provided the stripping factor, ie, the group mG_M/L_M, is known for each component on each tray of the column (165). This requires knowledge of the temperature and total flow (G_M and L_M) profiles through the tower. There are many suggestions about how to estimate these profiles (165,166,171).

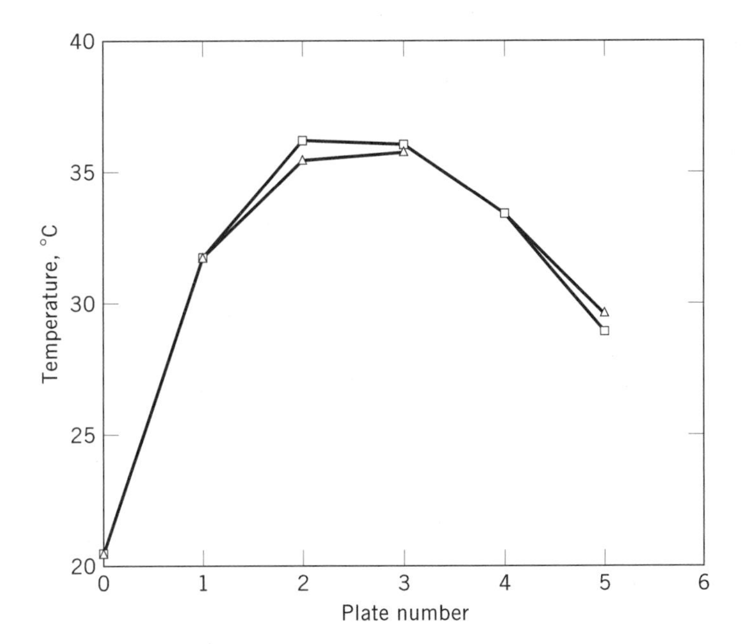

Fig. 28. Computed and experimental liquid temperature profiles in an ammonia absorber with five bubble cap trays (170). Water was used as a solvent. y_{N+1}=0.123; y_1 = 0.0242; L'_M/G'_M = 1.757; △ measured; □, calculated.

7.4. Plate Efficiency Estimation

Rate of Mass Transfer in Bubble Trays. The Murphree vapor efficiency, much like the height of a transfer unit in packed absorbers, characterizes the rate of mass transfer in the equipment. The value of the efficiency depends on a large number of parameters not normally known, and its prediction is therefore difficult and involved. Correlations have led to widely used empirical relationships, which can be used for rough estimates (172,173). The classical approach for tray efficiency estimation, however, remains the old Bubble Tray Design Manual (174), summarizing intensive research on this topic.

In large plates 0.6 m or more in diameter, the efficiency of the tray as a whole may differ from the efficiency observed at some particular point of the tray because the liquid is not uniformly mixed in the direction of the flow on the whole tray. The point value of the efficiency called E_{OG} is thus more closely related to interphase diffusion than E_{MV}. As the gas passes upward through the liquid covering a small area of the plate, mass transfer from gas to liquid occurs in a manner similar to a packed tower of height h_B, the depth of the bubbling area. Under the assumption that the liquid is completely mixed in the vertical direction, and that the gas travels through that minicolumn in a plug-flow-like fashion, the number of transfer units of the bubbling area may be calculated in terms of the gas concentrations above and below the area under consideration by applying the definition of N_{OG}, equation 64. This equation may be integrated by taking y^* as constant and equal to y_n^* because of the well-mixed nature of the liquid phase. By comparing the result with the definition of the plate efficiency, equation 92, formulated for a single point on the plate, the following relationship between the point efficiency and the number of transfer units arises:

$$E_{OG} = 1 - e^{-N_{OG}} \tag{97}$$

If resistance to transfer is present in both phase, N_{OG} may be expressed as an addition of resistances using equations 63, 56, 50, 53:

$$\frac{1}{N_{OG}} = \frac{1}{N_G} + \frac{mG_M}{L_M}\frac{1}{N_L} \tag{98}$$

Hence, the point efficiency E_{OG} may be computed if both N_G and N_L in the bubbling area are known. These parameters are determined by the prevailing transfer coefficients, the interfacial area, and by h_B: $N_G = k_G a P h_B / G_M$ and $N_L = k_L a \rho h_B / L_M$.

To estimate the number of transfer units for design, the following empirical correlations which were derived from efficiency measurements employing a variety of trays and operating conditions under the aforementioned assumptions are recommended (174):

$$N_G = (0.776 + 4.63h_w - 0.238F + 0.0712L')\mathrm{Sc}^{-0.5} \tag{99}$$

where h_w is the weir height in m; $F = \rho_G (u_G)^{1/2}$ in m/s [(kg/m^3)$^{1/2}$]; L' is the liquid flow rate per average width of stream in m^3/(s.m); and

$$N_L = 3050\sqrt{D_L}\,(68h_w + 1)t_L \tag{100}$$

where $t_L = h_L z_L / L'$ is the liquid residence time in s. The recommended correlation for h_L is

$$h_L = 0.0419 + 0.19h_w - 0.0135F + 2.46L' \tag{101}$$

Effect of Different Degrees of Mixing. Once E_{OG} is evaluated on the basis of N_G and N_L using equation 83, it has to be translated into E_{MV} by considering the degree of mixing on the tray. It is obvious that for a small plate with completely backmixed liquid,

$$E_{MV} = E_{OG} \tag{102}$$

If, on the other hand, the liquid flows in a plug-flow-like manner over the tray, but the vapor may be assumed to mix between the trays so that it enters each tray in uniform composition, the result may be calculated according to (175).

$$E_{MV} = \frac{L_M}{mG_M}\left[e^{(mG_M/L_M)E_{OG}} - 1\right] \tag{103}$$

In the case of unmixed vapors between the plates, the equations, being implicit in E_{MV}, have also been solved numerically (175). The results depend on the arrangement of the downcomers and are not too different numerically from equation 103. In reality, however, the liquid is neither completely backmixed nor can the tray be considered as a plug-flow device.

Many theories have been put forth to handle partial liquid backmixing on plates. Early calculations (139,176,177) used the so-called tanks-in-series model. Recycle models have been derived by assuming partial or complete backmixing of the liquid, or the vapor, or both (178–180). The suggested procedure (174) is based on the eddy diffusion model when the parameter characterizing the degree of liquid backmixing is the dimensionless group known as the Peclet number:

$$\mathrm{Pe} = \frac{z_\mathrm{L}^2}{D_\mathrm{E} t_\mathrm{L}} \tag{104}$$

The eddy dispersion coefficient D_E has been measured and correlated empirically as

$$\sqrt{D_\mathrm{E}} = 0.00378 + 0.0179 u_\mathrm{g} + 3.69 L' + 0.18 h_\mathrm{w} \tag{105}$$

The nomenclature is the same as that used in equation 99 and 100.

The relationship between E_MV, E_OG, and Pe has been calculated according to the recommended formulas (173) and presented in tabular form (4). A plot of this data is shown in Figure 29. Complete backmixing is characterized by $\mathrm{Pe} = 0$ ($D_\mathrm{E} = \infty$), where $E_\mathrm{MV} = E_\mathrm{OG}$. In larger columns, in which the liquid is not completely mixed horizontally, E_MV can be seen from Figure 29 to become larger than the point efficiency and may thus even exceed 100% (181). Entrainment, as well as by-passing, are always detrimental to the plate efficiency. Analytical expressions to correct for entrainment and by-passing have been developed (182,183).

7.5. Capacity Limitations. The fluid flow capacity of a bubble tray may be limited by any of three principal factors.

1. Flooding, often the most restrictive of the limitations, occurs when the clear liquid height in the downcomer, H_dc, exceeds a certain fraction of the tray spacing. During operation, the liquid level in the downcomer builds up as a result of the head necessary to overcome the various resistances to liquid flow, including the friction in the downcomer itself and the hydraulic gradient across the plate. A significant portion in the liquid backup is caused by the need for the liquid to overcome the difference in pressure between the inside and the outside of the downcomer, which in turn is caused by the pressure drop of the vapors through the next higher plate. If the diameter of the column is made too small for a given flow, the vapor pressure drop and thus the liquid backup in the downcomers will increase to the point where some liquid spills onto the next higher tray, and flooding sets in. In principle, the condition may be corrected by increasing the diameter or the tray spacing. A conservative design requires a clear liquid head in the downcomer of no more than half the tray spacing to allow for froth entrapped in the downcomer. The maximum allowable superficial velocity based on the column cross section u_G may be roughly estimated

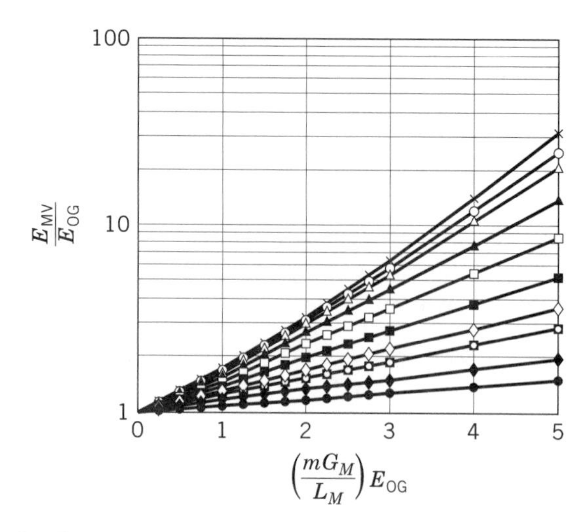

Fig. 29. Relationship between E_{OG} and E_{MV} at different degrees of liquid backmixing (4). Curves represent different Peclet numbers. From top to bottom Pe = ∞, +; Pe = 100, ○; Pe = 50, △; Pe = 20, ▲; Pe = 10, □; Pe = 5, ■; Pe = 3, ◇; Pe = 2, ■; Pe = 1, ◆; Pe = 0.5, ●.

for different tray spacings (6,184,185). A more reliable design would consider each pressure drop contributing to H_{dc} separately.

2. Entrainment occurs when spray or froth formed on one tray enters the gas passages in the tray above. In moderate amounts, entrainment will impair the countercurrent action and hence drastically decrease the efficiency. If it happens in excessive amounts, the condition is called priming and will eventually flood the downcomers.
3. At high liquid flow, the hydraulic gradient may become so large that the caps near the liquid feed point will stop bubbling, and the efficiency will suffer.

8. Nomenclature

Symbol	Definition	Units
A	L_M/mG_M, absorption factor	
a	effective interfacial area per packed volume	m^2/m^3
c_{Ai}	concentration of solute at interface	mol/m^3 or mol/L
C_P	specific heat of gas mixture	kJ/(mol·K)
C_q	specific heat of liquid mixture	kJ/(mol·K)
D_{ax}	axial dispersion coefficient	m^2/s
D_{AB}	diffusion coefficient of A in B	m^2/s
D_{Aj}	binary diffusion coefficient of A in the jth species other than A	m^2/s
$D_{A_{eff}}$	effective diffusion coefficient of A in a mixture, see Eq. 34	m^2/s
D_E	eddy dispersion (or diffusion) coefficient on a tray	m^2/s
D_L	liquid diffusion coefficient	m^2/s
E_{MV}	Murphree vapor plate efficiency, see equation 96	

Symbol	Definition	Units
E_{OG}	point value of plate efficiency	
F	$\mu_G\sqrt{\rho_G}$, F-factor for bubble tray gas load	$kg^{1/2}/(m^{1/2}{\cdot}s)$
g_j	moles of component j	mol
G_M	gas flow rate	$kmol/(m^2{\cdot}s)$
G'_M	flow rate of inert gas	$kmol/(m^2{\cdot}s)$
G^0_M	generalized molar gas flow rate, see equation 73	
H	Henry's constant	Pa
HETP	height equivalent to a theoretical plate	m
H_G	height of a gas-phase transfer unit, see equation 51	m
H_L	height of a liquid-phase transfer unit, see equation 54	m
H_{OG}	overall gas-phase height of a transfer unit, see equation 57	m
H_{OS}	integral heat of solution for the solute	kJ/mol
H_v	latent heat of pure solvent	kJ/mol
h	total height of packing	m
h_L	clear height of liquid on a tray	m
h_w	weir height	m
h_B	height of bubble layer on a tray	m
J_A	flux of component A due to diffusion	$mol/(m^2{\cdot}s)$
K	partition coefficient for gas–liquid equilibrium	
K_l	constant in bond stretching potential	$J/kmol \cdot m^2$
K_{OG}	overall gas-phase mass transfer coefficient	$mol/(s{\cdot}m^2{\cdot}Pa)$
K_θ	constant in bond bending potential	$J/kmol \cdot rad^2$
k_G	gas-phase mass transfer coefficient	$mol/(s{\cdot}m^2{\cdot}Pa)$
k_L	liquid-phase mass transfer coefficient	m/s
k_{sol}	thermal conductivity of the solution	$J/m \cdot K \cdot s$
k^r_L	mass transfer coefficient including effect of chemical reaction	m/s
k^0	mass transfer coefficient for equimolar counterdiffusion	m/s
k'	mass transfer coefficient for unidirectional molecular diffusion through a stagnant gas	m/s
k	mass transfer coefficient for general multicomponent diffusion situation	m/s
k_{Aj}	binary mass transfer coefficient of A in the jth species other than A (see equation 39)	m/s
L	mass flow rate of liquid	$kg/(s{\cdot}m^2)$
L_M	liquid flow rate	$kmol/(s{\cdot}m^2)$
L^0_M	generalized liquid flow rate, see equation 73	$kmol/(s{\cdot}m^2)$
L'_M	flow rate of solute-free solvent	$kmol/(s{\cdot}m^2)$
L'	liquid flow rate on a plate per width of stream	$m^3/(s{\cdot}m)$
l	bond length	m
l_0	equilibrium bond length	m
m	slope of equilibrium line	
N_A	flux of solute A through phase interface	$mol/(s{\cdot}m^2)$
N_G	number of gas-phase transfer units, see equation 52	
N_j	flux of component j through phase interface	$mol/(s{\cdot}m^2)$
N_l, N_s	length of linear hydrocarbon chains	
N_L	number of liquid-phase transfer units, see equation 55	
N_{OG}	number of overall gas-phase transfer units, see equation 58	
N_P	number of actual plates in the column	
N_T	N_{OG} for equimolar counterdiffusion, see equation 66	
N_{TP}	number of theoretical plates	
NTU_{ap}	N_{OG} calculated by equation 65, see also equation 82	
NTU	actual value of N_{OG} under the influence of axial dispersion, see equation 83	

Symbol	Definition	Units
P	pressure	Pa
p_A	partial pressure of A	Pa
p_{A_i}	partial pressure of A at the interface	Pa
P_s	vapor pressure of a pure component	Pa
Pe	uh/D_{ax}, dimensionless Peclet number	
Pe_i	Peclet number specifically defined for phase i, $i = x$ or y	
Pe_{HTU}	uH_{OG}/D_{ax}, dimensionless HTU Peclet number, see equation 84	
$Pe_{HTU,i}$	Peclet number specifically defined for phase i, $i = x$ or y, see equation 86	
R	L_M/G_M, *flow ratio*	
r	distance between interaction sites (atoms)	m
r_{ij}	relation velocity for species i in reaction j	$kmol/m^3s$
s	surface renewal rate, equation 41	s^{-1}
Sc	$\mu/\rho D$, the dimensionless Schmidt number	
T	temperature	°C or K
t_A	$\Sigma_j N_j/N_A$, parameter indicating the degree of counterdiffusion	
t_L	residence time of liquid on plate, see equation 104	s
U	total (intramolecular and intermolecular) interaction energy	J/kmol
U_A	linear velocity of diffusion of solute A	m/s
U_j	linear velocity of diffusion of the jth species other than the solute A	m/s
u	superficial velocity of gas or liquid phase	m/s
V^{LJ}	Lennard–Jones interaction potential	J/kmol
$V_{stretching}$	bond stretching potential	J/kmol
$V_{bending}$	bond bending potential	J/kmol
$V_{torsion}$	bond torsion potential	J/kmol
X_f	film factor, see equation 24	
X_A^0	$x_A/(1-t_Ax_A)$, generalized liquid concentration, see equation 73	
X'_A	$x_A/(1-x_A)$, liquid mole ratio	
x_A	liquid mole fraction of solute	
x_j	mole fraction of the jth species other than A	
x_A	arithmetic mean of mole fraction of A over diffusion path	
x_j	arithmetic mean of x_j over diffusion path	
x_{Ab}	mole fraction in bulk of the liquid phase	
x_{Ai} or x_i	liquid mole fraction of solute at the phase interface	
x_{BM}	logarithmic mean of solvent concentration between the phase interface and the bulk of the liquid, see equation 18	
Y	moles absorbed component per mole of entering vapor, equation 98	
Y'	$y/(1-y)$, mole ratio	
$Y^0{}_A$	$Y_A/(1-t_Ay_A)$, generalized gas concentration, see equation 64	
Y_f	film factor, see equation 23	
y_A^*	mx_A, concentration in equilibrium with bulk liquid	
y_{Ab}	mole fraction in bulk of gas phase	
y_{A_i} or y_i	gas mole fraction of solute at phase interface	
y_{BM}	logarithmic mean of inert gas concentration between the phase interface and the bulk of the gas phase, see equation 15	
y_{BM}^*	$[(1-y_A)(1-y_A^*)]/\ln[(1-y_A)/(1-y_A^*)]$, logarithmic mean of inert gas concentration between the equilibrium concentration and the bulk of the gas phase	

Symbol	Definition	Units
Y_f^*	$[(1-t_Ay_A)-(1-t_Ay_A^*)]/\ln[(1-t_Ay_A)/(1-t_Ay_A^*)]$, overall gas-phase film factor	
z_0	thickness of diffusion layer	m
z_L	length of liquid travel on tray	m
Δg_j	total amount of component j absorbed	$kmol/(s{\cdot}m^2)$
$\Delta H_{Rij}\,(T)$	molar enthalpy of reaction	J/kmol
ΔT_{max}	temperature associated with internal temperature maximum °C or K	
δ	film thickness	m
ε_{ij}	Lennard–Iones first interaction constant (potential well depth parameter)	J/kmol
θ	bond angle	rad
θ_0	equilibrium bond angle	rad
μ	viscosity	mPa·s(=cP)
μ	chemical potential	J/kmol
ϕ	dihedral or torison angle	rad
Γ	Correction factor for thermodynamic nonidealities in diffusion equations, see equation 33	
ρ	density	mol/m^3
$\bar{\rho}$	mean density of liquid phase	mol/m^3, mol/L, or g/cm^3
σ	surface tension	N/m
σ_{ij}	Lennard–Jones second interaction constant (atom size parameter)	m
ϕ	packing parameter for equation 93	m
ψ	packing parameter for equation 94	m
SUBSCRIPTS		
av	average	
A	component A, solute	
B	component B, usually nondiffusing	
G	gas	
L	liquid	
b	in the bulk; this symbol is normally omitted	
i	at gas–liquid interface	
0	liquid feed for plate columns	
x	pertaining to liquid phase	
y	pertaining to gas phase	
M	molar quantity	
n	referring to stream leaving the nth tray	
1	bottom of a packed column	
2	top of a packed column	

BIBLIOGRAPHY

"Absorption" in *ECT* 1st ed., Vol. 1, pp. 14–32, by E. G. Scheibel, Hoffmann-La Roche, Inc.; in *ECT* 2nd ed., Vol. 1, pp. 44–77, by F. A. Zenz, Squires International, Inc.; in *ECT* 3rd ed., Vol. 1, pp. 53–96, by C. W. Wilke and Urs von Stockar, University of California, Berkeley; in *ECT* 4th ed., Vol. 1, pp. 38–93, by Urs von Stockar, École Polytechnique Fédérale, Lausanne and C. W. Wilke, University of California, Berkeley; "Absorption" in *ECT* (online), posting date: December 4, 2000, by Urs von Stockar, École Polytechnique Fédérale, Lausanne and C. W. Wilke, University of Berkeley.

CITED PUBLICATIONS

1. F. A. Zenz, *Chem. Eng.* **120**, 1972 (Nov. 13, 1972).
2. R. E. Treybal, *Mass Transfer Operations*, 3rd ed., McGraw-Hill Book Co., Inc., New York, 1980.
3. W. S. Norman, Absorption, *Distillation and Cooling Towers*, Longmans, Green & Co., Ltd. (Wiley) New York, 1962.
4. T. K. Sherwood, R. L. Pigford, and C. R. Wilke, *Mass Transfer*, McGraw-Hill Book Co., Inc., New York, 1975.
5. A. Mersmann, H. Hofer, and J. Stichlmair, *Ger. Chem. Eng.* **2**, 249 (1979).
6. R. H. Perry, D. Green and J. O. Maloney, eds., *Perry's Chemical Engineer's Handbook*, 7th ed., McGraw-Hill Book Co., Inc., New York, 1997.
7. J. M. Prausnitz, R. N. Lichtenhaler, and E. Gomes de Azevedo, *Molecular Thermodynamics of Fluid-Phase Equilibria*, 3rd ed., Prentice-Hall, Englewood Cliffs, N. J., 1986.
8. R. C. Reid, J. M. Prausnitz, and B. E. Poling, *The Properties of Gases and Liquids*. 4th ed., McGraw-Hill Book Co., New York, 1988.
9. A. S. Kertes and co-workers, *Solubility Data Series*, Pergamon Press, Oxford, U.K., 1979.
10. W. C. Edmister, *Applied Hydrocarbon Thermodynamics*, Gulf Publishing Co., Houston, Tex., 1961.
11. N. B. Vargaftik, *Tables on the Thermophysical Properties of Liquids and Gases*, John Wiley & Sons, Inc., New York, 1975.
12. R. Battino, *Solubility Data Series*, Pergamon Press, Oxford, 1981.
13. J. Gmehling and U. Onken, *Vapor-Liquid Equilibrium Data Collection*. Chemistry Data Series, Dechema, Frankfurt, 1977 ff.
14. M. P. Allen and D. J. Tildesley, *Computer Simulation of Liquids*, Oxfort University Press, 1987.
15. D. Frenkel and B. Smit,*Understanding of Molecular Simulation From Algorithms to Applications*, Academic Press, 1999.
16. R. J. Sadus, *Molecular Simulation of Fluids: Theory, Algorithms and Object-Orientation*, Elsevier, New York, 1999.
17. W. J. Ward and J. A. Quinn, *AIChE J* **11**, 1005 (1965).
18. C. K. Brooks, III, M. Karplus, and B. M. Pettitt, Proteins: a Theoretical Perspective of Dynamics, Structure and Thermodynamics, in *Advances in Chemical Physics*, Vol. LXXI, John Wiley & Sons, Inc., New York, 1988.
19. P. Van der Ploeg and H. J. C. Berendsen, *J. Chem. Phys.* **76**, 3271 (1982).
20. J. P. Ryckaert and A. Bellemans, *Chem. Phys. Lett.* **30**, 123 (1975).
21. P. J. Flory, *Statistical Mechanics of Chain Molecules*, Interscience, New York, 1969.
22. A. R. Leach, *Molecular Modeling. Principles and Applications*, Longman, 1996.
23. W. B. Kay, *Ind. Eng. Chem.* **30**, 459 (1938).
24. F. Favari, A. Bertucco, N. Elvassore, and M. Fermeglia, *Chem. Eng. Sci.* **55**, 2379 (2000).
25. P. Paricaud, A. Galindo, and G. Jackson, *Fluid Phase Equil.* **194**, 87 (2002).
26. F. Tumakaka, J. Gross, and G. Sadowski, *Phase Equil.* **194**, 541 (2002).
27. M. Laso, J. J. de Pablo, and U. W. J. Suter, *Chem. Phys.* **97**, 2817 (1992).
28. N. F. A. van der Vegt, *J. Memb. Sci.* **205**, 125 (2002).
29. K. S. Shing, K. E. Gubbins, and K. Lucas, *Mol. Phys.* **65**, 1235 (1988).
30. M. Laso, J. J. De Pablo, and U. W. Suter, *J. Chem. Phys.* **97**(4), 2817 (1992).
31. S. Murad and S. Gupta, *Fluid Ph. Equil.* **187**, 29 (2001).
32. W. G. Whitman, *Chem. Metall. Eng.* **29**, 147 (1923).

33. A. F. Ward and L. H. Brooks, *Trans. Faraday Soc.* **48**, 1124 (1952).
34. E. J. Cullen and J. F. Davidson, *Trans. Faraday Soc.* **53**, 113 (1957).
35. J. L. Duda and J. S. Vrentas, *AIChE J* **14**, 286 (1968).
36. S. Lynn, J. R. Straatemeier, and H. Kramers, *Chem. Eng. Sci.* **4**, 58 (1955).
37. R. W. Schrage, *A Theoretical Study of Interface Mass Transfer*, Columbia University, New York, 1953.
38. B. Paul, *J. Am. Rocket Soc.* 1321 (Sept. 1962).
39. L. V. Delaney and L. C. Eagleton, *AIChE J* **8**, 418 (1962).
40. R. Cartier, D. Pindzola, and P. E. Bruins, *Ind. Eng. Chem.* **51**, 1409 (1959).
41. N. A. Clontz, R. T. Johnson, W. L. McCabe, and R. W. Rousseau, *Ind. Eng. Chem. Fundam.* **11**, 368 (1972).
42. J. T. Davies and E. K. Rideal, *Adv. Chem. Eng.* **4**, 1 (1963).
43. C. J. King, *AIChE J* **10**, 671 (1964).
44. R. B. Bird, W. E. Stewart, and E. N. Lightfoot, *Transport Phenomena*, John Wiley & Sons, Inc., New York, 1960.
45. C. R. Wilke, *Chem. Eng. Prog.* **46**, 95 (1950).
46. Curtiss and Hirschfelder, *J. Chem. Phys.* **17**, 552 (1949).
47. J. A. Wesselingh and R. Krishna, "Mass Transfer" in *Multicomponent Mixtures*, Delft University Press, Delft, The Netherlands, 2000.
48. R. Krishna and R. Taylor, *Multicomponent Mass Transfer*, John Wiley & Sons, Inc., New York, 1993.
49. R. E. Cunningham and R. J. J. Williams, *Diffusion in Gases and Porous Media*, Plenum Press, 1980.
50. E. L. Cussler, *Diffusion: Mass Transfer in Fluid Systems*, 2nd ed., Cambridge University Press, Cambridge, UK, 1997.
51. R. Zarzycki, A. Chacuk, *Absorption fundamentals & applications*, Pergamon Press, New York, 1993.
52. G. Ackermann, *Forschungsheft V.D.I.* **382**, 1 (1937).
53. W. K. Lewis and W. G. Whitman, *Ind. Eng. Chem.* **16**, 1215 (1924).
54. J. L. Bravo, J. A. Rocha and J. R. Fair, *Hydrocarbon Processing.* **1**, 91 (1985).
55. Y. Djebbar and R. M. Narbaitz, *Wat. Sci. Tech.* **38**, 295 (1998).
56. K. Onda, H. Takeuchi, and Y. Okumoto, *J. Chem. Eng. Jpn.* **1**, 56 (1968).
57. R. Higbie, *Trans. Am. Inst. Chem. Eng.* **31**, 365 (1935).
58. P. V. Danckwerts, *Ind. Eng. Chem.* **43**, 1460 (1951).
59. M. J. W. Frank, J. A. M. Kuipers, W. P. M. Van Swaaij, *Chem. Eng. Sci.* **55**, 3739 (2000).
60. H. L. Toor and J. M. Marchelo, *AIChEJ* **4**, 97 (1958).
61. H. López-Arjona, R. Lobo and T. Viveros-García, *Chem. Eng. Sci.* **55**, 6897 (2000).
62. M. Brinkmann, M. Schafer, H.-J. Warnecke, and J. Pruss, *Comp.Chem. Eng.* **22**, 515 (1998).
63. T. H. Chilton and A. P. Colburn, *Ind. Eng. Chem.* **26**, 1183 (1934).
64. G. F. Froment, in *Chemical Reaction Engineering* (Advances in Chemistry Series 109), American Chemical Society, Washington, D. C., (1972) p. 19.
65. P. V. Danckwerts, *Gas-Liquid Reactions*, McGraw-Hill Book Co., Inc., New York, 1970.
66. D. W. van Krevelen and P. J. Hoftijzer, *Rec. Trav. Chim.* **67**, 563 (1948).
67. K. F. Loughlin, M. A. Abul-Hamayel, and L. C. Thomas, *AIChE* **31**, 1614 (1985).
68. M. P. Pradhan, N. J. Suchak, P. R. Walse, and J. B. Joshi, *Chem. Eng. Sci.* **24**, 4569 (1997).
69. M. P. Pradhan, J. B. Joshi, *Chem. Eng. Sci.* **55**, 1269 (2000).
70. Ph. Leckner, R. O. Pearson and R. T. Wood, *Chem. Eng. Progr.* **78**, 65 (1982).
71. A. L. Kohl and R. B. Nielsen, *Gas Purification*, 5th ed., Gulf Professional Publishing, Houston, 1997.

72. A. M. Goldstein, E. C. Brown, F. J. Heinzelmann and G. R. Say, *Energy Prog.* **7**, 67 (1986).
73. B. E. Roberts and Alan E. Mather, *Chem. Eng. Comm.* **65**, 105 (1988).
74. T. T. Teng and A. E. Mather, *J. Chem. Eng. Data*, **35**, 410 (1990).
75. D. Zhang, H.-J. Ng, and R. Veldman, GPA 78th Annual Convention, March, 1999.
76. J. Carroll, J. Maddocks, and H.-J. Ng, GPA 78th Annual Convention, March, 1999.
77. R. Baur, R. Taylor and R. Krishna, *Chem. Eng. Sci.* **56**, 2085 (2001).
78. D. H. Miles and G. M. Wilson, Annual Report to the API for 1974, October, 1975.
79. H. L. Clever, S. A. Johnson, and M. E. Derrick, *J. Phys. Chem. Ref. Data*, **14**, 631 (1985).
80. M. Luckas, K. Lucas and H. Roth, *AIChEJ* **40**, 1892 (1994).
81. G. Sartori, W. Ho, D. Savage, G. Chludzinski, and S. Wiechert, *Sep. Purif. Meth.* **16**, 171 (1987).
82. G. Astarita, D. W. Savage and A. Bisio, *Gas Treating with Chemical Solvents*, John Wiley & Sons, New York, 1983.
83. P. M. M. Blauwhoff, G. F. Versteegand, and W. P. M. van Swaaij, *Chem. Eng. Sci.* **38**, 1411 (1983).
84. A. K. Saha, S. S. Bandyopadhyay, and A. K. Biswas, *J. Chem. Eng. Data* **38**, 78 (1993).
85. G. Astarita, *Mass Transfer with Chemical Reaction*, Elsevier, Amsterdam, the Netherlands, 1966.
86. A. P. Colburn, *Trans. Am. Inst. Chem. Eng.* **35**, 211 (1939).
87. J. H. Wiegand, *Trans. Am. Inst. Chem. Eng.* **36**, 679 (1940).
88. J. D. Raal and M. K. Khurana, *Can. J. Chem. Eng.* **51**, 162 (1973).
89. C. R. Wilke and U. v. Stockar, *Ind. Eng. Chem. Fundam.* **16**, 88 (1977).
90. J. Stichlmair, *Chem. Ind. Technol.* **44**, 411 (1972).
91. J. R. Bourne, U. v. Stockar, and G. C. Coggan, *Ind. Eng. Chem. Process Des. Dev.* **13**, 124 (1974).
92. C. R. Wilke and U. von Stockar, *Ind. Eng. Chem. Fundam.* **16** (2), 94 (1977).
93. I. A. Furzer, *Ind. Eng. Chem. Fundam.* **23**, 159 (1984).
94. N. Kolev and Kr. Semkov, *Chem. Eng. Prog.* **19**, 175 (1985).
95. N. Kolev and Kr. Semkov, *Vt Verfahrenstechnik* **17**, 474 (1983).
96. W. E. Dunn, T. Vermeulen, C. R. Wilke, and T. T. Word, Report UCRL 10394, University of California Radiation Laboratory as cited in Ref. 4, 1962.
97. W. E. Dunn, T. Vermeulen, C. R. Wilke, and T. T. Word, *Ind. Eng. Chem. Fundam.* **16**, 116 (1977).
98. E. T. Woodburn, *AIChE J* **20**, 1003 (1974).
99. M. Richter, *Chem. Tech.* **30**, 294 (1978).
100. U. von Stockar and P. F. Cevey, *Ind. Eng. Chem. Process Res. Dev.* **23**, 717 (1984).
101. A. K. Bi, B. Duczmal, and P. Machniewski, *Chem. Eng. Sci.*, **56**, 6233 (2001).
102. S. Moustiri, G. Hebrard, S. S. Thakre, and M. Roustan, *Chem. Eng. Sci.* **56**, 1041 (2001).
103. F. H. Yin, A. Afacan, K. Nandakumar, and K. T. Chuang, 50th Canadian Chemical Engineering Conference, Montreal, 2000.
104. B. J. Vinci, B. J. Watten, and M. B. Timmons, *Aquac. Eng.* **15**, 1 (1996).
105. R. A. Davis, *Chem. Eng. Edu.* **27**, 20 (1993).
106. T. Miyauchi and T. Vermeulen, *Ind. Eng. Chem. Fundam.* **2**, 113 (1963).
107. C. A. Sleicher, *AIChE J* **5**, 145 (1959).
108. S. Stemerding and F. J. Zuiderweg, *Chem. Eng. CE* 156 (May 1963).
109. J. C. Mecklenburgh and S. Hartland, *The Theory of Backmixing*, Wiley-Interscience, New York, Chapt. 10, 1975.
110. J. S. Watson and H. D. Cochran, *Ind. Eng. Chem. Process Res. Dev.* **10**, 83 (1971).

111. U. von Stockar and Xiao-Ping Lu, *Ind. Eng. Process Chem. Res. Dev.* **30**, 673 (1991).
112. U. von Stockar, ACHEMASIA, (International Meeting on Chemical Engineering & Biotechnology, Beijing, China), Dechema & Ciesc, (1989). p. 43.
113. T. K. Sherwood and F. A. L. Holloway, *Trans. Inst. Am. Chem. Eng.* **36**, 39 (1940).
114. F. F. Rixon, *Trans. Instn. Chem. Eng.* **26**, 119 (1948).
115. H. A. Koch, L. F. Stutzman, H. A. Blum, and L. E. Hutchings, *Chem. Eng. Prog.* **45**, 677 (1949).
116. E. L. Knoedler and C. F. Bonilla, *Chem. Eng. Prog.* **50**, 125 (1954).
117. J. E. Vivian and C. J. King, *AIChE J* **120**, 221 (1964).
118. D. Cornell, W. G. Knapp, and J. R. Fair, *Chem. Eng. Prog.* **56**, 68 (1960).
119. W. L. Bolles and J. R. Fair, *Chem. Eng.* **89**(July 12), 109 (1982).
120. P. H. Au-Yeung and A. B. Ponter, *Can. J. Chem. Eng.* **61**, 481 (1983).
121. D. M. Mohunta, A. S. Vaidyanathan, and G. S. Laddha, *Ind. Chem. Eng.* **11**, 73 (1969).
122. M. De Brito, U. von Stockar, A. Menéndez, P. Bomio and M. Laso, *Ind. Eng. Chem. Res.* **33**, 647 (1994).
123. R. Baur, A. P. Higler, R. Taylor, R. Krishna, *Chem. Eng. J.* **76**, 33 (2000).
124. M. de Brito, U. von Stockar, *IChE Symp. Ser.* **128**, B137 (1991).
125. J. B. Zech and A. B. Mersmann, *IChE Symp. Ser.* **56**, 39 (1979).
126. R. Mangers and A. B. Ponter, *Ind. Eng. Chem. Process Des. Dev.* **19**, 530 (1980).
127. M. S. Shi and A. B. Mersmann, *Ger. Chem. Eng.* **8**, 87 (1985).
128. R. Billet and M. Schultes, Paper given at AIChE Annual Meeting, Washington, D.C., 1988.
129. D. W. van Krevelen and P. J. Hoftijzer, *Rec. Trav. Chim. Pays-Bas* **66**, 49 (1947).
130. H. L. Schulman, C. F. Ulrich, A. Z. Proulx, and J. O. Zimmerman, *AIChE J* **1**, 253 (1955).
131. R. Billet, *Packed Column Analysis and Design*, Ruhr-Universität, Bochum (1989).
132. M. Laso, M. de Brito, P. Bomio, U. von Stockar, *Chem. Eng. J.* **58**, 251 (1995).
133. J. F. Davidson, *Trans. Inst. Chem. Eng.* **37**, 131 (1959).
134. J. Bridgewater and A. M. Scott, *Trans. Inst. Chem. Eng.* **52**, 317 (1974).
135. A. B. Ponter and P. H. Au-Yeung, *Can. J. Chem. Eng.* **60**, 94 (1982).
136. R. Echarte, H. Campana, and E. A. Brignole, *Ind. Eng. Chem. Process Des. Dev.* **23**, 349 (1984).
137. E. J. Lynch and C. R. Wilke, *AIChE J* **1**, 9 (1955).
138. H. L. Shulman, C. F. Ullrich, and N. Wells, *AIChE J* **1**, 247 (1955).
139. L. Spiegel and W. Meier, *Chem. Eng. Symp. Ser.* **104**, A203 (1987).
140. B. W. Gamson, G. Thodos, and O. A. Hougen, *Trans. Am. Inst. Chem. Eng.* **39**, 1 (1943).
141. R. G. Eckert and O. A. Hougen, *Chem. Eng. Prog.* **45**, 188 (1949).
142. C. R. Wilke and O. A. Hougen, *Trans. Am. Inst. Chem. Eng.* **41**, 445 (1945).
143. M. Hobson and G. Thodos, *Chem. Eng. Prog.* **47**, 370 (1951).
144. L. Fellinger, Ph.D., Dissertation, Massachusetts Institute of Technology, Boston, 1941 (see also Ref. 186).
145. J. L. Bravo and J. R. Fair, *Ind. Eng. Chem. Process Des. Dev.* **21**, 162 (1982).
146. L. Chunxi and W. Fürst, *Chem. Eng. Sci.* **55**, 2975 (2000).
147. F. Y. Jou, J. J. Carroll, A. E. Mather, and F. D. Otto, *Can. J. of Chem. Eng.* **71**, 264 (1993).
148. R. L. Kent and B. Eisenberg, *Hydrocarbon Proc.* **55**, 87 (1976).
149. G. Kuranov, B. Rumpf, G. Maurer, and N. A. Smirnova, *Fluid Phase Equilib.* **136**, 147 (1997).
150. M. L. Posey and G. T. Rochelle, *Ind. Chem. Eng. Res.* **36**, 3944 (1997).

151. E. B. Rinker, S. S. Ashour, and O. C. Sandall, *Chem. Eng. Sci.* **50**, 755 (1995).
152. D. P. Hagewiesche, S. S. Ashour, H. A. Al-Ghawas, and O. C. Sandall, *Chem. Eng. Sci.* **50**, 1071 (1995).
153. C. H. Liao and M. H. Li, *Chem. Eng. Sci.* **57**, 4569 (2002).
154. H. S. Fogler, *Elements of Chemical Reaction Engineering*, 3rd ed., Prentice Hall, 1998.
155. W. H. Press, A. Teukolsky, W. T. Vetterling, and B. P. Flannery, *Numerical Recipes in C++, The Art of Scientific Computing*, 2nd ed., Cambridge University Press, 2002.
156. T. K. Sherwood, G. H. Shipley, and F. A. L. Holloway, *Ind. Eng. Chem.* **30**, 765 (1938).
157. W. E. Lobo, L. Friend, F. Hashmall, and F. Zenz, *Trans. Am. Inst. Chem. Eng.* **41**, 693 (1945).
158. F. A. Zenz and R. A. Eckert, *Pet. Refiner.* **40**, 130 (1961).
159. R. A. Eckert, *Chem. Eng. Prog.* **66**, 39 (1970).
160. J. R. Fair, D. E. Steinmeyer, W. R. Penney, and B. B. Crocker, in Ref. 6, pp. 18–23.
161. E. V. Murphree, *Ind. Eng. Chem.* **17**, 474 (1925).
162. A. Kremser, *Nat. Pet. News* **22**(21), 42 (1930).
163. G. G. Brown and M. Souders, *Oil Gas J.* **31**(5), 34 (1932).
164. M. Souders and G. G. Brown, *Ind. Eng. Chem.* **24**, 519 (1932).
165. G. Horton and W. B. Franklin, *Ind. Eng. Chem.* **32**, 1384 (1940).
166. W. C. Edmister, *Ind. Eng. Chem.* **35**, 837 (1943).
167. G. G. Brown and M. Souders, *The Science of Petroleum*, Vol. 2, Oxford University Press, New York, Sect. 25, p. 1557 (1938).
168. J. R. Bourne, U. v. Stockar, and G. C. Coggan, *Ind. Eng. Chem. Process Des. Dev.* **13**, 115 (1974).
169. J. Stichlmair and A. Mersmann, *Chem. Ing. Technol.* **43**, 17 (1971).
170. U. von Stockar, *Gasabsorption mit Wärmeeffekten*, Diss. Nr 4917, ETH-Zurich (1973).
171. W. R. Owens and R. N. Maddox, *Ind. Eng. Chem.* **60**, 14 (1968).
172. H. G. Drickamer and J. R. Bradford, *Trans. Am. Inst. Chem. Eng.* **39**, 319 (1943).
173. A. E. O'Connell, *Trans. Am. Inst. Chem. Eng.* **42**, 741 (1946).
174. Research Committee, *Bubble Tray Design Manual*, American Institute of Chemical Engineers, New York, 1958.
175. W. K. Lewis, *Ind. Eng. Chem.* **28**, 399 (1936).
176. M. Nord, *Trans. Am. Inst. Chem. Eng.* **42**, 863 (1946).
177. M. F. Gautreaux and H. E. O'Connell, *Chem. Eng. Prog.* **51**, 232 (1955).
178. E. D. Oliver and C. C. Watson, *AIChE J* **2**, 18 (1956).
179. L. A. Warzel, Ph.D. Dissertation, University of Michigan, Ann Arbor, Mich., 1955.
180. V. M. Ramm, *Absorption of Gases*, Israel Program for Scientific Translations, Jerusalem, 1968 Chapt. 3.
181. G. G. Brown, M. Souders, H. V. Nyland, and W. H. Hessler, *Ind. Eng. Chem.* **27**, 383 (1935).
182. A. P. Colburn, *Ind. Eng. Chem.* **28**, 526 (1936).
183. C. P. Strand, *Chem. Eng. Prog.* **59**, 58 (1963).
184. M. Souders and G. G. Brown, *Ind. Eng. Chem.* **26**, 98 (1934).
185. J. R. Fair, *Petro/Chem Eng.* **33**(10), 45 (Sept. 1961).
186. T. K. Sherwood and R. L. Pigford, *Absorption and Extraction*, McGraw-Hill Book Co. Inc., New York, 1952.

GENERAL REFERENCES

References 2–4, 6–8, 21, 22, 37, 44, 47, 48, 50, 58, 65, 85, 109, and 154 are general references.
A. H. P. Skelland, *Diffusional Mass Transfer*, John Wiley & Sons, Inc., New York, 1974.

Manuel Laso
Universidad Politécnica de Madrid, ETSII (Spain)
Urs von Stockar
École Polytechnique Fédérale de Lausanne

ACETALDEHYDE

1. Introduction

Acetaldehyde [75-07-0] (ethanal), CH_3CHO, was first prepared by Scheele in 1774, by the action of manganese dioxide and sulfuric acid on ethanol. The structure of acetaldehyde was established in 1835 by Liebig from a pure sample prepared by oxidizing ethyl alcohol with chromic acid. Liebig named the compound "aldehyde" from the Latin words translated as al(cohol) dehyd(rogenated). The formation of acetaldehyde by the addition of water to acetylene was observed by Kutscherow in 1881.

Acetaldehyde, first used extensively during World War I as a starting material for making acetone [67-64-1] from acetic acid [64-19-7], is an important intermediate in the production of acetic acid, acetic anhydride [108-24-7], ethyl acetate [141-78-6], peracetic acid [79-21-0], pentaerythritol [115-77-5], chloral [302-17-0], glyoxal [107-22-2], alkylamines and pyridines. Commercial processes for acetaldehyde production include the oxidation or dehydrogenation of ethanol, the addition of water to acetylene, the partial oxidation of hydrocarbons, and the direct oxidation of ethylene [74-85-1].

Acetaldehyde is a product of most hydrocarbon oxidations. It is an intermediate product in the respiration of higher plants and occurs in trace amounts in all ripe fruits that have a tart taste before ripening. The aldehyde content of volatiles has been suggested as a chemical index of ripening during cold storage of apples. Acetaldehyde is also an intermediate product of fermentation (qv), but it is reduced almost immediately to ethanol. It may form in wine (qv) and other alcoholic beverages after exposure to air imparting an unpleasant taste; the aldehyde reacts to form diethyl acetal [105-57-7] and ethyl acetate [141-78-6]. Acetaldehyde is an intermediate product in the decomposition of sugars in the it body and hence occurs in trace quantities in blood.

2. Physical Properties

Acetaldehyde is a colorless, mobile liquid having a pungent, suffocating odor that is somewhat fruity and quite pleasant in dilute concentrations. Its physical

Table 1. **Physical Properties of Acetaldehyde**

Properties	Values
formula weight	44.053
melting point, °C	−123.5
boiling point at 101.3 kPa[a] (1 atm), °C	20.16
density, g/mL	
d_4^{0}	0.8045
d_4^{11}	0.7901
d_4^{15}	0.7846
d_4^{20}	0.7780
coefficient of expansion per °C (0–30°C)	0.00169
refractive index, n_D^{20}	1.33113
vapor density (air = 1)	1.52
surface tension at 20°C, mN/m (= dyn/cm)	21.2
absolute viscosity at 15°C, mPa·s (= cP)	0.02456
specific heat at 0°C, J/(g·K)[b]	
15°C	2.18
25°C	1.41
$\alpha = C_p/C_v$ at 30°C and 101.3 kPa[a] (1 atm)	1.145
latent heat of fusion, kJ/mol[b]	3.24
latent heat of vaporization, kJ/mol[b]	25.71
heat of solution in water, kJ/mol[b]	
at 0°C	−8.20
at 25°C	−6.82
heat of combustion of liquid at constant pressure, kJ/mol[b]	12867.9
heat of formation at 273 K, kJ/mol[b]	−165.48
free energy of formation at 273 K, kJ/mol[b]	−136.40
critical temperature, °C	181.5
critical pressure, MPa[c]	6.40
dipole moment, C·m[d]	8.97×10^{-30}
ionization potential, eV	10.50
dissociation constant at 0°C, K_a	0.7×10^{-14}
flash point, closed cup, °C	−38
ignition temperature in air, °C	165
explosive limits of mixtures with air, vol % acetaldehyde	4.5–60.5

[a] To convert kPa to psi, multiply by 0.14503.
[b] To convert J to cal, divide by 4.187.
[c] To convert MPa to psi, multiply by 145.
[d] To convert C·m to debyes, multiply by 2.998×10^{29}.

properties are given in Table 1; the vapor pressure of acetaldehyde and its aqueous solutions appear in Tables 2 and 3, respectively; and the solubilities of acetylene, carbon dioxide and nitrogen in liquid acetaldehyde are given in Table 4. Acetaldehyde is miscible in all proportions with water and most common organic solvents, eg, acetone, benzene, ethyl alcohol, ethyl ether, gasoline, paraldehyde, toluene, xylenes, turpentine, and acetic acid. The freezing points of aqueous solutions of acetaldehyde are 4.8 wt %, −2.5°C; 13.5 wt %, −7.8°C; and 31.0 wt % −23.0°C.

Given in the literature are: vapor pressure data for acetaldehyde and its aqueous solutions (1–3); vapor–liquid equilibria data for acetaldehyde–ethylene oxide (1), acetaldehyde–methanol (4), sulfur dioxide–acetaldehyde–water (5), acetaldehyde–water–methanol (6); the azeotropes of acetaldehyde–butane and

Table 2. **Vapor Pressure of Acetaldehyde**

Temperature, °C	Vapor pressure, kPa[a]	Temperature, °C	Vapor pressure, kPa[a]
−50	2.5	20	100.6
−20	16.4	20.16	101.3
0	44.0	30	145.2
5	54.8	50	279.4
10	67.7	70	492.6
15	82.9	100	1,014

[a] To convert kPa to mm Hg, multiply by 7.5.

Table 3. **Vapor Pressure of Aqueous Solutions of Acetaldehyde**

Temperature, °C	Mol %	Total vapor pressure, kPa[a]
10	4.9	9.9
10	10.5	18.6
10	46.6	48.4
20	5.4	16.7
20	12.9	39.3
20	21.8	57.7

[a] To convert kPa to mm Hg, multiply by 7.5.

acetaldehyde–ethyl ether (7); solubility data for acetaldehyde–water–methane (8), acetaldehyde–methane (9); densities and refractive indexes of acetaldehyde for temperatures 0–20°C (2); compressibility and viscosity at high pressure (10); thermodynamic data (11–13); pressure-enthalpy diagram for acetaldehyde (14); specific gravities of acetaldehyde–paraldehyde and acetaldehyde–acetaldol mixtures at 20/20°C vs composition (7); boiling point vs composition of acetaldehyde–water at 101.3 kPa (1 atm) and integral heat of solution of acetaldehyde in water at 11°C (7).

Table 4. **Solubility of Gases in Liquid Acetaldehyde at Atmospheric Pressure**

	Volume of gas (STP) dissolved in 1 volume acetaldehyde		
Temperature, °C	Acetylene	Carbon dioxide	Nitrogen
−16	54		
−6	27	11	
0	17	6.6	
12	7.3	14.5	0.15
16	5	1.5	
20	3		

3. Chemical Properties

The limits and products of the various combustion zones for acetaldehyde–oxygen and acetaldehyde–air have been described (15–18); the effect of pressure on the explosive limits of acetaldehyde–air mixtures has been investigated (19). In a study of the spontaneous ignition of fuels injected into hot air streams, it was found that acetaldehyde was the least ignitable of the aldehydes examined (20,21). The influence of surfaces on the ignition and detonation of fuels containing acetaldehyde has been reported (22,23). Ignition data have been published for the systems acetaldehyde–oxygen–peroxyacetic acid–acetic acid (24), acetaldehyde–oxygen–peroxyacetic acid (25), and ethylene oxide–air–acetaldehyde (26).

Acetaldehyde is a highly reactive compound exhibiting the general reactivity ofaldehydes (qv). Acetaldehyde undergoes numerous condensation, addition, and polymerization reactions; under suitable conditions, the oxygen or any of the hydrogens can be replaced.

3.1. Decomposition. Acetaldehyde decomposes at temperatures above 400°C, forming principally methane and carbon monoxide. The activation energy of the pyrolysis reaction is 97.7 kJ/mol (408.8 kcal/mol) (27). There have been many investigations of the photolytic and radical-induced decomposition of acetaldehyde and deuterated acetaldehyde (28–30).

3.2. The Hydrate and Enol Form. In aqueous solutions, acetaldehyde exists in equilibrium with the acetaldehyde hydrate [4433-56-1], ($CH_3CH(OH)_2$). The degree of hydration can be computed from an equation derived by Bell and Clunie (31). Hydration, the mean heat of which is −21.34 kJ/mol (−89.29 kcal/mol), has been attributed to hyperconjugation (32). The enol form, vinyl alcohol [557-75-5] ($CH_2{=}CHOH$) exists in equilibrium with acetaldehyde to the extent of approximately one molecule per 30,000. Acetaldehyde enol has been acetylated with ketene [463-51-4] to form vinyl acetate [108-05-4] (33).

3.3. Oxidation. Acetaldehyde is readily oxidized with oxygen or air to acetic acid, acetic anhydride, and peracetic acid (see ACETIC ACID AND DERIVATIVES). The principal product depends on the reaction conditions. Acetic acid [64-19-7] may be produced commercially by the liquid-phase oxidation of acetaldehyde at 65°C using cobalt or manganese acetate dissolved in acetic acid as a catalyst (34). Liquid-phase oxidation in the presence of mixed acetates of copper and cobalt yields acetic anhydride [108-24-7] (35). Peroxyacetic acid or a perester is believed to be the precursor in both syntheses. There are two commercial processes for the production of peracetic acid [79-21-0]. Low temperature oxidation of acetaldehyde in the presence of metal salts, ultraviolet irradiation, or ozone yields acetaldehyde monoperacetate, which can be decomposed to peracetic acid and acetaldehyde (36). Peracetic acid can also be formed directly by liquid-phase oxidation at 5–50°C with a cobalt salt catalyst (37) (see PEROXIDES AND PEROXY COMPOUNDS). Nitric acid oxidation of acetaldehyde yields glyoxal [107-22-2] (38,39). Oxidations of *p*-xylene to terephthalic acid [100-21-0] and of ethanol to acetic acid are activated by acetaldehyde (40,41).

3.4. Reduction. Acetaldehyde is readily reduced to ethanol (qv). Suitable catalysts for vapor-phase hydrogenation of acetaldehyde are supported nickel (42) and copper oxide (43). The kinetics of the hydrogenation of acetaldehyde over a commercial nickel catalyst have been studied (44).

3.5. Polymerization. Paraldehyde, 2,4,6-trimethyl-1,3-5-trioxane [123-63-7], a cyclic trimer of acetaldehyde, is formed when a mineral acid, such as sulfuric, phosphoric, or hydrochloric acid, is added to acetaldehyde (45). Paraldehyde can also be formed continuously by feeding liquid acetaldehyde at 15–20°C over an acid ion-exchange resin (46). Depolymerization of paraldehyde occurs in the presence of acid catalysts (47); after neutralization with sodium acetate, acetaldehyde and paraldehyde are recovered by distillation. Paraldehyde is a colorless liquid, boiling at 125.35°C at 101 kPa (1 atm).

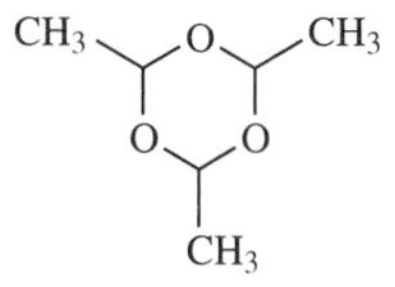

paraldehyde

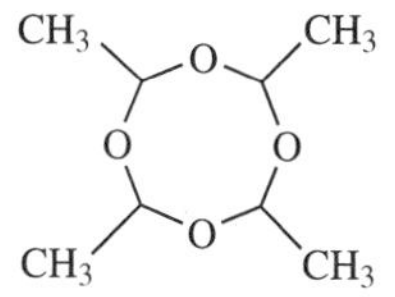

metaldehyde

Metaldehyde [9002-91-9], a cyclic tetramer of acetaldehyde, is formed at temperatures below 0°C in the presence of dry hydrogen chloride or pyridine–hydrogen bromide. The metaldehyde crystallizes from solution and is separated from the paraldehyde by filtration (48). Metaldehyde melts in a sealed tube at 246.2°C and sublimes at 115°C with partial depolymerization.

Polyacetaldehyde, a rubbery polymer with an acetal structure, was first discovered in 1936 (49,50). More recently, it has been shown that a white, nontacky, and highly elastic polymer can be formed by cationic polymerization using BF_3 in liquid ethylene (51). At temperatures below −75°C using anionic initiators, such as metal alkyls in a hydrocarbon solvent, a crystalline, isotactic polymer is obtained (52). This polymer also has an acetal [poly(oxymethylene)] structure. Molecular weights in the range of 800,000–3,000,000 have been reported. Polyacetaldehyde is unstable and depolymerizes in a few days to acetaldehyde. The methods used for stabilizing polyformaldehyde have not been successful with polyacetaldehyde and the polymer has no practical significance (see ACETAL RESINS).

3.6. Reactions with Aldehydes and Ketones. The base-catalyzed self-addition of acetaldehyde leads to formation of the dimer, acetaldol [107-89-1], which can be hydrogenated to form 1,3-butanediol or dehydrated to form crotonaldehyde [4170-30-3]. Crotonaldehyde (qv) can also be made directly by the vapor-phase condensation of acetaldehyde over a catalyst (53).

Acetaldehyde forms aldols with other carbonyl compounds containing active hydrogen atoms. Kinetic studies of the aldol condensation of acetaldehyde and deuterated acetaldehydes have shown that only the hydrogen atoms bound to the carbon adjacent to the CHO group take part in the condensation reactions and hydrogen exchange (54,55). A hexyl alcohol, 2-ethyl-1-butanol [97-95-0], is produced industrially by the condensation of acetaldehyde and butyraldehyde in dilute caustic solution followed by hydrogenation of the enal intermediate (see ALCOHOLS, HIGHER ALIPHATIC). Condensation of acetaldehyde in the presence of dimethylamine hydrochloride yields polyenals which can be hydrogenated to a mixture of alcohols containing from 4 to 22 carbon atoms (56).

The base-catalyzed reaction of acetaldehyde with excess formaldehyde is the commercial route to pentaerythritol [115-77-5]. The aldol condensation of three moles of formaldehyde with one mole of acetaldehyde is followed by a crossed Cannizzaro reaction between pentaerythrose, the intermediate product, and formaldehyde to give pentaerythritol (57). The process proceeds to completion without isolation of the intermediate. Pentaerythrose [3818-32-4] has also been made by condensing acetaldehyde and formaldehyde at 45°C using magnesium oxide as a catalyst (58). The vapor-phase reaction of acetaldehyde and formaldehyde at 475°C over a catalyst composed of lanthanum oxide on silica gel gives acrolein [107-02-8] (59).

Ethyl acetate [141-78-6] is produced commercially by the Tischenko condensation of acetaldehyde using an aluminum ethoxide catalyst (60). The Tischenko reaction of acetaldehyde with isobutyraldehyde [78-84-2] yields a mixture of ethyl acetate, isobutyl acetate [110-19-0], and isobutyl isobutyrate [97-85-8] (61).

3.7. Reactions with Ammonia and Amines. Acetaldehyde readily adds ammonia to form acetaldehyde–ammonia. Diethylamine [109-89-7] is obtained when acetaldehyde is added to a saturated aqueous or alcoholic solution of ammonia and the mixture is heated to 50–75°C in the presence of a nickel catalyst and hydrogen at 1.2 MPa (12 atm). Pyridine [110-86-1] and pyridine derivatives are made from paraldehyde and aqueous ammonia in the presence of a catalyst at elevated temperatures (62); acetaldehyde may also be used but the yields of pyridine are generally lower than when paraldehyde is the starting material. The vapor-phase reaction of formaldehyde, acetaldehyde, and ammonia at 360°C over oxide catalyst was studied; a 49% yield of pyridine and picolines was obtained using an activated silica–alumina catalyst (63). Brown polymers result when acetaldehyde reacts with ammonia or amines at a pH of 6–7 and temperature of 3–25°C (64). Primary amines and acetaldehyde condense to give Schiff bases: $CH_3CH{=}NR$. The Schiff base reverts to the starting materials in the presence of acids.

3.8. Reactions with Alcohols, Mercaptans, and Phenols. Alcohols add readily to acetaldehyde in the presence of trace quantities of mineral acid to form acetals; eg, ethanol and acetaldehyde form diethyl acetal [105-57-7] (65). Similarly, cyclic acetals are formed by reactions with glycols and other polyhydroxy compounds; eg, ethylene glycol [107-21-1] and acetaldehyde give 2-methyl-1,3-dioxolane [497-26-7] (66):

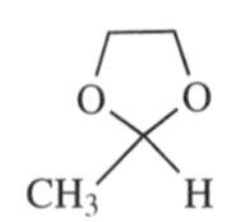

Mercaptals, $CH_3CH(SR)_2$, are formed in a like manner by the addition of mercaptans. The formation of acetals by noncatalytic vapor-phase reactions of acetaldehyde and various alcohols at 35°C has been reported (67). Butadiene [106-99-0] can be made by the reaction of acetaldehyde and ethyl alcohol at temperatures above 300°C over a tantala–silica catalyst (68). Aldol and crotonaldehyde are believed to be intermediates. Butyl acetate [123-86-4] has been prepared by the catalytic reaction of acetaldehyde with 1-butanol [71-36-3] at 300°C (69).

Reaction of one mole of acetaldehyde and excess phenol in the presence of a mineral acid catalyst gives 1,1-bis(*p*-hydroxyphenyl)ethane [2081-08-5]; acid catalysts, acetaldehyde, and three moles or less of phenol yield soluble resins. Hardenable resins are difficult to produce by alkaline condensation of acetaldehyde and phenol because the acetaldehyde tends to undergo aldol condensation and self-resinification (see PHENOLIC RESINS).

3.9. Reactions with Halogens and Halogen Compounds. Halogens readily replace the hydrogen atoms of the acetaldehyde's methyl group: chlorine reacts with acetaldehyde or paraldehyde at room temperature to give chloroacetaldehyde [107-20-0]; increasing the temperature to 70–80°C gives dichloroacetaldehyde [79-02-7]; and at a temperature of 80–90°C chloral [302-17-0] is formed (70). Catalytic chlorination using antimony powder or aluminum chloride–ferric chloride has also been described (71). Bromal [115-17-3] is formed by an analogous series of reactions (72). It has been postulated that acetyl bromide [506-96-7] is an intermediate in the bromination of acetaldehyde in aqueous ethanol (73). Acetyl chloride [75-36-5] has been prepared by the gas-phase reaction of acetaldehyde and chlorine (74).

Acetaldehyde reacts with phosphorus pentachloride to produce 1,1-dichloroethane [75-34-3] and with hypochlorite and hypoiodite to yield chloroform [67-66-3] and iodoform [75-47-8], respectively. Phosgene [75-44-5] is produced by the reaction of carbon tetrachloride with acetaldehyde in the presence of anhydrous aluminum chloride (75). Chloroform reacts with acetaldehyde in the presence of potassium hydroxide and sodium amide to form 1,1,1-trichloro-2-propanol [7789-89-1] (76).

3.10. Miscellaneous Reactions. Sodium bisulfite adds to acetaldehyde to form a white crystalline addition compound, insoluble in ethyl alcohol and ether. This bisulfite addition compound is frequently used to isolate and purify acetaldehyde, which may be regenerated with dilute acid. Hydrocyanic acid adds to acetaldehyde in the presence of an alkali catalyst to form cyanohydrin; the cyanohydrin may also be prepared from sodium cyanide and the bisulfite addition compound. Acrylonitrile [107-13-1] (qv) can be made from acetaldehyde and hydrocyanic acid by heating the cyanohydrin that is formed to 600–700°C (77). Alanine [302-72-7] can be prepared by the reaction of an ammonium salt and an alkali metal cyanide with acetaldehyde; this is a general method for the preparation of α-amino acids called the Strecker amino acids synthesis. Grignard reagents add readily to acetaldehyde, the final product being a secondary alcohol. Thioacetaldehyde [2765-04-0] is formed by reaction of acetaldehyde with hydrogen sulfide; thioacetaldehyde polymerizes readily to the trimer.

Acetic anhydride adds to acetaldehyde in the presence of dilute acid to form ethylidene diacetate [542-10-9]; boron fluoride also catalyzes the reaction (78).

Ethylidene diacetate decomposes to the anhydride and aldehyde at temperatures of 220–268°C and initial pressures of 14.6–21.3 kPa (110–160 mm Hg) (79), or upon heating to 150°C in the presence of a zinc chloride catalyst (80). Acetone (qv) [67-64-1] has been prepared in 90% yield by heating an aqueous solution of acetaldehyde to 410°C in the presence of a catalyst (81). Active methylene groups condense acetaldehyde. The reaction of isobutylene [115-11-7] and aqueous solutions of acetaldehyde in the presence of 1–2% sulfuric acid yields alkyl-*m*-dioxanes; 2,4,4,6-tetramethyl-*m*-dioxane [5182-37-6] is produced in yields up to 90% (82).

4. Manufacture

Since 1960, the liquid-phase oxidation of ethylene has been the process of choice for the manufacture of acetaldehyde. There is, however, still some commercial production by the partial oxidation of ethyl alcohol and hydration of acetylene. The economics of the various processes are strongly dependent on the prices of the feedstocks. Acetaldehyde is also formed as a coproduct in the high temperature oxidation of butane. A more recently developed rhodium catalyzed process produces acetaldehyde from synthesis gas as a coproduct with ethyl alcohol and acetic acid (83–94).

4.1. Oxidation of Ethylene. In 1894 F. C. Phillips observed the reaction of ethylene [74-85-1] in an aqueous palladium(II) chloride solution to form acetaldehyde.

$$C_2H_4 + PdCl_2 + H_2O \longrightarrow CH_3CHO + Pd + 2\ HCl$$

The direct liquid phase oxidation of ethylene was developed in 1957–1959 by Wacker-Chemie and Farbwerke Hoechst in which the catalyst is an aqueous solution of $PdCl_2$ and $CuCl_2$ (86).

Studies of the reaction mechanism of the catalytic oxidation suggest that a *cis*-hydroxyethylene–palladium π-complex is formed initially, followed by an intramolecular exchange of hydrogen and palladium to give a *gem*-hydroxyethyl-palladium species that leads to acetaldehyde and metallic palladium (88–90).

The metallic palladium is reoxidized to $PdCl_2$ by the $CuCl_2$ and the resultant cuprous chloride is then reoxidized by oxygen or air as shown.

$$Pd + 2\ CuCl_2 \longrightarrow PdCl_2 + 2\ CuCl$$

$$2\ CuCl + 1/2\ O_2 + 2\ HCl \longrightarrow 2\ CuCl_2 + H_2O$$

Thus ethylene is oxidized continuously through a series of oxidation–reduction reactions (87,88). The overall reaction is

$$C_2H_4 + 1/2\ O_2 \longrightarrow CH_3CHO \quad \Delta H = 427\ \text{kJ}\ (102\ \text{kcal})$$

There are two variations for this commercial production route: the two-stage process developed by Wacker-Chemie and the one-stage process developed

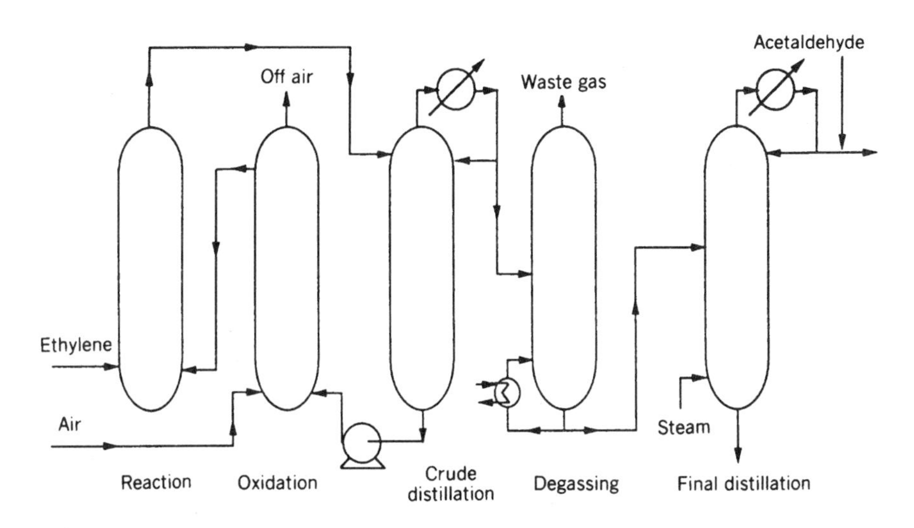

Fig. 1. The two-stage acetaldehyde process.

by Farbwerke Hoechst (91–92). In the two-stage process shown in Figure 1, ethylene is almost completely oxidized by air to acetaldehyde in one pass in a tubular plug-flow reactor made of titanium (93,94). The reaction is conducted at 125–130°C and 1.13 MPa (150 psig) using the palladium and cupric chloride catalysts. Acetaldehyde produced in the first reactor is removed from the reaction loop by adiabatic flashing in a tower. The flash step also removes the heat of reaction. The catalyst solution is recycled from the flash-tower base to the second stage (or oxidation reactor) where the cuprous salt is oxidized to the cupric state with air. The high pressure off-gas from the oxidation reactor, mostly nitrogen, is separated from the liquid catalyst solution and scrubbed to remove acetaldehyde before venting. A small portion of the catalyst stream is heated in the catalyst regenerator to destroy any undesirable copper oxalate. The flasher overhead is fed to a distillation system where water is removed for recycle to the reactor system and organic impurities, including chlorinated aldehydes, are separated from the purified acetaldehyde product. Synthesis techniques purported to reduce the quantity of chlorinated by-products generated have been patented (95).

In the one-stage process (Fig. 2), ethylene, oxygen, and recycle gas are directed to a vertical reactor for contact with the catalyst solution under slight pressure. The water evaporated during the reaction absorbs the heat evolved, and make-up water is fed as necessary to maintain the desired catalyst concentration. The gases are water-scrubbed and the resulting acetaldehyde solution is fed to a distillation column. The tail-gas from the scrubber is recycled to the reactor. Inert materials are eliminated from the recycle gas in a bleed-stream which flows to an auxiliary reactor for additional ethylene conversion.

This oxidation process for olefins has been exploited commercially principally for the production of acetaldehyde, but the reaction can also be applied to the production of acetone from propylene and methyl ethyl ketone [78-93-3] from

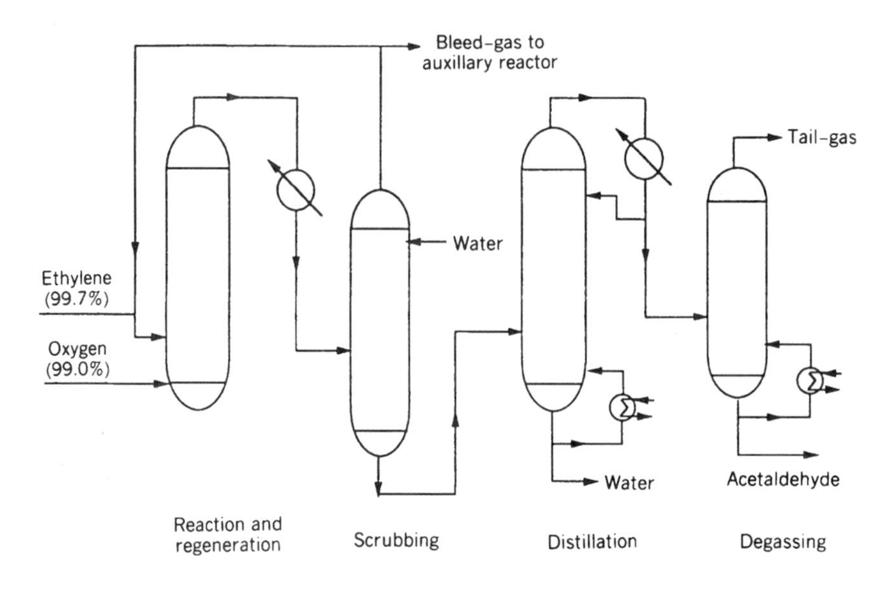

Fig. 2. The one-stage acetaldehyde process.

butenes (87,88). Careful control of the potential of the catalyst with the oxygen stream in the regenerator minimizes the formation of chloroketones (94). Vinyl acetate can also be produced commercially by a variation of this reaction (96,97). A method for preparing acetaldehyde from ethylene and oxygen has been patented (98). It is an improved process for the separation of catalyst.

4.2. From Ethyl Alcohol. Some acetaldehyde is produced commercially by the catalytic oxidation of ethyl alcohol. The oxidation is carried out by passing alcohol vapors and preheated air over a silver catalyst at 480°C (99).

$$CH_3CH_2OH + 1/2\ O_2 \longrightarrow CH_3CHO + H_2O$$

$$\Delta H = -242\ \text{kJ/mol}\ (-57.84\ \text{kcal/mol})$$

With a multitubular reactor, conversions of 74–82% per pass can be obtained while generating steam to be used elsewhere in the process (100).

4.3. From Acetylene. Although acetaldehyde has been produced commercially by the hydration of acetylene since 1916, this procedure has been almost completely replaced by the direct oxidation of ethylene. In the hydration process, high purity acetylene under a pressure of 103.4 kPa (15 psi) is passed into a vertical reactor containing a mercury catalyst dissolved in 18–25% sulfuric acid at 70–90°C (see ACETYLENE-DERIVED CHEMICALS).

$$HC{\equiv}CH + H_2O \xrightarrow[H_2SO_4(70-90^\circ C)]{Hg^{2+}} CH_3CHO$$

Fresh catalyst is fed to the reactor periodically; the catalyst may be added in the mercurous form but the catalytic species has been shown to be a mercuric ion complex (101). The excess acetylene sweeps out the dissolved acetaldehyde, which is condensed by water and refrigerated brine and then scrubbed with water; this crude acetaldehyde is purified by distillation; the unreacted acetylene is recycled. The catalytic mercuric ion is reduced to catalytically inactive mercurous sulfate and metallic mercury. Sludge, consisting of reduced catalyst and tars, is drained from the reactor at intervals and resulfated. The rate of catalyst depletion can be reduced by adding ferric or other suitable ions to the reaction solution. These ions reoxidize the mercurous ion to the mercuric ion; consequently, the quantity of sludge which must be recovered is reduced (80,102). In one variation, acetylene is completely hydrated with water in a single operation at 68–73°C using the mercuric–iron salt catalyst. The acetaldehyde is partially removed by vacuum distillation and the mother liquor recycled to the reactor. The aldehyde vapors are cooled to about 35°C, compressed to 253 kPa (2.5 atm), and condensed. It is claimed that this combination of vacuum and pressure operations substantially reduces heating and refrigeration costs (103).

4.4. From Synthesis Gas. A rhodium-catalyzed process capable of converting synthesis gas directly into acetaldehyde in a single step has been reported (83,84).

$$CO + H_2 \longrightarrow CH_3CHO + \text{other products}$$

This process comprises passing synthesis gas over 5% rhodium on SiO_2 at 300°C and 2.0 MPa (20 atm). Principal coproducts are acetaldehyde, 24%; acetic acid, 20%; and ethanol, 16%. Although interest in new routes to acetaldehyde has fallen as a result of the reduced demand for this chemical, one possible new route to both acetaldehyde and ethanol is the reductive carbonylation of methanol (85).

$$CH_3OH + CO + H_2 \longrightarrow CH_3CHO + H_2O$$

The catalyst of choice is cobalt iodide with various promotors from Group 15 elements. The process is run at 140–200°C, 28–41 MPa (4,000–6,000 psi), and gives an 88% conversion with 90% selectively to acetaldehyde. Neither of these acetaldehyde syntheses have been commercialized.

4.5. Other. A method for preparing acetaldehyde from acetic aid has been patented. The object of this preparation is to provide a method of producing acetylene that avoids the dangers associated with mercury and acetylene and the handling problems associated with reacting ethylene and oxygen, and for easy recovery of acetaldehyde (104).

5. Economic Aspects

The production pattern for acetaldehyde has undergone significant changes since the principal industrial routes to acetaldehyde were hydration of acetylene and oxidation of ethyl alcohol. First, increasing acetylene costs made this feedstock economically unattractive. Then the two aldehyde Wacker-Hoechst GmbH processes for the liquid-phase oxidation of ethylene to acetaldehyde began commercial operation. By 1968 more acetaldehyde was produced by the direct oxidation of ethylene using the Wacker process than from ethanol. Union Carbide discontinued its annual production of 90,700 t of acetaldehyde from ethanol in late 1977 (105). The percentage of ethanol consumed for acetaldehyde in the United States dropped from 20% in 1971 to zero in 1984.

Demand for acetaldehyde has decreased primarily as a result of less consumption for acetic acid production. At year-end 2000, all North American manufacture of acetic acid from acetaldehyde had been discontinued. Only China has increased capacity. Western Europe and China are the largest consumers of acetaldehyde. Since 1995 360 X 10^3 t capacity has been shut down in Western Europe and Mexico. More plant closures are expected (106).

6. Specifications, Analytical and Test Methods

Commercial acetaldehyde has the following typical specifications: assay, 99% min; color, water-white; acidity, 0.5% max (acetic acid); specific gravity, 0.790 at 20°C; bp, 20.8°C at 101.3 kPa (1 atm). It is shipped in steel drums and tank cars bearing the ICC red label. In the liquid state, it is noncorrosive to most metals; however, acetaldehyde oxidizes readily, particularly in the vapor state, to acetic acid. Precautions to be observed in the handling of acetaldehyde have been published (107).

Analytical methods based on many of the reactions common to aldehydes have been developed for acetaldehyde determination. In the absence of other aldehydes, it can be detected by the formation of a mirror from an alkaline silver nitrate solution (Tollens' reagent) and by the reduction of Fehling's solution. It can be determined quantitatively by fuchsin–sulfur dioxide solution (Schiff's reagent), or by reaction with sodium bisulfite, the excess bisulfite being estimated iodometrically. Acetaldehyde present in mixtures with other carbonyl compounds, such as organic acids, can be determined by paper chromatography of 2,4-dinitrophenylhydrazones (108), polarographic analysis either of the untreated mixture (109) or of the semicarbazones (110), the color reaction with thymol blue on silica gel (detector tube method) (111), mercurimetric oxidation (112), argentometric titration (113), microscopic (114) and spectrophotometric methods (115), and gas–liquid chromatographic analysis (116). However, gas–liquid chromatographic techniques have superseded most chemical tests for routine analyses.

Acetaldehyde can be isolated and identified by the characteristic melting points of the crystalline compounds formed with hydrazines, semicarbazides, etc; these derivatives of aldehydes can be separated by paper and column chromatography (108,117). Acetaldehyde has been separated quantitatively from other carbonyl compounds on an ion-exchange resin in the bisulfite form; the aldehyde is then eluted from the column with a solution of sodium chloride (118). In larger quantities, acetaldehyde may be isolated by passing the vapor into ether, then saturating with dry ammonia; acetaldehyde–ammonia crystallizes from the solution. Reactions with bisulfite, hydrazines, oximes, semicarbazides, and 5,5-dimethyl-1,3-cyclohexanedione [126-81-8] (dimedone) have also been used to isolate acetaldehyde from various solutions.

7. Health and Safety Factors

Acetaldehyde appears to paralyze respiratory muscles, causing panic. It has a general narcotic action which prevents coughing, causes irritation of the eyes and mucous membranes, and accelerates heart action. When breathed in high concentration, it causes headache and sore throat. Carbon dioxide solutions in acetaldehyde are particularly pernicious because the acetaldehyde odor is weakened by the carbon dioxide. Prolonged exposure causes a decrease of both red and white blood cells; there is also a sustained rise in blood pressure (119–121). IARC rates acetaldehyde as 2B (possible carcinogen to humans, NTP rates acetaldehyde as reasonably anticipated to be a human carcinogen. OSHA PEL = 200 ppm (122). In normal industrial operations there is no health hazard in handling acetaldehyde provided normal precautions are taken. Mixtures of acetaldehyde vapor and air are flammable; they are explosive if the concentrations of aldehyde and oxygen rise above 4 and 9%, respectively. Reference 107 discusses handling precautions.

8. Uses

Acetaldehyde production is linked with the demand for acetic acid, acetic anhydride, cellulose acetate, vinyl acetate resins, acetate esters, pentaerythritol, synthetic pyridine derivatives, terephthalic acid, and peracetic acid. In 1976 acetic acid production represented 60% of the acetaldehyde demand. That demand has diminished as a result of the rising cost of ethylene as feedstock and methanol carbonylation as the preferred route to acetic acid (qv).

Synthetic pyridine derivatives, peracetic acid, acetate esters by the Tischenko route, and pentaerythritol account for most (106) of acetaldehyde demand. This sector appears secure, but no new significant applications are expected for 2001–2005.

BIBLIOGRAPHY

"Acetaldehyde" in *ECT* 1st ed., Vol. 1, pp. 32–43, by M. S. W. Small, Shawinigan Chemicals Limited; in *ECT* 2nd ed., Vol. 1, pp. 77–95, by E. R. Hayes, Shawinigan Chemicals Limited (1963); in *ECT* 3rd ed., Vol. 1, pp. 97–112, by H. J. Hagemeyer, Jr., Texas Eastman Company (1978), in *ECT* 4th ed., Vol. 1, pp. 94–109, by H. J. Hagemeyer, Texas Eastman Company; "Acetaldehyde" in *ECT* (online), posting date: December 4, 2000, by H. J. Hagemeyer, Texas Eastman Company.

CITED PUBLICATIONS

1. K. F. Coles and F. Popper, *Ind. Eng. Chem.* **42**, 1434 (1950).
2. T. E. Smith and R. F. Bonner, *Ind. Eng. Chem.* **43**, 1169 (1951).
3. A. A. Dobrinskaya, V. C. Markovich, and M. B. Nelman, *Izv. Akad. Nauk SSR Ser. Khim.* **434**, (1953); *Chem. Abstr.* **48**, 4378 (1955).

4. R. P. Kirsanova and S. Sh. Byk, *Zh. Prikl. Khim. Leningrad* **31**, 1610 (1958).
5. A. E. Rabe, University Microfilm, Ann Arbor, Mich., L. C. Card Mic. 58–1920.
6. D. S. Tsiklis and A. M. Kofman, *Zh. Fiz. Khim.* **31**, 100 (1957).
7. W. H. Horsley, *Adv. Chem. Ser.* **35**, 27 (1962).
8. D. S. Tsiklis, *Zh. Fiz. Khim.* **32**, 1367 (1958).
9. D. S. Tsiklis and Ya. D. Shvarts, *Zh. Fiz. Khim.* **31**, 2302 (1957).
10. P. M. Chaudhuri and co-workers, *J. Chem. Eng. Data* **13**, 9 (1968).
11. C. F. Coleman and T. DeVries, *J. Am. Chem. Soc.* **71**, 2839 (1949).
12. K. A. Kobe and H. R. Crawford, *Pet. Refiner* **37**(7), 125 (1958).
13. K. S. Pitzer and W. Weltner, Jr., *J. Am. Chem. Soc.* **71**, 2842 (1949).
14. L. D. Christensen and J. M. Smith, *Ind. Eng. Chem.* **42**, 2128 (1950).
15. D. M. Newitt, L. M. Baxt, and V. V. Kelkar, *J. Chem. Soc.* **1703**, 1711 (1939).
16. J. H. Burgoyne and R. F. Neale, *Fuel* **32**, 5 (1953).
17. J. Chamboux and M. Lucquin, *Compt. Rend.* **246**, 2489 (1958).
18. A. G. White and E. Jones, *J. Soc. Chem. Ind. London* **69**, 2006, 209 (1950).
19. F. C. Mitchell and H. C. Vernon, *Chem. Metall. Eng.* **44**, 733 (1937).
20. B. P. Mullins, *Fuel* **32**, 481 (1953).
21. S. S. Penner and B. P. Mullins, *Explosions, Detonations, Flammability and Ignition*, Pergamon Press, Inc., New York, 1959, pp. 199–203.
22. R. O. King, S. Sandler, and R. Strom, *Can. J. Technol.* **32**, 103 (1954).
23. S. Ono, *Rev. Phys. Chem. Jpn.* **1946**, 42; *Chem. Abstr.* **44**, 4661 (1950).
24. N. M. Emanuel, *Dokl. Akad. Nauk SSSR* **59**, 1137 (1948); *Chem. Abstr.* **42**, 7142 (1948).
25. T. E. Pavlovskaya and N. M. Emanuel, *Dokl. Akad. Nauk SSSR* **58**, 1693 (1947); *Chem. Abstr.* **46**, 4231 (1952).
26. E. M. Wilson, *J. Am. Rocket Soc.* **23**, (1953).
27. H. Nilsen, *Tidsskr. Kjemi Bergves. Metall.* **17**, 149 (1957); *Chem. Abstr.* **53**, 10916 (1959).
28. F. P. Lossing, *Can. J. Chem.* **35**, 305 (1957).
29. R. K. Brinton and D. H. Volman, *J. Chem. Phys.* **20**, 1053, 1054 (1952).
30. P. D. Zemany and M. Burton, *J. Am. Chem. Soc.* **73**, 499, 500 (1951).
31. R. P. Bell and J. C. Clunie, *Trans. Faraday Soc.* **48**, 439 (1952).
32. R. P. Bell and M. Rand, *Bull Soc. Chim. Fr.* 115 (1955).
33. H. J. Hagemeyer, Jr., *Ind. Eng. Chem.* **41**, 766 (1949).
34. A. F. Cadenhead, *Can. Chem. Process.* **39**(7), 78 (1955).
35. G. Benson, *Chem. Metall. Eng.* **47**, 150, 151 (1940).
36. B. Phillips, F. C. Frostick, Jr., and P. S. Starcher, *J. Am. Chem. Soc.* **79**, 5982 (1957).
37. U.S. Pat. 2,8309,080 (reissued RE-25057) (Apr. 8, 1958), H. B. Stevens (to Shawinigan Chemicals Ltd.).
38. U.S. Pat. 3,290,378 (Dec. 6, 1966), K. Tsunemitsu and Y. Tsujino.
39. Ger. Pat. 573,721 (Apr. 5, 1933), M. Mugdam and J. Sixt (to Consortium für electrochemische Industrie G.m.b.H.).
40. U.S. Pat. 3,240,803 (Mar. 15, 1966), B. Thompson and S. D. Neeley (to Eastman Kodak Co.).
41. U.S. Pat. 2,287,803 (June 30, 1943), D. C. Hull (to Eastman Kodak Co.).
42. T. Sasa, *J. Soc. Org. Syn. Chem. Jpn.* **12**, 60 (1954); *Chem. Abstr.* **51**, 2780 (1957).
43. Fr. Pat. 973,322 (Feb. 9, 1951), H. M. Guinot (to Usines de Melle).
44. C. C. Oldenburg and H. F. Rase, *Am. Inst. Chem. Eng. J.* **3**, 462 (1957).
45. U.S. Pat. 2,318,341 (May 1943), B. Thompson (to Eastman Kodak Co.).
46. U.S. Pat. 2,479,559 (Aug. 1949), A. A. Dolnick and co-workers (to Publicker Ind. Inc.).
47. T. Kawaguchi and S. Hasegawa, *Tokyo Gakugei Daigaku Kiyo Dai 4 Bu* **21**, 64 (1969).
48. U.S. Pat. 2,426,961 (Sept. 2, 1947), R. S. Wilder (to Publicker Ind. Inc.).

49. M. W. Travers, *Trans. Faraday Soc.* **32**, 246 (1936).
50. M. Letort, *Compt. Rend.* **202**, 767 (1936).
51. O. Vogl, *J. Poly. Sci. A* **2**, 4591 (1964).
52. J. T. Furukawa, *Macromol. Chem.* **37**, 149 (1969).
53. U.S. Pat. 2,810,6 (Oct. 22, 1957), J. F. Gabbet, Jr. (to Escambia Chemical Corp.).
54. Y. Pocker, *Chem. Ind.* **1959**, 599 (1959).
55. R. P. Bell and M. J. Smith, *J. Chem. Soc.* 1691 (1958).
56. W. Langenbeck, J. Alm, and K. W. Knitsch, *J. Prakt. Chem.* **8**, 112 (1959).
57. E. Berlow, R. H. Barth, and J. E. Snow, *The Pentaerythritols*, Reinhold Publishing Corporation, New York, 1959, pp. 4–24.
58. Fr. Pat. 962,381 (June 8, 1950) (to Etablissements Kuhlmann).
59. U.S. Pat. 3,701,798 (Oct. 31, 1972), T. C. Snapp, A. E. Blood, and H. J. Hagemeyer, Jr. (to Eastman Kodak Co.).
60. A. S. Hester and K. Himmler, *Ind. Eng. Chem.* **51**, 1428 (1959).
61. U.S. Pat. 3,714,236 (Jan. 30, 1973), H. N. Wright and H. J. Hagemeyer, Jr. (to Eastman Kodak Co.).
62. M. S. Astle, *Industrial Organic Nitrogen Compounds*, Reinhold Publishing Corporation, New York, 1961, pp. 134–136.
63. S. L. Levy and D. F. Othmer, *Ind. Eng. Chem.* **47**, 789 (1955).
64. J. F. Carson and H. S. Olcott, *J. Am. Chem. Soc.* **76**, 2257 (1954).
65. J. Deschamps, M. Paty, and P. Pineau, *Compt. Rend.* **238**, 911 (1954).
66. Neth. Pat. Appl. 6,510,968 (Feb. 21, 1966) (to Lummus Co.); W. G. Lloyd, *J. Org. Chem.* **34**, 3949 (1969).
67. U.S. Pat. 2,691,684 (Oct. 12, 1954), L. K. Frevel and J. W. Hedelund (to The Dow Chemical Company).
68. B. B. Corson and co-workers, *Ind. Eng. Chem.* **42**, 359 (1950).
69. S. L. Lel'chuk, *Khim. Prom.* **1946**(9), 16 (1946); *Chem. Abstr.* **41**, 3756 (1947).
70. W. T. Cave, *Ind. Eng. Chem.* **45**, 1854 (1953).
71. Jpn. Pat. 4713 (Nov. 14, 1952), J. Imamura (to Bureau of Industrial Technics); *Chem. Abstr.* **47**, 11224 (1953).
72. M. N. Shchukina, *Zh. Obshch. Khim.* **18**, 1653 (1948); *Chem. Abstr.* **43**, 2575 (1949).
73. N. N. Lichton and F. Granchelli, *J. Am. Chem. Soc.* **76**, 3729 (1954).
74. Jpn. Pat. 153,599 (Nov. 2, 1942), Y. Kato; *Chem. Abstr.* **43**, 3027 (1949).
75. G. Illari, *Gazz. Chim. Ital.* **81**, 439 (1951); *Chem. Abstr.* **46**, 5532 (1952).
76. R. Lombard and R. Boesch, *Bull. Soc. Chim. Fr.* **1953**(10), C23 (1953).
77. K. Sennewald and K. H. Steil, *Chem. Ing. Tech.* **30**, 440 (1958).
78. E. H. Man, J. J. Sanderson, and C. R. Hauser, *J. Am. Chem. Soc.* **72**, 847 (1950).
79. C. C. Coffin, *Can. J. Res.* **5**, 639 (1931).
80. P. W. Sherwood, *Pet. Refiner* **34**(3), 203 (1955).
81. St. Grzelczyk, *Przem. Chem.* **12**(35), 696 (1956); *Chem. Abstr.* **52**, 12753 (1958).
82. M. I. Farberov and K. A. Machtina, *Uch. Zap. Yarosl. Tekhnol. Inst.* **2**, 5 (1957); *Chem. Abstr.* **53**, 18041 (1959).
83. W. Ger. Offen. 2,503,204 (Jan. 28, 1974), M. M. Bhasin (to Union Carbide Co.).
84. Belg. Pat. 824,822 (July 28, 1975) (to Union Carbide Co.).
85. U.S. Pat. 4,337,365 (June 29, 1982), W. E. Wakler (to Union Carbide Co.).
86. R. Jira and W. Freiesleben, *Organomet. React.* **3**, 22 (1972).
87. J. Smidt and co-workers, *Angew. Chem.* **71**, 176 (1959).
88. J. Smidt, *Chem. Ind. London* **1962**(2), 54 (1962).
89. R. Jira, J. Sedlmeier, and J. Smidt, *Justus Liebigs Ann. Chem.* **6993**, 99 (1966).
90. R. Jira, W. Blan, and D. Grimm, *Hydrocarbon Process.* **55**(3), 97 (1976).
91. Can. Pat. 625,430 (Aug. 8, 1961), J. Smidt and co-workers (to Consortium fur elektrochemische Industrie (G.m.b.H.).

92. *Pet. Refiner* **40**(11), 206 (1961).
93. R. P. Lowry, *Hydrocarbon Process.* **53**(11), 105 (1974).
94. R. Jira, in S. A. Miller, ed., *Ethylene and Its Industrial Derivatives*, Ernest Benn Ltd., London, 1969, pp. 639–553.
95. U.S. Pat. 4,720,474 (Sept. 24, 1985), J. Vasilevskis and co-workers (to Catalytica Associates).
96. Brit. Pat. 1,109,483 (Apr. 10, 1968), H. J. Hagemeyer, Jr., and co-workers (to Eastman Kodak Co.).
97. K. R. Bedell and H. A. Rainbird, *Hydrocarbon Process.* **51**(11), 141 (1972).
98. U.S. Pat. 6,140,544 (April. 19, 1999), B. Rinne and E. Franken-Stellamans (to Celanese GmbH).
99. W. L. Faith, D. B. Keyes, and R. L. Clark, *Industrial Chemicals*, 2nd ed., John Wiley & Sons, Inc., New York, 1957, pp. 2–3.
100. U.S. Pat. 3,284,170 (Nov. 8, 1966), S. D. Neeley (to Eastman Kodak Co.).
101. K. Schwabe and J. Voigt, *Z. Phys. Chem. Leipzig* **203**, 383 (1954).
102. *Pet. Refiner* **40**(11), 207 (1961).
103. D. F. Othmer, K. Kon, and T. Igarashi, *Ind. Eng. Chem.* **48**, 1258 (1956).
104. U.S. Pat. 6,121,498 (April. 30, 1998), G. C. Tuskin, L. S. Depew, and N. A. Collins (to Eastman Chemical Co.).
105. *Chem. Mark. Rep.* **3** (Dec. 27, 1976).
106. "Acetaldehyde," *Chemical Economics Handbook*, SRI International, Menlo Park, Calif., 2001.
107. Chemical Safety Data Sheet SD-43, *Properties and Essential Information for Safe Handling and Use of Acetaldehyde*, Manufacturing Chemists Association, Inc., Washington, D.C., 1952.
108. R. Ellis, A. M. Gaddis, and G. T. Currie, *Anal. Chem.* **30**, 475 (1958).
109. S. Sandler and Y. H. Chung, *Anal. Chem.* **30**, 1252 (1958).
110. D. M. Coulson, *Anal. Chim. Acta.* **19**, 284 (1958).
111. Y. Kobayashi, *Yuki Gosei Kagaku Kyokai Shi* **16**, 625 (1958); *Chem. Abstr.* **53**, 984 (1959).
112. J. E. Ruch and J. B. Johnson, *Anal. Chem.* **28**, 69 (1956).
113. H. Siegel and F. T. Weiss, *Anal. Chem.* **26**, 917 (1954).
114. R. E. Dunbar and A. E. Aaland, *Microchem. J.* **2**, 113 (1954).
115. J. H. Ross, *Anal. Chem.* **25**, 1288 (1953).
116. R. Stevens, *Anal. Chem.* **33**, 1126 (1961).
117. L. Nebbia and F. Guerrieri, *Chim. Ind. Milan* **39**, 749 (1957).
118. G. Gabrielson and O. Samuelson, *Sven. Km. Tidskr.* **64**, 150 (1952) (in English); *Chem. Abstr.* **46**, 9018 (1952).
119. E. W. Page and R. Reed, *Am. J. Physiol.* **143**, 122 (1945).
120. E. Skog, *Acta Pharmacol. Toxicol.* **6**, 299 (1950).
121. H. F. Smyth, Jr., C. P. Carpenter, and C. S. Weil, *J. Ind. Hyg. Toxicol.* **31**, 60 (1949).
122. R. L. Melnick, in E. Bingham, B. Cohrssen, and C. H. Powell, eds., *Patty*'s Toxicology, 5th ed., John Wiley & Sons, Inc., New York, 2001.

H. J. Hagemeyer
Texas Eastman Company

ACETIC ACID

1. Introduction

Acetic acid [64-19-7], CH_3COOH, is a corrosive organic acid having a sharp odor, burning taste, and pernicious blistering properties. It is found in ocean water, oilfield brines, rain, and at trace concentrations in many plant and animal liquids. It is central to all biological energy pathways. Fermentation of fruit and vegetable juices yields 2–12% acetic acid solutions, usually called vinegar (qv). Any sugar-containing sap or juice can be transformed by bacterial or fungal processes to dilute acetic acid.

Theophrastos (272–287 BC) studied the utilization of acetic acid to make white lead and verdigris [52503-64-7]. Acetic acid was also well known to alchemists of the Renaissance. Andreas Libavius (AD 1540–1600) distinguished the properties of vinegar from those of icelike (glacial) acetic acid obtained by dry distillation of copper acetate or similar heavy metal acetates. Numerous attempts to prepare glacial acetic acid by distillation of vinegar proved to be in vain, however.

Lavoisier believed he could distinguish acetic acid from acetous acid, the hypothetical acid of vinegar, which he thought was converted into acetic acid by oxidation. Following Lavoisier's demise, Adet proved the essential identity of acetic acid and acetous acid, the latter being the monohydrate, and in 1847, Kolbe finally prepared acetic acid from the elements.

Worldwide demand for acetic acid in 1999 was 2.8×10^6 t (6.17×10^9 lb). Estimated demand for 2003 is 3.1×10^6 t (6.84×10^9 lb) (1). Uses include the manufacture of vinyl acetate [108-05-4] and acetic anhydride [108-24-7]. Vinyl acetate is used to make latex emulsion resins for paints, adhesives, paper coatings, and textile finishing agents. Acetic anhydride is used in making cellulose acetate fibers, cigarette filter tow, and cellulosic plastics.

2. Physical Properties

Acetic acid, fp 16.635°C (2), bp 117.87°C at 101.3 kPa (3), is a clear, colorless liquid. Water is the chief impurity in acetic acid although other materials such as acetaldehyde, acetic anhydride, formic acid, biacetyl, methyl acetate, ethyl acetoacetate, iron, and mercury are also sometimes found. Water significantly lowers the freezing point of glacial acetic acid as do acetic anhydride and methyl acetate (4). The presence of acetaldehyde [75-07-0] or formic acid [64-18-6] is commonly revealed by permanganate tests; biacetyl [431-03-8] and iron are indicated by color. Ethyl acetoacetate [141-97-9]may cause slight color in acetic acid and is often mistaken for formic acid because it reduces mercuric chloride to calomel. Traces of mercury provoke catastrophic corrosion of aluminum metal, often employed in shipping the acid.

The vapor density of acetic acid suggests a molecular weight much higher than the formula weight, 60.06. Indeed, the acid normally exists as a dimer (5), both in the vapor phase (6) and in solution (6). This vapor density anomaly has important consequences in engineering computations, particularly in distillations.

Table 1. **Acetic Acid–Water Freezing Points**

Acetic acid, wt %	Freezing point, °C
100	16.635
99.95	16.50
99.70	16.06
99.60	15.84
99.2	15.12
98.8	14.49
98.4	13.86
98.0	13.25
97.6	12.66
97.2	12.09
96.8	11.48
96.4	10.83
96.0	10.17

Acetic acid containing <1% water is called glacial. It is hygroscopic and the freezing point is a convenient way to determine purity (8). Water is nearly always present in far greater quantities than any other impurity. Table 1 shows the freezing points for acetic acid–water mixtures.

The Antoine equation for acetic acid has recently been revised (3)

$$\ln(P) = 15.19234 + (-3654.622)/T + (-45.392)$$

The pressure P is measured in kelopascal (kPa) and the temperature T in kelvin (K). The vapor pressure of pure acetic acid is tabulated in Table 2. Precise liquid density measurements are significant for determining the mass of tank car quantities of acid. Liquid density data (9) as a function of temperature are given in Table 3.

Acetic acid forms a monohydrate containing ~23% water; thus the density of acetic acid–water mixtures goes through a maximum between 77 and 80 wt %

Table 2. **Acetic Acid Vapor Pressure**

Temperature, °C	Pressure, kPa[a]	Temperature, °C	Pressure, kPa[a]
0	4.7	110	776.7
10	8.5	118.2	1013
20	15.7	130	1386.5
30	26.5	140	1841.1
40	45.3	150	2461.1
50	74.9	160	3160
60	117.7	170	4041
70	182.8	180	5091
80	269.4	190	6333
90	390.4	200	7813
100	555.3	210	9612

[a] To convert kPa to psi, multiply by 0.145.

Table 3. **Density of Acetic Acid (Liquid)**

Temperature, °C	Density, kg/m^3
20	1049.55
25	1043.92
30	1038.25
47	1019.19
67	996.46
87	973.42
107	949.90
127	925.60
147	900.27
167	873.56
187	845.04
197	829.88
207	814.07
217	797.44

acid at 15°C. When water is mixed with acetic acid at 15–18°C, heat is given off. At greater acetic acid concentrations, heat is taken up. The measured heat of mixing is consistent with dimer formation in the pure acid. The monohydrate, sometimes called acetous acid, was formerly the main article of commerce. Data on solidification points of aqueous acetic acid mixtures have been tabulated, and the eutectic formation mapped (10). The aqueous eutectic temperature is about −26°C. A procedure for concentrating acetic acid by freezing, hampered by eutectic formation, has been sought for some time. The eutectic can be decomposed through adding a substance to form a compound with acetic acid, eg, urea or potassium acetate. Glacial acetic acid can then be distilled. The densities of acetic acid–water mixtures at 15°C are given in Table 4.

A summary of the physical properties of glacial acetic acid is given in Table 5.

Table 4. **Density of Aqueous Acetic Acid**

Acetic acid, wt %	Density, g/cm^3
1	1.007
5	1.0067
10	1.0142
15	1.0214
20	1.0284
30	1.0412
40	1.0523
50	1.0615
60	1.0685
70	1.0733
80	1.0748
90	1.0713
95	1.0660
100	1.0550

Table 5. **Properties of Glacial Acetic Acid**

Property	Value	Reference
freezing point, °C	16.635	2
boiling point, °C	117.87	5
density, g/mL at 20°C	1.0495	9
refractive index, $n^{25}{}_{D}$	1.36965	11
heat of vaporization ΔH_v, J/g[a] at bp	394.5	12
specific heat (vapor), J/(g·K)[a] at 124°C	5.029	12
critical temperature, K	592.71	3
critical pressure, MPa[b]	4.53	3
enthalpy of formation, kJ/mol[a] at 25°C		
liquid	−484.50	13
gas	−432.25	13
normal entropy, J/(mol·K)[a] at 25°C		
liquid	159.8	14
gas	282.5	14
liquid viscosity, mPa (=cP)		
20°C	11.83	15
40°C	8.18	15
surface tension, mN/m (=dyn/cm) at 20.1°C	27.57	16
flammability limits, vol % in air	4.0 to 16.0	16
autoignition temperature, °C	465	
flash point, °C		17
closed cup	43	
open cup	57	

[a] To convert J to cal, divide by 4.184.
[b] To convert MPa to psi, multiply by 145.

3. Chemical Properties

3.1. Decomposition Reactions. Minute traces of acetic anhydride are formed when very dry acetic acid is distilled. Without a catalyst, equilibrium is reached after ~7 h of boiling, but a trace of acid catalyst produces equilibrium in 20 min. At equilibrium, ~4.2 mmol of anhydride is present per liter of acetic acid, even at temperatures as low as 80°C (18). Thermolysis of acetic acid occurs at 442°C and 101.3 kPa (1 atm), leading by parallel pathways to methane [72-82-8] and carbon dioxide [124-38-9], and to ketene [463-51-4] and water (19). Both reactions have great industrial significance.

Single pulse, shock tube decomposition of acetic acid in argon involves the same pair of homogeneous, molecular first-order reactions as thermolysis (20). Platinum on graphite catalyzes the decomposition at 500–800 K at low pressures (21). Ketene, methane, carbon oxides, and a variety of minor products are obtained. Photochemical decomposition yields methane and carbon dioxide and a number of free radicals, which have complicated pathways (22). Electron impact and gamma rays appear to generate these same products (23). Electron cyclotron resonance plasma made from acetic acid deposits a diamond [7782-40-3] film on suitable surfaces (24). The film, having a polycrystalline structure, is a useful electrical insulator (25) and widespread industrial exploitation of diamond films appears to be on the horizon (26).

3.2. Acid–Base Chemistry. Acetic acid dissociates in water, $pK_a = 4.76$ at 25°C. It is a mild acid that can be used for analysis of bases too weak to detect in water (27). It readily neutralizes the ordinary hydroxides of the alkali metals and the alkaline earths to form the corresponding acetates. When the crude material pyroligneous acid is neutralized with limestone or magnesia the commercial acetate of lime or acetate of magnesia is obtained (8). Acetic acid accepts protons only from the strongest acids, such as nitric acid and sulfuric acid. Other acids exhibit very powerful, superacid properties in acetic acid solutions and are thus useful catalysts for esterifications of olefins and alcohols (28). Nitrations conducted in acetic acid solvent are effected because of the formation of the nitronium ion, NO_2^+. Hexamethylenetetramine [100-97-0] may be nitrated in acetic acid solvent to yield the explosive cyclotrimethylenetrinitramine [121-82-4], also known as cyclonit or RDX.

3.3. Acetylation Reactions. Alcohols may be acetylated without catalysts by using a large excess of acetic acid.

$$CH_3COOH + ROH \longrightarrow CH_3COOR + H_2O$$

The reaction rate is increased by using an entraining agent such as hexane, benzene, toluene, or cyclohexane, depending on the reactant alcohol, to remove the water formed. The concentration of water in the reaction medium can be measured, either by means of the Karl-Fischer reagent, or automatically by specific conductance and used as a control of the rate. The specific electrical conductance of acetic acid containing small amounts of water is given in Table 6.

Nearly all commercial acetylations are realized using acid catalysts. Catalytic acetylation of alcohols can be carried out using mineral acids, eg, perchloric acid [7601-90-3], phosphoric acid [7664-38-2], sulfuric acid [7664-93-9], benzenesulfonic acid [98-11-3], or methanesulfonic acid [75-75-2], as the catalyst. Certain acid-reacting ion-exchange resins may also be used, but these tend to decompose in hot acetic acid. Mordenite [12445-20-4], a decationized Y-zeolite, is a useful acetylation catalyst (29) and aluminum chloride [7446-70-0], Al_2Cl_6, catalyzes n-butanol [71-36-3] acetylation (30).

Table 6. **Specific Conductance of Aqueous Acetic Acid**

Acetic acid, wt %	Specific conductance κ, S/cm × 10^7
100	0.060
99.9515	0.065
99.746	0.103
99.320	0.261
98.84	0.531
97.66	2.19
96.68	5.45
94.82	20.1
92.50	59.9
90.75	111
82.30	688

Olefins add anhydrous acetic acid to give esters, usually of secondary or tertiary alcohols: propylene [115-07-1] yields isopropyl acetate [108-21-4]; isobutylene [115-11-7] gives *tert*-butyl acetate [540-88-5]. Minute amounts of water inhibit the reaction. Unsaturated esters can be prepared by a combined oxidative esterification over a platinum group metal catalyst. For example, ethylene-air-acetic acid passed over a palladium–lithium acetate catalyst yields vinyl acetate.

Acetylation of acetaldehyde to ethylidene diacetate [542-10-9], a precursor of vinyl acetate, has long been known (8), but the condensation of formaldehyde [50-00-0] and acetic acid vapors to furnish acrylic acid [97-10-7] is more recent (31). These reactions consume relatively more energy than other routes for manufacturing vinyl acetate or acrylic acid, and thus are not likely to be further developed. Vapor-phase methanol–methyl acetate oxidation using simultaneous condensation to yield methyl acrylate is still being developed (29). A vanadium–titania phosphate catalyst is employed in that process.

4. Manufacture

Commercial production of acetic acid has been revolutionized in the decade 1978–1988. Butane–naphtha liquid-phase catalytic oxidation has declined precipitously as methanol [67-56-1] or methyl acetate [79-20-9] carbonylation has become the technology of choice in the world market. Most commercial production of virgin synthetic acetic acid is based on methanol carbonylation (1). By-product acetic acid recovery in other hydrocarbon oxidations, eg, in xylene oxidation to terephthalic acid and propylene conversion to acrylic acid, has also grown. Production from synthesis gas is increasing and the development of alternative raw materials is under serious consideration following widespread dislocations in the cost of raw material (see CHEMURGY).

Ethanol fermentation is still used in vinegar production. Research on fermentative routes to glacial acetic acid is also being pursued. Thermophilic, anaerobic microbial fermentations of carbohydrates can be realized at high rates, if practical schemes can be developed for removing acetic acid as fast as it is formed. Under usual conditions, ~5% acid brings the anaerobic reactions to a halt, but continuous separation produces high yields at high production rates. Heat for the reaction is provided by the metabolic activity of the microorganisms. Fermentative condensation of CO_2 is another possible route to acetic acid.

Currently, almost all acetic acid produced commercially comes from acetaldehyde oxidation, methanol or methyl acetate carbonylation, or light hydrocarbon liquid-phase oxidation. Comparatively small amounts are generated by butane liquid-phase oxidation, direct ethanol oxidation, and synthesis gas. Large amounts of acetic acid are recycled industrially in the production of cellulose acetate, poly(vinyl alcohol), aspirin peracetic acid, and in a broad array of other proprietary processes. (These recycling processes are not regarded as production and are not discussed herein.)

4.1. Acetaldehyde Oxidation.

Ethanol [64-17-5] is easily dehydrogenated oxidatively to acetaldehyde (qv) using silver, brass, or bronze catalysts. Acetaldehyde can then be oxidized in the liquid phase in the presence of cobalt or manganese salts to yield acetic acid. Peracetic acid [79-21-0] formation is prevented by the transition metal catalysts (8). (Most transition metal salts decompose any peroxides that form, but manganese is uniquely effective.) Kinetic system models are useful for visualizing the industrial operation (32, 33). Stirred-tank and sparger reactor rates have been compared for this reaction and both are so high that they are negligible in the reaction's mathematical description.

Figure 1 is a typical flow sheet for acetaldehyde oxidation. The reactor is an upright vessel, fitted with baffles to redistribute and redirect the air bubbles. Oxygen is fully depleted by the time a bubble reaches the first baffle and bubbles above the first baffle serve mainly for liquid agitation. Such mechanical contacting decomposes transitory intermediates and stabilizes the reactor solution. Even though the oxidizer-reactor operates under mild pressure, sufficient aldehyde boils away to require an off-gas scrubber. Oxidate is passed into a column operated under a positive nitrogen pressure, hence to an acetaldehyde recovery column where unreacted aldehyde is recycled. More importantly, many dangerous peroxides are decomposed in this column, some into acetic acid, while traces of ethanol are esterified to ethyl acetate. Crude acid is taken off at the bottom and led to a column for stripping off the low boiling constituents other than aldehyde.

Crude oxidate is passed to a still where any remaining unreacted acetaldehyde and low boiling by-products, eg, methyl acetate and acetone [67-64-1], are removed as are CO, CO_2, and N_2. High concentration aqueous acetic acid is obtained. The main impurities are ethyl acetate, formaldehyde, and formic acid although sometimes traces of a powerful oxidizing agent, possibly diacetyl peroxide, are present. If the acetaldehyde contains ethylene oxide, then ethylene glycol diacetate is present as an impurity. Formic acid can be entrained using hexane or heptane, ethyl acetate, or a similar azeotroping agent. Often the total contaminant mass is low enough to permit destruction by chemical oxidation. The oxidizing agent, such as sodium dichromate, is fed down the finishing column as a concentrated solution. Potassium permanganate solution is also effective, but it often clogs the plates of the distillation tower.

Final purification is effected by distillation giving high purity acid. Some designs add ethyl acetate to entrain water and formic acid overhead in the finishing column. The acid product is removed as a sidestream. Potassium permanganate has been employed to oxidize formaldehyde and formic acid because the finished acid must pass a permanganate test. The quantity of water in the chemical oxidizer solution is important for regulating the corrosion rate of the finishing column: Acetic acid having a purity of 99.90–100% corrodes stainless steel SS-316 or SS-320. Lowering the acid concentration to 99.75–99.80% with distilled water in the permanganate solution diminishes the corrosion rate dramatically. Residues containing manganese acetate or chromium acetate are washed with a two-phase mixture of water–butyl acetate or water–toluene. The organic solvent removes high boiling materials, tars and residual acid, and the metallic acetates remain in the water layer (34).

Alternative purification treatments have been explored but have no industrial application. Nitric acid or sodium nitrate causes the oxidation of formic acid and formaldehyde, but provokes serious corrosion problems. Schemes have been devised to reduce rather than oxidize the impurities; eg, injecting a current of

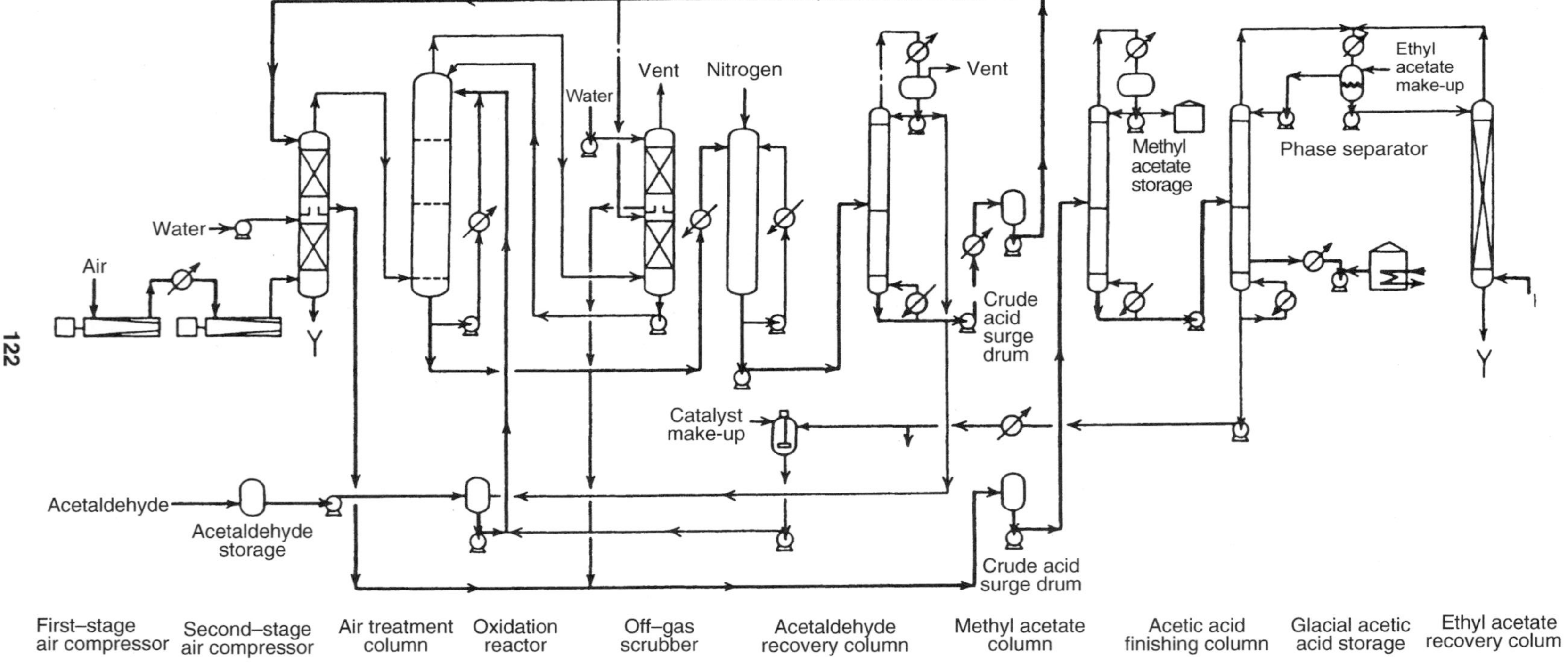

Fig. 1. A typical acetaldehyde oxidation flow sheet.

hydrogen and passing the acid over a metallic catalyst such as nickel or copper turnings. Since reduction occurs at 110–120°C, the reaction can be run in the final column. The risk of an explosion from the hydrogen passing through the column and venting at the top probably discourages the use of this treatment. Certain simple salts, eg, $FeSO_4$ or $MnSO_4$, may be introduced in the same way as the permanganate or dichromate discussed earlier. These serve to eliminate most of the quality-damaging impurities.

Conversion of acetaldehyde is typically >90% and the selectivity to acetic acid is higher than 95%. Stainless steel must be used in constructing the plant. This established process and most of the engineering is well understood. The problems that exist are related to more extensively automating control of the system, notably at start-up and shutdown, although even these matters have been largely solved. This route is the most reliable of acetic acid processes.

4.2. Methanol Carbonylation. Several processes were patented in the 1920s for adding carbon monoxide to methanol to produce acetic acid (35). The earliest reaction systems used phosphoric acid at 300–400°C under high CO pressures. Copper phosphate, hydrated tungstic oxide, iodides, and other materials were tried as catalysts or promoters. Nickel iodide proved to be particularly valuable. At that time, only gold and graphite were recognized as adequate to resist temperatures of 300–320°C and pressures of 20 MPa (2900 psi). In 1945–1946, when German work was disclosed by capture of the Central Research Files at Badische Anilin, a virtually complete plant design became public. Although this high pressure methanol carbonylation system suffered, many of the difficulties experienced in earlier processes, eg, loss of iodine, corrosive conditions, and dangerously high pressures, new alloys such as Hastelloy C permitted the containment of nearly all the practical problems. Experimental and pilot-plant units were operated successfully and, by 1963 BASF opened a large plant at Ludwigshafen, Germany providing license to Borden Chemical Company for a similar unit in Louisiana in 1966.

In 1968, a new methanol carbonylation process using rhodium promoted with iodide as catalyst was introduced by a modest letter (36). This catalyst possessed remarkable activity and selectivity for conversion to acetic acid. Nearly quantitative yields based on methanol were obtained at atmospheric pressure and a plant was built and operated in 1970 at Texas City, Tex. The effect on the world market has been exceptional (37).

Low pressure methanol carbonylation transformed the market because of lower cost raw materials, gentler, lower cost operating conditions, and higher yields. Reaction temperatures were 150–200°C and the reaction was conducted at 3.3–6.6 MPa (33–65 atm). The chief efficiency loss was conversion of carbon monoxide to CO_2 and H_2 through a water–gas shift as shown.

$$CO + H_2O \longrightarrow CO_2 + H_2$$

The subject has been reviewed (38, 39). Water may be added to the feed to suppress methyl acetate formation, but is probably not when operating on an industrial scale. Water increase methanol conversion, but it is involved in the unavoidable loss of carbon monoxide. A typical methanol carbonylation flow sheet is given in Figure 2.

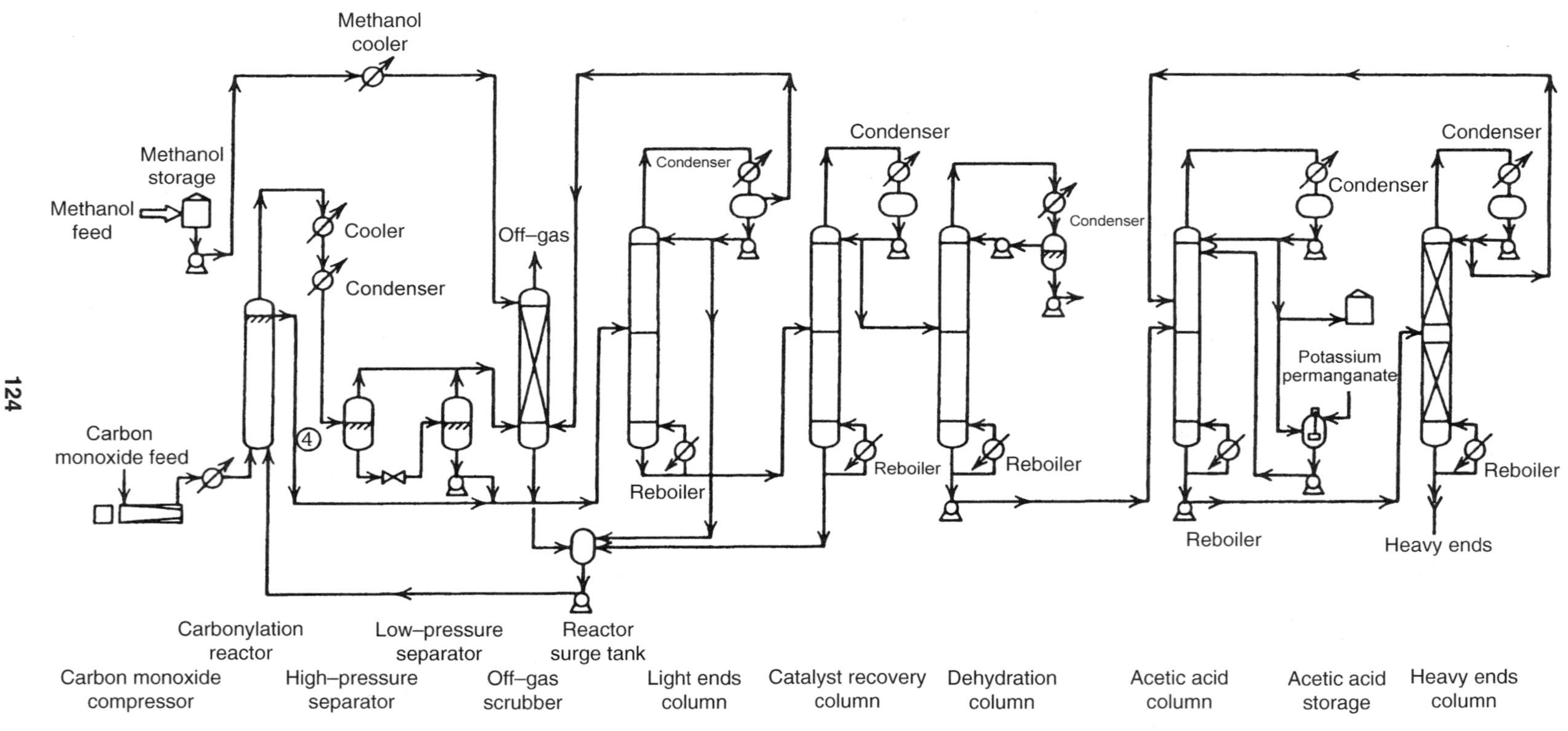

Fig. 2. A typical methanol carbonylation flow sheet.

Low boiling substances are removed from the chilled reactor product by distilling up to a cut point of 80°C. These low boilers are gaseous dimethyl ether, methyl acetate, acetaldehyde, butyraldehyde, and ethyl acetate. The bottoms are flash distilled to recover the rhodium catalyst. Flash distilled acid is azeotropically dehydrated. In the final distillation, glacial acid is obtained. Traces of iodine that may remain in the finished acid may be removed by fractional crystallization or by addition of a trace of methanol followed by distillation of the methyl iodide that forms. Somewhere in the carbonylation reaction, a minute amount of propionic acid seems to be made. It typically is found in the residues of the acetic acid finishing system and can be removed by purging the finishing column bottoms.

Vapor-phase methanol carbonylation over a supported metal catalyst has been described (40, 41). Methanol itself is obtained from synthesis gas, so the possibility of making acetic acid directly without isolating methanol has been explored (42) in both vapor and liquid phases. Alcohols are generated from CO and H_2 using halide promoted ruthenium, but acetic acid can be produced by addition of $Ru_3(CO)_{12}$ (43). A complex metallic catalyst containing rhodium, manganese, iridium, and lithium supported on silica has been used to provide selective synthesis (44). Ruthenium melt catalyst has been patented (45). Catalysts can be improved by running them in by stages to the optimum operating temperature over prolonged time periods, eg, 100–1000 h (46). A rhodium–nickel–silver catalyst for this reaction has been developed (47). An iridium catalyzed carbonylation process for the production of acetic acid has been patented (48).

Synthesis gas is obtained either from methane reforming or from coal gasification (see COAL CONVERSION PROCESSES). Telescoping the methanol carbonylation into an esterification scheme furnishes methyl acetate directly. Thermal decomposition of methyl acetate yields carbon and acetic anhydride,

$$2\,CH_3COCH_3 \longrightarrow (CH_3CO)_2O + C(soot)$$

but a pyrolytic route is not attractive because of excessive energy consumption. Methyl acetate carbonylation yields both anhydride and acetic acid, controllable in part by the conditions. A plant based on this process was put in operation in October 1983 (49).

4.3. Butane–Naphtha Catalytic Liquid-Phase Oxidation. Direct liquid-phase oxidation of butane and/or naphtha [8030-30-6] was once the most favored worldwide route to acetic acid because of the low cost of these hydrocarbons. Butane [106-97-8], in the presence of metallic ions, eg, cobalt, chromium, or manganese, undergoes simple air oxidation in acetic acid solvent (50). The peroxidic intermediates are decomposed by high temperature, by mechanical agitation, and by action of the metallic catalysts, to form acetic acid and a comparatively small suite of other compounds (51). Ethyl acetate and butanone are produced, and the process can be altered to provide larger quantities of these valuable materials. Ethanol is thought to be an important intermediate (52); acetone forms through a minor pathway from isobutane present in the hydrocarbon feed. Formic acid, propionic acid, and minor quantities of butyric acid are also formed.

The theoretical explanation of the butane reaction mechanism is as fully developed as is that of acetaldehyde oxidation (53). The theory of the naphtha oxidation reaction is more troublesome, however, and less well understood, largely because of a back-biting reaction that leads to cyclic products (54).

Liquid-phase butane oxidation is realized in a sparged column, fabricated of high alloy stainless steel. Cobalt, chromium, and manganese acetate catalyst is dissolved in acetic acid and introduced with the butane–acetic acid solution. Air or O_2-enriched air may be used. The temperature is kept just below the critical temperature of butane, 152°C. Pressure is ~5.6 MPa (812 psi) (55). The reactor product is cooled and the pressure slowly lowered. In stripping the low boiling constituents away from the reactor product, the first obtained is unreacted butane, which is often led through an expansion turbine that powers air compressors for the reactor and cools the product. The butane is then recycled to the reactor. After cooling, the reactor oxidate appears as two phases: a hydrocarbon-rich phase and a denser aqueous phase. The former is decanted and led back into the reactor; the latter is distilled to obtain the boiling oxygenates. Butanone [78-93-3] and ethyl acetate [141-78-6] are the chief constituents of the aqueous layer, but there are traces of methyl vinyl ketone (an unpleasant lacrimator), aldehydes, and esters. Formic acid, also present, forms a maximum-boiling azeotrope that boils higher than either of the chief constituents.

Although acetic acid and water are not believed to form an azeotrope, acetic acid is hard to separate from aqueous mixtures. Because a number of common hydrocarbons such as heptane or isooctane form azeotropes with formic acid, one of these hydrocarbons can be added to the reactor oxidate permitting separation of formic acid. Water is decanted in a separator from the condensate. Much greater quantities of formic acid are produced from naphtha than from butane, hence formic acid recovery is more extensive in such plants. Through judicious recycling of the less desirable oxygenates, nearly all major impurities can be oxidized to acetic acid. Final acetic acid purification follows much the same treatments as are used in acetaldehyde oxidation. Acid quality equivalent to the best analytical grade can be produced in tank car quantities without difficulties.

Two explosions, on November 14, 1987, at the largest butane liquid-phase oxidation plant resulted in 3 deaths and 37 people injured (56). The plant, which had operated since December 1952 free of such disasters, was rebuilt (57).

4.4. Prospective Processes. There has been much effort invested in examining routes to acetic acid by olefin oxidation or from ethylene, butenes, or *sec*-butyl acetate. Showa Denko brought a 100×10^3 t/year plant on stream based on direct oxidation of ethylene in 1997 (58). A process for a one-step gas-phase production of acetic acid from ethylene (59) from ethane (60), and a process from ethane–ethylene have been described (61).

Recovery of acid from wood distillate has long been viewed as a desirable prospect. Nearly all common lumber woods can be destructively distilled to furnish pyroligneous acid containing ~5–8% acetic acid. Coupled with azeotropic or extractive distillation, good quality glacial acetic acid could be prepared (62, 63). Indeed, glacial acetic acid used to be prepared from gray acetate of lime, formed by the reaction of pyroligneous acid and limestone, by reaction of the acetate with bicarbonate followed by sulfuric acid:

$$Ca(CH_3COO)_2 + Na_2CO_3 \longrightarrow 2\,Na(CH_3COO) + CaCO_3$$

$$NaOOCCH_3 + H_2SO_4 \longrightarrow NaHSO_4 + CH_3COOH$$

The acetic acid was distilled to give soda acetic acid and upon further purification, it was often used for food or pharmaceutical applications.

Mixtures of trioctylamine and 2-ethylhexanol have been employed to extract 1–9% by volume acetic acid from its aqueous solutions. Reverse osmosis for acid separation has been patented and solvent membranes for concentrating acetic acid have been described (64, 65). Decalin and trioctylphosphine were selected as solvents (66). Liquid–liquid interfacial kinetics is an especially significant factor in such extractions (67).

The fermentative fixing of CO_2 and water to acetic acid by a species of acetobacterium has been patented; acetyl coenzyme A is the primary reduction product (68). Different species of clostridia have also been used. Pseudomonads (69) have been patented for the fermentation of certain C_1 compounds and their derivatives, eg, methyl formate. These methods have been reviewed (70). The manufacture of acetic acid from CO_2 and its dewatering and refining to glacial acid has been discussed (71, 72).

The autotropic pathway for acetate synthesis among the acetogenic bacteria has been examined (73). Quantitative fermentation of 1 mol of glucose [50-99-7], $C_6H_{12}O_6$, yields 3 mol of acetic acid, while 2 mole of xylose [58-86-6], $C_5H_{10}O_5$, yields 5 mole. The glucose reaction is

$$C_6H_{12}O_6 + 2\,H_2O \longrightarrow 2\,CH_3COOH + 2\,CO_2 + 8\,H^+ + 8\,e^-$$

$$2\,CO_2 + 8\,H^+ + 8\,e^- \longrightarrow CH_3COOH + 2\,H_2O$$

Simply by passing gaseous H_2–CO_2 through an aqueous sugar mixture, the carbon dioxide is fixed into acetic acid:

$$2\,CO_2 + 4\,H_2 \longrightarrow CH_3COOH + 2\,H_2O$$

Using carbon monoxide, the reaction becomes

$$4\,CO + 2\,H_2O \longrightarrow CH_3COOH + 2\,CO_2$$

A number of C_1 compounds act as surrogates for reduction. A microbial process for the preparation of acetic acid as well as solvent for its extraction from the fermentation broth has been patented (74).

The possibility of using fermentation to generate a safe, noncorrosive road deicing composition has been studied (75). Calcium magnesium acetate [76123-46-1] is readily prepared from low concentration acetic acid produced from glucose or other inexpensive sugars. The U.S. Federal Highway Administration has financed development of anaerobic, thermophilic bacteria in an industrial process to manufacture calcium magnesium acetate from hydrolyzed corn starch and dolime [50933-69-2]. It is equally possible to utilize CH_3OH–CO_2 or H_2–CO_2 mixtures. These fermentative processes lead to dilute acid, often no more than 5 wt%, so that a concentration procedure is essential to recovery. Another new method for acetic acid production is electrodialytic conversion of dilute sodium acetate into concentrated acid (76). Electrodialytic fermentation systems have been subjected to computerized control and the energy cost for recovery is rapidly being diminished.

5. Shipping and Handling

Acetic acid, providing the concentration is greater than 99%, may be stored and shipped in aluminum. Aluminum slowly corrodes forming a layer of basic aluminum acetate that prevents further corrosion. Some of this basic oxide coating is suspended in the acid and the heels of tank cars often have a white or gray, cloudy appearance. Water increases the corrosion rate significantly; hence every effort must be devoted to maintaining a high acid concentration. Mercury in minute quantities catalyzes corrosion of aluminum by acetic acid so that a single broken thermometer can provoke catastrophic and dangerous corrosion in 99.6–99.7% glacial acid. Mercurial thermometers are often absolutely prohibited near aluminium tank cars or barrels. Acid can also be stored and shipped in stainless steel, glass carboys, and polyethylene drums.

Because glacial acetic acid freezes at $\sim$16°C, exceptional care must be taken for melting the product in cool weather. Electrical or steam heaters may be employed. Tank cars or tank wagons must be fitted with heating coils, which can be attached to a steam line and trap. Tank vents must be traced with electrical or steam lines to prevent crystallization. Acetic acid sublimes so that a single, large crystal can appear and completely fill an otherwise adequate vent.

According to the U.S. DOT regulations, acetic acid is a corrosive material (77). It may be shipped in metal or plastic packaging when no more than 0.45 kg is involved. Greater quantities may be shipped in boxed glass carboys, kegs or plywood drums, wooden barrels lined with asphalt or paraffin, earthenware containers in protective boxes, or plastic drums. It may not be shipped in plastic bags. Steel drums having polyethylene liners or polyethylene drums having steel overpacks are acceptable. Polyethylene drums do not appear to cause trace contamination after 12-months storage. Nonreturnable containers ought to be emptied and rinsed with fresh water. No other chemical ought to be shipped or stored in acetic acid containers.

Tank wagons are used for acetic acid deliveries in amounts intermediate between drum and tank car shipments, or to destinations not served by railways. Tank wagons are usually fabricated of stainless steel or aluminum alloy. Most shipments are carried in railway tank cars with nominal capacities of 38–76 m^3 (10,000–20,000 gal). For bookkeeping, capacities are determined by weight instead of volume measurements. Some tank cars are unloaded by suction, others by applying a positive nitrogen pressure and then siphoning the acetic acid into an education tube. Siphoning is frequently required by local laws.

Acetic acid is also transported in barges, sometimes in amounts of 1500–1750 tons. Acetic acid is not as hygroscopic as some other anhydrous organic substances, but barge shipments occasionally have specification problems because of wave splashing into the tanks or other careless handling.

6. Economic Aspects

Acetic acid has a place in organic processes comparable to sulfuric acid in the mineral chemical industries and its movements mirror the industry. Growth of synthetic acetic acid production in the United States was greatly affected by the

Table 7. **North American Producers of Acetic Acid and Their Capacities**[a]

Producer	Capacity × 10^3 t (×10^6 lb)/year
Virgin acid	
Celanese, Clear Lake, Tex.	998 (2,200)
Celanese, Pampa, Tex.	267 (590)
Eastman, Kingsport, Tenn.	254 (560)
Millennium, LaPorte, Tex.	453 (1000)
Sterling, Texas City, Tex.	453 (1000)
Total	*2,425 (5,350)*
Recovered Acid	
Air Products, Calvert City, Ky.	57 (125)
Air Products, Pasadena, Tex.	36 (80)
Celanese, Narrows, Va.	150 (330)
Celanese, Rock Hill, S.C.	150 (330)
DuPont, LaPorte, Tex.	82 (180)
Eastman, Kingsport, Tenn.	284 (625)
Primester, Kingsport, Tenn.	73 (160)
Other, small producers (9 locations)	98 (215)
Total	*930 (2045)*
U.S. Total	*3,355 (7395)*
Virgin Acid	
Celanese Canada, Edmonton, Alberta	91 (200)
Industrias Monfel, San Luis Potosi, Mexico	26 (45)
Total	*111 (245)*
Recovered Acid	
Celanese Canada, Edmonton, Alberta	36 (80)
Total	*36 (80)*
Other North American Total	*147 (325)*
Total	*3502 7720*

[a] Ref. 1, as of Feb. 26, 2001.

dislocations in fuel resources of the 1970s. The growth rate for 1994–1999 was 26% with an estimate of 2.6% through 2003.

North American producers and their capacities as of Feb. 2001 are listed in Table 7 (1).

Most commercial production of virgin acetic acid is based on methanol carbonylation. Celanese produces acetic acid by the liquid-phase oxidation of n-butane.

World producers of acetic acid and their production are listed in Table 8 (58) Major customers for acetic acid are listed below (58).

United States: Amoco Chemical, DuPont, Union Carbide.

Europe: Elf Atochem, DuPont, Courtaulds Fibres.

Asia/Pacific: Kohap Petrochemical, Samnam Petrochemical, Samsung Petrochemical, Capco, Dupont.

Historical prices for acetic acid for 1995–2000 was a high of $0.36/lb tech. tank, delivered, E., and a low of $0.23/lb, same basis. Prices in of Feb. 2001 were from $0.44 to $0.47/lb.

Table 8. **Other International Producers of Acetic Acid and Their Capacities**

Producer	Capacity × 10^3 t
South America	38
Western Europe	
U.K.	
BP Chemicals,	
Hull	675
France	
Acetex,	
Pardies	400
Germany	
Celanese Chemical,	
Frankfurt	180
Hurth-Knapsack	70
Wacker-Chemie,	
Burghausen	80
BASF,	
Ludwigshafen	50
Switzerland	
Lonza,	
Visp	30
Eastern Europe	207
Asia Pacific	
China	795
Japan	
Kyodo Sakussan,	
Himeji, Hyogo	400
Showa Denko,	
Oita	250
Daicel,	
Otake	36
Korea	
Samsung-BP Chemical,	
Ulsan	350
India	230
Taiwan	
China Petrochemical Development,	
Ta-sheh Hsian	100
Chang Chun Petrochemical,	
Miaoli City	60
Indonesia	
Indo Acidtama,	
Solo	33

7. Specifications and Analysis

Most specifications and analytical methods have been given (78). Most of the standards have remained unchanged for the past half-century. They were designed for acid recovered from wood tar condensates. All acid of commerce easily passes these tests.

Acetic acid made by methanol carbonylation sometimes has traces of iodine or bromine if the acid comes from the high pressure route. Qualitatively, these may be quickly detected by the Beilstein test for halogens: A copper wire is heated in a gas burner until no color can be seen and the coil plunged into the acetic acid, then brought into the gas flame again. Any trace of green or blue-green flame shows the presence of halogen. The lower identification threshold is ~0.7 ppm for chloride, ~0.65 ppm for bromide, and ~0.55 for iodide.

Super-pure acid is often specified by performance tests. Acetic acid to be used for Wijs Reagent must be very highly purified, otherwise the reagent deteriorates quickly. The dichromate–sulfuric acid test, made by dissolving potassium dichromate in concentrated H_2SO_4, and then mixing with an equal volume of acetic acid, is a sensitive test for minute quantities of certain oxidizing agents. These substances can be removed only by such special treatments as refluxing the acid with dichromate or permanganate, followed by redistillation. In commercial practice, instrumental methods are used to monitor quality but these methods are seldom given officially as standards. Gas chromatographic and mass spectrometric methods are capable of very high sensitivity.

Glacial acetic acid is considered to be 99.50 wt % or higher. A different grade has a minimum concentration of 99.70 wt %. Specialty users require water solutions of 86 and 36%. Such grades are prepared on special order. Only minor quantities of these grades are marketed, and their use is vanishing.

8. Health and Safety Factors

Acetic acid has a sharp odor and the glacial acid has a fiery taste and will penetrate unbroken skin to make blisters. Prolonged exposure to air containing 5–10 mg/m^3 does not seem to be seriously harmful, but there are pronounced, undesirable effects from constant exposure to as high as 26 mg/m^3 over a 10-day period (9).

Humans exude ~90 mg/day of volatile fatty acids in exhaled breath and perspiration, 80% of which is acetic acid (74). In a confined environment, as much as 15–20 mg/m^3 can accumulate and such concentrations can become serious in submarines or space capsules.

Concentrated aqueous or organic solutions can be strongly damaging to skin. Any solution containing >50% acetic acid should be considered a corrosive acid. Acetic acid can irreparably scar delicate tissues of the eyes, nose, or mouth. The acid penetrates the mucosa of the tongue that is near the p*K* of the acid (88). The action of acetic acid is insidious. There is no quick burning sensation when applied to the unbroken skin. Blisters appear within 30 min to 4 h. Little or no pain is experienced at first but when sensory nerve receptors are attacked, severe and unremitting pain results. Once blistering occurs, washing with water or bicarbonate seldom relieves the pain. Medical care should be sought immediately.

Care ought to be taken in handling acetic acid to avoid spillage or otherwise breathing vapors. Wash any exposed areas with large amounts of water. Once the odor of acetic acid vapors is noticeable, the area should be abandoned

Table 9. **United States and International Standards for Acetic Acid**[a]

Country	Exposure limits
U.S. OSHA PEL	TWA 10 ppm (25 mg/m^3)
U.S. NIOSH REL	TWA 10 ppm (25 mg/m^3)
	STEL/CEIL (C) 15 ppm (37 mg/m^3)
U.S. ACGIH TLV	TWA 10 ppm (25 mg/m^3)
	STEL/CEIL (C) 15 ppm (37 mg/m^3)
Australia	TWA 10 ppm (25 mg/m^3); STEL/CEIL (C) 15 ppm (37 mg/m^3)
Austria	TWA 10 ppm (25 mg/m^3)
Belgium	TWA 10 ppm (25 mg/m^3); STEL 15 ppm (37 mg/m^3)
Czechoslovakia	TWA 25 mg/m^3; STEL 15 ppm (37 mg/m^3)
Denmark	TWA 10 ppm (25 mg/m^3)
Finland	TWA 10 ppm (25 mg/m^3); STEL 15 ppm (37 mg/m^3)
France	STEL 10 ppm (25 mg/m^3)
Germany (DFG MAK)	TWA 10 ppm (25 mg/m^3)
Hungary	TWA 10 mg/m^3; STEL 20 mg/m^3
India	TWA 10 ppm (25 mg/m^3); STEL 15 ppm (37 mg/m^3)
Ireland	TWA 10 ppm (25 mg/m^3); STEL 15 (37 mg/m^3)
Japan (JSOH)	TWA 10 ppm (25 mg/m^3)
The Netherlands	TWA 10 ppm (25 mg/m^3)
The Philippines	TWA 10 ppm (25 mg/m^3)
Poland	TWA 15 mg/m^3; STEL 30 mg/m^3
Russia	TWA 10 ppm; STEL 5 mg/m^3, skin
Sweden	TWA 10 ppm (25 mg/m^3); STEL 15 ppm (35 mg/m^3)
Switzerland	TWA 10 ppm (25 mg/m^3); STEL 20 ppm (50 mg/m^3)
Thailand	TWA 10 ppm (25 mg/m^3)
Turkey	TWA 10 ppm (25 mg/m^3)
United Kingdom (HSE OES)	TWA 10 ppm (25 mg/m^3)
	STEL/CEIL (C) 15 ppm (37 mg/m^3)

[a] Ref. 81

immediately. Table 9 gives U.S. and international exposure limits for acetic acid (58).

Glacial acetic acid is dangerous, but its precise toxic dose is not known for humans. The LD_{50} for rats is said to be 3310 mg/kg, and for rabbits 1200 mg/kg (82). Ingestion of 80–90 g must be considered extraordinarily dangerous for humans. Vinegar, on the other hand, which is dilute acetic acid, has been used in foods and beverages since the most ancient of times. Although vinegar is subject to excise taxation in many countries of the world other than the United States, acetic acid for nonfood applications is commonly exempted (83).

Industrial plants for acetic acid production sometimes have waste streams containing formic and acetic acid. The quantity of acids in these streams is not significant to the overall plant efficiency, but from a safety viewpoint such material must either be recycled, treated with alkaline substances, or consumed microbially. Recycling is probably the most expensive route because the cost of processing is not repaid by an increase in efficiency. In Europe, these streams must be neutralized or degraded biologically. In Germany, no more than 3 kg/h may be emitted in vent gases, with a maximum of 150 mg/m^3 (83). The acid must be removed from the vent gas by scrubbing or chilling.

9. Uses

The uses for acetic acid are as follows: 41% is used in the production of vinyl acetate, 36% for the production of cellulose acetate and acetic anhydride, 11% for the production of acetate esters. 7% for the production of terephthalic acid, 5% miscellaneous including textiles and chloroacetic acid.

Acetic acid is used as solvent for the production of terephthalic acid. This use is growing at the rate of 6.5% per year because of the demand for terephthalic acid in the production of poly(ethylene trephthalate) resins.

Some uses for acetic acid described in recent patent applications include the following: as a signal enhancing agent in fluorescence spectroscopy (84); in the production of fruit vinegar from citrus juice (85); in a dilute cleaning composition for use in semiconductor fabrication (86); in a formulation used to control diseases in beehives (87); and in etching a substrate (88).

BIBLIOGRAPHY

"Acetic Acid" in *ECT* 1st ed., Vol. 1, pp. 56–74, by W. F. Schurig, The College of the City of New York; "Ethanoic Acid" in *ECT* 2nd ed., Vol. 8, pp. 386–404, by E. Le Monnier, Celanese Corporation of America; "Acetic Acid" in *ECT* 3rd ed., Vol. 1, pp. 124–147, by F. S. Wagner, Jr., Celanese Chemical Company; in *ECT* 4th ed., Vol. 1, pp. 131–139, by Frank S. Wagner, Jr, Nandina Corporation; "Acetic Acid" in *ECT* (online), posting date: December 4, 2000, by Frank S. Wagner, Jr., Nandina Corporation.

CITED PUBLICATIONS

1. "Acetic Acid", *Chemical News and Data, Chemical Profile*, http: www.chemexpo.com/news/Profile0102226.cfm, updated Feb. 26, 2001.
2. K. Hess and H. Haber, *Ber. Dtsh. Chem. Ges.* **70**, 2205 (1937).
3. D. Ambrose and N. B. Ghiassee, *J. Chem. Thermodyn.* **19**, 505 (1987).
4. D. D. Perrin and co-workers, *Purification of Laboratory Chemicals*, Pergamon Press, New York, 1966, p. 56.
5. I. Malijevska, *Collect. Czech. Chem. Commun.* **48**(8), 2147 (1983); O. K. Mikhailova and N. P. Markuzin, *Zh. Obshch. Khim.* **52**(10), 2164 (1982); *Zh. Obshch. Khim.* **53**(4), 713 (1983).
6. R. Buettner and G. Maurer, *Ber. Bunsenges. Phys. Chem.* **87**(10), 877 (1983).
7. Y. Fujii, H. Yamada, and M. Mizuta, *J. Phys. Chem.* **92**, 6768 (1988).
8. J. F. Thorpe and M. A. Whiteley, *Thorpe's Dictionary of Applied Chemistry*, 4th ed., Longmans, Green & Co., London, 1937, Vol. **1**, p. 51.
9. J. L. Hales, H. A. Gundry, and J. H. Ellender, *J. Chem. Thermodyn.* **15**, 211 (1983).
10. R. S. Barr and D. M. T. Newsham, *Chem. Eng. J. (Lausanne)* **33**(2), 79 (1986).
11. K. S. Howard, *J. Phys. Chem.* **62**, 1597 (1958).
12. W. Weltner, *J. Am. Chem. Soc.* **77**, 3941 (1955).
13. R. J. W. LeFévre, *Trans. Faraday Soc.* **34**, 1127 (1938).
14. D. D. Wagman and co-workers, *J. Phys. Chem. Ref. Data* **11**, Suppl. 2.
15. S. P. Miskidzh'yan and N. A. Trifonov, *Zh. Obshch. Khim.* **17**, 1033 (1947).
16. A. I. Vogel, *J. Chem. Soc.* **1948**, 1814 (1948).

17. R. J. Lewis, Sr., *Sax's Dangerous Properties of Industrial Materials*, 10th ed., John Wiley & Sons, Inc., New York, 2000, p. 15.
18. L. W. Hessel and E. C. Kooyman, *Pharm. Weekbl.* **104**, 687 (1969).
19. P. G. Blake and G. E. Jackson, *J. Chem. Soc.* **1968B**, 1153 (1968); **1969B**, 94 (1969).
20. J. C. Mackie and K. R. Doolan, *Int. J. Chem. Kinet.* **16**(5), 525 (1984).
21. J. J. Vajo, Y. K. Sun, and W. H. Winberg, *J. Phys. Chem.* **91**, 1153 (1987).
22. J. G. Calvert and J. N. Pitts, *Photochemistry*, 428 (1966).
23. A. S. Newton, *J. Chem. Phys.* **26**, 1764 (1957).
24. Jpn. Pat. JP 62-96,397 (May 2, 1987), S. Kawachi and K. Nakamura (to Ashai Chemical Industries).
25. Jpn. Pat. JP 62-113,797 (May 2, 1987), S. Kawachi and K. Katsuyuki (to Asahi Chemical Industries).
26. Y. Hirose, *Hyomen*, **25**(12), 734 (1987); *Chem. Abstr.* **108**, 189283g. Y. Hirose, *Seimitsu Kogaku Kaishi* **53**(10), 1507 (1987); *Chem. Abstr.* **108**, 61009e. N. Koshino, M. Kawarada, and K. Kurihara, *Denshi Zairyo* **27**(1), 49 (1988); *Chem. Abstr.* **109**, 16121c. H. Kawarada, J. Suzuki, and A. Hiraki, *Kagaku Kogyo* **39**(9), 784 (1988); *Chem. Abstr.* **110**, 41327v. Y. Hirose and F. Akatsuka, *Ibid.* **39**(8), 673 (1988); *Chem. Abstr.* **110**, 10257t. Y. Namba, *Ibid.* **39**(8), 666, 689(1988); *Chem. Abstr.* 110:10258u. F. Akatsuka, Y. Hirose, and K. Komaki, *Jpn. J. Appl. Phys. Pt 2* **27**(9),L1600 (1988).
27. A. Popoff, J. J. Lagowski, ed., *Chemistry of Nonaqueous Solvents*, Vol. 3, Academic Press, New York, 1970.
28. "Kohlenstoff," in K. von Baczko, ed., *Gmelins Handbuch der Anorganischen Chemie*, 8th ed., Teil C4, Frankfurt, 1975, 141–197.
29. M. Ai, *J. Catal.* **112**(1), 194 (1988).
30. P. S. T. Sai, *Reg. J. Energy, Heat Mass Transfer* **10**(2), 181 (1988).
31. J. F. Vitcha and V. A. Sims, *Ind. Eng. Chem. Prod. Res. Dev.* **5**, 50 (1966).
32. H. Hartig, *Chem. Ing. Tech.* **45**, 467 (1973).
33. A. Y. Yau, A. Manielec, and A. I. Johnson, in E. Rhodes and D. S. Scott, eds., *Proceedings of International Symposium on Research in Cocurrent Gas-Liquid Flow*, Plenum Press, New York, 1969, 607–632.
34. Ger. Pat. 2,153,767 (May 3, 1973), H. Schaum and H. Goessell (to Farbwerke Hoechst A. G.).
35. F. J. Weymouth and A. F. Millidge, *Chem. Ind. (London)* **1966**, 887 (May 28, 1966).
36. F. E. Paulik and J. F. Roth, *Chem. Commun.* **1968**, 1578 (1968).
37. *Chem. Eng. News*, 24 (Feb. 13, 1984); U.S. Pat. 4,690,912 (Sept. 1, 1987), F. E. Paulik, A. Hershman, W. R. Knox, and J. F. Roth (to Monsanto Company).
38. F. E. Paulik, *Catal. Rev.* **6**, 49 (1972).
39. R. S. Dickson, *Homogeneous Catalysis with Compounds of Rhodium and Iridium*, Reidel, Dordrecht, The Netherlands, 1985.
40. K. Omata and co-workers, *Stud. Surf. Sci. Catal.* **86**, 245 (1988); *Chem. Abstr.* **109**, 8355j (1988).
41. K. Fujimoto and co-workers, *Ind. Eng. Chem. Prod. Res. Dev.* **22**, 436 (1983).
42. A. F. Borowski, *Wiad. Chem.* **39**(10–12), 667(1985); J. Ogonowski, *Chemik* **40**(2), 38 **40**(3), 67 **40**(7), 199 (1987).
43. H. Ono and co-workers, *J. Organomet. Chem.* **331**, 387 (1987).
44. T. Nakajo, K. Sano, S. Matsuhira, and H. Arakawa, *Chem. Commun. (London)* **1987**(9), 647 (1987).
45. U.S. Pats. 4,440,570 (Apr. 3, 1984), 4,442,304 (Apr. 10, 1984), 4,557,760 (Dec. 10, 1985), H. Erpenbach and co-workers (to Hoechst A. G.).
46. J. F. Knifton, *Platinum Met. Rev.* **29**(2), 63 (1985); Jpn. Pat. Kokai Tokkyo Koho JP 59–25340 (Feb. 9, 1984) (to Agency of Industrial Sciences and Technology).

47. U.S. Pat. 4,351,908 (Sept. 28, 1984), H. J. Schmidt and E. I. Leupold (to Hoechst A. G.).
48. U.S. Pat. 6,140,535 (Oct. 31, 2000), B. L. Williams (to BP Chemicals Ltd.).
49. V. H. Agreda, *CHEM-TECH* **18**(4), 250(1988); *Chem. Eng. News* 30–32 (May 21, 1990).
50. F. Broich, *Chem. Ing. Tech.* **36**(5), 417 (1964).
51. H. Höfermann, *Chem. Ing. Tech.* **36**(5), 422 (1964).
52. C. C. Hobbs and co-workers, *Ind. Eng. Chem. Prod. Res. Dev.* **9**, 497 (1970); **11**, 220 (1972); *Ind. Eng. Chem. Process Des. Dev.* 59 (1972).
53. J. B. Saunby and B. W. Kiff, *Hydrocarbon Process*, **55**(11), 247 (1976).
54. R. K. Jensen and co-workers, *J. Am. Chem. Soc.* **101**, 7574 (1979).
55. N. M. Emanuel, E. T. Denisov, and Z. K. Maizus, *Liquid Phase Oxidation of Hydrocarbons*, Plenum Press, New York, 1967.
56. *New York Times Sect. I*, 15 (Nov. 16, 1987).
57. M. S. Reisch, *Chem. Eng. News* **65** (Nov. 23, 1987); **66**, 9 (Feb. 8, 1988); **66**, 5 (May 23, 1988).
58. "Acetic Acid", *http:www.chemweek.com/marketplace/product_focus/2000?*acetic_Acid.html.
59. U.S. Pat. 6,274,764 (Aug. 14, 2001), K. Karim and K. Sheikh (to Saudi Basic Industries Corp.)
60. U.S. Pat. 6,310,241 (Oct. 30, 2001), K. Karim and co-workers (to Saudi Basic Industries Corp.)
61. U.S. Pat. 6,274,765 (Aug. 14, 2001), H. Borchert, U. Dingerdissen, and R. Ranier (to Hoechst Research and Technology, Deutschland GmbH & Co. KG).
62. D. F. Othmer, *DECHEMA Monogr.* **33**, 9 (1959).
63. D. F. Othmer, R. E. White, and E. Trueger, *Ind. Eng. Chem.* **33**, 1240 (1941).
64. Jpn. Pat. Kokai Tokkyo Koho, JP 61-176,552 (Aug. 8, 1986), M. Tanaka, N. Kawada, and T. Morinaga (to Agency of Industrial Sciences and Technology).
65. Y. Kuo and H. P. Gregor, *Sepn. Sci. Technol.* **18**(5), 421 (1983).
66. U.S. Pat. 3,980,701 (Apr. 21, 1975), R. R. Grinstead (to The Dow Chemical Company); R. W. Hellsell, *Chem. Eng. Prog.* **73**, 55 (May 1977).
67. G. J. Hanna and R. D. Noble, *Chem. Rev.* **85**, 583 (1985).
68. T. Morinaga, *Hakko to Kogyo* **43**(11), 1015 (1985); Jpn. Pat. JP 63-84,495 (Apr. 15, 1988), T. Morinaga (to Agency of Industrial and Scientific Technology).
69. Jpn. Pat. Kokai Tokkyo Koho JP 59-179,089 (Oct. 11, 1984) (to Agency of Industrial Sciences and Technology).
70. H. Wood, H. L. Drake, and S. Hu in E. E. Snell, ed., *Some Historical and Modern Aspects of Amino Acid, Fermentations, and Nucleic Acids*, (symposium: June 3, 1981, St. Louis, Mo.), Annual Reviews Inc., Palo Alto, Calif., 1982, 29–56.
71. J. G. Zeikus, R. Kerby, and J. A. Krzycki, *Science* **227**(4691), 1167 (1985).
72. L. G. Ljungdahl, *Ann. Rev. Microbiol.* **40**, 415 (1986).
73. L. G. Ljungdahl and co-workers, *Biotechnology and Bioengineering, Symp. No. 15*, 207–223 (1985).
74. U.S. Pat. 6,368.819 (April 9, 2002), J. L. Gaddy and co-workers (to Bioengineering Resources Inc. and Celanese International Corp.)
75. L. G. Ljungdahl and co-workers, *CMA Manufacture (II) Improved Bacterial Strain for Acetate Production*, Final Report, FHWA/RD-86/117, U.S. Dept. of Transportation, Washington, D.C.
76. Y. Nomura and co-workers, *Appl. Environ. Microbiol.* **54**, 137 (1988).
77. *Code of Federal Regulations*, Title 49, 173.244–173.245, U.S. Dept. of Transportation, Washington, D.C.
78. E. F. Joy and A. J. Barnard, Jr., *Encyclopedia of Industrial Chemical Analysis*, Vol. 4, John Wiley & Sons, Inc., New York, 1967, 93–101.

79. H. S. Christensen and T. Luginbyl, eds., *Registry of Toxic Effects of Chemical Substances*, U.S. Dept. of Health, Education, Welfare, Rockville, Md., 1975.
80. V. P. Savina and B. V. Anisimov, *Kosm. Biol. Aviakosmicheskaya. Med.* **22**(1), 57–61 (1988).
81. E. Bingham, B. Cohrssen, C. H. Powell, eds., *Patty*'s Toxicology, Vol. 8, John Wiley & Sons, Inc., New York, 2001, pp. 1260–1261.
82. American Conference Governmental Industrial Hygienists, *Threshold Limit Values*, Cincinnati, Ohio, 1982; cf. Deutsche Forschungsgemeinschaft, *Maximale Arbeitsplatzkonzentrationen, 1983*, Verlag Chemie, Weinheim, 1983, p. 15.
83. *Vinegar Tax*, Bundesrepublik Deutschland, (April 25, 1972).
84. U.S. Pat. 6,241,662 (June, 5, 2001), R. Richards-Kortum and co-workers (to Lifespex, Inc.)
85. U.S. Pat. Appl. 20,010,038,869 (Nov. 8, 2001),Y. Kato and co-workers (to Pokka Corp.; Maruboshi Vinegar Co., Ltd, and Sakamoto Koryo Co. Ltd.).
86. U.S. Pat. Appl. 20020,032,136 (March 14, 2002), M. Hineman and G. T. Blalock (to Micron Technology, Inc.).
87. U.S. Pat. Appl. 20010,014,346 (Aug. 16, 2001), M. Watkins.
88. U.S. Pat. Appl. 20010,015,343 (Aug. 23, 2001), H. Sprey and co-workers.

FRANK S. WAGNER, JR.
Nandina Corporation

ACETIC ACID, HALOGENATED DERIVATIVES

1. Introduction

The most important of the halogenated derivatives of acetic acid is chloroacetic acid. Fluorine, chlorine, bromine, and iodine derivatives are all known, as are mixed halogenated acids. For a discussion of the fluorine derivatives see FLUORINE COMPOUNDS, ORGANIC.

2. Chloroacetic Acid

2.1. Physical Properties. Pure chloroacetic acid [79-11-8] ($ClCH_2COOH$), mol wt 94.50, $C_2H_3ClO_2$, is a colorless, white deliquescent solid. It has been isolated in three crystal modifications: α, mp 63°C, β, mp 56.2°C, and γ, mp 52.5°C. Commercial chloroacetic acid consists of the α form. Physical properties are given in Table 1.

Chloroacetic acid forms azeotropes with a number of organic compounds. It can be recrystallized from chlorinated hydrocarbons such as trichloroethylene, perchloroethylene, and carbon tetrachloride. The freezing point of aqueous chloroacetic acid is shown in Figure 1.

Table 1. **Physical Properties of Chloroacetic Acid**

Property	Value
boiling point, °C	189.1
density, at 25°C, g/mL	1.4043
dielectric constant at 60°C	12.3
free energy of formation, at 100°C, ΔG_f, kJ/mol[a]	−368.7
heat capacity, J/(mol·k)[b] at 100°C	181.0
heat of formation, at 100°C, ΔH_f, kJ/mol[a]	−490.1
heat of sublimation, at 25°C, ΔH_s, kJ/mol[a]	88.1
vapor pressure, kPa[c]	
at 25°C	8.68×10^{-3}
at 100°C	3.24
viscosity, at 100°C, mPa·s (= cP)	1.29
refractive index at 55°C	1.435
surface tension, at 100°C, mN/m (= dyn/cm)	35.17
dissociation constant K_a	1.4×10^{-3}
solubility (g/100 g solvent)	
water	614
acetone	257
methylene chloride	51
benzene	26
carbon tetrachloride	2.75

[a] To convert kJ/mol to kcal/mol, divide by 4.184.
[b] To convert J/(mol·K) to cal/(mol·K), divide by 4.184.
[c] To convert kPa to mm Hg, multiply by 7.5.

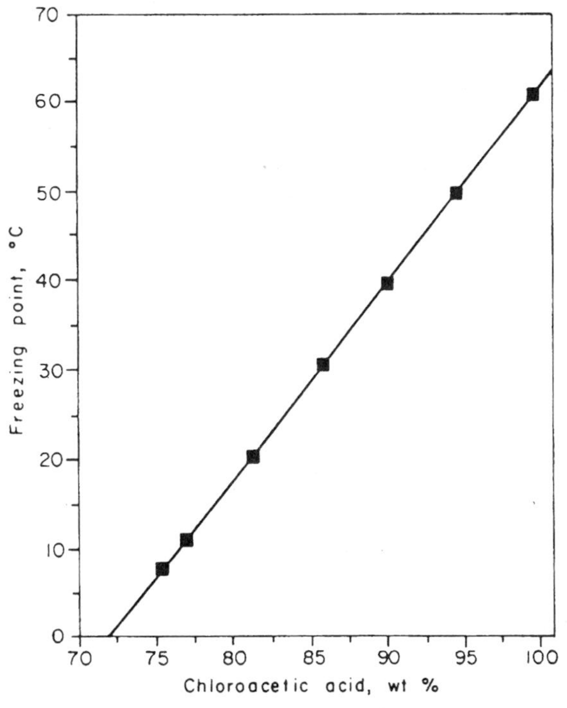

Fig. 1. The freezing point of monochloroacetic acid (MCAA)–water mixtures. For wt % acid >75%: fp(°C) = 2.17 × (wt % MCAA) − 156.5. Data courtesy of the Dow Chemical Company.

2.2. Chemical Properties. Chloroacetic acid has wide applications as an industrial chemical intermediate. Both the carboxylic acid group and the α-chlorine are very reactive. It readily forms esters and amides, and can undergo a variety of α-chlorine substitutions. Electron withdrawing effects of the α-chlorine give chloroacetic acid a higher dissociation constant than that of acetic acid.

The reaction of chloroacetic acid with alkaline cellulose yields carboxymethylcellulose which is used as a thickener and viscosity control agent (1) (see Cellulose ethers). Sodium chloroacetate reacts readily with phenols and cresols. 2,4,-Dichlorophenol and 4-chloro-2-cresol react with chloroacetic acid to give the herbicides 2,4-D and MCPA (2-methyl-4-chlorophenoxy acetic acid), respectively (2,3) (see Herbicides). Thioglycolic acid [68-11-1], $C_2H_4O_2S$, is prepared commercially by treating chloroacetic acid with sodium hydrosulfide. Sodium and ammonium salts of thioglycolic acid (qv) have been used for cold-waving human hair; the calcium salt is used as a depilating agent (see Cosmetics). Thioglycolic acid derivatives are also used as stabilizers in poly(vinyl chloride) (PVA) products.

Chloroacetic acid can be esterified and aminated to provide useful chemical intermediates. Amphoteric agents suitable as shampoos have been synthesized by reaction of sodium chloroacetate with fatty amines (4,5). Reactions with amines (6) such as ammonia, methylamine, and trimethylamine yield glycine [56-40-6], sarcosine [107-97-1], and carboxymethyltrimethylammonium chloride,respectively. Reaction with aniline forms *N*-phenylglycine [103-01-5], a starting point for the synthesis of indigo (7).

Reaction of chloroacetic acid with cyanide ion yields cyanoacetic acid [372-09-8], $C_3H_3NO_2$, (8) which is used in the formation of coumarin, malonic acid and esters, and barbiturates. Reaction of chloroacetic acid with hydroxide results in the formation of glycolic acid [79-14-1].

2.3. Manufacture. Most chloroacetic acid is produced by the chlorination of acetic acid using either a sulfur or phosphorus catalyst (9–12). The remainder is produced by the hydrolysis of trichloroethylene with sulfuric acid (13,14) or by reaction of chloroacetyl chloride with water.

A major disadvantage of the chlorination process is residual acetic acid and overchlorination to dichloroacetic acid. Although various inhibitors have been tried to reduce dichloroacetic acid formation, chloroacetic acid is usually purified by crystallization (15–17). Dichloroacetic acid can be selectively dechlorinated to chloroacetic acid with hydrogen and a catalyst such as palladium (18–20). Extractive distillation (21) and reaction with ketene (22) have also been suggested for removing dichloroacetic acid. Whereas the hydrolysis of trichloroethylene with sulfuric acid yields high purity chloroacetic acid, free of dichloracetic acid, it has the disadvantage of utilizing a relatively more expensive starting material and producing a sulfur containing waste stream.

Corrosive conditions of the chlorination process necessitate the use of glass-lined or lead-lined steel vessels in the manufacture of chloroacetic acid. Process piping and valves also are either glass-lined, or steel lined with a suitable polymer, eg, polytetrafluoroethylene (PTFE). Pumps, heat exchangers, and other process equipment can be fabricated from ceramic, graphite composite, tantalum, titanium, or certain high performance stainless steels. Chloroacetic acid can be stored as a molten liquid in glass-lined tanks for a short period of time, but

Table 2. **Producers of Chloroacetic Acid and Their Capacities**[a]

Producer	Capacity $\times 10^3$ t ($\times 10^6$ lb)
Dow, Midland, Mich.	22.7 (50)
Hercules, Hopewell, VA.	11.3 (25)
Niacet, Niagara Falls, N. Y.	9 (20)
Total	*43 (95)*

[a] Ref. 23, as of Jan. 2001.

develops color on aging. For long-term storage, the solid, flaked form of the acid is held in a polyethylene-lined fiber drum. The drum is constructed so that no chloroacetic acid vapors contact the fiber or metallic portions. Stainless steels are acceptable for the shipment of 80% solutions of chloroacetic acid provided the temperature is maintained $< 50°C$. However, long-term or continuous storage of the 80% solution results in significant pickup of iron from stainless steel.

2.4. Economic Aspects. U.S. producers of chloroacetic acid and their capacities are listed in Table 2. No new capacity is currently planned. Demand for chloroacetic acid in 1999 was 41.5×10^3 t (91.5×10^6 lb). Estimated demand for 2003 is 44.8×10^3 t (98.7×10^6 lb) (23).

Prices for the period 1994–1999 ranged from $0.24/kg ($0.53/lb) to $0.34/kg ($0.75 lb). Both prices are for high purity flake in bags, t.1., fob works. Current prices range from $0.33/kg ($0.73/lb) to $0.34/kg ($0.75/lb) (23).

2.5. Analytical and Test Methods. Gas or liquid chromatography (24) are commonly used to measure impurities such as acetic, dichloroacetic, and trichloroacetic acids. High purity 99 + % chloroacetic acid will contain $< 0.5\%$ of either acetic or dichloroacetic acid. Other impurities that may be present in small amounts are water and hydrochloric acid.

2.6. Health and Safety Aspects. Chloroacetic acid is extremely corrosive and will cause serious chemical burns. It also is readily absorbed through the skin in toxic amounts. Contamination of 5–10% of the skin area is usually fatal (25). The symptoms are often delayed for several hours. Single exposure to accidental spillage on the skin has caused human fatalities. The toxic mechanism appears to be blocking of metabolic cycles. Chloroacetic acid is 30–40 times more toxic than acetic, dichloroacetic, or trichloroacetic acid (26). When handling chloroacetic acid and its derivatives, rubber gloves, boots, and protective clothing must be worn. In case of skin exposure, the area should immediately be washed with large amounts of water and medical help should be obtained at once. Oral LD_{50} for chloroacetic acid is 76 mg/kg in rats (27).

2.7. Uses. Major industrial uses for chloroacetic acid are in the manufacture of thioglycolic acid (29%), cellulose ethers (mainly carboxymethylcellulose,CMC) (24%) and herbicides (18%). Other industrial uses (29%) include manufacture of glycine, amphoteric surfactants, cyanoacetic acid, phenoxyacetic acid, and chloroacetic acid esters (23).

3. Sodium Chloroacetate

Sodium chloroacetate [3926-62-3], mol wt 116.5, $C_2H_2ClO_2Na$, is produced by reaction of chloroacetic acid with sodium hydroxide or sodium carbonate. In many applications, chloroacetic acid or the sodium salt can be used interchangeably. As an industrial intermediate, sodium chloroacetate may be purchased or formed *in situ* from free acid. The sodium salt is quite stable in dry solid form, but is hydrolyzed to glycolic acid in aqueous solutions. The hydrolysis rate is a function of pH and temperature (28).

4. Dichloroacetic Acid

Dichloroacetic acid [79-43-6] ($Cl_2CHCOOH$), mol wt 128.94, $C_2H_2Cl_2O_2$, is a reactive intermediate in organic synthesis. Physical properties are mp 13.9°C, bp 194°C, density 1.5634 g/mL, and refractive index 1.4658, both at 20°C. The liquid is totally miscible in water, ethyl alcohol, and ether. Dichloroacetic acid ($K_a = 5.14 \times 10^{-2}$) is a stronger acid than chloroacetic acid. Most chemical reactions are similar to those of chloroacetic acid, although both chlorine atoms are susceptible to reaction. An example is the reaction with phenol to form diphenoxyacetic acid (29). Dichloroacetic acid is much more stable to hydrolysis than chloroacetic acid.

Dichloroacetic acid is produced in the laboratory by the reaction of chloral hydrate [302-17-0] with sodium cyanide (30). It has been manufactured by the chlorination of acetic and chloroacetic acids (31), reduction of trichloroacetic acid (32), hydrolysis of pentachloroethane [76-01-7] (33), and hydrolysis of dichloroacetyl chloride. Due to similar boiling points, the separation of dichloroacetic acid from chloroacetic acid is not practical by conventional distillation. However, this separation has been accomplished by the addition of azeotrope-forming hydrocarbons such as bromobenzene (34) or by distillation of the methyl or ethyl ester.

Dichloroacetic acid is used in the synthesis of chloramphenicol [56-75-7] and allantoin [97-59-6]. Dichloroacetic acid has virucidal and fungicidal activity. It was found to be active against several staphylococci (35). The oral toxicity is low: the LD_{50} in rats is 4.48 g/kg. It can, however, cause caustic burns of the skin and eyes and the vapors are very irritating and injurious (27).

5. Trichloroacetic Acid

Trichloroacetic acid [76-03-9] (Cl_3CCOOH), mol wt 163.39, $C_2HCl_3O_2$, forms white deliquescent crystals and has a characteristic odor. Physical properties are given in Table 3.

Trichloroacetic acid ($K_a = 0.2159$) is as strong an acid as hydrochloric acid. Esters and amides are readily formed. Trichloroacetic acid undergoes decarboxylation when heated with caustic or amines to yield chloroform. The decomposition of trichloroacetic acid in acetone with a variety of aliphatic and aromatic amines has been studied (36). As with dichloroacetic acid, trichloroacetic acid can be converted to chloroacetic acid by the action of hydrogen and palladium on carbon (17).

Trichloroacetic acid is manufactured in the United States by the exhaustive chlorination of acetic acid (37). The patent literature suggests two alternative

Table 3. **Physical Properties of Trichloroacetic Acid**

Property	Value
melting point, °C	59
boiling point, °C	197.5
density, at 64°C, g/mL	1.6218
refractive index at 61°C	1.4603
heat of combustion, kJ/g[a]	3.05
solubility, at 25°C, g/100 g solvent	
water	1306
methanol	2143
ethyl ether	617
acetone	850
benzene	201
o-xylene	110

[a] To convert kJ/g to kcal/g, divide by 4.184.

methods of synthesis: hydrogen peroxide oxidation of chloral (38) and hydrolytic oxidation of tetrachloroethene (39).

Sodium trichloroacetate [650-51-1], $C_2Cl_3O_2Na$, is used as a herbicide for various grasses and cattails (2). The free acid has been used as an astringent, antiseptic, and polymerization catalyst. The esters have antimicrobial activity. The oral toxicity of sodium trichloroacetate is quite low (LD_{50} rats, 5.0 g/kg). Although very corrosive to skin, trichloroacetic acid does not have the skin absorption toxicity found with chloroacetic acid (27).

6. Chloroacetyl Chloride

Chloroacetyl chloride [79-04-9] ($ClCH_2COCl$) is the corresponding acid chloride of chloroacetic acid (see Acetyl chloride). Physical properties include mol wt 112.94, $C_2H_2Cl_2O$, mp −21.8°C, bp 106°C, vapor pressure 3.3 kPa (25 mm Hg) at 25°C, 12 kPa (90 mm Hg) at 50°C, and density 1.4202 g/mL and refractive index 1.4530, both at 20°C. Chloroacetyl chloride has a sharp, pungent, irritating odor. It is miscible with acetone and benzene and is initially insoluble in water. A slow reaction at the water–chloroactyl chloride interface, however, produces chloroacetic acid. When sufficient acid is formed to solubilize the two phases, a violent reaction forming chloroacetic acid and HCl occurs.

Since chloroacetyl chloride can react with water in the skin or eyes to form chloroacetic acid, its toxicity parallels that of the parent acid. Chloroacetyl chloride can be absorbed through the skin in lethal amounts. The oral LD_{50} for rats is between 120 and 250 mg/kg. Inhalation of 4 ppm causes respiratory distress. A TLV of 0.05 ppm is recommended (27, 40).

Chloroacetyl chloride is manufactured by reaction of chloroacetic acid with chlorinating agents such as phosphorus oxychloride, phosphorus trichloride, sulfuryl chloride, or phosgene (41–43). Various catalysts have been used to promote the reaction. Chloroacetyl chloride is also produced by chlorination of acetyl chloride (44–46), the oxidation of 1,1-dichloroethene (47, 48), and the addition of chlorine to ketene (49, 50). Dichloroacetyl and trichloroacetyl chloride are produced by oxidation of trichloroethylene or tetrachloroethylene, respectively.

Much of the chloroacetyl chloride produced is used captively as a reactive intermediate. It is useful in many acylation reactions and in the production of adrenalin [51-43-4], diazepam [439-15-5], chloroacetophenone [532-27-4], chloroacetate esters, and chloroacetic anhydride [541-88-8]. A major use is in the production of chloroacetamide herbicides (3) such as alachlor [15972-60-8].

7. Chloroacetate Esters

Two chloroacetate esters of industrial importance are methyl chloroacetate [96-34-4], $C_3H_5ClO_2$, and ethyl chloroacetate [105-39-5], $C_4H_7ClO_2$. Their properties are given in Table 4.

Both esters have a sweet pungent odor and present a vapor inhalation hazard. They are rapidly absorbed through the skin and hydrolyzed to chloroacetic acid. The oral LD_{50} for ethyl chloroacetate is between 50 mg/kg (51).

Chloroacetate esters are usually made by removing water from a mixture of chloroacetic acid and the corresponding alcohol. Reaction of alcohol with chloroacetyl chloride is an anhydrous process that liberates HCl. Chloroacetic acid will react with olefins in the presence of a catalyst to yield chloroacetate esters. Dichloroacetic and trichloroacetic acid esters are also known. These esters are useful in synthesis. They are more reactive than the parent acids. Ethyl chloroacetate can be converted to sodium fluoroacetate by reaction with potassium fluoride (see FLUORINE COMPOUNDS, ORGANIC). Both methyl and ethyl chloroacetate are used as agricultural and pharmaceutical intermediates, specialty solvents, flavors, and fragrances. Methyl chloroacetate and β-ionone undergo a Darzens reaction to form an intermediate in the synthesis of Vitamin A. Reaction of methyl chloroacetate with ammonia produces chloroacetamide [79-07-2], C_2H_4ClNO (52).

8. Bromoacetic Acid

Bromoacetic acid [79-08-3] ($BrCH_2COOH$), mol wt 138.96, $C_2H_3BrO_2$, occurs as hexagonal or rhomboidal hygroscopic crystals, mp 49°C, bp 208°C, d^{50} 1.9335,

Table 4. **Physical Properties of Chloroacetate Esters**

Property	Methyl ester	Ethyl ester
molecular weight	108.52	122.55
melting point, °C	−32.1	−26
boiling point, °C	129.8	143.3
density, at 20°C, g/mL	1.2337	1.159
flash point, °C	57	66
vapor pressure, kPa[a]		
at 25°C	0.96	0.61
at 50°C	4.3	2.6
structural formula	$ClCH_2COOCH_3$	$ClCH_2COOC_2H_5$

[a] To convert kPa to mm Hg, multiply by 7.5.

$n^{50}{}_D$ 1.4804. It is soluble in water, methanol, and ethyl ether. Bromoacetic acid undergoes many of the same reactions as chloroacetic acid under milder conditions, but is not often used because of its greater cost. Bromoacetic acid must be protected from air and moisture, since it is readily hydrolyzed to glycolic acid. The simple derivatives such as the acid chloride, amides, and esters are well known. Esters of bromoacetic acid are the reagents of choice in the Reformatsky reaction, which is used to prepare β-hydroxy acids or α,β-unsaturated acids. Similar reactions with chloroacetate esters proceed slowly or not at all (53).

Bromoacetic acid can be prepared by the bromination of acetic acid in the presence of acetic anhydride and a trace of pyridine (54), by the Hell-Volhard-Zelinsky bromination catalyzed by phosphorus, and by direct bromination of acetic acid at high temperatures or with hydrogen chloride as catalyst. Other methods of preparation include treatment of chloroacetic acid with hydrobromic acid at elevated temperatures (55), oxidation of ethylene bromide with fuming nitric acid, hydrolysis of dibromovinyl ether, and air oxidation of bromoacetylene in ethanol.

9. Dibromoacetic Acid

Dibromoacetic acid [631-64-1] ($Br_2CHCOOH$), mol wt 217.8, $C_2H_2Br_2O_2$, mp 48°C, bp 232–234°C (decomposition), is soluble in water and ethyl alcohol. It is prepared by adding bromine to boiling acetic acid, or by oxidizing tribromoethene [598-16-3] with peracetic acid.

10. Tribromoacetic Acid

Tribromoacetic acid [75-96-7] (Br_3CCOOH), mol wt 296.74, $C_2HBr_3O_2$, mp 135°C, bp 245°C (decomposition), is soluble in water, ethyl alcohol, and diethyl ether. This acid is relatively unstable to hydrolytic conditions and can be decomposed to bromoform in boiling water. Tribromoacetic acid can be prepared by the oxidation of bromal [115-17-3] or perbromoethene [79-28-7] with fuming nitric acid and by treating an aqueous solution of malonic acid with bromine.

11. Iodoacetic Acid

Iodoacetic acid [64-69-7] (ICH_2COOH), mol wt 185.95, $C_2H_3IO_2$, is commercially available. The colorless, white crystals (mp 83°C) are unstable upon heating. It has a K_a of 7.1×10^{-4}. Iodoacetic acid is soluble in hot water and alcohol, and slightly soluble in ethyl ether. Iodoacetic acid can be reduced with hydroiodic acid at 85°C to give acetic acid and iodine (56). Iodoacetic acid cannot be prepared by the direct iodination of acetic acid (57), but has been prepared by iodination of acetic anhydride in the presence of sulfuric or nitric acid (58). Iodoacetic acid can also be prepared by reaction of chloroacetic or bromoacetic acid with sodium or potassium iodide (59).

12. Diiodoacetic Acid

Diiodoacetic acid [598-89-0] ($I_2CHCOOH$), mol wt 311.85, $C_2H_2I_2O_2$, mp 110°C, occurs as white needles and is soluble in water, ethyl alcohol, and benzene. It has been prepared by heating diiodomaleic acid with water (60) and by treating malonic acid with iodic acid in a boiling water solution (61).

13. Triiodoacetic Acid

Triiodoacetic acid [594-68-3] (I_3CCOOH), mol wt 437.74, $C_2HO_2I_3$, mp 150°C (decomposition), is soluble in water, ethyl alcohol, and ethyl ether. It has been prepared by heating iodic acid and malonic acid in boiling water (62). Solutions of triiodoacetic acid are unstable as evidenced by the formation of iodine. Triiodoacetic acid decomposes when heated above room temperature to give iodine, iodoform, and carbon dioxide. The sodium and lead salts have been prepared.

BIBLIOGRAPHY

"Acetic Acid Derivatives" under "Acetic Acid," in *ECT* 1st ed., Vol. 1, pp. 74–78, by L. F. Berhenke, F. C. Amstutz, and U. A. Stenger, The Dow Chemical Company; "Ethanoic Acid (Halogenated)" in *ECT*, 2nd ed., Vol. 8, pp. 415–422, by A. P. Lurie, Eastman Kodak Company; "Acetic Acid Derivatives (Halogenated)" in *ECT* 3rd ed., Vol. 1, pp. 171–178, by E. R. Freiter, The Dow Chemical Company; in *ECT* 4th ed., Vol. 1, pp. 165–175, by Earl D. Morris and John C. Bost, The Dow Chemical Company;"Acetic Acid, Halogenated Derivatives" in *ECT* (online), posting date: December 4, 2000, by Earl D. Morris and John C. Bost, The Dow Chemical Company.

CITED PUBLICATIONS

1. E. Ott, *Cellulose and Cellulose Derivatives*, 2nd ed., Wiley-Interscience, New York, 1955.
2. *Herbicide Handbook*, 5th ed., Weed Science Society of America, Champaign, Ill., 1983, p. 128.
3. M. Sittig, ed., *Pesticide Manufacturing and Toxic Materials Control Encyclopedia*, Noyes Data Corp., Park Ridge, N.J., 1980.
4. U.S. Pat. 2,961,451 (Nov. 22, 1960), A. Keough (to Johnson & Johnson Co., Inc.).
5. Jpn. Pat. 70/23,925 (Aug. 11, 1970), H. Marushige (to Lion Fat & Oil Co.); *Chem. Abstr.* **74**, 12622v, 1971.
6. N. D. Cheronis and K. H. Spitzmueller, *J. Org. Chem.* **6**, 349 (1941).
7. K. Venkataraman, *The Chemistry of Synthetic Dyes*, Vol. 2, Academic Press, Inc., New York, 1952, p. 1013.
8. L. Baker, *Org. Synth. Coll. Vol.* **1**, 181 (1941).
9. G. Koenig, E. Lohmar, and N. Rupprich, "Chloroacetic Acids," in *Ullman*'s Encyclopedia of Industrial Chemistry, Vol. A6, VCH Publishers, New York, 1986.
10. U.S. Pat. 2,503,334 (Apr. 11, 1950), A. Hammond (to Celanese Corporation); *Chem. Abstr.* **44**, 5901g (1950).
11. U.S. Pat. 2,539,238 (Jan. 23, 1951), C. Eaker (to Monsanto Corporation); *Chem. Abstr.* **45**, 4739i (1951).

12. G. Sioli, *Hydrocarbon Process* **58** (part 2), 111(1979).
13. U.S. Pat. 1,304,108 (May 20, 1919), J. Simon; *Chem. Abstr.* **13**, 2039 (1919).
14. Eur. Pat. Appl. 4,496 (Oct. 3, 1979), Y. Correia (to Rhône-Poulenc); *Chem. Abstr.* **92**, 58239a (1980).
15. U.S. Pat. 3,365,493 (Jan. 23, 1968); D. D. De Line (to The Dow Chemical Company); *Chem. Abstr.* **66**, 46104c (1967).
16. Brit. Pat. 949,393 (Feb. 12, 1964) (to Uddeholms Aktiebolag); *Chem. Abstr.* **60**, 15739h (1964).
17. Eur. Pat. Appl. EP 32,816 (July 29, 1981), R. Sugamiya (to Tsukishima Kikai); *Chem. Abstr.* **96**, 34575p (1982).
18. Neth. Pat. 109,768 (Oct. 15, 1964), G. van Messel (to N. V. Koninklijke Nederlandse Zoutindustrie); *Chem. Abstr.* **62**, 7643e (1965).
19. Ger. Offen. 2,240,466 (Feb. 21, 1974), A. Ohorodnik (to Knapsack Company); *Chem. Abstr.* **80**, 132809 (1974).
20. Can. Pat. 757,667 (Apr. 25, 1967), A. B. Foster (to Shawinigan Chemicals); *Chem. Abstr.* **67**, 53692s (1967).
21. U.S. Pat. 4,246,074 (Jan. 20, 1981), E. Fumaux (to Lonza Company); *Chem. Abstr.* **94**, 139253u (1981).
22. Ger. Offen. 2,640,658 (Feb. 9, 1978), E. Greth (to Lonza Company); *Chem. Abstr.* **88**, 152045y (1978).
23. "Chloroacetic Acid," *ChemExpo, Chemical Profiles, http://63.23666.84.14/news/PROFILE00010101.cfm.*
24. B. De Spiegeleer, *J. Liq. Chromatogr.* **11**, 863 (1988).
25. "Chloroacetic Acid—Toxicity Profile," The British Industrial Biological Research Association (BIBRA), Surrey, England, 1988.
26. G. Woodward, *J. Ind. Hyg. Toxicol.* **23**, 78 (1941).
27. G. Weiss, ed., *Hazardous Chemicals Data Book*, Noyes Data Corp., Park Ridge, N.J., 1980, pp. 245, 424, 628.
28. L. F. Berhenke and E. C. Britton, *Ind. Eng. Chem.* **38**, 544 (1946).
29. J. van Alphen, *Rec. Trav. Chim.* **46**, 144 (1927); *Chem. Abstr.* **21**, 1641 (1927).
30. A. C. Cope, J. R. Clark, and R. Connor, *Org. Synth. Coll. Vol.* **2**, 181 (1950).
31. U.S. Pat. 1,921,717 (Aug. 8, 1933), F. C. Amstutz (to The Dow Chemical Company); *Chem. Abstr.* **27**, 5084 (1933).
32. Ger. Pat. 246,661 (Apr. 27, 1911), K. Brand; *Chem. Abstr.* **6**, 2496 (1912).
33. Fr. Pat. 773,623 (Nov. 22, 1934), Alias, Froges, and Camargue (to Compagnie de Produits Chimiques et Electrometalluriques); *Chem. Abstr.* **29**,1437 (1935).
34. U.S. Pat 3,772,157 (Nov. 13, 1973), L. H. Horsley (to The Dow Chemical Company); *Chem. Abstr.* **80**, 36713a (1974).
35. Jpn. Kokai 74/109,525 (Oct. 18, 1974), Y. Momotari (to Hardness Chemical Industries Ltd.); *Chem. Abstr.* **82**, 165885y (1975).
36. T. Jasinkski and Z. Pawlak, *Zesz. Nauk. Wyzsz. Szk. Pedagog. Gdansku. Mat. Fiz. Chem.* **8**, 131 (1968); *Chem. Abstr.* **71**, 12355s (1969).
37. U.S. Pat. 2,613,220 (Oct. 7, 1952), C. M. Eaker (to Monsanto Company); *Chem. Abstr.* **47**, 8773c (1953).
38. Jpn. Kokai 72/42,619 (Dec. 16, 1972), H. Miyamori (to Mitsubishi Edogawa Chemical Co.); *Chem. Abstr.* **78**, 83833h (1973).
39. B. G. Yasnitskii, E. B. Dolberg, and C. I. Kovelenka, *Metody Poluch. Khim. Reakt. Prep.* **21**, 106 (1970); *Chem. Abstr.* **76**, 85321x (1972).
40. *Threshold Limit Values and Biological Exposure Indicies*, 5th ed., American Conference of Government Industrial Hygienists, Cincinnati, Ohio, 1986, p. 122.
41. Ger. Offen. 2,943,433 (May 7, 1981), G. Rauchschwalbe (to Bayer A.G./FRG); *Chem. Abstr.* **95**, 61533j (1981).

42. Ger. Offen. 2,943,432 (May 21, 1981), W. Heykamp (to Bayer A.G./FRG); *Chem. Abstr.* **95**, 80511s (1981).
43. Brit. Pat. 1,361,018 (July 24, 1974), O. Hertel (to Bayer A.G./FRG); *Chem. Abstr.* **77**, 74841m (1972).
44. Eur. Pat. Appl. 22,185 (Jan. 14, 1981), A. Ohorodnik (to Hoechst Corporation); *Chem. Abstr.* **94**, 174374f (1981).
45. U.S. Pat. 3,880,923 (Apr. 29, 1975), U. Bressel (to BASF A.G./FRG); *Chem. Abstr.* **81**, 120007b (1974).
46. U.S. Pat. 4,169,847 (Oct. 2, 1979), G. Degisher (to Saeurefabrik Schweizerhall); *Chem. Abstr.* **88**, 104701n (1978).
47. U.S. Pat. 3,674,664 (July 4, 1972), E. R. Larson (to The Dow Chemical Company).
48. K. Shinoda, *Kagaku Kaishi*, 1973, p. 527; *Chem. Abstr.* **79**, 4644s, 4649x, 4650r (1973).
49. U.S. Pat. 3,883,589 (May 13, 1975), V. W. Gash (to Monsanto Company). Also see related earlier patents by the same author: 3,758,569; 3,758,571; 3,794,679; 3,812,183.
50. F. B. Erickson and E. J. Prill, *J. Org. Chem.* **23**, 141(1958).
51. M. S. Bisesi in E. Bingham, B. Cohrssen, and C. H. Powell, eds., *Patty's* Toxicology, 5th ed.,Vol. 6, John Wiley & Sons, Inc., New York, 2001, Chapt. 81, p. 972.
52. U.S. Pat. 2,321,278 (June 8, 1941), E. C. Britton (to The Dow Chemical Company); *Chem. Abstr.* **37**, 6677 (1943).
53. E. E. Royals, *Advanced Organic Chemistry*, Prentice-Hall, Englewood Cliffs, N.J., 1954, p. 706.
54. S. Natelson and C. Gottfried, *Org. Synth. Coll. Vol.* **3**, 381 (1955).
55. Ger. Offen. 2,151,565 (Apr. 19, 1973), H. Jenker and R. Karsten (to Chemische Fabrik Kalk); *Chem. Abstr.* **79**, 18121f (1973).
56. K. Ichikawa and E. Miura, *J. Chem. Soc. Jpn.* **74**, 798 (1953).
57. F. Fieser and M. Fieser, *Organic Chemistry*, 3rd. ed., D. C. Heath & Co., Lexington, Mass., 1958, p. 171.
58. USSR Pat. 213,014 (Mar. 12, 1968), A. N. Novikou (to Tornsk Polytechnic Institute); *Chem. Abstr.* **69**, 51528k (1968).
59. Czech. Pat. 152,947 (Apr. 15, 1974), E. Prochaszka; *Chem. Abstr.* **81**, 135465y (1974).
60. L. Clarke and E. K. Bolton, *J. Am. Chem. Soc.* **36**, 1899 (1914).
61. R. A. Fairclough, *J. Chem. Soc.* **1938**, 1186 (1938).
62. R. L. Cobb, *J. Org. Chem.* **23**, 1368 (1958).

Earl D. Morris
John C. Bost
The Dow Chemical Company

ACETIC ANHYDRIDE

1. Introduction

Acetic anhydride [108-24-7], $(CH_3CO)_2O$, is a mobile, colorless liquid that has an acrid odor and is a more piercing lacrimator than acetic acid [64-19-7]. It is the largest commercially produced carboxylic acid anhydride: U.S. production capacity is over 1×10^6 t yearly. Its chief industrial application is for acetylation reactions; it is also used in many other applications in organic synthesis, and it has some utility as a solvent in chemical analysis.

First prepared by C. F. Gerhardt from benzoyl chloride and carefully dried potassium acetate (1), acetic anhydride is a symmetrical intermolecular anhydride of acetic acid; the intramolecular anhydride is ketene [463-51-4]. Benzoic acetic anhydride [2819-08-1]undergoes exchange upon distillation to yield benzoic anhydride [93-97-0] and acetic anhydride.

2. Physical and Chemical Properties

No dimerization of acetic anhydride has been observed in either the liquid or solid state. Decomposition, accelerated by heat and catalysts such as mineral acids, leads slowly to acetic acid (2). Acetic anhydride is soluble in many common solvents, including cold water. As much as 10.7 wt% of anhydride will dissolve in water. The unbuffered hydrolysis rate constant k at 20°C is 0.107 min^{-1} and at 40°C is 0.248 min^{-1}. The corresponding activation energy is ~31.8 kJ/mol (7.6 kcal/mol) (3). Although aqueous solutions are initially neutral to litmus, they show acid properties once hydrolysis appreciably progresses. Acetic anhydride ionizes to acetylium, CH_3CO^+, and acetate, $CH_3CO_2^-$, ions in the presence of salts or acids (4). Acetate ions promote anhydride hydrolysis. A summary of acetic anhydride's physical properties is given in Table 1.

Acetic anhydride acetylates free hydroxyl groups without a catalyst, but esterification is smoother and more complete in the presence of acids. For example, in the presence of p-toluenesulfonic acid [104-15-4], the heat of reaction for

Table 1. **Physical Properties of Acetic Anhydride**

Property	Value	Reference
freezing point, °C	−73.13	5
boiling point, °C at 101.3 kPa[a]	139.5	5
density, d_4^{20}, g/cm^3	1.0820	5
refractive index, n_D^{20}	1.39038	6
vapor pressure (Antoine equation), P in kPa[a] and T in K	$\ln(P) = \frac{14.6497 - 3467.76}{T - 67.0}$	7
heat of vaporization, ΔH_v, at bp, J/g[b]	406.6	5
specific heat, J/kg[c] at 20°C	1817	8
surface tension, mN/m (= dyn/cm)		
25°C	32.16	9
40°C	30.20	9
viscosity, mPa·s (= cP)		
15°C	0.971	10
30°C	0.783	10
heat conductivity, mW/(m·K)[d] at 30°C	136	11
electric conductivity, S/cm	2.3×10^{-8}	12

[a] To convert kPa to mm Hg, multiply by 7.5.
[b] To convert J/g to Btu/lb, multiply by 0.4302.
[c] To convert J to cal divide by 4.184.
[d] To convert mW/(m·K) to (Btu·ft)/(h·ft^2·°F), multiply by 578.

ethanol and acetic anhydride is −60.17 kJ/mol (−14.38 kcal/mol) (13):

$$ROH + (CH_3CO)_2O \longrightarrow CH_3COOR + CH_3COOH$$

Amines undergo an analogous reaction to yield acetamides, the more basic amines having the greater activity:

$$RNH_2 + (CH_3CO)_2O \longrightarrow CH_3CONHR + CH_3COOH$$

Potassium acetate, rubidium acetate, and cesium acetate are very soluble in anhydride in contrast to the only slightly soluble sodium salt. Barium forms the only soluble alkaline earth acetate. Heavy metal acetates are poorly soluble.

Triacetylboron [4887-24-5], $C_6H_9BO_6$, is generated when boric acid is added to acetic anhydride and warmed. Although explosions have resulted from carrying out this reaction, slowly adding the boric acid to a zinc chloride solution in acetic anhydride, and maintaining a temperature <60°C, gives a good yield (14). Acetic acid is also formed

$$H_3BO_3 + 3\,(CH_3CO)_2O \longrightarrow (CH_3CO)_3BO_3 + 3\,CH_3COOH$$

Heating triacetylboron at temperatures above its melting point, 123°C, causes a rearrangement to $B_2O(OCCH_3)_4$ (15). An explosive hazard is also generated by dissolving BF_3 in anhydride (see BORON COMPOUNDS).

Hydrogen peroxide undergoes two reactions with anhydride:

$$(CH_3CO)_2O + H_2O_2 \longrightarrow CH_3COOOH + CH_3COOH + H_2O$$

$$(CH_3CO)_2O + 2\,H_2O_2 \longrightarrow 2\,CH_3COOOH + H_2O$$

Peroxyacetic acid [79-21-0] is used in many epoxidations (16) where ion-exchange resins, eg, Amberlite IR-1180M, serve as catalysts. Pinene is epoxidized to sobrerol [498-71-5],$C_{10}H_{18}O_2$, using peroxyacetic acid at −5 to −10°C (17). Care must be taken, however, to avoid formation of the highly explosive diacetyl peroxide [110-22-5]:

$$(CH_3CO)_2O + CH_3COOOH \longrightarrow CH_3CO{-}OO{-}COCH_3 + CH_3COOH$$

Acetic anhydride can be used to synthesize methyl ketones in Friedel-Crafts reactions. For example, benzene [71-43-2] can be acetylated to furnish acetophenone [98-86-2]. Ketones can be converted to their enol acetates and aldehydes to their alkylidene diacetates. Acetaldehyde reacts with acetic anhydride to yield ethylidene diacetate [542-10-9] (18):

$$CH_3CHO + (CH_3CO)_2O \longrightarrow CH_3CH(O_2CCH_3)_2$$

Isopropenyl acetate [108-22-5], which forms upon reaction of acetone [67-64-1] with anhydride, rearranges to acetylacetone [123-54-6] in the presence of BF_3 (19):

$$CH_3COOC(CH_3){=}CH_2 \xrightarrow{BF_3} CH_3C({=}O){-}CH_2{-}C({=}O){=}CH_3$$

Unsaturated aldehydes undergo a similar reaction in the presence of strongly acid ion-exchange resins to produce alkenylidene diacetates. Thus acrolein [107-02-8] or methacrolein [78-85-3] react with equimolar amounts of anhydride at $-10^\circ C$ to give high yields of the *gem*-diacetates from acetic anhydride, useful for soap fragrances.

Acids react with acetic anhydride to furnish higher anhydrides (20). An acid that has a higher boiling point than acetic acid is refluxed with acetic anhydride until an equilibrium is established. The low boiling acetic acid is distilled off and the anhydride of the higher acid is left. Adipic polyanhydride is obtained in this manner (21).

3. Manufacture

3.1. The Acetic Acid Process.

Prior to the energy crisis of the 1970s, acetic anhydride was manufactured by thermal decomposition of acetic acid at pressures of 15–20 kPa (2.2–2.9 psi) (22), beginning with the first step:

$$CH_3COOH \longrightarrow CH_2{=}C{=}O + H_2O$$

The heat of reaction is approximately 147 kJ/mol (35.1 kcal/mol) (23). Optimum yields of ketene [463-51-4] require a temperature of $\sim$730–750°C. Low pressure increases the yield, but not the efficiency of the process. Competitive reactions are

$$CH_3COOH \longrightarrow CH_4 + CO_2 \quad \text{(endothermic)}$$

$$CH_3COOH \longrightarrow 2\,CO + 2\,H_2 \quad \text{(exothermic)}$$

$$2\,CH_2{=}C{=}O \longrightarrow CH_4 + C + 2\,CO \quad \text{(exothermic)}$$

$$2\,CH_2{=}C{=}O \longrightarrow CH_2{=}CH_2 + 2\,CO \quad \text{(exothermic)}$$

The second step is the liquid-phase ketene and acetic acid reaction

$$CH_2{=}C{=}O + CH_3COOH \longrightarrow (CH_3CO)_2O \quad \text{(exothermic)}$$

Triethyl phosphate is commonly used as dehydration catalyst for the water formed in the first step. It is neutralized in the exit gases with ammonia. Aqueous 30% ammonia is employed as solvent in the second step because water facilitates the reaction, and the small amount of water introduced is not significant overall. Compression of ketene using the liquid-ring pump substantially improves the formation of anhydride. Nickel-free alloys, eg, ferrochrome alloy, chrome–aluminum steel, are needed for the acetic acid pyrolysis tubes, because nickel promotes the formation of soot and coke, and reacts with carbon monoxide yielding a highly toxic metal carbonyl. Coke formation is a serious efficiency loss. Conventional operating conditions furnish 85–88% conversion, selectivity to ketene 90–95 mol%. High petroleum energy costs make these routes only marginally economic.

Acetone cracks to ketene, and may then be converted to anhydride by reaction with acetic acid. This process consumes somewhat less energy and is a popular subject for chemical engineering problems (24,25). The cost of acetone works against widespread application of this process, however.

3.2. The Acetaldehyde Oxidation Process. Liquid-phase catalytic oxidation of acetaldehyde (qv) can be directed by appropriate catalysts, such as transition metal salts of cobalt or manganese, to produce anhydride (26). Either ethyl acetate or acetic acid may be used as reaction solvent. The reaction proceeds according to the sequence

$$CH_3CHO + O_2 \longrightarrow CH_3COOOH$$

$$CH_3COOOH + CH_3CHO \longrightarrow CH_3COOOCH(OH)CH_3$$

$$CH_3COOOCH(OH)CH_3 \longrightarrow (CH_3CO)_2O + H_2O$$

$$CH_3COOOCH(OH)CH_3 \longrightarrow CH_3COOH + CH_3COOH$$

Acetaldehyde oxidation generates peroxyacetic acid, which then reacts with more acetaldehyde to yield acetaldehyde monoperoxyacetate [7416-48-0], the Loesch ester (26). Subsequently, parallel reactions lead to formation of acetic acid and anhydride plus water.

Under sufficient pressure to permit a liquid phase at 55–56°C, the acetaldehyde monoperoxyacetate decomposes nearly quantitatively into anhydride and water in the presence of copper. Anhydride hydrolysis is unavoidable, however, because of the presence of water. When the product is removed as a vapor, an equilibrium concentration of anhydride higher than that of acetic acid remains in the reactor. Water is normally quite low. Air entrains the acetic anhydride and water as soon as they form.

High purity acetaldehyde is desirable for oxidation. The aldehyde is diluted with solvent to moderate oxidation and to permit safer operation. In the liquid take-off process, acetaldehyde is maintained at 30–40 wt% and when a vapor product is taken, no $>$ 6 wt% aldehyde is in the reactor solvent. A considerable recycle stream is returned to the oxidation reactor to increase selectivity. Recycle air, chiefly nitrogen, is added to the air introducted to the reactor at 4000–4500 times the reactor volume per hour. The customary catalyst is a mixture of three parts copper acetate to one part cobalt acetate by weight. Either salt alone is less effective than the mixture. Copper acetate may be as high as 2 wt% in the reaction solvent, but cobalt acetate ought not rise $>$0.5 wt%. The reaction is carried out at 45–60°C under 100–300 kPa (15–44 psi). The reaction solvent is far above the boiling point of acetaldehyde, but the reaction is so fast that little escapes unoxidized. This temperature helps oxygen absorption, reduces acetaldehyde losses, and inhibits anhydride hydrolysis.

Product refining is quite facile, following the same general pattern for acetic acid (qv) recovery from acetaldehyde liquid-phase oxidation. Low boilers are stripped off using ethyl acetate or acetic acid, then anhydride is distilled. Residues, largely ethylidene diacetate and certain higher condensation products with the catalyst salts, can be recycled (27).

3.3. Methyl Acetate Carbonylation. Anhydride can be made by carbonylation of methyl acetate [79-20-9] (28) in a manner analogous to methanol carbonylation to acetic acid. Methanol acetylation is an essential first step in anhydride manufacture by carbonylation. (see Fig. 1). The reactions are

$$CH_3COOH + CH_3OH \longrightarrow CH_3COOCH_3 + H_2O \qquad -4.89\ \text{kJ/mol}\ (-1.2\ \text{kcal/mol})$$

$$CH_3COOCH_3 + CO \longrightarrow (CH_3CO)_2O \qquad -94.8\ \text{kJ/mol}\ (-22.7\ \text{kcal/mol})$$

Surprisingly, there is limited nonproprietary experimental data on methanol esterification with acetic acid (29). Studies have been confined to liquid-phase systems distant from equilibrium (30), in regions where hydrolysis is unimportant. A physical study of the ternary methanol–methyl acetate–water system is useful for design work (31). Methyl acetate and methanol form an azeotrope that boils at 53.8°C and contains 18.7% alcohol. An apparent methanol–water azeotrope exists, boiling at 64.4°C and containing ~2.9% water. These azeotropes seriously complicate methyl acetate recovery. Methyl acetate is quite soluble in water, and very soluble in water–methanol mixtures, hence two liquid phases suitable for decanting are seldom found.

The reaction mechanism and rates of methyl acetate carbonylation are not fully understood. In the nickel-catalyzed reaction, rate constants for formation of methyl acetate from methanol, formation of dimethyl ether, and carbonylation of dimethyl ether have been reported, as well as their sensitivity to partial pressure of the reactants (32). For the rhodium chloride [10049-07-7] catalyzed reaction, methyl acetate carbonylation is considered to go through formation of ethylidene diacetate (33):

$$2\ CH_3COOCH_3 + 2\ CO \longrightarrow 2\ (CH_3CO)_2O$$

$$2\ (CH_3CO)_2O + H_2 \longrightarrow CH_3CH(OOCCH_3)_2 + CH_3COOH$$

The role of iodides, especially methyl iodide, is not known. The reaction occurs scarcely at all without iodides. Impurities and coproducts are poorly reported in the patent literature on the process.

The catalyst system for the modern methyl acetate carbonylation process involves rhodium chloride trihydrate [13569-65-8], methyl iodide [74-88-4], chromium metal powder, and an alumina support or a nickel carbonyl complex with triphenylphosphine, methyl iodide, and chromium hexacarbonyl (34). The use of nitrogen-heterocyclic complexes and rhodium chloride is disclosed in one European patent (35). In another, the alumina catalyst support is treated with an organosilicon compound having either a terminal organophosphine or similar ligands and rhodium or a similar noble metal (36). Such a catalyst enabled methyl acetate carbonylation at 200°C under ~20 MPa (2900 psi) carbon

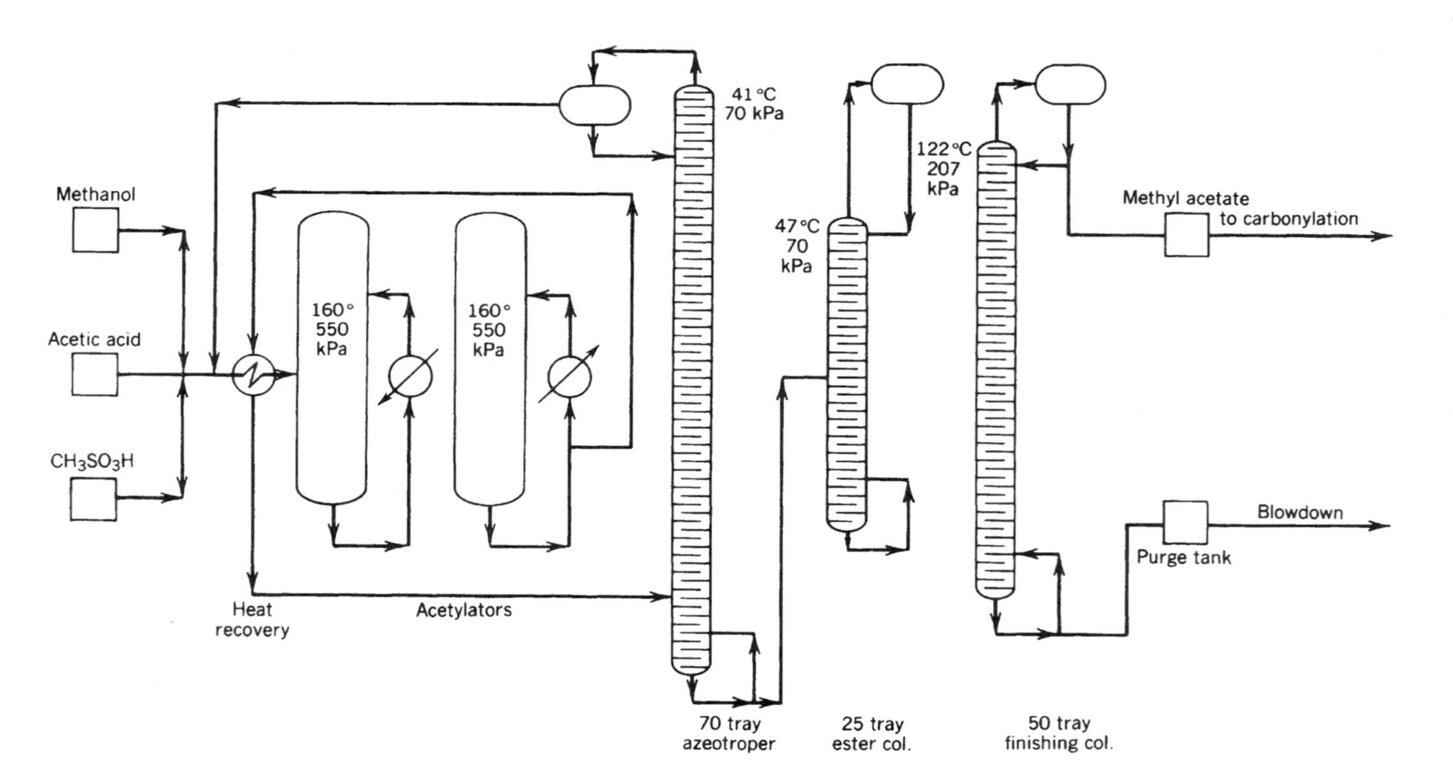

Fig. 1. Flow sheet for methyl acetate manufacture. To convert kPa to psi multiply by 0.145.

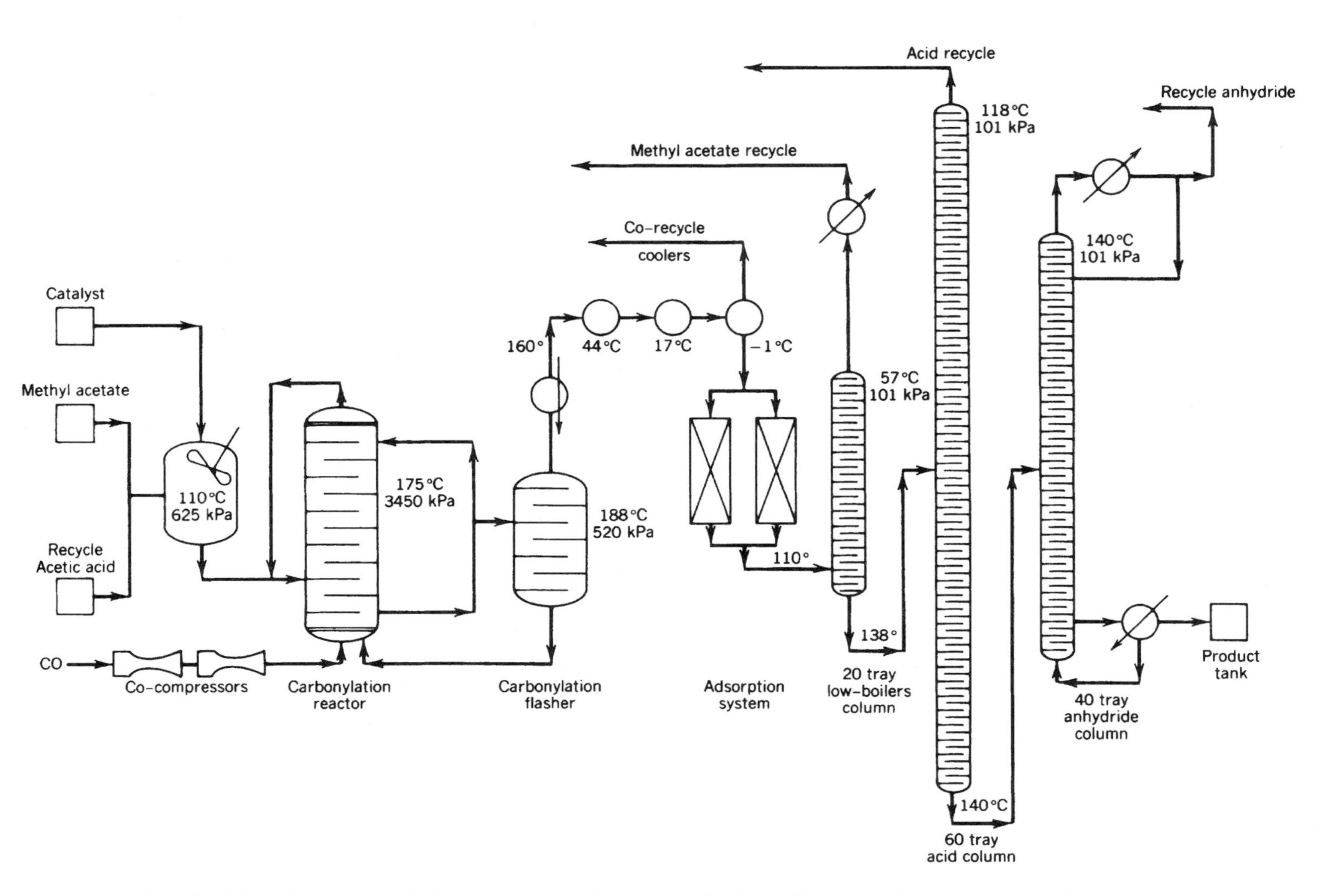

Fig. 2. Flow sheet for methyl acetate carbonylation to anhydride. To convert kPa to psi multiply by 0.145.

monoxide, with a space–time yield of 140 g anhydride per g rhodium per hour. Conversion was 42.8% with 97.5% selectivity. A homogeneous catalyst system for methyl acetate carbonylation has also been disclosed (37). A description of another synthesis is given where anhydride conversion is ~30%, with 95% selectivity. The reaction occurs at 445 K under 11 MPa partial pressure of carbon monoxide (37). A process based on a montmorillonite support with nickel chloride coordinated with imidazole has been developed (38). Other related processes for carbonylation to yield anhydride are also available (39,40).

The first anhydride plant in actual operation using methyl acetate carbonylation was at Kingsport, Tennessee (41). Infrared spectroscopy has been used to follow the apparent reaction mechanism (42).

The unit has virtually the same flow sheet (see Fig. 2) as that of methanol carbonylation to acetic acid (qv). Any water present in the methyl acetate feed is destroyed by recycle anhydride. Water impairs the catalyst. Carbonylation occurs in a sparged reactor, fitted with baffles to diminish entrainment of the catalyst-rich liquid. Carbon monoxide is introduced at ~15–18 MPa from centrifugal, multistage compressors. Gaseous dimethyl ether from the reactor is recycled with the CO and occasional injections of methyl iodide and methyl acetate may be introduced. Near the end of the life of a catalyst charge, additional rhodium chloride, with or without a ligand, can be put into the system to increase anhydride production based on net noble metal introduced. The reaction is exothermic, thus no heat need be added and surplus heat can be recovered as low pressure steam.

Catalyst recovery is a major operational problem because rhodium is a costly noble metal and every trace must be recovered for an economic process. Several methods have been patented (43–45). The catalyst is often reactivated by heating in the presence of an alcohol. In another technique, water is added to the homogeneous catalyst solution so that the rhodium compounds precipitate. Another way to separate rhodium involves a two-phase liquid such as the immiscible mixture of octane or cyclohexane and aliphatic alcohols having 4–8 carbon atoms. In a typical instance, the carbonylation reactor is operated so the desired products and other low boiling materials are flash-distilled. The reacting mixture itself may be boiled, or a sidestream can be distilled, returning the heavy ends to the reactor. In either case, the heavier materials tend to accumulate. A part of these materials is separated, then concentrated to leave only the heaviest residues, and treated with the immiscible liquid pair. The rhodium precipitates and is taken up in anhydride for recycling.

By-products remain unmentioned in most patents. Possibly there are none other than methyl acetate, acetic acid, and ethylidene diacetate, which are all precursors of anhydride.

In anhydride purification, iodide removal is of considerable significance; potassium acetate has been suggested for this procedure (46). A method for removing iodine from crude acetic anhydride has been patented (47). Because of the presence of iodide in the reaction system, titanium is the most suitable material of construction. Conventional stainless steel, SS-316, can be used for the acetylation of methanol, but it is unsuited for any part of the plant where halogen is likely to be present. Although such materials of construction add substantially to the capital cost of the plant, savings in energy consumed are expected to compensate. No authentic data on process efficiency has been published, but a rough estimate suggests a steam consumption of 1.9–2.0 kg/kg product anhydride.

3.4. Prospective Routes to Acetic Anhydride. Methyl acetate–dimethyl ether carbonylation seems to be the leading new route to acetic anhydride production (48). The high energy costs of older routes proceeding through ketene preclude their reintroduction. Thermolysis of acetone, methyl acetate, and ethylidene diacetate suffers from the same costly energy consumption. Acetic acid cracking, under vacuum and at atmospheric pressure, continues to be used for anhydride manufacture in spite of the clear obsolescence of the processes. The plants have high capital investment, and dismantling them demands a large economic advantage from any supplanting process.

Acetaldehyde oxidation to anhydride does not consume great amounts of energy. The strongly exothermic reaction actually furnishes energy and the process is widely used in Europe. Acetaldehyde must be prepared from either acetylene or ethylene. Unfortunately, use of these raw materials cancels the other advantages of this route. Further development of more efficient acetaldehyde oxidation as well as less expensive materials of construction would make that process more favorable.

Copper acetate, ferrous acetate, silver acetate [563-63-3], basic aluminum acetate,nickel acetate [373-02-4],cobalt acetate, and other acetate salts have been reported to furnish anhydride when heated. In principle, these acetates could be obtained from low concentration acetic acid. Complications of solids processing and the scarcity of knowledge about these thermolyses make industrial development of this process expensive. In the early 1930s, Soviet investigators discovered the reaction of dinitrogen tetroxide [10544-72-6] and sodium acetate [127-09-3] to form anhydride:

$$2\,NaOOCCH_3 + 2\,N_2O_4 \longrightarrow (CH_3CO)_2O + 2\,NaNO_3 + N_2O_3$$

Yields on the order of 85% were secured in the dry reaction (49). (Propionic anhydride and butyric anhydride can be obtained similarly from their sodium salts.) In as much as dinitrogen tetroxide can be regenerated, the economic prospects of this novel way of making anhydride are feasible.

Sodium acetate reacts with carbon dioxide in aqueous solution to produce acetic anhydride and sodium bicarbonate (50). Under suitable conditions, the sodium bicarbonate precipitates and can be removed by centrifugal separation. Presumably, the cold water solution can be extracted with an organic solvent, eg, chloroform or ethyl acetate, to furnish acetic anhydride. The half-life of aqueous acetic anhydride at 19°C is said to be no >1 h (2) and some other data suggests a 6 min half-life at 20°C (51). The free energy of acetic anhydride hydrolysis is given as −65.7 kJ/mol (−15.7 kcal/mol) (51) in water. In wet chloroform, an extractant for anhydride, the free energy of hydrolysis is strangely much lower, −50.0 kJ/mol (−12.0 kcal/mol) (52). Half-life of anhydride in moist chloroform may be as much as 120 min. Ethyl acetate, chloroform, isooctane, and *n*-octane may have promise for extraction of acetic anhydride. Benzene extracts acetic anhydride from acetic acid–water solutions (53).

Table 2. **North American Producers of Acetic Anhydride and Their Capacities**[a]

Producer	Capacity[a], $\times 10^3$ t ($\times 10^6$ lb)
Celanese, Narrows, Va.	118 (260)
Celanese, Pampa, Tex.	113 (250)
Celanese, Rock Hill, S.C.	113 (250)
Eastman, Kingsport, Tenn.	816 (1,800)
U.S. Total	*1,160 (2,560)*
Celanese, Cangrejera, Mexico	41 (90)
Celanese, Edmonton, Alberta	25 (55)
Other N. American Total	*66 (145)*
Total	*1,226 (2,705)*

[a] From Ref. 56, as of Feb. 19, 2001.

Ketene can be obtained by reaction of carbon oxides with ethylene (54). Because ketene combines readily with acetic acid, forming anhydride, this route may have practical applications.

4. Economic Aspects

Acetic anhydride is a mature commodity chemical in the United States. Exports are strong and have grown at the rate of 4.2% annually in recent years. Estimated demand for 2003 is 1.964×10^9 lb. Table 2 lists North American producers of acetic anhydride and their capacities. Prices for the period 1994–1999 were a high of 49.5 ¢/lb and a low of 47.5 ¢/lb. The price in February of 2001 was 49.5 ¢/lb. All prices quoted are per pound, tanks, delivered, East (55).

5. Analysis and Specifications

Analytical and control methods for acetic anhydride are fully discussed in reference 56. Performance tests are customarily used where the quality of the product is crucial, as in food or pharmaceutical products. Typical specifications are:

assay	99.0 wt% as anhydride
specific gravity$^{20}{}_{20}$	1.080–1.085
color, Pt–Co	10 max
$KMnO_4$ time	5 min needed to reduce 2 mL having no more than 0.1 mL of 0.1 N $KMnO_4$.
trace ions	$>$ 1 ppm Al, Cl, PO_4^{3-}, SO_4^{2-}, and Fe.
heavy metals	none
nitrates	none

A technical quality anhydride, assay ~97% maximum, often contains color bodies, heavy metals, phosphorus, and sulfur compounds. Anhydride manufactured by acetic acid pyrolysis sometimes contains ketene polymers, eg, acetylacetone, diketene, dehydroacetic acid, and particulate carbon, or soot, is occasionally encountered. Polymers of allene, or its equilibrium mixture, methylacetylene–allene, are reactive and refractory impurities, which if exposed to air, slowly autoxidize to dangerous peroxidic compounds.

Anhydride has been used for the illegal manufacture of heroin [561-27-3] (acetylmorphine) and certain other addictive drugs. Regulations on acetic anhydride commerce have long been a feature of European practice. After passage in 1988 of the Chemical Diversion and Trafficking Act, there is also U.S. control. Orders for as much as 1023-kg acetic anhydride, for either domestic sale or export, require a report to the Department of Justice, Drug Enforcement Administration (57).

6. Health and Safety Aspects

Acetic anhydride penetrates the skin quickly and painfully forming burns and blisters that are slow to heal. Anhydride is especially dangerous to the delicate tissues of the eyes, ears, nose, and mouth. The odor threshold is 0.49 mg/m^3, but the eyes are affected by as little as 0.36 mg/m^3 and electroencephalogram patterns are altered by only 0.18 mg/m^3. When handling acetic anhydride, rubber gloves that are free of pinholes are recommended for the hands, as well as plastic goggles for the eyes, and face-masks to cover the face and ears.

Acetic anhydride is dangerous in combination with various oxidizing substances and strong acids. Chromium trioxide [1333-82-0]and anhydride react violently to burn (58); mixtures containing nitric acid [7692-37-2] are said to be more sensitive than nitroglycerin (59). Thermal decomposition of nitric acid in acetic acid solutions is accelerated by the presence of anhydride (60). The critical detonation diameter for a nitric acid–acetic anhydride mixture has been subject to much study in the Soviet Union (61). The greatest explosion involving anhydride took place when a mixture of 568 L perchloric acid [7601-90-3] were admixed with 227 L of acetic anhydride. The mixture detonated, killing 17 people and destroying 116 buildings over several city blocks (62). The plant was an illegal metal-treating facility where the mixture was used to finish aluminum surfaces. Acetyl perchlorate is probably present in such solutions. These perchloric acid solutions are useful in metal finishing, but the risks in using them must be recognized. Such solutions are used commonly in metallography.

U.S. exposure standards for acetic anhydride are as follows: Primary bases, Current OSHA PEL, 5 ppm TWA; 1989 OSHA proposed PEL, and NIOSH REL; both 5 pp ceiling (63).

7. Uses

The biggest use of acetic anhydride is in the preparation of cellulose acetates (86%). Acetates produced include cellulose acetate, cellulose diacetate, cellulose triacetate, cellulose acetate propionate, and cellulose butyrate. The remaining 14% is consumed in miscellaneous uses including coatings, pesticides, aspirin, and acetaminophen (55).

Acetic anhydride is used in acetylation processes. There has been some diversification of anhydride usage in recent years. Acetylation of salicylic acid

[69-72-7] using anhydride to furnish acetylsalicyclic acid [50-78-2] (aspirin) is a mature process, although *N*-acetyl-*p*-aminophenol (acetaminophen) is currently making inroads on the aspirin market. Acetic anhydride is used to acetylate various fragrance alcohols to transform them into esters having much higher unit value and vitamins are metabolically enhanced by acetylation. Anhydride is also extensively employed in metallography, etching, and polishing of metals, and in semiconductor manufacture. Starch acetylation furnishes textile sizing agents.

Acetic anhydride is a useful solvent in certain nitrations, acetylation of amines and organosulfur compounds for rubber processing, and in pesticides. Though acetic acid is unexceptional as a fungicide, small percentages of anhydride in acetic acid, or in cold water solutions are powerful fungicides and bactericides. There are no reports of this application in commerce. It is possible that anhydride may replace formaldehyde for certain mycocidal applications.

BIBLIOGRAPHY

"Acetic Anhydride" in *ECT* 1st ed., Vol. 1, pp. 78–86, by Gwynn Benson, Shawinigan Chemicals Limited; in *ECT* 2nd ed., Vol. 8, pp. 405–414, by Ernest Le Monnier, Celanese Corporation of America; in *ECT* 3rd ed., Vol. 1, pp. 151–161, by Frank S. Wagner, Jr., Celanese Chemical Corporation; in *ECT* 4th ed., Vol. 1, pp. 142–154, by Frank S. Wagner, Jr., Nandina Corporation;"Acetic Anhydride" in *ECT* (online), posting date: December 4, 2000, by Frank S. Wagner, Jr., Nandina Corporation.

CITED PUBLICATIONS

1. C. F. Gerhardt, *Ann. Chim. Phys.* **37**(3), 285 (1852).
2. A. Lumière, L. Lumière, and H. Barbier, *Bull. Soc. Chim.* **33**, [iii], 783 (1905).
3. M. S. Sytilin, A. I. Morozov, and I. A. Makolkin, *Zh. Fiz. Khim.* **46**, 2266 (1972).
4. H. Schmidt, I. Wittkopf, and G. Jander, *Z. Anorg. Chem.* **256**, 113 (1948).
5. R. A. McDonald, S. A. Shrader, and D. R. Stull, *J. Chem. Eng. Data* **4**, 311 (1959).
6. Luck and co-workers, *J. Am. Chem. Soc.* **81**, 2784 (1959).
7. D. Ambrose and N. B. Ghiassee, *J. Chem. Thermodyn.* **19**, 911 (1987).
8. F. J. Wright, *J. Chem. Eng. Data* **6**, 454 (1961).
9. K. N. Kovalenko and co-workers, *Zh. Obshch. Khim.* **26**, 403 (1956).
10. T. V. Malkova, *Zh. Obshch. Khim.* **24**, 1157 (1954).
11. N. V. Tsederberg, *The Thermal Conductivity of Gases and Liquids*, Massachusetts Institute of Technology Press, Cambridge, Mass., 1965.
12. G. Jander and H. Surawski, *Z. Elektrochem.* **65**, 469 (1961).
13. Eur. Pat. 170,173 (Feb. 12, 1986), G. W. Stockton, D. H. Chidester, and S. J. Ehrlich, (to American Cyanamid Company).
14. L. M. Lerner, *Chem. Eng. News* **51**(34), 42 (1973).
15. I. G. Ryss and V. N. Plakhotnik, *Ukr. Khim. Zh. (Russ. ed.)* **36**, 423 (1970).
16. Ger. Pat. 3,447,864 (Jul. 10, 1986), K. Eckwert, and co-workers (to Henkel K.-G.a.A.).
17. Span. Pat. 548,516 (Dec. 1, 1986), J. L. Diez Amez; *Chem. Abstr.* **108**, 22107h (1988).
18. N. Rizkalla and A. Goliaszewski, in D. A. Fahey, ed., *Industrial Chemicals via C1 Processes* (ACS Symposium Series, 328), American Chemical Society, Washington, D.C., 1987, 136–153.
19. Ger. Pat. 1,904,141 (Sept. 11, 1969), W. Gay, A. F. Vellturo, and D. Sheehan.

20. Ger. Offen. 3,510,035 (Sept. 25, 1986), K. Bott and co-workers (to BASF A.G.); Jpn. Pats. JP 61-151,152, and JP 61-151,153 (July 9, 1986), and JP 61-161,241 (Jan. 8, 1985), K. Inoue, H. Takeda, and M. Kobayashi (to Mitsubishi Rayon Co. Ltd.); Ger. Pat. 3,644,222 (July 30, 1987), M. Hinenoya and M. Endo (to Daicel Chemical Industries Ltd.).
21. V. Zvonar, *Coll. Czech. Chem. Communs.* **38**(8), 2187 (1973).
22. Ger. Pat. 1,076,090 (Feb. 25, 1960), corresponding to U.S. 3,111,548 (Nov. 19, 1963), T. Alternschoepfer, H. Spes, and L. Vornehm (to Wacker-Chemie); H. Spes, *Chem. Ing. Tech.* **38**, 963 (1966).
23. W. Hunter, *BIOS Field Report 1050*, no. 22, Feb. 1, 1947; reissued as PB 68,123, U.S. Dept. of Commerce, Washington, D.C.
24. G. V. Jeffreys, *The Manufacture of Acetic Anhydride*, 2nd ed., The Institution of Chemical Engineers, London, 1964.
25. "Acetic Anhydride Design Problem" in J. J. McKetta, and W. A. Cunningham, eds., *Encyclopedia of Chemical Processing and Design*, Marcel Dekker, Inc., New York, Vol. 1, p. 271.
26. Ger. Pats. 699,709 (Nov. 7, 1940) and 708,822 (June 19, 1941), J. Loesch and co-workers (to A. G. für Stickstoffdünger); *Chem. Abstr.* **35**, 5133 (1941); **36**, 1955 (1942).
27. Fr. Pat. 1,346,360 (Dec. 20, 1963), G. Sitaud and P. Biarais (to Les Usines de Melle).
28. U.S. Pat. 2,730,546 (April 16, 1957), W. Reppe (to BASF A.G.).
29. H. Goldschmidt and co-workers, *Z. Physik. Chem.* **60**, 728 (1907); *Z. Physik. Chem.* **81**, 30 (1912); *Z. Physik. Chem.* **143**, 139 (1929); *Z. Physik. Chem.* **143**, 278 (1929); A. T. Williamson and co-workers, *Trans. Faraday Soc.* **34**, 1145 (1934); H. A. Smith, *J. Am. Chem. Soc.* **61**(1), 254 (1939).
30. A. G. Crawford and co-workers, *J. Chem. Soc.* **1949**, 1054 (1949).
31. Ger. Pat. 1,070,165 (Dec. 3, 1959), K. Kummerle (to Farbwerke Hoechst); Br. Pat. 2,033,385 (May 21, 1980), C. G. Wan (to Halcon).
32. K. Fujimoto and co-workers, *Am. Chem. Soc. Div. Pet. Chem., Prepr.* **31**(1), 85, 91 (1986).
33. E. Drent, *Am. Chem. Soc. Div. Pet. Chem. Prepr.* **31**(1), 97 (1986).
34. U.S. Pats. 4,002,678 and 4,002,677 (Jan. 11, 1977), A. N. Naglieri and co-workers (to Halcon International).
35. Eur. Pat. 8396 (March 5, 1980), H. Erpenbach and co-workers (to Hoechst).
36. Eur. Pat. 180,802 (Oct. 10, 1986), G. Luft and G. Ritter (to Hoechst A.G.).
37. G. Ritter and G. Luft, *Chem. Ing. Tech.* **59**(6), 485(1987); **58**(8), 668(1986).
38. Jpn. Pat. 62-135,445 (June 16, 1987), N. Okada and O. Takahashi (to Idemitsu Kosan K.K.).
39. U.S. Pat. 4,690,912 (Sept. 1, 1987), F. E. Paulik, A. Hershman, W. R. Knox, and J. F. Roth (to Monsanto Company).
40. Jpn. Pat. 62-226,940 (Oct. 5, 1987), H. Koyama, N. Noda, and H. Kojima (to Daicel Chemical Industry K.K.).
41. V. H. Agreda, *CHEMTECH* 250 (April 1988); T. H. Larkins Jr., *Am. Chem. Soc. Div. Pet. Chem. Prepr.* **31**(1), 74(1986); G. G. Mayfield and V. H. Agreda, *Energy Progress* **6**(4), 214 (Dec. 1986); *Chem. Eng. News*, 30 (May 21, 1990).
42. S. W. Polichnowski, *J. Chem. Educ.* **63**(3), 206 (1986).
43. U.S. Pat. 4,605,541 (Aug. 12, 1986), J. Pugach (to The Halcon SD Group, Inc.).
44. U.S. Pat. 4,650,649 (Mar. 17, 1987) and U.S. Pat. 4,578,368 (Mar. 25, 1986), J. R. Zoeller (to Eastman Kodak Company).
45. U.S. Pat. 4,440,570 (April 3, 1984), U.S. Pat. 4,442,304 (April 10, 1984), U.S. Pat. 4,557,760 (Dec. 1985), H. Erpenbach and co-workers (to Hoechst A.G.).
46. Brit. Pat. 2,033,385 (Ma. 21, 1980), P. L. Szecsi (to Halcon Research & Development).

47. U.S. Pat. 5,653,853 (Aug. 5, 1997), M. Kagotani and Y. Tsuji (to Diacel Chemical Industries, Ltd.).
48. U.S. Pat. Appl. 20,010,007,912 (July 12, 2001), R. E. Allan and J. D. Watson, East Yorkshire, U.K.; U.S. Pat. 5,554,790 (Sept. 10, 1996), Y. Harano and Y. Morimoto (to Daicel Chemical Industries, Ltd.).
49. V. M. Rodionov, A. I. Smarin, and T. A. Obletzove, *Chem. Abstr.* **30**, 1740, 4149 (1936); V. M. Rodionov and T. A. Oblitseva, *Chem. Zentr.* **1938**(II), 4054; *Chem. Abstr.* **34**, 6572 (1940).
50. Fr. Pat. 47,873 addn. to Fr. Pat. 809,731 (Aug. 14, 1937), A. Consalvo; Brit. Pat. 480,953 (Feb. 28, 1938); Brit. Pat. 486,964, addn. to 480,953 (June 14, 1938).
51. W. P. Jencks and co-workers, *J. Am. Chem. Soc.* **88**, 4464 (1966).
52. R. Wolfenden and R. Williams, *J. Am. Chem. Soc.* **107**, 4345 (1985).
53. Jpn. Pat. 60-204,738 (Oct. 16, 1985), M. Ichino and co-workers (to Daicel Chemical Industries Ltd.).
54. Fr. Pat. 973,160 (Feb. 8, 1951), G. H. van Hoek; Fr. Pat. 1,040,934 (Oct. 20, 1953), J. Francon.
55. "Acetic Anhydride," Chemical Profiles, *http://www.chemexpo.com/news/PROFILE 010219.cfm*, revised, Feb. 19, 2001.
56. E. F. Joy and A. J. Barnard Jr., in F. D. Snell and C. L. Hilton, eds., *Encyclopedia of Industrial Chemical Analysis*, Vol. 4, John Wiley & Sons Inc., New York, 1967, pp. 102–107.
57. D. Hanson, *Chem. Eng. News*, 17 (Feb. 27, 1989).
58. G. A. P. Tuey, *Chem. Ind. London*, **1948**, 766 (1948); D. A. Peak, *Chem. Ind. London* **1949**, 14 (1949).
59. J. Dubar and J. Calzia, *C.R. Acad. Sci., Ser. C.* **266**, 1114 (1968).
60. V. I. Semenikhim and co-workers, *Tr. Mosks. Khim. Tekhnol. Inst. im. D. I. Mendeleeva* **112**, 43–47 (1980); *Chem. Abstr.* **97**, 99151n (1982).
61. V. M. Raikova, *Tr. Mosks. Khim. Tekhnol. Inst.* **112**, 97–102 (1980); *Chem. Abstr.* **98**, 5958b (1983).
62. J. H. Kuney, *Chem. Eng. News* **25**, 1658 (1947).
63. K. Boyes, in E. Bingham, B. Cohrssen, and C. H. Powell, eds., *Patty*'s Toxicology, 5th ed., Vol. 2, John Wiley & Sons, Inc., New York, 2001, Chapt. 25, p. 58.

FRANK S. WAGNER, JR.
Nandina Corporation

ACETONE

1. Introduction

Acetone [67-64-1] (2-propanone, dimethyl ketone, CH_3COCH_3), molecular weight 58.08 (C_3H_6O), is the simplest and most important of the ketones. It is a colorless, mobile, flammable liquid with a mildly pungent, somewhat aromatic odor, and is miscible in all proportions with water and most organic solvents. Acetone is an excellent solvent for a wide range of gums, waxes, resins, fats, greases, oils, dyestuffs, and cellulosics. It is used as a carrier for acetylene, in the manufacture of a variety of coatings and plastics, and as a raw material for the chemical synthesis of a wide range of products such as ketene, methyl methacrylate, bisphenol A, diacetone alcohol, methyl isobutyl ketone, hexylene glycol (2-methyl-2,4-pentanediol), and isophorone. World production of acetone in 1999 was about 4.27 million metric tons per year, of which about 1.4 million was made in the United States. Most of the world's manufactured acetone is obtained as a coproduct in the process for phenol from cumene and most of the remainder from the dehydrogenation of isopropyl alcohol. Numerous natural sources of acetone make it a normal constituent of the environment. It is readily biodegradable.

2. Physical and Thermodynamic Properties

Selected physical properties are given in Table 1 and some thermodynamic properties in Table 2. Vapor pressure (P) and enthalpy of vaporization (H) over the temperature range 178.45–508.2 K can be calculated with an error of $<3\%$

Table 1. **Physical Properties**[a]

Property	Value		
melting point, °C	−94.6		
boiling point at 101.3 kPa[b], °C	56.29		
refractive index, n_D			
at 20°C	1.3588		
at 25°C	1.35596		
electrical conductivity at 298.15 K, S/cm	5.5×10^{-8}		
critical temperature, °C	235.05		
critical pressure, kPa[b]	4701		
critical volume, L/mol	0.209		
critical compressibility	0.233		
triple point temperature, °C	−94.7		
triple point pressure, Pa[b]	2.59375		
acentric factor	0.306416		
solubility parameter at 298.15 K, $(J/m^3)^{1/2}$[c]	19773.5		
dipole moment, C·m[d]	9.61×10^{-30}		
molar volume at 298.15 K, L/mol	0.0739		
molar density, mol/L			
solid at −99°C	16.677		
liquid at 298.15 K	13.506		
Selected physical properties as a function of temperature			
temperature, °C	0	20	40
surface tension, mN/m (= dyn/cm)	26.2	23.7	21.2
vapor pressure, kPa[b]	9.3	24.7	54.6
specific gravity at 20°C	0.807	0.783	0.759
viscosity, mPa·s (= cP)	0.40	0.32	0.27

[a] Extensive tables and equations are given in (1) for viscosity, surface tension, thermal conductivity, molar density, vapor pressure, and second virial coefficient as functions of temperature.
[b] To convert kPa to mmHg, multiply by 7.501.
[c] To convert $(J/m^3)^{1/2}$ to $(cal/m^3)^{1/2}$, divide by 2.045.
[d] To convert C·m to debyes, divide by 3.336×10^{-30}.

Table 2. **Thermodynamic Properties**[a–c]

Property	Value
specific heat of liquid at 20°C, J/g	2.6
specific heat of vapor at 102°C, J/(mol·K)	92.1
heat of vaporization at 56.1°C, kJ/mol	29.1
enthalpy of vaporization, kJ/mol	30.836
enthalpy of fusion at melting point, J/mol	5691.22
heat of combustion of liquid, kJ/mol	1787
enthalpy of combustion, kJ/mol	−1659.17
entropy of liquid, J/(mol·K)	200.1
entropy of ideal gas, J/(mol·K)	295.349
Gibbs energy of formation, kJ/mol	−152.716
enthalpy of formation, kJ/mol	
ideal gas	−217.15
gas	−216.5
liquid	−248

[a] Extensive tables and equations are given in (1) for enthalpy of vaporization and heat capacity at constant pressure.
[b] At 298.15 K unless otherwise noted.
[c] To convert J to cal, divide by 4.184.

from the following equations wherein the units are P, kPa; H, mJ/mol; T, K; and T_r = reduced temperature, T/T_c (1):

$$\log(P) = 70.72 - 5685/T - 7.351 \ln(T) + 0.0000063T^2$$

$$\log(H) = \log(49{,}170{,}000) + \left(1.036 - 1.294T_r + 0.672\,T_r^2\right) \log(1 - T_r)$$

Spectral characterization data are given in Table 3 (2).

3. Chemical Properties

The closed cup flash point of acetone is −18°C and open cup −9°C. The autoignition temperature is 538°C, and the flammability limits are 2.6–12.8 vol% in air at 25°C (3).

Table 3. **Spectral Parameters for Acetone**[a]

Method	Property
Absorption peaks, cm^{-1}	
infrared: SADG 77	3000, 1715, 1420, 1360, 1220, 1090, 900, 790, 530
Raman: SAD 162	3010, 2930, 2850, 2700, 1740, 1710, 1430, 1360, 1220, 1060, 900, 790, 520, 490, 390
Absorption peaks, nm	
ultraviolet: SAD 89	270 (in methanol)
Chemical shift, ppm	
^{1}H nmr: SAD 9228	2.1 ($CDCl_3$)
^{13}C nmr: JJ 28 FT	30.6, 206.0 ($CDCl_3$)
m/e (relative abundance)	
mass spec: Wiley 30	43(100), 58(42), 15(14), 42(6), 27(4), 39(3), 26(3), 29(2) molecular ion = 58.04

[a] Ref. 2.

Acetone shows the typical reactions of saturated aliphatic ketones. It forms crystalline compounds such asacetone sodium bisulfite [540-92-1], $(CH_3)_2C(OH)SO_3Na$, with alkali bisulfites. The highly reactive compound ketene [463-51-4], $CH_2{=}C{=}O$, results from the pyrolysis of acetone. Reducing agents convert acetone to pinacol [76-09-5], isopropyl alcohol [67-63-0], or propane [74-98-6]. Reductive ammonolysis produces isopropyl amines. Acetone is stable to many of the usual oxidants such as Fehling's solution, silver nitrate, cold nitric acid, and neutral potassium permanganate, but it can be oxidized with some of the stronger oxidants such as alkaline permanganate, chromic acid, and hot nitric acid. Metal hypohalite, or halogen in the presence of a base, oxidizes acetone to the metal acetate and a haloform, eg, iodoform. Halogens alone substitute for the H atoms, yielding haloacetones. Acetone is a metabolic product in humans and some other mammals and is a normal constituent of their blood and urine. In diabetics, it is present in relatively large amounts.

Compounds with active hydrogen add to the carbonyl group of acetone, often followed by the condensation of another molecule of the addend or loss of water. Hydrogen sulfide forms hexamethyl-1,3,5-trithiane probably through the transitory intermediate thioacetone that readily trimerizes. Hydrogen cyanide forms acetone cyanohydrin [75-86-5] $(CH_3)_2C(OH)CN$, which is further processed to methacrylates. Ammonia and hydrogen cyanide give $(CH_3)_2C(NH_2)CN$ [19355-69-2] from which the widely used polymerization initiator,azobisisobutyronitrile [78-67-1] is made (4).

Primary amines form Schiff bases, $(CH_3)_2C{=}NR$. Ammonia induces an aldol condensation followed by 1,4-addition of ammonia to produce diacetone amine (from mesityl oxide), 4-amino-4-methyl-2-pentanone [625-04-7], $(CH_3)_2C(NH_2)CH_2COCH_3$, and triacetone amine (from phorone), 2,2,6,6-tetramethyl-4-piperidinone [826-36-8]. Hydroxylamine forms the oxime and hydrazine compounds ($RNHNH_2$) form hydrazones ($RNHN{=}C(CH_3)_2$). Acetone and nitrous acid give the isonitroso compound that is the monoxime of pyruvaldehyde [306-44-5], $CH_3COCH{=}NOH$. Mercaptans form hemimercaptols by addition and mercaptols, $(CH_3)_2C(SR)_2$, by substitution following the addition.

With aldehydes, primary alcohols readily form acetals, $RCH(OR')_2$. Acetone also forms acetals (often called ketals), $(CH_3)_2C(OR)_2$, in an exothermic reaction, but the equilibrium concentration is small at ambient temperature. However, the methyl acetal of acetone, 2,2-dimethoxypropane [77-76-9], was once made commercially by reaction with methanol at low temperature for use as a gasoline additive (5). Isopropenyl methyl ether [116-11-0], useful as a hydroxyl blocking agent in urethane and epoxy polymer chemistry (6), is obtained in good yield by thermal pyrolysis of 2,2-dimethoxypropane. With other primary, secondary, and tertiary alcohols, the equilibrium is progressively less favorable to the formation of ketals, in that order. However, acetals of acetone with other primary and secondary alcohols, and of other ketones, can be made from 2,2-dimethoxypropane by transacetalation procedures (7,8). Because they hydrolyze extensively, ketals of primary and especially secondary alcohols are effective water scavengers.

Acetone has long been used as an agent to block the reactivity of hydroxyl groups in 1,2- and 1,3-diols, especially in carbohydrate chemistry. The equilibrium for the formation of acetals with hydroxyls in these compounds is more favorable because the products are five- and six-membered ring compounds,

1,3-dioxolanes and dioxanes, respectively. With glycerol the equilibrium constant for formation of the dioxolane is ~0.50 at 23°C and 0.29 at 48°C in a mixture resulting from acidification of equal volumes of acetone and glycerol at ambient temperature. The equilibrium can be displaced toward acetal formation by the use of a water scavenger, eg, an anhydrous metal salt such as copper sulfate.

Acetone undergoes aldol additions,

$$R^1R^2CH-\overset{\overset{O}{\|}}{C}-R^3 + CH_3-\overset{\overset{O}{\|}}{C}-CH_3 \longrightarrow R^1R^2C(OH)-C(CH_3)_2-\overset{\overset{O}{\|}}{C}-R^3, \quad R = H \text{ or alkyl}$$

and further reacts with the products, forming aldol chemicals, diacetone alcohol (4-hydroxy-4-methyl-2-pentanone [123-42-2]), mesityl oxide (4-methyl-3-penten-2-one [141-79-7]), isophorone (3,5,5-trimethyl-2-cyclohexenone [78-59-1]), phorone (2,6-dimethyl-2,5-heptadien-4-one [504-20-1]), and mesitylene (1,3,5-trimethylbenzene [108-67-8]). From these are produced the industrial solvents methyl isobutyl ketone (MIBK, 4-methyl-2-pentanone [108-10-1]), methylisobutylcarbinol (MIBC, 4-methyl-2-pentanol [108-11-2]), hexylene glycol (2-methyl-2,4-pentanediol [107-41-5]), and others. Acetone enters the typical nucleophilic addition and condensation reactions of ketones both at its carbonyl group and at its methyl groups, with aldehydes, other ketones, and esters. The Claisen reaction with ethyl acetate gives acetylacetone (2,4-pentanedione [123-54-6]); Mannich reaction with secondary amines gives R_2NCH_2 substituted acetones; and the Reformatzky reaction gives β-hydroxy esters. Glycidic esters (esters with a 2-epoxy group) can be made by condensation of acetone and chloroacetic esters with a metal alkoxide.

The para and ortho positions of phenols condense at the carbonyl group of acetone to make bisphenols, eg, bisphenol A, 4,4′-(1-methylethylidene)bisphenol [80-05-07]). If the H atom is activated, ClCH– compounds add to the carbonyl group in the presence of strong base; chloroform gives chloretone (1,1,1-trichloro-2-methyl-2-propanol [57-15-8]).

4. Manufacture

Acetone was originally observed ~1595 as a product of the distillation of sugar of lead (lead acetate). In the nineteenth century, it was obtained by the destructive distillation of metal acetates, wood, and carbohydrates with lime, and pyrolysis of citric acid. Its composition was determined by Liebig and Dumas in 1832.

Until World War I, acetone was manufactured commercially by the dry distillation of calcium acetate from lime and pyroligneous acid (wood distillate) (9). During the war processes for acetic acid from acetylene and by fermentation supplanted the pyroligneous acid (10). In turn, these methods were displaced by the process developed for the bacterial fermentation of carbohydrates (cornstarch and molasses) to acetone and alcohols (11). At one time, Publicker Industries, Commercial Solvents and National Distillers had combined biofermentation capacity of 22,700 metric tons of acetone per year. Biofermentation became noncompetitive ~1960 because of the economics of scale of the isopropyl alcohol dehydrogenation and cumene hydroperoxide processes.

Production of acetone by dehydrogenation of isopropyl alcohol began in the early 1920s and remained the dominant production method through the 1960s. In the mid-1960s, virtually all United States acetone was produced from propylene. A process for direct oxidation of propylene to acetone was developed by Wacker Chemie (12), but is not believed to have been used in the United States. However, by the mid-1970s 60% of United States acetone capacity was based on cumene hydroperoxide [80-15-9], which accounted for ~65% of the acetone produced.

Acetone was a coproduct of the Shell process for glycerol [56-8-5]. Propylene was hydrated to isopropyl alcohol. Some of the alcohol was catalytically oxidized to acrolein and some was oxidized to give hydrogen peroxide and acetone. Some more of the isopropyl alcohol and the acrolein reacted to give allyl alcohol and acetone. The allyl alcohol was then treated with the peroxide to give glycerol. About 1.26 kg of acetone resulted per kilogram of glycerol.

Direct oxidation of hydrocarbons and catalytic oxidation of isopropyl alcohol have also been used for commercial production of acetone.

Most of the world's acetone is now obtained as a coproduct of phenol by the cumene process. More then 90% of U.S. acetone is produced by this process. Cumene is oxidized to the hydroperoxide and cleaved to acetone and phenol. The yield of acetone is believed to average about 94%, and ~0.60–0.62 unit weight of acetone is obtained per unit of phenol (13).

Dehydrogenation of isopropyl alcohol accounts for most of the acetone production not obtained from cumene. The vapor is passed over a brass, copper, or other catalyst at 400–500°C, and a yield of ~95% is achieved (1.09 unit weight of alcohol per unit of acetone) (13).

Minor amounts of acetone are made by other processes. Until mid-1980 Shell Chemical Company obtained acetone and hydrogen peroxide as coproducts of noncatalytic oxidation of isopropyl alcohol with oxygen in the liquid phase. Yield to acetone was ~90%. In a process analogous to the cumene process, Eastman Chemical Products, Inc., and The Goodyear Tire & Rubber Company produce hydroquinone and acetone from diisopropylbenzene [25321-09-9] in the United States. Similarly, Mitsui Petrochemical Industries, Ltd., in Japan produces coproduct acetone with cresol from cymene [25755-15-1]. BP Chemicals, Ltd., in the United Kingdom recovers by-product acetone from the manufacture of acetic acid by the oxidation of light petroleum distillate.

Producers of acetone in the United States and their capacities and feedstocks are given in Table 4 (14).

4.1. Cumene Hydroperoxide Process for Phenol and Acetone. Benzene is alkylated to cumene, and then is in oxidized to cumene hydroperoxide, which in turn is cleaved to phenol and acetone.

$$C_6H_5CH(CH_3)_2 \xrightarrow[\text{oxygen}]{\text{air or}} C_6H_5\overset{\displaystyle OOH}{\overset{|}{C}}(CH_3)_2 \xrightarrow[\text{heat}]{\text{acid}} C_6H_5OH + (CH_3)_2C{=}O$$

One kilogram of phenol production results in ~0.6 kg of acetone or about ~0.40–0.45 kg of acetone per kilogram of cumene used.

Table 4. **U.S. Producers of Acetone and Their Capacities**

Company	Location	Annual[a] capacity, 10^3 t
Cumene feedstock		
Aristech Chemical Corporation	Haverhill, Ohio	203
JLM Chemicals	Blue Island, Ill.	30
Dow Chemical U.S.A.	Oyster Creek, Tex.	181
General Electric Company	Mount Vernon, Ind.	197
Georgia Gulf Corporation	Plaquemine, La.	45
Shell Oil Company	Deer Park, Tex.	185
Texaco Corporation	El Dorado, Kans.	38.6
Isopropyl alcohol feedstock		
Shell Oil Company	Deer Park, Tex.	190
Union Carbide Corporation	Institute, W.Va.	77
Diisopropylbenzene feedstock		
Eastman Kodak Company	Kingsport, Tenn.	11.3[b]
The Goodyear Tire and Rubber Company, Chemical Division	Bayport, Tex.	6.8[b]

[a] As of April 5, 1999 (14).
[b] Small amounts, in manufacture of hydroquinone.

There are many variations of the basic process and the patent literature is extensive. Several key patents describe the technology (15). The process steps are oxidation of cumene to a concentrated hydroperoxide, cleavage of the hydroperoxide, neutralization of the cleaved products, and distillation to recover acetone.

In the first step, cumene is oxidized to cumene hydroperoxide with atmospheric air or air enriched with oxygen in one or a series of oxidizers. The temperature is generally between 80 and 130°C and pressure and promoters, such as sodium hydroxide, may be used (16). A typical process involves the use of three or four oxidation reactors in series. Feed to the first reactor is fresh cumene and cumene recycled from the concentrator and other reactors. Each reactor is partitioned. At the bottom there may be a layer of fresh 2–3% sodium hydroxide if a promoter (stabilizer) is used. Cumene enters the side of the reactor, overflows the partition to the other side, and then goes on to the next reactor. The air (oxygen) is bubbled in at the bottom and leaves at the top of each reactor.

The temperatures decline from a high of 115°C in the first reactor to 90°C in the last. The oxygen ratio as a function of consumable oxygen is also higher in the later reactors. In this way, the rate of reaction is maintained as high as possible, while minimizing the temperature-promoted decomposition of the hydroperoxide.

This procedure may result in a concentration of cumene hydroperoxide of 9–12% in the first reactor, 15–20% in the second, 24–29% in the third, and 32–39% in the fourth. Yields of cumene hydroperoxide may be in the range of 90–95% (17). The total residence time in each reactor is likely to be in the range of 3–6 h. The product is then concentrated by evaporation to 75–85% cumene hydroperoxide. The hydroperoxide is cleaved under acid conditions with agitation in a vessel at 60–100°C. A large number of nonoxidizing inorganic acids are useful for this reaction, eg, sulfur dioxide (18).

After cleavage, the reaction mass is a mixture of phenol, acetone and a variety of other products such as cumylphenols, acetophenone, dimethyl-phenylcarbinol, α-methylstyrene, and hydroxyacetone. It may be neutralized with a sodium phenoxide solution (19) or other suitable base or ion-exchange resins. Process water may be added to facilitate removal of any inorganic salts. The product may then go through a separation and a wash stage, or go directly to a distillation tower.

A crude acetone product is recovered by distillation from the reaction mass. One or two additional distillation columns may be required to obtain the desired purity. If two columns are used, the first tower removes impurities such as acetaldehyde and propionaldehyde. The second tower removes undesired heavies, the major component being water.

The yield of acetone from the cumene–phenol process is believed to average 94%. By-products include significant amounts of α-methylstyrene [98-83-9] and acetophenone [98-86-2] as well as small amounts of hydroxyacetone [116-09-6], and mesityl oxide [141-79-7]. By-product yields vary with the producer. The α-methylstyrene may be hydrogenated to cumene for recycle or recovered for monomer use. Yields of phenol and acetone decline by 3.5–5.5% when the α-methylstyrene is not recycled (20).

4.2. Purification. A process for the purification of acetone containing cumene as an impurity using solvent extraction with triethylene glycol has been patented (21).

4.3. Dehydrogenation of Isopropyl Alcohol. Isopropyl alcohol is dehydrogenated in an endothermic reaction.

$$CH_3CHOHCH_3 + 66.5\ \text{kJ/mol (at 327°C)} \longrightarrow CH_3COCH_3 + H_2$$

The equilibrium is more favorable to acetone at higher temperatures. At 325°C, 97% conversion is theoretically possible. The kinetics of the reaction has been studied (22). A large number of catalysts have been investigated, including copper, silver, platinum, and palladium metals, as well as sulfides of transition metals of groups 4(IVB), 5(VB), and (VIB)6 of the periodic table. These catalysts are made with inert supports and are used at 400–600°C (23). Lower temperature reactions (315–482°C) have been successfully conducted using zinc oxide-zirconium oxide combinations (24), and combinations of copper–chromium oxide and of copper and silicon dioxide (25).

It is usual practice to raise the temperature of the reactor as time progresses to compensate for the loss of catalyst activity. When brass spelter is used as a catalyst, the catalyst must be removed at intervals of 500–1000 h and treated with a mineral acid to regenerate a catalytically active surface (26). When 6–12% zirconium oxide is added to a zinc oxide catalyst and the reaction temperatures are not excessive, the catalyst life is said to be a minimum of 3 months (24). The dehydrogenation is carried out in a tubular reactor. Conversions are in the range of 75–95 mol %. A process described by Shell International Research (23) is a useful two-stage reaction to attain high conversion, with lower energy cost and lower capital cost. The first stage uses a tubular reactor at 420–550°C to convert up to 70% of the alcohol to acetone. The second stage employs an unheated fixed-bed reactor with the same catalyst used in the tube reaction to complete the conversion at ~85%.

Although the selectivity of isopropyl alcohol to acetone via vapor-phase dehydrogenation is high, there are a number of by-products that must be removed from the acetone. The hot reactor effluent contains acetone, unconverted isopropyl alcohol, and hydrogen, and may also contain propylene, polypropylene, mesityl oxide, diisopropyl ether, acetaldehyde, propionaldehyde, and many other hydrocarbons and carbon oxides (24,27).

The mixture is cooled and noncondensable gases are scrubbed with water. Some of the resultant gas stream, mainly hydrogen, may be recycled to control catalyst fouling. The liquids are fractionally distilled, taking acetone overhead and a mixture of isopropyl alcohol and water as bottoms. A caustic treatment may be used to remove minor aldehyde contaminants prior to this distillation (28). In another fractionating column, the aqueous isopropyl alcohol is concentrated to ~88% for recycle to the reactor.

A yield of ~95% of theoretical is achieved using this process (1.09 units of isopropyl alcohol per unit of acetone produced). Depending on the process technology and catalyst system, such coproducts as methyl isobutyl ketone and diisobutyl ketone can be produced with acetone (29).

5. Production and Shipment

Acetone is produced in large quantities and is usually shipped by producers to consumers and distributors in drums and larger containers. Distributors repackage the acetone into containers ranging in size from small bottles to drums or even tank trucks. Specialty processors make available various grades and forms of acetone such as high purity, specially analyzed, analytical reagent grade, chromatography and spectrophotometric grades, and isotopically labeled forms, and ship them in ampoules, vials, bottles, or other containers convenient for the buyers.

The Department of Transportation (DOT) hazard classification for acetone is Flammable Liquid, identification number UN1090. DOT regulations concerning the containers, packaging, marking, and transportation for overland shipment of acetone are published in the *Federal Register* (30). Regulations and information for transportation by water in the United States are published in the *Federal Register* (31) and by the U.S. Coast Guard (32). Rules and regulations for ocean shipping have been published by the International Maritime Organization (IMO), a United Nations convention of nations with shipping interests, in the IMOBCH Code (33). The IMO identification number is 3.1. Because additions and changes to the regulations appear occasionally, the latest issue of the regulations should be consulted.

Small containers up to 4–5 L (~1 gal) are usually glass. Acetone is also shipped by suppliers of small quantities in steel pails of 18 L. Depending on the size of the container, small amounts are shipped by parcel delivery services or truck freight. Quantities that can be accepted by some carriers are limited by law and special "over-pack" outer packaging may be required. Usual materials for larger containers are carbon steel for 55-gal (0.21 m^3) drums, stainless steel or aluminum for tank trucks, and carbon steel, lined steel, or aluminum for rail tank cars. The types of tank cars and trucks that can be used are specified by law, and shippers may have particular preferences. Barges and ships are usually steel, but may have special inner or deck-mounted tanks. Increasing in use, especially for international shipments, are intermodal (IM) portable containers, tanks suspended in frameworks suitable for interchanging among truck, rail, and ship modes of transportation.

Containers less than bulk must bear the red diamond-shaped FLAMMABLE LIQUID label. Bulk containers must display the red FLAMMABLE placard in association with the UN1090 identification. Fire is the main hazard in emergencies resulting from spills. Some manufacturers provide transportation emergency response information. A listing of properties and hazard response information for acetone is published by the U.S. Coast Guard in its CHRIS manual (34). Two books on transportation emergencies are available (35). Immediate information can be obtained from CHEMTREC (36). Interested parties may contact their suppliers for more detailed information on transportation and transportation emergencies.

Tank cars contain up to 10, 20, or 30 thousand gal (10,000 gal = 38 m^3) of material, tank trucks 6000 gal (22.7 m^3), and barges 438,000 gal (~1270 tons). International shipments by sea are typically ~2000 tons.

6. Economic Aspects

The economics of acetone production and its consequent market position are unusual. Traditional laws of supply and demand cannot be applied because supply depends on the production of phenol and demand is controlled by the uses of acetone. Therefore, coproduct acetone from the cumene to phenol process will continue to dominate market supply. Deliberate production of acetone from isopropyl alcohol accommodates demand in excess of that supplied by the phenol process. More than 75% of world and 90% of U.S. production comes from the cumene to phenol process.

Current U.S. and World production data are shown in Tables 5 and 6 (37). Consumption of acetone is expected to grow at a rate of ~3% through 2003 annually, but phenol demand and consequent coproduct acetone production are expected to grow at a rate of 4%, thus resulting in excess supplies. The fastest growing outlet for acetone is for bisphenol A, mainly for growth in polycarbonate. Although bisphenol A production consumes 1 mol of acetone, it yields a net amount of 1 mol of acetone production because 2 mol of acetone accompany the production of the required phenol. Production of "on-purpose" acetone will probably decline as supplies of by-product acetone increase.

In 1998, demand for acetone was 1.26×10^9 kg (2.78×10^9 lb). Estimated demand for 2003 is 1.46×10^9 kg (3.22×10^9 lb) (14).

Prices as of 1999, were for the United States, 9.5–10.5 cents/lb (4.3–4.7 cents/kg) fob Gulf Coast; 11 cents/lb (4.9 cents/kg) fob first quarter contract. In Europe, \$209–231/metric ton,: \$264.50/metric ton; contract. In Asia/Pacific, \$270–310/metric ton spot (37).

Table 5. **United States Production of Acetone, 10^3 t**[a]

Company	Production, 10^3 t
Sun	
Frankford, Pa	310
Mount Vernon Phenol[b]	
Mount Vernon, Ind.	197
Aristech	
Haverhill, Ohio	197
Shell Chemical	
Deer Park, Texas	185
Dow Chemical	
Freeport, Texas	176
Georgia Gulf	
Plaquemine, La.	140
Pasadena, Texas	44
Union Carbide	
Institute, W. Va.	77
Texaco[c]	
El Dorado, Kan.	30
JLM Industries	
Blue Island, Ill.	26

[a] As of Feb. 1999, from Ref. 37.
[b] Joint Venture of G.E. Plastics. JLM Marketing and Citgo Group.
[c] Refinery: Output marketed by Plaza Group, Houston.

Table 6. **World Production of Acetone, 10^3 t**[a]

Country	Production, 10^3t
North America	
Canada	16
Mexico	
Fenoquimia	
Cosoleacaque, Veracruz	83
South America	
Argentina	18
Brazil	
Rhodia	
Paulinia	78
Western Europe	
Belgium	
Phenolchemie	
Antwerp	252
Finland	
Borealis Polymers	
Porvoo	78
France	
Rhône-Poulenc	
Roussillon	96
Shell Chimie	
Berre	75
Germany	
Phenolchemie	
Gladbeck	318

Table 6 (*Continued*)

Country	Production, 10^3t
DOMO Group	
Leuna	57
Italy	
EniChem	
Mantova	174
Porto Torres	108
Netherlands	
Shell Nederland	
Rotterdam	80
DSM Chemicals	
Botlek	66
Spain	
Ertisa	
Huelva	96
U.K	
BP Amoco	
Hull	62
Shell Chemicals	
Ellesmere Port	40
Eastern Europe	
Russia	
Salavat	
Salvat	96
Orgesteklo	
Dzerzhinsk	82
Others	322
Middle East	18
South Africa	17
Asia / Pacific	
China	126
India	36
Japan	
Mitsui Toatsu	120
Mitsui Sekka	120
Mitsubishi Kaggaku	108
Korea	
Kumho	132
Taiwan	
Formosa Chemical	221
Taiwan Prosperity Chemical	60

[a] Ref. 37

7. Specifications, Standards, and Quality Control

The ASTM "Standard Specification for ACETONE," D329, requires 99.5% grade acetone to conform to the following: apparent specific gravity 20/20°C, 0.7905–0.7930; 25/25°C, 0.7860–0.7885 (ASTM D891); color, not more than No. 5 on the platinum–cobalt scale (ASTM D1209); distillation range, 1.0°C, which shall include 56.1°C (ASTM D1078); nonvolatile matter, not >5 mg/100 mL (D1353); odor, characteristic, nonresidual (ASTM D1296); water, not >0.5 wt %

(ASTM D1364); acidity (as free acetic acid), <0.002 wt %, equivalent to 0.019 mg of KOH per gram of sample (ASTM D1613); water miscibility, no turbidity or cloudiness at 1:10 dilution with water (ASTM D1722); alkalinity (as ammonia), not >0.001 wt % (ASTM D1614); and permanganate time, color of added $KMnO_4$ must be retained at least 30 min at 25°C in the dark (ASTM D1363).

Higher or lower quality at more or less cost will meet the needs of some consumers. Acetone is often produced under contract to meet customer specifications that are different from those of ASTM D329. Some specialty grades are analyzed reagent, isotopically labeled, clean room, liquid chromatography, spectroscopic, ACS reagent (38), semiconductor (low metals), and Federal Specification O-A-51G.

Specification tests are performed on plant streams once or twice per worker shift, or even more often if necessary, to assure the continuing quality of the product. The tests are also performed on a sample from an outgoing shipment, and a sample of the shipment is usually retained for checking on possible subsequent contamination. Tests on specialty types of acetone may require sophisticated instruments, eg, mass spectrometry for isotopically labeled acetone.

8. Analytical and Test Methods, Storage

In current industrial practice, gas chromatographic analysis (gc) is used for quality control. The impurities, mainly a small amount of water (by Karl–Fischer) and some organic trace constituents (by gc), are determined quantitatively, and the balance to 100% is taken as the acetone content. Compliance to specified ranges of individual impurities can also be assured by this analysis. The gas chromatographic method is accurately correlated to any other tests specified for the assay of acetone in the product. Contract specification tests are performed on the product to be shipped. Typical wet methods for the determination of acetone are acidimetry (39), titration of the liberated hydrochloric acid after treating the acetone with hydroxylamine hydrochloride; and iodimetry (40), titrating the excess of iodine after treating the acetone with iodine and base (iodoform reaction).

Carbon steel tanks of welded construction, as specified in the American Petroleum Institute Standard 650 (41), are recommended for acetone storage. Gaskets should be ethylene–propylene rubber or Viton rubber. An inert gas pad should be used. Provisions should be made to prevent static charge buildup during filling. Design considerations of the National Fire Prevention Association Code 30 and local fire codes should be followed. Tank venting systems should comply with local vapor emission standards and conform with National Fire Prevention Association recommendations. Where the purity of the acetone is to be optimized, an inorganic zinc lining is recommended (42). One such lining is Carbozinc 11, metallic zinc in an ethyl silicate binder, available from Carboline Co., St. Louis, Missouri.

9. Health and Safety Factors

Acetone is among the solvents of comparatively low acute and chronic toxicity. High vapor concentrations produce anesthesia, and such levels may be irritating to the eyes, nose, and throat, and the odor may be disagreeable. Acetone does not have sufficient warning properties to prevent repeated exposures to concentrations that may cause adverse effects. In industry, no injurious effects have been reported other than skin irritation resulting from its defatting action, or headache from prolonged inhalation (43). Direct contact with the eyes may produce irritation and transient corneal injury.

Material Safety Data Sheets (MSDS) issued by suppliers of acetone are required to be revised within 90 days to include new permissible exposure limits (PEL). Current OSHA PEL is 1000 ppm TWA (44) and ACGIH threshold limit values (TLV) (44) are 500 ppm TWA and 750 ppm STEL. A report on human experience (44) concluded that exposure to 1000 ppm for an 8-h day produced no effects other than slight, transient irritation of the eyes, nose, and throat.

There are many natural sources of acetone including forest fires, volcanoes, and the normal metabolism of vegetation, insects, and higher animals (45). Acetone is a normal constituent of human blood, and it occurs in much higher concentrations in diabetics. Its toxicity appears to be low to most organisms. Acetone is ubiquitous in the environment, but is not environmentally persistent because it is readily biodegraded. In general, acetone is an environmentally benign compound, widely detected but in concentrations that are orders of magnitude below toxicity thresholds.

Acetone can be handled safely if common sense precautions are taken. It should be used in a well-ventilated area, and because of its low flash point, ignition sources should be absent. Flame will travel from an ignition source along vapor flows on floors or bench tops to the point of use. Sinks should be rinsed with water while acetone is being used to clean glassware, to prevent the accumulation of vapors. If prolonged or repeated skin contact with acetone could occur, impermeable protective equipment such as gloves and aprons should be worn.

Compatibility of acetone with other materials should be carefully considered, especially in disposal of wastes. It reacts with chlorinating substances to form toxic chloroketones, and potentially explosively with some peroxy compounds and a number of oxidizing mixtures. Mixed with chloroform, acetone will react violently in the presence of bases. Other incompatibilities are listed in the Sax handbook (43).

Vapor flammability range in air (2.6–12.8 vol %) and low flash point (−18°C, 0°F) make fire the major hazard of acetone. Quantities larger than laboratory hand bottles should be stored in closed metal containers. Gallon glass bottles should be protected against impacts. Areas where acetone is in contact with the ambient air should be free of ignition sources (flames, sparks, static charges, and hot surfaces above the autoignition temperature, ~500–600°C depending on the reference consulted). Fires may be controlled with carbon dioxide or dry chemical extinguishers. Recommended methods of handling, loading, unloading, and storage can be obtained from Material Safety Data Sheets and inquiries directed to suppliers of acetone.

10. Uses

Acetone is used as a solvent and as a reaction intermediate for the production of other compounds that are mainly used as solvents and/or intermediates for consumer products. Forty five percent is used as acetone cyanohydrin for methyl methacrylate, 20% for bisphenol A, solvent uses, 17% MIBK and MIBC 8%, and miscellaneous 10% (14).

10.1. Direct Solvent Use. A large volume, direct solvent use of acetone is in formulations for surface coatings and related washes and thinners, mainly for acrylic and nitrocellulose lacquers and paints. It is used as a solvent in the manufacture of pharmaceuticals and cosmetics (~7000 metric tons in nail polish removers), in spinning cellulose acetate fibers, in gas cylinders to store acetylene safely, in adhesives and contact cements, in various extraction processes, and in the manufacture of smokeless powder. It is a wash solvent in fiberglass boat manufacturing, a cleaning solvent in the electronics industry, and a solvent for degreasing wool and degumming silk.

10.2. Acrylics. Acetone is converted via the intermediate acetone cyanohydrin to the monomer methyl methacrylate (MMA) [80-62-6]. The MMA is polymerized to poly(methyl methacrylate) (PMMA) to make the familiar clear acrylic sheet. PMMA is also used in molding and extrusion powders. Hydrolysis of acetone cyanohydrin gives methacrylic acid (MAA), a monomer that goes directly into acrylic latexes, carboxylated styrene–butadiene polymers, or ethylene–MAA ionomers. As part of the methacrylic structure, acetone is found in the following major end-use products: acrylic sheet molding resins, impact modifiers and processing aids, acrylic film, ABS and polyester resin modifiers, surface coatings, acrylic lacquers, emulsion polymers, petroleum chemicals, and various copolymers (see METHACRYLIC ACID AND DERIVATIVES; METHACRYLIC POLYMERS).

10.3. Bisphenol A. One mole of acetone condenses with 2 mol of phenol to form bisphenol A [80-05-07], which is used mainly in the production of polycarbonate and epoxy resins. Polycarbonates (qv) are high strength plastics used widely in automotive applications and appliances, multilayer containers, and housing applications. Epoxy resins (qv) are used in fiber-reinforced laminates, for encapsulating electronic components, and in advanced composites for aircraft–aerospace and automotive applications. Bisphenol A is also used for the production of corrosion- and chemical-resistant polyester resins, polysulfone resins, polyetherimide resins, and polyarylate resins.

10.4. Aldol Chemicals. The aldol condensation of acetone molecules leads to the group of aldol chemicals that are themselves used mainly as solvents. The initial condensation product is diacetone alcohol (DAA), which is dehydrated to mesityl oxide. Because of its toxicity effects, mesityl oxide is no longer produced for sale, but is used captively to make methyl isobutyl ketone (MIBK) and methylisobutylcarbinol (MIBC) by hydrogenation. DAA is hydrogenated to hexylene glycol. Three molecules of acetone give isophorone and phorone, which is hydrogenated todiisobutyl ketone (DIBK) [108-83-8] and diisobutylcarbinol (DIBC) [108-82-7].

MIBK is a coatings solvent for nitrocellulose lacquers and vinyl and acrylic polymer coatings, an intermediate for rubber antioxidants and specialty surfactants, and a solvent for the extraction of antibiotics. MIBC is used mainly for the production of zinc dialkyl dithiophosphates, which are used as lubricating oil additives. It is a flotation agent for minerals and a solvent for coatings. Besides its use as a chemical intermediate, DAA is used as a solvent for nitrocellulose, cellulose acetate, oils, resins, and waxes, and in metal cleaning compounds.

Hexylene glycol is a component in brake fluids and printing inks. Isophorone is a solvent for industrial coatings and enamels. DIBK is used in coatings and leather finishes.

10.5. Other Uses. More than 70 thousand metric tons of acetone is used in small volume applications some of which are to make functional compounds such as antioxidants, herbicides, higher ketones, condensates with formaldehyde or diphenylamine, and vitamin intermediates.

BIBLIOGRAPHY

"Acetone" in *ECT* 1st ed., Vol. 1, pp. 88–95, by C. L. Gabriel and A. A. Dolnick, Publicker Industries; in *ECT* 2nd ed., Vol. 1, pp. 159–167, by R. J. Miller, California Research Corporation; in *ECT* 3rd ed., Vol. 1, pp. 179–191, by D. L. Nelson and B. P. Webb, The Dow Chemical Company. "Acetone" in *ECT* 4th ed., Vol. 1, pp. 176–194, by William L. Howard, The Dow Chemical Company; "Acetone" in *ECT* (online), posting date: December 4, 2000, by William L. Howard, The Dow Chemical Company.

CITED PUBLICATIONS

1. American Institute of Chemical Engineers, *Design Institute for Physical Property Data*, (DIPPR File), University Park, Pa., 1989. For other listings of properties, see *Beilsteins Handbuch der Organischen Chemie*, Springer-Verlag, Berlin, Vol. 1 and supplement; and J. A. Riddick, W. B. Bunger, and T. K. Sakano, "Organic Solvents, Physical Properties, and Methods of Purification," in *Techniques of Organic Chemistry*, Vol. 2, John Wiley & Sons, Inc., New York, 1986.
2. R. C. Weast and J. G. Grasselli, eds., *Handbook of Data on Organic Compounds*, 2nd ed., Vol. 6, CRC Press, Inc., Boca Raton, Fla., Compound No. 21433, p. 3731.
3. *Fire Hazard Properties of Flammable Liquids, Gases, and Volatile Solids, Report 325M-1984, National Fire Codes*, Vol. 8, National Fire Protection Association, Batterymarch Park, Quincy, Mass.
4. R. A. Smiley, "Nitriles" in M. Grayson, ed., *Kirk-Othmer Encyclopedia of Chemical Technology*, 3rd ed., Vol. 15, Wiley-Interscience, New York, 1981, p. 901.
5. U.S. Pat. 2,827,494 (Mar. 18, 1958), J. H. Brown, Jr., and N. B. Lorette (to The Dow Chemical Company); *Chem. Abstr.* **52**, 14655i (1958). U.S. Pat. 2,827,495 (March 18, 1958), G. C. Bond and L. A. Klar (to The Dow Chemical Company); *Chem. Abstr.* **52**, 14656a (1958). N. B. Lorette, W. L. Howard, and J. H. Brown, Jr., *J. Org. Chem.* **24**, 1731 (1959); *Chem. Abstr.* **55**, 12275g (1961).
6. U.S. Pat. 3,804,795 (Apr. 16, 1974), W. O. Perry, M. W. Sorenson, and T. J. Hairston, (to The Dow Chemical Company); *Chem. Abstr.* **81**, 65384v (1974). Ger. Offen. 2,424,522 (Dec. 12, 1974) and U.S. Pat. 3,923,744 (Dec. 2, 1975), M. V. Sorenson, R. C. Whiteside, and R. A. Hickner (to The Dow Chemical Company); *Chem. Abstr.* **82**, 141725v (1975).
7. N. B. Lorette and W. L. Howard, *J. Org. Chem.* **25**, 521 (1960); *Chem. Abstr.* **54**, 19531c (1960). W. L. Howard and N. B. Lorette, *J. Org. Chem.* **25**, 525 (1960); *Chem. Abstr.* **54**, 19528f (1960).
8. U.S. Pat. 3,127,450 (March 31, 1964), W. L. Howard and N. B. Lorette (to The Dow Chemical Company); *Chem. Abstr.* **60**, 15737f (1964). U.S. Pat. 3,166,600 (January 19, 1965), N. B. Lorette and W. L. Howard (to The Dow Chemical Company); *Chem. Abstr.* **62**, 7656i (1965).

9. E. G. R. Ardah, A. D. Barbour, G. E. McClellan, and E. W. McBride, *Ind. Eng. Chem.* **16**, 1133 (1924).
10. J. M. Weiss, *Chem. Eng. News* **36**, 70 (June 9, 1958).
11. U.S. Pat. 1,329,214 (Jan. 27, 1920), C. Weizmann and A. Hamlyn; *Chem. Abstr.* **14**, 998 (1920).
12. Brit. Pat. 876,025 (Aug. 30, 1961) and Ger. Pat. 1,080,994 (to Consortium Fuer Elektrochemische Industrie G.m.b.H.). Brit. Pat. 884,962 (Dec. 20, 1961) (to Consortium Fuer Elektrochemische Industrie G.m.b.H.); *Chem. Abstr.* **59**, 5024h (1963). Brit. Pat. 892,158 (Mar. 21, 1962) (to Consortium Fuer Elektrochemische Industrie G.m.b.H.); *Chem. Abstr.* **59**, 13826d (1963).
13. C. S. Read with T. Gibson and Z. Sedaghat-Pour, "Acetone" in *Chemical Economics Handbook*, SRI International, Menlo Park, Calif., 1989, p. 604.5000 H.
14. Acetone, Chemical Profile, Chem Expo, revised April 5, 1999, www.chemexpo.com, searched Nov. 29, 2001.
15. Brit. Pat. 1,257,595 (Dec. 22, 1971) and Fr. Pat. 2,050,175, R. L. Feder and co-workers (to Allied Chemical); *Chem. Abstr.* **76**, 3548q (1972). U.S. Pat. 2,632,774 (Mar. 24, 1953), J. C. Conner, Jr., and A. D. Lohr (to Hercules, Inc.). U.S. Pat. 2,744,143 (Sept. 2, 1953), L. J. Filar (to Hercules, Inc.). U.S. Pat. 3,365,375 (Jan. 23, 1968) and Brit. Pat. 1,193,119, J. R. Nixon, Jr. (to Hercules, Inc.). Brit. Pat. 999,441 (July 28, 1965) (to Allied Chemical); *Chem. Abstr.* **63**, 14764g (1965).
16. Brit. Pat. 1,257,595 of ref. 15. U.S. Pat. 2,799,711 (July 16, 1957), E. Beati and F. Severini (to Montecatini). Brit. Pat. 895,622 (May 2, 1962) (to Societa Italiana Resine); *Chem. Abstr.* **57**, 11108h (1962).
17. Brit. Pat. 1,257,595 of ref. 15.
18. Brit. Pat. 970,945 (Sept. 23, 1964) (to Societa Italiana Resine); *Chem. Abstr.* **61**, 14586a (1964). U.S. Pat. 2,757,209 (July 31, 1956), G. C. Joris (to Allied Chemical).
19. U.S. Pat. 2,632,774 of ref. 15.
20. Ref. 13, p. 604.5000 F. Detailed process information is available in *Phenol, Report No. 22B*, Process Economics Program, SRI International, Menlo Park, Calif., December 1977.
21. U. S. Pat. 5,788,818 (Aug. 4, 1998), L. Larengo and co-workers (Erichem SpA, Italy).
22. C. Sheely, Jr., *Kinetics of Catalytic Dehydrogenation of Isopropanol*, University Microfilms, Ann Arbor, Mich., 1953, p. 3.
23. Brit. Pat. 938,854 (Oct. 9, 1953) and Belg. Pat. 617965, J. B. Anderson, K. B. Cofer, and G. E. Coury (to Shell Chemical); *Chem. Abstr.* **59**, 13826a (1963).
24. Brit. Pat. 665,376 (Jan. 23, 1952), H. O. Mottern (to Standard Oil Development Co.); *Chem. Abstr.* **46**, 6142h (1952); this work is also U.S. Pat. 2,549,844; *Chem. Abstr.* **46**, 524c (1952).
25. Brit. Pat. 804,132 (Nov. 5, 1958) (to Knapsack-Griesheim Aktiengesellschaft); *Chem. Abstr.* **53**, 7990c (1959).
26. Brit. Pat. 817,622 (Aug. 6, 1959), W. Edyvean (to Shell Research Limited); *Chem. Abstr.* **54**, 7562e (1960).
27. Brit. Pat. 1,097,819 (Jan. 3, 1968) (to Les Usines DeMelle); *Chem. Abstr.* **68**, 63105n (1968). Brit. Pat. 610,397 (Oct. 14, 1948) (to Universal Oil Products); *Chem. Abstr.* **43**, 4287f (1949).
28. Brit. Pat. 742,496 (Dec. 30, 1955), W. G. Emerson, Jr., and J. R. Quelly (to Esso Research & Engineering Co.); *Chem. Abstr.* **50**, 8710a (1956); this work is also U.S. Pat. 2,662,848; *Chem. Abstr.* **48**, 8815a (1954).
29. *Acetone, Methyl Ethyl Ketone, and Methyl Isobutyl Ketone, Report No. 77*, May 1972, Process Economics Program, SRI CEH, SRI International, Menlo Park, Calif. p. 604.5000 F. Contains detailed process information.

30. "Code of Federal Regulations, Title 49, pt. 100 to 177," *Federal Register*, Washington, D.C., October 1, 1988, paragraphs 173.118, 173.119, 173.32C, and 172.101. Because of occasional changes, the latest issue of *Federal Register* should be consulted.
31. "Code of Federal Regulations, Titles 33 and 46," *Federal Register*, Washington, D.C. See also ref. 30.
32. *Chemical Data Guide for Bulk Shipment by Water*, United States Coast Guard, Washington, D.C.
33. *Code for Construction and Equipment of Ships Carrying Dangerous Chemicals in Bulk*, International Maritime Organization, Publications Section, London, England.
34. *Chemical Hazard Response Information System, Commandant Instruction M.16465.12A*, U.S. Coast Guard, U.S. Department of Transportation, Washington, D.C.
35. *Guidebook for Initial Response to Hazardous Materials Incidents, DOT P 5800.4*, U.S. Department of Transportation, Washington, D.C., 1987. *Emergency Handling of Hazardous Materials in Surface Transportation*, Bureau of Explosives, Association of American Railroads, Washington, D.C., 1981.
36. Chemical Transportation Emergency Center, a public service of the Chemical Manufacturers' Association, 2501 M Street, N.W., Washington, D.C. 20037-1303.
37. *Chem. Week*, **37** (Feb. 24, 1999).
38. J. A. Riddick, W. B. Bunger, and T. K. Sakano, "Organic Solvents, Physical Properties, and Methods of Purification," in *Techniques of Organic Chemistry*, Vol. 2, John Wiley & Sons, Inc., New York, 1986, p. 954.
39. M. Morosco, *Ind. Eng. Chem.* **18**, 701 (1926).
40. L. F. Goodwin, *J. Am. Chem. Soc.* **42**, 39 (1920).
41. *American Petroleum Institute Standard 650*, 1977 ed., American Petroleum Institute, Washington, D.C., paragraphs 3.5.2e1, 3.5.2e3.
42. *Acetone*, Form No. 115-598-84, product bulletin of The Dow Chemical Company, Midland, Mich., 1984.
43. N. I. Sax and R. J. Lewis, Sr., *Dangerous Properties of Industrial Materials*, 9th ed., Wiley, New York, 1998.
44. D. A. Morgott, "Acetone" in E. Bingham, B. Cohrssen, and C. H. Powell, eds., *Patty's* Toxicology, 5th ed., Vol. 6, Wiley-Interscience, New York, 2001, pp. 1–116.
45. T. E. Graedel, D. T. Hawkins, and L. D. Claxton, *Atmospheric Chemical Compounds*, Academic Press, Orlando, Fla., 1986, p. 263, cited in *Hazardous Substances Data Bank, Acetone from Toxicology Data Network (TOXNET)*, National Library of Medicine, Bethesda, Md., Jan. 1990, NATS section in the review.

William L. Howard
The Dow Chemical Company

ACETYLENE

1. Introduction

Acetylene, C_2H_2, is a highly reactive, commercially important hydrocarbon. It is used in metalworking (cutting and welding) and in chemical manufacture. Chemical usage has been shrinking due to the development of alternative routes to the same products based on cheaper raw materials. The reactivity of acetylene is related to its triple bond between carbon atoms and, as a consequence, its high positive free energy of formation. Because of its explosive nature, acetylene is generally used as it is produced without shipping or storage.

2. Physical Properties

The physical properties of acetylene [74-86-2] have been reviewed in detail (1). The triple point is at −80.55°C and 128 kPa (1.26 atm). The temperature of the solid under its vapor at 101 kPa (1 atm) is −83.8°C. The vapor pressure of the liquid at 20°C is 4406 kPa (43.5 atm). The critical temperature and pressure are 35.2°C and 6190 kPa (61.1 atm). The density of the gas at 20°C and 101 kPa is 1.0896 g/L. The specific heats of the gas, C_p and C_v (at 20°C and 101 kPa) are 43.91 and 35.45 J/mol·°C (10.49 and 8.47 cal/mol, °C), respectively. The heat of formation ΔH_f at 0°C is 227.1 kJ/mol (54.3 kcal/mol).

Tables and diagrams of thermodynamic properties have been given (1) and data on the solubility of acetylene in organic liquids in relation to temperature and pressure have been reviewed and correlated (1). The dissolving powers of some of the better solvents are compared in Table 1. The solubility in water at 20°C is 16.6 g/L at 1520 kPa (15.0 atm) and 1.23 g/L at 101 kPa. Acetylene forms a hydrate of approximate stoichiometry $C_2H_2{\cdot}6H_2O$. The dissociation pressure of the hydrate is 582 kPa (5.75 atm) at 0°C and 3343 kPa (33 atm) at 15°C. Its heat of formation at 0°C is 64.4 kJ/mol (15.4 kcal/mol) (1).

3. Chemical Properties

Acetylene is highly reactive due to its triple bond and high positive free energy of formation. Extensive reviews of acetylene chemistry are available (2–8). Important reactions involving acetylene are hydrogen replacements, additions to the triple bond, and additions by acetylene to other unsaturated systems. Moreover, acetylene undergoes polymerization and cyclization reactions. The formation of a metal acetylide is an example of hydrogen replacement, and hydrogenation, halogenation, hydrohalogenation, hydration, and vinylation are important addition

Table 1. **Solubility of Acetylene in Some Organic Liquids**

Solvent	CAS Registry Number	bp, °C	Acetylene solubility[a]
acetone	[67-64-1]	56.5	237
acetonitrile	[75-05-8]	81.6	238
N,N-dimethylformamide (DMF)	[68-12-2]	153	278
dimethyl sulfoxide	[67-68-5]	189	269
N-methyl-2-pyrrolidinone	[872-50-4]	202	213
γ-butyrolactone[b]	[96-48-0]	206	203

[a] g/L of solution at 15°C and 1520 kPa (15.0 atm) total pressure.
[b] Butanoic acid, 4-hydroxy-, lactone.

reactions. In the ethynylation reaction, acetylene adds to a carbonyl group (see ACETYLENE-DERIVED CHEMICALS).

Many of the reactions in which acetylene participates, as well as many properties of acetylene, can be understood in terms of the structure and bonding of acetylene. Acetylene is a linear molecule in which two of the atomic orbitals on the carbon are *sp* hybridized and two are involved in π bonds. The lengths and energies of the C—H σ bonds and C≡C$\sigma + 2\pi$ bonds are as follows:

Bond	Bond length, nm	Energy, kJ/mol
≡C—H	0.1059	506
—C≡C—	0.1205	837

The two filled π orbitals result in a greater concentration of electron density between the carbon atoms than exists in ethylene. The resulting diminution of electron density on the carbon atom makes acetylene more susceptible to nucleophilic attack than is ethylene. The electron-withdrawing power of the triple bond polarizes the C—H bond and makes the proton more acidic than the protons of ethylene or ethane. The pK_a of acetylene is 25 (9), and the acidic nature of acetylene accounts for its strong interaction with basic solvents in which acetylene is highly soluble (10,11). Acetylene forms hydrogen bonds with basic solvents (12), and as a result, the vapor pressures of such solutions deviate greatly from Raoult's law (13). The high concentration of electrons in the triple bond enables acetylene to behave as a Lewis base toward strong acids; it forms an adduct with HCl (14).

3.1. Metal Acetylides. The replacement of a hydrogen atom on acetylene by a metal atom under basic conditions results in the formation of metal acetylides that react with water in a highly exothermic manner to yield acetylene and the corresponding metal hydroxide. Certain metal acetylides can be prepared by reaction of the finely divided metal with acetylene in inert solvents such as xylene, dioxane, or tetrahydrofuran (THF) at temperatures of 38–45°C (15).

Acetylides of the alkali and alkaline-earth metals are formed by reaction of acetylene with the metal amide in anhydrous liquid ammonia.

$$C_2H_2 + MNH_2 \longrightarrow MC{\equiv}CH + NH_3$$

Aluminum triacetylide [61204-16-8] is formed from $AlCl_3$ and sodium acetylide [1066-26-8] in a mixture of dioxane and ethylbenzene at 70–75°C (16).

$$3\,NaC{\equiv}CH + AlCl_3 \longrightarrow Al(C{\equiv}CH)_3 + 3\,NaCl$$

Copper acetylides form under a variety of conditions (17–19). Cuprous acetylides are generally explosive, but their explosiveness is a function of the formation conditions and increases with the acidity of the starting cuprous solution. They are prepared by the reaction of cuprous salts with acetylene in liquid ammonia or by the reaction of cupric salts with acetylene in basic solution in the presence of a reducing agent such as hydroxylamine. Acetylides also form from copper oxides and salts produced by exposing copper to air, moisture, and acidic or basic conditions. For this reason, copper or brasses containing >66% copper or brazing materials containing silver or copper should not be used in an acetylene system. Silver and mercury form acetylides in a manner similar to copper.

Acetylene Grignard reagents, which are useful for further synthesis, are formed by the reaction of acetylene with an alkylmagnesium bromide.

$$C_2H_2 + 2\,RMgBr \longrightarrow BrMgC{\equiv}CMgBr + 2\,RH$$

With care, the monosubstituted Grignard reagent can be formed and it reacts with aldehydes and ketones to produce carbinols (see GRIGNARD REACTIONS).

3.2. Hydrogenation. Acetylene can be hydrogenated to ethylene and ethane. The reduction of acetylene occurs in an ammoniacal solution of chromous chloride (20) or in a solution of chromous salts in H_2SO_4 (20). The selective catalytic hydrogenation of acetylene to ethylene, which proceeds over supported Group 8–10% (VIII) metal catalysts, is of great industrial importance in the manufacture of ethylene by thermal pyrolysis of hydrocarbons (21–23). Nickel and palladium are the most commonly used catalysts. Partial hydrogenation to ethylene is possible because acetylene is adsorbed on the catalyst in preference to ethylene.

3.3. Halogenation and Hydrohalogenation. Halogens add to the triple bond of acetylene. $FeCl_3$ catalyzes the addition of Cl_2 to acetylene to form 1,1,2,2-tetrachloroethane, which is an intermediate in the production of the industrial solvents 1,2-dichloroethylene, trichloroethylene, and perchloroethylene (see CHLOROCARBONS AND CHLOROHYDROCARBONS). Acetylene can be chlorinated to 1,2-dichloroethylene directly using $FeCl_3$ as a catalyst and a large excess of acetylene. The compound *trans*-$C_2H_2Cl_2$ is formed from acetylene in solutions of $CuCl_2$, CuCl, and HCl (24–26). Bromine in solution or as a liquid adds to acetylene to form first 1,2-dibromoethylene and finally tetrabromoethylene. Iodine adds less readily and the reaction stops at 1,2-diiodoethylene. Hydrogen halides react with acetylene to form the corresponding vinyl halides. An example is the formation of vinyl chloride that is catalyzed by mercuric salts.

3.4. Hydration. Water adds to the triple bond to yield acetaldehyde via the formation of the unstable enol (see ACETALDEHYDE). The reaction has been carried out on a commercial scale using a solution process with $HgSO_4/H_2SO_4$ catalyst (27,28). The vapor-phase reaction has been reported at 250–400°C using a wide variety of catalysts (28) and even with no catalyst (29). Vapor-phase catalysts capable of converting acetic acid to acetone directly convert the steam–acetylene mixture to acetone (28,30,31).

$$2\,C_2H_2 + 3\,H_2O \longrightarrow CH_3COCH_3 + CO_2 + 2\,H_2$$

3.5. Addition of Hydrogen Cyanide. At one time the predominant commercial route to acrylonitrile was the addition of hydrogen cyanide to acetylene. The reaction can be conducted in the liquid (CuCl catalyst) or gas phase (basic catalyst at 400–600°C). This route has been completely replaced by the ammoxidation of propylene (SOHIO process) (see ACRYLONITRILE).

3.6. Vinylation. Acetylene adds weak acids across the triple bond to give a wide variety of vinyl derivatives. Alcohols or phenols give vinyl ethers and carboxylic acids yield vinyl esters (see VINYL POLYMERS).

$$ROH + C_2H_2 \longrightarrow ROCH{=}CH_2$$

$$RCOOH + C_2H_2 \longrightarrow RCOOCH{=}CH_2$$

Vinyl ethers are prepared in a solution process at 150–200°C with alkali metal hydroxide catalysts (32–34), although a vapor-phase process has been reported (35). A wide variety of vinyl ethers are produced commercially. Vinyl acetate has been manufactured from acetic acid and acetylene in a vapor-phase process using zinc acetate catalyst (36,37), but ethylene is the currently preferred raw material. Vinyl derivatives of amines, amides, and mercaptans can be made similarly. *N*-Vinyl-2-pyrrolidinone is a commercially important monomer prepared by vinylation of 2-pyrrolidinone using a base catalyst.

3.7. Ethynylation. Base-catalyzed addition of acetylene to carbonyl compounds to form -yn-ols and -yn-glycols (see ACETYLENE-DERIVED CHEMICALS) is a general and versatile reaction for the production of many commercially useful products. Finely divided KOH can be used in organic solvents or liquid ammonia. The latter system is widely used for the production of pharmaceuticals and perfumes. The primary commercial application of ethynylation is in the production of 2-butyne-1,4-diol from acetylene and formaldehyde using supported copper acetylide as catalyst in an aqueous liquid-filled system.

3.8. Polymerization and Cyclization. Acetylene polymerizes at elevated temperatures and pressures that do not exceed the explosive decomposition point. Beyond this point, acetylene explosively decomposes to carbon and hydrogen. At 600–700°C and atmospheric pressure, benzene and other aromatics are formed from acetylene on heavy-metal catalysts.

Cuprous salts catalyze the oligomerization of acetylene to vinylacetylene and divinylacetylene (38). The former compound is the raw material for the production of chloroprene monomer and polymers derived from it. Nickel catalysts with the appropriate ligands smoothly convert acetylene to benzene (39) or 1,3,5,7-cyclooctatetraene (40–42). Polymer formation accompanies these transition-metal catalyzed syntheses.

4. Explosive Behavior

4.1. Gaseous Acetylene. Commercially pure acetylene can decompose explosively (principally into carbon and hydrogen) under certain conditions of pressure and container size. It can be ignited, ie, a self-propagating decomposition flame can be established, by contact with a hot body, by an electrostatic spark, or by compression (shock) heating. Ignition is generally more likely the higher the pressure and the larger the cross-section of the container. The wire temperature required for ignition decreased from 1252 to 850°C with increasing

Table 2. **Ignition Energy of Gaseous Acetylene at Various Pressures**

pressure, kPa[a]	65	100	150	200	300	1000	2000
energy, J[b]	1200	100	10	2	0.3	0.002	0.0002

[a] To convert kPa to atm, divide by 101.3.
[b] To convert J to cal, divide by 4.184.

pressure from 170 to 2000 kPa (~20 atm) (43) when a platinum or nickel resistance wire of 0.25-mm diameter and 25 or 75 mm length was heated gradually in pure acetylene; the pure acetylene was initially at room temperature in a tube 50 mm in diameter and 256 mm in length.

When the wall of the container is heated, ignition occurs at a temperature that depends on the material of the wall and the composition of any foreign particles that may be present. In clean steel pipe, acetylene at 235–2530 kPa (2.3–25 atm) ignites at 425–450°C (44,45). In rusted steel pipe, acetylene at 100–300 kPa ignites at 370°C (46). In steel pipe containing particles of rust, charcoal, alumina, or silica, acetylene at 200–2500 kPa ignites at 280–300°C (44). Copper oxide causes ignition at 250°C (47) and solid potassium hydroxide causes ignition at 170°C (44).

For local, short-duration heat sources, such as electrostatic sparks, the reported ignition energies for different pressures are on the orders of magnitude given in Table 2 (48–50).

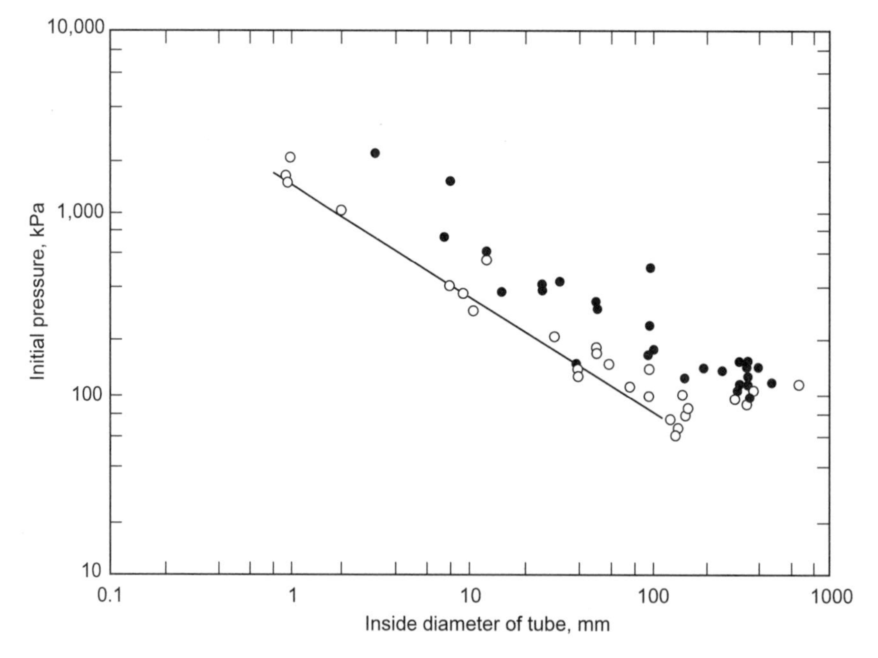

Fig. 1. Pressure required for propagation of decomposition flame through commercially pure acetylene free of solvent and water vapor in long horizontal pipes. Gas initially at room temperature; ignition by thermal nonshock sources. Curve shows approximate least pressure for propagation: (●), detonation, (○), deflagration (46,51–59). To convert kPa to atm, divide by 101.3.

Once a decomposition flame has formed, its propagation through acetylene in a pipe is favored by large diameter and high pressure. In a long pipe of given diameter, there is a pressure below which continued propagation of a flame, even though temporarily established, is very unlikely and may be impossible. In a pipe that is so short that heating by the ignition source raises the pressure, the required initial pressure is less than in a long pipe. Figure 1 shows pressures at which a flame travels through room temperature acetylene in long horizontal pipes of various diameters as the result of thermal (nonshock) ignition. The plotted points represent values reported in the literature (51–59) and the results of unpublished work (46). Many points representing detonation at higher pressures have been omitted. Propagation at pressures below atmospheric (101 kPa) require high energy ignition sources (100–1200 J or 24–287 cal). Pressure–diameter conditions near the curve, drawn at approximately the minimum pressure for propagation, tend to lead to deflagration rather than detonation, as indicated. However, the firing of a high explosive charge can cause a detonation wave to propagate at pressures even lower than those plotted as deflagrations, eg, 47 kPa in a 76-mm diameter tube and 80 kPa in a 13-mm diameter tube (60).

Deflagration flames in vessels and relatively short pipes usually propagate at an increasing velocity without becoming detonation waves, and develop pressures about ten times the initial pressure. In long pipes these flames usually become detonation waves. A slowly propagating deflagration flame in a long pipe occasionally neither accelerates nor dies out, but continues indefinitely. Decomposition flames in acetylene at pressures of 160–200 kPa in pipes of 50–150-mm diameter and 1500–6400 diameters in length have been observed to travel the full length at average velocities of 0.2 – 1 m/s (46). Generally, the velocity increases with increasing diameter. The pressure rises were <7 kPa (1 psi).

The calculated detonation velocity in room temperature acetylene at 810 kPa is 2053 m/s (61). Measured values are ~1000 – 2070 m/s, independent of initial pressure but generally increasing with increasing diameter (46, 60–64). In a time estimated to be ~6 s (65), an accidental fire-initiated decomposition flame in acetylene at ~200 kPa in an extensive piping system traveled successively through 1830 m of 76–203-mm pipe, 8850 m of 203-mm pipe, and 760 m of 152-mm pipe.

The predetonation distance (the distance the decomposition flame travels before it becomes a detonation) depends primarily on the pressure and pipe diameter when acetylene in a long pipe is ignited by a thermal, nonshock source. Figure 2 shows reported experimental data for quiescent, room temperature acetylene in closed, horizontal pipes substantially longer than the predetonation distance (44,46,52,56,58,64,66,67). The predetonation distance may be much less if the gas is in turbulent flow or if the ignition source is a high explosive charge.

The pressure developed by decomposition of acetylene in a closed container depends not only on the initial pressure (or more precisely, density), but also on whether the flame propagates as a deflagration or a detonation, and on the length of the container. For acetylene at room temperature and pressure, the calculated explosion pressure ratio, $P_{\text{final}}/P_{\text{initial}}$, is ~12 for deflagration and ~20 for detonation (at the Chapman-Jouguet plane). At 800 kPa (7.93 atm) initial

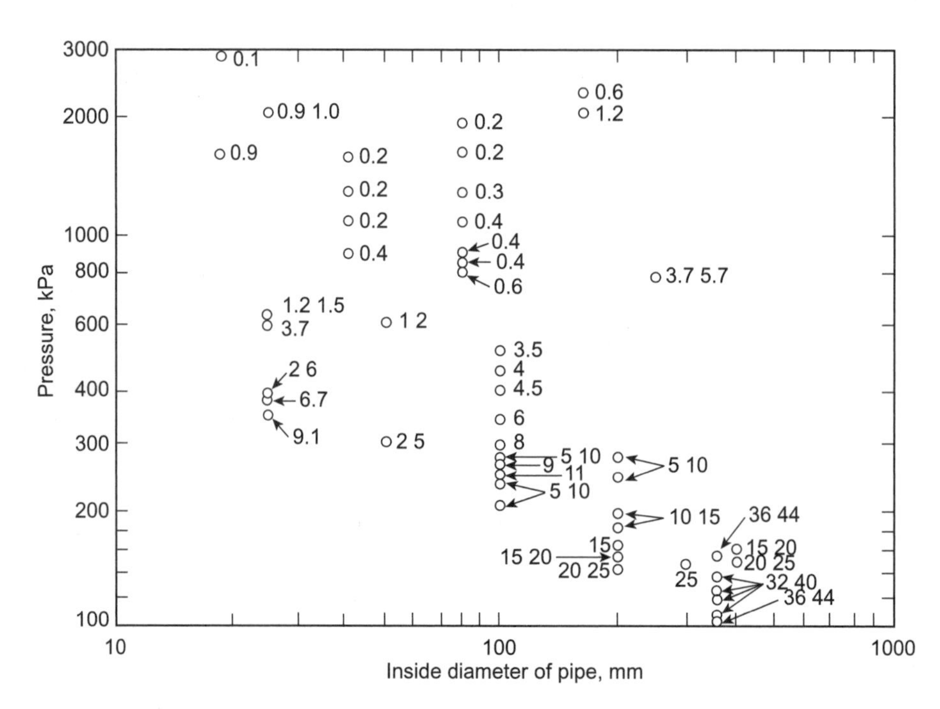

Fig. 2. Predetonation distances (in m) observed in acetylene at various pressures in horizontal pipes of various diameters. Gas quiescent, at room temperature, ignition by thermal nonshock sources (44,46,52,56,58,64,66,67). To convert kPa to atm, divide by 101.3.

pressure, the ratio is about the same for deflagration and 21.6 for detonation (61). The explosion pressure ratio for detonation refers to the pressure at the Chapman-Jouguet plane (of an ideal one-dimensional detonation wave) at which the heat-producing reaction has just come to equilibrium. This is the maximum effective pressure at the side wall of the pipe through which an established wave travels except at and near the end struck by the wave, where the pressure is increased by reflection. The calculated ratio, $P_{\text{reflected}}/P_{\text{initial}}$, for pressure developed by reflection of a detonation wave at the rigid end of a pipe is 48.5 for room temperature and pressure, and 52.5 for room temperature and 800-kPa initial pressure (46).

The measured explosion pressure ratio for deflagration in a container only a few diameters in length approaches the theoretical value; often it is ~10. However, in a pipe hundreds or thousands of diameters in length, deflagration may cause very little pressure rise because only a small fraction of the contents is hot at any time.

Explosion pressure ratios that have been measured in experiments involving detonation vary over a wide range, and depend not only on the density of the acetylene through which the detonation wave travels, but on the location of the pressure sensor and its dynamic response. Unpublished Union Carbide work, using bursting diaphragms and bonded-strain gauge sensors, records values ranging from the theoretical reflected pressure ratio of ~50–238. Russian work using crusher gauges indicates ratios as high as 658 (56,58). The Russian workers do not state whether the doubling effect of rapid rise inherent in the crusher gauge (68) was taken into account. The higher ratios have been indicated when the acetylene pressure has been only slightly above the minimum required for development of detonation. Evidently in these cases the acetylene in the far end of the pipe is compressed substantially while the flame moves at subsonic velocity toward it. The flame moves through acetylene at a density that is higher than the initial density after transition to detonation, particularly when the predetonation distance is a large fraction of the pipe length.

Flame Arresters. Propagation of a decomposition flame through acetylene in a piping system (by either deflagration or detonation) can be stopped by a hydraulic back pressure valve in which the acetylene is bubbled through water (65,69). It can also be stopped by filling the pipe with parallel tubes of smaller diameter, or randomly oriented Raschig rings (54,70–72). The small tubes should have a diameter less than that indicated (for the pressure to be used) by the curve of Figure 1, and the packed section should be long enough so that any decomposition products that are pushed through will not ignite the gas downstream. The presence of water or oil (on the walls or as mist) increases the effectiveness of the arrangement. Beds of granular ceramic material are effective with acetylene at cylinder pressure.

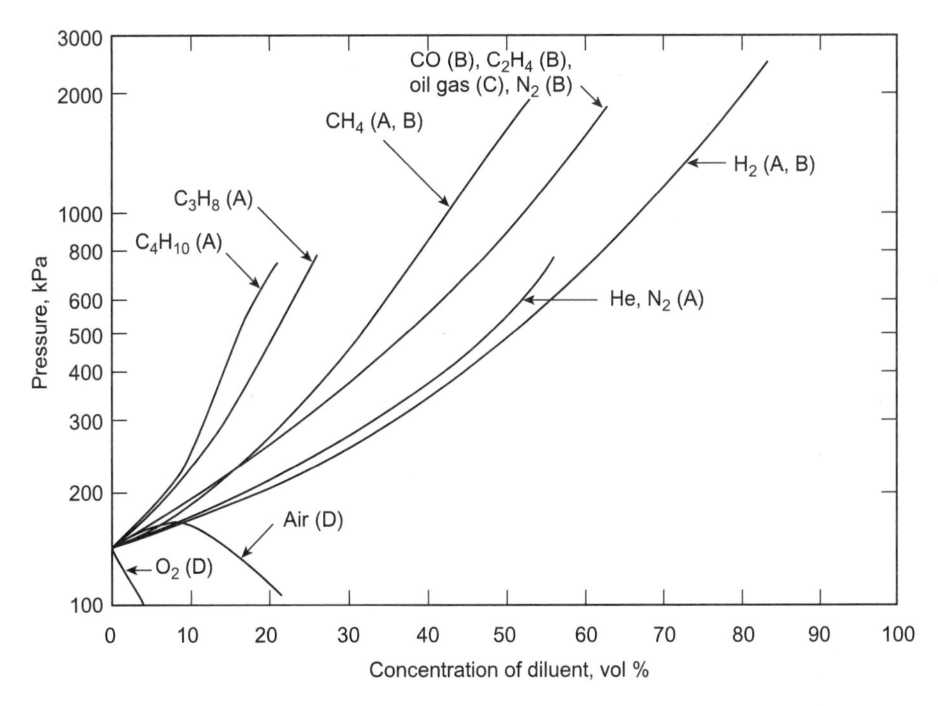

Fig. 3. Pressure required for ignition of mixtures of acetylene and a diluent gas (air, oxygen, butane, propane, methane, carbon monoxide, ethylene, oil gas, nitrogen, helium, or hydrogen) at room temperature. Initiation: fused resistance wire. Container: A, 50 mm dia × 305 mm length (73); B, 269 mm dia × 269 mm length (52); C, 102 mm dia × 254 mm length (74); and D, 120 mm dia sphere (75). To convert kPa to atm, divide by 101.3.

Ignition of Gaseous Acetylene Mixtures. The initial pressure required for ignition by fused resistance wire is given for several acetylene-diluent mixtures in Figure 3. Air in concentrations of less than ~13% inhibits ignition, but oxygen in any concentration promotes it. The data were obtained with relatively small containers and low ignition energies. With larger containers and higher ignition energies, the minimum pressure for ignition may be somewhat lower.

4.2. Acetylene—Air and Acetylene—Oxygen Mixtures. The flammability range for acetylene–air at atmospheric pressure is ~2.5–80% acetylene in tubes wider than 50 mm. The range narrows to ~8–10% as the diameter is reduced to 0.8 mm. Ignition temperatures as low as 300°C have been reported for 30–75% acetylene mixtures with air and for 70–90% mixtures with oxygen. Ignition energies are lower for the mixtures than for pure acetylene; a spark energy of 0.02 mJ (200 erg) has been found sufficient to ignite a 7.7% acetylene–air mixture at atmospheric pressure and room temperature (76,77).

In acetylene–air mixtures, the normal mode of burning is deflagration in relatively short containers and detonation in pipes. In oxygen mixtures, detonation easily develops in both short and long containers. Measurements of predetonation distances of acetylene–oxygen at 100 kPa and 40°C in 25- and 50-mm diameter tubes 2.9 and 3.6 m long gave values of 25–35 mm for mixtures of 25–50% acetylene (78). For gas mixtures at atmospheric temperature and pressure, the maximum detonation velocity has been calculated to be 2020 m/s for acetylene–air (at 15% acetylene) (79) and 2944 – 2960 m/s for acetylene–oxygen (at 50% acetylene) (79,80). The corresponding explosion pressure ratios for detonations in mixtures of these compositions have been calculated to be 21.9–22.4 for acetylene–air (79,81) and 43.5–50.2 for acetylene–oxygen (79,80), both referring to the Chapman-Jouguet plane. The $P_{\text{reflected}}/P_{\text{initial}}$ ratios for the same mixtures are estimated to be ~ 50 for acetylene–air and 110 for acetylene–oxygen (46). Except for the detonation pressures of acetylene–air mixtures, these quantities have been measured and the results agree approximately with the calculated values (46).

Several studies of spherical and cylindrical detonation in acetylene–oxygen and acetylene–air mixtures have been reported (82,83). The combustion and oxidation of acetylene are reviewed extensively in Ref. 84. A study of the characteristics and destructive effects of detonations in mixtures of acetylene (and other hydrocarbons) with air and oxygen-enriched air in earthen tunnels and large steel pipe is reported in Ref. 81.

4.3. Liquid and Solid Acetylene. Both the liquid and the solid have the properties of a high explosive when initiated by detonators or by detonation of adjoining gaseous acetylene (85). At temperatures near the freezing point neither form is easily made to explode by heat, impact, or friction, but initiation becomes easier as the temperature of the liquid is raised. Violent explosions result from exposure to mild thermal sources at temperatures approaching room temperature.

The minimum ignition energy of liquid acetylene under its vapor, when subjected to electrostatic sparks, has been found to depend on the temperature as indicated in Table 3 (86). Ignition appears to start in gas bubbles within the liquid.

Table 3. **Minimum Ignition Energy of Liquid Acetylene**

temperature, °C	−78	−50	−40	−35	−30	−27
vapor pressure, kPa[a]	145	537	779	931	1103	1234
minimum ignition energy, J[b]	>11	1.5	0.98–4.1	0.98	0.68	0.13

[a] To convert kPa to atm, divide by 101.3.
[b] To convert J to cal, divide by 4.184.

5. Manufacture from Hydrocarbons

Although acetylene production in Japan and Eastern Europe is still based on the calcium carbide process, the large producers in the United States and Western Europe now rely on hydrocarbons as the feedstock. Now >80% of the acetylene produced in the United States and Western Europe is derived from hydrocarbons, mainly natural gas or as a coproduct in the production of ethylene. In Russia ~40% of the acetylene produced is from natural gas.

Development of the modern processes for the manufacture of acetylene from hydrocarbons began in the 1920s when Badische Anilin- und Soda-Fabrik (BASF) initiated an intensive research program based on Berthelot's early (1860) laboratory investigations on the conversion of low molecular weight aliphatic hydrocarbons to acetylene by means of thermal cracking. BASFs development of the electric arc process led to the first commercial plant for the manufacture of acetylene from hydrocarbons. This plant was put into operation

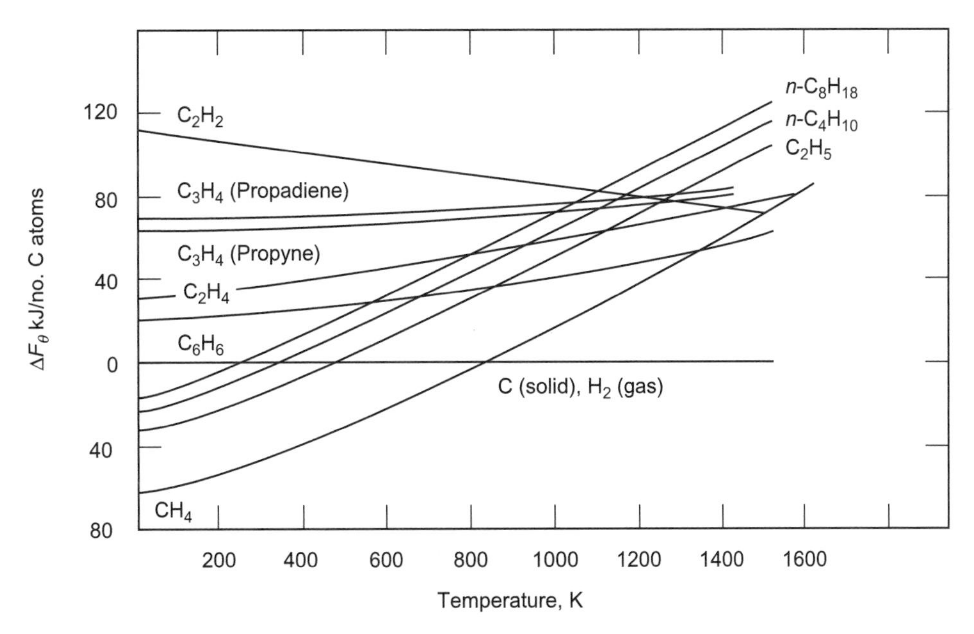

Fig. 4. Free energy of formation of several hydrocarbons. To convert kJ to kcal, divide by 4.184.

at Chemische Werke Hüls in Germany in 1940. In the United States, commercial manufacture of acetylene from hydrocarbons began in the early 1950s; expansion was rapid until the mid to late 1960s, when acetylene was gradually supplanted by cheaper ethylene as the main petrochemical intermediate.

5.1. Theory. The hydrocarbon to acetylene processes that have been developed to commercial or pilot-plant scale must recognize and take advantage of the unique thermodynamic properties of acetylene. As the free energy data shown in Figure 4 indicate, the common paraffinic and olefinic hydrocarbons are more stable than acetylene at ordinary temperatures. As the temperature is increased, the free energy of the paraffins and olefins become positive while that of the acetylene decreases, until at >1400 K acetylene is the most stable of the common hydrocarbons. However, it is also evident that, although it has the lowest free energy of the hydrocarbons at high temperature, it is still unstable in relation to its elements C and H_2. Thus it is necessary to heat the feedstock extremely fast to minimize its decomposition to its elements and, for a similar reason, the quench must be extremely rapid to avoid the decomposition of the acetylene product. Numerous acetylene production processes have been

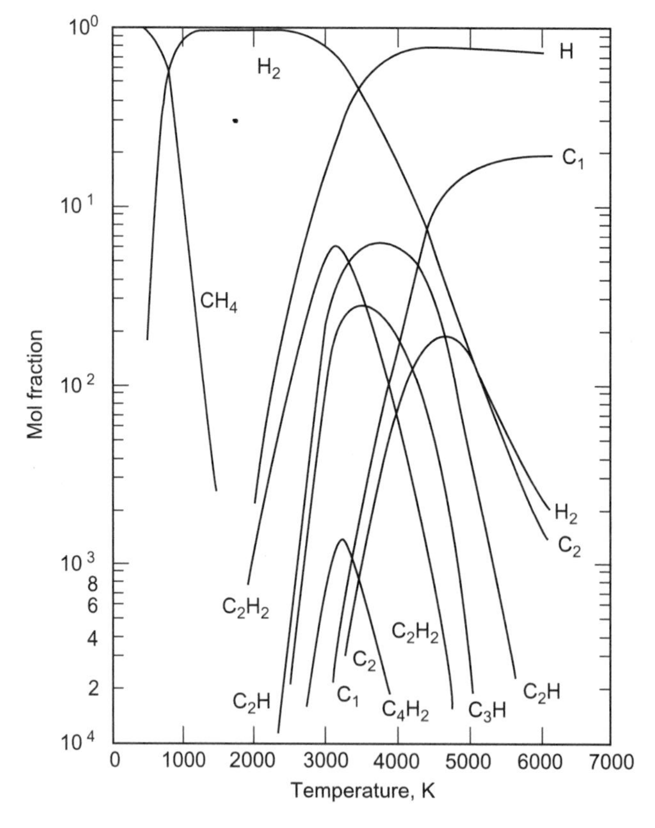

Fig. 5. Equilibrium diagram for carbon–hydrogen system at 101.3 kPa (1 atm). (C/H = 1/4).

developed, each in its own way and with varying degrees of success, accommodating the unique thermodynamics and pyrolysis kinetics of acetylene.

Examination of the equilibrium composition of the product gas mixture under relevant reactor conditions indicates the restrictive process conditions required to optimize the production process. Figure 5 illustrates the equilibrium composition for the carbon–hydrogen system with a C/H ratio of 1–4 at 101.3 kPa (1 atm) and at temperatures to 7000 K. This diagram is relevant to the pyrolysis of methane at atmospheric pressure. It is immediately evident that the hydrocarbon feedstock, CH_4, decomposes into its thermodynamically preferred state of C and H_2 at well <1000 K, whereas appreciable amounts of acetylene are not present until ~3000 K. Fortunately, the rate of the formation of acetylene is greater than the CH_4 decomposition rate. Thus it is important to heat the reactant as rapidly as possible to avoid decomposition of the feedstock to C and H_2 and to maximize the C_2H_2 formation. In a study to design an electric arc reactor (87) for producing acetylene from coal, it was found that the acetylene reaches equilibrium concentrations in <1 ms. Thus with rapid mixing and heating, it is possible to attain appreciable concentrations of acetylene with relatively little degradation of the feedstock to carbon.

Optimum reactor design and operating conditions can be further explored through equilibrium diagrams and computer models. The effects of feedstock composition pressures and temperature on the product composition have been explored in this manner in a study (87) in which in the added constraint that the heating would be rapid enough to preempt the degradation of the feedstock to carbon was imposed. Thus it was shown (Fig. 6) that at a C/H of 1–6 and at 51 kPa (0.5 atm), the concentration of acetylene at equilibrium conditions was as high as 16% and the temperature of peak concentration was 2000 K. Thus the process conditions of methane pyrolysis in excess hydrogen at reduced pressure are more promising than pyrolysis at atmospheric conditions as depicted in Figure 5.

Addressing the second step of the reaction, ie, the quench step, it is most important to quench the equilibrium mixture as quickly as possible in order to preserve the high acetylene concentration. In a study of the quenching mechanism (88), the acetylene-forming step was separated from the acetylene-preserving step by injecting known amounts of acetylene into a carbon-free plasma stream. The effect of various gases injected into the stream not only indicated the effectiveness of the quenching medium, but also revealed a great deal about the dynamics of the high temperature equilibrium composition. Hydrogen is much more effective in preserving the acetylene than the inert gases argon, helium, or nitrogen. Thus hydrogen injection allows recovery of as much as 90% of the acetylene, whereas with the other gases <50% of the acetylene was recovered. The observed effect of pressure was that in the 25–50 kPa (0.25–0.50 atm) range 85–90% of the acetylene is recovered, but at pressures between 50 and 100 kPa recovery decreases to 70%.

Because it was not possible to explain the differences in the effectiveness of hydrogen as compared to other gases on the basis of differences in their physical properties, ie, thermal conductivity, diffusivity, or heat capacity differences, their chemical properties were explored. To differentiate between the hydrogen

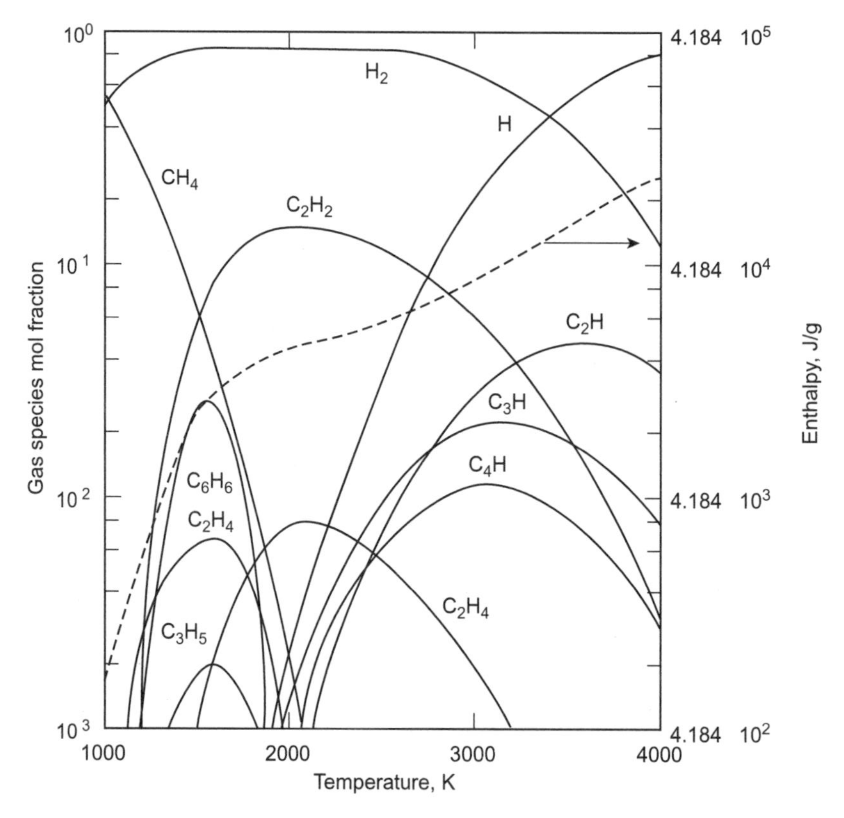

Fig. 6. Equilibrium diagram for carbon-hydrogen system at 51 kPa (0.5 atm). (C/H = 1/6). To convert J to cal, divide by 4.184.

atoms in the C_2H_2 molecules and those injected as the quench, deuterium gas was used as the quench. The data showed that although 90% of the acetylene was recovered, over 99% of the acetylene molecules had exchanged atoms with the deuterium quench to form C_2HD and C_2D_2.

To extend the study of the apparent decomposition recombination reaction, and specifically to determine if the carbon atoms exchange with other atoms in other acetylene molecules, tests using carbon isotopes were conducted. A mixture of 50% regular acetylene, $^{12}C_2H_2$, and 50% heavy acetylene, $^{13}C_2H_2$, was injected into the plasma stream. The results showed that, as before, 90% of the acetylene was recovered and that 97% of the acetylene molecules had exchanged carbon atoms.

These isotope exchange reactions not only provide an explanation for the effectiveness of hydrogen as a quench for the acetylene mixture, but also provide insight into the nature of the dynamic equilibrium present in hydrocarbon–hydrogen mixtures at high temperatures. The data indicate that essentially all of the acetylene molecules underwent total atom exchange, but because only a fraction of the energy required for decomposition of the C_2H_2 was supplied by the electric plasma, a chain or shuffle reaction is implied. The shuffle is initiated by the fragmentation of a relatively few C_2H_2 molecules into C_2H, C_2, CH, and H species. These fragments collide with C_2H_2 molecules, exchanging atoms and splitting off additional fragments. Allowing a residence time of 0.1 ms at an average plasma temperature of 3000 K, it can be estimated that each molecule undergoes $\sim 2 \times 10^4$ collisions. If an efficiency of 10% is assumed, each molecule experiences 2000 viable collisions in the first 2.5 cm of residency in the plasma stream. As the reaction mixture cools downstream, the number of collisions decrease and the chain reaction terminates as two CH fragments or a C_2H and H fragment collide to reform an acetylene molecule. Thus 90% of the acetylene is recovered but each C_2H_2 molecule has undergone some 2000 atom exchanges. As long as the exchanges occur between C and H species, high acetylene yield can be preserved. The introduction of inert species, A, He, or N_2, however, terminates the chain reaction and leads to acetylene degradation.

5.2. Process Technology. The processes designed to produce acetylene as the main product of a hydrocarbon feedstock are generally classified according to their energy source, ie, electricity or combustion. Using this classification, several processes that are now or have been operated commercially are listed in Table 4 and are described in the subsequent text. Two special cases, the production of acetylene by steam hydrocracking in oil refineries and the potentially commercial process of producing acetylene from coal, are also discussed.

Electric Discharge Processes. The synthetic rubber plant built by the I.G. Farbenindustrie during World War II at Hüls, contained the first successful commercial installation for the electric arc cracking of lower hydrocarbons to acetylene. The plant, with a capapcity of 200 t/day, was put into operation in August 1940.

The electric discharge processes can supply the necessary energy very rapidly and convert more of the hydrocarbons to acetylene than in regenerative or partial combustion processes. The electric arc provides energy at a very high

Table 4. **Acetylene Process Technology**

Energy source	Process designation	Feedstock	Typical cracked gas concentrations, mol%	
			Acetylene	Ethylene
		Electricity		
electric arc	Hüls	natural gas	15	0.9
arc plasma	Hüls	crude oil	14	7
	Hoechst	naphtha	14	7
		Combustion		
partial comb.	BASF, SBA Montecatini	natural gas, naphtha	8	0.2
pyrolysis	Hoechst HTP, SBA	natural gas	11	15
	BASF submerged	naphtha, bunker C	6	6
	Flame Wulff	range of hydrocarbons	14	8
	Kureha	crude oil		8[a]

[a] Concentrations depend on severity of pyrolysis. At a high severity (~2000°C) acetylene/ethylene ratio is 1, but at lower severity acetylene concentration is reduced and ethylene is increased.

flux density so that the reaction time can be kept to a minimum (see FURNACES, ELECTRIC, ARC FURNACES).

There have been many variations in the design of electric arc reactors but only three have been commercialized. The most important is the installation at Hüls. The other commercial arc processes were those of Du Pont (89) (a high speed rotating arc) and a Romanian process that produced both ethylene and acetylene. However, the Du Pont process has been shut down since 1969.

Hydrocarbon, typically natural gas, is fed into the reactor to intersect with an electric arc struck between a graphite cathode and a metal (copper) anode. The arc temperatures are in the vicinity of 20,000 K inducing a net reaction temperature of ~1500°C. Residence time is a few milliseconds before the reaction temperature is drastically reduced by quenching with water. Just under 11 kW·h of energy is required per kilogram of acetylene produced. Low reactor pressure favors acetylene yield and the geometry of the anode tube affects the stability of the arc. The maximum theoretical concentration of acetylene in the cracked gas is 25% (75% hydrogen). The optimum obtained under laboratory conditions was 18.5 vol % with an energy expenditure of 13.5 kW·h/kg (90).

Hüls Arc Process. The design of the Hüls arc furnace is shown in Figure 7. The gaseous feedstock enters the furnace tangentially through a turbulence chamber, E, and passes with a rotary motion through pipe H (length ~1.5 m, diameter 85–105 mm). The arc, G, burns between the bell-shaped cathode, C, and the anode pipe, H (grounded). Due to the rotary motion of the gas, the starting points of the arc rotate within the hollow electrodes. The cathodic or anodic starting point of the arc can move upward or downward freely. With the exception of the insulator, D, all parts of the furnace are made of iron. The wall thickness of the electrodes is 10–20 min.

The arc is ~100 cm long and extends ~40–50 cm into the anode pipe. About 20 cm below the anodic starting point of the arc, cold hydrocarbons ($C \geq 2$) are introduced into the tube through several nozzles to prequench the hot (~1750 K) reaction gases. The quench feed becomes partly cracked (mainly to ethylene). Immediately below the anode pipe the hot reaction mixture is cooled to a temperature of ~450 K by means of a water spray, I. The electrodes are water jacketed. The cathode is insulated from the other parts of the furnace which are grounded (insulator D). The arc is started by means of an ignition electrode. The arc is operated at 8000 kW, 7000 V, and a direct current of 1150 A. Off-peak power is generally used (91).

The feed to the arc consists of a mixture of fresh hydrocarbons and recycle gas. Table 5 indicates the composition of a typical feedstock as well as the composition of the gas leaving the arc furnace.

Taking into account the purification losses, the following operating requirements are necessary in order to obtain 100 kg of purified acetylene: 200 kg hydrocarbons (feedstock plus quench), 1030 kW·h electric energy for the arc, 250 kW·h electric energy for the separation unit, and 150 kg steam.

The by-products amount to 49.5-kg ethylene, 29-kg carbon black, 15-kg residual oil, and 280-m^3 hydrogen.

A considerable amount of carbon is formed in the reactor in an arc process, but this can be greatly reduced by using an auxiliary gas as a heat carrier.

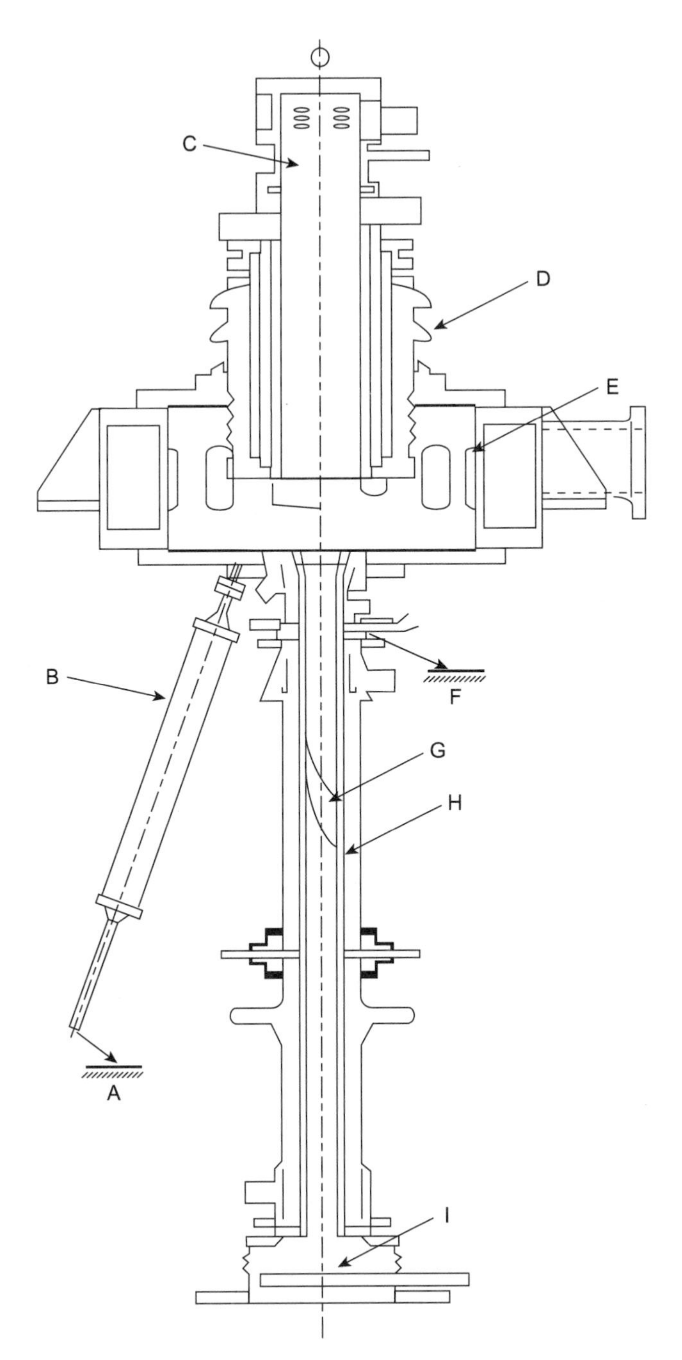

Fig. 7. Schematic drawing of a Hüls arc furnace. A, ground; B, ignition electrode; C, bell-shaped cathode; D, insulator; E, turbulence chamber; F, ground; G, arc; H, anode pipe; and I, water spray. Courtesy of Hüls AG.

Table 5. **Composition of Feedstock and Reaction Product, Arc Process**

Component	Feed gas, including recycle, vol %	Cracked gas,[a] vol %
C_2H_2	1.2	15.9
C_3H_4	1.0	1.0
C_4H_2	0.8	0.5
C_4H_4	0.7	0.5
C_2H_4	1.7	7.1
C_3H_6	2.3	0.9
C_4H_8	1.0	0.4
C_4H_6	0.4	0.3
CH_4	53.4	17.0
C_2H_6	10.2	1.2
C_3H_8	7.9	0.8
C_4H_{10}	12.5	2.1
C_5H_6	0.2	0.2
C_6H_6	0.4	0.4
C_7H_8		0.1
H_2	2.8	50.1
CO	0.8	0.7
N_2	2.7	0.8

[a] The cracked gas contains the products produced in the arc from the feedstock as well as the products obtained from the quench hydrocarbons. The liquid quench feed amounts to 120 kg/1000 kW·h and is composed of 25 kg C_3H_8 , 60 kg n-C_4H_{10}, and 35 kg *iso*-C_4H_{10}.

Hydrogen is a most suitable vehicle because of its ability to dissociate into very mobile reactive atoms. This type of processing is referred to as a plasma process and it has been developed to industrial scale, eg, the Hoechst WLP process. A very important feature of a plasma process is its ability to produce acetylene from heavy feedstocks (even from crude oil), without the excessive carbon formation of a straight arc process. The speed of mixing plasma and feedstock is critical (92).

Farbwerke Hoechst AG and Hüls AG have cooperated in the development of industrial-scale plasma units up to 10,000 kW (93). Yields of acetylene of 40–50 wt% with naphtha feedstock, and ~27 wt% with crude oil feedstock, have been obtained. Acetylene concentration in the cracked gas is in the 10–15 vol% range.

Hoechst WLP Process. The Hoechst WLP process uses an electric arc-heated hydrogen plasma at 3500–4000 K; it was developed to industrial scale by Farbwerke Hoechst AG (94). Naphtha, or other liquid hydrocarbon, is injected axially into the hot plasma and 60% of the feedstock is converted to acetylene, ehtylene, hydrogen, soot, and other by-products in a residence time of 2–3 ms. Additional ethylene may be produced by a secondary injection of naphtha (Table 6, Case A), or by means of radial injection of the naphtha feed (Case B). The oil quenching also removes soot.

Hüls Plasma Process. In the Hüls plasma process, the hydrocarbon is injected tangentially into the hot hydrogen. In crude oil cracking, a residence time of 2–4 ms converts 20–30% of the crude (94). Crude oil data are given

Table 6. **Characteristic Data of Electric Plasma Processes**[a]

Data	Hoechst WLP process[b] Case A	Case B	Hüls[c]
output, kW	10,000	9,000	8,500
naphtha input/100 kg acetylene, kg	192	250	
crude oil/100 kg acetylene, kg			367
quenching oil, kg	53	63	
products/100 kg acetylene,			
ethylene, kg	50	95	48
C_1–C_6 hydrocarbons, kg			82
hydrogen, m^3 (kg)	145 (13)	150 (13.5)	112 (10)
quenching oil, 20% carbon, kg	75	100	
crude oil residue with 20% carbon, kg			127
energy consumption/100 kg acetylene, kW·h	930	1,095	980
analysis of cracking gases, vol%			
C_2H_2	13.7	10.8	14.5
C_2H_4	6.4	9.8	6.5
yield ($C_2H_2 + C_2H_4$), wt %	78	78	56

[a] Ref. 93.
[b] Hydrogen plasma process using naphtha. Case A: secondary injection of naphtha; Case B: radial injection of the naphtha feed.
[c] Hydrogen plasma process using crude oil.

in Table 6, and data for naphtha and light hydrocarbon feeds are given in Table 7. In general, the arc processes achieve high temperatures easily, produce high yields of acetylene and few by-products, but can be handicapped by excessive carbon formation. On the strongly negative side are the high power consumption and the difficulty of controlling the arc geometry. Preheating the feed gas is one method to reduce cost in arc processes.

Electric arcs have been struck between grains of coal submerged in liquid hydrocarbons, such as kerosene and crude oil (95,96), to produce a gas with 30 vol% acetylene and 5–11 vol% ethylene (97). The energy consumption in those cases is about 9 kW · h/kg acetylene.

Flame or Partial Combustion Processes. In the combustion or flame processes, the necessary energy is imparted to the feedstock by the partial

Table 7. **Operational Results of the Hüls Plasma Process in the Cracking of Light Hydrocarbons**[a,b]

Data	Propane	n-Butane	Benzene	Naphtha
acetylene in the cracking gas, vol %	13.7	14.6	18.1	14.8
energy consumption, kW · h/100 kg C_2H_2	960	960	900	990
acetylene to ethylene ratio	2.2	1.7	18.0	1.8
carbon (rust) formation, kg/100 kg C_2H_2	2.3	3.1	44.5	6.1
yield ($C_2H_2 + C_2H_4$), wt %	61	61	56	54

[a] Courtesy of Applied Science Publishers Ltd.
[b] Ref. 93.

combustion of the hydrocarbon feed (one-stage process), or by the combustion of residual gas, or any other suitable fuel, and subsequent injection of the cracking stock into the hot combustion gases (two-stage process). A detailed discussion of the kinetics for the pyrolysis of methane for the production of acetylene by partial oxidation, and some conclusions as to reaction mechanism have been given (98).

There are several commercial versions of this partial combustion technique, including the widely used BASF process (formerly called the Sachsse process) and its various modifications with an overall similar design (99). Natural gas or other methane-rich feedstock is mixed with a limited amount of oxygen (insufficient for complete combustion), and fed through a specially designed distributor or burner to a single reaction zone in which ignition occurs. Preheating of the oxygen and methane, which is usually carried up to 500°C or above, supplies part of the energy and thus, by using less oxygen, reduces dilution of the acetylene by carbon oxides and hydrogen.

The design of the burner is of considerable importance (see COMBUSTION SCIENCE AND TECHNOLOGY). Combustion must be as brief and uniform as possible across the reaction chamber. Preignition, stability and blow-off of the flame, the possibility of backfiring through the ports of the burner head, and the deposition of carbon on the burner walls depend on the burner design and the velocities of the gas and the flame. The feasibility of partial combustion processes results from the high rates of reaction together with the relatively slow rate of decomposition of acetylene and hydrocarbon to carbon and hydrogen.

So-called tonnage oxygen, with a purity of 95–98%, is normally used as the oxidant. Although more expensive than air, its use gives several economic advantages, including a higher acetylene concentration in the cracked gas which results in lower purification costs. In addition, the plant off-gas obtained after separation of the acetylene contains high concentrations of hydrogen and carbon monoxide which, after further treatment, can be used for the synthesis of methanol or ammonia. The utilization of the off-gas is of considerable importance in establishing satisfactory economics for the partial combustion processes.

BASF Process. The basic design of the BASF process converter is shown in Figure 8. The burner is made of mild steel and is water cooled. The hydrocarbon feed can be methane, LPG, or naphtha, and these are separately preheated and mixed with oxygen. Self-ignition occurs if methane is preheated to 650°C and naphtha to 320°C. The oxygen and hydrocarbon feed are mixed in a venturi and passed to a burner block with >100 channels. The gas mixture speed in the channels is kept high enough to avoid backfiring but low enough to avoid blowout. The flame stability is enhanced by the addition of small amounts of oxygen flowing downward from the spaces between the channels. About one-third of the methane feed is cracked to acetylene and the remainder is burned.

The reaction gas is rapidly quenched with injected water at the point of optimum yield of acetylene, which happens to correspond with the point of maximum soot production. Coke will deposit on the walls of the burner and must be removed from time to time by a scraper.

The composition of the cracked gas with methane and naphtha and the plant feed and energy requirements are given in Table 8. The overall yield of acetylene based on methane is ~24% (100). A single burner with methane

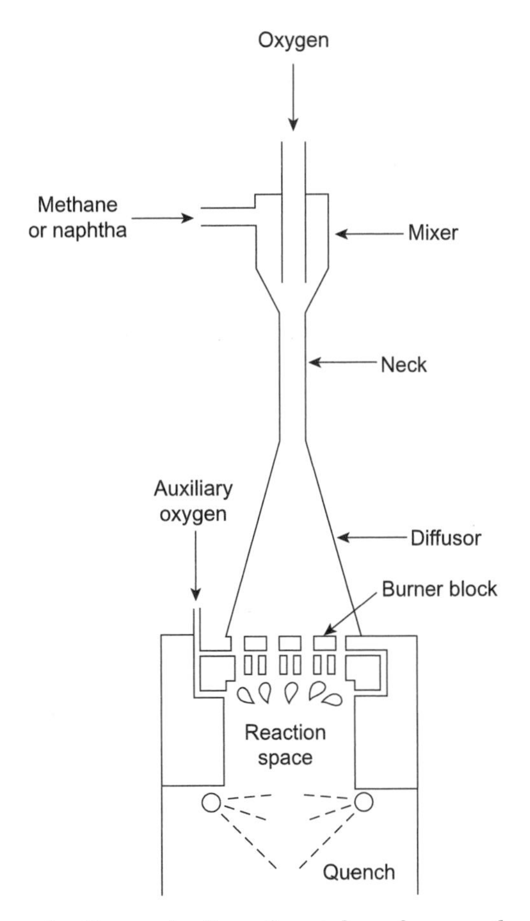

Fig. 8. BASF burner for the production of acetylene from methane or light naphtha (100). Courtesy of Verlag Chemie GmbH, Weinheim.

produces 25 t/d and with naphtha or LPG produces 30 t/d. The acetylene is purified by means of *N*-methylpyrrolidinone.

SBA Process. Two partial combustion processes have been developed by the Société Belge de l'Azote et des Produits Chimiques de Marly (located near Brussels). The first is a single-stage process using an entirely metallic converter. It produces 20–25 t/day of acetylene from methane, with an oxygen consumption of 4.6 kg/kg of acetylene produced. Methane and oxygen are preheated separately to 700°C and mixed. The oxygen is mixed with the methane through a series of holes in the internal shell. The volume flow rate of the mixed gas is set, using an inverted cone-shaped device, to that required at the point of ignition.

The flame-space walls are stainless steel and are water cooled. No mechanical coke scraper is required. A water quench cools the cracked gas stream rapidly at the point of maximum acetylene and this is followed by a secondary water quench. The primary quench point can be adjusted for variation in throughput, to accommodate the dependence of acetylene yield on residence time in the flame space.

Table 8. **BASF Process Consumptions and By-Product Yields and Cracked Gas Composition**[a,b]

Component	Methane	Naphtha
Feed and energy requirements		
hydrocarbon, kg/100 kg C_2H_2	410	430
oxygen, kg	490	430
N-methylpyrrolidinone, kg	0.5	0.5
electric energy consumption, kW·h	230	210
steam requirement, kg	450	450
residual gas, m^3	850	760
carbon, kg	5	30
Cracked gas, vol%		
C_2H_2	8.0	9.5
C_2H_4	0.2	0.2
CH_4	4.2	5.0
CO_2	3.4	3.8
CO	25.9	36.9
H_2	56.8	43.2
N_2	0.8	0.7
O_2	0.2	0.2
other hydrocarbons	0.5	0.7

[a] Courtesy of Verlag Chemie GmbH, Weinheim.
[b] Ref. 100.

Purification of the cracked gas is accomplished by water scrubbing, an electrostatic precipitator, and liquid ammonia absorption.

The SBA two-stage converter (Fig. 9) consists of two superimposed chambers. In the first (combustion) chamber, the combustion in oxygen of a hydrogen-rich gas is effected in the presence of superheated steam. By means of a special design, the combustion takes place with the formation of a ring of short flames, surrounded by steam. The energy required for pyrolysis is highly concentrated and thermal losses are reduced to a minimum (101). In the second (pyrolysis) chamber, the hydrocarbon feedstock is injected into the hot combustion gases. The reaction products are thoroughly quenched to avoid all parasitic reactions.

With this type of burner, a wide variety of raw materials, ranging from propane to naphtha, and heavier hydrocarbons containing 10–15 carbon atoms, can be used. In addition, the peculiar characteristics of the different raw materials that can be used enable the simultaneous production of acetylene and ethylene (and heavier olefins) in proportions which can be varied within wide limits without requiring basic modifications of the burner.

Montecatini Process. This partial combustion process operates at higher pressure, 405–608 kPa (4–6 atm), than the BASF and SBA processes. The burner dimensions are proportionately smaller. Because of the higher pressure, the danger of premature ignition of the methane–oxygen mixture is higher so that 2 vol% of steam is added to the gas mixture to alter the flammability limits.

The cracked gas composition is shown in Table 9 for the water quench operation (102). To produce 1800 m^3 of cracked gas, 1000 m^3 methane and 600 m^3 of oxygen are needed. If a naphtha quench is used, additional yields are

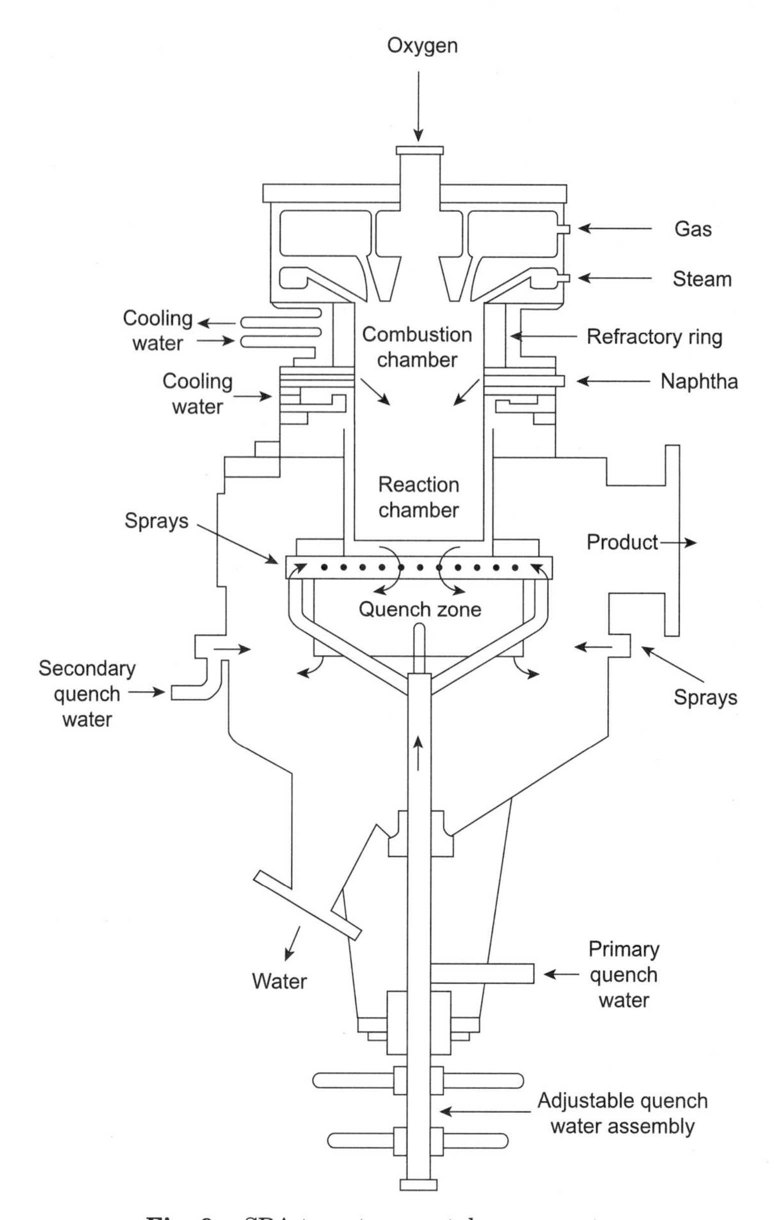

Fig. 9. SBA two-stage acetylene converter.

produced, consuming 130 kg of naphtha/1000 m^3 of methane (103). Purification of the acetylene is by methanol absorption.

Hoechst HTP Process. The two-stage HTP (high temperature pyrolysis) process was operated by Farbwerke Hoechst in Germany. The cracking stock for the HTP process can be any suitable hydrocarbon. With hydrocarbons higher than methane, the ratio of acetylene to ethylene can be varied over a range of 70:30–30:70. Total acetylene and ethylene yields, as wt% of the feed, are noted in Table 10.

Table 9. **Montecatini Process Cracked Gas Composition**[a]

Component	Composition, vol%	STP[b] m^3
C_2H_2	8.5	19.2
CO_2	3.8	2.6
CH_4	6.5	19.7
H_2	54.3	57.8
CO	25.2	9.8
C_2H_4 and higher hydrocarbons	1.7	31.2
Total	*100.0*	*140.3*

[a] Ref. 102.
[b] Per 100 kg of naphtha added.

The Hoechst burner is a water-cooled unit of all-metal construction. Fuel gas, which may be hydrogen, hydrocarbons, or off-gas from the process, is burned with oxygen in near stoichiometric amount in the combustion chamber. The hot combustion gases (tempered with dilution steam), together with the hydrocarbon feedstock injected, preferably as a vapor, enter the reaction zone where cracking of the feedstock takes place. Residence time in the reaction zone is very short, of the order of 1 ms. A rapid quench in specially designed equipment is effected to reduce the gas temperature below cracking temperatures.

BASF Submerged-Flame Process. This process can make acetylene from a wide range of feedstocks (naphtha to Bunker C oil) and, of course, crude oil itself. Oil is burned below the surface in an electrically ignited, oxygen-fed flame and quenching is immediate by the surrounding oil. The operating pressure is 900 kPa (9 bars) (100). The temperature of the oil is regulated at 200–250°C by circulation to a waste-heat boiler. The soot content of the oil is purged by burning it in the reactor. Crude oils with 12.4 wt% hydrogen can be cracked with a resulting soot level in the oil of 30%. Lower hydrogen content crudes can be handled by a separate purge of the oil to remove excess soot. An average composition of the cracked gas is shown in Table 11; it does not vary much with feedstock changes. The capcity of the commercial burner is 25 tons of acetylene and 30 tons of ethylene per day.

In summary, the bad features of partial combustion processes are the cost of oxygen and the dilution of the cracked gases with combustion products. Flame stability is always a potential problem. These features are more than offset by the inherent simplicity of the operation, which is the reason that partial combustion is the predominant process for manufacturing acetylene from hydrocarbons.

Table 10. **High Temperature Pyrolysis Process Yields**

Feed	Yield, wt%	Product
methane[a]	40.0	acetylene
butane	54.8	acetylene + ethylene (50:50)
light naphtha	54.0	acetylene + ethylene (30:70)
	50.0	acetylene + ethylene (70:30)

[a] Methane recycled.

Table 11. **BASF Submerged-Flame Process-Average Cracked Gas Composition**[a,b]

Component	Vol%
CH	43
H_2	29
CO_2	7
CH_4	4
C_2H_4	6.7
C_2H_2	6.2
C_3–plus higher hydrocarbons	4.0
H_2S	003–03

[a] Courtesy of Verlag Chemie GmbH, Weinheim.
[b] Ref. 100.

Regenerative Furnace Processes. The regenerative furnace processes supply the necessary energy for the cracking reaction by heat exchange with a solid refractory material. An alternating cycle operation is employed whereby the hydrocarbon feed is heated by the hot refractory mass to produce acetylene. Following this period, during which carbon and tars are deposited on the refractories, the process employs a combustion step in which the refractory mass is heated in an oxidizing atmosphere and the carbon and tar deposits are removed by burning. The refractories must resist both reducing and oxidizing atmosphere at ~1200°C. The refractories must also withstand the frequent and rapid heating and cooling cycles and abrasion in the case of moving refractory beds (pebbles).

Wulff Process. The regenerative technique is best exemplified by the Wulff process, licensed by Union Carbide Corp. The furnace consists basically of two masses of high purity alumina refractory tile having cylindrical channels for gas flow and separated by a central combustion space as shown in Figure 10. Its cyclic operation has four distinct steps, each of approximately 1 min in duration, the sequence being pyrolysis and heat in one direction followed by pyrolysis and heat in the other direction. Continuity of output is achieved by paired installations.

The regenerative nature of the Wulff operation permits the recovery of most of the sensible heat in the cracked gas. The gases leave the furnace at temperatures <425°C, thus obviating the need for special high temperature alloys in the switch valve and piping system.

This type of regenerative process runs at low pressure (just below atmospheric) and uses a considerable amount of dilution steam (two to three times the hydrocarbon feed). To crack methane, a reaction temperature of 1500°C must be reached, but higher hydrocarbons can be pyrolyzed to acetylene at lower temperatures, eg, 1200°C. Up to 15 vol% acetylene can be obtained in the cracked gas, but ethylene can also be produced at lower average cracking temperature and with lower acetylene yields. When cracking propane to acetylene and ethylene, the acetylene concentration in the cracked gases ranges from 14 to >16 mol%, and the ethylene concentration ranges from 8–13 mol% (104). Typical yields for acetylene plus ethylene (once-through cracking) on

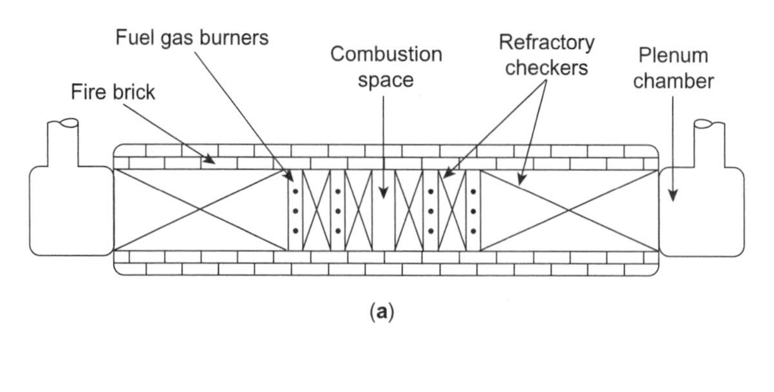

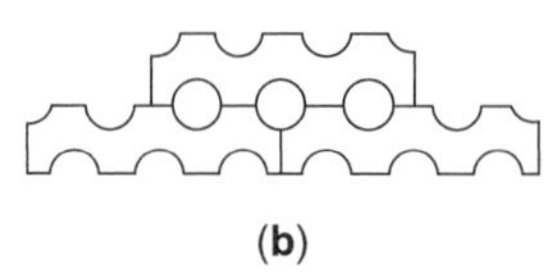

Fig. 10. (**a**) Wulff furnace design. (**b**) Checker detail of Wulff furnace refractory.

propane feed range from 51 to 59 wt% for acetylene to ethylene ratios of 3.5:1 and 1:3.5, respectively. Dimethylformamide (DM) is the purification solvent used (105).

Regenerative pyrolysis processing is very versatile; it can handle varied feedstocks and produce a range of ethylene to acetylene. The acetylene content of the cracked gases is high and this assists purification. On the other hand, the plant is relatively expensive and requires considerable maintenance because of the wear and tear on the refractory of cyclic operation.

Pyrolysis by Direct Firing. Pyrolysis of hydrocarbon in direct-fired tubes with steam dilution is practised extensively to make ethylene (qv). This technique is operated generally at the limits of metallurgy and at the maximum severity permissible (combination of time and temperature), while avoiding excessive coking rates inside the cracking tubes. The manufacture of acetylene requires even higher cracking temperatures. Such severe conditions normally induce an extreme rate of coking and an inoperable situation, If, however, the requisite high temperature can be reached without a high cracking severity and without excessively hot reactor walls, catastrophic coking rates can be avoided and useful operation is possible. Kureha Chemical Industries (Japan) developed a process based on this principle that operates at a level of pyrolysis and allows acceptable levels of acetylene production.

The unit Kureha operated at Nakoso to process 120,000 metric tons per year of naphtha produces a mix of acetylene and ethylene at a 1:1 ratio. Kureha's development work was directed toward producing ethylene from crude oil. Their work showed that at extreme operating conditions, 2000°C and short residence time, appreciable acetylene production was possible. In the process, crude oil or naphtha is sprayed with superheated steam into the specially designed reactor. The steam is superheated to 2000°C in refractory lined, pebble bed regenerative-type heaters. A pair of the heaters are used with countercurrent flows of combustion gas and steam to alternately heat the refractory and produce the superheated steam. In addition to the acetylene and ethylene products, the process produces a variety of by-products including pitch, tars, and oils rich in naphthalene. One of the important attributes of this type of reactor is its ability to produce variable quantities of ethylene as a coproduct by dropping the reaction temperature (106–108).

5.3. Separation and Purification of Hydrocarbon-Derived Acetylene. The pyrolysis of methane results in a cracked gas that is relatively low in acetylene content and that contains predominantly a mixture of hydrogen, nitrogen, carbon monoxide, carbon dioxide, unreacted hydrocarbons, acetylene, and higher homologues of acetylene. In cases where a higher hydrocarbon than methane is used as feedstock, the converter effluent also contains olefins (ethylene, propylene, propadiene, butadiene), aromatics (benzene, naphthalene), and miscellaneous higher hydrocarbons. Most acetylene processes produce significant amounts of carbon black and tars that have to be removed before the separation of acetylene from the gas mixture.

The isolation of the acetylene from the various converters presents a complicated problem. The unstable, explosive nature of acetylene imposes certain restrictions on the use of the efficient separation techniques developed for other hydrocarbon systems. The results of decomposition and detonation studies on acetylene and its mixtures with other gases indicate that operating conditions where the partial pressure of acetylene exceeds 103–207 kPa (15–30 psi) should be avoided. Similar limitations apply to the operating temperatures that should not exceed 95–105°C. Low temperatures may lead to the appearance of liquid or solid acetylene or its homologues with concomitant danger of unexpected decompositions. In view of these severe operating restrictions, it is not surprising that all commercial processes for the recovery of hydrocarbon-derived acetylene are based on absorption–desorption techniques using one or more selective solvents.

Of the many solvents proposed, only a few have found commercial application, including water (Hüls), anhydrous ammonia (SBA), chilled methanol (Montecatini), *N*-methylpyrrolidinone (BASF), butyrolactone, acetone, dimethylformamide, and hydrocarbon fractions.

The separation and purification of acetylene is further complicated by the presence in the pyrolysis gas of higher acetylenes which polymerize rather easily in solution. The removal of these constituents is a necessity, particularly in view of the utilization of the acetylene in chemical synthesis. This can be accomplished by scrubbing with small amounts of a suitable mineral oil or other organic solvent (SBA, Wulff) or by low temperature fractionation (Hüls). In the latter case, the concentrated, dry acetylene is cooled to the freezing point (195 K), whereby the higher acetylenes are liquefied and removed as a solution in methanol or benzene.

The carbon black (soot) produced in the partial combustion and electrical discharge processes is of rather small particle size and contains substantial amounts of higher (mostly aromatic) hydrocarbons which may render it hydrophobic, sticky, and difficult to remove by filtration. Electrostatic units, combined with water scrubbers, moving coke beds, and bag filters, are used for the removal of soot. The recovery is illustrated by the BASF separation and purification system (109). The bulk of the carbon in the reactor effluent is removed by a water scrubber (quencher). Residual carbon clean up is by electrostatic filtering in the case of methane feedstock, and by coke particles if the feed is naphtha. Carbon in the quench water is concentrated by flotation, then burned.

The BASF process uses *N*-methylpyrrolidinone as the solvent to purify acetylene in the cracked gas effluent. A low pressure prescrubbing is used to remove naphthalenes and higher acetylenes. The cracked gas is then compressed to 1 MPa (10 atm) and fed to the main absorption tower for acetylene removal. Light gases are removed from the top of this tower.

Stripping of acetylene from the solvent takes place at atmospheric pressure. Pure acetylene is removed from the side of the stripper; light impurities are removed overhead and recycled to the compressor. Higher acetylenes are removed from the side of a vacuum stripper with the acetylene overheads being recycled to the bottom of the acetylene stripper.

The gases leaving the purification system are scrubbed with water to recover solvent and a continuous small purge of solvent gets rid of polymers. The acetylene purity resulting from this system is 99%. The main impurities in the acetylene are carbon dioxide, propadiene, and a very small amount of vinylacetylene.

6. Manufacture From Calcium Carbide

Acetylene is generated by the chemical reaction between calcium carbide [75-20-7] and water with the release of 134 kJ/mol (900 Btu/lb of pure calcium carbide).

$$CaC_2 + 2\,H_2O \longrightarrow Ca(OH)_2 + C_2H_2$$

Because of the exothermic reaction and the evolution of gas, the most important safety considerations in the design of acetylene generators are the avoidance of excessively high temperatures and high pressures. The heat of reaction must be dissipated rapidly and efficiently in order to avoid local overheating of the calcium carbide which, in the absence of sufficient water, may become incandescent and cause progressive decomposition of the acetylene and the development of explosive pressures. Maintaining temperatures $<150°C$ also minimizes polymerization of acetylene and other side reactions that may form undesirable contaminants. For protection against high pressures, industrial acetylene generators are equipped with pressure relief devices which do not allow the pressure to exceed 204.7 kPa (15 psig). This pressure is commonly accepted as a safe upper limit for operating the generator.

Most carbide acetylene processes are wet processes from which hydrated lime, $Ca(OH)_2$, is a by-product. The hydrated lime slurry is allowed to settle in a pond or tank after which the supernatant lime-water can be decanted and reused in the generator. Federal, state, and local legislation restrict the methods of storage and disposal of carbide lime hydrate and it has become increasingly important to find consumers for the by-product. The thickened hydrated lime is marketed for industrial wastewater treatment, neutralization of spent pickling acids, as a soil conditioner in road construction, and in the production of sand-lime bricks.

6.1. Carbide-to-Water Generation. This process is the one most widely used in the United States for generating acetylene from calcium carbide. Standards for the design and construction of acetylene-generating equipment using this technique have been developed over the years by the acetylene industry. Underwriters Laboratories, Inc. have generally accepted design criteria for acetylene generating equipment (110). A water capacity of 3.78 L (1 gal) per 0.454 kg of carbide and a gas-generating rate of 0.028 m^3 (1 ft^3) per hour per 0.454 kg of carbide hopper capacity is considered normal. These design criteria apply to gravity feed generators where it is possible to have an uncontrolled release of the entire carbide hopper contents into the generating chamber. Other high capacity generators (up to 283 m^3/h) are designed so that it is impossible to have uncontrolled feed of carbide (screw-feed type); therefore, the chamber water capacity can be reduced, the carbide hopper capacity can be increased, and the gas production capacity can be raised. These high capacity generators also must pass prescribed safety tests before sale.

There are two classes of acetylene generators: the low pressure generator which operates <108.2 kPa (15.7 psi), and the medium pressure generator which operates between 108.2 and 204.7 kPa (29.7 psi). The latter is more prevalent in the United States.

There are numerous variations in the design of commercially available carbide-to-water acetylene generators. Basically, however, they are practically identical in that they consist of a water vessel or reaction chamber, a carbide feed mechanism, and a carbide storage container that empties into the feed mechanism. The water vessel is equipped with a means of filling with water and draining the lime slurry. Agitation, either hand or power driven, is provided for keeping the lime hydrate and reacting carbide in suspension. Pressure gauges and relief devices are also incorporated in the generating chamber. The water shell or generating chamber in the more common generator supports the feed mechanism and the carbide hopper, which is also fitted with pressure gauges and safety relief valves. The valved gas outlet from the generator leads directly to a flash arrester that protects the equipment against flashbacks originating from acetylene-consuming equipment downstream. The gas in the high capacity generator, owing to the higher operating temperatures, is first cooled in a water scrubber, which is an integral part of the generator, before passing through the flash arrester. The continuous supply of cooling water in the scrubber is also a source of water for the reaction with carbide. Most commercial acetylene generators are fitted with either mechanical, pneumatic, or electrical interference mechanisms which for safety reasons enforce a prescribed sequence of procedures for the operator using the equipment. Some of these safety mechanisms are high and low water level shutdown, high temperature cut out, high acetylene pressure shutdown, and shutdown on loss of either electric or pneumatic power to the generator controls. All safety conditions must be satisfied before start-up or during operation.

The gas demand dictates the rate of acetylene generation that is satisfied by the rate of carbide feed. One method of feed of properly sized carbide, which is used in medium pressure generators, employs gravity flow controlled by a valve activated by a spring-loaded rubber diaphragm. The motion of the diaphragm reflects the change of internal generator pressure and is transmitted to the carbide feed valve which, in turn, either opens or closes as the pressure decreases or increases; the generator operating pressure is set by the spring load applied to the rubber diaphragm. High capacity generators are equipped with screw conveyors for carbide feed with either constant speed on–off operation or of the variable-speed close-pressure control type. The on–off type is controlled by a simple electric pressure switch which pneumatically signals the required feed-screw operation to match the gas demand and can have pressure fluctuations up to 128.9 kPa (18.7 psi) or less.

Carbide-to-water generators exhibit a noticeable temperature rise in the course of normal operation. In properly designed units employing the recommended ratio of water to carbide, the temperature rises 21–27°C and thus attains a temperature on the order of 60–65°C. In high capacity continuous generators, where water addition is through sprays in the cooling tower or the generating chamber, temperatures in the slurry are allowed to rise to 82.2°C. With this type of generator the lime slurry must be dumped periodically to avoid flooding and there is a certain amount of gas lost each time. Since the solubility of acetylene in lime slurry is greatly diminished at 71.1–82.2°C, the gas loss through solubility is minimized by operation at these elevated temperatures.

Carbide of proper size, such as 14 ND or 6×2 mm, is important to the trouble-free operation of carbide-to-water generators. Larger size carbide interferes with the proper closing of the carbide valve causing uncontrolled feed. In screw feeds, the carbide size is limited only to the clearance between the screw and its housing. Generators that do not have continuous paddle agitation or water sprays to submerge and wet the freshly introduced carbide must use oiled carbide to reduce the reaction rate. Otherwise, small size (6×2 mm) carbide, which reacts rapidly with water, can be carried to the surface by the evolved gas where, without sufficient dissipation of heat, it can become incandescent and may initiate explosive decomposition of acetylene. Low pressure generators use a large grade of carbide, 35×9 mm. The low pressure generator is no longer made in the United States; however, some are still in use after >40 years of service. Any generator equipped with effective water sprays or paddles can use mixtures of dust carbide. It is imperative that the dust be completely wetted; otherwise, islands of dust will float on the surface where they can become incandescent and initiate explosive acetylene decomposition.

6.2. Water-to-Carbide Generation. This method of acetylene production has found only limited acceptance in the United States and Canada but has been used frequently in Europe for small-scale generation. The rate of generation is regulated by the rate of water flow to the carbide. Hazardous hot spots may occur and overheating may lead to the formation of undesirable polymer by-products. This method is, therefore, used mainly in small acetylene generators such as portable lights or lamps where the generation rate is slow and the mass of carbide is small.

6.3. Dry Generator. This water-to-carbide acetylene generation method is used in certain large-scale operations. The dry process uses about a kilogram of water per kilogram of carbide and the heat of reaction is dissipated by the vaporization of the water. Absolute control of the addition of water is critical and the reacting mass of dry lime and unreacted carbide must be continuously mixed to prevent hazardous localized overheating and formation of undesirable polymer by-products. The gas stream is filtered to remove lime dust. The dry lime by-product is considered to be advantageous compared to the wet lime by-product. Lime from dry generators is very fine and requires storage in silos or protection from scattering by wind currents. Transport must be in closed containers, such as bags.

6.4. Purification of Carbide Acetylene. The purity of carbide acetylene depends largely on the quality of carbide employed and, to a much lesser degree, on the type of generator and its operation. Carbide quality in turn is affected by the impurities in the raw materials used in carbide production, specifically, the purity of the metallurgical coke and the limestone from which the lime is produced. The nature and amounts of impurities in carbide acetylene are shown in Table 12.

The maximum amount of impurities in U.S. Grade B acetylene (111) (Carbide Generated Acetylene) is 2% on a dry basis. This gas meets commercial requirements for acetylene used in cutting and welding (qv). Production of U.S. Grade A acetylene (111) used in sensitive chemical reactions requires further purification to reduce impurities to 0.5%. There are four main impurities: phosphine [7803-51-2], ammonia [7664-41-7], hydrogen sulfide [7783-06-4], and organic sulfides. The purification involves oxidation of phosphine to phosphoric acid, the neutralization and absorption of ammonia, and the oxidation of hydrogen sulfide and organic sulfur compounds. Many processes are employed depending on the type and amount of impurities and the end use of the gas. These wet or dry processes range from simply passing the gas over purifying media to multistep chemical treatments.

The most commonly used dry methods employ oxidizing agents such as chromic acid or chromates, hypochlorite, permanganate, and ferric salts deposited on solid carriers such as diatomaceous earth arranged in beds or layers through which the gas is passed at ambient temperature. Some of the purifying media can be regenerated several times with diminishing effectiveness until they eventually lose their activity. Because of the high material and labor requirements, dry purification of acetylene is not practiced where large volumes of

Table 12. **Impurities in Carbide Acetylene**

Type	Amount, approx
PH_3	a few hundred ppm
$(CH_2{=}CH)_2S$	100 ppm (as H_2S)
NH_3	a few hundred ppm
O_2	250 ppm or less
N_2	few tenths of a percent (<1.0)
ArH_3	3 ppm or less
CH_4, CO_2, CO, H_2	a few hundred ppm
SiH_4	10 ppm or less
$CH_2{=}CH{-}C{\equiv}CH$	50 ppm
$CH_2{=}CH{-}C{\equiv}C{-}CH{=}CH_2$	50 ppm
$HC{\equiv}C{-}C{\equiv}H$	a few hundred ppm
$(CH_2{=}C{=}CH_2)^a$	traces (variable according to carbide quality)

[a] And other dienes, eg, hexadiene and butadienylacetylene; also methylacetylene.

gas have to be treated. Large-scale acetylene installations exclusively employ continuous, wet purification processes. Elaborate continuous purification methods have been developed in Europe, where at certain locations relatively low grade carbide is used to generate acetylene of low purity; these involve successive contact with water, dilute caustic solution, and chlorine–water or hypochlorite solution, followed in certain locations by a final treatment with activated carbon (112–114). Such an intensive purification of acetylene is beneficial in cases where the gas is to be used in processes employing sensitive catalytic systems.

7. Coproduct Acetylene From Steam Cracking

In the steam cracking of petroleum liquids to produce olefine, mainly ethylene, small concentrations of acetylene are produced. Although the concentrations are small, the large capacities of the olefin plants result in appreciable quantities of coproduct acetylene which, in many cases, are sufficient to satisfy the modest growth in acetylene demand due to specialty chemicals such as 1,4-butanediol (see ACETYLENE-DERIVED CHEMICALS). Because specifications for polymer-grade ethylene limit the acetylene contamination to <5 ppm, the refinery operator must decide on whether to hydrogenate the contaminte acetylene to ethylene or to separate it as a by-product. The decision is influenced by the concentration of the acetylene, which is sensitive to the composition of the feedstock and the severity of the cracking conditions, and the availability of an over-the fence use because acetylene by its nature cannot be economically transported any distance or stored. If a convenient use is available, it is generally cost-effective to recover the acetylene and sell it as a by-product, since it generally attracts a higher price than the ethylene.

The quantity of coproduct acetylene produced is sensitive to both the feedstock and the severity of the cracking process. Naphtha, for example, is cracked at the most severe conditions and thus produces appreciable acetylene; up to 2.5 wt% of the ethylene content. On the other hand, gas oil must be processed at lower temperature to limit coking and thus produces less acetylene. Two industry trends are resulting in increased acetylene output: (*1*) the ethylene plant capacity has more than doubled, and (*2*) furnace operating conditios of higher temperature and shorter residence times have increased the cracking severity.

7.1. Acetylene Recovery Process. A process to recover coproduct acetylene developed by Linde AG (Fig. 11), and reduced to practice in 11 commercial plants, comprises three sections: acetylene absorption, ethylene stripper, and acetylene stripper.

Acetylene Absorption. The gaseous feedstock containing the C_2 hydrocarbons is introduced into the acetylene absorption tower at a pressure range of 0.6–3 MPa (6–30 bar) depending on availability within the process design of the olefin plant. The absorption takes place in a countercurrent lean solvent flow, which is preferably DMF and with less frequency *N*-methylpyrrolidinone (NMP). The overhead gas fraction is partially condensed against refrigerant to avoid any solvent losses. The acetylene absorption tower is designed from thermodynamic and hydraulic points of view to minimize the acetylene content in the

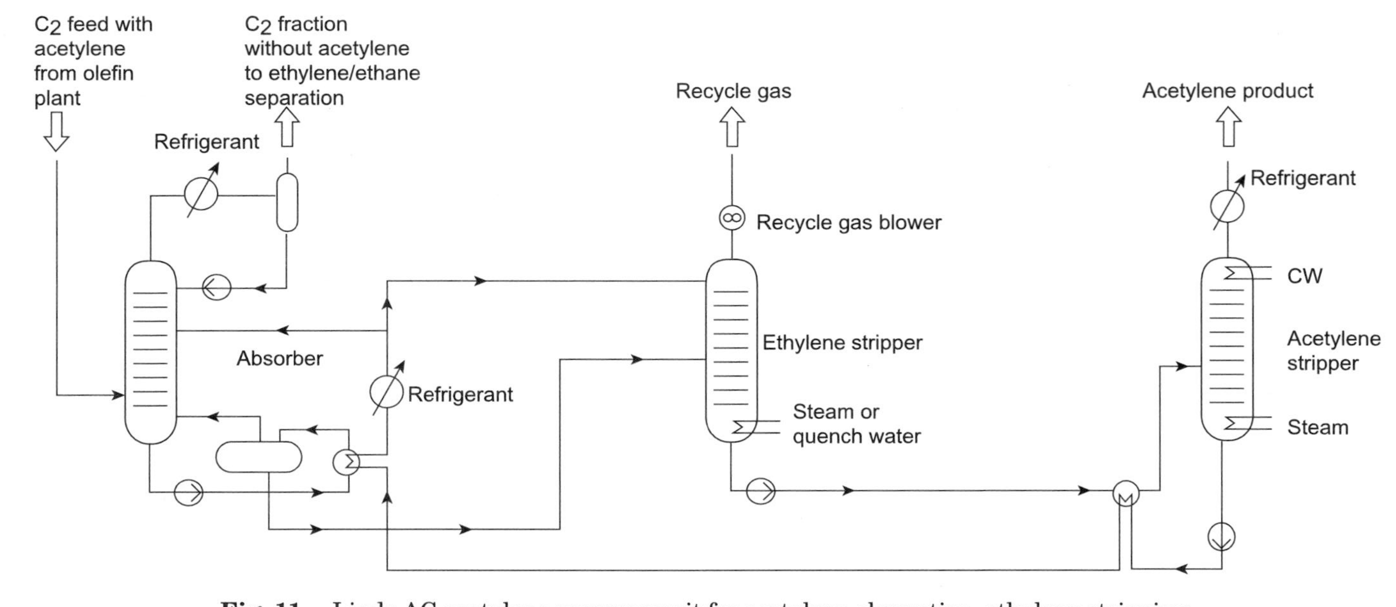

Fig. 11. Linde AG acetylene recovery unit for acetylene absorption, ethylene stripping, and acetylene stripping. CW = cooling water. Courtesy of P. Cl. Haehn, Linde AG.

overhead product (<0.2 ppm) and the recycle flow rate to the cracked gas compressor in the olefin plant (<2%).

Elthylene Stripping. The acetylene absorber bottom product is routed to the ethylene stripper, which operates at low pressure. In the bottom part of this tower, the loaded solvent is stripped by heat input according to the purity specifications of the acetylene product. A lean dimethylformamide fraction is routed to the top of the upper part for selective absorption of acetylene. This feature reduces the acetylene content in the recycle gas to its minimum (typically 1%). The overhead gas fraction is recycled to the cracked gas compression of the olefin plant for the recovery of the ethylene.

Acetylene Stripping. The loaded solvent with acetylene and traces of other basically olefinic components is pumped to the acetylene stripper tower for the delivery of the acetylene product in the overhead (typically 99.9% purity minimum). Solvent traces can be eliminated by chilling or water washing depending on downstream process requirements. The bottom product (lean solvent) is pumped back to the acetylene absorber and ethylene stripper towers after exchanging the maximum possible amount of its energy within the recovery process of economic reasons. The recovery process uses commercial solvents ithout the addition of an antifoaming agent. The applied solvents are not corrosive or fouling.

8. Acetylene from Coal

Coal, considered a solid hydrocarbon with a generic formula of $CH_{0.8}$, was explored by numerous workers (115–122) as a feedstock for the production of acetylene. Initially, the motivation for this work was to expand the market for the use of coal in the chemical process industry, and later when it was projected that the cost of ethylene would increase appreciably if pretroleum resources were depleted or constrained.

Acetylene traditionally has been made from coal (coke) via the calcium carbide process. However, laboratory and bench-scale experiments have demonstrated the technical feasibility of producing the acetylene by the direct pyrolysis of coal. Researchers in Great Britain (115,119), India (116), and Japan (118) reported appreciable yields of acetylene from the pyrolysis of coal in a hydrogen-enhanced argon plasma. In subsequent work (120), it was shown that the yields could be dramatically increased through the use of a pure hydrogen plasma.

Based on the bench-scale data, two coal-to-acetylene processes were taken to the pilot-plant level. These were the AVCO and Hüls arc-coal processes. The Avco process development centered on identifying fundamental process relationships (120). Preliminary data analysis was simplified by first combining two of three independent variables, power and gas flow, into a single enthaply term. The variation of the important criteria, specific energy requirements (SER), concentration, and yield with enthalpy are indicated in Figure 12. As the plots show, minimum SER is achieved at an enthalpy of ~5300 kW/(m^3/s) (2.5 kW/cfm), whereas maximum acetylene concentrations and yield are obtained at about 7400 kW/(m^3/s) (3.5 kW/cfm). An operating enthalpy between these two values

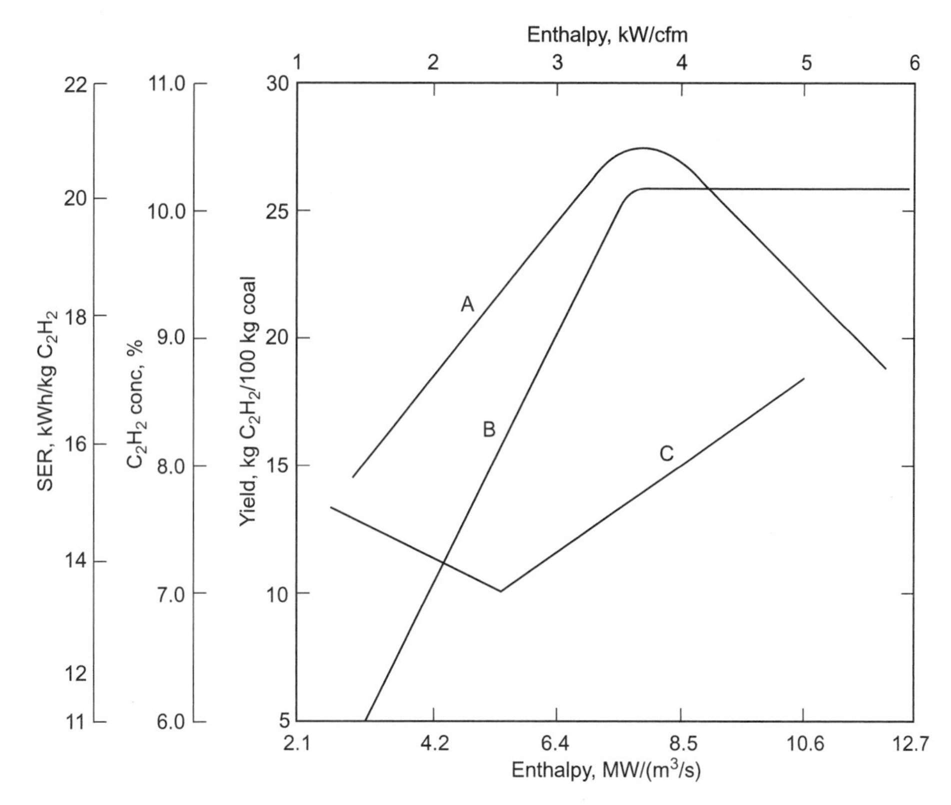

Fig. 12. Critical process parameters as a function of gas enthalpy where A is yield, B is concentration, and C is the specific energy requirement (SER).

should, therefore, be optimum. Based on the results of this work and the need to demonstrate the process at sufficient size to judge industrial applicability, AVCO built and operated a 1 MW reactor in 1979–1982. This project was jointly funded by the U.S. Department of Energy (121).

The AVCO reactor design is called a rotating arc reactor. In this design (Fig. 13), the arc is spread out radially from a center cathode to the walls that serve as the anode. In order to ensure temperature uniformity of the gas as well as the reactor wall, the arc is rotated using a magnetic field. The coal, which has been ground to conventional power plant grind, ie, 80% through a 74 μm (200 mesh) screen, is suspended in a hydrogen carrier gas and is fed through the top of the arc-coal reactor. The acetylene formed in the arc region is stabilized by rapidly quenching the gas stream to below 1400 K using a variety of quench media (hydrogen, methane, coal, hydrocarbons, water, etc).

The overall energy efficiency of the arc reactor is greatly enhanced by using a two-stage approach, ie, by using a chemically active quench in which further acetylene is produced. The active quench takes advantage of the latent heat of the gas below the arc to form additional acetylene. Two quench materials inverstigated were additional coal and a hydrocarbon (propane). In each case, the additional feed was injected below the arc reactor zone. Included in these experiments was a secondary water quench to freeze the acetylene yield.

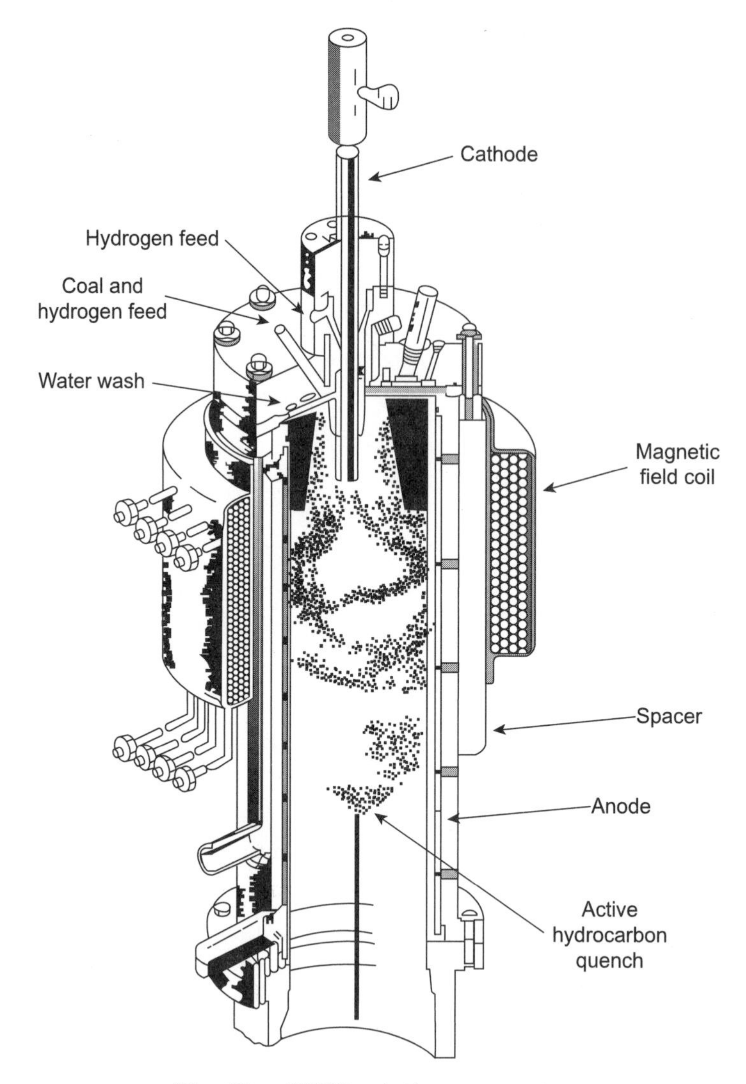

Fig. 13. AVCO rotating arc reactor.

An analytical model of the process has been developed to expedite process improvements and to aid in scaling the reactor to larger capacities. The theoretical results compare favorably with the experimental data, thereby lending validity to the application of the model to predicting directions for process improvement. The model can predict temperature and compositional changes within the reactor as functions of time, power, coal feed, gas flows, and reaction kinetics. It therefore can be used to project optimum residence time, reactor size, power level, gas and solid flow rates, and the nature, composition, and position of the reactor quench stream.

The economics of the arc-coal process is sensitive to the electric power consumed to produce a kilogram of acetylene. Early plant economic assessments indicated that the arc power consumption ($SER = kW \cdot h/kgC_2H_2$ must be below

13.2. The coal feedcoal quench experiments yielded a 9.0 SER with data that indicated a further reduction to below 6.0 with certain process improvements. In the propane quench experiment, ethylene as well as acetylene is produced. The combined process SER was 6.2 with a C_2H_2/C_2H_4 production ratio of 3:2. Economic analysis was completed utilizing the achieved acetylene yields, and an acetylene price ~35% lower than the price of ethylene was projected.

In subsequent work at Hüls (122) similar results were obtained. That is, using German coals it was also found that the magnetically rotated arc was the preferred reactor design and that the product mixture could be enriched through the use of a hydrocarbon quench. In this two-stage reactor, a SER of 11.5 kW·h/kg C_2H_2 was achieved, but it was projected that this could be reduced through further development work.

9. Shipment and Handling

The design of equipment for the handling and use of acetylene must take into consideration the possibility of acetylene decompositions. The design parameters must consider various factors, namely, pressure, temperature, source of ignitions, and ultimate pressures which may result from a decomposition. Decompositions do not occur spontaneously but must have a source of ignition. Decompositions in small vessels and short piping systems used at moderate pressures of 103 kPa (15 psi) result in a maximum pressure not greater than ~12 times the initial pressure (1240 kPa or 180 psi), in this example. Theoretically, during constant volume deflagration at ~100 – 500 kPa (1–5 atm) without loss of heat, the pressure rises to 11.5–12 times the initial pressure. The minimum acetylene pressure at which a deflagration flame can propagate throughout a long tube of any diameter has been determined experimentally. As a typical example, acetylene deflagrates in a 2.54-cm inside diameter tube at pressures above 241 kPa (35 psi) and a detonation does not develop. In pipelines of considerable lengths and of diameter sufficient to permit a true detonation to develop, the maximum pressure developed is ~50 times the initial pressure. Thus, the maximum pressure expected in a 2.54-cm inside diameter tubing at 277 kPa (40 psi) is 13.8 MPa (277 kPa × 50). In long pipelines, an effect called cascading may develop in which case the maximum pressure may be several hundred times the initial pressure.

Exact design criteria on equipment for handling acetylene is not readily available because of the great number of factors involved. However, recommendations have been made concerning the equipment, piping, compressors, flash arresters, and proper materials (69,123–125).

9.1. Acetylene in Cylinders. Acetylene cylinders are constructed to stabilize acetylene and, thereby, safely avoid the hazard of a detonation (126). Cylinders constructed for other gases do not have the same features and it is extremely important that such cylinders not be charged with acetylene. Likewise, acetylene cylinders should not be charged with other gases, even though they are capable of containing those gases up to the service pressure of the cylinder. The basic feature of an acetylene cylinder that is different from all other cylinders is that it is entirely filled with a monolithic porous mass. It is this monolithic mass that stabilizes the acetylene and permits its safe shipment. The monolithic mass is a unique technical development because it must have high porosity (up to 92%) to be economical and yet possess sufficient strength that it will not break down, crack, or crumble during many years of service. Thus whereas the porous mass completely fills the interior of the cylinder, it occupies only ~10% of the total volume. There is a slight clearance between the outside surface of the filler mass and the inside surface of the pressure shell, through which the acetylene flows to the valve for discharge. After the cylinder has been manufactured, a specified quantity of solvent is added, usually acetone. Acetone dissolves many times its own volume of acetylene and its purpose is to increase the amount of acetylene that may be safely charged and shipped. The acetylene is, therefore, not a free gas but is in solution. The solubility of acetylene in acetone increases with rising pressure and with diminishing temperature. Thus a pressure gauge attached to an acetylene cylinder reads the solution pressure and is not a direct measure of the amount of acetylene contained. The pressure is greatly affected by changes in cylinder temperature. For example, the gauge pressure of a cylinder may be 1590 kPa (230 psi) when the temperature is 21°C, and <692 kPa (100 psi) at −17.7°C without any acetylene having been withdrawn. It is therefore obvious that the contents of an acetylene cylinder, unlike oxygen or nitrogen cylinders, cannot be determined accurately by pressure gauge readings alone. Acetylene cylinder contents can, however, be accurately measured by weight, and it is on this basis that cylinder charging operations are conducted. Weight of acetylene can be converted into standard volume (atmospheric pressure, and 21°C) by the factor of 0.906 L/kg (14.7 ft^3/lb).

Acetylene cylinders are fitted with safety devices to release the acetylene in the event of fire. Cylinders manufactured in the United States are equipped with safety devices which contain a fusible metal that melts at 100°C. In large cylinders the safety devices are in the form of a replaceable, threaded steel plug with a core of fusible metal. Small cylinders (0.28 and 1.12 m^3; 10 and 40 ft^3, respectively) may have the fusible metal in passages in the cylinder valve.

The most common sizes of acetylene cylinders are those with nominal capacities of 0.28, 1.12, 2.80, 8.4, and 11.2 m^3 (10, 40, 100, 300, and 400 ft^3). The largest size produced in any quantity is about 28.0 m^3 (1000 ft^3) capacity and is commonly referred to as a lighthouse cylinder because of its use in lighting buoys, etc, for marine navigation purposes.

The manufacture and shipping of acetylene cylinders in the United States are in compliance with the specifications and regulations of the Department of Transportation (127). The specifications are verified by the Bureau of Explosives of the American Association of Railroads. DOT-8 and DOT-8AL specify the requirements for manufacture and testing of steel shell, porous mass, and quantity of acetone (therefore, acetylene) which may be charged. These regulations specify such things as the chemical analysis and physical properties of the shell material, certain critical fabrication limits, heat treatment and the tests which must be conducted on each cylinder, as well as destructive sampling tests. To be a legal article of commerce, each cylinder must bear the following markings: DOT Specification (for 8AL), serial number, registered symbol, inspector symbol, date of manufacture, and tare weight. The cylinder must be registered with the Bureau of Explosives through a manufacturing report which also includes the test results for that group of cylinders.

The tare weight (sometimes called stencil weight because it is cut into the cylinder metal) is the total weight of the cylinder and contents, but does not include a removable valve protection cap, if such is used. The saturation gas part of the tare weight is a calculated number which allows for the 11.4 g of acetylene required to saturate each 453.6 g of contained acetone at atmospheric pressure. The correct tare weight is an absolute necessity to the safe charging of acetylene cylinders.

The specifications set the maximum vol% of solvent that may be added to the cylinder shell (measured by its water capacity). The volume of solvent also varies with the capacity of the cylinder. Cylinders in the 90–92% porosity range with a capacity >9.1 kg of water may contain a maximum acetone charge of 43.4%, whereas those with 9.1 kg or less water capacity may contain up to 41.8 vol %. The first category of cylinders are normally referred to as welding cylinders and the latter as small tanks (those with 0.28 and 1.12 m^3 acetylene capacity).

The volume of acetylene that may be charged is limited by the DOT regulations to the amount that would produce an equilibrium pressure of 1833 kPa (266 psi) at 21°C. This maximum ratio by weight is 0.58 units of acetylene per unit of acetone. Welding cylinders designed and charged to these limits have a liquid-full temperature of ~65.5°C and the small tanks have a liquid-full temperature of ~79.4°C. The liquid-full temperature represents the point at which the liquid (acetylene–acetone) has expanded to completely fill the container. The DOT regulations specify that an acetylene cylinder shall not be filled and shipped at a pressure that would exceed the equivalent pressure of 1833 kPa at 21°C. A cylinder designed to the limits indicated above and charged to 1833 kPa at 21°C would have a pressure of ~2142 kPa (311 psi) at 32°C.

The DOT regulations also stipulate that prior to manufacture of cylinders of new design, or with substantially modified design, prototype cylinders must pass a set of Designed Qualification Tests administered by the Bureau of Explosives. These tests include a drop test to show that the porous mass will not compact, break down, or disintegrate during normal use; a flash test to show that the porous mass–acetone–shell combination will extinguish a flash that enters the cylinder (this requirement is the heart of acetylene cylinder design); a fire test to show the adequacy of the safety devices; and an impact test to show that the full charge of acetylene is stable under conditions such as falling off a fast-moving truck.

After a design has proved to be satisfactory by passing these tests, the regulations require inspection during manufacture to make certain that the cylinders produced are of equal quality.

Filling Cylinders with Acetylene. The filling and shipping of acetylene cylinders are subject to the regulations of the U.S. Department of Transportation. To completely charge acetylene cylinders in a reasonable period of time requires compression of acetylene to pressures >1833 kPa, usually in the range of 2074–2419 kPa (300–350 psi). Because acetylene at these pressures detonates, if a source of ignition is present, the cylinder charging plant must be carefully designed and constructed taking into account all of the safety hazards. The acetylene industry has prepared a basic set of guidelines (125).

Many factors must be taken into account in the charging operation, such as the rate of charging, cooling during charging, and interstage cooling of the gas to remove as much water as possible prior to the final drying. Water reduces the quantity of acetylene which may safely be carried, because the solubility of acetylene in water is less than in acetone. Other important factors include the mechanical reliability of the cylinder, valves and safety devices, the residual acetylene, and the presence of sufficient acetone to maintain the 0.58 ratio of acetylene/solvent. The charging of acetylene cylinders can be hazardous and should only be undertaken with the consent of the owner and by persons having full knowledge of the subject (128).

10. Economic Aspects

The relative economics of acetylene for chemical uses from calcium carbide and from hydrocarbon partial combustion or arc processes have swung rather clearly in favor of the hydrocarbon-based processes. Even more economically attractive is the acetylene produced as an unavoidable by-product in the manufacture of ethylene (qv). The economics apply to chemical uses, not industrial gases where calcium carbide does have advantages of scale that overcome its higher production cost. However, the key economic factor in the use of acetylene is the lower price of alternative materials which have decreased or eliminated some of the largest outlets for acetylene. Acetylene's triple bond inherently consumes more energy of formation than olefins; thus acetylene is more expensive. There seems no likelihood of reversing the decline in acetylene usage unless there is a change in raw material costs or more by-product acetylene is recovered.

Most by-product acetylene from ethylene production is hydrogenated to ethylene in the course of separation and purification of ethylene. In this process, however, acetylene can be recovered economically by solvent absorption instead of hydrogenation. Commercial recovery processes based on acetone dimethylformamide, or *N*-methylpyrrolidinone have a long history of successful operation. This represents a small volume for an economically scale derivatives unit.

The average price of pipeline shipments of acetylene in the United States in 2001 ranged from \$0.63 to \$0.80/kg. The range of prices due to the different sources of the acetylene can only be roughly estimated since each process has by-products and coproducts which may be credited or debited in more than one way. However (129), the relative prices of acetylene by the three primary categories of process may be calculated to be 2:1.5:1 g from calcium carbide, from hydrocarbon partial combustion, and for by-product acetylene, respectively. Pricing of the by-product acetylene from ethylene production at a value equivalent to ethylene plus recovery costs could reverse the trend away from acetylene-based processes. There is a commercial swing toward this objective as high cost calcium carbide sources are shut down.

Acetylene from calcium carbide can be advantageous in that calcium carbide may be shipped to the point of acetylene usage and acetylene generated on the spot. This avoids the necessity for low pressure, low pressure-drop gaseous acetylene pipelines, or high pressure cylinders for shipping acetylene. The carbide route is the preferred method of operation for most industrial gas

Table 13. **U.S. Producers of Acetylene and their Capacitities**[a]

Producer	Capacity, $\times 10^6$ kg ($\times 10^6$ lb)
BASF, Geismar, La.	45.4 (100)
Borden Chemicals, Geismar, La.	90.7 (200)
Carbide-Graphite Group, Calvert City, Ky, and Louisville, Ky.	34.0 (75)
ChevronPhillips, Cedar Bayou, Tex.	9.1 (20)
Dow, Seadrift, Tx.	5.4 (12)
Dow, Taft, La.	11.3 (25)
Dow, Texas City, Tex.	6.8 (15)
Equistar, LaPorte, Tex.	11.3 (25)
Rohm and Haas, Deer Park, Tex.	27.2 (60)
Total	*241.2 (532)*

[a] From Ref. 130.

operations. It is well suited to small-scale consumers. The high cost of acetylene in industrial gas applications reflects these scale, handling, and shipping factors.

Table 13 lists the producers of acetylene in the United States and their capacities. The capacity does not include industrial gas producers manufacturing acetylene from calcium carbide. The majority of the capacity is targeted for chemical use (~96%). This use includes acetylene black and supply to the industrial gas market. Most of the U.S. capacity, ~68%, is based on natural gas, 18% is based on ethylene coproduct, and 14% on calcium carbide (130).

In 2000, consumption of acetylene in the United States and Western Europe was 400×10^3t. Thirty-six percent was used for producing acetylenic chemicals (1,4-butanediol, vinyl ether, and *N*-vinyl-2-pyrrolidone), 27% was used to produce vinyl acetate monomer and vinyl chloride monomer. Consumption is expected to decline at a rate of 5% until 2005 in the U.S. Consumption is expected to decline slightly in Western Europe and to remain unchanged in Japan (131). For the year 2004, a demand of 98.4×10^6 kg (217×10^6 lb) is expected (130). Nearly all acetylene is consumed at the production site (131).

1,4-Butanediol accounts for ~90% of the demand for acetylene for producing acetylene chemicals. This demand is growing at a rate of 5%. Strong demand is for downstream intermediates, tetrahydrofuran and γ-butyolactone. A major use for butanediol is in resins for the automotive industry. The market for vinyl chloride is weak due to a soft market for poly(vinyl chloride) (130).

11. Analytical Methods

11.1. General Methods. Traces of acetylene can be detected by passing the gas through Ilosvay's solution that contains a cuprous salt in ammoniacal solution. The presence of acetylene is indicated by a pink or red coloration caused by the formation of cuprous acetylide, Cu_2C_2. The same method can be used for the quantitative determination of acetylene in parts per billion concentrations; the copper acetylide is measured colorimetrically (132).

The preferred quantitative determination of traces of acetylene is gas chromatography, which permits an accurate analysis of quantities <1 ppm. This procedure has been highly developed for air pollution studies (133) (see AIR POLLUTION CONTROL METHODS). Other physical methods, such as infrared and mass spectroscopy, have been widely used to determine acetylene in various mixtures.

Acetylene can be determined volumetrically by absorption in fuming sulfuric acid (or more conveniently in sulfuric acid activated with silver sulfate); or by reaction with silver nitrate in solution and titration of the nitric acid formed:

$$HC{\equiv}CH + 3\,AgNO_3 \longrightarrow (AgC{\equiv}CAg)AgNO_3 + 2\,HNO_3$$

The precipitated acetylide must be decomposed with hydrochloric acid after the titration as a safety measure. Concentrated solutions of silver nitrate or silver perchlorate form soluble complexes of silver acetylide (134). Ammonia and hydrogen sulfide interfere with the silver nitrate method, which is less accurate than the sulfuric acid absorption method. Acetylene and monosubstituted acetylenes may also be determined by means of potassium mercuric iodide and potassium hydroxide in methanol solution by back-titration of the excess potassium hydroxide (135).

$$2\,RC{\equiv}CH + K_2HgI_4 + 2\,KOH \longrightarrow (RC{\equiv}C)_2Hg + 4\,KI + 2\,H_2O$$

11.2. Acetylene Derived from Calcium Carbide. The analysis of acetylene derived from calcium carbide includes the determination of phosphorus, sulfur, and nitrogen compounds that are always present in the crude gas. Gas chromatographic methods that are accurate and convenient have been developed for phosphine, arsine, hydrogen sulfide, and ammonia (136). Chemically, the quantitative determination of phosphorus and sulfur can be achieved by oxidation with calcium or sodium hypochlorite. Phosphine and hydrogen sulfide present in the gas are oxidized to phosphate and sulfate and are measured gravimetrically as phosphomolybdate and barium sulfate, respectively. Ammonia is determined by the Nessler method after absorption in dilute hydrochloric acid. Oxygen, if present, can be determined by gas chromatography or by paramagnetic oxygen analyzer. It can also be determined chemically by absorption in alkaline pyrogallol after the acetylene has been removed by fuming sulfuric acid (20%) or it can be determined electrometrically after removal of interfering impurities. Qualitatively, hydrogen sulfide and phosphine can be detected in concentrations as low as 10 ppm by the brown to black discoloration of moist silver nitrate paper.

11.3. Acetylene Derived from Hydrocarbons. The analysis of purified hydrocarbon-derived acetylene is primarily concerned with the determination of other unsaturated hydrocarbons and inert gases. Besides chemical analysis, physical analytical methods are employed such as gas chromatography, infrared (ir), ultraviolet (uv), and mass spectroscopy. In industrial practice, gas chromatography is the most widely used tool for the analysis of acetylene. Satisfactory separation of acetylene from its impurities can be achieved using 50–80 mesh Porapak N programmed from 50–100°C at 4°C/min.

12. Health and Safety Aspects

Acetylene has a faint ethereal odor. Commercial-grade acetylene has a garliclike odor. The odor threshold is 657.2 mg/m^3 (137).

The NIOSH REL for acetylene is 2500 ppm as a ceiling. OSHA and ACGIH treat acetylene as a simple asphyxiant (137).

13. Uses

Acetylene is used primarily as a raw material for the synthesis of a variety of organic chemicals (see ACETYLENE-DERIVED CHEMICALS). In the United States, this accounts for ~90% of acetylene usage and most of the remainder is used for metal welding or cutting. The chemical markets for acetylene are shrinking as ways are found to substitute lower cost olefins and paraffins for the acetylene, with some products now completely derived from olefinic starting materials.

13.1. Chemical Uses. In Europe, products such as ethylene, acetaldehyde, acetic acid, acetone, butadiene, and isoprene have been manufactured from acetylene at one time. Wartime shortages or raw material restrictions were the basis for the choice of process. Coking coal was readily available in Europe and acetylene was easily accessible via calcium carbide.

The principal chemical markets for acetylene at present are its uses in the preparation of vinyl chloride, and 1,4-butanediol. Polymers from these monomers reach the consumer in the form of surface coatings (paints, film, sheets, or textiles), containers, pipe, electrical wire insulation, adhesives, and many other products that total billions of kilograms. The acetylene routes to these monomers were once dominant but have been largely displaced by newer processes based on olefinic starting materials.

Vinyl chloride (chloroethene) was a significant market for acetylene (see VINYL POLYMERS), but is soft at present. The reaction of acetylene and hydrogen chloride is carried out in the vapor phase at 150–250°C over a mercuric chloride catalyst. The acetylene route is usually coupled with an ethylene chlorination unit so that the hydrogen chloride derived from cracking dichloroethane can be consumed in the reaction with acetylene. Thus one mole each of ethylene, acetylene, and chlorine give 2 mol of vinyl chloride with a minimum of by-products. The oxychlorination of ethylene, however, eliminates by-product hydrogen chloride and thus much of the incentive for using the acetylene-based process. Hard cracking of hydrocarbons to a 1:1 molar mixture of ethylene and acetylene for use as feedstock for vinyl chloride production is done primarily outside the United States.

Vinyl acetate (ethnyl acetate) is produced in the vapor-phase reaction at 180–200°C of acetylene and acetic acid over a cadmium, zinc, or mercury acetate catalyst. However, the palladium-catalyzed reaction of ethylene and acetic acid has displaced most of the commercial acetylene-based units (see ACETYLENE-DERIVED CHEMICALS; VINYL POLYMERS).

Minor amounts of acetylene are used to produce chlorinated ethylenes. Trichloroethylene (trichloroethane) and perchloroethylene (tetrachloroethene) are prepared by successive chlorinations and dehydrochlorinations (see

CHLOROCARBONS AND CHLOROHYDROCARBONS). The chlorinations take place in the liquid phase using uv radiation and the dehydrochlorinations use calcium hydroxide in an aqueous medium at 70–100°C. Dehydrochlorination can also be carried out thermally (330–700°C) or catalytically (300–500°C).

Tetrachloroethylene can be prepared directly from tetrachloroethane by a high temperature chlorination or, more simply, by passing acetylene and chlorine over a catalyst at 250–400°C or by controlled combustion of the mixture without a catalyst at 600–950°C (138). Oxychlorination of ethylene and ethane has displaced most of this use of acetylene.

Acetylene is condensed with carbonyl compounds to give a wide variety of products, some of which are the substrates for the preparation of families of derivatives. The most commercially significant reaction is the condensation of acetylene with formaldehyde. The reaction does not proceed well with base catalysis which works well with other carbonyl compounds and it was discovered by Reppe (139) that acetylene under pressure [304 kPa (3 atm), or above] reacts smoothly with formaldehyde at 100°C in the presence of a copper acetylide complex catalyst. The reaction can be controlled to give either propargyl alcohol or butynediol (see ACETYLENE-DERIVED CHEMICALS). 2-Butyne-1, 4-diol, its hydroxyethyl ethers, and propargyl alcohol are used as corrosion inhibitors. 2,3-Dibromo-2-butene-1,4-diol is used as a flame retardant in polyurethane and other polymer systems (see BROMINE COMPOUNDS; FLAME RETARDANTS).

Much more important is the hydrogenation product of butynediol, 1, 4-butanediol [110-63-4]. The intermediate 2-butene-1,4-diol is also commercially available but has found few uses. 1,4-Butanediol, however, is used widely in polyurethanes and is of increasing interest for the preparation of thermoplastic polyesters, especially the terephthalate. Butanediol is also used as the starting material for a further series of chemicals including tetrahydrofuran, γ-butyrolactone, 2-pyrrolidinone, NMP, and *N*-vinylpyrrolidinone (see ACETYLENE-DERIVED CHEMICALS). The 1,4-butanediol market essentially represents the only growing demand for acetylene as a feedstock. This demand accounted for 90% demand for acetylene in 2001 (130).

A small amount of acetylene is used in condensations with carbonyl compounds other than formaldehyde. The principal uses for the resulting acetylenic alcohols are as intermediates in the synthesis of vitamins (qv).

Another small-scale use for acetylene is in the preparation of vinyl ethers from alcohols, including polyols and phenols. A base such as sodium or potassium hydroxide is used as catalyst in a liquid-phase high pressure reaction at 120–180°C. This general reaction is also a product of the acetylene research done at I. G. Farbenindustries by J. W. Reppe. A wide variety of alcohols can be vinylated, but only a few have achieved any commercial use. The most important is methyl vinyl ether (methoxyethene), which is used as a monomer and comonomer with maleic anhydride (Gantrez resins, GAF) for the preparation of adhesives, coatings, and detergents, as well as starting materials for further synthesis.

Acetylene black is prepared by the partial combustion of acetylene and has specialty uses in batteries.

Vinyl fluoride (fluoroethene), is manufactured from the catalyzed addition of hydrogen fluoride to acetylene. It is used to prepare poly(vinyl fluoride) which was found use in highly weather-resistant films (Tedlar film, Du Pont). Poly(vinylidene fluoride) also is used in weather-resistant coatings (see FLUORINE COMPOUNDS, ORGANIC). The monomer can be prepared from acetylene, hydrogen fluoride, and chlorine but other nonacetylenic routes are available.

At one time, the only commercial route to 2-chloro-1,3-butadiene (chloroprene), the monomer for neoprene, was from acetylene (see ELASTOMERS, SYNTHETIC, SURVEY). In the United States, Du Pont operated two plants in which acetylene was dimerized to vinylacetylene with a cuprous chloride catalyst and the vinyl acetylene reacted with hydrogen chloride to give 2-chloro-1, 3-butadiene. This process was replaced in 1970 with a butadiene-based process in which butadiene is chlorinated and dehydrochlorinated to yield the desired product (see CHLOROCARBONS AND CHLOROHYDROCARBONS).

13.2. Fuel Uses. At one time acetylene was widely used for home, street, and industrial lighting. These applications disappeared with the advent of electrical lighting during the 1920s. However, one of the first fuel uses for acetylen, metalworking with the oxyacetylene flame, continues to consume a significant amount of acetylene (140).

Fusion welding is the process of uniting metallic parts by heating the surfaces of the portions to be joined until the metal flows. Many electrical and chemical means are used to provide the heat for various welding processes, but the oxyacetylene flame remains the preferred choice of the gas welding processes. Cheaper fuels are available, such as propane and butane, but they do not reach the high flame temperature (3200°C) or achieve acetylene's combustion intensity (product of the burning velocity and the heating value of the fuel). The cheaper fuels are reserved for use in specialized applications where their properties are applicable. The oxyacetylene flame can be used in joining most meals and thus has a versatility advantage (see WELDING).

Large quantities of acetylene are used in metal cutting, which involves the combustion and melting of the metal; the oxyacetylene flame supplies the heat to initiate the process. Acetylene seems to have the advantage over other fuels because of the need for less oxygen and a shorter preheat time. Although oxyacetylene cutting is used in the field for construction and demolition, most cutting operations are performed in steel mills or fabricating shops.

Other uses of oxyacetylene flames in mill operations are in building up or hardfacing metal, lancing (piercing a hole in a metal mass), and a variety of metal cleaning procedures. A minor but interesting fuel use of acetylene is in flame spectrophotometry where oxygen and nitrous oxide are used as oxidants in procedures for a wide variety of the elements.

An internal combustion engine adapted to use an environmentally clean multifuel composition, comprising acetylene as a primary fuel and a combustible fuel, such as one or ore fluids selected from an alcohol such as ethanol, methanol or any other alcohol or alcohols from the group comprising C_1-C_{12} carbon chains, ethers such as from the group comprising dimethyl ether, diethyl ether, methyl *tert*-butyl ether, ethyl *tert*-butyl ether, *tert*-amyl methyl ether, diisopropyl ether, etc., low molecular weight esters such as from the group comprising methyl formate, methyl acetate, ethyl acetate, methyl propionate, ethyl propionate, ethyl malate, and butyl malate, or other suitable combustible fluid such as mineral spirits, as a secondary fuel for operatively preventing early ignition and knock arising from the primary fuel has been described (141).

13.3. Other. In a new strategy called "click" chemistry recently developed ar Scripps Research Institute, reactive molecular building blocks are designed to click together selectively and covalently. The expanded strategy is the use of chemical and biological receptor structures as templates to guide the formation of click products.

The reactive building blocks that are focused on at present have azide and acetylene groups that combine readily with each other (when in close proximity to each other) to from triazoles. Triazoles have been prepared that can inhibit the disease-associated enzyme, acetylcholinesterase. Acetylcholinesterase is a crucial enzyme in the mammalian nervous system and a current target for drugs to alleviate Alzheimer's disease (142).

BIBLIOGRAPHY

"Acetylene" in *ECT* 1st ed., Vol. 1, pp. 101–123, by G. R. Webster, Carbide and Carbon Chemicals Corp., R. L. Hasche, Tennessee Eastman Co., and K. Kaufman, Shawinigan Chemicals Ltd.; in *ECT* 1st ed., Suppl. 2, pp. 1–35, by H. B. Sargent, Linde Co., Div. of Union Carbide Co., and W. G. Schepman, Union Carbide Olefins Co.; In *ECT* 2nd ed., Vol. 1, pp. 171–211, by H. Beller and J. M. Wilkinson, Jr., GAF Corp.; "From Hydrocarbons," pp. 211–237, by D. A. Duncan, Institute of Gas Technology; "Economic Aspects," pp. 237–243, by R. M. Manyik, Union Carbide Corp.; "Properties" under "Acetylene" in *ECT* 3rd ed., Vol. 1, pp. 192–203, by C. M. Dietz and H. B. Sargent, Union Carbide Corp.; "Handling," pp. 203–210, by R. O. Tribolet, Union Carbide Corp.; "Manufacturing," pp. 203–210, by R. P. Shaffer, Union Carbide Corp.; "Hydrocarbons, Acetylene" *ECT* 4th ed., Vol. 13, pp. 760–811, by Robert M. Manyik, Union Carbide Corp., C. M. Dietz, H. B. Sargent, R. O. Thribolet, R. P. Schaffer, Consultants, and Richard E. Gannon, Textron Systems; "Acetylene, Properties and Manufacturing from Calcium Carbide" in *ECT* (online), posting date: December 4, 2000, by Robert M. Manyik, Union Carbide Corporation, C. M. Dietz, H. B. Sargent, R. O. Thribolet, R. P. Schaffer, Consultants.

CITED PUBLICATIONS

1. S. A. Miller, *Acetylene—Its Properties, Manufacture and Uses*, Vol. 1, Academic Press, Inc., New York, 1965.
2. J. W. Copenhaver and M. H. Bigelow, *Acetylene and Carbon Monoxide Chemistry*, Reinhold Publishing Co., New York, 1949.
3. S. A. Miller, *Acetylene—Its Properties, Manufacture and Uses*, Vol. 2, Academic Press Inc., New York, 1966.
4. R. A. Raphael, *Acetylene Compounds in Organic Synthesis*, Butterworths, London, 1955.
5. T. F. Rutledge, *Acetylenic Compounds—Preparation and Substitution Reactions*, Reinhold Book Corp., New York, 1968.
6. T. F. Rutledge, *Acetylenes and Allenes*, Reinhold Book Corp., New York, 1969.
7. H. G. Viehe, *Chemistry of Acetylenes*, Marcel Dekker, Inc., New York, 1969.
8. R. J. Tedeschi, *Acetylene-based Chemicals from Coal and Other Natural Resources*, Marcel Dekker, Inc., New York, 1982.

9. Ref. 6, p. 8.
10. A. C. McKinnis, *Ind. Eng. Chem.* **27**, 2928 (1962).
11. Ref. 1, p. 74.
12. R. C. West and C. S. Kraihanzel, *J. Am. Chem. Soc.* **83**, 765 (1961).
13. H. J. Copley and C. E. Holley Jr., *J. Am. Chem. Soc.* **61**, 1599 (1939).
14. D. Cook, Y. Lupien, and W. G. Schneider, *Can. J. Chem.* **34**, 957 (1956).
15. T. Mole and J. R. Suertes, *Chem. Ind. London,* 1727 (1963).
16. U.S. Pat. 3,321,487 (May 23, 1967) P. Chini and A. Baradal.
17. V. F. Bramfield, M. T. Clark, and A. P. Seyfang, *J. Soc. Chem. Ind. London* **65**, 346 (1947).
18. W. Feitknecht and L. Hugi-Carmes, *Schweiz. Arch. Angew. Wiss. Tech.* **23**, 328 (1957).
19. W. Pakulat, *Schweisstechnik Berlin* **19**, 374 (1964).
20. Ref. 3, p. 1.
21. G. Bond and P. B. Wells, in D. D. Eley, H. Pines, and P. B. Weisz, eds., *Advances in Catalysis*, Vol. 15, Academic Press, Inc., New York, 1964, p. 155.
22. Ref. 3, 1–22.
23. *Proceedings of the 4th Ethylene Producers Conference*, Vol. 1, American Institute of Chemical Engineers, New York, 1992, pp. 177–211.
24. U.S. Pat. 2,440,997 (May 4, 1948), E. Adler.
25. A. L. Klebanskii, A. S. Vol'kenshtein, and N. A. Orlova, *Zh. Obshch. Khim.* **5**, 1255 (1935); *J. Prakt. Chem.* **1**, 145 (1936).
26. Ger. Pat. 968,921 (Apr. 10, 1958); 969,191 (May 8, 1959); and 1,097,977 (Jan. 26, 1961), A. Jacabowsky and K. Sennewald.
27. J. A. Nieuwland and R. R. Vogt, *The Chemistry of Acetylene*, Reinhold, New York, 1945, pp. 115–123.
28. Ref. 3, pp. 134–145.
29. U.S. Pat. 3,291,839 (Dec. 13, 1966), R. W. Carney and G. A. Renberg.
30. Yu. A. Gorin, I. K. Gorin, and N. A. Rozenberg, *Zh. Prikl. Khim. (Leningrad)* **40**, 399 (1967).
31. Yu. A. Gorin, I. K. Gorin, and A. E. Kalaus, *Chem. Abstr.* **57**, 12305a (1962).
32. Ref. 2, pp. 37–40.
33. Ref. 3, pp. 198–207.
34. U.S. Pat. 3,370,095 (Feb. 20, 1968), J. F. Vitcha.
35. V. A. Sims and J. F. Vitcha, *I&*EC Product Res. Dev. **2**, 293 (1963).
36. Brit. Pat. 1,125,055 (Aug. 28, 1968), C. H. A. Borsboom, P. A. Gautier, and D. Medema.
37. Brit. Pat. 1,100,038 (Jan. 24, 1968), L. G. Smith.
38. J. A. Nieuwland and co-workers, *J. Am. Chem. Soc.* **53**, 4197 (1931).
39. U. Rosenthal and W. Schulz, *J. Organomet. Chem.* **321**, 103 (1987).
40. Ref. 2, p. 189.
41. W. Reppe and co-workers, *Justus Liebigs Ann. Chem.* **560**, 1 (1948).
42. G. N. Schrauzer, P. Glockner, and S. Eichler, *Angew. Chem. Internat. Edit.* **3**, 185 (1964).
43. C. M. Detz, *Combust. Flame* **26**, 45 (1976).
44. Ref. 1, p. 485.
45. W. Rimarski and M. Konschak, *Forschungsarb. Schweissens Schneidens* **4**, 43 (1929).
46. Technical data, Union Carbide Corp., South Charleston, W. Va., unpublished.
47. W. Rimarski and M. Konschak, *Forschungsarb. Schweissens Schneidens* **5**, 100 (1930).
48. B. A. Ivanov and S. M. Kogarko, *Zh. Prikl, Mekh. Tek. Fiz.* **59**(3) (1963).
49. B. A. Ivanov and S. M. Kogarko, *Nauch, Tekh. Probl. Goreniya Vzryva* **105**(2) (1965).

50. Y. Hashiguchi and T. Fujisaki, *Kogyo Kagaku Zasshi* **61**(6), 515 (1958).
51. P. Hõlemann, R. Hasselmann, and G. Dix, *Forschungsber. Landes Nordrhein Westfalen* **102** (1954).
52. N. A. Copeland and M. A. Youker, "German Techniques for Handling Acetylene in Chemical Operations," *FIAT Final Report No. 720 (PB 20078)*, 1946; Includes translations of reports by Boesler, Rimarski, and Weissweiler.
53. M. Gugger, Chemische Werke Hüls, private communication, 1954.
54. H. Schmidt and K. Haberl, *Tech. Ueberwach.* **7**(12), 423 (1955).
55. W. Rimarski and M. Konschak, *Forschungsarb. Schweissens Schneidens* **9**, 105 (1934).
56. S. M. Kogarko, A. G. Lyamin, and V. A. Mikhailov, *Khim. Prom. Moscow* **41**(8), 621 (1965); *Dokl. Akad. Nauk SSSR* **162**(4), 857 (1965).
57. S. M. Kogarko and B. A. Ivanov, *Dokl. Phys. Chem.* **140**(1), 676 (1961).
58. S. M. Kogarko and co-workers, *Khim. Prom. Moscow* **7**, 496 (1962).
59. E. Barsalou, *Mem. Poudres* **43**, 63 (1961).
60. R. E. Duff, H. T. Knight, and H. R. Wright, *J. Chem. Phys.* **22**, 1618 (1954).
61. E. Penny, *Discuss. Faraday Soc.* **22**, 157 (1956).
62. M. Berthelot and H. LeChatelier, *C. R. Acad. Sci.* **129**, 427 (1899).
63. Ref. 1, p. 495.
64. P. Hölemann, R. Hasselmann, and G. Dix, *Forschungsber. Landes Nordrhein Westfalen,* **382** (1957).
65. M. E. Sutherland and H. W. Wegert, *Chem. Eng. Prog.* **69**(4), 48 (1973).
66. W. Rimarski and M. Konschak, *Forschungsarb. Schweissens Schneidens* **6**, 92 (1931).
67. Ref. 1, p. 496.
68. A. B. Arons, *Underwater Explosion Research*, Vol. 1, U.S. Office of Naval Research, 1950.
69. *Acetylene Transmission for Chemical Synthesis, Pamphlet G 1.3*, Compressed Gas Association, Arlington, Va., 1984.
70. H. Beller, *Weld. J. (Miami, Fla.)* **37**, 1090 (1958).
71. S. M. Kogarko, A. G. Lyamin, and V. A. Mikhailov, *Khim. Prom. Moscow* **4**, 275 (1964).
72. B. A. Ivanov and S. M. Kogarko, *Int. Chem. Eng. Process. Ind.* **4**(4), 670 (1964).
73. G. W. Jones, R. E. Kennedy, and I. Spolan, *U.S., Bur. Mines Rep. Invest.* **4196** (1948).
74. W. Rimarski and M. Konschak, *Forschungsarb. Schweissens Schneidens* **8**, 113 (1933).
75. Ya. M. Landesman, L. M. Savichkaya, and M. A. Glikin, *Khim. Prom. Moscow* **47**(5), 347 (1971).
76. H. F. Calcote and co-workers, *Ind. Eng. Chem.* **44**, 2656 (1952).
77. J. B. Fenn, *Ind. Eng. Chem.* **43**, 2865 (1951).
78. L. E. Bollinger, M. C. Fong, and R. Edse, *Amer. Rocket Soc. J.* **31**, 588 (1961).
79. N. Manson, *Propagation des Détonations et des Déflagrations dans les Mélanges Gazeux*, Office National Etudes Recherches Aéronautiques et Institut Français Pétroles, Paris, 1947.
80. G. B. Kistiakowsky, H. T. Knight, and M. E. Malin, *J. Chem. Phys.* **20**, 884 (1952).
81. D. S. Burgess and co-workers, *U.S. Bur. Mines Rep. Invest.* **7196**, (1968).
82. G. A. Carlson, *Combust. Flame* **21**(3), 383 (1973).
83. J. H. Lee, B. H. K. Lee, and I. Shanfield, *10th Symposium Combustion, University Cambridge, 1964*, pp. 805–813, discussion 813–815 (Pub. 1965).
84. A. Williams and D. B. Smith, *Chem. Rev.* **70**, 267 (1970).
85. H. A. Mayes and H. J. Yallop, *Chem. Eng. London* **1965** (185).
86. D. W. Breck, H. R. Gallisdorfer, and R. P. Hamlen, *J. Chem. Eng. Data* **7**, 281 (1962).

87. C. S. Kim, R. E. Gannon, and S. Ubhayakar, *Proceedings of 87th Annual Meeting*, American Institute of Chemical Engineering, Boston, Mass., Aug. 1979.
88. V. J. Krukonis and R. E. Gannon, *Advances in Chemistry Series.* V 131, American Chemical Society, Washington, D.C., 1974, pp. 29–41.
89. *Chem. Week* **94**, 64 (Jan. 18, 1964).
90. Ref. 1, p. 402.
91. Ref. 1, p. 395.
92. J. E. Anderson and L. K. Case, *Ind. Eng. Chem. Process Des. Dev.* **1**(3), 161 (1962).
93. K. Gehrmann and H. Schmidt, *World Pet. Congr. Proc. 8th* **4**, 379 (1971).
94. H. Hoefermann and co-workers, *Chem. Ind.* **21**, 863 (1969).
95. U.S. Pat. 2,632,731 (Mar. 24, 1953), W. von Ediger (to Technical Assets Inc.).
96. L. Andrussow, *Erdoel Kohle* **12**, 24 (1959).
97. H. Kroepelin and co-workers, *Chem. Ing. Technol.* **28**, 703 (1956).
98. P. J. Leroux and P. M. Mathieu, *Chem. Eng. Prog.* **57**(11), 54 (1961).
99. E. Bartholome, *Chem. Ing. Technol.* **26**, 245 (1954).
100. H. Friz. *Chem. Ing. Technol.* **40**, 999 (1968).
101. U.S. Pat. 3,019,271 (May 18, 1959), F.F.A. Braconier (to Société Belge de l'Azote).
102. G. Fauser, *Chim. Ind. (Milan)* **42**(2), 150 (1960).
103. Brit. Pat. 932,429 (July 24, 1963) (to Montecatini).
104. U.S. Pat. 2,796,951 (June 25, 1957), M. S. P. Bogart (to The Lummus Co.).
105. G. H. Bixler and C. W. Coberly, *Ind. Eng. Chem.* **45**(12), 2596 (1953).
106. J. Happel and L. Kramer, *Ind. Eng. Chem.* **59**(1), 39 (1967).
107. A. Holman, O. A. Rokstad, and A. Solbakken, *Ind. Eng. Chem. Process Des. Dev.* **15**(3), 439 (1976).
108. J. M. Reid and H. R. Linden, *Chem. Eng. Prog.* **56**(1), 47 (1960).
109. *Hydrocarbon Process.* (Nov. 1971).
110. Underwriters Laboratories Inc., *Standards for Acetylene Generators No. 297 Portable Medium Pressure,* May 1973; *No. 408 Stationary Medium Pressure,* May 1973.
111. Federal Specification, *Acetylene Technical Dissolved BB-A-106B*, Washington, D.C.
112. Grimm, "Large-Scale Purification of Carbide Acetylene with Special Regard to the Production of Acetaldehyde and Ethylene," *OTS Report, PB 35209*, U.S. Department of Commerce, 1944.
113. Merkel, "Results of Acetylene Purification at Ludwigshafen," *OTS Report, PB 35211*, U.S. Department of Commerce, 1944.
114. W. E. Alexander, "Purification and Drying of Acetylene for Chemical Use," *OTS Report, PB 44943*, U.S. Department of Commerce, Washington, D.C.
115. R. L. Bond and co-workers, *Nature (London)* **200**(4913), 1313 (Dec. 28, 1963).
116. S. C. Chakravartty, D. Dutta, and A. lahiri, *Fuel* **55**(1), 43 (1976).
117. R. L. Coates, C. L. Chen, and B. J. Pope, *Adv. Chem. Ser.* **131**, 92 (1974).
118. Y. Kawana, *Chem, Econ. Eng. Rev.* **4**(1), (45), 13 (1972).
119. W. R. Ladner and R. Wheatley, *Fuel* **50**(4), 443 (Oct. 1971).
120. R. E. Gannon and V. Krukonis, "Arc-Coal Process Development," *R&*D Report No. 34—Final Report, Contract No. 14-01-0001-493, prepared for Office of Coal Research by AVCO Corp., 1972.
121. *Avco Arc-Coal Acetylene Process Development Program, final report*, contract DE-ACO2-79-C-S40214, prepared for U.S. Dept. of Energy by Avco Systems Division, Apr. 1981.
122. R. Muller and co-workers, *Proceedings of 8th International Symposium on Plasma Chemistry International Union of Pure and Applied Chemistry*, Vol. 2, Tokyo, 1987.
123. W. G. Schepman, paper presented at the *International Symposium Acetylene Association Meeting*, Philadelphia, Pa., 1958.

124. H. Schmidt and K. Haberl, *Tech. Ueberwach.* **423**(12), (1955); A. Ebert, *Explosivstoffe* **44**, 245 (1956).
125. *Acetylene Cylinder Charging Plants*, Pamphlet NFPA SIA, National Fire Prevention Association, Boston, Mass., 1974.
126. *Guidelines for Periodic Visual Inspection and Requalification of Acetylene Cylinders, Pamphlet C-13*, Compressed Gas Association, Arlington, Va., 1992; *Acetylene, Pamphlet G-1*, Compressed Gas Association, Arlington, Va., 1990.
127. Department of Transportation, *Code of Federal Regulations 49*, items 173.303; 173.306; 178.59; and 179.60, Washington, D.C., Oct. 1, 1973.
128. Department of Transportation, *CFR 49(10-1-73)* item 173.30(b), p. 207.
129. O. Horn, *Erdoel Kohle Erdgas Petrochem. Brennst. Chem.* **26**(3), 129 (1973).
130. "Acetylene, Chemical Profile," *Chem. Market Reporter*, Nov. 10, 2001.
131. J. Larson, U. Loechner, and G. Tok, *Chemical Economics Handbook*, SRI International, Menlo Park, Calif., Aug. 2001.
132. E. E. Hughes and R. Gorden Jr., *Anal. Chem.* **31**, 94 (1959).
133. *Test for C_1 through C_5 Hydrocarbons in the Atmosphere by Gas Chromatography*, ASTM Analysis D2820-72, Vol. 26.
134. L. Barnes, Jr. and I. J. Molinini, *Anal. Chem.* **27**, 1025 (1955).
135. J. G. Hanna and S. Siggia, *Anal. Chem.* **21**, 1469 (1949).
136. L. Chelmu, *Chim. Anal. Bucharest* **2**, 212 (1972).
137. T. Carreon, in E. Bingham, B. Cohrssen, and C. H. Powell, eds., *Patty's Toxicology*, 5th ed., Vol. 4, John Wiley & Sons, Inc., New York, 2001, pp. 119–120.
138. U.S. Pat. 2,538,723 (Jan. 16, 1951), O. Fruhwirth and co-workers (to Donau Chemie A. G.).
139. W. Reppe and co-workers, *Justus Liebigs Ann. Chem.* **596**, 1 (1955).
140. *Oxyacetylene Handbook*, 3rd ed., Union Carbide Corp., Linde Division, New York.
141. U.S. Pat. Appl. 20020014226 (Feb. 7, 2002), J. W. Wulff, M. Hulett, and S. Lee (to Lathrop and Gage, LC).
142. S. Borman, *Chem. Eng. News*, **80**(6), 29 (Feb. 11, 2002).

GENERAL REFERENCES

H. K. Kamptner, W. R. Krause, and H. P. Schilken, "Acetylene from Naphtha Pyrolysis," *Chem. Eng. N.Y.*, 80 (Feb. 28, 1966).

V. J. Krukonis and R. E. Gannon, "Deuterium and Carbon 13 Tagging Studies of the Plasma Pyrolysis of Coal," *Adv. Chem. Ser.* **131**, 29 (1974).

R. F. Goldstein and A. L. Waddams, *The Petroleum Chemicals Industry*, 3rd ed., E. & F. N. Spon Ltd., London, 1967, pp. 303–316.

"Acetylene Production Using Hydrogen Plasma," *Oil Gas J.* 82 (Mar. 12, 1973).

"Acetylene: Winning With Wulff?" *Chem. Week*, 89 (Apr. 16, 1966).

"Acetylene and Ethylene Processes—Conference Report," *Chem. Process Eng. (London)*, 101 (May 1968).

R. B. Stobaugh, W. C. Allen Jr., and Van R. H. Sterberg, "Vinyl Acetate: How, Where, Who Future," *Hydrocarbon Process.* **51**(5), 153 (1972).

R. J. Parsons, "Progress Review No. 61: The Use of Plasmas in Chemical Synthesis," *J. Inst. Fuel* **43**(359), 524 (Dec. 1970).

G. Duembgen and co-workers, "Untersuchungen zur Acetylen-Herstellund durch Methanund Leicht-benzin Spaltung," *Chem. Ing. Technol.* **40**, 1004 (1968).

L. S. Lobo and D. L. Trimm, "Carbon Formation from Light Hydrocarbons on Nickel," *J. Catal.* **29**(1), 15 (Apr. 1973).

D. T. Illin and co-workers, "Production of Acetylene by Electrocracking of Natural Gas in a Coaxial Reactor," translated from *J. Appl. Chem. USSR* **42**(3), 648 (1969).

H. K. Kamptner, W. R. Krause, and H. P. Schilken, "HTP: After Five Years," *Hydrocarbon Process Pet. Refiner* **45**(4), 187 (1966).
H. Bockhorn and co-workers, "Production of Acetylene in Premixed Flames and of Acetylene Ethylene Mixtures," *Chem. Ing. Technol.* **44**(14), 869 (1972).
"Thermal Decomposition of Ethane in a Plasma Jet," *Kogyo Kagaku Zasshi* **74**(9), 83 (1971).
"Production of Acetylene and Ethylene by Submerged Combustion," *Khim. Promst. (Moscow)* **49**(5), 330 (1973).
"Wulff Furnaces Make Acetylene, Ethylene," *Oil Gas J.*, 81 (Mar. 12, 1973).
K. Gerhard Baur, "Acetylene From Crude Oil Makes Debut in Italy," *Chem. Eng.*, 82 (Feb. 10, 1969).
"Acetylene Badische Anilin & Soda–Fabrik AG," *Hydrocarbon Process.* 118 (Nov. 1971).
"Acetylene—Wulff Process," *Hydrocarbon Process.* **46**(11), 139 (1967).
"Process Costs, Wulff Acetylene," *Chem. Process. Eng.* **47**(2), 71 (1966).
"Procedeul de obtinere a acetilenei si etenei cu flacara imersata," *Rev. Chim. (Bucharest)* **22**(12), 715 (Dec. 1971).
H. K. Kamptner, W. R. Krause, and H. P. Schilken, "High Temperature Cracking," *Chem. Eng.*, 93 (Feb. 28, 1966).
K. L. Ring and co-workers, *Chem. Ec. Handbook*, SRI International, Menlo Park, Calif., 1994.
Acetylene, report no. 76-2, Chem Systems, Inc., 1976.
H. Witcoff, *Chem. Systems, Inc.*, private communication, Apr. 1994.
T. Wett, "Marathon Tames the Wulffs at Burghausen," *Oil Gas J.* **70**, 101 (Sept. 4, 1972).
H. Gladisch, "Acetylen-Herstellung im elektrishen Lichtbogen," *Chem. Ing. Technol.* **41**(4), 204 (1969).
E. A. Schultz, "Das Marathon-Werk Burghausen," *Erdoel Kohle Erdgas Petrochem.* **21**, 481 (1968).
R. B. Stobaugh, *Petrochemical Manufacturing* & Marketing Guide, Gulf Publishing Co., Houston, Tex., 1966, pp. 1–17.

Richard E. Gannon
Textron Defense Systems
Robert M. Manyik
Union Carbide Corporation
C. M. Dietz
H. B. Sargent
R. O. Thribolet
R. P. Schaffer
Consultants

ACETYLENE-DERIVED CHEMICALS

1. Introduction

Acetylene [74-86-2], C_2H_2, is an extremely reactive hydrocarbon, principally used as a chemical intermediate (see Acetylene). Because of its thermodynamic instability, it cannot easily or economically be transported for long distances. To avoid large free volumes or high pressures, acetylene cylinders contain a porous solid packing and an organic solvent. Acetylene pipelines are severely restricted in size and must be used at relatively low pressures. Hence, for large-scale operations, the acetylene consumer must be near the place of acetylene manufacture.

Historically, the use of acetylene as a raw material for chemical synthesis has depended strongly on the availability of alternative raw materials. The United States, which until recently appeared to have limitless stocks of hydrocarbon feeds, has never depended on acetylene to the same extent as Germany, which had more limited access to hydrocarbons (1). During World War I the first manufacture of a synthetic rubber was undertaken in Germany to replace imported natural rubber, which was no longer accessible. Acetylene derived from calcium carbide was used for preparation of 2,3-dimethyl-1,3-butadiene by the following steps:

$$\text{acetylene} \longrightarrow \text{acetaldehyde} \longrightarrow \text{acetic acid} \longrightarrow \text{acetone} \longrightarrow$$

$$\text{2,3-dimethyl-2,3-butanediol} \longrightarrow \text{2,3-dimethyl-1,3-butadiene}$$

Methyl rubber, obtained by polymerization of this monomer, was expensive and had inferior properties, and its manufacture was discontinued at the end of World War I. By the time World War II again shut off access to natural rubber, Germany had developed better synthetic rubbers based upon butadiene [106-99-0] (see ELASTOMERS, SYNTHETIC SURVEY).

In the United States butadiene was prepared initially from ethanol and later by cracking four-carbon hydrocarbon streams (see BUTADIENE). In Germany, butadiene was prepared from acetylene via the following steps: acetylene ⟶ acetaldehyde ⟶ 3-hydroxybutyraldehyde ⟶ 1,3-butanediol ⟶ 1,3-butadiene.

Toward the end of the war, an alternative German route to butadiene was introduced, which required much less acetylene:

$$\underset{\text{acetylene}}{HC{\equiv}CH} + \underset{\text{formaldehyde}}{2\,CH_2O} \longrightarrow \underset{\text{1,4-butynediol}}{HOCH_2C{\equiv}CCH_2OH} \longrightarrow$$

$$\underset{\text{1,4-butanediol}}{HOCH_2CH_2CH_2CH_2OH} \longrightarrow \underset{\text{1,3-butadiene}}{CH_2{=}CHCH{=}CH_2}$$

Because of its relatively high price, there have been continuing efforts to replace acetylene in its major applications with cheaper raw materials. Such efforts have been successful, particularly in the United States, where ethylene has displaced acetylene as raw material for acetaldehyde, acetic acid, vinyl acetate, and chlorinated solvents. Only a few percent of U.S. vinyl chloride production is still based on acetylene. Propylene has replaced acetylene as feed for acrylates and acrylonitrile. Even some recent production of traditional Reppe acetylene chemicals, such as butanediol and butyrolactone, is based on new raw materials.

2. Reaction Products

2.1. Acetaldehyde. Acetaldehyde [75-07-0], C_2H_4O, (qv) was formerly manufactured principally by hydration of acetylene.

$$HC{\equiv}CH + H_2O \longrightarrow CH_3CHO$$

Many catalytic systems have been described; acidic solutions of mercuric salts are the most generally used. This process has long been superseded by more economical routes involving oxidation of ethylene or other hydrocarbons.

2.2. Acrylic Acid, Acrylates, and Acrylonitrile. Acrylic acid [79-10-7], $C_3H_4O_2$, and acrylates were once prepared by reaction of acetylene and carbon monoxide with water or an alcohol, using nickel carbonyl as catalyst. In recent years this process has been completely superseded in the United States by newer processes involving oxidation of propylene (2). In western Europe, however, acetylene is still important in acrylate manufacture (see ACRYLIC ACID AND DERIVATIVES; ACRYLIC ESTER POLYMERS).

In the presence of such catalysts as a solution of cuprous and ammonium chlorides, hydrogen cyanide adds to acetylene to give acrylonitrile [107-13-1], C_3H_3N (qv).

$$HC{\equiv}CH + HCN \longrightarrow CH_2{=}CHCN$$

Since the early 1970s this process has been completely replaced by processes involving ammoxidation of propylene (3).

2.3. Chlorinated Solvents. Originally, successive chlorination and dehydro-chlorination of acetylene was the route to trichloroethylene [79-01-6], C_2HCl_3, and perchloroethylene [127-18-4], C_2Cl_4.

$$HC{\equiv}CH + 2\,Cl_2 \longrightarrow CHCl_2CHCl_2 \longrightarrow CHCl{=}CCl_2 + HCl$$

$$CHCl{=}CCl_2 + Cl_2 \longrightarrow CHCl_2CCl_3 \longrightarrow CCl_2{=}CCl_2 + HCl$$

This route has been completely displaced, first by chlorination and dehydrochlorination of ethylene or vinyl chloride, and more recently by oxychlorination of two-carbon raw materials (2) (see CHLOROCARBONS AND CHLOROHYDROCARBONS SURVEY).

2.4. Cyclooctatetraene (COT). Tetramerization of acetylene to cyclooctatetraene [629-20-9], C_8H_8, although interesting, does not seem to have been used commercially. Nickel salts serve as catalysts. Other catalysts give benzene. The mechanism of this cyclotetramerization has been studied (4).

2.5. Ethylene. During World War II the Germans manufactured more than 60,000 t/year of ethylene [74-85-1], C_2H_4, by hydrogenation of acetylene, using palladium on silica gel as catalyst. Subsequently, cracking of hydrocarbons displaced this process. However, it is still utilized for purification of ethylene containing small amounts of acetylene as contaminant (5) (see ETHYLENE).

2.6. Vinyl Acetate. Vinyl acetate [108-05-04], $C_4H_6O_2$, used to be manufactured by addition of acetic acid to acetylene.

$$HC{\equiv}CH + CH_3COOH \longrightarrow CH_2{=}CHOOCCH_3$$

Liquid- and vapor-phase processes have been described; the latter appear to be advantageous. Supported cadmium, zinc, or mercury salts are used as catalysts. In 1963 it was estimated that 85% of U.S. vinyl acetate capacity was based on acetylene, but it has been completely replaced since ~1982 by newer technology using oxidative addition of acetic acid to ethylene (2) (see VINYL POLYMERS). In western Europe production of vinyl acetate from acetylene still remains a significant commercial route.

2.7. Vinylacetylene and Chloroprene. In the presence of cuprous salt solutions, acetylene dimerizes to vinylacetylene [689-97-4], C_4H_4. Yields of 87% monovinylacetylene, together with 10% of divinylacetylene, have been described (6).

$$2\,HC{\equiv}CH \longrightarrow HC{\equiv}CCH{=}CH_2$$

Using cuprous chloride as catalyst, hydrogen chloride adds to acetylene, giving 2-chloro-1,3-butadiene [126-99-8], chloroprene, C_4H_5Cl, the monomer for neoprene rubber.

$$HC{\equiv}CCH{=}CH_2 + HCl \longrightarrow CH_2{=}CClCH{=}CH_2$$

Manufacture via this process has been completely replaced by chlorination of butadiene (3) (see CHLOROPRENE; POLYCHLOROPRENE).

2.8. Vinyl Chloride and Vinylidene Chloride. In the presence of mercuric salts, hydrogen chloride adds to acetylene giving vinyl chloride [75-01-4], C_2H_3Cl.

$$HC{\equiv}CH + HCl \longrightarrow CH_2{=}CHCl$$

Once the principal route to vinyl chloride, in all but a few percent of current U.S. capacity this has been replaced by dehydrochlorination of ethylene dichloride. A combined process in which hydrogen chloride cracked from ethylene dichloride was added to acetylene was advantageous but it is rarely used because processes to oxidize hydrogen chloride to chlorine with air or oxygen are cheaper (7) (see VINYL POLYMERS).

In similar fashion, vinylidene chloride [75-35-4], $C_2H_2Cl_2$, has been prepared by successive chlorination and dehydrochlorination of vinyl chloride (see VINYLIDENE CHLORIDE MONOMER AND POLYMERS).

$$CH_2{=}CHCl + Cl_2 \longrightarrow CH_2ClCHCl_2$$

$$CH_2ClCHCl_2 \longrightarrow CH_2{=}CCl_2 + HCl$$

2.9. Vinyl Fluoride. Vinyl fluoride [75-02-5], C_2H_3F, the monomer for poly(vinyl fluoride), is manufactured by addition of hydrogen fluoride to acetylene (see FLUORINE CONTAINING COMPOUNDS, POLY(VINYL FLUORIDE)).

$$HC{\equiv}CH + HF \longrightarrow CH_2{=}CHF$$

3. Ethynylation Reaction Products

The name ethynylation was coined by Reppe to describe the addition of acetylene to carbonyl compounds (8).

$$HC{\equiv}CH + RCOR' \longrightarrow HC{\equiv}CC(OH)RR'$$

Although stoichiometric ethynylation of carbonyl compounds with metal acetylides was known as early as 1899 (9), Reppe's contribution was the development of catalytic ethynylation. Heavy metal acetylides, particularly cuprous acetylide, were found to catalyze the addition of acetylene to aldehydes. Although ethynylation of many aldehydes has been described (10), only formaldehyde has been catalytically ethynylated on a commercial scale. Copper acetylide is not effective as catalyst for ethynylation of ketones. For these, and for higher aldehydes, alkaline promoters have been used.

The following series of reactions illustrates the manufacture of the principal Reppe acetylene chemicals.

$$HC{\equiv}CH + 1 \text{ or } 2\ HCHO \longrightarrow \underset{\text{propargyl alcohol}}{HC{\equiv}CCH_2OH} + \underset{\text{2-butyne-1,4-diol}}{HOCH_2C{\equiv}CCH_2OH}$$

$$HOCH_2C{\equiv}CCH_2OH + H_2 \longrightarrow \underset{\text{2-butene-1,4-diol}}{HOCH_2CH{=}CHCH_2OH}$$

$$HOCH_2C{\equiv}CCH_2OH + 2\ H_2 \longrightarrow \underset{\text{1,4-butanediol}}{HOCH_2CH_2CH_2CH_2OH}$$

$$HOCH_2CH_2CH_2CH_2OH \longrightarrow \underset{\gamma\text{-butyrolactone}}{\text{(cyclic } -O-C{=}O)} + 2H_2$$

γ-butyrolactone

$$\text{γ-butyrolactone} + RNH_2 \longrightarrow \text{(cyclic } -N(R)-C{=}O) + H_2O$$

2-pyrrolidinone (R = H)
1-methyl-2-pyrrolidione (R = CH_3)

$$\text{2-pyrrolidinone (N-H)} + HC{\equiv}CH \longrightarrow \text{(cyclic } -N(CH{=}CH_2)-C{=}O)$$

N-vinyl-2-pyrrolidinone

Except for the pyrrolidinones (see PYRROLE AND PYRROLE DERIVATIVES), these products are discussed in the following.

3.1. Propargyl Alcohol. Propargyl alcohol [107-19-7],2-propyn-1-ol, C_3H_4O, is the only commercially available acetylenic primary alcohol. A colorless, volatile liquid, with an unpleasant odor that has been described as "mild geranium," it was first prepared in 1872 from β-bromoallyl alcohol (11). Propargyl alcohol is miscible with water and with many organic solvents. Physical properties are listed in Table 1.

Reactions. Propargyl alcohol has three reactive sites—a primary hydroxyl group, a triple bond, and an acetylenic hydrogen—making it an extremely versatile chemical intermediate.

The hydroxyl group can be esterified with acid chlorides, anhydrides, or carboxylic acids and it reacts with aldehydes (12) or vinyl ethers (13) in the presence of an acid catalyst to form acetals.

$$RCHO + 2\,HC{\equiv}CCH_2OH \longrightarrow RCH(OCH_2C{\equiv}CH)_2 + H_2O$$

$$CH_2{=}CHOR + HC{\equiv}CCH_2OH \longrightarrow \underset{\displaystyle OR}{CH_3\overset{}{C}HOCH_2C{\equiv}CH}$$

At low temperatures, oxidation with chromic acid gives propynal [624-67-9], C_3H_2O (14), or propynoic acid [471-25-0], $C_3H_2O_2$ (15), which can also be prepared in high yields by anodic oxidation (16).

$$HC{\equiv}CCH_2OH \longrightarrow HC{\equiv}CCHO$$

$$HC{\equiv}CCH_2OH \longrightarrow HC{\equiv}CCOOH$$

Various halogenating agents have been used to replace hydroxyl with chlorine or bromine. Phosphorus trihalides, especially in the presence of pyridine, are particularly suitable (17,18). Propargyl iodide is easily prepared from propargyl bromide by halogen exchange (19).

Table 1. **Physical Properties of Propargyl Alcohol**

Property	Value
melting point, °C	−52
boiling point, °C	114
specific gravity, $d^{20}{}_4$	0.948
refractive index, $n^{20}{}_D$	1.4310
viscosity at 20°C, mPa s (= cP)	1.65
dielectric constant, ε	24.5
specific heat, $C^{20}{}_p$, J/(g K)[a]	2.577
heat of combustion at constant vol, kJ/mol[b]	1731
heat of vaporization at 112°C, kJ/mol[b]	42.09
flash point, Tagliabue open cup, °C	36

[a] To convert J/(g·K) to cal/(g·°C) divide by 4.184.
[b] To convert kJ/mol to kcal/mol divide by 4.184.

Hydrogenation gives allyl alcohol [107-18-6], C_3H_6O, its isomer propanal [123-38-6] (20), or propanol, C_3H_8O [71-23-8] (21). With acidic mercuric salt catalysts, water adds to give acetol, hydroxyacetone, $C_3H_6O_2$ [116-09-6] (22).

$$HC{\equiv}CCH_2OH + H_2O \longrightarrow CH_3COCH_2OH$$

Using alcohols instead of water under similar conditions gives cyclic ketals (23), which can be hydrolyzed to acetol.

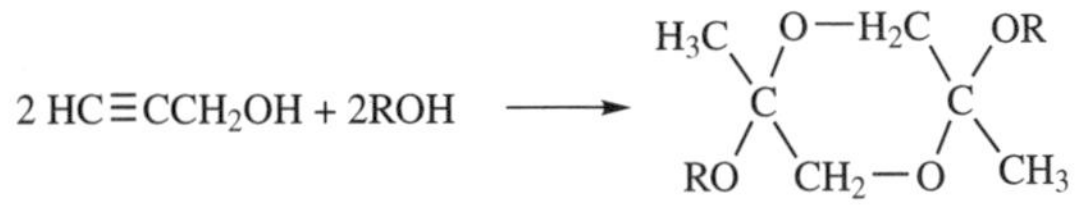

$$2\,HC{\equiv}CCH_2OH + 2ROH \longrightarrow$$

H₃C O—H₂C OR
C C
RO CH₂—O CH₃

Halogens add stepwise, giving almost exclusively dihaloallyl alcohols (24,25).

$$HC{\equiv}CCH_2OH + X_2 \longrightarrow CHX{=}CXCH_2OH$$

A second mole of halogen adds with greater difficulty; oxidative side reactions can be minimized by halogenating an ester instead of the free alcohol (26).

$$(HC{\equiv}CCH_2O)_3PO + 6\,Br_2 \longrightarrow (HCBr_2CBr_2CH_2O)_3PO$$

With mercuric salt catalysts, hydrogen chloride adds to give 2-chloroallyl alcohol, 2-chloroprop-2-en-1-ol [5976-47-6] (27).

$$HC{\equiv}CCH_2OH + HCl \longrightarrow CH_2{=}CClCH_2OH$$

In the presence of suitable nickel or cobalt complexes, propargyl alcohol trimerizes to a mixture of 1,3,5-benzenetrimethanol [4464-18-0] and 1,2,4-trimethanol [25147-76-6] benzene (28).

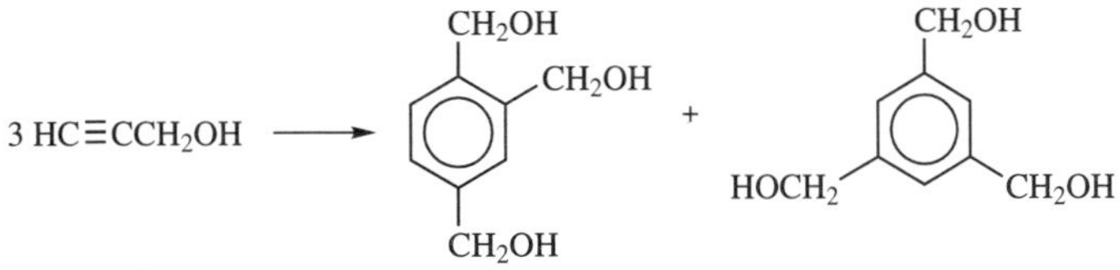

$$3\,HC{\equiv}CCH_2OH \longrightarrow$$

CH₂OH, CH₂OH, CH₂OH (1,2,4-isomer) + CH₂OH, HOCH₂, CH₂OH (1,3,5-isomer)

Cyclization with various nickel complex catalysts gives up to 97% selectivity to a mixture of cyclooctatetraene derivatives, with only 3% of benzene derivatives. The principal isomer is the symmetrical 1,3,5,7-cyclooctatetraene-1,3,5,7-tetramethanol (29).

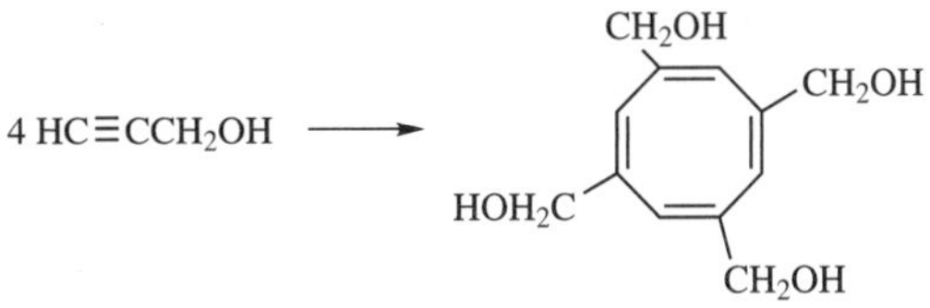

$$4\,HC{\equiv}CCH_2OH \longrightarrow$$

CH₂OH, CH₂OH, HOH₂C, CH₂OH

Nickel halide complexes with amines give mixtures of linear polymer and cyclic trimers (30). Nickel chelates give up to 40% of linear polymer (31).

When heated with ammonia over cadmium calcium phosphate catalysts, propargyl alcohol gives a mixture of pyridines (32).

In the presence of copper acetylide catalysts, propargyl alcohol and aldehydes give acetylenic glycols (33). When dialkylamines are also present, dialkylaminobutynols are formed (34).

$$HC{\equiv}CCH_2OH + RCHO \longrightarrow RCH(OH)C{\equiv}CCH_2OH$$

$$HC{\equiv}CCH_2OH + HCHO + R_2NH \longrightarrow R_2NCH_2C{\equiv}CCH_2OH + H_2O$$

With two equivalents of an organomagnesium halide, a Grignard reagent is formed, capable of use in further syntheses (35,36). Cuprous salts catalyze oxidative dimerization of propargyl alcohol to 2,4-hexadiyne-1,6-diol [3031-68-3] (37).

$$2\,HC{\equiv}CCH_2OH + \frac{1}{2}\,O_2 \longrightarrow HOCH_2C{\equiv}C{-}C{\equiv}CCH_2OH + H_2O$$

Manufacture. Propargyl alcohol is a by-product of butynediol manufacture. The original high pressure butynediol processes gave ~5% of the by-product; newer lower pressure processes give much less. Processes have been described that give much higher proportions of propargyl alcohol (38,39).

BASF produces propargyl alcohol in its plants in New Jersey and Germany. Processes for producing propargyl alcohol using (*1*) 1,2,3-trichloropropane and (*2*) 2, 3-dichloro-1-propanol in an industrially advantageous manner have been described (40).

Shipment, Storage, and Price. Propargyl alcohol is available in tank cars, tank trailers, and drums. It is usually shipped in unlined steel containers and transferred through standard steel pipes or braided steel hoses; rubber is not recommended. Clean, rust-free steel is acceptable for short-term storage. For longer storage, stainless steel (types 304 and 316), glass lining, or phenolic linings (Lithcote LC-19 and LC-24, Unichrome B-124, and Heresite) are suitable. Aluminum, epoxies, and epoxy-phenolics should be avoided.

Specifications and Analytical Methods. The commercial material is specified as 97% minimum purity, determined by gas chromatography or acetylation. Moisture is specified at 0.05% maximum (Karl-Fischer titration). Formaldehyde content is determined by bisulfite titration.

Health and Safety Factors. Although propargyl alcohol is stable, violent reactions can occur in the presence of contaminants, particularly at elevated temperatures. Heating in undiluted form with bases or strong acids should be avoided. Weak acids have been used to stabilize propargyl alcohol prior to distillation. Since its flash point is low, the usual precautions against ignition of vapors should be observed.

Propargyl alcohol is a primary skin irritant and a severe eye irritant and is toxic by all means of ingestion; all necessary precautions must be taken to avoid contact with liquid or vapors. The LD_{50} is 0.07 mL/kg for white rats and 0.06 mL/kg for guinea pigs.

Uses. Propargyl alcohol is a component of oil-well acidizing compositions, inhibiting the attack of mineral acids on steel (see CORROSION AND CORROSION CONTROL). It is also employed in the pickling and plating of metals.

It is used as an intermediate in preparation of the miticide Omite [2312-35-8], 2-(4′-*tert*-butylphenoxy)cyclohexyl 2-propynyl sulfite (41); sulfadiazine [68-35-9] (42); and halogenated propargyl carbonate fungicides (43).

3.2. Butynediol. Butynediol, 2-butyne-1,4-diol, [110-65-6] was first synthesized in 1906 by reaction of acetylene bis(magnesium bromide) with paraformaldehyde (44). It is available commercially as a crystalline solid or a 35% aqueous solution manufactured by ethynylation of formaldehyde. Physical properties are listed in Table 2.

Table 2. **Physical Properties of Butynediol, Butenediol, and Butanediol**

Property	Butynediol	Butenediol	Butanediol
molecular formula	$C_4H_6O_2$	$C_4H_8O_2$	$C_4H_{10}O_2$
CAS Registry Number	[110-65-6]	[110-64-5]	[110-63-4]
melting point, °C	58	11.8	20.2
boiling point, °C at kPa[a]			
0.133	101	84	86
1.33	141	122	123
13.3	194	176	171
101.3	248	234	228
specific gravity	$d_4^{20} = 1.114$	$d_{15}^{25} = 1.070$	$d_4^{20} = 1.017$
refractive index $n^{25}{}_D$	$\alpha = 1.450 \pm 0.002$	1.4770	1.4445
	$\beta = 1.528 \pm 0.002$		
heat of combustion,[b] kJ/mol[c]	2204		
flash point, Tagliabue open cup, °C	152	128	121
viscosity at 20°C, mPa s (=cP)		22	84.9
at 38°C		10.8	
at 99°C		2.5	
surface tension at 20°C, mN/m (=dyn/cm)			44.6
dielectric constant at 20°C, ε			31.5
solubility, g/100 mL solvent at 25°C			
water (0°C)	2		
water	374		miscible
ethanol	83		miscible
acetone	70		miscible
ether	2.6		3.1
benzene	0.04		0.3
hexane			<0.1

[a] To convert kPa, to mm Hg, multiply by 7.5.
[b] At constant volume.
[c] To convert kJ to kcal, divide by 4.184.

Reactions. Butynediol undergoes the usual reactions of primary alcohols. Because of its rigid, linear structure, many reactions forming cyclic products from butanediol or *cis*-butenediol give only polymers with butynediol.

Both hydroxyl groups can be esterified normally (45). The monoesters are readily prepared as mixtures with diesters and unesterified butynediol, but care must be taken in separating them because the monoesters disproportionate easily (46).

The hydroxyl groups can be alkylated with the usual alkylating agents. To obtain aryl ethers a reverse treatment is used, such as treatment of butynediol toluenesulfonate or dibromobutyne with a phenol (45). Alkylene oxides give ether alcohols (47).

In the presence of acid catalysts, butynediol and aldehydes (48) or acetals (49) give polymeric acetals, useful intermediates for acetylenic polyurethanes suitable for high energy solid propellants.

$$HOCH_2C{\equiv}CCH_2OH + CH_2O \longrightarrow HO(CH_2C{\equiv}CCH_2OCH_2O)_nH$$

Electrolytic oxidation gives acetylene dicarboxylic acid [142-45-0] (2-butynedioic acid) in good yields (50); chromic acid oxidation gives poor yields (51). Oxidation with peroxyacetic acid gives malonic acid [141-82-2] (qv) (52).

Butynediol can be hydrogenated partway to butenediol or completely to butanediol.

$$HOCH_2C{\equiv}CCH_2OH \longrightarrow HOCH_2CH{=}CHCH_2OH \longrightarrow HOCH_2CH_2CH_2CH_2OH$$

Noble metal containing hydrogenation catalyst for selective hydrogenation of 1,4-butynediol to 1, 4-butanediol and a process for preparation has been described (53).

Dichlorobutyne [821-10-3] and dibromobutyne [2219-66-1] are readily prepared by treatment with thionyl or phosphorus halides. The less-stable diiodobutyne is prepared by treatment of dichloro- or dibromobutyne with an iodide salt (54).

Addition of halogens proceeds stepwise, sometimes accompanied by oxidation. Iodine forms 2,3-diiodo-2-butene-1,4-diol (55). Depending on conditions, bromine gives 2,3-dibromo-2-butene-1,4-diol, 2,2,3,3-tetrabromobutane-1,4-diol, mucobromic acid, or 2-hydroxy-3,3,4,4-tetrabromotetrahydrofuran (56). Addition of chlorine is attended by more oxidation (57–59), which can be lessened by esterification of the hydroxyl groups.

Uncatalyzed addition of hydrochloric acid is accompanied by replacement of one hydroxyl group, giving high yields of 2,4-dichloro-2-buten-1-ol (60); with mercuric or cupric salt catalysts, addition occurs without substitution (61, 62).

$$HOCH_2C{\equiv}CCH_2OH + 2\ HCl \longrightarrow ClCH_2CH{=}CClCH_2OH + H_2O$$

$$HOCH_2C{\equiv}CCH_2OH \xrightarrow[\text{catalyst}]{\text{HCl}} HOCH_2CCl{=}CHCH_2OH \xrightarrow[\text{catalyst}]{\text{HCl}} HOCH_2CCl_2CH_2CH_2OH$$

When aqueous solutions of sodium bisulfite are heated with butynediol, one or two moles add to the triple bond, forming sodium salts of sulfonic acids (63).

In the presence of mercuric salts, butynediol rapidly isomerizes to 1-hydroxy-3-buten-2-one (64).

$$HOCH_2C{\equiv}CCH_2OH \longrightarrow CH_2{=}CH\overset{O}{\overset{\|}{C}}CH_2OH$$

This adds compounds with active hydrogen such as water, alcohols, and carboxylic acids (65), to give 1,4-dihydroxy-2-butanone or its derivatives.

$$CH_2{=}CH\overset{O}{\overset{\|}{C}}CH_2OH + H_2O \longrightarrow HOCH_2CH_2\overset{O}{\overset{\|}{C}}CH_2OH$$

Butynediol is more difficult to polymerize than propargyl alcohol, but it cyclotrimerizes to hexamethylolbenzene [2715-91-5] (benzenehexamethanol) with a nickel carbonyl–phosphine catalyst (66); with a rhodium chloride–arsine catalyst a yield of 70% is claimed (67).

When heated with acidic oxide catalysts, mixtures of butynediol with ammonia or amines give pyrroles (68) (see PYRROLE AND PYRROLE DERIVATIVES).

$$HOCH_2C{\equiv}CCH_2OH + RNH_2 \longrightarrow \text{N-R pyrrole}$$

Manufacture. All manufacturers of butynediol use formaldehyde ethynylation processes. The earliest entrant was BASF, which, as successor to I. G. Farben, continued operations at Ludwigshafen, FRG, after World War II. Later BASF also set up a U.S. plant at Geismar, La. The first company to manufacture in the United States was GAF in 1956 at Calvert City, Ky., and later at Texas City, Tex., and Seadrift, Tex.

At the end of World War II, the butynediol plant and process at Ludwigshafen were studied extensively (69, 70). Variations of the original high pressure, fixed-bed process, which is described below, are still in use. However, all of the recent plants use low pressures and suspended catalysts (71–77).

The hazards of handling acetylene under pressure must be considered in plant design and construction. Although means of completely preventing acetylene decomposition have not been found, techniques have been developed that prevent acetylene decompositions from becoming explosive. The original German plant was designed for pressures up to 20.26 MPa (200 atm), considered adequate for deflagration (nonexplosive decomposition), which could increase pressures approximately 10-fold. It was not practical to design for control of a detonation (explosive decomposition), which could increase pressure nearly 200-fold (78).

The reactors were thick-walled stainless steel towers packed with a catalyst containing copper and bismuth oxides on a siliceous carrier. This was activated by formaldehyde and acetylene to give the copper acetylide complex that functioned as the true catalyst. Acetylene and an aqueous solution of formaldehyde were passed together through one or more reactors at ~90–100°C and an acetylene partial pressure of ~500–600 kPa (5–6 atm) with recycling as required. Yields of butynediol were over 90%, in addition to 4–5% propargyl alcohol.

Shipment, Storage, and Price. Butynediol, 35% solution, is available in tank cars, tank trailers, and drums. Stainless steel, nickel, aluminum, glass, and various plastic and epoxy or phenolic liners have all been found satisfactory. Rubber hose is suitable for transferring. The solution is nonflammable and freezes at about −5°C.

Butynediol solid flakes are packed in polyethylene bags inside drums. The product is hygroscopic and must be protected from moisture.

Specifications and Analytical Methods. The commercial aqueous solution is specified as 34% minimum butynediol, as determined by bromination or refractive index. Propargyl alcohol is limited to 0.2% and formaldehyde to 0.7%.

The commercial flake is specified as 96.0% minimum butynediol content, with a maximum of 2.0% moisture. Purity is calculated from the freezing point (at least 52°C).

Health and Safety Factors. Although butynediol is stable, violent reactions can take place in the presence of certain contaminants, particularly at elevated temperatures. In the presence of certain heavy metal salts, such as mercuric chloride, dry butynediol can decompose violently. Heating with strongly alkaline materials should be avoided.

Butynediol is a primary skin irritant and sensitizer, requiring appropriate precautions. Acute oral toxicity is relatively high: LD_{50} is 0.06 g/kg for white rats.

Uses. Most butynediol produced is consumed by the manufacturers in manufacture of butanediol and butenediol. Small amounts are converted to ethers with ethylene oxide.

Butynediol is principally used in pickling and plating baths. Small amounts are used in the manufacture of brominated derivatives, useful as flame retardants. It was formerly used in a wild oat herbicide, Carbyne (Barban), 4-chloro-2-butynyl-*N*-(3-chlorophenyl)carbamate [101-27-9], $C_{11}H_9Cl_2NO_2$ (79).

3.3. Butenediol. 2-Butene-1,4-diol [110-64-5] is the only commercially available olefinic diol with primary hydroxyl groups. The commercial product consists almost entirely of the cis isomer.

trans-2-Butene-1,4-diol diacetate was prepared from 1,4-dibromo-2-butene in 1893 (80) and hydrolyzed to the diol in 1926 (81). The original preparation of the cis diol utilized the present commercial route, partial hydrogenation of butynediol.

Physical properties are listed in Table 2. Butenediol is very soluble in water, lower alcohols, and acetone. It is nearly insoluble in aliphatic or aromatic hydrocarbons.

Reactions. In addition to the usual reactions of primary hydroxyl groups and of double bonds, *cis*-butenediol undergoes a number of cyclization reactions.

The hydroxyl groups can be esterified normally: the interesting diacrylate monomer (82) and the biologically active haloacetates (83) have been prepared in this manner. Reactions with dibasic acids have given polymers capable of being cross-linked (84) or suitable for use as soft segments in polyurethanes (85). Polycarbamic esters are obtained by treatment with a diisocyanate (86) or via the bischloroformate (87).

$$HOCH_2CH{\equiv}CHCH_2OH + R(NCO)_2 \longrightarrow H(OC_4H_6O\overset{O}{\overset{\|}{C}}{-}NHRNH\overset{O}{\overset{\|}{C}}{-})_nOC_4H_6OH$$

$$HOCH_2CH{=}CHCH_2OH + COCl_2 \longrightarrow ClOCH_2CH{=}CHCH_2OCCl + \xrightarrow{R(NH_2)_2} H_2NRNH(\overset{O}{\overset{\|}{C}}OC_4H_6O\overset{O}{\overset{\|}{C}}NHRNH)_nH$$

The hydroxyl groups can be alkylated in the usual manner. Hydroxyalkyl ethers may be prepared with alkylene oxides and chloromethyl ethers by reaction with formaldehyde and hydrogen chloride (88). The terminal chlorides can be easily converted to additional ether groups.

$$HOCH_2CH{=}CHCH_2OH + HCHO + HCl \longrightarrow$$

$$ClCH_2OCH_2CH{=}CHCH_2OCH_2Cl \xrightarrow{NaOR} ROCH_2OCH_2CH{=}CHCH_2OCH_2OR$$

cis-Butenediol reacts readily with aldehydes (89), vinyl ethers (90), or dialkoxyalkanes (91) in the presence of acidic catalysts to give seven-membered cyclic acetals (4,7-dihydro-1,3-dioxepins).

$$HOCH_2CH{=}CHCH_2OH + RCHO \longrightarrow \overline{RCHOCH_2CH{=}CHCH_2O} + H_2O$$

$$HOCH_2CH{=}CHCH_2OH + ROCH{=}CH_2 \longrightarrow CH_3\overline{CHOCH_2CH{=}CHCH_2O} + ROH$$

$$HOCH_2CH{=}CHCH_2OH + RR'C(OR'')_2 \longrightarrow RR'\overline{COCH_2CH{=}CHCH_2O} + R''OH$$

The hydroxyl groups of butenediol are replaced by halogens by treatment with thionyl chloride or phosphorus tribromide (92, 93); by stopping short of total halogenation, mixtures can be obtained containing 4-halobutanols as the major constituent (94). The hydroxyl groups undergo typical allylic reactions such as being replaced by cyanide with cuprous cyanide as catalyst (95).

With a palladium chloride catalyst, butenediol is carbonylated by carbon monoxide, giving 3-hexenedioic acid [4436-74-2], $C_6H_8O_4$ (96).

$$HOCH_2CH{=}CHCH_2OH + 2\,CO \longrightarrow HOOCCH_2CH{=}CHCH_2COOH$$

An early attempt to hydroformylate butenediol using a cobalt carbonyl catalyst gave tetrahydro-2-furanmethanol (97), presumably by allylic rearrangement to 3-butene-1,2-diol before hydroformylation. Later, hydroformylation of butenediol diacetate with a rhodium complex as catalyst gave the acetate of 3-formyl-3-buten-1-ol (98). Hydrogenation in such a system gave 2-methyl-1,4-butanediol (99).

Heating with cuprous chloride in aqueous hydrochloric acid isomerizes 2-butene-1,4-diol to 3-butene-1,2-diol (100)] Various hydrogen-transfer catalysts isomerize it to 4-hydroxybutyraldehyde [25714-71-0], $C_4H_8O_2$ (101), acetals of which are found as impurities in commercial butanediol and butenediol.

$$HOCH_2CH{=}CHCH_2OH \longrightarrow HOCH_2CH_2CH_2CHO$$

Treatment with acidic catalysts dehydrates *cis*-butenediol to 2,5-dihydrofuran [1708-29-8], C_4H_6O (102). Cupric (103) or mercuric (104) salts give 2,5-divinyl-1,4-dioxane [21485-51-8], presumably via 3-butene-1,2-diol.

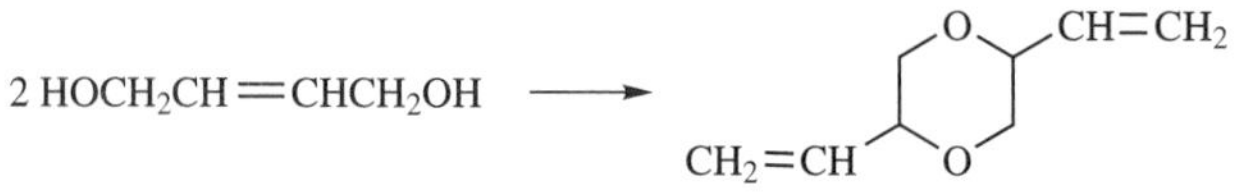

$HOCH_2CH{=}CHCH_2OH \longrightarrow$ [2,5-dihydrofuran ring structure]

$2\ HOCH_2CH{=}CHCH_2OH \longrightarrow$ [1,4-dioxane ring with CH=CH$_2$ and CH$_2$=CH substituents]

Mixtures of butenediol and ammonia or amines cyclize to pyrrolines when heated with acidic catalysts (68).

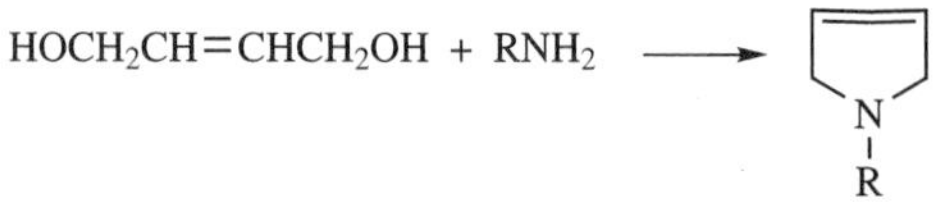

$HOCH_2CH{=}CHCH_2OH + RNH_2 \longrightarrow$ [N-R pyrroline ring structure]

Halogens add to butenediol, giving 2,3-dihalo-1,4-butanediol (92, 93). In a reaction typical of allylic alcohols, hydrogen halides cause substitution of halogen for hydroxyl (105).

When butenediol is treated with acidic dichromate solution, dehydration and oxidation combine to give a high yield of furan [110-00-9], $C_4H_4O_2$ (106) (see FURAN DERIVATIVES).

Treatment with hydrogen peroxide converts butenediol to 2,3-epoxy-1,4-butanediol (107) or gives hydroxylation to erythritol [149-32-6], $C_4H_{10}O_4$ (108). Under strongly acidic conditions, tetrahydro-3,4-furanediol is the principal product (109).

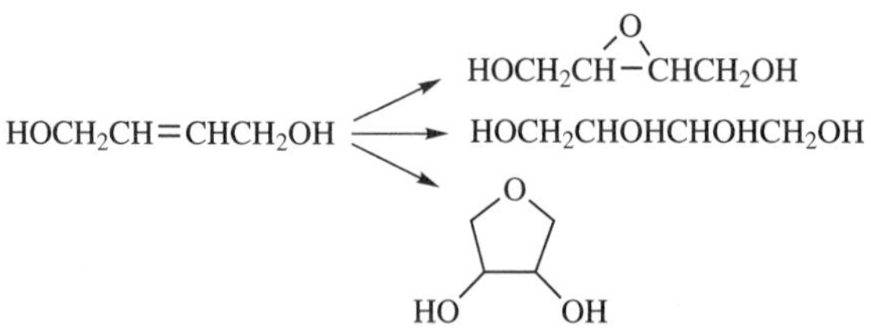

$HOCH_2CH{=}CHCH_2OH \longrightarrow$ $HOCH_2CH\!-\!CHCH_2OH$ (epoxide, O bridging) ; $HOCH_2CHOHCHOHCH_2OH$; [tetrahydrofuran ring with HO and OH substituents]

Butenediol is a weak dienophile in Diels-Alder reactions. Adducts have been described with anthracene (110) and with hexachlorocyclopentadiene (111).

Butenediol does not undergo free-radical polymerization. A copolymer with vinyl acetate can be prepared with a low proportion of butenediol (112).

Manufacture. Butenediol is manufactured by partial hydrogenation of butynediol. Although suitable conditions can lead to either cis or trans isomers (113), the commercial product contains almost exclusively *cis*-2-butene-1,4-diol. Trans isomer, available at one time by hydrolysis of 1,4-dichloro-2-butene, is unsuitable for the major uses of butenediol involving Diels-Alder reactions. The liquid-phase heat of hydrogenation of butynediol to butenediol is 156 kJ/mol (37.28 kcal/mol) (114).

The original German process used either carbonyl iron or electrolytic iron as hydrogenation catalyst (115). The fixed-bed reactor was maintained at 50–100°C and 20.26 MPa (200 atm) of hydrogen pressure, giving a product containing substantial amounts of both butynediol and butanediol. Newer, more selective processes use more active catalysts at lower pressures. In particular, supported palladium, alone (50) or with promoters (116, 117), has been found useful.

Shipment, Storage, and Price. Butenediol is available in unlined steel tank cars, tank trailers, and various sized drums. Because of its relatively high freezing point, tank cars are fitted with heating coils.

Specifications and Analytical Methods. Purity is determined by gas chromatography (gc). Technical grade butenediol, specified at 95% minimum, is typically 96–98% butenediol. The cis isomer is the predominant constituent; 2–4% is trans.

Principal impurities are butynediol (specified as 2.0% maximum, typically <1%), butanediol, and the 4-hydroxybutyraldehyde acetal of butenediol. Moisture is specified at 0.75% maximum (Karl-Fischer titration). Typical technical grade butenediol freezes at ~8°C.

Health and Safety Factors. Butenediol is noncorrosive and stable under normal handling conditions. It is a primary skin irritant but not a sensitizer; contact with skin and eyes should be avoided. It is much less toxic than butynediol. The LD_{50} is 1.25 mL/kg for white rats and 1.25–1.5 mL/kg for guinea pigs.

Uses. Butanediol is used to manufacture the insecticide Endosulfan, other agricultural chemicals, and pyridoxine (vitamin B_6) (see VITAMINS) (118). Small amounts are consumed as a diol by the polymer industry.

3.4. Butanediol. 1,4-Butanediol [110-63-4], tetramethylene glycol, 1,4-butylene glycol, was first prepared in 1890 by acid hydrolysis of *N,N′*-dinitro-1,4-butanediamine (119). Other early preparations were by reduction of succinaldehyde (120) or succinic esters (121) and by saponification of the diacetate prepared from 1,4-dihalobutanes (122). Catalytic hydrogenation of butynediol, now the principal commercial route, was first described in 1910 (123). Other processes used for commercial manufacture are described in the section on Manufacture. Physical properties of butanediol are listed in Table 2.

Reactions. The chemistry of butanediol is determined by the two primary hydroxyls. Esterification is normal. It is advisable to use nonacidic catalysts for esterification and transesterification (124) to avoid cyclic dehydration. When carbonate esters are prepared at high dilutions, some cyclic ester is formed; more concentrated solutions give a polymeric product (125). With excess phosgene the useful bischloroformate can be prepared (126).

$$HO(CH_2)_4OH + COCl_2 \longrightarrow \text{(cyclic carbonate, 7-membered ring: }O{-}C(=O){-}O\text{)} + H[O(CH_2)_4OC(=O)]_nOH$$

$$HO(CH_2)_4OH + 2COCl_2 \longrightarrow ClC(=O)O(CH_2)_4OC(=O)Cl$$

Ethers are formed in the usual way (127). The bis(chloromethyl) ether is obtained using formaldehyde and hydrogen chloride (88).

$$HO(CH_2)_4OH + 2\,HCHO + 2\,HCl \longrightarrow ClCH_2O(CH_2)_4OCH_2Cl + 2\,H_2O$$

With aldehydes or their derivatives, butanediol forms acetals, either seven-membered rings (1,3-dioxepanes) or linear polyacetals; the rings and chains are easily intraconverted (128, 129).

$$HO(CH_2)_4OH + RCHO \longrightarrow \text{(7-membered ring: }O{-}CH(R){-}O\text{)} + H[O(CH_2)_4OCHR]_nOH$$

Heating butanediol with acetylene in the presence of an acidic mercuric salt gives the cyclic acetal expected from butanediol and acetaldehyde (130).

A commercially important reaction is with diisocyanates to form polyurethanes (131) (see URETHANE POLYMERS).

$$HO(CH_2)_4OH + R(NCO)_2 \longrightarrow H[O(CH_2)_4OC(=O)NHRNHC(=O)]_nO(CH_2)_4OH$$

Thionyl chloride readily converts butanediol to 1,4-dichlorobutane [110-56-5] (132) and hydrogen bromide gives 1,4-dibromobutane [110-52-1] (133). A procedure using 48% HBr with a Dean-Stark water trap gives good yields of 4-bromobutanol [33036-62-3], free of diol and dibromo compound (134).

With various catalysts, butanediol adds carbon monoxide to form adipic acid. Heating with acidic catalysts dehydrates butanediol to tetrahydrofuran [109-99-9], C_4H_8O (see FURAN DERIVATIVES). With dehydrogenation catalysts, such as copper chromite, butanediol forms butyrolactone (135). With certain cobalt catalysts both dehydration and dehydrogenation occur, giving 2,3-dihydrofuran (136).

$$HO(CH_2)_4OH \xrightarrow{-H_2O} \text{(tetrahydrofuran)}$$

$$HO(CH_2)_4OH \xrightarrow{-H_2} \text{(butyrolactone, ring }O{-}C{=}O\text{)}$$

$$HO(CH_2)_4OH \xrightarrow{-H_2O,\ +H_2} \text{(2,3-dihydrofuran)}$$

Heating butanediol or tetrahydrofuran with ammonia or an amine in the presence of an acidic heterogeneous catalyst gives pyrrolidines (137,138). With a dehydrogenation catalyst, one or both of the hydroxyl groups are replaced by amino groups (139).

$$HO(CH_2)_4OH + R_2NH \longrightarrow \text{mixture of } R_2N(CH_2)_4OH \text{ and } R_2N(CH_2)_4NR_2$$

With an acidic catalyst, butanediol and hydrogen sulfide give tetrahydrothiophene [110-01-0], C_4H_8S (140).

$$HO(CH_2)_4OH + H_2S \longrightarrow \text{(tetrahydrothiophene)}$$

Vapor-phase oxidation over a promoted vanadium pentoxide catalyst gives a 90% yield of maleic anhydride [108-31-6] (141). Liquid-phase oxidation with a supported palladium catalyst gives 55% of succinic acid [110-15-6] (142).

Manufacture. Most butanediol is manufactured in Reppe plants via hydrogenation of butynediol. Recently, an alternative route involving acetoxylation of butadiene has come on stream and, more recently, a route based upon hydroformylation of allyl alcohol.

Another process, involving chlorination of butadiene, hydrolysis of the dichlorobutene, and hydrogenation of the resulting butenediol, was practiced by Toyo Soda in Japan until the mid-1980s (143).

Reppe Process

$$HC{\equiv}CH + 2\ HCHO \longrightarrow HOCH_2C{\equiv}CCH_2OH \xrightarrow{H_2} HOCH_2CH_2CH_2CH_2OH$$

Acetoxylation Process

$$H_2C{\equiv}CHCH{=}CH_2 + 2\ CH_3COOH \xrightarrow{O_2}$$

$$CH_3\overset{O}{\overset{\|}{C}}OCH_2CH{=}CHCH_2O\overset{O}{\overset{\|}{C}}CH_3 + H_2O \xrightarrow{\text{hydrogenation}} \xrightarrow{\text{hydrolysis}} HOCH_2CH_2CH_2CH_2OH$$

Hydroformylation Process

$$CH_3\overset{O}{CH{-}CH_2} \longrightarrow HC_2{=}CHCH_2OH \xrightarrow{H_2,\ CO}$$

$$HOCH_2CH_2CH_2CHO \xrightarrow{\text{hydrogenation}} HOCH_2CH_2CH_2CH_2OH$$

Economic Aspects. Table 3 lists the U.S. producers of 1,4-butanediol and their capacities. All current production except for Lyondell and BP Amoco is

Table 3. **U.S. Producers of 1,4-Butanediol and their Capacities**[a]

Producer	Capacity, $\times 10^6$ kg ($\times 10^6$ lb)
BASF, Geismar, La.	127 (280)
BP Amoco, Lima, Ohio	64 (140)
DuPont, LaPorte, Tex.	102 (225)
ISP, Texas City, Tex.	29 (65)
Lyondel, Channelview, Tex.	54 (120)
Total	*376* (*830*)

[a] From Ref. 144.

based on the traditional method of reacting acetylene and formaldehyde (Reppe process). Lyondell uses propylene oxide. BPAmoco's process is based in butane oxidation.

BASF is planning a plan to produce 90×10^6 kg (200×10^6 lb) per year. It would use butane/maleic anhydride as feed.

Demand for butanediol in 1998 was 293×10^6 kg (645×10^6 lb). The projected demand for 2003 is 363×10^6 kg (800×10^6 lb). Growth rate is expected at the rate of 4.5% through 2003 (144).

Current high price is $0.90/lb tank, fob, frt, equald. Current low price is $0.65/lb same basis. Current spot pricing is $0.50–0.60/lb, same basis.

Shipment and Storage. Tank cars and tank trailers, selected to prevent color formation, are of aluminum or stainless steel, or lined with epoxy or phenolic resins; drums are lined with phenolic resins. Flexible stainless steel hose is used for transfer. Because of butanediol's high freezing point (~20°C) tank car coil heaters are provided.

Specifications and Analytical Methods. Butanediol is specified as 99.5% minimum pure, determined by gc, solidifying at 19.6°C minimum. Moisture is 0.04% maximum, determined by Karl-Fischer analysis (directly or of a toluene azeotrope). The color is APHA 5 maximum, and the Hardy color (polyester test) is APHA 200 maximum. The carbonyl number is 0.5 mg KOH/g maximum; the acetal content can also be measured directly by gc.

Health and Safety Factors. Butanediol is much less toxic than its unsaturated analogous. It is neither a primary skin irritant nor a sensitizer. Because of its low vapor pressure, there is ordinarily no inhalation problem. As with all chemicals, unnecessary exposure should be avoided. The LD_{50} for white rats is 1.55 g/kg.

Uses. The largest uses of butanediol are internal consumption in manufacture of tetrahydrofuran (45%) and butyrolactone (22%). The largest merchant uses are for poly(butylene terephthalate) resins 24% (see POLYESTERS, THERMOPLASTIC) and in polyurethanes 5%, both as a chain extender and as an ingredient in a hydroxyl-terminated polyester used as a macroglycol. miscellaneous uses account for 4% of consumption and include uses as a solvent, as a coating resin raw material, and as an intermediate in the manufacture of other chemicals and pharmaceuticals.

3.5. Butyrolactone.

γ-Butyrolactone [96-48-0], dihydro-2(3*H*)-furanone, was first synthesized in 1884 via internal esterification of 4-hydroxybutyric acid

Table 4. **Physical Properties of Butyrolactone**

Property	Value
freezing point, °C	−44
boiling point, °C at kPa[a]	
0.133	35
1.33	77
13.3	134
101.3	204
specific gravity, $d^{20}{}_4$	1.129
$d^{25}{}_4$	1.125
refractive index, $n^{20}{}_D$	1.4362
$n^{25}{}_D$	1.4348
viscosity at 25°C, mPa s (= cP)	1.75
dielectric constant at 20°C, ε	39.1
heat capacity at 20°C, J/(g K)[b]	1.60
critical temperature, °C	436
critical pressure, MPa[c]	3.43
flash point, Tagliabue open cup, °C	98

[a] To convert kPa to mm Hg multiply by 7.5.
[b] To convert J to cal divide by 4.184.
[c] To convert MPa to atm divide by 0.1013.

(145). In 1991 the principal commercial source of this material is dehydrogenation of butanediol. Manufacture by hydrogenation of maleic anhydride (146) was discontinued in the early 1980s and resumed in the late 1980s. Physical properties are listed in Table 4.

Butyrolactone is completely miscible with water and most organic solvents. It is only slightly soluble in aliphatic hydrocarbons. It is a good solvent for many gases, for most organic compounds, and for a wide variety of polymers.

Reactions. Butyrolactone undergoes the reactions typical of γ-lactones. Particularly characteristic are ring openings and reactions in which ring oxygen is replaced by another heteroatom. There is also marked reactivity of the hydrogen atoms alpha to the carbonyl group.

Hydrolysis in neutral aqueous solutions proceeds slowly at room temperature and more rapidly at acidic conditions and elevated temperatures. The hydrolysis–esterification reaction is reversible. Under alkaline conditions hydrolysis is rapid and irreversible. Heating the alkaline hydrolysis product at 200–250°C gives 4,4′-oxydibutyric acid [7423-25-8] after acidification (147).

$$\text{butyrolactone} + NaOH \longrightarrow HO(CH_2)_3COONa \xrightarrow{\text{heat}} \xrightarrow{\text{acid}} O(CH_2CH_2CH_2COOH)_2$$

With acid catalysts, butyrolactone reacts with alcohols rapidly even at room temperature, giving equilibrium mixtures consisting of esters of 4-hydroxybutyric acid [591-81-1] with unchanged butyrolactone as the main component. Attempts to distill such mixtures ordinarily result in complete reversal to butyrolactone and alcohol. The esters can be separated by a quick flash distillation at high vacuum (148).

$$\text{butyrolactone} + ROH \rightleftharpoons HO(CH_2)_4COOR$$

When butyrolactone and alcohols are heated for long times and at high temperatures in the presence of acidic catalysts, 4-alkoxybutyric esters are formed. With sodium alkoxides, sodium 4-alkoxybutyrates are formed (149).

$$\text{butyrolactone} + 2\,ROH \longrightarrow RO(CH_2)_3COOR + H_2O$$

$$\text{butyrolactone} + NaOR \longrightarrow RO(CH_2)_3COONa$$

Butyrolactone and hydrogen sulfide heated over an alumina catalyst result in replacement of ring oxygen by sulfur (150).

$$\text{butyrolactone} + H_2S \longrightarrow \text{thiobutyrolactone}$$

Heating butyrolactone with bromine at 160–170°C gives a 70% yield of α-bromobutyrolactone (151). With phosphorus tribromide as catalyst, bromination is accelerated, giving 2,4-dibromobutyric acid, which dehydrobrominates to α-bromobutyrolactone when distilled (152). Chlorination gives α-position monochlorination at 110–130°C and α-dichlorination at 190–200°C (153).

$$\text{butyrolactone} + Br_2 \xrightarrow{PBr_3} BrCH_2CH_2CHBrCOOH \xrightarrow[\text{dist.}]{-HBr} \text{α-bromobutyrolactone}$$

The α-methylene group of butyrolactone condenses easily with a number of different types of carbonyl compounds; eg, sodium alkoxides catalyze self-condensation to α-dibutyrolactone (154), benzaldehyde gives α-benzylidenebutyrolactone (155), and ethyl acetate gives α-acetobutyrolactone (156).

$$2\,\text{butyrolactone} \longrightarrow \text{α-dibutyrolactone (C=C linked)}$$

$$\text{butyrolactone} + C_6H_5CHO \longrightarrow C_6H_5CH{=}C\text{(butyrolactone ring)}$$

$$C{=}O + CH_3COOC_2H_5 \longrightarrow \overset{O}{\overset{\|}{C}}{-}CH_3,\ C{=}O$$

The α-acetobutyrolactone, with or without isolation, can be used in the preparation of various 5-substituted 2-butanone derivatives, presumably by decarboxylation of the acetoacetic acid obtained by ring hydrolysis. Simple hydrolysis gives 5-hydroxybutan-2-one (157) and acidolysis with hydrochloric acid gives 5-chlorobutan-2-one in good yields (158).

The α-methylene groups also add to double bonds; eg, 1-decene at 160°C gives up to 80% of α-decylbutyrolactone (159). With photochemical initiation similar additions take place at room temperature (160).

$$C{=}O + CH_2{=}CHR \longrightarrow CH_2CH_2R,\ C{=}O$$

With Friedel-Crafts catalysts, butyrolactone reacts with aromatic hydrocarbons. With benzene, depending on experimental conditions, either phenylbutyric acid or 1-tetralone can be prepared (161).

$$C{=}O + C_6H_6 \longrightarrow C_6H_5(CH_2)_3COOH \quad \text{or} \quad C{=}O$$

Carbonylation of butyrolactone using nickel or cobalt catalysts gives high yields of glutaric acid [110-94-1] (162).

$$O + CO + H_2O \longrightarrow HOOC(CH_2)_3COOH$$

A series of ring-opening reactions are frequently unique and synthetically useful. Butyrolactone and anhydrous hydrogen halides give high yields of 4-halobutyric acids (163).In the presence of alcohols, esters are formed.

$$O + HX \longrightarrow X(CH_2)_3COOH \qquad X = Cl, Br, \text{ or } I$$

Phosgene (164) or thionyl chloride in the presence of an acid catalyst (165) gives good yields of 4-chlorobutyryl chloride. Heating butyrolactone and thionyl chloride in an alcohol gives good yields of 4-chlorobutyric esters (166).

Butyrolactone with sodium sulfide or hydrosulfide forms 4,4′-thiodibutyric acid (167); with sodium disulfide, the product is 4,4′-dithiodibutyric acid (168).

$$\text{(butyrolactone)} + Na_2S_x \longrightarrow S_x(CH_2CH_2CH_2COONa)_, \quad x = 1 \text{ or } 2$$

Salts of thiols (169) or of sulfinic acids (170) react like the alkoxides, giving 4-alkylthio- or 4-alkylsulfono-substituted butyrates. Alkali cyanides give 4-cyanobutyrates (171), hydroxylamine gives a hydroxamic acid (172), and hydrazine a hydrazide (173).

$$\text{(butyrolactone)} + NaCN \longrightarrow NC(CH_2)_3COONa$$

$$\text{(butyrolactone)} + HONH_2 \longrightarrow HO(CH_2)_3CONHOH$$

$$\text{(butyrolactone)} + NH_2NH_2 \longrightarrow HO(CH_2)_3CONHNH_2$$

Butyrolactone reacts rapidly and reversibly with ammonia or an amine forming 4-hydroxybutyramides (174), which dissociate to the starting materials when heated. At high temperatures and pressures the hydroxybutyramides slowly and irreversibly dehydrate to pyrrolidinones (175). A copper-exchanged Y-zeolite (176) or magnesium silicate (177) is said to accelerate this dehydration.

$$\text{(butyrolactone)} + RNH_2 \rightleftharpoons HO(CH_2)_3CONHR \longrightarrow \text{(N-R pyrrolidinone)} + H_2O$$

Manufacture. Butyrolactone is manufactured by dehydrogenation of butanediol. The butyrolactone plant and process in Germany, as described after World War II (178), approximates the processes presently used. The dehydrogenation was carried out with preheated butanediol vapor in a hydrogen carrier over a supported copper catalyst at 230–250°C. The yield of butyrolactone after purification by distillation was ∼90%. Preparation of γ-butyrolactone by catalytic hydrogenation of maleic anhydride has been described (179).

Shipment and Storage. Butyrolactone is shipped in unlined steel tank cars and plain steel drums. Plain steel, stainless steel, aluminum, and nickel are suitable for storage and handling; rubber, phenolics, and epoxy resins are not suitable. Butyrolactone is hygroscopic and should be protected from moisture. Because of its low freezing point (−44°C), no provision for heating storage vessels is needed.

γ-Butyrolactone is produced by BASF, Alfa Resan, The chemical Co., Penta Manufacturing Co., Quaked City Chemicals, and Vopak U.S.A.

Specifications and Analytical Methods. Purity is specified as 99.5% minimum, by gc area percentage, with a maximum of 0.1% moisture by Karl-Fischer titration. Color, as delivered, is 40 APHA maximum; samples may darken on long storage.

Health and Safety Factors. Butyrolactone is neither a skin irritant nor a sensitizer; however, it is judged to be a severe eye irritant in white rabbits. The acute oral LD_{50} is 1.5 mL/kg for white rats or guinea pigs. Subacute oral feeding studies were carried out with rats and with dogs. At levels up to 0.8% of butyrolactone in the diet there were no toxicologic or pathologic effects in the three months of the test.

Because of its high boiling point (204°C), it does not ordinarily represent a vapor hazard.

Uses. Butyrolactone is principally consumed by the manufacturers by reaction with methylamine or ammonia to produce *N*-methyl-2-pyrrolidinone [872-50-4] and 2-pyrrolidinone [616-45-5], C_4H_7NO, respectively. Considerable amounts are used as a solvent for agricultural chemicals and polymers, in dyeing and printing, and as an intermediate for various chemical syntheses.

4. Other Alcohols and Diols

Secondary acetylenic alcohols are prepared by ethynylation of aldehydes higher than formaldehyde. Although copper acetylide complexes will catalyze this reaction, the rates are slow and the equilibria unfavorable. The commercial products are prepared with alkaline catalysts, usually used in stoichiometric amounts.

Ethynylation of ketones is not catalyzed by copper acetylide, but potassium hydroxide has been found to be effective (180). In general, alcohols are obtained at lower temperatures and glycols at higher temperatures. Most processes use stoichiometric amounts of alkali, but true catalytic processes for manufacture

Table 5. **Commercial Secondary and Tertiary Acetylenic Alcohols and Glycols**

Alcohol or glycol	Molecular formula	CAS Registry No.	Starting material
1-hexyn-3-ol	$C_6H_{10}O$	[105-31-7]	butyraldehyde
4-ethyl-1-octyn-3-ol	$C_{10}H_{18}O$	[5877-42-9]	2-ethylhexanal
2-methyl-3-butyn-2-ol	C_5H_8O	[115-19-5]	acetone
3-methyl-1-pentyn-3-ol	$C_6H_{10}O$	[77-75-8]	methyl ethyl ketone
2,5-dimethyl-3-hexyne-2,5-diol	$C_8H_{14}O_2$	[142-30-3]	acetone
3,6-dimethyl-4-octyne-3,6-diol[a]	$C_{10}H_{18}O_2$	[78-66-0]	methyl ethyl ketone
2,4,7,9-tetramethyl-5-decyne-4,7-diol[a]	$C_{14}H_{26}O_2$	[126-86-3]	methyl isobutyl ketone

[a] These glycols are commercially available as mixtures of diastereoisomers.

Table 6. **Physical Properties of Acetylenic Alcohols**

Property	Hexynol	Ethyl-octynol	Methyl-butynol	Methyl-pentynol
molecular weight	98	154	84	98
freezing point, °C	−80	−45	2.6	−30.6
boiling point, °C	142	197.2	103.6	121.4
specific gravity, d^{20}_{20}	0.882	0.873	0.8672	0.8721
refractive index, n^{20}_{D}	1.4350	1.4502	1.4211	1.4318
viscosity at 20°C, mPa s (=cP)			3.79	2.65[a]
flash point, Tagliabue open cup, °C		83	25	38
water solubility (20°C), wt %	3.8	<0.1	miscible	9.9

[a]At 31°C.

of the alcohols have been described; the glycols appear to be products of stoichiometric ethynylation only.

Table 5 lists the principal commercially available acetylenic alcohols and glycols; Tables 6 and 7 list the physical properties of acetylenic alcohols and glycols, respectively.

4.1. Methylbutynol. 2-Methyl-3-butyn-2-ol [115-19-5], prepared by ethynylation of acetone, is the simplest of the tertiary ethynols, and serves as a prototype to illustrate their versatile reactions. There are three reactive sites, ie, hydroxyl group, triple bond, and acetylenic hydrogen. Although the triple bonds and acetylenic hydrogens behave similarly in methylbutynol and in propargyl alcohol, the reactivity of the hydroxyl groups is very different.

Reactions. As with other tertiary alcohols, esterification with carboxylic acids is difficult and esters are prepared with anhydrides (181), acid chlorides (182), or ketene (183). Carbamic esters may be prepared by treatment with an isocyanate (184) or with phosgene followed by ammonia or an amine (185).

The labile hydroxyl group is easily replaced by treatment with thionyl chloride, phosphorous chlorides, or even aqueous hydrogen halides. At low temperatures aqueous hydrochloric (186) or hydrobromic (187) acids give good yields of 3-halo-3-methyl-1-butynes. At higher temperatures these rearrange, first to 1-halo-3-methyl-1,2-butadienes, then to the corresponding 1,3-butadienes (188,189).

$$HC{\equiv}C\overset{OH}{\overset{|}{C}}(CH_3)_2 \xrightarrow{HX} HC{\equiv}C\overset{X}{\overset{|}{C}}(CH_3)_2 \longrightarrow XCH{=}C{=}C(CH_3)_2 \longrightarrow XCH{=}CH{-}\underset{CH_3}{\underset{|}{C}}{=}CH_2$$

Table 7. **Physical Properties of Acetylenic Glycols**

Property	Dimethyl-hexynediol	Dimethyl-octynediol	Tetramethyl-decynediol
molecular weight	142	170	226
melting point, °C	96–97	49–51	37–38
boiling point, °C	206	222	260
surface tension, mN/m (=dyn/cm) 0.1% in water at 25°C	60.9	55.3	31.6
water solubility (20°C), wt %	27.0	10.5	0.12

With acid catalysts in the liquid (190) or vapor (191) phase, methylbutynol is dehydrated to isopropenylacetylene.

$$HC{\equiv}C\overset{OH}{\overset{|}{C}}(CH_3)_2 \longrightarrow HC{\equiv}C\overset{CH_3}{\overset{|}{C}}{=}CH_2$$

Hydrogenation of methylbutynol gives 2-methyl-3-buten-2-ol and then 2-methylbutan-2-ol in stepwise fashion (192).

$$HC{\equiv}CC(OH)(CH_3)_2 \longrightarrow CH_2{=}CHC(OH)(CH_3)_2 \longrightarrow CH_3CH_2C(OH)(CH_3)_2$$

Acidic mercury salts catalyze hydration to form a ketone (193).

$$HC{\equiv}C\overset{OH}{\overset{|}{C}}(CH_3)_2 + H_2O \longrightarrow CH_3\overset{O}{\overset{\|}{C}}{-}\underset{OH}{\underset{|}{C}}(CH_3)_2$$

Bromination in polar solvents usually gives *trans*-3,4-dibromo-2-methyl-3-buten-2-ol; in nonpolar solvents, with incandescent light, the cis isomer is the principal product (194). Chlorine adds readily up to the tetrachloro stage, but yields are low because of side reactions (195).

$$HC{\equiv}C\overset{OH}{\overset{|}{C}}(CH_3)_2 \longrightarrow CHX{=}CX\overset{OH}{\overset{|}{C}}(CH_3)_2 \longrightarrow CHX_2CX_2\overset{OH}{\overset{|}{C}}(CH_3)_2$$

Upon treatment with suitable cobalt complexes, methylbutynol cyclizes to a 1,2,4-substituted benzene. Nickel complexes give the 1,3,5-isomer (196), sometimes accompanied by linear polymer (25) or a mixture of tetrasubstituted cyclooctatetraenes (26).

$$3\ HC{\equiv}C\overset{OH}{\overset{|}{C}}(CH_3)_2 \longrightarrow \text{1,2,4-}C_6H_3R_3 \text{ or } \text{1,3,5-}C_6H_3R_3$$

$$4\ HC{\equiv}C\overset{OH}{\overset{|}{C}}(CH_3)_2 \longrightarrow C_8H_4R_4 \qquad R = \alpha\text{-hydroxyisopropyl}$$

When bis(π-allyl)nickel is used, only small amounts of cyclic product are obtained and the principal product is formed by addition of one triple bond to another (197).

$$2\ HC{\equiv}C\overset{OH}{\overset{|}{C}}(CH_3)_2 \longrightarrow (CH_3)_2\overset{OH}{\overset{|}{C}}CH{=}CHC{\equiv}C\overset{OH}{\overset{|}{C}}(CH_3)_2$$

With a nickel carbonyl catalyst, hydrochloric acid, and an alcohol the initially formed allenic ester cyclizes on distillation (198).

$$(CH_3)_2C(OH)C{\equiv}CH \longrightarrow (CH_3)_2C{=}C{=}CHCOOR \longrightarrow \text{(5,5-dimethyl-2(5H)-furanone; } H_3C, H_3C \text{ on ring carbon adjacent to O, C=O)}$$

With palladium chloride catalyst, carbon monoxide, and an alcohol the labile hydroxyl is alkylated during carbonylation (199).

$$(CH_3)_2C(OH)C{\equiv}CH \longrightarrow (CH_3)_2C(OR)CH{=}CHCOOR$$

Copper salts catalyze oxidative dimerization to conjugated diynediols in high yields (200).

$$2\,(CH_3)_2C(OH)C{\equiv}CH \longrightarrow (CH_3)_2C(OH)C{\equiv}CC{\equiv}CC(OH)(CH_3)_2$$

Glycols are obtained by treatment with a ketone using alkali as catalyst or with an aldehyde using alkali or copper acetylide as catalyst (201,202).

$$(CH_3)_2C(OH)C{\equiv}CH + RCOR' \longrightarrow (CH_3)_2C(OH)C{\equiv}CC(OH)RR'$$

Hypohalites replace the acetylenic hydrogen with chlorine, bromine, or iodine (203).

$$(CH_3)_2C(OH)C{\equiv}CH + NaOX \longrightarrow (CH_3)_2C(OH)C{\equiv}CX$$

Ethynyl carbinols rearrange to conjugated unsaturated aldehydes. Copper or silver salts catalyze isomerization of the acetate to an allenic acetate, which can be hydrolyzed to an unsaturated aldehyde (204).

$$(CH_3)_2C(OOCCH_3)C{\equiv}CH \xrightarrow{Ag_2CO_3} (CH_3)_2C{=}C{=}CHOOCCH_3$$

Manufacture. In general, manufacture is carried out in batch reactors at close to atmospheric pressure. A moderate excess of finely divided potassium hydroxide is suspended in a solvent such as 1,2-dimethoxyethane. The carbonyl compound is added, followed by acetylene. The reaction is rapid and exothermic. At temperatures $< 5°C$ the product is almost exclusively the alcohol. At 25–30°C the glycol predominates. Such synthesis also proceeds well with non-complexing solvents such as aromatic hydrocarbons, although the conversion is usually lower (205).

Continuous processes have been developed for the alcohols, operating under pressure with liquid ammonia as solvent. Potassium hydroxide (206) or anion exchange resins (207) are suitable catalysts. However, the relatively small manufacturing volumes militate against continuous production. For a while a continuous catalytic plant operated in Ravenna, Italy, designed to produce about 40,000 t/year of methylbutynol for conversion to isoprene (208,209).

A number of secondary and tertiary acetylenic alcohols and glycols are manufactured by Air Products and Chemicals Co.

Health and Safety Factors. Under normal conditions acetylenic alcohols are stable and free of decomposition hazard. The more volatile alcohols present a fire hazard.

The alcohols are toxic orally, through skin absorption, and through inhalation. The secondary alcohols are more toxic than the tertiary. The glycols are relatively low in toxicity.

Compound	LD_{50}, mL/kg (mice)
hexynol	0.175
ethyloctynol	2.1
methylbutynol	2.2
methylpentynol	0.7
tetramethyldecynediol	4.6

Uses. The secondary acetylenic alcohols hexynol and ethyloctynol are used as corrosion inhibitors in oil-well acidizing compositions (see CORROSION AND CORROSION CONTROL). The tertiary alcohols methylbutynol and methylpentynol are used as chemical intermediates, for manufacture of Vitamin A and other products, and in metal plating and pickling operations. Dimethylhexynediol can be used in manufacture of fragrance chemicals and peroxide catalysts. Higher acetylenic glycols and ethoxylated acetylene glycols are useful as surfactants and electroplating additives.

5. Vinylation Reaction Products

Unlike ethynylation, in which acetylene adds across a carbonyl group and the triple bond is retained, in vinylation a labile hydrogen compound adds to acetylene, forming a double bond.

In early work, vinyl chloride had been heated with stoichiometric amounts of alkali alkoxides in excess alcohol as solvent, giving vinyl ethers as products (210). Supposedly, this involved a Williamson ether synthesis, where alkali alkoxide and organic halide gave an ether and alkali halide. However, it was observed that small amounts of acetylene were formed by dehydrohalogenation of vinyl chloride, and that this acetylene was consumed as the reaction proceeded. Hence, acetylene was substituted for vinyl chloride and only catalytic amounts of alkali were used. Vinylation proceeded readily with high yields (211).

Catalytic vinylation has been applied to a wide range of alcohols, phenols, thiols, carboxylic acids, and certain amines and amides. Vinyl acetate is no longer prepared this way in the United States, although some minor vinyl esters such as stearates may still be prepared this way. However, the manufacture of vinyl-pyrrolidinone and vinyl ethers still depends on acetylene.

5.1. *N*-Vinylcarbazole. Vinylation of carbazole proceeds in high yields with alkaline catalysts (212,213). The product, 9-ethenylcarbazole, $C_{14}H_{11}N$ [1484-13-5], forms rigid high melting polymers with outstanding electrical properties.

$$\text{N–H} + HC{\equiv}CH \longrightarrow \text{N–}CH_2{=}CH_2$$

5.2. Neurine. Neurine is trimethylvinylammonium hydroxide, $C_5H_{13}NO$ [463-88-7]. Tertiary amines and their salts vinylate readily at low temperatures with catalysis by free tertiary amines.

$$(CH_3)_3N + HC{\equiv}CH + H_2O \longrightarrow [(CH_3)_3NCH{=}CH_2]^+[OH]^-$$

Above ~50°C tetramethylammonium hydroxide is formed as a by-product; it is the sole product above 100°C (214).

5.3. *N*-Vinyl-2-pyrrolidinone. 1-Ethenyl-2-pyrrolidinone [88-12-0], C_6H_9NO, *N*-vinylpyrrolidinone, was developed by Reppe's laboratory in Germany at the beginning of World War II and patented in 1940 (215).

The major use of vinylpyrrolidinone is as a monomer in manufacture of poly(vinylpyrrolidinone) (PVP) homopolymer and in various copolymers, where it frequently imparts hydrophilic properties. When PVP was first produced, its principal use was as a blood plasma substitute and extender, a use no longer sanctioned. These polymers are used in pharmaceutical and cosmetic applications, soft contact lenses, and viscosity index improvers. The monomer serves as a component in radiation-cured polymer compositions, serving as a reactive diluent that reduces viscosity and increases cross-linking rates (see VINYL POLYMERS, *N*-VINYLAMIDE POLYMERS).

5.4. Vinyl Ethers. The principal commercial vinyl ethers are methyl vinyl ether (methoxyethene, C_3H_6O) [107-25-5]; ethyl vinyl ether (ethoxyethene, C_4H_8O) [104-92-2]; and butyl vinyl ether (1-ethenyloxybutane, $C_6H_{12}O$) [111-34-2]. (See Table 8 for physical properties.) Others such as the isopropyl, isobutyl, hydroxybutyl, decyl, hexadecyl, and octadecyl ethers, as well as the divinyl ethers of butanediol and of triethylene glycol, have been offered as development chemicals (see ETHERS).

Ethyl vinyl ether was the first to be prepared, in 1878, by treatment of diethyl chloroacetal with sodium (216). Methyl vinyl ether was first listed in Reppe patents on vinylation in 1929 and 1930 (210,211).

Reactions. Vinyl ethers undergo all of the expected reactions of olefinic compounds plus a number of reactions that are both useful and unusual.

With a suitable catalyst, usually a Lewis acid, many labile hydrogen compounds add across the vinyl ether double bond in the Markovnikov direction.

Table 8. **Physical Properties of Vinyl Ethers**[a]

Property	Methyl	Ethyl	Butyl
molecular weight	58	72	100
freezing point, °C	−122.8	−115.4	−91.9
boiling point, °C	5.5	35.7	93.5
vapor pressure at 20°C, kPa[b]	156.7	57	5.6
specific gravity, $d^{20}{}_4$	0.7511	0.7541	0.7792
refractive index, $n^{20}{}_D$	1.3730 (0°C)	1.3767	1.4020
flash point, °C	−56	<−18	−1
water solubility at 20°C, wt %	1.5	0.9	0.2

[a] Lower vinyl ethers are miscible with nearly all organic solvents.
[b] To convert kPa to mm Hg, multiply by 7.5.

Alcohols give acetals. This reaction has been frequently used to provide blocking groups in organic synthesis. The acetals are stable under neutral or alkaline conditions and are easily hydrolyzed with dilute acid after other desired reactions have occurred (217,218). Water gives acetaldehyde and the corresponding alcohol, presumably via disproportionation of the hemiacetal (219). Carboxylic acids give 1-alkoxyethyl esters (220). Thiols give thioacetals (221).

$$CH_2{=}CHOR + R'OH \longrightarrow CH_3CH(OR')OR$$

$$CH_2{=}CHOR + H_2O \longrightarrow [CH_3CH(OH)OR] \longrightarrow CH_3CHO + ROH$$

$$CH_2{=}CHOR + R'COOH \longrightarrow CH_3CH(OOCR')OR$$

Hydrogen halides react vigorously to give 1-haloethyl ethers, which are reactive intermediates for further synthesis (222). Conditions must be carefully selected to avoid polymerization of the vinyl ether. Hydrogen cyanide adds at high temperature to give a 2-alkoxypropionitrile (223).

$$CH_2{=}CHOR + HX \longrightarrow CH_3CH(X)OR \quad (X = Cl \text{ or } Br)$$

$$CH_2{=}CHOR + HCN \longrightarrow CH_3CH(CN)OR$$

Chlorine and bromine add vigorously, giving, with proper control, high yields of 1,2-dihaloethyl ethers (224). In the presence of an alcohol, halogens add as hypohalites, which give 2-haloacetals (225,226). With methanol and iodine this is used as a method of quantitative analysis, titrating unconsumed iodine with standard thiosulfate solution (227).

$$CH_2{=}CHOR + X_2 \longrightarrow XCH_2CH(X)OR$$

$$CH_2{=}CHOR + X_2 + R'OH \longrightarrow XCH_2CH(OR')OR$$

With Lewis acids as catalysts, compounds containing more than one alkoxy group on a carbon atom add across vinyl ether double bonds. Acetals give 3-alkoxyacetals; since the products are also acetals, they can react further with excess vinyl ether to give oligomers (228–230). Orthoformic esters give diacetals of malonaldehyde (231). With Lewis acids and mercuric salts as catalysts, vinyl ethers add in similar fashion to give acetals of 3-butenal (232,233).

$$CH_2{=}CHOR + R'CH(OR'')_2 \longrightarrow R'CH(OR'')CH_2CH(OR'')OR$$

$$CH_2{=}CHOR + CH(OR')_3 \longrightarrow CH(OR')_2CH_2CH(OR')OR$$

$$2\,CH_2{=}CHOR \longrightarrow CH_2{=}CHCH_2CH(OR)_2$$

Vinyl ethers and α,β-unsaturated carbonyl compounds cyclize in a hetero-Diels-Alder reaction when heated together in an autoclave with small amounts of hydroquinone added to inhibit polymerization. Acrolein gives 3,4-dihydro-2-methoxy-2*H*-pyran (234,235), which can easily be hydrolyzed to glutaraldehyde (236) or hydrogenated to 1,5-pentanediol (237). With 2-methylene-1,3-dicarbonyl compounds the reaction is nearly quantitative (238).

$$CH_2{=}CH{-}CH{=}O + CH_2{=}CH{-}OR \longrightarrow \text{2-alkoxy-3,4-dihydro-2}H\text{-pyran}$$

$$C_6H_5C(=O){-}C(=CH_2){-}C(=O)CH_3 + CH_2{=}CH{-}OCH_3 \longrightarrow \text{5-benzoyl-6-methyl-2-alkoxy-3,4-dihydro-2}H\text{-pyran}$$

Vinyl ethers cyclize with ketenes to cyclobutanones (239).

$$CH_2{=}CHOR + CH_2{=}C{=}O \longrightarrow \text{3-alkoxycyclobutanone}$$

Vinyl ethers serve as a source of vinyl groups for transvinylation of such compounds as 2-pyrrolidinone or caprolactam (240,241).

Compounds such as carbon tetrachloride (242) or trinitromethane (243) can add across the double bond.

$$CH_2{=}CHOR + CCl_4 \longrightarrow ROCHClCH_2CCl_3$$

$$CH_2{=}CHOR + CH(NO_2)_3 \longrightarrow ROCH_2CH_2C(NO_2)_3$$

With thionyl chloride as catalyst, hydrogen peroxide adds to vinyl ethers in anti-Markovnikov fashion, as do monothioglycols with amine catalysts (244).

$$2\,CH_2{=}CHOR + HOOH \longrightarrow ROCH_2CH_2OOCH_2CH_2OR$$

$$CH_2{=}CHOR + HSCH_2CH_2OH \longrightarrow ROCH_2CH_2SCH_2CH_2OH$$

Substances that form carbanions, such as nitro compounds, hydrocyanic acid, malonic acid, or acetylacetone, react with vinyl ethers in the presence of water, replacing the alkyl group under mild conditions (245).

$$CH_2{=}CHOR + CH_3NO_2 + H_2O \longrightarrow CH_3\underset{}{\overset{OH}{\overset{|}{C}}}HCH_2NO_2 + ROH$$

$$CH_2{=}CHOR + HCN + H_2O \longrightarrow CH_3\overset{OH}{\overset{|}{C}}HCN + ROH$$

The reaction of a vinyl ether with carbon dioxide and a secondary amine gives a carbamic ester (246).

$$CH_2{=}CHOR + CO_2 + (CH_3)_2NH \longrightarrow (CH_3)_2NCOO\overset{OR}{\overset{|}{C}}HCH_3$$

Manufacture. The principal manufacturers of vinyl ethers are BASF, Kowa American Corp. and Monomer Payment Dafoe Labs, Inc.

German vinyl ether plants were described in detail at the end of World War II and variations of these processes are still in use. Vinylation of alcohols from methyl to butyl was carried out under pressure: typically 2–2.3 MPa (20–22 atm) and 160–165°C for methyl, and 0.4–0.5 MPa (4–5 atm) and 150–155°C for isobutyl. An unpacked tower, operating continuously, produced about 300 t/month, with yields of 90–95% (247).

High boiling alcohols were vinylated at atmospheric pressure. The Germans used a tower packed with Raschig rings and filled with an alcohol containing 1–5% of KOH at 160–180°C. Acetylene was recycled continuously up through the tower. The heat of reaction, ~125 kJ/mol (30 kcal/mol), was removed by cooling coils. Fresh alcohol and catalyst were added continuously at the top and withdrawn at the bottom. Yields of purified, distilled product were described as quantitative (248).

Shipment, Storage, and Prices. Methyl vinyl ether is available in tank cars or cylinders, while the other vinyl ethers are available in tank cars, tank wagons, or drums. Mild steel, stainless steel, and phenolic-coated steel are suitable for shipment and storage. If protected from air, moisture, and acidic contamination, vinyl ethers are stable for years.

Specifications and Analytical Methods. Vinyl ethers are usually specified as 98% minimum purity, as determined by gas chromatography. The principal impurities are the parent alcohols, limited to 1.0% maximum for methyl vinyl ether and 0.5% maximum for ethyl vinyl ether. Water (by Karl-Fischer titration) ranges from 0.1% maximum for methyl vinyl ether to 0.5% maximum for ethyl vinyl ether. Acetaldehyde ranges from 0.1% maximum in ethyl vinyl ether to 0.5% maximum in butyl vinyl ether.

Health and Safety Factors. Because of their high vapor pressures (methyl vinyl ether is a gas at ambient conditions), the lower vinyl ethers represent a severe fire hazard and must be handled accordingly. Contact with acids can initiate violent polymerization and must be avoided. Although vinyl ethers form peroxides more slowly than saturated ethers, distillation residues must be handled with caution.

Inhalation should be avoided. A group of six rats that were exposed to 64,000 ppm of methyl vinyl ether in air for 4 h were anesthetized. All recovered and appeared normal after 72 h. One died after 96 h. The others survived the 2-week observation period without noticeable effect.

The lower vinyl ethers do not appear to be skin irritants or sensitizers. Oral toxicity is very low: Isobutyl vinyl ether has LD_{50} of 17 mL/kg for white rats.

Uses. Union Carbide consumes its vinyl ether production in the manufacture of glutaraldehyde [111-30-8]. BASF and GAF consume most of their production as monomers (see VINYL POLYMERS). In addition to the homopolymers, the copolymer of methyl vinyl ether with maleic anhydride is of particular interest.

BIBLIOGRAPHY

"Acetylene-Derived Chemicals" in *ECT* 3rd ed., Vol. 1, pp. 244–276, by Eugene V. Hort, GAF Corporation; in *ECT* 4th ed., Vol. 1, pp. 195–231, by Eugene V. Hort and Paul Taylor, GAF Corporation; "Acetylene-Derived Chemicals" in *ECT* (online), posting date: December 4, 2000, by Eugene V. Hort, Paul Taylor, GAF Corporation.

CITED PUBLICATIONS

1. S. A. Miller, *Acetylene, Its Properties, Manufacture and Uses*,Vol. 1,Academic Press, Inc., New York, 1965, pp. 24–28, 42–44.
2. M. J. Haley with T. Ball and S. Yoshikawa, *Acetylene, CEH Product Review, Chemical Economics Handbook*,SRI International, Menlo Park, Calif., Oct. 1988, p. 300.5000x.
3. Ref. 2, p. 300.5000y.
4. R. E. Colborn and K. P. C. Vollhardt, *J. Am. Chem. Soc.* **108**, 5470 (1986).
5. U.S. Pat. 4, 241, 230 (Dec. 23, 1980), B. M. Drinkard (to Mobil Oil Corp.).

6. Jpn. Kokai 78 59,605 (May 29, 1978), Y. Nambu and C. Fujii (to Denki Kaguku Kogyo K.K.).
7. Ref. 2, p. 300.5000s.
8. W. Reppe and co-workers, *Ann.* **596**, 2 (1955).
9. J. V. Nef, *Ann.* **308**, 277 (1899).
10. Ref. 8,p. 29.
11. L. Henry, *Berichte* **5**, 449 (1872).
12. U.S. Pat. 2,563,325 (Aug. 7, 1951), F. Fahnoe (to GAF Corp.).
13. U.S. Pat. 2,641,615 (June 9, 1953), R. F. Kleinschmidt (to GAF Corp.).
14. J. C. Sauer, *Org. Synth. Coll. Vol.* **4**, 813 (1963).
15. V. Wolf, *Chem. Berichte* **86**, 735 (1953).
16. V. Wolf, *Chem. Ber.* **87**, 668 (1954).
17. L. Henry, *Ber.* **6**, 728 (1873); **8**, 398 (1875).
18. A. Kirrmann, *Bull. Soc. Chim. Fr.* **39** (4), 698 (1926).
19. L. Henry, *Berichte* **17**, 1132 (1884).
20. Ref. 8, pp. 57–59.
21. F. J. McQuillin and W. O. Ord, *J. Chem. Soc.*, 2906 (1959).
22. Ref. 8, p. 61.
23. G. F. Hennion and W. S. Murray, *J. Am. Chem. Soc.* **64**, 1220 (1942).
24. U.S. Pat. 3,637,813 (Jan. 25, 1972), G. F. D'Allelio.
25. E. Cherbuliez, M. Gowhari, and J. Rabinowitz, *Helv. Chim. Acta* **47**, 2098 (1964).
26. U.S. Pat. 3,783,016 (Jan. 1, 1974), D. I. Randall and C. Vogel (to GAF Corp.).
27. Ref. 8, p. 69.
28. P. Chini, A. Santambrogio, and N. Palladino, *J. Chem. Soc. C*, 830 (1967).
29. W. Schulz, U. Rosenthal, D. Braun, and D. Walther, *Z. Chem.* **27**, 264 (1987).
30. W. E. Daniels, *J. Org. Chem.* **29**, 2936 (1964).
31. L. A. Akopyan and co-workers, *Polym. Sci. USSR* (in English) **17**(5), 1231 (1975).
32. M. G. Akhmerov, D. Usupov, and A. Kuchkarov, *Uzb. Khim. Zh.* **1979**(4), 59 (1979); *Chem. Abstr.* **92**, 58871 (1980).
33. U.S. Pat. 2,238,471 (Apr. 15, 1941), E. Keyssner and E. Eichler (to GAF Corp.).
34. R. L. Salvador and D. Simon, *Can. J. Chem.* **44**(21), 2570 (1966).
35. R. Lespieau and P. L. Viguier, *C. R. Acad. Sci.* **146**, 294 (1908).
36. I. G. Ali-Zade and co-workers, *Dokl. Akad. Nauk. SSSR* **173**, 89 (1967); *Chem. Abstr.* **67**, 32723 (1967).
37. L. Brandsma, *Preparative Acetylenic Chemistry*,Elsevier, Amsterdam, The Netherlands, 1971, pp. 166–168.
38. U.S. Pat. 3,257,465 (June 21, 1966), M. W. Leeds and H. L. Komarowski (to Cumberland Chemical Corp.).
39. Brit. Pat. 968,928 (Sept. 9, 1964), M. E. Chiddix and O. F. Hecht (to GAF Corp.).
40. U.S. Pat. Appl. 20,020,010,377 (Jan. 24, 2002), T. Aoki and co-workers (to Kawasaki-shi, Japan).
41. U.S. Pat. 3,272,854 (Sept. 13, 1966), R. A. Covey, A. E. Smith, and W. L. Hubbard (to Uniroyal Corp.).
42. U.S. Pat. 2,778,830 (Jan. 22, 1957), H. Pasedach and M. Seefelder (to BASF A.G.).
43. U.S. Pat. 3,923,870 (Dec. 2, 1975), W. Singer (to Troy Chemical Corp.).
44. G. Y. Yositsch, *Zh. Russ. Fiz. Khim. Ova.* **38**, 252 (1906).
45. A. W. Johnson, *J. Chem. Soc.*, 1009 (1946).
46. G. Dupont, R. Dulou, and G. Lefebvre, *Bull. Soc. Chim. Fr.*, 816 (1954).
47. Ref. 8, p. 56.
48. U.S. Pat. 3,083,235 (Mar. 26, 1963), D. J. Mann, D. D. Perry, and R. M. Dudak (to Thiokol Chemical Corp.).

49. U.S. Pat. 2,941,010 (June 14, 1960), D. J. Mann, D. D. Perry, and R. M. Dudak (to Thiokol Chemical Corp.).
50. V. Wolf, *Chem. Ber.* **88**, 717 (1955).
51. I. Heilbron, E. R. Jones, and F. Sondheimer, *J. Chem. Soc.*, 604 (1949).
52. V. Franzen, *Ann.* **587**, 131 (1954).
53. U.S. Pat. Appl. 20,020,099244 (July 25, 2002), R. V. Chaudhari and co-workers, Pune, India.
54. A. W. Johnson, *J. Chem. Soc.*, 1011 (1946).
55. A. W. Johnson, *J. Chem. Soc.*, 1014 (1946).
56. U.S. Pat. 3,746,726 (July 17, 1973), F. Reicheneder and K. Dury (to BASF A.G.).
57. U.S. Pat. 3,054,739 (Sept. 18, 1962), F. Reicheneder and K. Dury (to BASF A.G.).
58. K. Dury, *Angew. Chem.* **72**, 864 (1960).
59. H. Kleinert and H. Fuerst, *J. Pract. Chem.* **36**, 252 (1967).
60. L. H. Smith, *Synthetic Fiber Developments in Germany*,Textile Research Institute, New York, 1946, 534–541.
61. Ger. Pat. 1,074,569 (Feb. 4, 1960), H. Pasedach and D. Ludsteck (to BASF A.G.).
62. Ger. Pat. 32,828 (May 15, 1965), H. Kleinert; *Chem. Abstr.* **63**, 17901 (1965).
63. Ref. 8, p. 51.
64. Ref. 8, pp. 45–47.
65. Y. K. Yur'ev, I. K. Korobitsyna, and E. G. Brige, *Dokl. Akad. Nauk SSSR* **62**, 625 (1948); *Chem. Abstr.* **43**, 5003 (1949).
66. U.S. Pat. 2,542,417 (Feb. 20, 1951), R. F. Kleinschmidt (to GAF Corp.).
67. Ger. Pat. 229,689 (Nov. 13, 1985), H. Drevs and R. S. Koernig; *Chem. Abstr.* **104**, 225336 (1986).
68. U.S. Pat. 2,421,650 (June 3, 1947), W. Reppe, C. Schuster, and E. Weiss (to GAF Corp.).
69. C. J. S. Appleyard and J. F. C. Gartshore, "Manufacture of 1,4-Butynediol at I. G. Ludwigshafen," *BIOS Report 367, Item 22; OTS Report PB 28556*,U.S. Department of Commerce.
70. D. L. Fuller, A. O. Zoss, and H. M. Weir, "The Manufacture of Butynediol from Acetylene and Formaldehyde," *FIAT Report No. 926; OTS Report PB 80334*, U.S. Department of Commerce, 1946.
71. U.S. Pat. 3,560,576 (Feb. 2, 1971), J. R. Kirschner (to E.I. du Pont de Nemours & Co., Inc.).
72. Ger. Offen. 2,357,751 (Aug. 7, 1975), K. Baer and co-workers (to BASF A.G.)
73. Ger. Offen. 2,314,693 (Oct. 10, 1974), W. Reiss and co-workers (to BASF A.G.).
74. Ger. Offen. 2,240,401 (Mar. 7, 1974), H. Pasedach and H. Kroesche (to BASF A.G.).
75. U.S. Pat. 3,920,759 (Nov. 18, 1975), E. V. Hort (to GAF Corp.).
76. U.S. Pat. 4,117,248 (Sept. 26, 1978), J. L. Prater and R. L. Hedworth (to GAF Corp.).
77. U.S. Pat. 4,119,790 (Oct. 10, 1978), E. V. Hort (to GAF Corp.).
78. Ref. 1, pp. 476–542.
79. U.S. Pat. 2,906,614 (Sept. 29, 1959), T. R. Hopkins and J. W. Pullen (to Spencer Chemical Co.).
80. G. Griner, *C. R. Acad. Sci.* **116**, 723 (1893).
81. C. Prevost, *C. R. Acad. Sci.* **183**, 1292 (1926).
82. U.S. Pat. 2,877,205 (Mar. 10, 1959), J. Lal (to Justi and Son, Inc.).
83. U.S. Pat. 2,840,598 (June 24, 1958), H. Schwartz (to Vineland Chemical Co.).
84. U.S. Pat. 2,980,649 (Apr. 18, 1961), J. R. Caldwell and R. Gilkey (to Eastman Kodak Co.).
85. O. Bayer and co-workers, *Angew. Chem.* **62**, 61 (1950).
86. C. S. Marvel and C. H. Young, *J. Am. Chem. Soc.* **73**, 1066 (1951).

87. Jpn. Pat. 3164 (Apr. 25, 1958), T. Haya, M. Sato, and M. Yoshida (to Mitsubishi Co.); *Chem. Abstr.* **53**, 6083 (1959).
88. G. Lefebvre, G. Dupont, and R. Dulou, *C. R. Acad. Sci.* **229**, 222 (1949).
89. K. C. Brannock and G. R. Lappin, *J. Org. Chem.* **21**, 1366 (1956).
90. Ger. Pat. 855,864 (Nov. 17, 1952), A. Seib (to BASF A.G.).
91. U.S. Pat. 3,240,702 (Mar. 15, 1966), R. F. Monroe (to The Dow Chemical Company).
92. J. M. Bobbit, L. H. Amundsen, and R. I. Steiner, *J. Org. Chem.* **25**, 2230 (1960).
93. A. Valette, *Ann. Chim.* **3**(12), 644 (1948).
94. Ger. Pat. 857,369 (Nov. 27, 1952), H. Krzikalla and E. Woldan (to BASF A.G.).
95. P. Kurtz, *Ann.* **572**, 49, 69 (1951).
96. U.S. Pat. 4,633,015 (Dec. 30, 1986), A. S. C. Chan and D. E. Morris (to Monsanto Co.).
97. L. E. Craig, R. M. Elofson, and I. J. Ressa, *J. Am. Chem. Soc.* **72**, 3277 (1960).
98. U.S. Pat. 3,661,980 (May 9, 1972), W. Himmele, W. Qulla, and R. Prinz (to BASF A.G.).
99. U.S. Pat. 3,859,369 (Jan. 7, 1975), H. B. Copelin (to E. I. du Pont de Nemours & Co., Inc.).
100. Ger. Offen. 3,334,589 (Apr. 4, 1985), R. Schalenbach and H. Waldmann (to Bayer A.G.).
101. M. F. Abidova, A. S. Sultanov, and N. A. Savel'eva, *Katal. Pererab. Uglevodorodnogo. Syr'ya* **1971**(5), 175 (1971); *Chem. Abstr.* **79**, 125755 (1973).
102. N. O. Brace, *J. Am. Chem. Soc.* **77**, 4157 (1955).
103. Ger. Pat. 961,353 (Apr. 4, 1957), H. Friederich (to BASF A.G.).
104. U.S. Pat. 2,912,439 (Nov. 10, 1959), R. H. Hasek and J. E. Hardwicke (to Eastman Kodak Co.).
105. Ger. Pat. 1,094,732 (Dec. 15, 1960), H. Frensch (to Hoechst A.G.).
106. N. Clauson-Kaas, *Acta Chem. Scand.* **15**, 177 (1961); *Chem. Abstr.* **57**, 4619 (1962).
107. U.S. Pat. 2,833,787 (May 6, 1958), G. J. Carlson and co-workers (to Shell Oil Co.).
108. U.S. Pat. 3,284,419 (Nov. 8, 1966), F. G. Helfferich (to Shell Oil Co.).
109. Ger. Pat. 833,963 (Mar. 13, 1952), E. Bauer (to BASF A.G.).
110. Ref. 8, p. 142.
111. U.S. Pat. 2,779,700 (Jan. 29, 1957), P. Robitschek and C. T. Bean (to Hooker Chemical Co.).
112. U.S. Pat. 2,740,771 (Apr. 3, 1958), R. I. Longley, Jr., E. C. Chapin, and R. F. Smith (to Monsanto Chemical Co.).
113. F. J. McQuillin and W. O. Ord, *J. Chem. Soc.*, 2906 (1959).
114. I. A. Makolkin and co-workers, *Tr. Mosk. Inst. Nar. Khoz.* **1968**(46), 3 (1968); *Chem. Abstr.* **71**, 54383 (1969).
115. J. W. Copenhaver and M. H. Bigelow, *Acetylene and Carbon Monoxide Chemistry*, Reinhold, New York,1949, pp. 131–133.
116. P. W. Feit, *Chem. Ber.* **93**, 116 (1960).
117. U.S. Pat. 2,961,471 (Nov. 22, 1960), E. V. Hort (to GAF Corp.).
118. H. Tieckelmann, in R. A. Abromovitch, ed., *Pyridine and Its Derivatives*, Suppl. Part III, John Wiley & Sons, Inc., New York, 1974, pp. 670–673.
119. P. J. Dekkers, *Recl. Trav. Chim. Pays-Bas* **9**, 92 (1890).
120. C. Harries, *Berichte* **35**, 1187 (1902).
121. J. Baeseken, *Recl. Trav. Chim. Pays-Bas* **34**, 100 (1915).
122. J. Hamonet, *C. R. Acad. Sci.* **132**, 632 (1905).
123. R. Lespieau, *C. R. Acad. Sci.* **150**, 1761 (1910).
124. W. Griehl and G. Schnock, *Faserforsch. Textiltech.* **8**, 408 (1957); *Chem. Abstr.* **52**, 11781 (1958).
125. S. Sarel, L. A. Pohoryles, and R. Ben-Shoshan, *J. Org. Chem.* **24**, 1873 (1959).
126. Ger. Pat. 800,662 (Nov. 27, 1950), H. Krzikalla and K. Merkel (to BASF A.G.).

127. R. Riemschneider and W. M. Schneider, *Monatsh. Chem.* **90**, 510 (1959).
128. J. W. Hill and W. H. Carothers, *J. Am. Chem. Soc.* **57**, 925 (1935).
129. U.S. Pat. 2,870,097 (Jan. 20, 1959), D. B. Pattison (to E. I. du Pont de Nemours & Co., Inc.).
130. Ger. Pat. 800,398 (Nov. 2, 1950), B. Christ (to BASF A.G.).
131. V. V. Korshak and I. A. Gribova, *Izv. Akad. Nauk SSSR Otd. Tekh. Nauk*, 670 (1954); *Chem. Abstr.* **49**, 10893 (1955).
132. U.S. Pat. 2,222,302 (Nov. 19, 1940), W. Schmidt and F. Manchen (to GAF Corp.).
133. A. Muller and W. Vane, *Berichte* **77B**, 669 (1944).
134. S. K. Kang,W. S. Kim, and B. H. Moon, *Synthesis* **1985**(12), 1161 (1985).
135. Ref. 8, p. 178.
136. I. Geiman and co-workers, *Khim Geterotsickl Soedin.* **1981**(4), 448 (1981); *Chem. Abstr.* **95**, 80604 (1981).
137. G. A. Kliger and co-workers, *Otkrytiya Izobret.* **1988**(23), 106 (1988); *Chem. Abstr.* **109**, 212804 (1988).
138. R. E. Walkup and S. Searles, Jr., *Tetrahedron* **41**(1), 101 (1985).
139. Ger. Offen. 2,824,908 (Dec. 20, 1979), W. Mesch (to BASF A.G.).
140. Y. K. Yur'ev and N. G. Medovshchikov, *J. Gen. Chem. USSR* **9**, 628 (1939); *Chem. Abstr.* **33**, 7779 (1939).
141. *Chem. Eng.* **82**(7), 55 (1975).
142. *Chem. Mark. Rep.* **207**(11), 5 (1975).
143. M. J. Haley withW. E. Cox and S. Yoshikawa, *1,4-Butanediol*,CEH Product Review, Chemical Economics Handbook, SRI International, Menlo Park, Calif., Oct. 1988, pp. 621.5030f,g.
144. "1,4-Butandiol, Chemical Profile," *Chemical Market Reporter*, June 2000.
145. R. Fittig and M. B. Chanlaroff, *Berichte* **226**, 331 (1884).
146. S. Minoda and M. Miyajima, *Hydrocarbon Process.* **49**(11), 176 (1970).
147. K. Saotome and K. Sato, *Bull. Chem. Soc. Jpn.* **39**(3), 485 (1966).
148. H. C. Brown and K. A. Keblys, *J. Org. Chem.* **31**(2), 485 (1966).
149. Ref. 8, pp. 191–194.
150. Y. K. Yur'ev, E. G. Vendel'shtein, and L. A. Zinov'eva, *Zhur. Obshch. Khim.* **22**, 509 (1952); *Chem. Abstr.* **47**, 2747 (1953).
151. Swiss Pat. 264,598 (Jan. 16, 1950), G. Bischoff; *Chem. Abstr.* **45**, 1622 (1951).
152. J. E. Livak and co-workers, *J. Am. Chem. Soc.* **67**, 2218 (1945).
153. Ger. Pat. 810,025 (Aug. 6, 1951), C. Schuster and A. Simon (to BASF A.G.).
154. H. Hart and O. E. Curtis, Jr., *J. Am. Chem. Soc.* **78**, 112 (1956).
155. U.S. Pats. 2,993,891 (July 25, 1961), 3,030,361 (Apr. 17, 1962), 3,031,446 (Apr. 24, 1962),H. W. Zimmer and J. M. Holbert (to Chattanooga Medicine Co.).
156. Jpn. Pat. 8,271 (Sept. 24, 1956), M. Ohta (to Mitsubishi Co.); *Chem. Abstr.* **52**, 11904 (1958).
157. V. M. Markovich and co-workers, *Khim. Farm. Zh.* **16**, 1491 (1982); *Chem. Abstr.* **98**, 125367 (1983).
158. Indian Pat. 160,027 (June 20, 1987), A. Prakosh and R. S. Prasad (to Reckitt and Colman of India Ltd.).
159. G. I. Nikishin and co-workers, *Izv. Akad. Nauk SSSR Otd. Tekh. Nauk*, 146 (1962); *Chem. Abstr.* **57**, 16390 (1962).
160. D. Elad and R. D. Youssefyeh, *Chem. Commun.* **1965**(1), 7 (1965).
161. W. E. Truce and C. E. Olson, *J. Am. Chem. Soc.* **74**, 4721 (1952).
162. Ger. Pat. 1,026,297 (Mar. 20, 1958), N. von Kutepow (to BASF A.G.); *Chem. Abstr.* **54**, 9768 (1960).
163. D. J. Cram and H. Steinberg, *J. Am. Chem. Soc.* **76**, 3630 (1954).
164. U.S. Pat. 2,778,852 (Jan. 22, 1957), K. Adam and H. G. Trieschmann (to BASF A.G.).

165. Ger. Pat. 804,567 (Apr. 26, 1951), H. Kaltschmitt and A. Tartter (to BASF A.G.).
166. V. Y. Kortun, Z. M. Kol'tsova, and V. G. Yashunskii, *Zh. Prikl. Khim. (Leningrad)* **51**, 1919 (1978); *Chem. Abstr.* **89**, 196931 (1978).
167. U.S. Pat. 2,819,304 (Jan. 7, 1958), W. Reppe,H. Friederich, and H. Laib (to BASF A.G.).
168. Ger. Pat. 917,665 (Oct. 14, 1954), H. Haussmann and G. Grafinger (to BASF A.G.).
169. H. Plieninger, *Chem. Ber.* **83**, 265 (1950).
170. U.S. Pat. 2,603,658 (July 15, 1952), F. Hanusch (to BASF A.G.).
171. W. Reppe and co-workers, *Ann.* **596**, 198 (1955).
172. T. C. Bruice and J. J. Bruno, *J. Am. Chem. Soc.* **83**, 3494 (1961).
173. A. L. Dounce, R. H. Wardlow, and R. Connor, *J. Am. Chem. Soc.* **57**, 2556 (1935).
174. S. M. McElvain and J. F. Vozza, *J. Am. Chem. Soc.* **71**, 896 (1949).
175. E. Spath and J. Lintner, *Ber.* **69**, 2727 (1936).
176. K. Hatada and Y. Ono, *Bull. Chem. Soc. Jpn.* **50**, 2517 (1977) (in English).
177. U.S. Pat. 4,824,967 (Apr. 25, 1989), K. C. Liu and P. D. Taylor (to GAF Corp.).
178. A. O. Zoss and D. L. Fuller, "The Manufacture of γ-Butyrolactone," PB 60902, U.S. Department of Commerce Office of Technical Services, Oct. 1946.
179. U.S. Pat. Appl. 20,010,029,302 (Oct. 11, 2001), S.-H. Cho and co-workers (to Yoosung-ku, Seo-ku, and Daeduck-ku).
180. A. E. Favorskii and M. Skossarewsky, *Zh. Russ. Fiz. Khim. Ova.* **32**, 652 (1900); *Bull. Soc. Chim. Fr.* **26**, 284 (1901).
181. G. F. Hennion and co-workers, *J. Org. Chem.* **21**, 1142 (1956).
182. U.S. Pat. 2,882,287 (Apr. 14, 1959), D. C. Rowlands and W. H. Gillen (to Air Reduction Co.).
183. C. D. Hurd and W. D. McPhee, *J. Am. Chem. Soc.* **71**, 398 (1949).
184. U.S. Pat. 2,798,885 (July 9, 1957), H. Ensslin and K. Meier (to Ciba Pharmaceutical Products Inc.).
185. I. N. Nazarov and G. A. Schvekhgeimer, *Zh. Obshch. Khim.* **29**, 463 (1959); *Chem. Abstr.* **53**, 21661 (1959).
186. G. F. Hennion and K. W. Nelson, *J. Am. Chem. Soc.* **79**, 2142 (1957).
187. Y. Pasternak, *Bull. Soc. Chim. Fr.* **1963**(8–9), 1719 (1963).
188. T. A. Favorskaya, *J. Gen. Chem. USSR* **9**, 386 (1939); *Chem. Abstr.* **33**, 9281 (1939).
189. T. A. Favorskaya, *J. Gen. Chem. USSR* **10**, 461 (1940); *Chem. Abstr.* **34**, 7845 (1940).
190. U.S. Pat. 2,250,558 (July 29, 1941), T. H. Vaughn (to Union Carbide Corp.).
191. U.S. Pat. 3,388,181 (June 11, 1968), H. D. Anspon (to GAF Corp.).
192. R. J. Tedeschi and G. Clark, Jr., *J. Org. Chem.* **27**, 4323 (1962).
193. N. C. Rose, *J. Chem. Educ.* **43**(6), 324 (1966).
194. I. N. Nazarov and L. D. Bergel'son, *Zh. Obshch. Khim.* **27**, 1540 (1957); *Chem. Abstr.* **52**, 3660 (1958).
195. G. F. Hennion and G. M. Wolf, *J. Am. Chem. Soc.* **62**, 1368 (1940).
196. Ger. Offen. 3,633,033 (Apr. 7, 1988), G. Thelen and H. W. Voges (to Huels A.G.).
197. G. A. Chukhadzhyan and co-workers, *Zh. Org. Khim.* **8**(3), 476 (1972); *Chem. Abstr.* **77**, 4739 (1972).
198. E. R. H. Jones, G. H. Whitham, and M. C. Whiting, *J. Chem. Soc.*, 4628 (1957).
199. T. Nogi and J. Tsuji, *Tetrahedron* **25**(17), 4099 (1969).
200. H. A. Stansbury and W. R. Proops, *J. Org. Chem.* **27**, 320 (1962).
201. S. S. Dehmlow and E. V. Dehmlow, *Justus Liebigs Ann. Chem.* **10**, 1753 (1973).
202. Ref. 8, p. 36.
203. R.-R. Lii and S. I. Miller, *J. Am. Chem. Soc.* **95**(5), 1602 (1973).
204. G. Saucy and co-workers, *Helv. Chim. Acta* **42**, 1945 (1959).
205. A. V. Shchelkunov, A. Ashirbekova, and V. M. Rofman, *Deposited Doc.* **1980**, 5311 (1980).

206. U.S. Pat. 3,082,260 (Mar. 19, 1963), R. J. Tedeschi, A. W. Casey, and J. P. Russell (to Air Reduction Co.).
207. M. A. Dzhragatspanyan, A. G. Mirzkhanyan, and L. A. Ustynyuk, *Arm. Khim. Zh.* **36**, 476, 547 (1983); *Chem. Abstr.* **100**, 22301, 22302 (1984).
208. U.S. Pat. 3,283,014 (Nov. 1, 1966), A. Balducci and M. de Malde (to SNAM S.p.A.).
209. A. Heath, *Chem. Eng.* **80**(22), 48 (1973).
210. U.S. Pat. 1,941,108 (Dec. 26, 1923), W. Reppe (to I. G. Farbenind. A.G.).
211. U.S. Pat. 1,959,927 (May 22, 1934), W. Reppe (to I.G. Farbenind. A.G.).
212. O. Solomon, C. Ionescu, and I. Ciuta, *Chem. Tech. Leipzig* **9**, 202 (1957); *Chem. Abstr.* **51**, 15493 (1957).
213. Shansi University, *Hua Hsueh Tung Pao*, 21 (1977); *Chem. Abstr.* **87**, 134899 (1977).
214. C. Gardner and co-workers, *J. Chem. Soc.*, 789 (1949).
215. Fr. Pat. 865,354 (May 3, 1940), H. Weese,G. Hecht, and W. Reppe (to I. G. Farbenind. A.G.).
216. J. Wislicenus, *Ann.* **192**, 106 (1878).
217. I. W. J. Still, J. N. Reed, and K. Turnbull, *Tetrahedron Lett.* **17**, 1481 (1979).
218. A. Franke, F. F. Frickel, R. Schlecker, and P. C. Theime, *Synthesis*, 712 (1979).
219. A. Skrabal and R. Skrabal, *Z. Phys. Chem. A* **181**, 449 (1938).
220. E. Levas, *C. R. Acad. Sci.* **228**, 1443 (1949).
221. F. Kipnis, H. Solonay, and J. Ornfelt, *J. Am. Chem. Soc.* **73**, 1783 (1951).
222. U.S. Pat. 2,061,946 (Nov. 24, 1936), E. Kuehn and H. Hopff (to I. G. Farbenind. A.G.).
223. W. Reppe and co-workers, *Ann.* **601**, 109 (1956).
224. L. Summers, *Chem. Rev.* **55**, 317 (1955).
225. U.S. Pat. 2,433,890 (Jan. 6, 1948), O. W. Cass (to E. I. du Pont de Nemours & Co., Inc.).
226. U.S. Pat. 4,489,011 (Dec. 18, 1984), S. S. M. Wang (to Merrell Dow Pharm., Inc.).
227. S. Siggia and R. L. Edsberg, *Anal. Chem.* **20**, 762 (1948).
228. U.S. Pat. 2,165,962 (July 1, 1939), M. Mueller-Conradi and K. Pieroh (to I. G. Farbenind. A.G.).
229. U.S. Pats. 2,487,525 (Apr. 8, 1949) and 2,502,433 (Apr. 4, 1950), J. W. Copenhaver (to GAF Corp.).
230. R. I. Hoaglin and D. H. Hirsch, *J. Am. Chem. Soc.* **71**, 3468 (1949).
231. U.S. Pat. 2,527,533 (Oct. 31, 1950), J. W. Copenhaver (to GAF Corp.).
232. R. I. Hoaglin, D. G. Kubler, and A. E. Montagna, *J. Am. Chem. Soc.* **80**, 5460 (1958).
233. M. S. Nieuwenhuizen, A. P. G. Kieboom, and H. Van Bekkum, *Synthesis*, 712 (1981).
234. R. I. Longley, Jr., and W. S. Emerson, *J. Am. Chem. Soc.* **72**, 3079 (1950).
235. Jpn. Kokai Tokkyo Koho 59,108,734 (June 23, 1984) (to Daicel Ind., Ltd.).
236. U.S. Pat. 2,546,018 (Mar. 20, 1951), C. W. Smith and co-workers (to Shell Development Co.).
237. U.S. Pat. 2,546,019 (Mar. 20, 1951), C. W. Smith and co-workers (to Shell Development Co.).
238. M. Yamauchi, S. Katayama, O. Baba, and T. Watanabe, *J. Chem. Soc. Chem. Commun.*, 281 (1983).
239. R. W. Aben and H. W. Scheeren, *J. Chem. Soc. Perkin Trans.* **1**, 3132 (1979).
240. U.S. Pat. 3,019,231 (Jan. 30, 1962), W. J. Peppel and J. D. Watkins (to Jefferson Chemical Co.).
241. W. H. Watanabe and L. E. Conlon, *J. Am. Chem. Soc.* **79**, 2828 (1957).
242. U.S. Pat. 2,560,219 (July 10, 1951), S. A. Glickman (to GAF Corp.).
243. U.S. Pat. 3,050,565 (Aug. 21, 1962), P. O. Tawney (to Uniroyal Corp.).
244. U.S. Pat. 2,768,975 (Oct. 30, 1956), R. S. Schiefelbein (to Jefferson Chemical Co.).
245. U.S. Pat. 2,736,743 (Feb. 28, 1956), C. J. Schmidle and R. C. Mansfield (to Rohm and Haas Corp.).

246. Y. Yoshida and S. Inoue, *Chem. Lett.* **11**, 1375 (1977).
247. S. A. Miller, *Acetylene, Its Properties, Manufacture and Uses*, Vol. 2,Academic Press, Inc., New York, 1965, pp. 199–202.
248. J. W. Copenhaver and M. H. Bigelow, *Acetylene and Carbon Monoxide Chemistry*, Reinhold Publishing Corp., New York, 1949, pp. 37–38.

EUGENE V. HORT
PAUL TAYLOR
GAF Corporation

ACROLEIN AND DERIVATIVES

Acrolein (2-propenal) (C_3H_4O) [107-02-8], is the simplest unsaturated aldehyde (CH_2=CHCHO). The primary characteristic of acrolein is its high reactivity due to conjugation of the carbonyl group with a vinyl group. Controlling this reactivity to give the desired derivative is the key to its usefulness. Acrolein now finds commercial utility in several major products as well as a number of smaller volume products. More than 80% of the refined acrolein that is produced today goes into the synthesis of methionine. Much larger quantities of crude acrolein are produced as an intermediate in the production of acrylic acid. More than 85% of the acrylic acid produced worldwide is by the captive oxidation of acrolein. Several review articles (1–8) and a book (9) have been published on the preparation, reactions, and uses of acrolein.

Acrolein is a highly toxic material with extreme lacrimatory properties. At room temperature, acrolein is a liquid with volatility and flammability somewhat similar to acetone; but unlike acetone, its solubility in water is limited. Commercially, acrolein is always stored with hydroquinone and acetic acid as inhibitors. Special care in handling is required because of the flammability, reactivity, and toxicity of acrolein.

The physical and chemical properties of acrolein are given in Table 1. Additional data are available (9–12).

1. Manufacture

Acrolein was first reported in 1843 but was not produced commercially until the late 1930s. The first commercial processes were based on the vapor-phase condensation of acetaldehyde and formaldehyde (1). In the 1940s, a series of catalyst developments based on cuprous oxide and cupric selenites led to a vapor-phase propylene oxidation route to acrolein (13,14). In 1959, Shell was the first to commercialize the propylene oxidation route to acrolein. These early propylene oxidation catalysts were capable of only low per pass propylene conversions (~15%) and therefore required significant recycle of unreacted propylene (15–17). In

Table 1. **Properties of Acrolein**

Property	Value
Physical properties	
molecular formula	C_3H_4O
molecular weight	56.06
specific gravity at 20/20°C	0.8427
coefficient of expansion at 20°C, vol/°C	0.00140
boiling point, °C	
at 101.3 kPa[a]	52.69
at 1.33 kPa[a]	−36
vapor pressure at 20°C, kPa[a]	29.3
heat of vaporization at 101.3 kPa[a], kJ/kg[b]	510
critical temperature, °C	233
critical pressure, MPa[c]	5.07
critical volume, mL/mol	189
freezing point, °C	−87.0
solubility at 20°C, % by wt	
in water	20.6
water in	6.8
refractive index, $n^{20}{}_D$	1.4013
viscosity at 20°C, mPa·s (=cP)	0.35
heat capacity (specific), kJ/(kg·K)[b]	
liquid (25°C)	2.16
gas (25°C)	1.42
liquid density at 20°C, kg/L[d]	0.8412
Chemical properties	
flash point, open cup	−18°C
closed cup	−26°C
flammability limits in air, vol %	
upper	31
lower	2.8
autoignition temperature in air, °C	234
heat of combustion of 25°C, kJ/kg[b]	− 27,589
heat of polymerization (vinyl), kJ/mol[b]	− 71–80
heat of condensation (aldol), kJ/mol[b]	− 42

[a] To convert kPa to mm Hg, multiply by 7.5.
[b] To convert kJ to kcal, divide by 4.184.
[c] To convert MPa to psi, multiply by 145.
[d] To convert kg/L to lb/gal, multiply by 8.345.

1957, Standard Oil of Ohio (Sohio) discovered bismuth molybdate catalysts capable of producing high yields of acrolein at high propylene conversions (>90%) and at low pressures (18). Over the next several decades, much industrial and academic research and development was devoted to improving these catalysts, which are used in the production processes for acrolein, acrylic acid, and acrylonitrile. All commercial acrolein manufacturing processes known today are based on propylene oxidation and use bismuth molybdate-based catalysts.

Many key improvements and enhancements to the bismuth molybdate based propylene oxidation catalysts have occurred since its discovery in 1957. Table 2 shows a chronological list of representative bismuth molybdate catalysts from the patent literature.

Table 2. **Propylene Oxidation Catalyst for Acrolein Production**

Year	Catalyst	Company	Reference
1960	BiMo	Sohio	18
1965	BiMoFe	Knapsack	19
1969	BiMoFeNiCo	Nippon Kayaku	20,21
1970	BiMoFeNiCrSn	Toa Gosei	22
1972	BiMoFeNiPTlMg	Sumitomo	23
1974	BiMoFeCoWSiK	Nippon Shokubai	24,25
1990	BiMoFeCoNiPKSmSi	Degussa	26
1992	BiMoFeCoNiNaCaBKSi	Mitsubishi Pet. Co.	27
1997	BiMoFeCoWSiKZrS	Nippon Shokubai	28
1999	BiMoFeCoWSbZnK	Mitsubishi Rayon	29

The most efficient catalysts are complex mixed-metal oxides that consist largely of Bi, Mo, Fe, Ni, and/or Co, K, and either P, B, W, or Sb (30). Many additional combinations of metals have been patented, along with specific catalyst preparation methods to adjust specific surface area, pore volume, and pore size distribution. Most catalysts used commercially today are extruded neat metal oxides as opposed to supported (coated) catalysts. Propylene conversions are generally >93%. Acrolein selectivities of 80–90% are typical. The acrolein yields depend not only on the chemical composition of the catalyst, but also on the shape of the catalyst and catalyst loading configurations.

With the maturing of the propylene oxidation catalyst area, attention in the 1980s and 1990s was more focused on reaction process related improvements. Alternate feedstocks such as propane have also been investigated but has not yet lead to a commercial process (31).

The catalytic vapor-phase oxidation of propylene is generally carried out in a fixed-bed multitube reactor at near atmospheric pressures and elevated temperatures (~350°C); molten salt or other heat exchange media is used for temperature control. Air is commonly used as the oxygen source and steam is added to suppress the formation of flammable gas mixtures. Operation can be single pass or a recycle stream may be employed. As catalyst technology matured, interest focused on improving process efficiency and minimizing process wastes by defining process improvements that use recycle of process gas streams and/or use of new reaction diluents (32–36).

The reaction is very exothermic. The heat of reaction of propylene oxidation to acrolein is 340.8 kJ/mol (81.5 kcal/mol); the overall reactions generate ~418 kJ/mol (100 kcal/mol). The principal side reactions produce acrylic acid, acetaldehyde, acetic acid, carbon monoxide, and carbon dioxide. A variety of other aldehydes and acids are also formed in small amounts. Proprietary processes for acrolein manufacture have been described (37, 38).

The reactor effluent gases are cooled to condense and separate the acrolein from unreacted propylene, oxygen, and other low boiling components (predominantly nitrogen). This is commonly accomplished in two absorption steps where (1) aqueous acrylic acid is condensed from the reaction effluent and absorbed in a water-based stream, and (2) acrolein is condensed and absorbed in water to separate it from the propylene, nitrogen, oxygen, and carbon oxides. Acrylic acid may

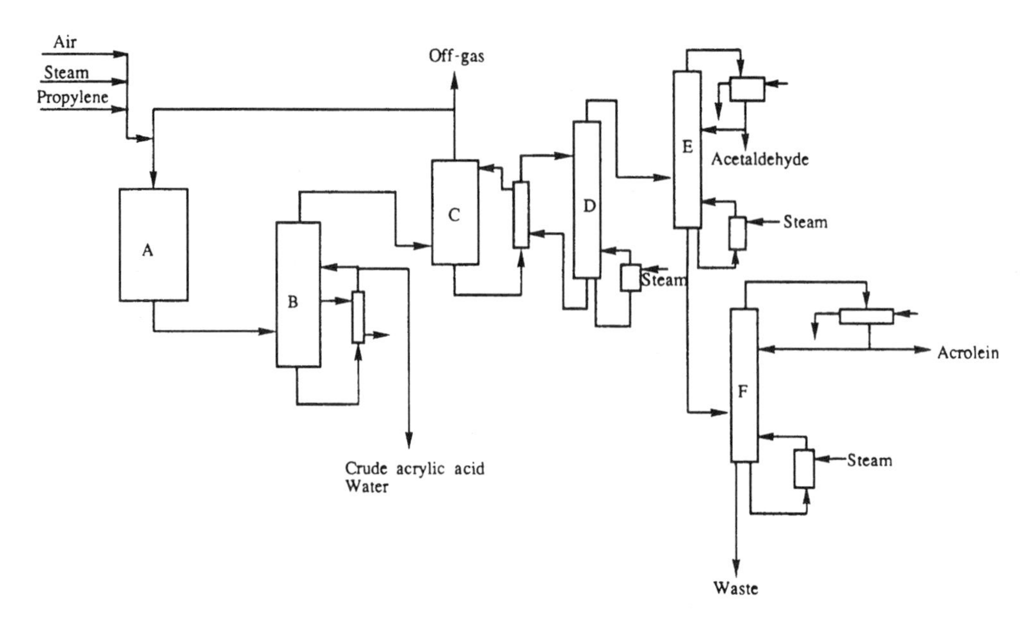

Fig. 1. A typical process flowsheet for acrolein manufacture. A, Fixed-bed or fluid-bed reactor; B, quench cooler; C, absorber; D, stripper; E and F, fractionation stills.

be recovered from the aqueous product stream if desired. Subsequent distillation separate water and acetaldehyde from the crude acrolein. In another distillation column, refined acrolein is recovered as an azeotrope with water. A typical process flow diagram is given in Figure 1.

2. Economic Aspects

Presently, the worldwide-refined acrolein nameplate capacity is ~250,000 t/yr. Both manufactures in the United States, Degussa and Dow have increased their production capacity in recent years to a total of ~122,000 t/yr. The key producers of refined acrolein are as noted in Table 3.

Of these producers, Elf Atochem, Degussa, and Daicel are reported to be in the merchant acrolein business. Dow supplies only to the acrolein derivative markets. Aventis also produces acrolein, primarily as a nonisolated intermediate to make methionine. A number of other small scale plants are located worldwide, which also produce acrolein as an intermediate to make methionine.

The significance of industrial acrolein production may be clearer if one considers the two major uses of acrolein—direct oxidation to acrylic acid and reaction to produce methionine via 3-methylmercaptopropionaldehyde (MMP). In acrylic acid production, acrolein is not isolated from the intermediate production stream. The 2000 acrylic acid production demand in the United States alone accounted for >1,300,000 t/yr (40), with worldwide capacity approaching 2,100,000 t/yr (41). Approximately 0.75 kg of acrolein is required to produce

Table 3. **Refined Acrolein Producers**

Producer	Annual nameplate capacity, 10^3 t/yr[a]
Degussa	110
Dow	72
Elf Atochem	30
Volzhskiy Orgsynthese	8
Daicel	9
Ohita	4.5
Sumitomo	15
(China)	4

[a]Estimated (39).

1 kg of acrylic acid. The methionine production process involves the reaction of acrolein with methyl mercaptan to produce the intermediate, MMP, which is further reacted with HCN to form methionine. Worldwide methionine capacity was estimated at ~570,000 t/yr in 2000 (42, 43) (see ACRYLIC ACID AND DERIVATIVES; AMINO ACIDS, SURVEY.)

3. Specifications and Analysis

Acrolein is produced according to the specifications in Table 4. Acetaldehyde and acetone are the principal carbonyl impurities in freshly distilled acrolein. Acrolein dimer accumulates at 0.50% in 30 days at 25°C. Analysis by two gas chromatographic (gc) methods with thermal conductivity detectors can determine all significant impurities in acrolein. The analysis with 0.91 mm × m 6.4 mm Porapak Q, 175–300 μm (50–80 mesh), programmed from 60 to 250°C at 10°C/min, does not separate acetone, propionaldehyde, and propylene oxide from acrolein. These separations are made with 3.66 mm × m 6.4 mm 20% Tergitol E-35 on 250–350 μm (45–60 mesh) Chromosorb W, kept at 40°C until acrolein elutes and then programmed rapidly to 190°C to elute the remaining components.

Alternatively, a bonded poly(ethylene glycol) capillary column held at 35°C for 5 min and programmed to 190°C at 8°C/min may be employed to determine all components but water. The Karl–Fischer method for water gives inaccurate results.

Table 4. **Specifications for Acrolein**

Requirement	Limit
acrolein, wt%, min	95.5
total carbonyl compounds other than acrolein, wt %, max	1.5
hydroquinone, wt %	0.10–0.25
water, wt %, max	3.0
specific gravity, 20°/20°C	0.842–0.846
pH of 10% solution in water at 25°C, max	6.0

Hydroquinone can be determined spectrophotometrically at 292 nm in methanol after a sample is evaporated to dryness to remove the interference of acrolein. An alternative method is high-performance liquid chromatography (hplc) on 10-μm LiChrosorb RP-2 at ambient temperature with 2.0 mL/min of 20% (v/v) 2,2,4-trimethylpentane, 79.20% chloroform, and 0.80% methanol with uv detection at 292 nm.

4. Reactions and Derivatives

Acrolein is a highly reactive compound because both the double bond and aldehydic moieties participate in a variety of reactions.

Acrolein is readily oxidized to acrylic acid, ($C_3H_4O_2$) [79-10-7], by passing a gaseous mixture of acrolein, air, and steam over a catalyst composed primarily of molybdenum and vanadium oxides (see ACRYLIC ACID AND DERIVATIVES). This process has been reviewed in a number of articles (44–47). Virtually all of the acrylic acid produced in the United States is made by the oxidation of propylene via the intermediacy of acrolein.

$$CH_2{=}CH{-}CHO \xrightarrow{\text{cat}} CH_2{=}CHCOOH$$

Direct formation of acrylic acid esters by oxidation of acrolein in the presence of lower alcohols has been studied (48). The intermediacy of acrylic acid is thereby avoided in the manufacture of these important acrylic acid derivatives.

$$CH_2{=}CH{-}CHO + O_2 + ROH \longrightarrow CH_2{=}CH{-}CO_2R$$

Because of a lack of discrimination between the double bond and carbonyl moieties, direct hydrogenation of acrolein leads to the production of mixtures containing propanol (C_3H_8O) [71-28-8], propionaldehyde (C_3H_6O) [123-38-6], and allyl alcohol, (C_3H_6O) [107-18-6]. However, proper selection of reaction conditions allows the carbonyl (49–62) and olefin (63–68) moieties to be selectively reduced to allyl alcohol (see ALLYL ALCOHOL AND DERIVATIVES) and propionaldehyde (see ALDEHYDES), respectively (69–89).

$$CH_2{=}CH{-}CHO \longrightarrow CH_2{=}CH{-}CH_2OH$$

$$CH_2{=}CH{-}CHO \longrightarrow CH_3CH_2CHO$$

The vapor-phase reduction of acrolein with isopropyl alcohol in the presence of a mixed-metal oxide catalyst yields allyl alcohol in a one-pass yield of 90.4%, with a selectivity (82) to the alcohol of 96.4%.

The addition of alcohols to acrolein may be catalyzed by acids or bases. By the judicious choice of reaction conditions the regioselectivity of the addition may be controlled and alkoxy-propionaldehydes (90), acrolein acetals, or alkoxypropionaldehyde acetals may be produced in high yields (91).

$$R'OH + CH_2{=}CH{-}CHO \longrightarrow CH_2{=}CH{-}CH(OR')_2 \;;\; R'OCH_2CH_2CHO \;;\; R'OCH_2CH_2CH(OR')_2$$

Table 5 lists a variety of alkoxypropionaldehydes and certain of their properties (92). Alcohols up to n-butyl have been added to acrolein in this fashion. Methyl, ethyl, and allyl alcohols react with ease, while the addition of hexyl or octyl alcohol proceeds in low yields. Although the alkoxypropionaldehydes have found only limited industrial utility, it is anticipated that they will find use as replacements for more toxic solvents. Furthermore, the alkoxypropionaldehydes may readily be reduced to the corresponding alkoxypropanols, which may also have desirable properties as solvents (93).

If the addition of alcohol to the olefin moiety of acrolein is carried out in the presence of a formaldehyde source, 2-(alkoxymethyl)acroleins are formed (94–97). 2-(Alkoxymethyl)acroleins are potential intermediates in the manufacture of substitued pyridines and quinolines.

$$CH_2{=}CH{-}CHO + \text{'}CH_2O\text{''} + ROH \longrightarrow CH_2{=}C(CHO)(CH_2OR)$$

Acrolein acetals have also been prepared in high yields (91). The formation of the acetal requires the careful control of reaction conditions to avoid additions to the double bond. Table 6 lists a variety of acrolein acetals that have been prepared and their boiling points (98).

The addition of certain glycols and polyols to acrolein leads to the production of cyclic acetal derivatives (99–103).

$$CH_2{=}CH{-}CHO + R{-}CH(OH){-}CH(OH){-}R \longrightarrow CH_2{=}CH{-}CH\langle O{-}CHR{-}CHR{-}O\rangle$$

Cyclic acrolein acetals are, in general, easily formed, stable compounds and have been considered as components in a variety of polymer systems. Table 7 lists a variety of previously prepared cyclic acrolein acetals and their boiling points (104).

Reactions of acrolein with alcohols producing high yields of alkoxypropionaldehyde acetals are also known. Examples of these are displayed in Table 8 (105). The alkoxypropionaldehyde acetals may be useful as solvents or as intermediates in the synthesis of other useful compounds (106).

A new and potentially significant use of acrolein is the manufacture of 1,3-propanediol ($C_3H_6O_2$) [504-63-2] (107–109). Addition of water to acrolein forms

Table 5. **Alkoxypropionaldehydes from Acrolein**

Compound added	Product				
	Structure	CAS Registry Number	Molecular formula	bp, °C	Pressure at bp, kPa[a]
CH_3OH	$CH_3OCH_2CH_2CHO$	[2806-84-0]	$C_4H_8O_2$	49	6.7
C_2H_5OH	$C_2H_5OCH_2CH_2CHO$	[2806-85-1]	$C_5H_{10}O_2$	57	5.3
n-C_3H_7OH	n-$C_3H_7OCH_2CH_2CHO$	[19790-53-5]	$C_6H_{12}O_2$	88	12
iso-C_3H_7OH	iso-$C_3H_7OCH_2CH_2CHO$	[39563-51-4]	$C_6H_{12}O_2$	45	2
n-C_4H_9OH	n-$C_4H_9OCH_2CH_2CHO$	[13159-34-2]	$C_7H_{14}O_2$	60	2
$CH_2{=}CHCH_2OH$	$CH_2{=}CHCH_2OCH_2CH_2CHO$	[44768-60-7]	$C_6H_{10}O_2$	55	1.9
$ClCH_2CH_2OH$	$ClCH_2CH_2OCH_2CH_2CHO$	[5422-33-3]	$C_5H_9ClO_2$	75.5	0.67
$CH_2{=}C(CH_3)CH_2OH$	$CH_2{=}C(CH_3)CH_2OHCH_2OCH_2CH_2CHO$	[76618-56-9]	$C_7H_{12}O_2$	62	1.2

[a] To convert kPa to mm Hg, multiply by 7.5.

Table 6. **Acrolein Acetals**

Compound added	Product				
	Structure	CAS Registry Number	Molecular formula	bp, °C	Pressure at bp, kPa[a]
C_2H_5OH	$CH{=}CHCH(OC_2H_5)_2$	[3054-95-3]	$C_7H_{14}O_2$	63	12.1
CH_3OH	$CH_2{=}CHCH(OCH_3)_2$	[6044-68-4]	$C_5H_{10}O_2$	40	16
n-C_3H_7OH	$CH_2{=}CHCH(OC_3H_7)_2$	[20615-55-8]	$C_9H_{18}O_2$	87.5–88	
iso-C_3H_7OH	$CH_2{=}CHCH(Oiso\text{-}C_3H_7)_2$	[14091-80-6]	$C_9H_{18}O_2$	54	1.6
n-C_4H_9OH	$CH_2{=}CHCH(OC_4H_9)_2$	[45094-50-6]	$C_{11}H_{22}O_2$	39	1.6
$C_6H_5CH_2OH$	$CH_2{=}CHCH(OCH_2C_6H_5)_2$	[40575-57-3]	$C_{17}H_{18}O_2$	120	6.7×10^{-4}
$CH_2{=}CHCH_2OH$	$CH_2{=}CHCH(OCH_2CH{=}CH_2)_2$	[3783-83-3]	$C_9H_{14}O_2$	75	3.7

[a] To convert kPa to mm Hg multiply by 7.5.

Table 7. **Acrolein Cyclic Acetals**

Compound added	Structure	Product: CAS Registry Number	Molecular formula	bp, °C	Pressure at bp, kPa[a]
$HOCH_2CH_2OH$	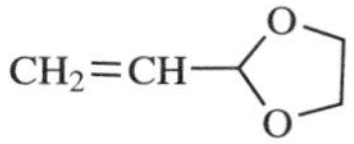 $CH_2=CH$–, O, O	[3984-22-3]	$C_5H_8O_2$	115.5–116.5	
$HOCH_2CH(OH)$-CH_2OH	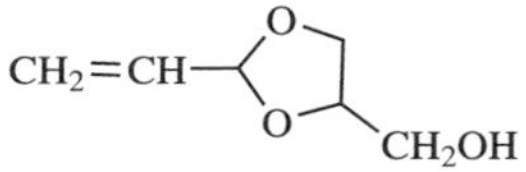 $CH_2=CH$–, O, O, CH_2OH	[4313-32-0]	$C_6H_{10}O_3$	80	0.4
$C(CH_2OH)_4$	 $CH_2=CH$–, O, O, O, O, –$CH=CH_2$	[78-19-3]	$C_{11}H_{16}O_4$	142–143 mp 41–42	1.6
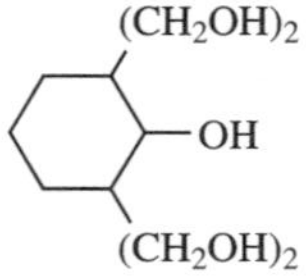 $(CH_2OH)_2$, –OH, $(CH_2OH)_2$	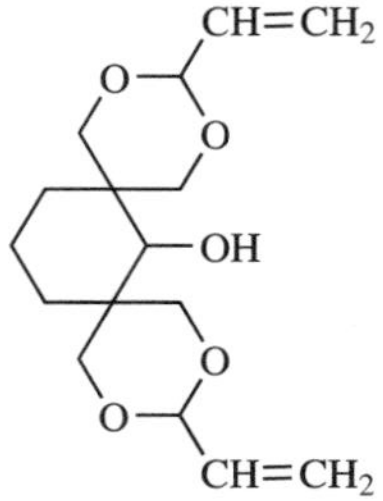 $CH=CH_2$, O, O, –OH, O, O, $CH=CH_2$	[b]	$C_{16}H_{24}O_5$	198–199	0.4
$HOCH_2CH_2CH_2OH$	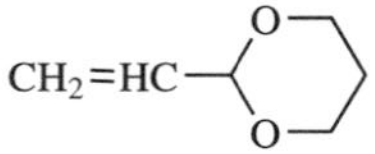 $CH_2=HC$–, O, O	[5935-25-1]	$C_6H_{10}O_2$	65–66	5.9

[a] To convert kPa to mm Hg, multiply by 7.5.
[b] No CAS Registry Number has been assigned.

Table 8. **Alkoxypropionaldehyde and Related Acetals**

Compound added	Product: Structure	CAS Registry Number	Molecular formula	bp, °C	Pressure at bp, kPa[a]
C_2H_5OH	$C_2H_5OCH_2CH_2CH(OC_2H_5)_2$	[7789-92-6]	$C_9H_{20}O_3$	69–70	1.3
CH_3OH	$CH_3OCH_2CH_2CH(OCH_3)_2$	[14315-97-0]	$C_6H_{14}O_3$	94–95	18.9
$n\text{-}C_3H_7OH$	$C_3H_7OCH_2CH_2CH(OC_3H_7)_2$	[53963-14-7]	$C_{12}H_{26}O_3$	109	1.6
$i\text{-}C_3H_7OH$	$C_3H_7OCH_2CH_2CH(O\textit{iso}\text{-}C_3H_7)_2$	[89769-16-4]	$C_{12}H_{26}O_3$	89	1.5
$C_6H_5CH_2OH$	$C_6H_5CH_2OCH_2CH_2CH(OCH_2C_6H_5)_2$	[b]	$C_{24}H_{26}O_3$	243–246	0.07
$CH_2{=}CHCH_2OH$	$CH_2{=}CHCH_2OCH_2CH_2$- $CH(OCH_2CH{=}CH_2)_2$	[8431-07-1]	$C_{12}H_{20}O_3$	113–114	1.3
$n\text{-}C_4H_9OH$	$C_4H_9OCH_2CH_2CH(OC_4H_9)_2$	[53963-15-8]	$C_{15}H_{32}O_3$	143–144	1.3
$n\text{-}C_5H_{11}OH$	$C_5H_{11}OCH_2CH_2CH(OC_5H_{11})_2$	[53963-17-6]	$C_{18}H_{38}O_3$	153–155	0.13
$ClCH_2CH_2OH$	$ClCH_2CH_2OCH_2CH_2CH(OCH_2CH_2Cl)_2$	[688-78-8]	$C_9H_{17}Cl_3O_3$	160–162	0.7
C_2H_5SH	$C_2H_5SCH_2CH_2CH(SC_2H_5)_2$	[19157-17-6]	$C_9H_{20}S_3$	143	1.3
$C_2H_5SH + HCl$	$ClCH_2CH_2CH(SC_2H_5)_2$	[19157-16-5]	$C_7H_{15}ClS_2$	115–117	1.5
$CH_2{=}CHCH_2OCH_2CH_2CHO +$ C_2H_5OH	$CH_2{=}CHCH_2OCH_2CH_2CH(OC_2H_5)_2$	[107023-55-2]	$C_{10}H_{20}O_3$	86	1.7
$C_2H_5OH + HCl$	$ClCH_2CH_2CH(OC_2H_5)_2$	[35573-93-4]	$C_7H_{15}ClO_2$	58–62	1.1
$CH_3OH + HCl$	$ClCH_2CH_2CH(OCH_3)_2$	[35502-06-8]	$C_5H_{11}ClO_2$	45	1.6
$C_2H_5OH + HBr$	$BrCH_2CH_2CH(OC_2H_5)_2$	[59067-07-1]	$C_7H_{15}BrO_2$	80–90	2.7
$n\text{-}C_2H_7OH + HCl$	$ClCH_2CH_2CH(OC_3H_7)_2$	[35502-07-9]	$C_9H_{19}ClO_2$	87	2.7
$i\text{-}C_4H_9OH + HCl$	$ClCH_2CH_2CH(O\textit{i}\text{-}C_4H_9)_2$	[35502-09-1]	$C_{11}H_{23}ClO_2$	105	0.6

[a] To convert kPa to mm Hg, multiply by 7.5.
[b] No CAS Registry Number has been assigned.

3-hydroxypropionaldehyde ($C_3H_6O_2$) [2134-29-4]. Hydrogenation of 3-hydroxypropionaldehyde forms 1,3-propanediol.

$$CH_2{=}CH{-}CHO + H_2O \longrightarrow HO{-}CH_2{-}CH_2{-}CHO \xrightarrow{H_2} HO{-}CH_2{-}CH_2{-}CH_2{-}OH$$

Competitive routes to 1,3-propanediol are ethylene oxide hydroformylation (110) and biofermentation of corn (111). The largest anticipated use of 1,3-propanediol is in the manufacture of polytrimethylene terephthalate (PTT). Shell and Dupont have announced commercial processes for this polyester.

One of the largest uses of acrolein is the production of 3-methylmercaptopropionaldehyde, (C_4H_8OS) [3268-49-3], which is an intermediate in the synthesis of D,L-methionine ($C_5H_{11}NO_2S$) [59-51-8], an important chicken feed supplement.

$$CH_2{=}CH{-}CHO + CH_3SH \longrightarrow CH_3SCH_2CH_2CHO$$

3-Methylmercaptopropionaldehyde is also used to make the methionine hydroxy analogue $CH_3SCH_2CH_2CH(OH)COOH$, ($C_5H_{10}O_2S$) [583-91-5], which is used commercially as an effective source of methionine activity (112). All commercial syntheses of methionine and methionine hydroxy analogues are based on the use of acrolein as a raw material. More than 570,000 t of this amino acid are produced yearly (43) (see AMINO ACIDS). One method for the preparation of methionine from acrolein via 3-methylmercaptopropionaldehyde is as follows:

$$H_3C{-}S{-}CH_2{-}CH_2{-}CHO \xrightarrow{HCN} H_3C{-}S{-}CH_2{-}CH_2{-}CH(OH){-}CN \xrightarrow[\text{2. hydrolysis}]{\text{1. }NH_3} H_3C{-}S{-}CH_2{-}CH_2{-}CH(NH_2){-}COOH$$

Methyl mercaptan adds to acrolein in nearly quantitative yields in the presence of a variety of basic catalysts (113–115). Other alkylmercaptopropionaldehydes produced by the reaction of acrolein with a mercaptan are known. Table 9 lists a variety of these and their boiling points (116).

Although the liquid-phase reaction of acrolein with ammonia produces polymers of little interest, the vapor-phase reaction, in the presence of a dehydration catalyst produces high yields of β-picoline, (C_6H_7N) [108-99-6] and pyridine, (C_5H_5N) [110-86-1] in a ratio of ~2:1.

$$CH_2{=}CH{-}CHO + NH_3 \xrightarrow[\text{cat}]{\Delta} \text{3-methylpyridine } (\beta\text{-picoline}) + \text{pyridine} + H_2O$$

β-Picoline may serve as an important source of nicotinic acid, ($C_6H_5NO_2$) [59-67-6] for dietary supplements. A variety of substituted pyridines may be prepared from acrolein (117–125).

Acrolein may participate in Diels–Alder reactions as the dieneophile or as the diene (126–130).

Table 9. **Alkylmercaptopropionaldehydes from Addition of Mercaptans to the Acrolein Double Bond**

	Product				
Compound added	Structure	CAS Registry Number	Molecular formula	bp, °C	Pressure at bp, kPa[a]
CH_3SH	$CH_3SCH_2CH_2CHO$	[3268-49-3]	C_4H_8OS	54–56	1.5
C_2H_5SH	$C_2H_5SCH_2CH_2CHO$	[5454-45-5]	$C_5H_{10}OS$	63–65	1.5
n-C_3H_7SH	n-$C_3H_7SCH_2CH_2CHO$	[44768-66-3]	$C_6H_{12}OS$	75	0.86
$C_6H_5CH_2SH$	$C_6H_5CH_2SCH_2CH_2CHO$	[16979-50-3]	$C_{10}H_{12}OS$	142.3	0.8
$CH_3C(O)SH$	$CH_3C(O)SCH_2CH_2CHO$	[53943-93-4]	$C_5H_8O_2S$	89	1.5
CF_3SH	$CF_3SCH_2CH_2CHO$	[58019-54-6]	$C_4F_3H_5OS$	46.5	2.7

[a] To convert kPa to mm Hg, multiply by 7.5.

Table 10. **Products of Dienes Added to Acrolein**[a]

THBA Product[b]	CAS Registry Number	Time, h	bp, °C	Pressure at bp, kPa[c]
THBA	[100-50-5]	1	51–52	1.7
2,5-*endo*-methylene-THBA	[5453-80-5]	several[d]	70–72	2.7
4-methyl-THBA	[7560-64-7]	3	70–71	1.9
2,5-*endo*-ethylene-THBA	[40570-95-4]	8	84–85	1.6

[a] Reaction at 100°C unless otherwise noted; yields are 90–95%.
[b] THBA from butadiene; 2,5-*endo*-methylene from 1,3-cyclopentadiene; 4-methyl-THBA from 2-methylbutadiene; 2,5-*endo*-ethylene from 1,3-cyclohexadiene.
[c] To convert kPa to mm Hg, multiply by 7.5.
[d] At 25°C.

The participation of acrolein as the dienophile in Diels–Alder reactions is, in general, an exothermic process. Dienes such as cyclopentadiene and 1-diethylamino-1,3-butadiene react rapidly with acrolein at room temperature.

Several Diels–Alder reactions in which acrolein participates as the dienophile are of industrial significance. These reactions involve butadiene or substituted butadienes and yield the corresponding 1,2,5,6-tetrahydrobenzaldehyde derivative (THBA); examples are given in Table 10 (131). These products have found use in the epoxy and perfume/fragrance industries.

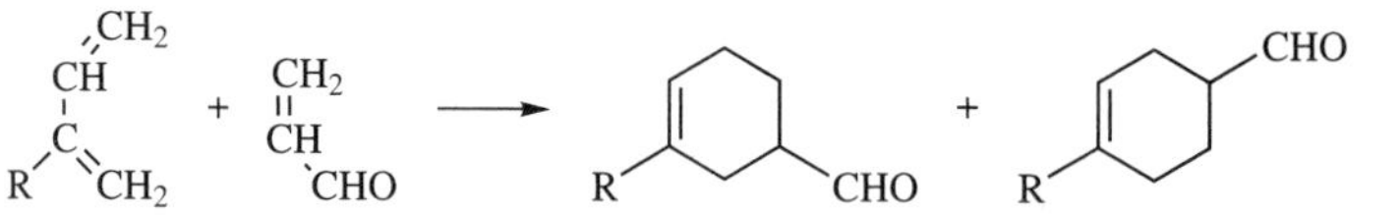

Many other acrolein derivatives produced via Diels–Alder reactions are classified as flavors and fragrances. Among those of commercial interest are lyral, ($C_{13}H_{22}O_2$), (1) [31906-04-4] (132, 133) and myrac aldehyde, ($C_{13}H_{20}O$), (2) [37677-14-8] (133, 134).

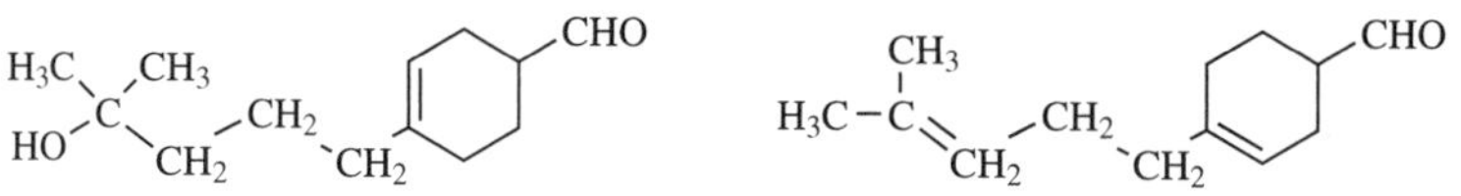

1 **2**

An industrially useful reaction in which acrolein participates as the diene is that with methyl vinyl ether. The product, methoxydihydropyran, ($C_6H_{10}O_2$) [4454-05-1] is an intermediate in the synthesis of glutaraldehyde, ($C_5H_8O_2$) [111-30-8].

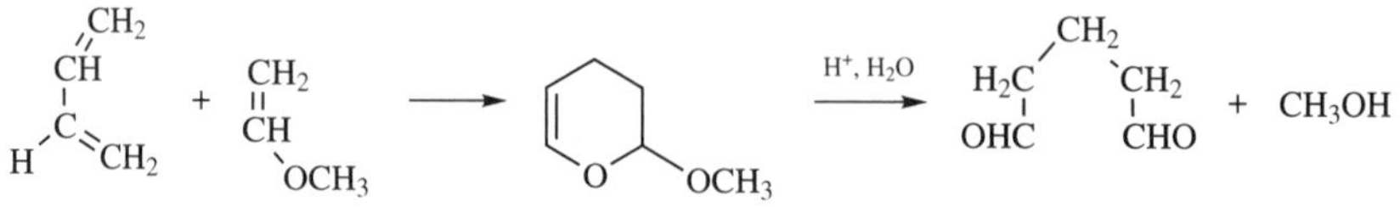

In addition to its principal use in biocide formulations (135), glutaraldehyde has been used in the film development and leather tanning industries (136). It may be converted to 1,5-pentanediol, ($C_5H_{12}O_2$) [111-29-5] or glutaric acid ($C_5H_8O_4$) [110-94-1].

Acting as both the diene and dienophile, acrolein undergoes a Diels–Alder reaction with itself to produce acrolein dimer, 3,4-dihydro-2-formyl-2*H*-pyran, ($C_6H_8O_2$) [100-73-2]. At room temperature, the rate of dimerization is very slow. However, at elevated temperatures (~200°C) and pressures the dimer may be produced in single-pass yields of 33% with selectivities >95%. Acrolein is efficiently regenerated from its dimer by a retro-Diels–Alder reaction at temperatures near 500°C. Use of this reaction has been proposed for delivery of high purity acrolein at remote locations (137).

CH2, CH, C, H, CH2 + CH2, CH, CHO ⇌ O, CHO

Acrolein dimer may be easily hydrated to α-hydroxyadipaldehyde, ($C_6H_{10}O_3$), [141-31-1] which may then be reduced to 1,2,6-hexanetriol, ($C_6H_{14}O_3$) [106-69-4]. Several uses for 1,2,6-hexanetriol, have been proposed including various polymer (in polyurethanes and polyesters), pharmaceutical, and cosmetics (as alternative to glycerol as a humectant) industry applications.

O, CHO —(H^+, H_2O)→ CH2, H2C, CH2, OHC, CH, HO, CHO —(H_2, cat)→ CH2, H2C, CH2, HOCH2, CH, HO, CH2OH

In the absence of inhibitors, acrolein polymerizes readily in the presence of anionic, cationic, or free-radical agents. The resulting polymers are insoluble, highly cross-linked solids with no known commercial use.

Copolymers, including one obtained by the oxidative copolymerization of acrolein with acrylic acid, a product of commercial interest, are known. There is a great variety of potential acrolein copolymers; however, significant commercial uses have not been developed. The possible application of polyacroleins or copolymers as polymeric reagents, polymeric complexing agents, and polymeric carriers has been recognized. Preparative methods as well as properties of the homopolymers and copolymers of acrolein have been reviewed (4, 9, 138).

5. Direct Uses of Acrolein

Because of its antimicrobial activity, acrolein has found use as an agent to control the growth of microbes in process feed lines, thereby controlling the rates of plugging and corrosion (see WASTES, INDUSTRIAL).

Acrolein at a concentration of <500 ppm is also used to protect liquid fuels against microorganisms. The dialkyl acetals of acrolein are also useful in this application. In addition, the growth of algae, aquatic weeds, and mollusks in recirculating process water systems is also controlled by acrolein. Currently, acrolein is used to control the growth of algae in oil fields and has also been used as an H_2S scavenger (139). The ability to use acrolein safely in these direct applications is a prime concern and is a deterrent to more widespread use.

In recent years, several acrolein derivatives have been proposed to provide a safer means of transport of acrolein to an application site. Acrolein dimer is thermally cracked to acrolein at temperatures near 500°C (137). Acetals of acrolein, most particularly, vinyl dioxolane, may revert to acrolein and alcohol in acidified aqueous solution.

6. Health and Safety Factors

The most frequently encountered hazards of acrolein are acute toxicity from inhalation and ocular irritation (140). Because of its high volatility, even a small spill can lead to a dangerous situation. Acrolein is highly irritating and a potent lacrimator. The odor threshold (50%), 0.23 mg/m^3 [0.09 ppm (v/v)], is close to the Occupational Safety and Health Administration (OSHA) permissible exposure limit (PEL) and ACGIH TLV-TWA8, 0.25 mg/m^3 (0.1 ppm). The odor threshold for 100% recognition, 0.5 mg/m^3 (0.2 ppm), is above the TLV (141) but the OSHA and ACGIH short-term exposure limit (STEL) is 0.8 mg/m^3 (0.3 ppm), so perception of acrolein will generally provide adequate warning.

Concentrations of acrolein vapor as low as 0.6 mg/m^3 (0.25 ppm) may irritate the respiratory tract, causing coughing, nasal discharge, chest discomfort or pain, and difficulty with breathing (142). A concentration of 5–10 mg/m^3 (2–4 ppm) is intolerable to most individuals in a minute or two (140) and is close to the concentration considered immediately dangerous to life and health (143). At higher concentrations there may be lung injury from inhaled acrolein, and prolonged exposure may be fatal. In a short time, exposure to 25 mg/m^3 (10 ppm) or more is lethal to humans (144).

Acrolein vapor is highly irritating to the eyes, causing pain or discomfort in the eye, profuse lacrimation, involuntary blinking, and marked reddening of the conjunctiva. Splashes of liquid acrolein will produce a severe injury to the eyelids and conjunctiva and chemical burns of the cornea.

A small amount of acrolein may be fatal if swallowed. It produces burns of the mouth, throat, esophagus, and stomach. Signs and symptoms of poisoning may include severe pain in the mouth, throat, chest, and abdomen; nausea; vomiting, which may contain blood; diarrhea; weakness and dizziness; and collapse and coma (142).

Acrolein is highly toxic by skin absorption. Brief contact may result in the absorption of harmful and possibly fatal amounts of material. Skin contact causes severe local irritation and chemical burns. Butyl rubber gloves should be used when handling acrolein (145).

There is no specific antidote for acrolein exposure. Treatment of exposure should be directed at the control of symptoms and the clinical condition. Most of the harmful effects of acrolein result from its highly irritating and corrosive properties.

Chronic human exposure is unlikely due to the lack of tolerance to acrolein. There is no evidence that acrolein is a human carcinogen (146), and inadequate animal data preclude any evaluation of its carcinogenicity (147). Acrolein has shown low, borderline, or moderate mutagenicity in bioassays, depending on the test system and frame of reference (148, 149). Animal studies that gave little indication of teratogenicity of acrolein are insufficient for determining whether acrolein is a teratogen (150). Some embryotoxicity over a narrow dose range with acrolein administered by injection indicates that acrolein is quite embryotoxic (150).

Acrolein is very flammable; its flash point is $<0^{\circ}C$, but a toxic vapor cloud will develop before a flammable one. The flammable limits in air are 2.8 and 31.0% lower and upper explosive limits, respectively, by volume. Acrolein is only partly soluble in water and will cause a floating fire, so alcohol type foam should be used in firefighting. The vapors are heavier than air and can travel along the ground and flash back from an ignition source.

Acrolein is a highly reactive chemical, and contamination of all types must be avoided. Violent polymerization may occur by contamination with either alkaline materials or strong mineral acids. Contamination by low molecular weight amines and pyridines such as α-picoline is especially hazardous because there is an induction period that may conceal the onset of an incident and allow a contaminant to accumulate unnoticed. After the onset of polymerization, the temperature can rise precipitously within minutes.

Acrolein reacts slowly in water to form 3-hydroxypropionaldehyde and then other condensation products from aldol and Michael reactions. Water dissolved in acrolein does not present a hazard. The reaction of acrolein with water is exothermic and the reaction proceeds slowly in dilute aqueous solution. This reaction will be hazardous in a two-phase adiabatic system in which acrolein is supplied from the upper layer to replenish that consumed in the lower, aqueous, layer. The rate at which these reactions occur will depend on the nature of the impurities in the water, the volume of the water layer, and the rate of heat removal. Thus a water layer must be avoided in stored acrolein.

Dimerization of acrolein is very slow at ambient temperatures but it can become a runaway reaction at elevated temperature ($\sim 90^{\circ}C$), a consideration in developing protection against fire exposure of stored acrolein.

7. Storage and Handling

The following cautions should be observed: Do not destroy or remove inhibitor. Do not contaminate with alkaline or strongly acidic materials. Do not store in the presence of a water layer. In the event of spillage or misuse that cause a release of product vapor to the atmosphere, thoroughly ventilate the area, especially near floor levels where vapors will collect.

Acrolein produced in the United States is stabilized against free-radical polymerization by 1000–2500 ppm of hydroquinone and is protected somewhat against base-catalyzed polymerization by ~ 100 ppm of acetic acid. To ensure stability, the pH of a 10% v/v solution of acrolein in water should be <6.

Since the principal hazard of contamination of acrolein is base-catalyzed polymerization, a "buffer" solution to shortstop such a polymerization is often employed for emergency addition to a reacting tank. A typical composition of this solution is 78% acetic acid, 15% water, and 7% hydroquinone. The acetic acid is the primary active ingredient. Water is added to depress the freezing point and to increase the solubility of hydroquinone. Hydroquinone (HQ) prevents free-radical polymerization. Such polymerization is not expected to be a safety hazard, but there is no reason to exclude HQ from the formulation. Sodium acetate may be included as well to stop polymerization by very strong acids. There is, however, a temperature rise when it is added to acrolein due to catalysis of the acetic acid–acrolein addition reaction.

Suitable materials of construction are steel, stainless steel, and aluminum 3003. Galvanized steel should not be used. Plastic tanks and lines are not recommended.

Storage tanks should have temperature monitoring with alarms to detect the onset of reactions. The design should comply with all applicable industry, federal, and local codes for a class 1B flammable liquid. The storage temperature should be $<37.8^\circ C$. Storage should be under an atmosphere of dry nitrogen and should vent vapors from the tank to a scrubber or flare.

In treatment of spills or wastes, the suppression of vapors is the first concern and the aquatic toxicity to plants, fish, and microorganisms is the second. Normal procedures for flammable liquids should also be carried out.

Even small spills and leaks (<0.45 kg) require extreme caution. Unless the spill is contained in a fume hood, do not remain in or enter the area unless equipped with full protective equipment and clothing. Self-contained breathing apparatus should be used if the odor of acrolein or eye irritation is sensed. Small spills may be covered with absorbant, treated with aqueous alkalies, and flushed with water.

Acrolein is very highly toxic to fish and to the microorganisms in a biological wastewater treatment plant. Avoid drainage to sewers or to natural waters. Safe, practical methods have been devised to handle contained spills of liquid acrolein. These entail covering the acrolein with 15 cm of 3M ATC foam to suppress evaporation followed by either (1) removing some of the foam and igniting the vapors to destroy most of the acrolein under controlled burning conditions, or (2) polymerizing the acrolein by the addition of a dilute (5–10%) aqueous sodium carbonate solution. In situations where the foam covering and controlled burning are not feasible, the acrolein spill may be covered uniformly with dry sodium carbonate amounting to 60–120 kg/m^3 (0.5–1 lb/gal) of acrolein, followed by dilution with 5–10 volumes of water per volume of acrolein. This procedure effectively destroys the acrolein by polymerization but leaves water-insoluble residue. More water ($\sim$20 volumes) is needed to get a solution or fine suspension of polymer. Other alkalies, such as dilute aqueous sodium hydroxide, will serve the same purpose but the polymerization is more violent than when the sodium carbonate is used.

8. Government Regulations

The Environmental Protection Agency (EPA) Risk Management Program (RMP), rule 40 CFR 68 lists acrolein as a toxic substance. Chemical processes containing acrolein in quantities of 5000 lb or more are subject to the RMP rule. The rule was promulgated to facilitate prevention and response planning for releases of extremely hazardous substances to the air and communication of these plans to the public. The rule strongly emphasizes prevention measures to minimize the consequences of a release to the surrounding community and environment.

OSHA Process Safety Management (PSM) rule 29 CFR 1910.119 lists acrolein as a toxic substance. Chemical processes containing acrolein quantities of 150 lb or more are subject to the rule. The purpose of the rule is to protect employees inside the workplace from process safety hazard.

The Comprehensive Environmental Response, Compensation, and Liability Act of 1980 (CERCLA) requires notification to the National Response Center of releases of quantities of hazardous substances equal to or greater than the reportable quantity (RQ) in 40 CFR 302.4. The reportable quantity for acrolein is 1 lb (0.454 kg).

The Superfund Amendments and Reauthorization Act of 1986 (SARA) Title III requires emergency planning based on threshold planning quantities (TPQ). The TPQ for acrolein is 500 lb (227 kg). SARA also requires submission of annual reports of release of toxic chemicals that appear on the list in 40 CFR 372.65 (for SARA 313). Acrolein appears on that list. This information must be included in all MSDSs that are copied and distributed for acrolein.

The Clean Air Act Amendments of 1990 list acrolein as a hazardous air pollutant. Processes involving acrolein may be subject to emission control requirements.

Acrolein in its "commercial product" (refined) form is a listed RCRA hazardous waste. Refined acrolein and wastes that contain or are derived from "commercial product" acrolein must be managed as RCRA hazardous waste with waste code P003.

Acrolein is a DOT Flammable Liquid having subsidiary DOT hazard classifications of Poison B and Corrosive Material. It is also an inhalation hazard that falls under the special packaging requirements of 49 CFR 173.3a.

BIBLIOGRAPHY

"Acrolein" in *ECT* 1st ed., Vol. 1, pp. 173–175, by R. L. Hasche, Tennessee Eastman Corporation; in *ECT* 1st ed; Suppl. 1, pp. 1–18, by H. R. Guest and H. A. Stansbury, Jr., Union Carbide Chemicals Company; "Acrolein and Derivatives" in *ECT* 2nd ed., Vol. 1, pp. 255–274, by H. R. Guest, B. W. Kiff, and H. A. Stansbury, Jr., Union Carbide Chemicals Company; in *ECT* 3rd ed., Vol. 1, pp. 277–297, by L. G. Hess, A. N. Kurtz, and D. B. Stanton, Union Carbide Corporation; *ECT* 4th ed., Vol. 1, pp. 232–251, by W. G. Etzkorn, J. J. Kurland, and W. D. Neilsen, Union Carbide Chemicals & Plastics Company Inc.; "Acrolein and Derivatives" in *ECT* (online), posting date: December 4, 2000, by W. G. Etzkorn, J. J. Kurland, and W. D. Neilsen, Union Carbide Chemicals & Plastics Company Inc.

CITED PUBLICATIONS

1. H. Schulz and H. Wagner, *Angew. Chem.* **62**, 105 (1950).
2. S. A. Ballard, H. de ViFinch, B. P. Geyer, G. W. Hearne, C. W. Smith, and R. R. Whetstone, *World Petroleum Congress Proceedings of the 4th Congress, Rome*, 1955, Sect. 4, Part C, pp. 141–154.

3. W. M. Weigert and H. Haschke, *Chem. Ztg.* **98**, 61 (1974).
4. R. C. Schulz in J. I. Kroschwitz, ed., *Encyclopedia of Polymer Science and Engineering*, 2nd ed., Vol. 1, Wiley-Interscience, New York, 1985, pp. 160–169.
5. T. Ohara, T. Sato, N. Shimizu, G. Prescher, H. Schwind, and O. Weiberg, *Ullman's* Encyclopedia of Industrial Chemistry, 5th ed., Vol. A1, 1985, pp. 149–160.
6. D. Arntz, M. Höpp, S. Jacobi, J. Sauer, T. Ohara, T. Sato, N. Shimizu, G. Prescher, H. Schwind, and O. Weiberg, *Ullmann's* Encyclopedia, Industrial Organic Chemicals, Starting Materials and Intermediates, Vol. 1, 1999, pp. 199–222.
7. S. Mourey, *Info. Chim. Mag.* **412**, 90 (1999).
8. P. H. M. Delaghe and M. Lautens, in L. A. Paquette, ed., *Encyclopedia of Reagent for Organic Synthesis*, Vol. 1, John Wiley & Sons Inc., Chichester, 1995, pp. 72–75.
9. C. W. Smith, ed., *Acrolein*, John Wiley & Sons, Inc., New York, 1962.
10. *Dictionary of Organic Compounds*, 5th ed., Vol. 5, Chapman and Hall, New York, 1982, p. 4784.
11. D. R. Lide and G. W. A. Milne, ed., *Handbook of Data on Organic Compounds*, 3rd ed., Vol. 5, CRC Press, Boca Raton, 1994, p. 4499.
12. *Dictionary of Organic Compounds*, 6th ed., Vol. 6, Chapman and Hall, London, 1996, p. 5432.
13. U.S. Pat. 2,383,711 (Aug. 28, 1945), A. Clark and R. S. Shutt (to Battelle Memorial Institute).
14. U.S. Pat. 2,593,437 (Apr. 22, 1952), E. P. Goodings and D. J. Hadley (to Distillers Co., Ltd.).
15. U.S. Pat. 2,451,485 (Oct. 19, 1948), G. W. Hearne and M. L. Adams (to Shell Development Co.).
16. U.S. Pat. 2,486,842 (Nov. 1, 1949), G. W. Hearne and M. L. Adams (to Shell Development Co.).
17. U.S. Pat. 2,606,932 (Aug. 12, 1952), R. M. Cole, C. L. Cunn, and G. J. Pierotti (to Shell Development Co.).
18. U.S. Pat. 2,941,007 (June 14, 1960), J. L. Callahan, R. W. Foreman, and F. Veatch (to Standard Oil Co.).
19. U.S. Pat. 3,171,859 (Mar. 2, 1965), K. Sennewald, K. Gehramann, W. Vogt, and S. Schaefer (to Knapsack-Griesheim, A.G.).
20. U.S. Pat. 3,454,630 (July 8, 1969), G. Yamaguchi and S. Takenaka (to Nippon Kayaku Co., Ltd.).
21. U.S. Pat. 3,576,764 (Apr. 27, 1971), G. Yamaguchi and S. Takenaka (to Nippon Kayaku Co., Ltd.).
22. Fr. Pat. 2,028,164 (1970), H. Ito, S. Nakamura and T. Nakano (to Toa Gosei Chem).
23. Ger. Pat. 2,133,110 (1972), T. Shiraishi, (to Sumitomo Chem.).
24. U.S. Pat. 3,825,600 (July 23, 1974), T. Ohara, M. Ueshima, and I. Yanagisawa (to Nippon Shokubai K.K.).
25. U.S. Pat. 3,907,712 (Sept. 23, 1975), T. Ohara, M. Ueshima, and I. Yanagisawa (to Nippon Shokubai K.K.).
26. Eur. Pat. 0417,722 (1990), W. Bock, D. Arntz, G. Prescher, and W. Burkhardt (to Degussa).
27. Eur. Pat. 0239,071 (1992), K. Sarumaru, E. Yamamoto, and T. Saito (to Mitsubishi Pet.).
28. U. S. Pat. 5,700,752 (1997) I. Kurimoto, T. Kawajiri, H. Onodera, M. Tanimoto, and Y. Aoki (to Nippon Shokubai).
29. U. S. Pat. 5,892,108 (1999), T. Shiotani, M. Sugiyama, T. Kuroda, and M. Oh-Kita (to Mitsubishi Rayon Co. Ltd.).
30. Y. Moro-Oka and W. Ueda, *Adv. Catal.* **40**, 233 (1994).
31. M. Baerns and O. V. Buyevskaya, *Erdöel, Erdgas, Kohle* **116**(1), 25 (2000).

32. U.S. Pat. 4,147,885 (Apr. 3, 1979), N. Shimizu, I. Yanagisawa, M. Takata, and T. Sato (to Nippon Shokubai K.K.).
33. U.S. Pat. 4,031,135 (June 21, 1977), H. Engelbach and co-workers (to BASF Aktiegesellschaft).
34. Eur. Pat. Appl. 253,409 (July 17, 1987), W. Etzkorn and G. Harkreader (to Union Carbide Corp.).
35. Eur. Pat. Appl. 257,565 (Aug. 20, 1987), W. Etzkorn and G. Harkreader (to Union Carbide Corp.).
36. Eur. Pat. Appl. 293,224 (May 27, 1988), M. Takata, M. Takamura, S. Uchida, and M. Sasaki (to Nippon Shokubai K.K.).
37. G. E. Schaal, *Hydrocarbon Process* **52**, 218 (1973).
38. W. Weigert, *Chem. Eng. News* **80**, 68 (1973).
39. *Chemical Week* **41** (Mar. 22, 2000).
40. *Chem. Mark. Rep.* (May 24,1999).
41. *Chemical Week* **34** (Aug. 5,1998).
42. *Chem. Mark. Rep.* **8** (Jan. 19, 1998).
43. *Chem. Mark. Rep.* **14** (June 15, 1998).
44. Jpn. Kokai Tokkyo Koho JP 63/146841 AZ [88/146841] (June 18, 1988), K. Sarumaru and T. Shibano (to Mitsubishi Petrochemical Co.).
45. J. B. Black, J. D. Scott, E. M. Serwicka, and J. B. Goodenough, *J. Catal.* **106**, 16 (1987).
46. E. M. Serwicka, J. B. Black, and J. B. Goodenough, *J. Catal.* **106**, 23 (1987).
47. T. V. Andrushkevich, *Catal. Rev. Sci. Eng.* **35**, 213 (1993).
48. Eur. Pat. 972 759 (2000), Y. Yoshida, H. Otake-shi and M. Oh-Kita (to Mitsubishi Rayon Co., Ltd.).
49. V. Kijenski, M. Glinski, and J. Reinhercs, *Stud. Surf. Sci. Catal.* **41** 231 (1988).
50. Y. Nagase and K. Wada, *Ibaraki Daigaku Kogakubu Kenkyu Shuho*, **33**, 223 (1985).
51. Y. Nagase, H. Hattori, and K. Tanabe, *Chem. Lett.* **10** 1615 (1983).
52. T. H. Vanderspurt, *Ann. N.Y. Acad. Sci.* **333**, 155 (1980).
53. U.S. Pat. 4,127,508 (Nov. 28, 1978), T. H. Vanderspurt (to Celanese Corp.).
54. Ger. Offen. DE2734811 (Feb. 9, 1978), T. H. Vanderspurt (to Celanese Corp.).
55. T. Nakano, S. Umano, Y. Kino, and Y. M. Ishii, *J. Org. Chem.* **53**, 3752 (1988).
56. Y. Nagase and T. Katou, *Chem. Lett.*, 436 (2000).
57. C. Mohr, H. Hofmeister, M. Lucas, and P. Claus, *Chem. Ing. Tech.* **71**, 869 (1999).
58. J. E. Bailie, C. H. Rochester, and G. J. Hutchings, *J. Mol. Catal. A: Chem.* **136**, 35 (1998).
59. M. A. Aramendía, V. Borau, C. Jiménez A. Marinas, J. M. Marinas, A. Porras, and F. J. Urbano, *Catal. Lett.*, **50** 173 (1998).
60. G. J. Hutchings, F. King, I. P. Okoye, M. B. Padley, and C. H. Rochester, *J. Catal.* **148**, 453 (1994).
61. B. Coq, F. Figuéras, C. Moreau, P. Moreau, and M. Warawdekar, *Catal. Lett.* **22**, 189 (1993).
62. U.S. Pat. 5,354,915 (Oct. 11, 1994), W. T. Reichle (to Union Carbide).
63. M. A. Aramendia, and co-workers, *React. Kinet. Catl. Lett.* **36**, 251 (1988).
64. L. M. Ryzhenko and A. D. Shebaldova, *Khim. Tekhnol. Elementoorg. Soedin. Polim.* 14 (1984).
65. D. L. Reger, M. M. Habib, and D. V. Fauth, *Tetrahedron Lett.*, 115 (1979).
66. M. Terassawa, K. Kaneda, T. Imanaka, and S. Tera, *J. Catal.* **51**, 406 (1978).
67. J. A. Cabello, and co-workers, *Bull. Soc. Chim. Belg.* **93**, 857 (1984).
68. H. M. Ali, A. A. Naiini, and C. H. Brubaker, Jr., *J. Mol. Catal.* **77**, 125 (1992).
69. K.-J. Yang and C. S. Chein, *Inorg. Chem.* **26** 2732 (1987).
70. Z. Poltarzewski, S. Galvagno, R. Pietropaolo, and P. Staiti, *J. Catal.* **102**, 190 (1986).

71. M. Funakoshi, H. Komiyama, and H. Inoue, *Chem. Lett.*, 245 (1985).
72. A. D. Shabaldova, and co-workers, in *Nukleofil'*nye Reacts. Karbonil'nykh Soedin. (conference proceedings, Saratov, USSR) 87–89 (1982).
73. G. P. Pez and R. A. Grey, *Fund. Res. Homogenous Catal.* **4**, 97 (1984).
74. J. M. Campello, A. Garcia, D. Luna, and J. M. Marinas, *React. Kinet. Catal. Lett.* **21**, 209 (1982).
75. Jpn. Pat. Jp 57/91743 AZ [82/917343] (June 8, 1982) (to Agency of Industrial Science & Technology).
76. G. V. Kudryavtsev, A. Yu Stakheev, and G. V. Lisichkin, *Zh. Vses. Khim. Ova.* **27**, 232 (1982).
77. J. M. Campello, A. Garcia, D. Luna, and J. M. Marinas, *Bull. Soc. Chim. Belg.* **91**, 131 (1982).
78. R. A. Grey, G. P. Pez, and A. Wallo, *J. Am. Chem. Soc.* **103**, 7536 (1981).
79. K. Murata and A. Matsuda, *Bull. Chem. Soc. Jpn.* **54**, 1989 (1981).
80. K. Kaneda, and co-workers, *Fund. Res. Homogenous Catal.* **3**, 671 (1979).
81. Y. Nagase, *Ibaraki Daigaku Kogakubu Kenkyu Shuho* **33**, 217 (1985).
82. Eur. Pat. Appl. EP 183225 Al (June 4, 1986), Y. Shimasaki, Y. Hino, and M. Ueshima (to Nippon Shokubai Kagaku Kogyo Co., Ltd.).
83. A. Alba, and co-workers, *React. Kinet. Catal. Lett.* **25**, 45 (1984).
84. V. P. Kukolev, N. A. Balyushima, and G. H. Chukhadzhyan, *Arm. Khim. Zh.* **35**, 688 (1982).
85. G. Horanyi and K. Torkos, *J. Electroanal. Chem. Interfacial Electrochem.* **136**, 301 (1982).
86. U. S. Pat. 4,292,452A (Sept. 29, 1981), R. J. Lee, D. H. Meyer, and D. M. Senneke (to Standard Oil Co.).
87. R. W. Hoffman and T. Herold, *Chem. Ber.* **114**, 375 (1981).
88. Y. Nagse and T. Washiyama, *Ibaraki Daigaku Kogakubu Kenkyu Shuho* **27**, 171 (1979).
89. Y. Ho and R. R. Squires, *Org. Mass Specrom.* **28**, 1658 (1993).
90. Jpn. Kokai Tokkyo Koho JP 2000 72,708 (March 7, 2000), H. Kawasaki and T. Yoshitome (to Idemitsu Petrochemical Co. Ltd., Japan).
91. G. V. Kryshtal, D. Dvorak, Z. Arnold, and L. A. Yanovskaya, *Isv. Akad. Nauk. SSSR, Ser. Khim.* **4**, 921 (1986).
92. Ref. 9, p. 140 and references cited therein.
93. PCT/WO 98 50,339 (Nov. 12, 1998), H. Kawasaki and T. Jintoku (to Idemitsu Petrochemical Co. Ltd., Japan).
94. Eur. Pat. Appl. EP 757,978 (Feb. 12, 1997), M. Hoepp, K. Koehler, and D. Arntz (to Degussa Aktiengesellschaft, Germany).
95. Eur. Pat. Appl. EP 548,520 (June 30, 1993), H. L. Strong, D. A. Cortes, and Z. Ahmed (to American Cyanamid Co.).
96. U. S. Pat. 5,177,266 (Jan. 5, 1993), H. L. Strong (to American Cyanamid Co.).
97. Jpn. Kokai Tokkyo Koho JP 2000 212,115 (Aug. 2, 2000), K. Maki (to Nippon Shokubai Kagaku Kogyo Co. Ltd., Japan).
98. Ref. 9, p. 122 and references cited therein.
99. Jpn. Kokai Tokkyo Koho JP 11 315,075 [99 315,075] (Nov. 16, 1999), K. Maki (to Nippon Shokubai Kagaku Kogyo Co. Ltd., Japan).
100. Eur. Pat. Appl. EP 704,441 (Apr. 3, 1996), M. Hoepp, D. Arntz, W. Boeck, A. Bosseplois, and K. Raible (to Degussa Aktiengesellschaft, Germany).
101. F. M. Bautista, J. M. Campelo, A. García, J. León, D. Luna, and J. M. Marinas, *J. Chem. Soc. Perkin Trans.*, 815 (1995).
102. PCT/WO 95 19,975 (July 27, 1995), B. V. Gregorovich (to du Pont de Nemours, E. I., and Co.).

103. Jpn. Kokai Tokkyo Koho JP 06 01,744 [94 01,744] (Jan. 11, 1994), S. Myazaki and H. Sonobe (to Mitsubishi Rayon Co.).
104. Ref. 9, pp. 124 and references cited therein.
105. Ref. 9, p. 130 and references cited therein.
106. PCT/WO 96 28,409 (Sep. 19, 1996), S. A. Brew, S. F. T. Froom, and S. R. Hodge (to BP Chemicals Limited).
107. D. Arntz, T. Haas, A. Müller, and N. Wiegand, *Chem. Ing. Tech.* **63** 733 (1991).
108. U. S. Pat. 5,284,979 (Feb. 8, 1994),. T. Haas, G. Böhme, and D. Arntz (to Degussa).
109. U. S. Pat. 5,364,984 (Nov. 15, 1994),. D. Arntz, T. Haas, and A. Schäfer-Sindlinger (to Degussa).
110. PCT/WO 96 10,550 (Apr. 4, 1996),. J. P. Arhancet, T. C. Forschner, J. B. Powell, T. C. Semple, L. H. Slaugh, T. B. Thomason, and P. R. Weider (to Shell Development, Co.).
111. PCT/WO 99 58,686 (Nov. 18, 1999),. G. M. Whited, B. Bulthuis, D, Trimbur, and A. A. Gatenby (to to du Pont de Nemours, E. I., and Co.; Genencor Int., Inc.).
112. U. S. Pat. 4,353,924 (Oct. 12, 1982), J. W. Beher, D. L. Mansfield, and D. J. Weinkauff (to Monsanto Company).
113. Rom. Pat. RO 85095B (Oct. 30, 1984), A. M. Pavlouschi, L. Levinta, and G. H. Gross (to Combinatul Petrochimic, Pitesti).
114. Jpn. Kokai Tokkyo Koho JP 55/16135 [80/16135] (Apr. 30, 1980), (to Asahi Chemical Industry Co., Ltd.).
115. PCT/WO 96 40,631 (Dec 19, 1996), T. F. Blackburn, P. F. Pellegrin, and A. H. Krantz (to Novus International, Inc., USA).
116. Ref. 9, p. 118 and references cited therein.
117. T. Y. Zhang, J. R. Stout, J. G. Keay, F. V. Scriven, J. E. Toomey, and G. L. Goe, *Tetrahedron* **51**, 13177 (1995)
118. Eur. Pat. Appl. EP 299362 A1 (Jan. 18, 1989), K. Nagao(to Osaka Soda Co., Ltd.).
119. Ger. Offen. DE 3634259 A1 (Apr. 21, 1988), W. Hoelderich, N. Goetz, and G. Fouquet (to BASF A.G.).
120. Ger. Offen. DE 3634975 A1 (Apr. 30, 1987), R. J. Doehner, Jr. (to American Cyanamid Co.).
121. Ger. Offen. DE 3337569 A1 (Apr. 25, 1985), T. Dockner, H. Hagen, and H. Krug (to BASF AG).
122. J. I. Grayson and R. Dinkel, *Helv. Chim. Acta*, 67, 2100 (1984).
123. C. Wang and Y. Li, *Yiyao Gongye* **6**, 1 (1984).
124. Eur. Pat. Appl. EP75727 A2 (Apr. 6, 1983), J. I. Grayson and R. Dinkel (to Lonza A.G.).
125. A. T. Soldatenkov and co-workers, *Zh. Org. Khim.* **16**, 188 (1980).
126. Eur. Pat. Appl. EP43507 A2 (Jan. 13, 1982), K. Bruns and T. N. Dang (to Henkel K. G. A. A.).
127. G. A. Trofimov, V. I. Lavrov, and L. N. Parshina, *Zh. Org. Khim.* **17**, 1716 (1981).
128. Jpn. Kokai Tokkyo Koho JP 61/161241 AZ [86/161241] (July 21, 1986), K. Inoue, H. Takeda, and M. Kobayashi (to Mitsubishi Rayon Co., Ltd.).
129. Jpn. Kokai Tokkyo Koho JP 62/141097 AZ [87/141097] (June 24, 1987), N. Tanaka, H. Takada, M. Oku, and A. Kimura (to Koa Corp.).
130. K. G. Akopyan, and co-workers, *Prom-st. Stroit. Arkhit. Arm.*, 34 (1988).
131. Ref. 9, pp. 216 and references cited therein.
132. U.S. Pat. 4,007,137 (Feb. 8, 1977), J. M. Sanders, W. L. Schreiber, and J. B. Hall (to International Flavors & Fragrances).
133. U.S. Pat. 4,107,217 (Aug. 15, 1978), W. L. Schreiber and A. O. Pittet (to International Flavors and Fragrances).
134. Ger. Offen., DE2,643,062 (Apr. 14, 1977), J. M. Sanders and co-workers (to International Flavors and Fragrances).

135. U. S. Pat. 4,244,876 (Jan. 13, 1981), G. H. Warner, L. F. Theiling, and M. G. Freid (to Union Carbide Corp.).
136. U. S. Pat. 2,941,859 (June 21, 1960), M. L. Fein and E. M. Filachione (to Union Carbide Corp.).
137. U. S. Pat. 5,243,082 (Sept 7, 1993), W. G. Etzkorn and W. D. Neilsen (to Union Carbide Corp.).
138. N. Yamashita in J. C. Salamone, ed., *Polymeric Materials Encyclopedia*, Vol. 1, CRC Press, Boca Raton, 1996, pp. 40–47.
139. Brit. Pat. Appl. GB 2023123 (Dec. 28, 1979), C. L. Kissel and F. F. Caserio (to Magna Corp.).
140. R. O. Beauchamp, D. A. Andjelkovich, A. D. Kligerman, K. T. Morgan, and H. d'A. Heck, *CRC Crit. Rev. Toxicol.* **14**, 309 (1985).
141. B. L. Carson, C. M. Beall, H. V. Ellis, L. H. Baker, and B. L. Herndon, Acrolein Health Effects, NTIS PB82-161282; EPA-460/3-81-034, *Gov. Rep. Announce.* Index **12**, 9–12 (1981). [A 121-page review of health effects literature primarily related to inhalation exposure.]
142. Acrolein, in *Material Safety Data Sheet, Union Carbide Chemicals and Plastics Company Inc.*, Specialty Chemicals Division, August 31, 1999.
143. Ref. 140, p. 339.
144. Syracuse Research Corporation, *Information Profiles on Potential Occupational Hazards*, Vol. 1, Single Chemicals Acrolein, NTIS PB81-147951, U. S. Department of Commerce, Springfield, Va., 1979, p. 11.
145. K. Forsberg and Z. F. Mansdorf, *Quick Selection Guide to Chemical Protective Clothing*, 3rd ed., John Wiley & Sons Inc., 1997, p 52.
146. Ref. 140, p. 342.
147. Acrolein, in *IARC Monographs on the Evaluation of Carcinogenic Risk of Chemicals to Humans. Some Monomers, Plastics and Synthetic Elastomers and Acrolein*, Vol. 19, International Agency for Research on Cancer, Lyon, France, 1979, pp. 479– 494.
148. Ref. 144, p. 9.
149. Ref. 141, p. 2.
150. Ref. 140, pp. 334–345.

W. G. Etzkorn
S. E. Pedersen
T. E. Snead
The Dow Company

ACRYLAMIDE

1. Introduction

Acrylamide [79-06-1] (NIOSH No: A533250) has been commercially available since the mid-1950s and has shown steady growth since that time, but is still considered a small volume commodity. Its formula,$H_2C{=}CHCONH_2$ (2-propeneamide), indicates a simple chemical, but it is by far the most important member of the series of acrylic and methacrylic amides. Water soluble polyacrylamides

(1) represent the most important applications. The largest use in this catagory is as a dewatering aid for sludges in the treatment of effluent from municipal wastewater treatment plants and industrial processes.

Other uses include flocculants in feed water treatment for industrial purposes, the mining industry and various other process industries, soil stabilization, papermaking aids, and thickeners. Smaller but none the less important uses include dye acceptors; polymers for promoting adhesion; additives for textiles, paints, and cement; increasing the softening point and solvent resistance of resins; components of photopolymerizable systems; and cross-linking agents in vinyl polymers.

2. Physical Properties

Acrylamide is a white crystalline solid that is quite stable at ambient conditions, and, even at temperatures as high as its melting point (for 1 day in the absence of light), no significant polymer formation is observed. Above its melting point, however, liquid acrylamide may polymerize rapidly with significant heat evolution. Precautions should be taken when handling even small quantities of molten material. In addition to the solid form, a 50% aqueous solution of acrylamide is a popular commercial product today. This solution is stabilized by small amounts of cupric ion (25–30 ppm based on monomer) and soluble oxygen. Several other stabilizers are also available for the aqueous monomer solution, such as ethylenediaminetetraacetic acid (EDTA) [60-00-4] (2), ferric ion (3), and nitrite (4,5). The only effect of oxygen is to increase the induction period for polymerization (6). Iron complexes of cyanogen or thiocyanogen have proven to be useful stabilizers for salt-containing acrylamide solutions (7). The physical properties of solid acrylamide monomer are summarized in Table 1. Solubilities of acrylamide in various solvents are given in Table 2, and typical physical properties of a 50% solution in water appear in Table 3.

3. Chemical Properties

Acrylamide, C_3H_5NO, is an interesting difunctional monomer containing a reactive electron-deficient double bond and an amide group, and it undergoes reactions typical of those two functionalities. It exhibits both weak acidic and basic properties. The electron-withdrawing carboxamide group activates the double bond, that consequently reacts readily with nucleophilic reagents, eg, by addition.

$$\mathrm{Nuc:H} + \mathrm{CH_2{=}CHCONH_2} \longrightarrow \mathrm{NucCH_2CH_2CONH_2}$$

Many of these reactions are reversible, and for the stronger nucleophiles they usually proceed the fastest. Typical examples are the addition of ammonia, amines, phosphines, and bisulfite. Alkaline conditions permit the addition of mercaptans, sulfides, ketones, nitroalkanes, and alcohols to acrylamide. Good examples of alcohol reactions are those involving polymeric alcohols such as poly(vinyl alcohol), cellulose, and starch. The alkaline conditions employed with

Table 1. **Physical Properties of Solid Acrylamide Monomer**[a]

Property	Value
molecular weight	71.08
melting point, °C	84.5 ±
vapor pressure, Pa[b]	
25°C	0.9
40°C	4.4
50°C	9.3
boiling point, °C	
0.27 kPa[b]	87
0.67 kPa[b]	103
1.4 kPa[b]	116.5
3.3 kPa[b]	136
heat of polymerization, kJ/mol[c]	−82.8
density, g/mL at 30°C	1.122
equilibrium moisture content, particle size 355 μm[d], at 22.8°C, 50% rh	1.7 g of water/kg of dry acrylamide
crystal system	monoclinic or triclinic
crystal habit	thin tabular to laminar
refractive indexes	
n_x	1.460 (calcd)
n_y	1.550 ± 0.003
n_z	1.581 ± 0.003
optic axial angles	2E 98°, 2V 58°
optic sign	(−)

[a] Ref. 5.
[b] To convert kPa to mm Hg, multiply by 7.5.
[c] To convert kJ/mol to kcal/mol, divide by 4.184.
[d] 45 mesh.

Table 2. **Solubilities of Acrylamide in Various Solvents at 30°C**

Solvent	g/100 mL
acetonitrile	39.6
acetone	63.1
benzene	0.346
ethylene glycol monobutyl ether	31
chloroform	2.66
1,2-dichloroethane	1.50
dimethylformamide	119
dimethyl sulfoxide	124
dioxane	30
ethanol	86.2
ethyl acetate	12.6
n-heptane	0.0068
methanol	155
pyridine	61.9
water	215.5
carbon tetrachloride	0.038

Table 3. **Physical Properties of 50% Aqueous Acrylamide Solution**[a]

Property	Value
pH	5.0–6.5
refractive index range, 25°C (48–52%)	1.4085–1.4148
viscosity, mPa (= cP) at 25°C	2.71
specific gravity, at 25°C	1.0412
density, 25/4°C	1.038
crystallization point, °C	8–13
partial phase diagram	
eutectic temperature, °C	−8.9
eutectic composition, wt%	31.2
boiling point at 101.3 kPa,[b] °C	99–104
vapor pressure	
at 23°C, kPa[b]	2.407
at 70°C, kPa[b]	27.93
specific heat (20–50°C), (20–50°C),J/(g · K)[c]	3.47
heat of dilution to 20 wt%, J/g soln[c]	−4.6
heat of polymerization, kJ/mol[c]	−85.4
heat of melting (solution), melting range −17.3 to + 19.7°C, J/g[c]	247.7
flammability	nonflammable

[a] Ref.5.
[b] To convert kPa to mm Hg, multiply by 7.5.
[c] To convert J to cal, divide by 4.184.

these reactions result in partial hydrolysis of the amide, yielding mixed carbamoylethyl and carboxyethyl products.

Some specific examples include the noncatalytic reaction of acrylamide with primary amines to produce a mono or bis product (5).

$$RNH_2 + CH_2{=}CHCONH_2 \longrightarrow RNHCH_2CH_2CONH_2 \longrightarrow RN(CH_2CH_2CONH_2)_2$$

Secondary amines give only a monosubstituted product. Both of these reactions are thermally reversible. The product with ammonia (3,3′,3″-nitrilotrispropionamide [2664-61-1], $C_9H_{18}N_4O_3$) (5) is frequently found in crystalline acrylamide as a minor impurity and affects the free-radical polymerization. An extensive study (8) has determined the structural requirements of the amines to form thermally reversible products. Unsymmetrical dialkyl hydrazines add through the unsubstituted nitrogen in basic medium and through the substituted nitrogen in acidic medium (9). Monoalkylhydroxylamine hydrochlorides react with preservation of the hydroxylamine structure (10). Primary nitramines combine in such a way as to keep the nitramine structure intact.

The reaction with sodium sulfite or bisulfite (5,11) to yield sodium-β-sulfopropionamide [19298-89-6] ($C_3H_7NO_4S{\cdot}Na$) is very useful since it can be used as a scavenger for acrylamide monomer. The reaction proceeds very rapidly even at room temperature, and the product has low toxicity. Reactions with phosphines and phosphine oxides have been studied (12), and the products are potentially useful because of their fire retardant properties. Reactions with sulfide and dithiocarbamates proceed readily but have no applications (5). However, the reaction with mercaptide ions has been used for analytical purposes (13).

Water reacts with the amide group (5) to form hydrolysis products, and other hydroxy compounds, such as alcohols and phenols, react readily to form ether compounds. Primary aliphatic alcohols are the most reactive and the reactions are complicated by partial hydrolysis of the amide groups by any water present.

Activated ketones react with acrylamide to yield adducts that frequently cyclize to lactams (14). The lactams can be hydrolyzed to yield substituted propionic acids. Chlorine and bromine react with acrylamide in aqueous solution to yield α,β-dihalopropionamide (5). Under acidic conditions, sizable quantities of acrylamide can be removed from water by chlorination (15). Hydrochloric and hydrobromic acids add to give β-halopropionamides. These adducts are also thermally reversible. A patent describes a procedure to prepare *N*-substituted acrylamide by direct transamidation of acrylamide (16). Dienes react with acrylamide to form Diels-Alder type adducts (17,18). Improved yields in the aza-annelation of cyclic ketones by the use of enamines and imines have been reported (19–21). Palladium reduced with borohydride (22), nickel boride (23), or rhodium carbonyl (24) reduces the double bond of acrylamide to yield propionamide, and acrylamide can be oxidized to a glycol with sodium hypochlorite using osmium tetroxide as a catalyst (25). In contrast, if osmium is not present, the attack occurs at the nitrogen to yield *N*-vinyl-*N*′-acryloylurea [19396-55-5] ($C_6H_8N_2O_2$) (26). When treated with a strong base in an aprotic solvent, acrylamide forms a head-to-tail dimer, 3-acrylamidopropionamide [21963-06-4]($C_6H_{10}N_2O_2$) (27). Electrolytic reductive dimerization of acrylamide proceeds through tail-to-tail addition to yield adipamide [628-94-4], ($C_6H_{12}N_2O_2$) (28).

The most important reactions of acrylamide are those that produce vinyl addition polymers (see ACRYLAMIDE POLYMERS). The initiation and termination mechanisms depend on the catalyst system, but the reaction can be started by any free-radical source. In practice, redox couples such as sodium persulfate and sodium bisulfite are commonly used, and the highest molecular weight polymers are obtained in aqueous solution, with molecular weights of several million prepared routinely. Acrylamide is remarkable for the very large value of $k_p/k_t^{1/2}$, $1.8 \times 10^4/(1.45 \times 10^7)^{1/2}$, which is a measure of chain length in the polymerization. However, it may be necessary to remove the inhibitor (cupric ions) from aqueous acrylamide solutions to obtain the desired polymerization results. Copolymers with acrylamide are also prepared with ease, although the molecular weights are consistently lower than that of polyacrylamide prepared under similar conditions. Acrylamide copolymerizes readily by a free-radical mechanism with other acrylates, methacrylates, and styrene. Acrylamide may be polymerized by a hydrogen-transfer mechanism catalyzed by strong base in basic or aprotic solvents. The product is poly(β-alanine) or nylon-3 (29), which has properties similar to natural silk. This polymer, on hydrolysis, yields β-aminopropionic acid. A hydrogen-transfer copolymer with acrolein has also been reported (30). A biocatalytic method of removing residual monomer from polymers that could become very important to acrylamide polymer users has been described (31).

The amide group is readily hydrolyzed to acrylic acid, and this reaction is kinetically faster in base than in acid solutions (5,32,33). However, hydrolysis of *N*-alkyl derivatives proceeds at slower rates. The presence of an electron-withdrawing group on nitrogen not only facilitates hydrolysis but also affects the polymerization behavior of these derivatives (34,35). With concentrated sulfuric acid, acrylamide forms acrylamide sulfate salt, the intermediate of the former sulfuric acid process for producing acrylamide commercially. Further reaction of the salt with alcohols produces acrylate esters (5). In strongly alkaline anhydrous solutions, a potassium salt can be formed by reaction with potassium *tert*-butoxide in *tert*-butyl alcohol at room temperature (36).

Several other interesting reactions include acrylamide transition metal complexes (37–40), complexes with nucleosides (41) in dimethyl sulfoxide solution, and also complexes with several inorganic salts (42–44). Dehydration of acrylamide by treatment with fused manganese dioxide (45) at 500°C or with phosphorus pentoxide (46) yields acrylonitrile. Aldehydes such as formaldehyde, glyoxal, and chloral hydrate react with acrylamide under neutral and alkaline conditions, producing the corresponding *N*-methylolacrylamide [924-42-5] (47,48). Under acidic conditions, *N,N*-methylenebisacrylamide [110-26-9] $[(H_2C{=}CHCONH)_2CH_2]$ is produced from formaldehyde and acrylamide (49). Under acidic conditions, methoylol ethers are formed from hydroxyl compounds and *N*-methylolacrylamide (50). Condensation products derived from *N*-methylolacrylamide and polyphenols have also been reported (51). By using *p*-toluenesulfonic acid as the catalyst in dioxane or ethyl acetate solvent, *N,N′*-oxydimethylenebisacrylamide [16958-71-7] ($C_8H_{12}O_3N_2$) has been obtained (52). These difunctional products have similar copolymerization parameters to acrylamide and are useful as cross-linking agents. Alcohols can be used to cap the methylol compound to provide the less reactive methylol ethers. Methanol is commonly employed, but, where increased compatibility with oleophilic systems is desired, one of the butanols is the preferred alcohol; oxalic acid is an example of a suitable catalyst (50). This reaction also occurs with cellulosic hydroxyls. Provided the system is not basic, the methylol derivative may also be condensed with carbamate esters (53), secondary amines (54), or phosphines (12) without involving the double bond. Acrylamido-*N*-glycolic acid and diacrylamidoacetic acid can be obtained from acrylamide and glyoxylic acid (55–58). *N*-Acylacrylamides are of minor interest industrially. One member of the series, diacrylamide [20602-80-6] ($C_6H_7NO_2$), is a suspected side-reaction product in the sulfuric acid process of manufacture. It may be prepared by the reaction of acrylamide with acrylic anhydride or acryloyl chloride. A specific preparation for *N*-acetylacrylamide [1432-45-7] ($C_5H_7NO_2$) is the addition of ketene to acrylamide (59).

4. Manufacture

The current routes to acrylamide are based on the hydration of inexpensive and readily available acrylonitrile [107-13-1] (C_3H_3N, 2-propenenitrile, vinyl cyanide, VCN, or cyanoethene) (see ACRYLONITRILE). For many years, the principal process for making acrylamide was a reaction of acrylonitrile with $H_2SO_4{\cdot}H_2O$ followed by separation of the product from its sulfate salt using a base neutralization or an ion exclusion column (60).

$$CH_2{=}CHCN + H_2SO_4{\cdot}H_2O \longrightarrow CH_2{=}CHCONH_2{\cdot}H_2SO_4$$

This process yields satisfactory monomer, either as crystals or in solution, but it also produces unwanted sulfates and waste streams. The reaction was usually run in glass-lined equipment at 90–100°C with a residence time of 1 h. Long residence time and high reaction temperatures increase the selectivity to impurities, especially polymers and acrylic acid, which controls the properties of subsequent polymer products.

The ratio of reactants had to be controlled very closely to suppress these impurities. Recovery of the acrylamide product from the acid process was the most expensive and difficult part of the process. Large scale production depended on two different methods. If solid crystalline monomer was desired, the acrylamide sulfate was neutralized with ammonia to yield ammonium sulfate. The acrylamide crystallized on cooling, leaving ammonium sulfate, which had to be disposed of in some way. The second method of purification involved ion exclusion (60), which utilized a sulfonic acid ion-exchange resin and produced a dilute solution of acrylamide in water. A dilute sulfuric acid waste stream was again produced, and, in either case, the waste stream represented a problem as well as an increased production cost. As far as can be determined, no commercial acrylamide is produced today via this process.

Even in 1960, a catalytic route was considered the answer to the pollution problem and the by-product sulfate, but nearly 10 years elapsed before a process was developed that could be used commercially. Some of the earlier attempts included hydrolysis of acrylonitrile on a sulfonic acid ion-exchange resin (61). Manganese dioxide showed some catalytic activity (62), and copper ions present in two different valence states were described as catalytically active (63), but copper metal by itself was not active. A variety of catalysts, such as Urushibara or Ullmann copper and nickel, were used for the hydrolysis of aromatic nitriles, but aliphatic nitriles did not react using these catalysts (64). Beginning in 1971, a series of patents were issued to The Dow Chemical Company (65) describing the use of copper metal catalysis. Full-scale production was achieved the same year. A solution of acrylonitrile in water was passed over a fixed bed of copper catalyst at 85°C, which produced a solution of acrylamide in water with very high conversions and selectivities to acrylamide. The heat of hydration is approximately −70 kJ/mol (−17 kcal/mol). This process usually produces no waste streams, but if the acrylonitrile feed contains other nitrile impurities, they will be converted to the corresponding amides. Another reaction that is prone to take place is the hydrolysis of acrylamide to acrylic acid and ammonia. However, this impurity can usually be kept at very low concentrations.

Mitsui Toatsu Chemical, Inc. disclosed a similar process using Raney copper (66) shortly after the discovery at Dow, and BASF came out with a variation of the copper catalyst in 1974 (67). Since 1971, several hundred patents have shown modifications and improvements to this technology, both homogeneous and heterogeneous, and reviews of these processes have been published (68). Nalco Chemical Company has patented a process based essentially on Raney copper catalyst (69) in both slurry and fixed-bed reactors and produces acrylamide monomer mainly for internal uses. Other producers in Europe, besides Dow and American Cyanamid, include Allied Colloids and Stockhausen, who are believed to use processes similar to the Raney copper technology of Mitsui Toatsu, and all have captive uses. Acrylamide is also produced in large quantities in Japan.

In 1985, Nitto Chemical Industry started using microorganisms for making acrylamide from acrylonitrile using an enzymatic hydration process (71,78). The reaction is catalyzed by nitrile hydralase, a nitrilasically active enzyme produced by organisms such as *Corynebacterium N-774* strain, *Bacillus*, *Bacteridium*, *Micrococcus*, *Nocardia*, and *Pseudomonas*. This is one of the initial uses of biocatalysis in the manufacture of commodity chemicals in the petrochemical industry. There are certainly other bioprocesses in use for fine chemicals in the amino acid area, as well as fermentation processes. Improved bacterial strains and cells immobilized in acrylamide gels as well as methods of concentrating the dilute product solutions are subjects of more recent patents (72–74). The most recent release indicates a switch to *Rhodococcus rhodochrous* bacteria, which will increase their capacity from 6000 to 20,000t/year (75). The reaction is run at 0–15°C and a pH 7–9 and gives almost complete conversions with very small amounts of by-products such as acrylic acid.

Acrylamide and its derivatives have been prepared by many other routes (4). The reactions of acryloyl chloride and acrylic anhydride with ammonia are classical methods. Primary and secondary amines may be used in place of ammonia to obtain *N*-substituted derivatives. Acryloyl isocyanate has been hydrolyzed to acrylamide but yields are poor. Exhaustive amination of methyl acrylate with ammonia yields 3,3′,3″-nitrilotrispropionamide [2664-61-1] ($C_9H_{18}N_4O_3$). This compound can be thermally decomposed to acrylamide by heating to 208–230°C at 2 kPa (15 mm Hg). Similarly, Michael-type addition products of alkylamines or aliphatic alcohols and methyl acrylate react with ammonia to give the corresponding β-substituted propionamides. These compounds may also be thermally decomposed to yield acrylamide. The Michael-type addition products of methyl acrylate and aliphatic amines may react further to give *N*-alkyl or *N,N*-dialkyl propionamide derivatives that can be thermally decomposed to monoalkyl- or dialkyl-substituted acrylamides, respectively. *N*-Substituted acrylamides may also be prepared from acetylene, carbon monoxide, and an amine using an iron or nickel carbonyl catalyst. However, the best route to mono-*N*-alkyl-substituted acrylamides is the Ritter reaction. This reaction is used to prepare diacetoneacrylamide [2873-97-4] ($C_9H_{15}NO_2$), 2-acrylamido-2-methylpropanesulfonic acid [15214-89-8] ($C_7H_{13}NO_4S$), *N*-isopropylacrylamide [2210-25-5] ($C_6H_{11}NO$), *N*-tert-butylacrylamide [107-58-4] ($C_7H_{13}NO$), and other *N*-alkyl acrylamides in which the carbon attached to the nitrogen is usually tertiary (76,77).

A process for manufacturing acrylamide microemulsified homopolymer has been disclosed (78).

5. Economic Aspects

Seventy percent of the world capacity of acrylamide is in the United States, Western Europe, and Japan. Western Europe and the United States consume ~48% of the supply or water management and paper production. For Japan and the rest of the world, the main uses are in paper production and enhanced oil recovery (79)

U. S. producers of acrylamide and their capacities are given in Table 4.

Prices for the period 1996–2001 were in the range of \$0.80/kg (\$1.76/lb)–\$0.84/kg (\$1.86/lb). Prices were for a 50% solution, 100% basis, bulk, and fob works. Current prices are the same and on the same basis (1).

Table 4. **U. S. Producers of Acrylamide and Their Capacities**[a]

Producer	Capacity $\times 10^3$t($\times 10^6$ lb)
Ciba Specialty Chemicals, Suffolk, Va.	15 (33)
Cytec Industries, Avondale, La.	41 (90)
Ondeo Nalco, Garyville, La.	16 (35)
SNF Floerger, Riceboro, Ga.	65 (143)
Total	*137 (301)*

[a] Ref. 1.

6. Specifications

The 50% aqueous acrylamide is the preferred form because it eliminates the handling of solids and because its cost is lower. This result is of the new manufacturing method put into effect in 1971. The aqueous form is applicable to nearly all the end uses of acrylamide when volume is taken into account. Aqueous acrylamide is shipped in tank trucks, rail cars, or drums, but small samples can also be obtained. The solution should be kept in stainless steel or in tanks coated with plastic resin (phenolic, epoxy, or polypropylene). All containers, including tank trucks and rail cars, must be rinsed prior to disposal or return. When shipping costs are an important consideration, solid acrylamide may be the desired form. Acrylamide should be stored in a well-ventilated area away from sunlight. The temperature should be <30°C, and under these conditions no change of quality should be noticed for at least 3 months. Typical specifications for the 50% aqueous solution are shown in Table 5 and for the solid monomer in Table 6.

Table 5. **Typical Specifications for 50% Aqueous Solutions**[a]

Property	Limit
assay, wt %	48–52
pH	5.0–6.5
polymer, ppm, max (BOM)[b]	100
Cu^{2+} inhibitor, ppm, max (BOM)[b]	25
color	water clear

[a] Refs.11, 81.
[b] Based on monomer = BOM.

Table 6. **Typical Specifications for Crystalline Acrylamide Monomer**

appearance	white, free flowing crystal
assay, %, min	98
water, %, max	0.8
iron, as Fe^0, ppm, max	15
color, 20% soln, max, APHA	50
water insoluble, %, max	0.2
butanol insoluble, %, max	1.5

Table 7. **Acrylamide Assay Techniques**[a]

Method	Approximate sensitivity, ppm[b]	Application	Interference	References
refractive index	50,000	quality control	anything affecting refractive index	11,81
bromate–bromide	1,000	assay product	unsaturated compounds	5
flame ionization	40	monomer in polymer		86
dc polarization	10	assay product	alkali cations, acrylic esters	84
differential pulse polarography	>1	environment concerns	alkali cations, acrylic esters, vinyl cyanide	82
spectrophotometry	0.1	urinalysis	aldehydes, ketones, pyrroles, indoles, hydrazine, aromatic amines	84,85
hplc	0.1	wipe and air		87
electron capture, gc	0.1 ppb	river water		83

[a] Ref. 85.
[b] Unless otherwise noted.

7. Analytical and Test Methods

The analysis of acrylamide monomer in water solutions containing at least 0.5% monomer is carried out by bromination (5). If the concentration is fairly high, in the 2–55% monomer range, then a refractive index method is easier (11,80,81). Polarography (82) and gas chromatography (83) can also be used to determine trace amounts of acrylamide monomer in other organic materials. For detecting small concentrations of polymer in aqueous acrylamides solutions, *n*-butanol addition will produce turbidity, which can then be compared to standards (5,81). Cupric ion inhibitor and other impurities in acrylamide samples can be determined by standard techniques (11,81,84). Other methods can also be employed to analyze the polyacrylamide content of monomer solutions, including turbidimeteric (Hach) and colorimetric (Klett) methods (11,81). A summary of various analytical techniques for assaying acrylamide monomer are listed in Table 7.

8. Health and Safety Factors

Contact with acrylamide can be hazardous and should be avoided. The most serious toxicological effect of exposure to acrylamide monomer is as a neurotoxin.

In contrast, polymers of acrylamide exhibit very low toxicity. Since the solid form sublimes, the solid or powder form of acrylamide is more likely to be a problem than the aqueous form because of possible exposure to dusts and vapors. An important characteristic of the toxicity of acrylamide monomer is that the signs and symptoms of exposure to toxic levels may be slow in developing and can occur after ingestion of small amounts over a period of several days or weeks. It is therefore important that people who have been exposed to acrylamide be monitored by a qualified physician. Signs and symptoms include increased sweating of hands and feet, numbness or tingling of the extremities, or even paralysis of the arms and legs. Acrylamide is readily absorbed through unbroken skin, and the signs are the same as with ingestion. Acute dermal LD_{50} is 2250 mg/kg for rabbits (77). Eye contact can produce conjunctival irritation and slight corneal injury and can lead to systemic exposure if contact is prolonged and/or repeated. Inhalation of vapors, dusts, and/or mists can result in serious injury to the nervous system, but again, symptoms may be slow in developing. Since the symptoms for minor repeated exposures over a long period of time are similar to those for gross human exposure, the development of such symptoms can be a signal that severe damage has already occurred. There are no reliable "early warning" signals of damaging exposure to toxic levels of acrylamide monomer, so it is imperative that all handling procedures be designed to prevent human contact. In a long-term study, rats that received relatively low concentrations of acrylamide monomer in the drinking water showed an increase in several types of malignant tumors (88). Suitable respirators and clothing that consists of a head covering, long-sleeved coverall, impervious gloves, and rubber footwear are recommended to avoid contact (81).

A large number of research studies have been published, many of which were released by government agencies (89–96). A threshold limit value (TLV) of 0.03 mg/m^3 (skin) has been set by the American Conference of Governmental Industrial Hygienists (ACGIH). ACGIH also categorizes acrylamide as A2 (suspect human carcinogen). Occupational Safety and Health Administration (OSHA) permissible exposure limit (PEL) is 0.3 mg/m^3 (98). Several studies demonstrate that acrylamide is biodegradable (97), and the hydrolysis of acrylamide proceeds readily both in rivers and in soils (99,100). Bioconcentration of acrylamide probably will not occur because of the ease of biodegradation and the high water solubility of this material. Acrylamide shows low acute toxicity to fish (61,101). Other derivatives, such as N-methylolacrylamide, are neurotoxins in their own right, but the LD_{50} is much higher than for acrylamide. Toxicity data for acrylamide and several derivatives are listed in Table 8.

Handling of dry acrylamide is hazardous primarily from its dust and vapor, and this is a significant problem, especially in the course of emptying bags and drums. This operation should be carried out in an exhaust hood with the operator wearing respiratory and dermal protection. Waste air from the above mentioned ventilation should be treated by a wet scrubber before purging to the open air, and the waste water should be fed to an activated sludge plant or chemical treatment facility. Solid acrylamide may polymerize violently when melted or brought into contact with oxidizing agents. Storage areas for solid acrylamide monomer should be clean and dry and the temperature maintained at 10–25°C, with a maximum of 30°C.

Table 8. **Toxicity of Acrylamide and Derivatives**

Compound	CAS Registry Number	Molecular formula	Animal	Oral LD_{50}, g/kg	References
acrylamide	[79-06-1]	C_3H_5NO	mouse	0.17	77
N-methylolacrylamide	[924-42-5]	$C_4H_7NO_2$	mouse	0.42	102
N,N′-methylenebisacrylamide	[110-26-9]	$C_7H_{10}N_2O_2$	rat	0.39	103
N-isobutoxymethylacrylamide	[16669-59-3]	$C_8H_{15}NO_2$	rat	1.0	104
N,N-dimethylacrylamide	[2680-03-7]	C_5H_9NO	rat	0.316	105
2-acrylamido-2-methylpropane-sulfonic acid	[15214-89-8]	$C_7H_{13}NO_4S$	rat	1.41	106

The 50% aqueous product is the most desirable where water can be tolerated in the process. Employees should not be permitted to work with acrylamide until thoroughly instructed and until they can practice the required precautions and safety procedures. Anyone handling acrylamide should practice strict personal cleanliness and strict housekeeping at all times. This should include wearing a complete set of clean work clothes each day and the removal of contaminated clothing immediately. If contact is made, the affected skin area should be washed thoroughly with soap and water and contaminated clothing should be replaced. When contact can occur, such as in maintenance and repair operations or connection and disconnection during transport, protective equipment should be used. This should include impervious gloves and footwear to protect the skin, and suitable eye protection such as chemical worker's goggles. If exposure to the face is possible, a face shield should be used in addition to the goggles. Food, candy, tobacco, and beverages should be banned from areas where acrylamide is being handled, and workers should wash hands and face thoroughly with soap and water before eating or drinking. The need for good personal hygiene and housekeeping to prevent exposure cannot be overemphasized.

Aqueous solutions of 50% acrylamide should be kept between 15.5 and 38°C with a maximum of 49°C. Below 14.5°C acrylamide crystallizes from solution and separates from the inhibitor. Above 50°C the rate of polymer buildup becomes significant. Suitable materials of construction for containers include stainless steel (304 and 316) and steel lined with plastic resin (polypropylene, phenolic, or epoxy). Avoid contact with copper, aluminum, their alloys, or ordinary iron and steel.

Disposal of small amounts of acrylamide may be done by biodegradation in a conventional secondary sewage treatment plant, but any significant amounts should be avoided. Such waste material should not be allowed to get into a municipal waste treatment or landfill operation unless all appropriate precautions have been taken. When the disposal of large quantities is necessary, the supplier should be contacted. Containers that have been used for acrylamide should be thoroughly rinsed and then disposed of in an appropriate manner. In any disposal of waste materials, all of the applicable federal, state, and local statutes, rules, and regulations should be followed. Persons contemplating large-scale use of acrylamide monomer should consult the manufacturers at an early stage in the planning to ensure that their facilities and operations are adequate. Many companies refuse to supply to operations that, in their opinion, are unsafe.

9. Uses

The largest use of acrylamide in the United States is for the production of polyacrylamides and consumes 94% of the total. In this category, the largest use of polyacrylamides is in water treatment, which accounts for 56%. This includes use as a dewatering aid for sludge in the treatment of effluent from municipal wastewater treatment plants (eg, sewage) and industrial processes (pulp and paper plant wastewater). Polyacrylamides are also used as flocculents for feed water treatment for industrial purposes. Other uses include in pulp and paper production (24%), mineral processing (10%), *N*-methylacrylamide and other monomers (6%), and miscellaneous (4%) (1).

BIBLIOGRAPHY

"Acrylamide" in *ECT* 2nd ed., Vol. 1, pp. 274–284, by Norbert M. Bikales and Edwin R. Kolodny, American Cyanamid Company; in *ECT* 3rd ed., Vol. 1, pp. 298–311, by D. C. MacWilliams, Dow Chemical USA; in *ECT* 4th ed., Vol. 1, pp. 251–266, by C. E. Hebermann, Dow Chemical, USA; "Acrylamide" in *ECT* (online), posting date: December 4, 2000, by C. E. Habermann, Dow Chemical, USA.

CITED PUBLICATIONS

1. "Acrylamide. ," *Chemical Profiles, Chem Expo, Http://63.236.84.14/news/profile.cfm.* May 6, 2002.
2. U.S. Pat. 2,917,477 (Dec. 15, 1959), T. J. Suen and R. L. Webb (to American Cyanamid Co.).
3. E. Collinson and F. S. Dainton, *Nature (London)* **177**, 1224 (1956).
4. U.S. Pat. 2,758,135 (Aug. 7, 1956), M. L. Miller (to American Cyanamid Co.).
5. *Chemistry of Acrylamide, Bulletin PRC 109*, Process Chemicals Department, American Cyanamid Co., Wayne, N.J., 1969.
6. J. P. Friend and A. E. Alexander, *J. Polym. Sci. Part A-1* **6**, 1833 (1968).
7. Ger. Pat. 1,030,826 (May 29, 1958), H. Wilhelm (to BASF A.G.).
8. A. LeBerre and A. Delaroix, *Bull. Soc. Chim. Fr.* **11**(2), 2639 (1974); *Chem. Abstr.* **82**, 97302 (1975); earlier paper in *Bull. Soc. Chim. Fr.* **2**(2), 640 (1973) is very significant.
9. A. LeBerre and C. Porte, *Bull. Soc. Chim. Fr.* **7–8**(2), 1627 (1975); *Chem. Abstr.* **84**, 58149 (1976).
10. U.S. Pat. 3,778,464 (Dec. 11, 1973), P. Klemchuck,
11. *Aqueous Acrylamide, Forms 260-951-88, Analytical Method PAA 46*, Chemicals and Metals Department, The Dow Chemical Company, Midland, Mich., 1976.
12. U.S. Pat. 3,699,192 (Oct. 17, 1972), P. Moretti (to U.S. Oil Company, Inc.).
13. The Dow Chemical Company, unpublished results.
14. D. Elad and D. Ginsberg, *J. Chem. Soc.* 4137 (1953).
15. B. T. Croll, G. M. Srkell, and R. P. J. Hodge, *Water Res.* **8**, 989 (1974).
16. Ger. Offen. 3,128,574 (Jan. 27, 1983), K. Laping, O. Petersen, K. H. Heinemann, H. Humbert, and F. Henn (to Deutsche Texaco A.G.).
17. J. S. Meek, R. T. Mernow, D. E. Ramey, and S. J. Cristol, *J. Am. Chem. Soc.* **73**, 5563 (1951).
18. A. I. Naimushin and V. V. Simonov, *Zh. Obshch. Khim.* **47**, 862 (1977); *Chem. Abstr.* **87**, 38678m (1977).
19. G. Stork, *Pure Appl. Chem.* **17**, 383 (1968).
20. I. Ninomiya, T. Naito, S. Higuchi, and T. Mori, *J. Chem. Soc. D* **9**, 457 (1971).
21. U.S. Pat. 4,198,415 (Apr. 15, 1980), N. J. Bach and E. C. Kornfeld (to Eli Lilly and Co.).
22. T. W. Russell and D. M. Duncan, *J. Org. Chem.* **39**, 3050 (1974).
23. T. W. Russell, R. C. Hoy, and J. E. Cornelius, *J. Org. Chem.* **37**, 3552 (1972).
24. T. Kitamura, N. Sakamoto, and T. Joh, *Chem. Lett.* **2**(4), 379 (1973).
25. U.S. Pat. 3,846,478 (Nov. 5, 1974), R. W. Cummins (to FMC Corp.).

26. U.S. Pat. 3,332,923 (July 25, 1967), L. D. Moore and R. P. Brown (to Nalco Chemical Co.).
27. A. Leoni and S. Franco, *Macromol. Synth.* **4**, 125 (1972).
28. U.S. Pats. 3,193,476 and 3,193,483 (July 6, 1965), M. M. Baizer (to Monsanto Co.).
29. D. S. Breslow, G. E. Hulse, and A. S. Matlack, *J. Am. Chem. Soc.* **79**, 3760 (1957).
30. N. Yamashita, M. Yoshihara, and T. Maeshima, *J. Polym. Sci. Part B* **10**, 643 (1972).
31. Eur. Pat. Appl. 272025 A2 (June 22, 1988), D. Byrom and M. A. Carver; Eur. Pat. Appl. EP 272026 A2 (June 22, 1988), M. A. Carver and J. Hinton (to Imperial Chemical Ind.).
32. Jpn. Kokai 76 86412 (July 29, 1976), F. Matsuda and T. Takazo (to Mitsui Toatsu Chem. Co.).
33. G. A. Chubarov, S. M. Danov, and V. I. Logutov, *Zh. Prikl. Khim. Leningrad* **52**, 2564 (1979); *Chem. Abstr.* **92**, 163293m (1980).
34. A. Conix, G. Smets, and J. Moens, *Ric. Sci. Suppl.* **25**, 200 (1954); *Chem. Abstr.* **54**, 11545e (1960).
35. T. Azuma and N. Ogata, *J. Polym. Sci. Polym. Chem. Ed.* **13**, 1959 (1975).
36. U.S. Pat. 3,084,191 (Apr. 2, 1963), J. R. Stephens (to American Cyanamid Co.).
37. M. F. Farona, W. T. Ayers, B. G. Ramsey, and J. G. Grasselli, *Inorg. Chim. Acta* **3**, 503 (1969).
38. J. Reedijk, *Inorg. Chim. Acta* **5**, 687 (1971).
39. A. Samantaray, P. K. Panda, and B. K. Mohapatra, *J. Indian Chem. Soc.* **57**, 430 (1980).
40. M. S. Barvinok and L. V. Mashkov, *Zh. Neorg. Khim.* **25**, 2846 (1980).
41. V. I. Bruskov and V. N. Bushuev, *Biofizika* **22**(1), 26 (1977); *Chem. Abstr.* **87**, 39783d (1977).
42. T. O. Osmanov, V. F. Gromov, P. M. Khomikovskii, and A. D. Abkin, *Polym. Sci. USSR* **22**, 739 (1980); **21**, 1948 (1979).
43. T. Asakara and N. Yoda, *J. Polym. Sci. Part A-1* **6**, 2477 (1968).
44. T. Asakara, K. Ikeda, and N. Yoda, *J. Polym. Sci. Part A-1* **6**, 2489 (1968).
45. U.S. Pat. 2,373,190 (Apr. 10, 1945), F. E. King (to B. F. Goodrich Co.).
46. C. Moureau, *Bull. Soc. Chim. Fr.* **9**, 417 (1973).
47. U.S. Pat. 3,064,050 (Nov. 13, 1962), K. W. Saunders and L. L. Lento, Jr. (to American Cyanamid Co.).
48. H. Fener and V. E. Lynch, *J. Am. Chem. Soc.* **75**, 5027 (1953).
49. U.S. Pat. 2,475,846 (July 12, 1949), L. A. Lindberg (to American Cyanamid Co.).
50. Ger. Offen. 2,310,516 (Sept. 19, 1974), K. Fischer and H. Petersen (to BASF A.G.).
51. T. Araki, C. Terunuma, K. Tanigawa, and N. Ando, *Kobunshi Ronbunshu* **31**, 309 (1974).
52. Jpn. Kokai 7,582,008 (July 3, 1975), K. Yamamoto and co-workers (to Mitsui Toatsu Chemicals, Inc.).
53. Jpn. Kokai 7,426,235 (Mar. 8, 1974), S. Kumi and co-workers (to Dainippon Ink and Chemicals, Inc.).
54. E. Mueller, K. Dinges, and W. Ganlich, *Makromol. Chem.* **57**, 27 (1962).
55. U.S. Pat. 3,185,539 (May 25, 1965), R. K. Madison and W. J. Van Loo Jr., (to American Cyanamid Co.).
56. Jpn. Pat. 15,816 (Aug. 5, 1964), T. Oshima and M. Suzuki (to Sumitomo Chemical Co., Ltd.).
57. U.S. Pat. 3,422,139 (Jan. 14, 1969), P. Talet and R. Behar (to Nobel-Bozel).
58. Fr. Pat. 1,406,594 (July 23, 1965), P. Talet and R. Behar (to Nobel-Bozel).
59. R. E. Dunbar and G. C. White, *J. Org. Chem.* **23**, 915 (1958).
60. U.S. Pat. 2,734,915 (Feb. 14, 1956), G. D. Jones (to The Dow Chemical Company).

61. U.S. Pat. 3,041,375 (June 26, 1962), S. N. Heiny (to The Dow Chemical Company).
62. M. J. Sook, E. J. Forbes, and G. M. Khan, *Chem. Commun.* (**5**), 121 (1966).
63. U.S. Pat. 3,381,034 (Apr. 30, 1968), J. L. Greene and M. Godfrey (to Standard Oil Co., Ohio).
64. K. Watanabe, *Bull. Chem. Soc. Jap.* **37**, 1325 (1964); *Chem. Abstr.* **62**, 2735b (1965).
65. U.S. Pat. 3,597,481 (Aug. 3, 1971), B. A. Tefertiller and C. E. Habermann (to The Dow Chemical Company); U.S. Pat. 3,631,104 (Dec. 28, 1971), C. E. Habermann and B. A. Tefertiller (to The Dow Chemical Company); U.S. Pat. 3,642,894 (Feb. 15, 1972), C. E. Habermann, R. E. Friedrich, and B. A. Tefertiller (to The Dow Chemical Company); U.S. Pat. 3,642,643 (Feb. 15, 1972), C. E. Habermann (to The Dow Chemical Company); U.S. Pat. 3,642,913 (Mar. 7, 1972), C. E. Habermann (to The Dow Chemical Company); U.S. Pat. 3,696,152 (Oct. 3, 1972), C. E. Habermann and M. R. Thomas (to The Dow Chemical Company); U.S. Pat. 3,758,578 (Sept. 11, 1973), C. E. Habermann and B. A. Tefertiller (to The Dow Chemical Company); U.S. Pat. 3,767,706 (Oct. 23, 1972), C. E. Habermann and B. A. Tefertiller (to The Dow Chemical Company).
66. Brit. Pat. 1,324,509 (July 25, 1973) (to Mitsui Toatsu Chemicals, Inc.).
67. Ger. Offen. 2,320,060 (Nov. 7, 1974), T. Dockner and R. Platz (to BASF A.G.).
68. E. Otsuka and co-workers, *Chem. Econ. Eng. Rev.* **7**(4), 29 (1975).
69. Brit. Pat. 2,018,240 (Oct. 17, 1979), I. Watanabe (to Nitto Chemical).
70. U.S. Pat. 4,343,900 (Aug. 10, 1982), I. Watanabe (to Nitto Chemical).
71. Fr. Demande 2,488,908 (Feb. 26, 1982), I. Watanabe and co-workers (to Nitto Chemical).
72. U.S. Pat. 4,390,631 (June 28, 1983), I. Watanabe and co-workers (to Nitto Chemical).
73. U.S. Pat. 4,414,331 (Nov. 8, 1983), I. Watanabe and co-workers (to Nitto Chemical).
74. *Chem. Eng.* **97**(7), 19–21 (July 1990).
75. J. J. Ritter and P. P. Minieri, *J. Am. Chem. Soc.* **70**, 4045 (1948).
76. H. Plant and J. J. Ritter, *J. Am. Chem. Soc.* **73**, 4076 (1951).
77. *Aqueous Acrylamide, Form No. 192-466-76*, Chemicals and Metals Department, The Dow Chemical Company, Midland, Mich., 1976.
78. U. S. Pat. 5,545,688 (Aug. 13, 1996), S.-Y. Huang (to Cytec Technology Corp.).
79. R. Will and G. Toki, "Acrylamides" in *Chemical Economics Handbook*, SRI International, Menlo park, Calif., 2002
80. *Acrylamide-50 Handling and Storage Procedures, PRC 22B*, American Cyanamid Co., Wayne, N.J., 1980.
81. *Aqueous Acrylamide, Forms 260-951-88, Analytical Method PAA 44*, Chemical and Metals Department, The Dow Chemical Company, Midland, Mich., 1976.
82. S. R. Betso and J. D. McLean, *Anal. Chem.* **48**, 766 (1976).
83. B. T. Croll and G. M. Simkins, *Analyst* **97**, 281 (1972).
84. M. V. Norris, in F. D. Snell and C. L. Hilton, eds., *Encyclopedia of Industrial Chemical Analysis*, Vol. 4, Wiley-Interscience, New York, 1967, pp. 160–168.
85. D. C. MacWilliams, D. C. Kaufman, and B. F. Waling, *Anal. Chem.* **37**, 1546 (1965); A. R. Mattocks, *Anal. Chem.* **40**, 1347 (1968).
86. B. T. Croll, *Analyst (London)* **96**, 67 (1971).
87. *HPLC Determinations of Acrylamide in Water and Air Samples, Analytical Method PAA 58,61 in Forms 260-951-88*, Chemicals and Metals Department, The Dow Chemical Company, Midland, Mich., 1981.
88. K. A. Johnson, S. J. Gorzinski, K. M. Bodner, R. A. Campbell, C. H. Wolf, M. A. Friedman, and R. W. Mast, *Toxicol. Appl. Pharmacol.* **85**, 154 (1986).
89. L. N. Davis, P. R. Durkin, P. H. Howard, and J. Saxena, *Investigation of Selected Potential Environmental Contaminants; Acrylamides*; EPA Report No. 560/2-76-008, 1976.

90. *Criteria for Recommended Standard Occupational Exposure to Acrylamide*, U.S. Department of Health, Education, and Welfare, Washington, D.C., 1976.
91. *Environmental and Health Aspects of Acrylamide, A Comprehensive Bibliography of Published Literature 1930 to April 1980*, EPA Report No. 560/7-81-006, 1981.
92. *Assessment of Testing Needs; Acrylamide*, EPA Report No. 560/11-80-016, 1980.
93. J. Going and K. Thomas, *Sampling and Analysis of Selected Toxic Substances; Task I Acrylamide*, EPA Report No. 560/13-79-013, 1979.
94. E. J. Conway, R. J. Petersen, R. F. Colingsworth, J. G. Craca, and J. W. Carter, *Assessment of the Need for a Character of Limitations on Acrylamide and Its Components*, EPA MRI Project No. 4308-N, 1979.
95. H. A. Tilson, *Neurobehav. Toxicol. Tetratol.* **3**, 445 (1981).
96. P. M. Edwards, *Br. J. Ind. Med.* **32**, 31 (1975).
97. B. T. Croll, G. M. Arkell, and R. P. J. Hodge, *Water Res.* **8**, 989 (1974).
98. R. L. Melnick, in E. Bingham, B. Cohrseen, and C. H. Powell, eds., *Patty's* Toxicology, 5th ed., Vol. 1, John Wiley & Sons, Inc., New York, 2001, p. 143.
99. M. J. Hynes and J. A. Pateman, *J. Gen. Microbiol.* **63**, 317 (1970).
100. H. M. Abdelmagid and M. A. Tabatabai, *J. Environ. Qual.* **11**, 701 (1982).
101. Krautter and co-workers, *Environ. Toxicol. Chem.* **5**, 373 (1986).
102. *N-Methylolacrylamide, PRC 14*, Process Chemicals Department, American Cyanamid Co., Wayne, N.J., 1972.
103. *N,N'*-Methylenebisacrylamide, PRT 47 A, American Cyanamid Co., Wayne, N.J., 1978.
104. *N-(iso-Butoxymethyl)acrylamide, PRT 126*, Process Chemicals Department, American Cyanamid Co., Wayne, N.J., 1977.
105. *N,N-Dimethylacrylamide, Technical Bulletin*, Alcolac, Inc., Baltimore, Md., 1977.
106. *AMPS Monomer*, Lubrizol Corp., Wickliffe, Ohio, 1981.

C. E. Habermann
Dow Chemical

ACRYLAMIDE POLYMERS

1. Introduction

The terminology used to describe acrylamide-containing polymers in the technical literature varies in its precision. In order to avoid confusion, throughout this article the term "poly(acrylamide)" will be reserved for the nonionic homopolymer of acrylamide, whereas the term "polyacrylamides" or "acrylamide polymers" will refer to acrylamide-containing polymers, including the homopolymer and copolymers. Specific nomenclature will be used for particular copolymers, for example, poly(acrylamide-co-sodium acrylate).

The diverse class of water-soluble and water-swellable polymers comprising polyacrylamides contains some of the most important synthetic polymeric materials used to improve the quality of life in our modern society. Acrylamide-containing polymers fall into three main categories: nonionic, anionic, and cationic. The projected annual sales growth rate of polyacrylamides between

1999 and 2002 is 4–7% (1). The multi-billion-dollar global market value of this class of materials makes it an economically important segment of the chemical industry.

Poly(acrylamide) is made by the free radical polymerization of acrylamide, which is derived from acrylonitrile by either catalytic hydrolysis or bioconversion. The unique chemistry of acrylamide, its favorable reactivity ratios with many comonomers, and the ability of poly(acrylamide) to be derivatized allows for a substantial variety of polymers to be tailor-made over a wide range of molecular weights (approximately 10^3–50×10^6 daltons), charge densities, and chemical functionalities.

The very large number of applications for acrylamide-containing polymers has been extensively reviewed (2–7). One major application area for polyacrylamides is in solid–liquid separations. The largest market segments therein are for use as flocculants and dewatering aids for municipal wastewater, thickening aids for industrial wastewater, secondary clarification and clarification of potable water, solids removal from biological broths, and animal feed recovery from waste. Because of major concern for the environment, the allowable suspended solids in most effluent streams are becoming more restricted by government regulations. New technologies for producing cationic polymers with a wide range of charge levels, novel structures, and very high molecular weights have addressed this need. These polymers have greatly improved the dewatering performances of centrifuges, screw presses, and belt presses used for such purposes. This has resulted in drier dewatered solids, which has translated into lower costs to either landfill or incinerate the solids.

The largest volume applications for polyacrylamides in paper mills are in on-machine wet-end processes. Paper retention aids and drainage aids are used to flocculate or bind fillers, fibers, and pigments. Glyoxalated cationic polyacrylamides are used as strengthening agents and promoters for paper sizing. Other papermaking applications include off-machine processes for recovering fiber from recycled paper waste and for deinking.

High-molecular-weight polyacrylamides have also traditionally been used in the minerals processing industry. Recent polymer technology developments, including-ultra-high-molecular-weight and novel anionic polyacrylamides, have yielded important materials. These products are used as flocculants in coal mining, the Bayer process for alumina recovery (red mud flocculants), precious metals recovery, and the solid–liquid separation of underflow streams in a variety of mining processes. Novel chemical modifications of low molecular weight polyacrylamides have resulted in materials that are used as modifiers in the selective separation of metal sulfides and magnetite and as depressants and flotation aids.

One large market segment for anionic polyacrylamides had traditionally been in enhanced oil recovery. However, low oil prices have resulted in a large decline in such applications. Since 1990, polymer flooding has virtually disappeared in the United States. However, during 1999 crude oil prices started to increase.

Other significant application areas for polyacrylamides include soil conditioning and erosion control, drag reduction, sugar processing, additives in cosmetics, and superabsorbents.

2. Physical Properties

2.1. Solid Polyacrylamides. Completely dry poly(acrylamide) is a brittle white solid. It is nontoxic, unlike the monomer. Dry polyacrylamides (including copolymers) are commercially available as nondusting powders and as spherical beads. These products can contain small amounts of additives that aid in both the stability and dissolution of the polymers in water. Commercially available acrylamide copolymer powders, which are typically dried under mild conditions, will usually contain about 5–15% water depending on their ionicity. The powders are hygroscopic, and generally become increasingly hygroscopic as the ionic character of the polymer increases. Cationic polymers are particularly hygroscopic.

Some physical properties of nonionic poly(acrylamide) are listed in Table 1. The tacticity and linearity of the polymer chain is claimed to be dependent on the polymerization temperature. Syndiotacticity is favored at low temperatures (8). Linear polymer chains are reportedly obtained below 50°C, but branching begins to occur as the temperature is increased above this level (9). A wide range of values of the glass transition temperature (T_g) of poly(acrylamide) have been published. This is because the measured value is highly sensitive to the presence of water, and also to the presence of nonacrylamide species along the polymer backbone. For example, small amounts of acrylate groups can arise from hydrolysis of the amide group during or after polymerization. This can dramatically change the T_g.

2.2. Solution Properties. The amide group (—$CONH_2$) in poly(acrylamide) provides for its solubility in water and in a few other polar solvents such as glycerol, ethylene glycol, and formamide. We can acquire a sense of poly(acrylamide)'s affinity for water by examining a few characteristic parameters. Theta (Θ) conditions for a polymer delineate a particular combination of solvent and temperature at which the polymer acts in an ideal manner (10), ie, the chains behave as random coils. The Θ temperature of poly(acrylamide) in water has been determined to be −8°C (11). Thus water at 25°C is a solvent of intermediate

Table 1. **Physical Properties of Solid Poly(acrylamide)**

Property	Value	Ref.
density	1.302 g/cm^3 (23°C)	250
glass-transition temperature (T_g)	195°C	251
critical surface tension (γ_c)	52.3 mN/m (20°C)	252
chain structure	mainly heterotactic linear or	253
	branched, some	25
	head-to-head addition	84, 254
crystallinity	amorphous (high molecular weight)	255
solvents	water, ethylene glycol, formamide	256
nonsolvents	ketones, hydrocarbons, ethers, alcohols	257
fractionation solvents	water–methanol	258
gases evolved on combustion in air	H_2, CO, CO_2, NH_3, nitrogen oxides	259

Table 2. **Physical Properties of Poly(acrylamide) in Solution**

Property	Value	Conditions	Ref.
steric hindrance parameter (σ)	2.72	water @ 30°C	260
characteristic ratio (C_∞)	14.8	water @ 30°C	260
persistence length (y)	15.2 Å	water @ 25°C	—[a]
partial specific volume (ν)	0.693 cm^3/g	water @ 20°C	261
theta temperature (Θ)	−8°C	water @ 25°C	11
theta conditions	0.40 v/v methanol/water	water @ 25°C	12
Flory χ parameter	0.48 ± 0.01	water @ 30°C	13
refractive index increment (dn/dc)	0.187 cm^3/g	$\lambda = 546.1$ nm	262
	0.185 cm^3/g	$\lambda = 632.8$ nm	

[a] Calculated from the values of K_0 (Mark-Houwink-Sakurada prefactor under Θ conditions) and Θ_0 (viscosity constant) found in Ref. 9, using the relationship: $y = \left(\frac{M_0}{2b}\right)\left(\frac{K_0}{\Theta_0}\right)^{2/3}$, where M_0 is the monomer molecular weight (71 g/mol), and b is the monomer length (2.5 Å).

quality for poly(acrylamide). Aqueous methanol (40 vol %), however, is a Θ solvent for poly(acrylamide) at 25°C (12). The Flory χ parameter, which is a measure of the relative affinity between the polymer segments with each other vs with the solvent, is 0.5 under Θ conditions. The Flory χ parameter of poly(acrylamide) has been determined to be 0.48 in water at 30°C (13). These and other properties of poly(acrylamide) in solution are collected in Table 2.

Poly(acrylamide) is soluble in liquid water at all concentrations, temperatures, and pH values. However, at high pH (> 10.5) the polymer will begin to hydrolyze on standing (14). Poly(acrylamide) is generally soluble in most salt solutions but can phase separate in some highly concentrated salt solutions, such as $(NH_4)_2SO_4$. Each amide group in poly(acrylamide) has roughly 2 strongly bound water molecules (15) associated with it, whereas the entire first hydration sheath contains a total of about 4–5 waters per monomer (16). This may be compared to poly(sodium acrylate), which has 4 strongly bound waters per repeat unit (17) and a total of 11 waters of hydration per repeat unit (18). Certain salts, however, can alter the hydrogen bonding between the primary amide groups and water in individual chains. For example, addition of potassium iodide [7681-11-0] to a poly(acrylamide) solution can increase the solution viscosity slightly (19). The inferred coil expansion involves a change in the hydration sheath of the polymer.

The amide group is capable of strong hydrogen bonding, which has effects on both the monomer and polymer properties. The relative rates of acrylamide polymerization in various organic solvents (20–22) are influenced by solvent–monomer interactions, which depend on the polarity and hydrogen bonding ability of acrylamide. Hydrogen bonding has been evidenced (with NMR) to occur mainly with the carbonyl oxygen in the acrylamide (23,24).

The hydrogen bonding ability of the amide group is also well worth considering when rationalizing the solution properties of polymers containing acrylamide. Two examples are presented here. The slow evolution of hydrogen-bonded aggregates (see the following) have been implicated in explaining the time dependence of the viscosities of poly(acrylamide) solutions in aqueous media (25,26). Second, it is well known that copolymers of acrylamide and sodium acrylate exhibit maximum values of the mean square radius (R_g), second virial coefficient (A_2), and intrinsic viscosity [η] at 60–70 mol % acrylate content (27). This can be rationalized from a consideration of intermolecular hydrogen bonding and electrostatic interactions.

Copolymers of acrylamide with ionic comonomers are also generally quite soluble in water. However, the solution properties of ionized copolymers of acrylamide are substantially different from those of the homopolymer. The incorporation of ionic comonomers leads to all of the traditional polyion effects such as chain expansion and viscosification at low ionic strength (polyelectrolyte effect), ionization-dependent dissociation constants, counterion condensation, ion exchange with charged surfaces, and specific binding of certain multi-valent ions. For example, anionic copolymers containing carboxylate groups will precipitate at certain multi-valent salt concentrations (28–30). Poly(sodium acrylate) can phase separate in the presence of divalent salts when there are about 0.8 equivalents of the divalent cations (31). The phase behavior of acrylamide–acrylic acid copolymers in mixtures of mono and divalent salts has been studied by François et al. (32). Trivalent cations (eg, Al^{3+} and Cr^{3+}) are even more efficient at precipitating polyions containing carboxylate groups (33). Under the right conditions, these physical cross-links can be used to form a reversible gel. This strategy has been employed in mobility control systems used in oil recovery (34).

The rate of dissolution of polyacrylamides can depend on the agitation conditions, dissolved salts, the material form of the polymer (eg, solid or emulsion), state of hydration, and the presence of other components. While salts only weakly affect the dissolution rate of poly(acrylamide), ionic copolymers tend to dissolve decidedly more slowly in salt solutions than in pure water. Increasing the mechanical energy input typically speeds up the dissolution process; however, mechanical degradation (chain scission) of very high-molecular-weight chains can occur. Flows with elongational components (eg, turbulent and porous media flows) are usually most egregious in this regard. The coil-to-stretched transition initiated at a critical elongation rate in these flows can be responsible for such phenomena as drag reduction (35) and apparent viscosity enhancement in porous media flow (36,37). However, chain scission can also occur at a second critical elongation rate. These critical elongation rates are functions of the degree of polymerization, the solvent quality, and the polymer concentration (38–42). Care is often taken to avoid mechanical degradation of the polymer during mixing [especially in impeller-type mixers (43), gear pumps, and orifice flow], filtration, and flow through a packed column (eg, in HPSEC analysis), all of which expose the polymer chains to some elongational flow kinematics.

Linear polyacrylamides in solution adopt nearly random coil configurations that are partially permeable (draining) to solvent. The coils are unassociated in dilute solution. The average shape of the isolated coils has been described as an ellipsoidal or bean-shaped structure (44). The individual chains are quite flexible, as is common with most vinyl polymers. This is indicated from several parameters shown in Table 2, such as the persistence length, steric hindrance parameter (σ), and characteristic ratio (C_∞). The persistence length of 15.2 Å for poly(acrylamide) in water is quite similar to the average intrinsic (bare) persistence length (~14 Å) of many vinyl polymers (45).

Table 3. **Suggested R_g–M_w Correlations for Polyacrylamides in Solution $R_G = K_r M_w^{a_r}$ (R_g in nm)**

Polymer	Solvent	Temp. (°C)	MW range (10^6 daltons)	$10^2 K_r$ (nm)	a_r	Ref.
poly(acrylamide)	water	25	0.83–13.4	0.725	0.64	11
poly(acrylamide)	0.1 *M* NaCl + 0.2% NaN_3	20	0.16–8.2	0.749	0.64	55
poly(Na acrylate$_{29}$–co-acrylamide$_{71}$)	0.1 *M* NaCl		0.96–6	2.50	0.60	30
poly(Na acrylate$_{20}$–co-acrylamide$_{80}$)	1 *M* NaCl pH 9	ambient	0.1–3.0	4.06	0.55	56
poly(acrylamide$_{70}$–co-AETAC$_{30}$)	1 *M* NaCl + biocide		0.5–2.7	3.30	0.54	58

One measure of the size of a polymer in solution is its mean square radius (R_g), sometimes referred to as its radius of gyration. The mean square radius scales with the weight-average molecular weight (M_w) to a fractional power (a_r) for a homologous series of polymers all within the same topology class (e.g., linear chains); $R_g = K_r M_w^{a_r}$. The K_r and a_r values depend on the polymer, solvent, and temperature. Suggested values derived from the literature for poly(acrylamide) and a few copolymers are listed in Table 3.

2.3. Solution Rheology. Solutions of polyacrylamides tend to behave as pseudoplastic fluids in viscometric flows. Dilute solutions are Newtonian (viscosity is independent of shear rate) at low shear rates and transition to pseudoplastic, shear thinning behavior above a critical value of the shear rate. This critical shear rate decreases with the polymer molecular weight, polymer concentration, and the thermodynamic quality of the solvent. A second Newtonian plateau at high shear rates is not readily seen, probably due to mechanical degradation of the chains (25). Viscometric data for dilute and semidilute poly (acrylamide) solutions can often be fit to a Carreau model (46,47). It is wise to remember the cautions that were cited previously about mechanical degradation of the high-molecular-weight components of a polyacrylamide sample when analyzing rheological data.

The viscosities of fully dissolved, high-molecular-weight poly(acrylamide)s in aqueous solutions have often, but not always, been seen to change with time over the periods of days to weeks. Typically, the solution viscosity decreases with time. Extensive studies of this instability phenomenon have been made by Kulicke and co-workers (25,26). They concluded that the evolution of intramolecular hydrogen bonds and the resulting change in macromolecular conformation were responsible for the time dependence, and not any molecular weight degradation. The instability can be avoided completely when the polymers are dissolved in formamide, aqueous ethylene glycol, or 2% 2-propanol in water. Competing viewpoints do exist about the interpretation of this solution aging (48,49).

The intrinsic viscosity $[\eta]$ of a polymer in solution is a measure of its molecular volume divided by its molecular weight. The $[\eta]$ value can be empirically correlated to the viscosity-average molecular weight (M_η) via the Mark-Houwink-Sakurada relationship (50): $[\eta] = K_\eta M_\eta^{a_\eta}$. Poly(acrylamide) and ionic

Table 4. **Suggested Mark-Houwink-Sakurada Correlations for Polyacrylamides in Solution $[\eta] = K_\eta M_\eta a_\eta$ ($[\eta]$ in cm^3/g)**

Polymer	Solvent	Temp. (°C)	MW range (10^6 daltons)	$10^2 K_\eta$ (cm^3/g)	a_η	Ref.
poly(acrylamide)	water	25	0.038–9	1.00	0.76	263
poly(acrylamide)	0.5 *M* NaCl	25	0.5–5.5	0.719	0.77	264
poly(acrylamide)	1.0 *M* NaCl	25	1.1–14.6	2.57	0.67	265
poly(Na acrylate$_{20}$–co-acrylamide$_{80}$)	0.5 *M* NaCl pH 9	25	0.12–3.0	1.40	0.75	56[a]
poly(Na acrylate$_{20}$–co-acrylamide$_{80}$)	0.5 *M* NaCl pH 9	25	0.12–3.0	1.09	0.78	56[b]
poly(Na acrylate$_{30}$–co-acrylamide$_{70}$)	0.5 *M* NaCl	25	0.77–5.5	1.12	0.79	264
poly(Na acrylate$_{20}$–co-acrylamide$_{80}$)	1.0 *M* NaCl pH 9	25	0.12–3.0	1.41	0.74	56[a]
poly(Na acrylate$_{20}$–co-acrylamide$_{80}$)	1.0 *M* NaCl pH 9	25	0.12–3.0	1.31	0.76	56[b]
poly(acrylamide$_{70}$–co-AETAC$_{30}$)	1.0 *M* NaCl + biocide	25	0.5–2.7	1.05	0.73	58

[a] Uncorrected for polydispersity.
[b] Corrected for polydispersity.

copolymers of acrylamide follow this empirical relationship, which is often used to estimate the polymer molecular weight. Table 4 lists suggested literature values of K_η and a_η for poly(acrylamide) and several copolymers in a variety of solvents.

At high concentrations of univalent ions ($\sim$ 1 molar) the solution properties of ionic copolymers of acrylamide tend to resemble those of the homopolymer, but they are not exactly the same. Considering the intrinsic viscosity data, the exponents a_η for anionic and cationic polyacrylamides in high salt concentrations tend to cluster between 0.7 and 0.8, which is similar to that for poly(acrylamide) (11, 25, 51–58). However, the prefactors (K_η) vary over a larger range, and this cannot be rationalized simply by considering only the degree of polymerization. This means that there are real differences in the short-range interactions along the chains, which depend on the copolymer composition.

The polymer concentration (c) dependence of the zero-shear-rate viscosity (η_0) for aqueous poly(acrylamide) solutions of various viscosity-average molecular weights does not seem to follow the entanglement model, wherein polymer chain interpenetration would dominate the viscometric behavior, and a master curve should result when η_0 is plotted against cM_η. Instead, the data can be better described using a suspension model, wherein $c[\eta]$ correlates the η_0 data on a master plot. Kulicke et al. (25) make a concise presentation of the relationship between η_0, M_η and c for poly(acrylamide) in water at 25°C.

Solutions of poly(acrylamide) that are well in excess of the overlap concentration can display viscoelastic properties. Viscoelasticity of these polymeric fluids can be observed in a variety of ways, including the presence of a normal stress and/or flow irregularities (eg, vortices) in steady-shear flow, stress overshoot during shear flow startup, a measurable storage modulus (G') in oscillatory flow, an apparent shear thickening in flows with an elongational component (eg, porous media flow), a measurable elongational viscosity, or the ability to pull a solution "fiber" (tubeless siphon effect). A simple means for qualitatively assessing the molecular weight of a linear poly(acrylamide) in solution is to see how long a thread one can pull out of a semidilute solution of the polymer using a rod.

3. Acrylamide Polymerization

Acrylamide [79-06-1] (2-propenamide, $C_{3J}H_5ON$) readily undergoes free-radical polymerization to high-molecular-weight poly(acrylamide) [9003-05-8]. The synthetic methods have been reviewed extensively (25). Free-radical initiation can be accomplished using organic peroxides, azo compounds, inorganic peroxides including persulfates, redox pairs, photoinduction, radiation-induction, electroinitiation, or ultrasonication. Several reasons account for the ultra-high-molecular-weights achievable. First, preparations of polyacrylamides are usually conducted in water, and the chain transfer constant to monomer and polymer appears to be zero in water (59). Second, the value of $k_p/k_t^{1/2}$, about 4.2, is unusually high (60) and is independent of the pH of the media. The rate of polymerization is proportional to the 1.2–1.5 power of the monomer concentration and to the square root of the initiator concentration (61–63). All this results in a high rate of propagation. Chain termination is primarily by disproportionation (64).

The large amount of heat (82.8 kJ/mol) that evolves during polymerization can result in a rapid temperature rise. One way in which this exotherm problem has been addressed in commercial high-solids and high-molecular-weight processes has been through the use of an adiabatic gel process in which the initiation temperature is 0°C. In another approach, controllable-rate redox polymerization of aqueous acrylamide-in-oil emulsions can be carried out at moderate temperatures of 40–60°C in order to accommodate the exotherm and to achieve very high molecular weights. At 70–100°C, a persulfate initiator can give a grafted or branched polymer (65). Additives greatly affect the rate and the kinetics of polymerization (66,67). These additives include metal ions, surfactants, chelating agents, and organic solvents. The high chain transfer constant of compounds such as 2-propanol, bisulfite ion, or persulfate ion to active polymer has been reported (68). Chain transfer agents have been used purposely to control molecular weight, minimize insoluble polymer, and control cross-linking and the degree of branching in commercial preparations.

4. Structural Modifications of Poly(acrylamide)

Poly(acrylamide) (PAM) is a relatively stable organic polymer. However, PAM can be degraded (eg, molecular weight decreases) under certain conditions. The amide functionality is acidic in nature and is capable of undergoing most of the chemical reactions of primary amides. Consequently, acrylamide polymers can be functionalized by post-polymerization chemical reactions. Examples illustrated in the following constitute the most-used chemical modifications. To obtain anionic derivatives, PAM can be hydrolyzed with caustic. Sulfomethylated PAM can be prepared by reacting PAM with formaldehyde and sodium bisulfite under acidic conditions. Reacting PAM with hydroxylamine under alkaline conditions can yield hydroxamated PAM. As an example of a cationic derivative, Mannich-base PAM can be obtained by reacting PAM with formaldehyde and dimethylamine to produce a cationic polymer with a charge that varies with pH. As an example of a nonionic derivative, PAM can be reacted with glyoxal to yield pendant aldehyde functionality. These structurally modified polyacrylamides are successful commercial products.

4.1. Degradation. Dry poly(acrylamide) is relatively stable. The onset of dry PAM decomposition occurs at 180°C (69). Inter- or intra-amide condensation (70) to an imide can occur in acidic media at high temperatures (140–160°C). At temperatures above 160°C, thermal degradation, imidization, nitrile formation, and dehydration take place. Polymer stability is very important in actual applications in order to maintain consistent and excellent performance. In most applications, polymer solutions are prepared and used at moderate temperatures; however, there are exceptions such as in the harsh reservoir conditions (high temperature and high salinity) found in some enhanced oil recovery operations. Impurities such as residual persulfate from batch manufacturing can degrade the polymer (71,72). A residual Fe^{2+} ETDA complex in the product can also enhance degradation at both ambient and elevated temperatures (73–75). Hydroxy radicals, which can form in the presence of oxygen (76,77), can attack the polymer backbone. In the absence of oxygen, anionic polyacrylamide solutions were stable at 90°C for 20 mo (78). Polyacrylamide in aqueous solution, in the presence of oxidizing agents such as $KMnO_4$, bromine, and $AgNO_3$, will degrade. The degraded polymer shows a reduced molecular weight, crosslinking and chain stiffening (79). Recently potassium peroxosulfate (80) was also reported to degrade hydrolyzed polyacrylamide. In the presence of ozone, very little degradation was found at low pH. However, random chain scission occurred at pH 10 (81). In the presence of sodium azide, a bactericide, a PAM in solution at room temperature showed no degradation for a long time (55). A combination of both pressure and elevated temperature can enhance polyacrylamide degradation. Polymer degradation can also occur under shear and elongational stresses (76,82). Backbone homolytic cleavage has been confirmed by a free radical trap technique. Under certain shear conditions one macroradical per 12 monomer units can be formed.

Numerous types of oxygen scavengers are used to inhibit and prevent oxidative degradation. These stabilizers have been reviewed (79) extensively. Effective compounds are thio compounds, hydroquinone, bisulfite, phenolic compounds, hydroxylamine, hydrazine, and others.

The biodegradability of PAM has not been definitively delineated in the literature (79). Recently, however, microorganisms (83), enterobacter agglomerans and azomonas macrocytogenes, were isolated from soil and the molecular weight of poly(acrylamide) in the presence of these microorganisms was found to undergo a 40-fold reduction as a result of chain degradation. The rate of biodegradation was equivalent to 20% of the carbon being consumed each day.

4.2. Hydrolyzed Polyacrylamide. Hydrolysis of poly(acrylamide) proceeds smoothly over a wide range of pH (see scheme below). Fundamental studies have been reviewed extensively (84–87). At alkaline pH, three reaction kinetics constants have been described, k_0, k_1 and k_2. The subscripts characterize the number of neighboring carboxylate groups next to the amide group being hydrolyzed. The rate constant k_0 is for no carboxylate neighbors, k_1 is for one carboxylate neighbor, and k_2 is for two carboxylate neighbors. Indirect evidence has shown that $k_0 > k_1 > k_2$. Under alkaline conditions, the rate of hydrolysis of poly(acrylamide) decreases with increasing conversion. The electrostatic repulsion from the increasing number of carboxylate groups in the backbone polymer opposes the approaching hydroxyl ion. Consequently, further hydrolysis will be severely retarded. Only about 80% of the amide groups (93) can be hydrolyzed by excess hydroxide ion even at elevated temperatures.

^{13}C-nmr studies (88–90) have shown that hydrolysis at high pH results in a nearly random distribution of carboxylate groups. In one industrial process, a polyacrylamide with about 30 mol % hydrolysis is prepared by heating an aqueous PAM solution containing excess sodium carbonate [497-19-8] (91). Polymerization of acrylamide in a water-in-oil emulsion in the presence of sodium hydroxide has also yielded a copolymer with about 30 mol % hydrolysis (92). A method of preparing a hydrolyzed PAM (93) with a viscosity-average molecular weight greater than 30×10^6 daltons was achieved in an inverse emulsion in the presence of caustic, ethoxylated fatty amine, and oil.

Hydrolysis of poly(acrylamide) proceeds slowly under acidic conditions. The undissociated carboxylic acid groups are protonated, neutral species under those conditions. The intramolecular catalysis by means of undissociated —COOH groups at low pH has been proposed as the main mechanism (75). An imide structure has been proposed to be an intermediate in the low-pH hydrolysis of poly(acrylamide), yielding short blocks of carboxyl groups distributed along the polymer chain (see scheme below). To date, there has been limited application of these block copolymer structures, and ones with high molecular weight have not been commercialized.

$$-(CH_2-\underset{\underset{NH_2}{|}}{\underset{C=O}{|}}{CH})_n- \xrightarrow{OH^-} -(CH_2-\underset{\underset{NH_2}{|}}{\underset{C=O}{|}}{CH}-CH_2-\underset{\underset{O^-}{|}}{\underset{C=O}{|}}{CH}-CH_2-\underset{\underset{NH_2}{|}}{\underset{C=O}{|}}{CH})_n-$$

$$\downarrow H^+$$

$$-(CH_2-\underset{O=C\diagdown}{\overset{}{CH}}-CH_2-\underset{\diagup C=O}{CH})_n- \;(\text{imide ring: } O=C-NH-C=O) \xrightarrow{OH^-} -(CH_2-\underset{\underset{NH_2}{|}}{\underset{C=O}{|}}{CH}-CH_2-\underset{\underset{O^-}{|}}{\underset{C=O}{|}}{CH}-CH_2-\underset{\underset{O^-}{|}}{\underset{C=O}{|}}{CH})_n-$$

Under neutral conditions, the observed mechanism of hydrolysis cannot be explained by a simple superposition of the retardation kinetics at high pH and intramolecular catalysis at low pH (75).

4.3. Cationic Carbamoyl Polymers. Poly(acrylamide) reacts with formaldehyde, [50-00-0], CH_2O, and dimethylamine, [124-40-3], C_2H_7N, to produce aminomethylated polyacrylamide (see the following scheme). This reaction has been studied extensively (94–98). A wide range of substitution can be produced in solution or in water-in-oil emulsion. ^{13}C-nmr studies (98) have verified that the Mannich substitution reaction follows second-order kinetics. The formation of the formaldehyde-dimethylamine adduct is very rapid. The high rate of Mannich substitution at high pH indicates a fast base-catalyzed condensation mechanism. The Mannich reaction is reversible and pH dependent. At low pH, the rate of substitution is very slow.

Because of the simplicity of the process, the small capital investment for manufacturing equipment and the low raw materials costs, this group of cationic water-soluble polymers constitutes a substantial percentage of commercially important flocculants. Solution Mannich PAMs are prepared and sold only at 4–6% solids, limited by the large solution viscosities and propensity for cross-linking on standing. The addition of formaldehyde scavengers such as guinadine compounds and dicyandiamide [461-58-5] (99) have improved the shelf stability. Aminomethylation of PAM is a reversible reaction. The reverse reaction can be retarded if the pendant amine groups are protonated by addition of an acid. Mannich base products protonated with organic acids or mineral acids have been patented (100). Low-charge-density quaternized aminomethylated products are also sold at polymer solids less than 3% due to very high solution viscosities at higher polymer concentrations.

Several disadvantages of solution Mannich PAMs are the problem of handling high solution viscosities, the added expense of shipping low-solids formulations, and the limitations to applications with low-pH substrates due to the decrease in cationic charge with increasing pH. Quaternized aminomethylated products in water-in-oil emulsions with greater than 20% solids have been developed. The charges in both high- and low-charge products were nearly independent of pH (97, 101–103). Microemulsion formulations have been developed and now replace certain polymer macroemulsions. In one such case, poly(acrylamide) was functionalized in a microdroplet (~100 nm in diameter) that contained only a few poly(acrylamide) molecules (104). Products based on this technology have been commercially successful as high performance cationic organic flocculants for municipal and industrial wastewater applications (105,106).

$$—(CH_2—\underset{\underset{NH_2}{|}}{\underset{C=O}{|}}{CH})_n— + (CH_3)_2NH + HCHO \longrightarrow —(CH_2—\underset{\underset{\underset{CH_2N(CH_3)_2}{|}}{\underset{NH}{|}}}{\underset{C=O}{|}}{CH})_n— + H_2O$$

$$\xrightarrow{CH_3Cl} —(CH_2—\underset{\underset{\underset{CH_2\overset{+}{N}(CH_3)_3\ Cl^-}{|}}{\underset{NH}{|}}}{\underset{C=O}{|}}{CH})_n—$$

4.4. Sulfomethylation. The reaction of formaldehyde and sodium bisulfite [7631-90-5] with polyacrylamide under strongly alkaline conditions at low temperature to produce sulfomethylated polyacrylamides has been reported many times (107–109). A more recent publication (110) suggests, however, that the expected sulfomethyl substitution is not obtained under the previously described strongly alkaline conditions of pH 10–12. This nmr study indicates that hydrolysis of polyacrylamide occurs and the resulting ammonia reacts with the sodium bisulfite and formaldehyde to form sulfomethyl amines and hexamethylenetetramine [100-97-0]. A recent patent describes a high-pressure, high-temperature process at slightly acid pH for the preparation of sulfomethylated polyacrylamide (111).

$$\text{—(CH}_2\text{—}\underset{\underset{\displaystyle NH_2}{|}}{\underset{\displaystyle C{=}O}{\underset{|}{\text{CH}}}}\text{)}_n\text{—} + NaHSO_3 + HCHO \longrightarrow \text{—(CH}_2\text{—}\underset{\underset{\displaystyle NHCH_2SO_3^-\ Na^+}{|}}{\underset{\displaystyle C{=}O}{\underset{|}{\text{CH}}}}\text{)}_n\text{—} + H_2O$$

4.5. Reaction with Other Aldehydes. Poly(acrylamide) reacts with glyoxal [107-22-2], $C_2H_2O_2$, under mild alkaline conditions to yield a polymer with pendant aldehyde functionality.

$$\text{—(CH}_2\text{—}\underset{\underset{\displaystyle NH_2}{|}}{\underset{\displaystyle C{=}O}{\underset{|}{\text{CH}}}}\text{)}_n\text{—} + O{=}\underset{\displaystyle H}{\underset{|}{C}}\text{—}\underset{\displaystyle H}{\underset{|}{C}}{=}O \longrightarrow \text{—(CH}_2\text{—}\underset{\underset{\displaystyle NH\text{—}\underset{\displaystyle OH}{\underset{|}{CH}}\text{—}CHO}{|}}{\underset{\displaystyle C{=}O}{\underset{|}{\text{CH}}}}\text{)}_n\text{—}$$

The rate of this reaction can be controlled by varying the pH and reaction temperature. Cross-linking is a competing reaction. The reaction rate increases rapidly with increasing pH and with increasing polymer concentration. In a typical commercial preparation a 10% aqueous solution of a low-molecular-weight polyacrylamide reacts with glyoxal at pH 8–9 at room temperature. As the reaction proceeds, solution viscosity increases slowly and then more rapidly as the level of functionalization and cross-linking increases. When the desired extent of reaction is achieved, before the gel point, the reaction is acidified to a pH below 6 to slow the reaction down to a negligible rate. These glyoxalated polyacrylamides are used as paper additives for improving wet strength (112).

A similar reaction occurs when poly(acrylamide) is mixed with glyoxylic acid [298-12-4], $C_2H_2O_3$, at pH about 8. This reaction produces a polymer with the CONHCH(OH)COOH functionality, which has found application in phosphate ore processing (113).

4.6. Transamidation. Poly(acrylamide) reacts with hydroxylamine [7803-49-8], H_2NOH, to form hydroxamated polyacrylamides with loss of ammonia (114).

$$\text{—(CH}_2\text{—}\underset{\underset{\displaystyle NH_2}{|}}{\underset{\displaystyle C{=}O}{\underset{|}{\text{CH}}}}\text{)}_n\text{—} + NH_2OH \longrightarrow \text{—(CH}_2\text{—}\underset{\underset{\displaystyle \underset{\displaystyle O^-\ Na^+}{\underset{|}{NH}}}{|}}{\underset{\displaystyle C{=}O}{\underset{|}{\text{CH}}}}\text{)}_n\text{—} + NH_3$$

This hydroxamation reaction occurs under alkaline conditions (115–117). Carboxyl groups can be produced due to hydrolysis of the amide (115,116). Acrylamide polymers can also be reacted with primary amines such as 2-aminoethanesulfonic acid (taurine) [107-35-7] at high temperature and acid pH to yield N—substituted copolymers containing sulfoethyl groups (118).

4.7. Hofmann Reaction. Polyacrylamide reacts with alkaline sodium hypochlorite [7681-52-9], NaOCl, or calcium hypochlorite [7778-54-3],

$Ca(OCl)_2$, to form a polymer with primary amine groups (119). Optimum conditions for the reaction include addition of a slight molar excess of sodium hypochlorite followed by addition of concentrated sodium hydroxide at low temperature (120). A two-stage addition of sodium hydroxide minimizes a side reaction between the pendant amine groups and isocyanate groups formed by the Hofmann rearrangement (121). Cross-linking sometimes occurs if the polymer concentration is high. High temperatures can result in chain scission. If long reaction times are used, NaOCl will cause chain scission and molecular weight decline. If very short reaction times are used at temperatures above 50°C, then polymers with high primary amine content can be obtained (122).

4.8. Reaction with Chlorine. Poly(acrylamide) reacts with chlorine under acid conditions or with NaOCl under mild alkaline conditions at low temperature to form reasonably stable *N*-chloropolyacrylamides. The polymers are water soluble and can provide good dry strength, wet strength, and wet web strength in paper (123).

5. Chemistry of Acrylamide Copolymers

5.1. Cationic Copolymers. The largest segment of the acrylamide polymer market has been dominated by cationic copolymers. The copolymers of acrylamide (AMD) and cationic quaternary ammonium monomers are manufactured by various commercial processes, which will be discussed in a later section. The most widely used of these cationic comonomers are cationic quaternary amino derivatives of (meth)acrylic acid esters or (meth)acrylamides, and diallydimethylammonium chloride.

The quaternary ammonium monomer contents in these copolymers are typically between 5 and 80 mol % for most applications. The composition actually employed depends on cost–performance relationships. Costs are largely dominated by the cationic monomer. Thus the cationic demand of the substrate for each application has to be optimized. Normally, low- to medium-charge copolymers are used for paper waste applications and medium- to high-charge copolymers are used for sludge dewatering. The molecular weights for flocculants are usually 5×10^6 daltons or greater. The higher-molecular-weight polymers often have the advantage of lower dosages in water treating and better fines capture in paper manufacture. Commercially important cationic comonomers, along with their reactivity ratios with acrylamide, are listed in Table 5.

Copolymers [69418-26-4] of acrylamide and AETAC (see Table 5, footnote abbreviation) are the most important flocculants because of a uniform sequence distribution of comonomers (124,125). Reactivity ratios obtained under very different free-radical copolymerization conditions can agree very well. For example, in one case, a free radical copolymerization was initiated using potassium persulfate (KPS) [7727-21-1] in aqueous solution at pH 6.1 (125), while in the other case the copolymerization was initiated using a TBHP/MBS redox pair in an inverse emulsion stabilized with sorbitan monooleate (SMO) at pH 3.5 (124). The surfactant in an inverse emulsion may alter the reactivity of both AMD and AETAC. For example, when SMO is utilized, in formulations made below the azeotropic monomer composition (ie, the copolymer composition is the same as the monomer

Table 5. **Acrylamide Monomer (M_1) Reactivity Ratios**

Comonomer M_2	CAS Registry Number	Molecular formula	r_1	r_2	Initiators[a]	Temp. (°C)	Ref.
Cationic Comonomer M[b]							
AETAC	[44992-01-0]	$C_8H_{16}NO_2Cl$	0.61	0.47	TBHP/MBS	40	124
AETAC	[44992-01-0]	$C_8H_{16}NO_2Cl$	0.61	0.47	KPS	40	125
METAC	[5039-78-1]	$C_9H_{18}NO_2Cl$	0.24	2.47	TBHP/MBS	40	124
METAC	[5039-78-1]	$C_9H_{18}NO_2Cl$	0.25	1.71	KPS	40	125
METAC	[5039-78-1]	$C_9H_{18}NO_2Cl$	0.57	1.11	KPS/NAS	26	127
DMAPAA	[3845-76-9]	$C_6H_{16}N_2O$	0.47	1.1	KPS	40	125
MAPTAC	[51410-72-1]	$C_{10}H_{21}N_2OCl$	0.57	1.13	KPS	40	125
DADMAC	[7398-69-8]	$C_8H_{16}NCl$	6.4–7.54	0.05–0.58	APS, ACV	20–60	125, 134, 140
Anionic Comonomer M[b]							
AA	[79-10-7]	$C_3H_4O_2$	0.25–0.95	0.3–0.95	KPS	30	139, 140
AA	[79-10-7]	$C_3H_4O_2$	0.89	0.92	AIBN	45	141
MAA	[79-41-4]	$C_4H_6O_2$	2.8–0.39	0.2–0.51	KPS	30	140
NaAMPS	[5165-97-9]	$C_6H_{12}O_4NSNa$	0.98	0.49	APS	30	142

[a] TBHP: *tert*-Butylhydroperoxide, MBS: sodium meta-bisulfite, KPS: potassium persulfate, APS: ammonium persulfate, ACV: azocyanovaleric acid, NAS: sodium sulfite, AIBN: 2,2′-azobisisobutyronitrile.

[b] AETAC: Acryloyloxyethyltrimethylammonium chloride, $CH_2{=}CHCO_2(CH_2)_2N^+(CH_3)_3Cl^-$; METAC: methacryloyloxyethyltrimethylammonium chloride, $CH_2{=}CCH_3CO_2(CH_2)_2N^+(CH_3)_3Cl^-$; DMAPAA: dimethylaminopropylacrylamide, $CH_2{=}CHCONH(CH_2)_3N(CH_3)_2$; MAPTAC: methacrylamidopropylacrylamide, $CH_2{=}CCH_3CONH(CH_2)_3N^+(CH_3)_3Cl^-$; DADMAC: diallyldimethylammonium chloride, $(CH_2{=}CHCH_2)_2N^+(CH_3)_2Cl^-$; AA: acrylic acid, $CH_2{=}CHCO_2H$; MAA: methacrylic acid, $CH_2{=}CCH_3CO_2H$; NaAMPS: 2-acrylamido-2-methylpropanesulfonic acid, Na salt, $CH_2{=}CHCONH(CH_3)_2CH_2SO_3\text{-}Na^+$.

feed composition, at about 58 mol % AETAC), AETAC is consumed slightly faster than AMD. On the other hand, if a block copolymeric surfactant, poly(ethylene oxide-b-12-hydroxysteric acid) (HB246), is utilized (126), then AMD is the faster reacting monomer. The results suggest that in the interfacial region near the discrete aqueous droplets, the AMD concentration is greater in the HB246 case than in the SMO case.

During AMD/MAETAC copolymerizations, MAETAC [5039-78-1] reacts with its own monomer significantly faster than with AMD. Consequently, copolymers [35429-19-7] can have severe compositional drift and often poor performance. Ha et al. (127) have studied how the sequence distribution can be improved if copolymerizations of AMD and MAETAC are conducted in water-in-oil microemulsion recipes. $K_2S_2O_8$-Na_2SO_3 redox initiator, the composite surfactants sorbitan monooleate, and octylphenol ethoxylate were used. They found reactivity ratios for AMD/MAETAC values were $r_{AMD} = 0.57$ and $r_{MAETAC} = 1.11$ (see Table 5).

Quaternary aminoester copolymers are very susceptible to base hydrolysis and are stable under very acidic conditions (124). In both manufacturing and in applications of these products great care is needed to control the pH in order to prevent hydrolysis. These products should possess sufficient buffering acid to maintain very acidic conditions. The hydrolytic instability of ester copolymers is primarily attributed to a base-catalyzed ester cleavage reaction that forms cyclic imides between neighboring AMD and AETAC groups. The loss in cationic charge is not due to direct ester hydrolysis (128). The chemistry of the six-membered imide ring is shown below (124,129).

—$(CH_2-CH(C{=}O\,NH_2)-CH_2-CH(C{=}O-O-CH_2-CH_2-N^+(CH_3)_2CH_3Cl))_n$— $\xrightarrow{OH^-}$ —$(CH_2-CH\,CH_2\,CH)_n$— (six-membered imide ring: $O{=}C-NH-C{=}O$) + $HO-CH_2-CH_2-N^+(CH_3)_2-CH_3\ Cl^-$

(not ↓)

—$(CH_2-CH(C{=}O\,NH_2)-CH_2-CH(C{=}O\,O^-))_n$— + $HO-CH_2-CH_2-N^+(CH_3)_2-CH_3\ Cl^-$

The effect of pH on hydrolytic stability of cationic ester–acrylamide copolymers has been long recognized (128). The decrease in viscosity and effectiveness, characteristic of this instability, do not take place in aqueous solutions at pH 2–5. Cationicity loss in AETAC and MAETAC copolymers depends on both pH and composition. For example, in a 43 mol % AETAC copolymer, at least 95% of the ester groups have at least one AMD neighbor. The effect of pH on cationicity loss in this copolymer was minimal at a pH of 2–3 at 90°C for 9 h. Above pH 3, ester loss increased dramatically. Prolonged heating (24 h) resulted in a greater degree of ester loss. It was found that the rate of isolated ester hydrolysis was first order in hydroxide concentration at 60°C at constant buffer pH of 5.5. If there were neighboring acrylamide groups in the chain, then there was a second-order dependence of ester disappearance on hydroxide concentration. This indicated that the imidization reaction was also first order in hydroxide ion concentration. The percentage of esters cleaved increased as the number of AETAC groups with neighboring AMD groups increased. Polymers with esters and no AMD neighbors such as the homopolymers of AETAC or MAETAC were found to have a low degree of hydrolysis. The rate of hydrolysis of AETAC and MAETAC copolymers were the same only when both had the same numbers of AMD neighbors.

Cationic copolymers derived from amide monomers, such as MAPTAC [51410-72-1] and APTAC [45021-77-0], are reasonably random and are hydrolytically stable. However, they are more expensive. The molecular weights of high-charge AMD/APTAC [75150-29-1] and AMD/MAPTAC [58627-30-8] copolymers typically do not reach the high molecular weights of AMD/AETAC copolymers because of impurities in the APTAC and MAPTAC. However, low-charge AMD/MAPTAC copolymers, containing ~3 mol % MAPTAC, are significant commercial products.

Diallydimethylammonium chloride (DADMAC) [7398-69-8] is the least expensive commercially available cationic monomer. This monomer has been successfully produced by reacting allyl chloride, dimethylamine, and sodium hydroxide in aqueous solution. (130,131). Monomer solutions with solids of 60–70% can be achieved and used directly for polymerization without further isolation and purification. DADMAC is a nonconjungated diene monomer that was found to homopolymerize to high-molecular-weight linear cationic polymer without cross-linking (130,132). Poly(diallydimethylammonium chloride) (PDADMAC) [26062-79-3] was the first synthetic organic flocculant approved for potable water clarification by the United State Public Health Service (133). The polymerization of DADMAC is known as kinetically favorable to give 98% of inter–intra cycloaddition and 2% pendant double bonds (134). The initiator radical attacks the terminal carbon on one allyl group, and the radical formed attacks the internal carbon on the other allyl group in the same molecule to form a five-membered pyrrodinium ring with a cis-to-trans ratio of 6:1 (135).

The rate law for DADMAC polymerization in an aqueous system, when persulfate is used, is not simple: $R_p = (S_2O_8{}^{2-})^{0.8}(\text{DADMAC})^{2.9}$. A combination of complicated initiation reactions and dimeric DADMAC interactions can account for the unusually high exponent of the DADMAC concentration (136). High monomer concentrations (>1.5 mol/L) used in commercial processes result in greater rates of polymerization and higher molecular weights. PDADMAC with low residual unreacted monomer can be manufactured in water using either persulfate addition or ammonium persulfate with sodium metabisulfite (137). Polymerization of DADMAC has also been studied in water-in-oil emulsion in a continuous stirred tank reactor (138). In that case, the oil-soluble initiator, 2,2′-azobis(2,4-dimethylvaleronitrile) (ADVN), and the surfactants sodium di-2-ethylhexylsulfosuccinate (AOT) and sorbitan monoleate (SMO) were used. The rate of polymerization was: $R_p = k\ (\text{ADVN})^{0.4}(\text{AOT})^{0.5}(\text{SMO})^{-0.4}(\text{DADMAC})^{3}$.

The negative order of SMO concentration was due to the fact that SMO is a radical scavenger. The influences of partitioning effects and ionic strength contributed to the third order in DADMAC concentration.

The molecular weight of PDADMAC is not as high as for acrylic polymers because of the large chain transfer constant of allylic radicals. However, a molecular weight of 5×10^5 daltons is sufficient for applications such as potable water clarification, color removal, and textile processing. These applications rely on the very high cationic charge of PDADMAC. This polymer is often used along with a high-molecular-weight anionic polyacrylamide in process-water clarification in paper deinking mills.

5.2. Anionic Copolymers. Anionic acrylamide copolymers such as poly(acrylamide-co-sodium acrylate) [25085-02-3] poly(acrylamide-co-ammonium acrylate) [26100-47-0], poly(acrylamide-co-sodium-2-acrylamido-2-methylpropanesulfonate [38193-60-1] (AMD/NaAMPS), and poly(acrylamide-co-2-acrylamido-2-methyl-1-propanesulfonic acid) [40623-73-2] (AMD/AMPS) have considerable practical importance. They can be prepared in solution, inverse emulsion (139,140), and inverse microemulsion (141). Comonomer reactivity ratios of AMD with acrylic acid or acrylic acid salts are given in Table 5. Reactivity ratios vary with pH. At high pH the reactivity ratio for AMD is higher, but at low pH the reactivity ratio of acrylic acid is higher. At a pH about 5 a random copolymer can be obtained. When AMD and sodium acrylate are copolymerized in a microemulsion at pH about 10, copolymer composition is independent of conversion and the reactivity ratios are equal. The copolymer chain composition conforms to Bernoullian statistics (141). These copolymers are used extensively as industrial flocculants for water treating, mining and paper manufacture, drag reduction agents, and in secondary and tertiary oil recovery. Reasons for their extensive use include their low cost and very high molecular weights. Their limitations include poor solubility at low pH and precipitation of the salt form in the presence of calcium ions.

Comonomer reactivity ratios for AMD and NaAMPS are given in Table 5 (142). AMD/AMPS copolymers and AMD/NaAMPS copolymers maintain their anionic charge at low pH and have a high tolerance to many divalent cations. They are used as flocculants for phosphate slimes, uranium leach residues, and coal refuse. There are also many oilfield applications.

6. Commercial Processes

There are numerous laboratory methods to prepare polyacrylamides. However, there are only a few viable commercial processes used to manufacture materials that meet the necessary performance standards. There are many requirements for commercial materials: very low to very high molecular weights, low insolubles content, low residual monomer content, fast dissolution rate, ease of handling, minimal dusting (for dry solids), product uniformity, long-term storage stability (to ensure performance consistency), high solids (to reduce shipping costs), and consistent performance characteristics. Several common commercial processes are summarized below.

6.1. Solution Polymerization. Commercial production of polyacrylamides by solution polymerization is conducted in aqueous solution, either adiabatically or isothermally. Process development is directed at molecular weight control, exotherm control, producing low levels of residual monomer, and control of the polymer solids to ensure that the final product is fluid and pumpable. A generic example of a solution polymerization follows.

An acrylamide monomer solution (2–30 wt % in water) is typically prepared, and deaerated by sparging it with an inert gas (eg, nitrogen) to reduce the oxygen content in solution. Stainless steel batch reactors or glass continuous stirred tank reactors are often used for solution polymerizations. A chelating agent is added to complex autopolymerization inhibitors such as copper or other metals if they are present. The polymerization is then initiated using one of several free-radical initiator systems (azo, peroxy, persulfate, redox, or combinations) at concentrations ranging from 0.001 to 10 wt % on monomer. The rate of polymerization depends on reaction conditions, but typically depends on the 1.2–1.6 order of monomer concentration and 0.5 order of initiator concentration. The heat evolved during polymerization (82.8 kJ/mol) can be removed by an external cooling system. For adiabatic processes, the temperature rise needs to be estimated and great care needs to be exercised to avoid exceeding the reflux temperature. Chain transfer agents and inorganic salts can be added to improve processing and to reduce insolubles. Monomer-to-polymer conversions of 99.5% are achievable in 4–6 h polymerization time. The products can have a molecular weight ranging from 1 thousand to 4 million. Polymer solids can be 2–30%. The process can be used in conjunction with thermal drying or precipitation methods in order to obtain products in either powder or granular form. Short residence times in drum drying have been used to avoid chain degradation and formation of insolubles. Precipitation in C_1–C_4 alcohols can be done to obtain nonsticky rubbery polymer gel that can be further extruded and then dried with hot air. The resulting granules can be milled and sieved to produce a uniform product. Care is taken to avoid very finely divided material that can cause dusting problems.

Some commercial low-molecular-weight polyacrylamides (LMPAM) are manufactured in solution and sold at 10–50% solids. For example, LMPAM containing DADMAC comonomer is made at 40% solids and can be reacted with glyoxal to produce a strengthening resin for paper. Furthermore, LMPAM hydrolyzed with sodium hydroxide to polyacrylate is manufactured at 30% solids and is used as an antiscalant. High-molecular-weight PAM is also prepared in solution at 2–6 wt % solids and is often further modified using, for example, the Mannich reaction.

6.2. Inverse Emulsion Process. A method of avoiding the high solution viscosities of high-molecular-weight water-soluble polymers comprises emulsifying the aqueous monomer solution in an oil containing surfactants, homogenizing the mixture to form a water-in-oil (inverse) emulsion, and then polymerizing the monomers in the emulsion. The resulting polymer latex can be inverted in water, releasing the polymer for use. A basic patent (143) illustrated this inverse emulsion process. Processes in which the inverse emulsion polymerization results in finely divided particles that are small enough to retard settling and can be sold without further modification have been developed

(144,145). Stability of the inverse emulsion to mechanical shear has recently been improved (146). Commercial production of inverse emulsion polymerization of AMD has been reviewed (147).

6.3. Polymerization on Moving Belts. Dry polyacrylamides are sometimes preferred, particularly when transportation distances are long. A variety of continuous processes has been developed for preparing dry polyacrylamides that consist of polymerizing aqueous acrylamide on a moving belt and drying the resulting polymer (148–150). In one such process (152) an aqueous solution of acrylamide and a photosensitizer is pumped onto a moving stainless steel belt, cooled on the underside by a water spray and covered on the upper side by a humid inert atmosphere. The belt passes under ultraviolet lamps that photoinitiate polymerization. The belt speed can be controlled so that the polymerization is complete when the polymer reaches the end of the belt. At the end of the belt the polymer gel that has formed can be sliced into small granules and dried in an oven. The dried polymer is then passed through a grinder to produce the desired particle size for handling and use.

Several recent patents describe improvements in the basic belt process. In one case a higher solids polymerization is achieved by cooling the starting monomer solution until some monomer crystallizes, and then introducing the resulting monomer slurry onto a belt. The latent heat of fusion of the monomer crystals absorbs some of the heat of polymerization, which otherwise limits the solids content of the polymerization (151). In another patent a concave belt that flattens near the end is described. This change is said to result in improved release of polymer from the belt (152).

6.4. Dry Bead Process. Dry polyacrylamides can also be prepared in the form of dry beads with bead sizes ranging from about 100 to 2000 μm (153,154). These beads are formed by azeotropically distilling water from inverse suspension polyacrylamides, collecting the beads by filtration, and further drying the beads in a fluid bed drier for short times. The resulting beads can be dissolved in water in a similar manner to other dry polyacrylamides. The size and shape of the beads prepared in the suspension polymerization process are a function of the types and amounts of surfactants and additives employed. Typically, 0.03–0.2 wt % (based on water plus polymer) of an oil-soluble polymeric surfactant is used to obtain the desired bead size. Greater amounts of surfactants lead to smaller beads (153). Certain water-soluble ionic organic compounds are said to be effective in improving the stability of the beads and providing a narrower bead size distribution when used in conjunction with the polymeric stabilizers (153). In the absence of the stabilizer, irregularly shaped, unstable particles can result. The choice of the stabilizer is considered to be dependent on the charge of the polyacrylamide being produced (153).

6.5. Microemulsion Polymerization. One inherent problem with water-in-oil emulsions of acrylamide-based polymers is the potential formation of unstable lattices both during production and in finished products. The coagulum that can form in the reactor can result in a time consuming cleanout (155). Technology has continuously improved reactor configuration, types of agitation, proper cooling (155), and a proper balance of aqueous, oil and emulsifier ratios (144,145). Microemulsion polymerization (156–166) can provide improvements to address these problems. Monomer microemulsions are thermodynamically stable systems comprising two liquids, insoluble in each other, and surfactant. They form spontaneously without homogenization. The resulting polymer microlattices are typically nonsettling, transparent, and about 100 nm in diameter. These systems can have high emulsifier levels: more than 8 wt %, which is about 4–5 times more than emulsifier levels in conventional inverse emulsions. Consequently, the cost of producing microemulsions becomes less attractive. However, further refinements to technology lead to the development of cost-effective microemulsified Mannich acrylamide polymers (167, 104–106). This technology was used to develop functionalized polyacrylamides (168). In one case, a PAM microlatex was reacted with formaldehyde and dimethylamine (Mannich reaction), and then quaternized with methyl chloride to yield a very highly charged cationic carbamoyl polymer (104–106, 169). These commercial products are widely used in many applications for solid–liquid separation. These products have been improved by treating them with buffer acid, a formaldehyde scavenger, and heat to produce a high-performance cationic polymer (170).

6.6. Environmentally Friendly Polyacrylamides. In recent years, commercial processes that use biodegradable oils to replace petroleum hydrocarbons have received a great deal of attention. Also, there has been a great deal of interest in polymerization in supercritical fluids. These future directions for the manufacture of polyacrylamides are summarized in the following.

6.7. Dispersion Polymerization. Water-in-oil emulsions contain at least 30% by weight of a petroleum-based hydrocarbon that is a valuable natural resource. By using such formulations, oils are consumed unnecessarily and can enter the world's waterways as a source of secondary pollution. An aqueous polymer dispersion is one environmentally responsible formulation that contains no oil or surfactant, and near-zero amounts of volatile organic compounds. Dispersion polymerization can be used to prepare cationic, anionic, and nonionic polyacrylamides.

6.8. Inverse Emulsions with Biodegradable Oils. Some examples of inverse emulsion polymerization processes employing biodegradable oils include materials with aqueous phase monomer mixtures, such as AMD and AETAC or AMD and MAETAC, dispersed in a biodegradable oil, such as bis-(2-ethylhexyl) adipate (171), containing a polymeric emulsifier that is a copolymer of dimethylaminoethylmethacrylate and mixtures of methacrylates. A buffering acid, such as a dicarboxylic acid, is used to stabilize cationic copolymers. Aliphatic dialkylethers are also used as biodegradable oils (172), in conjunction with SMO as an emulsifier, to produce high-molecular-weight cationic copolymers.

6.9. Inverse Emulsion Polymerization Acrylamide in Near-Critical and Supercritical Fluid Conditions. Supercritical fluids exhibit both liquid-like properties (eg, solubilizing power), and gaslike properties (eg, low viscosities). Aqueous AMD has been dispersed and even microemulsified in near-supercritical ethane–propane mixtures using nonionic surfactants such as ethoxylated alcohols (eg, Brij 30 and Brij 52). Emulsion polymerization of AMD was then conducted at 60°C for 5 h and 379 bar, at the near-supercritical condition of certain ethane–propane mixtures (173). 2,2′-azobis(isobutyronitrile) (AIBN) was used as the initiator. The resulting poly(acrylamide) had a low molecular weight, in the range of $(2.7–5.8) \times 10^5$ daltons. The ethane and propane can be easily recovered and recycled in a production plant.

Emulsion polymerization of AMD was also conducted at 60°C for 1 h and 345 bar in near-supercritical CO_2. AIBN was the initiator. An amide end-capped hexafluoropropylene oxide oligomer that has high solubility in the near-supercritical CO_2 was found to stabilize the dispersed particles (174–176). Only a few classes of polymers have good solubility in near-supercritical CO_2. The advantages of using carbon dioxide include very low viscosities during polymerization and ease of recovery.

7. Applications

Dewatering. Polyelectrolyte-assisted dewatering constitutes one of the most important application areas of polyacrylamides (3). Solid–liquid separations in aqueous media can be enhanced by the flocculation of small suspended particles into larger aggregates, which increases separation rates. Floc formation requires a destabilization and adherence of the smaller particles. This is usually accomplished by means of surface charge neutralization, charge-patch formation, and/or polymer bridging (177). Acrylamide-containing polymers make ideal candidates for such flocculants because of the large molecular weights achievable with them. High-molecular-weight cationic copolymers are typically employed in wastewater treatment. Solid–liquid separations in mining industries often benefit from the use of anionic copolymers, or in some cases dual polymer systems (cationic and anionic in sequence). Various dewatering processes in the papermaking industry regularly make use of cationic, anionic, or dual addition systems (178). Nonionic poly(acrylamide) finds less use in solid–liquid separations, save for some mining applications.

Mineral Processing. Both synthetic and natural hydrophilic polymers are used in the mineral processing industry as flocculants and flotation modifiers. Most synthetic polymers in use are polyacrylamides. Nonionic polymers are effective as flocculants for the insoluble gangue minerals in the acid leaching of copper and uranium (179,180), for thickening of iron ore slimes (181), and for thickening of gold flotation tailings (182). In some uranium leach operations, a cationic polymer with a relatively low charge density is used along with the nonionic polymer to improve supernatant clarity.

Anionic polyacrylamides are extensively used in the mining industry. They are used as flocculants for insoluble residues formed in cyanide leaching of gold (183). Acrylamide–acrylic acid copolymers are used for thickening copper, lead, and zinc concentrates in flotation of sulfide ores. These copolymers, containing from 50 to 100% carboxylate groups, are used to flocculate fine iron oxide particles in the manufacture of alumina from bauxite at high pH (184). Hydroxamated polyacrylamides, prepared by reaction of nonionic polyacrylamide or anionic polyacrylamide with hydroxylamine salts, are also effective in this Bayer process (185). Other uses for hydroxamated polyacrylamides include reduction of titanaceous and siliceous scale in Bayer alumina processes (186) and flocculation of titanium ore copper ore tailings in froth flotation processes (187). Copolymers [40623-73-2] of acrylamide and acrylamido-2-methylpropanesulfonic acid [15214-89-8] have been patented as phosphate slime dewatering aids (188).

Low-molecular-weight polyacrylamide derivatives with mineral specific functionalities have been developed as highly selective depressants for separation of valuable minerals from gangue minerals in froth flotation processes. These depressants have certain ecological advantages over natural depressants such as starches and guar gums. The depressants provide efficient mineral recovery without flocculation. They are often used along with hydrophobic mineral collectors (eg, sodium alkyl xanthates) and froth modifiers. Partially hydrolyzed polyacrylamides with molecular weights of 7,000–85,000 can be used in sylvanite (KCl) recovery (189). Polymers having the functionality —$CONHCH_2OH$ are efficient modifiers in hematite–silica separations (190). Polymers containing the —CONHCH(OH)COOH functionality provide excellent selectivity in separation of apatite from siliceous gangue in phosphate benefication. Valuable sulfide minerals containing copper and nickel can be separated effectively from gangue sulfide minerals such as pyrite in froth flotation processes when acrylamide–allylthiourea copolymers or acrylamide–allylthiourea–hydroxyethylmethacrylate terpolymers are added to depress the pyrite (191). Acrylamide copolymers can be used as iron ore pellet binders (192). When the ore slurry in water has a pH above 8, anionic polymers are effective. If the ore is acid washed to remove manganese, then a cationic polymer is effective.

Paper Manufacture. Polyacrylamides are used as wet-end additives to promote drainage of water from the cellulose web, to retain white pigments and clay fillers in the sheet, to promote sheet uniformity, and to provide dry tensile strength improvements (193). An important advance in papermaking technology has been the use of microparticle retention aids. Organic microparticles, prepared from acrylamide and anionic comonomers by microemulsion polymerization, provide good sheet formation characteristics and controlled drainage (194,195). Cationic polyacrylamides that have been reacted with glyoxal are used to promote wet strength (196). These wet strength resins have been used in paper towels. Recently, these glyoxalated polymers have been modified so that they can be used in toilet tissue. These polymers provide an initial high wet tensile strength with rapid tensile strength decay in water so that sewers may not become clogged (197,198). Anionic polyacrylamides have been used with alum to increase dry strength (199). Primary amide functionality promotes strong interfiber bonds between cellulose fibers. Sometimes paper mills use dry strength additives so that recycled fiber, groundwood, thermomechanical pulp, and other low cost fiber can be used to produce liner board and other paper grades which must meet ICC requirements for burst strength and crush strengths. Recently there has been an increasing demand for writing papers and copy paper that have good print characteristics. Print quality can be improved by use of surface sizes combined with acrylamide polymers. The acrylamide polymer gives the paper sheet better surface strength (200). Details on paper manufacture can be found in Reference 201. All additives used for manufacture of food-grade papers are subject to FDA regulations, and are listed in the *Code of Federal Regulations*, paragraphs 176.170, 176.180, 178.3400, and 178.3650 (1998).

Enhanced Oil Recovery. Polymer flooding is a potentially important use for anionic polyacrylamides having molecular weights greater than 5 million and carboxyl contents of about 30%. The ionic groups provide the proper viscosity and mobility ratio for efficient displacement. The anionic charge prevents excessive adsorption onto negatively charged pores in reservoir rock. Viscosity loss is observed in brines particularly when calcium ion is present. A primary advantage of anionic polyacrylamides is low cost (202). Profile modification is a process wherein flooding water is diverted from zones with high permeability to other zones of lower permeability containing oil. Polymeric hydrogels are used for this. Metals such as chromium and aluminum can be injected with anionic polyacrylamides to cross-link the polymers in more permeable reservoir zones prior to the water flood (203,204). The development of new more environmentally acceptable cross-linking systems has continued. A recent patent claims a composition consisting of hexamethylenetetramine [100-97-0] and 4-aminobenzoic acid [150-13-0] for this purpose (205).

Polyacrylamides are used in many other oilfield applications. These include cement additives for fluid loss control in well-cementing operations (206), viscosity control additives for drilling muds (207) and brines, and for fracturing fluids (208). Copolymers [40623-73-2] of acrylamide and acrylamidomethylpropanesulfonic acid do not degrade with the high concentrations of acids used in acid fracturing.

Hydrophobically Associating Polymers. Extensive research in the 1980s and 1990s focused on acrylamide copolymers containing small amounts of hydrophobic side chains. At zero or low shear rates, the apparent viscosity can be very large because of association of the hydrophobic groups between chains. In oil reservoir conditions, the polymers tolerate high salt concentrations while providing proper viscosifying properties (209). These associative thickeners are also used in coatings (210) and in oil spill cleanup (211). Reference 212 gives more information about associative polymers.

Hydrophobically associating acrylamide copolymers can be prepared by micellar polymerization. These copolymers have short blocks of hydrophobic groups randomly distributed in the backbone. A recent paper reviews the major advances in this area (213).

Superabsorbents. Water-swellable polymers are used extensively in consumer articles and for industrial applications. Most of these polymers are cross-linked acrylic copolymers of metal salts of acrylic acid and acrylamide or other monomers such as 2-acrylamido-2-methylpropanesulfonic acid. These hydrogel-forming systems can have high gel strength, as measured by the shear modulus (214). Sometimes, inorganic water-insoluble powder is blended with the polymer to increase gel strength (215). Patents describe processes for making cross-linked polyurethane foams that contain superabsorbent polymers (216,217). Recent patents describe grafted copolymers that are highly absorbent to aqueous electrolyte solutions (218).

8. Analytical Methods

Most of the traditional methods for polymer analysis (219) are applicable to polyacrylamides. We will only point out several special features regarding the use of some of these techniques for the analysis of polyacrylamides.

Oftentimes a preliminary step applied before many analytical methods is the isolation of the polymer. The isolation of polyacrylamides from the other components of the media in which they were prepared (eg, aqueous solution or inverse emulsion, with attendant surfactants and oil) is often readily accomplished by precipitation in short-chain alcohols or acetone. The individual solubilities of formulation components should be tested if there is any doubt. Experience shows that anionic copolymers are often best precipitated in the alcohols, and cationic copolymers in acetone (homopolyaminoesters are soluble in methanol). Since acrylamide is soluble in these organic solvents, it will also be separated from the polymer in this procedure.

Once the polymer is isolated, its chemical composition can be quantified using infrared or nmr spectroscopies (25,220). An nmr study can also give some information about chain architecture in the case of copolymers (221). Elemental analysis can be employed to confirm a composition. Ultraviolet spectroscopy is generally not used for compositional analysis per se; however, polymers containing acrylamide do absorb short-wavelength uv radiation, along with many other materials. A uv detector set around 215 nm is a common choice for measuring polymer concentration in the absence of interfering substances; many surfactants and some salts (eg, NO_3^-, SCN^-) are problematic in this regard. Differential refractometry is the logical alternative for concentration monitoring when uv-absorbing substances are present.

The extent of conversion during an acrylamide polymerization is most easily followed by determining the disappearance of the monomer. High-performance liquid chromatography (HPLC) is often found best for this purpose. An HPLC method in which poly(acrylamide) inverse emulsions can be used directly has been developed (222). In the case of copolymers, HPLC protocols that allow the simultaneous determination of all the unreacted monomers can be used to evaluate compositional drift as a function of conversion.

Global properties of the polymer chains (eg, molecular weight, coil dimensions, branching content) are most often evaluated using scattering, hydrodynamic, or viscometric techniques on dilute polymer solutions. The conventional methods appropriate for soluble polymeric materials typically apply equally well to polyacrylamides. In the case of high-molecular-weight polyacrylamides, the main difficulties in obtaining accurate information involve preparing a purified polymer solution that is in an equilibrium state, and passing it through the measurement device without substantially altering it in either process.

When using viscometric methods, experiment protocols can be designed to address any effects that instabilities may have on the viscosities of the polyacrylamide solutions, if not to alleviate the instabilities altogether. Mechanical degradation can also occur in high-molecular-weight polymers. This can happen during sample preparation (eg, mixing), purification (eg, filtration), or during the viscosity measurement itself (especially in elongational flows). In any case, one should estimate these handling effects for any set of protocols.

The presence of colloidal-size contamination ("dust") in polymer solutions can possibly affect either static or dynamic light scattering experiments (223,224). Neutron scattering is less afflicted by this kind of contamination (225). For high-molecular-weight polyacrylamides whose coil dimensions are roughly in the same size range as the colloidal contaminants, and which have a natural propensity to adsorb onto suspended materials (after all, many of these polymers are flocculants), any problems of sample purification should not be ignored. If one is simply looking for a clean sample, it is possible to exhaustively filter a solution in a recycle loop (226). Clarification of dilute polymer solutions by centrifugation is another method that can minimize mechanical degradation of the polymers. Centrifuging dilute solutions of high-molecular-weight linear polyacrylamides from 4 to 8 h in excess of $15{,}000 \times G$ is satisfactory in many instances. Good light scattering data can be acquired even in the case of a marginally clean polymer solution by attempting to "look through the dust." This is made easier by reducing the scattering volume, slowing (or stopping) the solution flow, and monitoring the scattering volume (either manually or with the aid of a computer algorithm) for periods that are free of point scatterers.

The effects of sample clarification must be gauged when trying to preserve and analyze the polymer in its original form. This includes situations when samples are passed through packed columns, as in HPSEC analysis. A uv absorbance study can be used to determine polymer loss if there are no interferences in the solution. A method that can monitor the molecular weight distribution (eg, HPSEC or dynamic light scattering), or one that is sensitive to the high-molecular-weight component (eg, elongational viscometry), can be used to assay for mechanical degradation.

The double extrapolation of light scattering data to zero polymer concentration and zero scattering angle yields an average property of the macromolecular ensemble: the weight-average molecular weight (M_w) from static light scattering and the z-average hydrodynamic radius ($\langle R_h \rangle_z$) from dynamic light scattering. In some cases, the details of the distribution of these quantities are also of interest. Dynamic light scattering data can be analyzed directly to give a distribution of the hydrodynamic size distribution of a sample. DiNapoli et al. (227) have demonstrated how to derive molecular weight distributions from dynamic light scattering data, but this involves knowing the correlation between the polymer diffusion coefficient and molecular weight, a relationship that is not always available.

Methods involving a physical separation of the components of the distribution, coupled with a method for measuring some feature of the macromolecules across this separated collection, find more use in determining molecular-weight (or size) distributions. Size exclusion chromatography (SEC) remains a popular way to separate macromolecular populations (228), including polyacrylamides (229). More recently, flow field flow fractionation (FFFF) (230) has been shown to have some advantages over SEC methods, especially for very high-molecular-weight polymers, including polyacrylamides. Since the fluid contact surface in FFFF is a membrane, as opposed to a packed bed of finely divided particles in SEC, there is less opportunity for altering the native distribution by means of polymer adsorption, retention, or mechanical degradation. High-molecular-weight cationic copolymers of acrylamide can be difficult to pass unaltered through commercial SEC columns. Ultracentrifugation (46,231,232) has also been used to separate the components of polyacrylamide samples for subsequent analysis, but this is currently a less popular method than either SEC or FFFF.

The early approaches to characterizing the molecular-weight distributions of samples separated using SEC or FFFF were based on retention time, requiring a correspondence to be made between the retention time and molecular weight. This was typically done by calibrating the separation device using fractionated (narrowly distributed) standards, which in some cases were only vaguely related chemically to the polymer of interest. More recently the use of in-line light scattering detectors for the purpose of directly determining M_w, R_g (static light scattering photometer), or R_h (dynamic light scattering photometer) for each "slice" of the separated distribution has been an alternative to these approaches (233). This has generally improved one's ability to characterize the details of the molecular-weight or size distributions for many acrylamide-containing polymers, for which standards consisting of narrow fractions are not readily available.

Titration methods are mostly applicable to ionic copolymers of acrylamide. Typically, potentiometric titrations are used for high salt concentrations, and conductometric titrations are used for low salt concentrations. This kind of information can be important since ionogenic groups with weak acid or base properties will have both dissociation-dependent and salt-dependent pK_as when they are in a polymer chain. In addition to titrating ionic acrylamide copolymers with low-molecular-weight titrants, their titration with other oppositely charged polyelectrolytes has proven useful (234). For example, poly(potassium vinylsulfonate) can be used to titrate cationic copolymers of acrylamide. This titration gives information about the available charge on the host macromolecule. Usually the macroion titrant is of lower molecular weight than the polymer of interest. In any case, the conditions of the polyelectrolyte–polyelectrolyte complexation reaction must allow for complete 1:1 complex formation. Care must be taken such that the kinetics of the complex formation does not influence the results. The end point can be detected in any one of several ways, including turbidometrically, or using a dye indicator. The color change of the dye at the end point can be determined visually or spectrophotometrically with an "optrode."

Detecting Polyacrylamides. In order to detect low concentrations of polyacrylamides as part of an analysis scheme (235), to optimize the use of, or to monitor the fate of these polymers in a variety of technological applications, an assay method for trace amounts of these polymers that remain in solution is usually needed. One approach, more appropriate for laboratory studies, has been to incorporate fluorescent groups in the polymer either by copolymerization or by post-polymerization derivatization. Acrylamide copolymers containing sodium fluorescein (236), various dyes [phenol red, or brilliant yellow (237)], *N*-2,4-dinitroaniline-acrylamide (238), or a pyrene-labeled monomer (239) have been described. Early methods based on chemical derivatizations describe coupling fluorescein isothiocyanate to amine groups on Hofmann-reacted poly(acrylamide) (240,241). Various other approaches have been developed to add fluorescein (242), dansyl (243), 9-xanthydrol (244), and other fluorescent groups to acrylamide-containing chains (245).

A completely different approach to detecting low levels of high-molecular-weight polyacrylamides in solution without recourse to prelabeling the polymers has been used in a number of instances. Methods based on the flocculation capacity of these polymers are surprisingly sensitive. Both the turbidity and the settling rate of a suspension can change measurably after exposure to even low concentrations (several ppm) of a flocculant. Lentz et al. (246) used such changes in the settling rate of kaolin suspensions to assay for low levels of anionic polyacrylamides in runoff water from a soil amendment application involving those polymers. They were able to reliably detect residual polymer in the water at the ppm level.

9. Specifications, Shipping, and Storage

The amount of residual acrylamide is usually determined for commercial polyacrylamides. In one method, the monomer is extracted from the polymer and the acrylamide content is determined by HPLC (247). A second method is based on analysis by cationic exchange chromatography (248). For dry products the particle size distribution can be quickly determined by use of a shaker and a series of test sieves. Batches with small particles can present a dust hazard. The percentage of insoluble material is determined in both dry and emulsion products.

Polyacrylamide powders are typically shipped in moisture-resistant bags or fiber packs. Emulsion and solution polymers are sold in drums, tote bins, tank trucks, and tank cars. The transportation of dry and solution products is not regulated in the United States by the Department of Transportation, but emulsions require a DOT NA 1693 label.

Under normal conditions, dry polymers are stable for 1 y or more. The emulsion and solution products have somewhat shorter shelf lives.

10. Health and Safety Factors

Commercial Polyacrylamides. Dry cationic polyacrylamides have been tested in subchronic and developmental toxicity studies in rats. No adverse effects were observed in either study. Chronic studies of polyacrylamides in rats and dogs indicated no chronic toxicity or carcinogenicity. Dry anionic and nonionic polyacrylamides (249) have acute oral (rat) and dermal (rabbit) LD_{50} values of greater than 2.5 and greater than 10.0 g/kg, respectively. Dry cationic polyacrylamides have acute oral (rat) and dermal (rabbit) LD_{50} values of greater than 5.0 and greater than 2.0 g/kg, respectively. Emulsion nonionic, anionic, and cationic polyacrylamides have both acute oral (rat) and dermal (rabbit) LD_{50} values of greater than 10 g/kg. Dry nonionic and cationic material caused no skin and minimal eye irritation during primary irritation studies with rabbits. Dry anionic polyacrylamide did not produce any eye or skin irritation in laboratory animals. Emulsion nonionic polyacrylamide produced eye irritation in rabbits, while anionic and cationic material produced minimal eye irritation in rabbits. Emulsion nonionic, anionic, and cationic polyacrylamide produced severe, irreversible skin irritation when tested in rabbits that had the test material held in skin contact by a bandage for 24 h. This represents an exaggeration of spilling the product in a boot for several hours. When emulsion nonionic, cationic, and anionic polyacrylamides were tested under conditions representing spilling of product on clothing, only mild skin irritation was noted. Polyacrylamides are used safely for numerous indirect food packaging applications, potable water, and direct food applications.

Experimental Polyacrylamides. It is wise to treat any laboratory-prepared "experimental" polyacrylamide as if it contains substantial amounts of unreacted monomer unless it has been isolated and purified as described above.

Acrylamide is commercially available as a 50% solution in water with a copper salt as a polymerization inhibitor. Polymerization is very exothermic and autopolymerization can occur under certain conditions. In the interest of safety, acrylamide solutions should be stored under the following conditions:

1. Maintain the storage temperature below 32°C (90°F) and above the solubility point.
2. Keep the solution free of contaminants.
3. Maintain the proper level of oxygen and Cu^{2+} inhibitors.
4. Maintain the pH at 5.2–6.0.
5. Store the solution in a container that is opaque to light.

It is recommended that these solutions be stored for no more than 3 mo due to depletion of the dissolved oxygen. All containers must be dated and no more than 93% full. Packaged acrylamide solutions should be consumed on a first-in, first-out basis.

11. Economic Aspects

Worldwide, there are many suppliers of polyacrylamides. Some of these are producers and some are repackagers. Suppliers are listed in Table 6. Selling prices for polyacrylamides vary considerably depending on the product form (solution, emulsion, dry), type (anionic, nonionic, cationic), and other factors. Prices on a polymer basis can range from as low as about $2/kg for simple dry nonionic polyacrylamides to $8/kg and more for highly charged cationic polymers. Prices in recent years have dropped due to price erosion and due to

Table 6. **Suppliers of Polyacrylamides**

Region	Companies
United States	Axchem
	Baker-Petrolite Co.
	BetzDearborn, Inc.
	Buckman Laboratories International, Inc.
	Calgon Corp.
	Callaway Chemical Co.
	Chemtall, Inc.
	CIBA Specialty Chemicals Corp. (Allied Colloids, Ltd.)
	Cytec Industries, Inc.
	Delta Chemical Corp.
	The Dow Chemical Co.
	Drew Chemical Corp. (Ashland Chemical, Inc.)
	Exxon Chemical Co.
	Hercules Inc.
	Nalco Chemical Co.
	Polydyne, Inc.
	S.N.F. Floerger SA
	Stockhausen, Inc.
Canada	Cytec Canada, Inc.
	Nalco Canada, Inc.
	Raisio Chemicals Canada, Inc.
	Rhodia Canada, Inc.
Mexico	BASF Mexicana, S.A. de C.V.
	Cytec, Atequiza Jalisco
	Nalco, Toluca

Table 6 *(Continued)*

Region	Companies
South America	Dispersol San Luis S.A. (Argentina)
	LaForestal Quimica S.A.I.C. (Argentina)
	Henkel Argentina S.A.
	Industrias Quimicas del Valle S.A. (Argentina)
	Proquima Productos Quimicos (Argentina)
	Adesol Produtos Quimicos Ltda. (Brazil)
	Quimicos Nacional Quiminasa S/A (Brazil)
	Cyquim de Columbia
	Quimicos Cyquim, C.A. (Venezuela)
Europe	BASF AG (Germany)
	Ciba Specialty Chemicals PLC (UK)
	Cytec Industries BV (Netherlands)
	Cytec Industries UK Ltd. (UK)
	Deutsche Nalco-Chemie GmbH (Germany)
	Kimira Oyj (Finland)
	Röhm GmbH (Germany)
	S.N.F. Floerger S.A. (France)
	Stockhausen GmbH (Germany)
Japan	Arakawa Chemical Industries, Ltd.
	Dai-Ichi Kogyo Seiyaku Company, Ltd.
	Diafloc Co., Ltd.
	Harima Chemicals, Inc.
	Hymo Corporation
	Kurita Water Industries, Ltd.
	Japan Polyacrylamide Ltd.
	Konan Chemical Industry Co., Ltd.
	Mitsubishi Chemical Industries Co., Ltd.
	Mitsui-Cytec, Ltd.
	Nippon Kayaku Company, Ltd.
	Sankyo Kasei Company, Ltd.
	Sanyo Chemical Industries, Ltd.
	Sumitomo Chemical Company, Ltd.
	Toa Gosei Chemical Industry Company, Ltd.
Republic of Korea	Cytec Korea, Inc.
	E-Yang Chemical Co., Ltd.
	Kolon Industries, Inc.
	Unico (Seoul)
Taiwan	Cytec Taiwan Corp.
	Taiwan Arakawa Chemical Ind., Ltd.
	Young Sun Chemical Works, Ltd.
China	China Petrochemical Corp.
India	Engineer's Poly-Chem
	Kaushal Aromatic Chemicals Pvt. Ltd.
	Somnath Products

lower manufacturing costs of AETAC and DADMAC cationic monomers. In many applications, such as sludge dewatering in waste treatment, the need for increased performance has lead to increased functionalization (eg, higher cationicity) and increased cost.

In the United States the major uses for polyacrylamides are in water treating and paper manufacturing. For water treating the best growth is expected to be for cationic copolymers due to use in dewatering equipment like belt presses that produce high-solids sludge cakes that can be more easily incinerated or disposed of in scarcer landfills. For paper manufacturing, glyoxalated cationic copolymers for paper wet strength, high-molecular-weight retention aids, and drainage aids are considered to grow in use. Increased use of PAM flocculants in recycled paper mills, particularly in deinking mills where better process water clarification is necessary because of closed water circuits, is also expected. Polyacrylamides are beginning to be used along with surface sizes in paper to improve the control of ink adsorption and print quality. In the mineral process industry in the United States (and Australia) there has been a great increase in the use of hydroxamated polyacrylamides in alumina manufacture. The market for polyacrylamides in enhanced oil recovery has decreased steadily in the United States In 1985, the total market exceeded 10 million metric tons, but in 1999 the use has been almost none. Total consumption of polyacrylamides is expected to increase about 4% per year during the next few years after 1999 in the United States. In Europe the use of water-treating chemicals on municipal sludge treatment will increase due to European Union legislation preventing sewage dumping. The treatment of wastewater is intensive, and many waste treatment plants and paper mills have closed circuits due to environmental concerns. The total consumption of PAMs is expected to increase about 3% per year in Europe. In Japan, the major consumption has been in paper manufacture (anionic and nonionic polyacrylamides) and water treatment (cationic and amphoteric copolymers). In municipal sludge treatment, highly cationic polyacrylamides are used for rapid flocculation in high-speed centrifuges. The use of cationic copolymers is expected to grow at higher rates. Japan exports a considerable amount of polyacrylamides (and acrylamide monomer) to Asian and other markets. Another potential market will be enhanced oil recovery in China. Table 7 gives an estimated breakdown of polyacrylamide consumption (1).

Table 7. **Consumption of Polyacrylamides**[a]

Year	Paper (%)	Water treating (%)	Mining (%)	EOR (%)	Other (%)	Total (10^3 t)
			United States			
1989	21.9	59	7.9	6	5.4	68.5
1993	18.3	63	13.6	0	5.1	86.0
1997	25	60	10.8	0	4.2	120.0
			Western Europe			
1997	38.4	53.5	5.8	0	2.3	86
			Japan			
1985	62	38				42
1989	57.7	42.3				52
1993	51	49				53
1997	52.4	47.6				63

Source: Ref. 1.

BIBLIOGRAPHY

"Acrylamide Polymers" in *ECT* 3rd ed., Vol. 1, pp. 312–330, by J. D. Morris and R. J. Penzenstadler, The Dow Chemical Company; "Acrylamide Polymers" in *ECT* 4th ed., Vol. 1, pp. 266–287, by David Lipp and Joseph Kozakiewicz, American Cynamid Company; "Acrylamide Polymers" in *ECT* (online), posting date: November 27, 2000, David Lipp and Joseph Kozakiewicz, American Cynamid Company.

CITED PUBLICATIONS

1. SRI Consulting, *Chemical Economics Handbook, Marketing Research Report on "Water Soluble Polymers"*, February 1999. Internet web site: *http://ceh.sri.com/*.
2. H. Dautzenberg, W. Jaeger, J. Kotz, B. Phillip, Ch. Seidel, and D. Stscherbina, *Polyelectrolytes: Formation, Characterization and Application*, Hanser Publishers, New York, 1994, p. 272.
3. R. S. Farinato, S.-Y. Huang, and P. Hawkins in R. S. Farinato and P. L. Dubin, eds., *Colloid–Polymer Interactions: From Fundamentals to Practice*, John Wiley & Sons, New York, 1999, p. 3.
4. S.-Y. Huang and D. W. Lipp in J. C. Salamone, ed., *Polymeric Materials Encyclopedia*, CRC Press, Inc., Boca Raton, Fla., 1996, p. 2427.
5. H. I. Heitner, *Kirk-Othmer: Encyclopedia of Chemical Technology*; 4th ed., Vol. 2 John Wiley & Sons, New York, 1994, p. 61.
6. H. I. Heitner, T. Foster, and H. Panzer, *Encyclopedia of Polymer Science and Engineering*, 2nd ed., Vol. 9, John Wiley & Sons, New York, 1987, p. 824.
7. D. A. Mortimer, *Polym. Int.* **25**, 29 (1991).
8. F. A. Bovey, *High Resolution NMR of Macromolecules*, Academic Press, New York, 1972, p. 147.
9. W.-M. Kulicke, R. Kniewske, and J. Klein, *Prog. Polym. Sci.* **8**, p. 379 (1982).
10. P. J. Flory, *Principles of Polymer Chemistry*, Cornell University Press, Ithaca, NY, 1953.
11. G. Medjahdi, D. Sarazin, and J. Francois, *Eur. Polym. J.* **26**(7), 823–829 (1990).
12. A. L. Izyumnikov, L. V. Mineyev, V. A. Maslennikov, L. S. Sidorina, O. S. Samsonova, and A. D. Abkin, *Polym. Sci. USSR* **30**(5), 1062–1071 (1988).
13. H. H. Hooper, J. P. Baker, H. W. Blanch, and J. M. Prausnitz, *Macromolecules* **23**, 1096–1104 (1990).
14. H. Volk and R. E. Friedrich in R.L. Davidson, ed., *Handbook of Water-Soluble Gums and Resins*, McGraw-Hill, New York, Ch. 16, 1980.
15. P. H. von Hippel, V. Peticolas, L. Schack, and L. Karlson, *Biochemistry* **12**(7), 1256 (1973).
16. B. P. Makogon and T. A. Bondarenko, *Polym. Sci. USSR* **27**(3), 630–634 (1986).
17. A. J. Begala and U. P. Strauss, *J. Phys. Chem.* **76**, 254 (1972).
18. K. Hiraoka and T. Yokoyama, *Polym. Bull.* **2**, 183 (1980).
19. M. Leca, *Polym. Bull.* **16**, 537–543 (1986).
20. A. Chapiro and L. Perec-Spritzer, *Eur. Polym. J.* **11**, 59 (1975).
21. T. O. Osmanov, V. F. Gromov, P. M. Khomikovskii, and A. D. Abkin, *Dokl. Akad. Nauk SSSR* **240**, 910–913 [Phys. Chem.] (1978).
22. V. F. Gromov, N. I. Galperina, T. O. Osmanov, P. M. Khomikovskii, and A. D. Abkin, *Eur. Polym. J.* **16**, 529–535 (1980).
23. Ye. V. Bune, I. L. Zhuravleva, A. P. Sheinker, Yu. S. Bogachev, and E. N. Teleshov, *Vysokomol. Soyed.* **A28**, 1279 (1986).

24. Yu. S. Bogachev, Ye. V. Bune, V. F. Gromov, I. L. Zhuravleva, E. N. Teleshov, and A. P. Sheinker, *Polym. Sci.* **33**(7), 1357–1360 (1991).
25. W.-M. Kulicke, R. Kniewske, and J. Klein, *Prog. Polym. Sci.* **8**, 373–468 (1982).
26. W.-M. Kulicke in B. Gampert, ed., *The Influence of Polymer Additives on Velocity and Temperature Fields—IUTAM Symposium Essen/Germany 1984*, Springer, Berlin, 163–172 (1985).
27. W.-M. Kulicke, and H. H. Horl *Colloid. Polym. Sci.* **263**, 530 (1985).
28. P. Albonico, and T. P. Lockhart, *J. Appl. Polym. Sci.* 55, 69–73 (1995).
29. V. Ya. Kabo, L. A. Itskovich, and V. P. Budtov, *Polym. Sci. USSR* **31**(10), 2217–2225 (1989).
30. T. Schwartz and J. François, *Makromol. Chem.* **182**, 2775–2787 (1981).
31. A. Ikegami and N. Imai, *J. Polym. Sci.* **56**, 133 (1962).
32. J. François, N. D. Truong, G. Medjahdi, and M. M. Mestdagh, *Polymer* **38**(25), 6115–6127 (1997).
33. R. Rahbari and J. François, *Polymer* **33**(7), 1449 (1992).
34. R. K. Prud'homme, J. T. Uhl, J. P. Poinsatte, and F. Halverson, *Soc. Petroleum Eng. J.* 804–808 (Oct. 1983).
35. W.-M. Kulicke, K. Kötter, and M. Gräger, *Adv. Poly. Sci.* **89**, 1 (1989).
36. F. Durst, R. Haas, and B. U. Kaczmar, *J. Appl. Polym. Sci.* **26**(9), 3125–3149 (1981).
37. G. Chauveteau, M. Moan, and A. Magueur, *J. Non-Newtonian Fluid Mech.* **16**, 315–327 (1984).
38. R. S. Farinato and W. S. Yen, *J. Appl. Polym. Sci.* **33**, 2353–2368 (1987).
39. W.-M. Kulicke and R. Haas, *Ind. Eng. Chem. Fundam.* **23**, 308–315 (1984).
40. S. A.-A. Ghoniem, *Rheol. Acta* **24**, 588–595 (1985).
41. K. A. Narh, J. A. Odell, A. J. Müller, and A. Keller, *Polym. Comm.* **31**, 2–5 (1990).
42. J. A. Odell and A. Keller, *J. Polym. Sci., Part B, Polym. Phys.* **24**, 1889–1916 (1986).
43. J. P. Scott, P. D. Fawell, D. E. Ralph, and J. B. Farrow, *J. Appl. Polym. Sci.* **62**, 2097–2106 (1996).
44. W. M. Kulicke, M. Koetter, and H. Graeger, *Advances in Polymer Science* **89**, Springer-Verlag Berlin, Heidelberg, 1989, p. 32.
45. Ref. 2, p. 224.
46. D. Vlassopoulos and W. R. Schowalter *J. Rheol.* **38**(5), 1427–1446 (1994).
47. A. Ait-Kadi, P. J. Carreau, and G. Chauveteau, *J. Rheol.* **31**(7), 537–561 (1987).
48. H. C. Haas and R. L. MacDonald, *J. Appl. Polym. Sci.* **16**, 2709–2713 (1972).
49. M. Chmelir, A. Kunschner, and E. Barthell, *Ang. Makromol. Chem.* **89**, 145–165 (1980).
50. J. Brandup and E. H. Immergut, eds., *Polymer Handbook*, 3rd ed., John Wiley & Sons, New York, p. VII-2.
51. P. Munk, T. M. Aminabhavi, P. Williams, D. E. Hoffman, and M. Chmelir, *Macromolecules* **13**, 871–875 (1980).
52. J. Klein and K. D. Conrad, *Makromol. Chem.* **179**, 1635–1638 (1978).
53. W.-M. Kulicke and J. Klein, *Ang. Makromol. Chem.* **69**, 169–188 (1978).
54. W. Scholtan, *Makromol. Chem.* **14**, 169 (1954).
55. J. Francois, D. Sarazin, T. Schwartz, and G. Weill, *Polymer* **20**, 969 (1979).
56. K. J. McCarthy, C. W. Burkhardt, and D. P. Parazak, *J. Appl. Polym. Sci.* **33**, 1699–1714 (1987).
57. T. Griebel and W.-M. Kulicke, *Makromol. Chem.* **193**, 811–821 (1992).
58. F. Mabire, R. Audebert, and C. Quivoron, *Polymer* **25**, 1317 (1984).
59. S. M. Shawki and A. E. Hamielec, *J. Appl. Polym. Sci.* **23**, 3341 (1979).
60. D. J. Currie, F. S. Dainton, and W. S. Watt, *Polymer* **6**, 451 (1965).
61. J. P. Riggs and F. Rodriguez, *J. Polym. Sci. A1* **5**, 3151 (1967).
62. C. Y. Chen and J. F. Kuo, *J. Polym. Sci., Chem.* **26**, 1115 (1988).

63. S. K. Ghosh and B. J. Mandal, *Polymer* **34**, 4287 (1993).
64. T. J. Suen, A. M. Schiller, and W. N. Russel, *J. Appl. Polym. Sci.* **3**, 126 (1960).
65. E. H. Gleason, M. L. Miller, and G. F. Sheats, *J. Polym. Sci.* **38**, 133 (1959).
66. V. F. Kurenkov and V. A. Myagchenkov, *Eur. Polym. J.* **16**, 229 (1980).
67. J. Barton, V. Jurvanicova and V. Vaskova, *Makromol. Chem.* **186**, 1935 (1985).
68. W. M. Thomas and D. W. Wang, *Encyc. of Polym. Sci. and Eng.*; 2nd ed., Vol. **1**, John Wiley & Sons, New York, 1985, p. 182.
69. P. Molyneux, *Water Soluble Synthetic Polymers: Properties and Behavior*, Vol. I, CRC Press, Inc., Boca Raton, Fla., 1983, p. 91.
70. S. J. Guerrero, P. B. Oldarino, and J. H. Zurimendi, *J. Appl. Polym. Sci.* **30**, 955 (1985).
71. H. C. Hass and R. L. McDonald, *Polym. Lett.* **10**, 401 (1973).
72. H. C. Hass and R. L. McDonald, *J. Appl. Polym. Sci.* **16**, 2709 (1972).
73. T. V. Stupnikova, B. P. Makopon, T. V. Vishnia, and I. L. Povh, *Visokomolecularnie Soedinenia* **A30** 1380 (1988).
74. D. K. Ramsden and K. McKay, *Polym. Degradation and Stability* **15**, 15 (1986).
75. D. K. Ramsden and K. McKay, *Polym. Degradation and Stability* **14**, 217 (1986).
76. W. Nagashiro, T. Tsunoda, M. Tanaka, and M. Oikawa, *Bull. Chem. Soc. Japan* **48**, 2597 (1975).
77. U. Grollman and W. Schnabel, *Polym. Degradation and Stability* **4**, 203 (1982).
78. R. G. Ryles, *Chemical Stability Limits of Water-Soluble Polymers Used in Oil Recovery Processes*, SPE **389**, 1359 (1985).
79. L. I. Tolstikh, N. I. Akimov, I. A. Golubeva, and I. A. Shvetsov, *Intern. J. Polym. Mater.* **17**, 177 (1992).
80. V. F. Kurenkov, R. V. Gerkin, V. S. Ivanov, and V. S. Luk'yanov, *Izv. Vyssh. Uchebn. Zaved., Khim. Tekhnol.* **41**(2), 73 (1998).
81. J. Suzuki, S. Iizuka, and S. Suzuki, *J. Appl. Polym. Sci.* **22**(8), 2109 (1978).
82. J. Klein and A. Westercamp, *J. Polym. Sci., Letters* **20**, 547 (1982).
83. K. Nakamiya and S. Kinoshita, *J. Fermentation and Bioengineering* **804**, 418 (1995).
84. S. Sawant and H. Morawetz, *Macromolecules* **17**, 2427 (1984).
85. H. Kharadmand, J. Francois, and V. Plazanet, *Polymer* **29**, 860 (1988).
86. K. Ysuda, K. Okajima, and K. Kamide, *Polymer J.* **20**, 1101 (1988).
87. H. Morawetz, ACS Symp. Series 364, Chapter 23 in *Chem. Reactions on Polymers*, 1988, p. 317.
88. F. Halverson, J. E. Lancaster, and M. N. O'Connor, *Macromolecules* **18**, 1139 (1984).
89. N. D. Truong, J. C. Galin, J. François, and Q. T. Pham, *Polymer* **27**, 459 (1986).
90. N. D. Truong, J. C. Galin, J. François, and Q. T. Pham, *Polymer* **27**, 467 (1986).
91. U. S. Pat. 3,022,279 (Feb. 20, 1962), A. C. Profitt (to The Dow Chemical Co.).
92. V. F. Kurenkov, A. S. Verihznikova, and V. A. Myagehenkov, *Vysokomol. Soyed.* **A28**, 488 (1986).
93. U. S. Pat. 5,286,806 (Feb. 15, 1994), R. E. Neff and R. G. Ryles (to American Cyanamid Co.).
94. U. S. Pat. 3,171,805 (Mar. 2, 1965), T. J. Suen and A. M. Schiller (to American Cyanamid Co.).
95. U. S. Pat. 3,539,535 (Nov. 10, 1970), R. Wisner (to Dow Chemical Co.).
96. A. G. Agababyan, G. A. Gevorgyan, and O. L. Mndzhoyan, *Russian Chem. Rev.* **51**(4), 387 (1982).
97. U. S. Pat. 4,179,424 (Dec. 18, 1979), K. J. Phillips, E. G. Ballweber and J. R. Hurlock (to Nalco Chemical Co.).
98. C. J. McDonald and R. H. Beaver, *Macromolecules* **12**, 203 (1979).
99. U. S. Pat. 3,988,277 (Feb. 20, 1976), C. R. Witschonke and R. Rabinowitz (to American Cyanamid Co.).

100. EP 0405712 A2 (Feb. 1, 1991), G. Nakra, J. P. Kane, and T. E. Sortwell (to Diatec Co.).
101. Ger. Offen. 2,333,927 (Jan. 23, 1975), K. Dahmen, W. H. Hübner, and E. Barthell (to Chemische Fabrik Stockhausen & Cie).
102. U. S. Pat. 4,230,608 (Dec. 18, 1980), L. A. Mura, (to Nalco Chemical Co.).
103. Can. Pat. CA 1204535 (Mar. 6, 1986), K. G. Srinivasan, (to Nalco Chemical Co.).
104. U. S. Pat. 4,956,399 (Sept. 11, 1990), J. J. Kozakiewicz and S.-Y. Huang, (to American Cyanamid Co.).
105. U. S. Pat. 5,037,881 (Aug. 6, 1991), J. J. Kozakiewicz and S.-Y. Huang, (to American Cyanamid Co.).
106. U. S. Pat. 5,132,023 (July 21, 1992), J. J. Kozakiewicz and S.-Y. Huang, (to American Cyanamid Co.).
107. U. S. Pat. 3,332,927 (July 25, 1967), M. F. Hoover (to Calgon Corp.).
108. A. M. Schiller and T. J. Suen, *Ind. Eng. Chem.* **48**, 2132 (1956).
109. U. S. Pat. 3,979,348 (Sept. 7, 1976), E. G. Ballweber and K. G. Phillips (to Nalco Chemical Co.).
110. D. P. Bakalik and D. J. Kowalski, *Polym. Mater. Sci. Eng.* **57**, 845 (1987).
111. U. S. Pat. 5,120,797 (Jun. 9, 1992), D. W. Fong and D. J. Kowalski (to Nalco Chemical Co.).
112. U. S. Pat. 3,556,932 (Jan. 19, 1971), A. T. Coscia and L. Williams (to American Cyanamid Co.).
113. D. R. Nagaraj, A. S. Rothenberg, D. W. Lipp, and H. P. Panzer, *Int. J. of Miner. Process.* **20**, 291 (1987).
114. U. S. Pat. 4,587,306 (May 6, 1986), L. Vio and G. Meunier (to Societe Nationale Elf Aquitaine).
115. A. J. Domb, E. G. Cravalho, and R. Langer, *J. Poly. Sci. Polym. Chem. Ed.* **26**, 2623 (1988).
116. U. S. Pat. 5,128,420 (July 7, 1992) A. J. Domb, R. S. Langer, E. G. Cravalho, G. Golomb, E. Mathiowitz, and C. T. Laurencin (to Massachusetts Institute of Technology).
117. U. S. Pat. 4,902,751 (Feb 20, 1990) M. E. Lewellyn and D. P. Spitzer (to American Cyanamid Co.).
118. D. W. Fong and D. J. Kowalski, *Polym. Mater. Sci. Eng.* **75**, 145 (1996).
119. H. Tanaka and R. Senju, *Bull. Chem. Soc. Jpn.* **49**(10), 2821 (1976).
120. H. Tanaka, *J. Polym. Sci. Polym. Chem. Ed.* **17**, 1239 (1979).
121. A. El Achari, X. Coqueret, A. Lablache-Combier, and C. Loucheaux, *Makromol. Chem.* **194**, 1879 (1993).
122. U. S. Pats. 5,039,757 (Aug. 13, 1991), 5,239,014 (Mar. 8, 1994), 5,292,821 (Mar. 8, 1994) T. Takaki, K. Tsuboi, H. Itoh, and A. Nitta (to Mitsui Toatsu Chemicals, Inc.).
123. Eur. Pat. Appl. 289,823A (May 5, 1987), Y. L. Fu, S.-Y. Huang, and R. W. Dexter (to American Cyanamid Co.).
124. D. R. Draney, S.-Y. Huang, J. J. Kozakiewicz, and D. W. Lipp, *Polymer Preprints* **31** (2), 500 (1990).
125. H. Tanaka, *J. Polym. Sci. Polym. Chem. Ed.* **24**, 29 (1986).
126. Joé Hernández Barajas, Ph.D. Thesis, Vanderbilt University, 1996, Chapter 4.
127. R. Ha and S. Hou, *Gaofenzi Xuebao*, No. 5, 570 (1993), CA. Vol. **122**: 161506x, 1995.
128. R. Aksberg and L. Wågberg, *Appl. Polym. Sci.* **38**, 297 (1989).
129. E. T. Hsieh, I. J. Westerman, and A. Moradi-Araghi, *Polym. Mat. Sci. Eng.* **56**, 700 (1986).
130. Y. Negi, S. Harada, and O. Ishizuka, *J. Polym. Sci., Chem.: Part A-1*, **5**, 1951 (1967).
131. C. Wandrey, José Hernández Barajas, and D. Hunkeler, *Adv. Poly. Sci.* **145**, 123–182 (1999).

132. U. S. Pat. 3,288,770 (Nov. 29, 1966), G. B. Butler, (to Peninsular Chem. Research, Inc.).
133. Anon., *Chem. Eng. News* **46** (Jan. 15, 1968).
134. Ch. Wandrey, W. Jaeger, and G. Reinisch, *Acta Polym.* **32** 197 (1981).
135. J. E. Lancaster, L. Baccei, and H. P. Panzer, *J. Polym. Sci., Polym. Lett. Ed.* **14**, 549 (1976).
136. M. Hahn and W. Jaeger, *Angew. Makromol. Chem.* **198**, 165 (1992).
137. U. S. Pat. 4,092,467 (May 30, 1978), R. P. Welcher, R. Rabinowitz, and A. S. Cibuslkas, (to American Cyanamid Co.).
138. M. Chen and K. H. Reichert, *Polym. React. Eng.* **1**(1), 145 (1993).
139. U. S. Pat. 4,439,332 (Mar. 27, 1984), S. Frank, A. T. Coscia, and J. M. Schmitt (to American Cyanamid Co.).
140. K. Plochocka, J. Macromol. *Sci. Rev. Macromol. Chem.* **C20** (1), 67 (1981).
141. F. Candau, Z. Zekhnini, and F. Heatley, *Macromolecules* **19**, 1895–1902 (1986).
142. C. L. McCormick and G. S. Chen, *J. Polym. Sci. Chem.* **20**, 817 (1982).
143. U. S. Pat. 3,284,393 (Nov. 8, 1966), J. W. Vanderhoff and R. M. Wiley (to The Dow Chemical Co.).
144. U. S. Pat. 3,826,771 (July 30, 1974), D. R. Anderson and A. J. Frisque (to Nalco Chemical Co.).
145. U. S. Pat. 5,376,713 (Dec. 27, 1994), M. N. O'Connor, L. J. Barker, and R. G. Ryles (to Cytec Technology Corp.).
146. U. S. Pat. 5,679,740 (Oct. 21, 1997), H. I. Heitner (to Cytec Technology Corp.).
147. David Hunkeler and José Hernández Barajas in J. C. Salamone, ed., *Polymeric Materials Encyclopedia*, Vol. 5, CRC Press, Inc., Boca Raton, Fla., 1996, p. 3330.
148. Fr. Pats. 2,428,053 and 2,428,054 (June 9, 1978), J. Boutin and J. Neel (to Rhône-Poulenc Industries).
149. U. S. Pat. 4,762,862 (Aug. 9, 1988), A. Yada, K. Shusaku, K. Matsumoto, K. Kawamori, Y. Adachi, and Y. Hatakawa (to Dai-Ichi Kogyo Seiyaku Co.).
150. Ger. Pat. DE 3,621,429 (Jan. 8, 1987), J. M. Lucas and A. C. Pericone (to Milchem, Inc.); *Chem. Abstr.* **106**, 138949e (1987).
151. Eur. Pat. EP 296331B (Feb. 2, 1995), W. B. Davies (to Cytec Technology Corp.).
152. U. S. Pat. 4,857,610 (Aug. 15, 1989), M. Chmelir and J. Pauen (to Chemische Fabrik Stockhausen GmbH).
153. U. S. Pat. 4,506,062 (Mar 19, 1985), P. Flesher and A. S. Allen (to Allied Colloids, Ltd.).
154. U. S. Pat. 4,962,150 (Oct. 9, 1990), A. S. Allen (to Allied Colloids, Ltd.).
155. David Hunkeler and José Hernández Barajas in Olagoke Olabisi, ed., *Handbook of Thermoplastics*, Vol. 10, Marcel Dekker, Inc., New York, 1997, p. 237.
156. F. Candau in M. A. El-Nokaly, ed., *Polymer Association Structures: Microemulsions and Liquid Crystals*; ACS Symp. Ser. **384**, 47–61 (1984).
157. U. S. Pat. 4,521,317 (Jun.4, 1985), F. Candau, Yee-Sing Leong, N. Kohler and F. Dawans, (to Inst. Franacais du Petrole, Rueil-Malmaison, France).
158. F. Candau, Y. S. Leong, and R. M. Fitch, *J. Polym. Sci., Chem. Ed.* **23**, 193 (1985).
159. U. S. Pat. 4,681,912 (Jul.21, 1987), J.-P. Durand, D. Nicolas, N. Kohler, F. Dawans, and F. Candau, (to Inst. Franacais du Petrole, Rueil-Malmaison, France).
160. WO 8810274 A1 (Dec. 29, 1988), F. Candau and P. Buchert (to Norsolor S.A.).
161. M. T. Carver, E. Hirsch, J. C. Wittmann, R. M. Fitch, and F. Candau, *J. Phys. Chem.* **93**, 4867 (1989).
162. F. Candau and P. Buchert, *Colloids Surf.* **48**(1–3), 107 (1990).
163. F. Candau, D. Collin, and F. Kern, *Makroml. Chem., Makroml. Symp.* 35–36 (*Copolym. Copolym. Disperse Media*), 105 (1990).

164. U. S. Pat. 5,093,009 (Jul.21, 1992), F. Candau, P. Buchert, and E. Esch (to Inst. Franacais du Petrole, Rueil-Malmaison, France).
165. D. Hunkeler, F. Candau, C. Pichot, A. E. Hamielec, T. Y. Xie, J. Barton, V. Vaskova, J. Guillot, M. V. Dimonie, and K. H. Reichert, *Adv. Polym. Sci.* **112**, 116 (1994).
166. J. Barton, *Prog. Polym. Sci.* **21**, 399 (1996).
167. U. S. Pat. 5,545,688 (Aug.13, 1996), S.-Y. Huang (to American Cyanamid Co.).
168. U. S. Pats. 4,956,400 (Sept. 11, 1990), 5,037,863 (Aug. 6, 1991), J. J. Kozakiewicz, D. L. Dauplaise, J. M. Schmitt, and S.-Y. Huang (to American Cyanamid Co.).
169. U. S. Pats. 5,883,181 (Dec. 10, 1996), 5,723,548 (Mar. 3, 1998), J. J. Kozakiewicz and S.-Y. Huang, (to Cytec Technology Corp.).
170. U. S. Pats. 5,627,260 (May. 6, 1997), 5,670,615 (Jan. 26, 1997), 5,863,982 (Jan. 26, 1999), S.-Y. Huang, A. Leone-Bay, J. M. Schmitt, and P. W. Waterman (to Cytec Technology Corp.).
171. U. S. Pat. 4,824,894 (April 25, 1989), R. Schnee, A. Scordialo, and J. Masanek (to Röhm GmbH, Chemische Fabrik, Germany).
172. WO 92/08744 (May 29, 1992), U. S. Pat. 5,358,988 (Oct. 25, 1994), L. Schieferstein, H. Fischer, H. Kroke, V. Wehle, and R. Jeschke, (to Henkel Co., Germany).
173. U. S. Pat. 4,933,404 (Feb. 20, 1990), E. J. Beckman, E. J. Smith, and J. L. Fulton (to Univ. of Pittsburgh).
174. E. J. Beckman, E. J. Smith, and J. L. Fulton, *J. Physical Chem.* **94**, 345 (1990).
175. E. J. Beckman, E. J. Smith, and J. L. Fulton, *Macromolecules* **27**, 312, (1994); **27**, 5238 (1994).
176. K. A. Shaffer and J. M. Desimone, *Trends in Polym. Sci.* **3**(5), 146 (1995).
177. J. Gregory in C. A. Finch, ed., *Industrial Water Soluble Polymers*, The Royal Society of Chemistry, Cambridge, UK, 1996, pp. 62–75.
178. G. Petzold in R. S. Farinato and P. L. Dubin, eds., *Colloid–Polymer Interactions: From Fundamentals to Practice*, John Wiley & Sons, New York, Ch. 3, 1999, pp. 83–100.
179. J. M. W. Mackenzie, *Eng. Min. J.* 80–87 (Oct. 1980).
180. J. P. MacDonald, P. L. Mattison, and J. M. W. MacKenzie, *J. S. Afr. Inst. Min. Metall.* **81**, 303 (1981).
181. *Superfloc 16 Plus Flocculant*, Cytec Industries, West Patterson, N. J., 1997.
182. *Mining Chemicals Handbook*, rev. ed., Cytec Industries, Stamford, Conn. 1986.
183. D. L. Dauplaise and M. F. Werneke, *Proceedings of the Consolidation and Dewatering of Fine Particles Conference*, University of Alabama, Tuscaloosa, Ala., 1982, pp. 90–113.
184. H. I. Heitner, T. Foster, and H. P. Panzer in J. I. Kroschwitz, ed., *Encyclopedia of Polymer Science and Engineering*, 2nd ed., Vol. **9**, Wiley-Interscience, New York, 1987, p. 830.
185. U. S. Pat. 4,767,540 (Aug. 30, 1988), D. P. Spitzer and W. S. Yen (to American Cyanamid Co.).
186. U. S. Pats. 5,733,459 and 5,733,460 (Mar. 31, 1998) A. S. Rothenberg, P. V. Avotins, R. Cole, and F. Kula (to Cytec Technology Corp.).
187. U. S. Pat 5,368,745 (Nov. 29, 1994), A. S. Rothenberg, R. G. Ryles, and P. So (to Cytec Technology Corp.).
188. U. S. Pat. 4,342,653 (Aug. 3, 1982), F. Halverson (to American Cyanamid Co.).
189. U. S. Pat. 4,533,465 (Aug. 6, 1985), R. M. Goodman and S. K. Lim (to American Cyanamid Co.).
190. D. R. Nagaraj, A. S. Rothenberg, D. W. Lipp, and H. P. Panzer, *Int. J. of Miner. Process.* **20**, 291 (1987), p. 293.
191. U. S. Pat. 5,756,622 (May 26, 1998), S. S. Wang and D. R. Nagaraj (to Cytec Technology Corp.).

192. Eur. Pat. EP 288,150B (Feb. 23, 1994), A. Allen (to Allied Colloids, Ltd.).
193. U. S. Pat. 4,824,523 (June 1, 1987), L. Wågberg and T. Lindström (to Svenska Traforskningsinstitutet).
194. U. S. Pat. 5,167,766 (Dec. 1, 1992), D. S. Honig and E. Harris, (to American Cyanamid Co.).
195. D. S. Honig, E. W. Harris, L. M. Pawlowska, M. P. O'Toole, and L. A. Jackson, *Tappi* **76**(9), 135–143 (1993).
196. C. E. Farley in *Wet Strength Resins and Their Application*, Tappi Press, 1994, p. 45.
197. U. S. Pat. 4,605,702 (Aug 12, 1986), G. J. Guerro, R. J. Proverb, and R. F. Tarvin (to American Cyanamid Co.).
198. U. S. Pat. 5,723,022 (Mar 3, 1998), D. L. Dauplaise and G. J. Guerro (to Cytec Technology Corp.).
199. W. F. Reynolds and R. B. Wasser in J. P. Casey, ed., *Pulp and Paper Chemistry and Chemical Technology*, John Wiley & Sons, Inc., New York, 1981, pp. 1447–1470.
200. U. S. Pat. 5,824,190 (Oct. 20, 1998), G. Guerro, D. Dauplaise, and R. Bazaj (to Cytec Technology Corp.).
201. J. P. Casey, ed., *Pulp and Paper Chemistry and Chemical Technology*, **4** vols., John Wiley & Sons, Inc., New York, 1987, pp. 1–2609.
202. J. K. Borchardt in J. I. Kroschwitz, ed., *Encyclopedia of Polymer Science and Engineering*, 2nd ed., Vol. 10, Wiley-Interscience, New York, 1987, p. 350.
203. U. S. Pat. 4,744,419 (May 17, 1988), R. Sydansk and P. Argabright (to Marathon Oil Co.).
204. U. S. Pat. 4,606,407 (Aug. 8, 1986), P. Shu (to Mobil Oil Corp.).
205. U. S. Pat. 5,905,100 (May 18, 1999), A. Moradi-Araghi (to Phillips Petroleum Co.).
206. U. S. Pat. 4,806,164 (Feb. 21, 1989), L. Brothers (to Halliburton Co.).
207. U. S. Pat. 4,423,199 (Dec. 27, 1983), C. J. Chang and T. E. Stevens (to Rohm and Haas Co.).
208. U. S. Pat. 4,417,989 (Nov. 29, 1983), W. D. Hunter (to Texaco Development Corp.).
209. J. Bock, P. L. Valint Jr., S. J. Pace, D. B. Siano, D. N. Schulz, and S. R. Turner in G. A. Stahl and D. N. Schulz, eds., *Water-Soluble Polymers for Petroleum Recovery*, Plenum Press, New York, 1988, pp. 147–160.
210. U. S. Pat. 4,722,962 (Feb. 2, 1988), G. D. Shay (to DeSoto, Inc.).
211. U. S. Pat. 4,734,205 (Mar. 29, 1988), D. F. Jacques and J. Bock (to Exxon Research and Engineering Co.).
212. A. Karunasena, R. G. Brown, and J. E. Glass in J. E. Glass, ed., *Polymers in Aqueous Media: Performance Through Association* (Adv. Chem. Ser. 223), American Chemical Society, Washington, D. C., 1989.
213. F. Candau and J. Selb, *Adv. Colloid Interface Sci.* **79**, 149–172 (1999).
214. U. S. Pat. 4,654,039 USRE 32649 (Apr. 19, 1988), K. Brandt, T. Inglin, and S. Goldman (to The Proctor and Gamble Co.).
215. U. S. Pat. 4,500,670 (Feb. 19, 1985), M. J. McKinley and D. P. Sheridan (to The Dow Chemical Co.).
216. U. S. Pat. 4,725,628 (Feb. 16, 1988), C. Garvey, J. Pazos, and G. Ring (to Kimberly Clark Corp.).
217. U. S. Pat. 4,725,629 (Feb. 16, 1988), C. Garvey and J. Pazos (to Kimberly Clark Corp.).
218. U. S. Pats. 5,439,983 (Aug 8, 1995) and 5,206,326 (Apr 27, 1993), I. Ahmed and H. L. Hsieh (to Phillips Petroleum Co.).
219. E. Schröder, G. Müller, and K.-F. Arndt, *Polymer Characterization*, Hanser, Munich (1989).
220. F. A. Bovey and G. V. D. Tiers, *J. Polym. Sci., Part A* **1**, 849 (163).

221. F. Halverson, J. E. Lancaster, and M. N. O'Connor, *Macromolecules* **18**, 1139–1144 (1985).
222. J. Hernandez-Barajas and D. J. Hunkeler, *J. Appl. Polym. Sci.* **61**, 1325 (1996).
223. P. Kratochvil, *Classical Light Scattering from Polymer Solutions*, Elsevier, New York, 1987.
224. B. Chu, *Laser Light Scattering: Basic Principles and Practice*, Academic Press, New York, 2nd ed., 1991.
225. J. S. Higgins and H. C. Benoit, *Polymers and Neutron Scattering*, Clarendon Press, Oxford, 1994.
226. W.-M. Kulicke and R. Kniewske, *Makromol. Chem., Rapid Commun.* **1**, 719–727 (1980).
227. A. DiNapoli, B. Chu, and C. Cha, *Macromolecules* **15**, 1174–1180 (1982).
228. M. G. Styring and A. E. Hamielec in A. R. Cooper, ed., *Determination of Molecular Weight*, John Wiley & Sons, New York, Ch. 10, 1989, p. 263.
229. L. A. Papazian, *J. Liquid Chromatog.* **13**(17), 3389–3398 (1990).
230. J. C. Giddings, K. D. Caldwell, and L. F. Kesner in A. R. Cooper, ed., *Determination of Molecular Weight*, John Wiley & Sons, New York, Ch. 12, 1989, p. 337.
231. G. Holzwarth, L. Soni, and D. N. Schulz, *Macromolecules* **19**, 422–426 (1986).
232. G. Holzwarth, L. Soni, D. N. Schulz, and J. Bock in G. A. Stahl and D. N. Schulz, eds., *Water-Soluble Polymers for Petroleum Recovery*, Plenum, New York, 1988, pp. 215–229.
233. P. J. Wyatt, *Anal. Chim. Acta* **272**, 1–40 (1993).
234. H. Terayama, *J. Polym. Sci.* **VIII**(2), 243 (1952).
235. M. A. Langhorst, F. W. Stanley, Jr., S. S. Cutié, J. H. Sugarman, L. R. Wilson, D. A. Hoagland, and R. K. Prud'homme, *Anal. Chem.* **58**, 2242–2247 (1980).
236. Y. Nishijima, A. Teramoto, M. Yamamoto, and S. Hiratsuka, *J. Polym. Sci., Part A-2*, **5**, 23–35 (1967).
237. U. S. Pat. 4,194,877 (Mar. 25, 1980), J. I. Peterson (to United States of America, Washington, D. C.)
238. P. Gramain and P. Myard, *Polym. Bull.* **3**, 627–631 (1980).
239. N. J. Turro and K. S. Arora, *Polymer* **27**, 783 (1986).
240. J. Rica, H. Gysel, J. Schneider, R. Nyffenegger, and T. Binkert, *Macromolecules* **20**, 1407–1411 (1987).
241. H. Tanaka and R. Senju, *Bull. Chem. Soc. Jpn.* **49**, 2821 (1976).
242. J. K. Inman and H. M. Dintzis, *Biochem.* **8**(10), 4074–4082 (1969).
243. H. Tanaka and L. Odberg, *J. Polym. Sci., Part A, Polym. Chem.* **27**, 4329–4339 (1989).
244. U. S. Pat. 4,813,973 (Mar. 21, 1989), M. A. Winnik and R. M. Borg (to Univ. Toronto Inventions Found.).
245. U. S. Pat. 5,389,548 (Feb. 14, 1995), J. E. Hoots, C. C. Pierce, and R. W. Kugel (to Nalco Chemical Co.).
246. R. D. Lentz, R. E. Sojka, and J. A. Foerster, *J. Environ. Qual.* **25**, 1015–1024 (1996).
247. *Methods Exam. Waters Assoc. Matl. 1988*, United Kingdom Dept. of the Environment; *Chem. Abstr.* **108**, 197583v (1988).
248. G. Schmoetzer, *Chromatographia* **4**, 391–395 (1971); *Chem. Abstr.* **76**, 25744c (1972).
249. D. D. Mc-Collister, C. L. Hake, S. E. Sadek, and V. K. Rowe, *Toxicol. Appl. Pharmacol.* **7**(5), 639 (1965).
250. O. G. Lewis, *Physical Constants of Linear Homopolymers*, Springer-Verlag, New York, 1968, p. 23.
251. I. Janigová, K. Csomorová, M. Stillhammerová, and J. Barton, *Macromol. Chem. Phys.*, **195**, 3609–3614 (1994).

252. J. Brandrup and E. H. Immergut, eds., *Polymer Handbook*, 3rd ed., John Wiley & Sons, Inc., New York, 1989, p. VI-416.
253. J. E. Lancaster and M. N. O'Connor, *J. Polym. Sci. Lett.* **20**, 547 (1982).
254. S. Sawant and H. Morawetz, *J. Polym. Sci. Polym. Lett. Ed.* **20**, 385–388 (1982).
255. Ref. 70, p. 962.
256. Ref. 25, p. 383.
257. J. Brandup and E. H. Immergut, eds., *Polymer Handbook*, 3rd ed., John Wiley & Sons, New York, 1989, p. VII-383.
258. Ref. 257, p. VII-245.
259. *SUPERFLOC 127 Flocculant, Material Safety Data*, Cytec Industries, West Paterson, N.J., 1996.
260. Ref. 257 p. VII-34.
261. Ref. 25, p. 386.
262. Ref. 25, p. 392.
263. Ref. 25, p. 407.
264. J. Klein and K. D. Conrad, *Makromol. Chem.* **179**, 1635–1638 (1978).
265. T. Griebel, W.-M. Kulicke, and A. Hashemzadeh, *Coll. Poly. Sci.* **269**, 113–120 (1991).

Sun-Yi Huang
David W. Lipp
Raymond S. Farinato
Cytec Industries

ACRYLIC ACID AND DERIVATIVES

1. Introduction

The term acrylates includes derivatives of both acrylic ($CH_2{=}CHCOOH$) and methacrylic acids ($CH_2{=}C(CH_3)COOH$). This article discusses the preparation, properties, and reactions of acrylic acid monomers only (see Methacrylic acid and derivatives). Acrylic acid (propenoic acid) was first prepared in 1847 by air oxidation of acrolein (1). Interestingly, after use of several other routes over the past half century, it is this route, using acrolein from the catalytic oxidation of propylene, that is currently the most favored industrial process. Polymerization of acrylic esters has been known for just over a century (2). However, it was not until 1930 that the technical difficulties of their manufacture and polymerization were overcome (3). The rate of consumption of acrylates grew between 10 and 20% annually during the late 1970s. Growth fluctuated with the economy in the early 1980s with some announced capacity increases being delayed until later in the decade. Although growth in the ester markets has dropped appreciably, new applications for polymers of acrylic acid in the superabsorbent and detergent fields surged in the late 1980s. U. S. demand for acrylic acid is expected to grow at a 4.0% annual rate, and reach 1.2×10^9 kg by 2004.

Acrylates are primarily used to prepare emulsion and solution polymers. The emulsion polymerization process provides high yields of polymers in a form suitable for a variety of applications. Acrylate polymer emulsions were first used as coatings for leather in the early 1930s and have found wide utility as coatings, finishes, and binders for leather, textiles, and paper. Acrylate emulsions are used in the preparation of both interior and exterior paints, floor polishes, and adhesives. Solution polymers of acrylates, frequently with minor concentrations of other monomers, are employed in the preparation of industrial coatings. Polymers of acrylic acid can be used as superabsorbents in disposable diapers, as well as in formulation of superior, reduced-phosphate-level detergents.

The polymeric products can be made to vary widely in physical properties through controlled variation in the ratios of monomers employed in their preparation, cross-linking, and control of molecular weight. They share common qualities of high resistance to chemical and environmental attack, excellent clarity, and attractive strength properties (see ACRYLIC ESTER POLYMERS). In addition to acrylic acid itself, methyl, ethyl, butyl, isobutyl, and 2-ethylhexyl acrylates are manufactured on a large scale and are available in better than 98–99% purity (4). They usually contain 10–200 ppm of hydroquinone monomethyl ether as polymerization inhibitor.

2. Physical Properties

Physical properties of acrylic acid and representative derivatives appear in Table 1. Table 2 gives selected properties of commercially important acrylate esters, and Table 3 lists the physical properties of many acrylic esters.

Acrylic acid is a moderately strong carboxylic acid. Its dissociation constant is 5.5×10^{-5}. Vapor pressure as a function of temperature is given in Table 4 for acrylic acid and four important esters (4,16–18). The lower esters form azeotropes both with water and with their corresponding alcohols.

Table 1. **Physical Properties of Acrylic Acid Derivatives**

Property	Acrylic acid	Acrolein	Acrylic anhydride	Acryloyl chloride	Acrylamide
molecular formula	$C_3H_4O_2$	C_3H_4O	$C_6H_6O_3$	C_3H_3OCl	C_3H_5ON
CAS Registry Number	[79-10-7]	[107-02-8]	[2051-76-5]	[814-68-6]	[79-06-1]
melting point, °C	13.5	−88			84.5
boiling point[a], °C	141	52.5	38[b]	75	125[c]
refractive index[d], n_D	1.4185[e]	1.4017	1.4487	1.4337	
flash point, Cleveland open cup, °C	68				
density[d], g/mL	1.045[e]	0.838		1.113	1.122[f]

[a] At 101.3 kPa = 1 atm unless otherwise noted.
[b] At 0.27 kPa.
[c] At 16.6 kPa.
[d] At 20°C, unless otherwise noted.
[e] At 25°C.
[f] At 30°C.

Table 2. **Properties of Commercially Important Acrylate Esters**[a]

Property	Methyl	Ethyl	*n*-Butyl	Isobutyl	2-Ethylhexyl
solubility at 23°C, parts per 100 of solvent					
in water	5	1.5	0.2	0.2	0.01
of water in ester	2.5	1.5	0.7	0.6	0.15
heat of vaporization, kJ/g[b]	0.39	0.35	0.19	0.30	0.25
specific heat, J/g·C[b]	2.00	1.97	1.93	1.93	1.93
boiling points[c] of azeotropes					
with water, °C	71	81.1	94.5		
water content, %	7.2	15	40		
with methanol, °C	62.5	64.5			
methanol content, %	54	84.4			
with ethanol, °C	73.5	77.5			
ethanol content, %	42.4	72.7			
with *n*-butanol, °C			119		
butanol content, %			89		
heat of polymerization, kJ/mol[b]	78.7	77.8	77.4		

[a] Refs. 4,5
[b] To convert J to cal, divide by 4.184.
[c] At 101.3 kPa = 1 atm.

3. Reactions

Acrylic acid and its esters may be viewed as derivatives of ethylene, in which one of the hydrogen atoms has been replaced by a carboxyl or carboalkoxyl group. This functional group may display electron-withdrawing ability through inductive effects of the electron-deficient carbonyl carbon atom, and electron-releasing effects by resonance involving the electrons of the carbon–oxygen double bond. Therefore, these compounds react readily with electrophilic, free-radical, and nucleophilic agents.

3.1. Carboxylic Acid Functional Group Reactions. Polymerization is avoided by conducting the desired reaction under mild conditions and in the presence of polymerization inhibitors. Acrylic acid undergoes the reactions of carboxylic acids and can be easily converted to salts, acrylic anhydride, acryloyl chloride, and esters (16,17).

Salts are made by reaction of acrylic acid with an appropriate base in aqueous medium. They can serve as monomers and comonomers in water-soluble or water-dispersible polymers for floor polishes and flocculants.

Acrylic anhydride is formed by treatment of the acid with acetic anhydride or by reaction of acrylate salts with acryloyl chloride. *Acryloyl chloride* is made by reaction of acrylic acid with phosphorous oxychloride, or benzoyl or thionyl chloride. Neither the anhydride nor the acid chloride is of commercial interest.

Esters. Most acrylic acid is used in the form of its methyl, ethyl, and butyl esters. Specialty monomeric esters with a hydroxyl, amino, or other functional group are used to provide adhesion, latent cross-linking capability, or different solubility characteristics. The principal routes to esters are direct esterification with alcohols in the presence of a strong acid catalyst such as sulfuric acid, a

Table 3. **Physical Properties of Acrylic Esters[a], $CH_2{=}CHCOOR$**

Compound	Molecular formula	CAS Registry Number	Boling point °C	Boling point kPa[b]	Refractive index n^{20}_D	Density d^{20}_4
n-Alkyl esters[c]						
methyl	$C_4H_6O_2$	[96-33-3]	80	101	1.4040	0.9535
ethyl	$C_5H_8O_2$	[140-88-5]	43	13.7	1.4068	0.9234
propyl	$C_6H_{10}O_2$	[925-60-0]	44	5.3	1.4130	0.9078
butyl	$C_7H_{12}O_2$	[141-32-2]	35	1.1	1.4190	0.8998
pentyl	$C_8H_{14}O_2$	[2998-23-4]	48	0.9	1.4240	0.8920
hexyl	$C_9H_{16}O_2$	[2499-95-8]	40	0.15	1.4280	0.8882
heptyl	$C_{10}H_{18}O_2$	[2499-58-3]	57	0.13	1.4311	0.8846
octyl	$C_{11}H_{20}O_2$	[2499-59-4]	57	0.007	1.4350	0.8810
nonyl	$C_{12}H_{22}O_2$	[2664-55-3]	76	0.03	1.4375	0.8785
decyl	$C_{13}H_{24}O_2$	[2156-96-9]	120	0.67	1.4400	0.8781
dodecyl	$C_{15}H_{28}O_2$	[2156-97-0]	120	0.11	1.4440	0.8727
tetradecyl	$C_{17}H_{32}O_2$	[21643-42-5]	138	0.05	1.4468	0.8700
hexadecyl	$C_{19}H_{36}O_2$	[13402-02-3]	170	0.20	1.4470 (30°C)	0.8620 (30°C)
Secondary and branched-chain alkyl esters[d]						
isopropyl	$C_6H_{10}O_2$	[689-12-3]	52	13.7	1.4060	0.8932
isobutyl	$C_7H_{12}O_2$	[106-62-8]	62	6.7	1.4150	0.8896
sec-butyl	$C_7H_{12}O_2$	[2998-08-5]	60	6.7	1.4140	0.8914
2-ethylhexyl	$C_{11}H_{20}O_2$	[103-11-7]	85	1.07	1.4365	0.8852
Esters of olefinic alcohols[e]						
allyl	$C_6H_8O_2$	[999-55-3]	47	5.33	1.4320	0.9441
2-methylallyl	$C_7H_{10}O_2$	[818-67-7]	68	6.67	1.4372	0.9285
Aminoalkyl esters[f]						
2-(dimethylamino)-ethyl	$C_7H_{13}O_2N$	[2439-35-2]	61	1.47	0.9434	1.4375
2-(diethylamino)ethyl	$C_9H_{17}O_2N$	[2426-54-2]	70	0.67	0.9251	1.4425
Esters of ether alcohols[c]						
2-methoxyethyl	$C_6H_{10}O_3$	[3121-67-7]	59	1.60	1.4272	1.0131
2-ethoxylethyl	$C_8H_{12}O_3$	[106-74-1]	78	3.07	1.4282	0.9819
Cycloalkyl esters[g]						
cyclohexyl	$C_9H_{14}O_2$	[3066-71-5]	75	1.47	1.4600	0.9796
4-methylcyclohexyl	$C_{10}H_{16}O_2$	[16491-65-9]	55	0.27	1.4550	0.9537
Esters of halogenated alcohols[h]						
2-bromoethyl	C_5H_7Br	[4823-47-6]	53	5	1.4770	1.4774
2-chloroethyl	C_5H_7Cl	[2206-89-5]	74	29	1.4477	
Glycol diacrylates[i]						
ethylene glycol (monoester)	$C_5H_8O_3$	[818-61-1]	40	0.001	1.4482 (25°C)	
ethylene glycol	$C_8H_{10}O_4$	[2274-11-5]	70	0.13	1.4529	
propylene glycol	$C_9H_{12}O_4$	[999-61-1]	63	0.04	1.4470	
1,3-propanediol	$C_9H_{12}O_4$	[25151-33-1]	65	<0.13	1.4529	
1,4-butanediol	$C_{10}H_{14}O$	[31442-13-4]	83	0.04	1.4538	
diethylene glycol	$C_{10}H_{14}O_5$	[4074-88-8]	94	0.03	1.4572	
1,5-pentanediol	$C_{11}H_{16}O_4$	[36840-85-4]	94	0.04	1.4551	
1,10-decanediol	$C_{16}H_{26}O_4$	[13048-45-5]	145	0.01		

[a] In most cases, the references include additional examples of the class alcohols. Nitroalkyl esters are known (6,7).
[b] To convert kPa to mm Hg, multiply by 7.5.
[c] Ref. 8
[d] Ref. 9.
[e] Ref. 10.
[f] Ref. 11.
[g] Ref. 12.
[h] Refs. 13–15.
[i] Ref. 5.

Table 4. **Vapor Pressures of Acrylic Acid and Important Esters[a], kPa[b]**

Temperature, °C	Acrylic acid	Methyl acrylate	Ethyl acrylate	Butyl acrylate	2-Ethylhexyl acrylate
−20		0.85	0.31		
−10		1.72	0.61		
0		3.12	1.16	0.15	
10		5.40	2.20	0.28	
20		9.09	3.93	0.53	
30	0.8	14.5	6.73	0.97	
40	1.4	22.7	10.9	1.71	
50	2.4	34.0	16.8	2.84	0.13
60	4.0	49.6	25.3	4.53	0.25
70	6.6	70.7	37.3	7.13	0.44
80	10		53.3	10.9	0.75
90	16		74.7	16.0	1.23
100	24			23.06	1.97
110	35			32.8	3.06
120	49			45.6	4.62
130	69			61.9	6.79
140	99			82.3	10.1
150					14.4
160					20.0
170					28.0
180					37.3
190					50.0
200					65.6
210					85.3

[a] Ref. 4.
[b] To convert kPa to mm Hg, multiply by 7.5.

soluble sulfonic acid, or sulfonic acid resins; addition to alkylene oxides to give hydroxyalkyl acrylic esters; and addition to the double bond of olefins in the presence of strong acid catalyst (19,20) to give ethyl or secondary alkyl acrylates.

$$CH_2{=}CHCOOH + ROH \xrightarrow{H^+} CH_2{=}CHCOOR + H_2O$$

$$CH_2{=}CHCOOH + \underset{\text{(ethylene oxide, } CH_2{-}CH_2 \text{ bridged by O)}}{CH_2{-}CH_2} \longrightarrow CH_2{=}CHCOOCH_2CH_2OH$$

$$CH_2{=}CHCOOH + CH_2{=}CH_2 \xrightarrow{H^+} CH_2{=}CHCOOCH_2CH_3$$

Acrylic esters may be saponified, converted to other esters (particularly of higher alcohols by acid catalyzed alcohol interchange), or converted to amides by aminolysis. Transesterification is complicated by the azeotropic behavior of lower acrylates and alcohols but is useful in preparation of higher alkyl acrylates.

Amides. Reaction of acrylic acid with ammonia or primary or secondary amines forms amides. However, acrylamide (qv) is better prepared by controlled hydrolysis of acrylonitrile (qv). Esters can be obtained by carrying out the nitrile hydrolysis in the presence of alcohol.

$$CH_2{=}CHC{\equiv}N \xrightarrow[H_2O]{H_2SO_4} CH_2{=}CHCO\overset{+}{N}H_3HSO_4^- \xrightarrow[ROH,\ H+]{H_2O} CH_2{=}CH\overset{O}{\overset{\|}{C}}OH(R)$$

3.2. Unsaturated Group Reactions. In addition to a comprehensive review of these reactions (16), there are excellent texts (17,18). Free-radical-initiated polymerization of the double bond is the most common reaction and presents one of the more troublesome aspects of monomer manufacture and purification.

Substituted ring compounds are formed readily by Diels-Alder reactions.

$$CH_2{=}CHCOOR \ + \ CH_2{=}CH{-}CH{=}CH_2 \longrightarrow \text{(cyclohexene ring)}{-}COOR$$

Additions. Halogens, hydrogen halides, and hydrogen cyanide readily add to acrylic acid to give the 2,3-dihalopropionate, 3-halopropionate, and 3-cyanopropionate, respectively (21).

On storage or at elevated temperatures, acrylic acid dimerizes to give 3-acryloxypropionic acid [24615-84-7], $C_6H_8O_4$.

$$2\,CH_2{=}CHCOOH \longrightarrow CH_2{=}CHCOOCH_2CH_2COOH$$

Although the reaction is second order in acrylic acid concentration, the rate of dimer formation for neat acrylic acid available commercially is quite adequately expressed by

$$\text{rate} = 3.58 \times 10^{17}\exp(-10,500/T)$$

over the first several percent conversion (5), where rate is in ppm/day and T is the Kelvin temperature. Since this rate is approximately 100 ppm/day at 20°C, significant dimer can build up on prolonged storage. The reaction is accelerated by addition of strong acids, bases, or large amounts of water (several wt %). However, water at concentrations as low as 0.1–0.2 wt %, present in commercially available acrylic acid, has negligible effect on dimer formation (5,22). Continuation of this reaction leads to a distribution of polyester oligomers formed by successive additions across the double bond (5). Acrylic acid can be regenerated by thermal or acid catalyzed cracking of these oligomers (23). Cracking of the corresponding esters gives acrylic acid and acrylic ester.

$$CH_2{=}CHCOOCH_2CH_2COOR \longrightarrow CH_2{=}CHCOOH + CH_2{=}CHCOOR$$

Michael condensations are catalyzed by alkali alkoxides, tertiary amines, and quaternary bases and salts. Active methylene compounds and aliphatic nitro compounds add to form β-substituted propionates. These addition reactions are frequently reversible at high temperatures. Exceptions are the tertiary nitro adducts which are converted to olefins at elevated temperatures (24).

$$CH_3CHNO_2CH_3 + CH_2{=}CHCOOC_2H_5 \longrightarrow CH_3{-}\underset{NO_2}{\overset{CH_3}{C}}HCH_2CH_2COOC_2H_5$$

$$CH_3{-}\underset{NO_2}{\overset{CH_3}{C}}HCH_2CH_2COOC_2H_5 \longrightarrow CH_2{=}\overset{CH_3}{C}CH_2CH_2COOC_2H_5 + HNO_2$$

The addition of alcohols to form the 3-alkoxypropionates is readily carried out with strongly basic catalyst (25). If the alcohol groups are different, ester interchange gives a mixture of products. Anionic polymerization to oligomeric acrylate esters can be obtained with appropriate control of reaction conditions. The 3-alkoxypropionates can be cleaved in the presence of acid catalysts to generate acrylates (26). Development of transition-metal catalysts for carbonylation of olefins provides routes to both 3-alkoxypropionates and 3-acryl-oxypropionates (27,28). Hence these are potential intermediates to acrylates from ethylene and carbon monoxide.

Additions of mercaptans with alkaline catalysts give 3-alkylthiopropionates (29). In the case of hydrogen sulfide, the initially formed 3-mercaptopropionate reacts with a second molecule of acrylate to give a 3,3′-thiodipropionate (30,31).

$$H_2S + 2\ CH_2{=}CHCOOR \longrightarrow S(CH_2CH_2COOR)_2$$

Polythiodipropionic acids and their esters are prepared from acrylic acid or an acrylate with sulfur, hydrogen sulfide, and ammonium polysulfide (32). These polythio compounds are converted to the dithio analogs by reaction with an inorganic sulfite or cyanide.

$$2\ CH_2{=}CHCOOCH_3 + S_x + H_2S \xrightarrow{(NH_4)_2S_x}$$

$$S_x(CH_2CH_2COOCH_3)_2 \xrightarrow{NaCN} CH_3OOCCH_2CH_2SSCH_2CH_2COOCH_3$$

Ammonia and amines add to acrylates to form β-aminopropionates, which add easily to excess acrylate to give tertiary amines. The reactions are reversible (33).

$$NH_3 + CH_2{=}CHCOOR \longrightarrow H_2NCH_2CH_2COOR \xrightarrow{CH_2{=}CHCOOR}$$

$$HN(CH_2CH_2COOR)_2 \xrightarrow{CH_2{=}CHCOOR} N(CH_2CH_2COOR)_3$$

Aqueous ammonia and acrylic esters give tertiary amino esters, which form the corresponding amide upon ammonolysis (34). Modern methods of molecular quantum modeling have been applied to the reaction pathway and energetics for several nucleophiles in these Michael additions (35,36).

Acrylic esters dimerize to give the 2-methylene glutaric acid esters catalyzed by tertiary organic phosphines (37) or organic phosphorous triamides, phosphonous diamides, or phosphinous amides (38). Yields of 75–80% dimer, together with 15–20% trimer, are obtained. Reaction conditions can be varied to obtain high yields of trimer, tetramer, and other polymers.

4. Manufacture

Various methods for the manufacture of acrylates are summarized in Figure 1, showing their dependence on specific raw materials. For a route to be commercially attractive, the raw material costs and utilization must be low, plant investment and operating costs not excessive, and waste disposal charges minimal.

After development of a new process scheme at laboratory scale, construction and operation of pilot-plant facilities to confirm scale-up information often require two or three years. An additional two to three years is commonly required for final design, fabrication of special equipment, and construction of the plant. Thus, projections of raw material costs and availability five to ten years into the future become important in adopting any new process significantly different from the current technology.

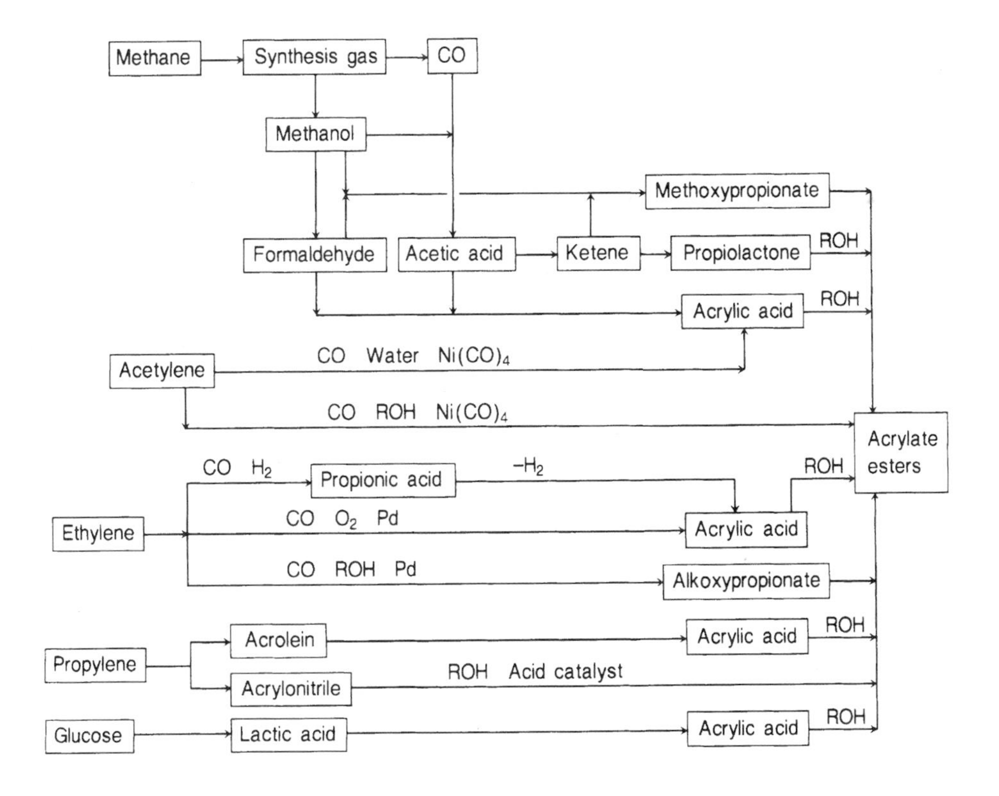

Fig. 1. Acrylate manufacturing technologies.

In the 1980s cost and availability of acetylene have made it an unattractive raw material for acrylate manufacture as compared to propylene, which has been readily available at attractive cost (see ACETYLENE-DERIVED CHEMICALS). As a consequence, essentially all commercial units based on acetylene, with the exception of BASF's plant at Ludwigshafen, have been shut down. All new capacity recently brought on stream or announced for construction uses the propylene route. Rohm and Haas Co. has developed an alternative method based on alkoxycarbonylation of ethylene, but has not commercialized it because of the more favorable economics of the propylene route.

Propylene requirements for acrylates remain small compared to other chemical uses (polypropylene, acrylonitrile, propylene oxide, 2-propanol, and cumene for acetone and phenol). Hence, cost and availability are expected to remain attractive and new acrylate capacity should continue to be propylene-based until after the turn of the century.

4.1. Propylene Oxidation. The propylene oxidation process is attractive because of the availability of highly active and selective catalysts and the relatively low cost of propylene. The process proceeds in two stages giving first acrolein and then acrylic acid (39) (see ACROLEIN AND DERIVATIVES).

$$CH_2{=}CHCH_3 + O_2 \longrightarrow CH_2{=}CHCHO + H_2O$$

$$CH_2{=}CHCHO + \frac{1}{2}\, O_2 \longrightarrow CH_2{=}CHCOOH$$

Single-reaction-step processes have been studied. However, higher selectivity is possible by optimizing catalyst composition and reaction conditions for each of these two steps (40,41). This more efficient utilization of raw material has led to two separate oxidation stages in all commercial facilities. A two-step continuous process without isolation of the intermediate acrolein was first described by the Toyo Soda Company (42). A mixture of propylene, air, and steam is converted to acrolein in the first reactor. The effluent from the first reactor is then passed directly to the second reactor where the acrolein is oxidized to acrylic acid. The products are absorbed in water to give about 30–60% aqueous acrylic acid in about 80–85% yield based on propylene.

Japan Catalytic Chemical Co. (43) and Mitsubishi Petrochemical Co. (44) offer licenses to their acrylate manufacturing technology (including high quality catalysts). Thus, although most manufacturers have also developed their own catalyst and process technologies, many have also taken licenses from these companies and either constructed entire plants based on those disclosures or combined that technology with their own developments for their operating plants.

Catalysts. Catalyst performance is the most important factor in the economics of an oxidation process. It is measured by activity (conversion of reactant), selectivity (conversion of reactant to desired product), rate of production (production of desired product per unit of reactor volume per unit of time), and catalyst life (effective time on-stream before significant loss of activity or selectivity).

Early catalysts for acrolein synthesis were based on cuprous oxide and other heavy metal oxides deposited on inert silica or alumina supports (39).

Later, catalysts more selective for the oxidation of propylene to acrolein and acrolein to acrylic acid were prepared from bismuth, cobalt, iron, nickel, tin salts, and molybdic, molybdic phosphoric, and molybdic silicic acids. Preferred second-stage catalysts generally are complex oxides containing molybdenum and vanadium. Other components, such as tungsten, copper, tellurium, and arsenic oxides, have been incorporated to increase low temperature activity and productivity (39,45,46).

Catalyst performance depends on composition, the method of preparation, support, and calcination conditions. Other key properties include, in addition to chemical performance requirements, surface area, porosity, density, pore size distribution, hardness, strength, and resistance to mechanical attrition.

Patents claiming specific catalysts and processes for their use in each of the two reactions have been assigned to Japan Catalytic (45,47–49), Sohio (50), Toyo Soda (51), Rohm and Haas (52), Sumitomo (53), BASF (54), Mitsubishi Petrochemical (56,57), Celanese (55), and others. The catalysts used for these reactions remain based on bismuth molybdate for the first stage and molybdenum vanadium oxides for the second stage, but improvements in minor component composition and catalyst preparation have resulted in yields that can reach the 85–90% range and lifetimes of several years under optimum conditions. Since plants operate under more productive conditions than those optimum for yield and life, the economically most attractive yields and productive lifetimes may be somewhat lower.

Oxidation Step. A review of mechanistic studies of partial oxidation of propylene has appeared (58). The oxidation process flow sheet (Fig. 2) shows equipment and typical operating conditions. The reactors are of the fixed-bed

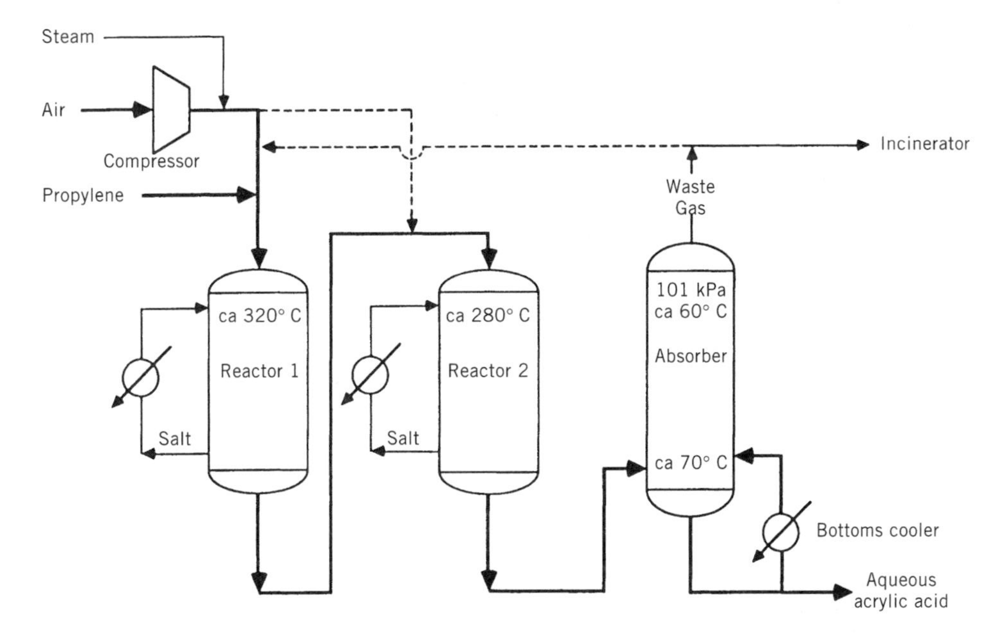

Fig. 2. Oxidation process. To convert kPa to mm Hg, multiply by 7.5.

shell-and-tube type (about 3–5 m long and 2.5 cm in diameter) with a molten salt coolant on the shell side. The tubes are packed with catalyst, a small amount of inert material at the top serving as a preheater section for the feed gases. Vaporized propylene is mixed with steam and air and fed to the first-stage reactor. The feed composition is typically 5–7% propylene, 10–30% steam, and the remainder air (or a mixture of air and absorber off-gas) (56,57,59,60).

The preheated gases react exothermically over the first-stage catalyst with the peak temperature in the range of 330–430°C, depending on conditions and catalyst selectivity. The conversion of propylene to waste gas (carbon dioxide and carbon monoxide) is more exothermic than its conversion to acrolein. At the end of the catalyst bed the temperature of the mixture drops toward that of the molten salt coolant.

If necessary, first-stage reactor effluent may be further cooled to 200–250°C by an interstage cooler to prevent homogeneous and unselective oxidation of acrolein taking place in the pipes leading to the second-stage reactor (56,59).

The acrolein-rich gaseous mixture containing some acrylic acid is then passed to the second-stage reactor, which is similar to the first-stage reactor, but packed with a catalyst designed for selective conversion of acrolein to acrylic acid. Here, the temperature peaks in the range of 280–360°C, again depending on conditions. The temperature of the effluent from the second-stage reactor again approximates that of the salt coolant. The heat of reaction is recovered as steam in external waste-heat boilers.

The process is operated at the lowest temperature consistent with high conversion. Conversion increases with temperature; the selectivity generally decreases only with large increases in temperature. Catalyst life also decreases with increasing temperatures. Catalysts are designed to give high performance over a range of operating conditions permitting gradual increase of salt temperature over the operating life of the catalysts to maintain productivity and selectivity near the initial levels, thus compensating for gradual loss of catalyst activity.

The gaseous reactor effluent from the second-stage oxidation is fed to the bottom of the aqueous absorber and cooled from about 250°C to less than 80°C by contact with aqueous acrylic acid. The gas passes through the absorber to complete the recovery of product. The water is fed to the top of the absorber at about 30–60°C to minimize acrylic acid losses and the absorber off-gas is sent to a flare or to a furnace to convert all residual organic material to waste gas. Some of the absorber off-gas may be recycled to the first-stage reactor feed to allow achievement of optimum oxygen-to-propylene ratio at reduced steam levels (59). If the resulting oxygen level is too low for best performance in the second-stage oxidation, an interstage feed of supplemental air (or air plus steam) may be introduced (56). The aqueous effluent from the bottom of the absorber is 30–60% acrylic acid depending on whether off-gas recycle with low steam feed or air feed with higher steam level is chosen. This is sent to the separations section for recovery. The overall yield of acrylic acid in the oxidation reaction steps is in the range of 75–86%, depending on the catalysts, conditions, and age of catalyst employed.

Acrylic Acid Recovery. The process flow sheet (Fig. 3) shows equipment and conditions for the separations step. The acrylic acid is extracted from the absorber effluent with a solvent, such as butyl acetate, xylene, diisobutyl ketone,

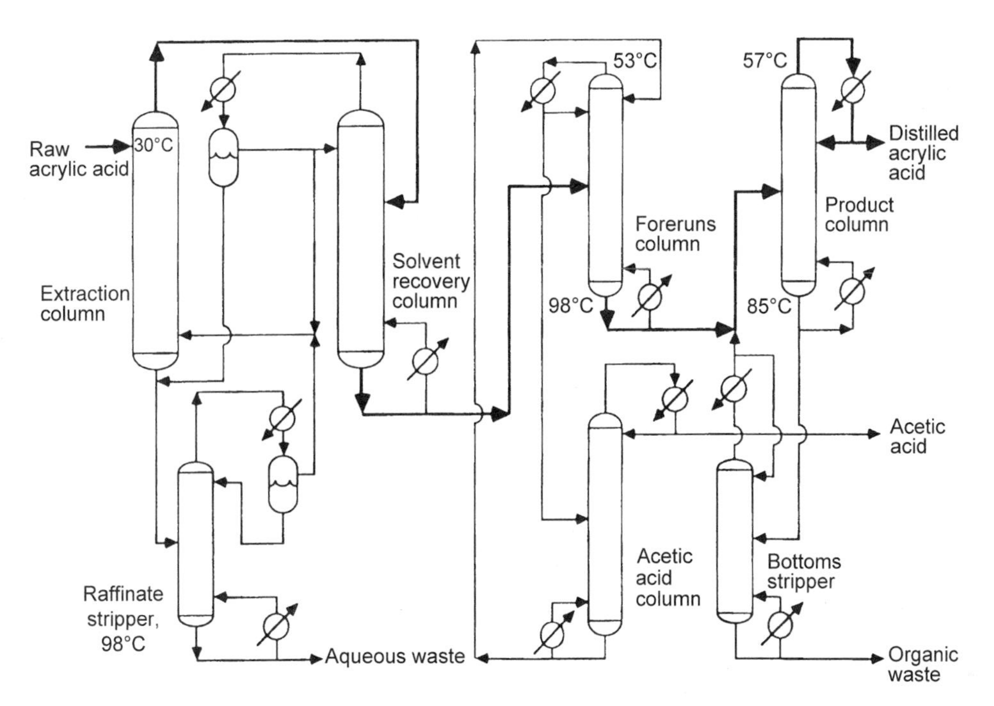

Fig. 3. Separations process.

or mixtures, chosen for high selectivity for acrylic acid and low solubility for water and by-products. The extraction is performed using 5–10 theoretical stages in a tower or centrifugal extractor (46,61–65).

The extract is vacuum-distilled in the solvent recovery column, which is operated at low bottom temperatures to minimize the formation of polymer and dimer and is designed to provide acrylic acid-free overheads for recycle as the extraction solvent. A small aqueous phase in the overheads is mixed with the raffinate from the extraction step. This aqueous material is stripped before disposal both to recover extraction solvent values and minimize waste organic disposal loads.

It is possible to dispense with the extraction step if the oxidation section is operated at high propylene concentrations and low steam levels to give a concentrated absorber effluent. In this case, the solvent recovery column operates at total organic reflux to effect azeotropic dehydration of the concentrated aqueous acrylic acid. This results in a reduction of aqueous waste at the cost of somewhat higher energy usage.

The bottoms from the solvent recovery (or azeotropic dehydration column) are fed to the foreruns column where acetic acid, some acrylic acid, and final traces of water are removed overhead. The overhead mixture is sent to an acetic acid purification column where a technical grade of acetic acid suitable for ester manufacture is recovered as a by-product. The bottoms from the acetic acid recovery column are recycled to the reflux to the foreruns column. The bottoms from the foreruns column are fed to the product column where the glacial acrylic acid of commerce is taken overhead. Bottoms from the product column are stripped to recover acrylic acid values and the high boilers are burned. The principal losses of acrylic acid in this process are to the aqueous raffinate and to the aqueous layer from the dehydration column and to dimerization of acrylic acid to 3-acryloxypropionic acid. If necessary, the product column bottoms stripper may include provision for a short-contact-time cracker to crack this dimer back to acrylic acid (60).

In any case, mild conditions and short residence times to minimize dimer formation are maintained throughout the separations section. In addition, free-radical polymerization inhibitors are fed to each unit to prevent polymer formation and resulting equipment failure. The glacial acrylic acid produced at this stage of the process is typically better than 99.5% pure, with the principal contaminants being water and acetic acid at about 0.1–0.2 wt %, acetic acid at about 0.1 wt %, and acrylic acid dimer at 0.1–0.5 wt % depending on storage time after distillation. Propionic acid is present at 0.02–0.04 wt %. The monomethyl ether of hydroquinone is added as storage and shipping stabilizer at 0.02 wt % (200 ppm). Low concentrations of aldehydes, primarily furfuraldehyde, but including acetaldehyde, acrolein, and benzaldehyde, may be present in commercial glacial acrylic acid. These impurities may have an adverse effect in achievement of very high molecular weight polymers. Further distillation or chemical treatment prior to final distillation is employed to remove carbonyl impurities. Effective agents include hydrazine, amino acids, and alkyl or aryl amines (66–68).

Esterification. The process flow sheet (Fig. 4) outlines the process and equipment of the esterification step in the manufacture of the lower acrylic esters

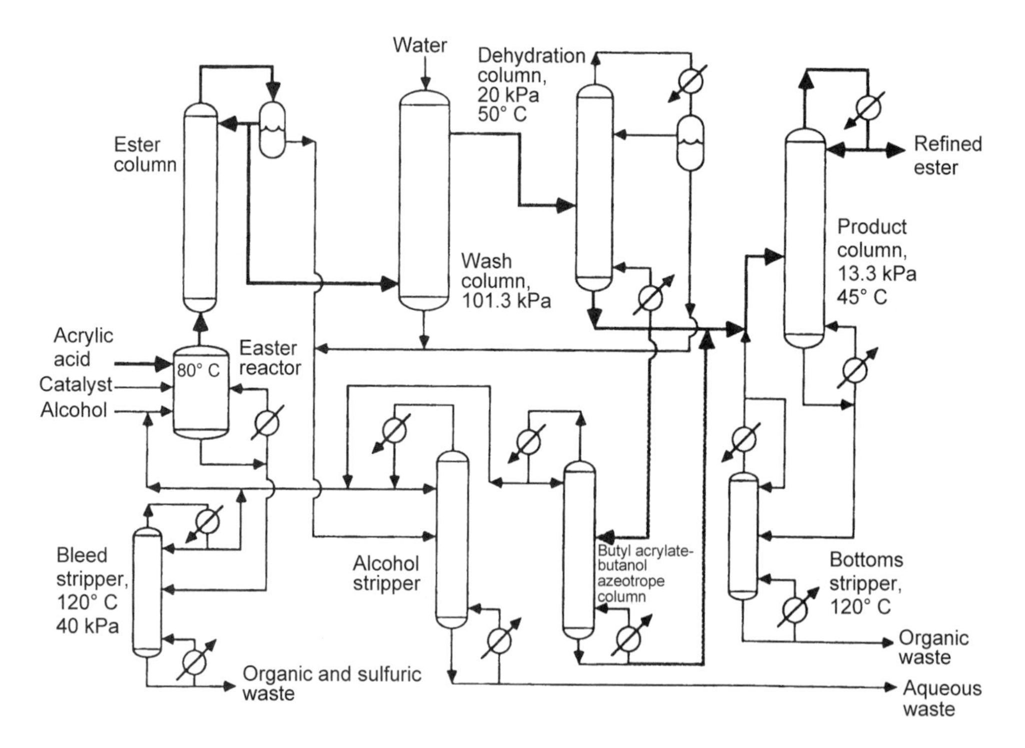

Fig. 4. Esterification process. To convert kPa to mm Hg, multiply by 7.5.

(methyl, ethyl, or butyl). For typical art, see References 69–74. The part of the flow sheet containing the dotted lines is appropriate only for butyl acrylate, since the lower alcohols, methanol and ethanol, are removed in the wash column. Since the butanol is not removed by a water or dilute caustic wash, it is removed in the azeotrope column as the butyl acrylate azeotrope; this material is recycled to the reactor.

Acrylic acid, alcohol, and the catalyst, eg, sulfuric acid, together with the recycle streams are fed to the glass-lined ester reactor fitted with an external reboiler and a distillation column. Acrylate ester, excess alcohol, and water of esterification are taken overhead from the distillation column. The process is operated to give only traces of acrylic acid in the distillate. The bulk of the organic distillate is sent to the wash column for removal of alcohol and acrylic acid; a portion is returned to the top of the distillation column. If required, some base may be added during the washing operation to remove traces of acrylic acid.

A continuous bleed is taken from the reactor to remove high boilers. Values contained in this bleed are recovered in the bleed stripper and the distillate from this operation is recycled to the esterification reactor. The bleed stripper residue is a mixture of high boiling organic material and sulfuric acid, which is recovered for recycle in a waste sulfuric acid plant.

If a waste sulfuric acid regeneration plant is not available, eg, as part of a joint acrylate–methacrylate manufacturing complex, the preferred catalyst for esterification is a sulfonic acid type ion-exchange resin. In this case the residue from the ester reactor bleed stripper can be disposed of by combustion to recover energy value as steam.

The wet ester is distilled in the dehydration column using high reflux to remove a water phase overhead. The dried bottoms are distilled in the product column to provide high purity acrylate. The bottoms from the product column are stripped to recover values and the final residue incinerated. Alternatively, the bottoms may be recycled to the ester reactor or to the bleed stripper.

Conventional polymerization inhibitors are fed to each of the distillation columns. The columns are operated under reduced pressure to give low bottom temperatures and minimize polymerization.

The aqueous layer from the ester column distillate, the raffinate from washing the ester, and the aqueous phase from the dehydration step are combined and distilled in the alcohol stripper. The wet alcohol distillate containing a low level of acrylate is recycled to the esterification reactor. The aqueous column bottoms are incinerated or sent to biological treatment. Biological treatment is common.

Process conditions for methyl acrylate are similar to those employed for ethyl acrylate. However, in the preparation of butyl acrylate the excess butanol is removed as the butanol–butyl acrylate azeotrope in the azeotrope column.

The esters are produced in minimum purity of 99.5%. The yield, based on acrylic acid, is in the range of about 95–98% depending on the ester and reaction conditions. Monomethyl ether of hydroquinone (10–100 ppm) is added as polymerization inhibitor and the esters are used in this form in most industrial polymerizations.

4.2. Acetylene-Based Routes. Walter Reppe, the father of modern acetylene chemistry, discovered the reaction of nickel carbonyl with acetylene and water or alcohols to give acrylic acid or esters (75,76). This discovery led to several processes which have been in commercial use. The original Reppe reaction requires a stoichiometric ratio of nickel carbonyl to acetylene. The Rohm and Haas modified or semicatalytic process provides 60–80% of the carbon monoxide from a separate carbon monoxide feed and the remainder from nickel carbonyl (77,78). The reactions for the synthesis of ethyl acrylate are

$$4\,C_2H_2 + 4\,C_2H_5OH + Ni(CO)_4 + 2\,HCl \longrightarrow 4\,CH_2{=}CHCOOC_2H_5 + H_2 + NiCl_2$$

$$C_2H_2 + C_2H_5OH + 0.05\,Ni(CO)_4 + 0.8\,CO + 0.1\,HCl \longrightarrow$$
$$CH_2{=}CHCOOC_2H_5 + 0.05\,NiCl_2 + 0.05\,H_2$$

The stoichiometric and the catalytic reactions occur simultaneously, but the catalytic reaction predominates. The process is started with stoichiometric amounts, but afterward, carbon monoxide, acetylene, and excess alcohol give most of the acrylate ester by the catalytic reaction. The nickel chloride is recovered and recycled to the nickel carbonyl synthesis step. The main by-product is ethyl propionate, which is difficult to separate from ethyl acrylate. However, by proper control of the feeds and reaction conditions, it is possible to keep the ethyl propionate content below 1%. Even so, this is significantly higher than the propionate content of the esters from the propylene oxidation route.

Other by-products formed are relatively easy to separate, including esters of higher unsaturated monobasic acids (alkyl 3-pentenoate and 3,5-heptadienoate) (5) and esters of multiply-unsaturated dibasic acids, eg, suberates.

The reaction is initiated with nickel carbonyl. The feeds are adjusted to give the bulk of the carbonyl from carbon monoxide. The reaction takes place continuously in an agitated reactor with a liquid recirculation loop. The reaction is run at about atmospheric pressure and at about 40°C with an acetylene:carbon monoxide mole ratio of 1.1:1 in the presence of 20% excess alcohol. The reactor effluent is washed with nickel chloride brine to remove excess alcohol and nickel salts and the brine–alcohol mixture is stripped to recover alcohol for recycle. The stripped brine is again used as extractant, but with a bleed stream returned to the nickel carbonyl conversion unit. The neutralized crude monomer is purified by a series of continuous, low pressure distillations.

The modified Reppe process was installed by Rohm and Haas at their Houston plant in 1948 and later expanded to a capacity of about 182×10^6 kg/yr. Rohm and Haas started up a propylene oxidation plant at the Houston site in late 1976. The combination of attractive economics and improved product purity from the propylene route led to a shutdown of the acetylene-based route within a year.

Reppe's work also resulted in the high pressure route which was established by BASF at Ludwigshafen in 1956. In this process, acetylene, carbon monoxide, water, and a nickel catalyst react at about 200°C and 13.9 MPa (2016 psi) to give acrylic acid. Safety problems caused by handling of acetylene are alleviated by the use of tetrahydrofuran as an inert solvent. In this process, the catalyst is a mixture of nickel bromide with a cupric bromide promotor. The liquid reactor effluent is degassed and extracted. The acrylic acid is obtained by distillation of the extract and subsequently esterified to the desired acrylic ester. The

BASF process gives acrylic acid, whereas the Rohm and Haas process provides the esters directly.

Nickel carbonyl is volatile, has little odor, and is extremely toxic. Symptoms of dangerous exposure may not appear for several days. Effective medical treatment should be started immediately. The plant should be designed to ensure containment of nickel carbonyl and to prevent operator contact.

All other organic waste-process and vent streams are burned in a flare, in an incinerator, or in a furnace where fuel value is recovered. Wastewater streams are handled in the plant biological treatment area.

A method for producing acrylic acid by acrolein oxidation using a metallic catalyst (M_0, W, V) has been described (79). The catalyst is broken into smaller pieces and suspended in water. This helps to prevent deterioration of the catalyst.

Although some very minor manufacturers of acrylic acid may still use hydrolysis of acrylonitrile (see below), essentially all other plants worldwide use the propylene oxidation process.

4.3. Acrylonitrile Route. This process, based on the hydrolysis of acrylonitrile (80), is also a propylene route since acrylonitrile (qv) is produced by the catalytic vapor-phase ammoxidation of propylene.

$$CH_2{=}CHCH_3 + NH_3 + \frac{3}{2}\,O_2 \longrightarrow CH_2{=}CHCN + 3\,H_2O$$

The yield of acrylonitrile based on propylene is generally lower than the yield of acrylic acid based on the direct oxidation of propylene. Hence, for the large volume manufacture of acrylates, the acrylonitrile route is not attractive since additional processing steps are involved and the ultimate yield of acrylate based on propylene is much lower. Hydrolysis of acrylonitrile can be controlled to provide acrylamide rather than acrylic acid, but acrylic acid is a by-product in such a process (81).

The sulfuric acid hydrolysis may be performed as a batch or continuous operation. Acrylonitrile is converted to acrylamide sulfate by treatment with a small excess of 85% sulfuric acid at 80–100°C. A hold-time of about 1 h provides complete conversion of the acrylonitrile. The reaction mixture may be hydrolyzed and the aqueous acrylic acid recovered by extraction and purified as described under the propylene oxidation process prior to esterification. Alternatively, after reaction with excess alcohol, a mixture of acrylic ester and alcohol is distilled and excess alcohol is recovered by aqueous extractive distillation. The ester in both cases is purified by distillation.

Important side reactions are the formation of ether and addition of alcohol to the acrylate to give 3-alkoxypropionates. In addition to high raw material costs, this route is unattractive because of large amounts of sulfuric acid–ammonium sulfate wastes.

4.4. Ketene Process. The ketene process based on acetic acid or acetone as the raw material was developed by B. F. Goodrich (82) and Celanese (83). It is no longer used commercially because the intermediate β-propiolactone is suspected to be a carcinogen (84). In addition, it cannot compete with the improved propylene oxidation process (see KETENES, KETENE DIMERS, AND RELATED SUBSTANCES).

4.5. Ethylene Cyanohydrin Process. This process, the first for the manufacture of acrylic acid and esters, has been replaced by more economical ones. During World War I, the need for ethylene as an important raw material for the synthesis of aliphatic chemicals led to development of this process (16) in both Germany, in 1927, and the United States, in 1931.

In the early versions, ethylene cyanohydrin was obtained from ethylene chlorohydrin and sodium cyanide. In later versions, ethylene oxide (from the direct catalytic oxidation of ethylene) reacted with hydrogen cyanide in the presence of a base catalyst to give ethylene cyanohydrin. This was hydrolyzed and converted to acrylic acid and by-product ammonium acid sulfate by treatment with about 85% sulfuric acid.

Losses by polymer formation kept the yield of acrylic acid to 60–70%. Preferably, esters were prepared directly by a simultaneous dehydration–esterification process.

The process has historic interest. It was replaced at the Rohm and Haas Company by the acetylene-based process in 1954, and in 1970 at Union Carbide by the propylene oxidation process.

4.6. Other Syntheses. Acrylic acid and other unsaturated compounds can also be made by a number of classical elimination reactions. Acrylates have been obtained from the thermal dehydration of hydracrylic acid (3-hydroxypropanoic acid [503-66-2]) (85), from the dehydrohalogenation of 3-halopropionic acid derivatives (86), and from the reduction of dihalopropionates (2). These studies, together with the related characterization and chemical investigations, contributed significantly to the development of commercial organic chemistry. Metallic oxide catalysts can produce acrylic acid by vapor phase catalytic oxidation of propane in high yield (87).

Vapor-Phase Condensations of Acetic Acid or Esters with Formaldehyde. Addition of a methylol group to the α-carbon of acetic acid or esters, followed by dehydration, gives the acrylates.

$$CH_3COOH(R) + CH_2O \longrightarrow CH_2{=}CHCOOH(R) + H_2O$$

The reaction is generally carried out at atmospheric pressure and at 350–400°C. A variety of catalysts, eg, bases and metal salts and oxides on silica or alumina–silicates, have been patented (88–93). Conversions are in the 30–70% range and selectivities in the 60–90% range, depending on the catalyst and the ratio of formaldehyde to acetate.

The procedure is technically feasible, but high recovery of unconverted raw materials is required for the route to be practical. Its development depends on the improvement of catalysts and separation methods and on the availability of low cost acetic acid and formaldehyde. Both raw materials are dependent on ample supply of low cost methanol.

Although the rapid cost increases and shortages of petroleum-based feedstocks forecast a decade ago have yet to materialize, shift to natural gas or coal may become necessary. Under such conditions, it is possible that acrylate manufacture via acetylene, as described above, could again become attractive.

It appears that condensation of formaldehyde with acetic acid might be preferred. A coal gasification complex readily provides all of the necessary intermediates for manufacture of acrylates (94).

Oxidative Carbonylation of Ethylene—Elimination of Alcohol from β-Alkoxypropionates. Spectacular progress in the 1970s led to the rapid development of organotransition-metal chemistry, particularly to catalyze olefin reactions (95,96). A number of patents have been issued (28,97–99) for the oxidative carbonylation of ethylene to provide acrylic acid and esters. The procedure is based on the palladium catalyzed carbonylation of ethylene in the liquid phase at temperatures of 50–200°C. Esters are formed when alcohols are included. Anhydrous conditions are desirable to minimize the formation of by-products including acetaldehyde and carbon dioxide (see ACETALDEHYDE).

During the reaction, the palladium catalyst is reduced. It is reoxidized by a co-catalyst system such as cupric chloride and oxygen. The products are acrylic acid in a carboxylic acid-anhydride mixture or acrylic esters in an alcoholic solvent. Reaction products also include significant amounts of 3-acryloxypropionic acid [24615-84-7] and alkyl 3-alkoxypropionates, which can be converted thermally to the corresponding acrylates (23,100). The overall reaction may be represented by:

$$CH_2{=}CH_2 + CO + HOR + \frac{1}{2}\,O_2 \longrightarrow CH_2{=}CHCOOR + H_2O$$

It is preferrable to carry out the reactions to give an intermediate β-alkoxypropionate. When the reaction is carried out in ethanolic ethyl β-ethoxypropionate as solvent, and a trace of mercury(II) is included in the catalyst system, the yield of intermediate ethyl β-ethoxypropionate rises to 95–97%; the principal by-product is the corresponding β-chloropropionate ester. Acid-catalyzed thermal cracking of the mixture at 120–150°C gives very high yields of ethyl acrylate (26). Although yields are excellent, the reaction medium is extremely corrosive, so high cost materials of construction are necessary. In addition, the high cost of catalyst and potential toxicity of mercury require that the inorganic materials be recovered quantitatively from any waste stream. Hence, high capital investment, together with continued favorable costs for propylene, have prevented commercialization of this route.

The elimination of alcohol from β-alkoxypropionates can also be carried out by passing the alkyl β-alkoxypropionate at 200–400°C over metal phosphates, silicates, metal oxide catalysts (101), or base-treated zeolites (98). In addition to the route via oxidative carbonylation of ethylene, alkyl β-alkoxypropionates can be prepared by reaction of dialkoxy methane and ketene (102).

Dehydrogenation of Propionates. Oxidative dehydrogenation of propionates to acrylates employing vapor-phase reactions at high temperatures (400–700°C) and short contact times is possible. Although selective catalysts for the oxidative dehydrogenation of isobutyric acid to methacrylic acid have been developed in recent years (see METHACRYLIC ACID AND DERIVATIVES) and a route to methacrylic acid from propylene to isobutyric acid is under pilot-plant development in Europe, this route to acrylates is not presently of commercial interest because of the combination of low selectivity, high raw material costs, and purification difficulties.

Liquid-Phase Oxidation of Acrolein. As discussed before, the most attractive process for the manufacture of acrylates is based on the two-stage, vapor-phase oxidation of propylene. The second stage involves the oxidation of acrolein. Considerable art on the liquid-phase oxidation of acrolein (17) is available, but this route cannot compete with the vapor-phase technology.

5. Specialty Acrylic Esters

Higher alkyl acrylates and alkyl-functional esters are important in copolymer products, in conventional emulsion applications for coatings and adhesives, and as reactants in radiation-cured coatings and inks. In general, they are produced in direct or transesterification batch processes (17,103,104) because of their relatively low volume.

Direct, acid catalyzed esterification of acrylic acid is the main route for the manufacture of higher alkyl esters. The most important higher alkyl acrylate is 2-ethylhexyl acrylate prepared from the available oxo alcohol 2-ethyl-1-hexanol (see ALCOHOLS, HIGHER ALIPHATIC). The most common catalysts are sulfuric or toluenesulfonic acid and sulfonic acid functional cation-exchange resins. Solvents are used as entraining agents for the removal of water of reaction. The product is washed with base to remove unreacted acrylic acid and catalyst and then purified by distillation. The esters are obtained in 80–90% yield and in excellent purity.

Transesterification of a lower acrylate ester and a higher alcohol (104,105) can be performed using a variety of catalysts and conditions chosen to provide acceptable reaction rates and to minimize by-product formation and polymerization.

$$CH_2{=}CHCOOR + R'OH \xrightarrow{H^+} CH_2{=}CHCOOR' + ROH$$

Pure dry reactants are needed to prevent catalyst deactivation; effective inhibitor systems are also desirable as well as high reaction rates, since many of the specialty monomers are less stable than the lower alkyl acrylates. The alcohol–ester azeotrope (8) should be removed rapidly from the reaction mixture and an efficient column used to minimize reactant loss to the distillate. After the reaction is completed, the catalyst may be removed and the mixture distilled to obtain the ester. The method is particularly useful for the preparation of functional monomers which cannot be prepared by direct esterification. Dialkylaminoethyl acrylic esters are readily prepared by transesterification of the corresponding dialkylaminoethanol (104,105). Catalysts include strong acids and tetraalkyl titanates for higher alkyl esters; and titanates, sodium phenoxides, magnesium alkoxides, and dialkyltin oxides, as well as titanium and zirconium chelates, for the preparation of functional esters. Because of loss of catalyst activity during the reaction, incremental or continuous additions may be required to maintain an adequate reaction rate.

Table 5. **Acrylic Acid Producers and Capacities, U.S., $\times 10^6$ kg ($\times 10^6$ lb)**

Producer	Capacity
BASF, Freeport, Tex.	220 (485)
American Acryl, Pasadena, Tex.	120 (265)
Celanese, Clear Lake, Tex.	290 (640)
Dow Chemical, Taft, La.	109 (240)
Rohm and Haas/StoHaas, Deer Park, Tex.	573 (1265)
Total	*1312 (2895)*

Hydroxyethyl and 2-hydroxypropyl acrylates are prepared by the addition of ethylene oxide or propylene oxide to acrylic acid (106,107).

The reactions are catalyzed by tertiary amines, quaternary ammonium salts, metal salts, and basic ion-exchange resins. The products are difficult to purify and generally contain low concentrations of acrylic acid and some diester which should be kept to a minimum since its presence leads to product instability and to polymer cross-linking.

6. Economic Aspects

Table 5 lists United States producers of acrylic acid and their capacities. Industrial production is by gas-phase oxidation of propylene. About three quarters of the production is converted directly into acrylate esters and the rest is purified into glacial acrylic acid (108).

Rohm and Haas formed a fifty-fifty joint venture with Stockhausen (Krefeld, Germany). The new company, StoHaas (Marl, Germany) has expanded capacity to 34×10^6 kg (75×10^6 lb) at Deer Park, Texas and plans to produce 100×10^6 kg (220×10^6 lb) in Marl in 2003. Acrylic acid and esters are marketed exclusively by Rohm and Haas.

Growth for acrylic acid is estimated at 4.0% per year and should reach 1.8×10^9 kg (2.6×10^9 lb) in 2004. *n*-Butyl acrylate is the most important of the esters and demand will grow at the rate of 4.2% (108).

Prices for the period 1995–2001 ranged from $0.37 to $0.39/kg ($0.81-0.87/lb), list, glacial acrylic acid, tanks, dil. Current prices: $0.39/kg ($0.87/lb), same basis. Market price: $0.28 to $0.32/kg ($0.62–0.70/lb) same basis (108).

7. Analytical Methods

Chemical assay is preferably performed by gas–liquid chromatography (glc) or by the conventional methods for determination of unsaturation such as bromination or addition of mercaptan, sodium bisulfite, or mercuric acetate.

Acidity is determined by glc or titration, and the dimer content of acrylic acid by glc or a saponification procedure. The total acidity is corrected for the dimer acid content to give the value for acrylic acid.

The relatively low flash points of some acrylates create a fire hazard. Also, the ease of polymerization must be borne in mind in all operations. The lower and upper explosive limits for methyl acrylate are 2.8 and 25 vol %, respectively. Corresponding limits for ethyl acrylate are 1.8 vol % and saturation, respectively. All possible sources of ignition of monomers must be eliminated.

8. Storage and Handling

Acrylic acid and esters are stabilized with minimum amounts of inhibitors consistent with stability and safety. The acrylic monomers must be stable and there should be no polymer formation for prolonged periods with normal storage and shipping (4,109). The monomethyl ether of hydroquinone (MEHQ) is frequently used as inhibitor and low inhibitor grades of the acrylate monomers are available for bulk handling. MEHQ at 10–15 ppm is generally adequate for the esters, but a higher concentration (200 ppm) is needed for acrylic acid.

The effectiveness of phenolic inhibitors is dependent on the presence of oxygen and the monomers must be stored under air rather than an inert atmosphere. Temperatures must be kept low to minimize formation of peroxides and other products. Moisture may cause rust-initiated polymerization.

Acrylic acid has a relatively high freezing point (13°C) and the inhibitor may not be distributed uniformly between phases when frozen acid is partially thawed. If the liquid phase is inadequately inhibited, it could polymerize and initiate violent polymerization of the entire mass. Provisions should be made to maintain the acid as a liquid. High temperatures should be avoided because of dimer formation. If freezing should occur, melting should take place at room temperature (25°C); material should not be withdrawn until the total is thawed and well-mixed to provide good distribution of the inhibitor and dissolved oxygen. No part of the mass should be subjected to elevated temperatures during the melting process.

Dimer formation, which is favored by increasing temperature, generally does not reduce the quality of acrylic acid for most applications. The term dimer includes higher oligomers formed by further addition reactions and present in low concentrations relative to the amount of dimer (3-acryloxypropionic acid). Glacial acrylic acid should be stored at 16–29°C to maintain high quality.

The acrylic esters may be stored in mild or stainless steel, or aluminum. However, acrylic acid is corrosive to many metals and can be stored only in glass, stainless steel, aluminum, or polyethylene-lined equipment. Stainless steel types 316 and 304 are preferred materials for acrylic acid.

For most applications, the phenolic inhibitors do not have to be removed. The low-inhibitor grades of acrylic monomers are particularly suitable for the manufacture of polymer without pretreatment. Removal of inhibitor from ester is best done by adsorption with ion-exchange resins or other adsorbents. Phenolic inhibitors may be removed from esters with an alkaline brine wash, generally a solution containing 5% caustic and 20% salt. Vigorous agitation during washing should be avoided to prevent the formation of emulsions. The washed monomers may be used without drying in emulsion processes. Washed uninhibited monomers are less stable and should be used promptly. They should be stored under refrigeration, but should not be permitted to freeze because of the danger of explosive polymerization during thawing. The heat of polymerization is approximately 75 kJ/mol (18 kcal/mol).

A method for transporting and storing pure acrylic acid comprises ensuring means of appropriate measures that the acrylic acid is partly crystalline during the entire duration of transport or storage (110).

9. Health and Safety Factors

The toxicity of common acrylic monomers has been characterized in animal studies using a variety of exposure routes. Toxicity varies with level, frequency, duration, and route of exposure. The simple higher esters of acrylic acid are usually less absorbed and less toxic than lower esters. In general, acrylates are more toxic than methacrylates. Data appear in Table 6.

With respect to acute toxicity, based on lethality in rats or rabbits, acrylic monomers are slightly to moderately toxic. Mucous membranes of the eyes, nose, throat, and gastrointestinal tract are particularly sensitive to irritation. Acrylates can produce a range of eye and skin irritations from slight to corrosive depending on the monomer.

Full eye protection should be worn whenever handling acrylic monomers; contact lenses must never be worn. Prolonged exposure to liquid or vapor can result in permanent eye damage or blindness. Excessive exposure to vapors causes nose and throat irritation, headaches, nausea, vomiting, and dizziness or drowsiness (solvent narcosis). Overexposure may cause central nervous system depression. Both proper respiratory protection and good ventilation are necessary wherever the possibility of high vapor concentration arises.

Table 6. **Acute Toxicity of Acrylic Acid and Esters**

Monomer	Methyl acrylate	Ethyl acrylate	Butyl acrylate	Acrylic acid
rat oral LD_{50}, g/kg	0.3	0.8–1.8	3.7–8.1	0.3–2.5
rabbit dermal LD_{50}, g/kg	1.3	1.2–3.0	1.7–5.7	0.3–1.6
rat inhalation LC_{50}, ppm	750–1350	2180	2370	1200–4000
rabbit eye irritation	severe to corrosive	severe to corrosive	slight to moderate	corrosive
rabbit skin irritator	severe irritation	severe irritation	moderate	corrosive
odor threshold, ppb	2.3–4.8	0.5	35[a]	94[b]
TLV/TWA[c,d], mg/m^3	35	20	52	5.9
ppm	10, skin	5	10	2, skin
STEL[d], mg/m^3		100		
ppm		25		
		A–2 suspect human carcinogen		

[a] May have delayed eye irritation.
[b] Vapor exposure can cause irreversible eye damage.
[c] Ref. 107.
[d] Ref. 83.

Swallowing acrylic monomers may produce severe irritation of the mouth, throat, esophagus, and stomach, and cause discomfort, vomiting, diarrhea, dizziness, and possible collapse.

Skin redness and from slight to corrosive irritation is caused by direct contact. Acrylic acid is more corrosive than esters. The monomers not only irritate the skin, but may also be absorbed through the skin. Therefore, gloves and protective clothing and shoes or boots should be used in addition to eye (or full face) protective equipment. Upon contact, the skin should be flushed with copious amounts of water; follow-up medical attention should be sought. Medical attention should also be obtained if any of the earlier mentioned symptoms appear.

Repeated exposures to acrylic monomers can produce allergic dermatitis (or skin sensitization) resulting in rash, itching, or swelling. After exposure to one monomer, this dermatitis may arise upon subsequent exposure to the same or even a different acrylic monomer.

Repeated exposures of animals to high (near-lethal) concentrations of vapors result in inflammation of the respiratory tract, as well as degenerative changes in the liver, kidneys, and heart muscle. These effects arise at concentrations far above those causing irritation. Such effects have not been reported in humans. The low odor threshold and irritating properties of acrylates cause humans to leave a contaminated area rather than tolerate the irritation.

Current TLV/TWA values are provided in *Material Safety Data Sheets* provided by manufacturers upon request. Values (83,111,112) appear in Table 5.

Acrolein, acrylamide, hydroxyalkyl acrylates, and other functional derivatives can be more hazardous from a health standpoint than acrylic acid and its simple alkyl esters. Furthermore, some derivatives, such as the alkyl 2-chloroacrylates, are powerful vesicants and can cause serious eye injuries. Thus, although the hazards of acrylic acid and the normal alkyl acrylates are moderate and they can be handled safely with ordinary care to industrial hygiene, this should not be assumed to be the case for compounds with chemically different functional groups (see INDUSTRIAL HYGIENE; PLANT SAFETY; TOXICOLOGY).

In 1983 the National Toxicology Program (NTP) reported that ethyl acrylate produced tumors in the rodent forestomach after gavage (forced feeding via stomach tube) for 2 yr. The response occurred only at the site of contact after lifetime exposure to levels that were both irritating and ulcerating to that tissue. Based on this study, both the NTP and the International Agency for Research on Cancer (IARC) concluded that there was sufficient evidence for carcinogenicity of ethyl acrylate in experimental animals and by extension classified ethyl acrylate as possibly carcinogenic in humans. Several other studies of simple acrylate esters using alternate exposure methods failed to show evidence of oncogenic response.

9.1. Regulation. Acrylic acid and certain esters may be used safely as a component of the uncoated or coated food-product surface of paper and paperboard intended for use in producing, manufacturing, packing, processing, preparing, treating, packaging, transporting and holding of dry food (21 CFR 176.180).

Acrylic acid polymers may be used as indirect food additives only for components of adhesives (21 CFR 175.105). Semirigid and rigid acrylic and modified acrylic plastics may be safely used as articles intended for use in contact with food and as components of articles intended for use in contact with food (21 CFR 177.1010).

10. Uses

Most acrylic acid is consumed in the form of the polymer. The dominant share of acrylic acid is converted to esters. Today growth is in the demand for superabsorbents (SAPs) for use in diapers and hygienic products. Acrylic acid accounts for 80–85% of raw materials used in the manufacture of SAPs.

Poly(acrylic acid) and salts (includes superabsorbent polymers, detergents, water treatment, and dispersants) account for 34% of total consumption; *n*-butyl acrylate, 32%; ethyl acrylate, 18%; 2-ethyl hexyl acrylate, 6%; methyl and specialty acrylates, 7%; and miscellaneous uses, 3% (108).

BIBLIOGRAPHY

"Acrylic and Methacrylic Acids" in *ECT* 1st ed., Vol. 1, pp. 176–179, by F. J. Glavis, Rohm & Haas Company; "Acrylic Resins and Plastics" in *ECT* 1st ed., Vol. 1, pp. 180–184, by E. H. Kroeker, Rohm & Haas Company; "Acrylic Acid and Derivatives" in *ECT* 2nd ed., Vol. 1, pp. 285–313,. by F. J. Glavis and E. H. Specht, Rohm & Haas Company; "Acrylic Acid and Derivatives" in *ECT* 3rd ed., Vol. 1, pp. 330–354, by J. W. Nemec and W. Bauer, Jr., Rohm and Haas Company, "Acrylic Acid and Derivatives" in *ECT* 4th ed., Vol. 1, pp. 287–314, by William Bauer, Jr., Rohm and Haas Company; "Acrylic Acid and Derivatives" in *ECT* (online), posting date: December 4, 2000, by William Bauer, Jr., Rohm and Haas Company.

CITED PUBLICATIONS

1. J. Redtenbacher, *Ann.* **47**, 125 (1843).
2. W. Caspary and B. Tollens, *Ann.* **167**, 240 (1873).
3. S. Hochheiser, *Rohm and Haas,* University of Pennsylvania Press, Philadelphia, 1986, pp. 31ff.
4. *Storage and Handling of Acrylic and Methacrylic Esters and Acids, Bulletin 84C7,* Rohm and Haas Co., Philadelphia, Pa., 1987; *Acrylic and Methacrylic Monomers—Specifications and Typical Properties, Bulletin 84C2,* Rohm and Haas Co., Philadelphia, Pa., 1986; *Rocryl Specialty Monomers—Specifications and Typical Properties, Bulletin 77S2,* Rohm and Haas Co., Philadelphia, Pa., 1989.
5. Rohm and Haas Company, Philadelphia, Pa., internal data.
6. N. S. Marans and R. P. Zelinski, *J. Am. Chem. Soc.* **72**, 2125 (1950).
7. U.S. Pat. 2,967,195 (Jan. 3, 1961), M. H. Gold (to Aerojet-General Corp.).
8. C. E. Rehberg and C. H. Fisher, *J. Am. Chem. Soc.* **66**, 1203 (1944); *Ind. Eng. Chem.* **40**, 1429 (1948).
9. C. E. Rehberg, W. A. Faucette, and C. H. Fisher, *J. Am. Chem. Soc.* **66**, 1723 (1944).
10. C. E. Rehberg and C. H. Fisher, *J. Org. Chem.* **12**, 226 (1947).
11. C. E. Rehberg and W. A. Faucette, *J. Am. Chem. Soc.* **71**, 3164 (1949).
12. C. E. Rehberg and W. A. Faucette, *J. Org. Chem.* **14**, 1094 (1949).

13. C. E. Rehberg, M. B. Dixon, and W. A. Faucette, *J. Am. Chem. Soc.* **72**, 5199 (1950).
14. D. W. Coddington, T. S. Reid, A. H. Ahlbrecht, C. H. Smith, Jr., and D. R. Usted, *J. Polym. Sci.* **15**, 515 (1955).
15. W. Postelnek, L. E. Coleman, and A. M. Lovelace, *Fortschr. Hochpolym. Forsch.* **1**, 75 (1958).
16. E. H. Riddle, *Monomeric Acrylic Esters,* Reinhold Publishing Co., New York, 1954.
17. H. Rauch-Puntigam and T. Volker, *Acryl- und Methacrylverbindungen,* Springer-Verlag, Berlin, 1967.
18. M. Sittig, *Vinyl Monomers and Polymers,* Noyes Development Corp., Park Ridge, N.J., 1966.
19. U.S. Pat. 3,703,539 (Nov. 21, 1962), B. A. Di Liddo (to B. F. Goodrich Co.).
20. U.S. Pat. 4,490,553 (Dec. 25, 1984), J. D. Chase and W. W. Wilkison (to Celanese Corporation).
21. R. Mozingo and L. A. Patterson, *Org. Synth. Coll. Vol. 3,* 576 (1955).
22. F. M. Wampler III, *Plant/Operations Progress* **1**(3), 183–189 (1988).
23. U.S. Pat. 3,888,912 (June 10, 1975), M. D. Burguette (to Minnesota Mining and Manufacturing Co.).
24. U.S. Pat. 3,642,843 (Feb. 15, 1972), J. W. Nemec (to Rohm and Haas Co.).
25. C. E. Rehberg, M. B. Dixon, and C. H. Fisher, *J. Am. Chem. Soc.* **68**, 544 (1946); **69**, 2966 (1947); C. E. Rehberg and M. B. Dixon, *J. Am. Chem. Soc.* **72**, 2205 (1950).
26. U.S. Pat. 3,227,746 (Jan. 4, 1966), F. Knorr and A. Spes (to Wacker-Chemie G.m.b.H.).
27. D. M. Fenton and K. L. Olivier, *Chem. Technol.* 220 (Apr. 1972).
28. U.S. Pat. 3,987,089 (Oct. 19, 1976), F. L. Slejko and J. S. Clovis (to Rohm and Haas Co.).
29. C. D. Hurd and L. L. Gershbein, *J. Am. Chem. Soc.* **69**, 2328 (1947).
30. L. L. Gershbein and C. D. Hurd, *J. Am. Chem. Soc.* **69**, 241 (1947).
31. E. A. Fehnel and M. Carmack, *Org. Syn.* **30**, 65 (1950).
32. U.S. Pat. 3,769,315 (Oct. 30, 1973), R. L. Keener and H. Raterink (to Rohm and Haas Co.).
33. S. M. McElvain and G. Stork, *J. Am. Chem. Soc.* **68**, 1049 (1946).
34. U.S. Pat. 2,580,832 (Jan. 1, 1952), E. W. Pietrusza (to Allied Chemical Co.).
35. C. B. Frederick and C. H. Reynolds, *Toxicol. Lett.* **47**, 241–247 (1989).
36. R. Osman, K. Namboodiri, H. Weinstein, and J. R. Rabinowitz, *J. Am. Chem. Soc.* **110**, 1701–1707 (1988).
37. U.S. Pat. 3,074,999 (Jan. 22, 1963), M. B. Rauhut and H. Currier (to American Cyanamid Co.).
38. U.S. Pats. 3,342,853 and 3,342,854 (Sept. 19, 1967), J. W. Nemec and co-workers (to Rohm and Haas Co.).
39. C. R. Adams, *Chem. Ind.* **26**, 1644 (1970).
40. S. Sakuyama, T. O'Hara, N. Shimizu, and K. Kubota, *Chem. Technol.* 350 (June 1973).
41. T. O'Hara and co-workers, "Acrylic Acid and Derivatives" in *Ullmanns Encyclopedia of Industrial Chemistry,* 5th ed., Vol. A1, VCH, Verlagsgesellschaft mbH, Weinheim 1985, 161–176.
42. *Hydrocarbon Process.* **48**(11), 145 (1969).
43. *Hydrocarbon Process.* **60**(11), 124 (1981).
44. *Hydrocarbon Process.* **68**(11), 91 (1989).
45. U.S. Pat. 3,475,488 (Oct. 28, 1969), N. Kurata, T. Ohara, and K. Oda (to Nippon Shokubai Kagaku Kogyo Co., Ltd.).
46. Brit. Pats. 915,799 and 915,800 (Jan. 16, 1963), D. J. Hadley and R. H. Jenkins (to Distillers Co., Ltd.).

47. U.S. Pats. 4,203,906 (May 20, 1980) 4,256,753 (Mar. 17, 1981), M. Takada, H. Uhara, and T. Sato (to Nippon Shokobai Kogaku Kogyo Co., Ltd.).
48. U.S. Pat. 4,537,874 (Aug. 27, 1985), T. Sato, M. Takata, M. Ueshima, and I. Nagai (to Nippon Shokubai Kogaku Co., Ltd.).
49. U.S. Pat. 4,438,217 (Mar. 20, 1984), M. Takata, R. Aoki, and T. Sato (to Nippon Shokubai Kogaku Co., Ltd.).
50. U.S. Pats. 2,881,212 (Apr. 7, 1959) and 3,087,964 (Apr. 30, 1963), J. D. Idol, J. L. Callahan, and R. W. Foreman (to Standard Oil Co., Ohio).
51. Jpn. Pat. 43-13609 (June 8, 1968), M. Izawa and co-workers (to Toyo Soda Manufacturing Co.).
52. U.S. Pats. 3,441,613 (Apr. 29, 1969) and 3,527,716 (Sept. 8, 1970), J. W. Nemec and F. W. Schlaefer (to Rohm and Haas Co.).
53. U.S. Pat. 4,092,354 (May 30, 1978), T. Shiraishi, T. Kechiwada, and Y. Nagaoka (to Sumitomo Chemical Co., Ltd.).
54. U.S. Pat. 3,527,797 (Sept. 8, 1970), R. Krabetz, H. Engelbach, and H. Zinke-Allmang (to Badische Anilin- und Soda-Fabrik A.G.).
55. U.S. Pats. 3,939,096 (Feb. 17, 1976) and 3,962,322 (June 8, 1976), P. C. Richardson (to Celanese Corp.).
56. U.S. Pat. 4,365,087 (Dec. 21, 1982), K. Kadowacki, K. Sarumaru, and T. Shibano (to Mitsubishi Petrochemical Co., Ltd.).
57. U.S. Pat. 4,356,114 (Oct. 26, 1982), K. Kadowacki, K. Sarumaru, and Y. Tanaka (to Mitsubishi Petrochemical Co., Ltd.).
58. T. P. Snyder and C. G. Hill, Jr., *Catal. Review—Sci. Eng.* **31**, 43–95 (1989). A current review with leading references to much of the literature of 1975–1990 in partial oxidation of propylene.
59. U.S. Pat. 4,147,885 (Apr. 3, 1979), N. Shimezur, I. Yonagisawa, M. Takata, and T. Sato (to Nippon Shokubai Kogaku Kogyo Co., Ltd.).
60. U.S. Pat. 4,317,926 (Mar. 2, 1982), T. Sato, M. Baba, and M. Okane (to Nippon Shokubai Kogaku Co., Ltd.).
61. Brit. Pat. 997,325 (July 7, 1965), F. C. Newman (to Distillers Co., Ltd.).
62. Jpn. Pat. 49-18728 (June 4, 1971) (to Toa Gosei Chemical Industry).
63. W. Krolikowski, *Soc. Plast. Eng. J.* 1031 (Sept. 1964).
64. U.S. Pat. 3,968,153 (July 6, 1976), T. Ohrui, T. Sakahibara, Y. Aono, M. Kato, H. Takao, and M. Ayano (to Sumitomo Chemical Co.).
65. U.S. Pat. 3,962,074 (June 8, 1976), W. K. Schropp (to Badische Anilin- und Soda-Fabrik A.G.).
66. U.S. Pat. 3,725,208 (Apr. 3, 1973), S. Maezawa, H. Yoshikawa, K. Sakamoto, J. Fugii, and M. Hashimoto (to Nippon Kayaku Co.).
67. U.S. Pat. 3,893,895 (July 8, 1975), J. Dehnert, A. Kleeman, T. Lussling, E. Noll, H. Schaefer, and G. Schreyer (to Deutsche Gold und Silber Scheideanstalt).
68. U.S. Pat. 4,358,347 (Nov. 9, 1982), B. Mettetal and R. Kolonko (to The Dow Chemical Company).
69. U.S. Pat. 3,914,290 (Oct. 21, 1975), S. Otsuki and I. Miyanohara (to Rohm and Haas Co.).
70. U.S. Pat. 2,916,512 (Dec. 8, 1959), G. J. Fischer and A. F. McLean (to Celanese Corp.).
71. U.S. Pat. 3,087,962 (Apr. 30, 1963), N. M. Bortnick (to Rohm and Haas Co.).
72. U.S. Pat. 3,882,167 (May 6, 1975), E. Lohmar, A. Ohorodnik, K. Gehrman, and P. Stutzke (to Hoechst A.G.).
73. U.S. Pat. 2,947,779 (Aug. 2, 1960), J. D. Idol, R. W. Foreman, and F. Veach (to Standard Oil Co. Ohio).
74. Brit. Pat. 923,595 (Apr. 18, 1963), F. J. Bellringer, C. J. Brown, and P. B. Brindley (to Distillers Co. Ltd.).

75. W. Reppe, *Justus Liebigs Ann. Chem.* **582**, 1 (1953).
76. U.S. Pat. 3,023,327 (Feb. 27, 1962), W. Reppe and R. Stadler (to Badische Anilin- und Soda Fabrik A.G.).
77. U.S. Pats. 2,582,911 (Jan. 15, 1952) and 2,613,222 (Oct. 7, 1952) and 2,773,063 (Dec. 4, 1956), H. T. Neher, E. H. Specht, and A. Neuman (to Rohm and Haas Co.).
78. M. Salkind, E. H. Riddle, and R. W. Keefer, *Ind. Eng. Chem.* **51**, 1232, 1328 (1959).
79. U.S. Pat. Appl. 20010004671 (June 21, 2001), W.-H. Lee.
80. *Hydrocarbon Process.* **44**(11), 169 (1965).
81. U.S. Pat. 2,734,915 (Feb. 14, 1956), G. D. Jones (to The Dow Chemical Company).
82. U.S. Pats. 2,356,459 (Aug. 22, 1944) and 2,361,036 (Oct. 24, 1944), E. F. King, and U.S. Pat. 3,002,017 (Sept. 26, 1961), N. Wearsch and A. J. De Paola (to B. F. Goodrich Co.).
83. U.S. Pat. 3,069,433 (Dec. 18, 1962), K. A. Dunn (to Celanese Corp.).
84. *Threshold Limit Values and Biological Exposure Indices for 1989–1990,* American Conference of Governmental Industrial Hygienists, Cincinnati, Ohio, 1989.
85. F. K. Beilstein, *Ann. Chem.* **122**, 372 (1862).
86. W. von Schneider and E. Erlenmeyer, *Ber.* **3**, 340 (1870).
87. U.S. Pat. Appl. 20010029234 (Oct. 11, 2001), X. Tu and co-workers.
88. U.S. Pat. 2,821,543 (Jan. 28, 1958), R. W. Etherington (to Celanese Corp. of America).
89. U.S. Pat. 3,014,958 (Dec. 26, 1961), T. A. Koch and I. M. Robinson (to E. I. du Pont de Nemours & Co., Inc.).
90. U.S. Pats. 3,578,702 (May 11, 1971) and 3,574,703 (Apr. 13, 1971), T. C. Snapp, Jr., A. E. Blood, and H. J. Hagemeyer, Jr. (to Eastman Kodak Co.).
91. U.S. Pats. 3,840,587 and 3,840,588 (Oct. 8, 1974), A. J. C. Pearson (to Monsanto Co.).
92. U.S. Pat. 3,933,888 (Jan. 20, 1976), F. W. Schlaefer (to Rohm and Haas Co.).
93. U.S. Pat. 4,490,476 (Dec. 25, 1984), R. J. Piccolini and M. J. Smith (to Rohm and Haas Co.).
94. J. Haggin, *Chem. Eng. News,* 7–13 (May 19, 1986).
95. R. F. Heck, *Organotransition Metal Chemistry,* Academic Press, Inc., New York, 1974, Chapt. IX.
96. E. I. Becker and M. Tsutsui, eds., *Organometallic Reactions,* Vol. 3, Wiley-Interscience, New York, 1972.
97. U.S. Pats. 3,346,625 (Oct. 10, 1967), D. M. Fenton and K. L. Olivier; 3,397,225 (Aug. 13, 1968), D. M. Fenton; 3,349,119 (Oct. 24, 1967); and 3,381,030 (Apr. 30, 1968), D. M. Fenton and K. L. Olivier (to Union Oil Co. of California).
98. U.S. Pats. 3,920,736 (Nov. 18, 1975) and 3,876,694 (Apr. 8, 1975), W. Gaenzler (to Rohm G.m.b.H. Chemische Fabrik).
99. U.S. 3,579,568 (May 18, 1971), R. F. Heck and P. M. Henry (to Hercules Inc.).
100. U.S. Pat. 4,814,492 (Mar. 21, 1989), E. C. Nelson (to Texaco, Inc.).
101. U.S. Pat. 3,022,339 (Feb. 20, 1962), E. Enk and F. Knoerr (to Wacker Chemie G.m.b.H.).
102. U.S. Pat. 4,827,021 (May 2, 1989), G. C. Jones, W. D. Nottingham, and P. W. Raynolds (to Eastman Kodak Co.).
103. U.S. Pat. 2,917,538 (Dec. 15, 1959), R. L. Carlyle (to The Dow Chemical Company).
104. P. L. De Beneville, L. S. Luskin, and H. J. Sims, *J. Org. Chem.* **23**, 1355 (1958).
105. U.S. Pat. 4,777,265 (Oct. 11, 1988), F. Merger and co-workers (to BASF AG).
106. U.S. Pat. 2,484,487 (Oct. 11, 1949), J. R. Caldwell (to Eastman Kodak Co.).
107. U.S. Pat. 3,059,024 (Oct. 16, 1962), A. Goldberg, J. Fertig, and H. Stanley (to National Starch and Chemical Corp.).
108. "Acrylic Acid" Chemical Profile, *Chemical Market Reporter* (April 1, 2002).
109. U.S. Pat. Appl. 20020165410 (November 7, 2002), H. Aichinger, G. Nestler, and P. L. Kageler (to BASF Akiengesellschaft).

110. L. S. Kirch, J. A. Kargol, J. W. Magee and W. S. Stuper, *Plant/Operations Progress* **1**(4) 270–274 (1988).
111. "Air Contaminants—Permissible Exposure Limit", *Title 29 Code of Federal Regulations Part CFR 1910.1000,* OSHA, 1989, p. 3112.
112. S. T. Cragg, in E. Bingham, B. Cohressen, and C. H. Powell, eds., *Patty's Toxicology*, 5th ed., Vol. 5, John Wiley & Sons, Inc., New York, 2001, p. 802.

WILLIAM BAUER, JR.
Rohm and Haas Company

ACRYLIC ESTER POLYMERS

1. Introduction

The usage of acrylic esters as building blocks for polymers of industrial importance began in earnest with the experimentation of Otto Rohm (1). The first recorded preparation of the basic building block for acrylic ester polymers, acrylic acid, took place in 1843; this synthesis relied on the air oxidation of acrolein (2,3). The first acrylic acid derivatives to be made were methyl acrylate and ethyl acrylate. Although these two monomers were synthesized in 1873, their utility in the polymer area was not discovered until 1880 when Kahlbaum polymerized methyl acrylate and tested its thermal stability. To his surprise, the polymerized methyl acrylate did not depolymerize at temperatures up to 320°C (4). Despite this finding of incredibly high thermal stability, the industrial production of acrylic ester polymers did not take place for almost another 50 years.

The commercial discovery of acrylic ester polymers took place while Otto Rohm was conducting his doctoral research in 1901. Rohm obtained a U.S. patent in 1912 covering the vulcanization of acrylates with sulfur (5). Commercial production of acrylic ester polymers by the Rohm and Haas Co. of Darmstadt, Germany, commenced in 1927 (6).

2. Physical Properties

The structure of the acrylic ester monomers is represented by the following:

$$\begin{matrix} H & & H \\ & C{=}C & \\ H & & COOR \end{matrix}$$

The R ester group dominates the properties of the polymers formed. This R side-chain group conveys such a wide range of properties that acrylic ester polymers are used in applications varying from paints to adhesives and concrete modifiers and thickeners. The glass-transition range for a polymer describes the temperature range below which segmental pinning takes place and the polymer takes on a stiff, rigid, inflexible nature. This range can vary widely among the acrylic ester polymers from −54°C for butyl acrylate ($R = C_4H_9$) to 103°C for acrylic acid (R = H). Film properties are dramatically influenced by this changing of the polymer flexibility.

When copolymerized, the actylic ester monomers typically randomly incorporate themselves into the polymer chains according to the percentage concentration of each monomer in the reactor initial charge. Alternatively, acrylic ester monomers can be copolymerized with styrene, methacrylic ester monomers, acrylonitrile, and vinyl acetate to produce commercially significant polymers.

Acrylic ester monomers are typically synthesized from the combination of acrylic acid and an alcohol. The properties of the polymers they form are dominated by the nature of the ester side chain as well as the molecular weight of the product. Acrylic ester polymers are similar to others in that they show an improvement in properties as a function of molecular weight until a certain threshold beyond which no further improvement is observed. This threshold is reached at a molecular weight value of 100,000–200,000 for acrylic polymers.

2.1. Glass-Transition Temperature. The glass-transition temperature (T_g) (qv) describes the approximate temperature below which segmental rigidity (ie, loss of rotational and translational motion) sets in. Although a single value is often cited, in reality a polymer film undergoes the transition over a range of temperatures. The reason for this range of temperatures for the glass transition is that segmental mobility is a function of both the experimental method used [dynamic mechanical analysis (dma) vs differential scanning calorimetry (dac)] as well as the experimental conditions. Factors such as hydroplasticization in varying degrees of humidity can skew T_g results. Most polymers experience an increase in the specific volume, coefficient of expansion, compressibility, specific heat, and refractive index. The T_g is typically measured as the midpoint of the range over which the discontinuity of these properties takes place. Care should be taken when analyzing T_g data, however, as some experimenters cite the onset of the discontinuity as the T_g value.

The rigidity upon cooling below T_g is manifested as an embrittlement of the polymer to the point where films are glass-like and incapable of handling significant mechanical stress without cracking. If, on the other hand, one raises the temperature to which a film is exposed above the glass-transition range, the polymer film becomes stretchable, soft, and elastic. For amorphous acrylic polymers, many physical properties show dramatic changes after passing through the glass-transition temperature range. Among these physical properties are diffusion chemical reactivity, mechanical and dielectric relaxation, viscous flow, load-bearing capacity, hardness, tack, heat capacity, refractive index, thermal expansivity, creep, and crystallization.

The most common thermal analyses used to determine the glass-transition temperature are dma and dsc. More information on these techniques and how to interpret the results are contained in References 7–9. The T_g values for the most Common homopolymers of acrylic esters are listed in Table 1.

The most common way of tailoring acrylic ester polymer properties is to copolymerize two or more monomers. In this fashion, the balance of hard (high T_g) and soft (low T_g) monomers used to make up the overall composition will

Table 1. **Physical Properties of Acrylic Polymers**

Polymer	Monomer molecular formula	CAS registry number	T_g, °C[a]	Density, g/cm³ [b]	Refractive index, n_D
methyl acrylate	$C_4H_6O_2$	[9003-21-8]	6	1.22	1.479
ethyl acrylate	$C_5H_8O_2$	[9003-32-1]	−24	1.12	1.464
propyl acrylate	$C_6H_{10}O_2$	[24979-82-6]	−45		
isopropyl acrylate	$C_6H_{10}O_2$	[26124-32-3]	−3	1.08	
n-butyl acrylate	$C_7H_{12}O_2$	[9003-49-0]	−50	1.08	1.474
sec-butyl acrylate	$C_7H_{12}O_2$	[30347-35-4]	−20		
isobutyl acrylate	$C_7H_{12}O_2$	[26335-74-0]	−43		
tert-butyl acrylate	$C_7H_{12}O_2$	[25232-27-3]	43		
hexyl acrylate	$C_9H_{16}O_2$	[27103-47-5]	−57		
heptyl acrylate	$C_{10}H_{18}O_2$	[29500-72-9]	−60		
2-heptyl acrylate	$C_{10}H_{18}O_2$	[61634-83-1]	−38		
2-ethylhexyl acrylate	$C_{11}H_{20}O_2$	[9003-77-4]	−65		
2-ethylbutyl acrylate	$C_9H_{16}O_2$	[39979-32-3]	−50		
dodecyl acrylate	$C_{15}H_{28}O_2$	[26246-92-4]	−30		
hexadecyl acrylate	$C_{19}H_{36}O_2$	[25986-78-1]	35		
2-ethoxyethyl acrylate	$C_7H_{12}O_3$	[26677-77-0]	−50		
isobornyl acrylate	$C_{13}H_{20}O_2$	[30323-87-6]	94		
cyclohexyl acrylate	$C_9H_{14}O_2$	[27458-65-7]	16		

[a] Refs. 7 and 10.
[b] Ref. 11

determine the overall hardness and softness of the polymer film. An estimate of the T_g, and therefore the film hardness, can be calculated using the Fox equation (eq. (1)) (12):

$$1/T_g = \Sigma W(i)/T_g(i) \qquad (1)$$

The factor W in this equation refers to the weight, or percent composition, of a given monomer with a given T_g value for the homopolymer.

As can be seen in Table 1, the most common acrylic ester polymers have low T_g values and, therefore, soften films in which they are copolymerized with other vinylic monomers. This effect results in an internal plasticization of the polymer. That is, the plasticization effect from acrylic esters, unlike plasticizer additives which are not covalently bound, will not be removed via volatilization or extraction.

Nondestructive techniques such as torsional modulus analysis can provide a great deal of information on the mechanical properties of viscoelastic materials (8,13–25). For this type of analysis, a higher modulus value is measured for those polymers which are stiffer, harder, or have a higher degree of cross-linking. The regions of elastic behavior are shown in Figure 1 with curve A representing a soft polymer and curve B a harder polymer. A copolymir with a composition

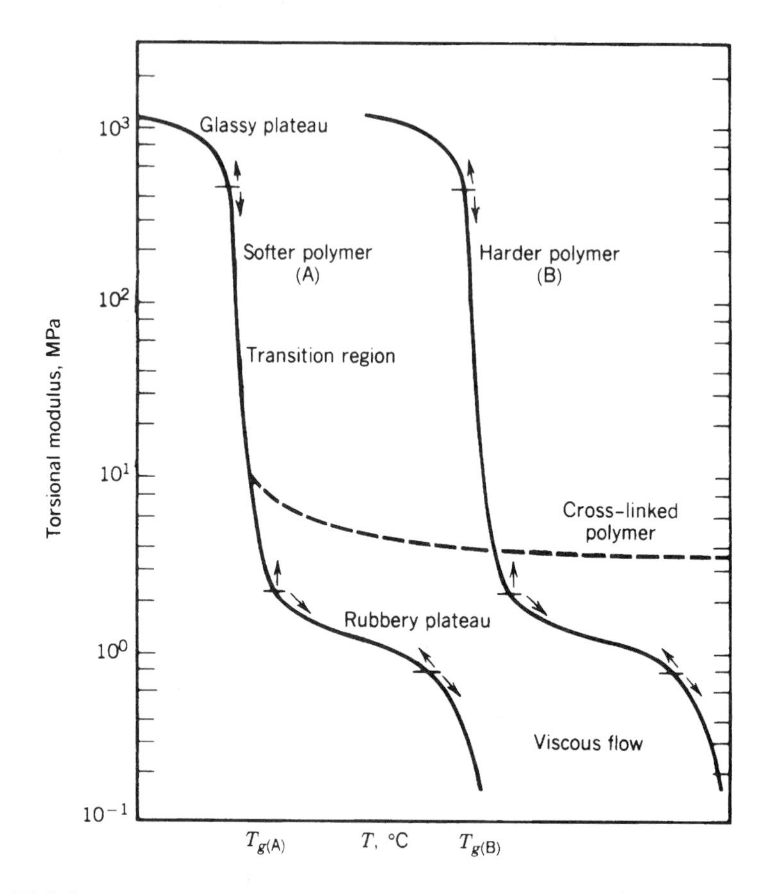

Fig. 1. Modulus-temperature curve of amorphous and cross-linked acrylic polymers. To convert MPa to kg/cm^2, multiply by 10.

between these two homopolymers would fall between the two depicted curves, with the relative distance from each curve determined by the similarity of the copolymer composition to one homopolymer or the other (26–28).

Acrylic ester polymers are susceptible to the covalent bonding of two or more polymer chains to form a cross-link (11,29–38). The above-described thermal analysis techniques are capable of distinguishing not only T_g but also varying degrees of cross-linking between polymers. A higher degree of cross-linking results in an elevation and extension of the rubbery plateau region. After a certain level of cross-linking is obtained, the segmental mobility of the polymer chains is impeded (23,25,28). This loss of mobility is measured as an increase in the T_g of the polymer. Further details on cross-linking within and between polymer chains can be found in References 11 and 29–38.

2.2. Molecular Weight. The properties of acrylic ester polymers (and most other types of polymers for that matter) improve as molecular weight increases. Beyond a certain level (100,000–200,000 for acrylic ester polymers) this improvement in polymer properties reaches a plateau. The glass-transition temperature can be described by the equation:

$$T_g = T_{gi} - k/M_n$$

where T_{gi} is the glass-transition temperature for a polymer of infinite molecular weight and M_n is the number average molecular weight. Typical values of k fall in the range of 2×10^5 (39). Reference 40 summarizes the effect of molecular weight on polymer properties.

2.3. Mechanical and Thermal Properties. The mechanical and thermal properties of a polymer are strongly dependent on the nature of the ester side-chain groups of its composite monomers. With H as a side chain, poly(acrylic acid) is a brittle material at room temperature, which is capable of absorbing large quantities of water. The first member of the acrylic ester family, poly (methyl acrylate), is a tough, rubbery, tack-free material at room temperature. The next higher chain length material, poly(ethyl acrylate), is softer, more rubbery, and more extensible. Poly(butyl acrylate) has considerable tack at room temperature and is capable of serving as an adhesive material. Information on these homopolymers is summarized in Table 2 (41). Softness of these polymers increases with increasing chain length until one reaches poly(*n*-nonyl acrylate). Beginning with this chain length, the side chains start to crystallize, which leads to a stiffening of the polymer. This stiffening translates into an embrittlement of the polymer (42); poly(*n*-hexadecyl acrylate), for example, is a hard, waxy material at room temperature.

Acrylic ester polymers are quite resilient to extreme conditions. This resilience gives finished products the durability that has earned acrylic polymers their reputation for value over time. In contrast to polymers of methacrylic esters, acrylic esters are stable when heated to high temperatures. Poly(methyl acrylate) can withstand exposure to 292–399°C in vacuo without generating significant quantities of monomer (43,44). Acrylic ester polymers are also resistant to oxidation. Hydroperoxides can be formed from polymer radicals and oxygen under forcing conditions (45–47), but by and large this is a minor concern.

2.4. Solublilty. Like most other properties, the side chain of acrylic ester polymers determines their solubility in organic solvents. Shorter side-chain polymers are relatively polar and will dissolve in polar solvents such as ether alcohols, ketones, and esters. With longer side-chain polymers, the solubility of a polymer shifts to the more hydrophobic solvents such as aromatic or aliphatic hydrocarbons. If a polymer is soluble in a given solvent, typically it is soluble in all proportions. Film formation occurs with the evaporation of the solvent, increase in solution viscosity, and the entanglement of the polymer chains.

Table 2. **Mechanical Properties of Acrylic Polymers**

Polyacrylate	Elongation, %	Tensile strength, kPa[a]
methyl	750	6895
ethyl	1800	228
butyl	2000	21

[a] To convert kPa to psi, multiply by 0.145.

Table 3. **Solubility Parameters of Acrylic Homopolymers Calculated by Small's Method**[a]

Polymer	$(J/cm^3)^{1/2}$ [b]
methyl acrylate	4.7
ethyl acrylate	4.5
n-butyl acrylate	4.3

[a] Refs. 23 and 53.
[b] To convert $(J/cm^3)^{1/2}$ to $(cal/cm^3)^{1/2}$, divide by 2.05.

Phase separation and precipitation are not usually observed for solution polymers.

Solubility is determined by the free energy equation (the Flory-Huggins equation) governing the mutual miscibility of polymers (eq. (2):

$$\Delta G_{\mathrm{Mix}} = kT\ (N_1 \ln \nu_1 + N_2 \ln \nu_2 + \chi_1 N_1 \nu_2) \qquad (2)$$

where k is the Boltzmann's constant, T the temperature, N_1 the number of solvent molecules, N_2 the number of polymer molecules, υ_1 the volume fraction of the solvent, υ_2 the volume fraction of the polymer, and χ_1 the Flory-Huggins interaction parameter.

With this equation, polymer dissolution takes place when the free energy of mixing is negative. A polymer in solution always has a much higher entropy level than undissolved polymer since it is free to move to a far greater extent. This means the change in entropy term will always have a large positive value. Therefore, the factor which determines whether or not a polymer will dissolve in a particular solvent is the heat term. If the difference in the solubility parameters for two substances is small, dissolution will occur since the heat of mixing will be small and the entropy difference will be large (this translates into a negative overall energy of mixing). A polymer will dissolve in a particular solvent if the solubility parameters and the polarities for the polymer and the solvent are comparable (38,48–53). Some relevant solubility parameters are given in Table 3.

Polymer solution viscosity is a function of the polymer molecular weight, concentration in solvent, temperature, polymer composition, and solvent composition (9,54–56).

3. Chemical Properties

Acrylic polymers and copolymers are highly resistant to hydrolysis. This property differentiates acrylic polymers from poly(vinyl acetate) and vinyl acetate copolymers. When exposed to highly extremely acidic or alkaline environments, acrylic ester polymers can hydrolyze to poly(acrylic acid) and the corresponding alcohol. Resistance to hydrolysis decreases in the order butyl acrylate > ethyl ethyl acrylate > methyl acrylate. Although it is the least hydrolytically stable, methyl acrylate is still far more resistant to hydrolysis than vinyl acetate (57,58).

Ultraviolet radiation is the other main stress encountered by polymers in the coatings arena. One hundred percent acrylic polymers are highly resistant to photodegradation because they are transparent to the vast majority of the solar spectrum (59). When uv-absorbing monomers, such as styrene, are incorporated into the polymer backbone, the uv-resistance of the resulting polymer decreases dramatically and a more rapid deterioration in polymer/coating properties is observed. On the other hand, a noncovalently bound uv absorber, such as hydroxybenzophenone further improves the uv stability of 100% acrylic polymers (59).

Higher energy radiation such as from gamma ray or electron beam sources results in the scission of both main and side chains (60). The ratio of backbone to side-chain scission is determined by the nature of the side chain (61,62).

4. Acrylic Ester Monomers

A wide variety of properties are encountered in the acrylic monomers area. This range of properties is made accessible by the variability of the side chain for acrylic monomers. Some of the key physical properties of the most commercially important monomers are included in Table 4. A more complete listing of both monomers and their properties is found in the article ACRYLIC ACID AND DERIVATIVES.

The two most common methods for production of acrylic ester monomers are (*1*) the semicatalytic Reppe process which utilizes a highly toxic nickel carbonyl catalyst and (*2*) the propylene oxidation process which primarily employs molybdenum catalyst. Because of its decreased cost and increased level of safety, the propylene oxidation process accounts for most of the acrylic ester production currently. In this process, acrolein is formed by the catalytic oxidation of propylene vapor at high temperature in the presence of steam. The acrolein intermediate is then oxidized to acrylic acid [79-10-7].

$$CH_2{=}CHCH_3 + O_2 \xrightarrow{\text{catalyst}} CH_2{=}CHCHO + H_2O$$

$$2\ CH_2{=}CHCHO + O_2 \xrightarrow{\text{catalyst}} 2\ CH_2{=}CHCOOH$$

Once the acrylic acid has been formed, the various acrylic ester monomers are synthesized by esterification of acrylic acid with the appropriate alcohol (63–66).

These monomers are then prevented from highly exothermic and hazardous autopolymerization processes during shipping and storage by the addition of a chemical inhibitor. The most common inhibitors currently used are hydroquinone, the methyl ether of hydroquinone (MEHQ) [150-76-5], and the newest member of the inhibitor family, 4-hydroxy TEMPO. 4-Hydroxy TEMPO, unlike the quinone inhibitors, does not require the presence of oxygen in order to be effective. Chemical inhibitors are only added at the <100 ppm level and are not typically removed prior to their commercial use. Finally, copper and its alloys can also function as inhibitors and should, therefore, be avoided when constructing a reactor for purposes of producing acrylic ester (co)polymers (67). With no inhibitor added, the monomers must be stored at temperatures below 10°C for

Table 4. **Physical Properties of Acrylic Monomers**

Acrylate	CAS registry number	Molecular weight	Bp, °C[a]	d^{25}, g/cm^3	Flash point, °C[b]	Water solubility, g/100 g H_2O	Heat of evaporation, J/g[c]	Specific heat, J/g·K[c]
methyl	[96-33-3]	86	79–81	0.950	10	5	385	2.01
ethyl	[140-88-5]	100	99–100	0.917	10	1.5	347	1.97
n-butyl	[141-32-2]	128	144–149	0.894	39	0.2	192	1.92
isobutyl	[106-63-8]	128	61–63[d]	0.884	42	0.2	297	1.92
t-butyl	[1663-39-4]	128	120	0.879	19	0.2		
2-ethylhexyl	[103-11-7]	184	214–220	0.880	90[e]	0.01	255	1.92

[a] At 101.3 kPa unless otherwise noted.
[b] Tag open cup unless otherwise noted.
[c] To convert J to cal, divide by 4.184.
[d] At 6.7 kPa = 50 mm Hg.
[e] Cleveland open cup.

Table 5. **Toxicities of Acrylic Monomers**

Monomer	Acute oral LD_{50} (rats), mg/kg	Acute Percutaneous LD_{50} (rabbits), mg/kg	Inhalation	
			LC_{50} (rats), mg/L	TLV, ppm
methyl acrylate	300	1235	3.8	10
ethyl acrylate	760	1800	7.4	25
butyl acrylate	3730	3000	5.3	

no longer than a few weeks. Failure to exercise these precautions can result in violent, uncontrolled, and potentially deadly polymerizations.

Common acrylic ester monomers are combustible liquids. Commercial acrylic monomers are shipped with DOT (Department of Transportation) red labels in bulk quantities, tank cars, or tank trucks. Mild steel is the usual material of choice for the construction of bulk storage facilities for acrylic monomers; moisture is excluded to avoid rusting of the storage tanks and contamination of the monomers.

A variety of methods are available for determining the purity of monomers by the measurement of their saponification equivalent and bromine number, specific gravity, refractive index, and color (68–70). Minor components are determined by iodimetry or colorimetry for hydroquinone or MEHQ, Karl-Fisher method for water content, and turbidimetry for measuring trace levels of polymer. Gas-liquid chromatography is useful in both the general measurement of monomer purity as well as the identification of minor species within a monomer solution.

Although toxicities for acrylic ester monomers range from slight to moderate, they can be handled safely and without difficulty by trained, personnel, provided that the proper safety instructions are followed (67,71). Table 5 contains animal toxicity data for common acrylic ester monomers under acute toxicity conditions.

Because of their higher vapor pressures, liquid methyl and ethyl acrylate are the two most potentially harmful acrylic ester monomers. Threshold limit values (TLV) for long-term low level exposures to these monomers in industrial situations have been established by OSHA (Table 5). Local regulations and classifications sometimes apply, however, to these monomers. Ethyl acrylate, for example, has been labeled a known carcinogen by the State of California (71).

5. Polymerization

5.1. Radical Polymerization. Free-radical initiators such as azo compounds, peroxides, or hydroperoxides are commonly used to initiate the polymerization of acrylic ester monomers. Photochemical (72–74) and radiation-initiated (75) polymerization are also possible. At constant temperature, the initial rate of polymerization is first order in monomer and one-half order in initiator. Rate data for the homopolymerization of several common acrylic ester monomers initiated by 2,2′-azobisisobutryonitrile (AIBN) [78-67-1] have been determined

Table 6. **Polymerization Data for Acrylic Ester Monomers in Solution**[a]

Acrylate	Concentration, solvent	k_{sp}, L/mol·h[b]	Heat, kJ/mol[c]	Shrinkage, vol%
methyl	3 M, methyl propionate	250	78.7	24.8
ethyl	3 M, benzene	313	77.8	20.6
butyl	1.5 M, toluene	324	77.4	15.7

[a] Ref. 76.
[b] At 44.1°C.
[c] To convert kJ to kcal, divide by 4.184.

and are contained in Table 6. Also included in this table are heats of polymerization and volume shrinkage data (76).

The polymerization of both acrylic and methacrylic ester monomers is accompanied by the release of a large quantity of heat as well as a substantial decrease in sample volume. Commercial processes must account for both these phenomena. Excess heat must be removed from industrial reactors by the use of high surface area heat exchangers. As for the shrinkage issue, the percent shrinkage encountered upon polymerization of the monomer is, in general, inversely proportional to the length of the monomer side chain. Mole for mole, the shrinkage amount is relatively constant (77).

The free-radical polymerization of acrylic monomers takes place through the classical stepwise chain-growth mechanism, which is described as the head-to-tail addition of individual monomer units through attack df the monomer double bond and formation of a single bond between the newly incorporated monomer units.

$$R'-CH_2\underset{\displaystyle COOR}{CH}\cdot + CH_2=\underset{\displaystyle COOR}{CH} \longrightarrow R'-CH_2\underset{\displaystyle COOR}{CH}-CH_2-\underset{\displaystyle COOR}{CH}\cdot$$

This stepwise growth continues until either termination or chain transfer of the radical chain end takes place. Termination can occur by combination or disproportionation, depending on the conditions of the polymerization (78,79).

The addition step typically takes place as a head-to-tail process although head-to-head addition has been observed as well (80). Oxygen has a strong inhibitory effect on the rate of polymerization of acrylic ester polymers. Oxygen is, therefore, excluded from commercial reactors primarily through the use of positive nitrogen flow. The nature of the oxygen inhibition is known: an alternating copolymer can be formed between oxygen and acrylic ester monomers (81,82).

$$R'-CH_2\underset{\displaystyle COOR}{CH}\cdot + O_2 \xrightarrow{\text{fast}} R'-CH_2\underset{\displaystyle COOR}{CH}OO-CH_2\underset{\displaystyle COOR}{CH}OO\cdot$$

The oxygen chain end is relatively unreactive when compared to the acrylic chain end and reduces the overall rate of polymerization. Additionally, the peroxy radical undergoes a faster rate of termination than the standard acrylic-based radical:

$$\underset{\displaystyle\text{COOR}}{\text{R}'\!-\!\text{CH}_2\text{CH}}\text{OO}-\underset{\displaystyle\text{COOR}}{\text{CH}_2\text{CH}}\text{OO}\cdot + \underset{\displaystyle\text{COOR}}{\text{CH}_2\!=\!\text{CH}} \xrightarrow{\text{slow}} \underset{\displaystyle\text{COOR}}{\text{R}'\!-\!\text{CH}_2\text{CH}}\text{OO}-\underset{\displaystyle\text{COOR}}{\text{CH}_2\text{CH}\cdot}$$

One will observe a drop in overall reaction rate, a change in polymer composition and properties, as well as a decrease in polymer molecular weight if oxygen is not excluded from a reactor when polymerizing acrylic ester monomers (83).

The wide variety of acrylic ester monomers dictates that a wide variety of homopolymers with radically different properties are accessible. An even wider variety of polymers can be formed through the copolymerization of two or more acrylic ester monomers (84,85).

Acrylic ester monomers are, in general, readily copolymerized with other acrylic and vinylic monomers. Table 7 presents data for the free-radical copolymerization of a variety of monomers 1:1 with acrylic ester monomers. These numbers are calculated through the use of reactivity ratios:

$$r_1 = k_{11}/k_{12}$$
$$r_2 = k_{22}/k_{21}$$

For a binary copolymer, the smaller reactivity ratio is divided by the larger r value and multiplied by 100. Values greater than 25 indicate that copolymerization proceeds smoothly; low values for the ease of copolymerization can be helped through the adjustment of comonomer composition as well as the monomer addition method (86).

A growing chain with monomer 1 as the chain-end radical has a rate constant for self-addition of k_{11}; the rate for addition of monomer 2 is k_{12}. The self-addition rate for a terminal monomer 2 radical is given as k_{22}; the rate for addition of monomer 1 is k_{21}. The reactivity ratios can also be calculated from the Price-Alfrey measures (87) of resonance stabilization (Q) and polarity (e) which are shown for common acrylic esters in Table 8. NMR can also be used to determine the composition distribution characteristics of acrylic copolymers (88,89).

Table 7. **Relative Ease of Copolymer Formation for 1:1 Ratios of Acrylic and Other Monomers, $\frac{r(\text{smaller})}{r(\text{larger})} \times 100$**

		Monomer 1		
Monomer 2	CAS registry number	Methyl acrylate	Ethyl acrylate	Butyl acrylate
acrylonitrile	[107-13-1]	53	46	74
butadiene	[106-99-0]	66	4.7	8.1
methyl methacrylate	[80-62-6]	50.3	30.6	14.6
styrene	[100-42-5]	21	16	26
vinyl chloride	[75-01-4]	2.7	2.1	1.6
vinylidene chloride	[75-35-4]	100	52	55
vinyl acetate	[108-05-4]	1.1	0.7	0.6

Table 8. **Q and e Values for Acrylic Monomers**[a]

Monomer	*Q*	*e*
methyl acrylate	0.44	+0.60
ethyl acrylate	0.41	+0.46
butyl acrylate	0.30	+0.74
isobutyl acrylate	0.41	+0.34
2-ethylhexyl acrylate	0.14	+0.90

[a] Ref. 88.

In addition to the standard side-chain variation discussed above, special functionality can be added to acrylic ester monomers by use of the appropriate functional alcohol. Through the use of small levels of functional monomers, one can allow an acrylic ester polymer to react with metalions, cross-linkers, or other types of resins. Table 9 contains information on some of the more common functional monomers.

Table 9. **Functional Monomers for Copolymerization with Acrylic Monomers**

Monomer	Structure	CAS registry number	Molecular formula
	Carboxyl		
methacrylic acid	$CH_2{=}C(CH_3)COOH$	[79-41-4]	$C_4H_6O_2$
acrylic acid	$CH_2{=}CHCOOH$	[79-10-7]	$C_3H_4O_2$
itaconic acid	$CH_2{=}C(CH_2COOH)COOH$	[97-65-4]	$C_5H_6O_4$
	Amino		
t-butylaminoethyl methacrylate	$CH_2{=}C(CH_3)COO(CH_2)_2NHC(CH_3)_3$	[24171-27-5]	$C_{10}H_{19}NO_2$
dimethylaminoethyl methacrylate	$CH_2{=}C(CH_3)COO(CH_2)_2N(CH_3)_2$	[2867-47-2]	$C_8H_{15}NO_2$
	Hydroxyl		
2-hydroxyethyl methacrylate	$CH_2{=}C(CH_3)COOCH_2CH_2OH$	[868-77-9]	$C_6H_{10}O_3$
2-hydroxyethyl acrylate	$CH_2{=}CHCOOCH_2CH_2OH$	[818-61-1]	$C_5H_8O_3$
	N-Hydroxymethyl		
N-hydroxymethyl acrylamide	$CH_2{=}CHCONHCH_2OH$	[924-42-5]	$C_4H_7NO_2$
N-hydroxymethyl methacrylamide	$CH_2{=}C(CH_3)CONHCH_2OH$	[923-02-04]	$C_5H_9NO_2$
	Oxirane		
glycidyl methacrylate	$CH_2{=}C(CH_3)COOCH_2CH{-}CH_2$ (epoxide, O bridging CH and CH_2)	[106-91-2]	$C_7H_{10}O_3$
	Multifunctional		
1,4-butylene dimethacrylate	$(CH_2{=}C(CH_3)COOCH_2)_2$	[2082-81-7]	$C_{12}H_{18}O_4$

5.2. Bulk Polymerization. Bulk polymerizations of acrylic ester monomers are characterized by the rapid formation of an insoluble network of polymers at low conversion with a concomitant rapid increase in reaction viscosity (90,91). These properties are thought to come from the chain transfer of the active radical via hydrogen abstraction from the polymer backbone. When two of these backbone radical sites propagate toward one another and terminate, a cross-link is formed (91).

5.3. Solution Polymerization. Of far greater commercial value than that of simple bulk polymerizations, solution polymerizations employ a co-solvent to aid in minimizing reaction viscosity as well as controlling polymer molecular weight and architecture. Lower polyacrylates are, in general, soluble in aromatic hydrocarbons, esters, ketones, and chlorohydrocarbons. Solubilities in aliphatic hydrocarbons, ethers, and alcohols are somewhat lower. As one moves to longer alcohol side-chain lengths, acrylics become insoluble in oxygenated organic solvents and soluble in aliphatic and aromatic hydrocarbons and chlorohydrocarbons. Solvent choices for acrylic solution polymerizations are made on the basis of cost, toxicity, flammability, volatility, and chain-transfer activity. The chain-transfer constants (C_s) for a variety of solvents in the solution polymerization of poly(ethyl acrylate) are shown in Table 10.

Initiators serve the dual role of beginning the polymerization of an individual chain as well as controlling the molecular weight distribution of a polymer sample. Initiators are chosen based on their solubility, thermal stability (rate of decomposition), and the end use for the polymer. Additionally, initiators can be used to control polymer architecture by cross-linking control; this property also allows initiators to serve a role in the regulation of molecular weight distribution. Levels of usage vary from hundredths of a percent to several percent by weight on the polymer formed. The types of initiators most commonly employed in solution polymerizations are organic peroxides, hydroperoxides, and azo compounds.

Molecular weight control can also be achieved through the use of a chain-transfer agent. The most commonly used species in this class are chlorinated aliphatic compounds and thiols (94). The chain-transfer constants (C_s at 60°C) for some of these compounds in the formation of poly(methyl acrylate) are as follows (87): carbon tetrabromide, 0.41: ethanethiol, 1.57; and butanethiol, 1.69.

Because of the volatile nature of the monomers used and high temperatures often employed, solution polymerizations are typically performed in reactors

Table 10. **Chain-Transfer Constants to Common Solvents for Poly(ethyl acrylate)**[a]

Solvent	$C_s \times 10^5$
benzene	5.2
toluene	26.0
isopropyl alcohol	260
isobutyl alcohol	46.5
chloroform	14.9
carbon tetrachloride	15.5

[a] Refs. 79, 92, and 93.

which can withstand pressures of at least 446 kPa (65 psi). Standard materials of construction include stainless steel (which may be glass-lined) or nickel. Anchor-type agitators are used for solution polymerizations with viscosities up to 1.0 Pa·s (1000 cP), but when viscosity levels move above this range, a slow ribbon-type agitator is used to sweep material away from the reactor walls. Improper agitation can result in the severe fouling of a reactor. Most industrial reactors are jacketed for steam heating and/or water cooling of a batch and contain a rupture disk to relieve pressure buildup. Additionally, there are numerous inlets in a typical industrial reactor as well as a thermocouple for monitoring temperatute. A valve is placed in the bottom of the reactor to release polymerized material to storage containers.

Cooling within a reactor is typically provided by a reflux condenser. Since polymerization is a highly exothermic process, temperature control is a safety concern as well as a product integrity issue. Temperature control is primarily obtained through the gradual addition of monomers into the reactor by gravity from storage containers close to the reactor. In this manner, the rate of monomer addition and reaction can be matched to the cooling capacity of the teactor so that temperatures remain relatively constant throughout the polymerization. If these measures fail to control the temperature of a particular batch, a chemical inhibitor, such as a hydroquinone, can be added to retard the rate of polymerization.

Oxygen can serve as an inhibitor of polymerization. Reactors typically maintain a blanket of nitrogen over the entire reactor kettle. In polymerizations with temperatures below reflux, nitrogen is used to purge the reaction solution; a nitrogen blanket is then placed over the reactor prior to the addition of the initiator. Total cycle times for solution polymerizations run in the range of 24 h (95).

A typical solution polymerization recipe is shown below:

Composition	Parts
Reactor charge	
Ethyl acetate	61.4
Benzoyl peroxide	0.1
Monomer charge	
Ethyl acrylate	36.5
Acrylic acid	2.0

This copolymer has an overall composition of 94.8% ethyl acrylate/5.2% acrylic acid with the monomer charged at a level of 39 wt% in a solution of ethyl acetate. Initially, the solvent and initiator, benzoyl peroxide in this case, are added to the reactor and heated to reflux (80°C). Forty percent of the monomer mixture is added to the reactor in one charge. Then, four equal aliquots of monomer are added 24,50, 79, and 110 min after the initial charge. Reflux is maintained within the reactor overnight to ensure complete reaction; the product is then cooled and packaged the next morning (96).

Storage and handling equipment are typically made from steel. In order to prevent corrosion and the transfer of rust to product, moisture is typically excluded from solution polymer handling and storage systems (97). Because of the temperature-sensitive nature of the viscosity of solution polymers, the temperature of the storage tanks and tranfer lines is regulated either through prudent location of these facilities or through the use of insulation, heating, and cooling equipment.

5.4. Emulsion Polymerization. Emulsion polymerization is the most industrially important method of polymerizing acrylic ester monomers (98,99). The principal ingredients within this type of polymerization are water, monomer, surfactant, and water-soluble initiator. Products generated by emulsion polymerization find usage as coatings or binders in paints, paper, adhesives, textile, floor care, and leather goods markets. Because of their film-forming properties at room temperature, most commercial acrylic ester Polymers are copolymers of ethyl acrylate and butyl acrylate with methyl methacrylate.

Lower acrylates are capable of polymerizing in water in the presence of an emulsifier and a water-soluble initiator. The polymeric product is typically a milky-white dispersion of polymer in water at a polymer solids content of 30–60%. Particle sizes for these latices fall in the range of 0.1–1.0 μm. Because of the compartmentalized nature of the process (99), high molecular weights are obtained with most emulsion polymerizations without the resulting viscosity build encountered with solution polymerizations. Additionally, the use of water as a dispersion medium provides attractive safety, environmental, and heat removal benefits when compared to other methods of polymerizing acrylic ester monomers. The emulsion polymerization of the higher (relatively water insoluble) acrylates can even be accomplished now through the use of a patented method for catalytically transferring monomer from droplets to the growing polymer particles (100).

The types of surfactants used in an emulsion polymerization span the entire range of anionic, cationic, and nonionic species. The most commonly used soaps are alkyl sulfates such as sodium lauryl sulfate, alkylaryl sulfates such as sodium dodecyl benzene sulfonate, and alkyl or aryl polyoxyethylene nonionic surfactants (87,101–104). Product stability and particle size control are the driving forces which determine the types of surfactants employed; mixtures of nonionic and anionic surfactants are commonly used to achieve these goals (105–108).

Water-soluble peroxides, such as sodium or ammonium persulfate, are commonly used in the industrial arena. The thermal dissociation of this initiator (109) results in the formation of sulfate radicals which initiate polymer chains in the aqueous phase. It is possible to use other oxidants, such as hydrogen peroxide [7722-84-1] or persulfates in the presence of reducing agents and/or polyvalent metal ions (87). In this manner, a redox initiator system is formed which allows the experimenter to initiate polymer chains over a much broader range of temperatures (25–90°C) than simple thermal initiation (75–90°C) (110). The primary disadvantage of this initiation method is that greater level of salt impurities are introduced to the reactor which could, perhaps, adversely influence final polymer properties such as stability.

Emulsion polymerization batches on the industrial scale are typically run in either stainless steel or glass-lined steel reactors which can safely handle internal pressures of 446 kPa (65 psi). Agitation within the reactors is controlled by use of a variable speed stirring shaft coupled at times with a baffling system within the reactor to improve mixing. Care must be taken to avoid excessive mixing forces being placed on the latex as coagulum will form under extreme conditions. Temperature control of batches is maintained through the use of either steam or cold waterjacketing. Multiple feed lines are necessary to provide for the addition of multiple streams of reactants such as initiators, monomer emulsions, inhibitors if necessary, and cooling water. Monitoring equipment for batches typically consists of thermocouples, manometers, sightglasses, as well as an emergency stack with a rupture disk in case of pressure buildup within the reactor. A typical industrial emulsion polymerization plant is shown in Figure 2.

There are numerous examples of typical industrial emulsion polymerization recipes available in the open literature (111,112). A process for the synthesis of a polymer with a 50% methyl methacrylate, 49% butyl acrylate, and 1% methacrylic acid terpolymer at a solids content of 45% is described below:

Charge	Parts
Monomer emulsion charge	
Deionized water	13.65
Sodium lauryl sulfate	0.11
Methyl methacrylate	22.50
Butyl acrylate	22.05
Methacrylic acid	0.45
Initiator charge	
Ammonium persulfate	0.23
Reactor charge	
Deionized water	30.90
Sodium lauryl sulfate	0.11

The monomer emulsion is first formed in a separate agitation tank by combination of the water, soap, and monomer with a proper level of mixing. Care must be taken to avoid excessive levels of agitation in the monomer emulsion tank to avoid incorporating air into the emulsion. The reactor water is heated under a nitrogen blanket to a temperature of at least 75°C prior to the addition of the initiator. Following the addition of the initiator, the monomer emulsion is fed into the reactor over the course of approximately 2.5 h. Temperature control is maintained during this time through both control of the monomer feed rate as well as use of the reactor jacket heating/cooling system. After the monomer emulsion feed is completed, the temperature is held above 75°C for at least 30 min to reduce the level of residual standing monomer within the system. The product is then cooled, filtered, and packaged.

Once packaged, the storage of acrylic latices is a nontrivial matter; problems commonly encountered with these polymer colloids include skinning (surface film formation), sedimentation, grit formation within the latex, formation of coagulum on storage container walls, and sponging (aerogel formation). Exposure of the material to extremes in temperature is avoided through prudent location of these facilities or the use of insulation, heating, and cooling equipment.

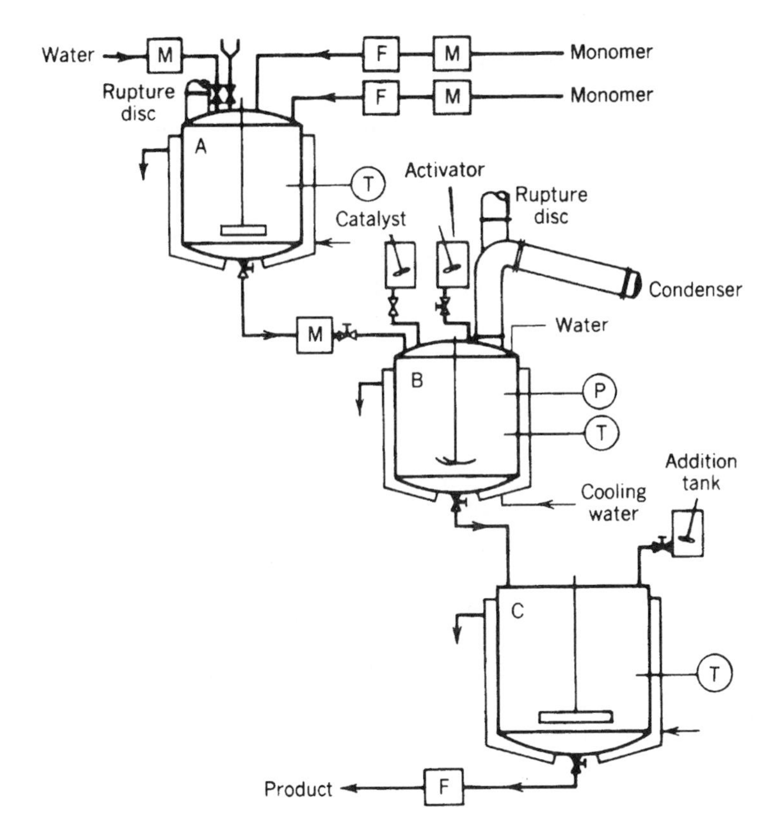

Fig. 2. Emulsion polymerization plant. A, Emulsion feed tank; B, polymerization reactor; C, drumming tank; F, filter; M, meter; P, pressure gauge; T, temperature indication.

Acrylic emulsion polymers, like many other types of polymers, are subject to bacterial attack. Proper adjustment of pH, addition of bactericides, and good housekeeping practices (95) can alleviate the problems associated with bacterial growth. Some advances in the industrial application of emulsion polymerization have been described in the open literature (113).

5.5. Suspension Polymerization. Suspension polymers of acrylic esters are industrially used as molding powders and ion-exchange resins. In this type of polymerization, monomers are dispersed as 0.1- to 5-mm droplets in water and are stabilized by protective colloids or suspending agents. In contrast to emulsion polymerization, initiation is accomplished by means of a monomer-soluble agent and occurs within the suspended monomer droplet. Water serves the same dual purpose as in emulsion (heat removal and polymer dispersion). The particle size of the final material is controlled through the control of agitation levels as well as the nature and level of the suspending agent. Once formed, the 0.1- to 5-mni polymer beads can be isolated through centrifugation or filtration.

The most commonly used suspending agents are cellulose derivatives, polyacrylate salts, starch, poly(vinyl alcohol), gelatin, talc, and clay derivatives (95).

The important function these agents must serve is to prevent the coalescence of monomer droplets during the course of the polymerization (114). Thickeners can also be added to improve suspension quality (95). Other additives such as lauryl alcohol, stearyl acid or cetyl alcohol lubricants and di- or trivinyl benzene, diallyl esters of dibasic acids, and glycol dimethacrylates cross-linkers are used to improve bead uniformity and bead performance properties.

Unlike emulsion polymerization, the initiators employed in suspension polymerization must not be water-soluble; organic peroxides and azo species are most commonly used. In similar fashion to bulk polymerization, the level of initiator used directly influences the molecular weight of the product (95,115,116). Developments in the method of suspension polymerization have been reviewed in the open literature (117,118).

5.6. Graft Copolymerization. Polymer chains can be attached to a preexisting polymer backbone of a similar or completely different composition to form what is termed a graft copolymer. Acrylic branches can be added to either synthetic (119,120) or natural (121–124) backbones. Attachment of graft polymer branches to preformed backbones is accomplished by chemical (125–127), photochemical (128,129), radiation (130), and mechanical (131) means. The presence of distinct compositions in this branched geometry often conveys properties which cannot otherwise be attained (132,133).

5.7. Living Polymerization. One of the most exciting areas currently in the radical polymerization of acrylic ester monomers is the field of living polymerization. Living polymers are defined in Reference 134 as "polymers that retain their ability to propagate for a long time and grow to a desired maximum size while their degree of termination or chain transfer is still negligible." Because of these properties, exceptional control can be exercised over the topology (ie, linear, comb), composition (ie, block, graft), and functional form (ie, telechelic, macromonomer) of these polymers (135).

Atom-transfer radical polymerization (ATRP) and nitroxide-mediated (136–138) polymerization both show promise in terms of the ability to fine tune polymer architecture using living radical methods. ATRP has been successfully used in the polymerization of methyl acrylate (139,140) as well as functional acrylates containing alcohol (141), epoxide (142), and vinyl groups (143) on the side chain. The main drawbacks to the ATRP method of creating acrylic ester homo- and copolymers are the relatively long reaction times and the high levels of metal-containing initiator required.

5.8. Radiation-Induced Polymerization. Coatings can be formed through the application of high energy radiation to either monomer or oligomer mixture. Ultraviolet curing is the most widely practiced method of radiation-based initiation (144–150); this method finds its main industrial applications in the areas of coatings, printing ink, and photoresists for computer chip manufacturing. The main disadvantage of the method is that uv radiation is incapable of penetrating highly pigmented systems.

To form a film via this method, a mixture of pigment, monomer, polymer, photoinitiator, and inhibitor are applied to a substrate and polymerized by controlled exposure to uv radiation. Polymers used as co-curing agents often have unsaturated methacrylate functionalities attached; higher order acrylates are often used as the solvent in photocure mixtures.

In order to avoid the problems associated with more highly pigmented systems, electron beam curing is employed (151). This high energy form of radiation is capable of penetrating through the entire coating regardless of the coating's pigment loading level.

5.9. Anionic Polymerization. The anionic polymerization of acrylic ester monomers, is accomplished by use of organometallic initiators in organic solvents. The main advantage to the use of anionic polymerization as opposed to other methods is its ability to generate stereoregular or block copolymers. Some examples of this type of polymerization include the anionic formation of poyl(*t*-butyl acrylate) (152–155), poly(isopropyl acrylate) (156), and poly(isobutyl acrylate) (157,158). Solvent conditions primarily determine tacticity of the resulting polymer product with nonpolar solvents generating isotactic product and polar solvents resulting in the formation of syndiotactic polymers. The strikingly different physical properties and mechanistic discussions on the formation of these two different types of polymer have been described in the polymer literature (159–162).

The initiation step for anionic polymerizations takes place via a Michael reaction:

$$R_3C^-M^+ + CH_2{=}CHCOOR \rightarrow R_3C{-}CH_2{-}\bar{C}H{-}COOR$$

A subsequent polymer growth occurs by head-to-tail addition of monomer to the growing polymer chain.

$$R_3C{-}CH_2{-}\bar{C}H{-}COOR \; + \; CH_2{=}CHCOOR \longrightarrow R_3C{-}CH_2{-}\underset{\displaystyle COOR}{CH}{-}CH_2{-}\underset{\displaystyle COOR}{\bar{C}H}$$

Because of cost constraints and toxicity issues involved with the organometallic initiators, anionic polymerization is of limited commercial significance. Both the living methods described above as well as DuPont's group-transfer polymerization method (163–167) are seen as alternative ways to achieve the same level of control over polymer architecture as that of anionic polymerization. All these methods offer the promise of narrow and controllable molecular weight distributions as well as the ability to form block copolymers through the sequential addition of monomers. Additionally, all the methods suffer from slow overall reaction rates and the difficulty of removing the specialty initiators after polymer formation has taken place.

6. Analytical Methods and Specifications

6.1. Emulsion Polymers. Current analytical methods allow for complete characterization of all crucial aspects of an acrylic latex (87). The main properties of interest are an acrylic latex's composition, percent solids content, viscosity, pH, particle size distribution (168,169), glass-transition temperature, minimum film-forming temperature (170), and surfactant type. In addition to these basic properties, the stability of a latex with respect to mechanical shear, freeze-thaw cycles, and sedimentation on standing for long periods of time are of interest in commercial products.

6.2. Solution Polymers. A solution polymer's composition, solids content, viscosity, molecular weight distribution, glass-transition temperature, and solvent are of interest. Standard methods allow for all of these properties to be readily determined (171,172).

7. Health and Safety Factors

Acrylic polymers are categorized as nontoxic and have been approved for the handling and packaging by the FDA. The main concerns with acrylic polymers deal with the levels of residual monomers and the presence of nonacrylic additives (primarily surfactants) which contribute to the overall toxicity of a material. As a result, some acrylic latex dispersions can be mild skin or eye irritants.

During the manufacture of an acrylic polymer, precautions are taken to maintain temperature control (173). In addition to these measures, polymerizations are run under conditions wherein the reactor are closed to the outside environment to prevent the release of monomer vapor into the local environment. As for final product properties, acrylic latices are classified as nonflammable substances and solution polymers are classified as flammable mixtures.

8. Uses

Because of their wide property range, clarity, and resistance to degradation by environmental forces, acrylic polymers are used in an astounding variety of applications that span the range from very soft adhesive materials to rigid non-film-forming products.

8.1. Coatings. Acrylic ester latex polymers are used widely as high quality paint binders because of their excellent durability, toughness, optical clarity, uv stability, and color retention. These properties allow acrylics to find use as binder vehicles in all types of paints (76): interior and exterior; flats, semi-gloss, and gloss; as well as primers to topcoats. Although all-acrylic compositions are most favored in exterior applications because of their excellent durability (174), other types of copolymers such as vinyl-acrylics and styrene-acrylics benefit in terms of performance properties from the acrylic portion of the composition; methods of manufacturing acrylic-based paints have been described previously in the literature (175). Acrylic emulsion polymers even find use in the protection of structural steel (176) (see COATING PROCESSES).

The industrial finishing area sees both acrylic emulsions as well as solution polymers utilized in a wide variety of applications including factory finished wood (177,178), metal furniture and containers (179), and can and coil coatings (180). In order to harden acrylic polymers for this type of demanding application, the polymers are often cross-linked with melamines, epoxies, and isocyanates. The coatings are applied via spraying, roll dipping, or curtain coating. Radiation curing using uv radiation or electron beam radiation (181–186), powder coating (187–190), electrode deposition of latices (191–193), and the use of higher solids level emulsion (194) represent newer methods for applying acrylic coatings to form industrial finishes. Excellent reviews on the use of water-based emulsions (195,196) and solution acrylics (197–199) can be found in the open literature (see COATINGS).

Hydrophobically modified acrylics are finding extensive usage as thickening agents in the paints marketplace as well as the area of industrial finishes (200). Flow and leveling improvements are observed when changing a formulation over from hydroxyethylcellulose to acrylic-based thickeners. Unlike cellulosic thickeners, the modified acrylics act through an associative thickening mechanism; they stabilize the dispersed polymer phase rather than thickening the aqueous phase of a polymer latex. Two main types of modified acrylics are of commercial value: HASE (hydrophobically modified alkali-soluble emulsions) and HEUR (hydrophobically modified ethylene oxide urethane block copolymers). These acrylics compete with hydrophobically modified hydroxyethylcellulose in the marketplace (201–203).

8.2. Textiles. Because of their durability, soft feel, and resistance to discoloration, acrylic emulsion polymers find a variety of uses in the textiles area including binders for fiberfill and nonwoven fabrics, textile bonding or laminating, flocking, backcoating and pigment printing applications. *N*-Methylolacrylamide is often used as a self-cross-linker in acrylic textile binders to improve washing and dry cleaning durability as well as overall binder strength (204).

Polyester (205–208), glass (209), and rayon (210) nonwoven and fiberfill mats have been manufactured using acrylic binders to hold the mats together. In this process, the acrylic emulsions are applied to a loose web or mat and are then heated to form a film at the fiber crossover points which maintains the structural integrity of the mat. The final products generated using this technology include quilting, clothing, disposable diapers, towels, filters, and roofing (see NONWOVEN FABRICS).

Acrylic polymers find use in applications that take advantage of their exceptional resistance to environmental assaults such as uv radiation, ozone, heat, water, dry cleaning, and aging (211). Acrylics are often used as the backing material for automotive and furniture upholstery to improve the dimensional handling properties, prevent pattern distortion, prevent unraveling, and minimize seam slippage. Strike-through problems are averted through the use of foamed or frothed acrylic coatings, which also yield a softer fabric and save on energy costs (212). Crushed acrylic latex foam are employed as backing materials for draperies. The foam protects the drapery from sun damage, mechanically stabilizes the fabric, improves drape, and gives a softer hand than conventional backing materials (213). Acrylics are also used as carpet-backings and to bond fabric-to-fabric, fabric-to-foam, and fabric-to-nonwoven materials (214).

The flocking process begins with the bonding of cut fibers to an adhesive-coated fabric to obtain a decorative and functional material (215). Acrylics can provide the softness and durability that are sought in flocked textiles; they also serve as binders for pigments in the printing of flocked fabrics (35,216,217).

The feel, soil release properties, and permanent-press behavior of a fabric can be finely tuned using acrylic as finishing polymers. Copolymers of acrylics with acrylic or methacrylic acid can be used as thickeners for textile coating formulation.

8.3. Adhesives. Acrylic emulsion polymers are used in a wide variety of adhesives. Pressure-sensitive Adhesives, which typically have T_g values less than 20°C, are the main type of acrylic adhesive. Acrylic polymers and copolymers find use as PSAs in tapes, decals, and labels. Along with their aforementioned superior chemical resistance properties, acrylics possess an excellent balance of tack, peel, and shear properties which is crucial in the adhesives market (218,219). Other types of adhesives that employ acrylics include construction formulations and film-to-film laminates.

8.4. Paper. Because of their excellent cost-performance balance, acrylic-vinyl acetate copolymer emulsion binders have been used as pigment binders for coated paper and boards. These binders provide higher brightness, opacity, coating solids, and improved adhesion versus styrene-butadiene copolymers (220,221). Acrylics also find usage as paper saturants with properties that compare favorably to natural rubber, butadiene-acrylonitrile, and butadiene-styrene (222). Finally, acrylic emulsion polymers are utilized in starch-latex-pigmented coatings (223) as well as size-press (224) and beater addition (225) applications.

8.5. Other Applications. The leather finishing area is a traditional stronghold of acrylic emulsion polymers (226). Acrylics are used throughout the entire process of pigskin leather production; the use of acrylics lends uniformity, break improvement, better durability, and surface resistance while preserving the natural appearance of the pigskin (227).

Acrylics have been used to impart impact strength and better substrate adhesion to cement (228). The ceramics industry uses both acrylic solution and emulsion polymers as temporary binders, deflocculants, and additives in ceramics bodies and glazes (229) (see CERAMICS).

Acrylics are used in the manufacture of aqueous and solvent-based caulks and sealants (230,231). Elastomeric acrylics are used in mastics to prevent uv radiation and chemical damage to the underlying polyurethane foam. Acrylics also impart hailstone resistance as well as flexibility over a broad temperature range (232). The manufacturing process for poly(vinyl chloride) uses acrylics as processing aids and plate-out scavengers in calendered and blown films. Acrylics allow for the manufacture of thick, smooth calendered vinyl sheets through modification of the melt viscosity of the vinyl sheet polymer (233). In the agricultural area, thin layers of acrylic emulsions have been applied to citrus leaves and fruit to control "Greasy Spot," a disease which causes leaf-spotting and eventually leaf loss (234). Acrylics have found a great deal of use in the floor polish area; a guide to formulating these coatings has been published (235).

Acrylic polymers have been used as alternatives to nitrile rubbers in some hydraulic and gasket applications because of their excellent heat-resistance properties (236,237). Ethylene-acrylate copolymers have been used as transmission seals, vibration dampeners, dust boots, and steering and suspension seals (238).

BIBLIOGRAPHY

"Acrylic Ester Polymers" in *ECT* 3rd ed., Vol. 1, pp. 386–408, by B. B. Kine and R. W. Novak, Rohm and Haas Co.; in *ECT* 4th ed., Vol. 1, pp. 314–343, by Ronald W. Novak,

Rohm and Haas Company. "Acrylic Ester Polymers" in *ECT* (online), posting date: December 4, 2000, by Ronald W. Novak, Rohm and Haas Company.

CITED PUBLICATIONS

1. J. Redtenbacher, *Ann.* **47**, 125 (1843).
2. Fr. Englehern, *Berichte* **13**, 433 (1880); L. Balbiano and A. Testa, *Berichte* **13**, 1984 (1880).
3. W. Caspary and B. Tollens, *Ann.* **167**, 241 (1873).
4. G. W. A. Kahlbaum, *Berichte* **13**, 2348 (1880).
5. U.S. Pat. 1,121,134 (Dec. 15, 1914), O. Rohm.
6. E. H. Riddle, *Monomeric Acrylic Esters*, Reinhold Publishing Corp., New York, 1954.
7. J. Brandrup and E. H. Immergut, *Polymer Handbook*, 2nd ed., Wiley-Interscience, New York, 1975.
8. H. B. Burrell, *Off. Dig. Fed. Soc. Paint Technol.* 131 (Feb. 1962).
9. D. W. van Krevelen, *Properties of Polymers*, Elsevier, Amsterdam, 1972.
10. J. L. Gardon, in N. M. Bikales, ed., *Encyclopedia of Polymer Science and Technology*, Vol. 3, Interscience Publishers, a division of John Wiley & Sons, Inc., New York, 1965, pp. 833–862.
11. T. G. Fox, Jr., *Bull. Am. Phys. Soc.* **1**(3), 123 (1956).
12. J. A. Shetter, *J. Polym. Sci., Part B* **1**, 209 (1963).
13. L. E. Nielsen, *Mechanical Properties of Polymers*, Van Nostrand Reinhold Co., Inc., New York, 1962, p. 122.
14. I. Williamson, *Br. Plast.* **23**, 87 (1950).
15. R. F. Clark, Jr. and R. M. Berg, *Ind. Eng. Chem.* **34**, 1218 (1942).
16. *ASTM Standards, ASTM D1043-61T*, Vol. 27, American Society for Testing and Materials, Philadelphia, Pa., 1964.
17. S. D. Gehman, D. E. Woodford, and C. S. Wikinson, Jr., *Ind. Eng. Chem.* **39**, 1108 (1947).
18. *ASTM Standards, ASTM D1053-61*, Vol. 19, American Society for Testing and Materials, Philadelphia, Pa., 1974.
19. A. C. Nuessle and B. B. Kine, *Ind. Eng. Chem.* **45**, 1287 (1953).
20. A. C. Nuessle and B. B. Kine, *Am. Dyest. Rep.* **50**(26), 13 (1961).
21. V. J. Moser, *Am. Dyest. Rep.* **53**(38), 11 (1964).
22. M. K. Lindemann, *Appl. Polym. Symp.* **10**, 73 (1969).
23. W. H. Brendley, Jr., *Paint Varn. Prod.* **63**(3), 23 (1973).
24. R. Zdanowski and G. L. Brown, Jr., *Resin Review*, Vol. 9, No. 1, Rohm and Haas Co., Philadelphia, Pa., 1959, p. 19.
25. R. Bakule and J. M. Blickensderfer, *Tappi* **56**(4), 70 (1973).
26. A. V. Tobolsky, *J. Polym. Sci., Polym. Symp.* **9**, 157 (1975).
27. L. J. Hughes and G. E. Britt, *J. Appl. Polym. Sci.* **5**(15), 337 (1961).
28. S. Krause and N. Roman, *J. Polym. Sci., Part A* **3**, 1631 (1965).
29. H. L. Gerhart, *Off. Dig. Fed. Soc. Paint Technol.* 680 (June 1961).
30. R. M. Christenson and D. P. Hart, *Off. Dig. Fed. Soc. Paint Technol.* 684 (June 1961).
31. H. A. Vogel and H. G. Brittle, *Off. Dig. Fed. Soc. Paint Technol.* 699 (June 1961).
32. J. D. Murdock and G. H. Segall, *Off. Dig. Fed. Soc. Paint Technol.* 709 (June 1961).
33. J. D. Petropoulous, C. Frazier, and L. E. Cadwell, *Off. Dig. Fed. Soc. Paint Technol.* 719 (June 1961).
34. D. G. Applegath, *Off. Dig. Fed. Soc. Paint Technol.* 737 (June 1961).
35. U.S. Pat. 2,886,474 (May 12, 1959), B. B. Kine and A. C. Nuessle (to Rohm and Haas Co.).

36. U.S. Pat. 2,923,653 (Feb. 2, 1960), N. A. Matlin and B. B. Kine (to Rohm and Haas Co.).
37. D. H. Klein, *J. Paint Technol.* **42**(545), 335 (1970).
38. S. H. Rider and E. E. Hardy, *Polymerization and Polycondensation Processes* (Adv. Chem. Ser. No. 34), American Chemical Society, Washington, D.C., 1962.
39. R. H. Wiley and G. M. Braver, *J. Polym. Sci.* **3**, 647 (1948).
40. J. R. Martin, J. F. Johnson, and A. R. Cooper, *J. Macrol. Sci. Rev. Macromol. Chem.* **8**, 57 (1972).
41. A. S. Craemer, *Kunststoffe* **30**, 337 (1940).
42. B. E. Rehberg and C. H. Fisher, *Ind. Eng. Chem.* **40**, 1429 (1948).
43. G. G. Cameron and D. R. Kane, *Makromol. Chem.* **109**, 194 (1967).
44. L. Gunawan and J. K. Haken, *J. Polym. Sci., Polym. Chem. Ed.* **23**, 2539 (1985).
45. A. R. Burgess, *Chem. Ind.*, 78 (1952).
46. R. Stelle and H. Jacobs, *J. Appl. Polym. Sci.* **2**(4), 86 (1959).
47. B. G. Achhammer, *Mod. Plast.* **35**, 131 (1959).
48. K. L. Hoy, *J. Paint Technol.* **43**, 76 (1970).
49. A. F. M. Martin, *Handbook of Solubility Parameters and Other Cohesion Parameters*, CRC Press, Inc., Boca Raton, Fla., 1983.
50. J. H. Hildebrand and R. L. Scott, *The Solubility of Non-Electrolytes*, 3rd ed., Rheinhold Publishing Corp., New York, 1949.
51. P. A. Small, *J. Appl. Chem.* **3**, 71 (1953).
52. H. Burrell, *Off. Dig. Fed. Soc. Paint Technol.* 726 (Oct. 1955).
53. J. L. Gardon, *J. Paint Technol.* **38**, 43 (1966).
54. A. Rudin and H. K. Johnston, *J. Paint Technol.* **43**(559), 39 (1971).
55. T. P. Forbarth, *Chem. Eng.* **69**(5), 96 (1962).
56. M. Salkind, E. H. Riddle, and R. W. Keefer, *Ind. Eng. Chem.* **51**, 1328 (1959).
57. H. Warson, *The Applications of Synthetic Resin Emulsions*, Ernest Benn Ltd., London, .
58. R. F. B. Davies and G. E. J. Reynolds, *J. Appl. Polym. Sci.* **12**, 47 (1968).
59. A. R. Burgess, *Chem. Ind.*, 78 (1952).
60. M. Tabata, G. Nilsson, and A. Lund, *J. Polym. Sci., Polym. Chem. Ed.* **21**, 3257 (1983).
61. R. K. Graham, *J. Polym. Sci.* **38**, 209 (1959).
62. W. Burlant, J. Minsch, and C. Taylor, *J. Polym. Sci., Part A* **2**, 57 (1964).
63. S. Sakuyama and co-workers, *Chem. Technol.*, 350 (June 1973).
64. F. T. Maler and W. Bayer, *Encycl. Chem. Process Des.* **1**, 401 (1976).
65. D. J. Hadley and E. M. Evans, *Propylene and Its Industrial Derivatives*, John Wiley & Sons, Inc., New York, 1973, pp. 416–497.
66. U.S. Pat. 3,875,212 (1975), T. Ohrui (to Sumitomo Chemical).
67. *Storage and Handling of Acrylic and Methacrylic Esters and Acids, CM-17*, Rohm and Haas Co., Philadelphia, Pa.
68. *Analytical Methods for the Acrylic Monomers, CM-18*, Rohm and Haas Co., Philadelphia, Pa.
69. *Acrylate Monomers, Bulletin F-40252*, Union Carbide Corp., New York.
70. *Celanese Acrylates, Product Manual N-70-1*, Celanese Chemical Co., New York.
71. *State of California Health and Welfare Agency Safe Drinking Water and Toxic Enforcement Act of 1986.*
72. M. H. Mackoy and H. W. Melville, *Trans. Faraday Soc.* **45**, 323 (1949).
73. G. M. Burnett and L. D. Loan, *International Symposium on Macromolecular Chemistry*, Prague, 1957, p. 113.
74. U.S. Pats. 2,367,660 and 2,367,661 (Jan. 23, 1945), C. L. Agre (to E. I. du Pont de Nemours & Co., Inc.).

75. A. Chapiro, *Radiative Chemistry of Polymeric Systems*, Wiley-Interscience, New York, 1972.
76. *Preparation, Properties and Uses of Acrylic Polymers*, CM-19, Rohm and Haas Co., Philadelphia, Pa.
77. T. G. Fox, Jr. and R. Loshock, *J. Am. Chem. Soc.* **75**, 3544 (1953).
78. J. L. O'Brien, *J. Am. Chem. Soc.* **77**, 4757 (1955).
79. E. P. Bonsall, *Trans. Faraday Soc.* **49**, 686 (1953).
80. P. G. Griffiths and co-workers, *Tetrahedron Lett.* **23**, 1309 (1982).
81. G. V. Schulz and G. Henrici, *Makromol. Chem.* **18/19**, 473 (1956).
82. F. R. Mayo and A. A. Miller, *J. Am. Chem. Soc.* **80**, 2493 (1956).
83. M. M. Mogilevich, *Russ. Chem. Rev.* **48**, 199 (1979).
84. D. A. Tirrell, in J. I. Kroschwitz, ed., *Encyclopedia of Polymer Science and Engineering*, Wiley-Interscience, New York, 1986, pp. 192–233.
85. O. Yoshiahi and T. Imato, *Rev. Phys. Chem. Jpn.* **42**(1), 34 (1972).
86. F. W. Billmeyer, Jr., *Textbook of Polymer Chemistry*, Interscience Publishers, New York, 1957.
87. *Emulsion Polymerization of Acrylic Monomers*, CM-104, Rohm and Haas Co., Philadelphia, Pa.
88. J. J. Uibel and F. J. Dinon, *J. Polym. Sci., Polym. Chem. Ed.* **21**, 1773 (1983).
89. F. A. Bovey, *High Resolution NMR of Macromolecules*, Academic Press, New York, 1972.
90. M. S. Matheson and co-workers, *J. Am. Chem. Soc.* **73**, 5395 (1951).
91. T. G. Fox and R. Gratch, *Ann. N. Y. Acad. Sci.* **57**, 367 (1953).
92. L. Maduga, *An. Quim.* **65**, 993 (1969).
93. P. G. Griffiths, E. Rizzardo, and D. H. Solomon, *J. Macromol. Sci., Chem.* **17**, 45 (1982).
94. J. G. Kloosterboer and H. L. Bressers, *Polym. Bull.* **2**, 205 (1982).
95. *The Manufacture of Acrylic Polymers, CM-107*, Rohm and Haas Co., Philadelphia, Pa.
96. *The Manufacture of Ethyl Acrylate-Acrylic Acid Copolymers, TMM-48*, Rohm and Haas Co., Philadelphia, Pa.
97. *Bulk Storage and Handling of Acryloid Coating Resins, C. 186*, Rohm and Haas Co., Philadelphia, Pa.
98. P. A. Lovell and M. S. El-Aasser *Emulsion Polymerization and Emulsion Polymers*, John Wiley & Sons, Inc., New York, 1997.
99. R. G. Gilbert, *Emulsion Polymerization: A Mechanistic Approach*, Academic Press. New York, 1995.
100. U.S. Pat. 5,521,266 (May 28, 1996), W. Lau (to Rohm and Haas Co.).
101. B. B. Kine and G. H. Redlich, in D. T. Wasan, ed., *Surfactants in Chemical/Process Engineering*, Marcel Dekker, Inc., New York, 1988, p. 163.
102. A. E. Alexander, *J. Oil Colour Chem. Assoc.* **42**, 12 (1962).
103. C. E. McCoy, Jr., *Off. Dig. Fed. Soc. Paint Technol.* **35**, 327 (1963).
104. A. F. Sirianni and R. D. Coleman, *Can. J. Chem.* **42**, 682 (1964).
105. G. L. Brown, *Off. Dig. Fed. Soc. Paint Technol.* **28**, 456 (1956).
106. Brit. Pat. 940,366 (Oct. 30, 1963), P. R. van Ess (to Shell Oil).
107. U.S. Pat. 3,080,333 (Mar. 5, 1963), R. J. Kray and C. A. Defazio (to Celenese).
108. A. R. M. Azad, R. M. Fitch, and J. Ugelstad, *Colloidal Dispersions Micellar Behavior* (ACS Symp. Ser. No. 9), American Chemical Society, Washington, D.C., 1980.
109. H. Fikentscher, H. Gerrens, and H. Schuller, *Angew. Chem.* **72**, 856 (1960).
110. A. S. Sarac, *Prog. Polym. Sci.* **24**, 1149 (1999).
111. U.S. Pat. 3,458,466 (July 29, 1969), W. J. Lee (to The Dow Chemical Company).

112. U.S. Pat. 3,344,100 (Sept. 26, 1967), F. J. Donat and co-workers (to B. F. Goodrich Co.).
113. H. Warson, *Makromol. Chem. Suppl.* **10/11**, 265 (1985).
114. G. F. D'Alelio, *Fundamental Principles of Polymerization*, John Wiley & Sons, Inc., New York, 1952.
115. G. S. Whitby and co-workers, *J. Polym. Sci.* **16**, 549 (1955).
116. B. N. Rutovshii and co-workers, *J. Appl. Chem. USSR* **26**, 397 (1953).
117. H. Warson, *Polym. Paint Colour J.* **178**, 625 (1988).
118. H. Warson, *Polym. Paint Colour J.* **178**, 865 (1988).
119. L. J. Hughes and G. L. Brown, *J. Appl. Polym. Sci.* **7**, 59 (1963).
120. B. N. Kishore and co-workers, *J. Polym. Sci., Polym. Chem. Ed.* **24**, 2209 (1986).
121. G. Graczyk and V. Hornof, *J. Macromol. Sci., Chem.* **12**, 1633 (1988).
122. O. Y. Mansour and A. B. Moustafa, *J. Polym. Sci., Polym. Chem. Ed.* **13**, 2795 (1975).
123. T. Nagabhushanam and M. Santoppa, *J. Polym. Sci., Polym. Chem. Ed.* **14**, 507 (1976).
124. L. Zhi-Chong and co-workers, *J. Macromol. Sci., Chem.* **12**, 1487 (1988).
125. C. E. Brockway and K. B. Moser, *J. Polym. Sci., Part A-1* **1**, 1025 (1963).
126. G. Smets, A. Poot, and G. L. Dunean, *J. Polym. Sci.* **54**, 65 (1961).
127. C. H. Bamford and E. F. T. White, *Trans. Faraday Soc.* **52**, 719 (1956).
128. J. A Hicks and H. W. Melville, *Nature (London)* **171**, 300 (1953).
129. I. Sahata and D. A. I. Goring, *J. Appl. Polym. Sci.*, **20**, 573 (1976).
130. R. K. Graham, M. J. Gluchman, and M. J. Kampf, *J. Polym. Sci.* **38**, 417 (1959).
131. D. J. Angier, E. D. Farlie, and W. F. Watson, *Trans. Inst. Rubber Ind.* **34**, 8 (1958).
132. W. J. Burlant and A. S. Hoffman, *Block and Graft Polymers*, Reinhold Publishing Corp., New York, 1960.
133. R. J. Ceresa, *Block and Graft Copolymers*, Butterworth, Inc., Washington, D.C., 1962.
134. M. Szwarc, *J. Polym. Sci., Part A: Polym. Chem.* **36**, ix, (1998).
135. T. E. Patten and K. Matyjaszewski, *Adv. Mater.* **10**, 901 (1998).
136. M. K. Georges and co-workers, *Macromolecules* **26**, 2987 (1993).
137. S. A. F. Bon and co-workers, *Macromolecules* **30**, 324 (1997).
138. C. J. Hawker and co-workers, *Macromolecules* **31**, 213 (1998).
139. J. S. Wang and K. Matyjaszewski, *Macromolecules* **28**, 7901 (1995).
140. T. E. Patten and co-workers, *Science* **272**, 866 (1996).
141. S. Coca and co-workers, *J. Polym. Sci., Part A: Polym. Chem.* **36**, 1417 (1998).
142. K. Matyjaszewski, S. Coca, and C. B. Jasiaczek, *Macromol. Chem. Phys.* **198**, 4001 (1997).
143. S. Coca and K. Matyjaszewski, *Polym. Prepr. (Am. Chem. Soc., Div. Polym. Chem.)* **38**, 691 (1997).
144. D. McGinniss, *Ultraviolet Light Induced Reactions in Polymers* (ACS Symp. Ser. No. 25), American Chemical Society, Washington, D.C., 1976.
145. J. W. Vanderhoff, *J. Polym. Sci.* **38** (1959).
146. S. P. Pappas, ed., *U.V. Curing: Science and Technology*, Technology Marketing Corp., Stamford, Conn., 1978.
147. U.S. Pat. 3,418,295 (Dec. 24, 1968), A. C. Schwenthaler (to E. I. du Pont de Nemours & Co., Inc.).
148. V. D. McGinnis, SME Technical Paper (Ser.) FC 76–486, 1976.
149. J. V. Koleche, *Photochemistry of Cycloaliphatic Epoxides and Epoxy Acrylates*, Vol. I, ASTM International, 1989.
150. *The Curing of Coatings with Ultra-Violet Radiation, D8667 G.D.*, Tioxide of Canada, Sorel, P.Q., Canada.
151. T. A. Du Plessis and G. De Hollain, *J. Oil Colour Chem. Assoc.* **62**, 239 (1979).

152. M. L. Miller and C. E. Rauhut, *J. Polym. Sci.* **38**, 63 (1959).
153. B. Garret, *J. Am. Chem. Soc.* **81**, 1007 (1959).
154. A. Kawasahi and co-workers, *Makromol. Chem.* **49**, 76 (1961).
155. G. Smets and W. Van Hurnbeeck, *J. Polym. Sci., Part A-1* **1**, 1227 (1963).
156. W. E. Goode, R. P. Fellman, and F. H. Owens, in C. G. Overberger, ed., *Macromolecular Synthesis*, Vol. 1, John Wiley & Sons, Inc., New York, 1963, p. 25.
157. J. Furukawa, T. Tsuruta, and T. Makimoto, *Makromol. Chem.* **42**, 162 (1960).
158. T. Makmoto, T. Tsuruta, and J. Furukawa, *Makromol. Chem.* **50**, 116 (1961).
159. Y. Heimei and co-workers, in N. A. J. Platzer, ed., *Appl. Polym. Symp.*, Vol. 26, 1975, p. 39.
160. D. E. Glusker and co-workers, *J. Polym. Sci.* **49**, 315 (1961).
161. D. J. Cram and K. R. Kopecky, *J. Am. Chem. Soc.* **81**, 2748 (1959).
162. H. Yuki and co-workers, in O. Vogl and J. Furukawan, eds., *Ionic Polymerization*, Marcel Dekker, Inc., New York, 1976.
163. M. T. Reetz, *Angew. Chem., Int. Ed. Engl.* **27**, 994 (1988).
164. V. V. Korshak and co-workers, *Makromol. Chem. Suppl.* **6**, 55 (1984).
165. C. P. Bosmyak, I. W. Parsons, and J. N. Hay, *Polymer* **21**, 1488 (1980).
166. F. L. Keohan and co-workers, *J. Polym. Sci., Polym. Chem. Ed.* **22**, 679 (1984).
167. Y. D. Lee and H. B. Tsai, *Makromol, Chem.* **190**, 1413 (1989).
168. E. A. Collines and co-workers, *J. Coat. Technol.* **47**, 35 (1975).
169. E. B. Bradford and J. W. Vanderhoff, *J. Polym. Sci. Part C* **1**, 41 (1963),
170. T. F. Protxman and G. L. Brown, *J. Appl. Polym. Sci.* 81 (1960).
171. P. W. Allen, *Technique of Polymers Characterization*, Butterworths, London, 1959.
172. *Dilute Solution Properties of Acrylic and Methacrylic Polymers, SP-160*, Rohm and Haas Co., Philadelphia, Pa.
173. M. Harmon and J. King, U.S. NTIS AD Rept. AD-A0I7443, p. 142, 1974.
174. R. E. Harren, A. Mercurio, and J. D. Scott, *Aust. Oil Colour Chem. Assoc. Proc. News* 17 (Oct. 1977).
175. *Resin Review*, Vol. 18, Rohm and Haas Co., Philadelphia, Pa., 1968.
176. R. N. Washburne, *Am. Paint Coatings J.* **67**, 40 (1983).
177. R. P. Hopkins, E. W. Lewandowski, and T. E. Purcell, *J. Paint Technol.* **44**, 85 (1972).
178. T. E. Purcell, *Am. Paint J.* (June 1972).
179. Technical Practices Committee, *Materials Performance*, Vol. 15, National Assoc. of Corrosion Engineers, Houston, Tex., 1976, pp. 4, 13.
180. K. E. Buffington, *Maint. Eng.* (Mar. 1976).
181. I. K. Schahidi, J. C. Trebellas, and J. A. Vona, *Paint Varn. Prod.* **64**, 39 (1974).
182. C. B. Rybyn and co-workers, *J. Paint Technol.* **46**, 60 (1974).
183. *Multifunctional Acrylates in Radiation Curing*, Celanese Co., New York.
184. C. H. Carder, *Paint Varn. Prod.* **64**, 19 (1974).
185. *Materials for Photocurable Coatings and Inks*, Union Carbide Corp., New York.
186. R. A. Hickner and L. M. Ward, *Paint Varn. Prod.* **64**, 27 (1974).
187. F. Wingler and co-workers, *Farbe Lack* **78**, 1063 (1972).
188. R. Muller, *Farbe Lack* **78**, 1070 (1972).
189. J. K. Rankin, *J. Oil Colour Chem. Assoc.* **56**, 112 (1973).
190. F. A. Kyrlova, *Lakokras. Mater. Ikh Primen.* **3**, 20 (1969).
191. J. B. Zicherman, *Am. Paint J.* **56**(46), 23–28 (1972).
192. U.S. Pat. 1,535,228 (Aug. 16, 1968), K. Shibayama (to Eiki Jidaito).
193. F. Beck and co-workers, *Farbe Lack* **73**, 298 (1967).
194. A. Mercurio and S. N. Lewis, *J. Paint Technol.* **47**, 37 (1975).
195. W. H. Brendley, Jr. and E. C. Carl, *Paint Varn. Prod.* **63**, 23 (1973).
196. M. R. Yunaska and J. E. Gallagher, *Resin Review*, Vol. 19, Rohm and Haas Co., Philadelphia, Pa., 1969, pp. 1–3.

197. G. Allyn, *Materials and Methods*, Reinhold Publishing Co., New York, 1956.
198. *Protecting the Surfaces of Copper and Copper-Base Alloys*, International Copper Re-search Association, New York.
199. R. R. Kuhn, N. Roman, and J. D. Whiteman, *Mod. Paint Coatings* **71**, 50 (1981).
200. J. C. Thiabault, P. R. Sperry, and E. J. Schaller, in J. E. Glass, ed., *Water-Soluble Polymers*, American Chemical Society, Washington, D.C., 1986, p. 375.
201. A. J. Reuvers, *Prog, Org. Coatings* **35**, 171 (1999).
202. T. L. Maver, *J. Coat. Technol.* **64**, 45 (1992).
203. T. Murakami, R. H. Fernando, and J. E. Glass, *Surf. Coat. Int.* **76**, 8 (1993).
204. N. Sutterlin, *Makromol. Chem. Suppl.* **10/11**, 403 (1985).
205. V. J. Moser, *Resin Review*, Vol. 14, Rohm and Haas Co., Philadelphia, Pa., 1964, p. 25.
206. D. I. Lunde, *Nonwoven Fabrics Forum*, Clemson University, June 15–17, 1982.
207. U.S. Pat. 3,157,562 (Nov. 17, 1964), B. B. Kine, V. J. Moser, and H. A. Alps (to Rohm and Haas Co.).
208. U.S. Pat. 3,101,292 (Aug. 20, 1963), B. B. Kine and N. A. Matlin (to Rohm and Haas Co.).
209. J. R. Lawrence, *Resin Review*, Vol. 6, Rohm and Haas Co., Philadelphia, Pa., 1956, p. 12.
210. U.S. Pat. 2,931,749 (Apr. 5, 1960), B. B. Kine and N. A. Matlin (to Rohm and Haas Co.).
211. G. C. Kantner, *Text. World* **132**, 89 (1982).
212. G. C. Kantner, *Am. Text.* **12**, 30 (Feb. 1983).
213. L. Thompson and H. Mayfield, *AATTC National Technical Conference*, 1974, p. 258.
214. F. X. Chancler and J. G. Brodnyan, *Resin Review*, Vol. 23, Rohm and Haas Co., Philadelphia, Pa., 1973, p. 3.
215. V. J. Moser and D. G. Strong, *Am. Dyest. Rep.* **55**, 52 (1966).
216. U.S. Pat. 2,883,304 (Apr. 21, 1959), B. B. Kine and A. C. Nuessle (to Rohm and Haas Co.).
217. Brit. Pat. 1,011,041 (1962), K. Craemer (to Badische Anilin).
218. K. F. Foley and S. G. Chu, *Adhes. Age* 24 (Sept. 1986).
219. D. Satas, *Adhes. Age*, 28 (Aug. 1988).
220. J. J. Latimer, *Pulp Pap. Can.* **82**, 83 (1981).
221. J. E. Young and J. J. Latimer, *South. Pulp Pap. J.* **45**, 5 (1982).
222. P. J. McLaughlin, *Tappi* **42**, 994 (1959).
223. L. Mlynar and R. W. McNamec, Jr., *Tappi* **55**, 359 (1972).
224. F. L. Schucker, *Resin Review*, Vol. 7, Rohm and Haas Co., Philadelphia, Pa., 1957, p. 20.
225. H. C. Adams, T. J. Drennen, and L. E. Kelley, *Tappi* **48**, 486 (1965).
226. J. A. Handscomb, *J. Soc. Leather Trades'* Chem. **43**, 237 (1959).
227. W. C. Prentiss, *J. Am. Leather Chem. Assoc.* **71**, 54 (1976).
228. J. A. Lavelle and P. E. Wright, *Resin Review*, Vol. 24, Rohm and Haas Co., Philadelphia, Pa., 1974, p. 3.
229. J. R. Johnson, *Resin Review*, Vol. 11, Rohm and Haas Co., Philadelphia, Pa., 1961, p. 3.
230. P. E. Wright and H. C. Young, *Resin Review*, Vol. 24, Rohm and Haas Co., Philadelphia, Pa., 1974, p. 17.
231. P. H. Dougherty and H. T. Freund, *Resin Review*, Vol. 17, Rohm and Haas Co., Philadelphia, Pa., 1967, p. 3.
232. L. S. Frankel, D. A. Perry, and J. J. Lavelle, *Resin Review*, Vol. 32, Rohm and Haas Co., Philadelphia, Pa., 1982, p. 1.
233. J. T. Lutz, Jr., in Ref. 197, p. 18.
234. M. Cohen, *Proc. Fl. State Hortic. Soc.* **72**, 56 (1959).

235. *Resin Review*, Vol. 16, Rohm and Haas Co., Philadelphia, Pa., 1966, p. 12.
236. T. M. Vial, *Rubber Chem. Technol.* **44**, 344 (1971).
237. P. Fram, in N. M. Bikales, ed., *Enclopedia of Polymer Science and Technology*, Vol. 1, Interscience Publishers, New York, 1964, pp. 226–246.
238. D. L. Schultz, *Rubber World* **182**, 51 (May 1980).

Robert V. Slone
Rohm and Haas Company

ACRYLONITRILE

1. Introduction

Prior to 1960, acrylonitrile [107-13-1] (also called acrylic acid nitrile, propylene nitrile, vinyl cyanide, propenoic acid nitrile) was produced commercially by processes based on either ethylene oxide and hydrogen cyanide or acetylene and hydrogen cyanide. The growth in demand for acrylic fibers, starting with the introduction of Orlon by Du Pont around 1950, spurred efforts to develop improved process technology for acrylonitrile manufacture to meet the growing market (see Fibers, acrylic). This resulted in the discovery in the late 1950s by Sohio (1) and also by Distillers (2) of a heterogeneous vapor-phase catalytic process for acrylonitrile by selective oxidation of propylene and ammonia, commonly referred to as the propylene ammoxidation process. Commercial introduction of this lower cost process by Sohio in 1960 resulted in the eventual displacement of all other acrylonitrile manufacturing processes. Today over 90% of the more than 4,000,000 metric tons produced worldwide each year are made using the Sohio-developed ammoxidation process. Acrylonitrile is among the top 50 chemicals produced in the United States as a result of the tremendous growth in its use as a starting material for a wide range of chemical and polymer products. Acrylic fibers remain the largest use of acrylonitrile; other significant uses are in resins and nitrile elastomers and as an intermediate in the production of adiponitrile and acrylamide.

2. Physical Properties

Acrylonitrile (C_3H_3N, mol wt = 53.064) is an unsaturated molecule having a carbon–carbon double bond conjugated with a nitrile group. It is a polar molecule because of the presence of the nitrogen heteroatom. There is a partial shift in the bonding electrons toward the more electronegative nitrogen atom, as represented by the following heterovalent resonance structures.

$$CH_2{=}CH{-}C{\equiv}N{:} \longleftrightarrow CH_2{=}CH{-}\overset{+}{C}{=}\overset{-}{\ddot{N}}{:} \longleftrightarrow \overset{+}{C}H_2{-}CH{=}C{=}\overset{-}{\ddot{N}}{:}$$

Tables 1 and 2 list some physical properties and thermodynamic information, respectively, for acrylonitrile (3–5).

Table 1. **Physical Properties of Acrylonitrile**

Property	Value
appearance/odor	clear, colorless liquid with faintly pungent odor
boiling point, °C	77.3
freezing point, °C	−83.5
density, 20°C, g/cm³	0.806
volatility, 78°C, %	>99
vapor pressure, 20°C, kPa[a]	11.5
vapor density (air = 1)	1.8
solubility in water, 20°C, wt%	7.3
pH (5% aqueous solution)	6.0–7.5
critical values	
temperature, °C	246
pressure, MPa[b]	3.54
volume, cm³/g	3.798
refractive index, n^{25}_{D}	1.3888
dielectric constant, 33.5 MHz	38
ionization potential, eV	10.75
molar refractivity (D line)	15.67
surface tension, 25°C, mN/m (=dyn/cm)	26.6
dipole moment, C·m[c]	
liquid	1.171×10^{-29}
vapor	1.294×10^{-29}
viscosity, 25°C, mPa·s (=cP)	0.34

[a] To convert kPa to mm Hg multiply by 7.5
[b] To convert MPa to psi multiply by 145.
[c] To convert C·m to debye, divide by 3.336×10^{-30}

Table 2. **Thermodynamic Data**[a]

Property	Value
flash point, °C	0
autoignition temperature, °C	481
flammability limits in air, 25°C, vol %	
lower	3.0
upper	17.0
free energy of formation, ΔG°_{g}, 25°C, kJ/mol	195
enthalpy of formation, 25°C, kJ/mol	
ΔH°_{g}	180
ΔH°_{l}	147
heat of combustion, liquid, 25°C, kJ/mol	1761.5
heat of vaporization, 25°C, kJ/mol	32.65
molar heat capacity, kJ/(kg·K)	
liquid	2.09
gas at 50°C, 101.3 kPa[b]	1.204
molar heat of fusion, kJ/mol	6.61
entropy, S, gas at 25°C, 101.3 kPa[b], kJ/(mol·K)	274

[a] To convert kJ to kcal divide by 4.184.
[b] 101.3 kPa = 1 atm.

Table 3. **Azeotropes of Acrylonitrile**

Azeotrope	Boiling point, °C	Acrylonitrile concentration, wt %
water	71.0	88
isopropyl alcohol	71.6	56
benzene	73.3	47
methanol	61.4	39
carbon tetrachloride	66.2	21
tetrachlorosilane	51.2	11
chlorotrimethylsilane	57.0	7

Table 4. **Solubilities of Acrylonitrile in Water**

Temperature, °C	Acrylonitrile in water, wt %	Water in acrylonitrile, wt %
−50		0.4
−30		1.0
0	7.1	2.1
10	7.2	2.6
20	7.3	3.1
30	7.5	3.9
40	7.9	4.8
50	8.4	6.3
60	9.1	7.7
70	9.9	9.2
80	11.1	10.9

Acrylonitrile is miscible in a wide range of organic solvents, including acetone, benzene, carbon tetrachloride, diethyl ether, ethyl acetate, ethylene cyanohydrin, petroleum ether, toluene, some kerosenes, and methanol. Compositions of some common azeotropes of acrylonitrile are given in Table 3. Table 4 presents the solubility of acrylonitrile in water as a function of temperature (6). Vapor–liquid equilibria for acrylonitrile in combination with acetonitrile, acrolein, HCN, and water have been published (6–9). Table 5 gives the vapor pressure of acrylonitrile over aqueous solutions.

Table 5. **Acrylonitrile Vapor Pressure over Aqueous Solutions at 25°C**

Acrylonitrile, wt %	Vapor pressure, kPa[a]
1	1.3
2	2.9
3	5.3
4	6.9
5	8.4
6	10.0
7	10.9

[a] To convert kPa to mm Hg multiply by 7.5.

Acrylonitrile has been characterized using infrared, Raman, and ultraviolet spectroscopies, electron diffraction, and mass spectroscopy (10–18).

3. Chemical Properties

Acrylonitrile undergoes a wide range of reactions at its two chemically active sites, the nitrile group and the carbon–carbon double bond. Detailed descriptions of specific reactions have been given (19, 20). Acrylonitrile polymerizes readily in the absence of a hydroquinone inhibitor, especially when exposed to light. Polymerization is initiated by free radicals, redox catalysts, or bases and can be carried out in the liquid, solid, or gas phase. Homopolymers and copolymers are most easily produced using liquid-phase polymerization (see ACRYLONITRILE POLYMERS). Acrylonitrile undergoes the reactions typical of nitriles, including hydration with sulfuric acid to form acrylamide sulfate ($C_3H_5NO{\cdot}H_2SO_4$ [15497-99-1]), which can be converted to acrylamide (C_3H_5NO [79-06-1]) by neutralization with a base; and complete hydrolysis to give acrylic acid ($C_3H_4O_2$ [79-10-7]). Acrylamide (qv) is also formed directly from acrylonitrile by partial hydrolysis using copper-based catalysts (21–24); this has become the preferred commercial route for acrylamide production. Industrially important acrylic esters can be formed by reaction of acrylamide sulfate with organic alcohols. Methyl acrylate ($C_4H_6O_2$ [96-33-3]) has been produced commercially by the alcoholysis of acrylamide sulfate with methanol. Reactions at the activated double bond of acrylonitrile include Diels-Alder addition to dienes, forming cyclic products; hydrogenation over metal catalysts to give propionitrile (C_3H_5N [107-12-0]) and propylamine (C_3H_9N [107-10-8]); and the industrially important hydrodimerization to produce adiponitrile ($C_6H_8N_2$ [111-69-3]) (25–27). Other reactions include addition of halogens across the double bond to produce dihalopropionitriles, and cyanoethylation by acrylonitrile of alcohols, aldehydes, esters, amides, nitriles, amines, sulfides, sulfones, and halides.

4. Manufacturing and Processing

Acrylonitrile is produced in commercial quantities almost exclusively by the vapor-phase catalytic propylene ammoxidation process developed by Sohio (28).

$$C_3H_6 + NH_3 + {}^3/_2\,O_2 \xrightarrow{\text{catalyst}} C_3H_3N + 3\,H_2O$$

A schematic diagram of the commercial process is shown in Figure 1. The commercial process uses a fluid-bed reactor in which propylene, ammonia, and air contact a solid catalyst at 400–510°C and 49–196 kPa (0.5–2.0 kg/cm^2) gauge. It is a single-pass process with about 98% conversion of propylene, and uses about 1.1 kg propylene per kg of acrylonitrile produced. Useful by-products from the process are HCN (about 0.1 kg per kg of acrylonitrile), which is used primarily in the manufacture of methyl methacrylate, and acetonitrile (about 0.03 kg per kg of acrylonitrile), a common industrial solvent. In the commercial operation the hot reactor effluent is quenched with water in a countercurrent

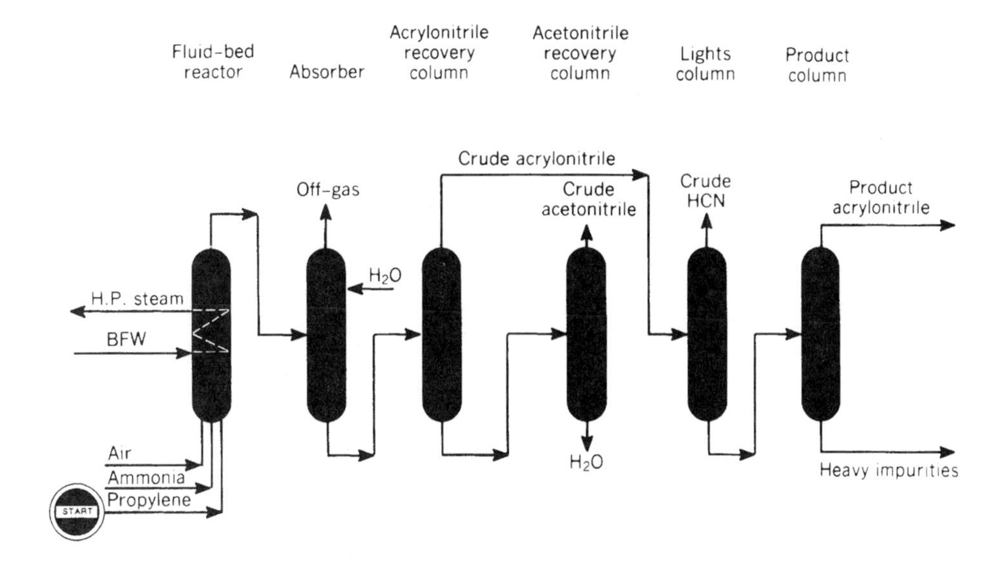

Fig. 1. Process flow diagram of the commercial propylene ammoxidation process for acrylonitrile. *BFW*, boiler feed water.

absorber and any unreacted ammonia is neutralized with sulfuric acid. The resulting ammonium sulfate can be recovered and used as a fertilizer. The absorber off-gas containing primarily N_2, CO, CO_2, and unreacted hydrocarbon is either vented directly or first passed through an incinerator to combust the hydrocarbons and CO. The acrylonitrile-containing solution from the absorber is passed to a recovery column that produces a crude acrylonitrile stream overhead that also contains HCN. The column bottoms are passed to a second recovery column to remove water and produce a crude acetonitrile mixture. The crude acetonitrile is either incinerated or further treated to produce solvent quality acetonitrile. Acrylic fiber quality (99.2% minimum) acrylonitrile is obtained by fractionation of the crude acrylonitrile mixture to remove HCN, water, light ends, and high boiling impurities. Disposal of the process impurities has become an increasingly important aspect of the overall process, with significant attention being given to developing cost-effective and environmentally acceptable methods for treatment of the process waste streams. Current methods include deep-well disposal, wet air oxidation, ammonium sulfate separation, biological treatment, and incineration (29).

Although acrylonitrile manufacture from propylene and ammonia was first patented in 1949 (30), it was not until 1959, when Sohio developed a catalyst capable of producing acrylonitrile with high selectivity, that commercial manufacture from propylene became economically viable (1). Production improvements over the past 30 years have stemmed largely from development of several generations of increasingly more efficient catalysts. These catalysts are multicomponent mixed metal oxides mostly based on bismuth–molybdenum oxide. Other types of catalysts that have been used commercially are based on iron–antimony oxide, uranium–antimony oxide, and tellurium–molybdenum oxide.

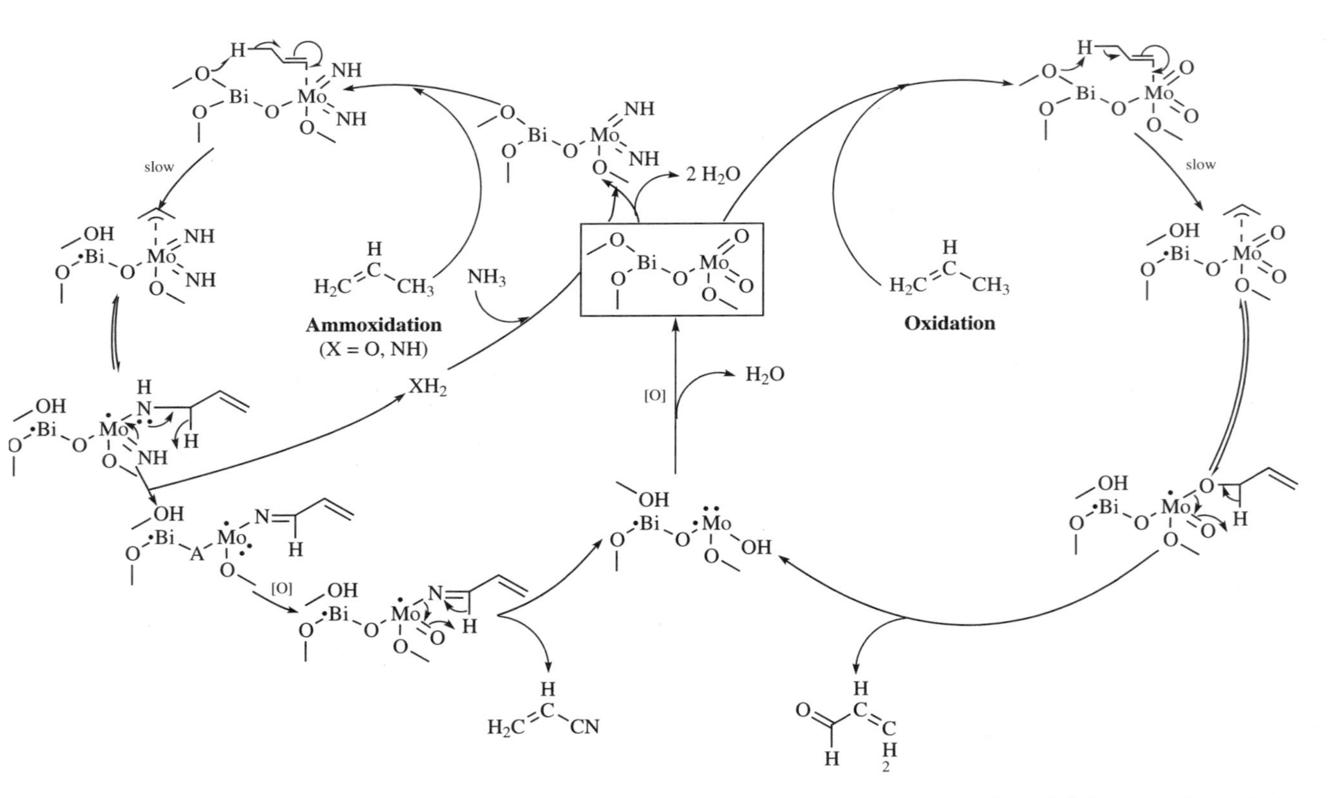

Fig. 2. Mechanism of selective ammoxidation and oxidation of propylene over bismuth molybdate catalysts (31).

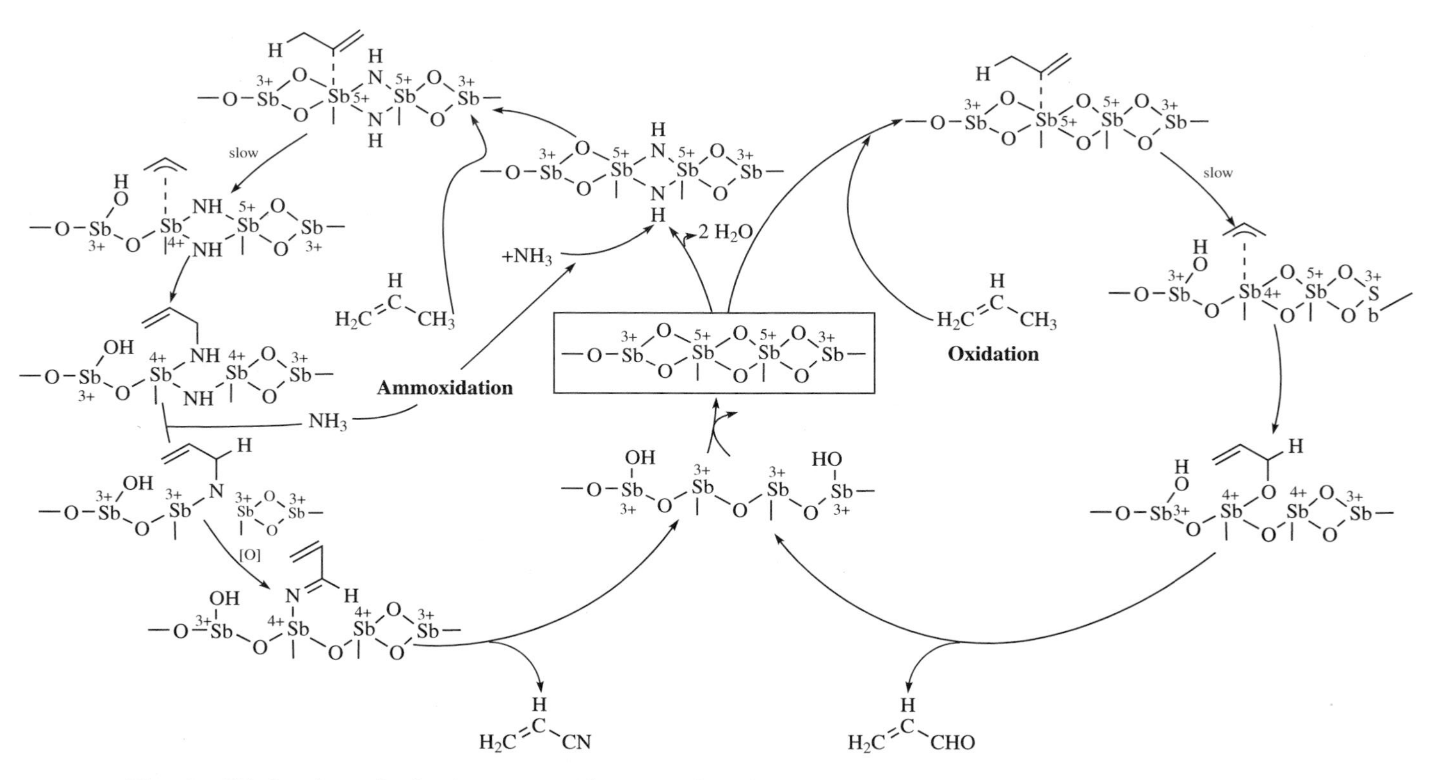

Fig. 3. Mechanism of selective ammoxidation and oxidation of propylene over antimonate catalysts (31).

Fundamental understanding of these complex catalysts and the surface-reaction mechanism of propylene ammoxidation has advanced substantially since the first commercial plant began operation. Mechanisms for selective ammoxidation of propylene over bismuth molybdate and antimonate catalysts are shown in Figures 2 and 3. The rate-determining step is abstraction of an α-hydrogen of propylene by an oxygen in the catalyst to form a π-allyl complex on the surface (31–33). Lattice oxygens from the catalyst participate in further hydrogen abstraction, followed by oxygen insertion to produce acrolein in the absence of ammonia, or nitrogen insertion to form acrylonitrile when ammonia is present (34–36). The oxygens removed from the catalyst in these steps are replenished by gas-phase oxygen, which is incorporated into the catalyst structure at a surface site separate from the site of propylene reaction. In the ammoxidation reaction, ammonia is activated by an exchange with O^{2-} ions to form isoelectronic NH^{2-} moieties according to the following:

$$NH_3 + O^{2-} \longrightarrow NH^{2-} + H_2O$$

These are the species inserted into the allyl intermediate to produce acrylonitrile.

The active site on the surface of selective propylene ammoxidation catalyst contains three critical functionalities associated with the specific metal components of the catalyst (37–39): an α-H abstraction component such as Bi^{3+}, Sb^{3+}, or Te^{4+}; an olefin chemisorption and oxygen or nitrogen insertion component such as Mo^{6+} or Sb^{5+}; and a redox couple such as Fe^{2+}/Fe^{3+} or Ce^{3+}/Ce^{4+} to enhance transfer of lattice oxygen between the bulk and surface of the catalyst. The surface and solid-state mechanisms of propylene ammoxidation catalysis have been determined using Raman spectroscopy (40, 41), neutron diffraction (42–44), x-ray absorption spectroscopy (45, 46), x-ray diffraction (47–49), pulse kinetic studies (36), and probe molecule investigations (50).

5. Obsolete Acrylonitrile Processes

Processes rendered obsolete by the propylene ammoxidation process (51) include the ethylene cyanohydrin process (52–54) practiced commercially by American Cyanamid and Union Carbide in the United States and by I. G. Farben in Germany. The process involved the production of ethylene cyanohydrin by the base-catalyzed addition of HCN to ethylene oxide in the liquid phase at about 60°C. A typical base catalyst used in this step was diethylamine. This was followed by liquid-phase or vapor-phase dehydration of the cyanohydrin. The liquid-phase dehydration was performed at about 200°C using alkali metal or alkaline earth metal salts of organic acids, primarily formates and magnesium carbonate. Vapor-phase dehydration was accomplished over alumina at about 250°C.

$$C_2H_4O + HCN \xrightarrow[\text{catalyst}]{\text{base}} HOC_2H_4CN \xrightarrow[-H_2O]{\text{catalyst}} C_3H_3N$$

A second commercial route to acrylonitrile used by Du Pont, American Cyanamid, and Monsanto was the catalytic addition of HCN to acetylene (55).

$$C_2H_2 + HCN \xrightarrow{\text{catalyst}} C_3H_3N$$

The reaction occurs by passing HCN and a 10:1 excess of acetylene into dilute hydrochloric acid at 80°C in the presence of cuprous chloride as the catalyst.

These processes use expensive C_2 hydrocarbons as feedstocks and thus have higher overall acrylonitrile production costs compared to the propylene-based process technology. The last commercial plants using these process technologies were shut down by 1970.

Other routes to acrylonitrile, none of which achieved large-scale commercial application, are acetaldehyde and HCN (56), propionitrile dehydrogenation (57, 58), and propylene and nitric oxide (59, 60):

$$CH_3CHO + HCN \longrightarrow CH_3CH(OH)CN \xrightarrow{-H_2O} C_3H_3N$$

$$CH_3CH_2CN \longrightarrow C_3H_3N + H_2$$

$$4\,C_3H_6 + 6\,NO \longrightarrow 4\,C_3H_3N + 6\,H_2O + N_2$$

Numerous patents have been issued disclosing catalysts and process schemes for manufacture of acrylonitrile from propane. These include the direct heterogeneously catalyzed ammoxidation of propane to acrylonitrile using mixed metal oxide catalysts (61–64).

$$C_3H_8 + NH_3 + 2\,O_2 \xrightarrow{\text{catalyst}} C_3H_3N + 4\,H_2O$$

A two-step process involving conventional nonoxidative dehydrogenation of propane to propylene in the presence of steam, followed by the catalytic ammoxidation to acrylonitrile of the propylene in the effluent stream without separation, is also disclosed (65).

$$C_3H_8 \xrightarrow{\text{catalyst}} C_3H_6 + H_2 \xrightarrow[+NH_3]{+\frac{3}{2}\,O_2} C_3H_3N + 3\,H_2O + H_2$$

Because of the large price differential between propane and propylene, which has ranged from \$155/t to \$355/t between 1987 and 1989, a propane-based process may have the economic potential to displace propylene ammoxidation technology eventually. Methane, ethane, and butane, which are also less expensive than propylene, and acetonitrile have been disclosed as starting materials for acrylonitrile synthesis in several catalytic process schemes (66, 67).

6. Economic Aspects

The propylene-based process developed by Sohio was able to displace all other commercial production technologies because of its substantial advantage in

Table 6. **Worldwide Acrylonitrile Production[a], 10^3 t**

Region	1998 (Estimated)	1997
Western Europe	1,112	1,073
Eastern Europe	182	189
United States	1,324	1,493
Japan	730	729
Far East/Asia	841	779
Africa/Middle East	152	147
Latin America/Mexico	246	232
Total	*4,587*	*4,642*

[a] Ref. 71.

overall production costs, primarily due to lower raw material costs. Raw material costs less by-product credits account for about 60% of the total acrylonitrile production cost for a world-scale plant. The process has remained economically advantaged over other process technologies since the first commercial plant in 1960 because of the higher acrylonitrile yields resulting from the introduction of improved commercial catalysts. Reported per-pass conversions of propylene to acrylonitrile have increased from about 65% to over 80% (28, 68–70).

More than half of the worldwide acrylonitrile production is situated in Western Europe and the United States (Table 6). In the United States, production is dominated by BP Amoco Chemicals, with more than a third of the domes-

Table 7. **U.S. Acrylonitrile Producers[a]**

Company	Approximate capacity[b], 10^3 t/y
BP Amoco Chemicals	640
Solutia, Inc.	260
Sterling Chemicals	360
E. I. du Pont de Nemours & Co., Inc.	185
Cytec Industries	220
Total	*1665*

[a] Ref. 71.
[b] As of 1997.

Table 8. **U.S. Acrylonitrile Exports[a], 10^3 t**

Destination	1997	1996
Far East/Asia	334	378
Japan	92	107
Western Europe	91	57
Canada	7	6
Latin America/Mexico	82	50
Middle East/Africa	91	57
Total export	*697*	*655*

[a] Ref. 71.

Table 9. **Worldwide Acrylonitrile Demand, 10^3 t/yr**

Region	1998 (Estimated)	1997	1995	1990	1986
Western Europe	1109	1116	1045	1136	1187
Eastern Europe	141	150	171	311	261
Japan	726	723	674	664	640
North America	781	800	756	641	638
Far East	1297	1264	1025	646	462
Africa/Middle East	261	257	223	135	142
Latin America/Mexico	302	281	244	206	213
Total	*4617*	*4591*	*4138*	*3739*	*3543*

tic capacity (Table 7). Nearly one-half of the U.S. production was exported in 1997 (Table 8), with most going to Far East Asia. The percentage of U.S. production exported grew from around 10% in the mid-1970s to about 42% in 1997 since Far East Asian producers, especially in the People's Republic of China (PRC), have not been able to satisfy the increasing domestic demand in recent years. In addition, the higher propylene costs relative to the United States generally makes it more economical to import acrylonitrile from the United States than to install new domestic production. Nevertheless, additions to Far East Asian acrylonitrile production capacity have been made in the 1990's, notably in South Korea. Table 9 provides a breakdown of worldwide demand between 1986 and 1998. Growth in demand has averaged about 3.6% per year.

7. Analytical Methods and Specifications

Standard test methods for chemical analysis have been developed and published (72). Included is the determination of commonly found chemicals associated with acrylonitrile and physical properties of acrylonitrile that are critical to the

Table 10. **Commercial Specifications for Acrylonitrile**

Parameter	Specification
acetone, ppm, max	300
acetonitrile, ppm, max	500
aldehydes, ppm, max	50
color, APHA, max	15
distillation range, °C, min, ibp	74.2
°C, max, 97%	78.8
HCN, ppm, max	5
inhibitor, hydroquinone monomethyl ether, ppm	35–45
iron, ppm, max	0.10
nonvolatile matter, ppm, max	100
peroxides, ppm, max	0.2
pH, 5% aqueous	6.0–7.5
refractive index, $n^{25}{}_{D}$	1.3891
water, wt %, max	0.2–0.5
purity, wt %, min	99.0

quality of the product (73–75). These include determination of color and chemical analyses for HCN, quinone inhibitor, and water. Specifications appear in Table 10.

8. Storage and Transport

Acrylonitrile must be stored in tightly closed containers in cool, dry, well-ventilated areas away from heat, sources of ignition, and incompatible chemicals. Storage vessels, such as steel drums, must be protected against physical damage, with outside detached storage preferred. Storage tanks and equipment used for transferring acrylonitrile should be electrically grounded to reduce the possibility of static spark-initiated fire or explosion. Acrylonitrile is regulated in the workplace by OSHA (29 CFR 1910).

Acrylonitrile is transported by rail car, barge, and pipeline. Department of Transportation (DOT) regulations require labeling acrylonitrile as a flammable liquid and poison. Transport is regulated under DOT 49 CFR 172.101. Bill of lading description is: Acrylonitrile, Inhibited, 3, 6.1, UN 1093, PGI, RQ.

9. Health and Safety Factors

Acrylonitrile is absorbed rapidly and distributed widely throughout the body following exposure by inhalation, skin contact or ingestion. However, there is little potential for significant accumulation in any organ with most of the compound being excreted primarily as metabolites in urine. Acrylonitrile is metabolized primarily by two pathways: conjugation with glutathione and oxidation. Oxidative metabolism leads to the formation of an epoxide, 2-cyanoethylene oxide, that is either conjugated with glutathione or directly hydrolyzed by epoxide hydrolase.

The acute toxicity of acrylonitrile is relatively high, with four-hour LC50s in laboratory animals ranging from 300 to 900 mg/m^3 and LD50s from 25 to 186 mg/kg (76, 77). Signs of acute toxicity observed in animals include respiratory tract irritation and two phases of neurotoxicity, the first characterized by signs consistent with cholinergic over-stimulation and the second being CNS dysfunction, resembling cyanide poisoning. In cases of acute human intoxication, effects on the central nervous system characteristic of cyanide poisoning and effects on the liver, manifested as increased enzyme levels in the blood, have been observed.

Acrylonitrile is a severe irritant to the skin, eyes, respiratory tract and mucous membranes. It is also a skin sensitizer.

Acrylonitrile is a potent tumorigen in the rat. Tumors of the central nervous system, ear canal, and gastrointestinal tract have been observed in several studies following oral or inhalation exposure. The mechanism of acrylonitrile's tumorigenesis in the rat and the relevance of these findings to humans is not clear. Available data are insufficient to support a consensus view on a plausible mode of action. There is evidence for weak genotoxic potential, but no evidence of DNA-adduct formation in target tissues. Recent work has provided indications that oxidative stress and resulting oxidative DNA damage may play a role.

There is extensive occupational epidemiology data on acrylonitrile workers. These investigations have not produced consistent, convincing evidence of an increase in cancer risk, although questions remain about the power of the database to detect small excesses of rare tumors. In 1998, The International Agency for Research on Cancer re-evaluated the cancer data for acrylonitrile and made a rare decision to downgrade the cancer risk classification (from probably carcinogen to humans to possibly carcinogenic to humans) based primarily on the growing epidemiology database (78).

Experimental evaluations of acrylonitrile have not produced any clear evidence of adverse effects on reproductive function or development of offspring at doses below those producing paternal toxicity.

The results of genotoxicity evaluations of acrylonitrile have been mixed. Positive findings *in vitro* have occurred mainly at exposures associated with cellular toxicity, and the most reliable *in vivo* tests have been negative.

Acrylonitrile will polymerize violently in the absence of oxygen if initiated by heat, light, pressure, peroxide, or strong acids and bases. It is unstable in the presence of bromine, ammonia, amines, and copper or copper alloys. Neat acrylonitrile is generally stabilized against polymerization with trace levels of hydroquinone monomethyl ether and water.

Acrylonitrile is combustible and ignites readily, producing toxic combustion products such as hydrogen cyanide, nitrogen oxides, and carbon monoxide. It forms explosive mixtures with air and must be handled in well-ventilated areas and kept away from any source of ignition, since the vapor can spread to distant ignition sources and flash back.

Federal regulations (40 CFR 261) classify acrylonitrile as a hazardous waste and it is listed as Hazardous Waste Number U009. Disposal must be in accordance with federal (40 CFR 262, 263, 264, 268, 270), state, and local regulations only at properly permitted facilities. It is listed as a toxic pollutant (40 CFR 122.21) and introduction into process streams, storm water, or waste water systems is in violation of federal law. Strict guidelines exist for clean-up and notification of leaks and spills. Federal notification regulations require that spills or leaks in excess of 100 lb (45.5 kg) be reported to the National Response Center. Substantial criminal and civil penalties can result from failure to report such discharges into the environment.

10. Uses

Worldwide consumption of acrylonitrile increased from 2.5×10^6 in 1976 to $\sim 4.6 \times 10^6$ t/yr in 1998. The trend in consumption over this time period is shown in Table 11 for the principal uses of acrylonitrile: acrylic fiber, acrylonitrile–butadiene–styrene (ABS) resins, adiponitrile, nitrile rubbers, elastomers, and styrene–acrylonitrile resins (SAN). Since the 1960s acrylic fibers have remained the major outlet for acrylonitrile production in the United States and especially in Japan and the Far East. Acrylic fibers always contain a comonomer. Fibers containing 85 wt % or more acrylonitrile are usually referred to as acrylics whereas fibers containing 35 to 85 wt% acrylonitrile are termed modacrylics (see FIBERS, ACRYLIC). Acrylic fibers are used primarily for the manufacture of apparel, including sweaters, fleece wear, and sportswear, as well as for home furnishings, including carpets, upholstery, and draperies. Acrylic fibers

Table 11. **Worldwide Acrylonitrile Uses and Consumption, 10^3 t**

Use	1998 (Estimated)	1997	1995	1990	1986
acrylic fibers	2615	2628	2313	2242	2350
ABS resins/SAN	1095	1079	996	781	598
adiponitrile	494	477	446	330	281
NB copolymers	144	143	134	143	125
misc.	269	264	249	243	189
Total	*4617*	*4591*	*4138*	*3739*	*3543*

consume about 57% of the acrylonitrile produced worldwide. Growth in demand for acrylic fibers in the 1990s is expected to be modest, between 2 and 3% per year, primarily from overseas markets. Domestic demand is expected to be flat.

ABS resins and adiponitrile are the fastest growing uses for acrylonitrile (see ACRYLAMIDE POLYMERS). ABS resins are second to acrylic fibers as an outlet for acrylonitrile. These resins normally contain about 25% acrylonitrile and are characterized by their chemical resistance, mechanical strength, and ease of manufacture. Consumption of ABS resins increased significantly in the 1980s and 1990s with its growing application as a specialty performance polymer in construction, automotive, machine, and appliance applications. Opportunities still exist for ABS resins to continue to replace more traditional materials for packaging, building, and automotive components. SAN resins typically contain between 25 and 30% acrylonitrile. Because of their high clarity, they are used primarily as a substitute for glass in drinking cups and tumblers, automobile instrument panels, and instrument lenses. The largest increase among the end uses for acrylonitrile has come from adiponitrile, which has grown to become the third largest outlet for acrylonitrile. It is used by Solutia as a precursor for hexamethylenediamine (HMDA, $C_6H_{16}N_2$ [124-09-4]) and is made by a proprietary acrylonitrile electrohydrodimerization process (25). HMDA is used exclusively for the manufacture of nylon-6,6. The growth of this acrylonitrile outlet in recent years stems largely from replacement of adipic acid ($C_6H_{10}O_4$ [124-04-9]) with acrylonitrile in HDMA production rather than from a significant increase in nylon-6,6 demand. The use of acrylonitrile for HMDA production should continue to grow at a faster rate than the other outlets for acrylonitrile, but it will not likely approach the size of the acrylic fiber market for acrylonitrile consumption.

Acrylamide (qv) is produced commercially by heterogeneous copper-catalyzed hydration of acrylonitrile (21–24). Acrylamide is used primarily in the form of a polymer, polyacrylamide, in the paper and pulp industry and in waste water treatment as a flocculant to separate solid material from waste water streams (see ACRYLONITRILE POLYMERS). Other applications include mineral processing, coal processing, and enhanced oil recovery in which polyacrylamide solutions were found effective for displacing oil from rock.

Nitrile rubber finds broad application in industry because of its excellent resistance to oil and chemicals, its good flexibility at low temperatures, high abrasion and heat resistance (up to 120°C), and good mechanical properties. Nitrile rubber consists of butadiene–acrylonitrile copolymers with an acrylonitrile content ranging from 15 to 45% (see ELASTOMERS, SYNTHETIC, NITRILE RUBBER). In addition to the traditional applications of nitrile rubber for hoses, gaskets, seals, and oil well equipment, new applications have emerged with the development of nitrile rubber blends with poly(vinyl chloride) (PVC). These blends combine the chemical resistance and low temperature flexibility characteristics of nitrile rubber with the stability and ozone resistance of PVC. This has greatly expanded the use of nitrile rubber in outdoor applications for hoses, belts, and cable jackets, where ozone resistance is necessary.

Other acrylonitrile copolymers have found specialty applications where good gas-barrier properties are required along with strength and high impact resistance. An example is BP Amoco Chemicals' Barex 210 acrylonitrile–methyl acrylate–butadiene copolymer and Monsanto's Lopac styrene-containing nitrile copolymer. These barrier resins compete directly in the alcoholic and other beverage bottle market with traditional glass and metal containers as well as with poly(ethylene terephthalate) (PET) and PVC in the beverage bottle market (see BARRIER POLYMERS). Other applications include food, agricultural chemicals, and medical packaging. Total acrylonitrile consumption for barrier resin applications is small, consuming less than about 1% of the total U.S. acrylonitrile production. Projections of a significant growth in demand for nitrile barrier resins remain unfulfilled because of an FDA ban in 1977 on the use of acrylonitrile-based copolymers in beverage bottles. Although the ban was lifted in 1982 and limits were set on acrylonitrile exposure in beverage and food packaging applications, it is uncertain that acrylonitrile copolymers can penetrate the current plastic bottle market dominated by PET.

A growing specialty application for acrylonitrile is in the manufacture of carbon fibers. They are produced by pyrolysis of oriented polyacrylonitrile fibers and are used to reinforce composites (qv) for high performance applications in the aircraft, defense, and aerospace industries. These applications include rocket engine nozzles, rocket nose cones, and structural components for aircraft and orbital vehicles where light weight and high strength are needed. Other small specialty applications of acrylonitrile are in the production of fatty amines, ion-exchange resins, and fatty amine amides used in cosmetics, adhesives, corrosion inhibitors, and water treatment resins. Examples of these specialty amines include 2-acrylamido-2-methylpropanesulfonic acid ($C_7H_{13}NSO_4$ [15214-89-8]), 3-methoxypropionitrile (C_4H_7NO [110-67-8]), and 3-methoxypropylamine ($C_4H_{11}NO$ [5332-73-0]).

BIBLIOGRAPHY

"Acrylonitrile" in *ECT* 1st ed., Vol. 1, pp. 184–189, by H. S. Davis, American Cyanamid Company; in *ECT* 2nd ed., Vol. 1, pp. 338–351, by W. O. Fugate, American Cyanamid Company; in *ECT* 3rd ed., Vol. 1, pp. 414–426, by Louis T. Groet, Badger, B. V.; "Acrylonitrile" in *ECT* 4th ed., Vol. 1, pp. 352–369, by J. F. Brazdil, BP Research; "Acrylonitrile" in *ECT* (online), posting date: December 4, 2000, by J.F. Brazdil, BP Research.

CITED PUBLICATIONS

1. U.S. Pat. 2,904,580 (Sept. 15, 1959), J. D. Idol (to The Standard Oil Co.).
2. Brit. Pat. 876,446 (Oct. 3, 1959) and U.S. Pat. 3,152,170 (Oct. 6, 1964), J. L. Barclay, J. B. Bream, D. J. Hadley, and D. G. Stewart (to Distillers Company Ltd.).

3. M. A. Dalin, I. K. Kolchin, and B. R. Serebryakov, *Acrylonitrile*, Technomic, Westport, Conn., 1971, 161–162.
4. R. M. Paterson, M. I. Bornstein, and E. Garshick, *Assessment of Acrylonitrile as a Potential Air Pollution Problem*, GCA-TR-75-32-G(6), GCA Corporation, 1976.
5. *IARC Monographs on the Evaluation of the Carcinogenic Risk of Chemicals to Humans*, International Agency for Research on Cancer, Vol. 19, Feb. 1979.
6. Ref. 3, p. 166.
7. N. M. Sokolov, *Rev. Chim.* **20**, 169 (1969).
8. N. M. Sokolov, *Proc. Int. Symp. Distill.* **3**, 110 (1969).
9. N. M. Sokolov, N. N. Sevryugova, and N. M. Zhavoronkor, *Theor. Osn. Khim. Tekhnol.* **3**, 449 (1969).
10. T. Fukuyama and K. Kuchitsu, *J. Mol. Struct.* **5**, 131 (1970).
11. *The Chemistry of Acrylonitrile*, 1st ed., American Cyanamid Company, New York, 1951, 14–15.
12. *EPA/NIH Mass Spectral Data Base*, Vol. 1, U.S. National Bureau of Standards, Washington, D.C., 1978, p. 5.
13. M. C. L. Gerry, K. Yamada, and G. Winnewisser, *J. Phys. Chem. Ref. Data* **8**, 107 (1979).
14. S. Suzer and L. Andrews, *J. Phys. Chem.* **93**, 2123 (1989).
15. A. R. H. Cole and A. A. Green, *J. Mol. Spectrosc.* **48**, 246 (1973).
16. V. I. Khvostenko, I. I. Furlei, V. A. Mazunov, and R. S. Rafikov, *Dokl. Akad. Nauk SSSR* **213**, 1364 (1973).
17. J. A. Nuth and S. Glicker, *J. Quant. Spectosc. Radiat. Trans.* **28**, 223 (1982).
18. G. Cazzoli and Z. Kisiel, *J. Mol. Spectrosc.* **130**, 303 (1988).
19. Ref. 3, 120–159.
20. Ref. 11, 21–51.
21. U.S. Pats. 3,597,481 (Aug. 3, 1971), 3,631,104 (Dec. 28, 1971), Re. 31,430 (Oct. 25, 1983), B. A. Tefertiller and C. E. Habermann (to The Dow Chemical Company).
22. U.S. Pat. 4,048,226 (Sept. 13, 1977), W. A. Barber and J. A. Fetchin (to American Cyanamid Co.).
23. U.S. Pat. 4,086,275 (Apr. 25, 1978), K. Matsuda and W. A. Barber (to American Cyanamid Co.).
24. U.S. Pat. 4,178,310 (Dec. 11, 1979), J. A. Fetchin and K. H. Tsu (to American Cyanamid Co.).
25. U.S. Pat. 3,193,480 (July 6, 1965), M. M. Baizer, C. R. Campbell, R. H. Fariss, and R. Johnson (to Monsanto Chemical Co.).
26. U.S. Pat. 3,529,011 (Sept. 15, 1970), J. W. Badham (to Imperial Chemical Industries Australia Ltd.).
27. Eur. Pat. Appl. E.P. 314,383 (May 3, 1989), G. Shaw and J. Lopez-Merono (to Imperial Chemical Industries PLC).
28. J. L. Callahan, R. K. Grasselli, E. C. Milberger, and H. A. Strecker, *Ind. Eng. Chem. Prod. Res. Dev.* **9**, 134 (1970).
29. *Chem. Eng. News* **67**(2), 23 (1989).
30. U.S. Pat. 2,481,826 (Sept. 13, 1949), J. N. Cosby (to Allied Chemical & Dye Corp.).
31. J. D. Burrington, C. T. Kartisek, and R. K. Grasselli, *J. Catal.* **87**, 363 (1984).
32. C. R. Adams and T. J. Jennings, *J. Catal.* **2**, 63 (1963).
33. C. R. Adams and T. J. Jennings, *J. Catal.* **3**, 549 (1964).
34. G. W. Keulks, *J. Catal.* **19**, 232 (1970).
35. G. W. Keulks and L. D. Krenzke, *Proceedings of the International Congress on Catalysis, 6th, 1976*, The Chemical Society, London, 1977, p. 806; *J. Catal.* **61**, 316 (1980).
36. J. F. Brazdil, D. D. Suresh, and R. K. Grasselli, *J. Catal.* **66**, 347 (1980).

37. R. K. Grasselli, J. F. Brazdil, and J. D. Burrington, *Proceedings of the International Congress on Catalysis, 8th, 1984*, Verlag Chemie, Weinheim, 1984, Vol. **V**, p. 369.
38. R. K. Grasselli, *Applied Catal.* **15**, 127 (1985).
39. R. K. Grasselli, *React. Kinet. Catal. Lett.* **35**, 327 (1987).
40. J. F. Brazdil, L. C. Glaeser, and R. K. Grasselli, *J. Catal.* **81**, 142 (1983).
41. L. C. Glaeser, J. F. Brazdil, M. A. Hazle, M. Mehicic, and R. K. Grasselli, *J. Chem. Soc. Faraday Trans. 1* **81**, 2903 (1985).
42. R. G. Teller, J. F. Brazdil, and R. K. Grasselli, *Acta Cryst.* **C40**, 2001 (1984).
43. R. G. Teller, J. F. Brazdil, R. K. Grasselli, R. T. L. Corliss, and J. Hastings, *J. Solid State Chem.* **52**, 313 (1984).
44. R. G. Teller, J. F. Brazdil, R. K. Grasselli, and W. Yelon, *J. Chem. Soc. Faraday Trans. 1*, **81**, 1693 (1985).
45. M. R. Antonio, R. G. Teller, D. R. Sandstrom, M. Mehicic, and J. F. Brazdil, *J. Phys. Chem.* **92**, 2939 (1988).
46. M. R. Antonio, J. F. Brazdil, L. C. Glaeser, M. Mehicic, and R. G. Teller, *J. Phys. Chem.* **92**, 2338 (1988).
47. J. F. Brazdil and R. K. Grasselli, *J. Catal.* **79**, 104 (1983).
48. J. F. Brazdil, L. C. Glaeser, and R. K. Grasselli, *J. Phys. Chem.* **87**, 5485 (1983).
49. A. W. Sleight, in J. J. Burton and R. L. Garten, eds., *Advanced Materials in Catalysis*, Academic Press, New York, 1977, 181–208.
50. R. K. Grasselli and J. D. Burrington, *Adv. Catal.* **30**, 133 (1981).
51. K. Weissermel and H. J. Arpe, *Industrial Organic Chemistry*, A. Mullen, trans., Verlag Chemie, New York, 1978, 266–267.
52. U.S. Pat. 2,690,452 (Sept. 28, 1954), E. L. Carpenter (to American Cyanamid Co.).
53. U.S. Pat. 2,729,670 (Jan. 3, 1956), P. H. DeBruin (to Stamicarbon N.V.).
54. *Chem. Eng. News*, **23**(20), 1841 (Oct. 25, 1945).
55. D. J. Hadley and E. G. Hancock, eds., *Propylene and Its Industrial Derivatives*, Halsted Press, a division of John Wiley & Sons, Inc., New York, 1973, p. 418.
56. K. Sennewald, *World Petroleum Congress Proceedings, 5th, 1959*, Section IV, Paper 19, 217–227.
57. U.S. Pat. 2,554,482 (May 29, 1951), N. Brown (to E. I. du Pont de Nemours & Co., Inc.).
58. U.S. Pat. 2,385,552 (Sept. 25, 1945), L. R. U. Spence and F. O. Haas (to Rohm and Haas Co.).
59. U.S. Pat. 2,736,739 (Feb. 28, 1956), D. C. England and G. V. Mock (to E. I. du Pont de Nemours & Co., Inc.).
60. U.S. Pat. 3,184,415 (May 18, 1965), E. B. Huntley, J. M. Kruse, and J. W. Way (to E. I. du Pont de Nemours & Co., Inc.).
61. U.S. Pats. 4,783,545 (Nov. 8, 1988), 4,837,233 (June 6, 1989), and 4,871,706 (Oct. 3, 1989), L. C. Glaeser, J. F. Brazdil, and M. A. Toft (to The Standard Oil Co.).
62. Brit. Pats. 1,336,135 (Nov. 7, 1973), N. Harris and W. L. Wood; 1,336,136 (Nov. 7, 1973), N. Harris (to Power-Gas Ltd.).
63. U.S. Pat. 3,833,638 (Sept. 3, 1974), W. R. Knox, K. M. Taylor, and G. M. Tullman (to Monsanto Co.).
64. U.S. Pats. 4,849,537 (July 18, 1989) and 4,849,538 (July 18, 1989), R. Ramachandran, D. L. MacLean, and D. P. Satchell, Jr. (to The BOC Group, Inc.).
65. U.S. Pat. 4,609,502 (Sept. 2, 1986), S. Khoobiar (to The Halcon SD Group, Inc.).
66. U.S. Pat. 3,751,443 (Aug. 7, 1973), K. E. Khchelan, O. M. Revenko, A. N. Shatalova, and E. G. Gelperina.
67. J. Perkowski, *Przem. Chem.* **51**, 17 (1972).
68. U.S. Pat. 4,746,753 (May 24, 1988), J. F. Brazdil, D. D. Suresh, and R. K. Grasselli (to The Standard Oil Co. (Ohio)).

69. U.S. Pat. 4,503,001 (Mar. 5, 1985), R. K. Grasselli, A. F. Miller, and H. F. Hardman (to The Standard Oil Co. (Ohio)).
70. U.S. Pat. 4,228,098 (Oct. 14, 1980), K. Aoki, M. Honda, T. Dozono, and T. Katsumata (to Asahi Kasei Kogyo Kabushiki Kaisha).
71. *World Acrylonitrile and Derivatives Supply/Demand Report*, PCI-Fibres & Raw Materials, 1998.
72. Ref. 3, 163–165.
73. *Annual Book of ASTM Standards, E 1178-97*, American Society for Testing and Materials, Philadelphia, Pa., 1999.
74. *Annual Book of ASTM Standards, E 203-96*, American Society for Testing and Materials, Philadelphia, Pa., 1999.
75. *Annual Book of ASTM Standards, E 299-97*, American Society for Testing and Materials, Philadelphia, Pa., 1999.
76. *Material Safety Data Sheet Number 1386*, BP Chemicals Inc., Cleveland, Ohio, 1989.
77. *Assessment of Human Exposures to Atmospheric Acrylonitrile*, SRI International, Menlo Park, Calif., 1979.
78. U.S. Pat. 3,489,789 (Jan. 13, 1970), R. A. Dewar and M. A. Riddolls (to Imperial Chemical Industries Australia Ltd.).
79. U.S. Pat. 3,549,685 (Dec. 22, 1970), J. W. Badham, P. J. Gregory, and J. B. Glen (to Imperial Chemical Industries Australia Ltd.).
80. International Agency for Research on Cancer (IARC). (1999, In Press) *In IARC Monographs on the Evaluation of the Carcinogenic Risk of Chemicals to Humans: Re-Evaluation of Some Organic Chemicals, Hydrazine and Hydrogen Peroxide*. Vol. 71.

James F. Brazdil
BP, Nitriles Catalysis Research

ACRYLONITRILE–BUTADIENE–STYRENE (ABS) POLYMERS

1. Introduction

Acrylonitrile–butadiene–styrene (ABS) polymers [9003-56-9] are composed of elastomer dispersed as a grafted particulate phase in a thermoplastic matrix of styrene and acrylonitrile copolymer (SAN) [9003-54-7]. The presence of SAN grafted onto the elastomeric component, usually polybutadiene or a butadiene copolymer, compatabilizes the rubber with the SAN component. Property advantages provided by this graft terpolymer include excellent toughness, good dimensional stability, good processability, and chemical resistance. Property balances are controlled and optimized by adjusting elastomer particle size, morphology, microstructure, graft structure, and SAN composition and molecular weight. Therefore, although the polymer is a relatively low cost engineering thermoplastic the system is structurally complex. This complexity is advantageous in that altering these structural and compositional parameters allows considerable versatility in the tailoring of properties to meet specific product requirements. This versatility may be even further enhanced by adding various monomers to raise the heat deflection temperature, impart transparency, confer flame retardancy, and, through alloying with other polymers, obtain special product features. Consequently, research and development in ABS systems is active and continues to offer promise for achieving new product opportunities.

2. Physical Properties

The range of properties typically available for general purpose ABS is illustrated in Table 1 (1). Numerous grades of ABS are available including new alloys and specialty grades for high heat, plating, flaming-retardant, or static dissipative product requirements (1,2). Reference 1 discusses stress–strain behavior, creep, stress relaxation, and fatigue in ABS materials.

2.1. Impact Resistance. Toughness is a primary consideration in the selection of ABS for many applications. ABS is structured to dissipate the energy of an impact blow through shear and dilational modes of deformation. Upon impact, the particulate rubber phase promotes both the initiation and termination of crazes. Crazes are regions of considerable strength that contain both voids and polymer fibrils oriented in the stress direction. Crazes are terminated by mutual interference or are stopped by other rubber particles, thereby dissipating energy without the formation of a crack, which would lead to catastrophic

Table 1. **Material Properties of General Purpose and Heat Distortion Resistant ABS**[a]

Properties	ASTM Method	High impact	Medium impact	Heat resistant
notched Izod impact at RT, J/m[b]	D256	347–534	134–320	107–347
tensile strength, MPa[c]	D638	33–43	30–52	41–52
tensile modulus, GPa[d]	D638	1.7–2.3	2.1–2.8	2.1–2.6
flexural modulus, GPa[d]	D790	1.7–2.4	2.2–3.0	2.1–2.8
elongation to yield, %	D638	2.8–3.5	2.3–3.5	2.8–3.5
Rockwell hardness	D785	80–105	105–112	100–111
heat deflection[e], °C at 1820 kPa[f]	D648	96–102	93–104	104–116
heat deflection[e], °C at 455 kPaf	D648	99–107	102–107	110–118
Vicat softening pt, °C	D1525	91–106	94–107	104–118
coefficient of linear thermal expansion, $\times 10^5$ cm/cm·°C	D696	9.5–11.0	7.0–8.8	6.5–9.2
dielectric strength, kV/mm	D149	16–31	16–31	14–35
dielectric constant, $\times 10^6$ Hz	D150	2.4–3.8	2.4–3.8	2.4–3.8

[a] Ref. 1.
[b] To convert J/m to ft·lb/in. divide by 53.4.
[c] To convert MPa to psi multiply by 145.
[d] To convert GPa to psi multiply by 145,000.
[e] Annealed.
[f] To convert kPa to psi multiply by 0.145.

failure. Shear deformation also contributes to stress relaxation. The behavior of the rubber phase is understood from analyses of the stress distribution surrounding the particulate rubber phase. The rubber component may exist in a state of triaxial tension due to the higher rate of thermal contraction of the rubber compared to styrene–acrylonitrile copolymer upon cooling after molding. Crazes, in general, are initiated at local points of stress concentration. Mechanisms and the factors affecting fracture toughness have been discussed in detail in the literature (3–12).

The inherent ductility of the matrix phase depends on the composition of the SAN copolymer and is reported to increase with increasing acrylonitrile content (3). Controlling rubber particle size, distribution, and microstructure are important in optimizing impact strength. Good adhesion between the rubber and the matrix phase is also essential and is achieved by an optimized graft structure (3,8,10,13). Typically, toughness is increased by increasing the rubber content and the molecular weight of the ungrafted SAN.

2.2. Rheology. Effects of structure of ABS on viscosity functions can be distinguished by considering effects at lower shear rates (<10/s) vs higher shear rates. At higher shear rates melt viscosity is primarily determined by ungrafted SAN structure and the percentage of graft phase. The modulus curves correspond in their shape to that of the ungrafted SAN component, and the rubber particle type and concentration have little effect on the temperature dependence of the viscosity function (14). The extrudate swell, however, becomes smaller with increasing rubber concentration (15).

By contrast, the graft phase structure has a marked effect on viscosity at small deformation rates. The long time relaxation spectra are affected by rubber particle–particle interactions (16,17), which are strongly dependent on particle size, grafting, morphology, and rubber content. Depending on particle surface area, a minimum amount of graft is needed to prevent the formation of three-dimensional networks of associated rubber particles (17). Thus at low shear rates ABS can behave similarly to a cross-linked rubber; the network structure, however, is dissolved by shearing forces. Extensive studies on the viscoelastic properties of ABS in the molten state have been reported (14–21). Effects of lubricants and other nonpolymeric components have also been described (22).

2.3. Gloss. Surface gloss values can be achieved ranging from a very low matte finish at <10% (60° Gardner) to high gloss in excess of 95%. Gloss is dependent on the specific grade and the mold or polishing roll surface.

2.4. Electrical Properties. (See Table 1.) A new family of ABS products exhibiting electrostatic dissipative properties without the need for nonpolymeric additives or fillers (carbon black, metal) is now also commercially available (2).

2.5. Thermal Properties. ABS is also used as a base polymer in high performance alloys. Most common are ABS–polycarbonate alloys which extend the property balance achievable with ABS to offer even higher impact strength and heat resistance (2).

2.6. Color. ABS is sold as an unpigmented powder, unpigmented pellets, precolored pellets matched to exacting requirements, and "salt-and-pepper" blends of ABS and color concentrate. Color concentrates can also be used for online coloring during molding.

3. Chemical Properties

The behavior of ABS may be inferred from consideration of the functional groups present within the polymer.

3.1. Chemical Resistance. The term chemical resistance is generally used in an applications context and refers to resistance to the action of solvents in causing swelling or stress cracking as well as to chemical reactivity. In ABS the polar character of the nitrile group reduces interaction of the polymer with hydrocarbon solvents, mineral and vegetable oils, waxes, and related household and commercial materials. Good chemical resistance provided by the presence of acrylonitrile as a comonomer combined with relatively low water absorptivity ($<1\%$) results in high resistance to staining agents (eg, coffee, grape juice, beef blood) typically encountered in household applications (23).

Like most polymers, ABS undergoes stress cracking when brought into contact with certain chemical agents under stress (23,24). Injection molding conditions can significantly affect chemical resistance, and this sensitivity varies with the ABS grade. Certain combinations of melt temperature, fill rate, and packing pressure can significantly reduce stress cracking resistance, and this effect is interactive in complex ways with the imposed stress level that the part is subjected to in service. Both polymer orientation and stress appear to be considerations; thus critical strains can be higher in the flow direction (25). Consequently, all media to be in contact with the ABS part during service should be evaluated under anticipated end-use conditions.

3.2. Processing Stability. Processing can influence resultant properties by chemical and physical means (26,27). Degradation of the rubber and matrix phases has been reported under very severe conditions (28). Morphological changes may become evident as agglomeration of dispersed rubber particles during injection molding at higher temperatures (28). Physical effects such as orientation and molded-in stress can have marked effects on mechanical properties. Thus the proper selection and control of process variables are important to maintain optimum performance in molded parts. Antioxidants (qv) added at the compounding step have been shown to help retention of physical properties upon processing (26).

Appearance changes evident under certain processing conditions include color development (26), changes in gloss (28), and splaying. Discoloration may be minimized by reducing stock temperatures during molding or extrusion. Splaying is the formation of surface imperfections elongated in the direction of flow and is typically caused by moisture, occluded air, or gaseous degradation products; proper drying conditions are essential to prevent moisture-induced splay.

Techniques for evaluating processing stability and mechanochemical effects include using a Brabender torque rheometer (29,30), injection molding (26,28), capillary rheometry (26,28), and measuring melt index as a function of residence time (26).

3.3. Thermal Oxidative Stability. ABS undergoes autoxidation and the kinetic features of the oxygen consumption reaction are consistent with an autocatalytic free-radical chain mechanism. Comparisons of the rate of oxidation of ABS with that of polybutadiene and styrene–acrylonitrile copolymer indicate that the polybutadiene component is significantly more sensitive to oxidation than the thermoplastic component (31–33). Oxidation of polybutadiene under these conditions results in embrittlement of the rubber because of cross-linking; such embrittlement of the elastomer in ABS results in the loss of impact resistance. Studies have also indicated that oxidation causes detachment of the grafted styrene–acrylonitrile copolymer from the elastomer which contributes to impact deterioration (34).

Examination of oven-aged samples has demonstrated that substantial degradation is limited to the outer surface (34), ie, the oxidation process is diffusion limited. Consistent with this conclusion is the observation that oxidation rates are dependent on sample thickness (32). Impact property measurements by high speed puncture tests have shown that the critical thickness of the degraded layer at which surface fracture changes from ductile to brittle is about 0.2 mm. Removal of the degraded layer restores ductility (34). Effects of embrittled surface thickness on impact have been studied using ABS coated with styrene–acrylonitrile copolymer (35).

Antioxidants have been shown to improve oxidative stability substantially (36,37). The use of rubber-bound stabilizers to permit concentration of the additive in the rubber phase has been reported (38–40). The partitioning behavior of various conventional stabilizers between the rubber and thermoplastic phases in model ABS systems has been described and shown to correlate with solubility parameter values (41). Pigments can adversely affect oxidative stability (32). Test methods for assessing thermal oxidative stability include oxygen absorption (31,32,42), thermal analysis (43,44), oven aging (34,45,46), and chemiluminescence (47,48).

3.4. Photooxidative Stability. Unsaturation present as a structural feature in the polybutadiene component of ABS (also in high impact polystyrene, rubber-modified PVC, and butadiene-containing elastomers) also increases liability with regard to photooxidative degradation (49–51). Such degradation only occurs in the outermost layer (52,53), and impact loss upon irradiation can be attributed to embrittlement of the rubber and possibly to scission of the grafted styrene–acrylonitrile copolymer (49,54). Oxidative degradation induced by prior processing may affect photosensitivity (49,55). Appearance changes such as yellowing are also induced by irradiation and caused by chromophore formation in both the polybutadiene and styrene–acrylonitrile copolymer components (49,56). Comparative data on ABS with other acrylic-based plastics have been reported (57).

Applications involving extended outdoor exposure, especially in direct sunlight, require protective measures such as the use of stabilizing additives, pigments, and protective coatings and film. Light stabilizers provide some measure of protection (58,59) as illustrated by the very successful use of ABS in interior automotive trim. Effects of polymer-bound stabilizers have been described (60). Pigments can significantly enhance stability. Paints are also highly effective in minimizing weather degradation (61). A particularly effective technique for sheet products is the lamination during extrusion of an acrylic film to ABS. Test methods for assessing light stability include outdoor exposure (62) and accelerated testing (62,63). Reactivity toward singlet oxygen has been reported (64). The current trend in accelerated light-aging for ABS in automotive applications is the use of xenon arc testing.

3.5. Flammability. The general purpose grades are usually recognized as 94 HB according to the requirements of Underwriters' Laboratories UL94 and also meet the requirements, dependent on thickness, of the Motor Vehicle Safety Standard 302. Flame-retardant (FR) grades (V0, V1, and V2) are also available which meet Underwriters' UL 94/94 5V and Canadian Standards Association (CSA) requirements beginning at a minimum thickness of 1.57 mm. Flame retardancy is achieved by utilizing halogen in combination with antimony oxide or by alloys with PVC or PC. A new FR grade utilizing polymer-bound bromine has been developed to avoid additive bloom and toxicity (65).

4. Manufacture

All manufacturing processes for ABS involve the polymerization of styrene and acrylonitrile monomers in the presence of an elastomer (typically polybutadiene or a butadiene copolymer) to produce SAN that has been chemically bonder or "grafted" to the rubber component termed the "substrate."

4.1. Rubber Chemistry. The rubber substrate is typically produced by the free-radical polymerization of butdiene. The radical source can be provided by either thermal decomposition or oxidation-reduction (redox) systems. The primary product is primarily 1,4-polybutadiene with some 1,2-polybutadiene, which contains a pendent vinyl group. Cross-linking of polymer occurs at high conversion through abstraction of reactive allylic sites or by copolymerization through double bonds (especially the double bonds in the more sterically accessibly pendent vinyl groups). Rubber cross-linking is controlled by the use of chain-transfer agents and the concentration and type of the initiator used; the reaction can also be affected by chain transfer to emulsifiers. For emulsion ABS, the rubber is typically both produced and subsequently used for grafting as a latex.

4.2. Graft Chemistry. Grafting of styrene and acrylonitrile onto a rubber substrate is the essence of the ABS process. Grafting is a free-radical process initiated by the abstraction of allylic hydrogens on the rubber substrate or by copolymerization through double bonds that are pendent or internal in the rubber substrate (66). Initiator level and type affects the extent of grafting (66–72) with oxyradicals yielding a higher degree of grafting than carbon radicals because of higher rates of abstraction from the rubber substrate. Chain-transfer agents are also used in controlling overall degree of grafting and grafting and graft molecular weight.

Ungrafted SAN is formed concurrently with grafted SAN, with the ratio controlled by factors that include temperature, chain-transfer agent, pendent vinyl content of rubber, initiator level, and initiator type (66–74). As previously described, occlusions of SAN can also form within the rubber particles with the mass process leading to significantly higher occulusion levels than the emulsion process (75,76). In the mass process, block copolymers of styrene and butadiene can be added to obtain unusual particle morphologies (eg, coil, rod, capsule, cellular) (75).

4.3. Emulsion Process. The emulsion process for making ABS has been commercially practiced since the early 1950s. Its advantage is the capability of producing ABS with a wide range of compositions, particularly higher rubber contents than are possible with other processes. Mixing and transfer of the heat of reaction in an emulsion polymerization is achieved more easily than in the mass polymerization process because of the low viscosity and good thermal properties of the water phase. The energy requirements for the emulsion process are generally higher because of the energy usage in the polymer recovery area. The emulsion polymerization process is typically a two-stage reaction process (77,78).

In the first stage, a rubber substrate, primarily composed of polybutadiene, is made using an emulsion polymerization process. The desired particle size of the rubber is either obtained by direct growth during polymerization or by an agglomeration process subsequent to polymerization. In a second-stage reaction, styrene and acrylonitrile are grafted onto the rubber substrate by emulsion polymerization. After the graft reaction is complete, the polymer can be recovered from the graft latex and compounded into a final pellet product (78–82).

4.4. Rubber Substrate Process. The rubber substrate can be made by a variety of different reaction processes including batch, semi-batch, and continuous (83). Butadiene monomer is primarily used in the substrate reaction, but comonomers such as styrene and acrylonitrile are common (81). The amount and type of comonomer employed will affect the glass transition of the rubber substrate and, thereby, influence the impact properties of the ABS polymer. Oxidation-reduction systems (eg, hydrogen peroxide and iron) or thermal initiators (eg, potassium persulfate or azobisisobutyronitrile) are used to initiate polymerization. Cross-link density is controlled by type and level of initiator, type and level of chain-transfer agent, reaction temperature, degree of conversion, or by the addition of comonomers. It is important to note that the graft process also can affect the cross-link density of the rubber. Various surfactant types can be employed to emulsify the monomer and stabilize the latex particles. Standard fatty acid soaps and derivatives are the most common emulsifiers employed; however, detergents such as sodium dobenzyl sulfonate and sodium lauryl sulfate can also be used. The use of nonionic surfactants has been reported (84). The "soap-free" emulsion polymerization of butadiene is possible using reactive surfactants (85), functional monomers such as acrylic acid (86), or high levels of potassium persulfate (87). The incorporation of surfactants into the polymer backbone provides the advantage of minimizing low molecular by-products in the final polymer that could result in mold buildup of juicing.

The incorporation of comonomers into the rubber substrate can be useful in achieving specialized performance of the final ABS polymer, such as adjusting the refractive index of the rubber phase to better match the continuous SAN phase to achieve a clear or more translucent ABS product (88). The incorporation of polymerizable antioxidants or uv stabilizers has also been reported (89). Typically, these modications increase the cost of ABS and are only employed for specialized applications.

Reactor productivity can be effected by various factors including initiator type, latex particle size, monomer purity, chain-transfer agents, and reaction temperature (83). As previously described, rubber particle size and distribution are important factors controlling the final properties of the ABS polymer. Large particles can be obtained by direct growth in the reactor, but much longer reaction times are needed. Comonomers such as AN can be added to speed the reaction rate and achieve relatively large particles in less time (90,91). Productivity can also be improved by the use of antifouling agents to minimize buildup of polymer on reactor heat-transfer surfaces (92–94). These antifouling agents improve heat transfer and minimize the time the reactor is down for oleaning.

4.5. Graft Process. Grafted SAN is critical to achieving effective dispersion of the rubber in the matrix phase, with key factors being SAN composition and rubber particle surface coverage. The composition of the grafted SAN depends on the monomer-feed composition and the monomer reactivity ratio. The composition of the polymer formed will equal the feed at the azeotropic composition, which occurs at ~3/1 mass ratio of styrene-to-acrylonitrile (76,84), and compositional drift will occur at monomer feed compositions other than the azeotropic concentration. Note that in aqueous systems, the difference in water phase solubility of acrylonitrile vs styrene can also perturb monomer concentrations at the reaction site and, thus, affect compositional drift. Polymerization techniques such as continuous vs batch processes and controlling pump rates can be used to control compositional drift (95–101). Surface coverage is controlled by rubber particle surface area and is effected by factors including initiator type, monomer feed to rubber level, and chain-transfer agents.

4.6. Resin Recovery Process. Typically, the polymer is recovered by the addition of coagulants which destabilize the ABS latex. Different coagulants are used depending on the surfactant. Thus, strong and weak acids work well with fatty acid soaps, and metal salts are used with acid stable soaps (102). The use of nonionic coagulants has also been reported (103,104). Acrylic latices have been used to control the coagulation process and obtain a narrow resin particle-size distribution (105).

Once coagulated, the resulting slurry can then be filtered or centrifuged to recover wet ABS resin, which is then dried to a low moisture content. A variety of dryers can be used for ABS, including tray, fluid bed, and rotary kiln-type dryers. Other methods of recovery have been employed such as spray drying (106) and extruder dewatering (107). Spray drying allows for good control of the final particle size of the resin, but uses a significant amount of energy in the drying process. In extruder dewatering, the latex is either directly fed into the extruder or is first coagulated and then fed into the extruder. Extruder dewatering allows for more efficient stripping and recovery of unreacted monomer than standard drying process.

4.7. Air and Water Treatment. The emulsion process exerts a greater demand on wastewater treatment than other processes (suspension or mass) because of the quantity of water used, and air emissions may be higher because of the types of process equipment employed. Recent federal and state EPA regulations governing air emission from ABS facilities affect the level of styrene, acrylonitrile, butadiene, and other volatile organic compounds that can be emitted into the air or sent to wastewater treatment facilities. In some cases, effluent water can be recycled and reused, but ultimately the water must be discharged, requiring treatment of the water prior to discharge. Air emissions from an emulsion ABS process can be reduced by improving the conversion of the monomers (108), the installation of equipment to strip and recover monomers, or the installation of end-of-pipe controls. End-of-pipe controls such as regenerative catalytic oxidation, regenerative thermal oxidation, fixed and fluid bed carbon absorption, and biofiltration are viable means of addressing air emission issues (109).

4.8. Mass Polymerization Process. In the mass (110–118) ABS process, the polymerization is conducted in a monomer medium rather than in water, usually employing a series of two or more continuous reactors. The rubber used in this process is most commonly a solution polymerized linear polybutadiene (or copolymer containing sytrene), although some mass processes utilize emulsion-polymerized ABS with a high rubber content for the rubber component (119). If a linear rubber is used, a solution of the rubber in the monomers is prepared for feeding to the reactor system. If emulsion ABS is used as the source of rubber, a dispersion of the ABS in the monomers is usually prepared after the water has been removed from the ABS latex.

In the mass process (120) using linear rubber, the rubber initially dissolved in the monomer mixture will phase separate, forming discrete rubber particles as SAN polymerization procedes. This process is referred to as phase inversion since the continuous phase shifts from rubber to SAN during the course of polymerization. Special reactor designs are used to control to phase inversion portion of the reaction (111,113–116). By controlling the shear rate in the reactor, the rubber particle size can be modified to optimize properties. Grafting of some of the SAN onto the rubber particles occurs as in the emulsion process. Typically, the mass-produced rubber particles are larger than those of emulsion-based ABS and contain much larger internal occlusions of SAN. The reaction recipe can include polymerization initiators, chain-transfer agents, and other additives. Diluents are sometimes used to reduce the viscosity of the monomer and polymer mixture to facilitate processing at high conversion. The product from the reactor system is devolatilized to remove the unreacted monomers and is then pelletized. Equipment used for devolatilization includes single- and twin-screw extruders, and flash and thin film/strand evaporators. Unreacted monomers are recovered for recycled back to the reactors to improve the process yield.

The mass ABS process was originally adapted from the mass polystyrene process (121). Mass produced ABS typically has very good unpigmented color and is usually somewhat more translucent because of the large rubber phase particle size and low rubber content. Increased translucency can reduce the concentration of colorants required. The extent of rubber incorporation is limited to approximately 20% because of viscosity limitations in the process; however, the mass-produced grafted rubber can be more efficient (on an equal percent rubber basis) at impact modification than emulsion-grafted rubber because of the presence of high occlusion levels in the rubber phase. The surface gloss of the mass-produced ABS is generally lower than that of emulsion ABS because of the presence of the larger rubber particles, but recent advances provide additional flexibility to achieve higher gloss (111–115).

4.9. Suspension Process. The suspension process utilizes a mass (122) or emulsion reaction (123,124) to produce a partially converted mixture of polymer and monomer and then employs a suspension process (125) to complete the polymerization. When the conversion of the monomers is approximately 15–30% complete, the mixture of polymer and unreacted monomers is suspended in water with the introduction of a suspending agent. The reaction is continued until a high degree of monomer conversion is attained and then unreacted monomers are stripped from the product before the slurry is centrifuged and dried producing product in the form of small beads. The morphology and properties of the mass suspension product are similar to those of the mass-polymerized product. The suspension process retains some of the process advantages of the water-based emulsion process, such as lower viscosity in the reactor and good heat removal capability.

4.10. Compounding. ABS either is sold as an unpigmented product, in which case the customer may add pigments during the forming process, or it is colored by the manufacturer prior to sale. Much of the ABS produced by the mass process is sold unpigmented; however, precolored resins provide advantages in color consistency. If colorants, lubricants, fire retardants, glass fibers, stabilizers, or alloying resins are added to the product, a compounding operation is required. ABS can be compounded on a range of equipment, including batch and continuous melt mixers, and both single- and twin-screw extruders. The device must provide sufficient dispersive and distributive mixing dependent on formulation ingredients for successful compounding, and low work or low shear counterrotating twin-screw extruders as used in PVC are not recommended. In the compounding step, more than one type of ABS may be employed (ie, emulsion and mass-produced) to obtain an optimum balance of properties for a specific application. Products can also be made in the compounding process by combining emulsion ABS having a high rubber content with mass- or suspension-polymerized SAN.

5. Processing

Good thermal stability plus shear thinning allow wide flexibility in viscosity control for a variety of processing methods. ABS exhibits non-Newtonian viscosity behavior. For example, raising the shear rate one decade from 100/s to 1000/s (typical in-mold shear rates) reduces the viscosity by 75% on a general purpose injection molding grade. Viscosity can also be reduced by raising melt temperature; typically increasing the melt temperature 20 to 30°C within the allowable processing range reduces the melt viscosity by about 30%. ABS can be processed by all the techniques used for other thermoplastics: compression and injection molding, extrusion, calendering, and blow-molding (see PLASTICS PROCESSING). Clean, undegraded regrind can be reprocessed in most applications (plating excepted), usually at 20% with virgin ABS. Postprocessing operations include cold forming; thermoforming; metal plating; painting; hot stamping; ultrasonic, spin, and vibrational welding; and adhesive bonding.

5.1. Material Handling and Drying. Although uncompounded powders are available from some suppliers, most ABS is sold in compounded pellet form. The pellets are either precolored or natural to be used for in-house coloring using dry or liquid colorants or color concentrates. These pellets have a variety of shapes including diced cubes, square and cylindrical strands, and spheroids. The shape and size affect several aspects of material handling such as bulk density, feeding of screws, and drying (qv). Very small particles called fines can be present as a carryover from the pelletizing step or transferring operations; these tend to congregate at points of static charge build up. Certain additives can be used to control static charges on pellets (126).

ABS is mildly hygroscopic. The moisture diffuses into the pellet and moisture content is a reversible function of relative humidity. At 50% relative humidity typical equilibrium moisture levels can be between 0.3 and 0.6% depending on the particular grade of ABS. In very humid situations moisture content can be double this value. Although there is no evidence that this moisture causes degradation during processing, drying is required to prevent voids and splay (127) and achieve optimum surface appearance. Drying down to 0.1% is usually sufficient for general purpose injection molding and 0.05% for critical applications such as plating. For nonvented extrusion and blow-molding operations a maximum of 0.02% is required for optimum surface appearance.

Desiccant hot air hopper dryers are recommended, preferably mounted on the processing equipment. Tray driers are not recommended, but if used the pellet bed should be no more than 5 cm deep. Many variables affect drying rates (128,129); the pellet temperature has a stronger effect than the dew point. Most pellet drying problems can be a result of actual pellet temperatures being too low in the hopper. Large particles dry much more slowly than pellets, thus regrind should be protected from moisture regain. Supplier data sheets should be consulted for specific drying conditions. Several devices are available commercially for analytically determining moisture contents in ABS pellets (130–132). Alternatives to pellet drying are vented injection molding (133) and cavity-air pressurization (counterpressure) (134).

5.2. Injection Molding. *Equipment.* Although plunger machines can be used, the better choice is the reciprocating screw injection machine because of better melt homogeneity. Screws with length-to-diameter ratios of 20:1 and a compression ratio of 2–3:1 are recommended. General purpose screws vary significantly in number and depth of the metering flights; long and shallow metering zones can create melt temperature override which is particularly undesirable with flame-retardent (FR) grades of ABS. Screws with a generous transition length perform best because of better melting rate control (135). Good results have been realized with a long transition "zero-meter" screw design (136). Some comments on the performance of general purpose and two-stage vented screws used for coloring with concentrates is given in reference 137. Guidelines for nozzle and nonreturn valve selection as well as metallurgy are given in references 138 and 139. Gas-nitrided components should be avoided; ion-nitrided parts are acceptable.

A variety of mold types can be used: two plate, three plate, stack, or runnerless. Insulated runner molds are not recommended. If heated torpedoes are used with hot manifold molds, they should be made from a good grade of stainless steel and not from beryllium copper. Molds are typically made from P-20, H-13, S-7, or 420 stainless; chrome or electroless plating is recommended for use with FR grades of ABS. Mold cavities should be well vented (0.05 mm deep) to prevent gas burns. Polished, full round, or trapezoidal runners are recommended; half or quarter round runners are not. Most conventional gating techniques are acceptable (138,139). On polished molds a draft angle of 0.5° is suggested to ease part ejection; side wall texturing requires an additional 1° per 0.025 mm of texture depth. Mold shrinkage is typically in the range of 0.5 to 0.9% (0.005

to 0.009 cm/cm) depending on grade, and the shrinkage value for a given grade can vary much more widely than this because of the design of parts and molding conditions.

Processing Conditions. Certain variables should be monitored, measured, and recorded to aid in reproducibility of the desired balance of properties and appearance. The individual ABS suppliers provide data sheets and brochures specifying the range of conditions that can be used for each product. Relying on machine settings is not adequate. Identical cylinder heater settings on two machines can result in much different melt temperatures. Therefore, melt temperatures should be measured with a fast response hand pyrometer on an air shot recovered under normal screw rpm and back-pressure. Melt temperatures range from 218 to 268°C depending on the grade. Generally, the allowable melt temperature range within a grade is at least 28°C. Excessive melt temperatures cause color shift, poor gloss control, and loss of properties. Similarly, a fill rate setting of 1 cm/s ram travel will not yield the same mold filling time on two machines of different barrel size. Fill time should be measured and adjusted to meet the requirements of getting a full part, and to take advantage of shear thinning without undue shear heating and gas burns. Injection pressure should be adjusted to get a full part free of sinks and good definition of gloss or texture. Hydraulic pressures of less than 13 MPa (1900 psi) usually suffice for most molding. Excessive pressure causes flash and can result in loss of some properties. Mold temperatures for ABS range from 27 to 66°C (60 to 82°C for high heat grades). The final properties of a molded part can be influenced as much by the molding as by the grade of ABS selected for the application (140). The factors in approximate descending order of importance are polymer orientation, heat history, free volume, and molded-in stress. Izod impact strength can vary severalfold as a function of melt temperature and fill rate because of orientation effects, and the response curve is ABS grade dependent (141). The effect on tensile strength is qualitatively the same, but the magnitude is in the range of 5 to 10%. Modulus effects are minimal. Orientation distribution in the part is very sensitive to the flow rate in the mold; therefore, fill rate and velocity-to-pressure transfer point are important variables to control (142). Dart impact is also sensitive to molding variables, and orientation and thermal history can also be key factors (143). Heat deflection temperature can be influenced by packing pressure (144) because of free volume considerations (145). The orientation on the very surface of the part results from an extensionally stretching melt front and can have deleterious effects on electro-plate adhesion and paintability. A phenomenon called the mold-surface-effect, which involves grooving the nonappearance half of the mold, can be employed to reduce unwanted surface orientation on the noncorresponding part surface (146–148). Other information regarding the influence of processing conditions on part quality are given in references 149–153.

Part Design. For optimum economics and production cycle time, wall thicknesses for ABS parts should be the minimum necessary to satisfy service strength requirements. The typical design range is 0.08 to 0.32 cm, although parts outside this range have been successfully molded. A key principle that guides design is avoiding stress concentrators such as notches and sharp edges. Changes in wall thickness should be gradual, sharp corners should be avoided, and generous radii (25% of the wall thickness) used at wall intersections with ribs and bosses. To avoid sinks, rib thickness should be between 50 and 75% of the nominal wall. Part-strength at weld lines can be diminished; thus welds should be avoided if possible or at least placed in noncritical areas of the part (154). Because of polymer orientation, properties such as impact strength vary from point to point on the same part and with respect to the flow direction (140). Locations of highest Izod impact strength can be points of lowest dart impact strength because of the degree and direction of orientation. ABS suppliers can provide assistance with design of parts upon inquiry and through design manuals (155). There are a number of special considerations when designing parts for metal plating to optimize the plating process, plate deposition uniformity, and final part quality (156). ABS parts can be also designed for solid–solid or solid–foam co-injection molding (157) and for gas-assisted-injection molding (158).

5.3. Extrusion. *Equipment.* Since moisture removal is even more critical with extrusion than injection molding, desiccant hot-air hopper drying of the pellets to 0.02% moisture is essential for optimum properties and appearance. The extruder requirements are essentially the same for pipe, profile, or sheet. Two-stage vented extruders are preferred since the improved melting control and volatile removal can provide higher rates and better surface appearance. Barrels are typically 24:1 minimum L/D for single-stage units and 24 or 36:1 for two-stage vented units. The screws are typically 2:1 to 2.5:1 compression ratio and single lead, full flighted with a 17.7° helix angle. Screen packs (20 – 40 mesh = 840 – 420 μm) are recommended.

For sheet, streamlined coat-hanger type dies are preferred over the straight manifold type. Typically, three highly polished and temperature controlled rolls are used to provide a smooth sheet surface and control thickness (159). Special embossing rolls can be substituted as the middle roll to impart a pattern to the upper surface of the sheet. ABS and non-ABS films can be fed into the polishing rolls to provide laminates for special applications, eg, for improved weatherability, chemical resistance, or as decoration. Two rubber pull rolls, speed synchronzied with the polishing rolls, are located far enough downstream to allow sufficient cooling of the sheet; finally, the sheet goes into a shear for cutting into lengths for shipping.

Pipe can be sized using internal mandrels with air pressure contained by a downstream plug or externally using a vacuum bushing and tank. Cooling can be done by immersion, cascade, or mist. Water temperatures of 41 to 49°C at the sizing zone reduce stresses. Foamcore pipe has increased in market acceptance significantly over the last few years, and cooling unit lengths must be longer than for solid pipe. Drawdown should not exceed 10 to 15%.

Profile dies can be flat plates or the streamline type. Flat plate dies are easy to build and inexpensive but can have dead spots that cause hang-up, polymer degradation, and shutdowns for cleaning. Streamlined, chrome-plated dies are more expensive and complicated to build but provide for higher rates and long runs. The land length choice represents a tradeoff; long lands give better quality profile and shape retention but have high pressure drops that affect throughput. Land length to wall thickness ratios are typically 10:1. Drawdown can be used to compensate for die swell but should not exceed 25% to minimize orientation.

Sizing jigs vary in complexity depending on profile design; water mist, fog, or air cooling can be used. The latter gives more precise sizing. Also, water immersion vacuum sizing can be used. Accurate, infinitely adjustable speed control is important to the takeoff end equipment to guarantee dimensional control of the profile.

With sheet or pipe, multilayer coextrusion can be used. Solid outer-solid core coextrusion can place an ABS grade on the outside that has special attributes such as color, dullness, chemical resistance, static dissipation, or fire-retardancy over a core ABS that is less expensive or even regrind. Composites can be created in which the core optimizes desired physical properties such as modulus, whereas the outer layer optimizes surface considerations not inherent in the core material. Solid outer–foam core can provide composites with significant reductions in specific gravity (0.7). Dry blowing agents can be "dusted" onto the pellets or liquid agents injected into the first transition section of the extruder.

Extrusion processing conditions vary depending on the ABS grade and application; vendor bulletins should be consulted for details. Information for assistance in troubleshooting extrusion problems can be found in Reference 160.

5.4. Calendering. The rheological characteristics of the sheet extrusion grades of ABS easily adapt them to calendering to produce film from 0.12 to 0.8 mm thick for vacuum forming or as laminates for sheet. The advantages of this process over extrusion are the capability for thinner gauge product and quick turnaround for short runs.

5.5. Blow Molding. Although ABS has been blow molded for over 20 years, this processing method has been gaining popularity recently for a variety of applications (161). Better blow-molding grades of ABS are being provided by tailoring the composition and rheological characteristics specifically to the process. Whereas existing polyolefin equipment can often be easily modified and adjusted to mold ABS, there are some key requirements that require attention.

Pellet predrying is required down to 0.02 to 0.03% moisture. High shear polyolefin screws must be replaced with low shear 2.0:1 to 2.5:1 screws with *L/D* ratios of 20:1 to 24:1 to keep the melt temperature in the 193 to 221°C optimum range. The land length of the tooling can be reduced to 3:1–5:1 because ABS shows less die swell; this also helps to reduce the melt pressure resulting from the higher viscosity. The accumulator tooling should be streamlined to reduce hang-up and improve re-knit, and be capable of handling the higher pressures required with large programmed parisons. Mold temperatures of 77 to 88°C provide good surface finish. It is recommended that the material vendor be consulted to confirm equipment capability and provide safety and processing information (162).

5.6. Secondary Operations. *Thermoforming.* ABS is a versatile thermoforming material. Forming techniques in use are positive and negative mold vacuum forming, bubble and plug assist, snapback and single- or twin-sheet pressure forming (163). It is easy to thermoform ABS over the wide temperature range of 120 to 190°C. As-extruded sheet should be wrapped to prevent scuffing and moisture pickup. Predrying sheet that has been exposed to humid air prevents surface defects; usually 1 to 3 h at 70–80°C suffices. Thick sheet should be heated slowly to prevent surface degradation and provide time for the core temperature to reach the value needed for good formability. Relatively inexpensive tooling can be made from wood, plaster, epoxies, thermoset materials, or metals. Tools should have a draft angle of 2° to 3° on male molds and ½° to 1° on female molds. More draft may be needed on textured molds. Vacuum hole diameters should not exceed 50% of the sheet thickness. Mold design should allow for 0.003 to 0.008 cm/cm mold shrinkage; exact values depend on mold configuration, the material grade, and forming conditions. Maximum depth of draw is usually limited to part width in simple forming, but more sophisticated forming techniques or relaxed wall uniformity requirements can allow greater draw ratios. Some definitions for draw ratios are given in Reference 164. Pressure forming, with well-designed tools, can make parts approaching the appearance and detailing obtained by injection molding. Additional information on pressure forming is given in Reference 165.

Cold Forming. Some ABS grades have ductility and toughness such that sheet can be cold formed from blanks 0.13–6.4 mm thick using standard metal-working techniques. Up to 45% diameter reduction is possible on the first draw; subsequent redraws can yield 35%. Either aqueous or nonaqueous lubrication is required. More details are available in Reference 166.

5.7. Other Operations. *Metallizing.* ABS can be metallized by electroplating, vacuum deposition, and sputtering. Electroplating (qv) produces the most robust coating; progress is being made on some of the environmental concerns associated with the chemicals involved by the development of a modified chemistry. An advantage to sputtering is that any metal can be used, but wear resistance is not as good as with electroplating. Attention must be paid to the molding and handling of the ABS parts since contamination can affect plate adhesion, and surface defects are magnified after plating. Also, certain aspects of part design become more important with plating; these are covered in References 147 and 156 (see also ELECTROLESS PLATING; METALLIC COATINGS).

Fastening, Bonding, and Joining. Often parts can be molded with various snap-fit designs (167) and bosses to receive rivets or self-tapping screws. Thermal-welding techniques that are easily adaptable to ABS are spin welding (168), hot plate welding, hot gas welding, induction welding, ultrasonic welding, and vibrational welding (169,170). ABS can also be nailed, stapled, and riveted. There are a variety of adhesives and solvent cements for bonding ABS to itself or other materials such as wood, glass, and metals; for more information, contact the material or adhesives suppliers. Joining ABS with materials of different coefficients of thermal expansion requires special considerations when wide temperature extremes are encountered. An excellent review of joining methods for plastics is given in Reference 171.

6. Economic Aspects

6.1. Capacity. Estimated ABS capacity worldwide in 2000 is given in Table 2 (172). Accurate ABS capacity figures are difficult to obtain because significant production capability is considered “swing” and can be used to manufacture polystyrene or SAN as well as ABS. From a regional standpoint, Asia-Pacific has the largest ABS nameplate production capability at 3977 t. The United States has approximately 17% of the world’s capacity at 1068 t. Most suppliers

Table 2. **Worldwide Capacity for ABS Plastics 1994–2000 by Region, 10^3 t**

Region	1994	1995	1996	1997	1998	1999	2000
Western Europe	841	838	865	882	1000	1000	990
Eastern Europe	80	80	81	81	82	82	82
Africa	0	0	0	0	0	0	0
North America	931	894	894	934	1083	1098	1098
Latin America	128	115	127	129	126	230	267
Middle East	0	0	0	0	0	0	0
Asia-Pacific	2130	2535	3158	3292	3712	3857	3977
Total	*4110*	*4462*	*5125*	*5318*	*6003*	*6267*	*6414*

have multiple facilities with the largest producers regionally being GE in North America, Bayer in Europe, and Chi Mei in the Pacific. As shown in Table 3, these three producers account for almost 50% of the world's capacity (172).

Table 4 lists the current U.S. producers of ABS and their capacities. The ABS resin is produced by either continuous mass (or bulk) suspension or emulsion polymerization. ABS contains over 50% styrene and varying amounts of butadiene and acrylonitrile. Some production is from swing capacity. ABS is the largest volume engineering thermoplastic. High impact polystyrene has become almost as expensive as ABS so some manufacturers have changed back to ABS resins (173).

ABS growth in 2002 was 2–4% due in part to new applications in recreational vehicles, solid automotive and appliance business, and increased activity in building and construction. The rise in the price of oil has impacted the

Table 3. **World Capacity of Leading ABS Producers**

Producer	2000 Capacity, 10^3 t	Largest producer in
GE Plastics	855	North America
Bayer	766	Europe
Chi Mei Industrial	1120	Pacific

Table 4. **Producers of ABS Resins and their Capacities**[a] **$\times 10^6$ kg ($\times 10^6$ lb)**

Producer	Capacity
Bayer, Addyston, Ohio	209 (450)
Diamond Polymers, Akron, Ohio	23 (50)
Dow, Allyn's Point, Conn.	25 (55)
Dow, Hanging Rock, Ohio	64 (140)
Dow, Midland, Mich.	88 (195)
GE Plastics, Bay St. Louis, Miss.	177 (280)
GE Plastics, Ottawa, Ill.	215 (475)
GE Plastics, Washington, W.Va.	73 (160)
Total	*819 (1805)*

[a] Ref. 174.

feedstock costs. The market has evolved into general purpose ABS and higher engineered products. Growth is expected at the rate of 2.4% annually.

Demand in 2001 was 0.58×10^9 kg (1.27×10^9 lb). Projected demand for 2005 is 0.63×10^9 kg (1.40×10^9 lb). In recent years U.S. exports have declined as new production in the Far East came on line. The Far East represents 70% of consumption, North America, 15% (173). Demand is expected to grow at an annual rate of 15% from 2000–2005 (174).

6.2. Price. Price history for 1996–2001: \$0.31–0.50/kg (\$0.69–\$1.04/lb) tl, delvd, market high impact injection molding grade. Current \$0.33–0.39/kg (\$0.72–0.85/lb same basis) (173).

7. Analytical Methods

Analytical investigations may be undertaken to identify the presence of an ABS polymer, characterize the polymer, or identify nonpolymeric ingredients. Fourier transform infrared (ftir) spectroscopy is the method of choice to identify the presence of an ABS polymer and determine the acrylonitrile–butadiene–styrene ratio of the composite polymer (175,176). Confirmation of the presence of rubber domains is achieved by electron microscopy. Comparison with available physical property data serves to increase confidence in the identification or indicate the presence of unexpected structural features. Identification of ABS via pyrolysis gas chromatography (177) and dsc (178) has also been reported.

Detailed compositional and molecular weight analyses involve: determining the percentage of grafted rubber; determining the molecular weight and distribution of the grafted SAN and the ungrafted SAN; and determining compositional data on the grafted rubber, the grafted SAN, and the ungrafted SAN. This information is provided by a combination of phase-separation and instrumental techniques. Separation of the ungrafted SAN from the graft rubber is accomplished by ultracentrifugation of ABS dispersions (179,180) which causes sedimentation of the grafted rubber. Cleavage of the grafted SAN from the elastomer is achieved using oxidizing agents such as ozone [10028-15-6] (180,181), potassium permanganate [7722-64-7] (182), or osmium tetroxide [20816-12-0] with *tert*-butylhydroperoxide [75-91-2] (183). Chromatographic and spectroscopic analyses of the isolated fractions provide structural data on the grafted and ungrafted SAN components (184). Information on the microstructure of the rubber is provided by analysis of the cleavage products derived from the substrate (94,96). The extraction of ungrafted rubber has also been reported (185).

Additional information on elastomer and SAN microstructure is provided by ^{13}C-nmr analysis (186). Rubber particle composition may be inferred from glass-transition data provided by thermal or mechanochemical analysis. Rubber particle morphology as obtained by transmission or scanning electron microscopy (187) is indicative of the ABS manufacturing process (77). (See Figs. 1 and 2.)

The isolation and/or identification of nonpolymerics has been described, including analyses for residual monomers (176,188,189) and additives (176,190–192). The determination of localized concentrations of additives within the phases of ABS has been reported; the partitioning of various additives

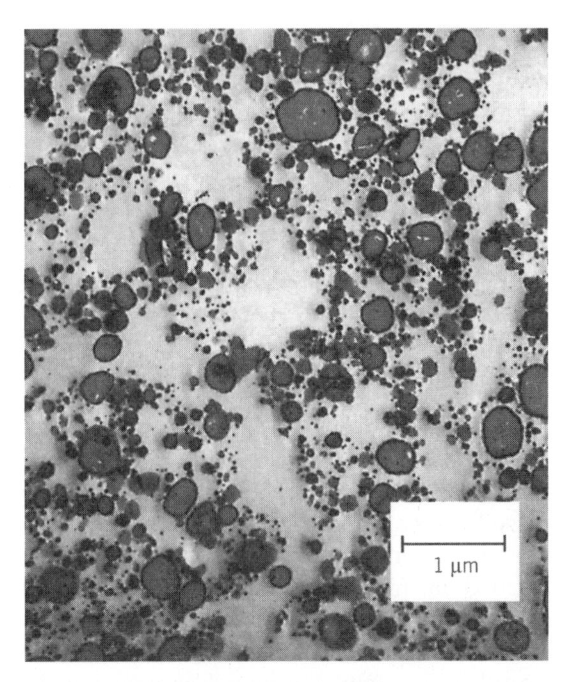

Fig. 1. Transmission electron micrograph of ABS produced by an emulsion process. Staining of the rubber bonds with osmium tetroxide provides contrast with the surrounding SAN matrix phase.

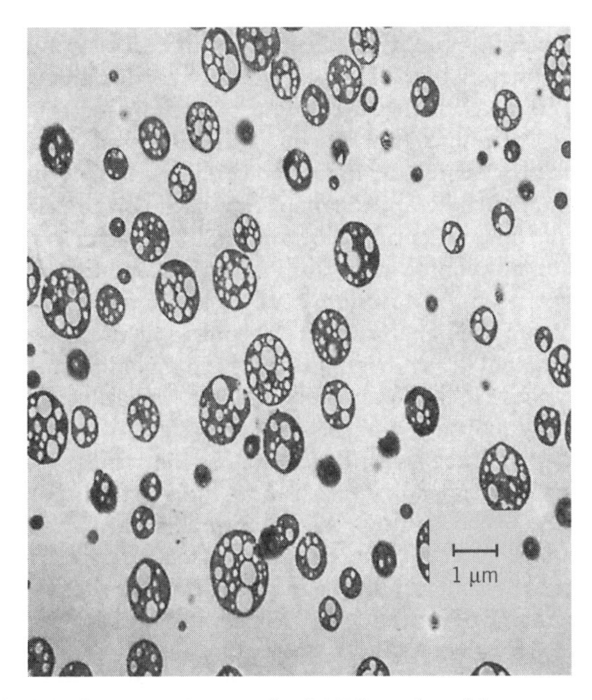

Fig. 2. Transmission electron micrograph of ABS produced by a mass process. The rubber domains are typically larger in size and contain higher concentrations of occluded SAN than those produced by emulsion technology.

between the elastomeric and thermoplastic phases of ABS has been shown to correlate with solubility parameter values (41).

8. Health and Safety Aspects

ABS is a questionable carcinogen. When heated to decomposition, it emits toxic vapors of nitrogen oxide and CN. These products are highly toxic when inhaled (193).

9. Uses

Its broad property balance and wide processing window has allowed ABS to become the largest selling engineering thermoplastic. ABS enjoys a unique position as a "bridge" polymer between commodity plastics and other higher performance engineering thermoplastics.

In 2002 the largest market for ABS resins worldwide was for appliances (23%). The majority of this consumption was for major appliances; extruded/thermoformed door and tank liners lead the way. Transparent ABS grades are also used in refrigerator crisper trays. Other applications in the appliance market include injection-molded housings for kitchen appliances, power tools, vacuum sweepers, sewing machines, and hair dryers. Transportation was the second largest market (21%). Uses are numerous and include both interior and exterior applications. Interior injection-molded applications account for the greatest volume. General purpose and high heat grades have been developed for automotive instrument panels, consoles, door post covers, and other interior trim parts. ABS resins are considered by many the preferred material for components situated above the "waistline" of the car. Exterior applications include radiator grilles, headlight housings, and extruded/thermoformed fascias for large trucks. ABS plating grades also account for significant ABS sales and include applications such as knobs, light bezels, mirror housings, grilles, and decorative trim. ABS in disks for brakes has been reported (194).

Pipe and fittings remain a significant market for ABS at 13%, particularly in North America. ABS foam core technology allows ABS resin to compete effectively with PVC in the primary drain-waste and vent (DWV) pipe market.

A large "value-added" market for ABS is business machines and other electrical and electronic equipment at 11%. Although general purpose injection-molding grades meet the needs of applications such as telephones and micro floppy disk covers, significant growth exists in more demanding flame-retardant applications such as computer housings and consoles.

The use of recycled ABS resin in electric and electronic applications has been reported (195).

Medical application accounted for 4% of use. Miscellaneous applications included toys, luggage, lawn and garden products, shower stalls, furniture and ABS resin blends with other polymers. Miscellaneous uses accounted for 28% of consumption (174).

BIBLIOGRAPHY

"Acrylonitrile Polymers, ABS Resins" in *ECT* 3rd ed., Vol. 1, pp. 442–456, by G. A. Morneau, W. A. Pavelich, and L. G. Roettger, Borg-Warner Chemicals; in *ECT* 4th ed., Vol. 2, pp. 391–411, by D. M. Kulich, J. E. Pace, L. W. Fritch, Jr., and A. Brisimitzakis, GE Plastics; "Acrylonitrile Polymers, ABS Resins" in *ECT* (online), posting date: December 4, 2000, by Donald M. Kulich, John E. Pace, Leroy W. Fritch, Jr., Angelo Brisimitzakis, GE Plastics.

CITED PUBLICATIONS

1. C. T. Pillichody and P. D. Kelley, in I. I. Rubin, ed., *Handbook of Plastic Materials and Technology*, John Wiley & Sons, Inc., New York, 1990, Chapt. 3.
2. R. D. Leaversuch, *Mod. Plast.* **66**(1), 77 (1989).
3. H. Kim, H. Keskkula, and D. R. Paul, *Polymer* **31**, 869 (1990).
4. G. H. Michler, *J. Mater. Sci.* **25**, 2321 (1990).
5. E. M. Donald and E. J. Kramer, *J. Mater. Sci.* **17**, 1765 (1982).
6. L. V. Newmann and J. G. Williams, *J. Mater. Sci.* **15**, 773 (1980).
7. C. B. Bucknall, *Toughened Plastics*, Applied Science Publishers, London, 1977.
8. M. Rink, T. Ricco, W. Lubert, and A. Pavan, *J. Appl. Polym. Sci.* **22**, 429 (1978).
9. J. Mann and G. R. Williamson, in R. N. Haward, ed., *The Physics of Glassy Polymers*, John Wiley & Sons, Inc., New York, 1973, Chapt. 8.
10. H. Keskkula, *Appl. Polym. Symp.* **15**, 51 (1970).
11. J. A. Schmitt, *J. Polym. Sci.* **C30**, 437 (1970).
12. S. L. Rosen, *Polym. Eng. Sci.* **7**, 115 (1967).
13. L. Bohn, *Angew. Makromol. Chem.* **20**, 129 (1971).
14. A. Zosel, *Rheol. Acta* **11**, 229 (1972).
15. H. Munstedt, *Polym. Eng. Sci.* **21**, 259 (1981).
16. Y. Aoki and K. Nakayama, *Polym. J.* **14**, 951 (1982).
17. Y. Aoki, *Macromolecules* **20**, 2208 (1987).
18. Y. Aoki, *J. Non-Newtonian Fluid Mech.* **22**, 91 (1986).
19. M. G. Huguet and T. R. Paxton, *Colloidal and Morphological Behavior of Block and Graft Copolymers*, Plenum, New York, 1971, pp. 183–192.
20. T. Masuda and co-workers, *Pure Appl. Chem.* **56**, 1457 (1984).
21. A. Casale, A. Moroni, and C. Spreafico, in *Copolymers, Polyblends, and Composites (Adv. in Chem. Ser. No. 142)*, American Chemical Society, Washington, D.C., 1975, p. 172.
22. L. L. Blyler, Jr., *Polym. Eng. Sci.* **14**(11), 806 (1974).
23. D. M. Kulich, P. D. Kelley, and J. E. Pace, in J. I. Kroschwitz, ed., *Encyclopedia of Polymer Science and Engineering*, 2nd ed., Vol. 1, Wiley-Interscience, New York, 1985, p. 396.
24. F. M. Smith, *Manufacture of Plastics*, Vol. 1, Reinhold Publishing Corp., New York, 1964, p. 443.
25. D. L. Fulkner, *Polym. Eng. Sci.* **24**, 1174 (1984).
26. J. M. Heaps, *Rubber Plast. Age*, 967 (1968).
27. T. H. Rogers and R. B. Roennau, *Chem. Eng. Progress* **62**(11), 94 (1966).
28. A. Casale and O. Salvatore, *Polym. Eng. Sci.* **15**, 286 (1975).
29. W. I. Congdon, H. E. Bair, and S. K. Khanna, *Org. Coatings Plast. Chem.* **40**, 739 (1979).
30. M. L. Heckaman, *Soc. Plast. Eng. Tech. Pap.* **18**, 512 (1972).

31. B. D. Gesner, *J. Appl. Polym. Sci.* **9**, 3701 (1965).
32. P. G. Kelleher, *J. Appl. Polym. Sci.* **10**, 843 (1966).
33. J. Shimada, *J. Appl. Polym. Sci.* **12**, 655 (1968).
34. M. D. Wolkowicz and S. Gaggar, *Polym. Eng. Sci.* **21**, 571 (1981).
35. P. So and L. J. Broutman, *Polym. Eng. Sci.* **22**, 888 (1982).
36. B. Gilg, H. Muller, and K. Schwarzenbach, Paper presented at *Advances in Stabilization and Controlled Degradation of Polymers*, New Paltz, N.Y., June 1982.
37. J. Shimada, K. Kabuki, and M. Ando, *Rev. Electr. Commun. Lab.* **20**, 564 (1972).
38. F. Gugumus, in G. Scott, ed., *Developments in Polymer Stabilization*, Vol. 1, Elsevier Applied Science Publishers, Ltd., London 1979, p. 319.
39. G. Scott, *Developments in Polymer Stabilization*, Vol. 4, Elsevier Applied Science Publishers, Ltd., London, 1981, p. 181.
40. Eur. Pats. 0109008 and 0108396, J. C. Wozny (to Borg-Warner Corp.).
41. D. M. Kulich and M. D. Wolkowicz, in *Rubber-Toughened Plastics* (Adv. in Chem. Ser. No. 222), American Chemical Society, Washington, D.C., 1989, p. 329.
42. B. D. Gesner, *SPE J.* **25**, 73 (1969).
43. J. Kovarova, L. Rosik, and J. Pospisil, *Polym. Mater. Sci. Eng.* **58**, 215 (1988).
44. F. Gugumus, in G. Scott, ed., *Developments in Polymer Stabilization*, Vol. 8., Elsevier, New York, 1987, p. 243.
45. D. M. Chang, *Org. Coatings Plast. Chem.* **44**, 347 (1981).
46. M. G. Wygoski, *Polym. Eng. Sci.* **16**, 265 (1976).
47. L. Zlatkevich, in P. Klemchuk, ed., *Polymer Stabilization and Degradation* (ACS Symp. Ser. No. 280), American Chemical Society, Washington, D.C., 1985, Chapt. 27.
48. L. Zlatkevich, *J. Polym. Sci. Polym. Phys. Ed.* **25**, 2207 (1987).
49. G. Scott and M. Tahan, *Eur. Polym. J.* **13**, 981 (1977).
50. M. Tahan, *Weathering of Plastics and Rubber, International Symposium of the Institute of Electrical Engineers, London, June 1976*, Chamelon Press, Ltd., London, 1976, p. A2.1.
51. J. B. Adeniyi, *Eur. Polym. J.* **20**, 291 (1984).
52. E. Priebe, P. Simak, and G. Stange, *Kunststoffe* **62**, 105 (1972).
53. T. Hirai, *Jpn. Plast.* 23, Oct. 1970.
54. M. Ghaemy and G. Scott, *Polym. Degrad. Stab.* **3**, 233 (1981).
55. J. B. Adeniyi and E. G. Kolawole, *Eur. Polym. J.* **20**, 43 (1984).
56. R. D. Deanin, I. S. Rabinovic, and A. Llompart, in *Multicomponent Polymer Systems* (Adv. in Chem. Ser. No. 99), American Chemical Society, Washington, D.C., 1971, p. 229.
57. A. Blaga and R. S. Yamasaki, *Durab. of Build. Mater.* **4**, 21 (1986).
58. J. Shimada and K. Kabuki, *J. Appl. Polym. Sci.* **12**, 671 (1968).
59. T. Kurumada, H. Ohsawa, and T. Yamazaki, *Polym. Degrad. Stab.* **19**, 263 (1987).
60. E. G. Kolawole and J. B. Adeniyi, *Eur. Polym. J.* **18**, 469 (1982).
61. T. R. Bullet and P. R. Mathews, *Plast. Polym.* **39**, 200 (1971).
62. A. Davis and D. Gordon, *J. Appl. Polym. Sci.* **18**, 1159 (1974).
63. P. G. Kelleher, D. J. Boyle, and R. J. Miner, *Mod. Plast.* 189 (Sept. 1969).
64. M. L. Kaplan and P. G. Kelleher, *J. Polym. Sci. Part A-1* **8**, 3163 (1970).
65. *Product Information*, GE Plastics, Pittsfield, Mass.
66. R. A. Hayes and S. Futamura, *J. Polym. Sci. Polym. Chem. Ed.* **19**, 985 (1981).
67. A. Brydon, G. M. Burnett, and C. G. Cameron, *J. Polym. Sci. Polym. Chem. Ed.* **12**, 1011 (1974).
68. P. W. Allen, G. Ayrey, and C. G. Moore, *J. Polym. Sci.* **36**, 55 (1959).
69. J. L. Locatelli and G. Riess, *Angew. Makromol. Chem.* **32**, 117 (1973).
70. N. J. Huang and D. C. Sundberg, *J. Polym. Sci., Polym. Chem. Ed.* **33**, 2551 (1995).
71. L. V. Zamoiskaya and co-workers, *Vysokomol. Soedin. Ser. A* **40**, 557 (1998).

72. C. S. Chern and G. W. Pehlein, *J. Polym. Sci., Polym. Chem. Ed.* **28**, 3073 (1990).
73. G. Reiss and J. L. Locatelli in *Copolymers, Polyblends, and Composites (Adv. in Chem. Ser. No. 142)*, American Chemical Society, Washington, D.C., 1975, p. 186.
74. B. Chauvel and J. C. Daniel, in *Copolymers, Polyblends, and Composites (Adv. in Chem. Ser. No. 142)*, American Chemical Society, Washington, D.C., 1975, p. 159.
75. A. Echte, in *Rubber-Toughened Plastics (Adv. in Chem. Ser. No. 222)*, American Chemical Society, Washington, D.C., 1989, p. 15.
76. E. Beati, M. Pegoraro, E. Pedemonte, *Angew. Makrom. Chem.* **149**, 55 (1987).
77. G. Odian, *Principles of Polymerization*, McGraw-Hill, Inc., New York, 1970, Chapt. 4.
78. P. J. Flory, *Principles of Polymer Chemistry*, Cornell University Press, Ithaca, N.Y., 1953, 203–217.
79. A. M. Aerdts, S. J. Theolen, and T. M. Smith, *Polymer* **35**, 1648 (1994).
80. R. Kuhn and co-workers, *Colloid Polym. Sci.* **271**, 133 (1993).
81. Jpn. Pat. 09286827 (Nov. 4, 1997), Y. Li and Y. Nakai (to Mitsubishi Rayon Co.); Jpn. Pat. 01259015 (Oct. 16, 1989), J. Sugiura and co-workers (to Mitsubishi Monsanto Chemical Co.).
82. U.S. Pat. 2,820,773 (Jan. 21, 1958), C. W. Childers and C. F. Fisk (to United States Rubber Co.).
83. U.S. Pat. 19970630 (1997), V. Lowry (to General Electric Co.).
84. E. S. Daniels and co-workers, *J. Appl. Polym. Sci.* **41**, 2463 (1990).
85. Jpn. Pat. 09296015 (Nov. 18, 1997), H. Kitayama, Y. Ishikawa, and M. Nomura (to Kao Corp.).
86. H. S. Chung and co-workers, *Komu Hakhoechi* **28**, 267 (1993).
87. H. S. Chung and Y. J. Shin, *Kongop Hsahak* **4**, 284 (1993).
88. Eur. Pat. 712894 (Apr. 2, 1997), V. J. Kuruganti, S. K. Gaggar, and R. A. St Jean (to General Electric Co.).
89. Eur. Pat. 479725 (Apr. 26, 1995), B. Gilg and co-workers (to Ciba Geigy).
90. Jpn. Pat. 05017508 (Jan. 26, 1993), A. Shichizawa and S. Ozawa (to Asahi Chemical Ind.).
91. Jpn. Pat. 05017506 (Jan. 26, 1993), A. Shichizawa and S. Ozawa (to Asahi Chemical Ind.).
92. Jpn. Pat. 04266902 (Sept. 22, 1992), M. Watanabe and co-workers (to Shin-Etsu Chemical Ind. Co.).
93. Eur. Pat. 496349 (Apr. 10, 1996), M. Watanabe and co-workers (to Shin-Etsu Chemical Ind. Co.).
94. Eur. Pat. 466141 (Nov. 8, 1995), M. Watanabe, S. Ueno, and M. Usuki (to Shin-Etsu Chemical Ind. Co.).
95. J. L. Locatelli and G. Riess, *Angew. Makromol. Chem.* **27**, 201 (1972).
96. C. F. Parsons and E. L. Suck, Jr., *Multicomponent Polymer Systems (Adv. in Chem. Ser. No. 99)*, American Chemical Society, Washington, D.C., 1971, p. 340.
97. J. Stabenow and F. Haaf, *Angew. Makromol. Chem.* **29–30**, 1 (1973).
98. J. L. Locatelli and G. Riess, *J. Polym. Sci. Polym. Chem. Ed.* **2**, 3309 (1973).
99. J. L. Locatelli and G. Riess, *Angew. Makromol. Chem.* **32**, 101 (1973).
100. P. Mathey and J. Guilot, *Polymer* **32**, 934 (1991).
101. R. Born and co-workers, *Dechema Mongr.* **134**, 409-418 (1998), 6th International Workshop on Polymer Reaction Engineering.
102. Jpn. Pat. 08027306 (Jan. 30, 1996), M. Motai and co-workers (to Japan Synthetic Rubber Co.).
103. Jpn. Pat. 03026705 (Feb. 5, 1991), M. Tsutsumi and M. Shidara (to Hitachi Chemical Co.).
104. Jpn. Pat. 08016126 (Feb. 21, 1996), H. Hayashi, T. Maruyama, and O. Murai (to Kao Corp.).

105. Jpn. Pat. 05163359 (June 29, 1993), F. Suzuki and co-workers (to Mitsubishi Rayon Co.).
106. Eur. Pat. 527605 (Feb. 17, 1993), M. C. Will (to Rohm and Haas Co.).
107. A. K. Ghosh and J. Lindt, *J. Appl. Polym. Sci.* **39**, 1553 (1990).
108. U.S. Pat. 3991136 (1976), William Dalton (to Monsanto Chemical Co.).
109. W. P. Flanagan, V. Lowry, T. E. Ludovico, A. P. Togna, and W. J. Fucich, *AICHE 1998, Annual Meeting, Session 16: Clean Air Technologies*, AICHE Publication, New York, 1998.
110. K. R. Sharma, *Polym. Mater. Sci. Eng.* **77**, 656 (1997).
111. U.S. Pat. 05569709 (Oct. 29, 1996), C. Y. Sue and co-workers (to General Electric Co.)
112. U.S. Pat. 05551859 (Sep. 3, 1996), J. E. Cantrill and T. R. Doyle (to Novacor Chemicals International).
113. U.S. Pat. 05550186 (Aug. 27, 1996), J. E. Cantrill and T. R. Doyle (to Novacor Chemicals International).
114. U.S. Pat. 055514750 (May. 7, 1996), J. E. Cantrill and T. R. Doyle (to Novacor Chemicals International).
115. U.S. Pat. 05414045 (May. 9, 1995), C. Y. Sue and co-workers (to General Electric Co.).
116. U.S. Pat. 04415708 (Nov. 15, 1983), S. Matsumura and co-workers (to Kanegafuchr Kagaku Kogyo Kabushiki Kaisha).
117. U.S. Pat. 04239863 (Dec. 16, 1980), C. Bredeweg (to The Dow Chemical Company.).
118. U.S. Pat. 03945976 (Mar. 23, 1976), J. L. McCurdy and N. Stein (to Standard Oil Co.).
119. U.S. Pat. 3,950,455 (Apr. 13, 1976), T. Okamoto and co-workers (to Toray Industries, Inc.).
120. U.S. Pat. 3,660,535 (May 2, 1972), C. R. Finch and J. E. Knutzsch (to The Dow Chemical Company).
121. U.S. Pat. 2,694,692 (Nov. 16, 1954), J. L. Amos, J. L. McCurdy, and O. R. McIntire (to The Dow Chemical Company).
122. U.S. Pat. 3,515,692 (June 2, 1970), F. E. Carrock and K. W. Doak (to Dart Industries, Inc.).
123. U. S. Pat. 1700011 (Dec. 23, 1991), A. Z. Zumer and co-workers (to USSR).
124. U.S. Pat. 04151128 (Apr. 24, 1979), A. J. Ackerman and F. E. Carrock (to Mobil Oil Corp.).
125. F. Rodriguez, *Principles of Polymer Systems*, McGraw-Hill, Inc., New York, 1970, Chapt. 5.
126. R. J. Pierce and J. W. Bozzelli, *Paper presented at the 45th Annual Technical Conference of the Society of Plastics Engineers*, May 1987, p. 19.
127. L. W. Fritch, *Paper presented at the 33rd Annual Technical Conference of the Society of Plastics Engineers*, May 1975, p. 70.
128. L. W. Fritch, *Plast. Technol.* 69 (1980).
129. *Cycolac Brand ABS Pellet Drying*, Technical Publication SR-601A, GE Plastics, 1989.
130. J. W. Bozzelli, B. J. Furches, and S. L. Janiki, *Mod. Plast.* 7 (1988).
131. *Product Bulletin: Micro Moisture II*, ZARAD Technology Inc., Cary, Ill.
132. *Product Bulletin E38855*, DuPont 903 Moisture Evolution Analyzer, DuPont Company, Instrument Systems, Wilmington, Del.
133. B. Miller, *Plast. World*, 51 (1987).
134. H. Lord, *Paper presented at the 36th Annual Technical Conference of the Society of Plastics Engineers*, May 1978, p. 83.
135. R. E. Nunn, *Injection Molding Handbook*, Van Nostrand Reinhold Co., New York, 1986, Chapt. 3.
136. B. Miller, *Plast. World* **40**(3), 34 (1982).

137. B. Furches and J. Bozzelli, *Paper presented at the 45th Annual Technical Conference of the Society of Plastics Engineers*, May 1987, p. 6.
138. *Cycolac Brand ABS Resin Injection Molding*, Technical Publication CYC-400, GE Plastics, Pittsfield, Mass., 1990.
139. *Molding Flame Retardent ABS*, Technical Publication P-408, GE Plastics, Pittsfield, Mass., 1989.
140. L. W. Fritch, *Injection Molding Handbook*, Van Nostrand Reinhold Co., New York, 1986, Chapt. 19.
141. L. W. Fritch, *Plast. Eng.*, 43 (1989).
142. L. W. Fritch, *Paper presented at the 45th Annual Technical Conference of the Society of Plastics Engineers*, May 1987, p. 218.
143. L. W. Fritch, *Paper presented at the 40th Annual Technical Conference of the Society of Plastics Engineers*, May 1982, p. 332.
144. L. W. Fritch, *Paper presented at the 5th Pacific Area Technical Conference of the Society of Plastics Engineers*, Feb. 1980.
145. S. Gaggar and J. Wilson, *Paper presented at the 40th Annual Technical Conference of the Society of Plastics Engineers*, May 1982, p. 157.
146. L. W. Fritch, *Paper presented at the 37th Annual Technical Conference of the Society of Plastics Engineers*, May 1979, p. 15.
147. L. W. Fritch, *Prod. Finish. Cincinnati* **48**(5), 42 (1984).
148. L. W. Fritch, *Plast. Machin. Equip.* 43 (1988).
149. R. M. Criens and H. Mosle, *Paper presented at the 42nd Annual Technical Conference of the Society of Plastics Engineers*, May 1984, p. 587.
150. J. W. Bozzelli and P. A. Tiffany, *Paper presented at the 44th Annual Technical Conference of the Society of Plastics Engineers*, May 1986, p. 120.
151. U. Wolfel and G. Menges, *Paper presented at the 45th Annual Technical Conference of the Society of Plastics Engineers*, May 1987, p. 292.
152. S. M. Janosz, *Paper presented at the 45th Annual Technical Conference of the Society of Plastics Engineers*, May 1987, p. 323.
153. H. Cox and C. Mentzer, *Polym. Eng. Sci.* **26**, 488 (1986).
154. G. Brewer, *Paper presented at the 45th Annual Technical Conference of the Society of Plastics Engineers*, May 1987, p. 252.
155. *Cycolac Brand ABS Resin Design Guide*, Technical Publication CYC-350, GE Plastics, Pittsfield, Mass., 1990.
156. *Cycolac Brand ABS Electroplating*, Technical Publication 402, GE Plastics, Pittsfield, Mass., 1990.
157. M. Snyder, *Plast. Machin. Equip.*, 50 (1988).
158. K. C. Rusch, *Paper presented at the 45th Annual Technical Conference of the Society of Plastics Engineers*, May 1987, p. 1014.
159. W. Virginski, *Paper presented at the 46th Annual Technical Conference of the Society of Plastics Engineers*, May 1988, p. 205.
160. *Plast. World*, 28 (1987).
161. L. E. Ferguson and R. J. Brinkmann, *Paper presented at the 45th Annual Technical Conference of the Society of Plastics Engineers*, May 1987, p. 866.
162. *Cycolac Brand ABS—General Purpose Blow Molding Grades*, Technical Publication SR-616, GE Plastics, Pittsfield, Mass., 1989.
163. *Thermoforming Cycolac Brand ABS*, Technical Publication P-406, GE Plastics, Pittsfield, Mass., 1989.
164. J. L. Throne, *Paper presented at the 45th Annual Technical Conference of the Society of Plastics Engineers*, May 1987, p. 412.
165. N. Nichols and G. Kraynak, *Plast. Technol.* 73 (1987).

166. R. Royer, *Paper presented at the Regional Conference of the Society of Plastics Engineers, Quebec Section*, 1968, p. 43.
167. G. Trantina and M. Minnicheli, *Paper presented at the 45th Annual Technical Conference of the Society of Plastics Engineers*, May 1987, p. 438.
168. T. L. La Bounty, *Paper presented at the 43rd Annual Technical Conference of the Society of Plastics Engineers*, May 1985, p. 855.
169. *Cycolac Brand ABS Resin—Assembly Techniques*, Technical Publication CYC-352, GE Plastics, Pittsfield, Mass., 1990.
170. H. Potente and H. Kaiser, *Paper presented at the 47th Annual Technical Conference of the Society of Plastics Engineers*, May 1989, p. 464.
171. V. K. Stokes, *Paper presented at the 47th Annual Technical Conference of the Society of Plastics Engineers*, May 1989, p. 442.
172. *International Trader Publications, ABS Global Capacity*, July 2000.
173. "ABS Resins, Chemical Profile," *Chemical Market Reporter* **263**, 27 (Jan. 13, 2003).
174. K.-L. Ring and H. Yoneyama, "ABS Resins," *Chemical Economics Handbook*, SRI, Menlo Park, CA, June 2001.
175. J. Haslam, H. A. Willis, and D. C. M. Squirrell, *Identification and Analysis of Plastics*, 2nd ed., Heyden Book Co. Inc., Philadelphia, 1980, Chapt. 8.
176. J. C. Cobbler and G. E. Stobbe, in F. D. Snell and L. S. Ettre, eds., *Encyclopedia of Industrial Chemical Analysis*, Vol. 18, Wiley-Interscience, New York, 1973, p. 332.
177. T. Okumoto and T. Tadaoki, *Nippon Kagaku Kaishi* **1**, 71 (1972).
178. K. Sircar and T. Lamond, *Thermochim. Acta* **7**, 287 (1973).
179. B. D. Gesner, *J. Polym. Sci. Part A* **3**, 3825 (1965).
180. L. D. Moore, W. W. Moyer, and W. J. Frazer, *Appl. Polym. Symp.* **7**, 67 (1968).
181. J. Tsurugi, T. Fukumoto, and K. Ogawa, *Chem. High Polym. (Tokyo)* **25**, 116 (1968).
182. H. Shuster, M. Hoffmann, and K. Dinges, *Angew. Makrom. Chem.* **9**, 35 (1969).
183. D. Kranz, K. Dinges, and P. Wendling, *Angew. Makrom. Chem.* **51**, 25 (1976).
184. D. Kranz, H. V. Pohl, and H. Baumann, *Angew. Makrom. Chem.* **26**, 67 (1972).
185. R. R. Turner, D. W. Carlson, and A. G. Altenau, *J. Elastomer Plast.* **6**, 94 (1974).
186. L. W. Jelinski and co-workers, *J. Polym. Sci. Polym. Chem. Ed.* **20**, 3285 (1982).
187. V. G. Kampf and H. Shuster, *Angew. Makrom. Chem.* **14**, 111 (1970).
188. D. Simpson, *Br. Plast.* 78 (1968).
189. L. I. Petrova, M. P. Noskova, V. A. Balandine, and Z. G. Guricheva, *Gig. Sanit.* **37**, 62 (1972).
190. T. R. Crompton, *Chemical Analysis of Additives in Plastics*, Pergamon Press, New York, 1971.
191. N. E. Skelly, J. D. Graham, and Z. Iskandarani, *Polym. Mater. Sci. Eng.* **59**, 23 (1988).
192. R. Yoda, *Bunseki* **1**, 29 (1984).
193. R. G. Lewis, *Sax's* Properties of Dangerous Industrial Materials, 10th ed., Vol. 2, John Wiley & Sons, Inc., New York, 2000, p. 71.
194. U.S. Pat. Appl. 20020029939 (March 14, 2002), K. J. Bunker (to Federal Mogul Technology Limited).
195. U.S. Pat. Appl. 2002013509 (Sept. 26, 2002), Y. Koike.

Donald M. Kulich
S. K. Gaggar
V. Lowry
R. Stepien
GE Plastics, Technology Center

ACRYLONITRILE POLYMERS, SURVEY AND STYRENE-ACRYLONITRILE (SAN)

1. Introduction

Acrylonitrile (AN), C_3H_3N, is a versatile and reactive monomer (1) that can be polymerized under a wide variety of conditions (2) and copolymerized with an extensive range of other vinyl monomers (3). It first became an important polymeric building block in the 1940s. Although acrylonitrile had been discovered in 1893, its unique properties were not realized until the development of nitrile rubbers during World War II (see ELASTOMERS, SYNTHETIC, NITRILE RUBBER) and the discovery of solvents for the homopolymer with resultant fiber applications (see FIBERS, ACRYLIC) for textiles and carbon fibers. Because of the difficulty of melt processing the homopolymer, acrylonitrile is usually copolymerized to achieve a desirable thermal stability, melt flow, and physical properties. As a comonomer, acrylonitrile (qv) contributes hardness, rigidity, solvent and light resistance, gas impermeability, and the ability to orient. These properties have led to many copolymer application developments since 1950.

The utility of acrylonitrile [107-13-1] in thermoplastics was first realized in its copolymer with styrene, C_8H_8 [100-42-5], in the late 1950s. Styrene is the largest volume of comonomer for acrylonitrile in thermoplastic applications. Styrene–acrylonitrile (SAN) copolymers [9003-54-7] are inherently transparent plastics with high heat resistance and excellent gloss and chemical resistance (4). They are also characterized by good hardness, rigidity, dimensional stability, and load-bearing strength (due to relatively high tensile and flexual strengths). Because of their inherent transparency, SAN copolymers are most frequently used in clear application. These optically clear materials can be readily processed by extrusion and injection molding, but they lack real impact resistance.

The subsequent development of acrylonitrile–butadiene–styrene (ABS) resins [9003-56-9], which contain an elastomeric component within a SAN matrix to provide toughness and impact strength, further boosted commercial application of the basic SAN copolymer as a portion of these rubber-toughened thermoplastics (see ACRYLONITRILE POLYMERS, ABS RESINS). When SAN is grafted onto a butadiene-based rubber, and optionally blended with additional SAN, the two-phase thermoplastic ABS is produced. ABS has the useful SAN properties of rigidity and resistance to chemicals and solvents, while the elastomeric component contributes real impact resistance. Because ABS is a two-phase system and each phase has a different refractive index, the final ABS is normally opaque. A clear ABS can be made by adjusting the refractive indexes through the inclusion of another monomer such as methyl methacrylate. ABS is a versatile material and modifications have brought out many specialty grades such as clear ABS and high-temperature and flame-retardant grades. Saturated hydrocarbon elastomers or acrylic elastomers (5,6) can be used instead of those based on butadiene, C_4H_6 [106-99-0] as weatherable-grade ABS.

In the late 1960s a new class of AN copolymers and multipolymers was introduced that contain more than 60% acrylonitrile. These are commonly known as barrier resins and have found their greatest acceptance where excellent barrier properties toward gases (7), chemicals, and solvents are needed. They may be processed into bottles, sheets, films, and various laminates, and have found wide usage in the packaging industry (see BARRIER POLYMERS).

Acrylonitrile has found its way into a great variety of other polymeric compositions based on its polar nature and reactivity, imparting to other systems some or all of the properties noted above. Some of these areas include adhesives and binders, antioxidants, medicines, dyes, electrical insulations, emulsifying agents, graphic arts, insecticides, leather, paper, plasticizers, soil-modifying agents, solvents, surface coatings, textile treatments, viscosity modifiers, azeotropic distillations, artificial organs, lubricants, asphalt additives, water-soluble polymers, hollow spheres, cross-linking agents, and catalyst treatments (8).

2. SAN Physical Properties and Test Methods

SAN resins possess many physical properties desired for thermoplastic applications. They are characteristically hard, rigid, and dimensionally stable with load-bearing capabilities. They are also transparent, have high heat distortion temperatures, possess excellent gloss and chemical resistance, and adapt easily to conventional thermoplastic fabrication techniques (9).

SAN polymers are random linear amorphous copolymers. Physical properties are dependent on molecular weight and the percentage of acrylonitrile. An increase of either generally improves physical properties, but may cause a loss of processability or an increase in yellowness. Various processing aids and

Table 1. **Physical/Mechanical Properties of Commercial Injection-Molded SAN Resins**[a]

	Bayer Lustran 31-2060	Dow Tyril 100	ASTM Method
specific gravity (23/23°C)	1.07	1.07	D 792
Vicat softening point (°C)	110	108	D 1525
tensile strength, MPa[b]	72.4	71.7	D 638
ultimate elongation @ breakage (%)	3.0	2.5	D 638
flexural modulus, GPa[c]	3.45	3.87	D 790
impact strength, notched Izod (J/m[d])	21.4 @ 0.125 in.	16.0 @ 0.125 in.	D 256
melt flow rate (g/10 min)	8.0	8.0	D 1238, cond. 1
refractive index, n_D	1.570	1.570	D 542
mold shrinkage (in./in.)	0.003–0.004	0.004–0.005	D 955
transmittance at 0.125-in. thickness (%)	89.0	89.0	D 1003
haze at 0.125-in. thickness (%)	0.8	0.6	D 1003

[a] Product literature.
[b] To convert MPa to psi, multiply by 145.
[c] To convert GPa to psi, multiply by 145,000.
[d] To convert J/m to ft lb/in., divide by 53.39.

Table 2. **Compositional Effects on SAN Physical Properties**[a]

AN, wt %	Tensile strength, MPa[b]	Elongation, %	Impact strength, J/m notch[c]	Heat distortion, temp., °C	Solution viscosity, mPa (=cP)
5.5	42.27	1.6	26.6	72	11.1
9.8	54.61	2.1	26.0	82	10.7
14.0	57.37	2.2	27.1	84	13.0
21.0	63.85	2.5	27.1	88	16.5
27.0	72.47	3.2	27.1	88	25.7

[a] Ref. 15.
[b] To convert MPa to psi, multiply by 145.
[c] To convert J/m to ft lb/in., divide by 53.39.

modifiers can be used to achieve a specific set of properties. Modifiers may include mold release agents, UV stabilizers, antistatic aids, elastomers, flow and processing aids, and reinforcing agents such as fillers and fibers (9). Methods for testing and some typical physical properties are listed in Table 1.

The properties of SAN resins depend on their acrylonitrile content. Both melt viscosity and hardness increase with increasing acrylonitrile level. Unnotched impact and flexural strengths depict dramatic maxima at ca 87.5 mol % (78 wt %) acrylonitrile (10). With increasing acrylonitrile content, copolymers show continuous improvements in barrier properties and chemical and UV resistance, but thermal stability deteriorates (11). The glass-transition temperature (T_g) of SAN varies nonlinearly with acrylonitrile content, showing a maximum at 50 mol % AN. The alternating SAN copolymer has the highest T_g (12,13). The fatigue resistance of SAN increases with AN content to a maximum at 30 wt %, then decreases with higher AN levels (14). The effect of acrylonitrile incorporation on SAN resin properties is shown in Table 2.

3. SAN Chemical Properties and Analytical Methods

SAN resins show considerable resistance to solvents and are insoluble in carbon tetrachloride, ethyl alcohol, gasoline, and hydrocarbon solvents. They are swelled by solvents such as benzene, ether, and toluene. Polar solvents such as acetone, chloroform, dioxane, methyl ethyl ketone, and pyridine will dissolve SAN (16). The interactions of various solvents and SAN copolymers containing up to 52% acrylonitrile have been studied along with their thermodynamic parameters, ie, the second virial coefficient, free-energy parameter, expansion factor, and intrinsic viscosity (17).

The properties of SAN are significantly altered by water absorption (18). The equilibrium water content increases with temperature while the time required decreases. A large decrease in T_g can result. Strong aqueous bases can degrade SAN by hydrolysis of the nitrile groups (19).

The molecular weight of SAN can be easily determined by either intrinsic viscosity or size-exclusion chromatography (SEC). Relationships for both multipoint and single point viscosity methods are available (20,21). The intrinsic viscosity and molecular weight relationships for azeotropic copolymers have been given (22,23):

$$[\eta] = 3.6 \times 10^{-4}\, M_w^{0.62} \text{ dL/g in MEK at } 30^\circ\text{C}$$

$$[\eta] = 2.15 \times 10^{-4}\, M_w^{0.68} \text{ dL/g in THF at } 25^\circ\text{C}$$

$$[\eta] = \frac{\eta_{sp}/c}{1 + k_\eta \eta_{sp}}$$

where k_η=0.21 for MEK at 30°C and 0.25 for THF at 25°C.

Chromatographic techniques are readily applied to SAN for molecular weight determination. Size-exclusion chromatography or gel permeation chromatography (GPC) (24) columns and conditions have been described for SAN (25). Chromatographic detector differences have been shown to be of the order of only 2–3% (26). High-pressure precipitation chromatography can achieve similar molecular weight separation (27). Liquid chromatography (LC) can be used with secfractioned samples to determine copolymer composition (28). Thin-layer chromatography will also separate SAN by compositional (monomer) variations (27).

Residual monomers in SAN have been a growing environmental concern and can be determined by a variety of methods. Monomer analysis can be achieved by polymer solution or directly from SAN emulsions (29) followed by "head space" gas chromatography (GC) (28,29). Liquid chromatography (LC) is also effective (32).

4. SAN Manufacture

The reactivities of acrylonitrile and styrene radicals toward their monomers are quite different, resulting in SAN copolymer compositions that vary from their monomer compositions (33). Further complicating the reaction is the fact that acrylonitrile is soluble in water (see ACRYLONITRILE) and slightly different behavior is observed between water-based emulsion and suspension systems and bulk or mass polymerizations (34). SAN copolymer compositions can be calculated from copolymerization equations (35) and published reactivity ratios (36). The difference in radical reactivity causes the copolymer composition to drift as polymerization proceeds, except at the azeotrope composition where copolymer composition matches monomer composition. Figure 1 shows these compositional variations (37). When SAN copolymer compositions vary significantly, incompatibility results, causing loss of optical clarity, mechanical strength, and moldability, as well as heat, solvent, and chemical resistance (38). The termination step has been found to be controlled by diffusion even at low conversions, and the termination rate constant varies with acrylonitrile content. The average half-life of the radicals increases with styrene concentration from 0.3 s at 20 mol % to 6.31 s with pure styrene (39). Further complicating SAN manufacture is the fact that both the heat (40,41) and rate (42) of copolymerization vary with monomer composition.

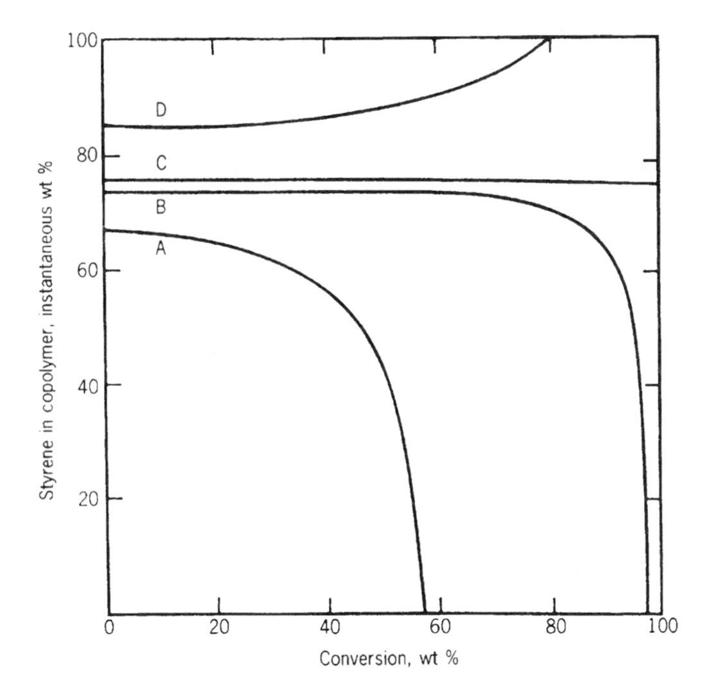

Fig. 1. Approximate compositions of styrene–acrylonitrile copolymers formed at different conversions starting with various monomer mixtures (35): S/AN = A, 65/36; B, 70/30; C, 76/24; D, 90/10.

The early kinetic models for copolymerization, Mayo's terminal mechanism (43) and Alfrey's penultimate model (44), did not adequately predict the behavior of SAN systems. Copolymerizations in DMF and toluene indicated that both penultimate and antepenultimate effects had to be considered (45, 46). The resulting reactivity model is somewhat complicated, since there are eight reactivity ratios to consider.

The first quantitative model, which appeared in 1971, also accounted for possible charge-transfer complex formation (47). Deviation from the terminal model for bulk polymerization was shown to be due to antepenultimate effects (48). The work with numerical computation and ^{13}C-nmr spectroscopy data on SAN sequence distributions indicates that the penultimate model is the most appropriate for bulk SAN copolymerization (49, 50). A kinetic model for azeotropic SAN copolymerization in toluene has been developed that successfully predicts conversion, rate, and average molecular weight for conversions up to 50% (51).

An emulsion model that assumes the locus of reaction to be inside the particles and considers the partition of AN between the aqueous and oil phases has been developed (52). The model predicts copolymerization results very well when bulk reactivity ratios of 0.32 and 0.12 for styrene and acrylonitrile, respectively, are used.

Commercially, SAN is manufactured by three processes: emulsion, suspension, and continuous mass (or bulk).

4.1. Emulsion Process. The emulsion polymerization process utilizes water as a continuous phase with the reactants suspended as microscopic particles. This low-viscosity system allows facile mixing and heat transfer for control purposes. An emulsifier is generally employed to stabilize the water-insoluble monomers and other reactants, and to prevent reactor fouling. With SAN the system is composed of water, monomers, chain-transfer agents for molecular weight control, emulsifiers, and initiators. Both batch and semibatch processes are employed. Copolymerization is normally carried out at 60 to 100°C to conversions of ~97%. Lower-temperature polymerization can be achieved with redox-initiator systems (53).

Figure 2 shows a typical batch or semibatch emulsion process (54). A typical semibatch emulsion recipe is shown in Table 3 (55).

The initial charge is placed in the reactor, purged with an inert gas such as N_2, and brought to 80°C. The initiator is added, followed by addition of the remaining charge over 100 min. The reaction is completed by maintaining agitation at 80°C for 1 h after monomer addition is complete. The product is a free-flowing white latex with a total solids content of 35.6%.

Compositional control for other than azeotropic compositions can be achieved with both batch and semibatch emulsion processes. Continuous addition of the faster reacting monomer, styrene, can be practiced for batch systems, with the feed rate adjusted by computer through gas chromatographic monitoring during the course of the reaction (56). A calorimetric method to control the monomer feed rate has also been described (10). For semibatch processes, adding the monomers at a rate that is slower than copolymerization can achieve

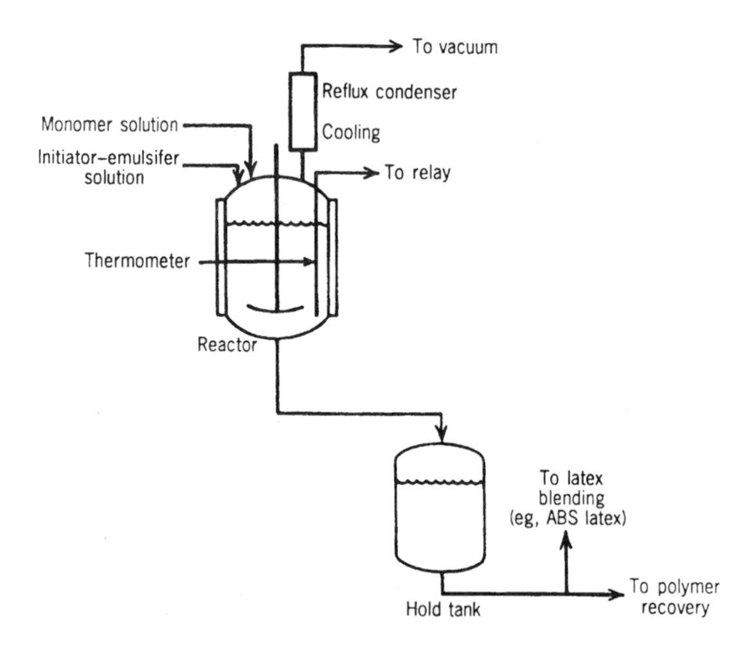

Fig. 2. Styrene–acrylonitrile batch emulsion process (54).

Table 3. **Semibatch Mode Emulsion Recipe for SAN Copolymers**

Ingredient	Parts
Initial reactor charge	
acrylonitrile	90
styrene	111
Na alkanesulfonate (emulsifier)	63
$K_2S_2O_8$ (initiator)	0.44
4-(benzyloxymethylene)cyclohexene (mol wt modifier)	1
water	1400
Addition charge	
acrylonitrile	350
styrene	1000
Na alkanesulfonate (emulsifier)	15
$K_2S_2O_8$ (initiator)	4
4-(benzyloxymethylene)cyclohexene (mol wt modifier)	10
water	1600

equilibrium. It has been found that constant composition in the emulsion can be achieved after ca 20% of the monomers have been charged (57).

Residual monomers in the latex are avoided either by effectively reacting the monomers to polymer or by physical or chemical removal. The use of *tert*-butyl peroxypivalate as a second initiator toward the end of the polymerization or the use of mixed initiator systems of $K_2S_2O_8$ and *tert*-butyl peroxybenzoate (58) effectively increases final conversion and decreases residual monomer levels. Spray devolatilization of hot latex under reduced pressure has been claimed to be effective (58). Residual acrylonitrile also can be reduced by postreaction with a number of agents such as monoamines (59) and dialkylamines (60), ammonium–alkali metal sulfites (61), unsaturated fatty acids or their glycerides (62, 63), their aldehydes, esters of olefinic alcohols, cyanuric acid (64,65), and myrcene (66).

The copolymer latex can be used "as is" for blending with other latexes, such as in the preparation of ABS, or the copolymer can be recovered by coagulation. The addition of electrolyte or freezing will break the latex and allow the polymer to be recovered, washed, and dried. Process refinements have been made to avoid the difficulties of fine particles during recovery (67,68).

The emulsion process can be modified for the continuous production of latex. One such process (70) uses two stirred-tank reactors in series, followed by insulated hold tanks. During continuous operation, 60% of the monomers are continuously charged to the first reactor with the remainder going into the second reactor. Surfactant is added only to the first reactor. The residence time is 2.5 h for the first reactor where the temperature is maintained at 65°C for 92% conversion. The second reactor is held at 68°C for a residence time of 2 h and conversion of 95%.

4.2. Suspension Process. Like the emulsion process, water is the continuous phase for suspension polymerization, but the resultant particle size is larger, well above the microscopic range. The suspension medium contains water, monomers, molecular weight control agents, initiators, and suspending

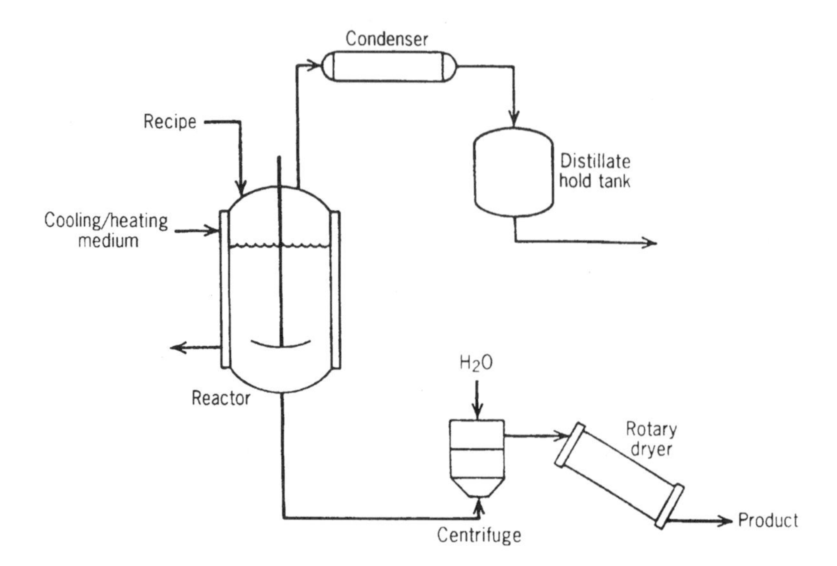

Fig. 3. Styrene–acrylonitrile suspension process (71).

aids. Stirred reactors are used in either batch or semibatch mode. Figure 3 illustrates a typical suspension manufacturing process while a typical batch recipe is shown in Table 4 (71). The components are charged into a pressure vessel and purged with nitrogen. Copolymerization is carried out at 128°C for 3 h and then at 150°C for 2 h. Steam stripping removes residual monomers (72), and the polymer beads are separated by centrifugation for washing and final dewatering.

Compositional control in suspension systems can be achieved with a corrected batch process. A suspension process has been described where styrene monomer is continuously added until 75–85% conversion, and then the excess acrylonitrile monomer is removed by stripping with an inert gas (73,74).

Elimination of unreacted monomers can be accomplished by two methods: dual initiators to enhance conversion of monomers to product (75–77) and steam stripping (72,78). Several process improvements have been claimed for dewatering beads (79), to reduce haze (80–83), improve color (84–88), remove monomer (89,90), and maintain homogeneous copolymer compositions (73,74,91).

Table 4. **Batch-Mode Recipe for SAN Copolymers**[a]

Ingredient	Parts
acrylonitrile	30
styrene	70
dipentene (4-isopropenyl-1-methylcyclohexene)	1.2
di-*tert*-butyl peroxide	0.03
acrylic acid–2-ethylhexyl acrylate (90:10) copolymer	0.03
water	100

[a] Ref. 71.

4.3. Continuous Mass Process. The continuous mass process has several advantages, including high space–time yield and good-quality products uncontaminated with residual ingredients such as emulsifiers or suspending agents. SAN manufactured by this method generally has superior color and transparency and is preferred for applications requiring good optical properties. It is a self-contained operation without waste treatment or environmental problems since the products are either polymer or recycled back to the process.

In practice, the continuous mass polymerization is rather complicated. Because of the high viscosity of the copolymerizing mixture, complex machinery is required to handle mixing, heat transfer, melt transport, and devolatilization. In addition, considerable time is required to establish steady-state conditions in both a stirred tank reactor and a linear flow reactor. Thus system start-up and product grade changes produce some off-grade or intermediate-grade products. Copolymerization is normally carried out between 100 and 200°C. Solvents are used to reduce viscosity or the conversion is kept to 40–70%, followed by devolatilization to remove solvents and monomers. Devolatilization is carried out from 120 to 260°C under vacuum at less than 20 kPa (2.9 psi). The devolatilized melt is then fed through a strand die, cooled, and pelletized.

A schematic of a continuous bulk SAN polymerization process is shown in Figure 4 (92). The monomers are continuously fed into a screw reactor, where copolymerization is carried out at 150°C to 73% conversion in 55 min. Heat of polymerization is removed through cooling of both the screw and the barrel walls. The polymeric melt is removed and fed to the devolatilizer to remove unreacted monomers under reduced pressure (4 kPa or 30 mm Hg) and high temperature (220°C). The final product is claimed to contain less than 0.7% volatiles. Two devolatilizers in series are found to yield a better quality product as well as better operational control (93,94).

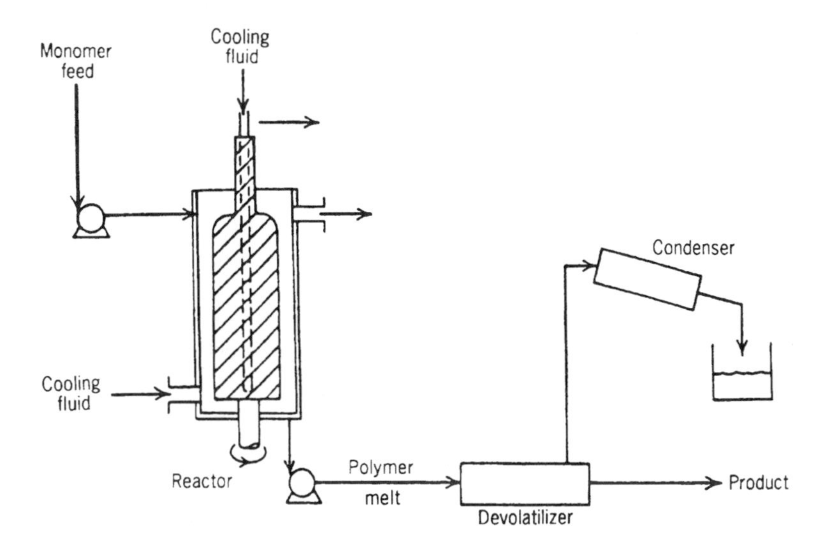

Fig. 4. Styrene–acrylonitrile continuous mass process (92).

Two basic reactor types are used in the continuous mass process, the stirred tank reactor (95) and the linear flow reactor. The stirred tank reactor consists of a horizontal cylinder chamber equipped with various agitators (96, 97) for mixing the viscous melt and an external cooling jacket for heat removal. With adequate mixing, the composition of the melt inside the reactor is homogeneous. Operation at a fixed conversion, with monomer make-up added at an amount and ratio equal to the amount and composition of copolymer withdrawn, produces a fixed composition copolymer. The two types of linear flow reactors employed are the screw reactor (92) and the tower reactor (97). A screw reactor is composed of two concentric cylinders. The reaction mixture is conveyed toward the outlet by rotating the inner screw, which has helical threads, while heat is removed from both cylinders. A tower reactor with separate heating zones has a scraper agitator in the upper zone, while the lower portion generates plug flow. In the linear flow reactors the conversion varies along the axial direction as does the copolymer composition, except where operating at the azeotrope composition. A stream of monomer must be added along the reactor to maintain SAN compositional homogeneity at high conversions. A combined stirred tank followed by a linear flow reactor process has been disclosed (97). Through continuous recycle copolymerization, a copolymer of identical composition to monomer feed can be achieved, regardless of the reactivity ratios of the monomers involved (98).

The devolatilization process has been developed in many configurations. Basically, the polymer melt is subjected to high temperatures and low pressures to remove unreacted monomer and solvent. A two-stage process using a tube and shell heat exchanger with enlarged bottom receiver to vaporize monomers has been described (94). A copolymer solution at 40–70% conversion is fed into the first-stage exchanger and heated to 120–190°C at a pressure of 20–133 kPa and then discharged into the enlarged bottom section to remove at least half of the unreacted acrylonitrile. The product from this section is then charged to a second stage and heated to 210–260°C at $<$20 kPa. The devolatilized product contains $\sim$1% volatiles. Preheating the polymer solution and then flashing it into a multipassage heating zone at lower pressure than the preheater produces essentially volatile-free product (93, 99). SAN can be steam stripped to quite low monomer levels in a vented extruder that has water injected at a pressure greater than the vapor pressure of water at that temperature (100).

A twin-screw extruder is used to reduce residual monomers from ca 50 to 0.6%, at 170°C and 3 kPa with a residence time of 2 min (96). In another design, a heated casing encloses the vented devolatilization chamber, which encloses a rotating shaft with specially designed blades (101, 102). These continuously regenerate a large surface area to facilitate the efficient vaporization of monomers. The devolatilization equipment used for the production of polystyrene and ABS is generally suitable for SAN production.

4.4. Processing. SAN copolymers may be processed using the conventional fabrication methods of extrusion, blow molding, injection molding, thermoforming, and casting. Small amounts of additives, such as antioxidants, lubricants, and colorants, may also be used. Typical temperature profiles for injection molding and extrusion of predried SAN resins are as follows (103).

Molding Temperatures	
Cylinder	193–288°C
Mold	49–88°C
Melt	218–260°C
Extrusion Temperatures	
Hopper zone	water-cooled
Rear zone	177–204°C
Middle zone	210–232°C
Torpedo zone and die	204–227°C

5. Other Copolymers

Acrylonitrile copolymerizes readily with many electron-donor monomers other than styrene and more than eight hundreds of acrylonitrile copolymers have been registered with *Chemical Abstracts* and a comprehensive listing of reactivity ratios for acrylonitrile copolymerizations is readily available (36, 104). Copolymerization mitigates the undesirable properties of acrylonitrile homopolymer, such as poor thermal stability and poor processability. At the same time, desirable attributes such as rigidity, chemical resistance, and excellent barrier properties are incorporated into melt-processable resins.

Barex (trademark of BP AMOCO Chemicals) resins, commercial high barrier resin produced by BP Amoco Chemicals, are copolymer of acrylonitrile and methyl acrylate [96-33-3]. This resin are excellent example of the use of acrylonitrile to provide gas and aroma/flavor barrier, chemical resistance, high tensile strength, and stiffness, and utilization of a comonomer to provide thermal stability and processability. In addition, modification with an elastomer provides toughness and impact strength. This material has a unique combination of useful packaging qualities, including transparency, and are excellent barriers to permeation by gases, organic solvents, and most essential oils. Barex resins also prevent the migration and scalping of volatile flavors and odors from packaged foods and fruit juice products (105,112). They also provide protection from atmospheric oxygen. Barex resins meet FDA compliance for direct food contact applications.

Barex resin extruded sheet and/or calendered sheet (112) can be easily thermoformed into lightweight, rigid containers (105,106). Packages can be printed, laminated, or metallized. Recent developments in extrusion and injection blow molding (105,107), laminated film structures (105,108), and coextrusion (105, 109) have led to packaging uses for a variety of products. Barex resins are especially well suited for bottle production. This acrylonitrile copolymer also provides a good example of the dependence of properties on the degree and temperature of orientation (110,111). Figure 5 illustrates the improvement in tensile strength, elongation, and the ability to absorb impact energy due to orientation (111) by Barex resins (for example, Barex 210). Tensile strength and impact strength increase with the extent of stretching, and decrease with the orientation temperature, and oxygen permeability decreases with orientation. These

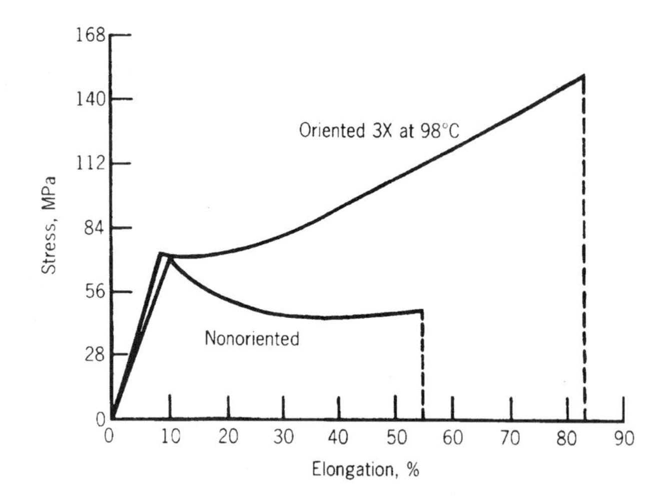

Fig. 5. Stress elongation of Barex 210 sheet (111). To convert MPa to psi, multiply by 145.

orientation properties have led to the commercialization Barex resins to fruit juice containers in France (112). Some typical physical properties of Barex resins are shown in Table 5.

Acrylonitrile–methyl acrylate–indene terpolymers, by themselves, or in blends with acrylonitrile–methyl acrylate copolymers, exhibit even lower oxygen and water permeation rates than the indene-free copolymers (113,114). Terpolymers of acrylonitrile with indene and isobutylene also exhibit excellent barrier

Table 5. **Physicial/Mechanical Properties of Commercial Barex Resins**[a]

	Barex 210[b]	Barex 218[b]	ASTM Method
specific gravity (23°C) (g/cm^3)	1.15	1.11	D 792
tensile strength, yield (MPa[c])	65.5	51.7	D 638
flexual modulus (GPa[d])	3.38	2.69	D 790
melt index (200c, 27.5 lbs)	3	3	D 1238
notched Izod impact (J/m[e])	267	481	D 790
heat deflection temperature (°C[f])	77	71	D 648
Gas permeability			
oxygen (23°C, 100% RH) $\frac{\text{nmol}}{\text{m·s·GPa}}$[f]	1.54	3.09	D 3985
carbon dioxide (23°C, 100% RH) $\frac{\text{nmol}}{\text{m·s·GPa}}$	2.32	3.09	D 3985
water vapor (38°C, 90% RH) $\frac{\text{nmol}}{\text{m·s·MPa}}$[g]	12.7	19.1	F 1249-90

Barex is a registered trademark of BP Amoco.
[a] Product literature.
[b] Extrusion grade.
[c] To convert MPa to psi, multiply by 145.
[d] To convert GPa to psi, multiply by 145,000.
[e] To convert J/m to ft lb/in., divide by 53.39.
[f] To convert from $\frac{\text{nmol}}{\text{m·s·GPa}}$ to $\frac{\text{cc-mm}}{\text{m}^2\text{·24h·bar}}$, divide by 5.145.
[g] To convert from $\frac{\text{nmol}}{\text{m·s·MPa}}$ to $\frac{\text{g-mm}}{\text{m}^2\text{·24h·atm}}$, divide by 6.35.

properties (115), and permeation of gas and water vapor through acrylonitrile–styrene–isobutylene terpolymers is also low (116,117).

Copolymers of acrylonitrile and methyl methacrylate (118) and terpolymers of acrylonitrile, styrene, and methyl methacrylate (119,120) are used as barrier polymers. Acrylonitrile copolymers and multipolymers containing butyl acrylate (121–124), ethyl acrylate (125), 2-ethylhexyl acrylate (121,124,126,127), hydroxyethyl acrylate (123), vinyl acetate (122,128), vinyl ethers (128,129), and vinylidene chloride (124,125,130–132) are also used in barrier films, laminates, and coatings. Environmentally degradable polymers useful in packaging are prepared from polymerization of acrylonitrile with styrene and methyl vinyl ketone (133).

Acrylonitrile multipolymers containing methyl methacrylate, α-methyl styrene, and indene are used as PVC modifiers to melt blend with PVC. These PVC modifiers not only enhance the heat distortion temperature but also improve the processability of the PVC compounds (134–138). The acrylonitrile multipolymers grafted on the elastomer phase provide the toughness and impact strength of the PVC compounds with high heat distortion temperature and good processability (139,140). Table 6 gives the structures, formulas, and CAS Registry Numbers for several comonomers of acrylonitrile.

Although the arrangement of monomer units in acrylonitrile copolymers is usually random, alternating or block copolymers may be prepared by using special techniques. For example, the copolymerization of acrylonitrile, like that of other vinyl monomers containing conjugated carbonyl or cyano groups, is changed in the presence of certain Lewis acids. Effective Lewis acids are metal compounds with nontransition metals as central atoms, including alkylaluminum halides, zinc halides, and triethylaluminum. The presence of the Lewis acid increases the tendency of acrylonitrile to alternate with electron-donor

Table 6. **Monomers Commonly Copolymerized with Acrylonitrile**

Monomer	Molecular formula	Structural formula	CAS Registry Number
methyl methacrylate	$C_5H_8O_2$	$CH_2{=}C(CH_3)COOCH_3$	[80-62-6]
methyl acrylate	$C_4H_6O_2$	$CH_2{=}CHCOOCH_3$	[96-33-3]
indene	C_9H_8		[95-13-6]
isobutylene	C_4H_8	$CH_2{=}C(CH_3)_2$	[115-11-7]
butyl acrylate	$C_7H_{12}O_2$	$CH_2{=}CHCOOC_4H_9$	[141-32-2]
ethyl acrylate	$C_5H_8O_2$	$CH_2{=}CHCOOC_2H_5$	[140-88-5]
2-ethylhexyl acrylate	$C_{11}H_{20}O_2$	$CH_2{=}CHCOOC_8H_{17}$	[103-11-7]
hydroxyethyl acrylate	$C_5H_8O_3$	$CH_2{=}CHCOOC_2H_4OH$	[818-61-1]
vinyl acetate	$C_4H_6O_2$	$CH_2{=}CHOOCCH_3$	[108-05-4]
vinylidene chloride	$C_2H_2Cl_2$	$CH_2{=}C(Cl)_2$	[75-35-4]
methyl vinyl ketone	C_4H_6O	$CH_2{=}CHCOCH_3$	[78-94-4]
α-methylstyrene	C_9H_{10}	$CH_2{=}C(CH_3)C_6H_5$	[98-83-9]
vinyl chloride	C_2H_3Cl	$CH_2{=}CHCl$	[75-01-4]
4-vinylpyridine	C_7H_7N	$CH_2{=}CHC_5H_4N$	[100-43-6]
acrylic acid	$C_3H_4O_2$	$CH_2{=}CHCOOH$	[79-10-7]

molecules, such as styrene, α-methylstyrene, and olefins (141–145). This alternation is often attributed to a ternary molecular complex or charge-transfer mechanism, where complex formation with the Lewis acid increases the electron-accepting ability of acrylonitrile, which results in the formation of a molecular complex between the acrylonitrile–Lewis acid complex and the donor molecule. This ternary molecular complex polymerizes as a unit to yield an alternating polymer. Cross-propagation and complex radical mechanisms have also been proposed (146).

A number of methods such as ultrasonics (147), radiation (148), and chemical techniques (149–151), including the use of polymer radicals, polymer ions, and organometallic initiators, have been used to prepare acrylonitrile block copolymers (152). Block comonomers include styrene, methyl acrylate, methyl methacrylate, vinyl chloride, vinyl acetate, 4-vinylpyridine, acrylic acid, and *n*-butyl isocyanate.

The living radical polymerization system (atom transfer radical polymerization, ATRP has been developed recently which allows for the controlled polymerization of acrylonitrile and comonomers and to produce the well-defined linear homopolymer, statistical copolymers, block copolymers, and gradient copolymers (153–156). The well-defined diblock copolymers comprising a polystyrene and an acrylonitrile–styrene (or isoprene) copolymer sequence are prepared by living radical polymerization system (157,158). The stereospecific acrylonitrile polymers are made by solid-state urea clathrate polymerization (159) and organometallic compounds of alkali and alkaline earth metals initiated polymerization (160).

Acrylonitrile has been grafted onto many polymeric systems. In particular, acrylonitrile grafting has been used to impart hydrophilic behavior to starch (161–163) and polymer fibers (164). Exceptional water absorption capability results from the grafting of acrylonitrile to starch, and the use of 2-acrylamido-2-methylpropanesulfonic acid [15214-89-8] along with acrylonitrile for grafting results in copolymers that can absorb over 5000 times their weight of deionized water (165). For example, one commercial product made by General Mills, Inc., Super Slurper, is a modified starch suitable for disposable diapers, surgical pads and paper towels applications. Acrylonitrile polymers also provide some unique applications. Hollow fibers of acrylonitrile polymers as ultrafiltration membrane materials are used in the pharmaceutical and bioprocessing industries (166). Polyacrylonitrile-based electrolyte with $Li/LiMn_2O_4$ salts is used for solid-state batteries (167). Polyacrylonitrile is also used as a binding matrix for composite inorganic ion exchanger (168).

6. Economic Aspects

SAN has shown steady growth since its introduction in the 1950s. The combined properties of SAN copolymers such as optical clarity, rigidity, chemical and heat resistance, high tensile strength, and flexible molding characteristics, along with reasonable price have secured their market position. Among the plastics with which SAN competes are acrylics, general-purpose polystyrene, and polycarbonate. SAN supply and demand are difficult to track because more than 75% of the resins produced are believed to be used captively for ABS compounding and in the production of acrylonitrile–styrene–acrylate (ASA) and acrylonitrile–EPDM–styrene (AES) weatherable copolymer (169). SAN is considered to be only an intermediate product and not a separate polymer in the production processes for these materials.

There are two major producers of SAN for the merchant market in the United States, Bayer Corporation and the Dow Chemical Company, which market these materials under the names of Lustran and Tyril, respectively. Bayer became a U.S. producer when it purchsed Monsanto's styrenics business in December 1995 (170). Some typical physical properties of these have been shown in Table 1. These two companies also captively consume the SAN for the production of ABS as well as SAN-containing weatherable polymers. The other two U.S. SAN producers either mainly consume the resin captively for ABS and ASA polymers (GE Plastics) or tool produce for a single client (Zeon Chemicals). BASF is expected to become a more aggressive SAN supplier in the United States when its Altamira, Mexico, stryenics plant comes on in early 1999. Overall, U.S. SAN consumption has been relatively stable for the last few years, ranging from $(43–44.5)\times10^3$ metric tons (95–98 million pounds) between 1994 and 1996. Most markets for SAN are growing at only GDP rates. Consumption growth for SAN in 1996–2001 is expected to continue at an average annual rate approximation that of GDP growth at 2%. Use for packaging will be flat and the automotive application may disappear altogether; other markets, however, are expected to increase at annual rates between 2.3% and 5.9%. Production and consumption figures for SAN resin in recent years are shown in Table 7 (170).

Table 7. **U.S. Production/Consumption of SAN (10^3 metric tons, dry-weight basis)**

	Production	Consumption[a]
1985	39.5	34.1
1986	41.8	35.9
1987	57.3	38.6[b]
1988	67.3	41.4
1989	51.4	34.1
1990	61.4	37.3
1991	49.5	37.7
1992	51.4	38.2
1993	47.7	40
1994	62.7	44.5[c]
1995	59.1	43.6[c]
1996	55.5	43.6[c]
1997	43.6	—[c]

[a] Includes captive consumption for uses other than ABS compounding and ASA/AES polymers production.
[b] According to the SPI, 45 metric tons of SAN resin were consumed domestically in 1987. Industry believes this figure to be incorrect. An estimate of 38.6 metric tons is believed to be more accurate.
[c] Reported SPI data for 1996–1997 includes both U.S. and Canadian information and, therefore, are not included in this table. The stated CEH statistics represent consumption only.

7. Health and Safety Factors

SAN resins themselves appear to pose few health problems in that SAN resins are allowed by FDA to be used by the food and medical for certain applications under prescribed conditions (171). The main concern over SAN resin use is that of toxic residuals, eg, acrylonitrile, styrene, or other polymerization components such as emulsifiers, stabilizers, or solvents. Each component must be treated individually for toxic effects and safe exposure level.

Acrylonitrile is believed to behave as an enzyme inhibitor of cellular metabolism (172), and it is classified as a probable human carcinogen of medium carcinogenic hazard (173) and can affect the cardiovascular system and kidney and liver functions (172). Direct potential consumer exposure to acrylonitrile through consumer product usage is low because of little migration of the monomer from such products; the concentrations of acrylonitrile in consumer products are estimated to be less than 15 ppm in SAN resins. OSHA's permissible exposure limit for acrylontrile is 2 ppm as an 8-h time-weighted average with no eye or skin contact; the acceptable ceiling limit is 10 ppm; and the action level, the concentration level that triggers the standard for monitoring, etc, is 1 ppm. Further information on the toxicology and human exposure to acrylonitrile is available (174–176) (see ACRYLONITRILE).

Styrene, a main ingredient of SAN resins, is a possible human carcinogen (IARC Group 2B/EPA-ORD Group C). It is an irritant to the eyes and respiratory tract, and while prolonged exposure to the skin may cause irritation and central nervous system effects such as headache, weakness, and depression, harmful amounts are not likely to absorbed through the skin. OSHA has set permissible exposure limits for styrene in an 8-h time-weighted average at 100 ppm; the acceptable ceiling limit (short-term, 15 min, exposure limit at 200 ppm) (177); and the acceptable maximum peak at 600 ppm (5-min max. peak in any 3 h). For more information on styrene environmental issues, see the CEH Styrene marketing research report (178,179).

In September 1996, the EPA issued a final rule requiring producers of certain thermoplastics to reduce emissions of hazardous air pollutants from their facilities. The final rule seeks to control air toxins released during the manufacture of seven types of polymers and resins, including SAN.

8. Uses

Acrylonitrile copolymers offer useful properties, such as rigidity, gas barrier, chemical and solvent resistance, and toughness. These properties are dependent upon the acrylonitrile content in the copolymers. SAN copolymers offer low cost, rigidity, processability, chemical and solvent resistance, transparency, and heat resistance in which the properties provide the advantages over other competing transparent/clear resins, such as: polymethyl methacrylate, polystyrene, polycarbonate, and styrene–butadiene copolymers. SAN copolymers are widely used in goods such as housewares, packaging, appliances, interior automotive lenses, industrial battery cases, and medical parts. U.S. consumption of SAN resins in major industrial markets are shown in Table 8.

Table 8. **SAN Copolymer Uses**[a]

Application	Articles
appliances	air conditioner parts, decorated escutcheons, washer and dryer instrument panels, washing machine filter bowls, refrigerator shelves, meat and vegetable drawers and covers, blender bowls, mixers, lenses, knobs, vacuum cleaner parts, humidifiers, and detergent dispensers
automotive	batteries, bezels, instrument lenses, signals, glass-filled dashboard components, and interior trim
construction	safety glazing, water filter housings, and water faucet knobs
electronic	battery cases, instrument lenses, cassette parts, computer reels, and phonograph covers
furniture	chair backs and shells, drawer pulls, and caster rollers
housewares	brush blocks and handles, broom and brush bristles, cocktail glasses, disposable dining utensils, dishwasher-safe tumblers, mugs, salad bowls, carafes, serving trays, and assorted drinkware, hangers, ice buckets, jars, and soap containers
industrial	batteries, business machines, transmitter caps, instrument covers, and tape and data reels
medical	syringes, blood aspirators, intravenous connectors and valves, petri dishes, and artificial kidney devices
packaging	bottles, bottle overcaps, closures, containers, display boxes, films, jars, sprayers, cosmetics packaging, liners, and vials
custom molding	aerosol nozzles, camera parts, dentures, disposable lighter housings, fishing lures, pen and pencil barrels, sporting goods, toys, telephone parts, filter bowls, tape dispensers, terminal boxes, toothbrush handles, and typewriter keys

[a] Refs. 9 and 103.

Acrylonitrile copolymers have been widely used in films and laminates for packaging (180–184) due to their excellent barrier properties. In addition to laminates (185–189), SAN copolymers are used in membranes (190–193), controlled-release formulations (194,195), polymeric foams (196,197), fire-resistant compositions (198,199), ion-exchange resins (200), reinforced paper (201), concrete and mortar compositions (202,203), safety glasses (204), solid ionic conductors (205), negative resist materials (206), electrophotographic toners (207), and optical recording as well (208). SAN copolymers are also used as coatings (209), dispersing agents for colorants (210), carbon-fiber coatings for improved adhesion (211), and synthetic wood pulp (212). SAN copolymers have been blended with aromatic polyesters to improve hydrolytic stability (213), with methyl methacrylate polymers to form highly transparent resins (214), and with polycarbonate to form toughened compositions with good impact strength (215–218). Table 8 lists the most common uses of SAN copolymers in major industrial markets (9,101). Some important modifications of SAN copolymers are listed in Table 9.

Acrylonitrile has contributed the desirable properties of rigidity, high-temperature resistance, clarity, solvent resistance, and gas impermeability to many polymeric systems. Its availability, reactivity, and low cost ensure a continuing market presence and provide potential for many new applications.

Table 9. **Modified SAN Copolymers**

Modifier	Remarks	Reference
polybutadiene	ABS, impact resistant	*a*
EPDM rubber[b]	impact and weather resistant	218, 219
polyacrylate	impact and weather resistant	220, 221
poly(ethylene–*co*-vinyl acetate) (EVA)	impact and weather resistant	222
EPDM + EVA	impact and weather resistant	223
silicones	impact and weather resistant	224
chlorinated polyethylene	impact and weather resistant and flame retardant	225
polyester, cross-linked	impact resistant	226
poly(α-methylstyrene)	heat resistant	227
poly(butylene terephthalate)	wear and abrasion resistant	228
ethylene oxide–propylene oxide copolymers	used as lubricants to improve processibility	229
sulfonation	hydrogels of high water absorption	230
glass fibers	high tensile strength and hardness	231

[a] See ACRYLONITRILE POLYMERS, ABS RESINS.
[b] Ethylene–propylene–diene monomer rubber.

BIBLIOGRAPHY

"Acrylonitrile Polymers (Survey and SAN)" in *ECT* 4th ed., Vol. 1, pp. 370–395, L. E. Ball and B. S. Curatolo BP Research; "Acrylonitrile Polymers, Survey and SAN" in *ECT* (online), posting date: December 4, 2000, by Lawrence & Ball, Benedicts, Curatolo, BP Research.

CITED PUBLICATIONS

1. C. H. Bamford, W. G. Barb, A. D. Jenkins, and P. F. Onyon, *The Kinetics of Vinyl Polymerizations by Radical Mechanisms*, Butterworths, London, 1958.
2. W. M. Thomas, *Fortschr. Hochpolymer. Forsch.* **2**, 401 (1961)
3. *The Chemistry of Acrylonitrile*, 2nd ed., American Cyanamid Co., Petrochemical Div., New York, 1959, p. ix.
4. Ref. 1, p. 29.
5. J. L. Ziska, "Olefin-Modified Styrene-Acrylonitrile," in R. Juran, ed., *Modern Plastics Encyclopedia 1989*, **65**(11), McGraw-Hill, Inc., New York, p. 105.
6. D. M. Bennett, "Acrylic–Styrene–Acrylonitrile," in R. Juran, ed., *Modern Plastics Encyclopedia*1989, **65**(11), McGraw-Hill, Inc., New York, p. 96.
7. S. P. Nemphos and Y. C. Lee, *Am. Chem. Soc., Div. Org. Coat. Plast. Chem. Pap.* **33**(2), 618 (1973).
8. Ref. 1, pp. 39–61.
9. F. L. Reithel, "Styrene–Acrylonitrile (SAN)," in R. Juran, ed., *Modern Plastics Encyclopedia 1989* **65**(11), McGraw-Hill, Inc., New York, p. 105.
10. B. N. Hendy, in N. A. J. Platzer, ed., *Copolymers, Polyblends, and Composites* (Adv. Chem. Ser. 142), American Chemical Society, Washington, D.C., 1975, p. 115.

11. W. J. Hall and H. K. Chi, *SPE National Technical Conference: High Performance Plastics*, Cleveland, Ohio, Oct. 5–7, 1976, pp. 1–5.
12. N. W. Johnston, *Polym. Prepr. Am. Chem. Soc.* **14**(1), 46 (1973).
13. N. W. Johnston, *J. Macromol. Sci. Rev. Macromol. Chem.* **C14**, 215 (1976).
14. J. A. Sauer and C. C. Chen, *Polym. Eng. Sci.* **24**, 786 (1984).
15. A. W. Hanson and R. L. Zimmerman, *Ind. Eng. Chem.* **49**, 1803 (1957).
16. Brit. Pat. 590,247 (July 11, 1947) (to Bakelite Ltd.).
17. R. F. Blanks and B. N. Shah, *Polym. Prepr. Am. Chem. Soc.* **17**(2), 407 (1976).
18. S. A. Jabarin and E. A. Lofgren, *Polym. Eng. Sci.***26**, 405 (1986).
19. M. Kopic, F. Flajsman, and Z. Janovic, *J. Macromol. Sci. Chem.* **A24**(1), 17 (1987).
20. H. U. Khan and G. S. Bhargava, *J. Polym. Sci., Polym. Lett. Ed.* **22**(2), 95 (1984).
21. H. U. Khan, V. K. Gupta, and G. S. Bhargava, *Polym. Commun.* **24**(6), 191 (1983).
22. V. H. Gerrens, H. Ohlinger, and R. Fricker, *Makromol. Chem.* **87**, 209 (1965).
23. Y. Shimura, I. Mita, and H. Kambe, *J. Polym. Sci., Polym. Lett. Ed.* **2**, 403 (1964).
24. A. R. Cooper, "Molecular Weight Determination," in J. I. Kroschwitz, ed., *Encyclopedia of Polymer Science and Engineering*, 2nd ed., Vol. 10, Wiley-Interscience, New York, 1987, pp. 1–19.
25. L. H. Garcia-Rubio, J. F. MacGregor, and A. E. Hamielec, *Polym. Prepr. Am. Chem. Soc.* **22**(1), 292 (1981).
26. K. Tsuchida, *Nempo—Fukui-ken Kogyo Shikenjo* **1980**, 74 (1981).
27. G. Gloeckner and R. Konigsveld, *Makromol. Chem. Rapid Commun.* **4**, 529 (1983).
28. G. Gloeckner, *Pure Appl. Chem.* **55**, 1553 (1983).
29. M. Alonso, F. Recasens, and L. Puiganer, *Chem. Eng, Sci.* **41**, 1039 (1986).
30. G. Hempel and U. Ruedt, *Dtsch. Lebensm.-Runsch.* **84**, 239 (1988).
31. I. Bruening, *Bol. Tec. PETROBRAS* **26**, 299 (1983).
32. K. Hidaka, *Shokuhin Eiseigaku Zasshi* **22**, 536 (1981).
33. R. G. Fordyce and E. C. Chapin, *J. Am. Chem. Soc.* **69**, 581 (1947).
34. W. V. Smith, *J. Am. Chem. Soc.* **70**, 2177 (1948).
35. P. J. Flory, *Principles of Polymer Chemistry*, Cornell Univ. Press, Ithaca, N.Y., 1957, Chapter V.
36. L. J. Young, in J. Brandrup and E. H. Immergut, *Polymer Handbook*, 2nd ed., John Wiley and Sons, Inc., New York, 1975, pp. 11-128–11-139; see also R. Z. Greenley, in J. Brandup and E. H. Immergut, eds., *Polymer Handbook*, 3rd ed., Wiley-Interscience, New York, 1989, pp. II-165–II-171.
37. C. H. Basdekis, *ABS Plastics*, Reinhold Publishing Corp., New York, 1964, p. 47.
38. Brit. Pat. 1,328,625 (Aug. 30, 1973) (to Daicel Ltd.).
39. R. V. Kocher, L. N. Anisimova. Yu. S. Zaitsev, N. B. Lachinov, and V. P. Zubov, *Polym. Sci. USSR*, **20**, 2793 (1978).
40. M. Suzuki, H. Miyama, and S. Fujimoto, *Bull. Chem. Soc. Jpn.* **35**, 60 (1962).
41. H. Miyama and S. Fujimoto, *J. Polym. Sci.* **54**, S32 (1961).
42. G. Mino, *J. Polym. Sci.* **22**, 369 (1956).
43. F. R. Mayo and F. M. Lewis, *J. Am. Chem. Soc.* **66**, 1594 (1944).
44. E. T. Mertz, T. Alfrey, and G. Goldfinger, *J. Polym. Sci.*, **1**, 75 (1946).
45. A. Guyot and J. Guillot, *J. Macromol. Scl. Chem.* **A1**, 793 (1967).
46. A. Guyot and J. Guillot, *J. Macromol. Sci. Chem.* **A2**, 889 (1968).
47. J. A. Seiner and M. Litt, *Macromolecules*, **4**, 308 (1971).
48. B. Sandner and E. Loth, *Faserforsch. Textiltech.* **27**, 571 (1976).
49. D. J. T. Hill, J. H. O'Donnell, and P. W. O'Sullivan, *Macromolecules*, **15**, 960 (1982).
50. G. S. Prementine and D. A. Tirrell, *Macromolecules* **20**, 3034 (1987).
51. C. C. Lin, W. Y. Chiu, and C. T. Wang, *J. Appl. Polym. Sci.*, **23**, 1203 (1978).
52. T. Kikuta, S. Omi, and H. Kubota, *J. Chem. Eng. Jpn.*, **9**, 64 (1976).
53. Brit. Pat. 1,093,349 (Nov. 29, 1967), T. M. Fisler (to Ministerul Industriei Chimice).

54. U.S. Pat. 3,772,257 (Nov. 13, 1973), G. K. Bochum, P. K. Hurth Efferen, J. M. Liblar, and H. J. K. Hurth (to Knapsack A. G.).
55. U.S. Pat. 4,439,589 (Mar. 27, 1984), H. Alberts, R. Schubart, and A. Pischtschan (to Bayer A.G.).
56. A. Guyot, J. Guillot, C. Pichot, and L. R. Gurerrero, in D. R. Bassett and A. E. Hamielec, eds., *Emulsion Polymers and Emulsion-Polymerization* (Symp. Ser. 165), American Chemical Society, Washington, D.C., 1981, p. 415.
57. J. Snuparek, Jr., and F. Krska, *J. Polym. Sci.* **21**, 2253 (1977).
58. Ger. Pat. 141,314 (Apr. 23, 1980), H. P. Thiele and co-workers (to VEB Chemische Werke Buna).
59. Pol. Pat. PL 129,792 (Nov. 15, 1985), P. Penezek, E. Wardzinska, and G. Cynkowska (to Instytut Chemii Przemyslowej).
60. U.S. Pat. 4,251,412 (Feb. 17, 1981), G. P. Ferrini (to B. F. Goodrich Co.).
61. U.S. Pat. 4,255,307 (Mar. 10, 1981), J. R. Miller (to B. F. Goodrich Co.).
62. U.S. Pat. 4,228,119 (Oct. 14, 1980), I. L. Gomez and E. F. Tokas (to Monsanto Co.).
63. U.S. Pat. 4,215,024 (July 29, 1980), I. L. Gomez and E. F. Tokas (to Monsanto Co.).
64. U.S. Pat. 4,215,085 (July 29, 1980), I. L. Gomez (to Monsanto Co.).
65. U.S. Pat. 4,275,175 (June 23, 1981), I. L. Gomez (to Monsanto Co.).
66. U.S. Pat. 4,252,764 (Feb. 24, 1981), E. F. Tokas (to Monsanto Co.).
67. U.S. Pat. 3,248,455 (Apr. 26, 1966), J. E. Harsch and co-workers (to U.S. Rubber Co.).
68. Brit. Pat. 1,034,228 (June 29, 1966) (to U.S. Rubber Co.).
69. U.S. Pat. 3,249,569 (May 3, 1966), J. Fantl (to Monsanto Co.).
70. U.S. Pat. 3,547,857 (Dec. 15, 1970), A. G. Murray (to Uniroyal, Inc.).
71. Brit. Pat. 971,214 (Sept. 30, 1964) (to Monsanto Co.).
72. U.S. Pat. 4,193,903 (Mar. 18, 1980), B. E. Giddings, E. Wardlow, Jr., and B. L. Mehosky (to The Standard Oil Co. (Ohio)).
73. U.S. Pat. 3,738,972 (June 12, 1973), K. Moriyama and T. Osaka (to Daicel Ltd.).
74. Brit. Pat. 1,328,625 (Aug. 30, 1973), K. Moriyama and S. Takahashi (to Daicel Ltd.).
75. Jpn. Pat. 80000,725 (Jan. 7, 1980), S. Kato and M. Astumi (to Denki Kagaku Kogyo K.K.).
76. Jpn. Pat. 78082,892 (July 21, 1978), S. Kato and M. Momoka (to Denki Kagaku Kogyo K.K.).
77. Jpn. Pat. 79020,232 (July 20, 1979), S. Kato and M. Momoka (to Denki Kagaku Kogyo K.K.).
78. Jpn. Pat. 79119,588 (Sept. 17, 1979), K. Kido, H. Wakamori, G. Asai, and K. Kushida (to Kureha Chemical Industry Co., Ltd.).
79. Jpn. Pat. 82167,303 (Oct. 15, 1982) (to Toshiba Machine Co.).
80. U.S. Pat. 3,198,775 (Aug. 3, 1965), R. E. Delacretaz, S. P. Nemphos, and R. L. Walter (to Monsanto Co.).
81. U.S. Pat. 3,258,453 (June 28, 1966), H. K. Chi (to Monsanto Co.).
82. U.S. Pat. 3,287,331 (Nov. 22, 1966), Y. C. Lee and L. P. Paradis (to Monsanto Co.).
83. U.S. Pat. 3,681,310 (Aug. 1, 1972), K. Moriyama and T. Moriwaki (to Daicel Ltd.).
84. U.S. Pat. 3,243,407 (Mar. 29, 1966), Y. C. Lee (to Monsanto Co.).
85. U.S. Pat. 3,331,810 (July 18, 1967), Y. C. Lee (to Monsanto Co.).
86. U.S. Pat. 3,331,812 (July 18, 1967), Y. C. Lee and S. P. Nemphos (to Monsanto Co.).
87. U.S. Pat. 3,356,644 (Dec. 5, 1967), Y. C. Lee (to Monsanto Co.).
88. U.S. Pat. 3,491,071 (Jan. 20, 1970), R. Lanzo (to Montecatini Edison SpA).
89. Jpn. Pat. 83103,506 (June 20, 1983) (to Mitsubishi Rayon, Ltd.).
90. Jpn. Pat. 88039,908 (Feb. 20, 1988) (to Mitsui Toatsu Chemicals, Inc.).
91. Belg. Pat. 904,985 (Oct. 16, 1986), S. Ikuma (to Mitsubishi Monsanto Chemical Co.).
92. U.S. Pat. 3,141,868 (July 21, 1964), E. P. Fivel (to Resines et Verms Artificiels).

93. U.S. Pat. 3,201,365 (Aug. 17, 1965), R. K. Charlesworth, W. Creck, S. A. Murdock, and K. G. Shaw (to The Dow Chemical Company).
94. U.S. Pat. 2,941,985 (June 21, 1960), J. L. Amos and C. T. Miller (to The Dow Chemical Company).
95. U.S. Pat. 3,031,273 (Apr. 24, 1962), G. A. Latinen (to Monsanto Co.).
96. U.S. Pat. 2,745,824 (May 15, 1956), J. A. Melchore (to American Cyanamid Co.).
97. Jpn. Pat. 73021,783 (Mar. 19, 1973), H. Sato, I. Nagai, T. Okamoto, and M. Inoue (to Toray, K. K. Ltd.).
98. R. L. Zimmerman, J. S. Best, P. N. Hall, and A. W. Hanso7aq in R. F. Gould, ed., *Polymerization and Polycondensation Processes* (Adv. Chem. Ser. 34), American Chemical Society, Washington, D.C., 1962, p. 225.
99. Jpn. Pat. 87179,508 (Aug. 6, 1987), N. Ito and co-workers (to Mitsui Toatsu Chemicals, Inc.).
100. Jpn. Pat. 83037,005 (Mar. 4, 1983) (to Mitsui Toatsu Chemicals, Inc.).
101. U.S. Pat. 3,067,812 (Dec. 11, 1962), G. A. Latinen and R. H. M. Simon (to Monsanto Co.).
102. U.S. Pat. 3,211,209 (Oct. 12, 1965), G. A. Latinen and R. H. M. Simon (to Monsanto Co.).
103. F. M. Peng, "Acrylonitrile Polymer," in J. I. Kroschwitz, ed., *Encyelopedia of Polymer Science and Engineering*, 2nd ed., Vol. 1, Wiley-Interscience Inc., New York, 1985, p. 463.
104. F. M. Peng, *J. Macromol, Sci. Chem.*, **A22**, 1241 (1985).
105. S. Woods, *Canadiqn. Packaging* **39**(9), 48 (1986).
106. P. R. Lund and T. J. Bond, *Soc. Plast. Eng. Tech. Pap.* **24**, 61 (1978).
107. R. C. Adams and S. J. Waisala, *TAPPI Pap. Synth. Conf. Prepr.*, Atlanta. Ga., **79** (1975).
108. Jpn. Pat. 85097,823 (May 31, 1985) (to Mitsui Toatsu Chemicals, Inc.).
109. U.S. Pat. 4,452.835 (June 5, 1984), G. Vasudevan (to Union Carbide Corp.).
110. Ger. Pat. 2,656,993 (June 22, 1978), R. E. Isley (to The Standard Oil Co. (Ohio)).
111. J. A. Carlson, Jr. and L. Borla, *Mod. Plast.* **57**(6), 117 (1980).
112. BP Amoco, Barex Resins Technical Data Sheets.
113. U.S. Pat. 3,926,871 (Dec. 16, 1975), L. W. Hensley and G. S. Li (to The Standard Oil Co. (Ohio)).
114. U.S. Pat. 4,195,135 (Mar. 25, 1980), G. S. Li and J. F. Jones (to The Standard Oil Co. (Ohio)).
115. U.S. Pat. 3,997,709 (Dec. 14, 1976), W. Y. Aziz, L. E. Ball, and G. S. Li (to The Standard Oil Co. (Ohio)).
116. Fr. Pat. 2,207,938 (June 21, 1974) (to Polysar Ltd.).
117. Can. Pat. 991,787 (June 22, 1976), M. H. Richmond and H. G. Wright (to Polysar Ltd.).
118. U.S. Pat. 4,301,112 (Nov. 17, 1981), M. M. Zwick (to American Cyanamid Co.).
119. U.S. Pat. 4,025,581 (May 24, 1977), J. A. Powell and A. Williams (to Rohm and Haas Co.).
120. Jpn. Pat. 79058,794 (May 11, 1979), H. Furukawa and S. Matsumura (to Kanegafuchi Chemical Industry Co., Ltd.).
121. Fr. Pat. 2,389,644 (Dec. 1, 1978), P. Hubin-Eschger (to ATO-Chimie S.A.).
122. Jpn. Pat. 75124,970 (Oct. 1, 1975), A. Kobayashi and M. Ohya (to Kureha Chemical Industry Co., Ltd.).
123. Jpn. Pat. 87013,425 (Jan. 22, 1987), M. Fujimoto, T. Yamashita, and T. Matsumoto (to Kanebo NSC K.K.).
124. U.S. Pat. 4,000,359 (Dec. 28, 1976), W. A. Watts and J. L. Wang (to Goodyear Tire and Rubber Co.).

125. U.S. Pat. 3,832,335 (Sept. 27, 1974), J. W. Bayer (to Owens-Illinois, Inc.).
126. Jpn. Pat. 74021,105 (May 29, 1974), M. Takahashi, K. Yanagisawa, and T. Mori (to Sekisui Chemical Co., Ltd.).
127. U.S. Pat. 3,959,550 (May 25, 1976), M. S. Guillod and R. G. Bauer (to Goodyear Tire and Rubber Co.).
128. Fr. Pat. 2,041,137 (Mar. 5, 1971), Q. A. Trementozzi (to Monsanto Co.).
129. Ger. Pat. 2,134,814 (May 31, 1972), T. Yamawaki, M. Hayashi, and K. Endo (to Mitsubishi Chemical Industries Co., Ltd.).
130. U.S. Pat. 3,725,120 (Apr. 3, 1973), C. A. Suter (to Goodyear Tire and Rubber Co.).
131. Ger. Pat. 1,546,809 (Jan. 31, 1974), D. S. Dixler (to Air Products and Chemicals, Inc.).
132. Jpn. Pat. 87256,871 (Nov. 9, 1987), H. Sakai and T. Kotani (to Asahi Chemical Industry Co., Ltd.).
133. Ger. Pat. 2,436,137 (Feb. 13, 1975), B. N. Hendy (to Imperial Chemical Industries Ltd.).
134. U.S. Pat. 4,596,856 (June 24, 1986), M. M. Wu, E. J. Dewitt and G. S. Li (to The Standard Oil C. (Ohio)).
135. U.S. Pat. 4,603,186 (Jul. 29, 1986), M. M. Wu, E. J. Dewitt and G. S. Li (to The Standard Oil C. (Ohio)).
136. U.S. Pat. 4,638,042 (Jan 20, 1987), G. S. Li, M. M. Wu and E. J. Dewitt (to The Standard Oil C. (Ohio)).
137. U.S. Pat. 4,681,916 (Jul. 21, 1987), M. M. Wu, E. J. Dewitt and G. S. Li (to The Standard Oil C. (Ohio)).
138. U.S. Pat. 4,681,917 (Jul. 21, 1987), G. S. Li, M. M. Wu and E. J. Dewitt (to The Standard Oil C. (Ohio)).
139. U.S. Pat. 4,761,455 (Aug. 2, 1988), M. M. Wu, E. J. Dewitt and G. S. Li (to The Standard Oil C. (Ohio)).
140. U.S. Pat. 4,677,164 (Jun. 30, 1987), G. S. Li, M. M. Wu and E. J. Dewitt (to The Standard Oil C. (Ohio)).
141. S. Yabumoto, K. Ishii, and K. Arita, *J. Polym. Sci.* **A1**, 1577 (1969).
142. N. G. Gaylord, S. S. Dixit, and B. K. Patnaik, *J. Polym. Sci. B* **9**, 927 (1971).
143. N. G. Gaylord, S. S. Dixit, S. Maiti, and B. K. Patnaik, *J. Macromol. Sci. Chem.* **6**, 1495 (1972).
144. K. Arita, T. Ohtomo, and Y. Tsurumi, *J Polym. Sci., Polym. Lett. Ed.* **19**, 211 (1981).
145. C. D. Eisenbach and co-workers. *Angew. Makromol. Chem.* **145/146**, 125 (1986).
146. H. Hirai, *J. Polym. Sci. Macromol. Rev.* **11**, 47 (1976).
147. A. Henglein, *Makromol. Chem.* **14**, 128 (1954).
148. A. Chapiro, *J. Polym. Sci.* **23**, 377 (1957).
149. A. D. Jenkins, *Pure Appl. Chem.* **46**, 45 (1976).
150. H. Craubner, *J. Polym. Sci., Polym. Chem. Ed.* **18**, 2011 (1980).
151. I. G. Krasnoselskaya and B. L. Erusalimskii, *Vysokomol. Soedin. Ser. B* **29**, 442 (1987).
152. A. Noshay and J. E. McGrath, *Block Copolymers: Overview and Critical Survey*, Academic Press, Inc., New York, 1977.
153. J. Chiefari and co-workers. *Macromol 31*, **16**, 5559 (1998).
154. K. Matyjaszewski, *J. Macromo. Sci. A.* **10**, 1785 (1997).
155. M. Baumert and co-wokers, ACS Polymeric Materials Sci and Eng, 79, Fall Meeting (1998).
156. K. Tharanikkarasu and G. Radhakrishnan, *J. Polym. Chem.* 34, **9**, 1723 (1996).
157. T. Fukuda and co-workers, *Macromol. 29* **8**, 3050 (1996).
158. I. Q. Li and co-workers, *Macromol 30*, **18**, 5195 (1997).
159. M. Minagawa and coworkers, *Polymer 36* **12**, 2343 (1995).
160. G. Optiz and coworkres, *Acta Polymerica, 47*, **2/3**, 67 (1996).

161. J. E. Turner, M. Shen, and C. C. Lin, *J. Appl. Polym. Sci.* **25**, 1287 (1980).
162. E. I. Stout, D. Trimnell, W. M. Doane, and C. R. Russell, *J Appl. Polym. Sci.* **21**, 2565 (1977).
163. W. P. Lindsay, *TAPPI. Annual Meeting Preprints*, Atlanta, Ga., 1977, p. 203.
164. F. Sundardi, *J. Appl. Polym. Sci.* **22**, 3163 (1978).
165. U.S. Pat. 4,134,863 (Jan. 16, 1979), G. F. Fanta, E. I. Stout, and W. M. Doane (to U.S. Dept. of Agriculture).
166. A. C. Orchard and L. J. Bates, *BHR Group Conf. Ser. Pub.* **3**, 59–66 (1993).
167. K. M. Abrham and M. Alamgir, *Pro. Int. Power Sources Symp*, **35th**, 264–6 (1992).
168. F. Sebesta and coworkers, *Acta Polyteg* **38**(3), 119 (1998).
169. *Chem. Eng. News* **67**(25), 45 (June 19, 1989).
170. SRI Report, July 1998.
171. United States Food and Drug Administration, *Fed. Regist.* **52**(173), 33802–33803 (Sept. 8, 1987).
172. United States Food and Drug Administration, *Fed. Regist.* 4510–26, **43**(11), 2586 (Jan. 17, 1978).
173. National Institute for Occupational Safety and Health, "A Recommended Standard for Occupational Exposure to Acrylonitrile," DHEW Publ. No. 78–116, U.S. Government Printing Office Washington, D.C., 1978.
174. N. I. Sax and R. J. Lewis, Sr., *Dangerous Properties of Industrial Materials*, 7th ed., Van Nostrand Reinhold, 1989.
175. R. E. Lenga, *The Sigma-Aldrich Library of Chemical Safety Data*, Sigma-Aldrich Corp., Milwaukee, Wis., 1985.
176. M. Sittig, ed., *Priority Toxic Pollutants*, Noyes Data Corporation, Park Ridge, N.J., 1980.
177. "Threshold Limit Values for Chemical Substances in the Work Environment Adopted by ACGIH for 1985–86," American Conference of Government Industrial Hygienists, Cincinnati, Ohio, 1985.
178. "Criteria for Recommended Standard Occupational Exposure to Styren," U.S. Department of Health and Human Services (NIOSH), Washington, D.C. Rep. 83–119, pp. 18, 227, 1983; available from NTIS, Springfield, Va.
179. J. Santodonato and co-workers, *Monograph on Human Exposure to Chemicals in the Work Place; Styrene, PB86-155132*, Syracuse, N.Y., July 1985.
180. Jpn. Pat. 76000,581 (Jan. 6, 1976), M. Nishizawa and co-workers (to Toray Industries, Inc.).
181. Jpn. Pat. 82185,144 (Nov. 15, 1982) (to Gunze Ltd.).
182. Jpn. Pat. 81117,652 (Sept. 16, 1981) (to Asahi Chemical Industry Co., Ltd.).
183. U.S. Pat. 4,389,437 (June 21, 1983), G. P. Hungerford (to Mobil Oil Corp.).
184. Jpn. Pat. 88041,139 (Feb. 22, 1988) (to Mitsui Toatsu Chemicals, Inc.).
185. Jpn. Pat. 83119,858 (July 16, 1983) (to Toyobo Co., Ltd.).
186. Jpn. Pat. 83183,465 (Oct. 26, 1983) (to Dainippon Printing Co., Ltd.).
187. Jpn. Pat. 84012,850 (Jan. 23, 1984) (to Asahi Chemical Industry Co., Ltd.).
188. Jpn. Pat. 85009,739 (Jan. 18, 1985) (to Toyo Seikan Kaisha, Ltd.).
189. Eur. Pat. Appl. EP138,194 (Apr. 24, 1985), J. H. Im and W. E. Shrum (to The Dow Chemical Company).
190. U.S. Pat. 4,364,759 (Dec. 21, 1982), A. A. Brooks, J. M. S. Henis, and M. K. Tripodi (to Monsanto Co.).
191. Jpn. Pat. 84 202,237 (Nov. 16, 1984) (to Toyota Central Research and Development Laboratories, Inc.).
192. Jpn. Pat. 85 202,701 (Oct. 14, 1985), T. Kawai and T. Nogi (to Toray Industries, Inc.).
193. H. Kawato, M. Kakimoto, A. Tanioka, and T. Inoue, *Kenkyu Hokoku–Asahi Garasu Kogyo Gijutsu Shoreikai* **49**, 77 (1986).

194. Eur. Pat. Appl. EP 141,584 (May 15, 1985), R. W. Baker (to Bend Research, Inc.).
195. Y. S. Ku and S. O. Kim, *Yakhak Hoechi* **31**(3), 182 (1987).
196. U.S. Pat. 4,330,635 (May 18, 1982), E. F. Tokas (to Monsanto Co.).
197. Ger. Pat. 3,523,612 (Jan. 23, 1986), N. Sakata and I. Hamada (to Asahi Chemical Industry Co., Ltd.).
198. Jpn. Pat. 84024,752 (Feb. 8, 1984) (to Kanebo Synthetic Fibers, Ltd.).
199. Ger. Pat. 3,512,638 (Feb. 27, 1986), H. J. Kress and co-workers (to Bayer A.G.).
200. Jpn. Pat. 88 089,403 (Apr. 20, 1988) (to Sumitomo Chemical Co., Ltd.).
201. Jpn. Pat. 82 059,075 (Dec. 13, 1982) (to Nichimen Co., Ltd.).
202. S. Milkov and T. Abadzhieva, *Fiz. Khim. Mekh.* **11**, 81 (1983).
203. A. M. Gadalla and M. E. El-Derini, *Polym. Eng. Sci.* **24**, 1240 (1984).
204. Ger. Pat. 2,652,427 (May 26, 1977), G. E. Cartier and J. A. Snelgrove (to Monsanto Co.).
205. Jpn. Pat. 82 137,359 (Aug. 24, 1982) (to Nippon Electric Co., Ltd.).
206. Jpn. Pat. 83 001,143 (Jan. 6, 1983) (to Fujitsu Ltd.).
207. Jpn. Pat. 88 040,169 (Feb. 20, 1988), Y. Takahashi, M. Nakamura, Y. Kitahata, and K. Maeda (to Sharp Corp.).
208. Jpn. Pat. 86 092,453 (May 10, 1986), Y. Ichihara and Y. Uratani (to Mitsubishi Petrochemical Co., Ltd.).
209. Braz. Pat. 87 818 (Dec. 22, 1987), T. P. Christini (to E. I. du Pont de Nemours & Co., Inc.).
210. Jpn. Pat. 84018,750 (Jan. 31, 1984) (to Sanyo Chemical Industries, Ltd.).
211. Jpn. Pat. 84 080,447 (May 9, 1984) (to Mitsubishi Rayon Co., Ltd.).
212. P. Albihn and J. Kubat, *Plast. Rubber Process. Appl.* **3**(3), 249 (1983).
213. U.S. Pat. 4,327,012 (Apr. 27, 1982), G. Salee (to Hooker Chemicals and Plastics Corp.).
214. Jpn. Pat. 83 217,536 (Dec. 17, 1983) (to Asahi Chemical Industry Co., Ltd.).
215. Eur. Pat. Appl. EP 96,301 (Dec. 21, 1983), H. Peters and co-workers (to Bayer A.G.).
216. Y. Fujita and co-workers, *Kobunshi Ronbzinshu* **743**(3), 119 (1986).
217. H. Takahashi and co-workers, *J. Appl. Polym. Sci.* **36**, 1821 (1988).
218. Ger. Pat. 2,830,232 (Mar. 29, 1979), W. J. Peascoe (to Uniroyal, Inc.).
219. Jpn. Pat. 79083,088 (July 2, 1979), S. Ueda. K. Tazaki, H. Kitayama, and I. Kuribayashi (to Asahi-Dow Ltd.).
220. Ger. Pat. 2,826,925 (Jan. 17, 1980), J. Swoboda, G. Lindenschmidt, and C. Bernhard (to BASF A.G.).
221. U.S. Pat. 3,944,631 (Mar. 16, 1976), A. J. Yu and R. E. Gallagher (to Stauffer Chemical Co.).
222. H. Bard and co-workers, Paper 15, *ACS National Meeting, Division of Industrial Engineering Chemistry*, Atlanta. Ga., Mar. 29–Apr. 3, 1981.
223. Jpn. Pat. 79083,049 (July 2, 1979), S. Ueda, K. Tazaki, H. Kitayama, and N. Asamizu (to Asahi-Dow Ltd.).
224. U.S. Pat. 4,071,577 (Jan. 31, 1978), J. R. Falender, C. M. Mettler, and J. C. Saam (to Dow Coming Corp.).
225. *Mod. Plast.* **60**(1), 92 (1983).
226. U.S. Pat. 4,224,207 (Sept. 23, 1980), J. C. Falk (to Borg-Warner Corp.).
227. U.S. Pat. 4,169,195 (Sept. 25, 1979), M. K. Rinehart (to Borg-Warner Corp.).
228. Jpn. Pat. 81016,541 (Feb. 17, 1981) (to Asahi-Dow Ltd.).
229. Ger. Pat. 2,916,668 (Nov. 13, 1980), J. Hambrecht, G. Lindenschmidt, and W. Regel (to BASF A.G.).
230. Jpn. Pat. 80157,604 (Dec. 8, 1980) (to Daicel Chemical Industries, Ltd.).
231. T. C. Wallace, *SPE Regional Technical Conference*, Las Vegas, Nev., Sept. 16–18, 1975.

GENERAL REFERENCES

C. H. Bamford and G. E. Eastmond, "Acrylonitrile Polymer," in N. M. Bikales, ed., *Encyclopedia, of Polymer Science and Technology*, Vol. 1, Interscience Publishers, a Division of John Wiley & Sons, Inc., New York, 1964, pp. 374–425.

Michael M Wu
BP Amoco Chemicals

ACTINIDES AND TRANSACTINIDES

1. Actinides

The actinide elements are a group of chemically similar elements with atomic numbers 89 through 103 and their names, symbols, atomic numbers, and discoverers are given in Table 1 (1–3) (see Thorium and thorium compounds; Uranium and uranium compounds; Plutonium and plutonium compounds; Nuclear reactors; and Radioisotopes).

Each of the elements has a number of isotopes (2,4), all radioactive and some of which can be obtained in isotopically pure form. More than 200 in number and mostly synthetic in origin, they are produced by neutron or charged-particle-induced transmutations (2,4). The known radioactive isotopes are distributed among the 15 elements approximately as follows: actinium, 29; thorium, 28; protactinium, 27; uranium, 22; neptunium, 20; plutonium, 18; americium, 14; curium, 15; californium, 20; einsteinium, 16; fermium, 18; berkelium, 12; mendelevium, 16 nobelium, 12; and lawrencium, 10. There is frequently a need for values to be assigned for the atomic weights of the actinide elements. Any precise experimental work would require a value for the isotope or isotopic mixture being used, but where there is a purely formal demand for atomic weights, mass numbers that are chosen on the basis of half-life and availability have customarily been used. A list of these is provided in Table 1.

Thorium and uranium have long been known, and uses dependent on their physical or chemical, not on their nuclear, properties were developed prior to the discovery of nuclear fission. The discoveries of actinium and protactinium were among the results of the early studies of naturally radioactive substances. The first transuranium element, synthetic neptunium, was discovered during an investigation of nuclear fission, and this event rapidly led to the discovery of the next succeeding element, plutonium. The realization that plutonium as ^{239}Pu undergoes fission with slow neutrons and thus could be utilized in a nuclear weapon supplied the impetus for its thorough investigation. This research has provided the background of knowledge and techniques for the production and identification of nine more actinide elements and three transactinide elements by 1974.

Table 1. **The Actinide Elements**

Atomic number	Element	CAS Registry Number	Symbol	Mass number[a]	Discoverers and date of discovery
89	actinium	[7440-34-8]	Ac	227	A. Debierne, 1899
90	thorium	[7440-29-1]	Th	232	J. J. Berzelius, 1828
91	protactinium	[7440-13-3]	Pa	231	O. Hahn and L. Meitner, 1917, and F. Soddy and J. A. Cranston, 1917
92	uranium	[7440-61-1]	U	238	M. H. Klaproth, 1789
93	neptunium	[7439-99-8]	Np	237	E. M. McMillan and P. H. Abelson, 1940
94	plutonium	[7440-07-5]	Pu	242	G. T. Seaborg, E. M. McMillan, J. W. Kennedy, and A. C. Wahl, 1940–1941
95	americium	[7440-35-9]	Am	243	G. T. Seaborg, R. A. James, L. O. Morgan, and A. Ghiorso, 1944–1945
96	curium	[7440-51-9]	Cm	248	G. T. Seaborg, R. A. James, and A. Ghiorso, 1944
97	berkelium	[744-40-6]	Bk	249	S. G. Thompson, A. Ghiorso, and G. T. Seaborg, 1949
98	californium	[7440-71-3]	Cf	249	S. G. Thompson, K. Street, Jr., A. Ghiorso, and G. T. Seaborg, 1950
99	einsteinium	[7429-92-7]	Es	254	A. Ghiorso, S. G. Thompson, G. H. Higgins, G. T. Seaborg, M. H. Studier, P. R. Fields, S. M. Fried, H. Diamond, J. F. Mech, G. L. Pyle, J. R. Huizenga, A. Hirsch, W. M. Manning, C. I. Browne, H. L. Smith, and R. W. Spence, 1952
100	fermium	[7440-72-4]	Fm	257	A. Ghiorso, S. G. Thompson, G. H. Higgins, G. T. Seaborg, M. H. Studier, P. R. Fields, S. M. Fried, H. Diamond, J. F. Mech, G. L. Pyle, J. R. Huizenga, A. Hirsch, W. M. Manning, C. I. Browne, H. L. Smith, and R. W. Spence, 1953
101	mendelevium	[7440-11-1]	Md	258	A. Ghiorso, B. G. Harvey, G. R. Choppin, S. G. Thompson, and G. T. Seaborg, 1955
102	nobelium	[10028-14-5]	No	259	A. Ghiorso, T. Sikkeland, J. R. Walton, and G. T. Seaborg, 1958
103	lawrencium	[22537-19-5]	Lr	262	A. Ghiorso, T. Sikkeland, A. E. Larsh, and R. M. Latimer, 1961

[a] Mass number of longest lived or most available isotope.

After a lapse of nearly seven years, a new production reaction and a new on-line separation technique were used to identify six more actinide elements (107–112) between 1981 and 1996. In 1999, evidence was reported for production and detection of the SuperHeavy Elements (SHEs) 114, 116, and 118. These elements are near the "island of nuclear stability" calculated in the mid-1960s to be at proton number 114 (or possibly 126) and neutron number 184 because of the extra stability resulting from the filling of the spherical shells predicted at those so-called "magic" numbers.

Thorium, uranium, and plutonium are well known for their role as the basic fuels (or sources of fuel) for the release of nuclear energy (5). The importance of the remainder of the actinide group lies at present, for the most part, in the realm of pure research, but a number of practical applications are also known (6). The actinides present a storage-life problem in nuclear waste disposal and consideration is being given to separation methods for their recovery prior to disposal (see WASTE TREATMENT, HAZARDOUS WASTE; NUCLEAR REACTORS, WASTE MANAGEMENT).

1.1. Source. Only the members of the actinide group through Pu have been found to occur in nature (2,3,7,8). Actinium and protactinium are decay products of the naturally occurring uranium isotope ^{235}U, but the concentrations present in uranium minerals are small and the methods involved in obtaining them from the natural source are very difficult and tedious in contrast to the relative ease with which the elements can be synthesized. Thorium and uranium occur widely in the earth's crust in combination with other elements, and, in the case of uranium, in significant concentrations in the oceans. The extraction of these two elements from their ores has been studied intensively and forms the basis of an extensive technology. Neptunium (^{239}Np and ^{237}Np) and plutonium (^{239}Pu) are present in trace amounts in nature, being formed by neutron reactions in uranium ores. Longer-lived ^{244}Pu, possibly from a primordial source, has been found in very small concentration (1 part in 10^{18}) in the rare earth mineral bastnasite [12172-82-6] (8). Mining these elements from these sources is not feasible because the concentrations involved are exceedingly small. Thus, with the exception of uranium and thorium, the actinide elements are synthetic in origin for practical purposes; ie, they are products of nuclear reactions. High neutron fluxes are available in modern nuclear reactors, and the most feasible method for preparing actinium, protactinium, and most of the actinide elements is through the neutron irradiation of elements of high atomic number (3,9).

Actinium can be prepared by the transmutation of radium,

$$^{226}\mathrm{Ra} + n \;\rightarrow\; ^{227}\mathrm{Ra} + \gamma; \qquad ^{227}\mathrm{Ra} \xrightarrow[41.2\ \mathrm{min}]{\beta^-} {}^{227}\mathrm{Ac}$$

and gram amounts have been obtained in this way. The actinium is isolated by means of solvent extraction or ion exchange.

Protactinium can be produced in the nuclear reactions

$$^{230}\mathrm{Th} + n \;\rightarrow\; ^{231}\mathrm{Th} + \gamma; \qquad ^{231}\mathrm{Th} \xrightarrow[25.6\,\mathrm{h}]{\beta^-} {}^{231}\mathrm{Pa}$$

However, the quantity of ^{231}Pa produced in this manner is much less than the amount (more than 100 g) that has been isolated from the natural source. The methods for the recovery of protactinium include coprecipitation, solvent extraction, ion exchange, and volatility procedures. All of these, however, are rendered difficult by the extreme tendency of protactinium(V) to form polymeric colloidal particles composed of ionic species. These cannot be removed from aqueous media by solvent extraction; losses may occur by adsorption to containers; and protactinium may be adsorbed by any precipitate present.

Kilogram amounts of neptunium (^{237}Np) have been isolated as a by-product of the large-scale synthesis of plutonium in nuclear reactors that utilize ^{235}U and ^{238}U as fuel. The following transmutations occur:

$$^{238}\mathrm{U} + n \rightarrow {}^{237}\mathrm{U} + 2n; \qquad {}^{237}\mathrm{U} \xrightarrow[6.75\,\mathrm{d}]{\beta^-} {}^{237}\mathrm{Np}$$

and

$$^{235}\mathrm{U} + n \rightarrow {}^{236}\mathrm{U} + \gamma; \qquad {}^{236}\mathrm{U} + n \rightarrow {}^{237}\mathrm{U} + \gamma; \qquad {}^{237}\mathrm{U} \xrightarrow{\beta^-} {}^{237}\mathrm{Np}$$

The wastes from uranium and plutonium processing of the reactor fuel usually contain the neptunium. Precipitation, solvent extraction, ion exchange, and volatility procedures (see DIFFUSION SEPARATION METHODS) can be used to isolate and purify the neptunium.

Plutonium as the important isotope ^{239}Pu is prepared in ton quantities in nuclear reactors. It is produced by the following reactions, wherein the excess neutrons produced by the fission of ^{235}U are captured by ^{238}U to yield ^{239}Pu.

$$^{235}\mathrm{U} + n \rightarrow \text{fission products} + 2.5\,n + 200\ \mathrm{MeV}$$

$$^{238}\mathrm{U} + n \rightarrow {}^{230}\mathrm{U} \xrightarrow[23.5\ \mathrm{min}]{\beta^-} {}^{239}\mathrm{Np} \xrightarrow[2.3\,\mathrm{d}]{\beta^-} {}^{239}\mathrm{Pu}$$

The plutonium usually contains isotopes of higher mass number (Fig. 1). A variety of industrial-scale processes have been devised for the recovery and purification of plutonium. These can be divided, in general, into the categories of precipitation, solvent extraction, and ion exchange.

The isotope ^{238}Pu, produced in kilogram quantities by the reactions

$$^{237}\mathrm{Np} + n \rightarrow {}^{238}\mathrm{Np} \quad \text{and} \quad {}^{238}\mathrm{Np} \xrightarrow[2.1\,\mathrm{d}]{\beta^-} {}^{238}\mathrm{Pu}$$

is an important fuel for isotopically powered energy sources used for terrestrial and extraterrestrial applications.

Kilogram quantities of americium as ^{241}Am can be obtained by the processing of reactor-produced plutonium. Much of this material contains an appreciable proportion of ^{241}Pu, which is the parent of ^{241}Am. Separation of the americium is effected by precipitation, ion exchange, or solvent extraction.

The nuclear reaction sequences of neutron captures and beta decays involved in the preparation of the actinide elements by means of the slow

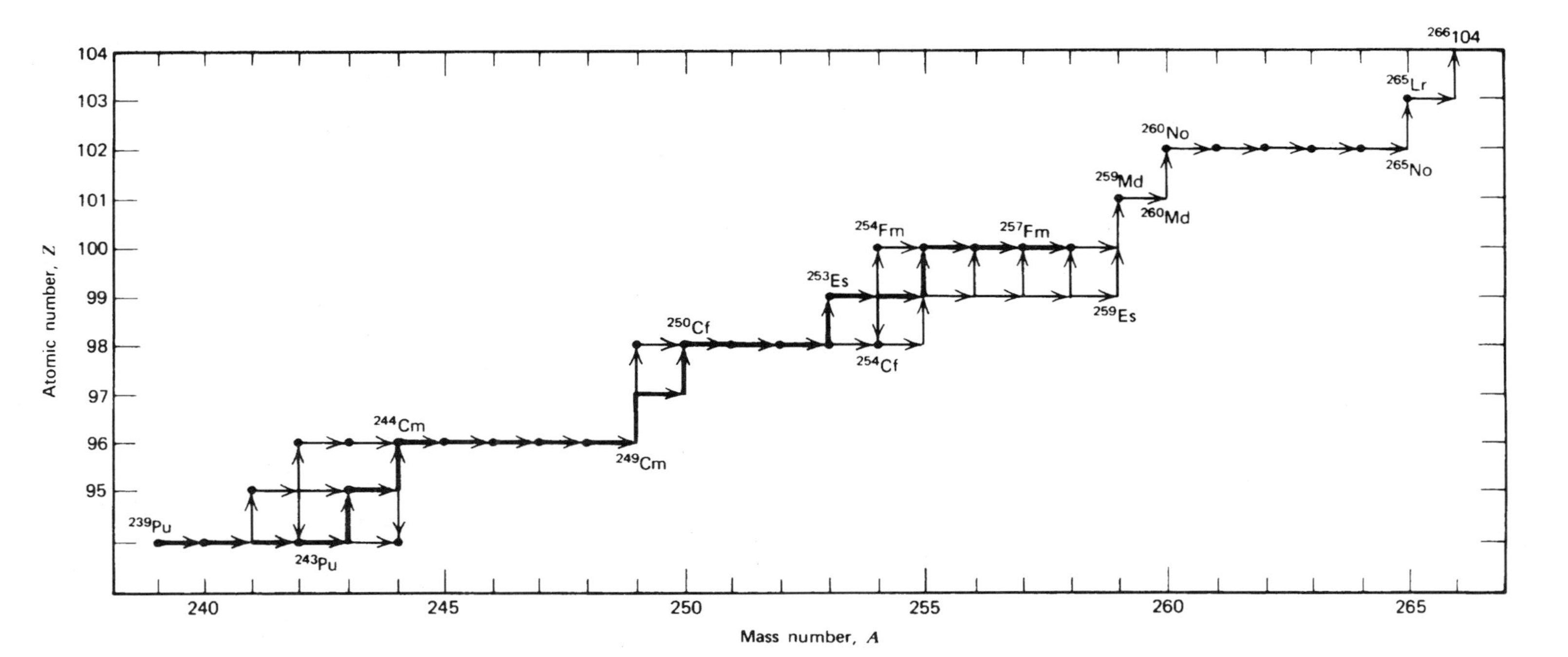

Fig. 1. Nuclear reactions for the production of heavy elements by intensive slow neutron irradiation. The main line of buildup is designated by heavy arrows. The sequence above ^{258}Fm represents predictions.

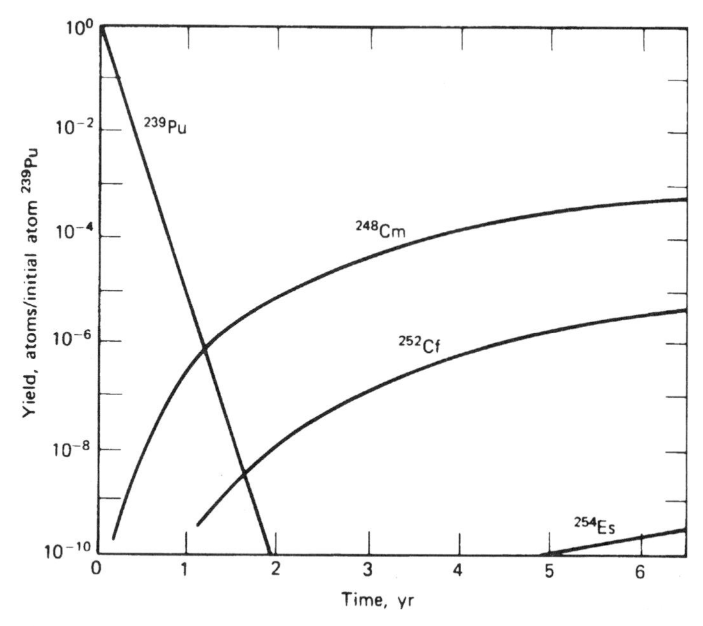

Fig. 2. Production of heavy nuclides by the irradiation of ^{239}Pu at a flux of 3×10^{14} neutrons/(cm$^2 \cdot$ s).

neutron irradiation of ^{239}Pu are indicated in Figure 1. The irradiations can be performed by placing the parent material in the core of a high-neutron-flux reactor where fluxes of neutrons in excess of 10^{14} may be available. Figure 2 gives an indication of the time required for typical preparation of various heavy isotopes from ^{239}Pu as the starting material. For example, beginning with 1 kg of ^{239}Pu, about 1 mg of ^{252}Cf would be present after 5–10 yr of continuous irradiation at a neutron flux 3×10^{14} neutrons/(cm^2·s). Much larger quantities can be produced by irradiating larger quantities of plutonium in production reactors, followed by irradiation of the curium thus produced in higher-neutron-flux reactors, ca 10^{15} neutrons/(cm^2·s), such as those at the Savannah River Plant in South Carolina and the High Flux Isotopes Reactor (HFIR) at the Oak Ridge National Laboratory (ORNL) in Tennessee. Such programs have led to the production of kilogram quantities of curium (^{244}Cm and heavier isotopes), gram quantities of californium, 100-mg quantities of berkelium, and milligram quantities of einsteinium (6,10). The elements 95 to 100 are also produced in increasing quantities by nuclear power reactors.

Ion exchange (qv; see also CHROMATOGRAPHY) is an important procedure for the separation and chemical identification of curium and higher elements. This technique is selective and rapid and has been the key to the discovery of the transcurium elements, in that the elution order and approximate peak position for the undiscovered elements were predicted with considerable confidence (9). Thus the first experimental observation of the chemical behavior of a new actinide element has often been its ion-exchange behavior—an observation coincident with its identification. Further exploration of the chemistry of the element often depended on the production of larger amounts by this method.

Solvent extraction is another useful method for separating and purifying actinide elements.

There are many similarities in the chemical properties of the lanthanide elements (see LANTHANIDES) and those of the actinides, especially with elements in the same oxidation state. A striking example of this resemblance is furnished by their ion-exchange behavior. Figure 3 shows the comparative elution data for tripositive actinide and lanthanide ions obtained by the use of the ion-exchange resin Dowex-50 (a copolymer of styrene and divinylbenzene with sulfonic acid groups) and the eluting agent ammonium α-hydroxyisobutyrate [2539-76-6]. In this system, which is used for illustration because of its historical importance,

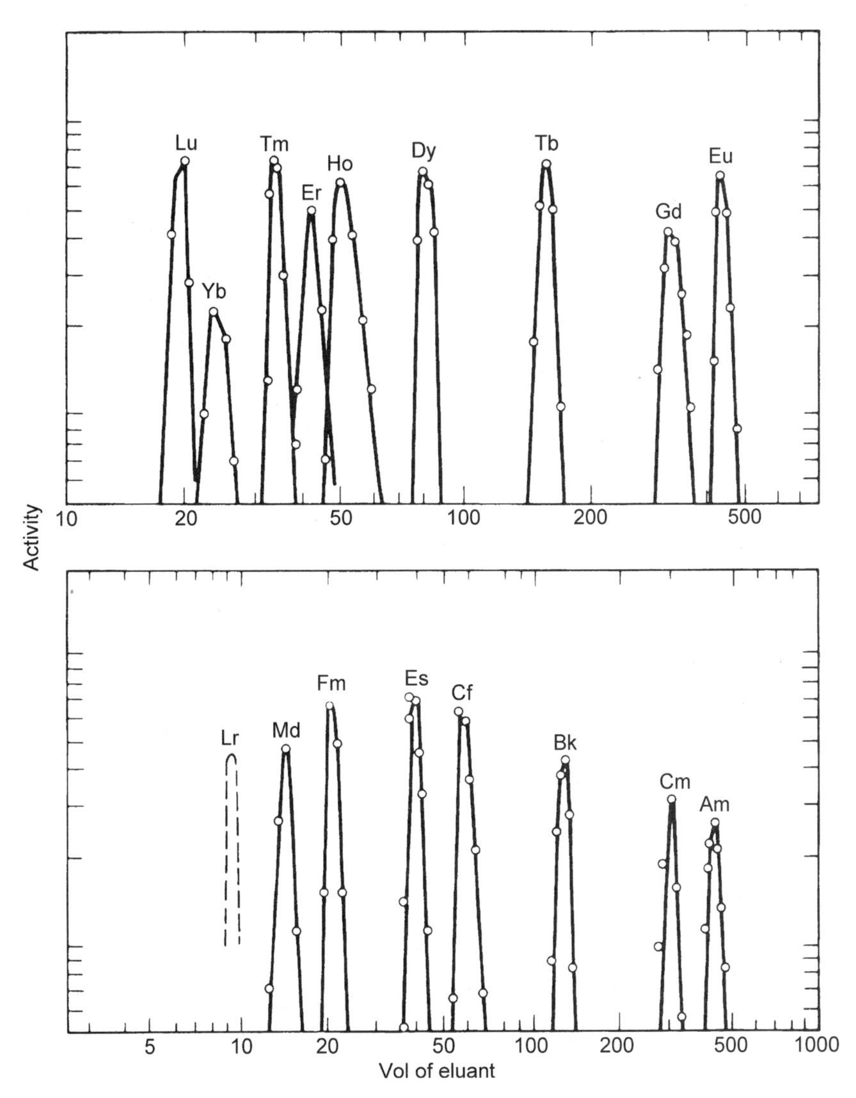

Fig. 3. The elution of tripositive actinide and lanthanide ions. Dowex-50 ion-exchange resin was used with ammonium α-hydroxyisobutyrate as the eluant. The position predicted for short-lived lawrencium is indicated by a broken line.

the elutions occur in the inverse order of atomic number. The elution sequence depends on a balance between the adherence to the resin and the stability of the complex ion formed with the eluting agent and may be correlated with the variation of ionic radius with atomic number.

Actinide ions of the III, IV, and VI oxidation states can be adsorbed by cation-exchange resins and, in general, can be desorbed by elution with chloride, nitrate, citrate, lactate, α-hydroxyisobutyrate, ethylenediaminetetraacetate, and other anions (11,12).

Ion-exchange separations can also be made by the use of a polymer with exchangeable anions; in this case, the lanthanide or actinide elements must be initially present as complex ions (11,12). The anion-exchange resins Dowex-1 (a copolymer of styrene and divinylbenzene with quaternary ammonium groups) and Amberlite IRA-400 (a quaternary ammonium polystyrene) have been used successfully. The order of elution is often the reverse of that from cationic-exchange resins.

Extraction chromatography (see ADSORPTION), in which the organic extractant is adsorbed on the surfaces of a fine, porous powder placed in a column, offers another excellent method for separating the actinide elements from each other. Useful cation extracting agents include bis(2-ethylhexyl)phosphoric acid [298-07-7], mono(2-ethylhexyl)phenylphosphonic acid ester [1518-07-6], and *n*-tributyl phosphate [126-73-8] (12). Excellent anion extracting agents include tertiary amines such as tricapryl amine [1116-76-3] or trilauryl amine [102-87-4], or quaternary amines such as tricaprylmethyl ammonium chloride [5137-55-3] (nitrate, thiocyanate) (12). Satisfactory supporting agents can be found in commercially available diatomaceous earths or silica microspheres.

It is possible to prepare very heavy elements in thermonuclear explosions, owing to the very intense, although brief (order of a microsecond), neutron flux furnished by the explosion (3,13). Einsteinium and fermium were first produced in this way; they were discovered in the fallout materials from the first thermonuclear explosion (the "Mike" shot) staged in the Pacific in November 1952. It is possible that elements having atomic numbers greater than 100 would have been found had the debris been examined very soon after the explosion. The preparative process involved is multiple neutron capture in the uranium in the device, which is followed by a sequence of beta decays. For example, the synthesis of ^{255}FM in the Mike explosion was via the production of ^{255}U from ^{238}U, followed by a long chain of short-lived beta decays,

$$^{255}\mathrm{U} \xrightarrow{-\beta^-} {}^{255}\mathrm{Np} \xrightarrow{-\beta^-} {}^{255}\mathrm{Pu} \xrightarrow{-\beta^-} \cdots \longrightarrow {}^{255}\mathrm{Fm}$$

all of which occur after the neutron reactions are completed.

The process of neutron irradiation in high-flux reactors cannot be used to prepare the elements beyond fermium (^{257}Fm), except at extremely high neutron fluxes, because some of the intermediate isotopes that must capture neutrons have very short half-lives that preclude the necessary concentrations. Transfermium elements are prepared in charged-particle bombardments (2,3). Such syntheses are characterized by the limited availability of target materials of high atomic number, the small reaction yields, and the difficulties intrinsic in the isolation of very short-lived substances. Nonchemical separations of short-lived isotopes from the target materials are carried out during bombardments. Numerous isotopes of mendelevium, nobelium, and lawrencium (the heaviest actinide elements) are produced by bombardment with heavy ions. Despite the fact that these are usually produced on a "one-atom-at-a-time" basis, the chemical properties of these elements have been studied using the tracer technique. Cyclotrons can be used to accelerate heavy ions; in addition, linear accelerators designed for this express purpose are in operation in several laboratories throughout the world.

Isotopes sufficiently long-lived for work in weighable amounts are obtainable, at least in principle, for all of the actinide elements through fermium (100); these isotopes with their half-lives are listed in Table 2 (4). Not all of

Table 2. **Long-Lived Actinide Nuclides Available in Weighable Amounts**

Element	Isotope	CAS Registry Number	Half-life
actinium-227	^{227}Ac	[14952-40-0]	21.8 yr
thorium-232	^{232}Th	[7440-29-1]	1.41×10^{10} yr
protactinium-231	^{231}Pa	[14331-85-2]	3.25×10^{4} yr
uranium-238	^{238}U[a]	[24678-82-8]	4.47×10^{9} yr
neptunium-236	^{236}Np[b]	[15770-36-4]	1.55×10^{5} yr
neptunium-237	^{237}Np	[13994-20-2]	2.14×10^{6} yr
plutonium-238	^{238}Pu	[13981-16-3]	87.8 yr
plutonium-239	^{239}Pu	[15117-48-3]	24,150 yr
plutonium-240	^{240}Pu	[14119-33-6]	6,540 yr
plutonium-241	^{241}Pu	[14119-32-5]	14.9 yr
plutonium-242	^{242}Pu	[13982-10-0]	3.87×10^{5} yr
plutonium-244	^{244}Pu	[14119-34-7]	8.3×10^{7} yr
americium-241	^{241}Am	[14596-10-2]	433 yr
americium-242	^{242}Am	[13981-54-9]	152 yr
americium-243	^{243}Am	[14993-75-0]	7,400 yr
curium-242	^{242}Cm	[15510-73-3]	163.0 d
curium-243	^{243}Cm	[15757-87-6]	30 yr
curium-244	^{244}Cm	[13981-15-2]	18.1 yr
curium-245	^{245}Cm	[15621-76-8]	8,540 yr
curium-246	^{246}Cm	[15757-90-1]	4,800 yr
curium-247	^{247}Cm	[15758-32-4]	1.6×10^{7} yr
curium-248	^{248}Cm	[15758-33-5]	3.6×10^{5} yr
curium-250	^{250}Cm[c]	[15743-88-6]	1.1×10^{4} yr
berkelium-247	^{247}Bkb	[15752-38-2]	1,380 yr
berkelium-249	^{249}Bk	[14900-25-5]	320 d
californium-249	^{249}Cf	[15237-97-5]	350 yr
californium-250	^{250}Cf	[13982-11-1]	13.1 yr
californium-251	^{251}Cf	[15765-19-2]	898 yr
californium-252	^{252}Cf	[13981-17-4]	2.6 yr
einsteinium-253	^{253}Es	[15840-02-5]	20.5 d
einsteinium-254	^{254}Es	[15840-03-6]	276 d
einsteinium-255	^{255}Es	[15840-04-7]	40 d
fermium-257	^{257}Fm	[15750-26-2]	100 d

[a] Natural mixture (^{238}U, 99.3%; ^{235}U, 0.72%; and ^{234}U, 0.006%). Half-life given is for the major constituent ^{238}U.

[b] Available so far only in trace quantities from charged particle irradiations.

[c] Available only in very small amounts from neutron irradiations in thermonuclear explosions.

these are available as individual isotopes. It appears that it will always be necessary to study the elements above fermium by means of the tracer technique (except for some very special experiments) because only isotopes with short half-lives are known.

1.2. Experimental Methods of Investigation. All of the actinide elements are radioactive and, except for thorium and uranium, special equipment and shielded facilities are usually necessary for their manipulation (9,14,15). On a laboratory scale, enclosed containers (gloved boxes) are generally used for safe handling of these substances. In some work, all operations are performed by remote control. Neptunium in the form of the long-lived isotope ^{237}Np is relatively convenient to work with in chemical investigations. Because of the existence of large quantities of the fissionable isotope ^{239}Pu, the physiological toxicity of plutonium deserves emphasis. In this form plutonium is a dangerous poison by reason of its intense alpha radioactivity [1.4×10^8 alpha particles/(mg · min)] and its physiological behavior. Ingested plutonium may be transferred to the bone and, over a period of time, give rise to neoplasms.

The study of the chemical behavior of concentrated preparations of short-lived isotopes is complicated by the rapid production of hydrogen peroxide in aqueous solutions and the destruction of crystal lattices in solid compounds. These effects are brought about by heavy recoils of high energy alpha particles released in the decay process.

Most chemical investigations with plutonium to date have been performed with ^{239}Pu, but the isotopes ^{242}Pu and ^{244}Pu (produced by intensive neutron irradiation of plutonium) are more suitable for such work because of their longer half-lives and consequently lower specific activities. Much work on the chemical properties of americium has been carried out with ^{241}Am, which is also difficult to handle because of its relatively high specific alpha radioactivity, about 7×10^9 alpha particles/(mg · min). The isotope ^{243}Am has a specific alpha activity about twenty times less than ^{241}Am and is thus a more attractive isotope for chemical investigations. Much of the earlier work with curium used the isotopes ^{242}Cm and ^{244}Cm, but the heavier isotopes offer greater advantages because of their longer half-lives. The isotope ^{248}Cm, which can be obtained in relatively high isotopic purity as the alpha-particle decay daughter of ^{252}Cf, is the most practical for chemical studies. Berkelium (as ^{249}Bk) and californium (as a mixture of the isotopes ^{249}Cf, ^{250}Cf, ^{251}Cf, and ^{252}Cf) are available as the result of intensive neutron irradiation of lighter elements. The best isotope for the study of californium is ^{249}Cf, which can be isolated in pure form through its beta-particle-emitting parent, ^{249}Bk. The isotope ^{253}Es (half-life, 20 d), also a product from such intensive neutron irradiation, is used to study the chemical properties of einsteinium. The isotope ^{254}Es (half-life, 276 d) is more useful for work with macroscopic quantities but is produced in much smaller amounts than ^{253}Es. Weighable amounts of berkelium, californium, and einsteinium are difficult to handle because of their intense radioactivity. Spontaneous fission is a mode of decay for ^{252}Cf (half-life, 2.6 yr), 1 μg of which emits approximately 2×10^8 neutrons/min, and the chief mode of decay of ^{254}Cf (half-life, 56 d), 1 μg of which emits approximately 8×10^{10} neutrons/min. Californium produced in the highest-flux reactors unfortunately contains ^{252}Cf, which makes it very difficult to handle. In work with more than a few micrograms of ^{252}Cf it is necessary to do all manipulations by remote control, which is exceedingly cumbersome on such a small scale; therefore, ^{249}Cf is generally used.

Special techniques for experimentation with the actinide elements other than Th and U have been devised because of the potential health hazard to the experimenter and the small amounts available (15). In addition, investigations are frequently carried out with the substance present in very low concentration as a radioactive tracer. Such procedures continue to be used to some extent with the heaviest actinide elements, where only a few score atoms may be available; they were used in the earliest work for all the transuranium elements. Tracer studies offer a method for obtaining knowledge of oxidation states, formation of complex ions, and the solubility of various compounds. These techniques are not applicable to crystallography, metallurgy, and spectroscopic studies.

Microchemical or ultramicrochemical techniques are used extensively in chemical studies of actinide elements (16). If extremely small volumes are used, microgram or lesser quantities of material can give relatively high concentrations in solution. Balances of sufficient sensitivity have been developed for quantitative measurements with these minute quantities of material. Since the amounts of material involved are too small to be seen with the unaided eye, the actual chemical work is usually done on the mechanical stage of a microscope, where all of the essential apparatus is in view. Compounds prepared on such a small scale are often identified by x-ray crystallographic methods.

Position in the Periodic Table and Electronic Structure. Prior to 1944 the location of the heaviest elements in the periodic table had been a matter of

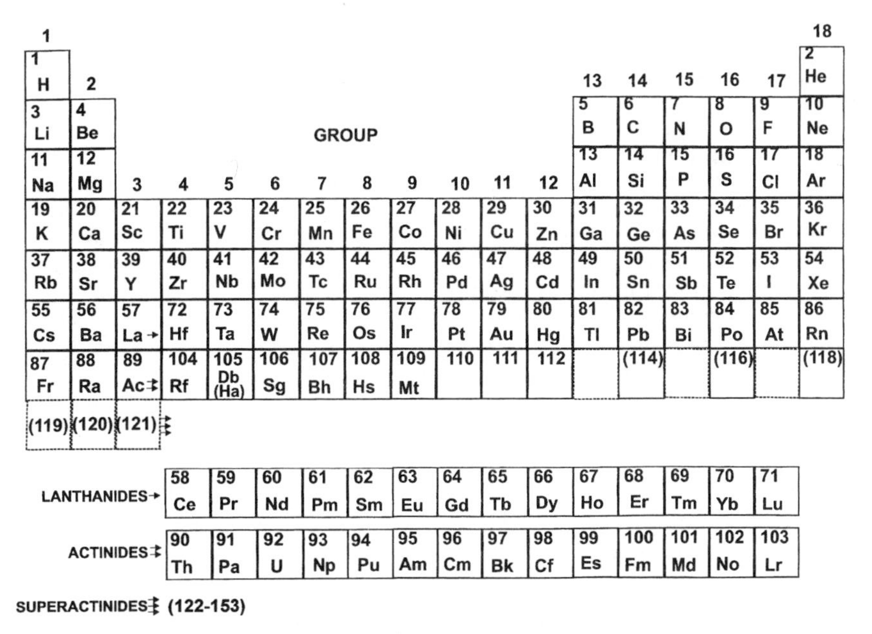

Fig. 4. Futuristic periodic table showing elements known as of early 2000. Predicted locations of undiscovered transuranium elements are shown in parentheses. Elements 114, 116, and 118 have been reported in 1999 but are not yet confirmed.

question, and the elements thorium, protactinium, and uranium were commonly placed immediately below the elements hafnium, tantalum, and tungsten. In 1944, on the basis of earlier chemical studies of neptunium and plutonium, the similarity between the actinide and the lanthanide elements was recognized (1, 7,14). The intensive study of the heaviest elements shows a series of elements similar to the lanthanide series, beginning with actinium (Fig. 4). Corresponding pairs of elements show resemblances in spectroscopic and magnetic behavior that arise because of the similarity of electronic configurations for the ions of the homologous elements in the same state of oxidation, and in crystallographic properties, owing to the near matching of ionic radii for ions of the same charge. The two series are not, however, entirely comparable. A difference lies in the oxidation states. The tripositive state characteristic of lanthanide elements does not appear in aqueous solutions of thorium and protactinium and does not become

Table 3. **Electronic Configurations for Gaseous Atoms of Lanthanide and Actinide Elements**

Atomic number	Element	CAS Registry Number4	Electronic configuration
57	lanthanum	[7439-91-0]	$5d6s^2$ [a]
58	cerium	[7440-45-1]	$4f5d6s^2$
59	praseodymium	[7440-10-0]	$4f^36s^2$
60	neodymium	[7440-00-8]	$4f^46s^2$
61	promethium	[7440-12-2]	$4f^56s^2$
62	samarium	[7440-19-9]	$4f^66s^2$
63	europium	[7440-53-1]	$4f^76s^2$
64	gadolinium	[7440-54-2]	$4f^75d6s^2$
65	terbium	[7440-27-9]	$4f^96s^2$
66	dysprosium	[7429-91-6]	$4f^{10}6s^2$
67	holmium	[7440-60-0]	$4f^{11}6s^2$
68	erbium	[7440-52-0]	$4f^{12}6s^2$
69	thulium	[7440-30-4]	$4f^{13}6s^2$
70	ytterbium	[7440-64-4]	$4f^{14}6s^2$
71	lutetium	[7439-94-3]	$4f^{14}5d6s^2$
89	actinium		$6d7s^2$ [b]
90	thorium		$6d^27s^2$
91	protactinium		$5f^26d7s^2$
92	uranium		$5f^36d7s^2$
93	neptunium		$5f^46d7s^2$
94	plutonium		$5f^67s^2$
95	americium		$5f^77s^2$
96	curium		$5f^76d7s^2$
97	berkelium		$5f^97s^2$
98	californium		$5f^{10}7s^2$
99	einsteinium		$5f^{11}7s^2$
100	fermium		$5f^{12}7s^2$
101	mendelevium		$(5f^{13}7s^2)$
102	nobelium		$(5f^{14}7s^2)$
103	lawrencium		$(5f^{14}6d7s^2$ or $5f^{14}7s^27p)$

[a] Beyond xenon.
[b] Beyond radon. The configurations enclosed in parentheses are predicted.

the most stable oxidation state in aqueous solution until americium is reached. The elements uranium through americium have several oxidation states, unlike the lanthanides. These differences can be interpreted as resulting from the proximity in the energies of the 7*s*, 6*d*, and 5*f* electronic levels.

Table 3 presents the actual or predicted electronic configurations of the actinide elements (2,14). Similar information for the lanthanide elements is given for purposes of comparison (14). As indicated, fourteen 4*f* electrons are added in the lanthanide series, beginning with cerium and ending with lutetium; in the actinide elements, fourteen 5*f* electrons are added, beginning, formally, with thorium and ending with lawrencium. In the cases of actinium, thorium, uranium, americium, berkelium, californium, and einsteinium the configurations were determined from an analysis of spectroscopic data obtained in connection with the measurement of the emission lines from neutral and charged gaseous atoms. The knowledge of the electronic structures for protactinium, neptunium, plutonium, curium, and fermium results from atomic beam experiments (15).

Measurements of paramagnetic susceptibility, paramagnetic resonance, light absorption, fluorescence, and crystal structure, in addition to a consideration of chemical and other properties, have provided a great deal of information about the electronic configurations of the aqueous actinide ions and of actinide compounds. In general, all of the electrons beyond the radon core in the actinide compounds and in aqueous actinide ions are in the 5*f* shell. There are exceptions, such as U_2U_3, and subnormal compounds, such as Th_2S_3, where 6*d* electrons are present.

1.3. Properties. The close chemical resemblance among many of the actinide elements permits their chemistry to be described for the most part in a correlative way (13,14,17,18).

Oxidation States. The oxidation states of the actinide elements are summarized in Table 4 (12–14). The most stable states are designated by bold face type and those which are very unstable are indicated by parentheses. These latter states do not exist in aqueous solutions and have been produced only in solid compounds. The IV state of curium is limited to CmO_2 and CmF_4 (solids) and a complex ion stable in highly concentrated cesium fluoride solution, whereas the IV state of californium is limited to CfO_2 and CfF_4 (solids), and double salts such as $7NaF{\cdot}6CfF_4$ (solid). In the second half of the series the II state first appears in the form of solid compounds at californium and becomes successively more stable in proceeding to nobelium. The II state is observed in aqueous solution for mendelevium (and presumably for fermium) and is the most stable state for nobelium. Americium(II), observed only in solid compounds, and berkelium(IV) show the stability of the half-filled 5*f* configuration ($5f^7$) and nobelium(II) shows the stability of the full 5*f* configuration ($5f^{14}$). The greater tendency toward the II state in the actinides, as compared to the lanthanides, is a result of the increasing binding of the 5*f* (and 6*d*) electrons upon approaching the end of the actinide series.

The actinide elements exhibit uniformity in ionic types. In acidic aqueous solution, there are four types of cations, and these and their colors are listed in Table 5 (12–14,17). The open spaces indicate that the corresponding oxidation states do not exist in aqueous solution. The wide variety of colors exhibited by

Table 4. **The Oxidation States of the Actinide Elements**

Atomic number and element														
89 Ac	90 Th	91 Pa	92 U	93 Np	94 Pu	95 Am	96 Cm	97 Bk	98 Cf	99 Es	100 Fm	101 Md	102 No	103 Lr
						(2)			(2)	(2)	2	2	**2**	
3	(3)	(3)	3	3	3	**3**	**3**	**3**	**3**	**3**	**3**	**3**	3	**3**
	4	4	4	4	**4**	4	4	4	(4)					
		5	5	**5**	5	5								
			6	6	6	6								
				7	(7)									

Table 5. **Ion Types and Colors for Actinide Ions**

Element	M^{3+}	M^{4+}	MO_2^+	MO_2^{2+}	MO_5^{3-}
actinium	colorless				
thorium		colorless			
protactinium		colorless	colorless		
uranium	red	green	color unknown	yellow	
neptunium	blue to purple	yellow-green	green	pink to red	dark green
plutonium	blue to violet	tan to orange-brown	rose	yellow to pink-orange	dark green
americium	pink or yellow	color unknown	yellow	rum-colored	
curium	pale green	color unknown			
berkelium	green	yellow			
californium	green	yellow			
einsteinium	colorless				

actinide ions is characteristic of transition series of elements. In general, protactinium(V) polymerizes and precipitates readily in aqueous solution, and it seems unlikely that ionic forms are present in such solutions.

Corresponding ionic types are similar in chemical behavior, although the oxidation–reduction relationships and therefore the relative stabilities differ from element to element. The ions MO_2^+ and MO_2^{2+} are stable with respect to their binding of oxygen atoms and remain unchanged through a great variety of chemical treatment. They behave as single entities with properties intermediate to singly or doubly charged ions and ions of similar size but of higher charge. The VII oxidation states found for neptunium and plutonium are probably in the form of ions of the type MO_5^{3-} in alkaline aqueous solution; in acid solution these elements in the VII oxidation state readily oxidize water.

The reduction potentials for the actinide elements are shown in Figure 5 (12–14,17,20). These are formal potentials, defined as the measured potentials corrected to unit concentration of the substances entering into the reactions; they are based on the hydrogen-ion–hydrogen couple taken as zero volts; no corrections are made for activity coefficients. The measured potentials were established by cell, equilibrium, and heat of reaction determinations. The potentials for acid solution were generally measured in 1 *M* perchloric acid and for alkaline solution in 1 *M* sodium hydroxide. Estimated values are given in parentheses.

The $M^{4+} \rightleftharpoons M^{3+}$ and $MO_2^{2+} \rightleftharpoons MO_2^+$ couples are readily reversible, and reactions are rapid with other one-electron reducing or oxidizing agents that involve no bond changes. The rate varies with reagents that normally react by two-electron or bond-breaking changes. The $MO_2^+ \rightleftharpoons M^{3+}$, $MO_2^{2+} \rightleftharpoons M^{3+}$, $MO_2^+ \rightleftharpoons M^{4+}$, and $MO_2^{2+} \rightleftharpoons M^{4+}$ couples are not reversible, presumably because of slowness introduced in the making and breaking of oxygen bonds.

Table 6 presents a summary of the oxidation–reduction characteristics of actinide ions (12–14,17,20). The disproportionation reactions of UO_2^+, Pu^{4+}, PuO_2^+, and AmO_2^+ are very complicated and have been studied extensively. In the case of plutonium, the situation is especially complex: four oxidation states

Actinium

Acid: $Ac^{3+} \xrightarrow{-4.9} (Ac^{2+}) \xrightarrow{-0.7} Ac$; $Ac^{3+} \rightarrow Ac$: -2.13

Base: $Ac(OH)_3 \xrightarrow{-2.5} Ac$

Thorium

Acid: $Th^{4+} \xrightarrow{-3.8} (Th^{3+}) \xrightarrow{-4.9} (Th^{2+}) \xrightarrow{+0.7} Th$; $Th^{4+} \rightarrow Th$: -1.83

Base: $ThO_2 \xrightarrow{-2.56} Th$

Protactinium

Acid: $PaOOH^{2+} \xrightarrow{-0.05} Pa^{4+} \xrightarrow{-1.4} (Pa^{3+}) \xrightarrow{-5.0} (Pa^{2+}) \xrightarrow{+0.3} Pa$; $Pa^{4+} \rightarrow Pa$: -1.47

Uranium

Acid: $UO_2^{2+} \xrightarrow{+0.17} UO_2^{+} \xrightarrow{+0.38} U^{4+} \xrightarrow{-0.52} U^{3+} \xrightarrow{-4.7} (U^{2+}) \xrightarrow{-0.1} U$; $UO_2^{2+} \rightarrow U^{4+}$: $+0.27$; $U^{3+} \rightarrow U$: -1.66; $U^{4+} \rightarrow U$: -1.38

Base: $UO_2(OH)_2 \xrightarrow{-0.3} UO_2 \xrightarrow{-2.6} U(OH)_3 \xrightarrow{-2.10} U$

Neptunium

Acid: $NpO_3^{+} \xrightarrow{+2.04} NpO_2^{2+} \xrightarrow{+1.24} NpO_2^{+} \xrightarrow{+0.64} Np^{4+} \xrightarrow{+0.15} Np^{3+} \xrightarrow{-4.7} (Np^{2+}) \xrightarrow{-0.3} Np$; $NpO_2^{2+} \rightarrow Np^{4+}$: $+0.94$; $Np^{3+} \rightarrow Np$: -1.79; $Np^{4+} \rightarrow Np$: -1.30

Base: $NpO_5^{3-} \xrightarrow{+0.6} NpO_2(OH)_2 \xrightarrow{+0.6} NpO_2OH \xrightarrow{+0.3} NpO_2 \xrightarrow{-2.1} Np(OH)_3 \xrightarrow{-2.23} Np$

Plutonium

Acid: $(PuO_2^{3+}) \xrightarrow{?} PuO_2^{2+} \xrightarrow{+1.02} PuO_2^{+} \xrightarrow{+1.04} Pu^{4+} \xrightarrow{+1.01} Pu^{3+} \xrightarrow{-3.5} (Pu^{2+}) \xrightarrow{-1.2} Pu$; $PuO_2^{2+} \rightarrow Pu^{4+}$: $+1.03$; $Pu^{3+} \rightarrow Pu$: -2.00; $Pu^{4+} \rightarrow Pu$: -1.25

Base: $PuO_5^{3-} \xrightarrow{+0.94} PuO_2(OH)_2 \xrightarrow{+0.3} PuO_2OH \xrightarrow{+0.9} PuO_2 \xrightarrow{-1.4} Pu(OH)_3 \xrightarrow{-2.46} Pu$

Fig. 5. Standard (or formal) reduction potentials of actinium and the actinide ions in acidic (pH 0) and basic (pH 14) aqueous solutions (values are in volts vs standard hydrogen electrode) (19).

of plutonium [(III), (IV), (V), and (VI)] can exist together in aqueous solution in equilibrium with each other at appreciable concentrations.

Hydrolysis and Complex Ion Formation. Hydrolysis and complex ion formation are closely related phenomena (13,14).

Of the actinide ions, the small, highly charged M^{4+} ions exhibit the greatest degree of hydrolysis and complex ion formation. For example, the ion Pu^{4+} hydrolyzes extensively and also forms very strong anion complexes. The hydrolysis of Pu^{4+} is of special interest in that polymers that exist as positive colloids can be produced; their molecular weight and particle size depend on the method of preparation. Polymeric plutonium with a molecular weight as high as 10^{10} has been reported.

The degree of hydrolysis or complex ion formation decreases in the order $M^{4+} > MO_2^{2+} > M^{3+} > MO_2^{+}$. Presumably the relatively high tendency toward

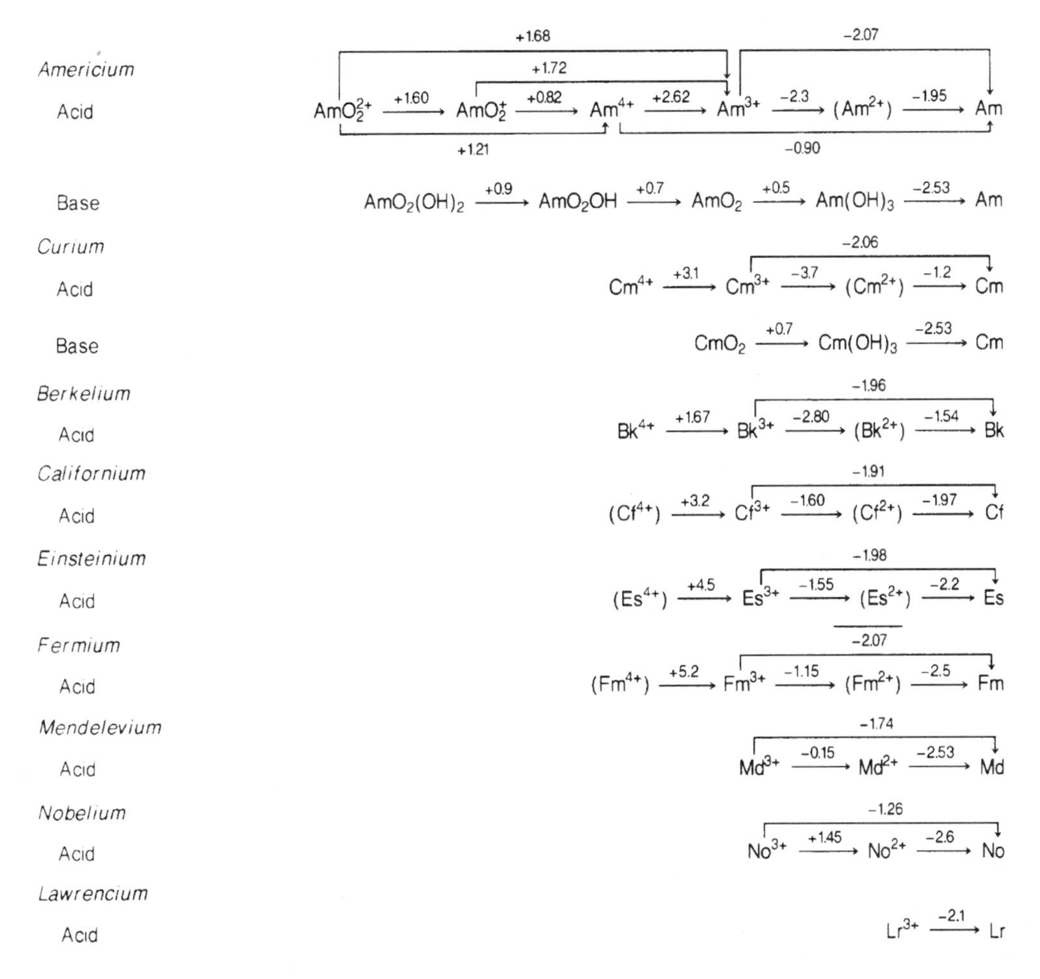

Fig. 5 *(Continued)*

hydrolysis and complex ion formation of MO_2^{2+} ions is related to the high concentration of charge on the metal atom. On the basis of increasing charge and decreasing ionic size, it could be expected that the degree of hydrolysis for each ionic type would increase with increasing atomic number. For the ions M^{4+} and M^{3+}, beginning at about uranium, such a regularity of hydrolytic behavior is observed, but for the remaining two ions, MO_2^+ and MO_2^{2+}, the degree of hydrolysis decreases with increasing atomic number, thus indicating more complicated factors than simple size and charge.

The extensive hydrolysis of protactinium in its V oxidation state makes the chemical investigation of protactinium extremely difficult. Ions of protactinium(V) must be held in solution as complexes, eg, with fluoride ion, to prevent hydrolysis.

The tendency toward complex ion formation of the actinide ions is determined largely by the factors of ionic size and charge. Although there is variation

Table 6. **Stability of Actinide Ions in Aqueous Solution**

Ion	Stability
Md^{2+}	stable to water, but readily oxidized
No^{2+}	stable
Ac^{3+}	stable
U^{3+}	aqueous solutions evolve hydrogen on standing
Np^{3+}	stable to water, but readily oxidized by air to Np^{4+}
Pu^{3+}	stable to water and air, but easily oxidized to Pu^{4+}; oxidizes slightly under the action of its own alpha radiation (in form of ^{239}Pu)
Am^{3+}	stable; difficult to oxidize
Cm^{3+}	stable
Bk^{3+}	stable; can be oxidized to Bk^{4+}
Cf^{3+}	stable
Es^{3+}	stable
Fm^{3+}	stable
Md^{3+}	stable, but rather easily reduced to Md^{2+}
No^{3+}	easily reduced to No^{2+}
Lr^{3+}	stable
Th^{4+}	stable
Pa^{4+}	stable to water, but readily oxidized
Pa^{5+}	stable; hydrolyzes readily
U^{4+}	stable to water, but slowly oxidized by air to UO_2^{2+}
Np^{4+}	stable to water, but slowly oxidized by air to NpO_2^{+}
Pu^{4+}	stable in concentrated acid, eg, 6 *M* HNO_3, but disproportionates to Pu^{3+} and PuO_2^{2+} at lower acidities
Am^{4+}	known in solution only as complex fluoride and carbonate ions
Cm^{4+}	known in solution only as complex fluoride ion
Bk^{4+}	marginally stable; easily reduced to Bk^{3+}
UO_2^{+}	disproportionates to U^{4+} and UO_2^{2+}; most stable at pH 2–4
NpO_2^{+}	stable; disproportionates only at high acidities
PuO_2^{+}	always tends to disproportionate to Pu^{4+} and PuO_2^{2+} (ultimate products); most stable at very low acidities
AmO_2^{+}	disproportionates in strong acid to Am^{3+} and AmO_2^{2+}; reduces fairly rapidly under the action of its own alpha radiation at low acidities (in form of ^{241}Am)
UO_2^{2+}	stable; difficult to reduce
NpO_2^{2+}	stable; easy to reduce
PuO_2^{2+}	stable; easy to reduce; reduces slowly under the action of its own alpha radiation (in form of ^{239}Pu)
AmO_2^{2+}	easy to reduce; reduces fairly rapidly under the action of its own alpha radiation (in form of ^{241}Am)
NpO_5^{3-}	observed only in alkaline solution
PuO_5^{3-}	observed only in alkaline solution; oxidizes water

within each of the ionic types, the order of complexing power of different anions is, in general, fluoride > nitrate > chloride > perchlorate for mononegative anions and carbonate > oxalate > sulfate for dinegative anions. The actinide ions form somewhat stronger complex ions than homologous lanthanide ions.

Actinide ions form complex ions with a large number of organic substances (12). Their extractability by these substances varies from element to element and depends markedly on oxidation state. A number of important separation procedures are based on this property. Solvents that behave in this way are tributyl phosphate, diethyl ether [60-29-7], ketones such as diisopropyl ketone [565-80-5] or methyl isobutyl ketone [108-10-1], and several glycol ether type solvents such as diethyl Cellosolve [629-14-1] (ethylene glycol diethyl ether) or dibutyl Carbitol [112-73-2] (diethylene glycol dibutyl ether).

A number of organic compounds, eg, acetylacetone [123-54-6] and cupferron [135-20-6], form compounds with aqueous actinide ions (IV state for reagents mentioned) that can be extracted from aqueous solution by organic solvents (12). The chelate complexes are especially noteworthy and, among these, the ones formed with diketones, such as 3-(2-thiophenoyl)-1,1,1-trifluoroacetone [326-91-0] ($C_4H_3SCOCH_2COCF_3$), are of importance in separation procedures for plutonium.

Metallic State. The actinide metals, like the lanthanide metals, are highly electropositive (13,14,17). They can be prepared by the electrolysis of molten salts or by the reduction of a halide with an electropositive metal, such as calcium or barium. Their physical properties are summarized in Table 7 (13,14, 17,21). Metallic protactinium, uranium, neptunium, and plutonium have complex structures that have no counterparts among the lanthanide metals. Plutonium metal has very unusual metallurgical properties. It is known to exist in six allotropic modifications between room temperature and its melting point. One of the most interesting features of plutonium metal is the contraction undergone by the δ and δ' phases with increasing temperature. Also noteworthy is the fact that for no phase do both the coefficient of thermal expansion and the temperature coefficient of resistivity have the conventional sign. The resistance decreases if the phase expands on heating. Americium is the first actinide to show resemblance in crystal structure to the lanthanide metals.

With respect to chemical reactivity, the actinide metals resemble the lanthanide metals more than metals of the 5*d* elements such as tantalum, tungsten, rhenium, osmium, and iridium. A wide range of intermetallic compounds has been observed and characterized, including compounds or alloys with members of groups IB, IIA, IIIA, IVA, VIII, VA, and the VIB chalcogenides (13,17). The 5*f* electrons in the lighter actinides are not as localized as the 4*f* electrons in the lanthanides and, with energies close to those of the 6*d* and 7*s* electrons, participate actively in bonding. The participation in bonding by 5*f* electrons is apparently more prominent in the early actinides than in the heavier actinides, where a localized behavior becomes evident. The behavior of the 5*f* electrons makes the actinide metals and their metallic compounds different in their behavior than the transition and lanthanide metals and their compounds.

Solid Compounds. The tripositive actinide ions resemble tripositive lanthanide ions in their precipitation reactions (13,14,17,20,22). Tetrapositive actinide ions are similar in this respect to Ce^{4+}. Thus the fluorides and oxalates are insoluble in acid solution, and the nitrates, sulfates, perchlorates, and sulfides are all soluble. The tetrapositive actinide ions form insoluble iodates and various substituted arsenates even in rather strongly acid solution. The MO_2^+ actinide ions can be precipitated as the potassium salt from strong carbonate solutions. In solutions containing a high concentration of sodium and acetate ions, the actinide MO_2^{2+} ions form the insoluble crystalline salt $NaMO_2(O_2CCH_3)_3$. The hydroxides of all four ionic types are insoluble; in the case of the MO_2^{2+} ions, compounds of the type exemplified by sodium diuranate

Table 7. **Properties of Actinide Metals**

Element	Melting point, °C	Heat of vaporization, ΔH_v, kJ/mol (kcal/mol)	Boiling point, °C	Crystal structure							
				Phase	Range of stability, °C	Symmetry	Lattice parameters, nm				Density, g/mL, at T, °C
							a_0	b_0		c_0	
actinium	1100 ± 50	293 (70)				FC cubic	0.5311				10.07, 25
thorium	1750	564 (130)	3850	α	RT–1360	FC cubic	0.5086				11.724, 25
				β	1360–1750	BC cubic	0.411				
protactinium	1575			α	RT–1170	tetragonal	0.3929			0.3241	15.37, 25
				β	1170–1575	BC cubic	0.381				
uranium	1132	446.4 (106.7)	3818	α	RT–668	orthorhombic	0.2854	0.5869		0.4956	18.97, 25
				β	668–774	tetragonal	1.0759			0.5656	18.11, 720
				γ	774–1132	BC cubic	0.3525				18.06, 805
neptunium	637 ± 2	418 (100)	3900	α	$RT - 280 \pm 5$	orthorhombic	0.4721	0.4888		0.4887	20.45, 25
				β	$280 \pm 5 - 577 \pm 5$	tetragonal	0.4895			0.3386	19.36, 313
				γ	$577 \pm 5 - 637 \pm 2$	BC cubic	0.3518				18.04, 600
plutonium	646	333.5 (79.7)	3235	α	RT–122	monoclinic	0.6183	0.4822	$\beta = 101.8°$	1.0963	19.86, 21
				β	122–207	BC monoclinic	0.9284	1.0463	$\beta = 92.13°$	0.7859	17.70, 190
				γ	207–315	orthorhombic	0.3159	0.5768		1.0162	17.13, 235
				δ	315–457	FC cubic	0.4637				15.92, 320
				δ′	457–479	tetragonal	0.3326			0.4463	16.01, 460
				ε	479–640	BC cubic	0.3636				16.48, 490
americium	1173	230 (55)	2011	α	RT–658	hexagonal	0.3468			1.1241	13.67, 20
				β	793–1004	FC cubic	0.4894				13.65, 20
				γ	1050–1173						
curium	1345	386 (92.2)	3110	α	below 1277	hexagonal	0.3496			1.1331	13.51, 25
				β	1277–1345	FC cubic	0.5039				12.9, 25
berkelium	1050			α	below 930	hexagonal	0.3416			1.1069	14.78, 25
				β	930–986	FC cubic	0.4997				13.25, 25
californium	900 ± 30			α	below 900	hexagonal	0.339			1.101	15.1, 25
				β		FC cubic	0.575				8.70, 25
einsteinium	860 ± 30			α	below 860	hexagonal					
				β		FC cubic	0.575				8.84

($Na_2U_2O_7$) can be precipitated from alkaline solution. The NpO_5^{3-} and PuO_5^{3-} anions, which seem to exist in alkaline solution, form insoluble compounds with several di- and tripositive cations. Peroxide solutions react with actinide ions, particularly M^{4+} ions, to form complex peroxy compounds in solution, and such compounds can be precipitated even from moderately acid solutions. Inorganic anions, eg, sulfate, nitrate, and chloride, are often incorporated in the solid peroxy compounds.

Thousands of compounds of the actinide elements have been prepared, and the properties of some of the important binary compounds are summarized in Table 8 (13,17,18,22). The binary compounds with carbon, boron, nitrogen, silicon, and sulfur are not included; these are of interest, however, because of their stability at high temperatures. A large number of ternary compounds, including numerous oxyhalides, and more complicated compounds have been synthesized and characterized. These include many intermediate (nonstoichiometric) oxides, and besides the nitrates, sulfates, peroxides, and carbonates, compounds such as phosphates, arsenates, cyanides, cyanates, thiocyanates, selenocyanates, sulfites, selenates, selenites, tellurates, tellurites, selenides, and tellurides.

Hundreds of actinide organic derivatives, including organometallic compounds, are known (12,19,23). A number of interesting actinide organometallic compounds of the π-bonded type have been synthesized and characterized. The triscyclopentadienyl compounds, although more covalent than the analogous lanthanide compounds, are highly ionic and include UCp_3, $NpCp_3$, $PuCp_3$, $AmCp_3$, $CmCp_3$, $BkCp_3$, and $CfCp_3$ ($Cp = C_5H_5$); each, except the uranium compound, is relatively stable and appreciably volatile but is sensitive to air (23). The tetrakiscyclopentadienyl complexes ($ThCp_4$, $PaCp_4$, UCp_4, and $NpCp_4$) are, like the Cp_3 complexes, soluble in organic solvents and moderately air-sensitive but not appreciably volatile. A number of triscyclopentadienyl actinide halides, of the general formula MCp_3X (M = Th, U, Np), are known, and these are soluble in a range of organic solvents, are more stable to heat than the tetrakis complexes, and are moderately air-sensitive. The pentamethylcyclopentadienides of thorium and uranium also exist. Of special interest are the cyclooctatetraene [629-20-1] (COT) complexes, including the bis compounds $Th(COT)_2$, $Pa(COT)_2$, $U(COT)_4$, $Np(COT)_4$, and $Pu(COT)_4$, and substituted derivatives of these (23). Characterized by monoclinic symmetry with a π-bonding sandwich structure involving 5*f* electron orbitals, the prototype compound involving uranium is known as uranocene, in view of the analogy to ferrocene. The compounds are air-sensitive, are only sparingly soluble in organic solvents, and can be sublimed in vacuum. A few are very stable to air. The actinides also form tetraallyl complexes, $M(C_3H_5)_4$, but these are stable only at low temperatures. Many organoactinide complexes with ς and π ligands of the type ($Cp_3M—R$(R = alkyl, aryl, or alkynyl) are known, as well as a number of borohydride compounds. Additional solid organoactinide complexes include alkoxides, dialkylamides, chelates such as β-diketones and β-ketoesters, β-hydroxyquinolines, *N*-nitroso-*N*-phenylhydroxylamines (cupferron type), tropolones, *N,N*-dialkyldithiocarbamates, and phthalocyanines; many of these are useful for separation schemes.

Crystal Structure and Ionic Radii. Crystal structure data have provided the basis for the ionic radii (coordination number = CN = 6), which are

Table 8. **Properties and Crystal Structure Data for Important Actinide Binary Compounds**

Compound	Color	Melting point, °C	Symmetry	Space group or structure type	Lattice parameters				Density, g/mL
					a_0, nm	b_0, nm	c_0, nm	Angle, deg	
AcH_2	black		cubic	fluorite ($Fm3m$)	0.5670				8.35
ThH_2	black		tetragonal	$F4/mmm$	0.5735		0.4971		9.50
Th_4H_{15}	black		cubic	$I\bar{4}3d$	0.911				8.25
α-PaH_3	gray		cubic	$Pm3n$	0.4150				10.87
β-PaH_3	black		cubic	β-W	0.6648				10.58
α-UH_3	?		cubic	$Pm3n$	0.4160				11.12
β-UH_3	black		cubic	β-W ($Pm3n$)	0.6645				10.92
NpH_2	black		cubic	fluorite	0.5343				10.41
NpH_3	black		trigonal	$P\bar{3}c1$	0.651		0.672		9.64
PuH_2	black		cubic	fluorite	0.5359				10.40
PuH_3	black		trigonal	$P\bar{3}c1$	0.655		0.676		9.61
AmH_2	black		cubic	fluorite	0.5348				10.64
AmH_3	black		trigonal	$P\bar{3}c1$	0.653		0.675		9.76
CmH_2	black		cubic	fluorite	0.5322				10.84
CmH_3	black		trigonal	$P\bar{3}c1$	0.6528		0.6732		10.06
BkH_2	black		cubic	fluorite	0.5248				11.57
BkH_3	black		trigonal	$P\bar{3}c1$	0.6454		0.6663		10.44
Ac_2O_3	white		hexagonal	La_2O_3 ($P\bar{3}m1$)	0.407		0.629		9.19
Pu_2O_3	?		cubic	$Ia3$ (Mn_2O_3)	1.103				10.20
Pu_2O_3	black	2085	hexagonal	La_2O_3	0.3841		0.5958		11.47
Am_2O_3	tan		hexagonal	La_2O_3	0.3817		0.5971		11.77
Am_2O_3	reddish brown		cubic	$Ia3$	1.103				10.57
Cm_2O_3	white to faint tan	2260	hexagonal	La_2O_3	0.3792		0.5985		12.17
Cm_2O_3			monoclinic	$C2/m$ (Sm_2O_3)	1.4282	0.3641	0.8883	$\beta = 100.29$	11.90
Cm_2O_3	white		cubic	$Ia3$	1.1002				10.80
Bk_2O_3	light green		hexagonal	La_2O_3	0.3754		0.5958		12.47
Bk_2O_3	yellow-green		monoclinic	$C2/m$	1.4197	0.3606	0.8846	$\beta = 100.23$	12.20
Bk_2O_3	yellowish brown		cubic	$Ia3$	1.0887				11.66
Cf_2O_3	pale green		hexagonal	La_2O_3	0.372		0.596		12.69
Cf_2O_3	lime green		monoclinic	$C2/m$	1.4121	0.3592	0.8809	$\beta = 100.34$	12.37

Cf_2O_3	pale green		cubic	$Ia3$	1.083				11.39
Es_2O_3	white		hexagonal	La_2O_3	0.37		0.60		12.7
Es_2O_3	white		monoclinic	$C2/m$	1.41	0.359	0.880	$\beta = 100$	12.4
Es_2O_3	white		cubic	$Ia3$	1.0766				11.79
ThO_2	white	ca 3050	cubic	fluorite	0.5597				10.00
PaO_2	black		cubic	fluorite	0.5509				10.45
UO_2	brown to black	2875	cubic	fluorite	0.5471				10.95
NpO_2	apple green		cubic	fluorite	0.5425				11.14
PuO_2	yellow-green to brown	2400	cubic	fluorite	0.53960				11.46
AmO_2	black		cubic	fluorite	0.5374				11.68
CmO_2	black		cubic	fluorite	0.5358				11.92
BkO_2	yellowish-brown		cubic	fluorite	0.5332				12.31
CfO_2	black		cubic	fluorite	0.5310				12.46
Pa_2O_5	white		cubic	fluorite-related	0.5446 or 0.5492				11.14
Np_2O_5	dark brown		monoclinic	$P2_1/c$	0.4183	0.6584	0.4086	$\beta = 90.32$	8.18
α-U_3O_8	black-green	1150 (dec)	orthorhombic	$C2mm$	0.6716	1.1960	0.4147		8.39
β-U_3O_8	black-green		orthorhombic	$Cmcm$	0.7069	1.1445	0.8303		8.32
γ-UO_3	orange	650 (dec)	orthorhombic	$Fddd$	0.981	1.993	0.971		7.80
$AmCl_2$	black		orthorhombic	$Pbnm$ ($PbCl_2$)	0.8963	0.7573	0.4532		6.78
$CfCl_2$	red-amber		?						
$AmBr_2$	black		tetragonal	$SrBr_2$ ($P4/n$)	1.1592		0.7121		7.00
$CfBr_2$	amber		tetragonal	$SrBr_2$	1.1500		0.7109		7.22
ThI_2	gold		hexagonal	$P6_3/mmc$	0.397		3.175		7.45
AmI_2	black	ca 700	monoclinic	EuI_2 ($P2_1/c$)	0.7677	0.8311	0.7925	$\beta = 98.46$	6.60
CfI_2	violet		hexagonal	CdI_2 ($P\bar{3}m1$)	0.456		0.699		6.63
CfI_2	violet		rhombohedral	$CdCl_2$ ($R\bar{3}m$)	0.743			$\alpha = 36$	6.58
AcF_3	white		trigonal	LaF_3 ($P\bar{3}c1$)	0.741		0.753		7.88
UF_3	black	>1140 (dec)	trigonal	LaF_3	0.718		0.7348		8.95
NpF_3	purple		trigonal	LaF_3	0.7129		0.7288		9.12
PuF_3	purple	1425	trigonal	LaF_3	0.7092		0.7254		9.33
AmF_3	pink	1393	trigonal	LaF_3	0.7044		0.7225		9.53
CmF_3	white	1406	trigonal	LaF_3	0.7014		0.7194		9.85
BkF_3	yellow-green		orthorhombic	YF_3 ($Pnma$)	0.670	0.709	0.441		9.70
BkF_3	yellow-green		trigonal	LaF_3	0.697		0.714		10.15

Table 8 (*Continued*)

Compound	Color	Melting point, °C	Symmetry	Space group or structure type	Lattice parameters				Density, g/mL
					a_0, nm	b_0, nm	c_0, nm	Angle, deg	
CfF_3	light green		orthorhombic	YF_3	0.6653	0.7039	0.4393		9.88
CfF_3	light green		trigonal	LaF_3	0.6945		0.7101		10.28
$AcCl_3$	white		hexagonal	UCl_3 ($P6_3/m$)	0.762		0.455		4.81
UCl_3	green	835	hexagonal	$P6_3/m$	0.7443		0.4321		5.50
$NpCl_3$	green	ca 800	hexagonal	UCl_3	0.7413		0.4282		5.60
$PuCl_3$	emerald green	760	hexagonal	UCl_3	0.7394		0.4243		5.71
$AmCl_3$	pink or yellow	715	hexagonal	UCl_3	0.7382		0.4214		5.87
$CmCl_3$	white	695	hexagonal	UCl_3	0.7374		0.4185		5.95
$BkCl_3$	green	603	hexagonal	UCl_3	0.7382		0.4127		6.02
α-$CfCl_3$	green	545	orthorhombic	$TbCl_3$ ($Cmcm$)	0.3859	1.1748	0.8561		6.07
β-$CfCl_3$	green		hexagonal	UCl_3	0.7379		0.4090		6.12
$EsCl_3$	white to orange		hexagonal	UCl_3	0.740		0.407		6.20
$AcBr_3$	white		hexagonal	UBr_3 ($P6_3/m$)	0.806		0.468		5.85
UBr_3	red	730	hexagonal	$P6_3/m$	0.7936		0.4438		6.55
$NpBr_3$	green		hexagonal	UBr_3	0.7916		0.4390		6.65
$NpBr_3$	green		orthorhombic	$TbCl_3$ ($Cmcm$)	0.4109	1.2618	0.9153		6.67
$PuBr_3$	green	681	orthorhombic	$TbCl_3$	0.4097	1.2617	0.9147		6.72
$AmBr_3$	white to pale yellow		orthorhombic	$TbCl_3$	0.4064	1.2661	0.9144		6.85
$CmBr_3$	pale yellow-green	625 ± 5	orthorhombic	$TbCl_3$	0.4041	1.2700	0.9135		6.85
$BkBr_3$	light green		monoclinic	$AlCl_3$ ($C2/m$)	0.723	1.253	0.683	$\beta = 110.6$	5.604
$BkBr_3$	light green		orthorhombic	$TbCl_3$	0.403	1.271	0.912		6.95
$BkBr_3$	yellow green		rhombohedral	$FeCl_3$ ($R\overline{3}$)	0.766			$\alpha = 56.6$	5.54
$CfBr_3$	green		monoclinic	$AlCl_3$	0.7214	1.2423	0.6825	$\beta = 110.7$	5.673
$CfBr_3$	green		rhombohedral	$FeCl_3$	0.758			$\alpha = 56.2$	5.77
$EsBr_3$	straw		monoclinic	$AlCl_3$	0.727	1.259	0.681	$\beta = 110.8$	5.62
PaI_3	black		orthorhombic	$TbCl_3$ ($Cmcm$)	0.433	1.40	1.002		6.69
UI_3	black		orthorhombic	$TbCl_3$	0.4328	1.3996	0.9984		6.76
NpI_3	brown		orthorhombic	$TbCl_3$	0.430	1.403	0.995		6.82
PuI_3	green		orthorhombic	$TbCl_3$	0.4326	1.3962	0.9974		6.92
AmI_3	pale yellow	ca 950	hexagonal	BiI_3 ($R\overline{3}$)	0.742		2.055		6.35

AmI_3	yellow		orthorhombic	$PuBr_3$	0.428	1.394	0.9974		6.95
CmI_3	white		hexagonal	BiI_3	0.744		2.040		6.40
BkI_3	yellow		hexagonal	BiI_3	0.7584		02.087		6.02
CfI_3	red-orange		hexagonal	BiI_3	0.7587		2.081		6.05
EsI_3	amber to light yellow		hexagonal	BiI_3	0.753		2.084		6.18
ThF_4	white	1068	monoclinic	UF_4 (C2/*c*)	1.300	1.099	0.860	$\beta = 126.4$	6.20
PaF_4	reddish-brown		monoclinic	UF_4	1.288	1.088	0.849	$\beta = 126.4$	6.38
UF_4	green	960	monoclinic	C2/*c*	1.2803	1.0792	0.8372	$\beta = 126.3$	6.73
NpF_4	green		monoclinic	UF_4	1.268	1.066	0.834	$\beta = 126.3$	6.86
PuF_4	brown	1037	monoclinic	UF_4	1.260	1.057	0.828	$\beta = 126.3$	7.05
AmF_4	tan		monoclinic	UF_4	1.256	1.058	0.825	$\beta = 125.9$	7.23
CmF_4	light gray-green		monoclinic	UF_4	1.250	1.049	0.818	$\beta = 126.1$	7.36
BkF_4	pale yellow-green		monoclinic	UF_4	1.2396	1.0466	0.8118	$\beta = 126.3$	7.55
CfF_4	light green		monoclinic	UF_4	1.2327	1.0402	0.8113	$\beta = 126.4$	7.57
α-$ThCl_4$	white		orthorhombic		1.118	0.593	0.909		4.12
β-$ThCl_4$	white	770	tetragonal	UCl_4 ($I4_1/amd$)	0.8473		0.7468		4.60
$PaCl_4$	greenish-yellow		tetragonal	UCl_4	0.8377		0.7481		4.72
UCl_4	green	590	tetragonal	$I4_1/amd$	0.8296		0.7481		4.89
$NpCl_4$	red-brown	518	tetragonal	UCl_4	0.8266		0.7475		4.96
α-$ThBr_4$	white		tetragonal	$I4_1/a$	0.6737		1.3601		5.94
β-$ThBr_4$	white		tetragonal	UCl_4	0.8931		0.7963		5.77
$PaBr_4$	orange-red		tetragonal	UCl_4	0.8824		0.7957		5.90
UBr_4	brown	519	monoclinic	2/*c*-/-	1.092	0.869	0.705	$\beta = 93.15$	
$NpBr_4$	dark red	464	monoclinic	2/*c*-/-	1.089	0.874	0.705	$\beta = 94.19$	
ThI_4	yellow	556	monoclinic	$P2_1/n$	1.3216	0.8068	0.7766	$\beta = 98.68$	6.00
PaI_4	black								
UI_4	black								
PaF_5	white		tetragonal	$I\bar{4}2d$	1.153		0.519		
α-UF_5	grayish white		tetragonal	$I4/m$	0.6512		0.4463		5.81
β-UF_5	pale yellow		tetragonal	$I\bar{4}2d$	1.1456		0.5196		6.47
NpF_5			tetragonal	$I4/m$	0.653		0.445		
$PaCl_5$	yellow	306	monoclinic	C2/*c*	0.800	1.142	0.843	$\beta = 106.4$	
α-UCl_5	brown		monoclinic	$P2_1/n$	0.799	1.069	0.848	$\beta = 91.5$	3.81

Table 8 (*Continued*)

Compound	Color	Melting point, °C	Symmetry	Space group or structure type	Lattice parameters				Density, g/mL
					a_0, nm	b_0, nm	c_0, nm	Angle, deg	
β-UCl_5	red-brown		triclinic	$P\bar{1}$	0.707	0.965	0.635	$\alpha = 89.10$ $\beta = 117.36$ $\gamma = 108.54$	
α-$PaBr_5$			monoclinic	$P2_1/c$	1.264	1.282	0.992	$\beta = 108$	
β-$PaBr_5$	orange-brown		monoclinic	$P2_1/n$	0.9385	1.2205	0.895	$\beta = 91.1$	
UBr_5	brown		monoclinic	$P2_1/n$					
PaI_5	black		orthorhombic		0.698	0.2160	2.130		
UF_6	white	64.02[a]	orthorhombic	$Pnma$	0.9900	0.8962	0.5207		5.060
NpF_6	orange	55	orthorhombic	$Pnma$	0.9909	0.8997	0.5202		5.026
PuF_6	reddish-brown	52	orthorhombic	$Pnma$	0.995	0.902	0.526		4.86
UCl_6	dark green	178	hexagonal	$P\bar{3}m1$	1.09		0.603		3.62

[a] At 151.6 kPa, to convert kPa to atm, divide by 101.3.

Table 9. **Ionic Radii of Actinide and Lanthanide Elements**

No. of 4*f* or 5*f* electrons	Lanthanide series				Actinide series			
	Element	Radius, nm	Element	Radius, nm	Element	Radius, nm	Element	Radius, nm
0	La^{3+}	0.1032			Ac^{3+}	0.112	Th^{4+}	0.094
1	Ce^{3+}	0.101	Ce^{4+}	0.087	(Th^{3+})	(0.108)	Pa^{4+}	0.090
2	Pr^{3+}	0.099	Pr^{4+}	0.085	(Pa^{3+})	0.104	U^{4+}	0.089
3	Nd^{3+}	0.0983			U^{3+}	0.1025	Np^{4+}	0.087
4	Pm^{3+}	0.097			Np^{3+}	0.101	Pu^{4+}	0.086
5	Sm^{3+}	0.0958			Pu^{3+}	0.100	Am^{4+}	0.085
6	Eu^{3+}	0.0947			Am^{3+}	0.0980	Cm^{4+}	0.084
7	Gd^{3+}	0.0938			Cm^{3+}	0.0970	Bk^{4+}	0.083
8	Tb^{3+}	0.0923	Tb^{4+}	0.076	Bk^{3+}	0.0960	Cf^{4+}	0.0821
9	Dy^{3+}	0.0912			Cf^{3+}	0.0950	Es^{4+}	0.081
10	Ho^{3+}	0.0901			Es^{3+}	0.0940		
11	Er^{3+}	0.0890						
12	Tm^{3+}	0.0880			Fm^{3+}	0.0924[a]		
13	Yb^{3+}	0.0868			Md^{3+}	0.0896[a]		
14	Lu^{3+}	0.0861			Lr^{3+}	0.0881[a]		

[a] These ionic radii were obtained by determining the elution position of these ions from cation exchange columns with alpha-hydroxyisobutyrate relative to those of Ho^{3+}, Er^{3+}, and Tm^{3+}. The ionic radius of Er^{3+} was taken from crystallographic data to be 0.0881. (D. H. Templeton and C. H. Dauben, J. Am. Chem. Soc. **76**, 5237 (1954).

summarized in Table 9 (13,14,17). For both M^{3+} and M^{4+} ions there is an actinide contraction, analogous to the lanthanide contraction, with increasing positive charge on the nucleus.

As a consequence of the ionic character of most actinide compounds and of the similarity of the ionic radii for a given oxidation state, analogous compounds are generally isostructural. In some cases, eg, UBr_3, $NpBr_3$, $PuBr_3$, and $AmBr_3$, there is a change in structural type with increasing atomic number, which is consistent with the contraction in ionic radius. The stability of the MO_2 structure (fluorite type) is especially noteworthy, as is shown by the existence of such compounds as PaO_2, AmO_2, CmO_2, and CfO_2 despite the instability of the IV oxidation state of these elements in solution. The actinide contraction and the isostructural nature of the compounds constitute some of the best evidence for the transition character of this group of elements.

Absorption and Fluorescence Spectra. The absorption spectra of actinide and lanthanide ions in aqueous solution and in crystalline form contain narrow bands in the visible, near-ultraviolet, and near-infrared regions of the spectrum (13,14,17,24). Much evidence indicates that these bands arise from electronic transitions within the $4f$ and $5f$ shells in which the $4f^n$ and $5f^n$ configurations are preserved in the upper and lower states for a particular ion. In general, the absorption bands of the actinide ions are some ten times more intense than those of the lanthanide ions. Fluorescence, for example, is observed in the trichlorides of uranium, neptunium, americium, and curium, diluted with lanthanum chloride (15).

1.4. Practical Applications. The practical use of three actinide nuclides, ^{239}Pu, ^{235}U, and ^{233}U, as nuclear fuel is well known (5,9). When a neutron of any energy strikes the nucleus of one of these nuclides, each of which is capable of undergoing fission with thermal (essentially zero-energy) neutrons, the fission reaction can occur in a self-sustaining manner. A controlled self-perpetuating chain reaction using such a nuclear fuel can be maintained in such a manner that the heat energy can be extracted or converted by conventional means to electrical energy. The complete utilization of nonfissionable ^{238}U (through conversion to fissionable ^{239}Pu) and nonfissionable ^{232}Th (through conversion to fissionable ^{233}U) can be accomplished by breeder reactors.

In addition, three other actinide nuclides (^{238}Pu, ^{241}Am, and ^{252}Cf) have other practical applications (6). One gram of ^{238}Pu produces approximately 0.56 W of thermal power, primarily from alpha-particle decay, and this property has been used in space exploration to provide energy for small thermoelectric power units (see THERMOELECTRIC ENERGY CONVERSION). The most noteworthy example of this latter type of application is a radioisotopic thermoelectric generator left on the moon. It produced 73 W of electrical power to operate the scientific experiments of the Apollo lunar exploration, and was fueled with 2.6 kg of the plutonium isotope in the form of plutonium dioxide, PuO_2. Similar generators powered the instrumentation for other Apollo missions, the Viking Mars lander, and the Pioneer and Voyager probes to Jupiter, Saturn, Uranus, Neptune, and beyond. Americium-241 has a predominant gamma-ray energy of 60 keV and a long half-life of 433 yr for decay by the emission of alpha particles, which makes it particularly useful for a wide range of industrial gaging applications, the diagnosis of thyroid disorders, and for smoke detectors. When mixed with beryllium it generates neutrons at the rate of 1.0×10^7 neutrons/(s·g) ^{241}Am. The mixture is designated ^{241}Am–Be, and a large number of such sources are in worldwide daily use in oil-well logging operations to find how much oil a well is producing in a given time span. Californium-252 is an intense neutron source: 1 g emits 2.4×10^{12} neutrons/s. This isotope is being tested for applications in neutron activation analysis, startup sources for nuclear reactors, neutron radiography, portable sources for field use in mineral prospecting and oil-well logging, and in airport neutron-activation detectors for nitrogenous materials (ie, explosives). Both ^{238}Pu and ^{252}Cf are being studied for possible medical applications: the former as a heat source for use in heart pacemakers and heart pumps and the latter as a neutron source for irradiation of certain tumors for which gamma-ray treatment is relatively ineffective.

2. Transactinides and Superheavy Elements

The elements beyond the actinides in the Periodic Table can be termed the transactinides. These begin with the element having atomic number 104 and extend, in principle, indefinitely. Nine such elements were definitely known by 1996. They were synthesized in bombardments of heavy targets with heavy-ion projectiles. In 1999, evidence was reported for the SHEs 114, 116, and 118, and 18 new isotopes of seaborgium (106) and heavier elements; these results have not yet been confirmed.

As indicated in Figure 4, the early transactinide elements find their place back in the main body of the Periodic Table as members of the new 6*d* transition series. The discoverers of the currently known transactinide elements, names and symbols approved by IUPAC in August 1997, and dates of discovery are listed in Table 10.

Fully relativistic calculations have indicated that the electronic configurations of the transactinide elements will be different than those based on simple extrapolation from their lighter homologues in the same group in the periodic table. There may be changes in the valence electron configurations, differences in ionic radii, complexing ability, and other chemical properties. Studies of the chemical properties (25) of elements 104 (rutherfordium) through 106 (seaborgium) in both aqueous and gas phase have now been conducted and confirm that these elements in general resemble the lighter members of groups 4 through 6 of the Periodic Table and should be placed under them in the Periodic Table as the heaviest members of those groups. In agreement with relativistic calculations, reversals in trends in properties in going down groups 4 and 5 have been observed. The behavior of the transactinides cannot be simply extrapolated from their lighter homologues, and in some cases they behave more like actinides of the same oxidation state.

Chemical studies are extremely difficult because the longest-lived isotopes of the transactinides are a minute or less as shown in Fig. 6 and the production rates for their synthesis are extremely small. The first chemical studies of bohrium (element 107) were conducted in 1999 using a longer-lived isotope, 17-s ^{267}Bh, discovered in early 1999. Studies of the volatility of its oxychloride based on detection of the decay of only four atoms showed that it was similar

Table 10. **The Transactinide Elements**

Atomic number	Element	Symbol	Mass number[a]	Discoverers and date of discovery
104	rutherfordium	Rf	261	A. Ghiorso, M. Nurmia, J. Harris, K. Eskola, and P. Eskola, 1969
105	(hahnium) dubnium	(Ha) Db[b]	262	A. Ghiorso, M. Nurmia, K. Eskola, J. Harris, and P. Eskola, 1970
106	seaborgium [54038-81-2]	Sg	266	A. Ghiorso, J. M. Nitschke, J. R. Alonso, C. T. Alonso, M. Nurmia, G. T. Seaborg, E. K. Hulet, and R. W. Lougheed, 1974
107	bohrium	Bh	267	G. Münzenberg, S. Hofmann, F. P. Hessberger, W. Reisdorf, K. H. Schmidt, J. H. R. Schneider, W. F. W. Schneider, P. Armbruster, C. C. Sahm, and B. Thuma, 1981
108	hassium	Hs	269	G. Münzenberg, P. Armbruster, H. Folger, F. P. Hessberger, S. Hofmann, J. Keller, K. Poppensieker, W. Reisdorf, K. H. Schmidt, H. J. Schött, M. E. Leino, and R. Hingmann, 1984
109	meitnerium	Mt	268	G. Münzenberg, P. Armbruster, F. P. Hessberger, S. Hofmann, K. Poppensieker, F. P. Hessberger, S. Hofmann, K. Poppensieker, W. Reisdorf, J. H. R. Schneider, K. H. Schmidt, C. C. Sahm, and D. Vermeulen, 1982

[a] Mass number of longest lived isotopes.
[b] The name dubnium (Db) has been approved by IUPAC, but the name hahnium (Ha) is commonly used in the US.
Note: 109 elements are listed here. In addition, discovery of elements 110–112 has been reported. Discovery of element 110 has been reported by three different groups between 1994–1996, but since they have reported different isotopes, none can be considered confirmation of the others. A Ghiorso *et al.*[1] at the Lawrence Berkeley National Laboratory (LBNL) in Berkeley, California reported evidence for element 110 with mass number 267; S. Hofmann *et al.*[2] at the Gesellschaft für Schwerionenforschung (GSI) at Darmstadt, Germany reported element 110 with mass numbers 269 and 271; Yu. Lazarev *et al.*[3] at the Flerov Laboratory for Nuclear Reactions, Dubna, Russia reported element 110 with mass 273. The group of S. Hofmann has the more convincing data for their discovery of 110, but the half-lives and cross sections of all appear reasonable. In 1995–1996 Hofmann et al.[4,5] also reported discovery of elements 111 and 112 at GSI with mass numbers of 272 and 277, respectively. These elements have not yet been named. In August 1999, scientists at LBNL published evidence for three new superheavy elements[6]: 118 with mass number 293, 116 with mass number 289, and 114 with mass number 285. In July 1999, a multinational group[7] working at Dubna, Russia published their evidence for observation of element 114 with mass number 287. A Dubna/Lawrence Livermore National Laboratory group[8] published evidence in Occtober 1999 for element 114 with mass number 289 and later reported[9] observation of element 114 with mass number 288. As of March 2000, none of these reports has been confirmed by another group.

(1) A. Ghiorso *et al., Nucl. Phys.* **A583**, 861 (1994); *Phys. Rev. C* **51**, R2293 (1995).
(2) S. Hofmann *et al., Z. Phys.* **A350**, 277 (1995).
(3) Yu. A. Lazarev *et al., Phys. Rev. C* **54**, 620 (1996).
(4) S. Hofmann *et al., Z. Phys.* **A350**, 281 (1995).
(5) S. Hofmann *et al., Z. Phys.* **A354**, 229 (1996).
(6) V. Ninov *et al., Phys. Rev. Lett.* **83** 1104 (1999).
(7) V. Ninov *et al., Phys. Rev. Lett.* **83** 1104 (1999).
(8) Yu. Ts. Oganessian *et al., Phys. Rev. Lett.* **83**, 3154 (1999).
(9) R. W. Lougheed *et al.*, 219th National ACS meeting, San Francisco, California, March 2000.

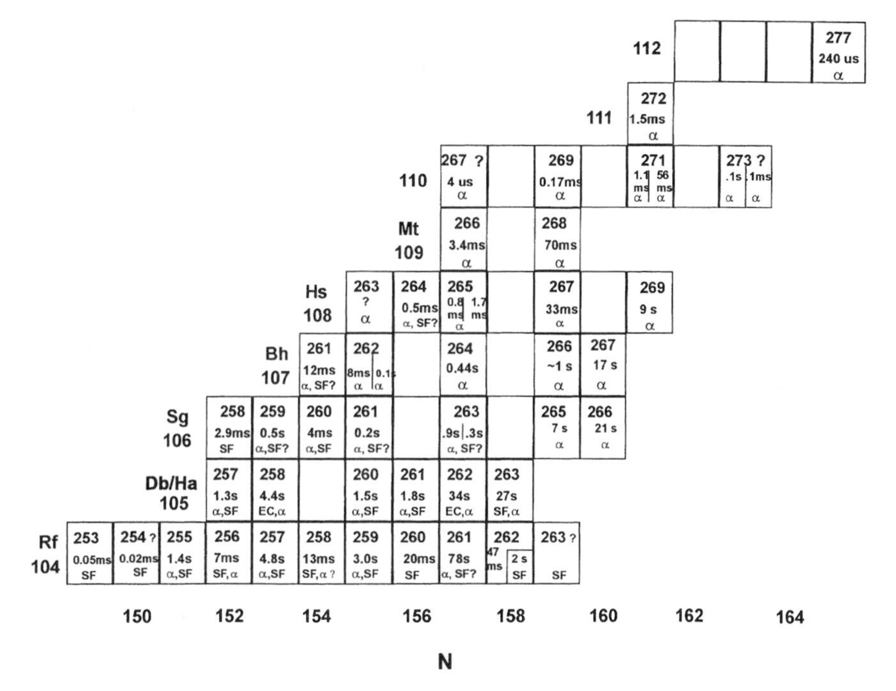

Fig. 6. All known isotopes of rutherfordium (element 104) through element 112 as of 1998.

to those of the lighter group 7 elements, technetium and rhenium, and different from the actinides. studies of hassium (element 108) probably will be possible in the near future using the known 9-s ^{269}Hs. It is expected that isotopes of meitnerium (109) can exist with half-lives long enought to permit studies of its chemical properties, and recent calculations suggest that ^{292}110 will decay by alpha emission with a half-life of about 50 yr. However, the number of atoms that can be produced for study is decreasing rapidly as we go to heavier and heavier elements. New synthesis reactions and improvements in production rates and in the efficiency and speed of chemical separations will be required in order to extend chemical studies to heavier elements.

Studies of the chemical properties of the heaviest elements are challenging for theorists as well as for experimentalists and fully relativistic calculations must be performed. Modern high-speed computers have made possible the calculation of electronic structures and the deviations that might be expected due to relativistic effects (13,26,27). Elements 104 through 112 are predicted to be members of a 6*d* transition series as shown in Figure 4, making them generally homologous in chemical properties with the elements hafnium through mercury as discussed earlier. The 7*p* shell is filled in elements 113–118 and they are expected to be similar to the elements thallium ($Z = 81$) through radon ($Z = 86$), although some calculations have suggested that elements 112 and 114 as well as 118 (eka-Rn) might be volatile, relatively inert gases, rather than similar to lead and mercury. If there is an island of stability at element 126, this element and its neightbors should have chemical properties similar to those of the actinide and lanthanide elements (26).

The 8*s* subshell should fill at elements 119 and 120, thus making these an alkali and alkaline earth metal, respectively. Next, the calculations point to the filling, after the addition of a 7*d* electron at element 121 of the inner 5*g* and 6*f* subshells, 32 places in all, which have been called the superactinide elements and which terminates at element 153. This is followed by the filling of the 7*d* subshell (elements 154 through 162) and 8*p* subshell (elements 163 through 168).

Actually, more careful calculations have indicated that the picture is not this simple, and that other electrons (8*p* and 7*d*) in addition to those identified in the above discussion enter prominently as early as element 121, and other anomalies may enter as early as element 103, thus further complicating the picture. These perturbations, caused by spin-orbit splitting, become especially significant beyond the superactinide series, and lead to predictions of chemical properties that are not consistent, element by element, with those suggested by Figure 4.

It should be pointed out that although these atomic calculations give some general guideance for experimental research, they do not predict the behavior of molecular species under actual experimental conditions. This even more difficult and complex theoretical problem has been addressed recently (28). systematic theoretical studies of heavy element compounds using relativistic quantum-chemical calculations are combined with fundamental physicochemical considerations to make detailed predictions relevant to both gas-phase volatility studies and partitioning studies between aqueous and organic phases.

On the basis of simple extrapolation of known half-lives, it would appear that the half-lives of the elements beyond element 112 would become ever shorter as the atomic number increases, even for the isotopes with the longest half-life for each element. This would make future prospects for the existence of heavier transuranium elements appear extremely unlikely, but new theoretical calculations and experimental observations have changed this outlook and led to optimism concerning the prospects for the synthesis and identification of elements beyond the observed upper limit of the periodic table, elements that have come to be referred to as SuperHeavy Elements.

The existence of superheavy elements was predicted as early as 1955. Theoretical calculations in the mid 1960s, based on filled shells (magic numbers) and other nuclear stability considerations, led to extrapolations to the far transuranium region (2,26,27). These suggestged the existence of closed nucleon shells at proton numbers (atomic numbers) 114 and 126 and neutron number 184 that helped stabilize the nucleus against decay by spontaneous fission, the main cause of instability of the heaviest elements. Table 11 shows the known closed proton and neutron shells and the predicted closed nuclear shells (in parantheses) that were originally predicted might be important in stabilizing the superheavy elements. Included by way of analogy are the long-known closed electron shells observed in the buildup of the electronic structure of atoms. These correspond to the noble gases, and the extra stability of these closed shells is reflected in the relatively small chemical reactivity of these elements. The predicted closed electron shells (in parentheses) occur at $Z = 118$ and 168.

Searches for long-lived superheavy elements in nature were initiated in the late 1960s sparked by the early predictions of half-lives as long as a billion years. If the half-life of a superheavy nucleus should happen to be as long as a few times

Table 11. **Closed Proton (*Z*) and Neutron (*N*) Shells with Closed Electron (Noble Gas) Shells for Comparison**[a]

Z	N	e^-
2(He)	2(^{4}He)	2(He)
8(O)	8(^{16}O)	10(Ne)
20(Ca)	20(^{40}Ca)	18(Ar)
28(Ni)	28(^{56}Ni)	36(Kr)
50(Sn)	50(^{88}Sr)	54(Xe)
82(Pb)	82(^{140}Ce)	86(Rn)
	126(^{208}Pb)	
(114)	(184)	(118)
(126)		
		(168)

[a] Predicted shells shown in parentheses.

10^8yr (now considered to be extremely unlikely), this would be long enough to allow the isotope to survive and still be present on the earth (as in the case of ^{235}U, which has a half-life of 7×10^8 yr), provided that it was initially present as a result of the cosmic nuclear reactions that led to the creation of the solar system. All evidence in nature, direct or indirect, for superheavy elements associated with the island of stability centered around element 114 has been inconclusive. Because of the physical limitations inherent in any experimental technique, it is not possible to say that such superheavy eleemnts do not exist in nature, but only to set limits on the amount that can be present. The results of such searches establish that their concentration is extremely small, eg, much less than one part in 10^{12} parts of ore. Searches have also been made in cosmic rays, meteorites, and moon rocks, with generally negative results except for some indirect evidence of possible previous presence in meteorites during the early history of the meteorite's life. Some evidence was reported for a now extinct superheavy element in the Allende meteorite, which fell in Mexico in 1969. The evidence for such an extinct superheavy element (now considered to be unlikely) was the observation of a unique composition of xenon isotopes that might have been formed from decay by spontaneous fission with a half-life of 10^7–10^9 yr (29). The postulated current synthesis of a broad range of chemical eleemnts, possibly even including superheavy elements, in stars might enhance the prospects for finding even shorter-lived superheavy elements in cosmic rays (30); elements as heavy as uranium have apparently been found in cosmic rays emanating from such stars.

The effects of a rather distinct deformed shell at $N = 152$ were clearly seen as early as 1954 in the alpha-decay energies and half-lives of isotopes of californium, einsteinium, and fermium. In fact, a number of authors have suggested that the entire transuranium region is stabilized by shell effects with an influence that increases markedly with atomic number. Thus the effects of shell substructure lead to an increase in spontaneous fission half-lives of up to about 15 orders of magnitude for the heavy transuranium elements, the heaviest of which would otherwise have half-lives of the order of those for a compound nucleus (10^{14} s or less) and not of milliseconds or longer, as found experimentally. But the influence of the N=152 subshell appeared to be nearly gone by element

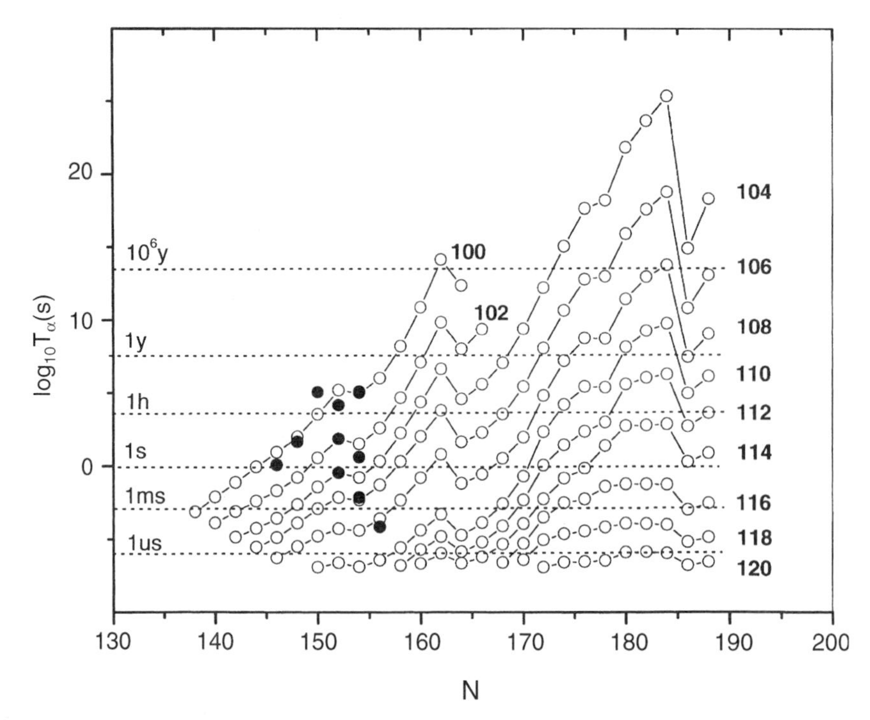

Fig. 7. Theoretical calculation of half-lives for alpha-decay of even-even isotopes of elements 100 through 120 plotted as a function of neutron number. Filled circles represent experimental data (31).

105. However, the recent theoretical calculations of alpha-decay half-lives as a function of neutron number shown in Figure 7 indicate even stronger deformed subshells around $N = 108$ (hassium) that lengthen the half-lives of nuclides in this region. The measurement of the unexpectedly long half-life of 21 s for the isotope of seaborgium with 160 neutrons, the identification of ^{266}Bh and ^{267}Bh with half-lives of $\sim$1 and $\sim$20 s, respectively, and the discovery that the elements bohrium (107) through 112 decay predominantly by alpha emission rather than spontaneous fission provide experimental support for these predictions. Furthermore, the calculations indicate that the half-lives continue to lengthen as neutron number 184 is approached so half-lives of milliseconds or more are predicted for isotopes of elements up through 116. The odd-neutron and odd-proton isotopes are expected to be considerably longer so it is reasonable to expect half-lives of milliseconds or longer for some isotopes of elements up through 120. It is now predicted that new isotopes with measurable half-lives can be produced all along the path to the "island" of nuclear stability at $Z = 114$ and $N = 184$. It is no longer a remote and inaccessible island and may extend as far as element 120!

Attempts at GSI in 1998 failed to produce element 113 using the same techniques as for elements 110–112. However, in 1999 evidence for the long-sought superheavy elements was reported by three other groups. (See footnote to Table 10.) In August, the Heavy Element Nuclear and Radiochemistry Group of the Lawrence Berkeley National Laboratory (LBNL)/University of California, Berkeley reported the observation of three six-member alpha-decay chains from

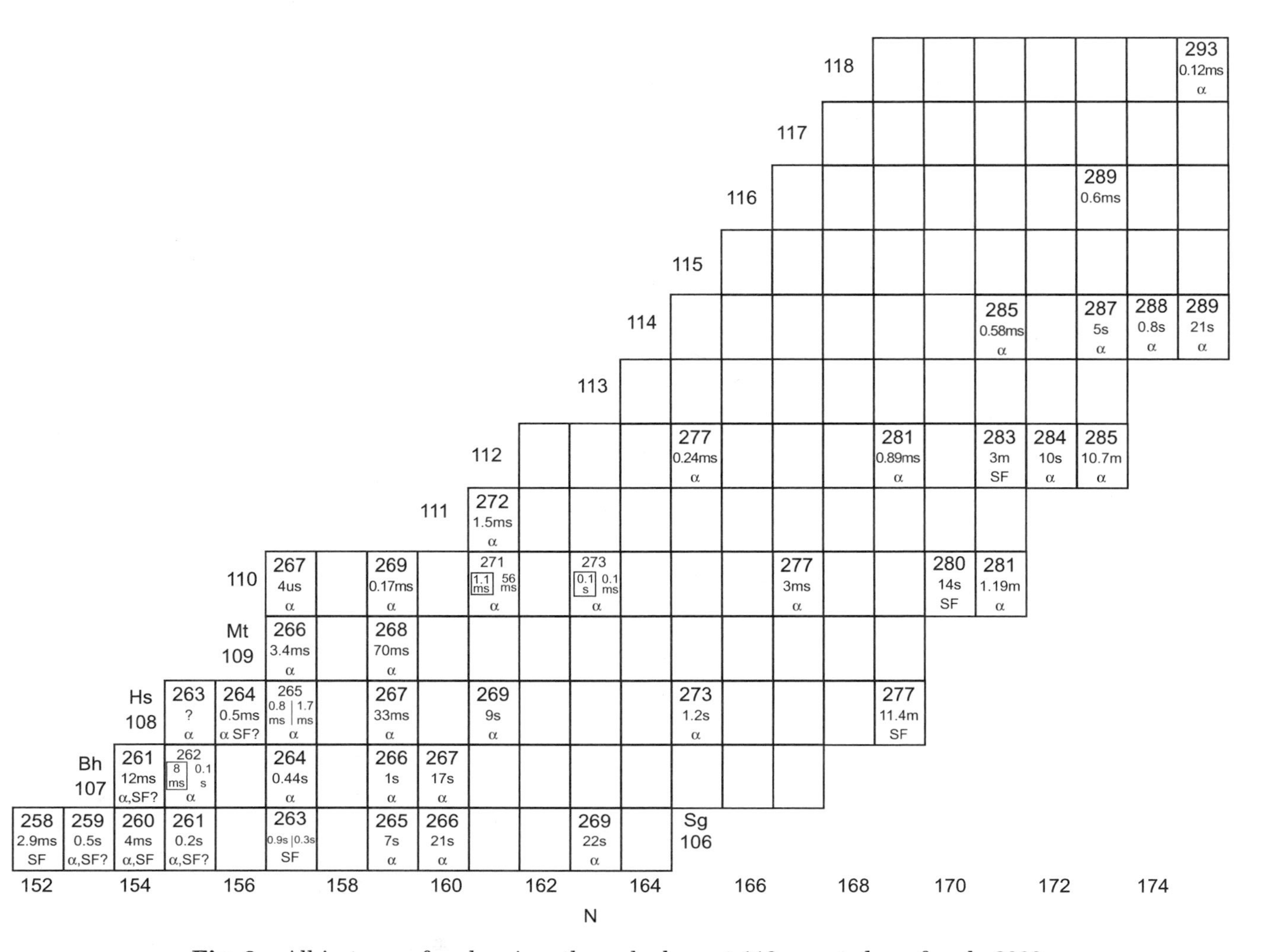

Fig. 8. All isotopes of seaborgium through element 118 reported as of early 2000.

$^{293}118$ produced in the reaction of ^{208}Pb targets with ^{86}Kr projectiles at the LBNL Cyclotron. The newly completed Berkeley Gas-filled Separator was used to separate and measure these decay chains. The sequence of successive alpha-decays from element 118 also produced the new elements 116 and 114 and new isotopes of 110, hassium, and seaborgium. The alpha-decay energies and half-lives of the all of the 118 decay chain members were in remarkable agreement with recent predictions (31). In july, a multinational group working at the U-400 cyclotron at the Flerov Laboratory for Nuclear Reactions in Dubna, Russia, reported using their vacuum separator to detect and measure two events in which alpha-decay was followed by spontaneous fission decay which they attributed to $^{287}114$ produced in the bombardment of ^{242}Pu targets with ^{48}Ca projectiles via emission of three neutrons from the compound nucleus. In October, a Dubna/Lawrence Livermore National Laboratory group reported using the Dubna gas-filled recoil separator to measure a single three-member alpha-decay chain followed by spontaneous fission, which was produced in the bombardment of ^{244}Pu with ^{48}Ca projectiles. They attributed this to single event to $^{289}114$ formed via a 3-neutron out reaction. These results were also in agreement with recent prodictions. In subsequent bombardments, they also recently reported observation of a two-member alpha-decay chain followed by spontaneous fission which they attribute to the decay of $^{288}114$ produced via a 4-neutron out reaction. None of these reports has yet been confirmed, but they are all included in Figure 8. If all of these and the reported new $^{266,267}Bh$ isotopes are confirmed, 18 new heavy element isotopes will have been added to the chart of the nuclides, nearly doubling the number known beyond element 105. The prospects for

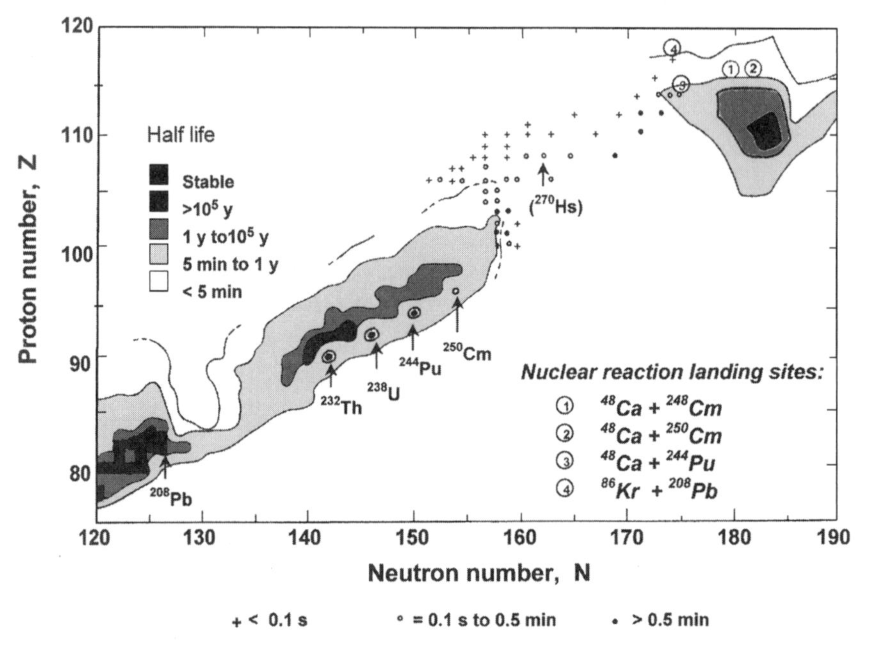

Fig. 9. Plot of heavy element topology from 1978. New heavy element isotopes reported as of 1999 are shown with the following symbols: + <0.1 s; o = 0.1 s to 0.5 min; • >0.5 min.

producing element 119 using similar reactions to those used for 118 appear promising. It is also expected to decay by a long sequence of high-energy alpha particles, which will produce the missing elements 117, 115, and 113.

A topological representation from about 1978 of the heavy element region and the spherical island of superheavy elements around $Z = 114$ and $N = 184$ is shown in Figure 9. The new heavy element isotopes discovered since then have been added. Possible reactions for making element 116 isotopes and their landing sites are shown as 1 and 2. The newly reported landing sites for $^{289}114$ and $^{293}118$ and the nuclear reactions used to make them are shown as 3 and 4. The doubly magic deformed nucleus ^{279}Hs ($Z = 108$, $N = 162$) is shown; it is expected to have a half-life of only about 5 s. Although the spherical doubly magic nucleus $^{298}114$ is now predicted to have a half-life of only about 12 min, the beta-stable nuclide $^{292}110$ is expected to alpha-decay with a half-life of about 50 yr. although it now appears that many long-lived superheavy elements can exist, new imaginative production reactions and techniques for increasing the overall yields and provision for "stockpiling" long-lived products for future studies must be developed in order to explore this exciting new landscape.

BIBLIOGRAPHY

"Actinides" in *ECT* 2nd ed., Vol. 1, pp. 351–371, by G. T. Seaborg, U.S. Atomic Energy Commission; "Actinides and Transactinides," in *ECT* 3rd ed., Vol. 1, pp. 456–488, by Glenn T. Seaborg, University of California, Berkeley; "Actinides and Transactinides" in *ECT* 4th ed., Vol. 1, pp. 412–445, by Glenn T. Seaborg; "Actinides and Transactinides" in *ECT* (online), posting date: December 4, 2000, by Glenn T. Seaborg, University of California, Berkeley.

CITED PUBLICATIONS

1. G. T. Seaborg and J. J. Katz, eds., *The Actinide Elements, National Nuclear Energy Series*, Div. IV, 14A, McGraw-Hill Book Co., Inc., New York, 1954; G. T. Seaborg, ed., *Transuranium Elements: Products of Modern Alchemy*, Benchmark Papers, Dowden, Hutchinson & Ross, Inc., Stroudsburg, Pa., 1978.
2. *Transurane, Gmelins Handbuch der anorganischen Chemie*, Part A, *The Elements*, Verlag Chemie GmbH, Weinheim/Bergstrasse, 1972–1973, A1,I, 1973, A1,II, 1974.
3. E. K. Hyde, I. Perlman, and G. T. Seaborg, *Nuclear Properties of the Heavy Elements*, Vol. **II**, *Detailed Radioactivity Properties*, Prentice-Hall, Inc., Englewood Cliffs, N.J., 1964.
4. C. M. Lederer, J. M. Hollander, and I. Perlman, *Table of Isotopes, Sixth Edition*, John Wiley & Sons, Inc., New York, 1967.
5. *Proceedings of the first United Nations International Conference on the Peaceful Uses of Atomic Energy, Geneva, 1955*, United Nations, New York, 1955; *Proceedings of the Second United Nations International Conference on the Peaceful Uses of Atomic Energy, Geneva, 1958*, United Nations, New York, 1958; *Proceedings of the Third United Nations International Conference on the Peaceful Uses of Atomic Energy, Geneva, 1964*, United Nations, New York, 1964; *Proceedings of the Fourth United Nations International Conference on the Peaceful Uses of Atomic Energy, Geneva, 1971*, United Nations, New York, and IAEA, Vienna, 1972.

6. G. T. Seaborg, *Nucl. Appl. Technol.* **9**, 830 (1970).
7. G. T. Seaborg, J. J. Katz, and W. M. Manning eds., *The Transuranium Elements: Research Papers, National Nuclear Energy Series*, Div. IV, 14B, McGraw-Hill Book Co., Inc., New York, 1949.
8. D. C. Hoffman, F. O. Lawrence, J. L. Mewherter, and F. M. Rourke, *Nature (London)* **234**, 132 (1971).
9. G. T. Seaborg, *Man-Made Transuranium Elements*, Prentice-Hall, Inc., Englewood Cliffs, N.J., 1963.
10. J. L. Crandall, *Production of Berkelium and Californium, Proceedings of the Symposium Commemorating the 25th Anniversary of Elements 97 and 98 held on January 20, 1975, Lawrence Berkeley Laboratory, Report LBL-4366*. Available as *TID 4500-R64* from National Technical Information Center, Springfield, Va., 1975.
11. *Series on Radiochemistry*, National Academy of Sciences. Reports available from Office of Technical Services, Department of Commerce, Washington, D.C.: P. C. Stevenson and W. E. Nervik, *Actinium* (with scandium, yttrium, rare earths), *NAS-NS-3020*; E. Hyde, *Thorium, NAS-NS 3004*; H. W. Kirby, *Protactinium, NAS-NS-3016*; J. E. Gindler, *Uranium, NAS-NS-3050*; G. A. Burney and R. M. Harbour, *Neptunium, NAS-NS-3060*; G. H. Coleman and R. W. Hoff, *Plutonium, NAS-NS-3058*; R. A. Penneman and R. K. Keenan, *Americium and Curium, NAS-NS-3006*; G. H. Higgins, *The Transcurium Elements, NAS-NS-3031*.
12. *Transurane, Gmelins Handbuch der anorganischen Chemie*, Part D, *Chemistry in Solution*, Springer-Verlag, Berlin, Heidelberg, New York, 1975. D1, D2.
13. C. Keller, *The Chemistry of the Transuranium Elements*, Verlag Chemie GmbH, 1971.
14. J. J. Katz and G. T. Seaborg, *The Chemistry of the Actinide Elements*, Methuen & Co., Ltd., London and John Wiley & Sons, Inc., New York, 1957.
15. Ref. 2, A2.
16. G. T. Seaborg, *The Transuranium Elements*, Yale University Press, New Haven, Conn., 1958.
17. J. C. Bailor, Jr., J. Emeleus, R. Nyholm, and A. F. Trotman-Dickenson, eds., *Comprehensive Inorganic Chemistry*, Vol. **5**, *Actinides*, Pergamon Press, Oxford, New York, 1973.
18. Ref. 2, Index, 1979.
19. J. J. Katz, G. T. Seaborg, and L. R. Morss, eds., *The Chemistry of the Actinide Elements*, 2nd ed., Chapman and Hall, London, 1986.
20. W. Müller and R. Lindner, eds., *Transplutonium 1975, 4th International Transplutonium Element Symposium, Proceedings of the Symposium at Baden Baden September 13–17, 1975*; W. Müller and H. Blank, eds., *Heavy Element Properties, 4th International Transplutonium Element Symposium, 5th International Conference on Plutonium and Other Actinides 1975, Proceedings of the Joint Session of the Baden Baden Meetings September 13, 1975*, North-Holland Publishing Co., Amsterdam, American Elsevier Publishing Co., Inc., New York.
21. Ref. 2, Parts B1, B2, B3, *The Metals and Alloys*, 1976, 1977.
22. Ref. 2, Part C, *The Compounds*, 1972.
23. E. C. Baker, G. W. Halstead, and K. N. Raymond, *Struct. Bonding (Berlin)* **25**, 23–68 (1976); T. J. Marks and A. Streitwieser, Jr., Chapt. 22 of ref. 19; T. J. Marks, Chapt. 23 of ref. 19.
24. W. T. Carnall and H. M. Crosswhite, Chapt. 16 of ref. 19.
25. D. C. Hoffman, and D. M. Lee, *J. Chem. Ed.* **76**, 331 (1999).
26. G. T. Seaborg, *Ann. Rev. Nucl. Sci.* **18**, 53 (1968); O. L. Keller, Jr., and G. T. Seaborg, *Ann. Rev. Nucl. Sci.* **27**, 139 (1977).
27. G. Hermann, *Superheavy Elements, International Review of Science, Inorganic Chemistry*, Series 2, Vol. **8**, Butterworths, London, and University Park Press,

Baltimore, Md., 1975; G. T. Seaborg and W. Loveland, *Contemp. Physics* **28**, 233 (1987).
28. V. Pershina, *Chem. Rev.* **96**, 1977 (1996). *Radiochim. Acta* **80**, 65 (1998).
29. E. Anders, H. Huguchi, J. Gros, H. Takahashi, and J. W. Morgan, *Science* **190**, 1262 (1975).
30. V. Trimble, *Rev. Mod. Phys.* **47**, 877 (1975).
31. R. Smolańczuk, *Phys. Rev. C* **56**, 812 (1997).

Glenn T. Seaborg
University of California, Berkeley
Darleance C. Hoffman
University of California, Berkeley
Diana M. Lee
Lawrence Berkeley National Laboratory

ADHESION

1. Introduction

Adhesion and adhesives play a role in many aspects of our daily lives. Secondary load-bearing structure in military and commercial aircraft is adhesively bonded to a large extent. The attainment of durable adhesion in these large structures is obviously tantamount in such applications. At the other end of the size scale, adhesion plays a role in the generation of modern electronics. Silicon chips are attached to lead frames by means of adhesives and electronic packages are often attached to circuit boards by means of adhesives. Indeed, the process of photolithography to generate microelectronic circuitry itself depends on adhesion. If the adhesion of the photoresist to the underlayers is not proper, etching solutions will undercut the resist and the edges of the circuitry will not be well defined. Life itself depends on adhesion. Insects can cling to walls by means of adhesion. Even small animals such as the gecko use adhesion as part of their life processes. These reptiles can reversibly adhere to a wide variety of surfaces and this allows them to prey upon their primary food source, insects. The formation of living organs is itself dependent on cell-to-cell adhesion. Nothing on this earth is independent of adhesion or the intermolecular forces that cause adhesion.

Adhesion is the physical attraction of the surface of one material for the surface of another. As such, it is dependent on the character of the physical forces that hold atoms and molecules together in each of the phases that are in contact and it is also dependent on how those forces match each other in the contact zone. An *adhesive* is a material that uses adhesion to effect an assembly between two other materials, which we call *adherends*. The assembly is known as an *adhesive bond* or an *adhesive joint*. The physical force necessary to break an adhesive joint is called practical *adhesion* (1). It is found that practical adhesion is dependent on adhesion but also that practical adhesion is primarily determined by the physical properties of the adhesive and adherends (2). When a polymeric adhesive is used, it is often found that practical adhesion exceeds adhesion by one or many orders of magnitude. Much of the science of adhesion concerns itself with the attempt to predict practical adhesion from fundamental forces at interfaces combined with the physical properties of the adhesive and the adherends. Another part of the science of adhesion is concerned with the improvement of practical adhesion by providing proper surfaces to which appropriately designed adhesives will display increased attraction. The final part of adhesion science concerns itself with the generation of improved adhesives and primers. The topic of adhesive chemistry is covered in another article in this volume.

We describe the phenomenon of adhesion from three aspects. First, our understanding of the phenomenon of making an adhesive bond is described. That is, we examine some explanations for how two materials interact at their surfaces. Second, we describe some methods by which we measure practical adhesion and examine several methods by which we determine how much force or energy it takes to break standard adhesive bonds. Third, we describe our present understanding of the connection between adhesion and practical adhesion.

This article is primarily concerned with adhesion and adhesive bonding rather than that of adhesion of coatings. However, essentially all of the principles discussed here are applicable to the adhesion of coatings to substrates, with the exception of the section on mechanical tests.

2. Fundamental Forces and the Strength of Materials

Our understanding of the properties of materials, including surface properties, extends from our understanding of the atomic and molecular structure of matter. Atoms have a central, positively charged nucleus surrounded by an electron cloud, while a molecule is a framework of positively charged nuclei around which the electron cloud exists. The forces of attraction between atoms and molecules are due to the relative motion of the electrons with respect to the nucleus or nuclear framework. The collection of forces that cause atoms and molecules to condense into liquids and solids are called *van der Waals forces* after the Dutch physicist J. D. van der Waals (3). Figure 1 provides a schematic representation of these van der Waals forces. Thus, the basis for dipole–dipole (Fig. 1**a**), dipole-induced dipole (Fig. 1**b**), and dispersion force interactions (Fig. 1**c**) are shown. All of these forces are due to the interaction of electron density distributions between atoms and molecules. The interactions can be quasistatic such as the case of the dipole–dipole and dipole–induced dipole interaction. Dispersion force interactions are due to instantaneous dipole–induced dipole interaction (hence the arrows.) Figure 1 also shows the approximate order of the strength of these interatomic or intermolecular interactions. It is important to note that despite the fact that dispersion force interactions are among the weakest, they occur between all atoms and molecules irrespective of molecular structure. Thus, dispersion force interactions are part of the basis for the strength of all materials and they play a role in all adhesion phenomena.

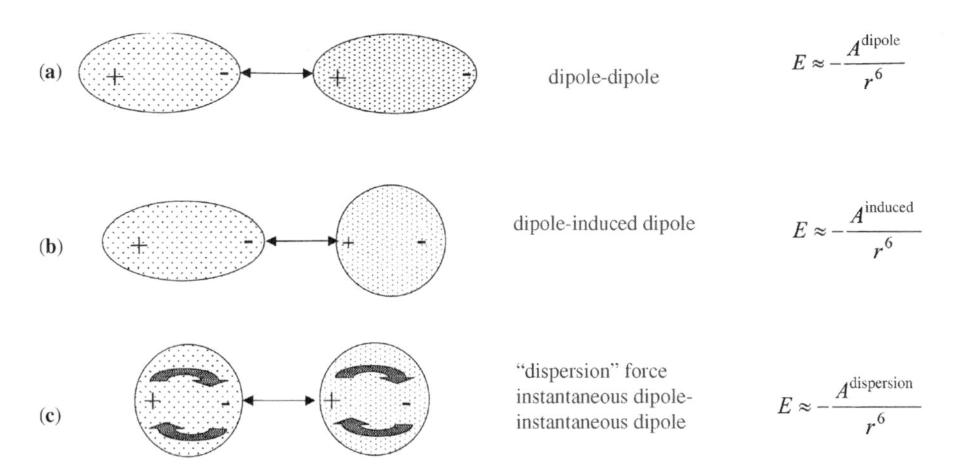

dipole-dipole > dipole-induced dipole > dispersion force

Fig. 1. Schematic representation of the basis for "van der Waals" forces.

There are additionally "chemical forces" that add to the "physical" forces discussed above. These chemical forces are due to chemical reactions such as those shown schematically in Figure 2. Acid–base interactions are shown in Figure 2**a**, examples of which are hydrogen bonding and donor–acceptor

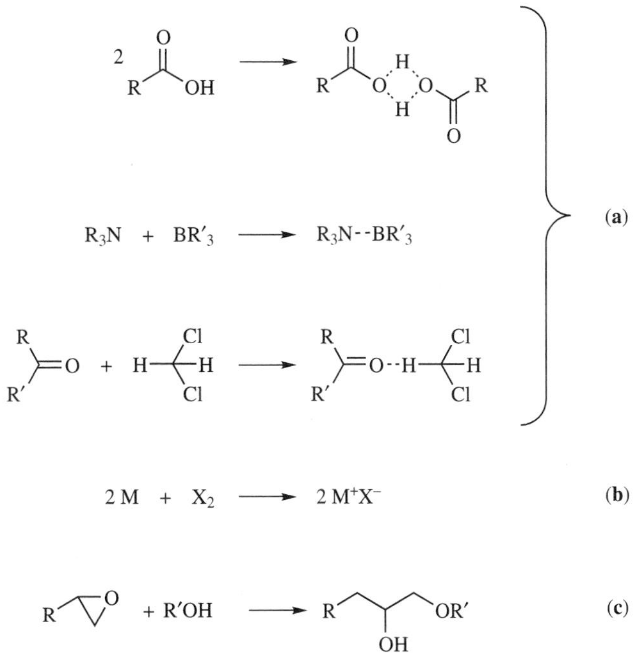

$2\ RC(=O)OH \longrightarrow RC(=O)O\cdots H\cdots O=C(R)O\cdots H$ (hydrogen-bonded dimer) (a)

$R_3N + BR'_3 \longrightarrow R_3N\text{--}BR'_3$

$RR'C{=}O + H\text{–}CCl_2\text{–}H \longrightarrow RR'C{=}O\cdots H\text{–}CCl_2\text{–}H$

$2\,M + X_2 \longrightarrow 2\,M^+X^-$ (b)

R-epoxide + R′OH ⟶ $R\text{–}CH_2\text{–}CH(OH)\text{–}CH_2\text{–}OR'$ (c)

Fig. 2. (**a**) Examples of "acid–base" interactions. (**b**) An example of ionic bond formation. (**c**) An example of covalent bond formation.

interactions. In these interactions, either a proton or electron is partially shared between two molecules. Figure 2**b** shows an example of an ionic interaction in which a metal, M, reacts with a halogen, X, transferring an electron from one to the other, to form a salt in which the resulting ions are attracted electrostatically. Figure 2**c** shows an example of covalent bond formation in which an epoxide reacts with an alcohol to yield an ether–alcohol. Such interactions are due to the sharing of an electron pair between two atoms or molecules. All of the forces and chemical reactions described in Figures 1 and 2 can play a role in the properties of a material, including surface properties. Chemical forces, in general, have a larger energy of interaction than do physical forces.

The simplest equation describing the potential energy of interaction between two interacting atomic or molecular species is the Lennard-Jones potential (4):

$$U^{L-J} = -\frac{A}{r^6} + \frac{B}{r^{12}} \tag{1}$$

In this equation, A contains all of the information regarding the forces of attraction between atoms and molecules while B contains the information regarding repulsive forces between atoms and molecules. The parameter r is the distance of separation of the atomic or molecular species. The fact that the denominators in equation 1 are a very high power of the distance indicates that these forces operate at very short ranges, in fact, the range of operation is on the order of nanometers or less.

The theoretical strength of a material in which the forces of attraction are due entirely to van der Waals forces can be determined by taking the derivative of the Lennard-Jones potential energy equation and setting that equal to zero (5). Thus

$$F_{max} = \frac{1}{9}\frac{\pi n^2 A}{\sqrt{3} r_0^3} \tag{2}$$

Where A is the attractive constant as defined in equation 2, n is the density of atoms or molecules in the material, and r_0 is the equilibrium distance of separation between the atoms or molecules. In a similar manner, the theoretical Young's modulus, E, of a material can be shown to be (6)

$$E = \frac{\pi n^2 A}{r_0^3} \tag{3}$$

We see, therefore, that both the theoretical strength and stiffness of a material are directly proportional to the strength of attraction between the atoms and molecules making up that material. If we invoke the presence of the other forces, as described in Figure 2, the strength and modulus increase proportionally to their contribution to the attraction. The equations describing these contributions are significantly more cumbersome and will not be included here.

3. Fundamental Forces and Surface Energy

We can understand the origin of surface energy by examination of the situation in Figure 3**a** where an assembly of atoms or molecules is depicted and a surface exists nearby (7). Let us imagine that the highlighted atom or molecule exists in a simple lattice having six nearest neighbors. If the energy of attraction between any pair of atoms or molecules is ϵ, then the total energy of attraction is

$$E_B = \frac{6\epsilon}{2} \tag{4}$$

where E_B is the energy of attraction in the bulk of the material and the two corrects for double counting. If, as shown in Figure 3**b**, the nearby surface is extended and the atom or molecule in question moves to the surface of the liquid, it finds itself in a higher energy state due to the lower number of nearest neighbors. The parameter E_S is the energy of the molecule at the surface. Thus

$$E_S = \frac{5\epsilon}{2} \tag{5}$$

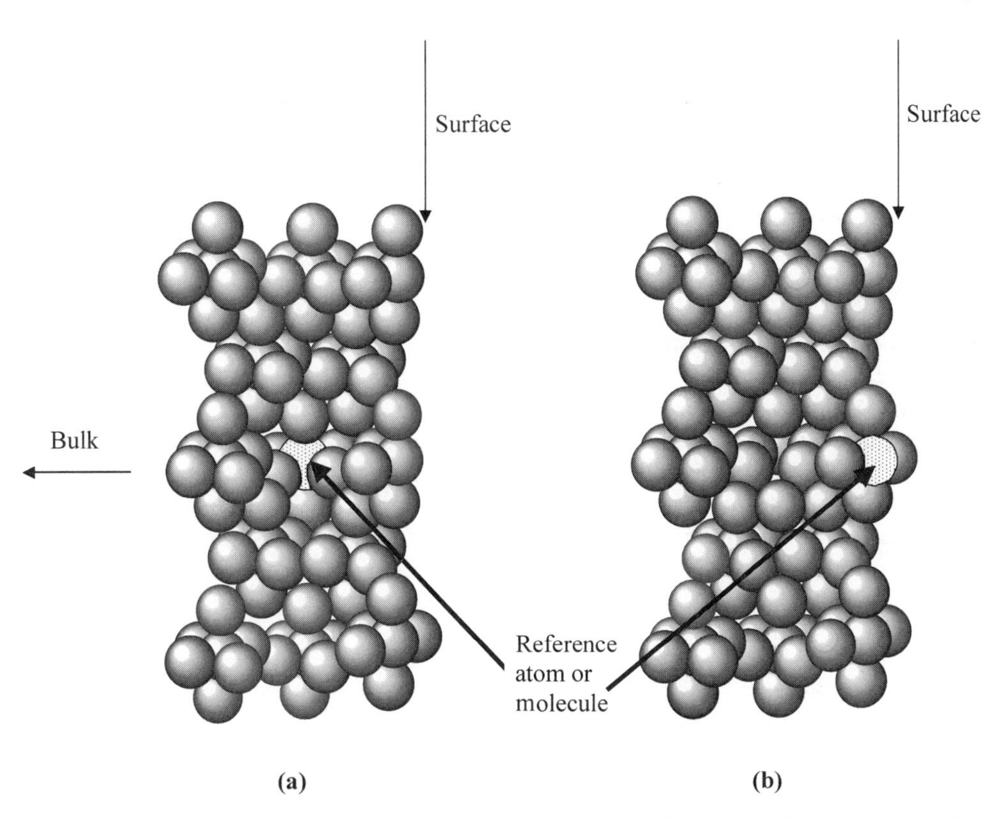

Fig. 3. Schematic representation of the difference in bonding state of an atom or molecule in the bulk of a material versus the surface of a material. The atom or molecule of interest is more lightly shaded than the majority.

Thus the energy expended to create surface per molecule is

$$E = \frac{1}{a_0}(E_S - E_B) = \frac{(5-6)\epsilon}{2a_0} = \gamma \quad (6)$$

where a_0 is the cross-sectional area on the surface occupied by the atom or molecule, Which is the amount of energy necessary to create a new surface per unit area and is given the symbol γ and is called the *surface energy*. Surface energy has units of millijoules per square meter (mJ/m^2). A typical surface energy of an organic liquid is on the order of 40 mJ/m^2 while water has a surface energy of 72 mJ/m^2 and a liquid metal like mercury has a surface energy of 480 mJ/m^2. When the substance in question is a liquid, the surface appears to exhibit a tension. This surface tension is due to the same forces as described above. For liquids, the surface tension and the surface energy are numerically identical.

Alternatively, one can use the method of Fowler and Guggenheim (8) to calculate the total energy of attraction between two van der Waals surfaces and one finds

$$\gamma = \frac{\pi n^2 A}{32 r_0^2} \quad (7)$$

which is functionally equivalent to the equation 6. Of primary importance, is that surface energy is also directly proportional to the attractive forces between atoms or molecules as described by the constant A. By comparing equation 7 and equations 2 and 3, we see surface energy, modulus, and the theoretical strength of a material are all related to the attractive constant, A, which is the basis for the statement above that adhesion is dependent on the same physical forces as the strength of materials.

4. The Work of Cohesion and the Work of Adhesion

Examine the situation in Figure 4. An imaginary plane is shown internal to an elastic material. The material can be the same on both sides of the imaginary plane or the two sides could be different. If we now split the material into two parts at the imaginary plane, two surfaces are created. If the material was the same on both sides of the plane, the amount of energy necessary to create this situation must be

$$W_C = 2\gamma \quad (8)$$

where W_C is the *work of cohesion*. If the process of separation creates two different materials, then we have

$$W_A = \gamma_1 + \gamma_2 - \gamma_{12} \quad (9)$$

where W_A is the *work of adhesion*. This equation is known as the Dupre equation (9). The $\gamma_i$values are the surface energies of the two materials and γ_{12} is the

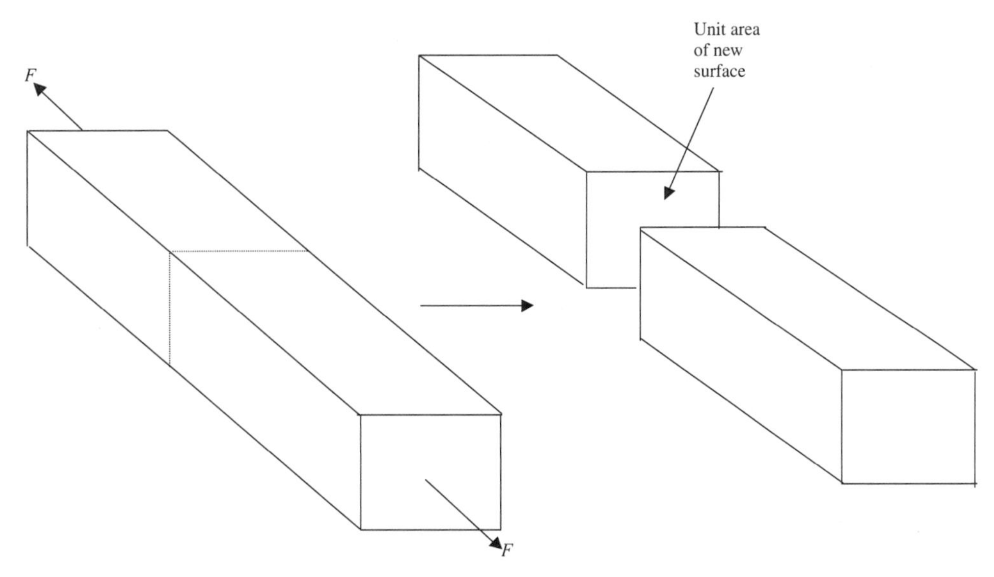

Fig. 4. The basis for the work of cohesion or adhesion. A brittle, elastic material with an imaginary plane in its interior is broken at that plane to present two new surfaces.

interfacial energy between the two materials. *Interfacial energy* is the amount of energy necessary to create a unit area of interface. Hence, when the interface disappears, that energy is released. The work of adhesion or cohesion is a thermodynamic parameter and is applicable only in reversible systems with elastic materials. The work of adhesion is a fundamental parameter that has formed the basis for much research in adhesion science. One can gather that adhesion will be the greatest for high values of the two surface energies in question and a small interfacial energy.

5. Wetting and Adhesion

To form an adhesive bond, a liquid adhesive must come into intimate contact with the surface of the adherend. If a drop of a liquid is placed on a smooth surface, it assumes a shape that is characteristic of the interaction of the liquid with the solid surface. (see Fig. 5). We say that when a liquid spreads on a surface, the liquid has *wetted the surface*. Young (10) derived an equation relating the angle of contact between the liquid and the solid to the surface tensions of the solid and liquid. Cherry (11) corrected Young's derivation and surface energies can be substituted for surface tensions. Thus,

$$\gamma_{LV} \cos\theta = \gamma_{SV} - \gamma_{SL} \tag{10}$$

Combining this equation with the Dupre equation provides

$$W_A = \gamma_{LV}\,(1 + \cos\theta) \tag{11}$$

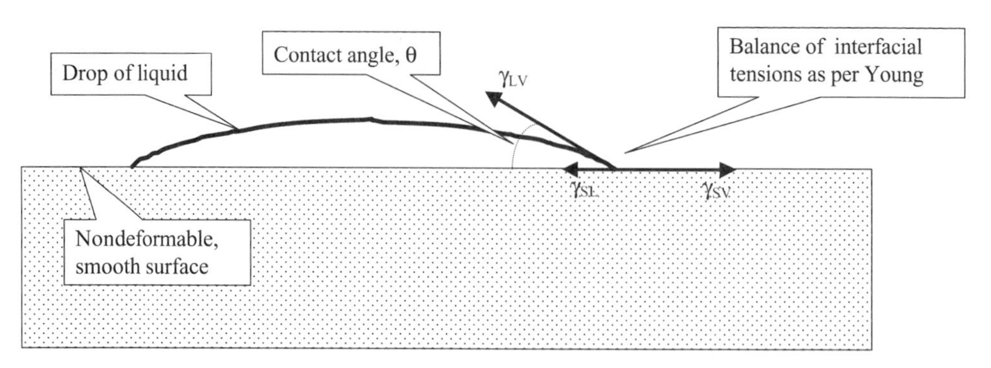

Fig. 5. Schematic representation of the contact angle measurement. The balance of forces, as proposed by Young is also shown.

This equation shows that the work of adhesion is maximized when $\cos\theta = 1$, ie, when $\theta = 0$. We also note that the work of adhesion can be, at most, twice the surface energy of the liquid doing the wetting. For water, this would be 144 mJ/m^2. This number should be kept in mind for the following discussion.

This deceptively simple equation relates the work of adhesion to a simply measured parameter related to wetting. Thus, we make the statement that a necessary (but not sufficient) criterion for adhesion is complete wetting of the surface by the liquid adhesive (ie, when $\cos\theta = 1$). The contact angle measurement has become a staple of adhesion scientists in that a nonwettable surface is also considered to be nonadhereable. Certainly, we are familiar with the use of poly(tetrafluoroethylene) as a nonadhesive material. The critical wetting tension (12) of poly(tetrafluoroethylene) is 18 mJ/m^2 and it is therefore nonwettable by materials such as water. We can also provide a potentially physically satisfying description. When the interaction forces between the liquid and the solid are so compatible that the atoms or molecules of the liquid would rather be in contact with the atoms or molecules in the surface of the adherend than to be in contact with themselves, wetting will be spontaneous with a contact angle of 0.

6. Adhesion and Interdiffusion

The ultimate in intimate contact occurs when the adherend and the adhesive are so compatible that they dissolve into one another. In molecular parlance, the molecules in the adhesive diffuse into the adherend and vice versa. We can examine this situation in terms of the intermolecular interactions that were discussed above. One means of measuring the degree of intermolecular interaction between molecules is to measure the enthalpy of vaporization, ΔH_{vap}. This quantity is the amount of energy necessary to separate the molecules in a liquid such that they no longer interact with one another. If we normalize the enthalpy of vaporization by the molar volume of the material, V, we obtain a quantity known as the cohesive energy density (CED):

$$\mathrm{CED} = \frac{\Delta H_{vap}}{V} \qquad (12)$$

We take the square root of this quantity and obtain an important parameter known as the *solubility parameter*

$$\delta = \sqrt{\text{CED}} \tag{13}$$

For any physical process to happen spontaneously, the change in the Gibb's free energy of the system must be negative. For two materials to be soluble in one another, we have

$$\Delta G_{\text{soln}} = \Delta H_{\text{soln}} - T\Delta S_{\text{soln}} < 0 \tag{14}$$

The Hildebrand theory of solutions (13) says that for a solute–solvent combination in which there is no interaction other than van der Waals interaction:

$$\Delta H_{\text{soln}} = \phi_1\phi_2(\delta_1 - \delta_2)^2 \tag{15}$$

where ϕ_1 and ϕ_2 are mole fractions and δ_1 and δ_2 are the solubility parameters of the 1 and 2 components of the mixture. Inspection of this equation shows that for this type of solution ΔH_{soln} cannot be negative but can be zero. Combining these two equations,

$$\Delta G_{\text{soln}} = \phi_1\phi_2(\delta_1 - \delta_2)^2 - T\Delta S_{\text{soln}} < 0 \tag{16}$$

we see that for ΔG_{soln} to be the most negative, $(\delta_1 - \delta_2) = 0$. This means that for a solution between two materials to form spontaneously, the solubility parameters must be equal. For solubility parameters to be the same, the molecular interactions in the two species have to be very similar. Alternatively, the enthalpy of solution could be negative due to chemical reaction such as acid–base interactions or hydrogen bonding. Thus, we can form another criterion for adhesion, *for diffusive adhesive bonding to occur, the molecular interactions in the two materials to be adhered must be very similar so that their solubility parameters are the same or the enthalpy of solution must be negative.*

An alternative view of solubility of polymers is found in the work of Flory that includes another parameter known as the *interaction parameter*, χ. This parameter is also related to the intermolecular interaction energy as used in equations 4–6.

7. Linear Elastic Fracture Mechanics

The strength of materials and interfaces as described in equation 2 is for a case in which the material or the interface is perfect, ie, it has no flaws. We know that such materials and interfaces are rare and difficult to manufacture. In most cases, materials available to us contain flaws and/or cracks. Fracture mechanics allows us to examine the strength of materials when flaws are present. Linear elastic fracture mechanics assumes that the material or interface in question follows Hooke's law (proportionality of stress and strain) for all extensions. If w is the work done on a sample and e is the strain energy released by the sample when it is unloaded, then the basis for linear elastic fracture mechanics is

$$\frac{1}{c}\frac{\delta}{\delta a}(w - e) \geq \mathcal{G}_C \tag{17}$$

c is the width of the specimen, a is the length of the crack in the specimen, and $\mathcal{G}_C$ is a materials parameter known as the *critical strain energy release rate*. This equation says that if the amount of work done on a sample less the amount of strain energy the sample can dissipate is greater than a certain amount of energy, then a crack will grow. Griffith (14) extended this formulation to say that if a crack grows, two new surfaces (the sides of the crack) are formed. Therefore, the smallest amount of energy necessary to generate a crack is

$$\frac{1}{c}\frac{\delta}{\delta a}(w - e) \geq 2\gamma \tag{18}$$

or if the crack is at an interface

$$\frac{1}{c}\frac{\delta}{\delta a}(w - e) \geq \gamma_1 + \gamma_2 - \gamma_{12} \tag{19}$$

We note that in the case of a linear elastic material,

$$W_A = \mathcal{G}_C \tag{20}$$

when the crack forms at an interface.

The discussion has so far been concerned with a linear elastic material. Materials, in general, are neither linear nor elastic and have many ways of dissipating mechanical energy other than cracking to generate surface. So in most cases,

$$\mathcal{G}_C >>>> W_A \tag{21}$$

This means that $\mathcal{G}_C$ for most materials is much larger than the properties of surfaces. It also brings up the fact that much of the stress that can be placed on an adhesively bonded assembly could be dissipated by various mechanisms in the adherends or in the adhesive before that stress can be transferred to the interface.

8. Covalent Bonding and Adhesion

The interfacial forces discussed thus far have been those involving the interactions between electron distributions surrounding atoms or molecules. With the exception of ionic bonds, these interactions typically have a very small potential energy. In terms of calories, these interactions are typically only on the order of 1–5 kcal/mol of interactions. Significantly higher potential energy of interaction can occur when protons or electron pairs are shared between atoms or molecules. Thus, hydrogen bonding can provide 10–25 kcal/mol of interactions while, when electron pairs are shared between atoms or molecules, the formation of a covalent bond provides a potential energy on the order of many tens to >100 kcal/mol of interaction. This level of interaction provides a level of adhesion of the order of joules per square meter (J/m^2), a significantly larger number than that obtained for purely van der Waals interactions (typically tens of mJ/m^2).

The potential for significantly higher levels of interfacial interaction due to hydrogen or covalent bonding at interfaces has become a technology for generating interfacial chemistry. The best example of interfacial chemistry for improvement of adhesion phenomena is that of silane coupling agents (15). An empirical formula for a silane coupling agent is

$$R - Si - (OR')_3$$

where R is a group that is reactive with an organic adhesive and the OR′ are alkoxy groups that are reactive with inorganic hydroxides on the surface of most inorganic materials. Thus, these silanes can be dissolved in water or water–alcohol mixtures in the presence of acidic or basic catalysts and applied to inorganic surfaces. An adhesive capable of reaction with R will then display improved adhesion to the surface through the formation of a covalent chemical bond. Although the primary reason for use of silanes is improvement in the durability of adhesive bonds, often one is able to find that initial bond strength can be improved over that of untreated interfaces.

9. Mechanical Roughness and Adhesion

In the absence of anything but van der Waals interactions at interfaces, an adhesive bond between an adhesive and a flat adherend will yield an adhesive joint that is likely to be mechanically weak. The reason for this situation becomes apparent when one considers the application of a force to the edge of such an adhesive bond. This situation is shown in Figure 6**a**. The only interaction holding the two surfaces together being van der Waals interactions, the work to open the bond will be very low. If there is no other means of absorbing mechanical energy, the interfacial crack will propagate easily between the two surfaces.

If instead of a planar interface, a mechanically rough interface exists between the adhesive and the adherend, there could exist two new means of absorbing mechanical energy. As a crack is propagating from the edge of the adhesive bond shown in Figure 6**b**, the crack can no longer follow a sharp interface. Instead, the crack must every so often encounter either the adherend or the adhesive. If the adherend has a higher stiffness than the adhesive, then the crack will divert into the adhesive where there will be a higher likelihood of energy dissipation due to plastic deformation. If the adherend surface is appropriately contoured, there is also a chance for the adhesive to "interlock" with the adherend (16). That is, if there are overhangs under which the adhesive can flow, then once it hardens, the adhesive cannot pass those overhangs without significant plastic deformation. The plastic deformation leads to significant absorption of mechanical energy before the bond can be parted.

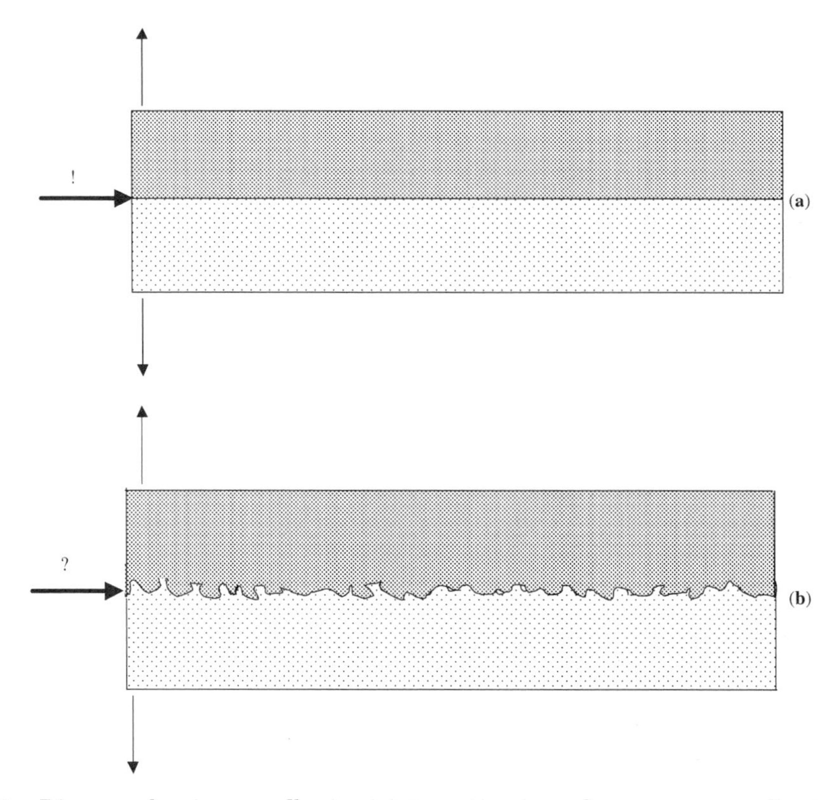

Fig. 6. Diagram showing an adhesive joint meeting in a plane versus an adhesive joint having microscopic morphology at its surface.

If a rough surface is to be bonded, there is another important consideration. Adhesives have non-zero viscosity and, in many cases, adhesive viscosity will increase with time after application. With a rough surface, the adhesive must penetrate into the pores of that surface during the application time. Packham (17) showed that the size of the pore on a mechanically rough surface can control the degree to which a viscous adhesive comes into intimate contact with that surface. Wicking of an adhesive into a porous surface is promoted by the proper adherend surface energy along with a pore size that is on the micron scale, or less.

10. Weak Boundary Layers and Surface Preparation

Not all surfaces are inherently amenable to adhesive bonding. Some surfaces are flat (as described in the last section) and some surfaces are contaminated with materials that subvert the possibility of intimate contact between the adhesive and the adherend. For example, when a metal sheet is formed into automobile parts by the process of deep drawing, a lubricant is used to aid in the process. That lubricant, if not removed, can act as a *weak boundary layer* between the metal adherend and the adhesive. Effective adhesive bonding cannot occur unless the adhesive subverts the lubricant. Weak boundary layers also occur in adhesive bonds to plastics since many plastics contain materials that bloom to surfaces.

To make effective adhesive bonds to adherends with flat surfaces or those contaminated with weak boundary layers, or both, one must modify the surface by means of a *surface preparation*. For metals, surface preparations usually involve some electrochemical method such as an etch (18) and/or an anodization procedure (19). In general, these treatments remove the weak boundary layer and provide a microscopically rough surface. For plastics, surface preparations include processes such as corona (20), flame (21), and plasma treatment (22). In general, these treatments oxidize the surface of the plastic (and thus increases its surface energy), removing or otherwise modifying the weak boundary layer material. Sometimes, the generation of a microscopic surface morphology occurs. For both metals and plastics, abrasion and simple wiping methods can aid in promoting adhesion but these are usually not as effective as the methods described above.

11. Priming

Priming is a chemistry and a process in which a chemically distinct layer is placed on an adherend before adhesive bonding to provide improved adhesion or improved durability of the resultant bond. The discussion on coupling agents (above) is one example of a primer but the technology of primers goes beyond silane coupling agents. In fact, many adhesive bonding primers contain silane coupling agents as one of their components. In general, primers are multicomponent mixtures, applied out of solvent or other carrier and are thus, low-viscosity materials. As discussed above, adhesion is dependent to a great extent on the degree of contact between the adhesive and the adherend. One way of making better contact between a high viscosity adhesive and a microscopically surface rough adherend is to apply a primer deposited from a low-viscosity carrier. Primers are applied to metals for adhesion improvement but are also applied for other purposes such as protection of the surface preparation or for corrosion protection of the metal in the bonded joint (23). One relatively new technology combines silane coupling agents with inorganic oxides to provide a water-based primer that provides improved durability (24).

Primers are also applied to wood as well as plastic surfaces. In this case, the primer is applied to create a graded interphase between the base plastic and the adhesive. For example, partially chlorinated polyolefins are used as primers for low-surface energy plastics. These materials are thought to diffuse into the low-energy substrate and leave a higher energy surface for the adhesive (25).

12. Adhesive Bond Breaking

Adhesive joints are mechanical systems. To evaluate the utility of an adhesive for a particular application, the adhesive bond must be examined mechanically. Specific tests have been developed to examine adhesives in their primary modes of loading. The primary modes of loading are shown schematically in Figure 7

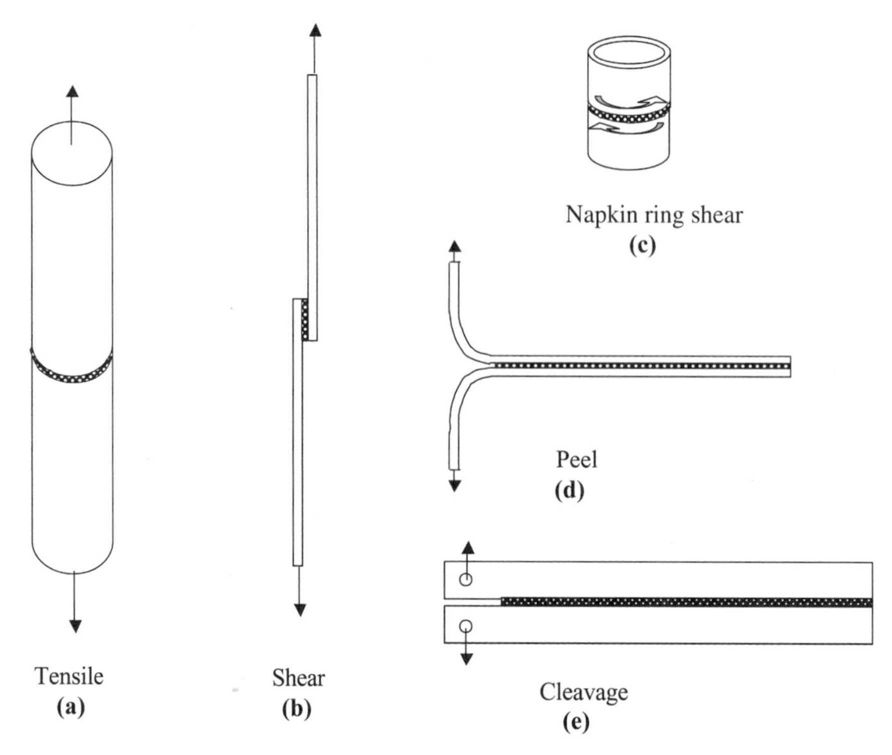

Fig. 7. Various specimens used to determine adhesive bond strength. (**a**) butt tensile joint, (**b**) lap shear specimen, (**c**) napkin ring shear specimen, (**d**) "T–peel" specimen, and (**e**) double cantilever beam specimen.

and are tensile, shear, peel, and cleavage. There are several agencies that specify adhesive bond tests (26). This article will discuss several methods described by the American Society for Testing and Materials (ASTM) (27).

12.1. Tensile Tests. There are several methods to test adhesive bonds in tension. One of these bonds, the butt tensile test, has been used on many occasions to examine adhesive bonds in tension. Figure 7**a** shows such a specimen and it is described in ASTM Method D2095. The specimen has some significant disadvantages, not the least of which is a nonuniform stress state. Adhesive bonds are seldom loaded in tension in practical adhesive bonds.

12.2. Shear Tests. The primary test specimen for loading adhesive bonds in shear is shown in Figure 7**b**. This specimen is described in ASTM Method D1002, the lap shear specimen. At first glance, it appears that this specimen does indeed load the adhesive in shear. During test, it is obvious that not only is the adhesive in shear, it is also loaded in peel (see below). This occurs at the ends of the lap and is demonstrated by plastic deformation of the adherends. The bending of the lap shear specimen has been investigated in a number of studies, in particular the classic work of Goland and Reissner (28). This specimen is used throughout the industry. Essentially all data sheets for adhesives contain lap shear data.

To determine the true shear properties of adhesives, two specimens are used. One is the napkin ring specimen (29) and the other is the thick adherend lap shear specimen (30). The napkin ring specimen, shown in Figure 7**c**, when carried out with proper instrumentation, can provide the true shear properties of an adhesive.

12.3. Fracture Tests. As described above, the strength of an adhesive bond is limited by the resistance of the assembly to the propagation of existing flaws in the structure. Fracture tests are meant to examine the resistance of the adhesive to the propagation of a flaw. The edge of the specimen provides the flaw.

12.4. Peel Tests. A peel test, such as that shown in Figure 7**d**, is a fracture test in which one or both of the adherends is capable of plastic deformation under the loads used to test the specimen. The specimen shown in Figure 7**d** is a T-peel specimen as described in ASTM D1876.

12.5. Cleavage Tests. A double cantilever beam specimen, such as that shown in Figure 7**e**, is a fracture test in which the adherends are thick enough that they do not significantly plastically deform during the test. Appropriately applied, this test can be used to measure the critical strain energy release rate of an adhesive. This test is described in ASTM D3433.

12.6. Contact Mechanics. A test method exists for the direct measurement of the forces of adhesion between two surfaces. The development of the method begins with the work of Hertz (31) who examined the mechanics of contact between two elastic spheres. Hertz was able to show that the contact radius between two elastic spheres was the following function of the load applied to the spheres.

$$a^3 = \left(\frac{3\pi}{4}\right)(k_1 + k_2)\left(\frac{R_1 R_2}{R_1 + R_2}\right)F \tag{22}$$

where many of the parameters are shown in Figure 8 and the k_i are

$$k_i = \frac{1 - v_i^2}{\pi E_i} \tag{23}$$

where v_i and E_i are the Poisson's ratio and Young's modulus of the ith material used in the measurement. The radius of contact between the two spheres is a. The Hertz theory is purely mechanical and takes no account of adhesion between the two spheres.

When the two spherical surfaces are made from materials that are elastic and have a modulus of 10^5–10^8 Pa, it is found that the radius of contact, a, is measureably larger than the radius predicted by Hertz theory (see Fig. 9). Using an approach based in fracture mechanics, Johnson, and co-workers (32) clearly demonstrated that the increase in contact area was due to adhesion between the two elastic bodies. The equation describing the contact between two adhering elastic bodies is

$$a_3 = \frac{R}{K}\left[F + 3W_A\pi R + \sqrt{6W_A\pi RF + (3W_A\pi R)^2}\right] \tag{24}$$

where $R = [R_1R_2/(R_1 + R_2)]$ and $K = (4/3\pi)(k_1 + k_2)$ and W_A is the work of adhesion between the two surfaces. Thus, if one knows the elastic constants of the

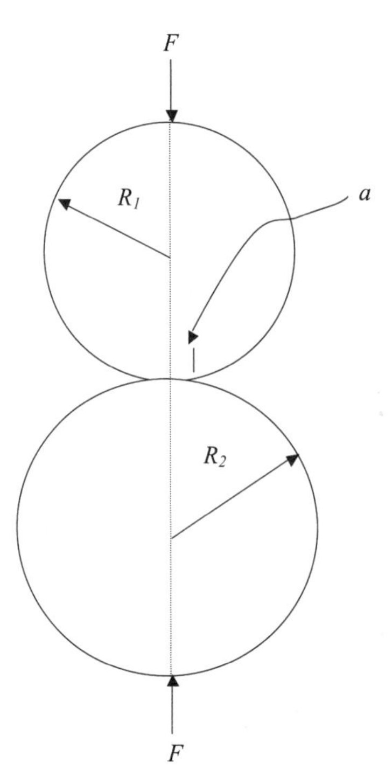

Fig. 8. The basis for the Hertz theory of contact mechanics.

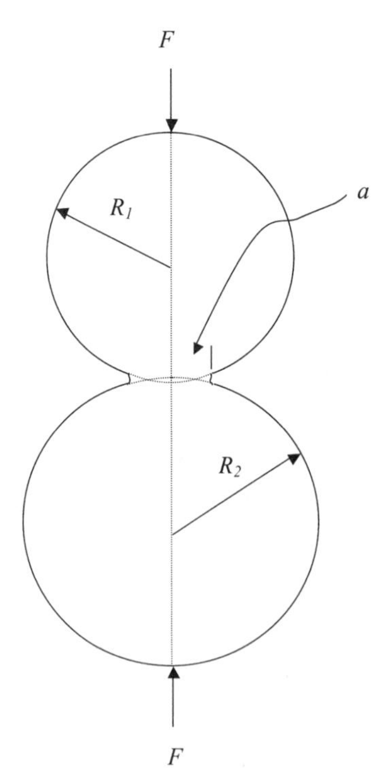

Fig. 9. The basis for the Johnson, Kendall, and Roberts (JKR) theory of contact mechanics that includes forces of adhesion at interfaces.

materials in contact, the radii of the spheres and if one measures the radius of contact as a function of load, F, then the work of adhesion between two solids can be determined directly from equation 24. Note that if $W_A = 0$, equation 24 becomes equation 22.

Two apparati have been developed to make use of this theory. One of these is the *surface forces apparatus* that has been extensively investigated by Israelachvilli and co-workers (33) and the other is the JKR apparatus described in the original work (34). In both cases, the data can be analyzed by means of equation 24 to yield the work of adhesion between the two surfaces. This theory and these techniques allow one to directly measure the forces of adhesion between two surfaces at very low rates. The use of this technique has increased significantly in recent years and it provides a means to connect the fundamental forces of adhesion to practical adhesion.

13. Adhesion and Practical Strength of Adhesive Bonds

Engineering has made significant advances in the analysis of structures through the use of computer-based models based in finite element analysis (35). It is tempting to apply such models to adhesive joints. Such analysis can be done for adhesive bonds if one makes the assumption that adhesion is perfect and that the forces in the interphase transfer forces perfectly (36). We know that in many cases, such transfer is not perfect and the models are thus limited. The previous two sections have described the surface aspects of adhesion and some mechanical tests of adhesive bonds. For adhesion science to become predictive and useful in modeling of adhesive bonds, a connection must be made between the mechanical strength of adhesive bonds and the fundamental forces of adhesion.

Several studies have attempted to connect fundamental adhesion with the practical strength of adhesive bonds. The studies have primarily been concerned with correlating the rate dependence of adhesive bond properties with some characteristic of the interface such as the work of adhesion. In particular, peel and cleavage measurements have been used.

Polymeric materials, including those used as adhesives, exhibit sensitivity to the rate of application of a mechanical stress. This phenomenon is known as time–temperature superposition. At low rates of application of mechanical stress, polymeric materials tend to be fully relaxed and act more liquid-like but at high rates of application of mechanical stress, polymeric materials tend to behave more like solids. The inverse is true as a function of temperature. If one measures the response of a polymer to an imposed mechanical stress as a function of temperature and rate, one can generate a "master curve" of that property as a function of a reduced rate or temperature. This process is described in the work of Williams and co-workers (37). This work describes the use of a quantity known as a shift factor, a_T, to generate a reduced rate or temperature that allows the data from the mechanical test to be placed on a single "master curve." Thus, one can measure the work necessary to break an adhesive bond as a function of the rate of application of mechanical stress. At low enough reduced rates (ie, high temperatures and low rates of application of mechanical

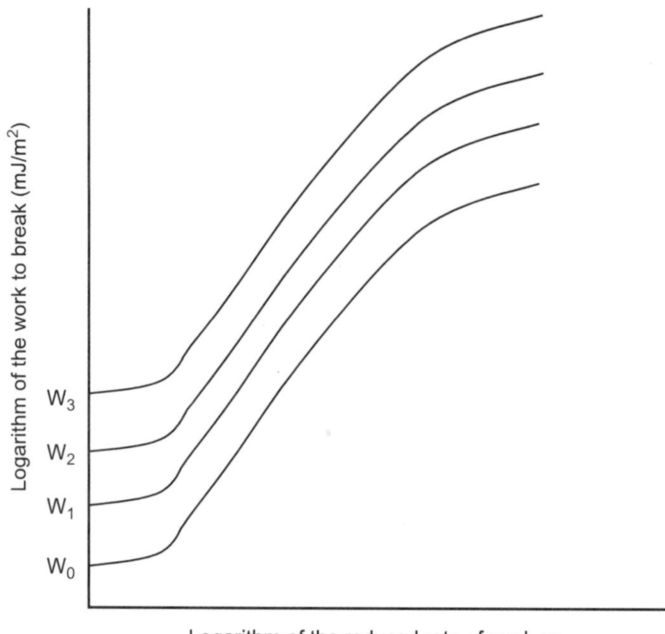

Fig. 10. Schematic representation of the results of Ahagon and Gent showing the dependence of practical adhesion on intrinsic adhesion and the rate at which the joint was peeled.

stress), it is possible that the work to break an adhesive bond becomes independent of rate. That is, the polymer can comply completely with the applied mechanical stress and the only force retaining the adhesive on the surface is adhesion.

Ahagon and Gent (38) did one such experiment in which they surface treated glass with silane coupling agents. These agents were chosen such that both were reactive with the glass but only one of which was reactive with the overlaying adhesive. They also used mixtures of these agents to treat the glass. Therefore, they created surfaces with increasing adhesion to the overlying adhesive. They peeled the adhesive from the treated glass surface as a function of rate and temperature. A schematic of the results of these measurements is shown in Figure 10. The logarithm of the work to break the adhesive bond is plotted as a function of a reduced rate of peel (the rate of peel, r, multiplied by the appropriate shift factor) of the adhesive from the treated glass surface. Each curve in Figure 10 corresponds to one of the treatments. The lowest curve is for the surface that was treated with the nonreactive silane and the top curve is for the surface treated with the reactive silane. The intermediate curves are for mixtures of the two. Note that the dependence of the logarithm of the work to break on ra_T for each level of adhesion is approximately the same as the other levels of adhesion. The dependence of work to break upon reduced rate of peel is governed by the mechanical properties of the adhesive. Each curve is translated upward by an amount dependent on the strength of the interface. Thus, the functional dependence of this data is

$$\log W_B = \log G_0 + \log \Phi(ra_T) \tag{25}$$

or, taking the antilog

$$W_B = G_0 \, \Phi(ra_T) \tag{26}$$

where G_0 is the bond strength at zero rate and $\Phi(ra_T)$ is a function that describes the response of the polymer to mechanical stress as a function of the reduced rate. The parameter Φ is 1 at sufficiently low rates and high temperatures. At that point $G_0 = W_B$.

We understand the polymer physics of these measurements and equation 26 by thinking about how polymers may respond to a stress when tethered to a surface. Polymers owe their rate-dependent mechanical characteristics to entanglements (due to their high molecular weight) or to cross-links or a combination of these. If $G_0 = 0$, then any force applied to an adhesive bond will cause the polymer to lift with no dissipated mechanical energy. Thus, when $G_0 = 0$, $W_B = 0$. If there is any modicum of adhesion of the polymer to the surface, then a slowly applied force will cause the polymer to begin disentanglement that absorbs mechanical energy. If enough mechanical energy is absorbed by disentanglement, there may not be enough mechanical energy to cause disbondment. As rate is increased, the amount of mechanical energy that the polymer can absorb increases up until the point that the material becomes stiff enough that it cannot disentangle on the timescale of the measurement. At that point, the polymer fails in a brittle manner or the polymer disbonds from the surface. Thus, adhesion becomes the key that unlocks the mechanisms of energy absorption of a polymeric adhesive in an adhesive bond.

Measurements similar to those of Ahagon and Gent have been carried out by Lake (39), Andrews and Kinloch (40) and Gent and Schultz (41). In all of these cases, results similar to that of Ahagon and Gent were observed.

More recently, researchers attempted to use the JKR method to examine the functional dependence of the work to break a contact adhesive bond. Maugis (42) found that the work necessary to separate a polyurethane elastomer from a glass surface on the rate of crack propagation at the interface was such that

$$\Phi \sim (ra_T)^{0.45}$$

Shull and co-workers (43) found a similar dependence when using materials similar to pressure sensitive adhesives.

Related studies by Li and co-workers (44) used the JKR apparatus and measurements of the adhesion energy (G) as a function of crack propagation rate between materials that mimic pressure sensitive adhesives. They also carried out standard peel tests as a function of rate using pressure sensitive adhesives chemically similar to those materials used in their JKR measurements. When plotted on the same axes, the results were similar to those shown in Figure 11. The bottom curve is for adhesion between two materials that have

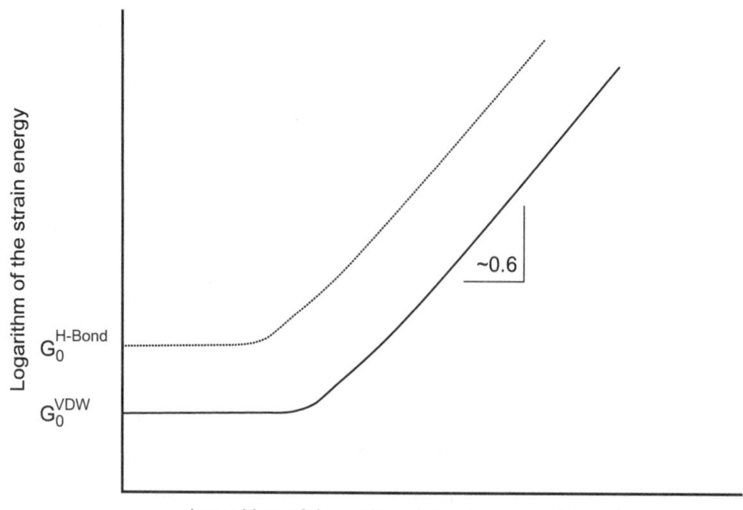

Fig. 11. Schematic representation of the results of Li, and co-workers (44), showing the dependence of practical adhesion on the intrinsic adhesion [based either in hydrogen bonding ("H-bond") or van der Waals interactions (VDW)] and the rate at which a crack propagates through the adhesive bond.

only van der Waals interactions possible between their surfaces. The second curve is for adhesion between two surfaces that have the ability to hydrogen bond with each other. The results have a dependence similar to that seen in Figure 10. These experiments are different from those of Ahagon and Gent, however, because both peel tests and JKR measurements are used to generate the curve, thus directly connecting fundamental adhesion (as measured in the JKR apparatus) and standard peel tests.

Models of the response of polymers at interfaces have been generated by a number of workers but Baljon and Robbins (45) have gone to the point of modeling polymers tethered in between two surfaces. Much of the physics described phenomenologically above was also demonstrated in their simulations. That is, if the adhesion of the polymer segments is high enough, unraveling of polymer chains followed by cavitation takes place internal to the adhesive.

Adhesion between glassy polymers has been investigated and has led to significant insights into the connection between adhesion, polymer physics, and fracture (46). This research used block copolymers as "primers" or "coupling agents" in an attempt to compatibilize the interface between incompatible polymers. As indicated in the previous section on diffusion and adhesion, very few high molecular weight polymer pairs are soluble in one another. A di-block copolymer is made of two segments. Each segment is made from one of the polymers to be adhered. A diagram showing the situation is provided in Figure 12. If the di-block polymer is of high enough molecular weight such that each of the blocks is sufficiently entangled, then the adhesive bond strength will be improved dramatically. In addition to the block length (molecular weight), the degree of

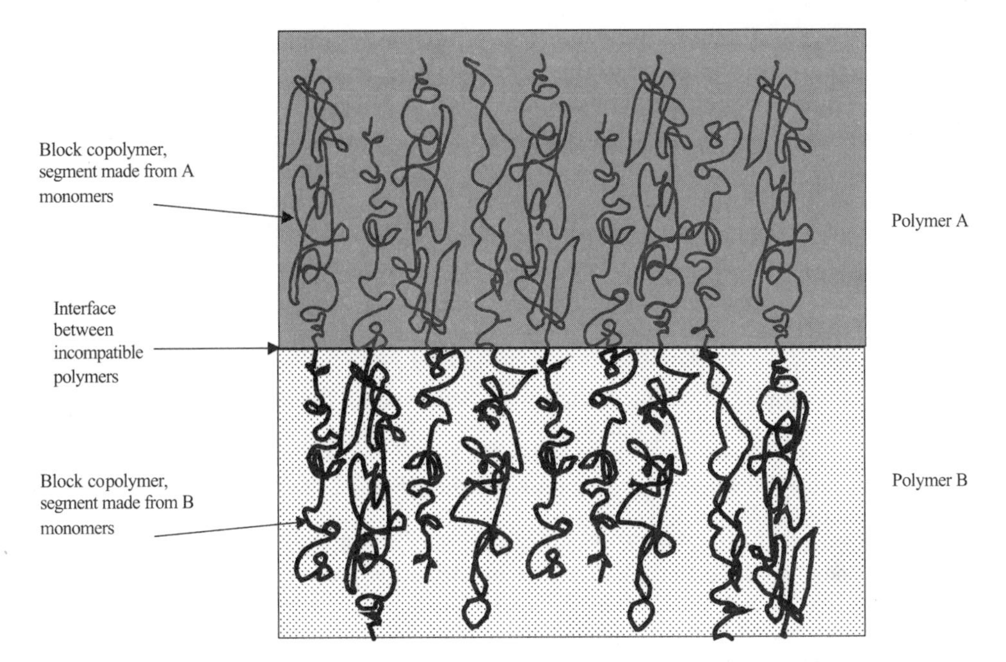

Fig. 12. Conceptual representation of a block copolymer at the interface between dissimilar glassy polymers.

improvement in adhesive bond strength is dependent on the number of block copolymer molecules crossing the interface as well as the crazing stress of either of the polymers in contact. Simply stated, if the interface becomes so strong due to the "stitching" action of the block copolymer, one or the other polymer will begin to yield before the interface yields. If the primary mechanism for yielding in that polymer is crazing (such as it is for polystyrene) then the crazing stress becomes the critical parameter. If the polymer crazes significantly, the energy to fracture the interface will be significantly higher than if no block copolymer is present. Thus, many of the concepts already presented for elastomeric materials, the concepts resulting from molecular modeling and these experiments using block copolymers between incompatible polymers, yield a story that relates fundamentals of interfaces and polymer physics to the strength of adhesive bonds.

14. Summary

Adhesion is a complex phenomenon whose understanding is addressed by three scientific disciplines: surface science, engineering mechanics, and polymer chemistry and physics. The phenomenon of adhesion is determined by the same intermolecular forces that provide the strength of all materials mitigated by the ability of adhesives to come into intimate contact with adherends. The strength of an adhesive bond is determined by a number of methods, most of which rely on destructive evaluation of an adhesive joint tailored to emphasize one or more modes of loading the joint. One method, based in contact mechanics, allows us to examine microscopic properties of adhesion at very slow rates. The measured strength of an adhesive bond results from a complex interweaving of adhesion with the physics of failure of the adhesive and the adherend. For elastomeric materials, a simple multiplicative relationship between intrinsic adhesion and the rheological properties of the adhesive and joint strength has been found.

BIBLIOGRAPHY

1. K. L. Mittal, in L. H. Lee, ed., *Adhesion Science and Technology*, Vol. 9A, Plenum Press, New York, 1975, p. 129.
2. K. L. Mittal, *Polymer Eng. Sci.* **17**, 46 (1977).
3. J. Israelachvili, *Intermolecular and Surface Forces*, 2nd ed., Academic Press, London, 1992, Chaps. 1, 4, 5, 6.
4. J. E. Lennard-Jones, *Proc. R. Soc. London, Ser. A* **196**, 463 (1924).
5. R. J. Good, in R. L. Patrick, *Treatise on Adhesion and Adhesives*, Vol. 1, Marcel Dekker, New York, 1967, Chap. 2.
6. R. J. Good, *op. cit.*
7. R. J. Stokes, and D. F. Evans, *Fundamentals of Interfacial Engineering*, Wiley-VCH, New York, 1997, pp. 53–55.
8. R. Fowler and E. A. Guggenheim, *Statistical Thermodynamics*, Cambridge University Press, Cambridge, U.K.
9. A. Dupre *Theorie Mechanique de la Chaleur*, Gauthier-Villars, Paris, 1869, p. 369.
10. T. Young, *Trans. R. Soc.* **95**, 65 (1805).
11. B. W. Cherry, *Polymer Surfaces*, Cambridge University Press, Cambridge, U.K., 1981, pp. 24–25.
12. (a) H. W. Fox and W. A. Zisman, *J. Colloid Sci.* **5**, 514 (1950). (b) W. A. Zisman, in L. H. Lee, ed., *Adhesion Science and Technology*, Vol. 9A, Plenum Press, New York, 1975, pp. 55–91.
13. J. Hildebrand and R. Scott, *The Solubility of Non-Electroytes*, 3rd ed., Reinhold, New York, 1950.
14. A. A. Griffith, *Philos. Trans. R. Soc. London Ser. A* **221**, 163 (1920).
15. E. P. Plueddemann, *Silane Coupling Agents*, Plenum Press, New York, 1982.
16. D. J. Arrowsmith, *Trans. Inst. Met. Finish.* **48**, 88 (1970).
17. D. E. Packham, in K. L. Mittal, ed., *Adhesion Aspects of Polymeric Coatings*, Plenum Press, New York, 1983, pp. 19–44.
18. ASTM Method D2674.
19. ASTM Method D3933.
20. T. Uehara, *Mater. Eng.*, **14** (Adhesion Promotion Techniques), 191 (1999).
21. D. Brewis and I. Mathieson, *Mater. Eng.* **14** (Adhesion Promotion Techniques), 175 (1999).
22. M. R. Wertheimer, L. Martinu, and J. E. Klember-Sapieha, *Mater. Eng.* **14** (Adhesion Promotion Techniques), 139 (1999).
23. R. H. Greer, and A. V. Pocius, *Electro-Deposited Primer Development and Low-Polluting Primer Evaluation*, Tech. Report. WRDC-TR-89-4069, August 1989, p. 2-2.
24. (a) R. A. Pike, *Int. J. Adhesion Adhesives* **5**, 3 (1985); (b) R. A. Pike, *Int. J. Adhesion Adhesives* **6**, 21 (1986); (c) U.S. Pat. 5,958,578, K. Y. Blohowiak, J. H. Osborne, and K. A. Krienke; (d) U.S. Pat. 6,037,060, K. Y. Blohowiak, J. H. Osborne, and K. A. Krienke; (e) U.S. Pat. 5,849,110, K. Y. Blohowiak, J. H. Osborne, and K. A. Krienke.

25. S. Waddington and D. Briggs, *Polym. Comm.* **32** 506 (1991).
26. (a) International Organization for Standardization, ISO, website at *www.iso.ch*; (b) Pressure Sensitive Tape Council, Chicago, Ill.
27. *ASTM Annual Book of Standards*, Volume 15.06, Adhesives.
28. M. Goland and E. Reissner, *J. Appl. Mech., Trans. Am. Soc. Eng.* **66**, A17 (1944).
29. N. K. Benson, in Houwinck and Salomon, eds., *Adhesion and Adhesives*, Elsevier, New York, 1967.
30. ASTM Method D5656.
31. H. Hertz, in Jones and Schott, eds., *Miscellaneous Papers*, Macmillan Publishing, London, 1896.
32. K. L. Johnson, K. Kendall, and A. D. Roberts, *Proc. R. Soc. London Ser. A* **324**, 301 (1971).
33. (a) J. N. Israelachvili, and D. Tabor., *Proc. Roy. Soc. London Ser. A* **331**, 19 (1972); (b) J. N. Israelachvili, and G. E. J. Adams, *J. Chem. Soc, Faraday Trans. 1* **74**, 975 (1978).
34. K. L. Johnson, K. Kendall and A. D. Roberts, *op cit.*
35. J. R. Brauer, *What Every Engineer Should Know About Finite Element Analysis*, Marcel Dekker, New York, 1993.
36. (a) J. A. Harris, and R. D. Adams, *Int. J. Adhesion Adhesives* **4**, 65 (1984); (b) R. D. Adams in L. H. Sharpe, and S. E. Wentworth, *The Science and Technology of Adhesive Bonding*, Gordon Breach, New York, 1990, pp. 321–344.
37. M. L. Williams, R. F. Landel, and J. D. Ferry, *J. Am. Chem. Soc.* **77**, 3701 (1955).
38. A. Ahagon and A. N. Gent, *J. Polym. Sci., Polym. Physics, Ed.* **13**, 1285 (1975).
39. G. J. Lake, in *Proceedings International Adhesion Conference*, The Plastics and Rubbers Institute., London UK, 1984, pp 22-1–22-4.
40. (a) E. H. Andrews, and A. J. Kinloch, *Proc. R. Soc. London Ser. A* **332**, 385 (1973); (b) E. H. Andrews, and A. J. Kinloch, *Proc. R. Soc. London Ser. A* **332**, 401 (1973).
41. A. N. Gent, and J. Schultz, *J. Adhesion* **3**, 281 (1972).
42. D. Maugis, in L. H. Lee, ed., *Adhesive Bonding*, Plenum Press, New York, 1991, p. 303.
43. K. R. Shull, D. Ahn, W.-L. Chen, C. M. Flanigan, and A. J. Crosby, *Macromol. Chem. Phys.* **199**, 489 (1998).
44. (a) L. Li, M. Tirrell, G. L. Korba, and A. V. Pocius, *J. Adhesion* **76**, 307 (2001); (b) L. Li, C. Macosko, A. V. Pocius, G. L. Korba, and M. Tirrell, in J. L. Emerson, ed., *Proceedings of the Adhesion Society 24th Annual Meeting*, Adhesion Society, Blacksburg Va, 2001, p. 270.
45. (a) A. Baljon and M. O. Robbins, *Macromolecules* **34**, 4200 (2001); (b) A. R. C and M. O. Robbins, *Science* **271**, 482 (1996).
46. (a) H. R. Brown, *Macromolecules* **22**, 2859 (1989); (b) C. Creton, E. J. Kramer, C.-Y. Hui, and H. R. Brown, *Macromolecules* **25**, 3075 (1992); (c) H. R. Brown, K. Char, V. R. Deline, and P. F. Green, *Macromolecules* **26**, 4155 (1993); (d) K. Char, H. R. Brown, and V. R. Deline, *Macromolecules* **26**, 4164 (1993); (e) H. R. Brown, *Macromolecules* **26**, 1666 (1993); (f) W. F. Reichert and H. R. Brown, *Polymer*, **34**, 2289 (1993); (g) H. R. Brown, *Science* **263**, 1411 (1994); (h) C. Creton, H. R. Brown, and K. R. Shull, *Macromolecules* **27**, 3174 (1994); (i) H. R. Brown, *Macromolecules* **24**, 2752 (1991).

GENERAL REFERENCES

I. M. Skeist, ed., *Handbook of Adhesives*, 3rd ed., Van Nostrand-Reinhold, New York, 1990.

R. D. Adams and W. C. Wake, *Structural Adhesives Joints in Engineering*, Elsevier Science, Inc., New York, 1984.
S. Wu, *Polymer Interfaces and Adhesion*, Marcel Dekker, Inc., New York, 1982.
A. J. Kinloch, *Adhesion and Adhesives: Science and Technology*, Chapman and Hall, New York, 1987.
A. V. Pocius, *Adhesion and Adhesives Technology: An Introduction*, Carl Hanser Verlag, Munich, January, 1997.
R. P. Wool, *Polymer Interfaces, Structure and Strength*, Carl Hanser Verlag, Munich, January, 1995.
R. J. Stokes, and D. F. Evans, *Fundamentals of Interfacial Engineering*, Wiley-VCH, New York, 1997.
J. Israelachvili, *Intermolecular and Surface Forces*, Academic Press, London, 1992.

Alphonsus V. Pocius
3M Company

ADHESIVES

1. Adhesive Compositions

An adhesive is a material that is used to join two objects through nonmechanical means. It is placed between the objects, which usually are called *adherends*, when it is part of a test piece or *substrate* that is part of an assembly, to create an adhesive joint. Although some adhesives form joints that almost immediately are as strong as they will be in actual use, other adhesives require further operations for the adhesive joint to reach its full strength. Adhesives can be made in several different physical forms, and the form of a given adhesive will define the possible methods of its application to the substrate.

An adhesive is comprised of a base chemical or a combination of chemicals that define its general chemical class. Most adhesives contain a curing agent or catalyst that will cause an increase in the molecular weight of the system and frequently the formation of a polymeric network. Nearly all adhesives also contain additives or modifiers that fine tune the adhesive and may significantly influence its behavior before and after formation of the adhesive joint. These additives include solvents, plasticizers, tackifiers, fillers, pigments, toughening agents, coupling agents, stabilizers, etc. Additives or modifiers increasingly are chosen for their ability to provide more than one benefit, eg, a pigment may not only color but also may reinforce an adhesive. In some cases, the process used to combine these diverse ingredients will strongly influence the properties of an adhesive. Although inorganic adhesives do exist, this article will be restricted to organic polymeric adhesives.

Consumers, designers, and engineers generally choose between adhesive bonding and mechanical or thermal methods when deciding how to join one object to another. Mechanical methods utilize bolts, screws, and rivets. Thermal methods include welding, soldering, and brazing. Adhesive bonding is the obvious choice for joining in cases in which the substrate is thin and relatively weak, eg, paper, or strong but relatively brittle, eg, glass. Use of adhesives in these situations avoids formation of stress concentration points and possible damage to the substrates. Even where the substrates will bear mechanical fastening, the geometry of certain parts sometimes makes welding or bolting more costly if not entirely impossible, as in the case of the aluminum honeycombs used in aerospace structures or tube-to-tube joining used in motor vehicle frame construction. Because they are usually applied so as to cover the entire joined surface in a continuous rather than point-by-point fashion, adhesives can provide a measure of environmental protection and mechanical reinforcement or stiffening well beyond the capabilities of mechanical fasteners. Stresses in adhesive joints are distributed over a relatively large area, which generally increases the mechanical and cosmetic integrity of joined parts. The energy damping capability of many polymeric adhesives contributes a mechanical damping component to joints that can increase their toughness and impact resistance. Adhesives are a great help in reducing the weight of structures because they add little weight and can facilitate the use of thinner substrates. Joining of dissimilar materials for reasons of economics, weight, or performance is frequently accomplished using adhesive bonding, providing properties already mentioned as well as electrical and thermal insulation, protection against galvanic corrosion, and acoustic damping. In some cases, adhesives are used in conjunction with joining methods such as welding and riveting, via weldbonding and rivet bonding, respectively, in order to maximize stiffness, strength, and fatigue resistance of joints.

Where an adhesive is the obvious choice, it is often the least expensive choice as well. In industrial situations, where the performance expected of the adhesive is high and broad and its cost is that of a specialty rather than a commodity material, it is common to see users take a systems approach to make the best choice of joining method or the best choice of adhesive, if adhesive bonding is seen to be the best joining method. The systems approach to choosing adhesives goes well beyond comparing the cost per gallon of adhesives. It considers the number of parts to be joined, the time and cost constraints of assembly, spatial limitations, the need for substrate surface cleaning or preparation, the cost of all application, fixturing, and curing equipment, environmental and safety requirements, disposal costs, and, finally, part performance, and lifetime.

2. Market Economics

In 1996, the global adhesive and sealant industry was estimated to have a size of ~7.5 million metric tons or 16.5 billion lb. The monetary value of this volume was considered to be ~US$20.0 billion (1). The value of the market was estimated to be $28.0 billion in 1998 led by North America with a 33% share followed by Europe (30%), the Far East (19%), and the rest of the world (18%) (2). By 2002, the same marketplace is expected to grow to 16.7 million metric tons or 36.8 billion lb (3). Adhesives make up >80% of the adhesives and sealants market. It has been estimated that the global use of adhesives will continue to grow annually 3–4% from 2000 through 2005, but for some types of adhesives and for some markets, the growth could be much larger.

In the United States, the adhesive and sealants business was producing ~US$1 billion from sales in 1972. By 1999, the U.S. adhesive business was estimated to be nearly 15.2 billion lb in size with an approximate value of US$9.5 billion. The size of the U.S. adhesive market is anticipated to grow to 17.4 billion lb by 2004 (4). The largest markets for adhesives in the United States are construction, primary wood-bonding, textiles, and packaging. Markets that command some of the highest prices for adhesives include dental, aerospace, and microelectronics.

The late 1990s were marked by significant numbers of consolidations and partnerships in the adhesive industry that are expected to continue into the 21st century. In 1999, only seven companies produced 49% of the adhesives sold in the world (5). The remainder of the adhesive industry is highly fragmented; in the United States alone there are ~500 adhesive companies. North American and European adhesive companies have partnered to serve the global operations of automotive OEMs expecting worldwide service. Several companies have formed joint ventures in the People's Republic of China (6) in anticipation of large market growth in that country. Such arrangements are expected to increase in number as makers of adhesives accelerate their pursuit of greater market share and opportunities in the most lucrative markets. Concurrently with these changes, many large resin suppliers have spun off their adhesive resin operations into new companies or sold off their adhesive raw materials divisions to established companies. An active adhesive formulator must keep track of raw materials sources and be prepared to trace older materials to their new sources.

The adhesives industry has been affected by environmental and regulatory concerns regarding health and safety issues of adhesive ingredients, use of solvents, and other issues. Less than 5% of the adhesives used in the United States in 1999 contained organic solvents. The use of adhesives with solvents is decreasing by ~2% annually. All other adhesives are waterborne or contain no carrier solvent. Recycling of adhesives has become more important as paper recycling has become very common, and the quality of recycled paper depends in part on the nature of adhesive residues present in recycle feedstock (7). Historically, certain adhesives have been based on natural products such as starch, natural rubber, and animal glue, and many adhesives still use as modifiers various tree-based rosins and terpenes, but there has been a strong shift away from naturally derived adhesives. Between 1972 and projected out to 2003, the value of U.S. adhesives made of synthetic resin and rubbers will have increased almost eight times while the value of U.S. adhesives made from natural bases will have increased only about five times (8). In the 1920s, nearly all primary wood bonding was done with adhesives produced from natural products (9). By the 1970s, that need was filled almost entirely by synthetic adhesives. As the price of crude oil rises and oil reserves dwindle, there is increasing interest in making more adhesives from renewable resources (10).

3. Classification of Adhesives

There are many ways to classify adhesives. These include chemical class, joint strength, bulk modulus, physical form, ultimate use, general market, method of application, and price. Another classification scheme involves considering the activation of an adhesive and the driving force for its change from a liquid-like system to a solid-like system. Each of these methods of classification provides a framework within which to understand adhesives.

The primary chemical classes from which adhesives are made include epoxies, acrylics, phenolics, urethanes, natural and synthetic elastomers, amino resins, silicones, polyesters, polyamides, aromatic polyheterocyclics, and the various natural products such as carbohydrates and their derivatives as well as plant- and animal-based proteins. Chemical class was once a relatively clean differentiator of adhesives, but so many adhesives now are hybrids, designed to take advantage of specific attributes of more than one chemical class or type of material. Hybridization can be accomplished by incorporating into an adhesive a nonreactive resin of a different chemical class; adding another type of reactive monomer, oligomer, or polymer; or chemically modifying an oligomer or polymer prior to adhesive compounding.

The measured overlap shear strength or peel strength of an adhesive joint is sometimes used to classify adhesives. The choice of substrate is a key element of such a comparison, certain aluminums or steels being most commonly chosen as standards, but glass, polyolefins, and other substrates are also used. Pressure-sensitive adhesives will be found at the low end of the bond strength spectrum, and structural adhesives will be found at the high end. In the middle will be found materials that are strong but not necessarily structural in nature; these are often called semistructural adhesives. Many sealants are strongly adherent, and some of these are referred to as adhesive sealants.

Market is a useful adhesive category for those interested in the buying and selling of adhesives, but market-based category can be very broad. Construction adhesives, eg, include joint compound, carpet glues, ceramic and vinyl tile adhesives, a variety of wood-bonding adhesives, and double-sided foam tapes for hanging architectural glass. Although the adhesives used in some of these product categories are relatively standardized, there are many choices in other product categories. Similar breadth and depth would be encountered among adhesives used in the automotive, medical, and electronics industries. Within each of these market areas and most other market areas there will be found both commodity and specialty adhesives.

Most of those who develop adhesive compositions consider the form and ultimate use of an adhesive to be the most useful categories because these guide and direct adhesive development. The ability of the adhesive formulator to satisfy an end use will be very much related to the completeness of the information available concerning performance attributes required or expected. Price is a category of immense interest to the adhesive developer as it helps to define the raw materials from which the formulator may choose. The adhesive development team often must work closely with the customer to learn what is really needed from an adhesive.

4. Forms and Types of Adhesives

As supplied, adhesives can be found in the form of low-viscosity liquids, viscous pastes, thin or thick films, semisolids, or solids. Before application to a substrate, an adhesive need not be sticky or otherwise particularly adherent. A distinct exception is the pressure-sensitive adhesive (PSA), which is inherently tacky when first made. Such an adhesive is applied as a thin film with or without a backing, the combination of the adhesive and the backing defining an adhesive tape. The PSA remains throughout its useful lifetime essentially the same material it was when first made. All other forms and types of adhesives undergo a transformation that is central to their function as an adhesive. This transformation is usually carried out through imposition of time, heat, or radiation, either actively or passively.

An adhesive applied as a true solution or a dispersion of solids will dry through loss of water or another solvent, leaving behind a film of adhesive. A reactive adhesive system will form internal chemical bonds through the processes of cross-linking, chemical reaction that joins dissimilar long-chain molecules, or polymerization, chemical reaction that joins similar monomer units. Solid adhesives are heated in order to be applied and then on cooling become functional adhesives. The transformation from a liquid, paste, or semisolid to a functional adhesive is loosely termed *curing*. Additional general terms that refer to this transformation include *setting up* and *hardening*. Adhesives may also cure in stages. The first stage of curing is sometimes referred to as the B stage, and adhesives that have undergone some level of precure in their manufacture are often said to have been *B-staged*. For many adhesive applications, the ability of an adhesive to gel, precure, or develop green strength or handling strength is a key characteristic, being most important for parts that will be bonded and then transported to the next step in their processing. *Adhesives* are referred to as such before and after cure.

5. Pressure-Sensitive Adhesives

PSAs are inherently and permanently soft, sticky materials that exhibit instant adhesion or tack with very little pressure to surfaces to which they are applied. The level of adhesion may build with time and be surprisingly high. PSAs generally have a high cohesive strength and often can be removed from substrates without leaving a residue. Some applications take advantage of a PSAs ability to quickly form a strong bond and under stress force failure elsewhere in a system, an attribute used to advantage in tamper-proof packaging and price stickers. At the other end of the spectrum lie PSAs that can be repeatedly repositioned. The primary characteristics used to describe the performance of PSAs are tack, adhesion strength in peel, and resistance to shear forces.

PSAs can be sold in bulk or solution for later coating by product manufacturers. Most PSAs, however, are sold as components of tapes or labels. PSAs are also used to make protective or masking films, some of which also function as conventional tape products, and others are sold in the form of aerosol sprays for graphic arts work. Tape products join one object to another, as when one wraps a gift, seals a box, or puts up a notice. They consist of a film or web carrier coated on one or both sides with a PSA. The carrier is usually a paper or synthetic polymer made in the form of a solid or a foamed film. The carrier is a key component of the tape. Such a construction is usually slit and wrapped on itself to form rolls of adhesive tape from which sections of the desired length can be removed. A release coating is sometimes added to the backside of the tape backing so that the tape can be removed from the roll cleanly, easily, and quietly without splitting the adhesive from the backing. Double-sided tapes that have no release liner effect release through opposite pairing of chemically different adhesives that are chemically incompatible or use of adhesives of different levels of cross-linking that are physically incompatible. There also may be a primer on one or both sides of the tape carrier to ensure better adhesion of the PSA or the release coating. Some tapes are sold with release liners that must be removed after the tape is taken off its roll. Tapes can be applied manually or via mechanized tape dispensers for packaging, splicing, and other applications. Labels are sold with the PSA already present for their attachment to a variety of surfaces. Transfer tapes are PSAs that are provided on a liner from which the adhesive film can be transferred to another surface. PSAs that are effectively sticky hot-melt adhesives can be applied in discrete lines, dots, or other shapes using manual and automated equipment. The convenience and adaptability of PSAs has gained them wide use in diverse applications in virtually every market served by adhesives.

Many PSA compositions contain a base elastomeric resin and a tackifier, which enhances the ability of the adhesive to instantly bond as well as its bond strength. The elastomer may be useful without cross-linking but will often require either chemical or physical cross-linking for establishment of sufficient cohesive strength. Heat, ultraviolet (uv), or radiation are usually the activators of the cross-linking, and suitable catalysts are used, with their choice depending on the base resin. Small amounts of epoxy or hydroxy functionality are sometimes added to allow uv cures if the base resins are not themselves uv-curable. Electron beam curing has received attention but tends to be more costly than uv curing. Elastomers used as the primary or base resin in tackified multicomponent PSAs include natural rubber, polybutadiene, polyorganosiloxanes, styrene–butadiene rubber, carboxylated styrene–butadiene rubber, polyisobutylene, butyl rubber, halogenated butyl rubber, and block polymers based on styrene with isoprene, butadiene, ethylene–propylene, or ethylene–butylene. Any of these resins may be blended with each other to alter or optimize properties. Polychloroprene, *cis*-polyisoprene, and some waxes are rarely used as the main components in PSAs but have found some use as modifiers. Natural rubber grafted with methyl methacrylate, styrene–acrylonitrile copolymers, and other elastomers have been found useful as components of primers for PSA products. Polymers that can be useful as PSAs without tackification but may be modified beneficially with their addition include poly(alkyl acrylate) homopolymers and copolymers, polyvinylethers, and amorphous polyolefins. Comonomers useful for acrylate PSAs include acrylic acid, methacrylic acid, lauryl acrylate, and itaconic acid.

Much of the art of making PSAs rests in the choice of tackifier and the balance between base resins and tackifiers, of which there are numerous choices (11). Tackifiers commonly used with natural rubber, butyl rubber, and polyacrylates include rosins and rosin derivatives manufactured from pine tree gums. The styrenic block polymer base resins respond well to tackification with aliphatic and partially aromatic materials miscible with their continuous nonstyrenic phase or phases. Materials useful as PSA tackifiers have a lower molecular weight than the base resin. They are useful because they lower the modulus of the bulk adhesive in the rubbery region of the modulus-temperature spectrum, that is, above the glass transition temperature. Tackifiers also tend to raise the glass-transition temperature of the system. Tackifiers that react with PSA resins have been introduced to counteract tendencies of tackifiers to migrate, bloom, or volatilize; these kinds of tackifiers are based on isocyanato-reactive or vinyl functional groups (12). Plasticizers are mentioned somewhat synonymously with tackifiers as modifiers for PSAs, but their use is recommended cautiously as any improvements they provide in tack can be quickly offset by losses in strength if the glass transition temperature of the material is lowered too much.

Silicone pressure-sensitive adhesives are blends or reaction products of the combination of a polyorganosiloxane, such as poly(dimethyl siloxane) or its copolymers with diphenylsiloxane or methylphenyl siloxane, with a polysiloxane resin, which is largely inorganic. Pendant vinyl groups may also be incorporated into silicone PSAs, making cross-linking possible with peroxide and other kinds of cures. These kinds of PSAs are most often tackified with additional silicone gums and siloxane resins of varying molecular weight. The silicone PSAs are unique in their resistance of temperatures up to 400°C; performance at elevated temperatures can be optimized using the siloxane resins and rare earth or transition metal esters (13).

The large bulk of PSAs are coated onto continuous webs or films to make pressure-sensitive tapes, labels, etc. While many PSAs continue to be coated out of organic solvents, many have been converted to water-based formulations or are extruded as hot-melt adhesives that upon cooling retain their tack. Aqueous emulsions of carboxylated styrene–butadiene and various acrylate copolymers are among the most useful as bases for water-based PSAs. The complexity of latex chemistry introduces additives such as chain-transfer agents and defoamers (14) into some emulsion-based PSAs. Proper coating of these kinds of PSAs can require addition of thickening agents based on water-soluble polymers. Other additives that may be found in PSAs include cross-linking agents, catalysts, heat stabilizers, antioxidants, photoinitiators, depolymerizers (or peptizers), and various fillers. Reinforcing agents such as phenolics and higher molecular weight relatives of the tackifiers are sometimes added to improve cohesive strength. As made, PSAs are generally colorless or off-white in appearance but are sometimes pigmented for color adjustment or become pigmented through addition of a colored filler such as titanium dioxide, talc, or silver.

6. Hot-Melt Adhesives

Hot-melt adhesives are solid adhesives that are heated to a molten liquid state for application to substrates, applied hot, and then cooled, quickly setting up a bond. The largest uses of hot-melt adhesives are in packaging, bookbinding, disposable paper products, wood-bonding, shoemaking, and textile binding. The advantages of hot-melt adhesives include their easy handling in solid form, almost indefinite shelf life, generally nonvolatile nature, and, most importantly, ability to form bonds quickly without supplementary processing. They are considered friendly to the environment and are expected to see expanded use on a worldwide basis as the market continues to move away from solvent-based adhesives. The disadvantages of hot-melts lie in their tendency to damage substrates that cannot withstand their application temperatures, limited high temperature properties, and only moderate strength.

Application temperatures typically used for hot-melts range from ~65–220°C. Although the industry still refers to most temperature-sensitive adhesives as hot-melts, one will see references to warm-melt adhesives that soften at ~121°C and cool-melt adhesives that soften below ~100°C, but these terms are somewhat arbitrarily applied. Decreases in the application temperatures for hot-melts have lessened safety concerns associated with this type of adhesive. While most hot-melts are supplied as sticks or pellets, they are also produced as flat films or sheets, rolls, fibrous nonwovens, powders, strings, bulk masses, or dots or lines on liners.

Hot-melts generally are based on one or more thermoplastic resin. The largest portion of commercial hot-melt adhesives has for many years been based on ethylene–vinyl acetate copolymers having a vinyl acetate content of ~20–40%. The styrenic block polymers that are thermoplastic elastomers also make up a large portion of hot-melts. Other resins that have been found useful as bases for hot-melts are synthetic elastomers, ethylene–ethyl acrylate copolymers, amorphous polyolefins, branched polyethylenes, polypropylene, polybutene-1, phenoxy resins, polyamides, polyesters, and polyurethanes. Combinations of these resins allows for property and cost adjustments. Tackifiers and plasticizers are commonly added to hot-melts to improve their flow and adhesion to substrates. Examples include synthetic hydrocarbons, natural terpenes, rosins, and various phthalates. Polybutene is occasionally used as the base resin for hot-melts having good cold flow and high wet-out characteristics, but it may also be used as a flexibilizer or plasticizer. Waxes are important hot-melt ingredients, lowering melt viscosity and improving wet out of the substrate. Reactive tackifiers exist to address migration. The polyamide, polyester, and polyurethane hot-melts are often classed separately from the other resins on which hot-melts are based. All are the result of condensation reactions. They are frequently used with few additives and their properties are instead adjusted by changing the starting ingredients of the polymers. They may, however, contain additives that make them better suited to specific uses. Adhesives based on these polymers are considered to deliver higher performance by virtue of better high-temperature resistance and higher strength and may provide better adhesion to polar substrates than the other largely hydrocarbon hot-melt adhesives (15).

Conventional hot-melt adhesives cool to set and do not chemically cross-link. Such systems have an open time of a few seconds to a few minutes. The need for more heat-stable adhesives and stronger bond strengths has driven the development of reactive hot-melts that undergo cross-linking. These are primarily based on polyurethane hot-melts with residual isocyanate groups that react with water after application to form a thermoset adhesive material. Water is provided by the surrounding air and substrate. Cure of these hot-melts is nearly complete within 24 h, but time for full cure will depend on temperature and ambient and substrate moisture content. An extension of the water-activated isocyanate cross-linking reaction is found in the use of polyurethanes that have been silylated to provide active hydrogens for reaction with residual isocyanates in polyurethanes (16). The acceptance of reactive polyurethane hot melts has led to development of reactive block polymer and acrylate hot-melts that rely on radiation cure through activation of epoxy or vinyl groups (17); these are used primarily as PSAs.

Hot-melt adhesives are usually clear, off-white, white, or amber. Colored versions are available for nonbonding decorative use, eg, arts and crafts. Good color retention with heat aging is an important feature of a heat-stable hot-melt system, and antioxidants and heat stabilizers are common ingredients in hot-melt adhesives. Photoinitiators are frequently present when uv or other radiation curing will be used. Other useful additives include fillers and reinforcing agents. When there is some lack of cohesiveness in blends of base resins, compatibilizers may be used to improve the apparent miscibility of these resins (18). Hot-melts can be based on either amorphous or semicrystalline resins. Particularly in the case of semicrystalline resins, the rate of cooling can dramatically affect adhesion to a substrate (19). To control the development of crystallinity, nucleating agents may be added to formulations based on crystallizable polymers such as polyesters.

7. Solution Adhesives

Adhesives delivered out of solutions are typically used for joining large areas destined for nonstructural or semistructural service. The solution may be made with an organic solvent or with water or may be an aqueous dispersion. It is important that the liquid carrier have some means of escaping from the bondline in order for the proper bond strength to develop. It should be appreciated that many PSAs are made by casting out of liquids, but when put into use as components of tapes or labels, these adhesives are soft solids containing virtually no liquid.

7.1. Solvent-Based Solution Adhesives. Contact adhesives, activatable dry film adhesives, and solvent-weld adhesives make up the solvent-based adhesives. Contact adhesives are solutions of high polymers that are applied to all surfaces to be joined via spray or brush, allowed to dry partially, and then given time under pressure to allow the adhesive layers to fuse. Heat is sometimes used to increase tack or accelerate drying. These adhesives are commonly used to join wood veneers to wood bases, synthetic laminates to particleboard countertops, and paper products to other materials. The major dry film adhesive is solvent-applied natural rubber, which is unique in its ability to adhere to itself without tackification and useful for self-sealing envelopes and similar employment. After being coated on to paper or another substrate, activatable dry film adhesives must be wiped or sprayed with a liquid to regain their adhesiveness; the activating liquid now is nearly always water. Solvent-weld adhesives are used to join plastic parts such as poly(vinyl chloride) (PVC) piping. The adhesive is usually a solution of PVC or chlorinated PVC that is applied to the outer surface of the pipe and the inner surface of a connector piece that are joined firmly together before the solvent has evaporated.

The most widely used contact adhesive is a solution of polychloroprene or modified polychloroprene in solvent blends of aromatic hydrocarbons, aliphatic hydrocarbons, esters, or ketones, eg, toluene–hexane–acetone. Viscosity, dry time needed before bonding, bond strength, and price are affected by the solvent. Using various combinations of the isomeric forms of polymerized 2-chlorobutadiene permits a fine-tuning of the crystallization rates of the dissolved polymer as the solvent evaporates. The polychloroprene may also be modified by the incorporation of methacrylic acid or mercaptans. Metal oxides (MgO and ZnO) that scavenge acids are often part of polychloroprene adhesives and also may act as cross-linking agents. Oxygen scavengers such as butylated hydroxytoluene (BHT) [128-37-0] or naphthylamines [25168-10-9] are added to prevent dehydrochlorination. To build initial handling strength, the solvent-based polychloroprene contact adhesives may be modified with alkyl phenolics, terpene phenolics, or phenolic-modified rosin esters, the first of these being the most effective and least deleterious (20). Chlorinated rubbers are sometimes added to these adhesives to improve their adhesion to plasticized PVC and other plastics. Added just before adhesive application, isocyanates are useful in modification of polychloroprene contact adhesives, reacting perhaps through hydrolysis of the pendant allylic groups present from the small number of 1,2 isomeric segments (21). The remainder of the solvent-based contact adhesives are comprised of polyurethane, SBR, styrene–butadiene–styrene block polymers, butadiene–acrylonitrile rubber, natural rubber, or various acrylic or vinyl resins in suitable solvents.

7.2. Water-Based Solution Adhesives. Solution adhesives based on water dispersions and aqueous emulsions are steadily gaining in use largely at the expense of solvent-based adhesives. These are rarely true solutions, with the exception of the viscosity modifiers often used to adjust flow characteristics. Dispersions of polyurethanes in water find use in bonding of plastic sheets and films, cloth, shoe parts, foams, PVC veneers, and carpets. Other water-dispersible resins can be added to the polyurethane dispersion to lower costs and modify performance characteristics. The largest group of water-dispersed or water-dissolved adhesives are made of natural products, which are covered separately. At one time, vegetable gums were used widely as water-activatable adhesives, but poly(vinyl alcohol) (PVA) has replaced them in envelope sealing and similar areas.

Poly(vinyl acetate) (PVAC) emulsions, the basis of the ubiquitous household white glues, are among the most familiar water-based adhesives. These are widely used for paper and wood bonding. They contain a substantial percentage of vinyl alcohol content, formed via partial hydrolysis from the vinyl acetate homopolymer as vinyl alcohol itself is not a stable molecule. Such latexes are stabilized through the use of surfactants, one choice being well-hydrolyzed PVAC. After application to the substrate, latex adhesives cure by the evaporation of water accompanied by the coalescence of the latex particles. On the porous substrates with which these are most frequently used, the water exits the bondline through the substrate as well as the adhesive, preventing voiding or foaming that might weaken the bond. Subtle changes in properties can be engineered through the use of other comonomers or the use of liquid plasticizers. Glyoxal [107-22-2] or other cross-linking agents can be added to PVAC latex adhesives to combat creep (22).

Polychloroprene latex adhesives have been available for many years. They are stable at pH values between $\sim$10 and 12. The latex particles are usually lightly cross-linked. Except for the substitution of water for the organic solvent, the ingredients in these kinds of adhesives are similar to those found in their solvent-based counterparts. Terpene-phenolics are particularly effective as tackifiers for contact adhesives based on polychoroprene latexes, but rosin acids, rosin esters, hydrocarbons, and coumarone-indenes are also useful, particularly where heat-assisted bonding is not possible. Dehydrochlorination leading to acid generation is particularly possible with the water-based polychloroprene adhesives. Like other water-based adhesives, these may require addition of biocides or preservatives to prevent the breeding of microorganisms (23).

8. Structural Adhesives

Structural adhesives are designed to bond structural materials. Most any adhesive giving shear strengths in excess of ~7 MPa (~1000 psi) may be called a structural adhesive. Structural adhesives are generally the first choice when bonding metal, wood, and high-strength composites to construct a load-bearing structure. Bonds formed with structural adhesives cannot be reversed without damaging one or the other substrate. They are the only kind of adhesive that might be expected to be able to sustain a significant percentage of its initial failure load in a hot and humid or hot and dry environment. Any one of these descriptors names structural adhesives the strongest and most permanent type of adhesive. For good reason, they are sometimes referred to as engineering adhesives. The strength and permanence of structural adhesives is largely achieved using reactive adhesives, a term that has become something of a synonym for structural adhesives. Epoxies are the most widely used class of structural adhesive chemistry, but acrylates, urethanes, phenolics, and other classes have been used to great advantage, and the combination of these different chemical classes to create hybrid adhesives propagates the best virtues of each. Reactive adhesive systems that are arguably not always considered structural adhesives but are conveniently grouped here are also reviewed in this section.

8.1. Epoxy Resins. Epoxy resins have a long and distinguished record as structural adhesives. Their use dates to 1950 or earlier, and their utility for adhesives was recognized upon their development. Most epoxy adhesives are resins based on what is commonly known as the diglycidyl ether of bisphenol A (DGEBPA). These resins are based on the reaction of 4,4′-isopropylidene diphenol (bisphenol A) ($C_{15}H_{16}O_2$) [80-05-7] and epichlorohydrin (C_3H_5ClO) [106-89-8]. The molecular weight of the commercial difunctional resins formed by this reaction will vary with the molar ratio of the reactants. At a molecular weight of ~400 or less, these resins are viscous liquids that are immensely useful in epoxy adhesives. Commercially viable solid resins based on DGEBPA have molecular weights ranging up to ~4000. Many epoxy adhesives will also contain a small amount of an epoxy diluent having low viscosity and a more flexible structure; this resin adjusts the flow of the system and also helps to wet out the fillers that are usually present.

A wide variety of epoxy resins are commercially available: monofunctional or polyfunctional, aliphatic, cyclic, or aromatic. Brominated epoxies may be useful where flammability is a concern. An oxirane functionality is all that is needed to make an epoxy resin, and structural adhesives are only one of over a dozen different uses for epoxy resins. Many epoxy resins on the market will not necessarily be suitable for adhesives, but their availability does expand the choices available for adhesive formulators. The specialty epoxy resins developed specifically for adhesive use sometimes will be more costly than the DGEBPA resins but may provide the basis for a specialty adhesive that can meet a unique need and therefore command a proportionally higher price. Examples of these are epoxy-functional dimer acids, urethanes, and various elastomers.

Epoxy resins based on DGEBPA usually are quite stable at temperatures up to 200°C. Curing agents, sometimes called hardeners, must be added to the epoxy to cause cross-linking and chain extension to occur and a bond to form. Certain types of curing agents will be favored over others for each of the three types of epoxy structural adhesives: one-part (1K) epoxy paste adhesives, 2K epoxy paste adhesives, and 1K epoxy film adhesives. The strained oxirane ring is reactive with functional groups having either nucleophilic (basic) or electrophilic (acidic) character. Acid anhydrides, carboxylic groups, and hydroxyl groups react very slowly with the oxirane ring and are usually used with catalysts that accelerate their reaction with epoxies. Those groups that readily react without catalysts but often benefit from their use include amines and mercaptans. Both the epoxy resin and the curative package (curing agent plus catalyst) will influence final cure speed.

One-part (1K) paste adhesives usually consist of a DGEBPA resin, a reactive diluent, and latent curing agents that are insoluble with the resin at room temperature but dissolve at elevated temperatures to trigger cure. These kinds of adhesives are in use in the aerospace, automotive, and electronics industries. Dicyanodiamide or dicyandiamide ($C_2H_4N_4$), [461-58-5], is the most frequently mentioned latent curing agent for cures occurring in the range 170–180°C; practitioners refer to this material as *dicy*. Also useful in this range are metal-complexed imidazoles, complexes of Lewis acids (eg, boron trifluoride with amines), and diaminodiphenylsulfone. Cure temperature can be lowered by using micronized dicyanodiamide ground to a particle size of 5–15 μm. Cure can be accelerated by use of aromatic tertiary amines, imidazole derivatives, and epoxy resin adducts with tertiary and other amines. Substituted ureas such as Monuron ($C_9H_{11}ClN_2O$), [150-68-5], and nonchlorinated substituted ureas such as 3-phenyl-1,1-dimethylurea ($C_9H_{12}N_2O$), [101-42-8], have also found use as accelerators in 1K epoxy adhesives. Dihydrazides offer a range of melting points depending on structure, their cure temperatures with epoxies beginning as low as 100–110°C. Adducts of dicyanodiamide that melt at temperatures in the 115–120°C range are available. Accelerated 1K epoxies show faster cures once heated but suffer from decreased shelf lives; after manufacture, they are usually stored in refrigerators or preferably freezers although this is usually impractical for drum quantities. For these same reasons, their manufacture is carried out at temperatures well below their activation temperatures and at low shear rates to avoid viscous heating.

The low viscosity two-part (2K) epoxy adhesives sold in hardware stores as 5-min epoxies are based on cure with polymercaptans regulated with amines to control worklife. The human nose can sense some mercaptans in air at the parts per billion (ppb) level, making them valuable as gas odorants, but they are tremendously useful as curing agents, particularly when used in thin films as for adhesives. Their low toxicity is also an advantage. Capcure 3-800 [101359-87-9] is a commonly found polymercaptan. Low odor polymercaptans have been developed that combine strategies of odor masking, odor counteracting, and absorbency to stabilize polymercaptans, reducing the level of odor by ~75% (24). Higher molecular weight versions of the polymercaptans are useful as the base resins of polysulfide sealants, which are sometimes categorized as adhesives. In full formulation, the polysulfide base resins are blended with curing agents such as manganese dioxide or sodium perborate, accelerators or retarders, fillers, plasticizers, thixotropes, adhesion promoters, and pigments (25). These materials are used primarily in the construction and aerospace industries.

Many 2K epoxies utilize curing agents that are the reaction products of amines of low molecular weight with fatty acids. These are variously known as polyamidoamines, polyamides, and amidoamines and sold in a range of molecular weights under trade names such as Versamid and Ancamide. The fatty acid portion of these amines gives them larger bulk than the lower molecular weight amine curing agents, which facilitates formulation of adhesives having mix ratios closer to 1:1 by volume. This volume ratio is of benefit for both packaging and off-ratio tolerance. Curing with polyamidoamines generally produces relatively flexible adhesives having good chemical resistance. Because they typically cure slowly, they are frequently used in combination with other amines such as diethylenetriamine (DETA), triethylenetriamine (TETA), tetraethylenepentamine, aminoethylpiperazine, modified imidazolines, and oligomeric amine-terminated polyethers. Some of the amines in this group are used as sole curing agents, and others, such as DETA and TETA, are used as epoxy adducts to reduce toxicity and increase stability. Aromatic amines, although useful for epoxy resin composite matrices, find little use in epoxy adhesives.

Another family of curing agents is based on substituted phenols such as tris(dimethylamino)phenol ($C_{12}H_{21}N_3O$), [31194-38-4], and tris[(dimethylamino)methyl]phenol ($C_{15}H_{27}N_3O$), [90-72-2]. These tertiary amines can produce rather brittle adhesives if used as sole curing agents, but are valuable as accelerators for other amines. They act as catalysts for dicarboxylic acid anhydride cures. Amines are also useful as accelerators for the oxirane-alcohol reaction, which is sluggish at room temperature but with catalysis will proceed above 120°C. Imidazoles are also generally useful as catalysts or cocuring accelerators for epoxy reactions with amines, hydroxyls, and thiols. Organic and inorganic salts sometimes find use in epoxy adhesives, coatings, and encapsulating compounds. Acid catalysts such as boron trifluoride–amine complexes find some use in epoxy adhesives but tend to require long cures, even at elevated temperatures, which normally works against their use in adhesives. Epoxy resins react slowly with acid anhydride curing agents but can be accelerated with acids or bases, imidazoles being used most often, however, anhydrides are not often used as curing agents in epoxy adhesives.

Epoxy film adhesives are 1K adhesives in film form. They are formulated much like 1K paste adhesives but often contain solid epoxy resins and additional resins that provide binding properties. These may be partially cured (B-staged) to provide a more dimensionally stable film. Epoxy film adhesives have been widely used in the aerospace industry where their relative stability accommodates the long build times needed for aircraft manufacture. Their cured properties can be outstanding in terms of strength, toughness, and durability. They can be supplied in film form and cut to size or provided as tapes in convenient slit widths. They may be made to be tacky using rubber resins and other mild tackifiers or they may be dry. Film adhesives of a more aggressive pressure-sensitive character have been developed by coating or laminating with pressure-sensitive formulations or formulating such that the bulk adhesive (26) is a pressure-sensitive adhesive in its own right but can be cured to a semistructural or structural strength. Epoxy film adhesives based on thermoplastic polyamide resins are very tough when cured but can be susceptible to moisture absorption.

In addition to resins and curing agents, epoxy adhesives will contain many functional additives and modifiers. Flexibilizers and tougheners such as polysulfides, epoxidized fatty acids, epoxidized polybutadiene, and amine- and carboxy-terminated acrylonitrile butadiene polymers react with the epoxy network. Flexibilizers remain in phase with the epoxy while tougheners typically phase separate to form domains, the result producing a tougher adhesive with more or less strength reduction relative to an unmodified system. Particulate tougheners may also be added to epoxy adhesives. These include core-shell resins, functionalized elastomeric particles, and ground reclaimed rubber. Positive aspects of structural adhesives based on epoxy resins include good adhesion to many substrates, no emission of volatiles upon cure, low shrinkage, and a broad formulating range based on a history of use dating to the 1940s. The lack of outgassing allows most curing to be done at ambient pressure although clamping till cure is standard protocol for any adhesive bonding operation. Shrinkage can be further decreased with use of appropriate fillers, with harder fillers by some reports providing the lowest shrinkage.

8.2. Acrylics. Historically, acrylics offer several useful characteristics as structural adhesives. Most well known is their relatively high speed of reaction via free radical polymerization. The details of their reaction provide a useful division of the different classes of acrylic structural adhesives into redox-activated adhesives, encompassing both anaerobic acrylics and nonaerobic structural acrylics, and cyanoacrylates. These will be considered in turn.

Oxygen inhibits the polymerization of acrylic monomers to a useful extent, and its exclusion kicks off polymerization of monomeric acrylates. Early versions of anaerobic acrylics relied solely on this mode of initiation and polymerization, containing little besides acrylate monomers and diacrylic esters (27). Later it was found that if hydroperoxides were incorporated into the acrylic monomer, small amounts of free metal ions from metal substrates could help to create free radicals that initiated polymerization of the acrylate monomers. Only small amounts of metal ions are needed, iron, nickel, zinc, and copper being some of those of major industrial interest. Even though a major alloying element, such as aluminum, eg, may not be capable of helping to generate free radicals via the redox reaction, minor alloying elements, such as copper, may be available that can act in this capacity. The speed of reaction is limited by the ability of the metal ion to reduce the peroxide. Free radical initiators used in anaerobic acrylics have included cumene hydroperoxide, *tert*-butyl hydroperoxide, and potassium persulfate ($K_2S_2O_8$), [7727-21-1]. Other useful initiators for this cure are combinations of saccharin [81-07-2] with aromatic amines such as *N,N′*-diisopropyl-*p*-toluidine [24544-09-0] or 1-acetyl-2-phenylhydrazine [114-83-0]; such combinations were originally thought to be accelerators useful only with peroxide initiators until it was found that they were themselves initiators (28). Various accelerators can be used with initiators to hasten cure of these adhesives; classes of compounds useful as accelerators include cyclic peroxides, amine oxides, sulfonamides, and triazines (29).

A key ingredient in anaerobic acrylic adhesives is the acrylate monomer or monomers. These include primarily acrylic acid and methacrylic acid and their many and various esters such as lauryl acrylate, cyclohexyl methacrylate, methyl methacrylate, hydroxyalkyl methacrylates, and tetrahydrofurfuryl methacrylate. These monomers vary in their volatility, reactivity, and cost, the less volatile monomers forming the basis of low odor acrylic adhesives. In addition to the monomer acrylates, there generally is also present a diacrylate that acts as a cross-linker, the alkyl glycol dimethacrylates being widely used in this function. Other ingredients used in these adhesives include stabilizers or polymerization inhibitors such as phenols or quinones; chelating agents that snatch up trace metals to prolong shelf life; and various modifiers such as inert fillers, inorganic and polymeric thickeners, elastomers to improve toughness, and bismaleimides that improve high temperature performance (30).

The low viscosities and good wetting properties of these adhesives allow them to penetrate and flow in tight spaces, which is taken advantage of in many of their uses. Threadlocking and sealing are primary applications. When applied to the threads of bolts or pipes, to flanges, and to other tight-fitting machine parts that are later screwed into or pressed against a mating surface, the adhesive cures due to the exclusion of air and the formation of free radicals via the reaction of metal ions with the initiator. Other applications include bonding of optical fibers, impregnation of porous parts, crimp-bonding of electrical parts, and fastening of press-fit parts. Anaerobic adhesives are one-part adhesives, usually packaged in small oxygen-permeable plastic containers that have not been entirely filled, this arrangement providing a sufficient supply of polymerization-inhibiting oxygen to ensure good shelf life.

The nonaerobic structural acrylic adhesives are two-part adhesive systems. They are generally less oxygen inhibited than the anaerobic acrylics and do not rely on metal surface activation in the same way as the anaerobics. These adhesives are very similar in formulation to the anaerobics, each borrowing technology from the other as it has developed. Lower oxygen sensitivity is accomplished through higher concentrations of accelerators and initiators. The accelerators and initiators are usually redox couples such as the commonly used hydroperoxide/amine–aldehyde condensates (oxidant–reductant), which react to form alkoxy radicals. The most widely used condensate is that resulting from reaction of *n*-butyraldehyde [123-72-8] with aniline [62-53-3] in the form of a polymeric resin that is available commercially [9003-37-6]. This material has a complex structure, the major component and active ingredient apparently being dihydropyridine [27790-75-6] (31). Another common redox couple is based on hydroperoxide coupled with an alkyl aromatic amine such as *N,N*-dimethylaniline [121-69-7]. A number of 2K acrylic formulations include metals, metal oxides, or metal salts (32).

The 2K nonaerobic acrylic adhesives can be used in any of three ways. The first is as a no-mix two-part, the use of which involves applying a thin layer of accelerator (in dilute solution) to one mating surface, flashing off the solvent, applying the adhesive to the second mating surface, and joining the two surfaces. It is perhaps a poor choice of terms, but the accelerator contains the initiator (eg, peroxide) or may contain a redox couple. As long as the bondline thickness is no more than ~500 μm (0.020 in.) for one-side activation or ~1000 μm for two-side activation, cure is expected to be adequate. The 2K acrylics that are meant to be mixed utilize a different kind of accelerator that contains the catalyst system in a carrier resin such as an epoxy and perhaps a diluent. These can be used in a fashion similar to the no-mix adhesives, but this approach may not produce optimal properties. Typically, the 2K acrylics are made by mixing the accelerator into the one-part acrylics and immediately applying this mixture to the substrate. Volume mix ratios will range from ~2:1 to ~20:1. Additional ingredients commonly found in these compositions include various elastomeric polymeric tougheners such as chlorosulfonated polyethylene, butadiene–acrylonitrile elastomers, and polyurethane acrylates. These tougheners are usually incorporated into the adhesives by dissolution in the acrylic monomers, creating adhesives sometimes referred to as second-generation acrylics. Their development by DuPont (33) and others marked the entry of acrylic structural adhesives into a large number of new applications.

Due to their high reactivity, these 2K acrylic adhesives are used in many situations where fast ambient cure is important. Since the incorporation of the redox couple catalysts, acrylic adhesives have advanced their use on metals as well as plastics, woods, and ceramic substrates. As a class, they tend to be fairly accommodating of oily metal and unprepared plastics and composites. Offensive odors often accompany the common forms that use the less expensive lower alkyl acrylates. Colors of these materials are clear, off-white, white, and amber. They are not often intentionally pigmented although they may be tinted with functional metal additives or aluminum powders.

A very important class of acrylic adhesives, the cyanoacrylates, are distinguished by their relative simplicity of formulation and their nearly instant bonding properties. The name recognition of "super glue" surpasses that of nearly any commercial adhesive though it is now known by a variety of other ungenericized trademarks. First discovered in the 1940s during World War II, cyanoacrylates were rediscovered and first truly appreciated in the 1950s and brought to the market in 1958. Then as now they are largely based on ethyl and methyl cyanoacrylate. Other monomers of interest have been the isopropyl, butyl, allyl, ethoxyethyl, methoxyethyl, methoxypropyl, and fluoroalkyl esters.

Cyanoacrylate adhesives cure by polymerizing anionically. They are catalyzed by mild nucleophiles (bases), such as an OH^- ion, which can readily be found in small quantities on many surfaces. Strong acids, found in many woods and acid-treated metals, can inhibit polymerization. As long as the adhesive film thickness is as low as possible, that is, practically zero, sufficient catalyst provided by the substrate will be available, hence the usual directive to apply the adhesive sparingly and to avoid using it as a void filler or to bond porous surfaces. Bond thicknesses higher than ~13 μm (0.005 in.) are not recommended unless appropriate surface activators are used. As the conversion to a cured adhesive is a polymerization, it passes through and is subject to the same stages as any addition polymerization: initiation, propagation, chain transfer, and chain termination. Like the anaerobic adhesives, these adhesives are conveniently initiated by coating onto surfaces suitable initiators such as alcohols, epoxides, various amines, caffeine, and other heterocyclic compounds (34). Compositions may also incorporate accelerators as well as inhibitors, the latter usually being either phenolics designed to inhibit premature polymerization due to heat or light or anionic polymerization inhibitors consisting of sulfur dioxide, other acid gases, or complexes of sulfur dioxide with organic or inorganic compounds. Normally quite brittle, cyanoacrylate adhesives can be flexibilized using monomers having longer alkyl side chains (2-octyl cyanoacrylate) or by incorporating plasticizers such as acetyl tri-butyl citrate (35). Various approaches have been taken to toughening the cyanoacrylates (36). As uncross-linked thermoplastic adhesives, the cyanoacrylates begin to soften and flow at $\sim$80°C and will also depolymerize. Their durability in hot moist environments is considered to be poor, especially on metals. This problem has been addressed through introduction of difunctional or bifunctional cross-linkers, addition of heat-resistant adhesion promoters, and various other strategies aimed at improving moisture resistance. The last important component of the cyanoacrylate adhesive is the thickener, which is usually polymeric in nature.

Cyanoacrylates have long been known to be effective adhesives for human skin and other soft human tissues. They are effective when used for sutureless wound closures and hemorrhage prevention, the butyl cyanoacrylate being most widely used (37) based on a good balance between biodegradability and inflammatory response. Flexibilizers as well as aids to biodegradation are added to make these more suitable for tissue bonding. In everyday use, the outstanding capability of cyanoacrylate adhesives to instantly bond human skin is seen as a negative feature. Skin-adhesion inhibitors that have been found useful include alkanols, carboxylic acid esters (38), and copolymers of maleic acid, vinyl chloride, and vinyl acetate (39). These slow the adhesive's reaction rates against human skin or at least lower adhesion to it.

8.3. Urethanes. The core of a urethane adhesive is an isocyanate compound. Isocyanates react with a variety of functional groups having active hydrogens to generate a variety of linkages that give the resulting polymers their names. These include reaction with alcohols to form urethanes [R–NH–CO–O–R′], with amines to form ureas [R–NH–CO–NH–R′], with thiols to form thiocarbamates [R–NH–CO–S–R′], with amides to form acylureas [R–NH–CO–N(R′)–CO–R″], with urethanes to form allophanates [R–NH–CO–N(R′)–CO–O–R″], and with ureas to form biurets [R–NH–CO–N(R′)–CO–NH–R″]. Isocyanates can also react with water, generating carbon dioxide through the degradation of the unstable carbamic acid [R-NH-COOH]. This last reaction is the basis for the making of polyurethane (PU) foams. To a great extent, what is classified as urethane chemistry encompasses the entire chemistry available to isocyanates.

Most polyurethane structural adhesives are two-part systems based on the reactions of isocyanates and polyisocyanates with oligomers or polymers having at least two hydroxyl groups, which are generically referred to as diols or polyols. Although part of many earlier adhesive formulations, toluene diisocyanate (TDI) is now decreasing in use while use of diphenylmethane diisocyanate (MDI) is growing. Other common diisocyanates include 1,6-hexamethylene diisocyanate (HMDI or HDI) and isophorone diisocyanate (IPDI). Also available are the modified MDIs, multifunctional isocyanates often termed polyisocyanates, polymeric polyisocyanates, and isocyanate-capped oligomers that are often referred to as urethane prepolymers (40). Materials now available that have very low monomeric isocyanate content are expected to bring about increased use of urethanes in adhesives (41). Hydroxyl-functional materials useful in urethane adhesives have molecular weights between ~500 and 3000 and functionalities between 2 and 3. The base oligomer is usually a polyester, polyether, polycarbonate, or polydiene such as polybutadiene. Cross-linked polyurethanes can be made with the use of trifunctional isocyanates and triols or through reactions of urethanes with urethanes, ureas, or isocyanates to yield the trimer isocyanurate.

In many cases, as polyurethanes are formed, long- and short-chain diols alternate along the chain to form segments that are either "soft" or "hard". On a microscope scale, the soft and hard segments coexist in a domain morphology characteristic of what are known as segmented polyurethanes. The very good impact and fatigue resistance of PUs is attributed to this phase-separated microstructure. Because it is the integral component of the soft segment, the particular diol or polyol chosen will greatly influence the rubbery and impact-resistance properties of the PU. Likewise, the isocyanate chosen will strongly influence the strength, modulus, and hardness of the PU. The domain morphology of segmented PUs is most pronounced for systems containing no chemical cross-linking. In contrast to most adhesive systems, low levels of cross-linking tend to degrade the properties of PU adhesives because of disruption of the domain morphology.

Because isocyanates react with so many different organic functional groups and can also react with water, which is found nearly everywhere, catalysts are very important for the control of isocyanate reactions. Many of the catalysts used may push one reaction over another, but they do not necessarily entirely block unwanted reactions. Tertiary amines, principally bis(dimethylaminoethyl)ether, are frequently used to promote the isocyanate–water reaction, producing a blowing or foaming that generally would not be desirable for adhesives. Compounds that drive the isocyanate–hydroxyl action without substantially encouraging the isocyanate–water reaction include organometallic complexes such as dibutyltin dilaurate and stannous octoate. At temperatures >100°C, urethanes and ureas will react with isocyanates to form the allophanates and biurets described previously, but above 130°C, these groups will decompose. Dimerization of isocyanates to form uretidiones is catalyzed by bases such as trialkylphosphines, pyridines, and tertiary amines. Formation of the trimer of isocyanates, isocyanurate, is favored through use of phosphines, amines, and various metal salts such as potassium acetate.

One-part urethane adhesives have been used for many years as high-performance sealants. In this capacity, they provide a useful combination of strength, flexibility, and elastic recovery. As adhesives, these systems have limited use unless formulated to overcome their inherent disadvantages. One-part polyurethane adhesives are typically moisture cured and rely on a multistep reaction sequence as follows: isocyanate reacts with water to form carbamic acid, the unstable carbamic acid loses carbon dioxide and generates an amine, the amine reacts with additional isocyanate to form a urea, and the urea reacts with additional isocyanate to form a biuret, which includes a cross-link. Unless it diffuses out of the system, the CO_2 can cause foaming. Formulators learn to minimize the isocyanate content (%NCO) of a system in order to balance cure speed with foam control. Cure speeds—and foaming rates—of these systems decrease from the outside in and vary with the amount of atmospheric moisture in the air, which changes hourly and seasonally.

A different kind of moisture-activated 1K urethane adhesive utilizes a moisture-activated curing agent such as oxazolidine (42). Oxazolidines are formed by dehydration and subsequent ring closure of aminoalcohols by aldehydes or ketones. When the presence of water causes that reaction to reverse, hydroxyl and amine groups are formed. These react readily and directly with isocyanates. Monooxazolidines are useful primarily as water scavengers, but bis (oxazolidines) can participate in the curing reactions of urethane adhesives.

More sophisticated 1K urethane adhesives use blocked isocyanates along with polyol curing agents. Useful blocking compounds include phenols, malonates, methylethylketoxime, and caprolactam. These react with isocyanates, but at high temperatures or in the presence of strong nucleophiles, the reaction reverses, freeing the isocyanate. Such systems do not rely on water for reaction, nor do they suffer from the detriments of CO_2 generation, but they do require heat for cure. Another approach to a stable 1K urethane is to use a solid polyol, such as pentaerythritol, that melts at elevated temperatures and then reacts with the isocyanate (43). Other schemes for 1K urethanes are described by Edwards (44).

As a class, urethane adhesives have somewhat poorer thermooxidative and moisture resistance than acrylic and epoxy structural adhesives. This finding has historically limited their expansion into certain areas of use. A 2K adhesive having the ability to survive automotive paint oven temperatures, which run as high as 205°C, uses polyols with high percentages of hydroxyl groups, an acrylonitrile-grafted triol, a phosphorus adhesion promoter, and a DABCO trimerization catalyst (45). The 1K adhesives made with blocked isocyanates tend to be unable to withstand high temperatures due to volatility of the blocking agents, and other approaches are also unsatisfactory for high-temperature stability. Use of micronized dicyanodiamide as a latent catalyst and curing agent for isocyanates has produced 1K urethane adhesives showing some capability to tolerate heating to well over 250°C while bonding well to FRP (46). Sensitivity to hydrolysis has been another of the historic disadvantages of traditional urethane structural adhesives.

Two-part polyurethane adhesives will usually contain fillers and may contain pigments that facilitate visual qualitative off-ratio mixing detection. To increase cure speed, polyamines are sometimes added to the polyol curative, which also contains the catalysts. In addition to their primary ingredients, one-part moisture-curing urethane adhesives will typically contain fillers and perhaps pigments. Arguably the largest user of urethane structural adhesives is the transportation industry, which uses urethane structural adhesives for bonding of automotive parts made of SMC, FRP, and RRIM composites and plastics. One-part urethanes are widely used for bonding of windshields to automotive vehicle frames. Though 1K urethanes are not conventionally considered to be structural in nature, automotive engineers hold that the windshield is part of the primary structure of the vehicle, conferring on these one-part urethanes the status of a structural adhesive. Wood bonding is another significant market for polyurethane structural adhesives.

As a group, polyurethane structural adhesives produce bond strengths on the lower end of the strength scale for structural adhesives, but their high flexibility, usually strong peel strength, and generally good impact and fatigue resistance recommend their use when these characteristics are important. A variety of adhesives have been developed that incorporate polyurethanes into acrylic or epoxy structural adhesives (47). Inclusion is done through use of isocyanate-functional ingredients or polyurethanes end-capped with a non-isocyanato functional group. The broad reactivity of isocyanates offers many other options for hybridization.

8.4. Phenolics. Phenolic resins were the basis of the first synthetic structural adhesives. They are formed by the reaction of phenol, (C_6H_6O) [108-95-2], and formaldehyde, (CH_2O) [50-00-0]. There are two types of phenolic resins, resoles and novolaks (or novolacs), the former being comprised of methylol-terminated resins and the latter, of phenol-terminated resins. Resoles result from use of basic reaction conditions and an excess of formaldehyde and will cure via self-condensation at 100–200°C with loss of water. Novolaks are produced using acidic reaction conditions and formaldehyde/phenol molar ratios of 0.5–0.8. They require addition of a curing agent for cure. Hexamethylenetetramine ($C_6H_{12}N_4$), [100-97-0], is a widely used novolak curing agent. Resoles and novolaks are sometimes referred to as one- and two-step resins, respectively.

Formulators can choose from a variety of commercially available phenolic compounds, including, in addition to phenol itself, the isomers of cresol, the isomers of xylenol, resorcinol, catechol, hydroquinone, bisphenol A, and various alkylphenols. Formaldeyhde is usually used as the second major component, but acetaldehyde, furfuraldehyde, and paraformaldehyde (the polymer of formaldehyde) have been used sometimes alone and sometimes along with formaldehyde. The reactions of these various components are complex but have been elucidated by painstaking research described by Robins (48) and others.

Like epoxies, the phenolics are very brittle unless modified by tougheners. The first successful tougheners were poly(vinyl formal) resins that were added as a powder sprinkled over a layer of resole phenolic applied out of solution. These Redux adhesives were the first toughened thermoset adhesives and were the basis of the first durable adhesive bonding technology for aerospace aluminum in the 1940s and 1950s. These were superseded in the 1960s by film adhesives formed from liquid phenolics filled with poly(vinyl formal) powders. Other tougheners followed: poly(vinyl butyral), nitrile rubbers, polyamides, acrylics, neoprenes, and urethanes. Epoxy–phenolics are important hybrid adhesives and offer an immensely useful combination of strength, toughness, durability, and heat resistance. Phenolic structural adhesives as a class of materials are highly resistant to most chemicals.

Phenolic adhesives are found as powders, liquids, pastes, and supported and unsupported films. Among the pastes, both 1K and 2K systems are available. Fillers are commonly used in paste adhesives. Support of film adhesives is provided by glass, cotton fabric, nylon, or polyester scrims. The novolaks are almost exclusively powders in pure form, but the resoles often are found as liquids. The resole systems are usually cured at temperatures exceeding 170°C. The condensation cure of the resole phenolics systems requires that they be cured under high pressures to minimize evolution of bubbles from water vapor, which is usually done in autoclaves or hot presses at pressures of ~200 to nearly 1400 kPa (29–203 psi) (49). Cure times range from 1 to 4 h depending on temperature. The cure conditions required for the resole phenolic adhesives have limited their use, and to a great extent they as well as the relatively brittle novolak phenolics have been displaced by epoxies for aerospace aluminum bonding applications for which they were once the first choice. Nitrile–phenolic adhesives have a long history of use not only in aerospace applications but also in automotive applications such as the bonding of brake linings and the friction materials used in transmissions. Resole phenolic resin adhesives are widely used in the making of plywood and particle-board as both binders and for laminating of veneers; resorcinol is frequently used along with phenol or as the sole hydroxyl compound. In wood bonding, the porosity of the wood allows escape of the water vapor generated during curing of the adhesive and is believed to facilitate mechanical anchoring of the adhesive in the wood. Phenolics are also widely used as foundry resins for making sand-shell molds.

8.5. Urea–Formaldehyde and Related Adhesives. Urea–formaldehydes (UF) are the most significant members of the class of materials known as the amino resins or aminopolymers. These are the polymeric condensation products of the reaction of aldehydes with amines or amides. A molar excess of formaldehyde is used, and this along with the temperature and the pH dictate the properties of the final product. The initial reactions of urea and formaldehyde to form mono- and dimethylolureas can be catalyzed by either acids or bases, but the final condensation reactions will proceed only under acid conditions. These adhesives are widely used to make plywood and particleboard in processes utilizing heated hydraulic presses with multiple outlets for water vapor release. Temperatures up to 200°C may be used. The UF adhesives in the first use contain hardeners composed of ammonium chloride or ammonium sulfate solutions or mixtures of urea and ammonium chloride plus fillers such as grain and wood flours. Particleboard adhesives, which are really binders, contain similar hardening agents, a worklife extender (ammonia solution), insecticides, wax emulsions, and fire-retarders. The slow hydrolysis of the methylenebisurea ($NH_2CONHCH_2NHCONH_2$), [13547-17-6], has been linked to the slow release of formaldehyde from UF adhesives (50). The wood industry has been under increasing pressure to reduce and eliminate unreacted and evolved formaldehyde from these products and has made great efforts to do so. Melamine–formaldehyde (MF) and the less expensive melamine–urea–formaldehyde (MUF) resins are the bases of high performing wood-bonding adhesives. Their resistance to water is superior to that of the UF resins, but their higher cost has limited their use. The urea in the MUF resins decreases the cost of the MF resins. Uses of these are similar to those for the UF resins with the addition of paper-laminates for wood panels. Melamine reacts more easily with formaldehyde than does urea, making possible full methylolation of melamine (51). Condensation of methylolated melamine with formaldehyde does occur under both acidic and slightly alkaline conditions, but acid catalysts or compounds generating acids are usually used in MF adhesives. Compounds such as acetoguanamine, ε-caprolactame, and *p*-toluenesulfonamide are often added to combat inherent brittleness and decrease stiffness. Ammonium salts are useful in making bulk wood products, but laminates can be adversely affected by these compounds; a complex of morpholine and *p*-toluenesulfonic acid is one hardener employed for this particular kind of MUF or MF adhesive. Defoamers and judicious amounts of release or wetting agents may also be used.

8.6. High Performance Adhesives. A number of adhesive needs exist that require resistance to very high temperatures and other environmental stressors such as certain gases, solvents, radiation, and mechanical loads. The upper temperature limits of the most durable epoxy and phenolic adhesives lie between ~200 and 250°C. The aerospace industry requires adhesives that are resistant to temperatures of nearly 400°C for hundreds of hours or ~150°C for much longer times. Heterocyclic polymers such as polyimides and polyquinoxalines have been the basis of most heat-resistant adhesives. Microelectronics adhesives sometimes also must deal with high heat, but they must also conduct heat away from heat-sensitive parts. This has been the inevitable result of increasing miniaturization. Epoxies continue to be the basis of many microelectronics adhesives, but adhesives based on stiff-chained thermoplastic resins such as polyethersulfone and polyetheretherketone have made some inroads. Electrical conductivity is most commonly enhanced with silver flake or powder, but nickel, copper, and metal-coated metals are also being used in this function (52). Thermal conductivity is usually adjusted through incorporation of aluminum, aluminum nitride, or other metals or ceramics (53).

9. Adhesives Made from Natural Products

The first adhesives developed by humans were based on naturally available materials such as bone, blood, milk, minerals, and vegetable matter. Beginning with the commercial development of Baekeland's phenolic resin adhesives by the General Bakelite Co. around 1910, synthetic adhesives began to replace natural product adhesives for existing applications. The use of adhesives by industry began to grow and diversify over the ensuing decades. In certain industries, among them furniture, food, bookbinding, and textiles, adhesives based on natural products continue to be used to a significant extent. These adhesives can be divided into those based on proteins, carbohydrates, and natural rubbers or oils. Historically, glue is a term used to refer to adhesives made from animal matter or vegetable-based protein.

9.1. Protein-Based Adhesives. The protein sources for these adhesives include mammals, fish, milk, soybeans, and blood. Animal and fish parts that yield useful proteins include hides, skins, bones, and collagen from cartilage and connective tissues. Most animal proteins are extracted using water and vary considerably in molecular weight, amino acid sequence, and inorganic impurities. For those proteins that are not already soluble in water, such as collagen, solubilization is accomplished by imposition of heat, pressure, or, most commonly, addition of acids or alkalis. Final molecular weights are in the range of

10,000–250,000 (54). Following solubilization, the protein solution is boiled down and dried to a final moisture content of 10–15%. Milk and cheese yield the relatively simple mixture of proteins called casein [9000-71-9]. Proteins are extracted from milk through direct acidification following decreaming and may also be generated through fermentation of lactose by bacteria to create lactic acid. Blood is almost entirely made up of proteins and after spray drying to remove water can be stored for an extended period of time. Soybeans are important sources of both proteins and triglyceride oils. Proteins for adhesives are obtained from harvested soybeans by extracting or pressing out oils and then heating the remaining matter no higher than 70°C lest its alkaline solubility be compromised. Soybean meal is ~45–55% protein, the balance consisting of carbohydrates (~30%) and ash (55).

Proteins are highly susceptible to changes in their structure through changes in pH, and the process of denaturation used when necessary to unfold protein molecules and break down their molecular weight to effect solubilization must only go far enough to obtain those effects but not deteriorate their adhesive qualities. Additional acids and bases are used in preparation of working adhesives made from proteins. Formulations of protein-based adhesives, in general, include the dried protein, water, an alkali compound that helps dispersion, and a hydrocarbon oil defoamer. Hydrated lime and sodium silicate solutions are usually added to modulate viscosity and to improve water resistance. Plasticizers are sometimes added as are fillers, biocides, preservatives, and fungicides. Protein-based adhesives are widely used for bonding of porous substrates such as wood, and as water is removed from the adhesive by absorption, air drying, and the optional application of heat, the proteins become fully denatured and the adhesive is set. A variety of denaturing and curing agents or cross-linkers can be used with protein-based adhesives, including hexamethylenetetramine, carbon disulfide [75-15-0], thiourea [62-56-6], dimethylolurea, and various metal salts. Blood glues may contain aldehydes and alkaline phenol–formaldehydes as cross-linkers. Although very strong, protein-based adhesives have been largely restricted to nonstructural interior wood bonding applications and other uses where their susceptibility to water and moisture do not jeopardize their stability, and the use of the various cross-linkers is targeted primarily at improving their water resistance. The most water-resistant protein-based adhesives are the blood or blood–soybean blends, but even they are not fully weatherproof. Casein or casein–soybean blends are next in line, and soybean and animal hide glues exhibit the least water resistance. The use of blood and casein adhesives is limited by the low yield of adhesive-grade dried blood from drying processes and the lack of appreciable suppliers of casein in the United States alongside a large number of diverse global sources. There has been a strong push from the soybean industry to have soy products more widely accepted in various industrial uses, but considerable work remains to be done in this area. Protein-based fibrin sealants have been the subject of considerable interest as medical adhesives and are considered by some to have many advantages when compared to cyanoacrylates and other types of adhesives (56), but their development has been limited due to human blood contamination issues.

9.2. Carbohydrate-Based Adhesives. Carbohydrates are available from a wide variety of plants, the shells of marine crustaceans, and bacteria.

The raw adhesive materials obtained from these sources include cellulose, starch, and gum. Cellulose [9004-34-6] is a semicrystalline polymeric form of glucose having a molecular weight of <1000 to nearly 30,000. It is present in plant matter at a level between ~ 30 and 90%. Like some of the naturally occurring proteins, cellulose must be chemically treated in order to be used as an adhesive. Reaction of its hydroxyl groups is used to convert cellulose to cellulose esters and ethers. Important cellulose esters include cellulose nitrate, cellulose propionate, cellulose butyrate, cellulose acetate propionate, and cellulose acetate butyrate (57). The most important cellulose ethers include carboxymethylcellulose (CMC), ethylcellulose, methylcellulose, and hydroxyethylcellulose. The cellulose adhesives are film formers having a thermoplastic nature. A typical adhesive formulation includes a few percent of the cellulose, less than a percent each of a plasticizer and a natural protein, and the great balance of water or another solvent. Methylcellulose is the basis of a common nonstaining waterbased wallpaper adhesive. Celluloses are very effective aqueous solution thickeners and are sometimes used in that capacity, so their solubility is limited by viscosity increases. Starches are the most significant class of carbohydrate adhesives. The source of the basic materials is broad and includes corn, wheat, rice, and potatoes as well as seeds, fruits, and roots from which starch is isolated by hot water leaching. Starch is a naturally occurring polymer of glucose. It occurs for the most part in either of two forms or something intermediate between the forms: amylose [9005-82-7], which is highly linear and has a degree of polymerization of 500–700, and amylopectin [9037-22-3], which is branched and has a degree of polymerization of $\sim 1500–2000$. Starch is also semicrystalline in nature. Its tightly packed granules must be opened to make it suitable for adhesive use, which is accomplished through heating, oxidation, or alkali or acid treatment. Colloidal suspensions of starches can be made by heating in water, but these have a tendency to solidify on cooling. Treatment with an alkali such as sodium hydroxide can lower the gelation temperature. Treatment with a mineral acid plus heat followed by neutralization with a base degrades the amorphous regions of the starch granule but does not disturb the crystalline regions, allowing a higher percentage of solids to be used in making an aqueous solution called a thin-boiling starch. Oxidation with alkaline hypochlorite produces a material similar to acid-treated starch but having better tack and adhesive properties. Dry roasting of starch in the presence of an acid catalyst produces dextrin [9004-53-9], which ranges in color from white to yellow to dark brown and shows different tendencies to repolymerize depending on the temperatures, times, and catalyst concentrations used. Additives used in dextrin adhesives include tackifiers such as borax [1303-96-4], viscosity stabilizers, fillers, plasticizers, defoamers, and preservatives. Formaldehyde precondensates and other compounds are added to improve water resistance. Starch-based adhesives are used in corrugated cardboard, paper bags, paper or paperboard laminates, carton sealing, tube winding, and remoistenable adhesives. Gums are naturally occurring polysaccharides obtained from various plants or microorganisms and usually prepared as adhesives by dispersion in either hot or cold water. Although they find use in applications similar to those mentioned for starches, they are more often found as additives in synthetic adhesives in which they act as rheology modifiers.

9.3. Other Nature-Based Adhesives. The use of natural rubber, an important adhesive component obtained from the rubber tree, is discussed in the section on Pressure-Sensitive Adhesives. Tannins are polymeric polyphenols isolated as one of two products from the bark of conifers and deciduous trees. Lignin is widely available as a waste material from pulp mills and has a complex structure. Tannin-based adhesives have attained some level of success in the marketplace. Despite considerable interest in and work toward more commercial use of lignins in adhesives for wood bonding, they have not yet succeeded in capturing market share. A vinyl-functionalized sugar has been developed for use in products including, most prominently, adhesives (58). Modification of sugars to make liquid epoxy resins has also been accomplished (59). Use of whey and whey by-products as adhesive components has been investigated (60). Modification of natural materials to make polyols and diisocyanates has been pursued in both the United States and the United Kingdom (61). It can be expected that additional plant-based monomers and polymers will be developed as the chemical industry comes to terms with the limited supply and rising costs of petrochemicals, making "green adhesives" a not-uncommon reality in the not-too-distant future (62).

10. Direct Bonding

Strictly speaking, direct bonding does not include the use of conventional adhesives or seemingly any adhesive at all. However, the joining of two extremely smooth solid surfaces into a spontaneous bond requires careful preparation and surface treatment that reflect the sophisticated use of chemistry, physics, and engineering. Practitioners of direct bonding consider its gluelessness to be a considerable benefit within its primary areas of applications, optics, electronics, and semiconductors, which benefit from minimal or no contamination (63). Such bonds are also considered jointless due to the atomic distances between the joined surfaces. The most prominent use of direct bonding may be wafer bonding, a key part of the silicon-on-insulator technology behind the making of integrated circuits, that is, computer chips (64). Direct bonding also is used in construction of waveguides for optical devices.

The inclusion of direct bonding among a list of adhesive types reflects the supposition that conventional adhesives of any composition are useful because they compensate for the shortcomings of most surfaces one might wish to join. Indeed, if smooth enough, even polytetrafluoroethylene will adhere to itself. In the case of what is called stiction, direct bonding is not seen as desirable, and steps are taken to prevent it from occurring (65). Redesign can be used to avoid material contact altogether. Surfaces can be roughened on a fine scale using chemical treatments.

11. Adhesive Formulation and Design

A 1999 compilation of chemicals used in adhesives listed 6300 materials (66), but the total number of compounds available for adhesive formulating is well in excess of this figure. Formulators of adhesives are in constant search of unique adhesive ingredients and their unusual combinations in order to satisfy the ever-increasing needs of their customers. In the interests of competition, many vendors of adhesive raw materials continue to protect the proprietary nature of their products by providing coded product names, a practice that though entirely understandable runs contrary to the need for the educated formulator to know the chemistry and structure of raw materials rather than relying on vague descriptions of the effects of a raw material in some standard formulation on some standard substrate.

Formulating adhesives is both a skill and an art. The novice formulator will find it invaluable to seek out other formulators in the same organization and learn from them as much as possible or at least whatever their time and patience allow. Maintaining such relationships over time can provide great benefit to the beginner as well as the veteran formulator, who will soon start learning from the former novice. The written and electronic literature of many vendors of adhesive raw materials includes information on formulating, including starting formulations. To the extent possible, one can also consult with vendor technical staff. The open technical literature, encompassing technical and trade journals, conference proceedings, and patents, provides considerable information on formulations, and its age should not discourage one from reading it as there is much to be learned from the older literature. The literature on nonadhesive polymer-based products, such as coatings, molding plastics, and composite matrix materials, may prove helpful in describing interesting raw materials not commonly used in adhesives. Likewise, components commonly used in one class of adhesives may be found to be useful in modifying adhesives of another class. The best teacher of formulating is experience, that is, trial and error.

There is more to adhesive formulation than the combining of various raw materials. The formulator must be a multidimensional technical professional able to juggle several different fields of science and engineering, legal issues, environmental considerations, computer hardware and software, and business concerns. It is not unusual to create a remarkable adhesive only to find that a key ingredient is unstable or too expensive for the intended market or poses unacceptable health and safety risks. Some customers have lists of ingredients that will not be allowed in items sold to them. Government entities require increasingly stricter labeling of adhesives and other chemical products, the requirements varying from country to country.

Better tools for adhesive formulation have been developed with the onset of the personal computer and computer workstations. These include software for design of experiments, databases used to track endless variations in adhesive recipes, mixtures design software for faster product optimization, and simple and complex spreadsheets used to determine cost at the front end of development. On-line searching of and access to the scientific and patent literature as well as the information on business trends and supplier's products available on the Internet have made information gathering easier. Adhesive development accelerates more each year, and the savvy formulator must keep pace.

BIBLIOGRAPHY

"Adhesives" in *ECT* 1st ed., Vol. 1, pp. 191–206, by V.N. Morris, C.L. Weidner, and N. St. Landau, Industrial Tape Corporation; "Adhesives" in *ECT* 1st ed., Suppl., pp. 18–32, by R.F. Blomquist, Forest Products Laboratory, Forest Service, U.S. Department of Agriculture; "Adhesives" in *ECT* 2nd ed., Vol. 1, pp. 371–405, by R.F. Blomquist, Forest Products Laboratory, Forest Service, U.S.D.A.; "Adhesion" in *ECT* 2nd ed., Suppl. Vol., pp. 16–27, by H. Schonhorn, Bell Telephone Laboratories; "Adhesives" in *ECT* 3rd ed., Vol. 1, pp. 488–508, by Fred A. Keimel, Bell Telephone Laboratories; "Adhesives" in *ECT* 4th ed., Vol. 1, pp. 445–466, by Alphonsus V. Pocius, The 3M Company; "Adhesives" in *ECT* (online), posting date: December 4, 2000, by Alphonsus V. Pocius, The 3M Company.

CITED PUBLICATIONS

1. *The Global Adhesive and Sealant Industry: An Executive Market Trend Analysis*, 2nd ed., CHEM Research GmbH and DPNA International, Frankfurt, Germany, 1997, p. 12.
2. *U. S. Industry* & Trade Outlook 2000, McGraw-Hill, New York, p. 11–11.
3. *Adhesives Age* **42**(3), 9 (1999).
4. *Adhesives, VII*, Skeist Incorporated, Whippany, N.J., 2000, p. 24.
5. J. Talmage, *Adhesives Sealants Ind.*, **7**(10), 20 (Dec. 2000/Jan. 2001).
6. L. H. Lee, L. Shi-Duo, and W. Zhi-Lu, *Proc. Annu. Meeting. Adhes. Soc.*, 371 (1996).
7. H. Onusseit, *Adhesives Sealants Ind.* **7**(7), 24 (2000).
8. *SBI Market Profile: Adhesives and Sealants*, FIND/SVP. Inc., New York, 1998, p. 18.
9. A. L. Lambuth, in R. W. Hemingway, A. H. Conner, S. J. Branham, Eds., *Adhesives from Renewable Resources, ACS Symp. Ser., 385*, American Chemical Society, Washington, D.C., 1989, Chap. 1, pp. 1–10.
10. C. W. Paul, M. L. Sharak, and M. Blumenthal, *Adhesives Age* **42**(7), 34 (1999).
11. I. Benedek and L. J. Heymans, *Pressure-Sensitive Adhesives Technology*, Marcel Dekker Inc., New York, 1997, pp. 142–145. Chap. 5 of this work is a good general reference on PSA compositions.
12. Jap. Pat. 2,000,256,639 A2 (Sept. 9, 2000), N. Watanabe, J. Nakamura, and Y. Mashimo, (to Toyo Ink Mfg. Co., Ltd.); U. S. Pat. 5,130,375 (July 14, 1992), M. M. Bernard and S. S. Plamthottam (to Avery Dennison Corp.).
13. S. B. Lin, in M. R. Tant, J. W. Connell, and H. L. N. McManus, eds., *High-Temperature Properties and Applications of Polymeric Materials*, ACS Symposium. Series, 603, American Chemical Society, Washington , D.C., 1995, Chap. 3, pp. 37–51.
14. A. J. DeFusco, K. C. Sehgal, and D. R. Bassett, in J. M. Asua, ed., *Polymeric Dispersions: Principles and Applications*, Kluwer Academic Publishers, the Netherlands, 1997, pp. 379–396.
15. A. Hardy, in *Synthetic Adhesives and Sealants, Critical Reports on Applied Chemistry*, Vol. 16, John Wiley & Sons, Inc., Chichester, England, 1987, pp. 31–58.
16. M. Huang, R. R. Johnston, P. Lehmann, N. Stasiak, and B. A. Waldman, *Adhesive Technology* **15**(2), 20 (1998); H. Mack, *Adhesives Age* **43**(8), 28 (2000).
17. For example, M. Dupont, *J. Adhesive Sealant Council*, 229 (1997); Jap. Pat. 1,114,041,0 A2 (May 25, 1999), Y. Nagai and Y. Ikegami, (to Mistubishi Rayon Co., Ltd.).
18. J. Piglowski, M. Trelinska-Wlazlak, and B. Paszak, *J. Macromol. Sci.—Phys.* **B38** (5–6), 515 (1999).
19. S. Ghosh, D. Khastgir, and A. K. Bhowmick, *J. Adhes. Sci. Technol.* **14**(4), 529 (2000).

20. R. S. Whitehouse, in W. C. Wake, ed., *Synthetic Adhesives and Sealants, Critical Reports on Applied Chemistry*, Vol. 16, John Wiley & Sons Inc., Chichester, England, 1987, pp.1–30.
21. J. Comyn, *Adhesion Science*, The Royal Society of Chemistry, Cambridge, England, 1997, p.56.
22. *Ibid.*, p. 62.
23. P. L. Wood, *Adhesive Technol.*, **15**(2), 8 (1998).
24. C. Frihart, A. Natesh, and U. Nagorny, *Adhesives Sealants Indu.* **8**(1), 26 (2001).
25. E. A. Peterson, and A. D. Yazujian, *Adhesives Age* **30**(6), 6 (1987).
26. Eur. Pat. Appl. (2000), A. Pahl (to Lohmann GmbH & Co).
27. W. C. Wake, *Adhesion and the Formulation of Adhesives*, Applied Science Publishers Limited, London, 1976, p.188.
28. C. W. Boeder, in S. R. Hartshorn, ed., *Structural Adhesives*, Plenum Press, New York, 1986, Chap. 5, pp. 225–226.
29. *Ibid.*, pp. 228–229.
30. *Ibid.*, p. 231.
31. *Ibid.*, p. 238.
32. For example, U. S. Pat. 4,855,001 (Aug. 8, 1989), D. J. Damico and R. M. Bennett (to Lord Corporation); U. S. Pat. 4,857,131 (Aug. 15, 1989), D. J. Damico, K. W. Mushrush, and R. M. Bennett (to Lord Corporation); Jap. Pat. 07,109,442 A2 (April 25, 1995), T. Fujisawa and O. Hara (to Three Bond Co. Ltd.); Eur. Pat. Appl. 540,098 A1 (May 5, 1993), V. DiRuocco, L. Gila, and F. Garbassi, (Ministero dell'Universita e della Ricerca Scientifica e Tecnologica).
33. U. S. Pat. 3,890,407 (June 17, 1975), P. C. Briggs and L. C. Muschiatti (to E. I. Du Pont de Nemours & Co. Inc.).
34. G. H. Millet, in S. R. Hartshorn, ed., *Structural Adhesives*, Plenum Press, New York, 1986, Chap. 6, pp. 262–263.
35. U. S. Pat. 5,981,621 (Nov. 9, 1999), J. G. Clark and J. C. Leung (Closure Medical Corporation).
36. Ref. 34, pp. 276–278.
37. A. C. Roberts, *Adhesive Technol.* **15**(2), 4, 6 (1998).
38. Ger. Pat. 4317886 (Dec. 2, 1993), S. Takahaski, A. Kaai, T. Okuyama, S. Tajima, and T. Horie (to Toagosei Chemical Industry Co., Ltd.).
39. U. S. Pat. 4,444,933 (Apr. 24, 1984), P. S. Columbus and J. Anderson (to Borden, Inc.).
40. See, for example, product literature from Bayer Corporation and BASF Corporation.
41. S. R. Hartshorn and K. C. Frisch, Jr., *Proc. 25th Anniv. Symp. of the Polymer Institute (Univ. of Detroit)*, Technomic Publishing Co., Inc., Lancaster, UK, 1994, pp. 1–10.
42. N. Weeks, *Adhesive Technol.* **17**(3), 19 (2000).
43. U. S. Pat. 4,390,678 (Jun. 28, 1983), S. B. LaBelle and J. E. Hagquist (to H. B. Fuller Company).
44. B. H. Edwards, in *Structural Adhesives*, S. R. Hartshorn ed., Plenum Press, New York, 1986, Chap. 4., pp. 197–200.
45. E. G. Melby, in *Advances in Urethane Science and Technology*, Vol. 14, K. C. Frisch and D. Klempner eds., Technomic Poblishing Co., Inc., Lancaster, UK, 1998, pp. 317–319.
46. *Ibid.*, pp. 319–325.
47. U. S. Pat. 3,525,779 (Aug. 25, 1970), J. M. Hawkins (to The Dow Chemical Co.); J. A. Clarke, *J. Adhesion*, **3**, 295–306 (1972); U. S. Pat. 5,232,996 (Aug. 3, 1993), D. N. Shah and T. H. Dawdy (to Lord Corporation); U. S. Pat. 5,278,257 (Jan. 11, 1994), R. Mulhaupt, J. H. Powell, C. S. Adderley, W. Rufenacht (to CIBA-GEIGY Corp.).

48. J. Robins, in *Structural Adhesives*, S. R. Hartshorn, ed., Plenum Press, New York, 1986, Chap. 2.
49. A. Higgins, *Int. J. Adhes. Adhes.* **20**(5), 367 (2000).
50. A. Pizzi, *Advanced Wood Adhesives Technology*, Marcel Dekker, Inc, New York, 1994, p. 21.
51. *Ibid.*, p. 68 and Ref. 1 therein.
52. S. K. Kang and S. Purushothaman, *J. Electron. Mater.* **28**(11), 1314 (1999).
53. K. Gilleo and P. Ongley, *Microelectronics Int.* **16**(2), 34 (1999).
54. R. Vabrik et al., *Prog. Rubber Plast. Technol.* **15**(1), 28 (1999).
55. A. L. Lambuth, in A. Pizzi and K. L. Mittal, eds., *Handbook of Adhesive Technology*, Marcel Dekker, Inc., New York, 1994, Chap. 13.
56. D. H. Sierra, *J. Biomater. Appl.* **7**(4), 309 (1993).
57. M. G. D. Bauman and A. H. Conner, in A. Pizzi and K. L. Mittal, eds., *Handbook of Adhesive Technology*, Marcel Dekker, Inc., New York, 1994, Chap. 15.
58. S. Bloembergen, I. J. McLennan and C. S. Schmaltz, *Adhesive Technology* **16**(3), 10 (1999).
59. J. Suszkiw, *Agri. Res.* **47**(6), 22 (1999).
60. T. Viswanathan, in R. W. Hemingway, A. H. Conner and S. J. Branham, eds., *Adhesives from Renewable Resources*, ACS Symposium. Series., 385, American Chemical Society, Washington, D.C., 1989, Chap. 28.
61. M. S. Holfinger, A. H. Conner, L. F. Lorenz and C. G. Hill, Jr., *J. Appl. Polym. Sci.* **49**(2), 337 (1993); J. L. Stanford, R. H. Still, J. L. Cawse and M. J. Donnelly, in R. W. Hemingway, A. H. Conner and S. J. Branham, Eds., *Adhesives from Renewable Resources*, ACS Symposium Series, 385, American Chemical Society, Washington, D.C., 1989, Chap. 30.
62. C. W. Paul, M. L. Sharak and M. Blumenthal, *Adhesives Age* **42**(7), 34 (1999).
63. J. Haisma et al., *Appl. Op.* **33**(7), 1154 (1994).
64. C. A. Desmond-Colinge and U. Gosele, *MRS Bull.* **23**(12), 30 (1998).
65. N. Tas et al., *J. Micromechan. Microengin.* **6**(4), 385 (1996).
66. M. Ash and I. Ash, *Handbook of Adhesive Chemicals and Compounding Ingredients*, Synapse Information Resources, Endicott, N.Y., 1999.

GENERAL REFERENCES

S. C. Temin, "Adhesive Compositions", in *Encyclopedia of Polymer Science and Technology*, 2nd ed., Vol.1, 1985.

J. Johnston, *Pressure Sensitive Adhesive Tapes: A Guide to their Function, Design, Manufacture, and Use*, Pressure Sensitive Tape Council, Northbrook, Ill, 2000.

I. Benedek, *Development and Manufacture of Pressure-Sensitive Products*, Marcel Dekker, Inc., New York, 1999.

A. Pizzi, *Advanced Wood Adhesive Technology*, Marcel Dekker, Inc., New York, 1994.

S. R. Hartshorn, ed., *Structural Adhesives: Chemistry and Technology*, Plenum Press, New York, 1986.

A. Pizzi and K. L. Mittal, *Handbook of Adhesive Technology*, Marcel Dekker, Inc., New York, 1994.

W. C. Wake, *Adhesion and the Formulation of Adhesives*, Applied Science Publishers, London, 1976.

A. V. Pocius, *Adhesion and Adhesives Technology: An Introduction*, Hanser Publishers, Munich, 1997.

K. J. Saunders, *Organic Polymer Chemistry*, Chapman and Hall, London, 1973.

G. Wypych, *Handbook of Fillers*, 2nd ed., ChemTec Publishers, Toronto, 1999. May be referenced under the name Jerzy Wypych.

McCutcheon's Functional Materials, North American ed., McCutcheon's Division, Manufacturing Confectioner Publishing Co., Glen Rock, New Jersey, multivolume, published annually. An International edition is also published annually.

Elaine M. Yorkgitis
3M Company

ADIPIC ACID

1. Introduction

Adipic acid, hexanedioic acid, 1,4-butanedicarboxylic acid, mol wt 146.14, $HOOCCH_2CH_2CH_2CH_2COOH$ [124-04-9], is a white crystalline solid with a melting point of ~152°C. Little of this dicarboxylic acid occurs naturally, but it is produced on a very large scale at several locations around the world. The majority of this material is used in the manufacture of nylon-6,6 polyamide [32131-17-2], which is prepared by reaction with 1,6-hexanediamine [124-09-4]. W. H. Carothers' research team at the Du Pont Company discovered nylon in the early 1930s (1), and the 50th anniversary of its commercial introduction was celebrated in 1989. Growth has been strong and steady during this period, resulting in an adipic acid demand of nearly 2 billion metric tons per year worldwide in 1999. The large scale availability, coupled with the high purity demanded by the polyamide process, has led to the discovery of a wide variety of applications for the acid.

2. Chemical and Physical Properties

Adipic acid is a colorless, odorless, sour tasting crystalline solid. Its fundamental chemical and physical properties are listed in Table 1. Further information may be obtained by referring to studies of infrared and Raman spectroscopy of adipic acid crystals (11,12), ultraviolet spectra of solutions (13), and specialized thermodynamic properties (4,14). Solubility and solution properties are described in Table 2. The crystal morphology is monoclinic prisms strongly influenced by impurities (21). Both process parameters (22) and additives (21) profoundly affect crystal morphology in the crystallization of adipic acid, an industrially significant process. Aqueous solutions of the acid are corrosive and their effect on various steel alloys have been tested (23). Generally, austenitic stainless steels containing nickel and molybdenum and >18% chromium are resistant. Data on 20 metals were summarized in one survey (24). Bulk and handling properties of adipic acid are summarized in Table 3.

Table 1. **Physical and Chemical Properties of Adipic Acid**

Property	Value	References
molecular formula	$C_6H_{10}O_4$	
molecular weight	146.14	
melting point, °C	152.1 ± 0.3	2
specific gravity	1.344 at 18°C (sol)	3
	1.07 at 170°C (liq)	4
coefficient of cubical	4.0×10^{-4} at 35–150°C (sol)	4
expansion, K^{-1}	10.3×10^{-4} at 155–168°C (liq)	5
vapor density, air = 1	5.04	
vapor pressure, Pa[a]		
solid at °C		6
18.5	9.7	
32.7	19.3	
47.0	38.0	
liquid at °C		7
205.5	1,300	
216.5	2,000	
244.5	6,700	
265.0	13,300	
specific heat, kJ/kg K[b]	1.590 (solid state)	8
	2.253 (liquid state)	9,8
	1.680 (vapor, 300°C)	
heat of fusion, kJ/kg[b]	115	
entropy of fusion, J/mol K[b]	79.8	4,10
heat of vaporization, kJ/kg[b]	549	
melt viscosity, mPa s(= cP)	4.54 at 160°C	
	2.64 at 193°C	
heat of combustion, kJ/mol[b]	2,800	10

[a] To convert Pa to mm Hg divide by 133.3.
[b] To convert J to cal divide by 4.184.

3. Chemical Reactions

Adipic acid undergoes the usual reactions of carboxylic acids, including esterification, amidation, reduction, halogenation, salt formation, and dehydration. Because of its bifunctional nature, it also undergoes several industrially significant polymerization reactions.

Esters and polyesters comprise the second most important class of adipic acid derivatives, next to polyamides. The acid readily reacts with alcohols to form either the mono- or diester. Although the reaction usually is acid catalyzed, conversion may be enhanced by removal of water as it is produced. The methyl ester is an industrially important material, because it is a distillable derivative which provides a means of separating or purifying acid mixtures. Recent modifications of adipic acid manufacturing processes have included methanol esterification of the dicarboxylic acid by-product mixture. Thus glutaric acid [110-94-1] and succinic acid [110-15-6] can be recovered upon hydrolysis, or disposed of as the esters (28). Monomethyl adipate can be electrolyzed as the salt to give dimethyl sebacate [106-79-6] (Kolbe synthesis) (29), an important 10-carbon

Table 2. **Solution Properties of Adipic Acid**

Property	Value	References
heat of solution in H_2O, kJ/kg[a]	214 at 10–20°C	15
	241 at 90–100°C	
dissociation constant in H_2O	$K_1 K_2$	16,17
at 25°C	3.7×10^{-5}	
	3.86×10^{-6}	
at 50°C	3.29×10^{-5} 3.22×10^{-6}	
at 74°C	2.90×10^{-5} 2.55×10^{-6}	
solubility in H_2O, g/100 g H_2O		18
at 15°C	1.42	
at 40°C	4.5	
at 60°C	18.2	
at 80°C	73	
at 100°C	290	
pH of aqueous solutions		19
0.1 wt %	3.2	
0.4 wt %	3.0	
1.2 wt %	2.8	
2.5 wt %	2.7	
solubility in organic solvents		
at 25°C		
very soluble in	methanol, ethanol	
soluble in	acetone, ethyl acetate	
very slightly soluble in	cyclohexane, benzene	
distribution coefficient		
organic solvents vs H_2O	$D, \frac{\text{wt \% in } H_2O}{\text{wt \% in solvent}}$	
CCl_4, $CHCl_3$, C_6H_6	>10	20
disopropyl ketone	4.8	
butyl acetate	2.9	
ethyl ether	2.2	
methyl isobutyl ketone	1.2	
ethyl acetate	0.91	
methyl propyl ketone	0.55	
methyl ethyl ketone	0.50	
cyclohexanone	0.32	
n-butanol	0.31	

[a] To convert kJ to kcal divide by 4.184.

Table 3. **Bulk Phase Handling Properties of Adipic Acid**

Property	Value	References
bulk density[a], kg/m^3	640–800	19
flash point, Cleveland open cup, °C	210	5
flash point, closed cup, °C	196	
autoignition temperature, °C	420	5
dust cloud ignition temperature, °C	550	
minimum explosive concentration (dust in air), kg/m^3	0.035	25,26
minimum dust cloud ignition energy, J[b]	6.0×10^{-2}	27
maximum rate of pressure rise, MPa[c]/s	18.6	

[a] A function of particle size.
[b] To convert J to cal divide by 4.184.
[c] To convert MPa to psi multiply by 145.

Table 4. **Esters of Adipic Acid**

Ester	CAS Registry Number	Pressure, kPa[a]	Boiling point, °C
monomethyl	[627-91-8]	1.3	158
dimethyl	[627-93-0]	1.7	115
monoethyl	[626-86-8]	0.9	160
diethyl	[141-28-6]	1.7	127
di-*n*-propyl	[106-19-4]	1.5	151
di-*n*-butyl	[105-99-7]	1.3	165
di-2-ethylhexyl	[103-23-1]	0.67	214
di-*n*-nonyl	[151-32-6]	0.67	230
di-*n*-decyl	[105-97-5]	0.67	244
di-tridecyl	[16958-92-2]	101.3	349
octyl decyl	[110-29-2]	0.67	235
di-(2-butoxyethyl)	[141-18-4]	0.53	215

[a] To convert kPa to mm Hg multiply by 7.5.

diacid. Diesters from moderately long-chain (8 or 10 carbon) alcohols are also an important group, finding use as plasticizers, eg, for poly(vinyl chloride) (PVC) resins. Table 4 lists the boiling points of several representative adipate esters. Reactions with diols (especially ethylene glycol) give polyesters, also important as plasticizers in special applications. In another important use of adipate esters, low molecular weight polyesters terminated in hydroxyl groups react with polyisocyanates to give polyurethane resins. Polyurethanes consumed ~48% of adipic acid production in the United States in 1999 (30).

Salt-forming reactions of adipic acid are those typical of carboxylic acids. Alkali metal salts and ammonium salts are water soluble; alkaline earth metal salts have limited solubility (see Table 5). Salt formation with amines and diamines is discussed in the next section.

Heating of the diammonium salt or reaction of the dimethyl ester with concentrated ammonium hydroxide gives adipamide [628-94-4], mp 228°C, which is relatively insoluble in cold water. Substituted amides are readily formed when amines are used. The most industrially significant reaction of adipic acid is its reaction with diamines, specifically 1,6-hexanediamine. A water-soluble polymeric salt is formed initially upon mixing solutions of the two materials; then heating with removal of water produces the polyamide, nylon-6,6. This reaction

Table 5. **Solubility of Adipic Acid Salts**

Salt	CAS Registry Number	Temperature, °C	Solubility, g/100 g H_2O
disodium (hemihydrate)	[7486-38-6]	14	59
dipotassium	[19147-16-1]	15	65
diammonium	[3385-41-9]	14	40
calcium			
(monohydrate)	[18850-78-7]	13	4
(anhydrous)	[22322-28-7]	100	1
barium			
(monohydrate)		12	12
(anhydrous)	[60178-88-0]	100	7

has been studied extensively, and the literature contains hundreds of references to it and to polyamide product properties (31).

$$n\,\mathrm{HOOC(CH_2)_4COOH} + n\,\mathrm{H_2N(CH_2)_6NH_2} \longrightarrow [^{-}\mathrm{OOC(CH_2)_4CO\overset{-}{O}\overset{+}{N}H_3(CH_2)_6\overset{+}{N}H_3}]_n \longrightarrow$$

$$(2n-1)\mathrm{H_2O} + \mathrm{HO[\overset{O}{\overset{\|}{C}}(CH_2)_4\overset{O}{\overset{\|}{C}}NH(CH_2)_6NH]_nH}$$

Hydrogenation of dimethyl adipate over Raney-promoted copper chromite at 200°C and 10 MPa produces 1,6-hexanediol [629-11-8], an important chemical intermediate (32). Promoted cobalt catalysts (33) and nickel catalysts (34) are examples of other patented processes for this reaction. An earlier process, which is no longer in use, for the manufacture of the 1,6-hexanediamine from adipic acid involved hydrogenation of the acid (as its ester) to the diol, followed by ammonolysis to the diamine (35).

Heating above the melting point results in elimination of water and formation of a linear or polymeric anhydride [2035-75-8], not the cyclic anhydride as produced in the case of glutaric anhydride [108-55-4] and succinic anhydride [108-30-5]. Decarboxylation occurs at temperatures >230–250°C, leaving cyclopentanone [120-92-3] as the chief product, bp 131°C. This reaction is catalyzed by metals such as calcium (36) or barium (37). Behavior of adipic acid upon Curie-point pyrolysis has been reviewed; mass spectroscopy was used to analyze the anhydrides, cyclic ketones, and rearranged fragments (38). Cyclization of the esters is accomplished by standard condensation chemistry with basic reagents. For example, cyclization via the acyloin condensation occurs in the presence of sodium metal, producing 2-hydroxycyclohexanone [533-60-8] (39).

Conversion of the acid to the acid chloride is accomplished using standard laboratory techniques. The resulting acid chloride frequently is used in subsequent synthesis reactions. An example is the laboratory synthesis of nylon-6,6 via the nylon rope trick, in which the diamine reacts with adipoyl chloride [111-50-2] in a two-phase system. Polyamide produced at the interface may be pulled continuously from the open vessel in a startling demonstration of polymerization chemistry (40). The acid–nitrile interchange is another unique reaction, in which a mixture of adipic acid and adiponitrile [111-69-3] are heated together, producing an equilibrium mixture containing significant amounts of 5-cyanopentanoic acid [5264-33-5]. This material is a precursor to caprolactam [105-60-2] and may be isolated from the reaction mixture by a number of methods, including esterification and hydrogenation (41).

4. Manufacture and Processing

Several general reviews of adipic acid manufacturing processes have been published since it became of commercial importance in the 1940s (42–46), including a very thorough report based on patent studies (47). Adipic acid historically has been manufactured predominantly from cyclohexane [110-82-7] and, to a lesser extent, phenol [108-95-2]. During the 1970s and 1980s, however, much research has been directed to alternative feedstocks, especially butadiene [106-99-0] and cyclohexene [110-83-8], as dictated by shifts in hydrocarbon pricing. All current industrial processes use nitric acid [7697-37-2] in the final oxidation stage. Growing concern with air quality may exert further pressure for alternative routes as manufacturers seek to avoid NO_x abatement costs, a necessary part of processes that use nitric acid.

Since adipic acid has been produced in commercial quantities for almost 50 years, it is not surprising that many variations and improvements have been made to the basic cyclohexane process. In general, however, the commercially important processes still employ two major reaction stages. The first reaction stage is the production of the intermediates cyclohexanone [108-94-1] and cyclohexanol [108-93-0], usually abbreviated as KA, KA oil, ol-one, or anone-anol. The KA (ketone, alcohol), after separation from unreacted cyclohexane (which is recycled) and reaction by-products, is then converted to adipic acid by oxidation with nitric acid. An important alternative to this use of KA is its use as an intermediate in the manufacture of caprolactam, the monomer for production of nylon-6 [25038-54-4]. The latter use of KA predominates by a substantial margin on a worldwide basis, but not in the United States.

4.1. Preparation of KA by Oxidation of Cyclohexane. There are three main variations to the basic cyclohexane oxidation process pioneered by Du Pont in the 1940s. The first, which can be termed metal-catalyzed oxidation, is the oldest process still in use and forms the base for the other two. It employs a cyclohexane-soluble catalyst, usually cobalt naphthenate [61789-51-3] or cobalt octoate [136-52-7], and moderate temperatures (150–175°C) and pressures (800–1200 kPa). Air is fed to each of a series of stirred tank reactors or to a column reactor that contains numerous reaction stages, along with cyclohexane. The catalyst, at 0.3–3 ppm based on cyclohexane feed, is usually premixed by injection into the feed stream, though it is not uncommon to divide the catalyst stream into many separate additions to each of the series reactors. The conversion of cyclohexane to oxidized products is 3–8 mol %, which is quite low compared to most important industrial processes. There are claims of commercial processes operating as low as 1 mol% conversion (48), which translates to 99% of the feed material being recovered and recycled to the oxidation reactors. Low conversion is the major factor in achieving high selectivities to ketone (K) and alcohol (A) (and to cyclohexylhydroperoxide [766-07-4] discussed below). This is so because the intermediates of interest (K, A, and cyclohexylhydroperoxide) are all much more easily oxidized than is cyclohexane (49,50). Selectivities vary inversely and linearly with conversion, ranging from ~90 mol% at 1–2 mol% conversion to 65–70 mol% at 8 mol% conversion. Table 6 illustrates the range of reaction conditions to be found in the patent literature.

Because the process operates at such low conversion of cyclohexane per pass through the oxidation reactors, large quantities of unreacted cyclohexane must be recovered by distillation of the oxidizer effluent. This, and the increase in energy prices in the 1970s, has resulted in considerable attention being given to the energy conservation schemes employed in recovering the cyclohexane. Examples of techniques used in energy conservation are process–process heat interchange, high efficiency packed distillation columns, and use of the "pinch-point" technique in designing recovery steps. Contacting the final crude KA oil with water or solutions of caustic soda, or both, for removal of mono- and dibasic

Table 6. **Reaction Conditions for Air Oxidation of Cyclohexane**

Process and company	Temperature, °C	Pressure, MPa[a]	Catalyst or additive	Reactor type	Cyclohexane conversion, mol %	KA yield, mol %	References
Metal-catalyzed							
Du Pont	170	1.1	Co	column	6	76	51
Stamicarbon	155	0.9	Co	tank	4	77	52
High peroxide							
BASF							
oxidation	145	1.1	none	tank	3	83	53–55
deperoxidation	125		Co/NaOH				
Du Pont							
oxidation	160	1.0	Co	column	4	82	56–58
deperoxidation	120		Co, Cr				
Rhône Poulenc							
oxidation	175	1.8	none	tank	4	84	59–62
deperoxidation	115		Cr, V, Mo				
Stamicarbon							
oxidation	160	1.3	none	tank	3	86	48,63,64
deperoxidation	100		Co/NaOH				
Boric acid							
Halcon	165	1.0	H_3BO_3	tank	3	87	65,66
ICI	165	1.0	H_3BO_3	tank	5	85	67
IFP	165	1.2	H_3BO_3		12	85	68,69
Solutia (formerly Monsanto)	165	1.0	H_3BO_3	tank	4	87	70

[a] To convert MPa to psi multiply by 145.

acid impurities also can be considered an energy conservation technique since this treatment can eliminate the final steam stripper often used to purify the crude KA oil.

Regardless of the techniques used to purify the KA oil, several waste streams are generated during the overall oxidation–separation processes and must be disposed of. The spent oxidation gas stream must be scrubbed to remove residual cyclohexane, but afterwards will still contain CO, CO_2, and volatile hydrocarbons (especially propane, butane, and pentane). This gas stream is either burned and the energy recovered, or it is catalytically abated. There are usually several aqueous waste streams arising from both water generated by the oxidation reactions and wash water. The principal hydrocarbon constituents of these aqueous wastes are the C_1–C_6 mono- and dibasic acids, but also present are butanol [71-36-3], pentanol [71-41-0], ε-hydroxycaproic acid [1191-25-9], and various lactones and diols (71,72). The spent caustic streams contain similar components in addition to the caustic values. These streams can be burned for recovery of sodium carbonate or sold directly as a by-product for use in the paper industry. The most concentrated waste stream is one often called still bottoms, heavy ends, or nonvolatile residue. It comes from the final distillation column in which the KA oil is steam-stripped overhead. The tails stream from this column contains most of the nonvolatile by-products, as well as metals and residues from the catalysts and from corrosion. Both the metals and acid content may be high enough to dictate that this stream be classified as a hazardous waste. It usually is burned and the energy used to generate steam (73). Much effort has gone into recovering valuable materials from it over the years, including adipic acid, which may be present in as much as 3–4% of the cyclohexane oxidized (74). It has potential as a feedstock in the production of monobasic acids, polyester polyols, butanediol, and maleic acid (75,76). The frequency of fugitive emissions from cyclohexane oxidation plants has been reviewed (77).

An alternative to maximizing selectivity to KA in the cyclohexane oxidation step is a process which seeks to maximize cyclohexylhydroperoxide, also called P or CHHP. This peroxide is one of the first intermediates produced in the oxidation of cyclohexane. It is produced when a cyclohexyl radical reacts with an oxygen molecule (78) to form the cyclohexylhydroperoxy radical. This radical can extract a hydrogen atom from a cyclohexane molecule, to produce CHHP and another cyclohexyl radical, which extends the free-radical reaction chain.

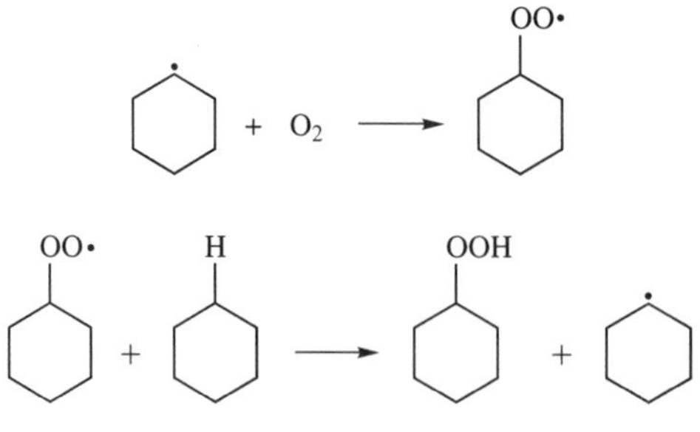

The peroxide can be converted to KA easily, and in high yield, in a number of ways; thus maximization of CHHP, at high yield, gives a process with high

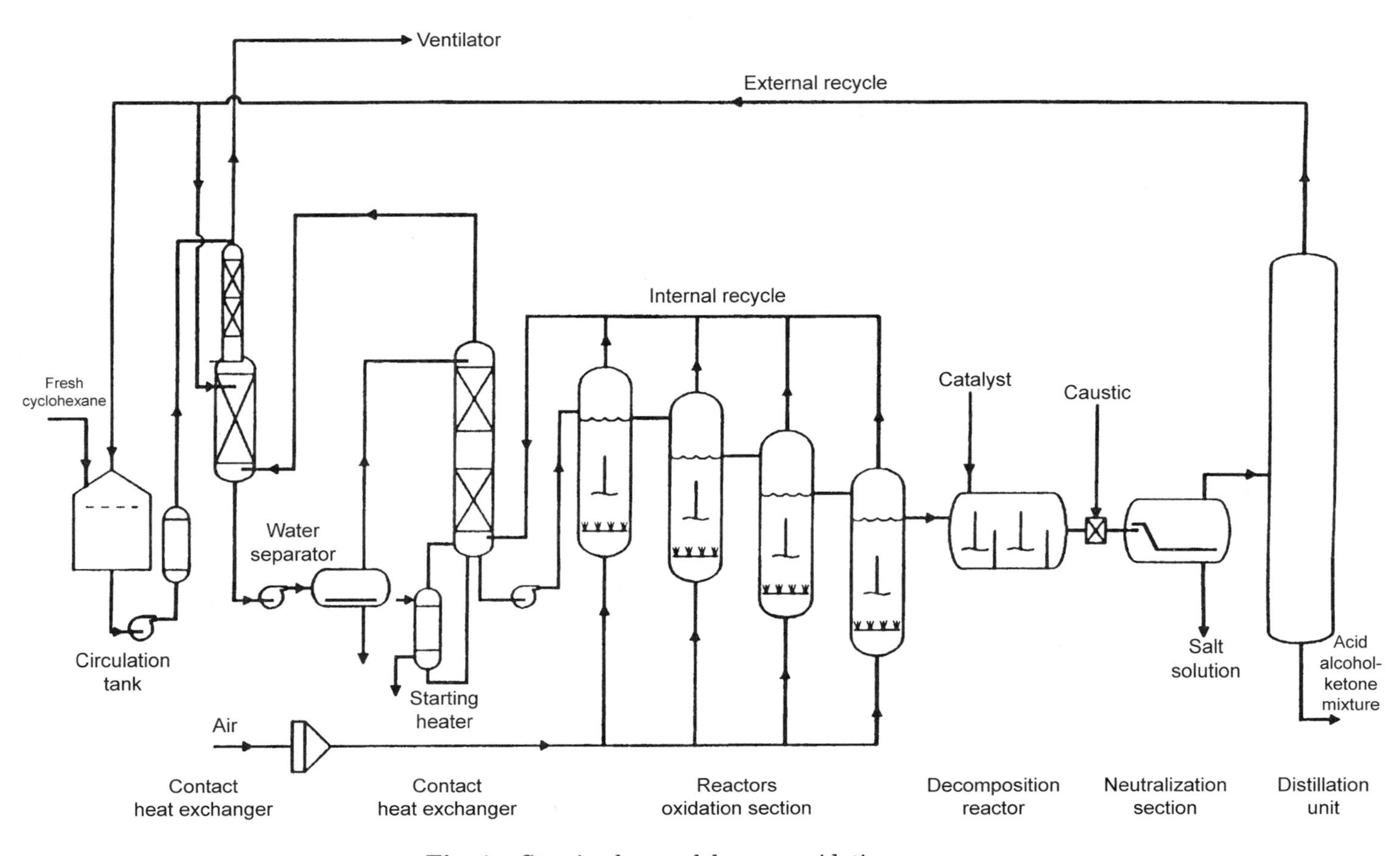

Fig. 1. Stamicarbon cyclohexane oxidation process.

yield to KA. Techniques employed to produce high CHHP yield include drastically cutting or eliminating metal catalysts in the oxidation step, minimizing cyclohexane conversion, passivating reactor walls, lowering reaction temperature (to as low as 140°C), adding water to the reaction mix to extract acid catalysts from the cyclohexane phase, and adding metal-chelating agents to the reaction mix. Optimization of this process can produce CHHP in a proportion as high as 75% of the reaction products (59). The CHHP then can be converted to KA by any of the following methods: decomposing it with homogeneous or heterogeneous catalysts from the group Co, Cr, Mo, V, Cu, or Ru; dehydrating it by treatment with caustic soda (which preferentially gives K); or hydrogenating it (which preferentially gives A). KA is separated from the reaction mixture in a manner similar to the conventional process. It may be possible, however, to avoid a final steam distillation of the KA overhead if the tails stream from the distillation train is sufficiently clean. This could result from a high yield process that employs thorough water and caustic washing. Figure 1 illustrates schematically the high peroxide process practiced by Stamicarbon (60).

Another alternative to the basic cyclohexane oxidation process is one which maximizes only the yield of A. This process uses boric acid as an additive to the cyclohexane stream as both a promoter and an esterifying agent for the A that is produced. Metaboric acid [10043-35-3] is fed to the first series oxidizer as a slurry in cyclohexane to give a molar ratio of boron:cyclohexane of around 1.5:100. No other metal catalyst is used. Esterifying the A effectively shields it from overoxidation and thus allows the attainment of very high yields (~90%) (65). The ratio A : K in the final product can exceed 10:1. The process was developed in the mid-1960s by a number of companies, including Halcon/Scientific Design (79,80), Institute Francais Petrole (68,81), and Stamicarbon (82). The process was licensed and commercialized by several companies in the decade following its development, including Solutia (formerly Monsanto), ICI, and Bayer. The major drawback to the process is the need to hydrolyze the borate ester in order to recover A. This is an energy-intensive step and can be quite a mechanical nuisance because of the requirement for handling boric acid solids. Without careful attention to energy conservation and engineering, the savings that accrue from the high yield can be more than offset. The process does, because of its high yield, offer advantages in waste minimization and product purity. It does, however, introduce boron into the waste streams.

4.2. Preparation of KA From Phenol. In past years, economics has dictated against the preparation of KA from phenol [108-95-2] because of the relatively high cost of this material compared to cyclohexane. However, given new routes to phenol and occasional periods of overcapacity for this commodity chemical, such technology has been revisited.

For example, the Solutia Benzene to Phenol process (83) has been touted as a step-change in adipic acid manufacturing technology. As shown in Figure 2, the reaction between benzene and nitrous oxide over an Fe- ZSM-5 catalyst (84) produces phenol in high yield and selectivity. Here, the nitrous oxide is obtained from removing the NO_x, oxygen and low-level organic compounds from the adipic acid offgas (85). Rather than destroying the N_2O (see Section 8 under N_2O Abatement Technology), the benzene to phenol process converts it to a value-added product which also serves as an adipic acid intermediate per the phenol to KA

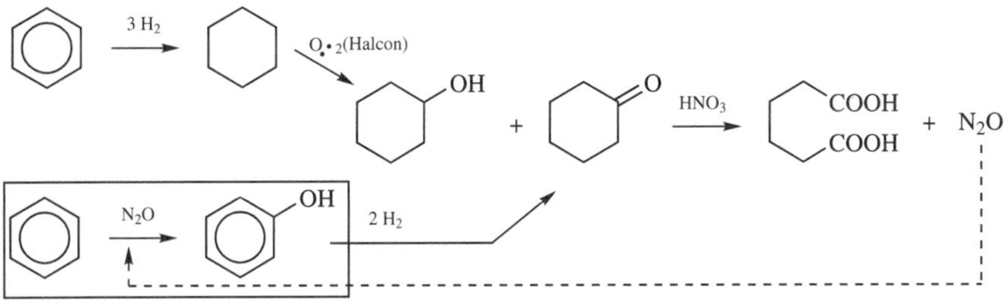

Fig. 2. Solutia benzene to phenol technology.

process, discussed below. The reaction runs at 3 atm and 300–500°C. Since it takes place in the gas phase with N_2O levels maintained below the minimum oxidant concentration (86), it is inherently safer than the high pressure, liquid phase cyclohexane to KA operation, which has a potential for flammable vapor cloud release. A moving bed design is used to provide a continuous operation and extend catalyst life.

Downstream, the vapor phase hydrogenation of phenol to KA is used. Its primary selling points over cyclohexane to KA routes include a lower benzene usage and a 99% KA yield (87,88). The catalyst is Pd/Al_2O_3. Reaction conditions are 130–160°C, 165 kPa (15 psi), and 4 H_2/phenol. A three-step distillation separates the product KA from the light ends, high boiler by-products (eg, cyclohexenyl and cyclohexanone) and unreacted phenol. Steam generated in the tubular isothermal reactor provides the energy required by the distillation section.

The liquid phase hydrogenation of phenol to KA also gives a very high yield, typically 97–99% (89). Just as in the vapor phase process, high KA yield leads to a simple purification section, consisting of an ion exchange step to remove the unreacted phenol (90). Typical reaction conditions are 140°C and 400 kPa (58 psi) using a heterogeneous nickel on silica catalyst (89).

In both liquid- and vapor-phase phenol to KA processes, one can obtain varying ratios of K to A. If the desired product is to be further oxidized with nitric acid to make adipic acid, then there is an optimal concentration of K that maximizes the tradeoff between adipic yield (91) and nitric acid usage. If the desired product is caprolactam, then a high K concentration would be required.

Cyclohexanol (A) can be manufactured from cyclohexene using a process developed and commercialized by Asahi (92,93). This process, illustrated in Figure 3 involves a selective partial hydrogenation of benzene to cyclohexene

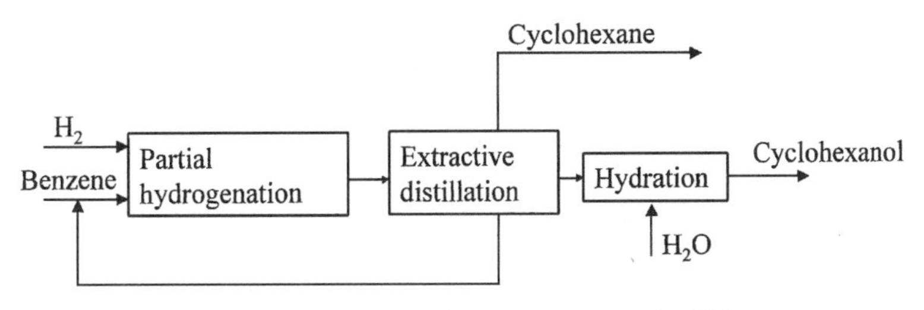

Fig. 3. Asahi cyclohexene process for KA.

using an aqueous ruthenium catalyst system containing metal oxide dispersants (eg, Al_2O_3) and zinc compound promoters (94). Reaction conditions of 120–180°C and 30–100 atm give a benzene conversion of 50–60% and a cyclohexene selectivity of 80%. Cyclohexane is the primary byproduct. The reactor contents are well agitated in order to create adequate mixing between gas (hydrogen), oil (benzene raw material and cyclohexene;cyclohexane products), and water (catalyst) phases. After hydrogenation, the oil (disperse phase) is separated from the water (continuous phase). Next, the unreacted benzene is recovered via extractive distillation using a polar solvent such as dimethylacetamide (95,96). The cyclohexene is then hydrolyzed in the presence of a ZSM-5 zeolite (97) suspended in water. The conversion is 10–15% and the cyclohexanol selectivity is 99%. This result is based on the nature of the extractive reaction step. Here, the cyclohexene is adsorbed on the catalyst introduced in the process as an aqueous phase slurry. Cyclohexanol, once formed, separates from the reaction medium and disperses back to the organic phase. Using proprietary multistep extractive distillation process, the cyclohexanol is separated from unreacted cyclohexene and purified (98). Provided that one has a use for the cyclohexane byproduct, this low waste process is an attractive alternative to other more traditional technology.

ARCO has developed a coproduct process that produces KA along with propylene oxide [75-56-9] (99–101). Cyclohexane is oxidized as in the high peroxide process to maximize the quantity of CHHP. The reactor effluent then is concentrated to about 20% CHHP by distilling off unreacted cyclohexane and cosolvent *tert*-butyl alcohol [75-65-0]. This concentrate then is contacted with propylene [115-07-1] in another reactor in which the propylene is epoxidized with CHHP to form propylene oxide and KA. A molybdenum catalyst is employed. The product ratio is ~2.5 kg of KA per kilogram of propylene oxide.

4.3. Nitric Acid Oxidation of Cyclohexanol(One). Although many variations of the cyclohexane oxidation step have been developed or evaluated, technology for conversion of the intermediate ketone–alcohol mixture to adipic acid is fundamentally the same as originally developed by Du Pont in the early 1940s (102,103). This step is accomplished by oxidation with 40–60% nitric acid in the presence of copper and vanadium catalysts. The reaction proceeds at high rate, and is quite exothermic. Yield of adipic acid is 92–96%, the major byproducts being the shorter chain dicarboxylic acids, glutaric and succinic acids, and CO_2. Nitric acid is reduced to a combination of NO_2, NO, N_2O, and N_2. The trace impurities patterns are similar in the products of most manufacturers since essentially all commercial adipic acid production uses this nitric acid oxidation process.

Papers addressing the mechanism of nitric acid oxidation began appearing in the mid-1950s (104). Then, a series of reports beginning in 1962 described the mechanism of the oxidation in considerable detail (105–109). The reaction pathway diagram shown in Figure 4 is based on these and other studies of nitric acid oxidation chemistry. A key intermediate in the reaction sequence is 2-oximinocyclohexanone [24858-28-4], produced via nitrosation of cyclohexanone. Nitrous acid [7782-77-6] is produced during the conversion of cyclohexanol to the ketone, and also upon oxidation of aldehyde and alcohol impurities usually accompanying the KA and arising in the cyclohexane oxidation step. The nitric acid oxidation chemistry is controlled by nitrous acid, which is in equilibrium with NO, NO_2, HNO_3, and H_2O in the reacting mixture. Total inhibition of reaction can be achieved by incorporating a small amount of urea [57-13-6], which effectively scavenges (110) nitrous acid from the mixture. Further nitration leads to 2-nitro-2-nitrosocyclohexanone [23195-89-3], which is converted via hydrolytic cleavage of the ring to 6-nitro-6-hydroximinohexanoic acid (nitrolic acid) [1069-46-1]). Of all the intermediates shown in Figure 4, the nitrolic acid is the only one of sufficient stability to be isolable under very mild conditions. It is hydrolyzed to adipic acid in one of the slowest steps in the sequence. Nitrous oxide (N_2O) is formed by further reaction of the nitrogen-containing products of nitrolic acid hydrolysis. The NO and NO_2 are reabsorbed and converted back to nitric acid, but N_2O cannot be recovered in this manner, and thus is the major nitric acid derived by-product of the process.

About 60–70% of the reaction occurs as in path 1 in Figure 4, the remainder by other pathways. About 20% of the reaction occurs by the vanadium oxidation of 1,2-dioxygenated intermediates (path 2 in Fig. 4). This chemistry has been discussed in detail (108,109). This path is noteworthy since it does not produce the nonrecoverable nitrous oxide. The other reactions shown in Figure 4 occur to varying degrees, depending on either an excess or deficiency of nitrous acid, arising from variations in reaction conditions. These lead to varying yields of the lower dicarboxylic acids. Yield of monobasic and dibasic acid by-products also is a function of the purity of the KA feed. A distinguishing characteristic for several of the commercial processes is the degree to which the intermediate KA is refined, prior to feeding it to nitric acid oxidation.

In a typical industrial adipic acid plant, as schematically illustrated in Figure 3, the KA mixture reacts in reactor A with 45–55% nitric acid containing copper (0.1–0.5%) and vanadium (0.02–0.1%) catalyst (111,112). Design of the oxidation reactor for optimum yield and heat removal has been the subject of considerable research and development over the years of use of this process (113). The reaction occurs at 60–90°C and 0.1-0.4 MPa (14–58 psi). It is very exothermic (6280 kJ/kg = 1500 kcal/kg), and can reach an autocatalytic runaway state at temperatures above ~150°C. Control is achieved by limiting the KA feed to a large excess of nitric acid in a stirred tank or circulating loop reactor. Two stages of oxidation are sometimes employed to achieve improved product quality (114). Oxides of nitrogen are removed by bleaching with air in column C, then water is removed by vacuum distillation in column E.

The concentrated stream, nominally adipic acid and lower dibasic acid coproducts in 35–50% HNO_3 (organic-free basis), is then cooled and crystallized (F). Crude adipic acid product is removed via filtration or centrifugation (G), and the mother liquor is returned to the oxidizer. Further refining is required to achieve polymer grade material, usually by recrystallization from water. Residual lower dibasic acids, nitrogen-containing impurities, and metals are removed in this step. Additional purification steps occasionally are employed, including slurry washing, further recrystallization, and charcoal treatment. The bleacher off-gas, containing NO and NO_2, is combined with air and absorbed in water generating nitric acid for reuse (D).

In order to control the concentration of lower dibasic acid by-products in the system, a portion of the mother liquor stream is diverted to a purge treatment

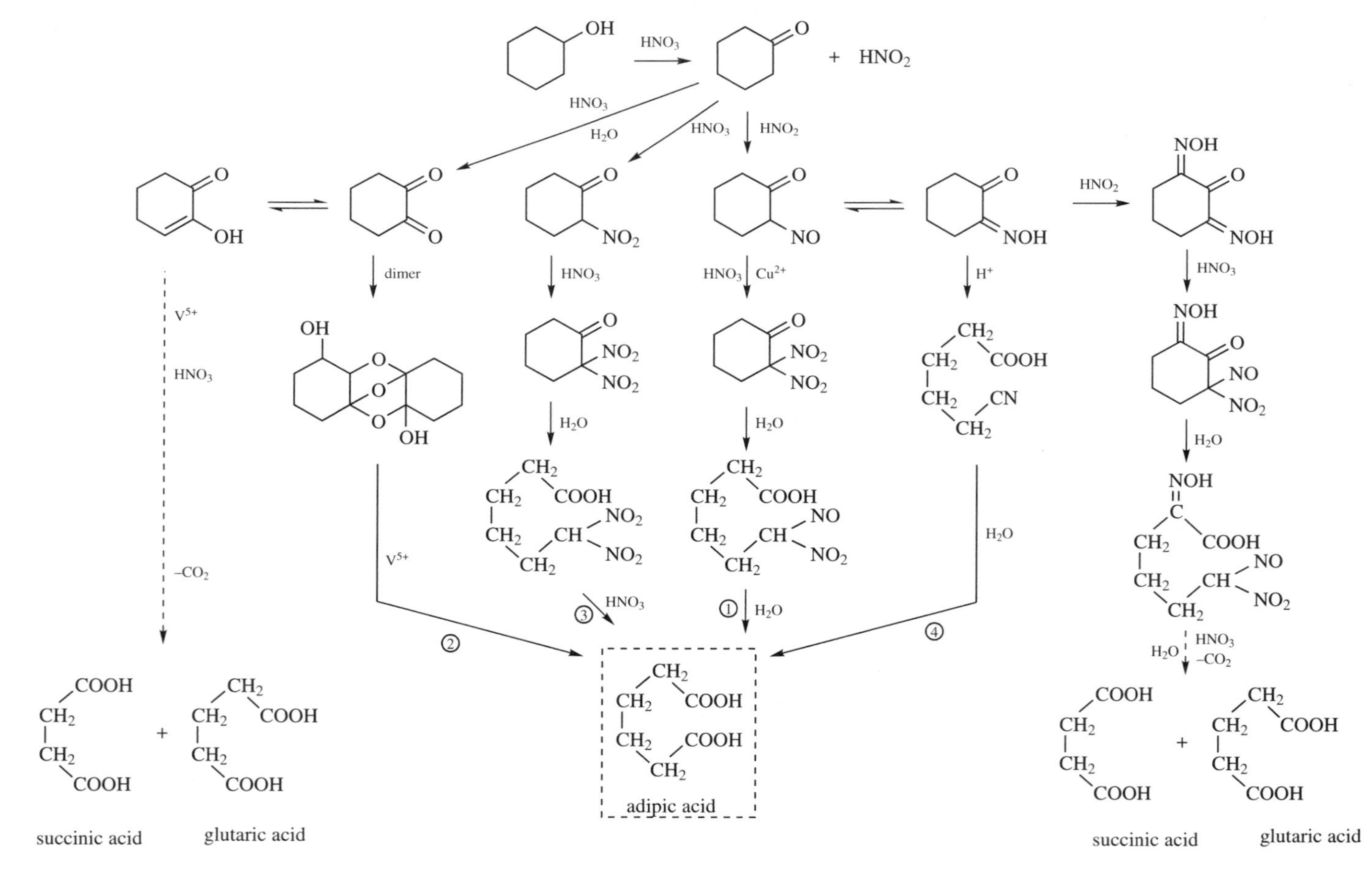

Fig. 4. Reaction paths to adipic acid. Encircled numbers label the pathways discussed in the text.

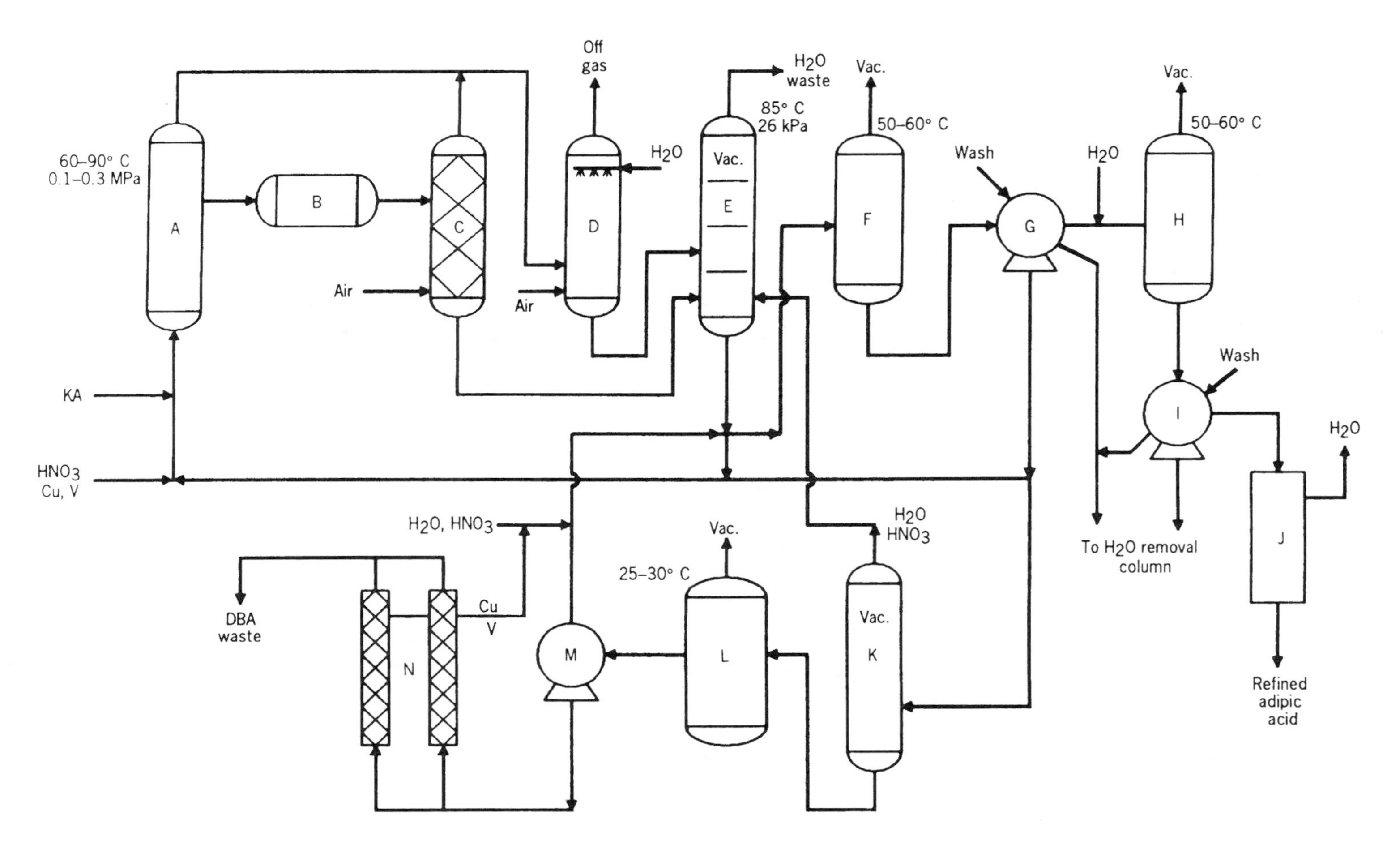

Fig. 5. Typical nitric acid oxidation process. A, reactor; B, optional cleanup reactor; C, bleacher; D, NO_x absorber; E, concentrating still; F, crude crystallizer; G, centrifuge or filter; H, refined crystallizer; I, centrifuge or filter; J, dryer; K, purge evaporator; L, purge crystallizer; M, centrifuge or filter; N, ion-exchange beds; DBA = dibasic acids.

process. Following removal of nitric acid by distillation (Fig. 5, K), copper and vanadium catalyst are recovered by ion-exchange treatment (Fig. 5, N). This area of the process has received considerable attention in recent years as companies strive to improve efficiency and reduce waste. Patents have appeared describing addition of SO_2 to improve ion-exchange recovery of vanadium (115), improved separation of glutaric and succinic acids by dehydration and distillation of anhydrides (116), formation of imides (117), improved nitric acid removal prior to dibasic acid recovery (118), and other claims (119).

Because of the highly corrosive nature of the nitric acid streams, adipic acid plants are constructed of stainless steel, or titanium in the more corrosive areas, and thus have high investment costs.

Nitric acid oxidation may be used to recover value from waste streams generated in the cyclohexane oxidation portion of the process, such as the water wash (120) and nonvolatile residue (76) streams. The nitric acid oxidation step produces three major waste streams: an off-gas containing oxides of nitrogen and CO_2; water containing traces of nitric acid and organics from the water removal column; and a dibasic acid purge stream containing adipic, glutaric, and succinic acids. The off-gas usually is passed through a reducing-flame burner to the atmosphere, or it may be oxidized back to NO_x at 1000–1300°C and recovered as nitric acid, as claimed in a patent (121). The overhead water stream usually is treated (eg, neutralization, biotreatment) and reused. This stream can also be burned or disposed of by deepwell injection or biotreatment along with the waste dibasic acids. However, as more uses for these acids are discovered, the necessity for their disposal diminishes. The principal emissions of concern from these processes are related to nitric acid, either as the various oxides of nitrogen or as a very dilute solution of the acid itself. The fate of these waste streams varies widely, subject to the usually very complex environmental and regulatory situations at each individual manufacturing site. These issues are now a prime consideration, equal to economics, in the design of chemical processing systems in the petrochemical industry (122).

4.4. Other Routes to Adipic Acid. A number of adipic acid processes rely on feedstocks other than cyclohexane and phenol and produce neither K nor A as intermediates. Although these have been investigated, none has been employed at a commercial scale. A one-step air oxidation process, first researched by Halcon (123,124) and Gulf (125) in the 1960s, and developed by Asahi and others (126–130) in the 1970s, uses an acetic acid [64-19-7] solvent for the cyclohexane. High concentrations of soluble cobalt catalyst (60–300 ppm) are used, along with cyclohexanone or acetaldehyde [75-07-0] promoter. Yields to adipic acid of 70–75% are reported at cyclohexane conversions of 50–75%. Reaction temperature is a moderate 70–100°C. References to air oxidation processes have continued to appear through the 1990s (131–136).

It has been known since the early 1950s that butadiene reacts with CO to form aldehydes and ketones that could be treated further to give adipic acid (137). Processes for producing adipic acid from butadiene and carbon monoxide [630-08-0] have been explored since around 1970 by a number of companies, especially ARCO, Asahi, BASF, British Petroleum, Du Pont, Monsanto, and Shell. BASF has developed a process sufficiently advanced to consider commercialization (138). There are two main variations, one a carboalkoxylation and the other a hydrocarboxylation. These differ in whether an alcohol, such as methanol [67-56-1], is used to produce intermediate pentenoates (139), or water is used for the production of intermediate pentenoic acids (140). The former is a two-step process which uses high pressure, >31 MPa (306 atm), and moderate temperatures (100–150°C) (138–141). Butadiene, CO, and methanol react in the first step in the presence of cobalt carbonyl catalyst and pyridine [110-86-1] to produce methyl pentenoates. A similar second step, but at lower pressure and higher temperature with rhodium catalyst, produces dimethyl adipate [627-93-0]. This is then hydrolyzed to give adipic acid and methanol (141), which is recovered for recycle. Many variations to this basic process exist. Examples are ARCOs palladium/copper-catalyzed oxycarbonylation process (142–144), and Monsanto's palladium and quinone [106-51-4] process, which uses oxygen to reoxidize the by-product hydroquinone [123-31-9] to quinone (145).

Other processes explored, but not commercialized, include the direct nitric acid oxidation of cyclohexane to adipic acid (146–149), carbonylation of 1,4-butanediol [110-63-4] (150), and oxidation of cyclohexane with ozone [10028-15-5] (151–154) or hydrogen peroxide [7722-84-1] (155–156). Production of adipic acid as a by-product of biological reactions has been explored in recent years (157–162).

5. Storage, Handling, and Shipping

When dispersed as a dust, adipic acid is subject to normal dust explosion hazards. See Table 7 for ignition properties of such dust–air mixtures. The material is an irritant, especially upon contact with the mucous membranes. Thus protective goggles or face shields should be worn when handling the material. Prolonged contact with the skin should also be avoided. Eye wash fountains, showers, and washing facilities should be provided in work areas. However, *MSDS Sheet 400* (5) reports that no acute or chronic effects have been observed.

The material should be stored in corrosion-resistant containers, away from alkaline or strong oxidizing materials. In the event of a spill or leak, nonsparking equipment should be used, and dusty conditions should be avoided. Spills should be covered with soda ash, then flushed to drain with large amounts of water (5).

Adipic acid is shipped in quantities ranging from 22.7 kg (50-lb bags) to 90.9 t (200,000-lb hopper cars). Upon long standing, the solid material tends to cake, dependent on such factors as initial particle size and moisture content. Shipping data in the United States are "Adipic Acid," *DOT-ID NA 9077, DOT Hazard Class ORM-E*. It is regulated only in packages of 2.3 t (5000 lb) or more (hopper cars and pressure-differential cars and trucks) (163).

6. Economic Aspects

The continuing pursuit of a wide variety of alternate manufacturing processes indicates an effort by competitors to position themselves to take advantage of potential shifts in petrochemical feedstock prices. A large number of the reports concern the use of C_4 feedstocks, notably butadiene, although several major

Table 7. **Worldwide Adipic Acid Capacities**[a]

Company	Location	Capacity, 10^3 t/year
North America		
Du Pont	Orange, Tex	220
Du Pont	Victoria, Tex	380
Du Pont	Maitland, Canada	190
Inolex Chemical Co. (formerly Allied)	Hopewell, va	30
Solutia, Inc. (formerly Monsanto)	Pensacola, Fla	830
Western Europe		
Rhodia Alsachimie SAS	Chalampe, France	355
BASF Aktiengeschaft	Ludwigshafen, Germany	260
Bayer AG	Leverkusen, Germany	55
Radici Chimica SpA	Zeitz, Germany	80[b]
Radici Chimica SpA	Novara, Italy	60
DuPont (U.K.) Ltd	Wilton, U.K.	260
Asia		
CNP Liaoyanf Petrochemical Fiber Co.	Liaoyang, China	100
Pingdingshan Petrochemical Co.	Pingdingshan, China	37
Taiyuan Chemical Industry Group	Taiyuan, China	2
Asahi Chemical Industry Co., Ltd	Nobeoka, Japan	120
Sumitomo Chemical Co., Ltd	Niihama, Japan	2
Kofran Chemical Co., Ltd (100% owned by Rhodia)	Ulsan, Korea	65
DuPont Singapore Pte. Ltd	Pulau Sakra, Sing	114
Nan Ya Plastics Corp.	Mailiao City, Taiwan	40[c]
Other Regions		
Rhodia S.A.	Paulina, Brazil	67
Remaining		15

[a] See Ref. 167.
[b] Scheduled for startup in 2001.
[c] Scheduled for startup in 2002.

modifications to cyclohexane- or benzene-based processes are included. The continued buildup of capacity in nylon-6,6 intermediates, especially in the Far East, attests to the confidence in continued growth by the major participants. Although the nylon-6 [25038-54-4] market currently is larger in Europe, both markets will share in the growth, especially in the developing areas of the world. The emergence of new polymers for specialized applications may tend to limit growth in certain areas. For example, polypropylene may take a significant share of the lower cost carpet market. Specialized polyamides such as nylon-4,6 [24936-71-8] have now appeared, although this one consumes adipic acid. Adipic acid is a very large volume organic chemical. Worldwide production in 1999 reached 2.1×10^6t (4.6×10^9 lb) (30) and by 2004, it is estimated to reach 2.45×10^6t (Table 7). It is one of the top 50 (164) chemicals produced in the United States in terms of volume, with 1999 production estimated at 918,000 t (165). Although the current economic climate has temporarily slowed the adipic acid growth rate, when these conditions improve, growth in demand in the United States is expected to match the 1996–1999 growth of 2.8%/year (166). Table 7 provides individual capacities for U.S. manufacturers. Western European capacity is essentially equivalent to that in the United States at 685,000 t/year. Demand is highly cyclic (167), reflecting the automotive and housing markets especially.

Prices usually follow the variability in crude oil prices. In 1998, adipic acid for was responsible for >50% of U.S. cyclohexane. In 1999 ~84% of U.S. adipic acid production was used in nylon-6,6 (66% fiber and 18% resin), 7.4% in polyurethanes, 3.5% in plasticizers, 2.7% miscellaneous, and 2.44% net exported (166).

Du Pont plans to expand its North American capacity by 150,000 t as a result of growth in the nylon 6,6, polyureathane and plasticizer markets (168).

7. Specifications and Analysis

Because of the extreme sensitivity of polyamide synthesis to impurities in the ingredients (eg, for molecular-weight control, dye receptivity), adipic acid is one of the purest materials produced on a large scale. In addition to food-additive and polyamide specifications, other special requirements arise from the variety of other applications. Table 8 summarizes the more important specifications. Typical impurities include monobasic acids arising from the air oxidation step in synthesis, and lower dibasic acids and nitrogenous materials from the nitric acid oxidation step. Trace metals, water, color, and oils round out the usual specification lists.

Standard methods for analysis of food-grade adipic acid are described in the Food Chemicals Codex (see Refs. in Table 8). Classical methods are used for assay (titration), trace metals (As, heavy metals as Pb), and total ash. Water is determined by Karl–Fisher titration of a methanol solution of the acid. Determination of color in methanol solution (APHA, Hazen equivalent, max. 10), as well as iron and other metals, are also described elsewhere (170). Other analyses frequently are required for resin-grade acid. For example, total nitrogen content is determined by chemiluminescence. Hydrolyzable nitrogen (NH_3, amides, nitriles, etc) is determined by distillation of ammonia from an alkaline solution. Reducible nitrogen (nitrates and nitroorganics) may then be determined by adding DeVarda's alloy and continuing the distillation. Hydrocarbon oil contaminants may be determined by gas chromatographic analysis.

Table 8. **Quality Specifications**

	Application		
Parameter	Food grade	Other	Page reference[a]
melting range	151.5–154.0°C		519
assay	99.6% min		11
water	0.2% max		552
residue on ignition	20.0 ppm max		11
arsenic (as As)	3.0 ppm max		464
heavy metals (as Pb)	10.0 ppm max		11,513
iron (as Fe)		2.0 ppm	
ICV color		5.0 max	
caproic acid		10.0 ppm	
succinic acid		50.0 ppm	
nitrogen		3 ppm	
hydrocarbon oil		10 ppm	

[a] Refers to pages in Ref. 169.

Monobasic acids are determined by gas chromatographic analysis of the free acids; dibasic acids usually are derivatized by one of several methods prior to chromatographing (171,172). Methyl esters are prepared by treatment of the sample with BF_3–methanol, H_2SO_4–methanol, or tetramethylammonium hydroxide. Gas chromatographic analysis of silylation products also has been used extensively. Liquid chromatographic analysis of free acids or of derivatives also has been used (173). More sophisticated high-performance liquid chromatography (hplc) methods have been developed recently to meet the needs for trace analyses in the environment, in biological fluids, and other sources (174,175). Mass spectral identification of both dibasic and monobasic acids usually is done on gas chromatographically resolved derivatives.

8. Health and Safety Factors

Adipic acid is relatively nontoxic; no OSHA PEL or NIOSH REL have been established for the material. Airborne exposure should be limited to 10 mg/m^3 (total dust), the ACGIH TLV-TWA for an organic nuisance dust (5). Toxicity in laboratory animals based on exposure to adipic acid has been reported (176).

eye, rabbit (eye irritant)	20 mg/24 h (SEV)
oral, rat	LDL_o: 3600 mg/kg
intraperitoneal, rat	LD_{50}: 275 mg/kg
oral, mouse	LD_{50}: 1900 mg/kg
intraperitoneal, mouse	LD_{50}: 275 mg/kg
intravenous, mouse	LD_{50}: 680 mg/kg

Adipic acid is excreted essentially unmetabolized in human urine, based on tests with a series of dicarboxylic acids (177). However, adipic acid may be produced via liver metabolism of longer chain diacids, as observed in a recent study with rats (178). The acid has achieved "generally recognized as safe" (GRAS) status from the U.S. Food and Drug Administration for use as a direct ingredient in food for such uses as acidulant, leavening agent, or pH control agent (179). The sodium salt [23311-84-4] has not achieved GRAS status. Maximum permissible usage of the acid in foods was studied with respect to toxicity and teratological and mutagenicity effects (180). No mutagenic or teratological activity was observed (181). Recommended maximum concentration in water reservoirs is 2 mg/L (5).

Adipic acid is an irritant to the mucous membranes. In case of contact with the eyes, they should be flushed with water. It emits acrid smoke and fumes on heating to decomposition. It can react with oxidizing materials, and the dust can explode in admixture with air (see Table 3). Fires may be extinguished with water, CO_2, foam, or dry chemicals.

Airborne particulate matter (182) and aerosol (183) samples from around the world have been found to contain a variety of organic monocarboxylic and dicarboxylic acids, including adipic acid. Traces of the acid found in southern California air were related both to automobile exhaust emission (184) and, indirectly, to cyclohexene as a secondary aerosol precursor (via ozonolysis) (185). Dibasic acids (eg, succinic acid) have been found even in such unlikely sources as the Murchison meteorite (186). Public health standards for adipic acid contamination of reservoir waters were evaluated with respect to toxicity, odor, taste, transparency, foam, and other criteria (187). Biodegradability of adipic acid solutions was also evaluated with respect to BOD/theoretical oxygen demand ratio, rate, lag time, and other factors (188).

8.1. N_2O Abatement Technology. During the 1990s, the adipic acid industry updated its offgas control technology (189). These process modifications were based on the disclosure that nitrous oxide (N_2O), a gas-phase byproduct of adipic acid manufacture, is the major source of strasospheric nitric oxide (NO) and thus has a global warming potential many times more than CO_2 (190). At 0.15–0.3 tons of N_2O per ton of adipic acid, these emissions were cited as the source of recent measured increases in atmospheric levels.

Reducing flame burner technology represents the high-temperature option to N_2O abatement (1200–1500°C). Here, natural gas reduces N_2O to nitrogen, CO_2 and water. Staged injection of fuel and air abates the N_2O and organics present in the vent gas. Flue gas recycle and/or a convection section provide additional residence time. With these two features, N_2O destruction efficiencies >99% are obtained without an increase in NO_x emissions.

Other options commercially practiced depend on the use of a catalyst. An intermediate temperature process (1000–1500°C) developed by DuPont and Rhone Poulenc is based on the catalytic reaction of N_2O to NO. This product is then followed by an air oxidation forming NO_2 which is subsequently absorbed in water. The final product, nitric acid, is then recycled back to the adipic acid process. The low-temperature catalytic process (400–700°C) is designed to destroy N_2O without the formation of NO_x. Such facilities can be installed with or without heat recovery depending on the value of steam. Since startup in the mid-1990s, they have been reported to achieve N_2O abatement efficiencies of 98% or better (191).

Table 9 lists the major adipic acid manufacturers and the N_2O Abatement Technology practiced at their respective site.

Table 9. **N_2O Abatement Technology Practiced at Major Adipic Acid Production Sites**[a]

Site	Technology	Start-up Date
Dupont Singapore	Thermal	1994
Dupont Orange, Tex	Catalytic	1996
Dupont Maitland, Ontario	Catalytic	1997
Dupont Victoria, Tex	Catalytic	1997
Dupont Wilton, U.K.	Thermal	1998
Asahi	Thermal	1999
BASF	Catalytic	1997
Bayer	Thermal	1993
Rhodia	Conversion to HNO_3	1998
Solutia	Thermal	1972

[a] See Ref. 189.

9. Uses

About 86% of U.S. adipic acid production is used captively by the producer, almost totally in the manufacture of nylon-6,6 (192). The remaining 14% is sold in the merchant market for a large number of applications. These have been developed as a result of the large scale availability of this synthetic petrochemical commodity. Prices for 1960–1989 for standard resin-grade material have paralled raw material and energy costs (petroleum and natural gas) growing at a rate of 1.7%/year. In the early 1990s, the price leveled off around 1.37 $/kg. By the late 1990s, it jumped to 1.53 $/kg (193).

In 1999, 66% of U.S. demand for adipic acid was for nylon-6,6 fiber, while 11% was used in nyon-6,6 resins (166). In Western Europe only about 40% was for polyamide. Nylon-6,6 resins were distributed between injection molding (90%) for such applications as automotive and electrical parts.

Less than 5% of the U.S. polyurethanes market in 1999 was derived from the condensation product of polyisocyanates with low molecular weight polyadipates having hydroxyl end groups (166). In 1999 this amounted to 68,000 t, or 84% of total adipic acid consumption. The percentage in Western Europe was closer to 20%. About 90% of these adipic acid containing polyurethanes are used in flexible or semirigid foams and elastomers, with the remainder used in adhesives, coatings, and spandex fibers.

About 4% of U.S. adipic acid consumed in 1999 was used in two basic types of adipic ester based plasticizers (194). Simple adipate esters prepared from C_8–C_{13} alcohols are used especially as PVC plasticizers (qv). For special applications requiring low volatility or extraction resistance, polyester derivatives of diols or polyols are preferred.

1,6-Hexanediamine, the second ingredient in the production of Nylon-6,6 polyamide, is prepared by hydrogenation of adiponitrile [111-69-3]. For many years, the nitrile was produced from adipic acid by dehydration of the ammonium salt (195); however, this process is no longer used in the United States or Western Europe. New processes based on propylene and butadiene have supplanted this technology in the United States. For several years, Du Pont operated a process based on the chlorination and cyanation of butadiene, but this was shut down in 1983 (196,197). Du Pont produces adiponitrile at two large U.S. plants and one French joint venture by direct nickel(0)-catalyzed homogeneous hydrocyanation of butadiene (198). Monsanto/Solutia and Asahi developed and practiced the electrolytic coupling of acrylonitrile process, used in the United States, Western Europe, and Japan (199,200).

About 2.4% of U.S. consumption in 1999 was distributed among several other applications, amounting to several thousand tons each (193). Wet-strength resins based on polyamide–epichlorohydrin products consumed about 16,000–18,000 t in 1998. Unsaturated polyester resins (4000 t in 1998) are used in surface coatings, flexible alkyd resins (qv), coil coatings, and other coatings because of their curing properties. Adipic acid also is used as a food acidulant in jams, jellies, and gelatins. Although it has only 2% of the acidulant market, 3200 t were used for this purpose in 1989 (201). The synthetic lubricant market consumed about 7000 t as the C_{8-13} adipate esters in 1998, for gas turbines, compressors, and military jet engines. An environmentally significant use of the acid, and especially its dibasic acid by-products, is as a buffer in the scrubbing operation of power plant flue gas desulfurization (202–206). Adipoyl chloride is occasionally used as a softening agent for leather.

BIBLIOGRAPHY

"Adipic Acid" under "Acids, Dicarboxylic" in *ECT* 1st ed., Vol. 1, pp. 153–154, by P. F. Bruins, Polytechnic Institute of Brooklyn; "Adipic Acid" in *ECT* 2nd ed., Vol. 1, pp. 405–421, by W. L. Standish and S. V. Abramo, E. I. du Pont Nemours & Co., Inc.; in *ECT* 3rd ed., Vol. 1, pp. 510–531, by D. C. Danly and C. R. Campbell, Monsanto Chemical Intermediates Company; in *ECT* 4th ed., Vol. 1. pp. 466–493, by Darwin D. Davis and Donald R. Kemp, E.I. du Pont de Nemours and Co., Inc.; "Adipic Acid" in *ECT* (online), posting date: December 4, 2000, by Darwin D. Davis and Donald R. Kemp, E. I. Du Pont de Nemours and Co.

CITED PUBLICATIONS

1. E. Bolton, *Ind. Eng. Chem.* **34**, 53 (1942).
2. H. Serwy, *Bull. Soc. Chim. Belg.* **42**, 483 (1933).
3. Armour Research Foundation, *Anal. Chem.* **20**, 385 (1948).
4. S. Khetarpal, L. Krishan, and H. Bhatnager, *Ind. J. Chem., Sect. A.* **19A**, 516 (1986).
5. P. Igoe, D. Wilson, and W. Silverman, *Material Safety Data Sheet No. 400* in *Genum's Reference Collection*, Genum Publishing Corp., Schenectady, N.Y., 1989.
6. A. Granovskaya, *Zh. Fiz. Khim.* **21**, 967 (1947).
7. F. Kraft and H. Noerdlinger, *Ber. Dtsch. Chem. Ges.* **22**, 818 (1889).
8. A. Van Dooren and B. Mueller, *Thermochim. Acta* **54**(1–2), 115 (1982).
9. P. E. Verkade, H. Hartman, and J. Coops, *Recl. Trav. Chim. Pays-Bas* **45**, 380 (1926).
10. I. Contineanu, E. Corlateanu, J. Herscovici, and I. Dumitri, *Rev. Chim. (Bucharest)* **31**, 763 (1980).
11. S. Kahnyakina and G. Puchkovskaya, *Zh. Prikl. Spektrosk.* **34**, 885 (1981).
12. Y. Morechal, *Can. J. Chem.* **63**, 1684 (1985).
13. K. Urano, K. Kawamoto, and K. Hayoshi, *Yosui To Haisui* **23**(2), 196 (1981).
14. A. Babinkov and co-workers, *Termodin. Organ. Sordin. Gor'kii* **1979**(8), 28 (1979).
15. A. Apelblat, *J. Chem. Thermodyn.* **18**, 351 (1986).
16. I. Jones and R. Soper, *J. Chem. Soc.* **1936**, 135 (1936).
17. J. Burgot, *Talanta* **25**, 233 (1978).
18. A. Apelblat and E. Manzurola, *J. Chem. Thermodyn.* **19**, 317 (1987).
19. *Adipic Acid, Product Bulletin E-99079-1*, E. I. du Pont de Nemours & Co., Inc., 1989.
20. C. S. Marvel and J. C. Richards, *Anal. Chem.* **21**, 1480 (1949).
21. K. Chow, J. Go, M. Mehdizadeh, and D. Grant, *Int. J. Pharm.* **20**(1–2), 3 (1984).
22. L. Hus, C. Chang, J. Beddow, and A. Vetter, *Proc. Tech. Prog. Int. Powder and Bulk Solids Handling and Processing, Atlanta, Georgia, May 24–26, 1983*, Books Demand UMI, Ann Arbor, Mich., 1983, pp. 52–66.
23. O. Georgescu, S. Ivascon, and M. Apostolescu, *Rev. Chim. (Bucharest)* **36**, 839 (1985).
24. *Corrosion Data Survey, Metals Section*, 5th and 6th ed., National Association of Corrosion Engineers, Houston, Tex., 1974 and 1985, pp. 6 and 4.
25. G. Lunn, *J. Hazard. Mater.* **17**(2), 207 (1988).
26. E. Scholl and co-workers, *Inst. Explos. Sprengtech. Bergbau-Versuchstrecke FRG, STF-Rep.* **1979**(2), 99 (1979).

27. D. Felstead, R. Rogers, and D. Young, *Conf. Ser.—Inst. Phys. (Electrostatics), ICI, United Kingdom* **66**, 105 (1983).
28. U.S. Pat. 4,375,552 (Mar. 1, 1983), V. Kuceski (to C. P. Hall Co.).
29. M. Seko, A. Yomiyama, and T. Isoya, *Chem. Econ. Eng. Rev.* **11**(9), 48 (1979).
30. R. T. Gerry, "Adipic Acid" in *Chemical Economics Handbook, Marketing Research Report*, SRI International, Menlo Park, Calif., 1987, p. 608.5000A, p. 4.
31. M. I. Kohan, *Nylon Plastics*, John Wiley & Sons, Inc., New York, 1973, pp. 14–73.
32. Ger. Offen. 3,510,876 (Oct. 2, 1986), W. Hoelderich and co-workers (to Badische Anilinund Soda-Fabrik A. G.).
33. Jpn. Kokai Tokkyo Koho 6105,036 (Jan. 10, 1986), K. Tsukada, N. Fukuoka, and I. Kinoshita (to Kao Corp.).
34. Jpn. Kokai Tokkyo Koho 80 04,090 (Jan. 29, 1980), J. Kanetaka and S. Mori (to Mitsubishi Petrochemical Co.).
35. Fr. Pat. 1,509,288 (Jan. 12, 1968), P. Volpe and W. Humphrey (to Celanese Corp.).
36. W. Hentzchel and J. Wislicenus, *Justus Leibigs Ann. Chem.* **275**, 312 (1983).
37. G. Vavon and A. Apchie, *Bull. Soc. Chim. Fr.* **43**, 667 (1928).
38. J. Dullinga, N. Nibbering, and A. Boerboom, *J. Chem. Soc. Perkin Trans.* **2**, 1065 (1984).
39. J. Sheehan, R. O'Neill, and M. White, *J. Am. Chem. Soc.* **72**, 3376 (1950).
40. P. Morgan, *J. Chem. Educ.* **36**(4), 182 (1959).
41. Ger. Offen. 3,235,938 (Apr. 21, 1983), K. Kimura and T. Isoya (to Asahi Chemical Industries, Ltd.).
42. I. V. Berezin, E. T. Denisov, and N. M. Emanuel, *The Oxidation of Cyclohexane*, Pergamon Press, Oxford, England, 1965.
43. S. A. Miller, *Chem. Process Eng.*, **1969**, 63 (June 1969).
44. Tamarapu Sridhar, *Mass Transfer in Cyclohexane Oxidation*, Ph.D. Thesis, Department of Chemical Engineering, Monash Univ., Australia, 1978.
45. V. D. Luedeke, "Adipic Acid", in *Encyclopedia of Chemical Process and Design*, J. McKetta and W. Cunningham, eds. Vol. 2, Marcel Dekker, Inc., New York, 1977, pp. 128–146.
46. A. K. Suresh, T. Shidhar, and O. E. Potter, *AIChE J.* **34** (1), 55 (1988).
47. Y. C. Yen and S. Y. Wu, *Nylon 6,6, PEP Report 54B*, SRI International, Menlo Park, Calif., 1987.
48. U.S. Pat. 4,238,415 (Dec. 9, 1980), W. O. Bryan (to Stamicarbon N. V.).
49. L. Bateman, H. Hughes, and A. L. Morris, *Faraday Discuss. Chem. Soc.* **1953**(14), 190 (1953).
50. D. G. Hendry and co-workers, *J. Org. Chem.* **41**, 2 (1976).
51. U.S. Pat. 3,530,185 (Sept. 22, 1970), K. Pugi (to E. I. du Pont de Nemours & Co., Inc.).
52. *Hydrocarbon Process.* **48**, 163 (1969).
53. U.S. Pat. 4,163,027 (July 31, 1979), P. Magnussen, G. Herrman, and E. Frommer (to Badische Anilin- und Soda-Fabrik A. G.).
54. Ger. Offen. 3,328,771 (Feb. 28, 1985), Stoessel and co-workers (to Badische Anilin-und Soda-Fabrik A. G.).
55. U.S. Pat. 4,704,476 (Nov. 3, 1987), J. Hartig, G. Herrman, and E. Lucas (to Badische Anilin- und Soda-Fabrik A. G.).
56. U.S. Pat. 3,957,876 (May 18, 1976), M. Rapoport and J. O. White (to E. I. du Pont de Nemours & Co., Inc.).
57. U.S. Pat. 3,987,100 (Oct. 19, 1976), W. J. Barnette, D. L. Schmitt, and J. O. White (to E. I. du Pont de Nemours Co., Inc.).
58. U.S. Pat. 4,465,861 (Aug. 14, 1984), J. Hermolin (to E. I. du Pont de Nemours & Co., Inc.).

59. U.S. Pat. 3,925,316 (Dec. 9, 1975), J. C. Brunie, N. Creene, and F. Maurel (to Rhône-Poulenc S. A.).
60. Adapted from L. L. van Dierendonck and J. A. de Leeuw den Bouter, *PT/Procestechniek* **39**(3), 44–48 (1984).
61. U.S. Pat. 3,923,895 (Dec. 2, 1975), M. Costantini, N. Creene, M. Jouffret, and J. Nouvel (to Rhône-Poulenc S. A.).
62. U.S. Pat. 3,927,105 (Dec. 16, 1975), J. C. Brunie and N. Creene (to Rhône-Poulenc S. A.).
63. U.S. Pat. 4,326,085 (Apr. 20, 1982), M. De Cooker (to Stamicarbon N. V.).
64. Eur. Pat. 092,867 (Nov. 2, 1983), J. G. Housmans and co-workers (to Stamicarbon N. V.).
65. U.S. Pat. 3,932,513 (Jan. 13, 1976), J. L. Russell (to Halcon International, Inc.).
66. U.S. Pat. 3,796,761 (Aug. 18, 1971), Marcell and co-workers (to Halcon International, Inc.).
67. Brit. Pat. 1,590,958 (June 10, 1981), J. F. Risebury (to Imperial Chemical Industries, Ltd.).
68. J. Alagy and co-workers, *Hydrocarbon Process.* **47**(12), 131 (1968).
69. H. Van Landeghem, *Ind. Eng. Chem. Process. Des. Dev.* **13**, 317 (1974).
70. U.S. Pat. 3,895,067 (Jan. 12, 1973), G. H. Mock and co-workers (to Monsanto Co.).
71. E. F. J. Duynstee and co-workers, *Recl. Trav. Chim. Pays-Bas* **89**, 769 (1970).
72. Ref. (50), pp. 1–5.
73. C. T. Chi and J. H. Lester, *Presentation to MCA Waste Minimization Workshop*, New Orleans, La., Nov. 11–13, 1987.
74. U.S. Pat. 3,260,743 (July 12, 1966), W. B. Hogeman (to Monsanto Co.); U.S. Pat. 3,365,490 (Jan. 23, 1968), W. J. Arthur and L. S. Scott (to E. I. du Pont de Nemours & Co., Inc.); U.S. Pat. 3,969,465 (July 13, 1976), J. K. Brunner (to Badische Anilin- und Soda-Fabrik A. G.); U.S. Pat. 4,105,856 (Aug. 8, 1978), C. A. Newton (to El Paso Products, Co.); K. J. Mehta and co-workers, *Chem. Eng. World* **24**(30), 63 (1989).
75. U.S. Pat. 3,993,691 (Nov. 23, 1976), J. K. Brunner (to Badische Anilin- und Soda-Fabrik A. G.); U.S. Pat. 4,166,056 (Aug. 28, 1979), K. P. Satterly and F. E. Livingston (to Witco Chemical Corp.); U.S. Pat. 4,233,408 (Nov. 11, 1980), K. P. Satterly and F. E. Livingston (to Witco Chemical Corp.).
76. Jpn. Pat. 82-041456-B (Sept. 3, 1982), (to Sumitomo Chemical Co.).
77. B. Harris and B. Tichenor, *Proceedings of 74th Annual Meeting*, Air Pollution Control Association, Pittsburgh, Pa., 1981, Vol. 3, paper 81–41.5.
78. Ref. 43, p. 69.
79. U.S. Pat. 3,243,449 (Mar. 29, 1966), C. N. Winnick (to Halcon International, Inc.)..
80. *Eur. Chem. News* **15**, 22 (May 2, 1969).
81. *Eur. Chem. News* **11**, 32 (June 9, 1967).
82. U.S. Pat. 3,287,423 (Nov. 22, 1966), J. Steeman and J. von den Hoff (to Stamicarbon, N. V.).
83. A. Uriarte, M. Rodkin, M. Gross, A. Kharitonov, and G. Panov, *Studies in surface Science and Catalysis* **110**, 857 (1997).
84. G. Panov, G. I. G. Sheveleva, G. A. Kharitonov, A. V. Romannikov, and L. Vostrikova, *Appl. Catal.* **82**(1), 31–6 (1992).
85. J. Oppenheim, *16th North American Catalysis Society Meeting*, Boston, Mass. June 2, 1999.
86. C. Buechler, J. Ebner, M. Gross, W. McGhee, J. Morries, E. Sall, and A. Uriarte, PCT Int. Appl. (1998).
87. I. Dodgson, K. Griffin, G. Barberis, F. Pignataro, and G. Tauszik, *Chem. Ind.* (London) **(24)**, 830–833 (1989).

88. H. Naumann, H. Schaefer, H. Oberender, D. Timm, H. Meye, and G. Pohl, *Chem. Tech.* (Leipzig) **29(1)**, 38 (1977) .
89. U.S. Pat. 2,794,056 (May 28, 1957), L. O. Winstrom (to Allied Chemical Co.).
90. Brit. Pat. 979,268 (Jan. 1, 1965), J. G. Mather and F. G. Webster (to Imperial Chemical Industries, Ltd.).
91. U.S. Pat. 3,987,100, (1976) Barnette, (to E.I. du Pont de Nemours & Co., Inc.).
92. *Jpn. Chem. Week*, 5 (Oct. 29, 1987).
93. *Comline Chemicals and Materials*, Comline News Service, Tokyo 107, Japan, Feb. 22, 1988.
94. Jpn. Pat. 62-205037 (1987) H. Nagahara, M. Konishi (to Asahi Chem. Ind. Co Ltd).
95. Jpn. Pat. 04-41442 (1992) S. Kodama, K. Nakagawa (to Asahi Chem. Ind. Co. Ltd).
96. Jpn. Pat. 04-41441 (1992) S. Kodama, K. Nakagawa (to Asahi Chem. Ind. Co. Ltd).
97. Jpn. Pat. 01-192.717 (1989) H. Ishida, K. Nakagawa (to Asahi Chem. Ind. Co. Ltd).
98. Jpn. Pat. 04-41448 (1993) S. Kodama, K. Nakagawa (to Asahi Chem. Ind. Co. Ltd).
99. U.S. Pat. 3,987,115 (Oct. 19, 1976), J. G. Zajacek and F. J. Hilbert (to Atlantic Richfield Co.).
100. U.S. Pat. 4,080,387 (Mar. 21, 1978), J. C. Jubin, I. E. Katz, and R. G. Tave (to Atlantic Richfield Co.).
101. T. T. Shih and W. J. Klingebiel, paper presented to *The First Shanghai International Symposium on Technology of Petroleum and Petrochemical Industry*, May 16–20, 1989, Shanghai, China.
102. U.S. Pat. 2,557,282 (1951), C. Hamblett and A. Mac Alevy (to E. I. du Pont de Nemours & Co., Inc.).
103. U.S. Pat. 2,703,331 (1953), M. Goldbeck and F. Johnson (to E. I. du Pont de Nemours & Co., Inc.).
104. H. Godt and J. Quinn, *J. Am. Chem. Soc.* **78**, 1461 (1956).
105. I. Lubyanitski, R. Minati, and M. Furman, *Russ. J. Phys. Chem. (Engl. trans.)*, **32**, 294 (1962).
106. I. Lubyanitski, *Zh. Obshch. Khim.* **36**, 343 (1962).
107. I. Lubyanitski, *Zh. Prikl. Khim. (Leningrad)* **36**, 819 (1963).
108. D. van Asselt and W. van Krevelen, *Chem. Eng. Sci.* **18**, 471 (1963).
109. D. van Asselt and W. van Krevelen, *Recl. Trav. Chim. Pays-Bas* **82**, 51, 429, 438 (1963).
110. U.S. Pat. 3,758,564 (Sept. 11, 1973), D. Davis (to E. I. du Pont de Nemours & Co., Inc.).
111. U.S. Pat. 3,564,051 (1971), E. Haarer and G. Wenner (to Badische Anilin- und Soda-Fabrik A. G.).
112. Brit. Pat. 1,092,603 (1969), G. Riegelbauer, A. Wegerich, A. Kuerzinger, and E. Haarer (to Badische Anilin- und Soda-Fabrik A. G.).
113. T. Hearfield, *Chem. Eng. (London)* **1980** (361), 625 (1980).
114. U.S. Pat. 3,359,308 (Dec. 19, 1967), O. Sampson (to E. I. du Pont de Nemours & Co., Inc.).
115. Eur. Pat. Appl. 122-249A1 (Oct. 17, 1984), C. Hsu and D. Laird (to Monsanto Co.).
116. U.S. Pat. 4,254,283 (Mar. 3, 1981), G. Mock (to Monsanto Co.).
117. Ger. Offen. 3,002,256 (July 30, 1981), W. Rebafka, G. Heilen, and W. Klink (to Badische Anilin- und Soda-Fabrik A. G.).
118. U.S. Pat. 4,014,903 (Mar. 29, 1977), W. Moore (to Allied Chemical Corp.).
119. Brit. Pat. 1,480,480 (July 20, 1977), A. Bowman (to Imperial Chemical Industries, Ltd.).
120. U.S. Pat. 4,227,021 (1980), O. Grosskinsky and co-workers (to Badische Anilin- und Soda-Fabrik A. G.).

121. Jpn. Kokai Tokkyo Koho, JP61-257940 (Nov. 15, 1986), T. Sakamoto, H. Suga, and T. Sakasegawa (to Asahi Chemical Industries Co., Ltd.).
122. S. Fathi-Afshar and J. Yang, *Chem. Eng. Sci.* **40**, 781 (1985).
123. Brit. Pat. 956,779 (Apr. 29, 1964) (to Halcon International, Inc.).
124. Brit. Pat. 956,780 (Apr. 29, 1964) (to Halcon International, Inc.).
125. U.S. Pat. 3,231,608 (Jan. 22, 1966), J. Kollar (to Gulf Research and Development Corp.).
126. Jpn. Pat. 45-16444 (June 8, 1970), G. Inoue and co-workers (to Asahi Chemical Industries, Ltd.).
127. Jpn. Pat. 50-116415 (Sept. 11, 1975), S. Furuhashi (to Asahi Chemical Industries, Ltd.).
128. Jpn. Pat. 51-29427 (Mar. 12, 1976), M. Nishino and co-workers (to Toray Industries).
129. U.S. Pat. 4,032,569 (June 28, 1977), A. Onopchenko and co-workers (to Gulf Research and Development Corp.).
130. K. Tanaka, *Chem. Technol.* **4**(9), 555 (1974).
131. Jpn. Pat. 58-021642 (Feb. 8, 1983), M. Suematsu and K. Nakaoka (to Toray Industries).
132. H. Shen and H. Weng, *Ind. Eng. Chem. Res.* **27**, 2254 (1988).
133. E. Sorribes, J. Navarro, A. Romero, and L. Jodra, *Rev. R. Acad. Cienc. Exactas, Fis. Nat. Madrid* **81**(1), 233 (1987).
134. D. Rao and R. Tirukkoyilur, *Ind. Eng. Chem. Process Des. Dev.* **25**(1), 299 (1986).
135. U.S. Pat. Appl. No.PCT/US97/10830 (2002) Aldrich, M. Sharon; Decoster, C. David; Vassiliou, Eustathious; Dassel, W. Mark; Rostami, M. Ader (RPC Inc., USA).
136. C. Xu, J. Oyloe; Richardson, and E. David, *Single-step synthesis of adipic acid by catalytic oxidation of cyclohexane in air*. 218th ACS National Meeting, New Orleans, August 22–26, 1999.
137. D. Forster and J. F. Roth, eds., *Homogeneous Catalysis II* (Advances in Chemistry Series 132), American Chemical Society, Washington, D.C. 1974; H. Adkins and co-workers, *J. Org. Chem.* **17**,980–987 (1952); U.S. Pat. 2,729,651 (Jan. 3, 1956), W. Reppe (to Badische Anilin-und Soda-Fabrik A. G.); USSR Pat. 198,324 (June 28, 1967), N. S. Imyanitov and co-workers; U.S. Pat. 3,509,209 (Apr. 28, 1970), D. M. Fenton (to Union Oil of California); U.S. Pat. 3,876,695 (Apr. 8, 1975), N. Von Kutepow (to Badische Anilin- und Soda-Fabrik A. G.).
138. *Chem. Eng. News*, 14 (May 25, 1987).
139. S. Hosaka and co-workers, *Tetrahedron* **27**, 3821 (1971); W. E. Billeys and co-workers, *Chem. Commun.* 1067 (1971).
140. Ger. Offen. 2,630,086 (Jan. 12, 1978), H. Schneider and co-workers (to Badische Anilin-und Soda-Fabrik A. G.); U.S. Pat. 4,316,047 (Feb. 16, 1982), R. Kummer and co-workers (to Badische Anilin- und Soda-Fabrik, A. G.).
141. U.S. Pat. 4,169,956 (Oct. 2, 1979), R. Kummer and co-workers (to Badische Anilin-und Soda-Fabrik A. G.); U.S. Pat. 4,171,451 (Oct. 16, 1979), R. Kummer and co-workers (to Badische Anilin- und Soda-Fabrik A. G.); U.S. Pat. 4,360,695 (Nov. 23, 1982), P. Magnussen and co-workers (to Badische Anilin- und Soda-Fabrik A. G.).
142. U.S. Pat. 4,171,450 (Oct. 16, 1979), H. S. Kesling, Jr., and co-workers (to Atlantic Richfield Co.).
143. U.S. Pat. 4,166,913 (Sept. 4, 1979), H. S. Kesling, Jr., and co-workers (to Atlantic Richfield Co.).
144. U.S. Pat. 4,195,184 (Mar. 25, 1980), H. S. Kesling, Jr., and co-workers (to Atlantic Richfield Co.).
145. U.S. Pat. 4,575,562 (Mar. 11, 1986), C. K. Hsu and co-workers (to Monsanto Co.).
146. U.S. Pat. 3,306,932 (Feb. 28, 1967), D. D. Davis (to E. I. du Pont de Nemours & Co., Inc.).

147. U.S. Pat. 3,654,355 (Nov. 19, 1969), W. H. Mueller and co-workers (to Monsanto Co.).
148. U.S. Pat. 3,636,100 (Jan. 18, 1972), W. H. Mueller and co-workers (to Monsanto Co.).
149. U.S. Pat. 3,636,101 (Jan. 18, 1972), T. F. Doumani (to Union Oil Co. of California).
150. Brit. Pat. 1,278,353 (June 21, 1972), H. Arnold and co-workers (to Monsanto Co.).
151. Brit. Pat. 1,239,224 (July 14, 1971), C. Gardner (to Imperial Chemical Industries, Ltd.).
152. U.S. Pat. 3,607,926 (Sept. 21, 1971), R. D. Smetana (to Texaco, Inc.).
153. Fr. Add. 96,191 (May 19, 1972), C. Gardner (to Imperial Chemical Industries, Ltd.); Jpn. Pat. 56-5374 (Feb. 4, 1981), S. Miyazaki (to Agency of Industrial Sciences and Technology).
154. Brit. Pat. 1,361,749 (July 31, 1974), S. D. Razumovskii and co-workers (to USSR).
155. Fr. Pat. 2,140,088 (Dec. 1, 1973), O. Grosskinsky, G. Herrmann, and R. Kaiser (to Badische Anilin- und Soda-Fabrik A. G.).
156. Jpn. Pat. 54-135720 (Oct. 22, 1979), Y. Ishii (to Yasutaka).
157. U.S. Pat. 3,912,586 (Oct. 14, 1975), H. Kaneyuki and co-workers (to Mitsui Petrochemical Industries).
158. Jpn. Pat. 57-129694 (Aug. 11, 1982), H. Nakano and co-workers (to Dainippon Ink and Chemicals).
159. U.S. Pat. 4,400,468 (Aug. 23, 1983), M. Faber (to Hydrocarbon Research).
160. Jpn. Pat. 58-149687 (Sept. 6, 1983), T. Minoda, T. Oomori, and H. Narishima (to Nissan Chemical Industries).
161. Eur. Pat. 74,169 (Mar. 16, 1983), P. C. Maxwell (to Celanese Corp.).
162. *Chem. Econ. Eng. Rev.* 35 (Jan./Feb. 1986).
163. Hazardous Materials Table, *Code of Federal Regulations 49CFR 172.101* (revised Nov. 1989).
164. *Chem. Eng. News* **67**(15), 12 (1989).
165. Ref. 30, p. 13.
166. Ref. 30, p. 16.
167. Ref. 30, p. 12, 31, 33, 37, 41, 43.
168. *Chem. Week*, **160**(12), 21 (1 Apr, 1998).
169. *Food Chemicals Codex*, 3rd ed., National Academy of Sciences, National Academy Press, Washington, D.C., 1981.
170. R. Keller in F. Snell and C. Hilton, eds., *Encyclopedia of Industrial Chemicals Analysis*, Vol. 4, Wiley-Interscience, New York, 1967, pp. 408–423.
171. "Chromatography" in R. Freis and J. Lawrence, eds., *Derivatization in Analytical Chemistry*, Vol. 1, Plenum Press, New York, 1981.
172. J. Drozd, *Chemical Derivatization in Gas Chromatography*, Elsevier, Amsterdam, The Netherlands, 1980.
173. R. Schwarzenbach, *J. Chromatogr.* **251**, 339 (1982).
174. M. Gennaro and co-workers, *Ann. Chim. (Rome)* **78**(3–4), 137 (1988).
175. G. Lippe and co-workers, *Clin. Biochem.* **20**(4), 275 (1987).
176. N. Sax, *Dangerous Properties of Industrial Materials*, Van-Nostrand Reinhold Co., New York, 1984, p. 141.
177. J. Svendsen, L. Sydnes, and J. Whist, *Spectrosc. Inst. J.* **3**(4–5), 380 (1984).
178. J. Vamecq, J. Draye, and J. Brison, *Am. J. Physiol.* **256**(4, pt. 1), G680–688 (1989).
179. *Federal Register* **47**(123) (June 25, 1982).
180. Y. Hirayama, *Shokuhin Eisei Kenkyu* **33**, 852 (1983).
181. H. Shimuzu and co-workers, *Sangyo Igaku* **27**, 400–419 (1985); see also D. Guest and co-workers, eds., *Patty's* Industrial Hygiene and Toxicology, 3rd ed., Vol. 2C, Wiley-Interscience, New York, 1982, p. 4945.
182. Y. Yokouchi and Y. Ambi, *Atmos. Environ.* **20**, 1727 (1986).

183. R. Ferek, A. Lazrus, P. Haagenson, and J. Winchester, *Environ. Sci. Technol.* **17**, 315 (1983).
184. K. Kawamura and I. Kaplan, *Environ. Sci. Technol.* **21**, 105 (1987).
185. B. Appel and co-workers, *Environ. Sci. Technol.* **13**, 98 (1979).
186. E. Pelzer, J. Bada, G. Schlesinger, and S. Miller, *Adv. Space. Res.* **4**(12), 69 (1984).
187. Y. Novikov and co-workers, *Gig. Sanit.* **1983**(9), 72 (1983).
188. K. Urano and Z. Kato, *J. Hazard Mater.* **13**(2), 147 (1986).
189. R. Reimer, C. Slaten, M. Seapan, T. Koch, and V. Triner, *Proceedings of the International Symposium, 2nd, Noordwijkerhout, Netherlands, September 8–10, 1999 (2000), Meeting Date 1999, 347–358.*
190. M. Thiemens, and W. Trogler, *Science*, **251**, 932 (1991).
191. Seapan et al, Proceedings of the AlChE Annual Meeting. Los Angeles. Calif. (nov. 1997). *Proceedings of the AlChE Annual Meeting. Los Angeles, Ca (Nov. 1997).*
192. Ref. 30, p. 6.
193. Ref. 30, p. 27.
194. Ref. 30, p. 23.
195. Ref. 30, p. 24.
196. J. Szymanowski and A. Sobczynska, *Przem. Chem.* **66**, 373 (1987).
197. U.S. Pat. 2,680,761 (1952), R. Halliwell (to E. I. du Pont de Nemours & Co., Inc.).
198. U.S. Pat. 2,518,608 (1947), M. Farlow (to E. I. du Pont de Nemours & Co., Inc.).
199. *Eur. Chem. News* **23**(2), 17 (1973); U.S. Pats. 3,496,217; 3,496,218; 3,766,237; 3,526,654; 3,542,847; 3,536,748; W. Drinkard and co-workers (to E. I. du Pont de Nemours & Co., Inc.).
200. M. M. Baizer and D. E. Danly, *Chem. Technol.* **10**(10), 161, 302 (1980).
201. Ref. 30, p. 26.
202. M. Kato, *Nikkakyo Geppo* **26**, 561 (1973).
203. *CPI Purchasing*, 31 (Aug. 1989).
204. U.S. Pat. 4,423,018 (Dec. 27, 1983), D. Danly and J. Lester (to Monsanto Co., now dedicated to the public).
205. S. Litherland and co-workers, *Energy Research Abstr.* **10**(8), Abstr. No. 12145 (1985).
206. C. Chi and J. Lester, Jr., *CHEMTECH*, 308 (May 1990).

GENERAL REFERENCES

Y. C. Yen and S. Y. Wu, *Nylon-6,6, Report No. 54B, Process Economics Program*, SRI International, Menlo Park, Calif., Jan. 1987, pp. 1–148.

V. D. Luedeke, "Adipic Acid" in *Encyclopedia of Chemical Processing and Design*, J. McKetta and W. Cunningham, eds., Vol. 2, Marcel Dekker, Inc., New York, 1977, 128–146.

Jpn. Pat. 59184138 (Oct. 19, 1984), O. Mitsui and Y. Fukuoka (to Asahi Chemical Industry, Ltd.).

Jpn. Pat. 59186929 (Oct. 23, 1984), O. Mitsui and Y. Fukuoka (to Asahi Chemical Industry, Ltd.).

Jpn. Pat. 61050930 (Mar. 13, 1986), H. Nagahara and Y. Fukuoka (to Asahi Chemical Industry, Ltd.).

Jpn. Pat. 62045541 (Feb. 27, 1987), H. Nagahara and M. Konishi (to Asahi Chemical Industry, Ltd.).

Jpn. Pat. 60104029 (June 8, 1985), Y. Fukuoka and O. Mitsui (to Asahi Chemical Industry, Ltd.).

Fr. Pat. 2,554,440 (May 10, 1985), O. Mitsui and Y. Fukuoka (to Asahi Chemical Industry, Ltd.).

Eur. Pat. 162,475 (Nov. 11, 1985), M. Tojo and Y. Fukuoka (to Asahi Chemical Industry, Ltd.).
Ger. Offen. 3,441,072 (May 23, 1985), O. Mitsui and Y. Fukuoka (to Asahi Chemical Industry, Ltd.).
Ref. 30, pp. 12, 33, 37, 41, 43.
Chem. Mark. Rep. **236**(15), 54 (1989).
Chem. Mark. Rep. **236**(17), 50 (1989).
Chem. Econ. Eng. Rev. **17**(6), 45 (1985).
Eur. Chem. News **50**(1329), 25 (1988).
Eur. Chem. News **52**(1370), 10 (1989).
Chem. Eng. News **65**(21), 14 (1987).
Chem. Week **142**(2), 11 (1990).
Agence Economique and Financiere, 7 (Nov. 1, 1987).
Chem. Eng. News **67**(49), 15 (1989).
Jpn. Chem. Week 2 (April 27, 1989).
Eur. Chem. News **50**(1329), 4 (1988).
U.S. Exports, EM546, U.S. Dept. of Commerce, Bureau of Census, data for 1986.
Ref. (30), p. 608.5031A.
Food Chemicals Codex, 3rd ed., National Academy of Sciences, National Academy Press, Washington, D.C., 1981.
Chem. Mark. Rep. **230**(14), 58 (1989).
Koon-Ling Ring, "Adipic Acid", in Chemical Economics Handbook, Marketing Research Report, SRI International, Menlo Park, Calif., 200 608.5001 G, p 6.

JUDITH P. OPPENHEIM
GARY L. DICKERSON
Solutia Inc.

ADSORPTION

1. Introduction

Adsorption is the term used to describe the tendency of molecules from an ambient fluid phase to adhere to the surface of a solid. This is a fundamental property of matter, having its origin in the attractive forces between molecules. The force field creates a region of low potential energy near the solid surface and, as a result, the molecular density close to the surface is generally greater than in the bulk gas. Furthermore, and perhaps more importantly, in a multicomponent system the composition of this surface layer generally differs from that of the bulk gas since the surface adsorbs the various components with different affinities. Adsorption may also occur from the liquid phase and is accompanied by a similar change in composition, although, in this case, there is generally little difference in molecular density between the adsorbed and fluid phases.

The enhanced concentration at the surface accounts, in part, for the catalytic activity shown by many solid surfaces, and it is also the basis of the application of adsorbents for low pressure storage of permanent gases such as methane.

However, most of the important applications of adsorption depend on the selectivity, ie, the difference in the affinity of the surface for different components. As a result of this selectivity, adsorption offers, at least in principle, a relatively straightforward means of purification (removal of an undesirable trace component from a fluid mixture) and a potentially useful means of bulk separation.

2. Fundamental Principles

2.1. Forces of Adsorption. Adsorption may be classified as chemisorption or physical adsorption, depending on the nature of the surface forces. In physical adsorption the forces are relatively weak, involving mainly van der Waals (induced dipole–induced dipole) interactions, supplemented in many cases by electrostatic contributions from field gradient–dipole or –quadrupole interactions. By contrast, in chemisorption there is significant electron transfer, equivalent to the formation of a chemical bond between the sorbate and the solid surface. Such interactions are both stronger and more specific than the forces of physical adsorption and are obviously limited to monolayer coverage. The differences in the general features of physical and chemisorption systems (Table 1) can be understood on the basis of this difference in the nature of the surface forces.

Heterogeneous catalysis generally involves chemisorption of the reactants, but most applications of adsorption in separation and purification processes depend on physical adsorption. Chemisorption is sometimes used in trace impurity removal since very high selectivities can be achieved. However, in most situations the low capacity imposed by the monolayer limit and the difficulty of regenerating the spent adsorbent more than outweigh this advantage. The higher capacities achievable in physical adsorption result from multilayer formation and this is obviously critical in such applications as gas storage, but it is also an important consideration in most adsorption separation processes since the process cost is directly related to the adsorbent capacity.

In very small pores the molecules never escape from the force field of the pore wall even at the center of the pore. In this situation the concepts of

Table 1. **Parameters of Physical Adsorption and Chemisorption**

Parameter	Physical adsorption	Chemisorption
heat of adsorption ($-\Delta H$)	low, <1.5 or 2 times latent heat of evaporation	high, >2 or 3 times latent heat of evaporation
specificity	nonspecific	highly specific
nature of adsorbed phase	monolayer or multilayer, no dissociation of adsorbed species	monolayer only may involve dissociation
temperature range	only significant at relative low temperatures	possible over a wide range of temperature
forces of adsorption	no electron transfer, although polarization of sorbate may occur	electron transfer leading to bond formation between sorbate and surface
reversibility	rapid, nonactivated, reversible	activated, may be slow and irreversible

monolayer and multilayer sorption become blurred and it is more useful to consider adsorption simply as pore filling. The molecular volume in the adsorbed phase is similar to that of the saturated liquid sorbate, so a rough estimate of the saturation capacity can be obtained simply from the quotient of the specific micropore volume and the molar volume of the saturated liquid.

2.2. Selectivity. Selectivity in a physical adsorption system may depend on differences in either equilibrium or kinetics, but the great majority of adsorption separation processes depend on equilibrium-based selectivity. Significant kinetic selectivity is in general restricted to molecular sieve adsorbents—carbon molecular sieves, zeolites, or zeolite analogues. In these materials the pore size is of molecular dimensions, so that diffusion is sterically restricted. In this regime small differences in the size or shape of the diffusing molecule can lead to very large differences in diffusivity. In the extreme limit one species (or one class of compounds) may be completely excluded from the micropores, thus giving a highly selective molecular sieve separation. The most important example of such a process is the separation of linear hydrocarbons from their branched and cyclic isomers using a 5A zeolite adsorbent. A second example, where the difference in diffusivities is less extreme but still large enough to produce an efficient separation, is air separation over carbon molecular sieve or 4A zeolite, in which oxygen, the faster diffusing component, is preferentially adsorbed.

A degree of control over the kinetic selectivity of molecular sieve adsorbents can be achieved by controlled adjustment of the pore size. In a carbon sieve this may be accomplished by adjusting the burn-out conditions or by controlled deposition of an easily crackable hydrocarbon. In a zeolite, ion exchange offers the simplest possibility but controlled silanation or boration has also been shown to be effective in certain cases (1).

Control of equilibrium selectivity is generally achieved by adjusting the balance between electrostatic and van der Waals forces. This may be accomplished by changing the chemical nature of the surface and also, to a lesser extent, by adjusting the pore size. In carbon adsorbents surface oxidation offers a simple and effective way of introducing surface polarity and thus modifying the selectivity. One example is shown in Figure 1. On an untreated carbon adsorbent n-hexane is adsorbed more strongly than sulfur dioxide, whereas on an oxidized surface the relative affinities are reversed. With zeolite adsorbents, changing the nature of the exchangeable cation by ion exchange or adjusting the silicon–aluminum ratio of the framework, which determines the cation density, are the most common approaches. In some instances the aluminum-free zeolite analogue (a porous crystalline silicate) may be prepared with the same channel geometry but with a nonpolar surface.

Adsorption on a nonpolar surface such as pure silica or an unoxidized carbon is dominated by van der Waals forces. The affinity sequence on such a surface generally follows the sequence of molecular weights since the polarizability, which is the main factor governing the magnitude of the van der Waals interaction energy, is itself roughly proportional to the molecular weight.

2.3. Hydrophilic and Hydrophobic Surfaces. Water is a small, highly polar molecular and it is therefore strongly adsorbed on a polar surface as a result of the large contribution from the electrostatic forces. Polar adsorbents such as most zeolites, silica gel, or activated alumina therefore adsorb

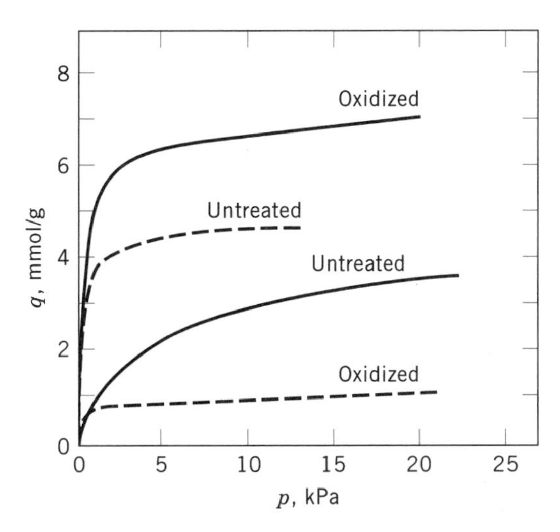

Fig. 1. Equilibrium isotherms for adsorption on activated carbon at 298 K showing the effect of surface modification (2). —, SO_2; – – –, *n*-hexane. To convert kPa to torr multiply by 7.5.

water more strongly than they adsorb organic species, and, as a result, such adsorbents are commonly called hydrophilic. In contrast, on a nonpolar surface where there is no electrostatic interaction water is held only very weakly and is easily displaced by organics. Such adsorbents, which are the only practical choice for adsorption of organics from aqueous solutions, are termed hydrophobic.

The most common hydrophobic adsorbents are activated carbon and silicalite. The latter is of particular interest since the affinity for water is very low indeed; the heat of adsorption is even smaller than the latent heat of vaporization (3). It seems clear that the channel structure of silicalite must inhibit the hydrogen bonding between occluded water molecules, thus enhancing the hydrophobic nature of the adsorbent. As a result, silicalite has some potential as a selective adsorbent for the separation of alcohols and other organics from dilute aqueous solutions (4).

2.4. Capillary Condensation. The equilibrium vapor pressure in a pore or capillary is reduced by the effect of surface tension. As a result, liquid sorbate condenses in a small pore at a vapor pressure that is somewhat lower than the saturation vapor pressure. In a porous adsorbent the region of multilayer physical adsorption merges gradually with the capillary condensation regime, leading to upward curvature of the equilibrium isotherm at higher relative pressure. In the capillary condensation region the intrinsic selectivity of the adsorbent is lost, so in separation processes it is generally advisable to avoid these conditions. However, this effect is largely responsible for the enhanced capacity of macroporous desiccants such as silica gel or alumina at higher humidities.

3. Practical Adsorbents

To achieve a significant adsorptive capacity an adsorbent must have a high specific area, which implies a highly porous structure with very small micropores.

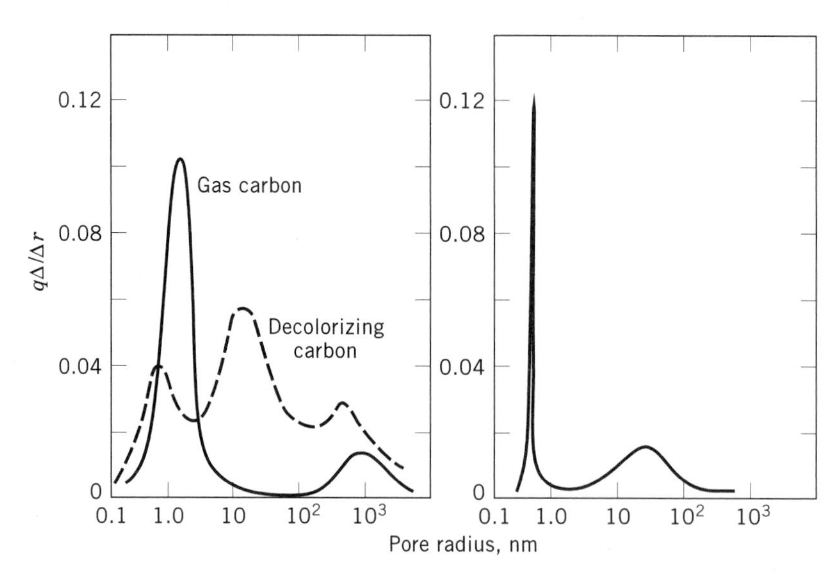

Fig. 2. Pore size distribution of typical samples of activated carbon (small pore gas carbon and large pore decolorizing carbon) and carbon molecular sieve (CMS). $\Delta v/\Delta r$ represents the increment of specific micropore volume for an increment of pore radius.

Such microporous solids can be produced in several different ways. Adsorbents such as silica gel and activated alumina are made by precipitation of colloidal particles, followed by dehydration (see ALUMINUM COMPOUNDS, ALUMINUM OXIDE (ALUMINA); SILICA, AMORPHOUS SILICA). Carbon adsorbents are prepared by controlled burn-out of carbonaceous materials such as coal lignite, and coconut shells (see CARBON, ACTIVATED CARBON). These procedures generally yield a fairly wide distribution of pore size (Fig. 2). The crystalline adsorbents (zeolite and zeolite analogues) are different in that the dimensions of the micropores are determined by the crystal structure and there is therefore virtually no distribution of micropore size (see MOLECULAR SIEVES). Although structurally very different from the crystalline adsorbents, carbon molecular sieves also have a very narrow distribution of pore size. The adsorptive properties depend on the pore size and the pore size distribution as well as on the nature of the solid surface. A simple classification of some of the common adsorbents according to these features is as follows:

	Pore size distribution	
Surface polarity	Narrow	Broad
polar	zeolites (Al rich)	activated alumina silica gel
nonpolar	carbon molecular sieves silicalite	activated carbon

Despite the difference in the nature of the surface, the adsorptive behavior of the molecular sieve carbons resembles that of the small pore zeolites. As their

Table 2. **Properties and Applications of Amorphous Adsorbents**

Adsorbent	Pore diameter, nm	Particle density, g/cm^3	Specific area, m^2/g	Applications
activated carbon (large pore)	1–10^3 (broad range)	0.6–0.8	200–600	water purification, sugar decolorizing
activated carbon (small pore)	1–10	0.5–0.9	400–1200	removal of light organics
carbon molecular sieve	0.4–0.5, 10–10^2 (bimodal)	0.9–1.0	100–300	air separation (N_2 production)
silica gel (high area)	2–10	1.09	800	general purpose
silica gel (low area)	10–50	0.62	300	general purpose
activated alumina	2–10	1.2–1.3	300–400	desiccants

name implies, molecular sieve separations are possible on these adsorbents based on the differences in adsorption rate, which, in the extreme limit, may involve complete exclusion of the larger molecules from the micropores.

Important properties and a number of applications of several commercial adsorbents are summarized in Tables 2–4.

3.1. Amorphous Adsorbents. The amorphous adsorbents (silica gel, activated alumina, and activated carbon) typically have specific areas in the 200–1000-m^2/g range, but for some activated carbons much higher values have been achieved (~1500 m^2/g). The difficulty is that these very high area carbons tend to lack physical strength and this limits their usefulness in many practical applications. The high area materials also contain a large proportion of very small pores, which renders them unsuitable for applications involving adsorption of large molecules. The distinction between gas carbons, used for adsorption of low molecular weight permanent gases, and liquid carbons, which are used for adsorption of larger molecules such as color bodies from the liquid phase, is thus primarily a matter of pore size.

Table 3. **Properties and Application of Polymeric Adsorbents**

Type of polymer	Representative commercial product	Properties	Applications
sulfonated styrene–divinylbenzene copolymers[a] with various degrees of cross-linking	Dowex-50 Amberlite IR120B	pore diameter, porosity, density, etc, vary with degree of hydration or dehydration	sugar separations[b], eg, various fructose, glucose
macroreticular sulfonated styrene–divinylbenzene	Diaion HPK-25	porosity ~0.33, microparticles ~80-μm diameter	removal of NH_3 or light amines[c]

[a] Ion exchanged to Ca^{2+} form.
[b] Liquid-phase operation.
[c] Gas or liquid phase.

Table 4. **Properties and Applications of Crystalline Adsorbents**[a]

Structure	Cation	Typical formula of unit cell or pseudocell	Window	Effective channel diameter, nm	Applications
4A	Na^+	$Na_{12}[(AlO_2)_{12}(SiO_2)_{12}]$	obstructed 8-ring	0.38	desiccant; CO_2 removal; air separation (N_2)
5A	Ca^{2+}	$Ca_5Na_2[(AlO_2)_{12}(SiO_2)_{12}]$	free 8-ring	0.44	linear paraffin separation; air separation (O_2)
3A	K^+	$K_{12}[(AlO_2)_{12}(SiO_2)_{12}]$	obstructed 8-ring	0.29	drying of reactive gases
13X	Na^+	$Na_{86}[(AlO_2)_{86}(SiO_2)_{106}]$	12-ring (free)	0.84	air separation (O_2),
10X	Ca^{2+}	$Ca_{43}[(AlO_2)_{86}(SiO_2)_{106}]$	12-ring (obstructed)	0.80	removal of mercaptans
SrBaX	Sr^{2+}, Ba^{2+}	$Sr_{21}Ba_{22}[(AlO_2)_{86}(SiO_2)_{106}]$	12-ring (obstructed)	0.80	separation of C_8 aromatics
KY	K^+	$K_{56}[(AlO_2)_{56}(SiO_2)_{136}]$	12-ring (free)	0.80	
Mordenite	H^+	$H_8[(AlO_2)_8(SO_2)_{40}]$	12-ring (free)	0.70	trapping of Kr from nuclear off-gas
	Ag^+	$Ag_8[(AlO_2)_8(SiO_2)_{40}]$	12-ring (free)	0.70	trapping of CH_3I from nuclear off-gas
AgX	Ag^+	$Ag_{86}[(AlO_2)_{86}(SiO_2)_{106}]$	12-ring (free)	0.84	
silicalite/HZSM5		$(SiO_2)_{96}$	10-ring	0.60	removal of organics in aqueous systems

[a] Structural details can be found in refs. 5 and 6. A simplified description is given in ref. 7.

In a typical amorphous adsorbent the distribution of pore size may be very wide, spanning the range from a few nanometers to perhaps one micrometer. Since different phenomena dominate the adsorptive behavior in different pore size ranges, IUPAC has suggested the following classification:

micropores, <2-nm diameter
mesopores, 2–50-nm diameter
macropores, >50-nm diameter

This division is somewhat arbitrary since it is really the pore size relative to the size of the sorbate molecule rather than the absolute pore size that governs the behavior. Nevertheless, the general concept is useful. In micropores (pores which are only slightly larger than the sorbate molecule) the molecule never escapes from the force field of the pore wall, even when in the center of the pore. Such pores generally make a dominant contribution to the adsorptive capacity for molecules small enough to penetrate. Transport within these pores can be severely limited by steric effects, leading to molecular sieve behavior.

The mesopores make some contribution to the adsorptive capacity, but their main role is as conduits to provide access to the smaller micropores. Diffusion in the mesopores may occur by several different mechanisms, as discussed below. The macropores make very little contribution to the adsorptive capacity, but they commonly provide a major contribution to the kinetics. Their role is thus analogous to that of a super highway, allowing the adsorbate molecules to diffuse far into a particle with a minimum of diffusional resistance.

3.2. Crystalline Adsorbents. In the crystalline adsorbents, zeolites and zeolite analogues such as silicalite and the microporous aluminum phosphates, the dimensions of the micropores are determined by the crystal framework and there is therefore virtually no distribution of pore size. However, a degree of control can sometimes be exerted by ion exchange, since, in some zeolites, the exchangeable cations occupy sites within the structure which partially obstruct the pores. The crystals of these materials are generally quite small (1–5 μm) and they are aggregated with a suitable binder (generally a clay) and formed into macroporous particles having dimensions large enough to pack directly into an adsorber vessel. Such materials therefore have a well-defined bimodal pore size distribution with the intracrystalline micropores (a few tenths of a nanometer) linked together through a network of macropores having a diameter of the same order as the crystal size (∼1 μm).

3.3. Desiccants. A solid desiccant is simply an adsorbent which has a high affinity and capacity for adsorption of moisture so that it can be used for selective adsorption of moisture from a gas (or liquid) stream. The main requirements for an efficient desiccant are therefore a highly polar surface and a high specific area (small pores). The most widely used desiccants (qv) are silica gel, activated alumina, and the aluminum rich zeolites (4A or 13X). The equilibrium adsorption isotherms for moisture on these materials have characteristically different shapes (Fig. 3), making them suitable for different applications.

The zeolites have high affinity and high capacity at low partial pressures, shown by the nearly rectangular form of the isotherm. This makes them useful desiccants where a very low humidity or dew point is required. The 3A zeolite is a

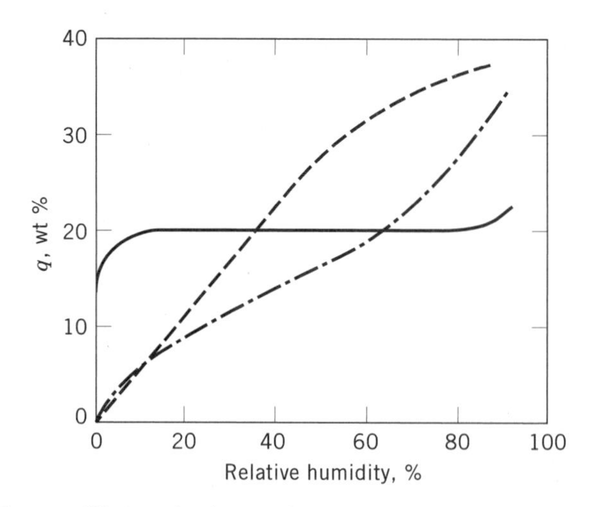

Fig. 3. Adsorption equilibrium isotherms for moisture on three commercial adsorbents: pelletized 4A zeolite (—), silica gel (– – –), and a typical activated alumina (— - —).

molecular sieve desiccant, since its micropores are small enough to exclude most molecules other than water. It is therefore useful for drying reactive gases. The major disadvantage of zeolite desiccants is that a high temperature is required for regeneration (>300°C), which makes their use uneconomic when only a moderately low dew point is required.

Considerable variation in the moisture isotherm for alumina can be obtained by different preparation and pretreatment. However, in general, the initial slope of the isotherm is not as steep as that of a zeolite, indicating a lower moisture affinity at low partial pressure, but the capacity at high humidities is often higher than that of a zeolitic adsorbent. Regeneration temperatures are typically in the 250–350°C range. Alumina adsorbents are also more robust than zeolites and less sensitive to deactivation by organics, but they are generally less suitable than the zeolites where very low humidity is the primary requirement.

The isotherm for silica gel is more nearly linear over a wide range of partial pressure, although the affinity for moisture is lower than that of either alumina or the zeolites. However, a correspondingly lower regeneration temperature is also required. This can be as low as 120°C, making silica gel the most suitable candidate for pressure swing driers, desiccant cooling systems (8), and other applications where low grade heat is used for regeneration of the adsorbent.

3.4. Loaded Adsorbents. Where highly efficient removal of a trace impurity is required it is sometimes effective to use an adsorbent preloaded with a reactant rather than rely on the forces of adsorption. Examples include the use of zeolites preloaded with bromine to trap traces of olefins as their more easily condensible bromides; zeolites preloaded with iodine to trap mercury vapor, and activated carbon loaded with cupric chloride for removal of mercaptans.

4. Adsorption Equilibrium

4.1. Henry's Law. Like any other phase equilibrium, the distribution of a sorbate between fluid and adsorbed phases is governed by the principles of thermodynamics. Equilibrium data are commonly reported in the form of an isotherm, which is a diagram showing the variation of the equilibrium adsorbed-phase concentration or loading with the fluid-phase concentration or partial pressure at a fixed temperature. In general, for physical adsorption on a homogeneous surface at sufficiently low concentrations, the isotherm should approach a linear form, and the limiting slope in the low concentration region is commonly known as the Henry's law constant. The Henry constant is simply a thermodynamic equilibrium constant and the temperature dependence therefore follows the usual van't Hoff equation:

$$\lim_{p \longrightarrow 0} \left(\frac{\partial q}{\partial p} \right)_T \equiv K' = K'_0 e^{-\Delta H_0/\mathrm{RT}} \tag{1}$$

in which $-\Delta H_0$ is the limiting heat of adsorption at zero coverage. Since adsorption, particularly from the vapor phase, is usually exothermic, $-\Delta H_0$ is a positive quantity and K' therefore decreases with increasing temperature. A corresponding dimensionless Henry constant (K) may also be defined, based on the ratio of adsorbed and fluid-phase concentrations:

$$\lim_{c \longrightarrow 0} \left(\frac{\partial q}{\partial c} \right)_T \equiv K = K_0 e^{-\Delta U_0/\mathrm{RT}} \tag{2}$$

Since, for an ideal vapor phase, $p = cRT$, these quantities are related by

$$K = RTK'; \qquad -\Delta H_0 = -\Delta U_0 + RT \tag{3}$$

Henry's law corresponds physically to the situation in which the adsorbed phase is so dilute that there is neither competition for surface sites nor any significant interaction between adsorbed molecules. At higher concentrations both of these effects become important and the form of the isotherm becomes more complex. The isotherms have been classified into five different types (9) (Fig. 4). Isotherms for a microporous adsorbent are generally of type I; the more complex forms are associated with multilayer adsorption and capillary condensation.

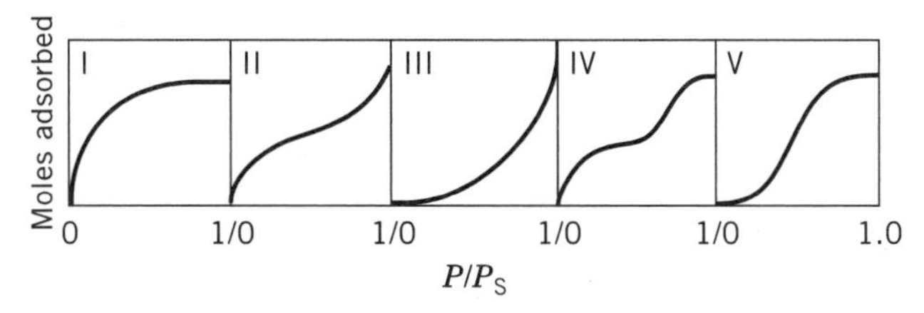

Fig. 4. The Brunaner classification of isotherms (I–V).

4.2. Langmuir Isotherm. Type I isotherms are commonly represented by the ideal Langmuir model:

$$\frac{q}{q_s} = \frac{bp}{1+bp} \tag{4}$$

where q_s is the saturation limit and b is an equilibrium constant which is directly related to the Henry constant ($K' = bq_s$). The Langmuir model was originally developed to represent monolayer adsorption on an ideal surface, for which q_s corresponds to the monolayer coverage. However, in applying this model to physical adsorption on a microporous solid, the saturation limit becomes the quantity of sorbate required to fill the micropore volume. This expression is of the correct form to represent a type I isotherm, since at low pressure it approaches Henry's law while at high pressure it tends asymptotically to the saturation limit. The equilibrium constant b ($= K'/q_s$) decreases with increasing temperature (eq. 1); therefore, for a given pressure range, the isotherm approaches rectangular or irreversible form at low temperatures (large b) and linear form at high temperatures (small b).

Although very few systems conform accurately to the Langmuir model, this model provides a simple qualitative representation of the behavior of many systems and it is therefore widely used, particularly for adsorption from the vapor phase. According to the Langmuir model the heat of adsorption should be independent of loading, but this requirement is seldom fulfilled in practice. Both increasing and decreasing trends are commonly observed (Fig. 5). For a polar sorbate on a polar adsorbent (ie, a system in which electrostatic forces are dominant) a decreasing trend is normally observed, since the relative importance of

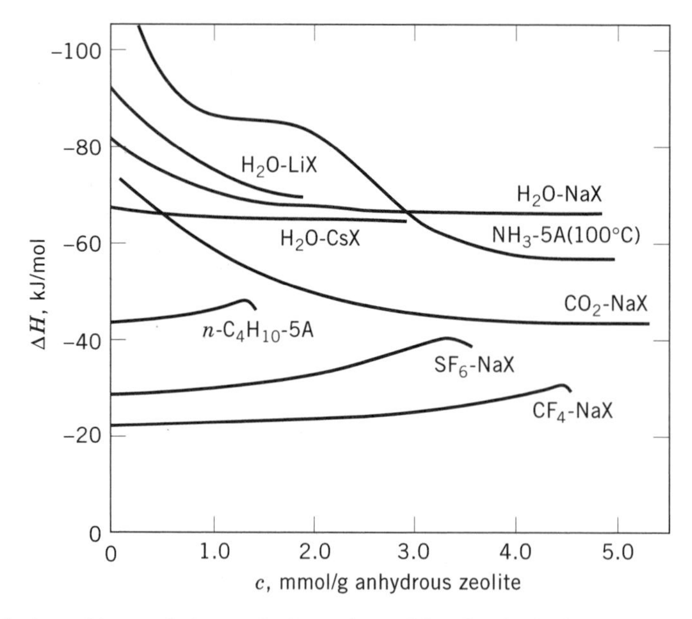

Fig. 5. Variation of isosteric heat of adsorption with adsorbed phase concentration. Reprinted from ref. 10, courtesy of Marcel Dekker, Inc. To convert kJ to kcal divide by 4.184.

the electrostatic contribution declines at high loadings as a result of preferential occupation of the most favorable sites and consequent screening of cations. In contrast, where van der Waals forces are dominant (nonpolar sorbates), a rising trend of heat of adsorption with loading is generally observed. This is commonly attributed to the effect of intermolecular attractive forces, but other explanations are also possible (11). In homologous series such as the linear paraffins the heat of adsorption increases linearly with carbon number (12).

4.3. Freundlich Isotherm. The isotherms for some systems, notably hydrocarbons on activated carbon, conform more closely to the Freundlich equation:

$$q = bp^{1/n} \qquad (n > 1.0) \tag{5}$$

Although the Freundlich expression does not reduce to Henry's law at low concentrations, it often provides a good approximation over a wide range of conditions. This form of equation can be explained as resulting from energetic heterogeneity of the surface. Superposition of a set of Langmuir isotherms with different b values (corresponding to sites of different energy) yields an expression of this form.

4.4. Adsorption of Mixtures. The Langmuir model can be easily extended to binary or multicomponent systems:

$$\frac{q_1}{q_{s1}} = \frac{b_1 p_1}{1 + b_1 p_1 + b_2 p_2 + \cdots}; \qquad \frac{q_2}{q_{s2}} = \frac{b_2 p_2}{1 + b_1 p_1 + b_2 p_2;\ + \cdots} \tag{6}$$

Thermodynamic consistency requires $q_{s1} = q_{s2}$, but this requirement can cause difficulties when attempts are made to correlate data for sorbates of very different molecular size. For such systems it is common practice to ignore this requirement, thereby introducing an additional model parameter. This facilitates data fitting but it must be recognized that the equations are then being used purely as a convenient empirical form with no theoretical foundation.

Equation 6 shows that the adsorption of component 1 at a partial pressure p_1 is reduced in the presence of component 2 as a result of competition for the available surface sites. There are only a few systems for which this expression (with $q_{s1} = q_{s2} = q_s$) provides an accurate quantitative representation, but it provides useful qualitative or semiquantitative guidance for many systems. In particular, it has the correct asymptotic behavior and provides explicit recognition of the effect of competitive adsorption. For example, if component 2 is either strongly adsorbed or present at much higher concentration than component 1, the isotherm for component 1 is reduced to a simple linear form in which the apparent Henry's law constant depends on p_2:

$$q_1 \simeq \left(\frac{b_1 q_{s1}}{1 + b_2 p_2}\right) p_1 \tag{7}$$

For an equilibrium-based separation, a convenient measure of the intrinsic selectivity of the adsorbent is provided by the separation factor (α_{12}), which is defined by analogy with the relative volatility as

$$\alpha_{12} = \frac{(X_1/Y_1)}{(X_2/Y_2)} \tag{8}$$

where X and Y refer to the mole fractions in the adsorbed and fluid phases, respectively, at equilibrium. For a system that obeys the Langmuir model (eq. 6) it is evident that $\alpha_{12} = b_1/b_2$ and is thus independent of concentration. The Langmuir isotherm is therefore sometimes referred to as the constant separation factor model.

The assumption of a constant separation factor is often a reasonable approximation for preliminary process design but this assumption is often violated in real systems, where some variation of the separation factor with composition is common and more extreme variations involving azeotrope formation ($\alpha = 1.0$ at a particular composition) and selectivity reversal (α varying from greater than 1.0 to less than 1.0 with changing composition) are not uncommon. There have been many attempts to improve the correlation of equilibrium data by using more complex expressions, one of the more widely used being the Langmuir-Freundlich or loading ratio correlation (13):

$$\frac{q_1}{q_s} = \frac{b_1 p_1^{n_1}}{1 + b_1 p_1^{n_1} + b_2 p_2^{n_1} + \cdots}; \qquad \frac{q_2}{q_s} = \frac{b_2 p_2^{n_2}}{1 + b_1 p_1^{n_1} + b_2 p_2^{n_2} + \cdots} \tag{9}$$

This has the advantage that the expressions for the adsorbed-phase concentration are simple and explicit, and, as in the Langmuir expression, the effect of competition between sorbates is accounted for. However, the expression does not reduce to Henry's law in the low concentration limit and therefore violates the requirements of thermodynamic consistency. Whereas it may be useful as a basis for the correlation of experimental data, it should be treated with caution and should not be used as a basis for extrapolation beyond the experimental range.

4.5. Ideal Adsorbed Solution Theory. Perhaps the most successful approach to the prediction of multicomponent equilibria from single-component isotherm data is ideal adsorbed solution theory (14). In essence, the theory is based on the assumption that the adsorbed phase is thermodynamically ideal in the sense that the equilibrium pressure for each component is simply the product of its mole fraction in the adsorbed phase and the equilibrium pressure for the pure component *at the same spreading pressure*. The theoretical basis for this assumption and the details of the calculations required to predict the mixture isotherm are given in standard texts on adsorption (7) as well as in the original paper (14). Whereas the theory has been shown to work well for several systems, notably for mixtures of hydrocarbons on carbon adsorbents, there are a number of systems which do not obey this model. Azeotrope formation and selectivity reversal, which are observed quite commonly in real systems, are not consistent with an ideal adsorbed phase and there is no way of knowing a priori whether or not a given system will show ideal behavior.

5. Adsorption Kinetics

5.1. Intrinsic Kinetics. Chemisorption may be regarded as a chemical reaction between the sorbate and the solid surface, and, as such, it is an activated process for which the rate constant (k) follows the familiar Arrhenius rate law:

$$k = k_0 e^{-E/RT} \tag{10}$$

Depending on the temperature and the activation energy (E), the rate constant may vary over many orders of magnitude.

In practice the kinetics are usually more complex than might be expected on this basis, since the activation energy generally varies with surface coverage as a result of energetic heterogeneity and/or sorbate-sorbate interaction. As a result, the adsorption rate is commonly given by the Elovich equation (15):

$$q = \frac{1}{k'} \ln(1 + k''t) \tag{11}$$

where k' and k'' are temperature-dependent constants.

In contrast, physical adsorption is a very rapid process, so the rate is always controlled by mass transfer resistance rather than by the intrinsic adsorption kinetics. However, under certain conditions the combination of a diffusion-controlled process with an adsorption equilibrium constant that varies according to equation 1 can give the appearance of activated adsorption.

As illustrated in Figure 6, a porous adsorbent in contact with a fluid phase offers at least two and often three distinct resistances to mass transfer: external film resistance and intraparticle diffusional resistance. When the pore size distribution has a well-defined bimodal form, the latter may be divided into macropore and micropore diffusional resistances. Depending on the particular system and the conditions, any one of these resistances may be dominant or the overall rate of mass transfer may be determined by the combined effects of more than one resistance.

5.2. External Fluid Film Resistance. A particle immersed in a fluid is always surrounded by a laminar fluid film or boundary layer through which an adsorbing or desorbing molecule must diffuse. The thickness of this layer, and therefore the mass transfer resistance, depends on the hydrodynamic conditions. Mass transfer in packed beds and other common contacting devices has been widely studied. The rate data are normally expressed in terms of a simple linear rate expression of the form

$$\frac{\partial q}{\partial t} = k_f a(c - c^*) \tag{12}$$

and the variation of the mass transfer coefficient (k_f) with the hydrodynamic conditions is generally accounted for in terms of empirical correlations of the general form

$$\mathrm{Sh} \equiv \frac{2k_f R}{D_m} = f(\mathrm{Re},\ \mathrm{Sc}) \tag{13}$$

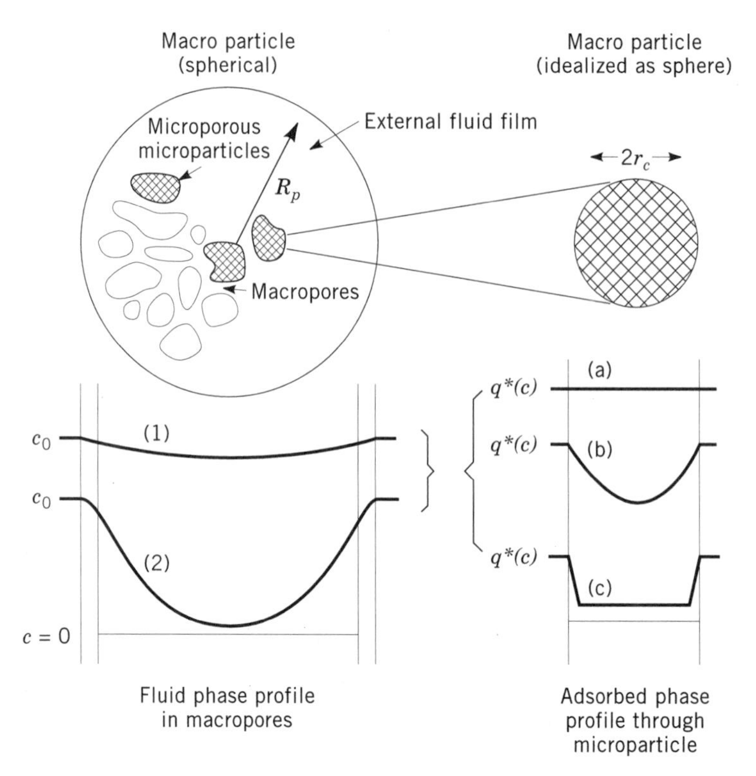

Fig. 6. Concentration profiles through an idealized biporous adsorbent particle showing some of the possible regimes. (1) + (a) rapid mass transfer, equilibrium throughout particle; (1) + (b) micropore diffusion control with no significant macropore or external resistance; (1) + (c) controlling resistance at the surface of the microparticles; (2) + (a) macropore diffusion control with some external resistance and no resistance within the microparticle; (2) + (b) all three resistances (micropore, macropore, and film) significant; (2) + (c) diffusional resistance within the macroparticle and resistance at the surface of the microparticle with some external film resistance.

where Re and Sc are the (particle-based) Reynolds and Schmidt numbers. One of the most widely used correlations, applicable to both gas and liquid systems over a wide range of conditions, is (16)

$$\mathrm{Sh} = \frac{2k_f R}{D_m} = 2.0 + 1.1\ \mathrm{Sc}^{1/3}\mathrm{Re}^{0.6} \tag{14}$$

5.3. Macropore Diffusion. Transport in a macropore can occur by several different mechanisms, the most important of which are bulk molecular diffusion, Knudsen diffusion, surface diffusion, and Poiseuille flow. In liquid systems bulk molecular diffusion is generally dominant, but in the vapor phase the contributions from Knudsen and surface diffusion may be large or even dominant. The contribution from Poiseuille flow, ie, forced flow through the pore under the influence of the pressure gradient, is generally relatively minor since pressure gradients are usually kept small. However, this is not true in the pressurization and blowdown steps of a pressure swing process, where the contribution from Poiseuille flow can be dominant.

A molecule colliding with the pore wall is reflected in a specular manner so that the direction of the molecule leaving the surface has no correlation with that of the incident molecule. This leads to a Fickian mechanism, known as Knudsen diffusion, in which the flux is proportional to the gradient of concentration of partial pressure. The Knudsen diffusivity (D_K) is independent of pressure and varies only weakly with temperature:

$$D_K = 9700\rho\sqrt{T/M} \qquad (\mathrm{cm^2/s}) \tag{15}$$

where ρ is the pore radius (cm) and M the molecular weight. Knudsen diffusion becomes dominant when collisions with the pore wall occur more frequently than collisions between diffusing molecules, ie, when the pore diameter is smaller than the mean free path. Since the mean free path varies inversely with pressure there is a gradual transition from the molecular to the Knudsen regime as the pressure is reduced, but the pressure at which this occurs depends on the pore size. At atmospheric temperature and pressure Knudsen diffusion is dominant in pores of less than about 10 nm diameter. In the intermediate region, which spans the range of the macropores in many commercial adsorbents, both mechanisms are of comparable significance.

The combined effects of Knudsen and molecular diffusion may be estimated approximately from the reciprocal addition rule:

$$\frac{1}{\epsilon_p D_p} = \frac{\tau}{\epsilon_p}\left(\frac{1}{D_K} + \frac{1}{D_m}\right) \tag{16}$$

The factor ε_p takes account of the fact that diffusion occurs only through the pore and not through the matrix; τ is a tortuosity factor which accounts for the increased path length and reduced concentration gradient arising from the random orientation of the pores as well as any other geometric effects. In a typical adsorbent $\tau \sim 3.0$ and $\varepsilon_p \sim 0.3$, so the effect of these two factors is to reduce the diffusivity by about one order of magnitude relative to the value for a straight cylindrical capillary.

5.4. Micropore Diffusion. In very small pores in which the pore diameter is not much greater than the molecular diameter the diffusing molecule never escapes from the force field of the pore wall. Under these conditions steric effects and the effects of nonuniformity in the potential field become dominant and the Knudsen mechanism no longer applies. Diffusion occurs by an activated process involving jumps from site to site, just as in surface diffusion, and the diffusivity becomes strongly dependent on both temperature and concentration.

The true driving force for any diffusive transport process is the gradient of chemical potential rather than the gradient of concentration. This distinction is not important in dilute systems where thermodynamically ideal behavior is approached. However, it becomes important at higher concentration levels and in micropore and surface diffusion. To a first approximation the expression for the diffusive flux may be written

$$J = -Bq\, \partial\mu/\partial z \tag{17}$$

where q is the concentration in the adsorbed phase. Assuming an ideal vapor phase, the expression for the chemical potential is

$$\mu = \mu^{\circ} + RT \ln p \tag{18}$$

where

$$\frac{\partial \mu}{\partial z} = RT \frac{d \ln p}{dq} \frac{\partial q}{\partial z} \tag{19}$$

Combining equations 17 and 18 yields, for the Fickian diffusivity (defined by $J = -D\partial q/\partial z$),

$$D = D_0 \frac{d \ln p}{d \ln q} \tag{20}$$

where $D_0 = BRT$. If the equilibrium relation is linear, $d \ln p/d \ln q = 1.0$ and $D \longrightarrow D_0$. At higher concentrations the equilibrium relationship is nonlinear, and as a result the diffusivity is generally concentration dependent. For the special case of a Langmuir isotherm (eq. 4), $d \ln p / d \ln q = (1 - q/q_s)^{-1}$ so

$$D = \frac{D_0}{1 - q/q_s} \tag{21}$$

A rapid increase in diffusivity in the saturation region is therefore to be expected, as illustrated in Figure 7 (17). Although the corrected diffusivity (D_0) is, in principle, concentration dependent, the concentration dependence of this quantity is generally much weaker than that of the thermodynamic correction factor ($d \ln p / d \ln q$). The assumption of a constant corrected diffusivity is therefore an acceptable approximation for many systems. More detailed analysis shows that the corrected diffusivity is closely related to the self-diffusivity or tracer diffusivity, and at low sorbate concentrations these quantities become identical.

The temperature dependence of the corrected diffusivity follows the usual Eyring expression

$$D_0 = D_{\infty} e^{-E/RT} \tag{22}$$

in which E is the activation energy or the energy barrier between adjacent sites. In small pore zeolites and carbon molecular sieves the dominant contribution to this energy is the repulsive interaction encountered by the molecule in penetrating the concentration in the pore. As a result, the activation energy shows a well-defined correlation with the molecule diameter and the size of the micropore, as illustrated in Figure 8.

More detailed information on micropore diffusion is available in the book by Kärger and Ruthven (18) and in the review article by Krishna and Wesselingh (19) which stresses the usefulness of the Stefan-Maxwell approach to the modelling of diffusion in binary and multicomponent systems.

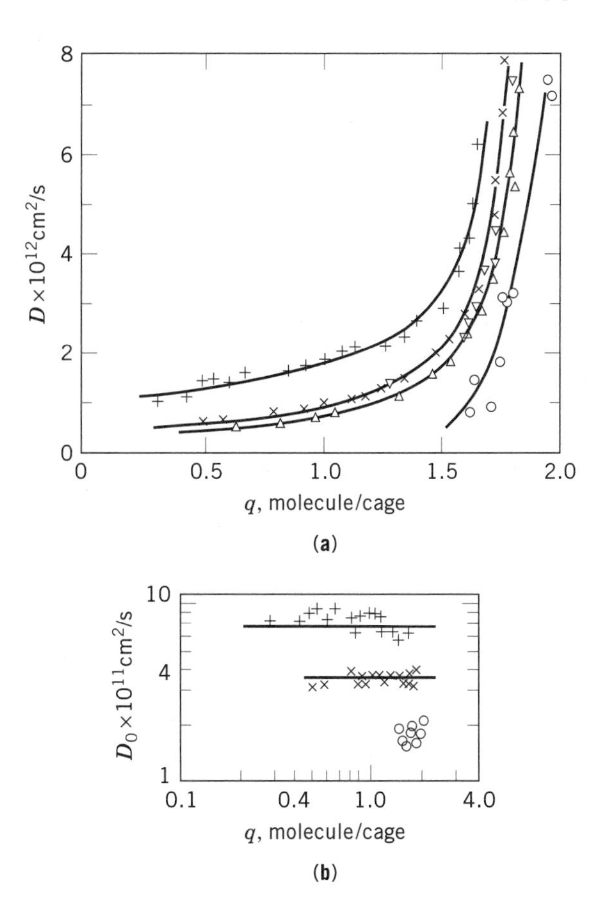

Fig. 7. Variation of (**a**) intracrystalline diffusivity and (**b**) corrected diffusivity (D_0) with sorbate concentration for n-heptane in a commercial sample of 5A zeolite crystals: ○ 409 K; △, ▽, 439 K (ads, des); ×, 462 K; +, 491 K. Reproduced by permission of National Research Council of Canada from ref. 17.

5.5. Sorption Rates in Batch Systems.

Direct measurement of the uptake rate by gravimetric, volumetric, or piezometric methods is widely used as a means of measuring intraparticle diffusivities. Diffusive transport within a particle may be represented by the Fickian diffusion equation, which, in spherical coordinates, takes the form

$$\frac{\partial q}{\partial t} = D\left(\frac{\partial^2 q}{\partial r^2} + \frac{2}{r}\,\frac{\partial q}{\partial r}\right) \tag{23}$$

For a step change in sorbate concentration at the particle surface ($r = R$) at time zero, assuming isothermal conditions and diffusion control, the expression for the uptake curve may be derived from the appropriate solution of this differential equation:

$$\frac{m_t}{m_\infty} = 1 - \frac{6}{\pi^2}\sum_{n=1}^{\infty}\frac{1}{n^2}\,e^{-n^2\pi^2 Dt/R^2} \tag{24}$$

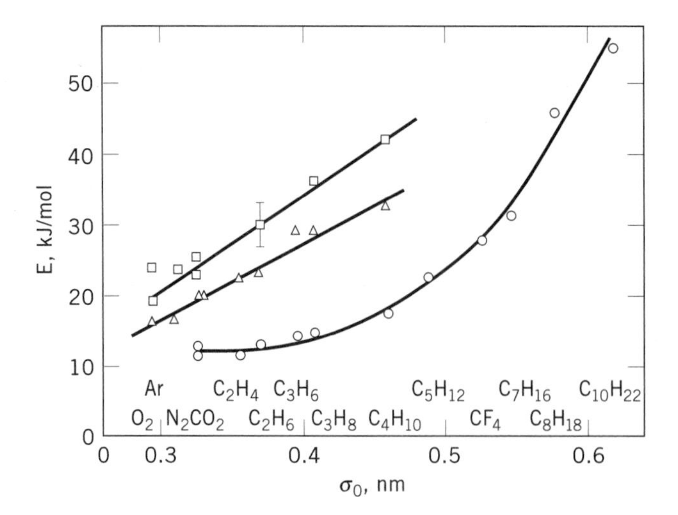

Fig. 8. Variation of activation energy with kinetic molecular diameter for diffusion in 4A zeolite (□), 5A zeolite (○), and carbon molecular sieve (MSC-5A) (△). Kinetic diameters are estimated from the van der Waals co-volumes. From ref. 7. To convert kJ to kcal divide by 4.184.

or, in the initial region,

$$\frac{m_r}{m_\infty} = \frac{6}{\sqrt{\pi}} \left(\frac{Dt}{R^2}\right)^{1/2} \tag{25}$$

The time constant R^2/D, and hence the diffusivity, may thus be found directly from the uptake curve. However, it is important to confirm by experiment that the basic assumptions of the model are fulfilled, since intrusions of thermal effects or extraparticle resistance to mass transfer may easily occur, leading to erroneously low apparent diffusivity values.

In certain adsorbents, notably partially coked zeolites and some carbon molecular sieves, the resistance to mass transfer may be concentrated at the surface of the particle, leading to an uptake expression of the form

$$\frac{m_t}{m_\infty} = 1 - e^{-k_s t} \tag{26}$$

in place of equation 24. The difference between surface resistance and intraparticle diffusion control is easily apparent from the form of the uptake curves (see Fig. 9). Since both D and k_s are generally concentration dependent, it is preferable to make differential measurements over small concentration steps in order to simplify the interpretation of the experimental data.

For a macroporous sorbent the situation is slightly more complex. A differential balance on a shell element, assuming diffusivity transport through the macropores with rapid adsorption at the surface (or in the micropores), yields

$$\epsilon_p \frac{\partial c}{\partial t} + (1 - \epsilon_p) \frac{\partial q}{\partial t} = \epsilon_p D_p \left(\frac{\partial^2 c}{\partial r^2} + \frac{2}{r} \frac{\partial c}{\partial r}\right) \tag{27}$$

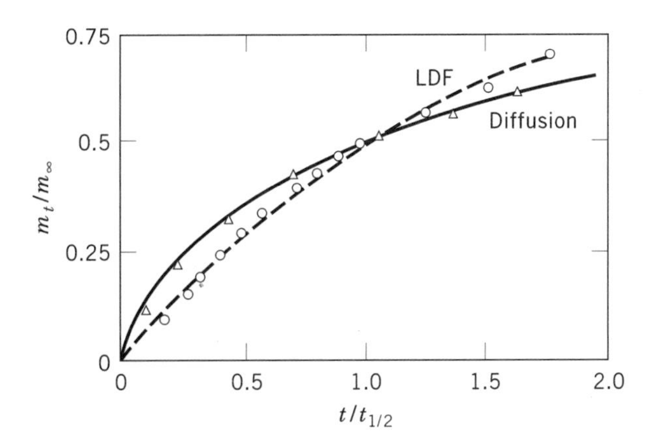

Fig. 9. Uptake curves for N_2 in two samples of carbon molecular sieve showing conformity with diffusion model (eq. 24) for sample 1 (△), and with surface resistance model (eq. 26) for example 2 (○); LDF = linear driving force. Data from ref. 20.

Assuming a linear equilibrium relationship (over the range of the small concentration step, $q^* = Kc$), this becomes

$$\frac{\partial c}{\partial t} = \frac{\epsilon_p D_p}{1 + (1 - \epsilon_p)K}\left(\frac{\partial^2 c}{\partial r^2} + \frac{2}{r}\frac{\partial c}{\partial r}\right) \tag{28}$$

which is of the same form as equation 23 but with an effective diffusivity given by

$$D = \frac{\epsilon_p D_p}{1 + (1 - \epsilon_p)K} \tag{29}$$

For adsorption from the vapor phase, K may be very large (sometimes as high as 10^7) and then clearly the effective diffusivity is very much smaller than the pore diffusivity. Furthermore, the temperature dependence of K follows equation 2, giving the appearance of an activated diffusion process with $E \approx (-\Delta U)$.

As a result of these difficulties the reported diffusivity data show many apparent anomalies and inconsistencies, particularly for zeolites and other microporous adsorbents. Discrepancies of several orders of magnitude in the diffusivity values reported for a given system under apparently similar conditions are not uncommon. Since most of the intrusive effects lead to erroneously low values, the higher values are probably the more reliable.

6. Adsorption Column Dynamics

In most adsorption processes the adsorbent is contacted with fluid in a packed bed. An understanding of the dynamic behavior of such systems is therefore needed for rational process design and optimization. What is required is a mathematical model which allows the effluent concentration to be predicted for any defined change in the feed concentration or flow rate to the bed. The flow pattern can generally be represented adequately by the axial dispersed plug-flow model, according to which a mass balance for an element of the column yields, for the basic differential equation governing the dynamic behavior,

$$-D_L \frac{\partial^2 c_i}{\partial z^2} + \frac{\partial}{\partial z}(vc_i) + \frac{\partial c_i}{\partial t} + \left(\frac{1-\epsilon}{\epsilon}\right)\frac{\partial \bar{q}_i}{\partial t} = 0 \qquad (30)$$

The term $\partial \bar{q}_i/\partial t$ represents the overall rate of mass transfer for component i (at time t and distance z) averaged over a particle. This is governed by a mass transfer rate expression which may be thought of as a general functional relationship of the form

$$\frac{\partial \bar{q}}{\partial t} = f(c_i, c_j, \ldots, q_i, q_j, \ldots) \qquad (31)$$

This rate equation must satisfy the boundary conditions imposed by the equilibrium isotherm and it must be thermodynamically consistent so that the mass transfer rate falls to zero at equilibrium. It may be a linear driving force expression of the form

$$\frac{\partial \bar{q}}{\partial t} = k_s(q_i^* - \bar{q}_i) \qquad (32)$$

where $q^*_i(c_i, c_j, \ldots)$ represents the equilibrium adsorbed phase concentration of component i, or it may be a set of diffusion equations with their associated boundary conditions.

For an isothermal system the simultaneous solution of equations 30 and 31, subject to the boundary conditions imposed on the column, provides the expressions for the concentration profiles $c_i(z,t)$, $\bar{q}_i(z,t)$ in both phases. If the system is nonisothermal, an energy balance is also required and since, in general, both the equilibrium concentration and the rate coefficients are temperature dependent, all equations are coupled. Analytical solutions are possible only for the simpler cases: single-component isothermal systems with linear or rectangular equilibrium isotherms. In the general case of a multicomponent nonisothermal system, numerical solutions offer the only practical approach.

The form of the response for an adiabatic three-component system (two adsorbable components in an inert carrier) is illustrated in Figure 10. In general, if there are n components (counting both heat and nonadsorbing species as components), the response contains $(n-1)$ transitions or mass transfer zones, separated by $(n-2)$ plateaus between the initial and final states. When the change imposed at the column inlet involves an increase in the concentration of the more strongly adsorbed species, the concentration at the intermediate plateau will exceed both its initial concentration and its final steady-state concentration. This phenomenon, known as roll-up, results from displacement by the more strongly adsorbed species, which travels more slowly through the column.

6.1. Equilibrium Theory. The general features of the dynamic behavior may be understood without recourse to detailed calculations since the overall

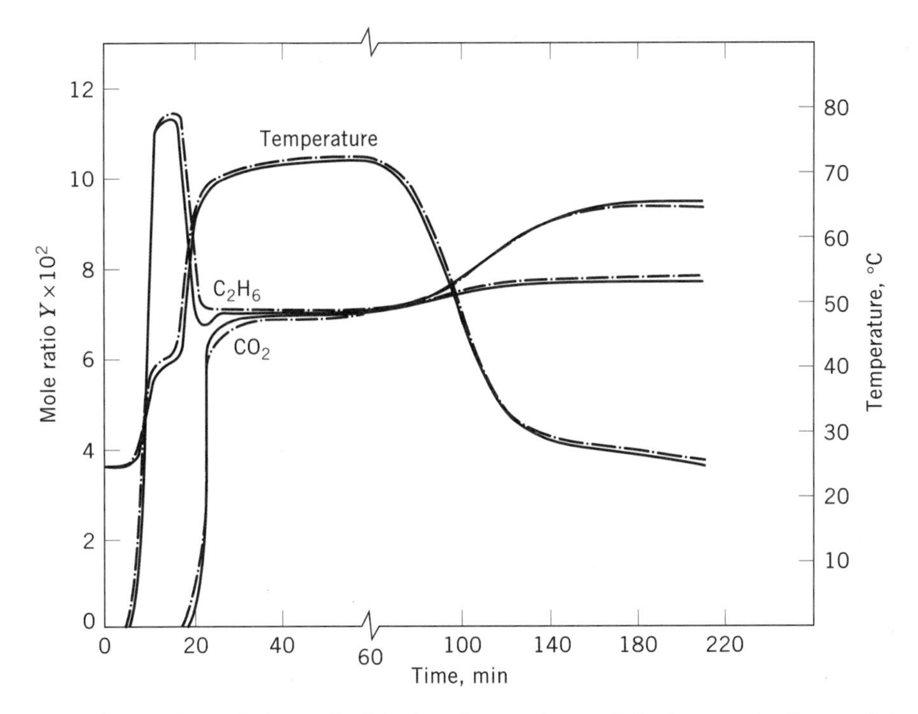

Fig. 10. Comparison of theoretical (—) and experimental (—·) concentration and temperature breakthrough curves for sorption of C_2H_6–CO_2 mixtures from a N_2 carrier on 5A molecular sieve. Feed: 10.5% CO_2, 7.03% C_2H_6 (molar basis) at 24°C, 116.5 kPa (1.15 atm). Column length, 48 cm. Theoretical curves were calculated numerically using the linear driving force model with a Langmuir equilibrium isotherm. Experimental data are from ref. 21. From ref. 22, courtesy of Pergamon Press.

pattern of the response is governed by the form of the equilibrium relationship rather than by kinetics. Kinetic limitations may modify the form of the concentration profile but they do not change the general pattern. To illustrate the different types of transition, consider the simplest case: an isothermal system with plug flow involving a single adsorbable species present at low concentration in an inert carrier, for which equation 30 reduces to

$$v\frac{\partial c}{\partial z} + \frac{\partial c}{\partial t} + \left(\frac{1-\epsilon}{\epsilon}\right)\frac{\partial \bar{q}}{\partial t} = 0 \tag{33}$$

Assuming local equilibrium, $\bar{q} = f(c)$ where this function represents the isotherm equation, this becomes

$$\frac{v}{1 + ((1-\epsilon)/\epsilon)f'(c)}\frac{\partial c}{\partial z} + \frac{\partial c}{\partial t} = 0 \tag{34}$$

where $f'(c) = dq^*/dc$ is simply the slope of the equilibrium isotherm at concentration c.

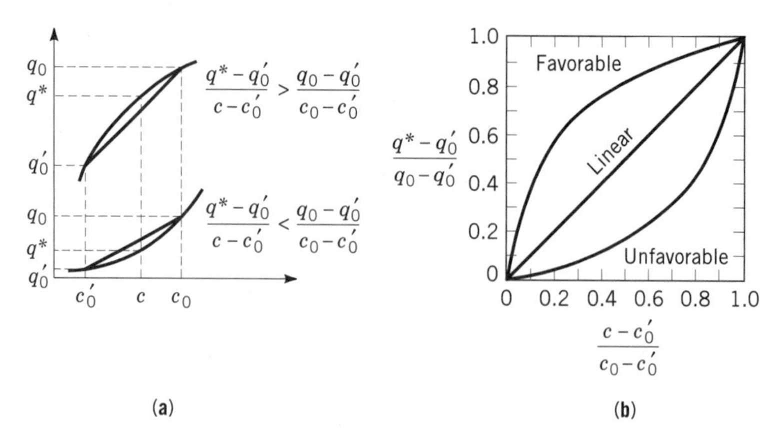

Fig. 11. (**a**) Equilibrium isotherm and (**b**) dimensionless equilibrium diagram showing favorable, linear, and unfavorable isotherms.

Equation 34 has the form of the kinematic wave equation and represents a transition traveling with the wave velocity w, given by

$$w = \left(\frac{\partial z}{\partial t}\right)_c = \frac{v}{1 + ((1-\epsilon)/\epsilon)f'(c)} \tag{35}$$

For a linear system $f'(c) = K$, so the wave velocity becomes independent of concentration and, in the absence of dispersive effects such as mass transfer resistance or axial mixing, a concentration perturbation propagates without changing its shape. The propagation velocity is inversely dependent on the adsorption equilibrium constant.

For a nonlinear system the behavior depends on the shape of the isotherm. If the isotherm is unfavorable (Fig. 11), $f'(c)$ increases with concentration so that w decreases with concentration. This leads to a spreading profile, as illustrated in Figure 12**b**. However, if the isotherm is favorable (in the direction of the concentration change), an entirely different situation arises. Then $f'(c)$ decreases with concentration so that w increases with concentration. This would lead to the physically unreasonable overhanging profiles shown in Figure 12**a**. In fact, what happens is that the continuous solution is replaced by the equivalent shock transition so that response becomes a shock wave which propagates at a steady velocity w' given by

$$w' = \frac{v}{1 + ((1-\epsilon)/\epsilon)(\Delta q/\Delta c)} \tag{36}$$

where $\Delta q/\Delta c$ represents the ratio of the concentration changes in the adsorbed and fluid phases.

6.2. Constant Pattern Behavior. In a real system the finite resistance to mass transfer and axial mixing in the column lead to departures from the idealized response predicted by equilibrium theory. In the case of a favorable isotherm the shock wave solution is replaced by a constant pattern solution. The

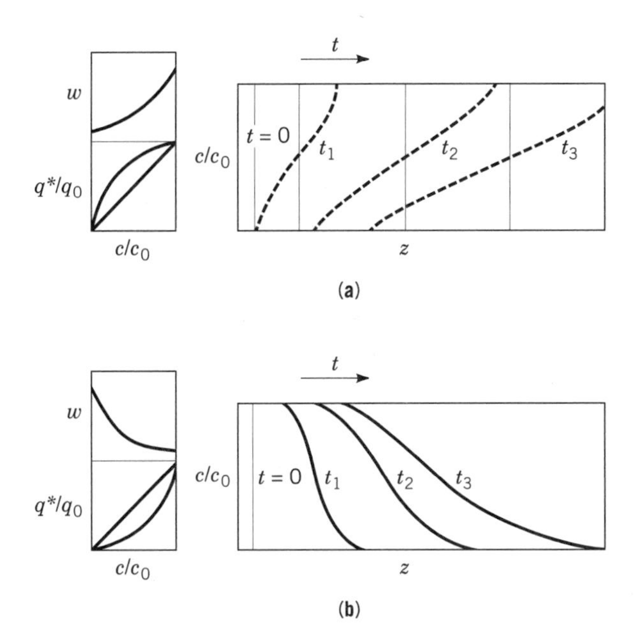

Fig. 12. (**a**) Development of the physically unreasonable overhanging concentration profile and the corresponding shock profile for adsorption with a favorable isotherm and (**b**) development of the dispersive (proportionate pattern) concentration profile for adsorption with an unfavorable isotherm (or for desorption with a favorable isotherm). From ref. 7.

concentration profile spreads in the initial region until a stable situation is reached in which the mass transfer rate is the same at all points along the wave front and exactly matches the shock velocity. In this situation the fluid-phase and adsorbed-phase profiles become coincident, as illustrated in Figure 13. This represents a stable situation and the profile propagates without further change in shape—hence the term constant pattern. The form of the concentration profile under constant pattern conditions may be easily deduced by integrating the mass transfer rate expression subject to the condition $c/c_0 = q/q_0$, where q_0 is the adsorbed phase concentration in equilibrium with c_0.

The distance required to approach the constant pattern limit decreases as the mass transfer resistance decreases and the nonlinearity of the equilibrium isotherm increases. However, when the isotherm is highly favorable, as in many adsorption processes, this distance may be very small, a few centimeters to perhaps a meter.

6.3. Length of Unused Bed. The constant pattern approximation provides the basis for a very useful and widely used design method based on the concept of the length of unused bed (LUB). In the design of a typical adsorption process the basic problem is to estimate the size of the adsorber bed needed to remove a certain quantity of the adsorbable species from the feed stream, subject to a specified limit (c') on the effluent concentration. The length of unused bed, which measures the capacity of the adsorber which is lost as a result of the spread of the concentration profile, is defined by

$$\text{LUB} = (1 - q'/q_0)L = (1 - t'/\bar{t})L \tag{37}$$

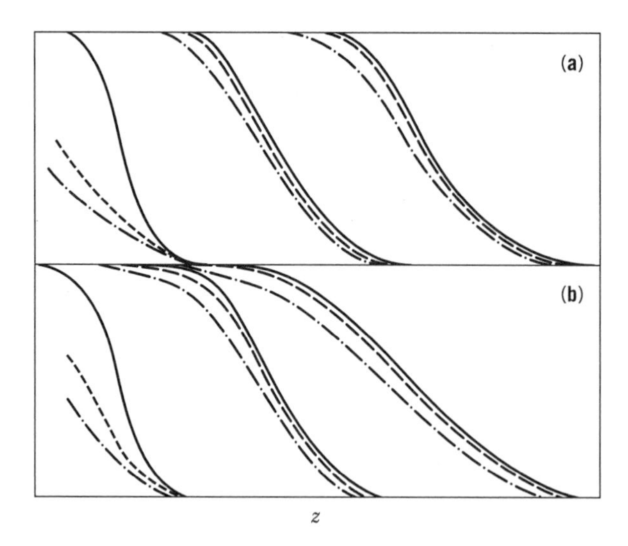

Fig. 13. Schematic diagram showing (**a**) approach to constant pattern behavior for a system with a favorable isotherm and (**b**) approach to proportionate pattern behavior for a system with an unfavorable isotherm. y axis: c/c_0, —; $\bar{q}/q_0$, -; c^*/c_0, —·. From ref. 7.

where q' is the capacity at the break time t' and $\bar{t}$; is the stoichiometric time (see Fig. 14). The values of t', $\bar{t}$, and hence the LUB are easily determined from an experimental breakthrough curve since, by overall mass balance,

$$\bar{t} = \frac{L}{v}\left[1 + \left(\frac{1-\epsilon}{\epsilon}\right)\left(\frac{q_0}{c_0}\right)\right] = \int_0^\infty \left(1 - \frac{c}{c_0}\right) dt \tag{38}$$

$$t' = \frac{L}{v}\left[1 + \left(\frac{1-\epsilon}{\epsilon}\right)\left(\frac{q'}{c_0}\right)\right] = \int_0^{t'} \left(1 - \frac{c}{c_0}\right) dt \tag{39}$$

Under constant pattern conditions the LUB is independent of column length although, of course, it depends on other process variables. The procedure is therefore to determine the LUB in a small laboratory or pilot-scale column packed with the same adsorbent and operated under the same flow conditions. The length of column needed can then be found simply by adding the LUB to

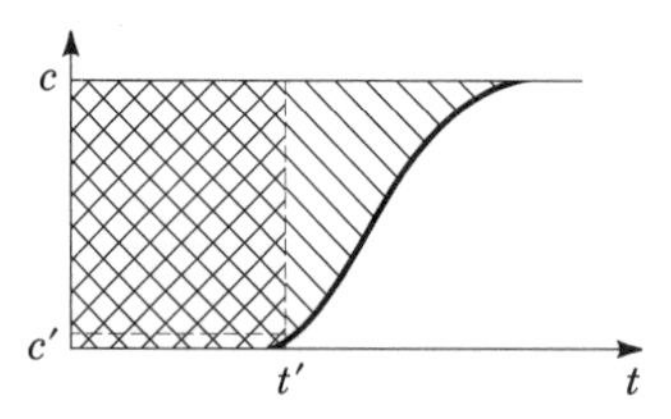

Fig. 14. Sketch of breakthrough curve showing break time t' and the method of calculation of the stoichiometric time $\bar{t}$; and LUB. From ref. 7. \\\\\\ = the integral of equation 38; ◇◇◇ = the integral of equation 39.

the length calculated from equilibrium considerations, assuming a shock concentration front.

One potential problem with this approach is that heat loss from a small scale column is much greater than from a larger diameter column. As a result, small columns tend to operate almost isothermally whereas in a large column the system is almost adiabatic. Since the temperature profile in general affects the concentration profile, the LUB may be underestimated unless great care is taken to ensure adiabatic operation of the experimental column.

6.4. Proportionate Pattern Behavior. If the isotherm is unfavorable, the stable dynamic situation leading to constant pattern behavior can never be achieved. The situation is shown in Figure 13**b**. The equilibrium adsorbed-phase concentration lies above rather than below the actual adsorbed-phase profile. As the mass transfer zone progresses through the column it broadens, but the limiting situation, which is approached in a long column, is simply local equilibrium at all points ($c = c^*$) and the profile therefore continues to spread in proportion to the length of the column. This difference in behavior is important since the LUB approach to design is clearly inapplicable under these conditions.

Favorable and unfavorable equilibrium isotherms are normally defined, as in Figure 11, with respect to an increase in sorbate concentration. This is, of course, appropriate for an adsorption process, but if one is considering regeneration of a saturated column (desorption), the situation is reversed. An isotherm which is favorable for adsorption is unfavorable for desorption and vice versa. In most adsorption processes the adsorbent is selected to provide a favorable adsorption isotherm, so the adsorption step shows constant pattern behavior and proportionate pattern behavior is encountered in the desorption step.

6.5. Detailed Modeling Results. The results of a series of detailed calculations for an ideal isothermal plug-flow Langmuir system are summarized in Figure 15. The solid lines show the form of the theoretical breakthrough curves for adsorption and desorption, calculated from the following set of model equations and expressed in terms of the dimensionless variables ζ, τ, and β:

Differential Balance for Column

$$v\frac{\partial c}{\partial z} + \frac{\partial c}{\partial t} + \left(\frac{1-\epsilon}{\epsilon}\right)\frac{\partial \bar{q}}{\partial t} = 0 \tag{40}$$

Rate Equation

$$\frac{\partial \bar{q}}{\partial t} = k(q^* - \bar{q}) \tag{41}$$

Equilibrium Isotherm

$$\frac{q^*}{q_s} = \frac{bc}{1+bc} \tag{42}$$

Initial Conditions

$$\begin{aligned} &\bar{q}(z,\,0) = c(z,\,0) = 0 \quad \text{(adsorption)} \\ &\bar{q}(z,\,0) = q_0,\; c(z,\,0) = c_0 \quad \text{(desorption)} \end{aligned} \tag{43}$$

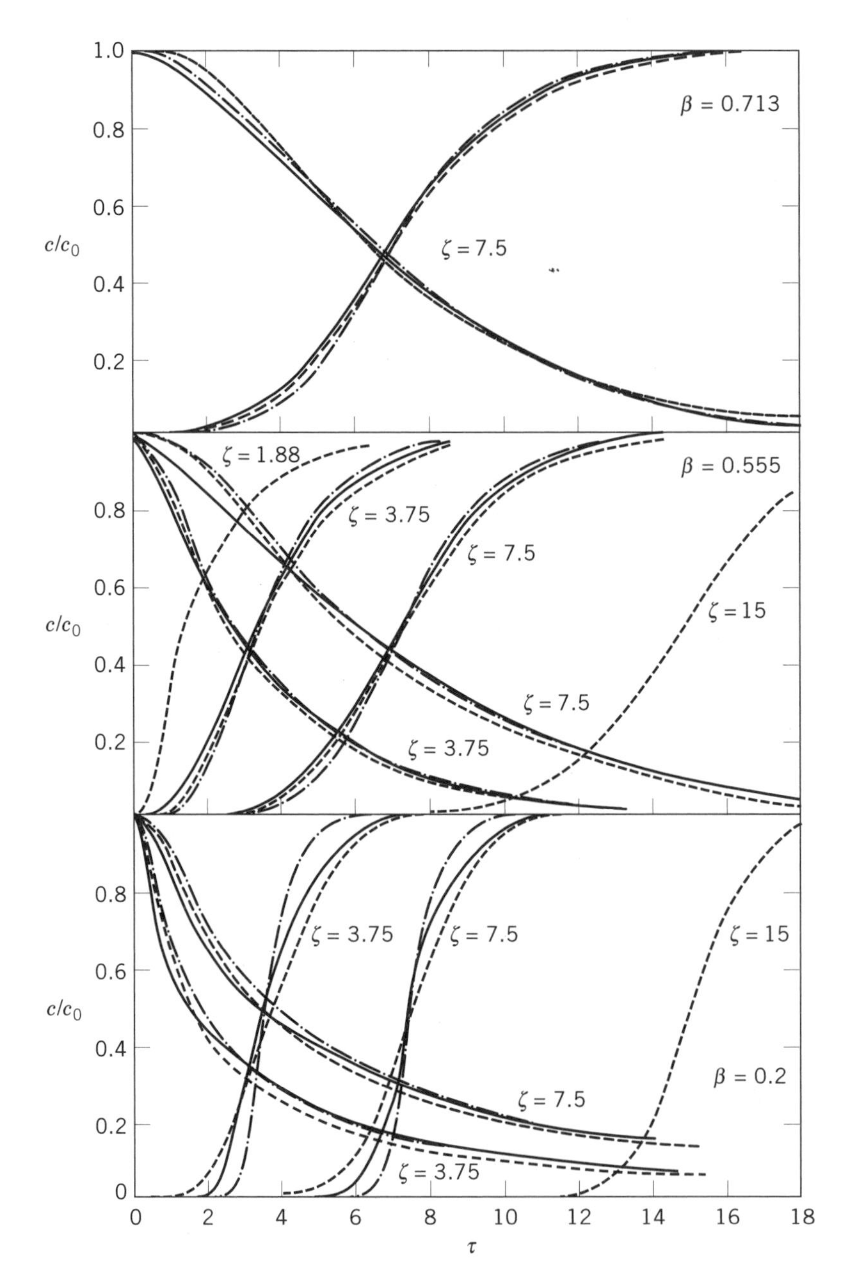

Fig. 15. Theoretical breakthrough curves for a nonlinear (Langmuir) system showing the comparison between the linear driving force (—), pore diffusion (-), and intracrystalline diffusion (—·) models based on the Glueckauf approximation (eqs. 40–45). From Ref. 7.

Boundary Conditions

$$\begin{aligned} c(0, t) &= c_0 \quad \text{(adsorption)} \\ c(0, t) &= 0 \quad \text{(desorption)} \end{aligned} \tag{44}$$

Bed Length Parameter

$$\zeta = \frac{z}{v}\left(\frac{q_0}{c_0}\right)\left(\frac{1-\epsilon}{\epsilon}\right)$$

Dimensionless Time

$$\tau = k\left(t - \frac{z}{v}\right) \tag{45}$$

Nonlinearity Parameter

$$\beta = 1 - \frac{q_0}{q_s} = \frac{1}{1 + bc_0}$$

Also shown are the corresponding curves calculated for the same system assuming a diffusion model in place of the linear rate expression. For intracrystalline diffusion $k = 15D_0/r_c^2$, whereas for macropore diffusion $k = (15\epsilon_p D_p/R_p^2)(c_0/q_0)$, in accordance with the Glueckauf approximation (23).

For linear or moderately nonlinear systems ($\beta \longrightarrow 1.0$) there is little difference in the response curves for all three models, thus verifying the validity of the Glueckauf approximation. Differences between the models, however, become more significant for a highly nonlinear isotherm ($\beta \longrightarrow 0$). For linear or near linear systems the adsorption and desorption curves are mirror images, but as the isotherm becomes more nonlinear the adsorption and desorption curves become increasingly asymmetric. The adsorption curve approaches its limiting constant pattern form whereas the desorption curve approaches the limiting proportionate pattern form. In the long-time region the desorption curve is governed entirely by equilibrium, so that the curves for all three rate models again become coincident.

The main conclusion to be drawn from these studies is that for most practical purposes the linear rate model provides an adequate approximation and the use of the more cumbersome and computationally time consuming diffusing models is generally not necessary. The Glueckauf approximation provides the required estimate of the effective mass transfer coefficient for a diffusion controlled system. More detailed analysis shows that when more than one mass transfer resistance is significant the overall rate coefficient may be estimated simply from the sum of the resistances (7):

$$\frac{1}{kK} = \frac{R}{3k_f} + \frac{R^2}{15KD_c} + \frac{R^2}{15\epsilon_p D_p} \tag{46}$$

6.6. Adsorption Chromatography. The principle of gas-solid or liquid-solid chromatography may be easily understood from equation 35. In a linear multicomponent system (several sorbates at low concentration in an inert carrier) the wave velocity for each component depends on its adsorption equilibrium constant. Thus, if a pulse of the mixed sorbate is injected at the column inlet, the different species separate into bands which travel through the column at their

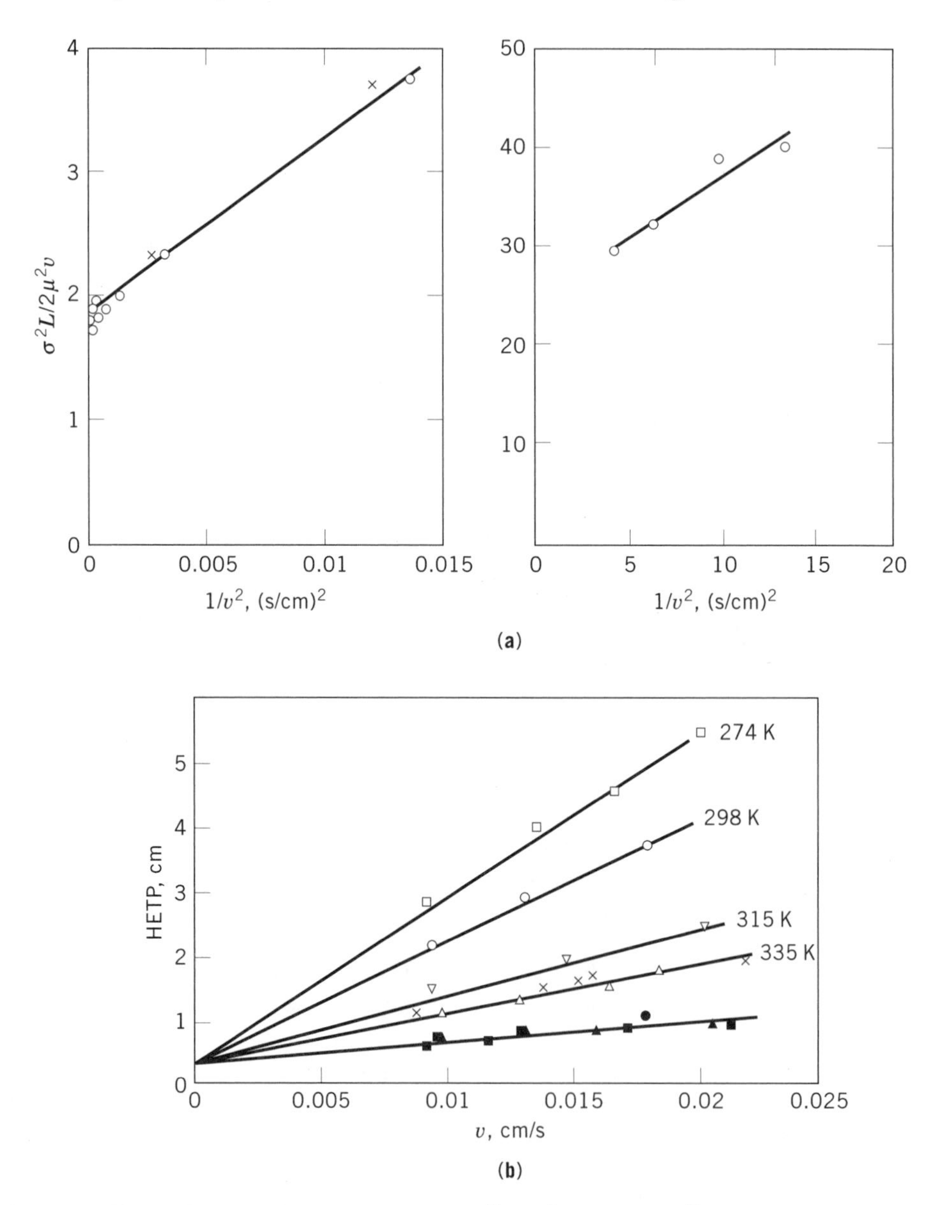

Fig. 16. Plots showing (**a**) variation of ($\varsigma^2 L/2\mu^2 v$) with $1/v^2$ for O_2 (left plot, ×, 0.84 – 0.72 mm = 20 – 25 mesh; ○, 0.42 – 0.29 mm = 40 – 50 mesh) and N_2 (right plot, on 3.2-mm pellets) in Bergbau-Forschung carbon molecular sieve and (**b**) variation of HETP with liquid velocity (interstitial) for fructose (solid symbols), and glucose (open symbols) in a column packed with KX zeolite crystals. From refs. 24 and 25.

characteristic velocities, and at the outlet of the column a sequence of peaks corresponding to the different species is detected. Measurement of the retention time ($\bar{t}$) under known flow conditions thus provides a simple means of determining the equilibrium constant (Henry constant):

$$\int_0^\infty \frac{ct\,dt}{\int_0^\infty c\,dt} = \bar{t} = \frac{L}{v}\left[1 + \left(\frac{1-\epsilon}{\epsilon}\right)K\right] \tag{47}$$

In an ideal system with no axial mixing or mass transfer resistance the peaks for the various components propagate without spreading. However, in any real system the peak broadens as it propagates and the extent of this broadening is directly related to the mass transfer and axial dispersion characteristics of the column. Measurement of the peak broadening therefore provides a convenient way of measuring mass transfer coefficients and intraparticle diffusivities. The simplest approach is to measure the second moments of the response peak over a range of flow rates:

$$\sigma^2 \equiv \int_0^\infty \frac{c(t-\bar{t})^2\,dt}{\int_0^\infty c\,dt} \tag{48}$$

Solution of the model equations shows that, for a linear isothermal system and a pulse injection, the height equivalent to a theoretical plate (HETP) is given by

$$H = \frac{\sigma^2}{\bar{t}^2}L = \frac{2D_L}{v} + \frac{2v}{kK}\left(\frac{\epsilon}{1-\epsilon}\right)\left[1 + \frac{\epsilon}{(1-\epsilon)K}\right]^{-2} \tag{49}$$

where $1/kK$ is the overall mass transfer resistance defined by equation 46.

For liquid systems D_L/v is approximately independent of velocity, so that a plot of H versus v provides a convenient method of determining both the axial dispersion and mass transfer resistance. For vapor-phase systems at low Reynolds numbers D_L is approximately constant since dispersion is determined mainly by molecular diffusion. It is therefore more convenient to plot H/v versus $1/v^2$, which yields D_L as the slope and the mass transfer resistance as the intercept. Examples of such plots are shown in Figure 16.

7. Applications

The applications of adsorbents are many and varied and may be classified as nonregenerative uses, in which the adsorbent is used once and discarded, and regenerative applications, in which the adsorbent is used repeatedly in a cyclic manner involving sequential adsorption and regeneration steps.

Nonregenerative uses
Desiccant in dual pane windows
Odor removal in health care products

Desiccant in refrigeration and air conditioning systems
Cigarette filters

Regenerative uses
Water purification (some systems)
Removal of trace impurities from gases or liquid streams
Bulk separations (gas or liquid)
Low pressure storage of methane
Desiccant cooling (open-cycle air conditioning)

In terms of tonnage usage, some of the nonregenerative applications, notably as desiccants in dual pane windows, in cigarette filters, and in water purification, are surprisingly important, but most of the important chemical engineering applications are regenerative since, with a few notable exceptions, the cost of the adsorbent is too great to allow nonregenerative use.

The application of adsorbents (generally high area activated carbon) as a means of storing methane fuel (natural gas) at relatively high density under moderate pressure is relatively new. With current technology, capacities of about 180 m^3 STP/m^3 at 3 MPa (30 atm) pressure are achievable (26) but somewhat higher capacities are needed to compete with liquid fuels for motor vehicles. However, depending on the cost of crude oil and the potential for improvement of the adsorbent, this technology could become important in the future.

Open-cycle desiccant cooling is another area of emerging technology (8). Rather than cooling and dehumidifying by mechanical work, as in a conventional air conditioning system, in an open-cycle desiccant system dehumidification is achieved directly, while cooling is achieved by controlled evaporation. The energy input is in the form of the heat required to regenerate the desiccant. A significant advantage of this system is that it can be designed to operate with a low regeneration temperature, thus making it possible to utilize low grade heat or even solar heat to drive the system.

7.1. Adsorption Separation and Purification Processes. The main area of current application of adsorption is in separation and purification processes. Many different ways of operating such processes have been devised and it is helpful to consider the various systems according to the mode of fluid-solid contact (see Fig. 17). In a cyclic batch process at least two beds are employed and each bed is successively saturated with the preferentially adsorbed species (or class of species) during the adsorption step and then regenerated during a desorption step in which the direction of mass transfer is reversed to remove the adsorbed species from the bed. In the continuous countercurrent process the adsorbent can be regarded as circulating continuously between the adsorption and desorption beds, in both of which fluid and solid contact in countercurrent flow. More commonly, as in the Sorbex type of process, the adsorbent is not physically circulated but the same effect is achieved in a fixed adsorbent bed equipped with multiple inlet and outlet ports to which the fluid streams are directed in sequence. Such systems can achieve a close approximation to counter-current flow without the problems inherent in circulating the solid adsorbent. However, the system is relatively expensive, so it is generally used only for difficult separations (low separation factor) which cannot be carried out efficiently in a simple batch process.

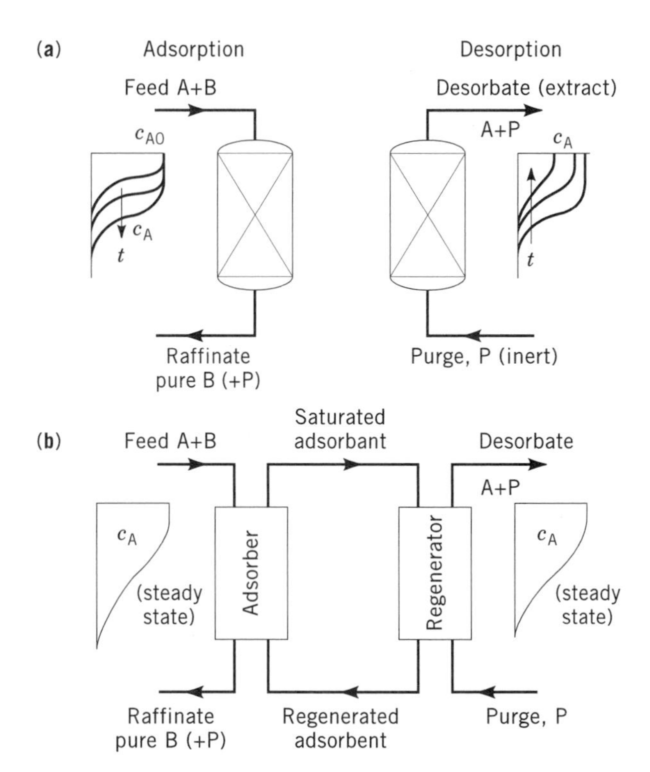

Fig. 17. The two basic modes of operation for an adsorption process: (**a**) cyclic batch system; (**b**) continuous countercurrent system with adsorbent recirculation. From ref. 7.

The other major difference between adsorption processes lies in the method by which the adsorbent bed is regenerated. The advantages and disadvantages of three different methods—temperature swing, pressure swing, and displacement—are summarized in Table 5. For efficient removal of trace impurities it is normally essential to use a highly selective adsorbent on which the sorbate is strongly held. Temperature swing regeneration is therefore generally used in such applications. However, in bulk separations all three regeneration methods are widely used.

7.2. Adsorbent Contactors. The randomly packed bed of adsorbent particles (with axial flow) is the workhorse of adsorption process technology. Such a contactor is cheap and robust and offers the important advantage of relatively low resistance to mass transfer. However, the pressure drop is relatively high so for applications such as VOC removal or desiccant cooling, which involve large flows of dilute gas streams, the energy losses can be prohibitive. It has been shown that a monolithic or parallel passage contactor with sufficiently small holes or spacing between the adsorbent layers has more favorable characteristics, in terms of the pressure drop per theoretical stage, and therefore, for such applications, offers an attractive alternative to the randomly packed bed (27). Processes based on parallel passage contactors, often constructed in the form of a slowly rotating wheel to allow continuous cycling between a cold adsorption

Table 5. **Factors Governing Choice of Regeneration Method**[a]

Method	Advantages	Disadvantages
thermal swing	good for strongly adsorbed species; small change in T gives large change in q^*	thermal aging of adsorbent
	desorbate may be recovered at high concentration	heat loss means inefficiency in energy usage
	gases and liquids	unsuitable for rapid cycling, so adsorbent cannot be used with maximum efficiency
		in liquid systems the latent heat of the interstitial liquid must be added
pressure swing	good where weakly adsorbed species is required at high purity	very low P may be required
	rapid cycling—efficient use of adsorbent	mechanical energy more expensive than heat
		desorbate recovered at low purity
displacement desorption	good for strongly held species	product separation and recovery needed (choice of desorbent is crucial)
	avoids risk of cracking reactions during regeneration	
	avoids thermal aging of adsorbent	

[a] Ref. 7.

zone and a hot regeneration zone, have been developed by several manufacturers including Dürr and Seibu-Geiken.

7.3. Process Design. As with any chemical engineering process, the choice of process type and the details of the design are dictated primarily by economic considerations, subject to the overriding requirements of safety and reliability. Although the principles of adsorption processes are well understood, most practical designs still rely on a good deal of empiricism since factors such as the aging and deterioration of an adsorbent under practical operating conditions are difficult to predict except from experience.

In general the sophistication of the design procedures is closely related to the level of sophistication of the process. In the simple thermal swing, batch type processes for removal of trace impurities the beds are generally sized on the basis of equilibrium capacity and LUB with a suitable allowance for aging of the adsorbent. The desorption or regeneration temperature is generally selected on the basis of equilibrium data as the minimum temperature which will allow the required specification of the purity of the raffinate product to be easily met. Use of a higher regeneration temperature generally gives a purer product but only at the cost of increased energy consumption and reduction of the service life of the adsorbent. The quantity of purge gas is estimated in two ways: from the overall heat balance with the required temperature rise and from the mass balance with the assumption of equilibrium at the bed outlet. Depending on the particular system and the process conditions, either of these considerations may be limiting. In general, when regeneration is carried out by purging at atmospheric pressure, the purge requirement is determined by the heat balance, but when the regeneration is carried out at elevated pressure, the mass balance is often the major constraint.

The design of pressure swing systems is also largely dependent on the scale-up of pilot-plant units, although, with such systems, the use of a detailed numerical simulation to guide the optimization of the operating conditions is more common. This is true also of countercurrent and simulated countercurrent processes where the initial design is commonly based on a simple McCabe-Thiele diagram (28) (see ADSORPTION, LIQUID SEPARATION).

7.4. Notation

a	ratio of external surface area to particle volume
b	Langmuir equilibrium constant
B	mobility
c	sorbate concentration in fluid phase
c_0	initial value of c
D	diffusivity
D_e	effective diffusivity
D_m	molecular diffusivity
D_K	Knudsen diffusivity
D_L	axial dispersion coefficient
D_0	corrected diffusivity
D_∞	preexponential factor
D_p	pore diffusivity
E	activation energy
$-\Delta H_0$	limiting heat of adsorption
K	dimensionless equilibrium constant
K_0	preexponential factor
K'	Henry's law constant
K'_0	preexponential factor
k	rate constant
k_0	preexponential factor
k', k''	constants in Elovich equation
k_f	fluid film mass transfer coefficient
k_s	surface mass transfer coefficient
L	bed length
M	molecular weight
m_t	mass adsorbed (or desorbed) at time t
m_∞	mass adsorbed (or desorbed) at $t \longrightarrow \infty$
p	partial pressure of sorbate
q	adsorbed phase concentration
q_0	initial value of q
$\overline{q}$	adsorbed phase concentration averaged over a particle
q^*	equilibrium value of q
q_s	saturation limit in Langmuir expression
r	radial coordinate
R	radius of adsorbent particle
R	gas constant
t	time
$\bar{t}$	mean residence time or stoichiometric time
t'	break time
T	temperature
$-\Delta U$	change of interval energy on adsorption
v	interstitial fluid velocity
w	wave velocity

w'	shock velocity
z	distance
α	separation factor
ε	voidage of adsorbent bed
ε_p	porosity of particle
μ	chemical potential
ρ	mean pore radius
σ^2	variance of pulse response
τ	tortuosity factor

BIBLIOGRAPHY

"Adsorptive separation, introduction" in *ECT* 3rd ed., Vol. 1, pp. 531–544, by Theodore Vermeulen, University of California, Berkeley; "Adsorption, theoretical" in *ECT* 2nd ed., Vol. 1, pp. 421–459, by Sydney Ross, Rensselaer Polytechnic Institute; "Adsorption, theoretical" in *ECT* 1st ed., Vol. 1, pp. 206–222, by P. H. Emmett, Mellon Institute of Industrial Research; "Adsorption" in *ECT* 4th ed., Vol. 1, pp. 493–528, by Douglas M. Ruthven, University New Brunswick, Canada; "Adsorption" in *ECT* (online), posting date: December 4, 2000, by Douglas M. Ruthven, University of New Brunswick, Canada.

CITED PUBLICATIONS

1. A. Thijs, G. Peters, E. F. Vansant, I. Verhaert, and P. deBievre, *J. Chem. Soc. Faraday Trans. 1* **79**, 2821 (1983).
2. Y. Matsumura, *Proceedings of the First Indian Carbon Conference, New Delhi*, December 1982, 99–106.
3. E. M. Flanigen and co-workers, *Nature* **271**, 512 (1978).
4. S. M. Klein and W. H. Abraham, *AIChE Symp. Ser.* **79**(230), 53 (1984).
5. D. W. Breck, *Zeolite Molecular Sieves*, John Wiley & Sons, Inc., New York, 1974.
6. W. M. Meier and D. H. Olson, *Atlas of Zeolite Structure Types*, Juris Druck and Verlag AG, Zurich, 1978.
7. D. M. Ruthven, *Principles of Adsorption and Adsorption Processes*, Wiley-Interscience, New York, 1984.
8. T. R. Penney and I. Maclaine-Cross, *Proceedings, Desiccant Cooling and Dehumidification Workshop, June 10–11, 1986, Chattanooga, Tenn.*, Sponsored by Electric Power Research Institute, Gas Research Institute, and Tennessee Valley Authority.
9. S. Brunauer, L. S. Deming, W. E. Deming, and J. E. Teller, *J. Am. Chem. Soc.* **62**, 1723 (1940).
10. D. M. Ruthven, *Sep. Purif. Methods* **5**(2), 189 (1976).
11. D. M. Ruthven and K. F. Loughlin, *J. Chem. Soc. Faraday Trans. 1* **68**, 696 (1972).
12. A. V. Kiselev and K. D. Shcherbakova in "Molecular Sieves," *Proceedings 1st International Zeolite Conference, London, 1967*, Society of Chemical Industry, London, 1968.
13. R. Sips, *J. Chem. Phys.* **16**, 490 (1948).
14. A. L. Myers and J. M. Prausnitz, *AIChE J.* **11**, 121 (1965).
15. P. G. Ashmore, *Catalysis and Inhibition of Chemical Reactions*, Butterworths, London, 1963, p. 164.
16. N. Wakao and T. Funazkri, *Chem. Eng. Sci.* **33**, 1375 (1978).
17. I. H. Doetsch, D. M. Ruthven, and K. F. Loughlin, *Can. J. Chem.* **52**, 2717 (1974).
18. J. Kärger and D.M. Ruthven, *Diffusion in Zeolities and other Micropores Solids*, John Wiley, New York, 1992.

19. R. Krishna and J. A. Wesselingh, *Chem. Eng. Sci.* **52**, 861 (1997)
20. J. A. Dominguez, D. Psaris, and A. I. La Cava, *AIChE Symp. Ser.* **84**(264), 73 (1988).
21. D. Basmadjian and D. W. Wright, *Chem. Eng. Sci.* **36**, 937 (1981).
22. A. I. Liapis and O. K. Crosser, *Chem. Eng. Sci.* **37**, 958 (1982).
23. E. Glueckauf, *Trans. Faraday Soc.* **51**, 1540 (1955).
24. D. M. Ruthven, N. S. Raghavan, and M. M. Hassan, *Chem. Eng. Sci.* **41**, 1325 (1986).
25. C. B. Ching and D. M. Ruthven, *Zeolites* **8**, 68 (1988).
26. S. S. Barton, J. A. Holland, and D. F. Quinn, *Proceedings of the 2nd International Conference on Adsorption, Santa Barbara, May 1986*, Engineering Foundation, New York, 1987, p. 99.
27. D. M. Ruthven and C. Thaeron *Gas. Sep. and Purif.* **10**, 63–73 (1996).
28. D. M. Ruthven and C. B. Ching, *Chem. Eng. Sci.* **44**, 1011 (1989).

GENERAL REFERENCES

D. M. Ruthven, *Principles of Adsorption and Adsorption Processes*, Wiley-Interscience, New York, 1984.

M. Suzuki, *Adsorption Engineering*, Kodansha-Elsevier, Tokyo, 1990.

R. T. Yang, *Gas Separation by Adsorption Processes*, Butterworths, Stoneham, Mass., 1987.

P. Wankat, *Large Scale Adsorption and Chromatography*, CRC Press, Boca Raton, Fla., 1986.

A. E. Rodrigues and D. Tondeur, eds., *Percolation Processes*, NATO ASI No. 33, Sijthoff & Noordhoff, Alpen aan den Rijn, 1980.

A. E. Rodrigues, M. D. Le Van, and D. Tondeur, *Adsorption: Science and Technology*, NATO ASI E158, Kluwer, Amsterdam, 1989.

N. Wakao, *Heat and Mass Transfer in Packed Beds*, Gordon & Breach, New York, 1982.

M. Smisek and S. Cerny, *Active Carbon*, Elsevier, Amsterdam, 1970.

T. Vermeulen, M. D. LeVan, N. K. Hiester, and G. Klein, "Adsorption and Ion Exchange," Section 16 of *Perry's Chemical Engineers' Handbook*, 6th ed., McGraw-Hill Book Co., New York, 1984.

D. Basmadjian, *The Little Adsorption*. CRC Press, Boca Raton, (1997).

D. D. Do *Adsorption Analysis; Kinetics and Equilibria* Imperial College Press, London (1998).

D. M. Ruthven, S. Farooq, and K. Kraebel, *Pressure Swing Adsorption* VCH, New-York (1994).

Douglas M. Ruthven
University of Maine

ADSORPTION, GAS SEPARATION

1. Introduction

Gas-phase adsorption is widely employed for the large-scale purification or bulk separation of air, natural gas, chemicals, and petrochemicals (Table 1). In these

Table 1. **Commercial Adsorption Separations**

Separation[a]	Adsorbent
gas bulk separations	
normal paraffins, isoparaffins, aromatics	zeolite
N_2/O_2	zeolite
O_2/N_2	carbon molecular sieve
CO, CH_4, CO_2, N_2, Ar, NH_3/H_2	zeolite, activated carbon
acetone/vent streams	activated carbon
C_2H_4/vent streams	activated carbon
H_2O/ethanol	zeolite
gas purifications	
H_2O/olefin-containing cracked gas, natural gas, air, synthesis gas, etc.	silica, alumina, zeolite
CO_2/C_2H_4, natural gas, etc.	zeolite
organics/vent streams	activated carbon, others
sulfur compounds/natural gas, hydrogen, liquefied petroleum gas (LPG), etc.	zeolite
solvents/air	activated carbon
odors/air	activated carbon
NO_x/N_2	zeolite
SO_2/vent streams	zeolite
Hg/chlor–alkali cell gas effluent	zeolite

[a] Ref. 1.

uses, it is often a preferred alternative to the older unit operations of distillation and absorption.

An adsorbent attracts molecules from the gas, the molecules become concentrated on the surface of the adsorbent, and are removed from the gas phase. Many process concepts have been developed to allow the efficient contact of feed gas mixtures with adsorbents to carry out desired separations and to allow efficient regeneration of the adsorbent for subsequent reuse. In nonregenerative applications, the adsorbent is used only once and is not regenerated.

Most commercial adsorbents for gas-phase applications are employed in the form of pellets, beads, or other granular shapes, typically ~1.5–3.2 mm in diameter. Most commonly, these adsorbents are packed into fixed beds through which the gaseous feed mixtures are passed. Normally, the process is conducted in a cyclic manner. When the capacity of the bed is exhausted, the feed flow is stopped to terminate the loading step of the process, the bed is treated to remove the adsorbed molecules in a separate regeneration step, and the cycle is then repeated.

The growth in both variety and scale of gas-phase adsorption separation processes, particularly since 1970, is due in part to continuing discoveries of new, porous, high-surface area adsorbent materials (particularly molecular sieve zeolites) and, especially, to improvements in the design and modification of adsorbents. These advances have encouraged parallel inventions of new process concepts. Increasingly, the development of new applications requires close cooperation in adsorbent design and process cycle development and optimization.

2. Adsorption Principles

The design and manufacture of adsorbents for specific applications involves manipulation of the structure and chemistry of the adsorbent to provide greater attractive forces for one molecule compared to another, or, by adjusting the size of the pores, to control access to the adsorbent surface on the basis of molecular size. Adsorbent manufacturers have developed many technologies for these manipulations, but they are considered proprietary and are not openly communicated. Nevertheless, the broad principles are well known.

This article is focused on physical adsorption, which involves relatively weak intermolecular forces, because most commercial applications of adsorption rely on this phenomenon alone. Chemisorption is discussed only briefly in some sections on specific applications.

2.1. Adsorption Forces. Coulomb's law allows calculations of the electrostatic potential resulting from a charge distribution, and of the potential energy of interaction between different charge distributions. Various elaborate computations are possible to calculate the potential energy of interaction between point charges, distributed charges, etc. See (2) for a detailed introduction.

An electric dipole consists of two equal and opposite charges separated by a distance. All molecules contain atoms composed of positively charged nuclei and negatively charged electrons. When a molecule is placed in an electric field between two charged plates, the field attracts the positive nuclei toward the negative plate and the electrons toward the positive plate. This electrical distortion, or polarization of the molecule, creates an electric dipole. When the field is removed, the distortion disappears, and the molecule reverts to its original condition. This electrical distortion of the molecule is called induced polarization; the dipole formed is an induced dipole.

The magnitude of the induced dipole moment depends on the electric field strength in accord with the relationship $\mu_i = \alpha F$, where μ_i is the induced dipole moment, F is the electric field strength, and the constant α is called the polarizability of the molecule. The polarizability is related to the dielectric constant of the substance. Group-contribution methods (2) can be used to estimate the polarizability from knowledge of the number of each type of bond within the molecule, eg, the polarizability of an unsaturated bond is greater than that of a saturated bond.

The total potential energy of adsorption interaction may be subdivided into parts representing contributions of the different types of interactions between adsorbed molecules and adsorbents. Adopting the terminology of Barrer and Vaughan (3), the total energy Φ_{Total} of interaction is the sum of contributions resulting from dispersion energy Φ_D, close-range repulsion Φ_R, polarization energy Φ_P, field–dipole interaction $\Phi_{F-\mu}$, field gradient–quadrupole interaction $\Phi_{\delta F-Q}$, and adsorbate–adsorbate interactions, denoted self-potential Φ_{SP}:

$$\Phi_{\text{Total}} = \underbrace{\Phi_D + \Phi_R + \Phi_P}_{\text{nonspecific}} + \underbrace{\Phi_{F-\mu} + \Phi_{\delta F-Q}}_{\text{specific}} + \underbrace{\Phi_{\text{SP}}}_{\text{adsorbate-adsorbent}}$$

The Φ_D and Φ_R terms always contribute, regardless of the specific electric charge distributions in the adsorbate molecules, which is why they are called nonspecific. The third nonspecific Φ_P term also always contributes, whether or not the adsorbate molecules have permanent dipoles or quadrupoles; however, for adsorbent surfaces that are relatively nonpolar, the polarization energy Φ_P is small.

The $\Phi_{F-\mu}$ and $\Phi_{\delta F-Q}$ terms are specific contributions, which are significant when adsorbate molecules possess permanent dipole and quadrupole moments. In the absence of these moments, these terms are zero, as is true also if the adsorbent surface has no electric fields, a completely nonpolar adsorbent.

Finally, the Φ_{SP} term is the contribution resulting from interactions between adsorbate molecules. At low coverages of the adsorbent by adsorbate molecules, this contribution approaches zero, and at high coverage it often causes a noticeable increase in the heat of adsorption.

The $\Phi_D + \Phi_R$ (dispersion plus repulsion) terms are known as the London or van der Waals forces. Spherical, nonpolar molecules are well described by the familiar Lennard-Jones 6–12 potential equation:

$$\Phi_D + \Phi_R = 4\varepsilon\left[-(\sigma/r)^6 + (\sigma/r)^{12}\right]$$

where r is the intermolecular separation distance, and σ (length units) and ε (energy units) are constants characteristic of the colliding molecules. Values of force constants σ and ε have been compiled (2).

These forces arise from the fact that each molecule contains atoms having a nucleus and surrounded by a cloud of electrons. The electron cloud fluctuates and is nonsymmetrical at various instants in time. Although a nonpolar neutral molecule has no net permanent charge or dipole, these fluctuating electron distributions provide fluctuating dipoles in each molecule. These fluctuating dipoles interact to generate forces between molecules or between adsorbed molecules and adsorbent surfaces. These contributions to the potential energy of adsorption are present even if the adsorbed molecules are nonpolar and even if the adsorbent structure contains no strong electrostatic fields.

The contribution Φ_P is due to the polarization of the molecules by electric fields on the adsorbent surface, eg, electric fields between positively charged cations and the negatively charged framework of a zeolite adsorbent. The attractive interaction between the induced dipole and the electric field is called the polarization contribution. Its magnitude is dependent on the polarizability α of the molecule and the strength of the electric field F of the adsorbent (4): $\Phi_P = -1/2\alpha F^2$.

The first of the two specific interaction terms $\Phi_{F-\mu}$ is due to the attractive interaction between the permanent dipole moment μ of a molecule and the electric field on the adsorbent surface (4):

$$\Phi_{F-\mu} = -F\mu\cos\Theta$$

where Θ is the dipole–axis/field angle.

The other specific interaction term $\Phi_{\delta F-Q}$ is due to the attractive interaction between the permanent quadrupole moment Q of the molecule and the electric field gradient on the adsorbent surface (4):

$$\Phi_{\delta F-Q} = \frac{1}{2} Q\, dF/dR$$

The final contribution, the self-potential term Φ_{SP}, is the sum of all the above interactions of adsorbed molecules with each other.

Finally, an analysis of the energies of adsorption on many practical polar and nonpolar adsorbents has shown not only that the magnitude of the Φ_P term depends directly on the polarizability α, but also that the sum of all of the nonspecific terms taken together, ie, $\Phi_D + \Phi_R + \Phi_P$, increases monotonically, with increasing α (4).

2.2. Adsorption Selectivities. For a given adsorbent, the relative strength of adsorption of different adsorbate molecules depends on the relative magnitudes of the polarizability α, dipole moment μ, and quadrupole moment Q of each. These properties for some common molecules are given in Table 2. Often, just the consideration of the values of α, μ, and Q allows accurate qualitative predictions to be made of the relative strengths of adsorption of given molecules on an adsorbent or of the best adsorbent type (polar or nonpolar) for a particular separation.

For example, the strength of the electric field F and field gradient ($\delta F = dF/dr$) of the highly polar cationic zeolites is strong. For this reason, nitrogen is more strongly adsorbed than is oxygen on such adsorbents, primarily because of the stronger quadrupole of N_2 compared to O_2.

In contrast, nonpolar activated carbon adsorbents lack strong electric fields and field gradients. Such adsorbents adsorb O_2 slightly more strongly than N_2, because of the slightly higher polarizability of O_2. Relative selectivities on nonpolar adsorbents often parallel the relative volatilities of the same compounds. Compounds with higher boiling points are more strongly adsorbed. In this

Table 2. **Electrostatic Properties of Some Common Molecules**

Molecule	Polarizability $\alpha \times 10^{40}$, $C^2{\cdot}m^2/J$ [a]	Dipole moment $\mu \times 10^{30}$, C·m [b]	Quadrupole moment $Q \times 10^{40}$, $C{\cdot}m^2$ [c]
Ar	1.83	0.00	0.00
H_2	0.90	0.00	2.09
N_2	0.78	0.00	−4.91
O_2	1.77	0.00	−1.33
CO	2.19	0.37	−6.92
CO_2	3.02	0.00	−13.71
CS_2	9.41	0.00	12.73
N_2O	3.32	0.54	−12.02
NH_3	2.67	5.10	−7.39
C_2H_6	4.97	0.00	−3.32
C_6H_6	11.49	0.00	−30.7
HCl	2.94	3.57	13.28

[a] To convert $C^2{\cdot}m^2/J$ to cm^3, divide by 1.113×10^{-16}.
[b] To convert C·m to debyes, divide by 3.336×10^{-30}.
[c] To convert $C{\cdot}m^2$ to Buckinghams, divide by 3.336×10^{-40}.

case, the higher boiling O_2 (bp ~ 90 K) is more strongly adsorbed than is N_2 (bp ~ 77 K).

The polarizabilities of molecules in a homologous series increase steadily with increasing numbers of atoms. Therefore, the relative strengths of adsorption also increase (along with the boiling points).

For a given adsorbate molecule, the relative strength of adsorption on different adsorbents depends largely on the relative polarizability and electric field strengths of adsorbent surfaces. On the one hand, water molecules, with relatively low polarizability but a strong dipole and moderately strong quadrupole moment, are strongly adsorbed by polar adsorbents (eg, cationic zeolites), but only weakly adsorbed by nonpolar adsorbents (eg, silicalite or nonoxidized forms of activated carbon). On the other hand, saturated hydrocarbons with low molecular weight have greater polarizabilities than does water, but no dipoles and only weak quadrupoles. These molecules are adsorbed less strongly than water on polar adsorbents, but more strongly than water on nonpolar adsorbents. Therefore, polar adsorbents are often called hydrophilic adsorbents and nonpolar adsorbents are called hydrophobic adsorbents.

2.3. Isotherms and Isobars. The graphical presentation of the equilibrium adsorbate loading vs adsorbate pressure (or concentration) at constant temperature (Fig. 1) is an adsorption isotherm. A graph of the adsorbate loading vs temperature at constant adsorbate pressure (Fig. 2) is an adsorption isobar. The greater the strength of adsorption, the greater is the adsorbate loading at a given temperature and partial pressure of the adsorbate up to the point where the maximum adsorption capacity of the adsorbent has been attained.

The strength of adsorption of unsaturated hydrocarbons by a polar adsorbent (zeolite) is much greater than for saturated hydrocarbons, and increases with increasing carbon number (Fig. 3). This observation may be understood as a consequence of the increasing polarizability of molecules with increasing

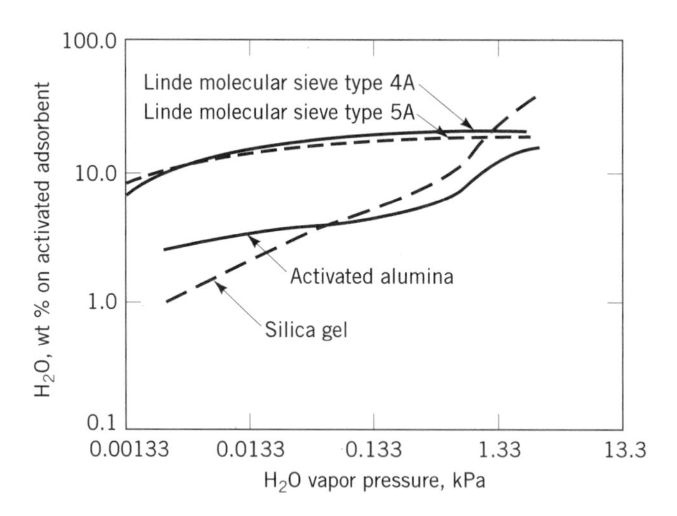

Fig. 1. Water isotherms for various adsorbents (1). Activation conditions: Linde molecular sieves, 350°C and <1.33 Pa; activated alumina, 350°C and <1.33 Pa; silica gel, 175°C and <1.33 Pa. To convert kPa to mm Hg, divide by 0.133.

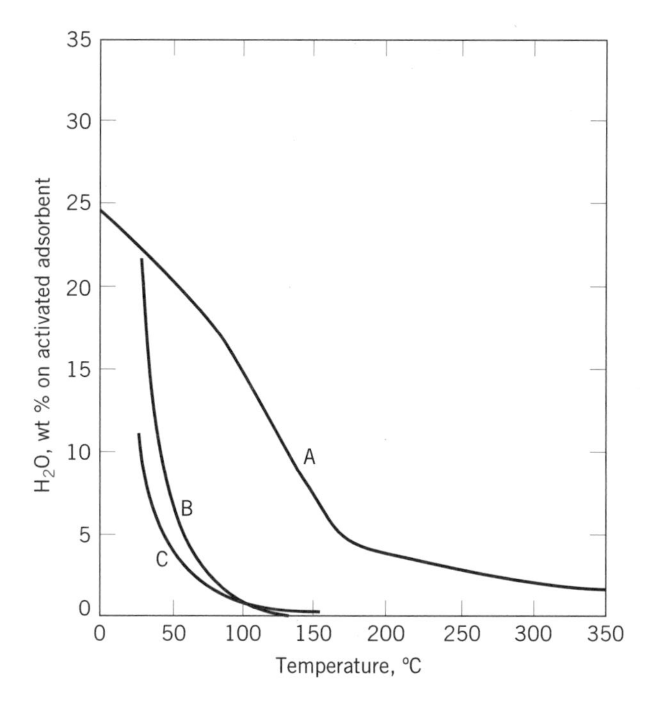

Fig. 2. Water isobars for various adsorbents: Equilibrium H_2O capacity vs temperature for three adsorbents (1); $p_{H_2O} = 1.33$ kPa (10 mm Hg) at 12°C and 101.3 kPa (1 atm). Activation conditions: A, Linde molecular sieve, 350°C and <1.33 Pa; B, activated alumina, 350°C and <1.33 Pa; C, silica gel, 175°C and <1.33 Pa. To convert Pa to mm Hg, multiply by 0.0075.

numbers of bonds and the presence of dipole and stronger quadrupole moments in the unsaturated hydrocarbons compared to the saturated hydrocarbons.

2.4. Heats of Adsorption. Physical adsorption processes are exothermic, ie, they release heat. Because the entropy change ΔS on adsorption is negative (adsorbed molecules are more ordered than in the gas phase) and the free energy change ΔG must be negative for adsorption to be favored, thermodynamics ($\Delta G = \Delta H - T\Delta S$) requires the enthalpy change ΔH on adsorption (heat of adsorption) to be negative (exothermic). Adsorption strengths thus decrease with increasing temperature.

The integral heat of adsorption is the total heat released when the adsorbate loading is increased from zero to some final value at isothermal conditions. The differential heat of adsorption δH_{iso} is the incremental change in heat of adsorption with a differential change in adsorbate loading. This heat of adsorption δH_{iso} may be determined from the slopes of adsorption isosteres (lines of constant adsorbate loading) on graphs of ln P vs $1/T$ (Fig. 4) through the Clausius-Clapeyron relationship:

$$\frac{d \ln P}{d\,(1/T)} = -\frac{\delta H_{\text{iso}}}{R}$$

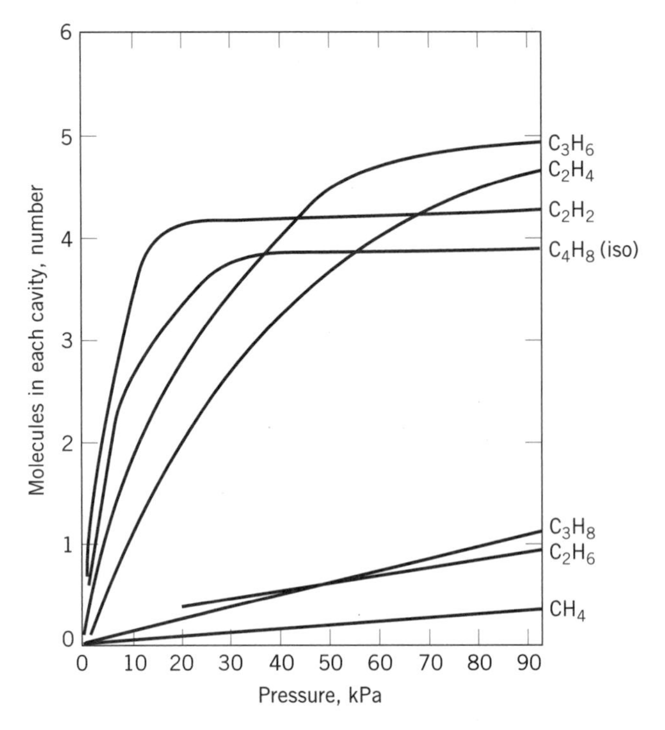

Fig. 3. Adsorption of hydrocarbons by zeolites is much greater for unsaturated hydrocarbons whose molecules contain double or triple bonds. From top to bottom, the curves show adsorption (at 150°C) of propylene, ethylene, acetylene, and isobutylene (unsaturated) and propane, ethane, and methane (saturated) (5). To convert kPa to mm Hg, multiply by 7.5. Courtesy of *Scientific American*.

where R is the gas constant, P the adsorbate absolute pressure, and T the absolute temperature.

Differential heats of adsorption for several gases on a sample of a polar adsorbent (natural zeolite chabazite) are shown as a function of the quantities adsorbed in Figure 5. Consideration of the electrical properties of the adsorbates, included in Table 2, allows the correct prediction of the relative order of adsorption selectivity:

$$\mathrm{Ar} < \mathrm{O_2} < \mathrm{N_2} < \mathrm{CO} \ll \mathrm{CO_2}$$

At low adsorbate loadings, the differential heat of adsorption decreases with increasing adsorbate loadings. This is direct evidence that the adsorbent surface is energetically heterogeneous, ie, some adsorption sites interact more strongly with the adsorbate molecules. These sites are filled first so that adsorption of additional molecules involves progressively lower heats of adsorption.

All practical adsorbents have surfaces that are heterogeneous, both energetically and geometrically (not all pores are of uniform and constant dimensions). The degree of heterogeneity differs substantially from one adsorbent type to another. These heterogeneities are responsible for many nonlinearities, both in single component isotherms and in multicomponent adsorption selectivities.

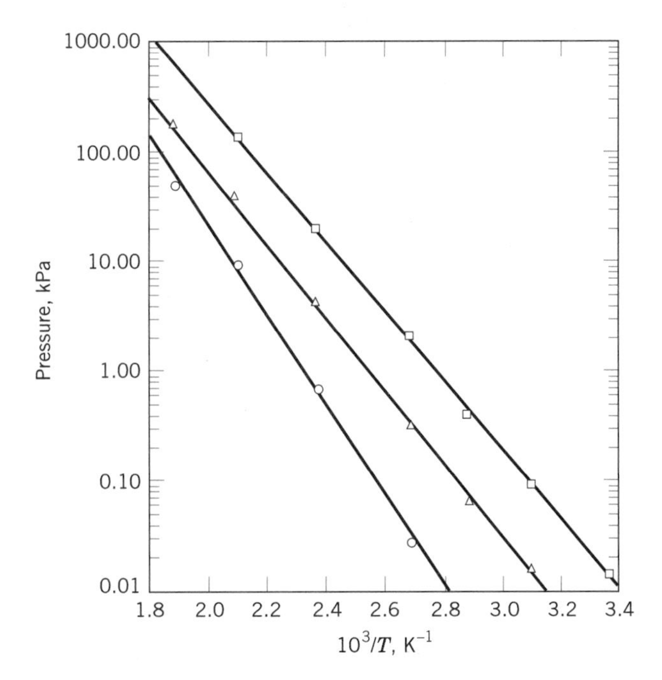

Fig. 4. Adsorption isosteres, water vapor on 4A (NaA) zeolite pellets (6). H_2O loading: □, 15 kg/100 kg zeolite; △, 10 kg/100 kg, ○, 5 kg/100 kg. To convert kPa to mm Hg, multiply by 7.5. Courtesy of Union Carbide.

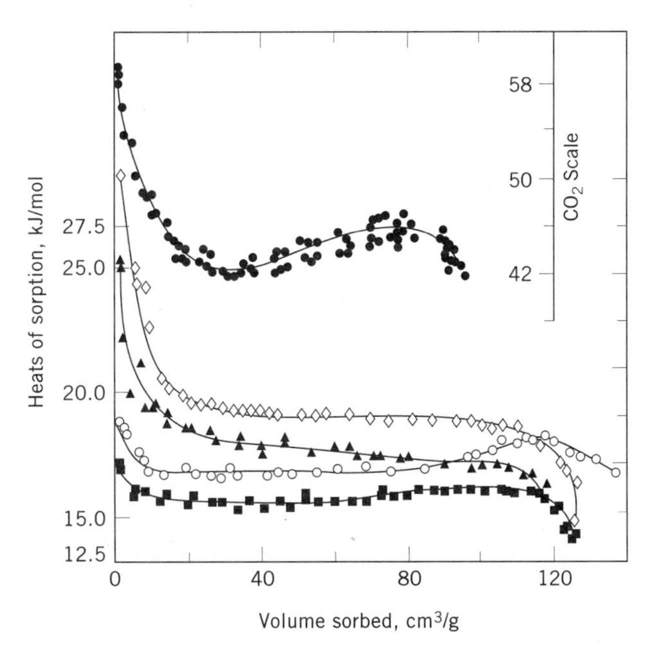

Fig. 5. Differential heats of sorption in nature chabazite (4). ▲ = N_2; ■ = Ar; ○ = O_2; ◇ = CO; ● = CO_2. See Table 2 for polarizability, dipole moment, and quadrupole moment values for the gases. Volume adsorbed is expressed as cm^3 of adsorbate as liquid. To convert kJ to kcal, divide by 4.184. Courtesy of Academic Press.

In Figure 5, the heat of adsorption of CO_2 increases slightly at the higher adsorbate loadings. This increase is due to the increasing self-potential contribution at the higher loadings.

2.5. Isotherm Models. Many efforts have been made over the years to develop isotherm models for data correlation and design predictions for both single component and multicomponent adsorption. Unfortunately, no single model is accurate over broad ranges of adsorbent and adsorbate types, pressures, temperatures, and loadings, especially for multicomponent systems. The reason is probably due to deficiencies in the models in adequately describing both the heterogeneities of the surface and the effects of the adsorbate on the properties of the adsorbent itself. Most models assume the adsorbent is inert, ie, not altered by the presence of the adsorbate molecules; however, partial changes in some adsorbent properties are commonly observed.

Nevertheless, each of the more popular isotherm models have been found useful for modeling adsorption behavior in particular circumstances. The following outlines many of the isotherm models presently available. Detailed discussions of derivations, assumptions, strengths, and weaknesses of these and other isotherm models are given in (4,7–16).

Not all of the isotherm models discussed next are rigorous in the sense of being thermodynamically consistent. For example, specific deficiencies in the Freundlich, Sips, Dubinin-Radushkevich, Toth, and vacancy solution models have been identified (14).

The Sips and related loading ratio correlation (LRC) models fail to properly predict Henry's law behavior (as required for thermodynamic consistency) at the zero pressure limit (8). Thermodynamic inconsistency of the LRC model had also been noted by the original authors (17); nevertheless, the model has been found useful in predicting multicomponent performance from single component data and correlating multicomponent data (18). However, users of models lacking thermodynamic consistency must take due care, particularly in extrapolation beyond the range of actual experimental data.

Because the Dubinin, Langmuir, and modified Langmuir (or LRC) do a very good job of fitting data over broad ranges of parameters, several authors had made modifications to them to make them thermodynmically consistent. The Dubinin equations have been modified to have a proper Henry's law region (19). A consisent dual site Langmuir model has been offered (20). The LRC equations have been extended to have correct Henry's law behavior and to be thermodynamically consistent (21).

Thermodynamically Consistent Isotherm Models. These models include both the statistical thermodynamic models and the models that can be derived from an assumed equation of state for the adsorbed phase plus the thermodynamics of the adsorbed phase, ie, the Gibbs adsorption isotherm,

$$\left(\frac{d\Phi}{dP}\right)_T = \frac{qRT}{P}$$

where Φ is the spreading pressure, P the partial pressure of adsorbate, q the adsorbate loading x per quantity w of adsorbent $= x/w$, T the temperature, and R the gas constant. In the following models, $\Theta = q/q_{\max}$ is the fractional surface coverage, where q_{max} is the maximum loading. Constants are q_{max}, all K's, all k's, all A's, b, c, n, s, t, β, and τ. The vapor pressure of pure adsorbate is P_0.

$$\begin{aligned}
\text{Henry's law:} \quad & q = KP_0 \\
\text{Langmuir:} \quad & K'P_0 = \Theta/(1-\Theta) \\
\text{Volmer:} \quad & bP_0 = [\Theta/(1-\Theta)] \exp[\Theta/(1-\Theta)] \\
\text{van der Waals:} \quad & K''P_0 = [\Theta/(1-\Theta)] \exp[\Theta/(1-\Theta)] \exp[B/RT] \\
\text{Virial:} \quad & K'''P_0/x = \exp\left[2A_1x + \left(\tfrac{3}{2}\right)A_2x^2 + \cdots\right]
\end{aligned}$$

The Langmuir model is discussed in (22); the Volmer in (23); and the van der Waals and virial equations in (8).

Statistical Thermodynamic Isotherm Models. These approaches were pioneered by Fowler and Guggenheim (24) and Hill (25). Examples of the application of this approach to modeling of adsorption in microporous adsorbents are given in (3,26–30). Excellent reviews have been written (4,31).

Semiempirical Isotherm Models. Some of these models have been shown to have some thermodynamic inconsistencies and should be used with due care. Nevertheless, they have each been found to be useful for data correlation and interpolation, as well as for the calculation of some thermodynamic properties.

Polanyi Adsorption Potential. These models are based on the adsorption potential

$$A = RT \ln(P_0/P)$$

The Dubinin-Radushkevich model (32) is the same as the more general Dubinin-Astakhov equation (33) (see below), with $n = 2$.

The Dubinin-Astakhov model is

$$\Theta = \exp[-(A/E)^n]$$

where n is generally between 1 and 3.

Radke-Prausnitz. This model (34) is also known as the Langmuir-Freundlich model:

$$\Theta = \frac{k'P}{[1 + (k'P)]^{\tau}}$$

Toth. This model (35) is represented as

$$\Theta = \frac{kP}{[1 + (kP^t)^{1/t}]}$$

UNILAN. The uniform distribution, Langmuir local isotherm model (12):

$$\Theta = \frac{1}{2s} \ln\left[\frac{(c + Pe^{s})}{(c + Pe^{-s})}\right]$$

BET. This model (36) estimates the coverage corresponding to one monolayer of adsorbate and is used to measure the surface areas of solids:

$$\Theta = \frac{b(P/P_0)}{[(1-P/P_0)\,(1-P/P_0+bP/P_0)]}$$

Isotherm Models for Adsorption of Mixtures. Of the following models, all but the ideal adsorbed solution theory (IAST) and the related heterogeneous ideal adsorbed solution theory (HIAST) have been shown to contain some thermodynamic inconsistencies. References to the limited available literature data on the adsorption of gas mixtures on activated carbons and zeolites have been compiled, along with a brief summary of approximate percentage differences between data and theory for the various theoretical models (16). In the following models the subscripts i and j refer to different adsorbates.

Markham and Benton. This model (37) is known as the extended Langmuir isotherm equation for two components, i and j:

$$\Theta_i = K_iP_i/\left(1+K_iP_i+K_jP_j\right)$$
$$\Theta_j = K_jP_j/\left(1+K_iP_i+K_jP_j\right)$$

Leavitt Loading Ratio Correlation (LRC) Method. The LRC model (17) for a single component i parallels Sips model (38):

$$\Theta_i = \left(K_iP_i\right)^{1/n_i} \Big/ \left[1+\left(K_iP_i\right)^{1/n_i}\right]$$

but with

$$-\ln K_i = A_{1i} + A_{2i}/T$$

For the binary system of components i and j, the LRC model (17) is

$$\Theta_i = \left(K_iP_i\right)^{1/n_i} \Big/ \left[1+\left(K_iP_i\right)^{1/n_i}+\left(K_jP_j\right)^{1/n_j}\right]$$

Ideal Adsorbed Solution (IAS) Model. For components i and j, assuming ideal gas behavior, this model (39) is

$$\frac{\Phi A}{RT} = \int_0^{P_i^o} \left[q_i^o(P)\right] d(\ln P) = \int_0^{P_j^0} \left[q_j^0(P)\right] d(\ln P)$$

$$PY_i = P_i^0X_i$$

$$PY_j = P_j^0X_j = P_j^0(1-X_i)$$

where P_i^o is the vapor pressure of component i, $q_i^o(P)$ the equilibrium loading of pure i at pressure P, Y_i the vapor-phase mole fraction of component i, and X_i the adsorbed phase mole fraction of component i. These equations are solved simultaneously to determine P_i^o, P_j^o, and X_i, and the following equations are used to calculate q_i, q_j, and q_{total}:

$$1/q_{\text{total}} = X_i/\lfloor q_i(P_i^0)\rfloor + X_j/\lfloor q_j(P_j^0)\rfloor$$

$$q_i = q_{\text{total}}X_i$$

$$q_j = q_{\text{total}}X_j$$

Heterogeneous Ideal Adsorbed Solution Theory (HIAST). This IAS theory has been extended to the case of adsorbent surface energetic heterogeneity and is shown to provide improved predictions over IAST (12).

Vacancy Solution Model (VSM). The initial model (40) considered the adsorbed phase to be a mixture of adsorbed molecules and vacancies (a vacancy solution) and assumed that nonidealities of the solution can be described by the two-parameter Wilson activity coefficient equation. Subsequently, it was found that the use of the three-parameter Flory-Huggins activity coefficient equation provided improved prediction of binary isotherms (41).

2.6. Molecular Modeling. Since the 1980s, significant advances have been made in molecular modeling of adsorption (42,43). The availability of high speed computers has made it practical to model adsorbate–adsorbent interactions in micro- and meso-pores using statistical thermodynamic principles. Many universities and research organizations have developed Monte Carlo computer programs; a commercial simulation program, Cerius2, is available from Molecular Simulation Inc. Adsorbent pores have been modeled in various ways to predict adsorption equilibrium and diffusivity phenomenon—as flat surfaces, as narrow slits, and as the regular crystalline structures of zeolites. The results vary greatly depending on such choices. Monte Carlo simulation has been used as an aid in the design of zeolites (44) and of activated carbons (45). Molecular modeling has been used to study pore size distribution by varying slit width (45) and to interpret NMR data (46).

2.7. Adsorption Dynamics. An outline of approaches that have been taken to model mass-transfer rates in adsorbents has been given (see ADSORPTION). Detailed reviews of the extensive literature on the interrelated topics of modeling of mass-transfer rate processes in fixed-bed adsorbers, bed concentration profiles, and breakthrough curves include (16,29). The related simple design concepts of WES, WUB, and LUB for constant-pattern adsorption are discussed later.

2.8. Reactions on Adsorbents. To permit the recovery of pure products and to extend the adsorbent's useful life, adsorbents should generally be inert and not react with or catalyze reactions of adsorbate molecules. These considerations often affect adsorbent selection and/or require limits be placed upon the severity of operating conditions to minimize reactions of the adsorbate molecules or damage to the adsorbents. However, even then, gradual reactions of trace impurities in a feed stream or slowly occurring reactions that modify the adsorbent may still cause a gradual decline in the adsorbent performance, as illustrated in Figure 6.

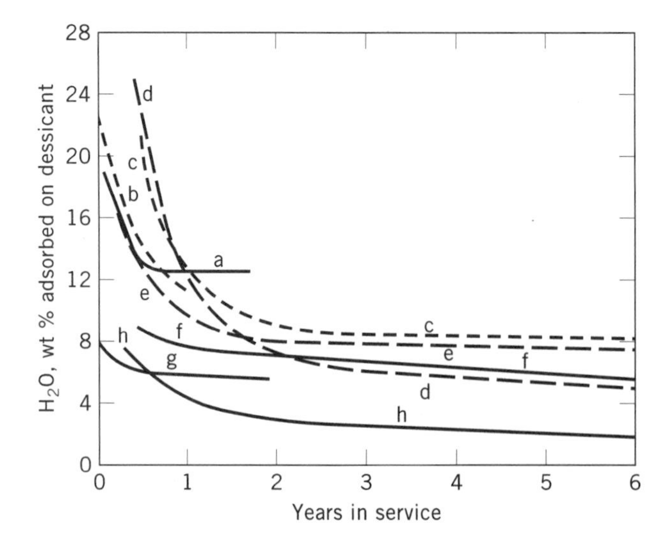

Fig. 6. Adsorption capacity of various dessicants vs years of service in dehydrating high pressure natural gas (47). (**a**) Alumina H-151, gas ~27°C and 123 kPa, from oil and water separators; (**b**) silica gel, gas ~38°C and 145 kPa, from oil absorption plant; (**c**) sorbead, 136-kPa gas from absorption plant; regeneration gas inlet temperature 243°C (maximum allowable dew point −6.7°C); (**d**) sorbead, 40-kPa gas containing propane; regeneration gas temperature 177°C (maximum allowable dew point −34°C); (**e**) sorbead, 1950–1956 data; (**f**) activated alumina, same gas as for Curve **d**; (**g**) activated bauxite(florite), residue gas from gasoline absorption plant; (**h**) activated alumina, same gas for for Curve **c**. Courtesy of Gulf Publishing Company.

3. Adsorbent Principles

3.1. Principal Adsorbent Types. Commercially useful adsorbents can be classified by the nature of their structure (amorphous or crystalline), by the sizes of their pores (micropores, mesopores, and macropores), by the nature of their surfaces (polar, nonpolar, or intermediate), or by their chemical composition. All of these characteristics are important in the selection of the best adsorbent for any particular application.

However, the size of the pores is the most important initial consideration because, if a molecule is to be adsorbed, it must not be larger than the pores of the adsorbent. Conversely, by selecting an adsorbent with a particular pore diameter, molecules larger than the pores may be selectively excluded, and smaller molecules can be allowed to adsorb.

Pore size is also related to surface area and thus to adsorbent capacity, particularly for gas-phase adsorption. Because the total surface area of a given mass of adsorbent increases with decreasing pore size, only materials containing micropores and small mesopores (nanometer diameters) have sufficient capacity to be useful as practical adsorbents for gas-phase applications. Micropore diameters are <2 nm; mesopore diameters are between 2 and 50 nm; and macropores diameters are >50 nm, by International Union of Pure and Applied Chemistry (IUPAC) classification (48).

The practical adsorbents used in most gas-phase applications are limited to the following types, classified by their amorphous or crystalline nature.

- *Amorphous*: silica gel, activated alumina, activated carbon, molecular sieve carbons and macroreticular resins.
- *Crystalline*: molecular sieve zeolites, and related molecular sieve materials that are not technically zeolites, eg, silicalite, $AlPO_4$s, and SAPOs, as well as mesoporous silicates/aluminosilicates, eg, MCM-41.

Typical pore size distributions for these adsorbents have been given (see ADSORPTION). Only molecular sieve carbons and crystalline molecular sieves have large pore volumes in pores <1 nm. Only the crystalline molecular sieves have monodisperse pore diameters because of the regularity of their crystalline structures (49).

Activated carbons are made by first preparing a carbonaceous char with low surface area followed by controlled oxidation in air, carbon dioxide, or steam. The

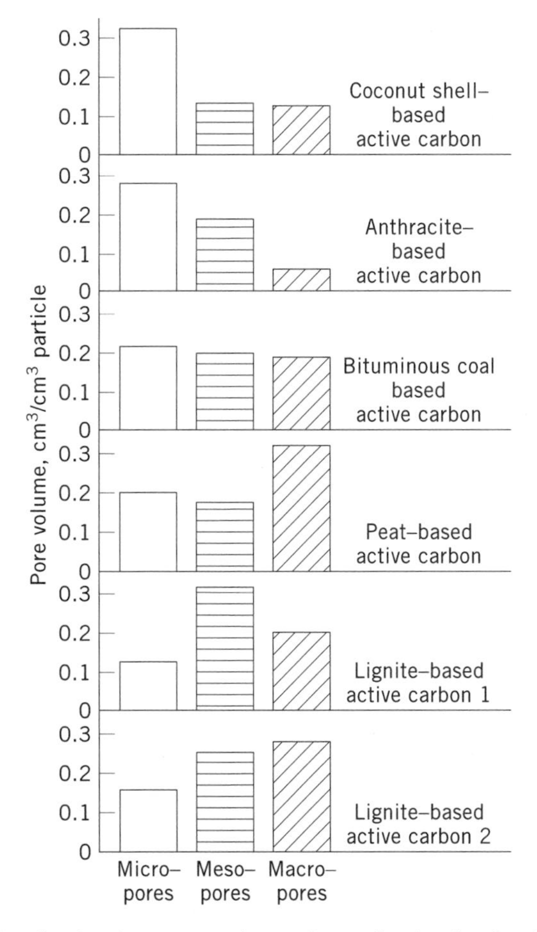

Fig. 7. Pore size distribution in some active carbons obtained using different precursors (50).Courtesy of Marcel Dekker Publishing Company.

pore-size distributions of the resulting products are highly dependent on both the raw materials and the conditions used in their manufacture, as may be seen in Figure 7.

Assuming the pores are large enough to admit the molecules of interest, the most important consideration is the nature of the adsorbent surface, because this characteristic controls adsorption selectivity.

Practical adsorbents may also be classified according to the nature of their surfaces.

- *Highly polar*: molecular sieve zeolites with high aluminum and cation contents.
- *Moderately polar*: crystalline molecular sieves with low aluminum and low cation contents, silica gel, activated alumina, activated carbons with highly oxidized surfaces, crystalline molecular sieve $AlPO_4$'s.
- *Nonpolar*: silicalite, F-silicalite, other high silica content crystalline molecular sieves, activated carbons with reduced surfaces.

3.2. Adsorption Properties. Typical adsorption isotherms for water on various adsorbents are given in Figure 1, and the corresponding isobars in Figure 2. Not only do the more highly polar molecular sieve zeolites adsorb more water at lower pressures than do the moderately polar silica gel and alumina gel, but they also hold onto the water more strongly at higher temperatures. For the same reason, temperatures required for thermal regeneration of water-loaded zeolites is higher than for less highly polar adsorbents.

Isotherms for H_2O and n-hexane adsorption at room temperature and for O_2 adsorption at liquid oxygen temperature on 13X (NaX) zeolite and on the crystalline SiO_2 molecular sieve silicalite are are shown in Figure 8. Silicalite adsorbs water very weakly. Further modification of silicalite by fluoride incorporation provides an extremely hydrophobic adsorbent, shown in Figure 9. These examples illustrate the broad range of properties of crystalline molecular sieves.

Activated carbons contain chemisorbed oxygen in varying amounts unless special care is taken to eliminate it. Desired adsorption properties often depend on the amount and type of chemisorbed oxygen species on the surface. Therefore, the adsorption properties of an activated carbon adsorbent depend on its prior temperature and oxygen-exposure history. In contrast, molecular sieve zeolites and other oxide adsorbents are not affected by oxidizing or reducing conditions.

This principle is illustrated in Figure 10. Water adsorption at low pressures is markedly reduced on a poly(vinylidene chloride)-based activated carbon after removal of surface oxygenated groups by degassing at 1000°C. Following this treatment, water adsorption is dominated by capillary condensation in mesopores, and the size of the adsorption–desorption hysteresis loop increases, because the pore volume previously occupied by water at the lower pressures now remains empty until the water pressure reaches pressures ($\sim$0.3 to 0.4 times the vapor pressure) at which capillary condensation can occur.

Typical adsorption isotherms for light hydrocarbons on activated carbon prepared from coconut shells are shown in Figure 11. The polarizabilities and

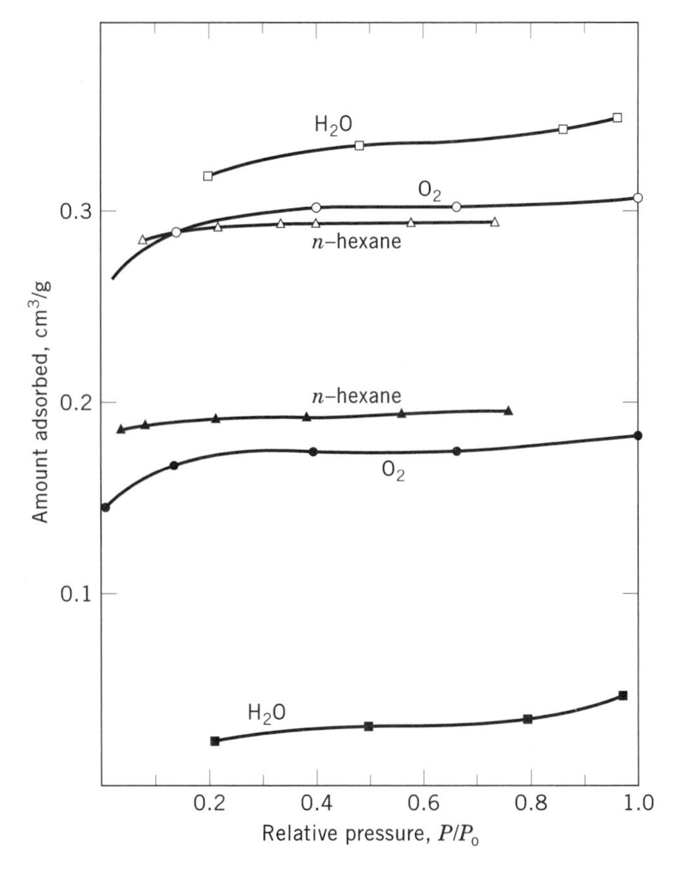

Fig. 8. Adsorption isotherms of H_2O, O_2, and *n*-hexane on zeolite NaX (open symbols) and silicalite (filled symbols). Oxygen is at −183°C and water and *n*-hexane (C_6H_{14}) at room temperature (RT). Volume adsorbed is expressed as cm^3 of adsorbate as liquid. Courtesy of *Nature, London* (51).

boiling points of these compounds increase in the order

$$CH_4 < C_2H_4 < C_2H_6 < C_3H_6 < C_3H_8$$

The relative strengths of adsorption of these compounds follow the same order, as expected for a nonpolar adsorbent, except that C_3H_6 was adsorbed more strongly than C_3H_8. This result indicates that the surface is weakly polar and that specific (dipole–field and quadrupole–field gradient) contributions to the adsorption potential alter the expected order slightly. This situation may also be due to chemisorbed oxygen species on the surface.

3.3. Physical Properties. Physical properties of importance include particle size, density, volume fraction of intraparticle and extraparticle voids when packed into adsorbent beds, strength, attrition resistance, and dustiness. Any of these properties can be varied intentionally to tailor adsorbents to specific applications (See ADSORPTION LIQUID SEPARATION; CARBON, ACTIVATED CARBON; ION EXCHANGE; MOLECULAR SIEVES; AND SILICON COMPOUNDS, SYNTHETIC INORGANIC SILICATES).

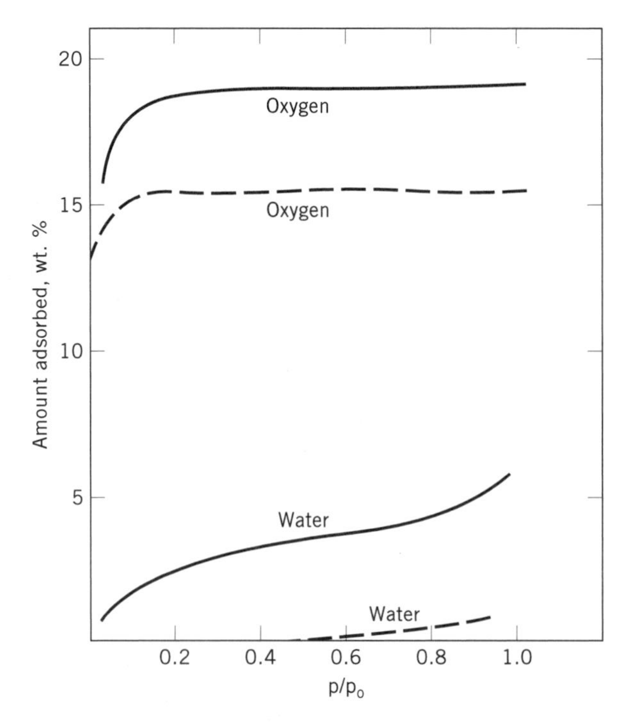

Fig. 9. Adsorption of oxygen (90 K) and water (RT) on silicalite (—) and F-silicalite (– – –) (52).

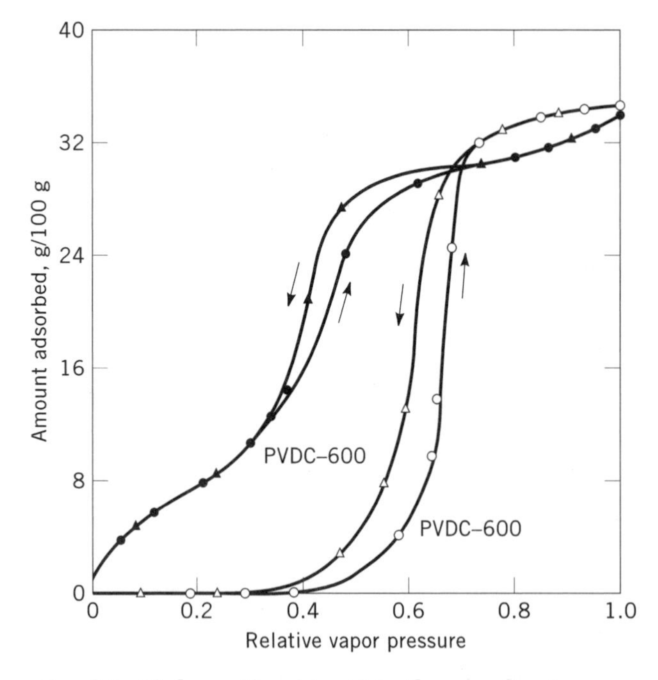

Fig. 10. Adsorption (●, ○)-desorption (▲, △) isotherms of water vapor on poly(vinylidene chloride) (PVDC) carbon before (filled symbols) and after (open symbols) outgassing at 1000°C (53). Courtesy of *Carbon*.

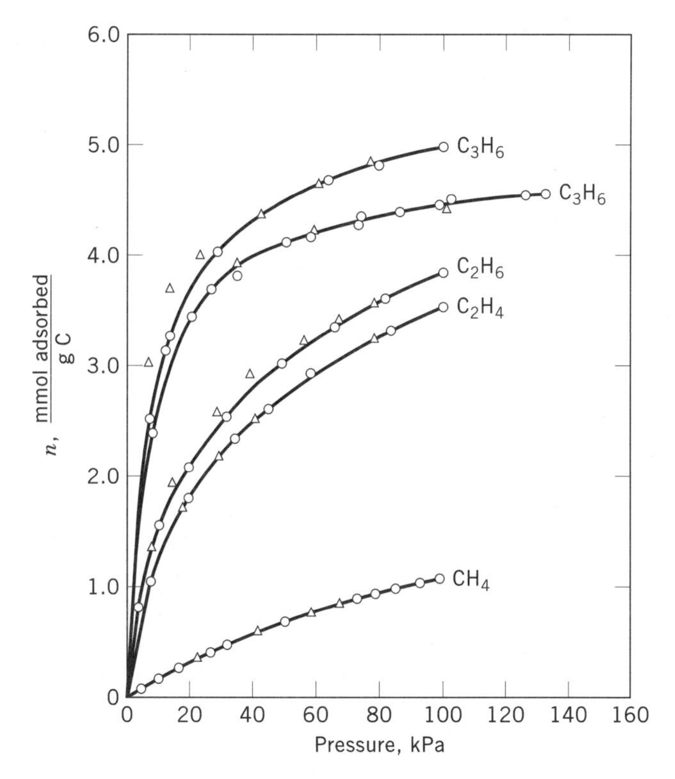

Fig. 11. Adsorption isotherms for hydrocarbons on activated coconut-shell carbon at 25°C (54). ○, Adsorption; △, desorption. To convert kPa to mm Hg, multiply by 7.5. Courtesy of *Industrial and Engineering Chemistry*.

Most commercial adsorbents for gas-phase separations are employed in the form of pellets, beads, or granular shapes, typically ~1.5–3.2 mm in size. However, a growing number of atmospheric presssure applications are employing adsorbents in the form of honeycomb monoliths and fabrics. These configurations have been introduced to reduce the pressure drop and thus utility consumption.

3.4. Deactivation. All adsorbents, no matter how inert, will be deactivated during extended usage by reaction with inpurities, reaction of adsorbates, or thermal damage. To compensate, adsorbent beds are sized to account for the gradual loss in capacity and to allow their use for a given period of time. Most commonly, at the end of its useful life, the adsorbent is dumped from the beds and replaced with fresh adsorbent.

The most common degradation when hydrocarbons are present is the formation of "coke" (high molecular weight, high carbon to hydrogen ratio material). Depending on conditions, coke formed from unsaturated hydrocarbons can be found on adorption sites reducing capacity, in pores decreasing mass transfer, or on the external surface (55). In some cases, the process equipment can be designed to allow periodic *in situ* rejuvenation of the adsorbent, eg, a periodic burning-off of coke accumulated on the adsorbent.

Adsorbent degradation by chemical attack or physical damage is not reversible. Acids or acid gases can react with adsorbents with alkaline surface chemistry, eg, low silica zeolites, and cause loss of adsorption capacity. Other adsorbents, such as silica gel high silica zeolites, are sensitive to alkalies. Oxidative or reductive conditions can change the adsorptive surface characteristics of activated carbon. The constant thermal expansion and contraction in temperature-swing adsorption (TSA) processes can cause damage to the internal pore and/or crystal structure. Activated alumina and silica gel can be dehydrated by excessive temperatures. When water is present, hydrothermal cycling can cause explosive steam release that physically damages some adsorbents. Some types of silica gel are susceptible to breakup caused by water droplets; special decrepitation-resistant grades are available.

4. Adsorption Processes

Adsorption processes are often identified by their method of regeneration. Pressure-swing adsorption (PSA) and TSA are the most frequently applied process cycles for gas separation. Purge-swing cycles and nonregenerative approaches are also applied to the separation of gases. Special applications exist in the nuclear industry. Others take advantage of reactive sorption. Most adsorption processes use fixed beds, but some use moving or fluidized beds, or rotary wheels.

4.1. Temperature Swing. A temperature-swing or TSA cycle is one in which desorption takes place at a temperature much higher than adsorption. The principal application is for separations in which contaminants are present at low concentration, ie, for purification. The TSA cycles are characterized by low residual loadings and high operating loadings. Figure 12 depicts the isotherms for the two temperatures of a TSA cycle. The available operating capacity is the difference between the loadings X_1 and X_2. These high adsorption capacities for low concentrations mean that cycle times are long, hours to days, for reasonably sized beds. This long cycle time is fortunate, because packed beds of adsorbent respond slowly to changes in gas temperature. A purge and/or vacuum removes the thermally desorbed components from the bed, and cooling returns the bed to adsorption condition. Systems in which species are strongly adsorbed are especially suited to TSA. Such applications include drying, sweetening, CO_2 removal, and pollution control.

Principles. In a TSA cycle, two processes occur during regeneration: heating and purging. Heating must provide adequate thermal energy to raise the adsorbate, adsorbent and adsorber temperature, desorb the adsorbate, and make up for heat losses. Regeneration is heating limited (or stoichiometric limited) when transfer of energy to the system is limiting. Equilibrium determines the maximum capacity of the purge gas to transfer the desorbed material away. Regeneration is stripping limited (or equilibrium limited) when transferring adsorbate away is limiting.

Heating occurs by either direct (external heat exchange to the purge gas) or, less commonly, indirect (heating elements, coils or panels, inside the adsorber) contact of the adsorbent by the heating medium. Direct heating is simpler and is invariably used for stripping-limited regeneration. Alternate methods for supplying indirect heat for desorption are microwave fields and joule's heat (electrothermal). Microwave energy has been successfully applied to regeneration of

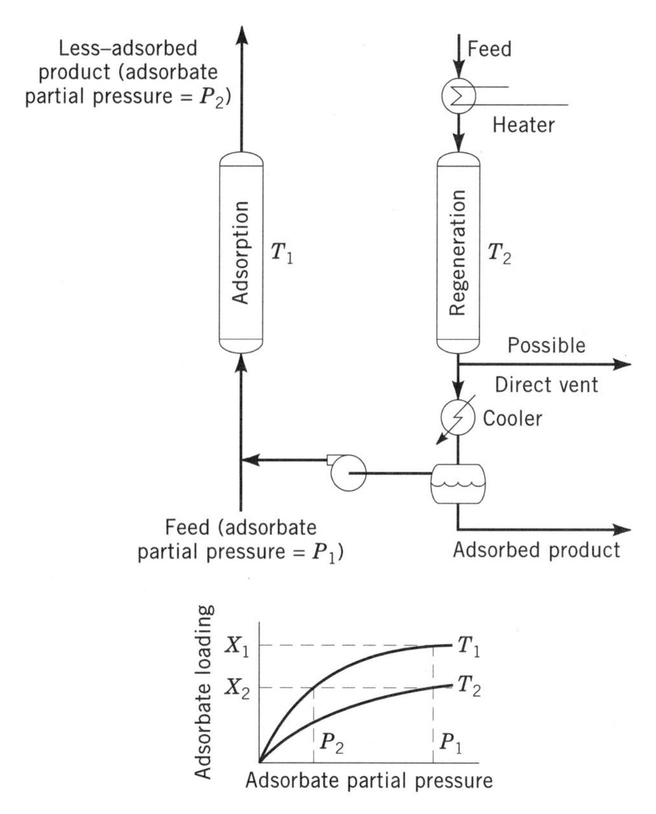

Fig. 12. Temperature-swing cycle (1). Loading X_1 at T_1 and feed partial pressure P_1; X_2 at the higher T_2 and the lower P_2 needed in the product.

low and high silica zeolites, activated carbon, and macroreticular resins (56). The electrical conductivity of carbon has been utilized to generate indirect heat in activated carbon adsorbers made of cloth (57) of granules (58), and of beads (59). However, the complexity of all of these indirect heating methods limits their use to heating-limited regeneration where purge gas is in short supply. Coils or panels can supply indirect cooling as well. The use of steam for the regeneration of activated carbon is a combination of thermal desorption and purge displacement; direct heating is supplied by water adsorption.

Steps. Thermal-swing cycles have at least two steps: adsorption and heating. A cooling step is also normally used after the heating step. A portion of the feed or product stream can be utilized for heating, or an independent fluid can be used. Easily condensable contaminants may be regenerated with noncondensable gases and recovered by condensation. Water-immiscible solvents are stripped with steam, which may be condensed and separated from the solvent by decantation. Fuel and/or air may be used when the impurities are to be burned or incinerated.

The highest regeneration temperatures are the most efficient for desorption. However, heater cost, metallurgy, and the thermal stability of the adsorbent and the fluids must be considered. Silica gel requires the lowest temperatures and the lowest amount of heat of any commercial adsorbent. Activated carbons, aluminas, and high silica molecular sieves can tolerate the highest temperatures. Although thermal-swing regeneration can be done at the same pressure as adsorption, lowering the pressure can achieve better desorption and is often used; such cycles are actually a hybrid of PSA and TSA referred to as pressure/thermal swing adsorption (PTSA). The heating gas is normally used for the cooling step. Rather than cooling the bed, adsorption can sometimes be started on a hot bed. If certain criteria are met (60), the dynamic adsorption performance does not depend on cooling.

Flow Sheet. The most common processing scheme is a pair of fixed-bed adsorbers alternating between the adsorption step and the regeneration steps. An example is given in Figure 12. However, the variations possible to achieve special needs are endless. The flow directions can be varied. Single beds provide interrupted flow, but multiple beds can ensure constant flow. Beds can be configured in lead-trim, parallel trains, series cool-heat, or closed-loop. Regeneration may even be *ex situ* rather than *in situ*.

The normal flow direction through a fixed bed is usually in a vertical direction. The mechanical complexities required for horizontal- or annular-flow beds often outweigh the decrease in pressure drop achieved. Because allowable velocities for crushing exceed those for lifting, the cycle step with the highest pressure drop should be downward. All other flows can then be in the same direction as the limiting flow (cocurrent) or in the opposite direction (countercurrent). Each combination of flow directions for heating and cooling produces a different residual of adsorbate (Fig. 13).

Although most applications of fixed beds have multiple adsorber beds to treat continuous streams, batch operation using a single adsorber bed is an alternative. For purification applications, where one vessel can contain enough adsorbent to provide treatment for days, weeks, or even months, the cost savings and simplicity often justify the inconvenience of stopping adsorption treatment periodically for a short regeneration.

When the mass-transfer zone is a major portion of an adsorbent bed, the equilibrium capacity is poorly utilized. A lead-trim configuration uses the adsorbent more fully. The feed flows successively through a lead bed and then a trim bed. The lead bed is nearly exhausted before it is taken out of service to be regenerated. When a lead bed is removed from adsorption, the trim bed becomes the lead, and a fully regenerated bed becomes the new trim bed.

When large flows are to be treated, designing and building a single adsorber vessel large enough to treat the entire stream is not practical. Instead, the feed flow is split equally between parallel beds and/or trains of adsorbers. This provides the additional advantage of a convenient method of turning down the process to save on utilities.

At the start of the cooling step, the adsorber vessel is a large heat sink containing valuable energy: the sum of all of the sensible heats of the adsorbent, the vessel, and any internals. If we use three adsorber beds—one on adsorption, one on heating, and one on cooling—the purge gas flows in series first to cool a hot bed and then to heat a spent bed. Thus all of the heat from the bed being cooled is recovered.

Thermal energy can also be conserved by using a thermal-pulse cycle. When desorption is heat limited, only a short soak time at temperature completes

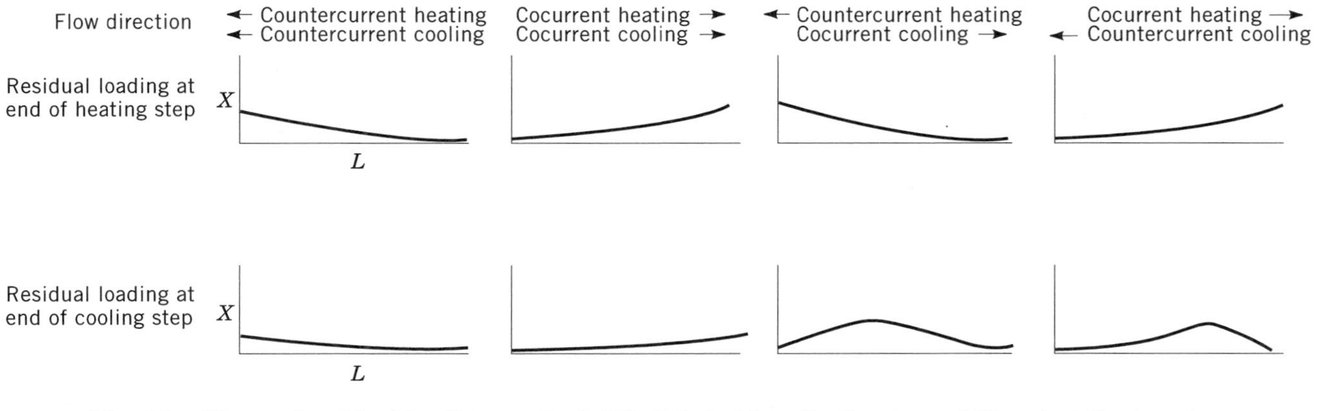

Fig. 13. Shape of residual loading gradient (61). L is bed length. Courtesy of *Chemical Engineering*.

regeneration. The entire adsorbent bed need not be at desorption temperature before beginning the cooling step. Only a pulse of heating gas that contains the heat of desorption is required to move through the bed, desorbing the adsorbate until it exhausts its thermal energy as it reaches the outlet. Because temperature fronts spread as they move through packed beds, a small excess of heat is added to the stoichiometric quantity to ensure that the outlet reaches the desired level before being cooled.

When the gas available for regeneration is in short supply, the regeneration steps are often carried out in a closed loop. This recycle of the bed effluent back to the inlet has the advantage of concentrating the impurity and making it easier to separate by condensation or other recovery means. Heating is usually accomplished with a semiclosed loop that has a constant fresh gas makeup and a bleed to draw off the desorbed material. However, contaminant is at a higher level than in an open loop and product purity is harder to achieve.

Drying. The single most common gas-phase application for TSA is drying. The natural gas, chemical, and cryogenics industries all use zeolites, silica gel, and activated alumina to dry streams.

Zeolites, activated alumina, and silica gel have all been used for drying of pipeline natural gas. Alumina and silica gel have the advantage of having higher equilibrium capacity and of being more easily regenerated with waste level heat (62). However, the much lower dewpoint and longer life attainable with 4A makes zeolites the predominant adsorbent. Special acid-resistant zeolites are used for natural gas containing large amounts of acid gases, such as CO_2 and H_2S.

The low dewpoint that can be achieved with zeolites is especially important when drying feed streams to cryogenic processes to prevent freeze-up at process temperatures. Natural gas is dried before liquefaction to liquefied natural gas (LNG), both in peak demand and in base load facilities. Zeolites have largely replaced silica gel and activated alumina in drying natural gas for ethane recovery utilizing the cryogenic turboexpander process, and for helium recovery. The air to be cryogenically distilled into N_2, O_2, and Ar must be purified of both H_2O and CO_2, and traces of mercury and hydrocarbon to protect the aluminum heat exchangers. This purification is accomplished with 13X zeolites or a compound bed of activated alumina and zeolite.

The 4A zeolite, silica gel, and activated alumina all find applications drying synthesis gas, inert gas, hydrocracker gas, rare gases, and reformer recycle H_2. Cracked gas before low temperature distillation for olefin production is a reactive stream. The 3A or pore-closed 4A zeolite size selectively adsorbs water but excludes the hydrocarbons, thus preventing coking (62). This molecular sieving also prevents coadsorption of hydrocarbons that would otherwise be lost during desorption with the water. Small pore zeolites are also applied to the drying of ethylene, propylene, and acetylene as they are drawn from salt cavern, or conventional, storage. When industrial gases containing Cl_2, SO_2, and HCl are dried, acid-resistant zeolites are used.

A recently developed drying application is the dehumification of air in buildings utilizing desiccant wheels. In an increasingly environmentally conscious world, adsorption is an alternative to the use of vapor compression air conditioning with its high electrical consumption and the potential for release of harmful refrigerants. In a typical system, the fresh air flows through the rotary dryer is partially cooled in a cross exchanger, and is brought to the desired temperature by an evaporative cooler. Exhaust air is used as a cooling source for the cross-exchanger and is heated by low level energy hot enough to remove the moisture from the desiccant wheel (63). The adsorbent can be silica gel, zeolite, or mixtures (64). The adsorbent can also remove some of the organic contaminants from the outside ambient air. The 5A zeolite was found to be superior to silica gel in removing CO, NO_2, and SO_2 in desiccant service, but more difficult to regenerate (65).

Sweetening. Another significant purification application area for adsorption is sweetening. Hydrogen sulfide, mercaptans, organic sulfides and disulfides, and COS need to be removed to prevent corrosion and catalyst poisoning. They are to be found in H_2, natural gas, deethanizer overhead, and biogas. Often adsorption is attractive because it dries the stream as it sweetens.

In the sweetening of wellhead natural gas to prevent pipeline corrosion, 4A zeolites allow sulfur compound removal without CO_2 removal (to reduce shrinkage), or the removal of both to upgrade low thermal content gas. When minimizing the formation of COS during desulfurization is desirable, bivalent cation-exchanged zeolites are commonly used because they are less catalytically active for the reaction of CO_2 with H_2S to form COS and water. Such calcium-, manganese-, and zinc-exchanged zeolites exhibit reduced reactivity while maintaining H_2S capacity with some loss of mass-transfer rate (66). Natural gas for steam–methane reforming in ammonia production must be sweetened to protect the sulfur sensitive, low-temperature, shift catalyst. Zeolites are better than activated carbon because mercaptans, COS, and organic sulfides are also removed. Many refinery H_2 streams require H_2S and water removal by 4A and 5A zeolites to prevent poisoning of catalysts such as those in catalytic reformers.

Other Separations. Other TSA applications range from CO_2 removal to hydrocarbon separations, and include removal of air pollutants and odors, and purification of streams containing HCl and boron compounds. Because of their high selectivity for CO_2 and their ability to dry concurrently, 4A, 5A, and 13X zeolites are the predominant adsorbents for CO_2 removal by temperature-swing processes. The 4A-type zeolite is used for CO_2 removal from baseload and peak-shaving natural gas liquefaction facilities.

The removal of volatile organic compounds (VOC) from air is most often accomplished by TSA, especially with steam. Air streams needing treatment can be found in most chemical and manufacturing plants, especially those using solvents. At concentrations from 500 to 15,000 ppm, recovery of the VOC from steam used to regenerate activated carbon adsorbent thermally is economically justified, or they may be recovered using PSA (see below). Concentrations $>$15,000 ppm are typically in the explosive range and require the use of inert gas rather than air for regeneration. Below $\sim$500 ppm, recovery is not economically justifiable, but environmental concerns often dictate adsorptive recovery followed by destruction. Activated carbon is the long-established adsorbent for these applications, which represent the second largest use for gas-phase carbons. Hydrophobic adsorbents such as high silica zeolites and macroreticular resins are finding increased useage for VOC. The high silica zeolites can be regenerated by air without the ignition threat posed by activated carbon. This stability allows

VOC destruction with adsorber/oxidizer hybrids systems (67). Honeycomb adsorbent wheels have been fabricated with activated carbon fibers (68), and with high silica zeolites (69). These rotary adsorbent beds reduce utility consumption due to lower pressure drop, reduce adsorbent inventory requirement, and provide continuous operation with nearly constant compositions. Activated carbon cloth has also been incorporated into fixed beds (70) and regenerated electrothermally (57). Wheels have been developed for cabin air cleanup in automobiles and airplanes (71).

A number of inorganic pollutants are removable by TSA processes. One of the major pollutants requiring removal is SO_2 from flue gases and from sulfuric acid plant tail gases. The Sulfacid and Hitachi fixed-bed processes, the Sumitomo and BF moving-bed processes, and the Westvaco fluidized-bed process all use activated carbon adsorbents for proven SO_2 removal (72). Zeolites with high acid resistance, such as mordenite and clinoptilolite, have proven to be effective adsorbents for dry SO_2 removal from sulfuric acid tail gas (73). Hydrophobic zeolite adsorbents (74), pillared interlayered clays (PILCs) (75), and resins (76) have all been shown to work in this application.

Zeolites have also proven applicable for removal of nitrogen oxides (NO_x) from wet nitric acid plant tail gas (73). The removal of NO_x from flue gases can also be accomplished by adsorption. The Unitaka process utilizes activated carbon with a catalyst for reaction of NO_x with ammonia, and activated carbon has been used to convert NO to NO_2, which is removed by scrubbing (72). Mercury is another pollutant that can be removed and recovered by TSA. Activated carbon impregnated with elemental sulfur is effective for removing Hg vapor from air and other gas streams; the Hg can be recovered by *ex situ* thermal oxidation in a retort (77). Mordenite and clinoptilolite zeolites are used to remove HCl from Cl_2, chlorinated hydrocarbons, and reformer catalyst gas streams (78). Activated aluminas are also used for such applications, and for the adsorption of fluorine and boron–fluorine compounds from alkylation (qv) processes.

4.2. Pressure Swing. A PSA cycle is one in which desorption takes place at a pressure much lower than adsorption. Its principal application is for bulk separations where contaminants are present at high concentration. The PSA cycles are characterized by high residual loadings and low operating loadings. Figure 14 shows the operating loading (X_1–X_2) that derives from the partial pressure at feed conditions and the lower pressure P_2 at the end of desorption. These low adsorption capacities for high concentrations mean that cycle times must be short, seconds to minutes, for reasonably sized beds. Fortunately, packed beds of adsorbent respond rapidly to changes in pressure. A purge usually removes the desorbed components from the bed, and the bed is returned to adsorption condition by repressurization. Applications may require additional steps. Systems with weakly adsorbed species are especially suited to PSA adsorption. The applications of PSA include drying, upgrading of H_2 and fuel gases, and air separation. Several broad reviews of PSA have been written (79–82).

Principles. In a PSA cycle, two processes occur during regeneration: depressurizing and purging. Depressurization must provide adequate reduction in the partial pressure of the adsorbates to allow desorption. Enough purge gas must flow through the adsorbent to transfer the desorbed material away. Equilibrium determines the maximum capacity of the gas to accomplish this. These

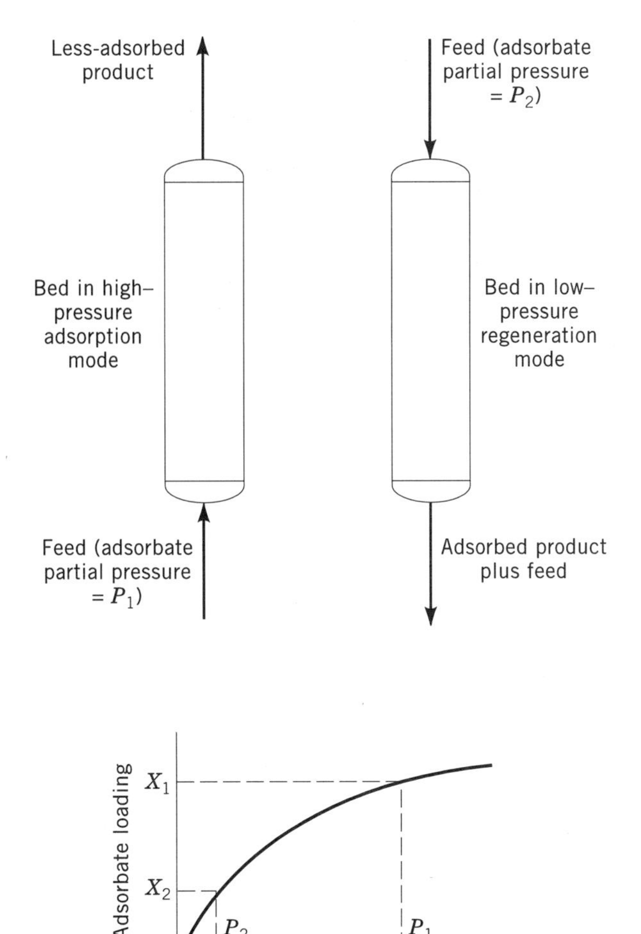

Fig. 14. Pressure-swing cycle (1).

cycles operate at nearly constant temperature and require no heating or cooling steps. They utilize the exothermic heat of adsorption remaining in the adsorbent to supply the endothermic heat of desorption. Pressure-swing cycles are classified as PSA, VSA (vacuum-swing adsorption), PSPP (pressure-swing parametric pumping) or RPSA (rapid pressure-swing adsorption). PSA swings between a high superatmospheric and a low superatmospheric pressure, and VSA swings from a superatmospheric pressure to a subatmospheric pressure. Otherwise, the principles involved are the same.

The other means of accomplishing pressure cycling of an adsorbent is parametric pumping, in which a single adsorbent bed is alternately pressurized with forward flow and depressurized with backward flow through the column from reservoirs at each end. Like TSA parametric pumping, one component concentrates in one reservoir and one in the other. As the name implies, pressure-swing parametric pumping embodies pressure changes that are more than pressuring and depressurizing a bed of adsorbent. Significant pressure gradients occur in the bed. These gradients are especially critical to the way that RPSA cycles operate and result in much smaller adsorbent beds and simpler processes.

In most applications of adsorption, the separation is carried out by adsorbing the more strongly adsorbed species from the less strongly adsorbed. These separations are thus equilibrium limited. However, an adsorptive separation can also be based on a rate or kinetically limited system. Slightly larger molecules diffuse more slowly through a microporous adsorbent with properly selected pore diameter. Therefore, in a rapidly cycling process such as PSA, smaller molecules can be preferentially adsorbed even in the absence of any equilibrium selectivity.

Steps. A PSA cycle has at least three steps: adsorption, blowdown, and repressurization. Although not always necessary, a purge step is normally used. In finely tuned processes, cocurrent depressurization and pressure-equalization steps are frequently added.

At the completion of adsorption, the less selectively adsorbed components have been recovered as product. However, a significant quantity of the weakly adsorbed species are held up in the bed, especially in the void spaces. A cocurrent depressurization step reduces the bed pressure by allowing flow out of the bed cocurrently to feed flow and thus reduces the amount of product retained in the voids (holdup), improving product recovery, and increases the concentration of the more strongly adsorbed components in the bed. The purity of the more selectively adsorbed species has been shown to depend strongly on the cocurrent depressurization step for some applications (83). A cocurrent depressurization step is optional because a countercurrent one always exists. Criteria have been developed to indicate when the use of both is justified (84).

None of the selectively adsorbed components is removed from the adsorption vessel until the countercurrent depressurization (blowdown) step. During this step, the strongly adsorbed species are desorbed and recovered at the adsorption inlet of the bed. The reduction in pressure also reduces the amount of gas in the bed. By extending the blowdown with a vacuum (ie, VSA), the productivity of the cycle can be greatly increased.

Additional stripping of the adsorbates from the adsorbent and purging of them from the voids is accomplished by the purge step. This step can occur concurrently with the end of the blowdown or be carried out afterward. This step is accomplished with a flow of product into the product end to provide a low residual of the selectively adsorbed components at the effluent end of the bed. Although the purge can be done at the same temperature as adsorption, raising the temperature can achieve better desorption and is sometimes used; such cycles are actually a hybrid of PSA and TSA referred to as PTSA. Because PSA cycles are shorter than those of TSA, a thermal assist of PSA does not gain as much as a vacuum assist of TSA.

The repressurization step returns the adsorber to feed pressure and completes the steps of a PSA cycle. Pressurization is carried out with product and/or feed. Pressurizing with product is done countercurrent to adsorption so that purging of the product end continues; indeed it may be merely a continuation of the purge step but with the bed exit valve closed. Pressurizing with feed cocurrent to adsorption in effect begins adsorption without producing any product.

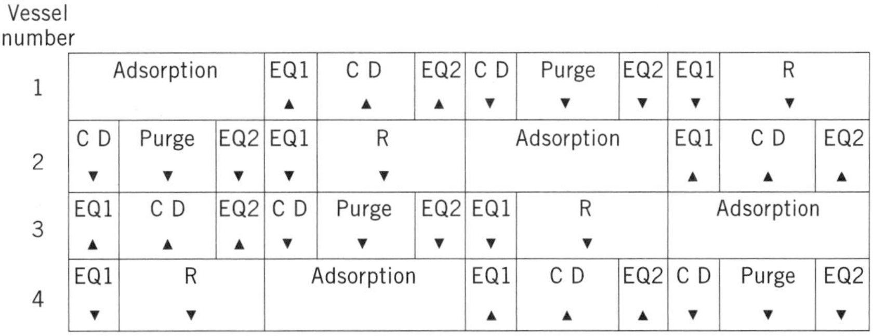

Vessel number												
1	Adsorption			EQ1 ▲	C D ▲	EQ2 ▲	C D ▼	Purge ▼	EQ2 ▼	EQ1 ▼	R ▼	
2	C D ▼	Purge ▼	EQ2 ▼	EQ1 ▼	R ▼		Adsorption			EQ1 ▲	C D ▲	EQ2 ▲
3	EQ1 ▲	C D ▲	EQ2 ▲	C D ▼	Purge ▼	EQ2 ▼	EQ1 ▼	R ▼		Adsorption		
4	EQ1 ▼	R ▼		Adsorption			EQ1 ▲	C D ▲	EQ2 ▲	C D ▼	Purge ▼	EQ2 ▼

Fig. 15. Four-bed PSA system cycle sequence chart (82). EQ, equalization; C D ▲, cocurrent depressurization; C D ▼, countercurrent depressurization; R, repressurization; ▲, cocurrent flow; ▼, countercurrent flow. Courtesy of American Institute of Chemical Engineers.

Pressure equalization steps are used to conserve gas and compression energy. They are applied to reduce the quantity of feed or product gas needed to pressurize the beds. Portions of the effluent gas during depressurization, blowdown, and purge can be used for repressurization.

Flow Sheet. The most common processing scheme has two or three fixed-bed adsorbers alternating between the adsorption step and the desorption steps. The simplest two-bed configuration is illustrated in Figure 14. However, the variations possible to achieve special separations are endless. Single beds with external surge vessels provide continuous flow; multiple beds are used to accommodate additional steps. An example of the bed sequencing needed for multiple steps in a four-bed PSA is shown in Figure 15. Beds can be configured in series or parallel to accomplish coproduction. The UOP Polybed PSA system uses five to ten beds to maximize the recovery of the less selectively adsorbed component and to extend the process to larger capacities (85).

The flow directions in a PSA process are fixed by the composition of the stream. The most common configuration is for adsorption to take place upflow. All gases with compositions rich in adsorbate are introduced into the adsorption inlet end, and so effluent streams from the inlet end are rich in adsorbate. Similarly, adsorbate-lean streams to be used for purging or repressurizing must flow into the product end.

Because RPSA is applied to gain maximum product rate from minimum adsorbent, single beds are the norm. In such cycles where the steps take only a few seconds, flows to and from the bed are discontinuous. Therefore, surge vessels are usually used on feed and product streams to provide uninterrupted flow. Some RPSA cycles incorporate delay steps unique to these processes. During these steps, the adsorbent bed is completely isolated; and any pressure gradient is allowed to dissipate (86).

Purifications. The major purification applications for PSA are for hydrogen, methane, and drying. One of the first commercial uses was for gas drying in which the original two-bed Skarstrom cycle was used. This cycle uses adsorption, countercurrent blowdown, countercurrent purge, and cocurrent repressurization to produce a dry air stream with <1 ppm H_2O (87). About one half of all dryers of instrument air use a PSA cycle similar to this one, most commonly using activated alumina or silica gel (88). Zeolites are used to obtain the lowest possible dewpoints. Some applications for drying air do not require a low level of H_2O, but only a significant lowering of the dew point. The pneumatic compressor systems used in vehicle air-brakes are an example; when a 10–30 K dew point depression is needed for higher discharge air temperatures in the presence of compressor oil, zeolites have been demonstrated to have an advantage over activated alumina and silica gel (89).

High purity H_2 is needed for applications such as hydrogenation, hydrocracking, and ammonia and methanol production. As a significant source of such gas, PSA is able to produce purities as high as 99.9999% using technologies such as the UOP Polybed approach (85). Most H_2 purification by PSA is associated with steam reforming of natural gas and with ethylene-plant and refinery off-gas streams. Hydrogen is also available in coke-oven gas, cracked ammonia, and coal-gasification gas. The contaminants that have to be removed by PSA include carbon oxides, N_2, O_2, Ar, NH_3, CH_4, and heavier hydrocarbons. To remove these components, adsorbent beds are compounded of activated carbon, zeolites, and carbon molecular sieves.

PSA can also be used to recover VOC from exhaust streams and from gasoline storage and loading facilities. Dow Chemical markets such systems using beads of activated carbon or resin under the trade name SORBATHENE (90).

Bulk Separations. Air separation, methane enrichment, and iso-/normal separations are the principal bulk separations for PSA. Others are the recovery of CO and CO_2.

The PSA process is used to separate air into N_2 and O_2. Many companies market systems for PSA O_2; zeolites 5A, 13X, clinoptilolite and mordenite, and carbon molecular sieves are commonly used in PSA, VSA, and RPSA cycles. The product purity ranges from 85 to 95% (limited by the argon, which normally remains with the O_2). The majority of this O_2 is employed for electric furnace steel, with lesser amounts for waste water treating and solid waste and kilns. Smaller production units, especially those based on RPSA cycles, are used for patients requiring respiratory inhalation therapy in the hospital and at home and for pilots on board aircraft. Enriched air, 25–55% O_2, used to enhance combustion, chemical reactions, and ozone production can be produced by tuning PSA processes.

Although air depleted in oxygen can be produced by an equilibrium-limited PSA, most PSA systems to produce N_2 are based on kinetic PSA. High purity, up to 99.99%, N_2 is produced when rate-limited PSA preferentially adsorbs oxygen from air on A-type zeolites or carbon molecular sieve even though the equilibrium selectivity favors N_2 (91). The major use for such N_2 is inert blanketing, such as in metal heat-treating furnaces; small units are used to purge aircraft fuel tanks and in the food and beverage industry.

The upgrading of methane natural gas pipeline quality is another significant PSA separation area. When high nitrogen causes the natural gas to be of poor quality, kinetic PSA is used to increase the energy content. The diffusion rate of nitrogen into clinoptilolite is several orders of magnitude faster than methane (92).

Methane is recovered from fermentation gases of landfills and wastewater purification plants and from poor-quality natural gas wells and tertiary oil recovery when CO_2 is the major bulk contaminant. Fermentation gases are saturated with water and contain "garbage" components such as sulfur and halogen compounds, alkanes, and aromatics. These impurities must first be removed by TSA using activated carbon or carbon molecular sieves. The CO_2 is then selectively adsorbed in a PSA cycle using either zeolites or silica gel in an equilibrium separation (93). Or, because the diffusivity for CO_2 is $\sim$200 times that of CH_4 on a properly selected carbon molecular sieve, the separation can be made in a kinetic-assisted equilibrium separation (94).

4.3. Purge Swing. A purge-swing adsorption cycle is one in which desorption takes place at the same temperature and total pressure as adsorption. Regeneration is accomplished either by partial-pressure reduction by an inert gas purge or by adsorbate displacement by an adsorbable gas. Its major application is for bulk separations when contaminants are at high concentration. Like PSA, purge cycles are characterized by high residual loadings, low operating loadings, and short cycle times (minutes). Mixtures of weakly adsorbed components are especially suited to purge-swing adsorption. Applications include the separation of normal from branched and cyclic hydrocarbons, and for gasoline vapor recovery.

Principles. Purging must provide adequate reduction in the partial pressure of the adsorbates to allow desorption. With enough purge volume, loadings as high as the loading X_1 in equilibrium with the feed partial pressure P_1 can be achieved, as shown in Figure 16. Reduction in partial pressure operates analogously to the reduction in system pressure in PSA cycles. Equilibrium determines the maximum capacity of the gas to purge the adsorbate. These cycles operate adiabatically at nearly constant inlet temperature and require no heating or cooling steps. As with PSA, purge processes utilize the exothermic heat of adsorption remaining in the adsorbent to supply the endothermic heat of desorption. Purge cycles are divided into two categories, inert purge and displacement purge. In inert-purge stripping, inert refers to the fact that the purge gas is not appreciably adsorbable at the cycle conditions. Inert purging desorbs the adsorbate solely by partial pressure reduction.

In displacement-purge stripping, displacement refers to the displacing action of the purge gas caused by its ability to adsorb at the cycle conditions. This competitive adsorption tends to desorb the adsorbate in addition to the partial pressure reduction of dilution. Displacement purging is not as dependent on the heat of adsorption remaining on the adsorbent, because the adsorption of purge gas can release much or all of the energy needed to desorb the adsorbate. The adsorbate must be more selectively adsorbed than the displacement purge so that it can desorb purge fluid during the adsorption step. The displacement purge gas composition must be carefully selected, because it contaminates both the product stream and the recovered adsorbate and requires distillation as illustrated in Figure 17. The displacement purge is more efficient for less selective adsorbate–adsorbent systems; systems with high equilibrium loading of adsorbate require more purging (95).

Steps. A purge-swing cycle usually has two steps: adsorption and purge. Sometimes, a cocurrent purge is added. After the adsorption step has been

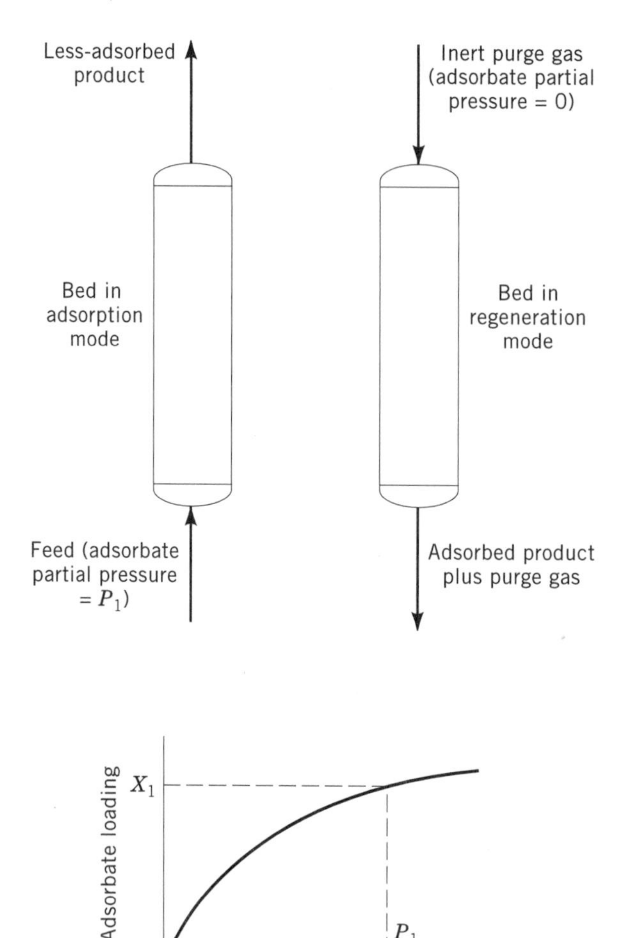

Fig. 16. Inert-purge cycle (1).

completed and the less selectively adsorbed components have been recovered, an appreciable amount of product is still stored in the bed. A purge cocurrent to feed can increase recovery by displacing the fluid held in the voids.

The more selectively adsorbed components are stripped from the adsorbent bed during the countercurrent purge step. By purging into the product end of the bed, a lower residual loading of the selectively adsorbed species can be achieved in the portion of the adsorber that determines product quality.

Flow Sheet. Most purge-swing applications use two fixed-bed adsorbers to provide a continuous flow of feed and product (Fig. 16). Single beds are used when the flow to be treated is intermittent or cyclic. Because the purge flow is invariably greater than the total feed treated, purge is carried out in the down-flow direction to prevent bed lifting, and adsorption is up-flow.

Applications. Several purge-swing processes for the separation of C10–C18 iso- from normal paraffins have been commercialized: Exxon's Ensorb,

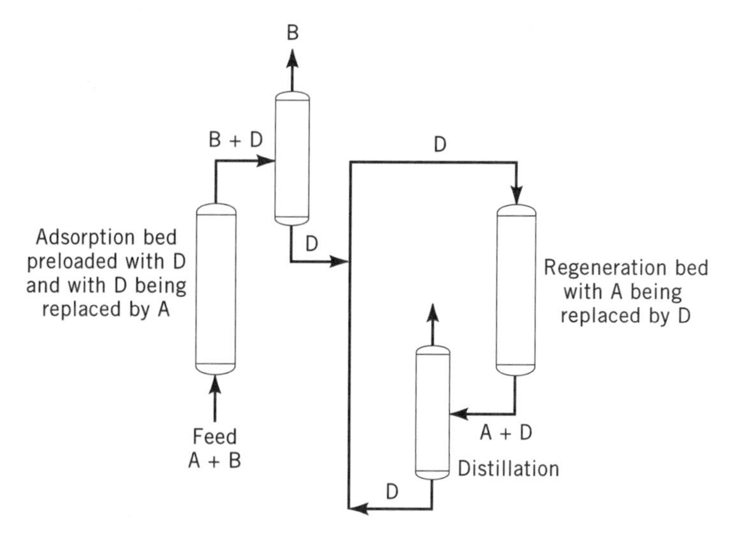

Fig. 17. Displacement-purge cycle (1).

Texaco Selective Finishing (TSF), Leuna Werke's Parex, and the Shell Process (62). All of these processes take advantage of the molecular size selectivity of 5A zeolite, but vary in the purge fluid. Ammonia is used in a displacement-purge cycle in Ensorb. Normal paraffins or light naphtha with a carbon number of two to four less than the feed stream are used for the displacement purge for TSF, Parex, and the Shell Process (96). Since 1971 all U.S. automobile models must have canisters of activated carbon to control gasoline vapors. Any gasoline vapors from the carburetor or the gas tank during running, from the tank during diurnal cycling, and from carburetor hot-soak losses are adsorbed by the carbon and held until they can be regenerated. The vapors are desorbed by an inert purge of air and are drawn into the carburetor as fuel when the engine is running (97). This gas-phase use for activated carbon is the third largest after solvent recovery and air purification.

4.4. Nonregenerative Processing. Gas-phase adsorption can also be used when regenerating the adsorbent is not practical. Most of these applications are used where the facilities to effect a regeneration are not justified by the small amount of adsorbent in a single unit. Nonregenerative adsorbents are used in packaging, dual-pane windows, odor removal, and toxic chemical protection.

Applications. Silica gel is the adsorbent most commonly used as a desiccant in packaging. Activated carbon is used in packaging and storage to adsorb other chemicals for preventing the tarnishing of silver, retarding the ripening or spoiling of fruits, "gettering" (scavenging) outgassed solvents from electronic components, and removing odors.

Adsorbents are used in dual-pane windows to prevent fogging between the sealed panes that could result from the condensation of water or the solvents used in the sealants. Synthetic zeolites (3A, 4A, 13X) or, less frequently, blends of zeolites with silica gel are installed in the spacing strips in double-glazed windows to adsorb water during initial dry-down and any in-leakage and to adsorb organic solvents emitted from the sealants during their cure. The adsorbent or mix of adsorbents applied depends on the sealant system and filling gas used.

The largest use of activated carbon is for the purification of air streams. Much of this carbon is used to treat recirculated air in large occupied enclosures, such as office buildings, apartments, and plants. The carbon is incorporated into thin filter-like frames to treat the large volumes of air with low pressure drop. Odors are also removed from smaller areas by activated carbon filters in kitchen hoods, air conditioners, and electronic air purifiers. On a smaller scale, gas masks containing carbon or carbon impregnated with promoters are used to protect wearers from odors and toxic chemicals. The smallest scale carbon filters are those used in cigarettes. In addition to protection from hazardous chemicals in industry, activated carbon gas masks can protect against gas-warfare chemicals. Activated carbon fibers have been formed into fabrics for clothing to protect against vesicant and percutaneous chemical vapors.

4.5. Reactive Adsorption. Although chemisorbents are not used as extensively as physical adsorbents, a number of commercially significant processes employ chemisorption for gas purification.

Desulfurization. An old method for removal of sulfur compounds involves contacting gases containing H_2S and H_2O with α- or γ-ferric oxide monohydrates at $\sim$38°C to adsorb the sulfur in the form of ferric sulfide, followed by periodic reoxidation of the surface to form elemental sulfur and to "revivify" the ferric oxides (98). The iron sponge is reused in this cycle until buildup of sulfur in its pores reduces its effectiveness and it is replaced with fresh adsorbent. The process is most efficient when the treated gases contain oxygen to allow continuous revivification. Spent adsorbents may be regenerated, e.g., by oxidation of the sulfur to SO_2 to be fed to a sulfuric acid plant or by solvent extraction with carbon disulfide, and reused.

When activated carbon is used to adsorb H_2S from air it undergoes similar reactions. Regeneration can be accomplished with water to form sulfuric acid (99), or with heated air to form SO_2 (100).

Mercury Removal. Trace amounts of mercury found in natural gas in some parts of the world are known to cause significant pinhole corrosion damage to aluminum heat exchanger surfaces in cryogenic coldboxes upstream of liquefied natural gas LNG plants. Mercury can be removed from such streams, and other industrial gases, down to low concentrations by treatment in an *ex situ* TSA regenerative process using an activated carbon adsorbent containing sulfur, and allowing reactions involving the formation of mercuric sulfide to remove mercury from the gas. Alternatively, the mercury can be removed by a newly developed adsorbent that may be employed in either nonregenerative or TSA regenerative process cycles (101).

Nuclear Waste Management. Separation of radioactive wastes provides a number of relatively small scale but vitally important uses of gas-phase purification applications of adsorption. Such applications often require extremely high degrees of purification because of the high toxicity of many radioactive elements.

Delay for Decay. Nuclear power plants generate radioactive xenon and krypton as products of the fission reactions. Although these products are trapped inside the fuel elements, portions can leak out into the coolant (through fuel cladding defects) and can be released to the atmosphere with other gases through an air ejector at the main condenser.

To prevent such release, off gases are treated in Charcoal Delay Systems, which delay the release of xenon and krypton, and other radioactive gases, such as iodine and methyl iodide, until sufficient time has elapsed for the short-lived radioactivity to decay. The delay time is increased by increasing the mass of adsorbent and by lowering the temperature and humidity; for a boiling water reactor (BWR), a typical system containing 21 t of activated carbon operated at 255 K, at 500 K dewpoint, and 101 kPa (15 psia) would provide $\sim$42 days holdup for xenon and 1.8 days holdup for krypton (102). Humidity reduction is typically provided by a combination of a cooler-condenser and a molecular sieve adsorbent bed.

If the spent fuel is processed in a nuclear fuel reprocessing plant, the radioactive iodine species (elemental iodine and methyl iodide) trapped in the spent fuel elements are ultimately released into dissolver off gases. The radioactive iodine may then be captured by chemisorption on molecular sieve zeolites containing silver (103).

Other Applications. Many applications of adsorption involving radioactive compounds simply parallel similar applications involving the same compounds in nonradioactive forms, eg, radioactive carbon-14, or deuterium- or tritium-containing versions of CO_2, H_2O, hydrocarbons. For example, molecular sieve zeolites are commonly employed for these separations, just as for the corresponding nonradioactive uses.

4.6. Moving/Fluidized Beds and Wheels. Most adsorption systems use stationary-bed adsorbers. However, efforts have been made over the years to develop moving-bed adsorption processes in which the adsorbent is moved from an adsorption chamber to another chamber for regeneration, with countercurrent contacting of gases with the adsorbents in each chamber. Union Oil's Hypersorption Process (104) is an example. However, this process proved uneconomical, primarily because of excessive losses resulting from adsorbent attrition.

The commercialization by Kureha Chemical Co. of Japan of a new, highly attrition-resistant, activated-carbon adsorbent as Beaded Activated Carbon (BAC) allowed development of a process employing fluidized-bed adsorption and moving-bed desorption for removal of volatile organic carbon compounds from air. The process has been marketed as GASTAK in Japan and is now marketed as SOLDACS by Daikin Industries, Ltd; such a system is offered in the United States by Carbon Resources. Nobel Chematur has developed a similar system based on resin beads marketed in the United States by American Purification and by Weatherly.

Another application for moving beds is in the treatment of flue or exhaust gases. Here the adsorbent flows downward in a cross-flow mode to minimize pressure drop. Copper oxide on alumina was used in such a configuration to remove SO_2 and NO_x (105), activated carbon to adsorb SO_2 (72), and activated carbon to remove VOC (106).

Recently wheels have been commercialized for ambient pressure applications in order to reduce costly pressure drop. Their major application has been TSA cycles for drying and for VOC removal (see above).

5. Design Methods

Design techniques for gas-phase adsorption range from empirical to theoretical, from simple to computationally intensive. Methods have been developed for equilibrium, mass transfer, and combined dynamic performance. Approaches are available for the regeneration methods of heating, purging, steaming, and pressure swing. Many tools exist to aid the adsorption system designer. Computer-based models have been presented in the literature, and AspenTech offers one commercially trademarked ADSIM, which includes the most popular equilibrium models, as well as allowing user-defined ones, ADSIM has modules for unsteady-state simulation to examine control strategies. ProSim has an adsorption column model in its ProSimPlus software based on (39,107,108). Several broad reviews have been published on analytical equations describing adsorption (109), on experimental adsorption equilibrium and kinetic data (110,111), on theoretical models for adsorption processes (112,113), on adsorption design considerations (1,114,115), and on molecular modeling (116). An extensive bibliographic listing of adsorption (and other separation processes) is maintained at, and available by subscription from, the School of Chemical Engineering, Curtin University of Technology, Perth, Australia (117).

5.1. Adsorption. In the design of the adsorption step of gas-phase processes, two phenomena must be considered: equilibrium and mass transfer. Sometimes adsorption equilibrium can be regarded as that of a single component, but more often several components and their interactions must be accounted for. Design techniques for each phenomenon exist as well as some combined models for dynamic performance.

Equilibrium. Among the aspects of adsorption, equilibrium is the most studied and published. Many different adsorption equilibrium equations are used for the gas phase; the more important have been presented (see the section on Isotherm Models). Equally important is the adsorbed phase mixing rule that is used with these other models to predict multicomponent behavior.

Many simple systems that could be expected to form ideal liquid mixtures are reasonably predicted by extending pure-species adsorption equilibrium data to a multicomponent equation. The potential theory has been extended to binary mixtures of several hydrocarbons on activated carbon by assuming an ideal mixture, and to O_2 and N_2 on 5A and 10X zeolites (118,119). Mixture isotherms predicted by IAST agree with experimental data for methane + ethane and for ethylene + CO_2 on activated carbon, and for $CO + O_2$ and for propane + propylene on silica gel (39). A statistical thermodynamic model has been successfully applied to equilibrium isotherms of several nonpolar species on 5A zeolite, to predict multicomponent sorption equilibria from the Henry constants for the pure components (29). A set of equations that incorporate surface heterogeneity into the IAST model (HIAST) provides a means for predicting near-ideal multicomponent equilibria (120).

For most models of adsorptive equilibrium, however, the coefficients derived from pure species are not adequate to predict multicomponent equilibrium for nonideal mixtures. Fitting the systems ethane + ethylene + propane on 5A zeolite and H_2S + CO_2 on H-mordenite required using binary parameters with the IAST or the real adsorbed solution theory models (121). A coalescing factor applied to the potential theory did collapse all isotherms to a single curve for activated carbon, zeolites, and silica gel. A binary interaction parameter that is a function of the coalescing factor was needed to gain agreement with binary data (122). For the multicomponent system of H_2, CO, CH_4, CO_2, and H_2S on activated carbon, an interaction parameter was required in the extended Langmuir equation to predict multicomponent equilibrium (123). Cross-correlation coefficients were necessary to apply a statistical model to three nonideal ternary zeolite systems (124). An activity coefficient whose composition dependence is described by the Wilson equation has been added to the VSM to fit data for hydrocarbons on activated carbon and $O_2 + N_2$ on 10X zeolite (40). Activity coefficients of the adsorbate–adsorbate interactions, or treatment of the surface as heterogeneous, are correlative methods that allow extension of the IAST to binary adsorption (125). Several models were tested with the addition of activity coefficients for the nonideal systems CO_2, ethylene and ethane on 5A zeolite, and carbon molecular sieve (126). A two-dimensional virial equation of state was needed to predict nonideal equilibrium of CO_2, H_2S and propane on H-mordenite and *n*-hexane and water on activated carbon (127).

Mass Transfer. The degree of approach to equilibrium that can be achieved in adsorption is determined by the mass-transfer rates. One useful design concept is the mass-transfer zone (MTZ), an extension of the ion-exchange zone method (128). Figure 18**b** is a depiction of the adsorbate loading in a fixed bed during adsorption. The ordinate is loading (X) and the abscissa is distance (L) from the inlet of the bed. Between the inlet and the exhaustion point (L_e), the loading is in equilibrium with the feed gas, and this section is called the equilibrium section. From the breakthrough point (L_b) to the outlet of the bed, the adsorbate loading is still at the residual loading level and is unused bed. Mass transfer between the gas and the adsorbent is occurring between the breakthrough and exhaustion points, and so this zone is called the MTZ. The length of the bed, L_b to L_e, is called the mass-transfer zone length (MTZL). The MTZL is usually correlated to flow rate or flow velocity (129).

Most dynamic adsorption data are obtained in the form of outlet concentrations as a function of time as shown in Figure 18**a**. The area iebai measures the removal of the adsorbate, as would the stoichiometric area idcai, and is used to calculate equilibrium loading. For constant pattern adsorption, the breakthrough time (Θ_b), and the stoichiometric time (Θ_s), are used to calculate LUB as $(1 - \Theta_b/\Theta_s)L_{\text{bed}}$ (130). This LUB concept is commonly used for drying and desulfurization design in the natural gas industry and for air prepurification before cryogenic distillation.

Another way of subdividing the bed is illustrated in Figure 18**b**. If the mass-transfer resistance were negligible, the MTZ would become a square or stoichiometric front along the line dsc. The area febgf represents used adsorbent loading, while the area ehbe between the potential loading and the actual loading curve eb is unused. By material balance, the area fdcgf up to the stoichiometric front would also represent the used capacity. Therefore, areas febgf and fdegf are equal. This portion of the bed up to the stoichiometric point, *s*, is called the weight of equivalent equilibrium section (WES). The rest of the adsorbent from the stoichiometric point to the breakthrough point is termed the weight of unused bed (WUB), because it is equivalent to a bed with no usable capacity in

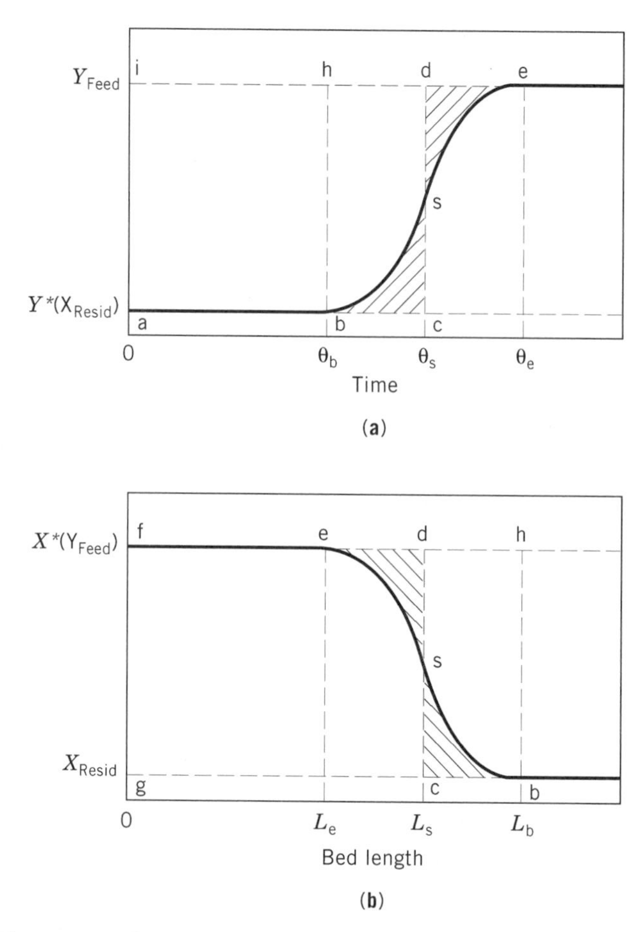

Fig. 18. (**a**) Time trace of adsorbate composition in an adsorber effluent during adsorption. (**b**) Adsorbate loading along the flow axis of an adsorber during adsorption (1).

the stoichiometric interpretation. Adsorption beds can thus be sized by combining a WES calculated from equilibrium data and a WUB derived from kinetic data.

MTZs can be designed by the HTU/NTU mass-transfer concept used in countercurrent liquid/liquid and gas/liquid absorption (131). When this technique is used, it has been found that the mass transfer (HTU) can be uncoupled from the axial dispersion (131). Theoretical stages form the essence of the discrete cell model graphical procedures, which are applied to flat isotherms and incorporate pore diffusivity and axial dispersion.

Dynamic Performance. More complex models do not attempt to separate the equilibrium behavior from the mass-transfer behavior. Rather they treat adsorption as one dynamic process with an overall dynamic response of the adsorbent bed to the feed stream. Although numerical solutions can be attempted for the rigorous partial differential equations, simplifying assumptions are often made to yield more manageable calculating techniques. For systems with a large number of components, the design can be simplified by combining the adsorbates into pseudospecies based on Freundlich exponent and on mass-transfer coefficients (133).

The *J*-function is a definite integral of an expression including I_0, the modified Bessel function of the first kind. *J*-function curves use stoichiometric time and the number of theoretical stages as the two parameters to fit breakthrough curves and extend to other conditions. These curves have been approximated for use on PC microcomputers (134). A phenomenological model requires the determination of two parameters, a transfer coefficient, and a linear isotherm constant, from a complete breakthrough curve. The solution to the model is in an infinite series form, which is calculable by a hand-held calculator or personal computer (135). Another method separates the equilibrium from the kinetic effects by constructing effective equilibrium curves. Because the solution to the model involves nonlinear algebraic or differential equations, graphs called solution charts are used to predict breakthrough fronts (136). Another solution technique is the use of fast Fourier transforms. Linear isotherms are required, but their applicability for predicting breakthrough curves has been demonstrated for isothermal and nonisothermal adsorbers (137). Another model, with a solution in infinite series form, incorporates separate mass-transfer coefficients for external film, macropore, and micropore resistances (138). Techniques have also been developed to predict breakthrough from fluidized beds. The behavior of organic solvents adsorbed from air on activated carbon was shown to exhibit breakthrough times that can be correlated to the adsorption capacity and the amount of bed expansion (139).

Some specific design methods have been developed for particular applications. Several procedures have been published for the design of gas dryers. The *J*-function has been applied to silica gel dryers after a correlated correction factor accounted for the nonisothermality (140). In other work on drying with activated alumina and silica gel, constant-pattern LUBs were shown applicable for designs at H_2O contents of <0.003 kg/kg air (141). Equilibrium and kinetic parameters for H_2O on activated alumina were determined for a more rigorous nonisothermal model that predicted adiabatic behavior (142). Breakthrough times for several organic vapors on activated carbon respirator cartridges have been found to be predictable by using a theory of statistical moments. In one study, equilibrium data was correlated by the Potential Theory and breakthrough was calculated using the normal probability distribution curve (143). In another, the equilibrium data was represented by a Freundlich equation (144). For heavy hydrocarbon recovery from natural gas on silica gel, the equilibrium data were fit to a Freundlich isotherm and the breakthrough composition was found to have a power dependence on the extent of adsorptive saturation of the adsorbent (137). A WUB design approach was found to predict breakthrough for several organics on activated carbon using a Potential Theory equilibrium curve (145). Correlations of equilibrium capacity and WUB were also found applicable in the removal of H_2S from natural gas using 5A zeolite (146).

5.2. Regeneration. In recent years, considerable effort has been expended to better understand and quantify the process of regeneration. Methods are available to predict thermal, purge, and steaming requirements. Models are available to simulate all of the regeneration types, temperature, pressure, and purge swings (115).

Thermal Requirements. When a TSA cycle is heating limited, the regeneration design is only concerned with transferring energy to the system. Charts for isothermal, linear-isotherm adsorption that were derived by Hougen and Marshall (147) from earlier work on heat transfer from a gas to fixed beds (148) can be reapplied to heat transfer for heating-limited regeneration when the heat of adsorption is negligible (61). This approach has been expanded to include both a correction of one dimensionless time per dimensionless bed length for heat losses and another correction particular to the H_2O–4A zeolite system studied (149). Because cooling is a heating-limited step, it can be calculated by the modified Hougen and Marshall method. As mentioned in the discussion of thermal swing, the cooling step can be performed under some process conditions by starting adsorption with a hot adsorbent; performance is not affected.

Purge Requirements. The amount of purge gas needed in stripping-limited regeneration is similar to that for purge regeneration, but it differs primarily in the temperature at which the isothermal desorption occurs. For a pressure-swing process, the theoretical minimum volumetric purge/feed ratio is the ratio of the purge pressure to the feed pressure (61), and one model shows the optimum ratio to be the minimum purge volume that can be used with the given cyclic steady-state conditions (150). For a thermal-swing process, the minimum purge/adsorbent ratio is the ratio of the heat capacity of the solid to that of the gas (151). Specific design purge data have been published for purge-swing activated-carbon automotive evaporative emissions control (97) and for pressure-swing drying of pneumatic system air (89). The pneumatic system process exhibits an optimum purge ratio for maximizing the attainable dewpoint depression. An isothermal purge-swing model that uses Langmuir equilibrium to simulate adsorbent performance has been presented (152).

Steaming Requirements. The steaming of fixed beds of activated carbon is a combination of thermal swing and displacement purge swing. The exothermic heat released when the water adsorbs from the vapor phase is much higher than is possible with heated gas purging. This cycle has been successfully modeled by equilibrium theory (153).

Temperature Swing. Several fairly comprehensive reviews of thermal-swing adsorption models are found in the literature (107,110,154). Many of these models have been used to carry out parametric analyses with a goal of energy minimization. A nonisothermal model for single components using equilibrium theory demonstrated that efficiency improves with increased purge contact time and high heat capacity purge gas but is minimally affected by initial bed loading (155), and the model defined conditions under which the desorption can be continued with a cold stream without additional overall purge gas. That work also introduced the concept that minimum thermal energy is required at the characteristic temperature, that temperature at which the slope of the equilibrium isotherm is equal to the ratio of the heat capacity at the adsorbent to that of the purge gas. A nonequilibrium, mechanistic model with a multicomponent VSM model of equilibrium (156) and a nonequilibrium nonadiabatic computer model with a Langmuir-like isotherm (107) reached similar conclusions on optimization. Another nonequilibrium nonadiabatic computer model with an Antoine equation isotherm was used with supporting data to demonstrate that significant energy can be saved by proper timing of the cooling step (157). Most modeling has assumed that the purge gas is clean even though solvent recovery processes use a recirculated stream when an inert purge gas is employed. Using the method of characteristics and a Freundlich isotherm, an equilibrium theory modeled the incorporation of closed-loop heating and cooling steps (158).

Pressure Swing. Design equations have been developed to predict temperature rise, minimum bed length to retain the heat front, minimum purge rate, and effluent composition (159). A nonequilibrium, nonisothermal simulation program with a Freundlich isotherm equation was found to agree with data for drying with silica gel (160). A somewhat simpler isothermal model using an isotherm approximated by two straight lines successfully calculated the volumetric purge/feed ratio needed to achieve varying product dryness using silica gel (161). An adiabatic equilibrium model with a Langmuir isotherm was used to study the blowdown step of a cycle removing CO_2 on activated carbon and 5A zeolite (162). Changing to an isothermal assumption introduced significant errors into the results. The countercurrent pressurization step was investigated with an isothermal equilibrium model using a Langmuir isotherm for O_2 production from air with 5A zeolite (163). The model predicted the dependence of O_2 concentration on countercurrent pressure and was used to study other parameters. An isothermal model with linear isotherms and component-specific pore diffusivity was used and compared to data for the kinetic-limited separation of air by RS-10 zeolite (164). The simulations agreed well with the experimental parametric studies of time and pressure of feed, blowdown, purge, and pressurization. An equilibrium model was formulated to simulate RPSA using a Freundlich isotherm for separation of N_2 and CH_4 (165). Pressure responses, flow rates, and compositions compared favorably as a function of feed pressure, cycle frequency, and product rate. A nonequilibrium, nonisothermal model for RPSA was developed using a linear isotherm and Darcy's law for pressure drop (166). The model predicted performance in agreement with previous data for air separation on 5A zeolite.

5.3. Pressure Drop. The prediction of pressure drop in fixed beds of adsorbent particles is important. When the pressure loss is too high, costly compression may be increased, adsorbent may be fluidized and subject to attrition, or the excessive force may crush the particles. As discussed previously, RPSA relies on pressure drop for separation. Because of the cyclic nature of adsorption processes, pressure drop must be calculated for each of the steps of the cycle. The most commonly used pressure drop equations for fixed beds of adsorbent are those of Ergun (167), Leva (108), and Brownell and co-workers (168). Each of these correlations uses a particle Reynolds number ($\mathrm{Re} = D_p G/\mu$) and friction factor (f) to calculate the pressure drop (ΔP) per unit length (L) by the equation

$$\frac{\Delta P}{L} = \frac{fG^2}{2g_c D_p \rho}$$

where D_p is the particle diameter, G the mass flux, μ the gas viscosity, and ρ the gas density. The methods differ in their definition of D_p and f. For up-flow in fixed-bed adsorbers, fluidization occurs when the pressure drop just balances the weight, corrected by any buoyancy:

$$\frac{\Delta P}{L} = \frac{(1-\epsilon)(\rho_s - \rho)g}{g_c}$$

where ρ_s is the density of the solid. For down-flow in packed beds, the potential for crushing the adsorbent must be checked. Two forces act to crush the particles, pressure drop and the weight of the bed. The sum of these two $(\Delta P + (1-\varepsilon)\rho_s L)$ should be kept less than that which is known to cause adsorbent damage.

6. Future Directions

Advances in fundamental knowledge of adsorption equilibrium and mass transfer will enable further optimization of the performance of existing adsorbent types. Continuing discoveries of new adsorbent materials will also provide adsorbents with new combinations of useful properties. New adsorbents and adsorption processes will be developed to provide needed improvements in pollution control, energy conservation, and the separation of high value chemicals. New process cycles and new hybrid processes linking adsorption with other unit operations will continue to be developed.

6.1. Fundamentals. Marked improvements in the prediction of multicomponent equilibrium from single-component data will be achieved by developing more realistic theoretical models that provide for nonideal adsorbate phases and heterogeneities of surface energetics and geometries, and that allow for the effect of adsorbates on adsorbent properties. Molecular modeling and molecular-dynamic simulations of adsorption phenomena on high speed computers will enable better prediction of multicomponent adsorption behavior and design of adsorbents with desired properties.

6.2. New Adsorbent Materials. Hydrophobic molecular sieves, mesoporous molecular sieves, macroreticular resins, and new carbon molecular sieves will continue to find new application. Carbon nanotubes and pillared interlayer clays (PILCS) will become more available for commercial applications, including adsorption. Adsorbents with enhanced performance, both highly selective physical adsorbents and easily regenerated, weak chemisorbents will be developed.

6.3. Process Concepts. More hybrid systems involving gas-phase adsorption coupled with catalytic processes and with other separations processes (especially distillation and membrane systems) will be developed to take advantage of the unique features of each. The roles of adsorption systems will be to efficiently achieve very high degrees of purification; to lower fouling contaminant concentrations to very low levels in front of membrane and other separations processes; or to provide unique separations of azeotropes, close-boiling isomers, and temperature-sensitive or reactive compounds.

6.4. Design Methods. Improvements in the ability to predict multicomponent equilibrium and mass-transfer rate performance will allow continued improvements in the design of new adsorption systems and in the energy efficiency of existing systems.

6.5. Computer Systems. Improved "smart" control systems based on new computer capabilities and control algorithms will be used increasingly in adsorption systems to provide more efficient operation. For example, the adjustment of the adsorption—regeneration cycle to account for changing feed compositions and flow rates can significantly reduce energy consumption by carrying out regenerations only when needed. Enhanced computer capabilities will also allow coupling of more sophisticated equilibrium models with more exact models for adsorption dynamics to provide improved design tools.

BIBLIOGRAPHY

"Adsorptive Separation, Gases" in *ECT* 3rd ed., Vol. 1, pp. 544–581, by D. B. Broughton, UOP Process Division, UOP Inc.; in *ECT* 4th ed., Vol. 1, pp. 529–573, by John D. Sherman and Carmen M. Yon, UOP; "Adsorption, Gas Separation" in *ECT* (online), posting date: December 4, 2000, by John D. Sherman, Carmen M. Yon, UOP.

CITED PUBLICATIONS

1. G. E. Keller, II, R. A. Anderson, and C. M. Yon, in R. W. Rousseau, ed., *Handbook of Separation Process Technology*, John Wiley & Sons, Inc., New York, 1987, pp. 644–696.
2. J. O. Hirschfelder, C. F. Curtiss, and R. B. Bird, *Molecular Theory of Gases and Liquids*, John Wiley & Sons, Inc., New York, 1954, pp. 215, 949, 1110.
3. R. M. Barrer and D. E. W. Vaughan, *J. Phys. Chem. Solids* **32**, 731 (1971).
4. R. M. Barrer, *Zeolites and Clay Minerals as Adsorbents and Catalysts*, Academic Press, London, 1978, pp. 164, 174, 185.
5. D. W. Breck and J. V. Smith, *Sci. Am.*, 8 (Jan. 1959).
6. *Data from Union Carbide Molecular Sieves*, UOP, Tarrytown, N.Y.
7. W. A. Steele, *Adv. Colloid Interface Sci.* **1**, 3 (1967). (Review with 360 refs.).
8. D. M. Ruthven, *Principles of Adsorption and Adsorption Processes*, John Wiley & Sons, Inc., New York, 1984, Chapts. 3, 4, p. 108.
9. W. Rudzinski, K. Nieszporek, H. Moon, and H-K Rhee, *Pol. Heterog. Chem. Rev.* **1**, 275 (1994) (in English); *Chem. Abstr.* **123**, 535166 (1995).
10. P. Selvam, S. K. Bhatia, and C. G. Sonwane, *Ind. Eng. Chem. Res.* **40**, 3237 (2001).
11. A. L. Myers, *NATO ASI Ser., Ser. E.* **158** (*Adsorpt. Sci. Technol.*), 15 (1989).
12. D. P. Valenzuela and A. L. Myers, *Adsorption Equilibrium Data Handbook*, Prentice Hall, Engelwood Cliffs, N.J., 1989.
13. D. P. Valenzuela and A. L. Myers, *Sep. Purif. Methods* **13**(2), 153 (1984).
14. O. Talu and A. L. Myers, *AIChE J.* **34**, 1887 (1988).
15. Y. K. Tovbin, *Langmuir* **13**, 979 (1997).
16. R. T. Yang, *Gas Separation by Adsorption Processes*, Butterworths, Boston, 1987, p. 86.
17. C. M. Yon and P. H. Turnock, *AIChE Symp. Ser.* **67**(117), 75 (1971).
18. R. T. Maurer, in J. R. Katzer, ed., *Molecular Sieves—II* (ACS Symp. Ser. 40) American Chemical Society, Washington, D.C., 1977, p. 379.
19. N. Sundaram, *Langmuir* **9**, 1568 (1993).
20. P. M. Mathias, R. Kumar, J. D. Moyer Jr., J. M. Schork, S. R. Srinivasan, S. R. Auvil, and O. Talu, *Ind. Eng. Chem. Res.* **35**, 2477 (1997).

21. R. T. Maurer, *AIChE J.* **43**, 388 (1997).
22. I. Langmuir, *J. Am. Chem. Soc.* **40**, 1361 (1918).
23. M. Volmer, *Z. Phys. Chem.* **115**, 253 (1925).
24. R. H. Fowler and E. A. Guggenheim, *Statistical Thermodynamics*, Cambridge University Press, Cambridge, 1939.
25. T. L. Hill, *Introduction to Statistical Thermodynamics*, Addison-Wesley, Reading, Mass., 1960.
26. V. A. Bakaev, *Dokl. Akad. Nauk SSSR* **167**, 369 (1967).
27. L. Riekert, *Adv. Catal.* **21**, 287 (1970).
28. P. Brauer, A. Lopatkin, and G. Ph. Stepanez, in E. M. Flanigen and L. B. Sand, eds., *Molecular Sieve Zeolites, Adv. in Chem 102*, American Chemical Society, Washington, D.C., 1971, p. 97.
29. D. M. Ruthven, K. F. Loughlin, and K. A. Holborrow, *Chem. Eng. Sci.* **28**, 701 (1973).
30. D. M. Ruthven, *Nat. Phys. Sci.* **232**(29), 10 (1971).
31. Ref. 8, p. 75ff.
32. M. M. Dubinin and L. V. Radushkevich, *Dokl. Akad. Nauk SSSR, Ser. Khim.* **55**, 331 (1947).
33. M. M. Dubinin and V. A. Astakhov, *Izv. Akad. Nauk. SSSR, Ser. Khim.* **71**, 5 (1971).
34. C. J. Radke and J. M. Prausnitz, *Ind. Eng. Chem. Fundam.* **11**, 445 (1972); *AIChE J.* **18**, 761 (1972).
35. J. Toth, *Acta. Chim. Acad. Sci. Hung.* **69**, 311 (1971).
36. S. Brunauer, P. H. Emmett, and E. Teller, *J. Am. Chem. Soc.* **60**, 309 (1938).
37. E. C. Markham and A. F. Benton, *J. Am. Chem. Soc.* **53**, 497 (1931).
38. R. Sips, *J. Chem. Phys.* **16**, 490 (1948).
39. A. L. Myers and J. M. Prausnitz, *AIChE J.* **11**, 121 (1965).
40. S. Suwanayuen and R. P. Danner, *AIChE J.* **26**, 68, 76 (1980).
41. T. W. Cochran, R. L. Kabel, and R. P. Danner, *AIChE J.* **31**, 268 (1985).
42. D. Nicholson and T. Stubos, *Membr. Sci. Technol. Ser.* **6**, 231 (2000); *Chem. Abstr.* **133**, 753508 (2000).
43. K. E. Gubbins, *NATO ASI Ser. Ser. C* **491**, 65 (1997); *Chem. Abstr.* **129**, 248956 (1998).
44. J. J. Low, J. D. Sherman, L. S. Cheng, R. L. Patton, A. Gupta and R. Q. Snurr, Paper Presented at the 7th International Conference on Fundamentals of Adsorption, Nagasaki, Japan, May 20–25, 2001.
45. G. M. Davies and N. A. Seaton, in F. Meunier, ed., *Fund. of Ads., Proc. VIth Int. Conf. of Fund. of Ads. (1998), LAS*, pp. 835–840 (1998).
46. R. L. Portsmouth and L. F. Gladden, *Chem. Eng. Res. Dev.* **70**, 186 (1992).
47. A. Kohl and F. Riesenfeld, *Gas Purification*, 4th ed., Gulf Publishing Co., Houston, Tex., 1985, p. 651.
48. K. S. W. Sing and co-workers, *Pure Appl. Chem.* **57**, 603 (1985).
49. D. W. Breck, *Zeolite Molecular Sieves—Structure, Chemistry, and Use*, John Wiley & Sons, Inc., New York, 1974.
50. R. C. Bansal, J.-B. Donnet, and F. Stoeckli, *Active Carbon*, Marcel Dekker, New York, 1988, p. ix.
51. E. M. Flanigen, J. M. Bennett, R. W. Grose, J. P. Cohen, R. L. Patton, R. M. Kirchner, and J. V. Smith, *Nature (London)* **271**, 512 (1978).
52. E. M. Flanigen and R. L. Patton, UOP, Tarrytown, N.Y., private communication.
53. R. C. Bansal, T. L. Dhami, and S. Parkash, *Carbon* **16**, 389 (1978).
54. W. K. Lewis, E. R. Gilliland, B. Chertow, and W. P. Cadogan, *Ind. Eng. Chem.* **42**, 1326 (1950).
55. D. Chen, H. P. Rebo, K. Moljord, and A Holmen, *Chem. Eng. Sci.* **51**, 2687 (1996).
56. D. W. Price and P. S. Schmidt, *J. Air Waste Mgmt. Ass.* **48**, 1135 (1998).

57. M. Lordgooei, M. J. Rood, and M. Rostam-Abadi, *Proc., 91st Annu. Meet.—Air Waste Manage. Assoc.*, WA47B02/1-12 (1998).
58. J. D. Snyder and J. G. Leesch, *Ind. Eng. Chem. Res.* **40**, 2925–2933 (2001).
59. S. Saysset, G. Grevillot, and A. S. Lamine, *Recents Prog. Genie Procedes* **13**, 389 (1999) (in English); *Chem. Abstr.* **135**, 399274 (2001).
60. D. Basmadjian, *Can. J. Chem. Eng.* **53**, 234 (1975).
61. G. M. Lukchis, *Chem. Eng.* **80**, (13), 111; (16), 83; (18), 83 (1973).
62. D. M. Ruthven, *Chem. Eng. Progr.* **84**, 42 (1988).
63. E. Van den Bulck, J. W. Mitchell, and S. A. Klein, *J. Heat Transfer* **108**, 684 (1986).
64. Y. Zhou, S. Y. Lee, and T. K. Ghosh, *Chem. Eng. Commun.* **169**, 57 (1998).
65. S. M. Reiwani and D. J. Moschandreas, *NTIS. Report No. PB87-114120/GAR* 1–96 (1986).
66. M. Buelow and A. Micke, in M. D. LeVan, ed., *Fund. of Ads., Proc. 5th Int. Conf. of Fund. of Ads.*, pp. 131–138 (1996).
67. A. Salden and G. Eigenberger, *Chem. Eng. Sci.* **56**, 1605 (2001).
68. M. Yates, J. Blanco, P. Avila, and M. P. Martin, *Micro. Meso. Mater.* **37**, 201 (2000).
69. Y. Mistsuma, Y. Ota, and T. Hirose, *J. Chem. Eng. Jpn.* **31**, 482 (1998).
70. C. C. Huang, F. C. Lin and F. C. Lu, *Sep. Sci. Tech.* **34**, 555 (1999).
71. K. Gadkaree, D. L. Hickman, and T. V. Johnson, Soc. Automot. Eng., [Spec. Publ] SP-1165 *(Aspects of Automotive Filtration)*, pp. 89–92 (1996).
72. H. Juentgen, *Carbon* **15**, 273 (1977).
73. J. R. Kiovsky, P. B. Koradia, and D. S. Hook, *Chem. Eng. Progr.* **72**, 98 (1976).
74. S. G. Deng and Y. S. Lin, *Ind. Eng. Chem. Res.* **34**, 4063 (1995).
75. M. S. A. Baksh and R. T. Yang, *AIChE J.* **38**, 1357 (1992).
76. E. S. Kikkinides and R. T. Yang, *Ind. Eng. Chem. Res.* **32**, 2365 (1993).
77. W. D. Lovett and F. T. Cunniff, *Chem. Eng. Progr.* **70**, 43 (1974).
78. A. Dyer, *An Introduction to Zeolite Molecular Sieves*, John Wiley & Sons, Inc., New York, 1988, pp. 102–105.
79. G. V. Baron, *Beig. Process Technol. Proc.* **11**(*Separation Technology*) pp. 201–20. (1994) (in English): *Chem. Abstr.* **121**, 582551 (1994).
80. R. V. Jasra, N. V. Choudary, and S. G. T. Bhat, *Sep. Sci. Technol.* **26**, 885 (1991).
81. S. Sircar, M. B. Rao, and T. C. Golden, *Stud. Surf. Sci. Catal.* **120A**, 395 (1999).
82. R. T. Cassidy and E. S. Holmes, *AIChE Symp. Ser.* **80**, 68 (1984).
83. P. Cen and R. T. Yang, *Ind. Eng. Chem. Fundam.* **25**, 758 (1986).
84. S. S. Suh and P. C. Wankat, *AIChE J.* **35**, 523 (1989).
85. R. T. Cassidy, in W. H. Flank, ed., *Adsorption and Ion Exchange with Synthetic Zeolites, Am. Chem. Soc. Symp. Ser. 135*, American Chemical Society, Washington, D.C., 1980, pp. 248–259.
86. G. E. Keller, II, in T. E. White, Jr., C. M. Yon, and E. H. Wagener, eds., *Industrial Gas Separations, Am. Chem. Soc. Symp. Ser. 223*, American Chemical Society, Washington, D.C., 1983, pp. 145–169.
87. C. W. Skarstrom, in N. N. Li, ed., *Recent Developments in Separation Science*, Vol. 2, CRC Press, Boca Raton, Fla., 1975, pp. 95–106.
88. J. W. Armond, in R. P. Townsend, ed., *The Properties and Applications of Zeolites*, The Chemical Society, London, 1980, pp. 92–102.
89. J. P. Ausikaitis in Ref. 18, pp. 681–695.
90. S. J. Collick, H. A. Johnson, and L. A. Robbins, *Environ. Prog.* **16**, 16 (1997).
91. M. Kawai and T. Kaneko, *Gas Sep. Purif.* **3**, 2 (1989).
92. T. C. Frankiewicz and R. G. Donnelly, in T. E. White, Jr., C. M. Yon, and E. H. Wagener, eds., *Industrial Gas Separations, Am. Chem. Soc. Symp. Ser. 223*, American Chemical Society, Washington, D.C., 1983, pp. 213–233.

93. E. Richter, *Erdoel Kohle, Erdgas, Petrochem.* **40**, 432 (1987).
94. A. Kapoor and R. T. Yang, *Chem. Eng. Sci.* **44**, 1723 (1989).
95. S. Sircar and R. Kumar, *Ind. Eng. Chem. Proc. Des. Dev.* **24**, 358 (1985).
96. R. T. Yang, *Gas Separation by Adsorption Processes*, Butterworths, Stoneham, Mass., 1987.
97. P. J. Clarke, J. E. Gerrard, C. W. Skarstrom, J. Vardi, and D. T. Wade, *SAE Trans.* **76**, 824 (1968).
98. Ref. 47, p. 421.
99. F. Adib, A. Bagreev, and T. J. Bandosz, *Ind. Eng. Chem. Res.* **39**, 2439 (2000).
100. A. Bagreev, H. Rahman, and T. J. Bandosz, *Carbon* **39**, 1319 (2001).
101. T. Y. Yan, *Ind. Eng. Chem. Res.* **33**, 3010 (1994).
102. J. T. Collins, M. J. Bell, and W. M. Hewitt, in A. A. Moghissi and co-workers, eds., *Nuclear Power Waste Technology*, American Society of Mechanical Engineers, New York, 1978, Chapt. 4.
103. D. W. Holladay, *A Literature Survey: Methods for the Removal of Iodine Species from Off-Gases and Liquid Waste Streams of Nuclear Power and Nuclear Fuel Reprocessing Plants, with Emphasis on Solid Sorbents*, Report ORNL/TM-6350 (Jan., 1979), p. 46, available from National Technical Information Service, Springfield, Va.
104. C. Berg, *Trans. Amer. Inst. Chem. Eng.* **42**, 665 (1946).
105. P. Cengiz, J. Abbasian, R. B. Slimane, K. K. Ho and N. R. Khalili, *Proc. Int. Tech. Conf. Coal Util. Fuel Syst. (2001), 26th*, pp. 893–897 (2001); Chem. Abstr. **135**, 517448 (2001).
106. E. S. Larsen and M. J. Pilat, *J. Air Waste Mgmt. Ass.* **41**, 1199 (1991).
107. J. M. Schork and J. R. Fair, *Ind. Eng. Chem. Res.* **27**, 457 (1988).
108. M. Leva, *Chem. Eng.* **56**, 115 (1949).
109. S. Sircar and A. L. Myers, *Ads. Sci. Technol.* **2**, 69 (1985).
110. M. S. Ray, *Sep. Sci. Tech.* **18**, 95 (1983).
111. M. Sakuth, A. Schweer, S. Sander, J. Meyer, and J. Gmehling, *Chem.-Ing.-Tech.* **70**, 1324 (1998) (in German).
112. M. S. Ray, *Stud. Surf. Sci. Catal.* **120A** (*Adsorption and Its Applications in Industry and Environmental Protection, Vol. 1*), pp. 977–1049 (1999).
113. D. M. Ruthven, *Ind. Eng. Chem. Res.* **39**, 2127 (2000).
114. K. Knaebel, D. Ruthven, J. L. Humphrey, R. Carr, J. R. Hufton, A. L. Myers, J. C. Crittenden and J. L. Bulloch, in P. P. Radecki, ed., *Emerging Sep. Sep. React. Technol. Process Waste Reduct*, American Institute of Chemical Engineers, New York, 1999, pp. 33–129.
115. A. Mersmann, A. , G. G. Borger, and S. Scholl, *Chem. Ing. Tech.* **63**, 892 (1991) (in German).
116. A. Dabrowski, in A. Dabrowski, ed., *Stud. Surf. Sci. Catal.* **120A**, 3 (1999).
117. M. S. Ray, *Ads. Sci. Tech.* **18**, 439 (2000).
118. E. Richter, W. Schutz and H. Juntgen, in A. B. Mersmann and co-workers, eds., *Fund. of Ads., Proc. 3rd Int. Conf. of Fund. of Ads.*, pp. 735–743 (1991).
119. S. J. Doong and R. T. Yang, *Ind. Eng. Chem. Res.* **27**, 630 (1988).
120. F. Karavias and A. L. Myers, *Langmuir* **7**, 3118 (1991).
121. G. Gamba, R. Rota, G. Storti, S. Carra, and M. Morbidelli, *AIChE J.* **35**, 959 (1989).
122. S. D. Mehta and R. P. Danner, *Ind. End. Chem. Fundam.* **24**, 325 (1985).
123. J. A. Ritter and R. T. Yang, *Ind. Eng. Chem. Res.* **26**, 1679 (1987).
124. R. Rota, G. Gamba, R. Paludetto, S. Carra, and M. Morbidelli, *Ind. Eng. Chem. Res.* **27**, 848 (1988).
125. M. Sakuth, J. Meyer, and J. Gmehling, *Chem. Eng. Proc.* **37**, 267 (1998).
126. G. Calleja, I. Pau, P. Perez, and J. A. Calles, in M. D. LeVan, ed., *Fund. of Ads., Proc. 5th Int. Conf. of Fund. of Ads.*, pp. 147–154 (1996).

127. J. Appel, *Surface Sci.* **39**, 237 (1973).
128. A. S. Michaels, *Ind. Eng. Chem.* **44**, 1922 (1952).
129. H. M. Barry, *Chem. Eng.* **67**, 105 (1960).
130. J. J. Collins, *Chem. Eng. Progr. Symp. Ser.* **63**, 31 (1967).
131. T. Vermeulen and G. Klein, *A.I.Ch.E. Symp. Ser.* **117**, 65 (1971).
132. A. Gorius, M. Bailly and D. Tondeur, *Chem. Eng. Sci.* **46**, 677 (1991).
133. S. Ramaswami, and C. Tien, *Ind. Eng. Chem. Proc. Des. Dev.* **25**, 133 (1986).
134. S. L. Forbes and D. W. Underhill, *JAPCA* **36**, 61 (1986).
135. R. Mohilla, J. Argelan, and R. Szolcsanyi, *Int. Chem. Eng.* **27**, 723 (1987).
136. D. Basmadjian and C. Karayannopoulos, *Ind. Eng. Chem. Proc. Des. Dev.* **24**, 140 (1985).
137. C. L. Humphries, *Hydrocarbon Process.* **45**, 88 (1966).
138. P. I. Cen and R. T. Yang, *AIChE J.* **32**, 1635 (1986).
139. H. Hori, I. Tanaka, and T. Akiyama, *JAPCA* **38**, 269 (1988).
140. H. Lee and W. P. Cummings, *Chem. Eng. Progr. Symp. Series* **63**(74), 42 (1967).
141. L. C. Eagleton and H. Bliss, *Chem. Eng. Progr.* **49**, 543 (1953).
142. J. W. Carter, *Br. Chem. Eng.* **14**, 303 (1969).
143. O. Grubner and W. A. Burgess, *Environ Sci. Technol.* **15**, 1346 (1981).
144. Y. E. Yoon and J. H. Nelson, *Am. Ind. Hyg. Assoc. J.* **45**, 517 (1984).
145. L. A. Jonas and J. A. Rehrmann, *Carbon* **11**, 59 (1973).
146. C. W. Chi and H. Lee, *AIChE Symp. Ser.* **69**, 95 (1973).
147. O. A. Hougen and W. K. Marshall, *Chem. Eng. Progr.* **43**, 197 (1947).
148. C. C. Furnas, *Trans. Am. Inst. Chem. Eng.* **24**, 142 (1930).
149. C. W. Chi, *AIChE Symp. Ser.* **74**, 42 (1977).
150. R. P. Underwood, *Chem. Eng. Sci.* **41**, 409 (1986).
151. R. Kumar and G. L. Dissinger, *Ind. Eng. Chem. Proc. Des. Dev.* **25**, 456 (1986).
152. I. Zwiebel, R. L. Gariepy, and J. J. Schnitzer, *AIChE J.* **18**, 1139 (1972); **20**, 915 (1974).
153. A. Jedrzejak and M. Paderewski, *Int. Chem. Eng.* **28**, 707 (1988).
154. J. L. Bravo, Report of DOE Contract No. DE-AS07-831D12473, 1984, pp. 150–181.
155. D. Basmadjian, K. D. Ha, and C. Y. Pan, *Ind. Eng. Chem. Proc. Des. Dev.* **14**, 328 (1975).
156. C. Huang and J. R. Fair, *AIChE J.* **35**, 1667 (1989).
157. M. M. Davis and M. D. LeVan, *Ind. Eng. Chem. Res.* **28**, 778 (1989).
158. A. Jedrzejak, *Chem. Eng. Technol.* **11**, 352 (1988).
159. D. H. White, Jr., and P. G. Barkley, *Chem. Eng. Progr.* **85**, 25 (1989).
160. K. Chihara and M. Suzuki, *J. Chem. Eng. Jpn.* **16**, 293 (1983).
161. J. W. Carter and M. L. Wyszynski, *Chem. Eng. Sci.* **38**, 1093 (1983).
162. R. Kumar, *Ind. Eng. Chem. Res.* **28**, 1677 (1989).
163. J. L. Liow and C. N. Kenney, *AIChE J.* **36**, 53 (1990).
164. H. Shin and K. S. Knaebel, *AIChE J.* **34**, 1409 (1988).
165. P. H. Turnock and R. H. Kadlec, *AIChE J.* **17**, 335 (1971).
166. S. J. Doong and R. T. Yang, *AIChE Symp. Ser.* **84**, 145 (1988).
167. S. Ergun, *Chem. Eng. Prog.* **48**, 89 (1952).
168. L. E. Brownell, H. S. Dombrowski, and C. A. Dickey, *Chem. Eng. Progr.* **46**, 415 (1950).

Carmen M. Yon
John D. Sherman
UOP

ADSORPTION, LIQUID SEPARATION

1. Introduction

Recovery and purification of the desired product are generally as important as the synthesis of the product itself. Although most of the value in chemical conversion is added via reaction, it is the separation that largely determines the capital cost of production. Nearly every chemical manufacturing operation requires the use of separation processes to recover and purify the desired product. In most circumstances, the efficiency of the separation process has a significant impact on both the quality and the cost of the product (1). Liquid-phase adsorption has long been used for the removal of contaminants present at low concentrations in process streams. In most cases, the objective is to remove a specific feed component; alternatively, the contaminants are not well defined, and the objective is to improve feed quality as defined by color, taste, odor, and storage stability (2). Deodorization of water, decolorization of sugar, ion exchange of fermentation broths are a few examples of processes in which trace impurities are removed. More recently the simulated moving bed (SMB) processes are finding applications in Biotechnology semi batch and in protein purification.

While most of the strategies to remove trace impurities are batch processes, bulk adsorptive separation processes are continuous or semicontinuous in operation, because in bulk separation processes, where the feed component may be present in large enough concentration, it is imperative to maximize utilization of the adsorbent. The Hypersorption process (3) developed by Union Oil Company in the early 1950s for the recovery of propane and heavier components from natural gas is the earliest example of large-scale countercurrent adsorption processes.

The first commercial operation of a liquid-phase simulated countercurrent adsorption process occurred in 1960 with the advent of the Molex process discovered by Universal Oil Products (UOP), for recovery of high purity n-paraffins (4–6). Since that time, bulk adsorptive separation of liquids has been used to solve a broad range of problems, including individual isomer separations and class separations. The commercial availability of synthetic molecular sieves and ion-exchange resins and the development of novel process concepts have been the two significant factors in the success of these processes.

This article is devoted mainly to the theory and practice of batch and continuous liquid-phase bulk adsorptive separation processes.

2. Batch versus Continuous Operation

Industrial-scale adsorption processes can be classified as batch or continuous (7,8). In a batch process, the adsorbent bed is saturated and regenerated in cyclic operation. In a continuous process, a countercurrent staged contact between the adsorbent and the feed and desorbent is established by either a true or a simulated recirculation of the adsorbent. The efficiency of an adsorption process is significantly higher in a continuous mode of operation than in a cyclic batch mode (9). In a batch chromatographic operation, the liquid composition at a given level in the bed undergoes a cyclic change with time, and large portions of the bed do not perform any useful function at a given time. In continuous operation, the composition at a given level is invariant with time, and every part of the bed performs a useful function at all times. The height equivalent of a theoretical plate (HETP) in a batch operation is roughly three times that in a continuous mode. For difficult separations, batch operation may require 25 times more adsorbent inventory and twice the desorbent circulation rate than does a continuous operation. In addition, in a batch mode, the four functions of adsorption, purification, desorption, and displacement of the desorbent from the adsorbent are inflexibly linked, whereas a continuous mode allows more degrees of freedom with respect to these functions, and thus a better overall operation.

3. Continuous Countercurrent Processes

The need for a continuous countercurrent process arises because the selectivity of available adsorbents in a number of commercially important separations is not high. In the *p*-xylene system, eg, if the liquid around the adsorbent particles contains 1% *p*-xylene, the liquid in the pores contains ~2% *p*-xylene at equilibrium. Therefore, one stage of contacting cannot provide a good separation, and multistage contacting must be provided in the same way that multiple trays are required in fractionating materials with relatively low volatilities. A number of commercial moving-bed designs exist mainly for ion exchange. A good review of these designs can be found in (10).

The multistage countercurrent contacting concept was originally used in a process developed and licensed by UOP under the name Sorbex (11,12). Other versions of the SMB system are also used commercially for industrial scale separations (13). Toray Industries built the Aromax process for the production of *p*-xylene (14–16). Illinois Water Treatment and Mitsubishi have commercialized SMB processes for the separation of fructose from dextrose (17–19). Institut Francais du Petrole have commercialized the Eluxyl process for production of *p*-xylene (21). More recently, SMB processes are finding increased application in the purification of specialty chemicals, enantiomers, vitamins and proteins (22,23). Particularly in the area of drug development, the advent of SMB has provided a high throughput, high yield, solvent efficient, safe and cost effective process option.

The following discussion involves the principles and practice of the SMB processes for liquid-phase separation.

4. Moving-Bed Operation

A hypothetical moving-bed system and a liquid-phase composition profile are shown in Figure 1. The adsorbent circulates continuously as a dense bed in a closed cycle and moves up the adsorbent chamber from bottom to top. Liquid

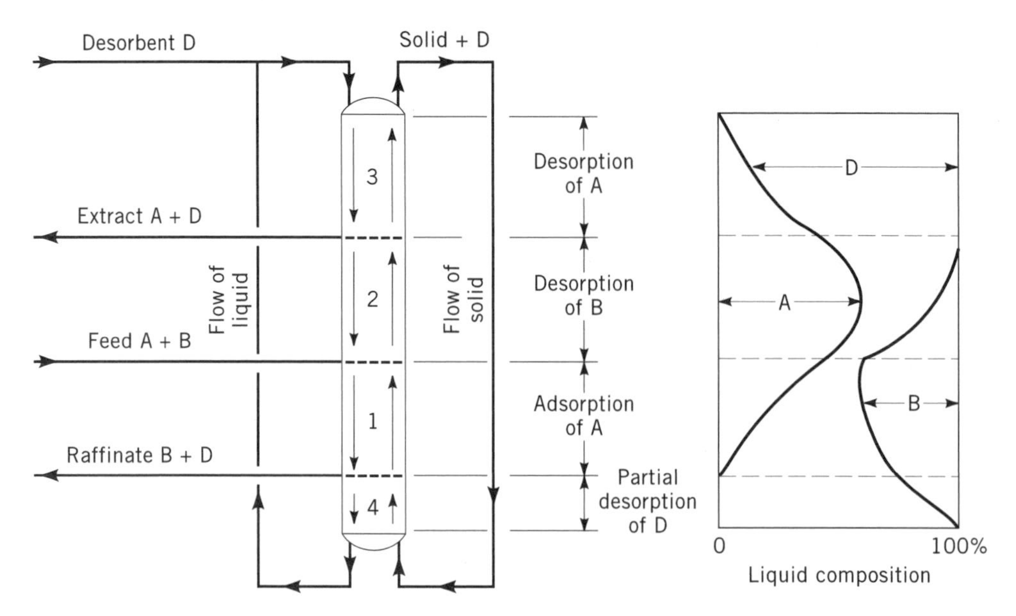

Fig. 1. Adsorptive separation with moving bed.

streams flow down through the bed countercurrently to the solid. The feed is assumed to be a binary mixture of A and B, with component A being adsorbed selectively. Feed is introduced to the bed as shown.

Desorbent D is introduced to the bed at a higher level. This desorbent is a liquid of different boiling point from the feed components and can displace feed components from the pores. Conversely, feed components can displace desorbent from the pores with proper adjustment of relative flow rates of solid and liquid.

Raffinate product, consisting of the less strongly adsorbed component B mixed with desorbent, is withdrawn from a position below the feed entry. Only a portion of the liquid flowing in the bed is withdrawn at this point; the remainder continues to flow into the next section of the bed. Extract product, consisting of the more strongly adsorbed component A mixed with desorbent, is withdrawn from the bed; again, only a portion of the flowing liquid in the bed is withdrawn, and the remainder continues to flow into the next bed section.

The positions of introduction and withdrawal of net streams divide the bed into four zones, each of which performs a different function as described below.

Zone 1. The primary function of this zone is to adsorb A from the liquid.

Zone 2. The primary function of this zone is to remove B from the pores of the solid.

Zone 3. The function of this zone is to desorb A from the pores.

Zone 4. The purpose of this zone is to act as a buffer to prevent component B, which is at the bottom of *Zone 1*, from passing into *Zone 3*, where it would contaminate extracted component A.

4.1. Difficulties of Moving-Bed Operation. The use of a moving bed introduces the problem of mechanical erosion of the adsorbent. Obtaining uniform flow of both solid and liquid in beds of large diameter is also difficult. The performance of this type of operation can be greatly impaired by nonuniform flow of either phase.

The use of a series of fluidized beds may be considered when solid overflows from each bed to the next. However, this arrangement involves a sacrifice in mass-transfer efficiency because the number of theoretical equilibrium trays cannot exceed the number of physical beds. In contrast, the flow through dense and fixed beds of adsorbent, as practiced in chromatography, can provide hundreds of theoretical trays in beds of modest length. In view of these difficulties, only a few fluidized-bed operations are practiced commercially. The Purasiv HR system, developed by Union Carbide Corporation, used beaded activated carbon for the recovery of solvent. This process used a staged fluidized bed for adsorption and a moving bed for regeneration (24).

4.2. Simulated Moving Bed Operation. In the moving-bed system of Figure 1, solid is moving continuously in a closed circuit past fixed points of introduction and withdrawal of liquid. The same results can be obtained by holding the bed stationary and periodically moving the positions at which the various streams enter and leave. A shift in the positions of the introduction of the liquid feed and the withdrawal in the direction of fluid flow through the bed simulates the movement of solid in the opposite direction.

Of course, moving the liquid feed and withdrawal positions continuously is impractical. However, approximately the same effect can be produced by providing multiple liquid-access lines to the bed and periodically switching each stream to the adjacent line. Functionally, the adsorbent bed has no top or bottom and is equivalent to a toroidal bed. Therefore, the four liquid-access positions can be moved around the bed continually, always maintaining the same distance between the various streams.

The commercial application of this concept (25) is portrayed in Figure 2, which shows the adsorbent as a stationary bed. A liquid circulating pump is provided to pump liquid from the bottom outlet to the top inlet of the adsorbent chamber. A fluid-directing device known as a rotary valve (26,27) is provided. The rotary valve functions on the same principle as a multiport stopcock in directing each of several streams to different lines. At the right-hand face of the valve, the four streams to and from the process are continuously fed and withdrawn. At the left-hand face of the valve, a number of lines are connected that terminate in distributors within the adsorbent bed. The rotary valve is the most widely used method in industrial scale applications for accomplishing the stepwise movement of external streams from bed to bed. This same function can be accomplished using manifolds of on–off valves (28,29). For a 24-bed process application, >100 such valves could be required. Small scale SMB applications, such as for chiral separations, make extensive use of switching valves (2,29).

At any particular moment, only four lines from the rotary valve to the adsorbent chamber are active. Figure 2 shows the flows at a time when lines 2, 5, 9, and 12 are active. When the rotating element of the rotary valve is moved to its next position, each net flow is transferred to the adjacent line;

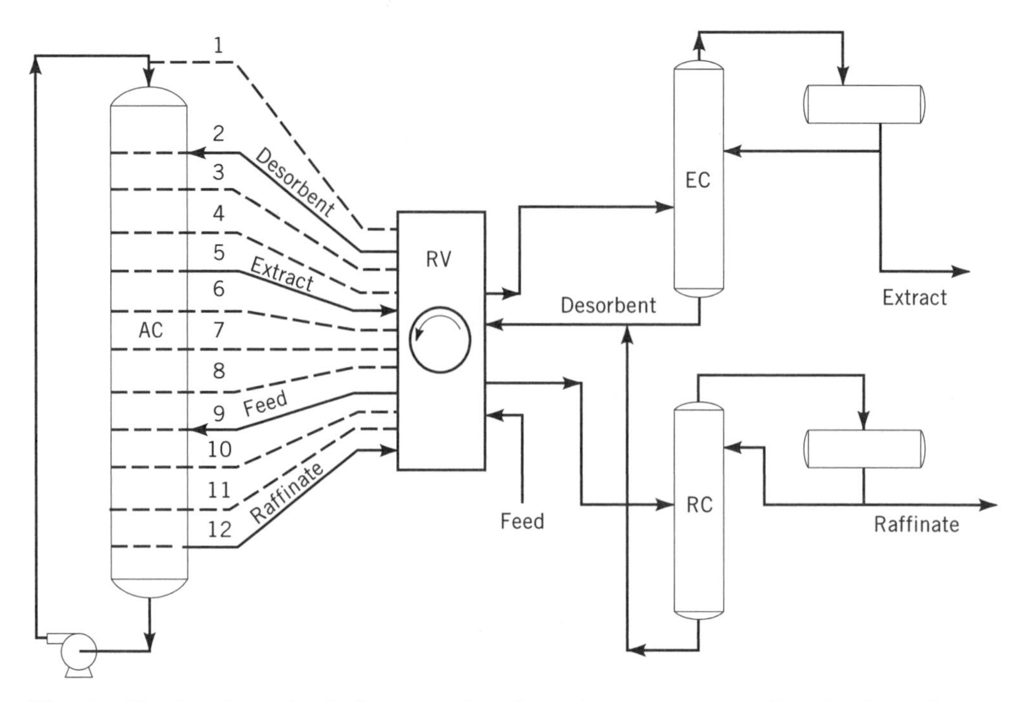

Fig. 2. Simulated moving-bed process for adsorptive separation. AC = adsorbent chamber, RV = rotary valve, EC = extract column, RC = raffinate column.

thus, desorbent enters line 3 instead of line 2, extract is drawn from 6 instead of 5, feed enters 10 instead of 9, and raffinate is drawn from 1 instead of 12.

Figure 1 shows that in the moving-bed operation, the liquid flow rate in each of the four zones is different because of the addition or withdrawal of the various streams. In the simulated moving-bed of Figure 2, the liquid flow rate is controlled by the circulating pump. At the position shown in Figure 2, the pump is between the raffinate and desorbent ports, and therefore should be pumping at a rate appropriate for *Zone 4*. However, after the next switch in position of the rotary valve, the pump is between the feed and raffinate ports, and should therefore be pumping at a rate appropriate for *Zone 1*. Stated briefly, the circulating pump must be programmed to pump at four different rates. The control point is altered each time an external stream is transferred from line 12 to line 1.

To complete the simulation, the liquid-flow rate relative to the solid must be the same in both the moving-bed and simulated moving-bed operations. Because the solid is physically stationary in the simulated moving-bed operation, the liquid velocity relative to the vessel wall must be higher than in an actual moving bed operation.

The primary control variable at a fixed feed rate, as in the operation pictured in Figure 2, is the cycle time, which is measured by the time required for one complete rotation of the rotary valve (this rotation is the analogue of adsorbent circulation rate in an actual moving-bed system), and the liquid flow rate in *Zones 2–4*. When these control variables are specified, all other net rates to and from the bed and the sequence of rates required at the liquid circulating pump are fixed. An analysis of sequential samples taken at the liquid circulating pump can trace the composition profile in the entire bed. This profile provides a guide to any changes in flow rates required to maintain proper performance before any significant effect on composition of the products has appeared. Various aspects of process control are described in the patent literature (30–38).

Temperature and pressure are generally not considered as primary operating variables: Temperature is set sufficiently high to achieve rapid mass-transfer rates, and pressure is sufficiently high to avoid vaporization. In liquid-phase operation, as contrasted to vapor-phase operation, the required bed temperature bears no relation to the boiling range of the feed, an advantage when heat-sensitive stocks are being treated.

4.3. Theoretical Modeling of SMB Systems. The McCabe–Thiele approach has been adapted to describe the SMB process (39). Two feed components, A and B, with a suitable adsorbent and a desorbent, C, are separated in an isothermal continuous countercurrent operation. If A is the more strongly adsorbed component and the system is linear and non-interacting, the flows in each section of the process must satisfy the following constraints for complete separation of A from B:

Section	*Condition*
IV	$SI(D+F-E-R) > K_{CB}$
I	$SI(D+F-E) > K_{BA}$
II	$SI(D-E) < K_{AB}$
III	$S/D < K_{CA}$

The required direction of the net flow of each component is illustrated in Figure 3. The SMB process has four flow-rate variables *(SIF, DIF, EIF*, and *RIF)* and four inequality constraints, one for each section of the bed. Once the equilibrium is fixed, the only remaining degree of freedom is the margin by which the inequality constraints are fulfilled. Once that is decided, the inequality constraints become four equations that define all flow-rate ratios for the system. Once the flow rates are fixed, a preliminary estimate of the number of theoretical stages in each section may be obtained by a McCabe–Thiele diagram, shown in Figure 4.

McCabe–Thiele diagrams for nonlinear and more practical systems with pertinent inequality constraints are illustrated in Figures 5 and 6. The convex isotherms are generally observed for zeolitic adsorbents, particularly in hydrocarbon separation systems, whereas the concave isotherms are observed for ion exchange resins used in sugar separations. Many types of adsorbents have been used in large scale SMB processes. These include amorphous inorganic materials such as silica, alumina; crystalline inorganic materials such as zeolites; organic materials such as activated carbon and polymeric materials such as strong acid, base resins. More recently, specialized adsorbents for chiral–enantiomer separations have been used in an SMB mode. The adsorbent plays a major role in determining the effectiveness of the separation processes. Selectivity, capacity and

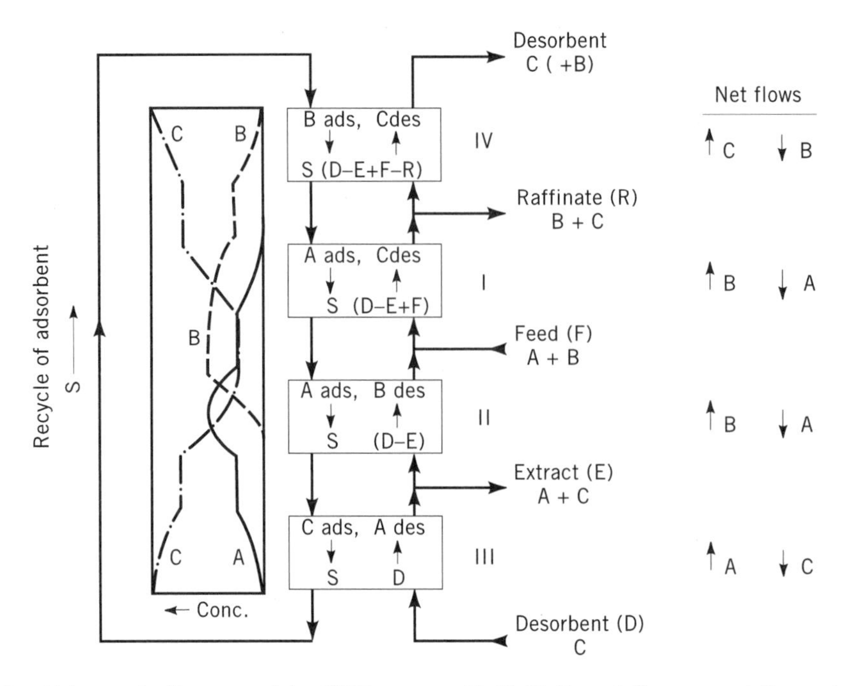

Fig. 3. Schematic diagram of the SMB process *D, E, F, R*, and *S* represent flow rates for desorbent, extract, feed, raffinate, and net solids, respectively.

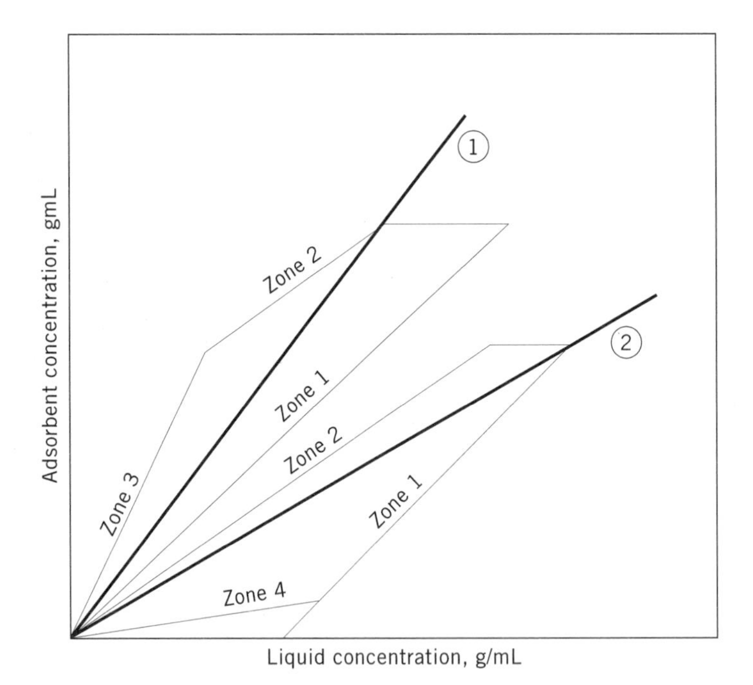

Fig. 4. SMB operation with linear isotherms. Slope of *(D KI;* slope of *M K2*. Conditions for separation: *K*, < *K2, L31S*~!! *K1K2 !~ L2IS:!,~K1, K2:!*LIIS:!~*K1, L4/S :!~–K2, L,* *–L2* = *F*; where *K* = adsorption coefficient, *L* = net liquid flow rate, *S* net solids flow rate, and *F* = feed flow rate.

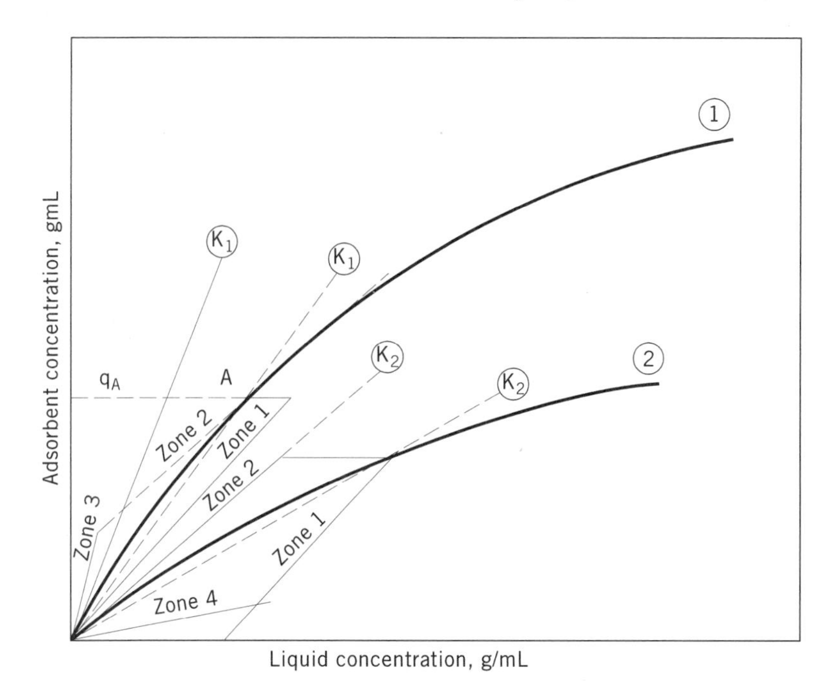

Fig. 5. SMB operation with convex isotherms. Conditions for separation: at point A, slope of (D *K2, K,* >*K2, L31S* ~t *K1, K2:!~ L21S : 5KT:!~K1, K~:!~ LIIS*K1, *L4IS K~, L, –L2 F*, where *K, L, S*, and *F* are as defined in Figure 4.

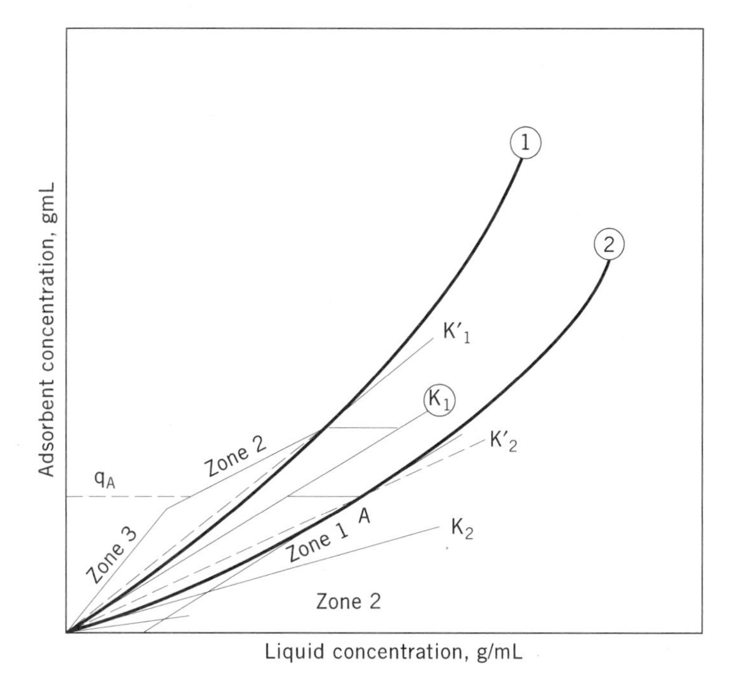

Fig. 6. Conditions for separation: At point A, slope of *K1, K,* >*K2, L31S* >*K'*j, *K2 :5KT* ~ *9 L11S* :!~*KI, K~* : *5 L21S :! Kj, L41S :!~ K2*; where *K, L*, and *S* are as defined in Figure 4. Adsorbent concentration of component 2 < q*A*.

mass transfer rate are determined largely by the adsorbate–adsorbent interactions.

5. Adsorbate–Adsorbent Interactions

An adsorbent can be visualized as a porous solid having certain characteristics. When the solid is immersed in a liquid mixture, the pores fill with liquid, which at equilibrium differs in composition from that of the liquid surrounding the particles of the adsorbent. These compositions can then be related to each other by enrichment factors that are analogous to relative volatility in distillation. The adsorbent is selective for the component that is more concentrated in the pores than in the surrounding liquid.

The choice of separation method to be applied to a particular system depends largely on the phase relations that can be developed by using various separative agents. Adsorption is usually considered to be a more complex operation than is the use of selective solvents in liquid–liquid extraction (see EXTRACTION, LIQUID–LIQUID), extractive distillation, or azeotropic distillation (see DISTILLATION, AZEOTROPIC AND EXTRACTIVE). Consequently, adsorption is employed when it achieves higher selectivities than those obtained with solvents.

An example of unique selectivities is the separation of olefins from paraffins in feed mixtures containing about five successive molecular sizes, eg, C_{10} to C_{14}. Liquid–liquid extraction might be considered for this separation. However, polar solvents give solubility patterns of the type shown in Figure 7. Each olefin is more soluble than the paraffin of the same chain length, but the solubility of both species declines as chain length increases.

Thus in a broad boiling mixture, solubilities of paraffins and olefins overlap and separation becomes impossible. In contrast, the relative adsorption of olefins and paraffins from the liquid phase on the adsorbent used commercially for this operation is shown in Figure 8. Not only is there selectivity between an olefin and paraffin of the same chain length, but also chain length has little effect on

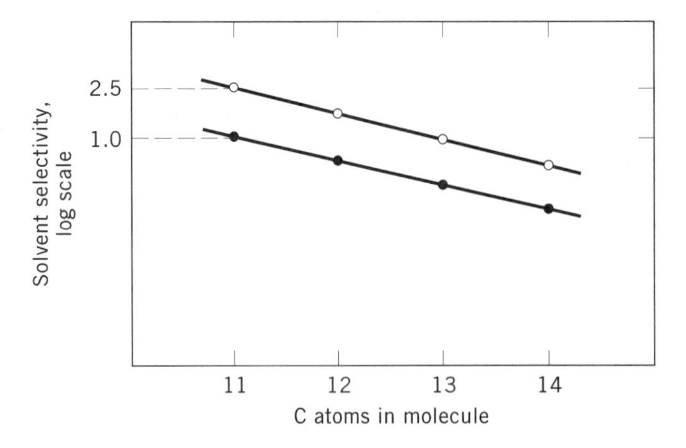

Fig. 7. Liquid–liquid extraction selectivity: olefins; paraffins.

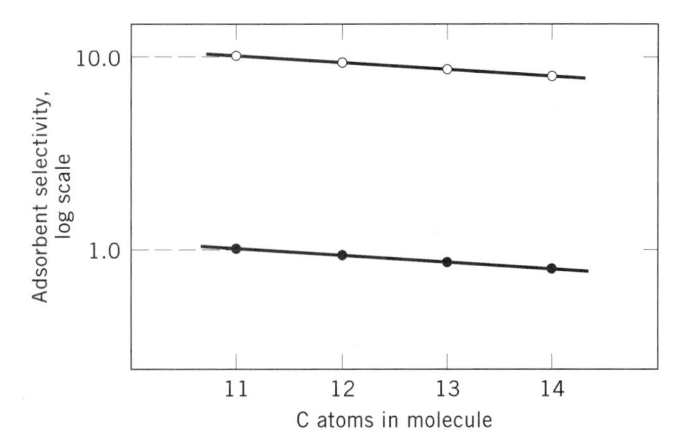

Fig. 8. Liquid-phase selectivity of UOPs Olex adsorbent; olefins; paraffins.

selectivity. Consequently, the complete separation of olefins from paraffins becomes possible.

Unique adsorption selectivities are employed in the separation of C_8 aromatic isomers, a classical problem that cannot be easily solved by distillation, crystallization, or solvent extraction (10). Although *p*-xylene [106-42-3] can be separated by crystallization, its recovery is limited because of the formation of eutectic with *m*-xylene [108-58-3]. However, either *p*-xylene, *m*-xylene, *o*-xylene [95-47-6], or ethylbenzene [100-41-4] can be extracted selectively by suitable modification of zeolitic adsorbents.

Literature dealing with adsorbent–adsorbate interactions in liquid phase is largely confined to patents (40–81). Although theoretical consistency tests exist for such data (82), the search for an adsorbent of suitable selectivity remains an art.

Recent publication by Pharmacia (83) illustrates the adsorbate–adsorbent interactions in extremely difficult separations of chiral molecules. Here the mobile phase or the desorbent is also employed to amplify selectivites via the use of phenomenon generally known as reversed phase chromatography.

6. Practical Adsorbents

The search for a suitable adsorbent is generally the first step in the development of an adsorption process. A practical adsorbent has four primary requirements: selectivity, capacity, mass-transfer rate, and long-term stability. The requirement for adequate adsorptive capacity restricts the choice of adsorbents to microporous solids with pore diameters ranging from a few tenths to a few tens of nanometers.

Traditional adsorbents such as silica [7631-86-9], SiO_2 activated alumina [1318-23-61], A1203; and activated carbon [7440-44-0], C, exhibit large surface areas and micropore volumes. The surface chemical properties of these adsorbents make them potentially useful for separations by molecular class. However, the micropore size distribution is fairly broad for these materials (84). This characteristic makes them unsuitable for use in separations in which steric hindrance can potentially be exploited (see ALUMINUM COMPOUNDS, ALUMINA; SILICON COMPOUNDS, SYNTHETIC INORGANIC SILICATES).

Typical polar adsorbents are silica gel and activated alumina. Equilibrium data have been published on many systems (40–47,85,86). The order of affinity for various chemical species is saturated hydrocarbons < aromatic hydrocarbons = halogenated hydrocarbons < ethers = esters = ketones < amines = alcohols ≤ carboxylic acids. In general, the selectivities are parallel to those obtained by the use of selective polar solvents; in hydrocarbon systems, even the magnitudes are similar. Consequently, the commercial use of these adsorbents must compete with solvent-extraction techniques.

The principal nonpolar-type adsorbent is activated carbon. Equilibrium data have been reported on hydrocarbon systems, various organic compounds in water, and mixtures of organic compounds (40,41,45,85,86). With some exceptions, the least polar component of a mixture is selectively adsorbed; eg, paraffins are adsorbed selectively relative to olefins of the same carbon number, but dicyclic aromatics are adsorbed selectively relative to monocyclic aromatics of the same carbon number (see CARBON, ACTIVATED CARBON).

Polymeric resins [81133-25-7] are widely used in the food and pharmaceutical industries as cation–anion exchangers for the removal of trace components and for some bulk separations, such as fructose from glucose (87). These resins are primarily attractive for aqueous-phase separations and offer a fairly wide potential range of surface chemistries to fit a number of separation needs. For example, polymeric resins are effective in partitioning by size and molecular weight and may also be effective in ion exclusion (see ION EXCHANGE).

In contrast to these adsorbents, zeolites offer increased possibilities for exploiting molecular-level differences among adsorbates. Zeolites are crystalline aluminosilicates containing an assemblage of SiO_4 and $A1O_4$ tetrahedral joined together by oxygen atoms to form a microporous solid, which has a precise pore structure (88). Nearly 40 distinct framework structures have been identified to date. The versatility of zeolites lies in the fact that widely different adsorptive properties may be realized by the appropriate control of the framework structure, the silica-to-alumina ratio (Si/Al), and the cation form. For example, zeolite A, shown in Figure 9, has a three-dimensional (3D) isotropic channel structure constricted by an eight-membered oxygen ring. Its effective pore size can be controlled at ~3–4 Å. The potassium form, with 3-Å pores, is used for removing water from olefinic hydrocarbons. The sodium form can be used to efficiently remove water from nonreactive hydrocarbons, such as alkanes. The substitution of calcium can provide a pore size that will admit *n*-paraffins and exclude other hydrocarbons.

Large-pore zeolites, X, Y, and mordenites, have pores defined by 12-membered oxygen rings with a free diameter of 7.4 Å. The framework structure of X and Y faujasites sketched in Figure 9 consists of a total of 192 SiO_2 and A102 units. The Si/Al (atomic) ratio for X is generally between 1.0 and 1.5, whereas for Y it is between 1.5 and 3.0. With suitable procedures, Y can be dealuminated to make ultrastable Y with Si/Al ratios exceeding 100. Adsorption properties of faujasites are strongly dependent on not only the cation form, but

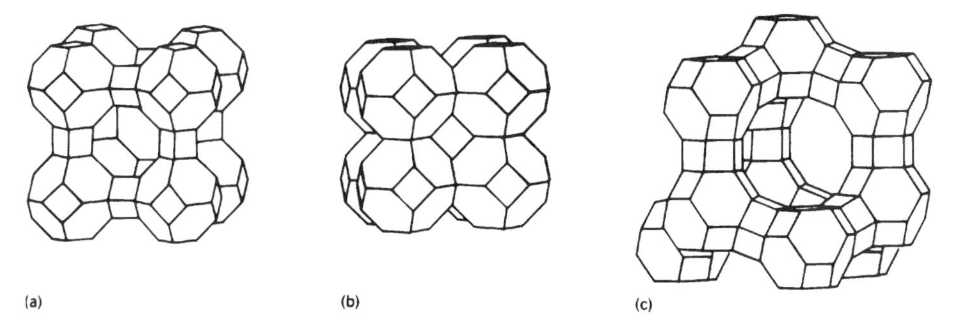

Fig. 9. Three zeolites with the same structural polyhedron, cubooctahedrons. (**a**) Type A, $Na_{12}[(AlO_2)_{12}(SiO_2)_{12}]\cdot 27H_2O$; (**b**) sodalite [1302-90-5]; (**c**) faujasite (Type X, Y), where $X = Na_{86}[(AlO_2)_{86}(SiO_2)_{106}]\cdot 264H_2O$; $Y = Na_{56}[(AlO_2)_{56}\text{-}(SiO_2)_{136}]\cdot 250H_2O$.

also the Si/Al ratio. The flexibility provided by faujasites in the adsorption of C8 aromatics is shown in Table 1. The selectivity order, from the most selectively adsorbed to the least selectively adsorbed, can be changed significantly by the choice of zeolite properties.

In addition to the fundamental parameters of selectivity, capacity, and mass-transfer rate, other more practical factors, namely, pressure drop characteristics and adsorbent life, play an important part in the commercial viability of a practical adsorbent (89).

6.1. Desorbent. In addition to adsorbent, the desorbent or the eluant plays an important role in the commercial viability of the SMB process. Desorbent is usually physically separable from the product, ie, its boiling point must be either higher or lower by sufficient degrees. Also desorbent selectivity must fall between the two key components which one wants to separate in an SMB mode. The third and equally important property is for the Desorbent to not hinder mass transfer. This can be very important in sterically hindered transfer processes such as in zeolites. For the separation of close boiling isomers the choice of desorbents is limited to the family of molecules which are similar to the key feed components.

7. SMB Applications

UOP has pioneered the use of SMB technology in numerous industrial-scale applications. A good example of the acceptance of SMB technology can be

Table 1. **Selectivity of Zeolites In C_8 Aromatic Systems**[a]

	Adsorbent			
	No. 1	No. 2	No. 3	No. 4
p-xylene	1	2	3	4
ethylbenzene	2	1	4	3
m-xylene	3	3	1	2
o-xylene	4	4	2	1

[a] Key: 1 = most selectively adsorbed, 4 = least selectively adsorbed.

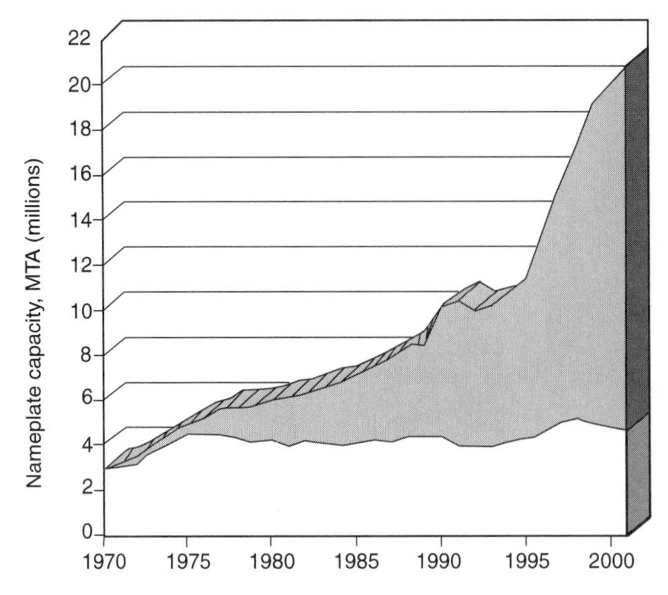

Fig. 10. Technologies for recovery of p-xylene: Production capacities since 1970.

found in the recovery of high purity p-xylene. Figure 10 (90) shows that liquid-phase adsorptive separation technology has significantly overshadowed crystallization as the preferred means of recovering high-purity p-xylene. This is due in large part to the success of the Parex process developed by UOP, but also includes capacity added by others. Hybrid technologies employing adsorption and crystallization can also be practiced (91–95).

8. Adsorptive Separation Technologies

The first commercial application of SMB was through the UOP Molex process to separate n-paraffins from branched paraffins, cyclic paraffins, and aromatics. This plant started up in 1960, and its products were used for the manufacture of biodegradable detergents. Since then, at UOP alone, $\sim$ 134 industrial-scale units have been put on-stream in a variety of applications that produce in excess

Table 2. **SMB Sorbex Processes for Commodity Chemicals**

UOP	Processes	units
Parex	p-xylene from C_8 aromatics	78
Molex	n-paraffins from branched and cyclic	37
Olex	olefins from paraffins	6
Cymex	p- or m-cymene from cymene isomers	1
Cresex	p- or m-cresol from cresol isomers	1
Sarex	fructose from dextrose plus polysaccharides	5
MX Sorbex	m-xylene/C_8 aromatics	6
total		*134*

Table 3. ***m*-Xylene Separation**

Component	Feed, wt %	Extract	Raffinate wt %
m-xylene	30.5	99.7	0.4
p-xylene	12.9	0.1	18.4
Ethyl benzene	35.8	0.1	51.3
o-xylene	20.8	0.1	29.9
total	*100.0*	*100.0*	*100.0*

of 20 million t/yr of products, worth >10 billion USD/yr. The extent of Sorbex technology commercialization, as of July 2001, is shown in Table 2 (96).

8.1. *m*-Xylene Separation. *m*-Xylene is another important component of the C_8 aromatics stream originating in refineries. It is used primarily to produce Isophthalic acid, which is an additive in the manufacture of PET resins. These "isopolyester" resins are being used in the rapidly growing applications for food and beverage packaging. *m*-Xylene has been traditionally produced via complexation with HF/BF_3. UOP has recently commissioned several MX Sorbex units to recover high purity *m*-xylene by adsorptive separation (Table 3), accounting for more than one-half the world's capacity for *m*-xylene. This technology was recently nominated for the prestigious Kirkpatrick award (97). A sixth unit has been designed.

In addition to the above hydrocarbon separation applications there have been >300 U.S. patents issued for application of SMB in fatty chemicals, carbohydrates, biochemicals, and pharmaceuticals, some of which are given here (98–117).

8.2. Ion-Exclusion Processes for Sucrose. Molasses, which is a by-product of raw cane or beet sugar manufacturing processes, is a heavy, viscous liquid that is separated from the final low grade massecuite from which no more sugar can be crystallized by the usual methods. Molasses has a reasonably high sugar content. The recovery of sucrose from molasses has been the object of intense investigation for >50 years. In 1953, the ion-exclusion process was introduced by the Dow Chemical Company (118). This process, which was developed to separate the ionic from the nonionic constituents of molasses, was based on the fact that, under equilibrium conditions, certain ion-exchange resins have a different affinity for nonionic species than for ionic species.

The ion-exclusion process for sucrose purification has been practiced commercially by Finn Sugar (119). This process operates in a cyclic-batch mode and provides a sucrose product that does not contain the highly molassogenic salt impurities and thus can be recycled to the crystallizers for additional sucrose recovery.

9. Liquid-Phase Adsorption and Separation of Pharmaceuticals and Related Compounds

While adsorption from the liquid phase was originally identified as an important process in biochemical analyses by Tswett at the turn of twentieth century (120), >70 years passed before the power of adsorption as a *process* for production of pharmaceutical substances and biologically active materials was realized. Since that time, various modes of adsorptive separations have acquired a key role in processes for the production of numerous biologically significant compounds, including:

- Sugars (monosaccharide resolution (121), oligio- (122), and disaccharide purification (123), and recovery from molasses (124,125).
- Amino acids (126) and peptides (127).
- Proteins (128), nucleic acids, and oligonucleotides (129).
- Antibodies [both monoclonal (130) and recombinant (131)].
- Pharmaceutically significant small organic molecules (132), including resolution of enantiomeric and diastereomeric compounds (133–135).

Although many similarities exist with petrochemical applications of adsorptive separations, there are two significant differences in liquid-phase adsorption processes for pharmaceuticals and related compounds. First, the sorbents, which are commonly referred to as "stationary phases" (136), are generally composed of different materials and take slightly different forms compared to the sorbents used in petrochemical applications. Second, the factors that influence the selection of a mode of operation are significantly different in the purification of pharmaceutical and biologically active materials. Therefore, while the available modes of operation are not different, the frequency of application of the various modes differs between the two areas.

In the following sections, we provide a brief overview of sorbents and modes of operation commonly encountered in pharmaceutical liquid-phase adsorptive separations. We illustrate the commercial status of the techniques through an overview of two specific example applications. As a means to provide some guidance for the selection when more than one mode of operation is available, we provide a short discussion of the economics of separation modes at various scales of operation. Finally, we provide a snapshot of the current challenges facing liquid-phase adsorptive separations of pharmaceutically significant compounds.

9.1. Sorbents and Eluents in Pharmaceutical Separations. As in any separation, the liquid-phase adsorptive separation of two or more compounds requires at least one physical property difference that can be exploited. In this manner, a mechanism can be chosen to exploit the desired physical property difference in the molecules to be separated and control the partitioning of the molecules between the liquid and adsorbed phases. The combination of the physical property being exploited and the mechanism through which partitioning occurs, in general, dictates the type of sorbent used to effect the separation.

Generally speaking, sorbents used in separation of pharmaceutical and biological substances fall into two broad categories: solids and gels. Solid packings are based on a polar or inert solid material, usually a metal oxide (eg, silica gel, zirconia, etc), which is typically spherical in shape and between 1 and 100 μ in diameter. Packings of this type are the most common and are routinely applied in the separation of small to moderately sized organic molecules of interest to the pharmaceutical and fine chemical communities. In addition, solid packings with larger pore sizes have become more widely used in recent years, particularly for the reverse-phase and/or size exclusion purification of peptides (137,138), proteins (138) and oligonucleotides (139). Gel-type packings are typically composed of carbohydrate matrices, with or without cross-linking with agarose or acrylamide (140). These gels are quite soft and can only be used in low- or moderate-pressure chromatography. Gel-type packings are often chosen for the separation of biologics, such as size exclusion-based purification of proteins (140).

The following sections describe most commonly utilized sorbents with a special focus on separation mechanisms used for adsorptive separation processes.

Normal Phase. The designation of normal phase sorbents typically implies the presence of a polar solid-phase material (136). The polarity of the solid phase can be obtained by the use of the solid base itself (eg, silica gel) or through the modification of the solid base with a chemically bonded (bonded-phase chromatography) or a physically adsorbed (liquid-partition chromatography) polar compound. Since the sorbent provides a source of polarity in the separation, the eluents used in normal-phase chromatography are typically non-polar substances slightly modified with components of varying polarity. Separations of this type have been termed "normal" because the solute molecules generally elute in order of increasing polarity.

Reverse Phase. As implied by the name, reverse-phase chromatography provides solute elution in order of decreasing polarity for small organic molecules, which is the opposite of the order observed in normal-phase chromatography. This elution order change is obtained by transposing the polarity of the sorbent and the eluent: In reverse-phase systems, the eluent is usually strongly polar (eg, water) while the sorbent has been modified by a relatively nonpolar compound. Silanes based on trimethyl, butyl, octyl, octadecyl, phenyl, and cyano ligands are the most commonly used. While reverse-phase chromatography packings also exist both as bonded-phase and physically adsorbed-phase materials, application of the bonded-phase packings has dominated since the early 1980s.

Ion Exchange. Electrostatic charge that exists on or near the surface of a solute molecule can be caused to interact with groups, clusters, ligands, or atoms containing the opposite charge on the surface of a solid through the proper choice of conditions and solids (128,141). As the term ion-exchange implies, the fundamental mechanism involves the exchange of a counterion on the solid material with the solute material possessing the same charge as the displaced counterion. Ion-exchange exists in two basic forms, anion exchange and cation exchange. Sorbents in common use for ion exchange are composed of base materials (eg, silica, cross-linked polymers) containing pore diameters ranging from 10 to $>$100 nm that have been derivitized with either positively or negatively charged ligands (142). These ligands can be separated into either strong or weak ion-exchange resins based on their pH susceptibility. In general, strong ion-exchange resins retain their exchange capacity over a wider range of pH, while weak ion-exchange resins have acceptable capacity over narrower windows of pH.

Size Exclusion. Also called gel permeation chromatography, size exclusion chromatography separates solutes solely on the basis of molecular size. As such, it is critical that sorbents used in size exclusion have no physical or chemical interaction with the solutes. Additionally, sorbents that have narrow, well-controlled and known pore distributions are required for practical and efficient application. Choice of a size exclusion sorbent is typically made by considering the desired pressure of operation as well as the molecular sizes of the solutes to be separated. As described above, low- and moderate-pressure applications can use either carbohydrate (soft-gels) or carbohydrates cross-linked with agarose or acrylamides, respectively. For high-pressure applications, size exclusion chromatography typically employs a spherical silica support of 5–10 μ in diameter possessing a controlled pore size distribution. The surface area of these silica packings is typically modified by a covalently bonded neutral molecule to eliminate surface silanol groups (140).

9.2. Recent Advances in Sorbent Materials. Hjerten and co-workers (143) showed that certain types of acrylamide gels, prepared directly within the column, can be compressed above the typical operating pressure yielding aggregated polymer gels with small (3–4 μ) channels between particles. These materials have good dynamic capacities that are maintained as flow rates are increased, allowing significant decreases in analysis time with no loss of resolution. While analytical columns of this type are available, preparative and process chromatographic application has yet to occur even though the promise of these materials for high-throughput separations is quite high.

Macroscopic rods of silica, formed by a reactive molding technique, have been created that contain micron-sized pores with shallow mesoporous "pockets" in the 10–30 nm range (144). The ready access of these shallow pores from the relatively small flow channels allows even the most rapid adsorption–desorption phenomena to occur in the absence of mass transfer limitations. Recently, these materials have been made available at the preparative or semiprocess scale by E. Merck, although published details of their application is presently limited.

9.3. Sorbents, Eluents, Solubility, Method Optimization, and Modeling. While limited sample solubility often can be overcome through a variety of techniques for analytical separations, the cost-effective application of most preparative- and process-scale separations requires moderate (>20 mg/mL) solubility in the eluent. As such, method development for process separations is more demanding, requiring good chromatographic performance with the least complex, most readily available eluents in which the solute components are readily soluble. It must be recognized that optimization of process chromatographic methods, regardless of the mode of operation, requires attention to many such details and, eventually, requires an iterative process often full of compromises. Given these important criteria, the significant advancements over the past decade made in the area of modeling the modes of preparative and process adsorptive separations are quite easy to understand (145–148). Nonetheless, it has become common-place, and nearly necessary, to employ either short-cut (149,150) or rigorous calculation models (151,152) to properly optimize adsorptive separations of pharmaceutical and biological interest.

10. Equilibrium Theory

An improvement over the traditional McCabe–Thiele analysis for adsorptive separations can be found within the equilibrium theory model (147). While this model does not include the effects of axial dispersion nor does it address the potential limitations imposed by mass-transfer resistances, the impact of adsorption equilibrium phenomena are adequately represented by this model. Analytical solutions have now been obtained for a variety of cases and subcases utilizing a variety of basic isotherm equations and assumptions (153–156). Details of the solution to these design problems have been ably covered by Morbidelli and coworkers (152–156) and Chaing (150,157). Only the basic aspects results of this theoretical analysis are presented here.

For a true moving-bed (TMB) countercurrent system at steady state, the equilibrium theory analysis identifies a series of j key design parameters defined by notation L_k, where j is the number of sections within the system. These design parameters, commonly referred to as zone ratios, are defined as the ratio of the net fluid phase flow rate and the net adsorbed phase flow rate within each section (147). Please note that various authors use various numbering schemes for the sections of the TMB/SMB process. Here, we number the zones of a TMB/SMB process in a manner that is consistent with Figure 1.

Equilibrium theory yields a specific set of key operating parameters that will accomplish the complete separation (two pure products) given a well-defined isotherm, the composition of the feedstock, and knowledge of certain physical properties of the sorbent, feedstock and desorbent (147). Provided that certain criteria with regard to L_3 and L_4 zone ratios are met, the complete separation region can be visualized in the two-dimensional (2D) space defined by the L_1 and L_2 design parameters as illustrated in Figure 11.

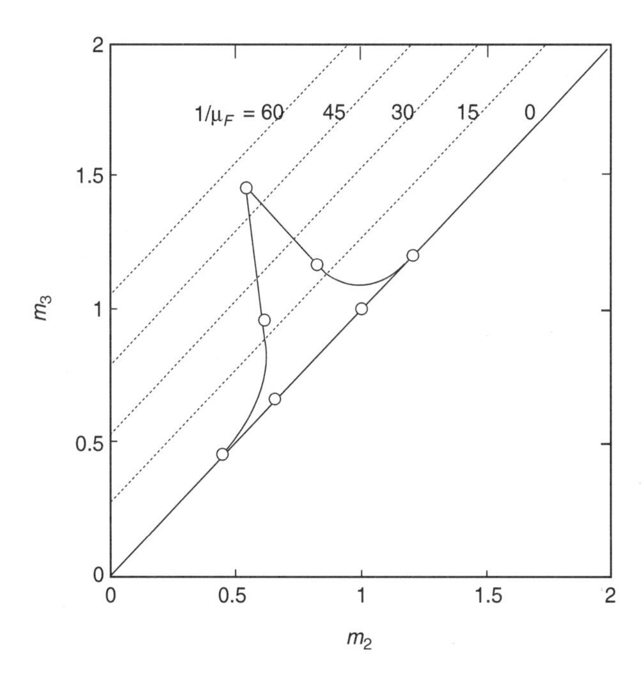

Fig. 11. Region of complete separation of in the L_1–L_2 plane using a two-component feed and a Gurvitsch isotherm. The dashed lines are lines of constant productivity, which increases as you move away from the diagonal. Reproduced with premission of the American Institute of Chemical Engineers.

Due to the shape of this allowable operating region, equilibrium theory has often been referred to as "triangle theory" within the literature. The shape of this operating region is governed by the type of adsorption isotherm selected as well as the relative strengths and, in some cases, adsorption capacities of the various components. Equilibrium theory has been successfully applied to several experimental systems [eg, see Migliorini and co-workers (155) and Francotte co-workers (155)].

11. Modes of Adsorptive Separation

While the applications of liquid-phase adsorptive separations in the petrochemical industry have tended toward the use of continuous processes, the same cannot be said for pharmaceutical and biomolecule separations. In fact, in contrast with the availability of continuous processes for >40 years, only within the past 10 years have modes of operation *other* than batch adsorptive separations been described in the literature. The potential reasons for the slow progress of continuous adsorptive separation processes (indeed, continuous process *in general*) in the pharmaceutical industry are potentially many. However, three particular hurdles to application can be readily identified.

First, and probably foremost, is the tremendous difference between the petrochemical and pharmaceutical applications "objective functions" that are used to determine the optimum process choice. Petrochemical process options tend toward minimization of the cost of manufacture (COM), while COM is rarely a substantial portion of the overall cost of development of a pharmaceutically important substance. This finding is especially true as the location of the separation in the production sequence moves further toward the active pharmaceutical ingredient (API). In addition to cost differences, different valuations exist between the two sectors for such fundamental parameters as time-to-market and risk-avoidance. Finally, while petrochemical processes tend to be carried out at high volume in dedicated facilities, the low volumes and product turnover experienced in the pharmaceutical industry favor manufacturing through general rather than specialized pieces of equipment.

A second governing issue is the generally greater complexity of the most common adsorptive separations in the pharmaceutical and bioprocess industries as compared to the general chemical industry. For example, most adsorptive separations in the petrochemical industry can be treated as binary separations, while the same is certainly not true for purification and separation of most pharmaceutical applications. A third seminal hurdle to be overcome by continuous adsorptive separations in pharmaceutical and biomolecule applications are the issues associated with validation and regulation, particularly in the application of current good manufacturing practices (cGMPs) to continuous separations.

These and other critical issues have led to the present situation in which batch adsorptive separation is preferred in pharmaceutical and biomolecule processes. This situation is, however, changing rapidly with a noticeable acceleration in the past 5 years. While batch adsorption processes will continue to be important for manufacturing of biologically important materials, there will continue to be cases which are particularly well suited to continuous processes, and these instances are likely to increase in frequency with each successful application.

The following sections briefly describe the different process modes of adsorptive separation as they are practiced in the pharmaceutical and biomolecule sector. As considerable overlap exists between the operating modes with the previously described petrochemical applications, only details that are fundamentally different are mentioned. Where relevant, mention has also been made of commercial or near-commercial examples of practical interest.

12. Batch Processes

Batch processes are relatively abundant in the purification and separation of pharmaceutically important substances, particularly in the areas of peptide (127), proteins (128), and oligonucelotides (129). In the most common applications, periodic injections are made on a column containing the appropriate sorbent media, and the eluent composition is either held constant (isocratic elution) or programmed (gradient elution) while the solute components are separated and the appropriate portion or portions of the purified effluent is collected. This can be followed, if required, by a column cleaning–equilibration step.

One commercial example is the purification of synthetic oligonucleotides for use as antisense drugs. Isis Pharmaceuticals, Inc. (Carlsbad, Calif.) has recently launched an antisense oligonucleotide drug that utilizes a single-column batchwise purification of a synthetic oligonucleotide late-stage intermediate (139), and three practical approaches to the postsynthesis purification have been recently described (158). One such scheme is a reverse-phase purification using either silica or polymer supports derivitized with octadecylsilanol (ODS) and weakly buffered eluents such as sodium- or ammonium acetate (158). In this purification scheme, the particular protected phosphothioate N mer of interest is separated from earlier eluting chain-growth failures, unattached protecting groups, unprotected oligonucleotides, and oligonucleotides that are smaller than the target (<N mers). In addition, some later eluting impurities are present and must be excluded from the purified material. An anion-exchange chromatography method may follow the reverse-phase method described above, resulting in >96% pure oligonucleotide (139,158). This combined process has been used by Isis Pharmaceuticals for the production of Vitravene (fomivirsen sodium), which was the first antisense drug approved for market by the U.S. Federal Drug Administration (FDA) (139,158,159). Annual cGMP production of antisense oligonucleotides is expected to reach 1 ton within the next few years (160).

13. Cyclic Processes

Similar to batch separation processes, the pharmaceutical and biomolecule separations industry defines cyclic processes as those in which solute material is recycled, with or without fresh solute material, back to the sorbent bed for further processing. Preparative and process separations have used eluent recycle closed-loop recycle (161–163) to improve eluent efficiency and/or the number of theoretical plates available for separation. If fresh solute were to be sent to the column along with the recycle solute, the two solute materials would be well mixed (162). In a recent advancement of this process, Grill (164) introduced a closed-loop recycle technique in which the chromatographic profile of the recycled solute is maintained and the fresh solute is injected at precise, optimized location while avoiding significant mixing. This technique has the advantage of further increasing the apparent number of theoretical plates for the separation through maintenance of the original profile. Although this technique holds promise, it has been shown to provide sorbent efficiencies that are superior to batch process, but inferior to SMB (165,166). Likewise the process flexibility is somewhat > SMB but inferior to that of batch processes.

Japan Organo Co. has introduced a "pseudo-SMB" process that uses column effluent recycle technology along with period introduction of both solute components and eluent (130). A recent introduction, the capabilities, advantages, and limitations of this process have not been fully explored or published, however, it appears to offer some of the same benefits of the steady-state recycle technique described above while maintaining some of the flexibilities of the batch processing, including the ability to perform ternary and higher order separations (167).

14. Continuous Processes

Although it had long been established as a viable, practical, and cost-effective liquid-phase adsorptive separation technique, the pharmaceutical and biomolecule separations community did not show considerable interest in SMB technology until the mid-1990s. In 1992, Daicel Chemical Industries, Ltd.(168,169) first published the resolution of optical isomers through SMB. Since that time, >600 articles, patents, books, and reviews on the application of SMB to pharmaceutically important compounds have been published. Though resolution of enantiomers is only one such application of SMB in pharmaceuticals and biopharmaceuticals, it has quickly become an area of significant focus. For example, ~25 companies, universities, and institutes worldwide had installed commercial or pilot-scale capacity for SMB applications in the pharmaceutical sector, with a particular emphasis on resolution of chiral products.

In early 2002, Forest Laboratories, Inc. (New York) announced that the U.S. FDA had issued an approval letter for the use of the compound escitalopram oxalate in treating major depressive disorder. Escitalopram oxalate, a selective serotonin reuptake inhibitor, is the single-isomer version of the successful Celexa antidepressant and is licensed from H. Lundbeck. It is well known within the industry (170) that a critical step in the production of escitalopram oxalate is the resolution of the (*S*)-enantiomer of citalopram from the racemic mixture by SMB adsorptive separation. This success is likely to raise awareness and application of SMB in pharmaceutical industries. Indeed, the Belgian pharmaceutical company UCB Pharma is presently exploring (170) the potential use of SMB for enantiomeric resolution and ultimate commercial production of a pharmaceutical compound.

Enantiomeric resolutions are only one area of potential application of SMB in pharmaceutical and biotechnology manufacture. Indeed, the voluminous literature is filled with other applications, including the purification of paclitaxel (171), betaine recovery from beet molasses (172), antibiotics (173), and azeotropic protein separations (174). The later two cases required the use of SMB in a relatively new mode, that of providing an eluent composition gradient, and theoretical treatments of this new method are now beginning to appear in the literature (174–176). It is therefore likely that the most significant contributions of SMB in the pharmaceutical and biotechnology sectors are yet to be discovered.

The dynamic field of continuous chromatography is itself evolving at a rapid pace. New hardware, new sorbents and new methods of operating are being identified and published at a significant rate. For example, a variation of SMB has been demonstrated in which the positions of the net fluid flows to and from the system are changed independently rather than simultaneously (177). It has been reported that this modification allows, in some cases, an increased sorbent and eluent efficiency while potentially requiring less hardware.

15. Economics of Pharmaceutical Separations

Schmidt-Traub and co-workers (178) recently performed a detailed economic comparison for the enantiomeric separation of a late-stage pharmaceutical intermediate through batch and SMB chromatography. The authors examined the cost of separation for optimized batch and SMB operations, and compared them on the basis of the cost of capital, operation, sorbent, eluent, and lost crude. The calculations were prepared at two different production levels, 1 and 5 MT/year. Although some of the cost factors assumed by the authors are quite low for operation in a cGMP environment, the results, shown in Table 4, are instructive.

As expected, costs for SMB separations are equivalent or superior to batch chromatography at these scales, with the benefit to SMB operation increasing as the production scale increases. Two unexpected results were obtained, however. First, the cost of the sorbent material was, on average, a lower cost than has been anticipated by other, less rigorous investigations. Second, the authors suggest that the cost contribution of eluent is actually higher in SMB operation than for batch operations. Although some of the authors' conclusions may be biased

Table 4. **Comaprison of SMB and Batch Chromatography for Enantiomeric Purification of a Pharmacuetical Intermediate**[a]

	SMB		Batch	
Contribution (% of total)	1 MT/a	5 MT/a	1 MT/a	5 MT/a
capital	18.9	9.2	8.3	4.2
operation	8.6	2.1	8.3	4.2
eluent	64.1	78.4	49.1	55.3
sorbent	8.4	10.3	21.0	23.6
lost crude			13.3	15.0
total cost ($/g)	*2.11*	*1.72*	*2.18*	*1.94*

[a] Ref. 178

by the particular separation example chosen, the conclusions offer a starting point for future, detailed investigations of the optimized cost comparisons of these operations.

16. Outlook

Liquid adsorption processes hold a prominent position in several applications for the production of high purity chemicals on a commodity scale. Many of these processes were attractive when they were first introduced to the industry and continue to increase in value as improvements in adsorbents, desorbents, and process designs are made. The Parex process alone has seen three generations of adsorbent and four generations of desorbent. Similarly, liquid adsorption processes can be applied to a much more diverse range of problems than those presented in Table 2.

A surprisingly large number of important industrial-scale separations can be accomplished with the relatively small number of zeolites that are commercially available. The discovery, characterization, and commercial availability of new zeolites and molecular sieves are likely to multiply the number of potential solutions to separation problems. A wider variety of pore diameters, pore geometries, and hydrophobicity in new zeolites and molecular sieves as well as more precise control of composition and crystallinity in existing zeolites will help to broaden the applications for adsorptive separations and likely lead to improvements in separations that are currently in commercial practice.

With the increased speed of discovery in the area of pharmaceuticals and biotechnology, SMB does offer an attractive alternative to conventional batch chromatography separation methods. SMB overcomes the general view of batch chromatography as low throughput, low yield, and high cost process and will likely find increased acceptance as a unit operation in both the discovery as well as production phase of drug development.

BIBLIOGRAPHY

"Adsorptive Separation, Liquids," in *ECT* 3rd ed., Vol. 1, pp. 563–581, D. B. Broughton, UOP Process Division, UOP, Inc; "Adsorption, Liquid Separation," *ECT* 4th ed., Vol. 1, pp. 573–600, by Stanley A. Gembicki, Anil R. Oraskav, and Jone A. Johnson, UOP LLC; "Adsorption, Liquid Separation" in *ECT* (online), posting date: December 4, 2000, by Stanley A. Gembicki, UOP LLC.

CITED PUBLICATIONS

1. *Separation and Purification: Critical Needs and Opportunities*, National Research Council Report, National Academy Press, 1987.
2. C. L. Mantell, *Adsorption*, 2nd ed., McGraw-Hill, Book Co., New York, 1951.
3. C. Berg, *Trans. Am. Inst. Chem. Eng.* **42**, 665 (1946).
4. D. B. Broughton, *Chem. Eng. Prog.* **64**, 60 (1968).

5. D. B. Broughton and A. G. Lickus, *Pet. Refiner* **40**(5), 173 (1961).
6. D. B. Carson and D. B. Broughton, *Pet. Refiner* **38**(4), 130 (1959).
7. D. M. Ruthven, *Principles of Adsorption and Adsorption Processes*, John Wiley & Sons, Inc., New York, 1984.
8. G. E. Keller 11, in T. E. Whyte and co-workers, eds., *Industrial Gas Separation* ACS Symposium Series No. 223, American Chemical Society, Washington, D.C., 1983.
9. D. B. Broughton, R. W. Neuzil, J. M. Pharis, and C. S. Brearly, *Chem. Eng. Prog.* **66**(9), 70 (1970).
10. P. C. Wankat, *Large Scale Adsorption and Chromatography*, CRC Press, Boca Raton, Fla., 1986.
11. D. B. Broughton, H. J. Bieser, and M. C. Anderson, *Pet. Int. (Milan)*, **23**(3), 91 (1976) (in English).
12. D. B. Broughton, H. J. Bieser, and R. A. Persak, *Pet. Int. (Milan)*, **23**(5), 36 (1976) (in English).
13. P. E. Barker and G. Gavelson, *Separation and Purification Methods* **17**, 1 (1988).
14. U.S. Pat. 3,761,533 (Sept. 25, 1973), S. Otani and co-workers (to Toray Industries Inc.).
15. S. Otani and co-workers, *Chem. Econ. Eng. Rev.* **3**(6), 56 (1971).
16. S. Otani, *Chem. Eng.* **80**(9), 106 (1973).
17. *Making Waves in Liquid Processing*, Illinois Water Treatment Company, IWT Adsep System, Rockford, Ill., 1984, VI (1).
18. Tetsua Hirota, *Sugar Azucar* (Jan. 1980).
19. Advertisement, *Sugar Azucar* (March 1980).
20. J. E. Rekoske and K. Zuckerman, *Spec. Chem.* **21**(1), 16 (2001).
21. U.S. Pat. 5,284,992 (Feb.8, 1994), G. Hotier, C. Roux, and T. Thanh (to Institut Francais du Petole).
22. J. E. Rekoske, *AIChE J.* **47**(1), 2 (2001).
23. J. E. Bauer, A. K. Chandhok, B. W. Scanlon, and S. A. Wilcher *A Comprehensive Look at Scaling-up SMB Chiral Separations*, Proceedings of Chiratech '97, Philadelphia, Catalyst Consultants Publishing, Inc., 1997.
24. *Chem. Eng.* 39, (Aug. 29, 1977).
25. U.S. Pat. 2,985,589 (May 23,1961), D. B. Broughton and C. G. Gerhold (to UOP Co.).
26. U.S. Pat. 3,040,777 (June 26, 1962), D. B. Carson (to UOP Co.).
27. U.S. Pat. 3,192,954 (July 6,1965), C. G. Gerhold and D. B. Broughton (to UOP Co.).
28. U.S. Pat. 4,434,051, (Feb. 28, 1984), M. W. Golem (to UOP) .
29. U.S. Pat. 5,565,104, (Oct. 15, 1996), J. W. Priegnitz (to UOP).
30. U.S. Pat. 3,268,604 (Aug. 23, 1966), D. M. Boyd (to UOP Co.).
31. U.S. Pat. 3,268,603 (Aug. 23, 1966), D. M. Boyd (to UOP Co.).
32. U.S. Pat. 3,131,232 (Apr. 28, 1964), D. B. Broughton (to UOP Co.).
33. U.S. Pat. 5,912,395, (June 15, 1999), J. L. Noe (to UOP LLC).
34. U.S. Pat. 5,470,482, (Nov. 298, 1995), R. E. Holt (to UOP LLC).
35. U.S. Pat. 5,457,260, (Oct. 10, 1995), R. E. Holt (to UOP LLC).
36. U.S. Pat. 6,284,134, (Sep. 4, 2001), G. Ferschneider, R. Huin, J. Viguie, and D. Humeau (to Institut Francais du Petrole).
37. U.S. Pat. 6,096,218, (Aug. 1, 2000), W. Hauck, and R. M. Nicoud (to Institute Francais du Petrole).
38. U.S. Pat. 5,569,808, (Oct. 29, 1996), F. Causell, G. Hotier, and P. Marteau, N. Zanier (to Institute Francais du Petrole).
39. C. Ho, C. B. Ching, and D. M. Ruthven, *Ind. Eng. Chem. Res.* **26**, 1407 (1987).
40. A. E. Herschler and T. S. Mertes, *Ind. Eng. Chem.* **47**, 193 (1955).
41. D. Haresnape, F. A. Fidler, and R. A. Lowry, *Ind. Eng. Chem.* **41**, 2691 (1949).
42. B. J. Mair and M. Shamaiengar, *Anal. Chem.* **30**, 276 (Feb. 1958).

43. B. J. Mair, A. L. Gaboriault, and F. D. Rossini, *Ind. Eng. Chem.* **39**, 1072 (1947).
44. S. Eagle and J. W. Scott, *Ind. Eng. Chem.* **42**, 1287 (1950).
45. A. E. Hirschler and S. Amon, *Ind. Eng. Chem.* **30**, 276 (Feb. 1958).
46. U.S. Pat. 3,133,126 (May 12, 1964), R. N. Fleck and C. G. Wright (to Union Oil Company).
47. Brit. Pat. 1,108,305 (Apr. 3, 1968), D. W. Peck, R. R. Gentry, and H. E. Frite (to Union Carbide Chemicals and Plastic Co.).
48. U.S. Pat. 3,843,518 (Oct. 22, 1974), E. M. Magee and F. J. Healy (to Esso Research & Engineering Company).
49. U.S. Pat. 3,686,343 (Aug. 22, 1972), R. Bearden and R. J. De Feo, Jr. (to Esso Research & Engineering Company).
50. U.S. Pat. 3,724,170 (Apr. 3, 1973), P. T. Allen, B. M. Drinkard, and E. H. Vager (to Mobil Research and Development Corp.).
51. U.S. Pat. 3,626,020 (Dec. 7, 1971), R. W. Neuzil (to UOP Co.).
52. U.S. Pat. 3,558,730 (Jan. 26, 1971), R. W. Neuzil (to UOP Co.).
53. U. S. Pat. 3,558,732 (Jan. 26, 1971), R. W. Neuzil (to UOP Co.).
54. U.S. Pat. 3,663,638 (May 16, 1972), R. W. Neuzil (to UOP Co.).
55. U.S. Pat. 3,686,342 (Aug. 22, 1972), R. W. Neuzil (to UOP Co.).
56. U.S. Pat. 3,734,974 (May 22, 1973), R. W. Neuzil (to UOP Co.).
57. U.S. Pat. 3,706,813 (Dec. 19, 1972), R. W. Neuzil (to UOP Co.).
58. U.S. Pat. 3,851,006 (Nov. 26, 1974), A. J. de Rosset and R. W. Neuzil (to UOP Co.).
59. U.S. Pat. 3,698,157 (Oct. 17, 1972), P. T. Allen and B. M. Drinkard (to Mobil Research and Development Corp.).
60. U.S. Pat. 3,917,734 (Nov. 4, 1975), A. J. de Rosset (to UOP Co.).
61. U.S. Pat. 3,665,046 (May 23, 1972), A. J. de Rosset (to UOP Co.).
62. U.S. Pat. 3,510,423 (May 5, 1973), R. W. Neuzil (to UOP Co.).
63. U.S. Pat. 3,723,561 (Mar. 27, 1973), J. W. Priegnitz (to UOP Co.).
64. U.S. Pat. 3,851,006 (Nov. 26, 1974), A. J. de Rosset (to UOP Co.).
65. F. Wolf and K. Pilchowski, *Chem. Technol.* **23**(11), (1971) (in German).
66. R. M. Moore and J. R. Katzer, *AIChE J.* **18**, 816 (1972).
67. C. N. Satterfield and C. S. Cheng, *AIChE J.* **18**, 710 (1972).
68. F. Wolf, K. Pilchowski, K. H. Mohrmann, and E. Hause, *Chem. Technol.* **27**(12), (1975) (in German).
69. S. K. Suri and V. Ramkrishna, *Trans. Faraday Soc.* **65**(6), 1960 (1969).
70. J. F. Walter and E. B. Stuart, *AIChE J.* **10**, 889 (1964).
71. U.S. Pat. 3,929,669 (Dec. 30, 1975), D. H. Rosback and R. W. Neuzil (to UOP Co.).
72. S. Sirear and A. L. Meyers, *AIChE J.* **17**, 186 (1971).
73. J. A. Johnson and co-workers. *Olex: A Process for producing High Purity Olefins*, presented at the AICHE Summer National Meeting, Minneapolis, Minn., August 1987.
74. R. W. Neuzil and R. H. Jensen, 85th National Meeting of the AICHE, Philadelphia, Pa, June 1978.
75. Discussion on Batch and SMB Chromatography, AICHE Chicago Section, March 14, 2001, Pharmacia.
76. R. T. Yang, *Gas Separation by Adsorption Processes*, Butterworth, London, 1986.
77. E. Heftmann, ed., *Chromatography*, Van Nostrand-Reinhold, New York, 1975.
78. J. J. Kipling, *Adsorption from Solutions of Non-Electrolytes*, Academic Press, Inc., New York, 1965.
79. F. C. Nachod and J. Schubert, *Ion-Exchange Technology*, Academic Press, Inc., New York, 1956.
80. R. M. Barrer, *Zeolites and Clay Minerals as Sorbents and Molecular Sieves*, Academic Press, Inc., London, 1978.

81. J. A. Johnson and A. R. Oroskar, "Sorbex Technology for Industrial Scale Separation," in H. G. Karge and J. Weitkamp, eds., *Zeolites as Catalysts, Sorbents, and Detergent Builders*, Elsevier Science Publishers BV, Amsterdam, The Netherlands, 1989.
82. UOP Communication.
83. U.S. Pat. 3,813,452, (May 28, 1974), H. J. Bieser, (to UOP).
84. U.S. Pat. 5,329,061, (July 12, 1994), J. D. Swift, (to UOP LLC).
85. U.S. Pat. 5,329,060, (July 12, 1994), J. D. Swift, (to UOP LLC).
86. U.S. Pat. 5,401,476, (May 28, 1995), G. Hotier, C. R. Guerraz, and T. N. Thanh, (to Institut Francais du Petrole).
87. U.S. Pat. 5,284,992, (Feb. 28, 1994), G. Hotier, C. R. Guerraz, and T. N. Thanh, (to Institut Francais du Petrole).
88. UOP Communication.
89. Kirkpatrick award.
90. U.S. Pat. 6,222,088, (Apr. 24, 2001), S. Kulprathipanja, (to UOP LLC).
91. U.S. Pat. 5,276,246, (Jan. 4, 1994), B. McCulloch, and J. R. Lansbarkis, (to UOP LLC).
92. U.S. Pat. 5,223,589, (June 29, 1993), S. Kulprathipanja, (to UOP LLC).
93. U.S. Pat. 5,220,102, (June 15, 1993), G. A. Funk, J. R. Lansbarkis, A. R. Oroskar, and B. Mcculloch, (to UOP LLC).
94. U.S. Pat. 5,177,300, (Jan. 5, 1993), S. Kulprathipanja, K. K. Kuhlne, M. S. Patton, and R. L. Fergin (to UOP LLC).
95. U.S. Pat. 5,177,295, (Jan. 5, 1993), A. R. Oroskar, R. E. Prada, J. A. Johnson, G. C. Anderson, and H. A. Zinnen, (to UOP LLC).
96. U.S. Pat. 5,159,131, (Oct. 27, 1992), H. A. Zinnen, (to UOP LLC).
97. U.S. Pat. 5,149,887, (Sept. 22, 1992), H. A. Zinnen, (to UOP LLC).
98. U.S. Pat. 5,143,586, (Sept. 1, 1992), B. McCulloch, (to UOP LLC).
99. U.S. Pat. 5,071,560, (Dec. 10, 1991), B. McCulloch and W. H. Goodman, (to UOP LLC).
100. U.S. Pat. 5,019,271, (May 28, 1991), H. A. Zinnen (to UOP LLC).
101. U.S. Pat. 5,012,039, (Apr. 30, 1991), T. J. Barder, (to UOP LLC).
102. U.S. Pat. 5,004,853, (Apr. 2, 1991), P. T. Barger, T. J. Barder, D. Y. Lin, and S. H. Hobbs, (to UOP LLC).
103. U.S. Pat. 4,992,621, (Feb. 26, 1991), B. McCulloch, and M. G. Gatter, (to UOP LLC).
104. U.S. Pat. 4,992,618, (Feb. 12, 1991), S. Kulprathipanja.
105. U.S. Pat. 4,977,243, (Dec. 11, 1990), T. J. Barder, B. W. Bedwell, and S. P. Johnson, (to UOP LLC).
106. U.S. Pat. 4,876,390, (Oct. 24, 1989), B. McCulloch, (to UOP LLC).
107. U.S. Pat. 4,797,233, (Jan. 10, 1989, H. A. Zinnen, (to UOP LLC).
108. U.S. Pat. 4,784,807, (Nov. 5, 1988), H. A. Zinnen, (to UOP LLC).
109. U.S. Pat. 4,770,819, (Sept. 13, 1988), H. A. Zinnen, (to UOP LLC).
110. M. Wheaton and W. C. Bauman, *I&EC* Eng. Proc. Dev. **45**, 228 (1953).
111. H. Hongisto and H. Heikkila, *Sugar Azucar* **56**, 60 (Mar. 1978).
112. M. S. Tswett, *Tr. Protok. Varshav. Obshch. Eststvoistpyt., Otd. Biol.* **14**, 20 (1905).
113. U.S. Pat. 4,837,315, (1989), S. Kulprathipanja, (to UOP).
114. PCT Int. Appl. WO 97/23511, M. B. Van Leeuwen, T. M. Slaghek, D. De Wit, H. C. Kuzee, and H. W. C. Raaymakers, 17 pp. (1997).
115. J. R. Kerns and R. J. Linhardt, *J. Chromatogr., A* **705**(2), 369 (1995).
116. B. W. Pynnonen, *Processes Sugar Ind., Proc. S.P.R.I. Workshop*, (1996) p. 120.
117. U.S. Pat. 6,200,390, (2001) M. M. Kearney, K. R. Peterson, and M. W. Mumm (to Amalgamated Research, Inc.).
118. Eur. Pat. Appl. EP 1,106,602, (2001) T. P. Binder, (to Archer Daniels Midland).

119. D. L. Husband, C. T. Mant, and R. S. Hodges, *J. Chromatogr. A* **893**, 81 (2000).
120. M. P. Nowlan and K. M. Gooding, in C. T. Mant and R. S. Hodges, eds., *High-Performance Liquid Chromatography of Peptides and Proteins* CRC Press, Boca Raton, Fla, 1991.
121. M. Gilar, *Anal. Biochem.* **298**, 196 (2001).
122. W. Schwartz, D. Judd, M. Wysocki, P. Santambien, L. Guerrier, and B. Oschetti Egisto, *Abstr. Pap. - Am. Chem. Soc.*, 221st BIOT-088, 2001.
123. R. L. Fahner, D. H. Whitney, M. Vanderlaan, and G. S. Blank, *Biotechnol. Appl. Biochem.* **30**, 128 (1999).
124. M. Agosot, N.-H. L. Wang, and P. C. Wankat, *Ind. Eng. Chem. Res.* **28**, 1358 (1989).
125. E. Kusters, G. Gerber and F. D. Antia, *Chromataographia* **40**, 387 (1995).
126. E. R. Francotte and P. Richert, *J. Chromatogr. A* **769**, 101 (1997).
127. E. R. Francotte, P. Richert, M. Mazzotti, and M. Morbidelli, *J. Chromatogr. A* **796**, 239 (1998).
128. L. R. Snyder and J. J. Kirkland, *Introduction to Modern Liquid Chromatography*, 2nd ed., John Wiley & Sons Inc., New York, 1979.
129. C. A. Hoeger, R. Galyean, R. A. McClintock, and J. E. Rivier, in C. T. Mant and R. S. Hodges, eds., *High-Performance Liquid Chromatography of Peptides and Proteins* CRC Press, Boca Raton, Fla., 1991.
130. M. I. Aguilar and M. T. W. Hearns, in M. T. W. Hearns, ed., *HPLC of Proteins, Peptides and Polynucleotides* VCH Publishers, New York 1991. p. 247.
131. Y. S. Sanghvi, M. Andrade, R. R. Deshmukh, L. Holmberg, A. N. Scozzari, and D. L. Cole, in G. Hartman, and S. Endres, eds., *Manual of Antisense Methodology* Kluwer Academic Publishers, New York, 1999. pp. 3–23.
132. R. L. Cunico, K. M. Gooding, and T. Wehr, *Basic HPLC and CE of Biomolecules*, Bay Bioanalytical Laboratory, Richmond, Calif., 1998.
133. F. E. Regnier and R. M. Chicz, in K. M. Gooding and F. E. Regnier, eds., *HPLC of Biological Macromolecules: Methods and Applications*, Marcel Dekker, Inc., New York, 1990.
134. G. Vanacek and F. E. Regnier, *Anal. Biochem* **109**, 345 (1980).
135. J.-L. Liao, R. Zhang, and S. Hjerten, *J. Chromatogr.* **586**, 21 (1991).
136. H. Minakuchi, K. Nakanishi, N. Soga, N. Ishizuka, and N. Tanaka, *Anal. Chem.* **68**, 3498 (1996).
137. D. M. Ruthven and C. B. Ching, *Chem. Eng. Sci.* **44**, 1011 (1989).
138. J. A. Breninger, R. D. Whitley, X. Zhang, and N.-H. L. Wang, *Comp. Chem. Eng.* **15**, 749 (1991).
139. G. Storti, M. Mazzotti, M. Morbidelli, and S. Carra, *AIChE J.* **39**, 471 (1993).
140. M. Mazzotti, G. Storti, and M. Morbidelli, *J. Chromatogr. A* **769**, 3 (1997).
141. C. Migliorini, M. Mazzotti, G. Zenoni and M. Morbidelli, *AIChE J.* **48**, 69 (2002).
142. A. S. T. Chiang, *AIChE J.* **44**, 2431 (1998).
143. H. Schmidt-Traub, J. Strube, H. I. Paul, and S. Michel, *Chem.-Ing.-Tech.* **67**, 323 (1995).
144. J. Strube and H. Schmidt-Traub, *Comput. Chem. Eng.* **22**, 1309 (1998).
145. M. Mazzotti, G. Storti, and M. Morbidelli, *AIChE J.* **40**, 1825 (1994).
146. A. Gentilini, C. Migliorini, M. Mazzotti, and M. Morbidelli, *J. Chromatogr. A* **805**, 37 (1998).
147. C. Migliorini, M. Mazzotti, and M. Morbidelli, *J. Chromatogr. A* **827**, 161 (1998).
148. G. Biressi, O. Lundemann-Hombourger, M. Mazzotti, R.-M. Nicoud, and M. Morbidelli, *J. Chromatogr. A* **876**, 3 (2000).
149. A. S. T. Chiang, *AIChE J.* **44**, 333 (1998).

150. R. R. Deshmukh, W. E. Leitch II, Y. S. Sanghvi, and D. L. Cole, in S. Ahuja, ed., *Separation Science and Technology*, Vol. 2 (Handbook of Bioseparations, Academic Press, San Diego, Calif., 2000, p. 511.
151. R. R. Deshmukh and Y. S. Sanghvi, "Recent Trends in Large-Sclae Purification of Antisense Oligonucleotides," from *IBC Conference: Large Scale Oligonucleotide Synthesis*, San Diego, Calif. (1997).
152. D. L. Cole, "GMP Manufacturing of Anitsense Oligonucleotides at Isis Pharmaceuticals for Clinical Trials and Marketplace: Yesterday, Today and Tomorrow," *IBC Conference: Oligonucleotide Pept. Manuf. Strategies*, San Diego, Calif. (1999).
153. R. S. Porter and J. F. Johnson, *Nature* (London) **183**, 391 (1959).
154. M. Bailly and D. Tondeur, *Chem. Eng. Sci.* **37**, 1199 (1982).
155. J. Dingenen and J. N. Kinkel, *J. Chromatogr., A* **666**, 627 (1994).
156. C. M. Grill, *J. Chromatogr., A* **796**, 101 (1998).
157. C. M. Grill and L. Miller *J. Chromatogr., A* **827**, 369 (1998).
158. C. M. Grill and L. Miller, Presentation at *SPICA 2000: Symposium on Preparative and Industrial Chromatography and Allied Techniques*, Zurich, Oct. 2000.
159. V. G. Mata and A. E. Rodrigues, *J. Chromatogr. A* **939**, 23 (2001).
160. M. Negawa and F. Shoji, *J. Chromatogr.* **590**, 113 (1992).
161. U.S. Pat. 5,126,055 (1992) A. Yamashita and F. Shoji, (to Daicel Chemical Industries, Ltd.).
162. M. McCoy, *Chem. Eng. News*, June 19 (2000).
163. D.-J. Wu , Z. Ma, and N.-H. L. Wang, *J. Chromatogr. A* **855**, 71 (1999).
164. S. Giacobello, G. Storti, and G. Tola, *J. Chromatogr. A* **872**, 23 (2000).
165. T. B. Jensen, T. G. P. Reijns, H. A. H. Billiet, and L. A. M. van der Wielen, *J. Chromatogr. A* **873**, 149 (2000).
166. J. Houwing, H. A. H. Billiet, and L.A.M. van der Wielen, *J. Chromatogr. A* **944**, 189 (2002).
167. S. Abel, M. Mazzotti, and M. Morbidelli, *J. Chromatogr. A* **944**, 23 (2002).
168. D. Antos and A. Seidel-Morgenstern, *J. Chromatogr. A* **944**, 77 (2002).
169. O. Lundemann-Hombourger, G. Pigorini, R.-M. Nicoud, D. S. Ross, and G. Terfloth, *J. Chromatogr, A* **947**, 59 (2002).
170. A. Jupke, A. Epping and H. Schmidt-Traub, *J. Chromatogr. A* **944**, 93 (2002).

STANLEY A. GEMBICKI
UOP LLC

ADVANCED MATERIALS, ECONOMIC EVALUATION

1. Introduction

The emergence of the advanced materials industry, beginning in the late 1970s, represents one of the most important and dynamic chapters in U.S. and international technological development. These materials possess new and different types of internal structures and exhibit a variety of novel physical and chemical properties that have a wide range of industrial and commercial applications.

The advanced materials industry encompasses such product areas as: biochemicals (including genetic-based materials); bioengineered materials; catalysts; ceramics and clays; coatings; composites; crystal materials; fuels; fullerenes; metal alloys; nanometarials (eg, nanotubes, nanopowders, nanospheres, nanofibers); optical and photonic materials; polymers (eg, plastics, rubber, fibers) and polymer matrices; powdered metals; sensor materials; superconducting materials; and thin films (1,2).

While advanced materials are highly diverse with respect to structure, physical and chemical characteristics, and applications, they form a coherent industry due to a number of criteria, including common processes and technical and economic interrelationships. The importance of the industry resides in the fact that its materials diffuse into and impact virtually all of the major industrial sectors, including aerospace, automotive, biomedical, construction, consumer electronics, defense, energy, food processing, healthcare, materials processing, mining, packaging, petrochemical, security, telecommunications and utilities.

Because the advanced materials industry impacts the economic well being of a growing range of businesses and industries within the United States and internationally, it will ultimately rival the traditional chemical, pharmaceutical, and metals industries in terms of revenue, diversity, and economic importance. The advanced materials field is fundamentally altering how, and even what, the world's leading industries produce, and how they structure themselves and perform their basic operations. Indeed, very few major new technologies can emerge without the application of advanced materials. Accordingly, an examination of the advanced materials industry is also a study of technological change in general in the modern age.

2. The Historical Context

Today's advanced material technology follows and expands upon what is generally considered the first material revolution that emerged during and following World War II. Beginning in the mid-nineteenth century, the German chemical industry dominated advanced chemical materials. German firms, including Bayer, Badische, and Hoechst, revolutionized industrial chemistry by synthesizing from coal tar materials a broad range of commercial organics (including dyes), pharmaceuticals, and plastics. In the years leading up to World War II, the United States began to compete with Germany in the commercialization of new materials. As the largest oil-producing country in the world, the United States substituted petroleum for coal as the dominant feedstock for its advanced material products. Some of these petroleum-based materials, notably synthetic rubber, toluene (for making explosives), and advanced fuels, served critical strategic purposes during World War II (3–5).

The postwar period, building upon these earlier achievements, ushered in the age of synthetic "materials by design." The large, established chemical companies and petroleum refiners—Dupont, Dow Chemical, Exxon (Jersey Standard), Sun Oil and, increasingly, the specialized process engineering firms—created and controlled this revolution because they had the means, engineering talent, and the mass production facilities to do so. The new technology at this time included the mass manufacture of the new synthetic rubbers, fibers, resins, and plastics, such as nylon, polyethylene, and the like. These novel materials percolated rapidly throughout the U.S. economy, transforming a wide range of industrial operations, including transportation, metals production, packaging, construction, electronics, biomedical, etc (6–8).

By the 1970s and early 1980s, shrinking returns from innovation due to overproduction, the global energy crisis, and declining technical possibilities from petroleum, put a brake on new material development. The large, integrated petrochemical producers, now shying away from new product development, focused on improving and modifying existing processes and products as their major competitive strategy. This retreat by the traditional chemical industry from more revolutionary technology provided an opportunity for newcomers to explore the commercial potential of a new generation of innovative materials. This encroachment by "upstart" R&D firms led the way to the new materials revolution of the 1980s and 1990s (6,9).

3. The New Materials Revolution: 1980s—Present

The current generation of new materials, in part, has evolved from earlier advanced material technology. For example, new types of "high performance" polymers derived from, and are an improvement over, such first generation materials as polyethylene (and polypropylene), nylon, etc. This is true as well for new types of catalysts, metal alloys, pharmaceuticals, and biochemicals.

But important technical and economic differences exist between earlier, post World War II technology and the newer generation of materials. Rather than relying on a single feedstock (ie, petroleum), the new and emerging processes consume a variety of raw materials: ceramics, clays, crystals, minerals, coal, petroleum, natural gas, graphite, and agricultural by-products. In contrast to earlier production technology, advanced materials technology does not depend on large, integrated plants and are not inexorably linked to mass production methods. This is so because they often become critical additives in a wide range of composite materials and therefore do not have to be made in the massive quantities typical of new petrochemical products in the past. For example, a small quantity of carbon nanotubes can be added to a relatively large volume of an already existing polymer material to make commercial quantities of a totally new type of composite material. Similarly, a relatively small amount of a particular element or compound, integrated into a mass of steel or aluminum, produces large quantities of advanced alloys (1,9).

These trends mean that the current generation of advanced materials depends to a greater extent than previously on creative laboratory investigation, venture capital funding, and extensive licensing arrangements, rather than on in-house, large-scale engineering capability. Accordingly, the established petrochemical firms, once the dominant player in creating new materials, increasingly find themselves competing with the university and government laboratories, and the start-up operations and spin-off firms that license promising inventions for advanced material markets.

4. Process Flow and Technology

While the advanced materials industry creates a wide range of products, common patterns emerge that characterize the process flow of the industry. First, the manufacturing processes that transform raw materials into advanced materials are generally chemical in nature. Typically, the processes employ thermal energy through various means—eg, thermal furnaces, laser technology, and high temperature reactors—To effect the requisite chemical reactions. Then too, an advanced material, once produced, generally requires subsequent processes to modify and prepare the material for market. Nanotubes, eg, need to be made into a composite materials, coatings have to be applied to the surfaces of components, and so forth in order for the advanced material to be useful in various applications. Depending on the industry involved, additional processing steps may include assembling components or parts containing an advanced material into a final product to be distributed and sold to an original equipment manufacturer (OEM) (1,6).

Currently, separate firms perform these processes for the advanced material companies. Increasingly, the latter resides near to the assorted downstream plants that serve it, thus forming tightly knit manufacturing complexes. This clustering of production provides essential economies in developing and manufacturing advanced materials through the regular movement of raw materials, energy, labor, knowledge and technology between companies. Over time, as they grow and diversify and integrate operations, the advanced material companies themselves undertake some or all of these operations in-house (9,10).

5. Industry Structure: Competition, Diversity, and Geography

The advanced materials industry exhibits a high degree of competitiveness. While one company may dominate a new material for a short time, companies offering competing technology—whether a new product, process, or both—emerge rapidly thereafter. Also, product lifecycles tend to be short as new companies and products regularly enter the market. New materials must also compete against older products that serve the same markets. The inevitable push of new technology is not always the case since markets, generally conservative in nature, do not readily accept novel materials over more familiar and tested products. Consumers of the older materials understand the advantages and limits of the established products and have adjusted their processes accordingly. They also are aware that adopting new materials can be costly, such as in retooling of plants and retraining of personnel. Further, the price of advanced materials cannot initially compete against the offerings of established producers, who attempt to maintain their markets by reducing unit prices. Such dynamic competition, coupled with the fact that production processes for advanced materials achieve greater economies over time, means that unit prices for these products tend to decline steadily over time (8,9).

Within the industry, there are "captive" producers and "open" producers. The former produces advanced materials—eg, nanotubes—for internal use. They may also purchase advanced materials on the open market. Captive producers include universities and government and industrial laboratories. They may also include the larger, multinational corporations, such as Honeywell. On the other hand, "open" companies produce and sell their materials on the open market. Purchasers of these materials may want to conduct experiments on them or incorporate them into proprietary production processes. Suppliers of advanced materials to the market may be small start up firms, large corporations, or academic and government laboratories.

The advanced materials industry encompasses a wide variety of companies in terms of size, diversity, and the degree of integration. These companies range from the large, integrated chemical, petrochemical, and process engineering firms who employ tens of thousands of people worldwide and post annual sales in the billion of dollars, to the small start up and spin-off companies who rarely employ >100 persons and who may record annual sales revenues <1 million dollars. The former type of company generally supports its research and development activity with internally generated funds and obtains patents on the technology it creates in-house. It may, in time, license its technology to outside companies in exchange for royalties. The large corporation offers a wide range of products, both established, well-know materials as well as new types of materials that often are modifications of existing commercial products. For its part, the small start up and spin off firm usually takes out licenses on promising material processes from university and government laboratories. In time, they then sublicense to other firms in order to expand manufacturing capacity or exploit locational advantages, ie, nearness to raw materials or markets. The start up firm typically obtains funds for development and commercialization activities from outside sources—venture capitalists, government grants, etc (9,11–13).

While certain of the large corporations, such as Dow, General Electric, Xerox, 3M, Motorola, and IBM, continue to develop new material technologies, the established corporation typically has turned away from developing the most radical technologies. This is so for a variety of reasons including increasing development costs; decreasing returns on investment in R&D (due in part to the need to utilize an extensive R&D department and the high costs of retooling and restructuring existing, large-scale plants); the growing likelihood that new technology must compete against the innovating company's own existing products; and the difficulties and costs involved in establishing new supply lines and distribution networks (and the possible alienation of existing suppliers and customers). Moreover, the growing professionalization of corporate management blunts the desire and capability of companies to undertake new technological development. This is, in part, due to the rise of intellectual and cultural boundaries between managers that reflect the greater degree of specialization that characterizes corporate management professionals. Such specialists, embracing their own particular professional goals, problems, and even language, hinder the close cooperation and free flow of information between different departments, which is often required in successfully undertaking major technological projects (5–8,14).

In contrast, the small, start-up and "spin-off" firms enjoy a greater degree of flexibility in pursuing new technology. They are not burdened with an existing system of plants nor do they have to support an extensive supply, distribution, and R&D infrastructure. They do not have established products that could compete with new material technologies nor do they need to cosset an established network of suppliers, distributors, or customers. Indeed, commercializing radically new materials is absolutely critical to their existence since doing so distinguishes them from their competitors, both large corporations and other smaller newcomers, and is the key to capturing new markets. In addition, the organizational structure in these firms is more loosely organized. As a result, the lines separating specialties and departments tend to be highly porous and even blurred. Often, individual executive managers undertake a wide range of functions, including engineering, procurement, and marketing. This informal organizational structure facilitates a free flow of information, know how, and insight throughout the company and permits decisions involving new technology development to be made quickly and efficiently (6–8,11,14).

The geographical distribution of advanced materials companies, whether in the United States or internationally, depends on a number of economic, political, intellectual, and demographic factors. These factors include the local availability of raw materials, potential markets, knowledge-based institutions (ie, universities, industrial and government laboratories, incubator facilities), government support and incentives (eg, tax breaks, R&D subsidies, technology development programs), and funding sources (eg, venture capital) (4,10–13).

The advanced materials industry is active in different parts of the United States and internationally. Within the United States, particularly dynamic areas include the Southeast, Southwest, and West Coast. Internationally, western Europe, and particularly Germany and the United Kingdom, conduct extensive research and development work. Now that much of eastern Europe is part of the European Union and can benefit from information, technology, and markets of the western European countries, this region is likely to expand its role in the field over the next few years. Within the Pacific area, Japan is a major participant in advanced materials, including advanced ceramics, since the 1980s, and Australia continues to increase its presence in the field. China, India, and certain Middle Eastern countries (eg, Israel), continue to develop as centers of advanced materials, a trend that will continue over the next 10–20 years. United States advanced materials companies continue to establish links with foreign firms through partnership arrangements, especially with European and Asian countries. These arrangements serve U.S. producers through information and technology transfer and ready access to foreign markets (9).

6. The Commercialization Process

Commercialization is the process by which a laboratory invention enters into the market arena. Commercialization involves a number of closely interrelated tasks, including product and process development, and marketing and distribution strategy. Patterns of commercialization for advanced materials differ, depending on whether the company is a large, integrated corporation or the small start-up or spin-off firm.

The large, integrated corporation generally controls all (or at least most) phases of commercialization. The corporation's research department investigates the technical and economic viability of new materials. The patent department, staffed by both technical and legal personnel, arranges the application of the patent with the U.S. Patent Office and determines the terms and conditions for licensing. The engineering department perfects the production process, which usually requires the development of continuous (semicontinuous) mass production systems, and often sees to the coordination between supply and production schedules. The marketing department then secures customers for the new material. These may be independent processors who further shape the material into some form (eg, molders), original equipment manufacturers (OEMs) that incorporate the material as a component in an assembled product (eg, automobile, airplane, computer, appliance) or, in certain cases, the final consumer who purchases the material directly from the manufacturers (or from distributors). The marketing department often works closely with—or is part of—technical services. The latter, a hybrid engineering and marketing service, instructs customers as to the proper use and possible application of new materials and the components derived from them. Technical services personnel, many of whom have engineering backgrounds, also examine possible ways in which materials require modification in order to attract potential customer sources. In doing so, technical services works together with the company's research, engineering, and marketing departments (6,9).

For the large, integrated corporation, commercialization proceeds in-house using the resources of the entire organization. With increasing frequency, however, commercialization of advanced materials takes place through the close interaction and cooperation of the university and government laboratory and the small start-up firm. The university and government technology transfer offices play an instrumental role in steering promising new materials from their organizations to the commercial sector. As part of their responsibilities, these offices work with in-house university and government researchers to patent inventions in a timely manner and to help license technologies to outside businesses. State (and federal) governments also provide assistance to the commercialization process. Federal R&D programs, such as the Small Business Research Initiative (SBIR), provide grants to firms commercializing promising new technology. The Departments of Energy (DOE), Commerce (DOC), and Defense (DOD) operate various programs to aid the development of advanced material products and processes. Also, certain states facilitate the transfer of advanced materials technology from universities and government laboratories to selected firms through such organizations as centers for innovation and technology, regional technology councils, and incubator facilities. Within Virginia, examples of such organizations include Virginia's Center for Innovative Technology (VACIT) and the Hampton Roads Technology Council. These agencies offer a number of functions, including the funding (eg, through grants) of research and development, providing research facilities, and identifying and bringing together potential strategic partners in joint technology development arrangements (9–13).

Once the start-up firm secures license rights to a patent, it continues to develop the technology through commercialization and, eventually, market development. The innovating university or government laboratory typically supplies technical assistance through much of the process. Often the original inventors retain some role in the new firm, possibly as technical consultant or lead

R&D person. To continue financing the project prior to the creation of significant sales revenues, the start up firm usually depends on government money in the form of federal and state grants and contracts. This money supports a small core staff of managers, engineers, and production crew. At some point, the firm approaches outside investors, usually in the form of venture capitalists. Investors participate in the commercialization process at a distance (ie, through lending funds only) or more intimately by taking part in the operations of the firm. In the latter case, they actively guide the progress of the enterprise by participating as members of the Board of Directors or Executive Committee (11–13,15).

Typically, the start-up firm does not generate revenue until it actually starts selling the advanced material on the open market. Reaching this stage typically requires 3–5 years from the time of first obtaining licensing rights since time is needed to undertake further research and development on the technology to improve the process, achieve commercial-grade material and lower production costs (9,14).

As the new material diffuses through markets, the firm expands production to meet the growing demand. With increased production, economies of scale come into play through the adoption of fully (or semi) continuous process technology and better coordination of the company's supply, processing, distribution, and marketing functions. These economies create a downward pressure on prices and, in turn, the further expansion in market demand. From this point, demand for the new material increasingly depends on price, as well as quality, as the process technology strives to dislodge the existing materials from their market positions. As production expands, the firm broadens the management team by hiring additional and diverse personnel to handle the increased volume and coordinate material flow, process technology, and market development. It is at this point that the material achieves economic prominence (9,14,16,17).

7. Evolutionary Patterns and Market Strategies

The evolution of an advanced material generally adheres to the classical product life-cycle model. In the first (or "creation") phase of an advanced material's life-cycle, the new material is not yet in its final commercial form. It undergoes frequent and significant alterations in its physical and chemical properties. Because the material is not yet standardized, the production process tends to be inefficient and generally makes the material in batches rather than continuously. As a consequence, unit prices remain high. Markets for the new material tend to focus on specialized, niche applications, such as for defense needs. Market strategies in this phase of the cycle hinge on the superior characteristics of the product rather than on competitive pricing. By necessity, inputs are limited to generally available materials and energy sources, facilities are often small scale and located near user or source of technology, and labor costs tend to be a large percentage of the total costs of production (9,14).

In the second (or "formative") phase of a new material, technical considerations and preferences of the marketplace determine which variation of the new material becomes the one, dominant design. This stabilization of design means that the production process can become fully continuous thus permitting the rapid scale up to high volume production. As the mass production process evolves, "islands of automation" appear in the production line. The production process then requires increasing volumes of specialized raw material and capital inputs. The production process becomes more rigid in that it is specifically designed for the large-scale production of the single type of material. In this second phase of the material's life cycle, the percentage of total costs consumed by labor decreases since the manufacturing facility generates an increasing output of materials per unit of man hours. As volume of output expands, other factors of production play a more important part in the cost equation. These include raw materials, capital expenditures (ie, such as on equipment and buildings), and energy usage (9,14).

During this period of initial production expansion, the advanced material company expands its present capacity, both by adding onto a current plant as well as building new facilities. Employees of the company begin to grow in number but not in proportion to output. Economies of scale and greater efficiencies come into play thus limiting the number of new production workers required. The firm continues to rely on the initial group of engineers and technical personnel since they possess a clear understanding of the nature of the material and its production technology. However, the company begins to bring in new engineering, marketing, and managerial talent in order to meet the increasing scale and diversity of operations (9,14).

In the third (or "ascendant") phase of the product life cycle, the market for the advanced material undergoes rapid growth as costs per unit of product decline. The company then undertakes aggressive marketing strategies. It looks to sublicense to other, secondary enterprises. These secondary companies may specialize in certain product categories (eg, nanospheres of certain diameter range, particular types of fullerenes, etc) or on particular markets and applications. The initial start-up firms then begin to earn income from both its own sales and from royalties collected from its secondary companies (ie, its licensees). Competition now focuses as much on economic—ie, low unit price—as well as product quality. As this third phase proceeds, patent protection and leaning curve advantages provide the advanced material firms and their licensees cover from competition. Nevertheless, prices continue to fall as production processes improve in order to maximize market growth (9,14).

The material eventually enters its fourth, or "developed", phase. At this time, competition from other firms increases as both patent protection and the learning curve advantage come to an end. Market strategies now strive toward incremental product and process changes and cumulative improvements as the original producer and his licensees struggle to retain market share. Firms now compete on the basis of product differentiation, marginal price reductions, and creative marketing techniques. They attempt to fully integrate supply, production, and marketing functions to gain the maximum economies as possible from their operations (9,14).

Finally, the material reaches its fifth, or "mature", phase. At this point, the material has become and established, "impacted" technology. Now, the process of "creative destruction" proceeds swiftly as the next generation of advanced materials comes to the fore and edges these older and once-dominant technologies out of critical markets (9,14).

At any one time within the industry, there exists a pool of advanced materials at differing phases of their product life cycles. Some materials, such as superconducting materials and molecular computers, remain at the research or laboratory phase and have not yet gone past the "critical divide" to commercialization. In contrast, other types of materials have just crossed over that barrier and have reached that point of initial production. These include such materials as piezoelectric ceramics and certain types of nanomaterials. Other materials—including thin films, ceramic composites, bioengineered synthetics, and sensors—are further along in the cycle and have begun to scale up for full mass production. More developed materials—eg, advanced structural alloys, ceramics, and polymer–metal composites—already in mass production mode, now pursue expanding markets and, thus can be considered fully formed commercial products. Finally, there are those materials, such as high performance polymers, in full commercial production for a number of years and well known in their markets. Now in their mature years, these materials have undergone a series of modifications in order to retain market share in the face of increasing competition from newer material technologies (9,14).

8. The Market: United States and International

As an advanced material proceeds along its product life cycle, its market grows in volume and diversity. Market volume expands as unit price contracts—due to the increased adoption of continuous, mass production methods—and the technical capabilities of the product increases. Moreover, market diversity broadens as researchers learn how to modify the material in order to take advantage of different commercial applications. Over the next few years, certain advanced materials, such as nanospheres and nanotubes, will find application across a wide range of industries including aerospace, appliances, biomedical, defense, electronics and telecommunications, automotive, construction, electrical products and systems, petrochemicals, portable energy systems (ie, batteries and fuel cells), security systems, power generation, electro-optical systems, and textiles (1,2,9,15–19).

This process of market entrance and diffusion ultimately determines the projected rate of growth of a new material. Overly optimistic forecasts must be adjusted to take into account the various difficulties that often arise when attempting to introduce a new material into the market place. While comprehensive data on the size of the U.S. advanced material industry are not currently

Table 1. **U.S. Market Trends for Advanced Materials, $\times 10^9$ \$[a,b]**

	2002	2005	2012
direct economic impact	10.0	12.75	21.5
total economic impact[c]	41.5	46.2	77.2

[a] Refs. 1,2,9,15–19.
[b] Includes all advanced materials; excludes sales of final products incorporating advanced materials.
[c] Includes direct + indirect + induced impact.

Table 2. **World Market Trends for Advanced Materials, $\times 10^9$ $[a,b]**

	2002	2005	2012
direct economic impact	38.50	45.36	65.6
total economic impact[c]	72.10	98.45	262.40

[a] Refs. 1–3,3,9,15–19.
[b] Includes all advanced materials; excludes sales of final products incorporating advanced materials.
[c] Includes direct + indirect + induced impact.

available, an in-depth analysis of the industry indicates that all sales of advanced materials in the United States can reach $21.5 billion annually by 2012. Further, the full economic impact of the industry—taking into account the direct, indirect and induced effects of new materials production—is likely to be between $77 and $78 billion by 2012 (1,2,9,15–19) (Table 1).

On a global scale, World markets are expected to closely shadow U.S. sales in terms of growth. By 2012, total sales of advanced materials will be ~$66 billion, representing a full economic impact worldwide—ie, the sum of direct, indirect, and induced impacts—of ~ $262 billion (Table 2).

9. The Emerging Advanced Materials

The following sections examine specific advanced material areas. Industry specialists consider these products as the most important advanced materials that have recently entered the market, are on the verge of commercialization, or may be commercialized over the next few years. As a group, these materials account for the bulk of current and projected advanced materials sales within the United States and worldwide for the 2002–2012 period.

9.1. Bioengineered Materials (9,20–24). Biochemicals play an increasingly critical role in the advanced materials industry. An important biochemical technology that is just beginning commercialization is the so-called bioengineered materials. These materials bridge the biochemical and synthetic organic fields and promise to provide large volumes of synthetic materials over the next few years.

Bioengineering technology involves the biochemical transformation —in so-called "biorefineries"—of agricultural feedstock, by-products, and wastes into useful synthetic materials. Biorefineries are directly comparable to the petrochemical plant in that both petroleum and biomass refineries use a particular raw material to produce a wide range of synthetic products varying greatly in their chemical properties and physical characteristics and serving a diversity of markets. These products include synthetic plastics and packaging, clothing, fuel additives, chemicals (eg, alcohols, polymers, ethylene, phenolics, acetic acid, etc), biologics, food products, adhesives and sealants, and a variety of commodity and industrial products (22).

An important advantage of bioengineered materials is that they conserve on energy and provide additional markets for the products and wastes of the farming industry. Also, processing costs are low because the technology uses known methods and does not require complex and expensive bioreactors and associated facilities for upstream production. Scale-up of biorefineries appears to be rapid and relatively inexpensive. From the environmental viewpoint, these materials degrade more readily than traditional synthetics. They also generate fewer greenhouse gases and require less energy, water, and raw materials to produce compared to petroleum-based materials (20–22).

In general, the production of bioengineered materials occurs near to agricultural areas. Recent advances in bioengineering technology involve cellulosic-based feedstocks, including vegetable crops, starch-producing crops, oil seeds, wood and other lignocellulosic biomass. Current research in the field focuses on three major production technologies: fermentation (including enzyme processing), pyrolysis, and low temperature technology. Of these possible processes, fermentation, in conjunction with such operations as distillation and polymerization, appears to offer the most promising commercial method (20–22).

Bioengineering research is international in scope. Besides the United States and Canada, such European countries as Austria, Denmark, Germany, Iceland, and Switzerland pursue particularly active programs in the field. In addition international companies are emerging that specialize in the construction and operation of advanced biorefineries.

Within the United States, a variety of agencies and organizations actively pursue bioengineering research and development. For example, the Department of Agriculture, through the Economic Research Service, has established a wide range of programs and grants to support the research and development of biorefineries for the purpose of converting biomass materials into commercial products (eg, fuels, chemicals) and energy sources (eg, electricity). Similarly, the Department of Energy—through the National Renewable Energy Laboratory (NREL), Oak Ridge National Laboratory (ORNL), and other DOE-affiliated laboratories—administers a variety of research and development programs related to the development of biorefineries for the purpose of creating a wide range of bioengineering materials. For example, NREL provided the company Genencor International with a $17 million contract to find a process to convert biomass to alcohol. Also, the U.S. DOEs Idaho National Engineering and Environmental Laboratory (INEEL) is partnering with universities and industry (eg, National Association of Wheat Growers) to find biorefining technology to efficiently convert wheat and other crops into chemicals, fuels, and industrial and consumer products (20–22).

The most significant U.S. commercial venture into bioengineering technology to date involves a several hundred million dollar facility built in Blair, Nebraska by Cargill Dow, a joint venture between Cargill Inc. (Minnetonka, Minnesota), an agricultural corporation, and Dow Chemical Company (Midland, Michigan). The facility, which began semi-commercial operations in 2002, is designed to eventually consume 40,000 bushels of locally produced corn to produce synthetic materials. Through fermentation, distillation, and polymerization processes, the Cargill's technology synthesizes polylactide (PLA) and related compounds from corn by-products. The polylactide (PLA) compounds, in turn, provides the building block for a variety of synthetic products, including fiber materials with superior characteristics, such as wear resistance and insulation. The costs of production of the PLA fibers are relatively low since the production

Table 3. **U.S. Market Trends for Advanced Bioengineered Products, $\times 10^9$ \$**[a,b]

	2002	2005	2012
direct economic impact	0.010	0.046	1.3
total economic impact[c]	0.058	2.3	5.5

[a] Refs. 20–22.
[b] Includes all advanced materials.
[c] Includes direct + indirect + induced impact.

process uses inexpensive agricultural waste, including corn stalks, wheat straw, rice hulls, and even sawdust and prairie grass. Dow Cargill plans to incorporate PLA materials in furnishings, containers, packaging, and numerous industrial applications (20–21,23).

A number of other chemical companies plan to enter into bioengineering of synthetics. Dupont in particular has formed a partnership with the sugar producer Tate & Lyle plc to build biorefineries to utilize genetically-engineered microbes in transforming naturally occurring sugars into a synthetic material useful in making clothing, packaging and plastics. Other chemical companies—including Celanese, BASF and Chevron—may soon follow Dupont and Dow in entering into the field (21–22).

The future prospect of bioengineered materials hinges on a number of factors. Most importantly, biorefineries need to achieve continuous, full-scale production to effectively compete in price against existing synthetic materials. Moreover, the companies involved in the technology must accelerate their technical services programs in order to locate and capture increased market share. This depends as well on finding new applications for the materials. Stricter environmental regulations at the federal and state levels promise to increase the interest in and demand for bioengineered materials (20–22,24).

Bioengineered materials face considerable competition from the synthetic materials traditionally made from petroleum and natural gas. It is expected that, with the backing of some of the large chemical companies, bioengineered materials production will find markets for their products over the next 10 years. By 2012, it is estimated that total U.S. sales of these materials could reach in excess of \$1 billion. This means that the total economic impact—direct, indirect, and induced—within the United States of bioengineered materials will be between \$5 and \$6 billion (Table 3). Much of this impact will be focused in the Midwest and other agricultural regions throughout the United States.

9.2. Advanced Ceramics (9,25–27). Ceramics are inorganic and nonmetallic materials. There are three major forms of ceramics: amorphous glasses, polycrystalline materials, and single crystals. Ceramics are generally made from powders and additives under high temperatures. Traditional types of ceramics includes bricks, tile, enamels, refractories, glassware, and porcelain. Advanced ceramic materials, developed more recently, possess superior physical, mechanical, and electrical properties. They are made from metal powders that undergo innovative processing methods.

Advanced ceramics increasingly enter into a wide variety of applications. In general, advanced ceramics extend equipment life, decrease fuel costs, and increase power and performance. One of their major markets is in electronics, which accounts of ~ 66% of the total demand for advanced ceramics. Important ceramic products in electronics applications include both the pure and mixed oxides—alumina, zirconia, silica, ferrites—and doped barium and lead titanates. The important electronic application of these materials includes their use in substrates and packaging, capacitors, transformers, inductors, and piezoelectric devices and sensors. Approximately two-thirds of total electronic ceramic consumption goes into integrated circuit packages and capacitors. Japan is particularly active in developing advanced ceramics for these applications. (9,26–27).

Advanced ceramics are also used in structural applications due to their resistance to corrosion and high temperatures. Ceramics perform very well as a material for equipment components or as an industrial coating. Such ceramics are important materials for infrastructural applications, such as power plants, construction, and bridges as well as in industrial equipment, eg, bearings, seals, cutting tools. The automotive industry employs advanced structural ceramics in catalytic converters and for certain under-the-hood components. Important advanced structural ceramics include various forms of aluminum oxide, zirconia, silicon carbide, and silicon nitride.

Recent developments involve the creation of new types of advanced ceramics and production processes. One of the most important lines of research within the United States and internationally includes the development of ceramic metal matrix composites incorporating reinforcing materials such as carbon fibers. These materials possess superior mechanical properties, excellent thermal stability and a low friction coefficient (to serve as a superior lubricant). Examples of such materials include silicon carbide fibers in silicon carbide matrix and aluminum oxide fibers in aluminum oxide matrix. Research in the field continues to find a wider range of new composites and to reduce unit costs (25–27).

A second major area of advanced ceramics research involves new techniques to make ceramic powders. For example, the use of thermal plasmas may prove a superior process to generate very fine powders, such as aluminum nitride, silicon carbide, titanium carbide, and aluminum oxide. In this process, the plasmas operate under very high temperatures that vaporize raw materials and speed reactions and, then cause rapid quenching of the particles, thus forming ultrafine powders. The process produces high purity powders as well as eliminating intermediate production steps currently employed in traditional manufacturing methods, thus conserving on time and costs, and minimizing wastes (9,25–27).

A third line of research centers on the development of advanced ceramic coatings technology that allows the deposition of a thin layer of ceramic on complex surfaces at low costs for improved resistance to corrosion, mechanical wear, and thermal shocks. These processes impart superior properties without industry needing to go through the time and expense to make entirely new parts and components. An important example of an advanced coating system is zirconium oxide coating for gas turbine engines. These coatings provide a thermal barrier that allows engines to run hotter by protecting underlying metal. In turn, the coating extends component life, increases engine efficiency, and reduces fuel consumption. Research in advanced ceramic coatings focuses on improving adhesion of the coating to surfaces, increasing the properties of the coatings, and reducing the costs of the coating process. The various coating processes currently employed or being investigated include plasma and flame spray, high velocity oxy-fuel deposition, and electron beam techniques (9,26).

Overall, the market for advanced ceramics within the United States and internationally is growing at between 3.0 and 4.0% annually. Advanced ceramics accounts for the largest portion of the total advanced materials market and will continue to dominate the advanced materials field throughout the 2002–2012 period. In 2002, the world market for advanced ceramics stood at \$21.4 billion. By 2005, the global market will be \$23.4 billion, and by 2012 nearly \$29 billion. In 2002, the U.S. market accounted for 40% of world demand for advanced ceramics, or \$8.6 billion. This translates into a total economic impact within the United States of > \$34 billion. By 2005, the U.S. market will stand at \$9.5 billion. In 2012, the U.S. advanced ceramics industry will see a market of \$12.2 billion, or 42% of total world demand. The total economic impact of advanced ceramics in the United States at this time will be \$48 billion (9,25–27) (Table 4).

Two advanced ceramic material areas appear particularly promising: nanoceramics and piezoelectric ceramics. While these materials currently account for a relatively small percentage of the total advanced ceramics market, they attract a disproportionate amount of research and development activity. As a result, they represent the cutting edge in the advanced ceramics field and promise further development, increased production, and a growing range of applications through 2012 and beyond.

Nanoceramics (9,28–30). One of the most promising developments in the development of advanced ceramics is in the area of nanoceramic materials. In this technology, free-flowing nanopowders, processed under intense thermal conditions and elevated pressures, form a variety of parts and components. Properties and applications of nanoceramic materials generally depend on the type and average particle size range of the metals.

Nanoceramics provide both cost savings and new material applications to industry. Traditional ceramics tend to be hard and brittle thus making parts made of the material difficult to machine, and, in turn, significantly limiting markets. In contrast, nanoceramics, characterized by very small internal grain size, provide to parts a mechanical flexibility that allows greater ease in forming, shaping, and finishing (eg, grinding and polishing) in lower temperature environments. Nanoceramics also possess superior structural characteristics exhibiting high strength and excellent abrasion, deformation, and wear resistance, even under high temperatures. Cost savings to industry result from lower energy use,

Table 4. **U.S. Market Trends for Advanced Ceramic Materials, $\times 10^9$ \$**[a,b]

	2002	2005	2012
direct economic impact	8.6	9.5	12.1
total economic impact[c]	34.4	38.0	48.4

[a] Refs. 9,25–27.
[b] Includes all advanced materials.
[c] Includes direct + indirect + induced impact.

reduced time to complete operations, and material savings from fewer damaged parts requiring replacement (28,29).

There are a variety of nanceramic materials. These include titanium nitride, silicon nitride, aluminum nitride, zirconia (and zirconia–aluminua), yittrium–aluminum compounds, and ceria and gallium oxides. Other countries, in addition to the United States, are actively engaged in developing and commercializing these compounds. Germany and Japan represent the most serious competitors to the United States in such materials as silicon nitride and electronic grade aluminum nitride. Other countries active in the filed include Russia, Austria, and Poland. Nevertheless, the United States dominates in most of the other nanoceramic materials, in part due to the greater economic growth of the United States in the 1990s compared to Europe and Asia. The added impetus from military spending after the 9/11 attacks has also aided nanoceramic development with the United States. Currently, the powders used for producing nanocermaics are available from a range of United States and international companies in laboratory, semicommercial, and commercial quantities (28–30).

Applications for nanoceramics include a wide range of structural and industrial uses, such as in machine tools, electroplated hard coatings, and thermal barrier coatings. In the automotive area, research focuses on the use of nanoceramics for "under-the-hood" automotive applications (eg, use in automotive engine cylinders providing greater retention of heat and more complete and efficient combustion of fuel). Small, light weight sensors that incorporate nanoceramics help to measure air/fuel ratios in exhaust gases. This in turn leads to more efficient cars and aids in curtailing environmental pollutants. Potential use for nanoceramics includes applications in appliances, industrial machinery, and petrochemical and power plants. Nanoceramics will see growing application as well in liners and components for appliances, heat exchange systems, industrial sensors, electric motor shafts, gears and spindles, high strength springs, ball bearings, and, potentially, thousands of additional structural parts and components (9,28,29).

Due to its unique internal structure—ie, possessing a large number of molecular-sized "holes"—nanoceramics can serve as advanced molecular sieves and catalytic carriers for a variety of chemical, refining, and biotechnology operations. Because of their generally nonhazardous nature, chemical inactivity, and biocompatibility, nanoceramics can supply materials for industrial, chemical, and biochemical ultrafiltration equipment, "delivery" systems for more efficiently and effectively introducing bioactive agents into the body, and equipment and apparatus for chemical and biochemical research and manufacture. Promising biotechnology applications include the use of nanoceramics in new bone implantation systems, and as implanted medical prostheses (29,30).

Nanoceramic materials also possesses superior electrooptical properties that have applications as materials for semiconductors, electronic components, and related technology and systems including optical filters, capacitors, floppy discs and tapes, magnetic media, fiber optics, and superconducting products (eg, flexible superconducing wire). Nanoceramics, in the form of alumina, ceria, zirconia, and titania oxides, can be used in industrial micropolishing operations since ultra fine abrasive particles provide for superior mechanical polishing of dielectric and metallic layers deposited on silicon wafers. Nanoceramic materials demonstrate unique translucent properties useful in advanced lighting systems. In particular, translucent alumina-based ceramic tubes can operate in high pressure sodium lamps and metal halide lamp tubes for indoor lighting. Nanoceramics also promise increased use in critical energy-related technology including advanced fuel and solar cells and new generation microbatteries (9,28,29).

In 2002, the largest share (53.4%) of the nanoceramic market went into electrooptical applications, followed by chemical/environmental applications (40.1%). The remaining 6.5% of the market for nanoceramics entered into a variety of structural applications. Over the next few years, it is expected that the electrooptical and structural areas will gain a little ground at the expense of chemical and environmental applications, as indicated in the anticipated average annual growth rates for the various nanoceramic markets: structural (9.6%); electrooptical (7.5%); and chemical/environmental (6.9%) (28,29).

Important issues related to the continued growth of ceramic materials for these applications include the development of new and improved nanoceramic materials and composites and advanced processing technology. For example, a key concern is to improve the strength and fracture toughness of nanoceramics since sudden structural failure hinders wider applications of the material. A novel type of structural composite being investigated involves placing nanosilicon carbide particles inside alumina and in developing silicon nitride composite matrices. These types of composites appear to possess superior wear resistance, chemical inertness, anticorrosion properties, and excellent thermal insulation.

Also of vital interest in the area is development of new processes to make such composites as well as to manufacture existing nanoceramic materials more efficiently and with higher quality (eg, improved compacting and pore size distribution). Innovative processes include new types of powder synthesis technologies involving advanced thermal plasma and laser-based methods. As critically, advances in nanoceramic processing and shaping include such activities as high pressure sintering, ultrasonic compacting and shaping, and advanced "wet" molding (9,28,29).

Overall, the U.S. market for nanoceramic materials is growing, and at a faster rate than for advanced ceramics as a whole. The expansion in the market for nanoceramics is in part due to the declining unit price of nanoceramic powders. In 2002, the average price per pound of nanoceramic powders stood at $6.6/lb. By 2007, it is estimated that the average price will be $5.1/lb and by 2012, <$4.0/lb. In addition to the price issue is the fact that nanaoceramics remains a developing field with a wide range of possible markets that are just beginning to be exploited (28,29).

In 2002, the market for nanoceramic powders stood at $154 million, or just 1.8% of the total value of the advanced ceramics market. By 2005, sales of nanoceramic powders will be $\sim$ $194 million, which represents 2.0% of the total advanced ceramics demand. By 2012, that percentage will increase to 2.7%, reflecting the growing importance of nanoceramics in the advanced ceramics markets. At that time, the market for nanoceramics will stand at $332 million, which translates into a full economic impact within the US of >$1 billion. After 2012, the market for, and economic impact of, nanoceramics will increase

Table 5. **U.S. Market Trends for Nanoceramic Powders, $\times 10^9$ \$**[a,b]

	2002	2005	2012
direct economic impact	0.154	0.194	0.332
total economic impact[c]	0.616	0.776	1.33

[a] Refs. 28–30.
[b] Includes all advanced materials.
[c] Includes direct + indirect + induced impact.

rapidly as the material continues to influence the advanced ceramics industry (28–30) (Table 5).

Piezoelectric Ceramics (9,31–38). Piezoelectric materials—composed of mixtures or complexes of zirconium, titanium, lead and other metals—create driving voltages when placed under mechanical stress (the "generator effect") and, contrariwise, undergo mechanical movement or deformation when subjected to electrical impulse (the "motor effect"). There are four types of piezoelectric materials: ceramics, crystals (eg, piezoelectric quartz), ceramic/polymer composites, and polymer films. Of the four types of piezoelectric materials, piezoelectric ceramics represents the largest and most mature market segment, accounting for ~90% of the total piezoelectric market. If piezoceramic/polyer composites are included, then piezoelectric ceramic materials have a presence in ~93.3% of the total piezoelectric market (31).

There are a number of piezoelectric ceramic materials, most composed of some form of lead and titanium. Currently, the most common piezoelectric ceramic is lead zirconate titanate (PZT). Other types of piezoelectric materials include barium titanate, bismuth titanate, lead titanate, and lead metaniobate. The various properties that these materials offer depend on the type of production process used and the chemical and operating conditions under which they are formed. The general process by which nonpiezoelectric ceramics are transformed into piezoelectric materials involves heating the ceramic materials in a dielectric oil bath by which the applied electric field aligns the dipole units existing in the material (9,31,34).

Piezoelectric ceramics come in "bulk" and "multilayered" form. The bulk form of the ceramic consists of a single ceramic block from which is produced various shapes: blocks, plates, disks, cylinders, rods, etc. In contrast, the multilayered variety consists of several thin layers of the ceramic material stacked up into rectangular and cylindrical shapes, such as bars, plates, and disks. The quality and performance level of these various forms of piezoelectric ceramics are measured by a number of variables including dielectric constant, dielectric loss factor, electromagnetic coupling factor, piezoelectric load constant, elastic compliance, elastic stiffness, electrical resistance, and thermal coefficient (34).

As with other advanced materials, the United States faces increasing competition from other countries in the piezoelectric ceramic field. Germany and Japan represent the major competition to the United States in piezoelectric ceramics. Other countries in Europe (Denmark, France) and Asia (Taiwan) actively pursue development of new piezoelectric ceramic technology.

Piezoelectric ceramics find application within the United States and internationally in a wide range of industries, including biomedical, aerospace, automotive, industrial, consumer, and marine. While the government dominates the demand for United States produced piezoelectric ceramics, the industrial and consumer markets continue to gain a presence in the field. Industry and government increasingly view these materials as useful in making such general components as electric circuit elements, transformers, actuators, transducers, and energy generators (eg, batteries). These components, such as actuators, allow a large force capability and short response time. This means that, in time, they provide rapid, precise, and carefully regulated displacement of devices, equipment, and systems in response to even small applied voltages. These piezoelectric ceramic components, in turn, find current and potential application in such devices and systems as sensors (medical, pressure, flow, acceleration), sonar equipment and hydrophones, laser positioning, industrial tools and hardware (valves, meters, cutting and polishing machines, displacement gauges), electrical devices (remote control switches, relay contact drivers, electroacoustic devices, microposition actuators, electrical appliances, security alarms, camera shutters), and security systems (9,31,33–38).

One area that is particularly promising for piezoelectric ceramics is their use in vibration control due to the general use of more powerful machinery and equipment in industry. Vibration control is especially important in such areas as aircraft manufacture, hospitals (eg, vibrations due to MRI equipment), and power plants. Another potential market for piezoelectric ceramics is in the manufacture of wireless switching equipment for both residential and business structures. The market in Europe and Asia appears particularly promising for these devices due to the higher cost of installing and replacing wired systems in these regions. The specific uses for piezoelectric ceramics in nonwired applications include switching and lighting systems, appliances, security systems, doorbells, and burglar alarms (9,31,33,35,38).

Future applications of piezoelectric ceramics hinges on their superior power density and cost and size advantage. As a result, they will enter into nonmagnetic transformer components in radiofrequency (RF) transmissions systems and remote control devices to power back lighting for computer screens and, in the form of ceramic fibers, as critical material in the monitoring of stresses and strains in aircraft bodies, automotive engines, and building structures (9).

Investigation into more advanced piezoelectric materials and production processes continues at a rapid rate within the United States and internationally. Attempts to find alternative materials that do not depend on lead result from stricter environmental policy. Denmark, eg, has been investigating the Alkaline Niobates as a possible substitute for PZT. Work is also underway in various countries on a new variety of porous piezoelectric ceramics that promise superior performance for transducers operating underwater (eg, in hydrophones). Finding new ways to fabricate piezoelectric ceramics also has a priority. Significant work in the United States, England, and other countries centers around "net shape fabrication" and the process of "plasticising" the powder-binder mixture in order to limit sintering and, in turn, the structural defects and high production costs associated with the sintering operation (31–33).

As with nanoceramics, the rate at which piezoelectric ceramics gain markets in the United States, Europe, and Asia depends on the ability of the production process to reduce unit costs. The cost of ceramic-based raw materials—which currently represents approximately one-fifth of total manufacturing costs—remains a bottleneck to commercialization. However, it is likely that his percentage will decline to 5% or less as the production process achieves mass production status Currently, the unit price for advanced ceramics is about $75 per wireless ceramic element. By 2005, this is expected to decline to ~$50 and by 2012 ~$20 per element (9).

As the unit price contracts, and the technology advances, new markets for piezoelectric ceramics will emerge. The annual rate of growth for piezoelectric ceramics is expected to remain at 8.9%. The material commands a somewhat greater role in the advanced ceramics market compared to nanoceramics, and will continue to do so throughout the 2002–2012 period. The U.S. market for piezoelectric ceramics stood at $237 million in 2002, representing 2.8% of the total advanced ceramics market. By 2005, sales of piezoelectric ceramics will be $306 million, or 3.2% of that market, and by 2012, $556 million, accounting for 4.6% of advanced ceramics demand. The increase in the direct sales of piezoelectric ceramics between 2002–2012 represents a growth in total economic impact in the United States l from $948 million to $2.2 billion (31,33) (Table 6).

9.3. Advanced Coatings. Advanced coatings provide a vital and growing area in the new materials field. These advances emerge from recent research in surface chemistry and solid state physics. Technical and economic growth of these materials continues to accelerate.

The advanced coatings field consists of an increasing number of smaller firms specializing in manufacturing particular types of coatings and coating application technology. Many of these firms are start up operations that license technologies from the university and government (eg, NASA, DOE, DOD) and who carryout their own R&D to further commercialization. These firms create novel coating materials that provide new ways to protect surfaces from the environment—ie, heat, impacts, erosion, and chemical degradation—thus increasing the life and performance of components, equipment, and systems across a wide range of industries and technologies. The more advanced coatings impart to surfaces and objects heightened ability to sense and respond to the full range of changes in the environment. Consequently, advanced coatings promises to be a central component in the development of the so-called "smart" materials that are expected to revolutionize twentieth century technology.

Table 6. **U.S. Market Trends for Piezoceramic Materials, $\times 10^9$ $**[a,b]

	2002	2005	2012
direct economic impact	0.237	0.306	0.556
total economic impact[c]	0.948	1.22	2.22

[a] Refs. 31,33.

[b] Includes all advanced materials.

[c] Includes direct + indirect + induced impact.

Thermal Barrier Coatings (9,39–42). Thermal barrier coatings protect surfaces in one of three ways: providing a simple physical barrier to thermal energy ("passive" heat control); dissipating, dispersing or reflecting heat; or minimizing heat-producing friction. A new generation of thermal barrier coatings promises to protect surfaces from very high temperature environments by affecting all three modes of heat management.

Thermal coatings are produced in a variety of forms, including paints, tapes, and vacuum deposited metals. Currently, thermal coatings tend to be passive in nature in that they simply act as a barrier to protect surfaces from a given thermal environment up to a certain temperature. Above this limit, degradation to the coating and, in turn, surface takes place. Advances in coating technology allows the materials to be "active" agents that sense changes in outside temperatures and, through electronic means, adjust their internal molecular structure to optimally reflect, absorb, or dissipate thermal energy and, in effect, be able to withstand far higher temperatures than currently possible. The ability to create "smarter" thermal coatings depends on an increasing understanding of the structure of coatings and their behavior in high-temperature environments (9).

A growing range of thermal coating materials is currently being developed. Such materials include a variety of aluminum alloys and metal matrix composites, ceramic-based materials, the aluminum oxides, titanium alloys, zirconia–yttria compounds, and molybdenum plasmas. One of the more promising passive thermal coating material that is emerging comes from advances made in polymer technology, and in particular, polyimide chemistry. Polyimide coatings withstand temperatures of up to 700°, or ~50% more than current coating materials. The polyimide materials, which are made in thermal reactors under relatively low pressures, also provide surfaces with superior resistance to corrosive agents (9,39,41).

An important part of thermal coating technology involves the method of applying the coating material to the surface. Thermal spray technique, in particular, represents the state of the art in applications technology. It employs a combination of thermal and kinetic energy to direct particles evenly and at desired thickness onto a surface. Improvements to the technology—including the development of high velocity oxygen fuel systems and incorporation of robotics—will increase productivity and lower overall costs (9,41,42).

One of the more advanced applications techniques currently being investigated and developed includes electron-beam–physical vapor deposition by which vaporized metal coating is directed onto a surface. This process allows the formation of tailored, composite coatings on surface substrates by coevaporating different coating materials and alternating layers of the different materials onto the surface. Additional processes that appear promising include vacuum plasma spray and arc and flame spray technologies (9,41,42).

The potential market for advanced thermal coatings range from the automotive, aerospace, and defense industries to high temperature microelectronic circuit boards, industrial motors, electric and nuclear power production, biomedical systems, chemical and petrochemical plants, and composite materials for construction applications and machine tools (9,39).

Currently (2002), prices for the more advanced coatings remain relatively high. For example, the unit price for polyimide thermal coatings is around

\$350/lb. Over time, the adoption of mass production techniques will force prices down. Thus, by 2005, the price for polyimde coating material will decline to \$172/lb and be at only \$20–25/lb in 2012. By this time, market penetration for the new generation of thermal barrier coatings will stand at ~20% (9).

Conductive Coatings (9,43–47). Conductive coatings consist of an electrically conductive material mixed into, or bonded onto, some nonconductive medium through such means as vapor-phase deposition or electroplating. In this sense, conductive coatings can be termed composite materials. Currently, conductive coatings exist commercially in three main forms, defined by the type of medium employed: conductive paint, metal plating (or cladding), and synthetic resin (eg, epoxy, urethane, acrylic). With respect to metal-based conducting coatings, current technology utilizes metals such as copper, nickel, and silver as the conducting agents, either separately or in combination. In terms of cost, silver paint tends to be the most expensive type of coating, nickel paint the least expensive, and copper-nickel cladding an intermediate cost conductive coating material (43).

A fertile area for recent and future research is in the area of conductive resins. This field evolved from Nobel Prize winning work that showed that plastic materials could be made electrically conductive. Researchers in the United States and internationally are investigating a number of potentially useful polymers and their particular conductive coating applications including: polythiopene derivatives (antistatic agents, photography), polypyrrole and polyaniline (electrostatic speakers, computer screens), polyphenylenevinylene (phone displays), and the polydialkylfluorenes (advanced color screens for video and TV) (9,43).

As the case of conducting polymer shows, emerging conductive coatings, both metal and polymer, find potential use in a wide variety of electrical and optoelectronic applications. Conducting coatings, when incorporated into a battery's current collector, enhance the power and life of batteries. The coatings impart portability, compactness, and lower costs as well. Adhesives made from conducting coatings (eg, epoxy medium) appear excellent for repairing printed circuits and replacing metallic solder. Conductive coatings embedded onto glass substrates promise commercial development of two-dimensional antenna systems for use in automobiles and telecommunications equipment. Transparent conducting films offer the prospect of increased application in optical systems, dielectric mirrors, and holographic devices (43–46).

One of the most important applications for metal-based conductive coatings is in electromagnetic interference (EMI) shielding, which is so because conductive coatings absorb, emit, or reflect certain optical and radio frequencies. Moreover, they create a magnetic field or three-dimensional geometry that scatters radar signals or reduce aircraft and ship signature. The coating protects equipment from interfering signals and sudden and potentially disruptive electromagnetic pulses. Conductive coatings can shield entire rooms containing electronic equipment or replace plastic as the packaging material for printed circuits and electronic components and devices (eg, computers mobile phones). Industries currently or potentially using conductive coatings in shielding systems include the aerospace, defense, electronics, security, health care, financial, and communications sectors (9,43,45).

Attempts by researchers to find superior alternatives to current conductive coatings result from limitations inherent in the existing technology and the need to extend the range of application. While silver is a superior conductor and anticorrosion agent, it is expensive, which restricts its use to only the most specialized applications. On the other hand, copper and nickel, while relatively inexpensive, do not offer the same level of conductivity as silver and, for example, cannot adequately shield entire rooms from electromagnetic interference. Moreover, copper tends to oxidize, which further reduces its conductivity, while nickel poses environmental problems. The increasing power required in electronic and telecommunications components, devices, and equipment also call for more advanced conductive materials in batteries, optical equipment, and EMI shielding systems. Advanced research within the United States and internationally seeks to develop new conductive coating materials to solve these problems and extend markets. These novel materials consist of special alloys, composites, and polymer compounds (43,45).

The current unit price of certain of these coatings—as high as \$1200/gal—prohibit a significant market for the more advanced conductive coatings. However, as production processes improve, the industry expects unit costs to contract significantly. By 2005, it is expected that the price of advanced conductive coating material will stand at ~\$650/gal and by 2012 will be as low as \$40/gal. As a result, by 2012 market penetration could reach between 6 and 7% (3).

Anti-Corrosion Metallic Coatings (9,48–51). Anticorrosion coatings prevent oxidation reactions from taking place at a metal's surface. Currently, anticorrosion coatings are solvent-based and employ both inorganic metals and organic resins. Specifically, coatings make use of metal-based primers (eg, zinc, aluminum, nickel) with polymer coatings used for intermediate (eg, epoxy) and top (eg, urethane) layers. A number of companies produce anticorrosion coating materials. In the United States alone, there are ~ 100 manufacturers of anticorrosion coatings providing specialized products to small geographical areas (48,50).

A variety of processes exist to apply coatings to surfaces. A typical application technology is thermal (flame) spray process by which a metal is melted in an oxy-acetylene flame and atomized under compressed air to form fine particles that are then sprayed onto a surface. On contact with metal, the metal spray solidifies to form a uniform coating. Other spray processes include powder and plasma thermal spray methods.

A number of problems continue to plague current anticorrosion coating technology. Current coatings tend to cause environmental problems and require expensive and time-consuming preparation of the surface. Also, coatings degrade over time, resulting in flaking and peeling. Current research in the field focuses on the development of coating materials to meet stricter environmental standards, do not require expensive surface preparation, and have good adhesion—superior bond strength and uniformity—and longer life under rigorous climatic conditions. Promising materials under investigation are nonsolvent-based coatings incorporating advanced polymer materials including polyester, polyaniline, and silcone and silicon–glycol resins (9,48–50).

A potentially revolutionary line of research being conducted in the United States and Europe concentrates on certain organic films that form tightly bound multilayers on a surface through electrolytic action. These materials are gel-like films of alternating layers of positively and negatively charged molecules. As opposing charges pair up, they hold adjacent layers together tightly while, at the same time, a positively charged bottom electrolyte layer adheres to negatively charged metal surface, thus avoiding degradation and flaking over a long period of time (51).

Also being investigated are purely metallic anticorrosion coatings, in particular a complex of aluminum, a rare earth metal (eg, cerium) and a transition metal (eg, iron or cobalt) combined in various proportions. The nature of the alloy itself, produced in an innovative thermal process, allows quenching of the molten metal at a relatively low rate (as measured in degrees cooled per second) compared to current aluminum alloy materials. This less radical quenching process, undertaken in a thermal furnace, produces an amorphous alloy without structural damage to the metal. This noncrystalline structure serves as both an anti-corrosion and anti-deformation coating (9).

Researchers within the United States and Europe are also examining new forms of applications technology. These efforts include finding ways to improve the more traditional thermal and plasma spray methods, especially in applications involving metal-based coatings. Since thermal techniques tend to weaken a metal's surface thus resulting in early degradation and coating failure, research work looks to reduce the need for heat during spraying. One alternative is the recently developed cold plasma spray, which offers an advanced alternative that retains the integrity of the host metal (9).

In addition, development efforts are now underway to commercialize totally new processes, notably a process based on the creation of ionic self-assembled coating layers. In this technology, a charged substrate is dipped into an aqueous solution of a cationic material followed by dipping in an anionic solution. Adsorption to the surface of the substrate results from electrostatic attraction of "interlayer charges", with each layer of uniform thickness. Multilayers of several microns thickness are easily fabricated via repeated dipping process and are rapidly dried and fixed at room temperatures. The low-cost process produces an ultra thin, impermeable, and tightly bound coating. The process produces specific coatings, depending on the applications involved, through molecular manipulation (9,48).

Continued innovation, and the decreasing costs over time in producing the coating material and in applying the coating to a surface, promise expanding markets for anticorrosion coating technology. In addition to making further inroads into such traditional markets as shipbuilding and repair, public infrastructure (bridges, buildings, etc), public utilities, machinery, buildings and construction, the new generation of anticorrosion coatings technology is expected to find increased applications in such industries as aerospace, automotive, electronics, industrial gases, telecommunications, and petrochemicals. (9,49,50).

Multifunctional Coatings (9,49,52,53). Multifunctional coatings, which first emerged in the 1990s, perform a number of operations—anticorrosion protection, conduction, electromagnetic shielding, thermal protection—simultaneously. Their development, both within the United States and internationally, results from exploiting the commercial potential of surface engineering. Typically, small R&D and start-up firms license multifunctional coating technology from the government and universities. In addition, certain large corporations (eg, Dow corning) look to expanding their product capability in advanced multi-functional coating technology (49).

Multi-functional coatings incorporate a number of new materials, either separately or in combination. These materials include fluoropolymer composites, the urea–formaldehyde resins, multicomponent pigments, and the carbides, nitrides, and borides of certain metals (eg, titanium). These materials can be incorporated into different media such as paint, ink, and adhesives. Multifunctional coatings may be composed of one material capable of performing different functions, or, more commonly, be a multilayered composite of different materials. Increasingly, work in the field focuses on the synthesis of nanocomposite coatings with multifunctional properties that impart a wide range of surface qualities, including transparency, surface hardness, reflectivity, etc. The use of diamond films doped with certain metals also offers a promising route for the custom-design of multifunctional coatings with desired electrical, mechanical, and optical properties (49,52,53).

A number of process technologies currently being developed in the creation of multifunctional coatings include thermal vaporization, electrospark alloying, laser heat processing, ion implantation, plasma deposition, magnetron sputtering, and solution-phase ionic self-assembly. In addition to these technologies, fluidization technique, typically used in refining or power generation, may provide a route to the mass encapsulation of anticorrosion agents for use in multifunctional coating systems (9,52).

Numerous applications exist for multifunctional coatings. In the automotive sector, multifunctional coatings meet increased demands for strength and thermal and corrosion protection of steel surfaces. In the textile industry, research in the United States and Germany looks to development of "smart" hybrid polymeric coatings for fibers that allow fibers and textiles to switch or tune their properties in response to external stimuli. In the metallurgical industries, multifunctional coatings provide superior hardness, anticorrosion properties, thermal protection, abrasion resistance, and chemical "inertness". Accordingly, they enter into a variety of metallurgical operations, such as pressure die casting operations. For its part, the defense and aerospace industries require multifunctional coatings in its defense systems and aircraft for sensing, conductivity, energy absorption, and thermal dissipation. Such coatings increase the performance capability and lifetime of components, equipment and systems (9,52,53).

One of the most promising areas for multifunctional coatings involves the development of new types of sensors. An emerging application for multi-functional sensors is in a new generation of micro electromechanical systems (MEMS) requiring the simultaneous detection of temperature, pressure, radiation, gas concentrations, electromagnetic fields, and so forth. Multifunctional coatings also act as sensors for detecting and monitoring structural defects in buildings, bridges, and aircraft, and to carry and deliver chemical agents to strengthen critical points in the structures. In one variation of the technology, the sensing system uses small synthetic spheres arrayed in a crystalline lattice and imbedded within a coating material. As the coating shifts or otherwise changes its configuration due to structural distortion, the internal lattice also

Table 7. **U.S. Market Trends for Advanced Coating Materials, $\times 10^{9a,b}$**

	2002	2005	2012
direct economic impact	0.085	0.170	1.80
total economic impact[c]	0.034	0.680	6.50

[a] Refs. 9,52,53.
[b] Includes all advanced materials.
[c] Includes direct + indirect + induced impact.

changes its structure. An optical system then monitors these changes over time. These spheres also contain various anticorrosion agents and deliver them to pivotal sites (9).

The market for advanced coatings—including thermal conductive, anticorrosion, and multifunctional—is increasing at a rapid rate. In the decade 2002–2012, thermal and anticorrosion coatings will continue to dominate the advanced coatings group, accounting for well $>$ 90% of sales throughout the period. However, during this time conductive and multifunctional coatings will gain ground as technology advances and unit costs decrease. These two coatings will exhibit the largest growth rate during the latter part of the period and in the decade to follow.

The production of advanced coatings as a whole in terms of dollar sales is expected to accelerate rapidly after 2005. By 2012, total U.S. production of advanced coatings will stand at $1.8 billion million, producing a total economic impact within the United States of $\sim$ $6.5 billion (9,52,53) (Table 7).

9.4. Nanopowders and Nanocomposites (9,54–60). In recent years, powder metal technology has emerged as a major segment of the metals industry. It is a growing presence in a variety of commercial and industrial applications. Most recently, this field has been advancing into the still new area of nanotechnology. Nanopowders are typically composed of metals or metal mixtures and complexes with particulate sizes in the micron ranges. The potential markets for these materials depend on the fact that they can be formed into diverse shapes and forms possessing a variety of important mechanical, electrical, and chemical characteristics. Nanopowders are composed of a broad range of metals and their compounds. These include (but are not limited to) the oxides of aluminum, magnesium, iron, zinc, cerium, silver, titanium, yttrium, vanadium, manganese, and lithium; the carbides and nitrides of such metals as tungsten and silicon; and metal mixtures, such as lithium/titanium, lithium manganese, silver/zinc, copper/tungsten, indium/tin, antimony/tin, and lithium vanadium (9,57–59).

As with a number of other advanced materials, including nanotubes, nanopowders come from a relatively small number of companies worldwide, including the United States, Europe, Asia, and the Mideast (eg, Israel). These companies range from large international corporations with diversified operations to small start-up firms, licensing technology from academic and government laboratories and specializing in a narrow range of nanopowder products. Nanopowder producers are either captive, ie, making powders for their own internal research and commercial use, or "open," ie, producing powders for sale to research organizations and commercial facilities (9,54).

Companies produce nanopowders through a number of processes, including furnace and laser-based technologies. Plasma chemical synthesis (PCS), eg, employs microwave methods to produce nanoparticulates through the creation and consequent rapid quenching of hot ionized gas plasma. Another process, a modification of the Xerox "emulsion aggregation" technology, utilizes emulsion polymerization technique. The resulting powders produced by these various processes possess narrow particle size distribution, high purity and energy efficiency, and superior metallic and ceramic properties. One of the more active areas in the nanopowder field is nano-based coatings. These coatings possess more tightly packed structures than exists in the case of traditional coating material. This structure, in turn, imparts to the surface a high degree of transparency, hardness, and abrasion and scuff resistance. These materials, when added to a resin base, produces superior paints and varnishes (9,54).

Additional applications for nanopowders include their use as abrasives for polishing silicon wafers and chips, hard disk drives (for higher data storage capability), and optical and fiber optical systems; an advanced catalyst for petroleum refining and petrochemicals production, as well as in automotive catalytic converters (providing more complete conversion of fuel to nontoxic gases); as pigments in paints and coatings; and as an additive to plastics in a new generation of semiconductor packaging (9,54,60).

One of the most important applications of nanopowders is in the manufacture of nanopowder–plastic composites. Typically, nanopowder composites contain under 6% by weight of nanometer-sized mineral particles embedded in resins. One of the first such composite used nylon as the plastic medium. More recently, other plastics have come to the fore, such as polypropylene and polyester resins. A new generation of plastics with superior properties is currently being developed for future application. In addition to a number of smaller R&D companies, a few of the larger corporations within the United States and internationally continue to develop advanced nanocomposites, including Bayer, Honeywell, and GE Plastics. In addition to the United States, Germany, China, Korea, and Japan are particularly active in the field (60).

Nanocomposites offer a range of beneficial properties including great strength and durability, shock resistance, electrical conductivity, thermal protection, gas impermeability, and flame retardancy. New and more sophisticated processes can manufacture composite powders with a uniform, nanolayer thick metallic or ceramic coating for high density parts with superior thermal, mechanical and electrical properties (54,60).

As a result of their superior properties and advancing manufacturing technology, nanocomposites face new market opportunities. For example, in the automotive area, General Motors recently entered into production of the first polymer nanocomposite part for the exterior of a car. The biomedical field also appears a particularly promising area for nanopowder composites, especially as delivery systems for the application of bioactive agents into the body, as materials for dental and medical micro abrasion applications, and for use in orthopedic implants (eg, artificial bones for hips) and heart valves (9,54,58–60).

Additional potential applications for nanopowders and nanopowder composites include electrodes for more efficient and longer lasting portable power sources (batteries, solar cells); superior materials for military weapons (eg, as armor and in projectiles); advanced instrumentation (eg, for automotive applications) and biomedical and environmental sensors; materials for stronger, lighter, and more flexible structural shapes and more durable, high performance cutting tools and industrial abrasives; advanced refractory material for chemical, metallurgical, and power generation; ceramic liners (made of zirconia and alumina) in more efficient internal combustion engine cylinders and ignition systems for automotive and aerospace applications; more powerful industrial magnets in magnetic resonance imaging (MRI) systems for medical applications; and a new generation of electrical and electronic components (eg, induction coils, piezoelectric crystals, oscillators) (9,58,59).

9.5. Nanocarbon Materials. Nanocarbon materials contain molecular-sized clusters composed of a number of carbon atoms arranged in various configurations. One such group falls into the category of fullerenes. In this case, a series of carbon atoms arranged spherically enclose one or more metal atoms. In the second type of material, the carbon atoms join together to form a tubular-like structure. These structures may or may not enclose metal atoms. These materials, known as nanotubes, have important applications in the advanced composites area. As a group, the nanocarbon materials have begun to enter the marketplace as commercial materials. A number of companies in the United States are particularly active in developing and commercializing nanocarbon materials and composites. These include Carbon Nanotechnologies Inc. (Houston, Tex.), Applied Nanotechnologies, Inc. (Chapel Hill, N.C.) and Luna Innovations (Blacksburg, V.A.).

Metal Fullerenes (9,61–65). Fullerenes in general refer to a group of materials composed of carbon structures of 60–90 carbon atoms, each enveloping a single metal atom. Currently, a number of U.S. companies manufacture fullerene materials in varying compositions and amounts. Since the 1990s, research undertaken in the United States has led to the creation of a particular type of fullerene—in which the carbon cage contains three distinct metallic atoms—that is of particular interest commercially. These triatomic fullerenes possess commercially useful properties now being explored by research and industry groups.

Two viable thermal processes produce advanced fullerenes. One technology involves application of the electric arc, using graphite to provide the carbon atoms. The major problem with the process is that it is highly energy-intensive and is therefore expensive if placed on a mass production basis. The second approach, referred to as the "soot-flame" process, is in fact currently utilized to manufacture certain traditional fullerenes. However, it is readily modified to generate commercial amounts of the more advanced (ie, triatomic) fullerene materials by burning a mixture of acetylene (or related hydrocarbon) and the required metals to be "encaged". The advantage of the soot-flame process is that it is relatively cost efficient and permits carefully controlled production, and therefore more precise product design (9).

Advanced fullerenes offer a variety of potentially important applications. For example, they are at the heart of new types of multi-functional catalyst systems for the petrochemical industry. In this application, the carbon structures encapsulate the different catalytically active metals (eg, iron, platinum, nickel), which are then released in tandem and in a controlled way as the external carbon structure disintegrates during reaction. The unique optical properties of the fullerene material offer additional applications in industrial photovoltaic sensing systems for incorporation into monitoring and automated control technology. The electromagnetic and optical properties of advanced fullerenes will find application in semiconductor, fiber optic, and microelectronic systems (9,61,64,65).

Within the biomedical field, advanced fullerenes appear to be superior "contrasting" agents for use in magnetic resonance imaging (MRI) systems. In this case, the fullerenes, ingested into the body orally, enhance MRI images 50–100 times more than current capability. As a consequence of this improved MRI performance, manufacturers incorporate smaller and less powerful magnets in their machines, resulting in more compact, portable, and cheaper equipment. This advantage, in turn, expands the applications for MRI technology in a number of markets, such as rural, less developed regions, in smaller to mid-sized clinics and hospitals, and in military field hospitals. Further, the smaller MRI equipment, because they operate with less powerful magnetic fields, reduce the costs to the larger hospitals and clinics of housing and maintaining large superconducting magnets. Industry expects the full-scale commercial production of advanced fullerenes for these applications by 2005.

Nanotubes (9,66–75). Nanotubes are carbon-based structures with cylindrical shapes and diameters between 0.8 and 300 nm. Nanotubes resemble small, rolled tubes of graphite. As such they possess high tensile strength and can act as an excellent conductor or semiconductor material. There are two main varieties of nanotubes: single-walled and multiwalled. Multiwalled structures are the less pure form of nanotubes and offer only a limited number of applications. The more advanced, purer form of nanotube, ie, defined by a single-walled structure, is the more promising material commercially, especially for incorporation into polymer materials in the synthesis of composites with superior structural, thermal, and electrical characteristics. (9,66,75).

Currently, a handful of companies in the United States and internationally produce advanced fullerenes and nanotubes. In 2002, it was estimated that there were between 20 and 30 captive producers of nanotubes worldwide. These companies included large multinationals (eg, Honeywell) as well as a number of small, independent research firms. In 2002, there were only five "open" companies that produced nanotubes solely for sale in the market. Both captive and open companies sell the bulk of their nanotube production to universities, R&D organizations and other companies, generally for research purposes. A third type of company designs and sells (or licenses) production equipment and systems to other firms to make nanotubes. A current trend in the industry is the formation of international partnerships between United States and foreign—especially Asian—companies to jointly develop and sell nanotubes materials and technologies. These partnership arrangements bring together complementary skills and knowledge and facilitate the sales of U.S.-produced nanotubes within Asian markets. For example, In 2002, the U.S. firm Carbon Nanotechnologies Inc. partnered with Sumitomo Corporation to market carbon nanotubes in Asia (9,66,70,74,75).

The production of nanotubes takes place using one of three major processes: gas-phase catalysis, chemical vapor deposition, and laser-based technology. Chemical vapor deposition involves heating a selected gas in a furnace and flowing the heated gas over a reactive metal surface. This process produces excellent yields with the production of a low concentration of contaminants, but with the nanotubes incurring a large number of defects.

The catalytic process requires acetylene gas to move over a catalyst located within a furnace at high temperatures (~700°C). In the process, the acetylene molecules decompose and rearrange themselves into nanotubes. This method, which operates on a semi-continuous basis, can generate a significant amount of nanotubes. However, these are typically of the less pure variety and therefore are of limited use commercially. Another problem with this approach is that metal particles from the catalyst tend to attach themselves to newly formed nanotubes. These particles magnetize the nanotubes thus limiting their use for applications in critical electronic components, such as transistors. More generally, both the chemical vapor deposition and catalytic process produce significant amounts of undesired byproducts, eg, carbon black and amorphous carbon. The removal of these impurities from nanotube yields is expensive and limits the economic feasibility of these processes (9,66,68,73).

The laser approach for making the purer nanotubes offers an alternative approach, albeit with its own set of problems. The process involves the use of free electron lasers (FEL). FELs operate at high energy levels and with very short pulses. They produce pure nanotubes by vaporizing graphite-catalyst mixtures. Removal of the impurities (such as spent catalyst and graphite materials), critical in the making of high grade catalysts, involves a solution-based purification process involving dissolution and precipitation of unwanted contaminants. The nanotubes, once purified, are mixed into a polymer host or matrix in various proportions of nanotubes to polymer. Examples of polymers used in nanotube composites include nylon, epoxy and polyester. The extent of dispersal of the nanotubes through the polymer determines the quality and ultimate commercial potential of the final composite. Ultrasonic technology offers one possible commercially viable dispersal technology. The FEL process for making advanced nanotubes is currently under development by government laboratories, universities, and start up companies. The technology produces fewer impurities than other techniques but is energy and capital intensive and offers as yet slow production rates (9).

Research is underway as well in developing other processes, such as improved electric arc technology. In 2002, IBM unveiled its new process for making single-walled nanotubes. The process involves a nanofabrication method centered on silicon crystal technique. The technology, which is still under development, promises minimum creation of by-products and contaminants and little damage to the nanotube structures (73).

Despite advances in nanotube production, process technology currently cannot accurately control the structure and distribution of nanotubes from one batch to the next. This limitation results in nanotube output possessing a high degree of variability in the material's physical and electronic properties. As a result, existing technology cannot as yet satisfactorily custom design nanotubes for particular applications.

One of the most promising large-scale markets for nanotubes involves electronic and optical applications. Because nanotubes have dimensions in the wavelength range of visible light, they can be used directly as active optoelectronic devices. For example, Motorola, Samsung, and other electronics companies are developing advanced electronic displays based on nanotubes. This work is leading to ultrathin screens and flat-panel displays capable of high resolution imaging and high power efficiency, and to a new generation of giant, low cost illuminated signs. A potential market for nanotube display technology is for 20–40 in. television screens since neither LCDs nor other existing display technology has as yet secured a dominant position in the field. A related area is the use of nanotubes in microelectronic devices. In particular, IBM recently succeeded in making microelectronic switches from nanotubes. This device is expected to find a large number of applications in computer and consumer electronic products. The aerospace and defense industries promise markets for advanced nanotube composites as well. These composite materials are both strong and light (20–30% lighter than carbon fibers) and consequently make excellent materials for aircraft components and structures. Current research suggests the possibility that advanced nanocomposite fibers may replace carbon fibers in many structural applications. Additional potential applications for carbon nanotubes include incorporation into thermally conductive fibers for clothing, carpets and fabrics, and into electrical conducting polymers and fibers for use as electromagnetic shielding materials and in various components for wireless communications, micro sensors and monitoring devices. Over the longer term, nanotube composites can provide a superior drug delivery system and advanced storage systems for hydrogen-based fuel cells (9,66,67,70–73,75).

Currently, the price of nanotubes prohibits their extensive commercial use. In 2002, the average price of nanotubes stood at ~\$40–\$50/g, or a number of times more expensive that gold. While the price of nanotubes has declined sharply over the last few years, nanotubes remain too expensive for other than their use in research work and in limited commercial applications (eg, microscope probe tips and membranes). It is estimated that the price of nanotubes would need to drop to ~\$15,000/lb (\$20–\$30/g) in order for their commercial use in flat panel displays for PCs and television sets. The price would then need to reach the \$10,000/lb level for nanotubes to be applied in such applications as microwave devices (eg, antenna) and radar-absorbing coatings for aircraft. A significant drop in price to \$200/lb or less would be required before nanotubes would be used in making fuel cells, batteries, drug delivery systems, and as commercial composites for fabrics, beams, structural members, shielding material for consumer electronics devices, and lightweight automotive and aerospace components (66,68).

The rate at which markets open up to the nanopowder, nanocomposite, and nanotube group of materials depends on the pace of development of production technology and its ability to manufacturer lower priced, high quality, customized materials. Currently, the U.S. market for these materials stands at ~\$25 million. By 2005, it is expected that, as a group, these materials will begin exploiting commercial markets to a significant degree, resulting in total sales of \$250 million. Between 2005 and 2012, the price of the nanomaterials will decline rapidly as their quality and ability to be customized for specific markets increases. By the

Table 8. **U.S. Market Trends for Nanopowders, Nanocomposites, and Nanotubes $\times 10^9$ \$[a,b]**

	2002	2005	2012
direct economic impact	0.025	0.250	2.50
total economic impact[c]	0.050	0.750	6.80

[a] Refs. 9,66,68,72,75.
[b] Includes all advanced materials.
[c] Includes direct + indirect + induced impact.

end of the period, U.S. sales will reach \$2.5 billion, representing a total economic impact accruing to the United States of ~\$6.8 billion (9,66,68,72,75) (Table 8).

9.6. Nanofibers (9,78–83). Nanofiber technology refers to the synthesis by various means of fiber materials with diameters less than 100 nm. Nanofiber technology remains a new but growing field with promising applications. In general, advantages of nanofibers depend on their high flexibility and therefore their ability to conform to a large number of three-dimensional configurations. They also have a very high surface area allowing a myriad of interactions with chemical and physical environments. Recent research suggests possible industrial applications as ceramic ultrafilters, gas separator membranes, electronic substrates, medical and dental composites, fiber reinforced plastics, electrical and thermal insulation, structural aerospace materials, and catalyst substrates for petrochemical synthesis. Nanofibers also may be applied in advanced optical systems, according to the shape, number, and composition of the fibers (9,81,82).

Nanofiber materials may also be incorporated into new types of textiles. In addition to the United States, South Korea is particularly active in this field. Nanofibers potentially can impart beneficial properties to both natural and synthetic fibers, such as superior thermal insulation, durability, strength, resilience, texture, wrinkle resistance, and flexibility. Nanofibers may become the fiber itself through polymerization or, in the form of ultra-thin whiskers, be added to a traditional fiber to modify its properties. Currently, Burlington Industries (Burlington, N.C.), partnering with Nano-Tex (Greensboro, N.C.), leads research in this latter approach. Formed into shirts, pants, and other forms, these whisker-modified fabrics just recently entered the market. Advanced research also points to the possibility of polymerizing textile-grade nanosized fibers through the self-assembling of acetylene-based molecules or through biosynthesis that effects various polymeric combinations of protein materials (9,78).

One of the most promising areas of nanofiber technology involves applications in the biomedical area. Nanofibers potentially can be integrated into advanced drug delivery systems. Even more importantly, nanofibers can produce three-dimensional collagen-based matrices or "scaffolds." When these scaffolds are "seeded" with specific types of human cells, blood vessels of small diameter are formed. These vessels can then be transplanted into a patient. This application of nanofiber technology offers one of the most promising routes to manmade blood vessels. The market for this technology continues to grow rapidly. Currently, nearly 1.5 million hospital operations requiring arterial prostheses are performed in the United States annually, including one-half million coronary by-pass operations. Since no acceptable synthetic arteries currently exist, implanted arteries need to be harvested from the patient, a procedure that often results in complications (and failure), extends recovery time, and is limited to certain patients with usable vessels (9,79,80,83).

The process of making nanofiber collagens involves the use of "electrospinning" technology, similar to the first such method used by DuPont in making its early synthetic fibers. The modern process produces nanofibers from collagen (polylactic and poly glycolic acids) and various human protein materials. An electrical charge is placed on a syringe containing the collagen. The electrical field forces the collagen liquid through the syringe in a thin stream which dries once in the air and is collected on a spool as a fibrous material (9,83).

For this application, the electrospinning process, in a redesigned and computerized form, creates fibers with the necessary "layering" and orientation so that the nanofiber matrix has similar properties to naturally occurring blood vessels. Following synthesis and spinning, the nanofibers are then weaved into a cloth matrix (ie, the scaffolding), which exhibit high porosity and large surface area. The process also uses an innovative bioreactor designed to maintain the matrix structure and hinder necrosis during the cell growth process. Because the fibers closely resemble naturally occurring tissue, cells readily grow in the man-made scaffold. Over time, the technology promises to find application in the synthesis of organs, nerves, muscles, and other tissues. In the near term, the synthetic nanofiber collagen mats may serve as an innovative bandage to stop bleeding during surgeries and to act as scaffolding in order to speed growth of new tissue at the wound site (9,79,80,83).

The company NanoMatrix, Inc. (Irving, Tex.), which licensed its process from Virginia Commonwealth University (Richmond, Va.), is currently developing a process to synthesize bandages and eventually arteries and organs from collagen-based nanofibers. Researchers anticipate that nanofiber bandages to be on the market by 2005 and nanofiber arteries (and other organs) by 2010 (9).

Production of nanofiber materials as a whole is growing as unit prices continue to decline. Total sales of nanofibers in the United States are likely to reach ~\$89 million by 2005. By 2012, estimates place sales at >\$350 million. This means that, by 2012, the total economic impact in the United States due to the production of these materials will be between \$1 and \$2 billion (9,78,79,83) (Table 9).

9.7. Thin Films (9,84–92). Advanced thin film materials represent one of the newest and most promising of the emerging material technologies. In general, thin film materials are composed of different advanced materials—polymers, metals, and polycrystals—layered a few tenths of an Angstrom deep onto a foundation or substrate, such as glass, acrylic, steel, ceramics, silica, and

Table 9. **U.S. Market Trends for Advanced Nanofibers, $\times 10^9$ \$**[a,b]

	2002	2005	2012
direct economic impact	0.033	0.089	0.350
total economic impact[c]	0.132	0.356	1.40

[a] Refs. 3,78–79,83.
[b] Includes all advanced materials.
[c] Includes direct + indirect + induced impact

plastics. Whereas coatings are applied to surfaces, thin films often operate as stand alone components in a variety of products and systems including consumer electronics and electronic components, telecommunications devices, optical systems (eg, reflective, antireflective, polarizing, and beam splitter coatings), biomedical technology, sensor systems, electromagnetic and microwave systems, and energy sources and products (eg, batteries, photovoltaic cells) (9,84,91,92).

While a variety of thin film systems remain to be commercialized, certain types are currently in production and have entered into particular markets. A number of companies within the United States and internationally produce various types and amounts of thin film materials. Thin-film companies tend to be relatively small (eg, <200 employees) and specialized (eg, concentrating in optical thin films). Some companies focus only on the process technology; others synthesize the coating materials themselves, as well as manufacture the thin film unit (ie, film materials-substrate composite). These companies typically sell their products to original equipment manufacturers for incorporation into final components and devices. A number of these companies continue to pursue R&D on new thin film technology. These efforts are often supported by government agencies in the form of grants. The Departments of Energy and Defense remain particularly active in the thin film field (9,84–86,88–90).

The future success of thin-film technology depends to a large extent on the viability of the production process. One possible approach involves a thermal laser-based deposition technique. Also known as pulsed laser deposition (PLD), this method involves hitting a target composed of the desired film material with a laser beam of short pulse. The laser's thermal energy causes single atoms or atom clusters to project up at a right angle to the beam onto the desired substrate, forming a homogeneous thin film. This process avoids the formation of unwanted particulates that can cause defects in, and hinder performance of, the final film. Adjusting various parameters of the laser, as well as modifying the target and substrate materials, permits the custom design of a broad range of thin films. The PLD process is often associated with metallic thin film materials. A similar process, called chemical vapor deposition, takes place in a vacuum and involves the diffusion and adsorption of the film material in the form of vapors onto the surface of the substrate. Variations of this process have yet to be fully developed. These processes include plasma enhanced chemical vapor deposition, ultraviolet injection liquid source chemical vapor deposition, and metallorganic chemical vapor deposition (9).

The second general type of process in the manufacture of thin films, known as electrostatic (or ionic) self-assembly (ESA), offers a superior technique in the production of organic, as well as metallic, thin-film materials. The process, using chemical solution (or liquid-phase) deposition, depends on repeatedly dipping the selected substrate— which is cleaned and left with an electrical charge—into alternate aqueous solutions containing anionic and cationic metallic materials. The ESA process conserves on costs because it does not require an ultraclean environment and is not energy intensive. The process also permits scale up for mass production manufacture through the use of automatic dipping machines and robot-controlled fabrication stations. In this process, the electrically charged substrate is repeatedly "dipped" into the solution. As this occurs, thin layers of materials form on the foundation. The material itself, as well as the number and types of layers, determine the optical, electrical, magnetic and mechanical properties of the final thin-film product. Other process technologies appear promising. These include magnetron sputtering, which allows the deposition of different materials only a few atoms deep. In this process, the magnetic field of a magnetron acts on a plasma material, which is transported to the substrate through the sputtering action to the substrate (9,86).

Following the formation of a thin film on a substrate, a variety of film patterning techniques come into play, especially to pattern electronic circuits onto the film. Recent advances in nanolithographic (etching) techniques in particular allow more rapid and precise thin film patterning. These techniques, including X-ray lithography and electron and ion beam lithography, are critical for the economic production of complex thin-film circuitry (9).

Thin-film technology brings together different advanced materials as the primary film substance. In general, these film materials can be organic polymers (eg, organic polymer electronic—OPE—synthetic resins), metals and alloys, or crystals of various sorts (eg, titanium dioxide, magnesium fluoride). The 3M Corporation, eg, is developing a new generation of fluoroacrylate polymers that possess both electronic and anticorrosive properties. These polymers can form ultra-thin transparent coatings on a number of substrates, including copper, aluminum, ceramic, steel, tin, or glass. Possible applications for these types of materials include their use in wireless telecommunications systems, liquid crystal and electrochromatic display technology, reflective or light-emitting (smart) windows, advanced sensors, magnetic and laser devices, piezoelectric products and systems, biomedical devices and implants, antistatic electronic packaging (eg, for use in packaging and protecting microchips), photovoltaic systems, corrosion protection, and xerographic applications, It is expected that, in time, OPE thin films may replace silicon in a number of the most important electronic applications. In part, this is because the process of producing the OPE thin film is potentially significantly cheaper than the vacuum-deposition processes required in silicon technology (9,92).

Metal-based thin films also appear close to achieving a reduction in the size of circuits and circuit components for electronic applications and may, in fact, compete against the polymer thin films in these markets. Metal-based thin films offer greater purity and durable interconnections between microcircuit components. These advantages allow future computers to be made much smaller and to operate faster than current technology. Metal-based thin films promise to advance a new generation of microelectronic and electromagnetic components including capacitors, resisters, thermistors, transducers, inductors, and related elements. Specific types of metals, metal compounds, and alloys used in advanced thin films include alumina, tantalum, nickel, nickel-aluminum alloys, copper, silver, silver-palladium alloys, platinum, and zinc (9,84,85).

Both polymer- and metal-based thin films also provide a route to "printed" low-cost antennas for attachment onto different surfaces. These antennas possess large surface areas for capacitive coupling and may compete against certain types of metallic conductive coatings. Additional potential markets for thin film materials include applications in more efficient photovoltaic systems, thermally and electrically conductive adhesives (for adhesives for chip-to-substrate bonding or for connecting materials in electronic enclosures), thin-film transistors, carpets

Table 10. **U.S. Market Trends for Advanced Thin Film Materials, $\times 10^9$ \$**[a,b]

	2002	2005	2012
direct economic impact	0.056	0.24	0.725
total economic impact[c]	0.112	0.96	2.9

[a] Refs. 9,84,88,90–92.
[b] Includes all advanced materials.
[c] Includes direct + indirect + induced impact.

and fabrics, wireless identification tags, electrodes for ultrasmall electronic devices (replacing traditional materials such as indium tin oxide), future fuel cell components, and less expensive, smaller, and more advanced microelectromechanical systems (MEMS) that combine computers with tiny mechanical devices such as sensors, valves, gears, mirrors and actuators embedded in semiconductor chips (9,84–88,91,92).

Despite a wide range of possible applications, a number of technical and economic risks exist as advanced thin film materials attempt to broaden their market base. These problems include uncertain interface control; physical degradation of the polymer material in the presence of high temperatures, high electric fields, and exposure to solvents used in the circuit printing process (which limits the types of circuits that can be designed); uncontrolled charge leakage between thin film-based devices and circuit elements resulting in lower operating life and increased signal interference; reduced electrical performance and mechanical degradation due to impurities in the polymer or metal (9).

Beyond these technical issues are industry and market barriers. For example, the suppliers of silicon represent a particularly difficult market in which to compete since they are well entrenched in the industry and enjoy strong customer loyalty. Moreover, they compete vigorously against new competition by improving quality and lowering the price of their technology. Costs remain a concern as well for thin film producers. Currently, thin-film materials vary widely in cost, from $<\$50/m^2$ to in excess of $\$1000/m^2$, depending on the type of thin-film system involved. The higher costs of thin film technology complicate fibers attempts by producers to compete against the more traditional materials. However, as production processes enter into full mass production, unit prices for thin films as a whole will continue to decline, resulting in increased market penetration and sales. While U.S. production of advanced thin films will remain relatively modest through 2005, production will expand rapidly after that. By 2012, total U.S. sales are estimated at ~\$700 million. At this time, the total economic impact of advanced thin films will be nearly \$3 billion (9,84,88,90,92) (Table 10).

BIBLIOGRAPHY

1. "Horizons in Advanced Materials," *The Update (Online)*, www.acq.osd.mil/bmdo/bmdolink/html/update/sum01/updhor1.htm (Summer 2001).
2. *Trends in Nanotechnology Newsletter* (2000–2002).
3. S. Moskowitz, "Synthetic Organic Chemicals, Economic Evaluation," *Kirk-Othmer Encyclopedia of Chemical Technology (Online Ed.)*, John Wiley & Sons, Inc., New York, 2001.

4. S. Moskowitz, "History of Refining," *Macmillan Encyclopedia of Energy*, J. Zumerchik, ed., Vol. 3, Macmillan Reference U.S.A., New York, 2001.
5. A. Arora and N. Rosenberg, "Chemicals: A U.S. Success Story," in *Chemicals and Long-Term Economic Growth: Insights from the Chemical Industry*, A. Arora, R. Landau, and N. Rosenberg, eds., John Wiley & Sons, Inc., New York, 1998, pp. 71–102.
6. P. Spitz, *Petrochemicals: The Rise of an Industry*, John Wiley & Sons, Inc., New York, 1988.
7. A. Arora and A. Gambardella, "Evolution of Industry Structure in the Chemical Industry," in *Chemicals and Long-Term Economic Growth: Insights from the Chemical Industry*, A. Arora, R. Landau, and N. Rosenberg, eds., John Wiley & Sons, Inc., New York, 1998, pp. 379–413.
8. A. Chandler, Jr., T. Hikino, and D. Mowery, "The Evolution of Corporate Capabilities and Corporate Strategy and Structure within the World's Largest Chemical Firms: The Twentieth Century in Perspective," in *Chemicals and Long-Term Economic Growth: Insights from the Chemical Industry*, A. Arora, R. Landau, and N. Rosenberg, eds., John Wiley & Sons, Inc., New York, 1998, pp. 415–457.
9. S. Moskowitz, "The US Advanced Materials Industry: A Regional Analysis," SLM International Group (2002).
10. Greater Richmond Technology Council, "Technology Industries Profile: 2000–2001" (2001).
11. J. Mason, "VC Firm Harris & Harris Specializes in Small Tech," *Smalltimes*, www.smalltimes.com (February 26, 2002).
12. "In the Arm of 'Angels, Raising Venture Capital in a world without VCs," *Biospace* (Online), www.biospace.com/articles/072099.cfm (July 20, 1999).
13. H. Gillespie, "A Venture Capital Idea," American Chemical Society Online, http://pubs.acs.org/subscribe/journals/mdd/v05/i02/html/02money.html (February 2002).
14. J. Utterback, "The Dynamics of Product and Process Innovation in Industry," in *Technological Innovation for a Dynamic Economy*, C. Hill and J. Utterback, eds., Pergamon Press, New York, 1979, pp. 40–65.
15. M. Schnabel, "Grants to Aid Research for Luna's "Buckyball"," *The Roanoke Times* (October 16, 2001).
16. NanoSonic (Website), www.nanosonic.com (April 10, 2002).
17. E. Gardner, "Firms Lay Financial Foundation Resting on Nanoclay Composites," *Smalltimes*, www.smalltimes.com (April 4, 2002).
18. American Chemical Society, "Nanotechnology," *Chemical &* Engineering News (Online), http://pubs.acs.org/cen/nanotechnology/7842/7842business.html (October 16, 2002).
19. Unitech (Website), www.rp46.com (April 23, 2002).
20. Chea, Terence, "From Fields to Factories: Plant-Based Materials Replace Oil-Based Plastics, Polyesters," *Washington Post* (May 3, 2002).
21. U.S. Department of Energy, Idaho National Engineering and Environmental Laboratory, "Bioenergy Research Starts on the Farm," *Press Release* (December 2002).
22. SC Johnston Associates Inc., "Biorefineries," *TechNewsNote* (May 22, 2000).
23. Cargill (Website), www.cargill.com (2003).
24. U.S. Department of Agriculture, "Farm Policy: Title IX - Energy," (September 2002).
25. United States Advanced Ceramics Association, "Advanced Ceramics Technology Roadmap," (December 2000).
26. United States Advanced Ceramics Association, "What are Advanced Ceramics," *Government Affairs Seminar* (March 6, 2003).
27. U. Fink, "Advanced Ceramics Materials," *SRI Consulting* (December 2001).
28. "Nanoceramics on the Rise," *Design Engineering* (April 24, 2003).

29. T. Abraham, "Healthy Prospects for Nanoceramic Powders," *Ceramic Ind.* (May 1999).
30. Global Information, Inc., "Opportunities in Nanostructured Materials: biomedical, Pharmaceutical and Cosmetic - Table of Contents," theinfoshop.com, www.theinfoshop.com/study/bc8285_nanostructured_materials_toc.html. (August 2001).
31. "U.S. Piezoelectric Market Continues to Boom," *Business Communications Co.* (April 2001).
32. N. J. Porch and co-workers, "Net Shape Fabrication of Piezoelectric Ceramics and Composites," *School of Engineering, University of Birmingham* (n.d.).
33. "Piezoelectric Ceramics Market and Materials," *Business Communications Co.* (2002).
34. EDO Corporation, "Piezoelectric Ceramic Technology." (www.edocorp.com).
35. NASA, "High-Performance, Durable Actuators for Demanding Applications," *Technology Opportunity* (June 30, 2000).
36. NASA, "Piezo-Electric, Active, Fluid Flow Control Valve," *Technology Profile* (Sept. 12, 2000).
37. NASA, "Macro-Fiber Composite Actuator," *Technology Profile* (July 7, 2000).
38. "Piezoelectric Products: Actuators, Sensors, and Transducers," EDO Corporation (www.edocorp.com).
39. R. Nathan Katz, "Advanced Ceramics: Thermal Barrier Coatings Beat the Heat" *Ceramic Ind.* (April 1, 2001).
40. National Paint and Coatings Association, Industry Information (On-Line), www.paint.org/ind_info/ffspecl.htm (April 25, 2002).
41. "Thermal Spray Application in the Automotive Sector," *Plasma and Thermal Coatings* (May 16, 2001).
42. "Tailor-Made Surface Properties by Thermal Spraying," *Plasma and Thermal Coatings* (June 20, 2000).
43. W. D. Kimmel and D. D. Gerke, "Conductive Coatings and Applications," *Conformity 2003: The Annual Guide* (2003).
44. NASA, "Unitech, LLC Produces High-Performance Polyimide Material," *NASA TechFinder*, http://technology.nasa.gov/scripts/nls_ax.dll/w3SuccItem (202389) (May 21, 2001).
45. "Conductive Coatings for EMI/RFI Shielding," *Applied Coating Technology* (March 7, 2001).
46. S. Schreiber, "Conductive Coatings and Future Requirements in Vehicle Glass Design," *International Glass Rev.*, Issue 3 (2000).
47. Dynaloy Inc. (Website). www.dynaloy.com (May 2003).
48. U.S. Environmental Protection Agency, "Environmentally Friendly Anti-Corrosion Coatings," *Final Report* (March 2001).
49. Dow Corning (Website). www.dowcorning.com (2002).
50. SRI, "High Performance Anticorrosion Coatings," *Abstract* (2002).
51. B. Harder, "Steely Glaze: Layered Electrolytes Control Corrosion," *Science News Online*, 161(15) (April 13, 2002).
52. J. Escarsega, "Multifunctional Protective Coatings for Weapon Systems," *Clean Air Symposium* (May 22, 2002).
53. H. A. Corne and co-workers, "Multifunctional Nanocomposite Coatings," *Mater. Res. Symp. Proc.*, **628** (2000).
54. K. Canning, "How Small Can You Go: Nanopowders Enhance Industry Applications," *Chemical Processing* (April 11, 2003).
55. Business Communications Co., "The Year in Review," *Nanoparticle News (Online)*, http://www.buscom.com/letters/fptnpromo/fptn/fptn.html (January 2002).

56. Business Communications Co., *Nanoparticle News (Online)*, http://www.buscom.com/letters/fptnpromo/fptn/fptn.html (October 2002).
57. AP Materials (Website). www.apmaterials.com (2001).
58. Nanopowders Industries (Website). www.nanopowders.com (2001).
59. Accumet Materials (Website). www.accumetmaterials.com (2002).
60. R. Leaversuch, "Nanocomposites Broaden Role in Automotive, Barrier Packaging," *Plastics Technology Online* (Oct. 2001).
61. C. Stuart, "You've Come A Long Way, Bucky! Fullerenes Finally Leave the Lab," *Small Times* (Online), www.smalltimes.com (November 8, 2001).
62. *Luna Innovations* (Website), www.lunainnovations.com. (March 26, 2002).
63. Luna Nanomaterials (Website), www.lunananomaterials.com (March 26, 2002).
64. G. Kranz, "Virginia Firm Hopes Buckyballs Will Strike Rich New Vein for MRI," *Small Times* (Online), www.smalltimes.com (November 19, 2001).
65. "Metallofullerenes: Fullerene Molecules With Three Metal Atoms," *Composites / Plastics* (Online), http://composite.about.com/library/PR/1999/blvatech4.htm (September 1, 1999).
66. S. Brauer, "Emerging Opportunities for Carbon Nanotubes," *Ceramic Ind.* (January, 2002).
67. L. Kalaugher, "Nanotubes Line Up to Make Photonic Crystals," nanotechweb.org (December 4, 2002).
68. I. Amato, "The Soot That Could Change the World," *Fortune Magazine* (June 25, 2001).
69. P. Ball, "Nanotubes Hang Tough," *Nature (London)* (Oct. 14, 2002).
70. Carbon Nanotechnologies, Inc. (Website). www.cnanotech.com (2002)
71. J. Gorman. "Fracture Protection: Nanotubes Toughen Up Ceramics," *Science News Online*, Vol. 163, No. 1 (January 4, 2003).
72. American Chemical Society, "Nanotech Offers Some There, There" *Chem. Eng. News*, **79**, 48; CENAR 79 48, 21–26 (Nov. 26, 2001).
73. M. Kanellos, "IBM Finds New Way to Make Nanotubes," CNET News.com (September 29, 2002).
74. NanoDevices (Website). www.nanodevices.com (2002).
75. "Nanoparticle Industry Review," Business Communications Company, Inc., Norwalk, Connecticut (June 2003).
76. Applied Nanotechnologies, Inc. (Website), www.applied-nanotech.com (2002).
77. M. Roco, "From Vision to the Implementation of the U.S. National Nanotechnology Initiative," *J. Nanoparticle Res.*, **3**(1) (Nov. 5, 2000).
78. K. Maney, "Nanotech Advances Nanew [sic] Fabric," *USA Today* (2002).
79. D. Tennant, "Researchers Able to Make Natural Blood Vessels in Lab," *The Virginia Pilot* (December 22, 2002).
80. National Institute of Standards and Technology, "Living Vascular Implant," *Advanced technology Program 2001 Competition* (June 2002).
81. NASA, "Polyimide Fibers," *Technology Profile* (September 12, 2000).
82. NASA, "Polyimide Fibers," *The NASA Langley Edge* (February 28, 2001).
83. "Electrospun Fibers Make Bandages That are Absorbed by the Body," *United Press International* (February 11, 2003).
84. K. Zweibel, "Thin Films: Past, Present, Future," *Progress in Photovoltaics*, Vol. 3, No. 5 (Sept/Oct 1995).
85. MTI Corporation (Website): www.mticrystal.com (2002).
86. Applied Thin Films (Website): www.atfinet.com (2002).
87. National Physical Laboratory (UK), "Sensors & Functional Materials," www.npl.co.uk (2003).

88. J. Jackson, "Building a Business Layer by Layer," *The Update Online*, www.acq.osd.mil/bmdo/bmdolink/html/update/fall00/updcov.htm (Fall 2000).
89. Bp Solar (Website), www.solarex.com (2002).
90. M. Schnabel, "Virginia Nanotechnology Firm Works on Energy Alternatives," *Richmond Times-Dispatch* (August 30, 2001). (NanoSonic).
91. NASA, "Novel Polymer Film Technologies," *Technology Opportunity* (May 4, 2000).
92. Sopheon Corp., *International Markets for Organic Polymer Electronics*, Minneapolis, MN (2000).

Sanford L. Moskowitz
Villanova University

AERATION, BIOTECHNOLOGY

1. Introduction

The supply of oxygen to a growing biological species, aeration, in aerobic bioreactors is one of the most critical requirements in biotechnology. It was one of the biggest hurdles that had to be overcome in designing bioreactors (fermenters) capable of turning penicillin from a scientific curiosity to the first major antibiotic (1). Aeration is usually accomplished by transferring oxygen from the air into the fluid surrounding the biological species, from where it is in turn transferred to the biological species itself. The rate at which oxygen is demanded by the biological species in a bioreactor depends very significantly on the species, on its concentration, and on the concentration of the other nutrients in the surrounding fluid (1,2) (see Cell culture technology). There is no unique set of units used to define this rate requirement, but some typical figures are given in Table 1. The very wide range is noteworthy; during the course of a batch bioreaction, oxygen demand often passes through a marked maximum when the species is most biologically active (1).

The main reason for the importance of aeration lies in the limited solubility of oxygen in water, a value that decreases in the presence of electrolytes and other solutes and as temperature increases. A typical value for the solubility of oxygen

Table 1. **Oxygen Demands of Biological Species**

Biological species	kg $O_2/(m^3{\cdot}h)$	References
bacteria/yeasts	1–7	1, 3
plant cells	0.03–0.3	4
seed priming[a]	$1–8 \times 10^{-2}$	5
mammalian cells[b]	$2–10 \times 10^{-3}$	6

[a] Based on a seed density of 100 kg/m^3.
[b] Based on a cell density of 10^{12} cells/m^3.

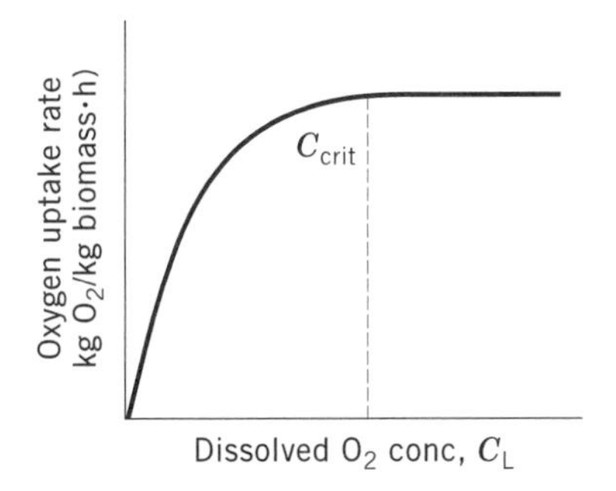

Fig. 1. The relationship between rate of oxygen uptake and dissolved oxygen, concentration where C_{crit} is the critical oxygen concentration.

(the equilibrium saturation concentration) in water in the presence of air at atmospheric pressure at 25°C is ~0.008 kg O_2/m^3 (=8 ppm=0.25 mmol/L). Thus, for a yeast or bacterial bioreaction demanding oxygen at the rates given in Table 1, all oxygen is utilized in ~10–40 s (3,7).

In addition to each bioreaction demanding oxygen at a different rate, there is a unique relationship for each between the rate of reaction and the level of dissolved oxygen (1,8). A typical generalized relationship is shown in Figure 1 for a particular species, eg, *Penicillium chrysogenum* or yeast. The shape of the curve is such that a critical oxygen concentration, C_{crit}, can be defined above which the rate of the bioreaction is independent of oxygen concentration, ie, zero order with respect to oxygen. The typical values given in Table 2 indicate that the critical concentration is usually on the order of 1–20% of the oxygen saturation value. Thus for each species, oxygen should be transferred rapidly enough to allow the oxygen demand to be met throughout the volume in which the bioreaction is occurring using a level of dissolved oxygen above the critical value. Failure to do so leads to a reduction in the overall rate and possibly a change of bioreaction to a different and unwanted metabolic pathway. If oxygen concentration falls low enough, an anaerobic reaction may develop. Yeast fermentations are a particularly good example of such possibilities (1,8).

In the cases of pellets (9), flocs (10), and immobilized cells and enzymes (11), aeration becomes more complex (1). Here it is the level of oxygen at the active biological site within the solid particle or aggregate rather than the level of dissolved oxygen in the surrounding fluid that determines the overall rate of reaction. Indeed, in certain cases such as penicillin fermentations, pellets form

Table 2. **Critical Oxygen Concentrations at 30°C**[a]

Organism	C_{crit}, (kg O_2/m^3) × 10^4
Azotobacter vinelandi	6–16
Pseudomonas denitrificans	3
Penicillium chrysogenum	3
Aspergillus oryzae	6.5

[a] Ref. 1.

of such a size that the center of the pellet becomes inactive (12). A similar effect is observed when biofilms form on cooling surfaces. At a certain thickness the oxygen concentration at the base of the film becomes zero, ie, that region becomes anaerobic, and the biological species, eg, *Pseudomonas fluorescens*, at the surface that is being fouled dies and sloughs off (13).

2. Principles of Oxygen Transfer

2.1. The Basic Mass-Transfer Steps. Figure 2 shows the steps through which oxygen must pass in moving from air (or oxygen-enriched air) to the reaction site in a biological species (1,14). The steps consist of transport through the gas film inside the bubble, across the bubble–liquid interface, through the liquid film around the bubble, across the well-mixed bulk liquid (broth), through the liquid film around the biological species, and finally transport within the species (eg, cell, seed, microbial floc) to the bioreaction site. Each step offers a resistance to oxygen transfer. In the last step, the resistance to transport is negligible for freely suspended bacteria or cells that are extremely small, but for immobilized cells and biological flocs and pellets the rate of oxygen diffusion through their structure may be rate limiting. In the more complex situation, the rate of oxygen diffusion to the active sites depends on mass transfer through the external boundary layer followed by diffusion through the solid as governed by Fick's law. The theory is essentially the same as that applied to chemical catalysis (qv.) and is described very well in (1), and other standard texts.

The rate-limiting step typically occurs at the air–liquid interface and, for biological species without diffusion limitations, the overall relationship can be simply written at steady state as

oxygen transfer rate from air (OTR) = oxygen uptake rate by biological species (OUR).

Provided this equality is satisfied and the dissolved oxygen concentration in the well-mixed liquid is greater than the critical concentration throughout the bioreactor, then the maximum oxygen demand of the species should be met

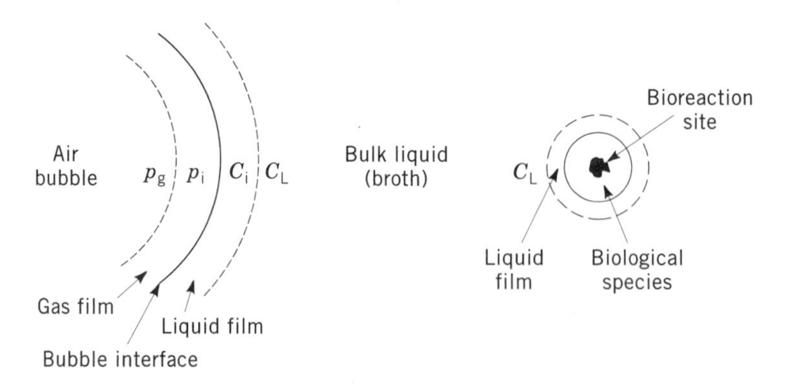

Fig. 2. Steps by which oxygen is transferred from the gas phase to the biological reaction site. Terms are defined in the text.

satisfactorily. Design of the bioreactor must ensure that the above requirements are achieved economically and without damaging the biological species.

When the oxygen demand is low, eg, at the start of a batch fermentation when the amount of biomass is small, the oxygen uptake rate at the bioreaction site is the limiting step and the level of dissolved oxygen is high (near saturation). When the oxygen demand becomes high, however, the rate-limiting step is that associated with transfer across the bubble–liquid interface and this rate of transfer is critically dependent on the fluid flow at the gas–liquid interface as well as the concentration of oxygen in both the gas and the liquid phases. The liquid-phase oxygen concentration is now low, but it should not be allowed to fall below the critical value. It is when this gas–liquid mass-transfer step is rate-limiting that the biggest demands are made on the bioreactor. Then there is the greatest difficulty in getting enough oxygen into the suspending fluid (broth) to satisfy the oxygen demand of the biological species.

2.2. The Basic Mass Transfer Relationship. The basic principles that underlie oxygenation (aeration) are exactly the same as those that determine the rate of transfer of any sparingly soluble gas (oxygen) from the gas stream (air) to the unsaturated liquid (broth). The rate at which this transfer takes place is dependent on four principal parameters (1,14,15). The first is the area of contact between the gas and the liquid. In most bioreactors, this is provided by dispersing air in the suspending fluid (medium or broth) to give a large specific area of contact as implied by Figure 2. However, there are also other methods, especially in wastewater treatment, by which large specific areas of contact are produced. Solid supports are used on which films of the irrigated, biologically active species grow, eg, trickling bed filters and rotating disk contactors (11,14). The other three mass-transfer rate parameters are the driving force available (ie, the difference in concentration of oxygen in the two phases); the two-phase fluid dynamics (including the effect of viscosity); and the chemical composition of the liquid.

The first assumption in all such physical mass transfer processes is that equilibrium exists at the interface between the two phases. This assumption implies that, at the interface, the concentration of the gas in the liquid, C_i, is equal to its solubility at its partial pressure in the gas phase, p_i. Since, for sparingly soluble gases such as oxygen, there is a direct proportionality between the two,

$$p_i = HC_i \tag{1}$$

where H is the Henry's law constant. The second assumption is that as the gas is absorbed, its concentration falls progressively, from a high in the bulk concentration in the gas phase, p_g, to that at the interface, p_i, then again from that on the liquid side of the interface, C_i, to a low in the bulk liquid, C_L (see Fig. 2).

The rate of mass transfer, J, is then assumed to be proportional to the concentration differences existing within each phase, the surface area between the phases, A, and a coefficient (the gas or liquid film mass transfer coefficient, k_g or k_L, respectively) which relates the three. Thus

$$J = k_g A(p_g - p_i) \tag{2}$$

$$= k_L A(C_i - C_L) \tag{3}$$

Equations 2 and 3 also stand as the definitions of k_g and k_L. For sparingly soluble gases, however, $p_i \simeq p_g$ so that from equation 1,

$$C_i = p_i/H = p_g/H = C_g^* \tag{4}$$

and

$$J = k_L A\left(C_g^* - C_L\right) \tag{5}$$

where C_g^* is the solubility of oxygen in the broth that is in equilibrium with the gas phase partial pressure of oxygen. Thus the aeration rate per unit volume of bioreactor, N, is given by

$$N = J/V = k_L\,(A/V)\left(C_g^* - C_L\right) = k_L a\left(C_g^* - C_L\right) \tag{6}$$

where a is the interfacial area per unit volume. At steady state, N must equal the OUR, which should be that demanded by the biological species in the range $C_L > C_{crit}$.

2.3. The Driving Force for Mass Transfer. The rate of mass transfer increases as the driving force, $C_g^* - C_L$, is increased. C_g^* can be enhanced as follows. From Dalton's law of partial pressures

$$p_g = P_g y \tag{7}$$

where y is the mole (volume) fraction of oxygen in the gas phase and P_g is the total pressure, ie, back pressure plus static head. C^*_g (eqs. 4 and 5) is also a function of composition and temperature, decreasing with both. C_L can be low provided it is above C_{crit}. However, in a very large scale bioreactor where circulation times are of the same order as the time for total oxygen depletion (3,7), the average value may need to be kept well above C_{crit} so that local values below it may be avoided. There are very few studies on the variations of C_L in any bioreactor (16,17), partly because of the difficulty of measuring it. C_L is usually measured by an oxygen electrode that gives C_L/C^*_g as a percentage and this value is sensitive to the local velocity over the probe, particularly if the liquid is viscous (18).

In a stirred bioreactor, the liquid is generally considered well-mixed, ie, C_L is spatially constant. The gas phase too may be well mixed (19) so that

$$p_g = \text{constant} = (p_g)_{out} \tag{8}$$

where $(p_g)_{out}$ is the partial pressure of oxygen in the exit gas and

$$C_g^* = (p_g)_{out}/H \tag{9}$$

This situation is most probable with fairly intense agitation and on the small scale. On the other hand, for ease of application, it is often assumed that no oxygen is utilized. This is the so-called no-depletion model. In this case

$$C_g^* = (p_g)_{in}/H \tag{10}$$

Again, it is most reasonable on the small scale, but only with low $k_{L}a$ values. Most of the data leading to $k_{L}a$ values in the literature [especially those in the review in (20)] were obtained by making this assumption.

For large-scale bioreactors (21), especially those of the air lift type (22), the gas phase is best considered as being in plug flow, so that a log mean value of driving force is obtained:

$$\Delta C_{\text{im}} = \frac{\Delta C_{\text{in}} - \Delta C_{\text{out}}}{\ln(\Delta C_{\text{in}}/\Delta C_{\text{out}})} \tag{11}$$

where

$$\Delta C_{\text{in}} = [(p_{\text{g}})_{\text{in}}/H] - C_{\text{L}} \tag{12}$$

and

$$\Delta C_{\text{out}} = [(p_{\text{g}})_{\text{out}}/H] - C_{\text{L}} \tag{13}$$

2.4. The Mass-Transfer Coefficient, $k_{L}a$. Because of the interaction between k_{L} and a when air is dispersed in the media, the two have not often been measured separately. When they have been measured, k_{L} has been found to be dependent on the relative velocity between the phases, the level of turbulence in the liquid, and the size of bubbles. However, it is a relatively weak function of all of those. In addition, it appears that impurities, whether in solution or suspension, reduce the ease with which oxygen can pass across the interface (23) and, consequently, reduce the value of k_{L}.

The Sauter mean bubble size, d_{B}, is also relatively insensitive to the fluid dynamics, especially in low viscosity broths. On the other hand, it is quite sensitive to the composition of the fluid, especially to the presence of substances that inhibit coalescence. In fact, coalescence and its inhibition plays a much larger role in controlling bubble size than bubble break-up. Theories of break-up suggest that bubble size increases with increasing surface tension, σ. However, in aqueous solutions, alcohols reduce σ whilst electrolytes increase it but both inhibit coalescence and reduce d_{B} (24). Unfortunately, as a result, a satisfactory equation is not available for calculating d_{B}. Nevertheless, those interested in CFD predictions of bioreactors generally do not appear to recognise this fact.

When the bubbles are smaller, especially in stirred bioreactors, bubbles very easily circulate with the broth. This increased recirculation when coalescence is inhibited leads to a significantly larger hold-up, ie, increased volume fraction of gas phase, ε_{H}. d_{B}, ε_{H}, and a are all related.

$$a = 6\epsilon_{\text{H}}/d_{\text{B}} \tag{14}$$

Therefore, a values in the presence of coalescence inhibitors are much higher than without them (25), the changes in a outweighing changes in k_{L}. Thus, it is found that coalescence inhibitors such as electrolytes and low molecular weight alcohols increase $k_{L}a$ by up to an order of magnitude compared to water (26). Coalescence promoters such as antifoams generally lower $k_{L}a$ by somewhat similar amounts (27,28). In addition, neither d_B nor k_L is a strong function of the fluid dynamics, but the amount of air recirculation is, so k_La and ε_H are both related to the fluid dynamics in similar ways (29). However, exceptions to this interrelationship have been reported (30) and therefore it should be treated with care. For example, if a certain agitator produces a higher hold-up without a commensurate increase in k_La, it represents a reduction in performance because productivity is linked to the volume of broth in the bioreactor.

Increases in broth viscosity significantly reduce k_La and cause bubble size distributions to become bimodal (31). Overall, k_La decreases approximately as the square root of the apparent broth viscosity (32). k_La can also be related to temperature by the relationship (33)

$$k_La = (k_La)_{20}\theta^{(T-20)} \tag{15}$$

where $(k_La)_{20}$ is the value at 20°C, $\theta \simeq 1.022$, and T is the temperature, °C; ie, k_La increases by ~2.5%/ °C.

2.5. The Measurement of k_La. There are two main methods for measuring k_La, the unsteady-state method, and the steady-state method. In the most common unsteady-state method, the level of dissolved oxygen is first reduced to zero, either by bubbling through nitrogen or by adding sodium sulfite (20). Then, the increase in dissolved oxygen concentration as a function of time is followed using an oxygen electrode. One of the assumptions set out earlier about the extent of gas-phase mixing has to be made and the choice can be very significant, giving k_La values when they are high differing by an order of magnitude for the same raw data (34). If two oxygen electrodes, one in the liquid phase and one in the gas, are employed, this difficulty can be eliminated, but the technique is quite difficult to use (35,36). Two other ways of circumventing these gas-phase mixing issues have also been proposed. In one, pure oxygen is used after degassing has been completed so that the driving force is always constant (37). In the other, a step change in backpressure is utilized to produce the transient (38). Both are rather specialized and it is not apparent that they are being used significantly.

In addition, the electrode response time requires consideration (18,20) as does the quality of liquid-phase mixing (22). Most significantly, it should again be emphasized that k_La is very dependent on composition. For example, the k_La of distilled water measured by nitrogen degassing is significantly less than that measured by sulfite deoxygenation. This difference occurs because the sulfite acts as a coalescence inhibitor, greatly enhancing hold-up, and therefore a and k_La.

For stirred bioreactors, the unsteady-state technique can be adapted to give k_La values for real bioreactions (39) including animal cell culture (40). Though the method has the same inherent weaknesses as all the dynamic techniques, its advantage is that the k_La is determined for the system of interest. These experimental values may well be very different from those predicted by literature correlations based on water or idealized liquids. This point is likely to be particularly valid for mycelial fermentations for example. Additionally, the presence of antifoam can cause a dramatic reduction (28).

Steady-state techniques are better and often can be carried out on real bioreactions. In most cases, however, and especially for batch systems, it is necessary to carry out an oxygen balance on the air in order to determine the mass of oxygen utilized and thus the mass flux represented in equation 6. Assumptions must again be made about the gas- and liquid-phase mixing. The significance of the assumptions is less than in the unsteady-state, except on the large scale where the well-mixed liquid assumption breaks down (16,17). No really satisfactory way of overcoming this problem has yet been proposed. Given suitable instrumentation, the steady-state method of measurement for the bioreaction of interest is the recommended technique. However, it is not suitable in systems having very low oxygen demands, where the extent of oxygen utilization is insufficient to give an accurate measure of the drop in oxygen concentration. In that case, the unsteady-state method (39,40) is best, as exemplified by its successful use with seed priming bioreactors (5).

Recently, a new steady-state technique has been introduced and it has become quite widely adopted. It involves the catalytic reduction of hydrogen peroxide either by catalase (41) or manganese dioxide (42). The former catalyst is less robust but is very good with polymer solutions simulating high viscosity fermentation broths where it has been utilised in fermenters up to 80 m^3 (43). However, the catalase is easily denatured especially by electrolytes and in that case, MnO_2 is better (42). The only data required is the constant flow rate of H_2O_2 and the dissolved oxygen concentration.

3. Aeration in Bioreactors

A huge variety of bioreactors has been developed and a thorough review is available (44). It is not feasible to consider them all and large numbers are only curiosities. A useful subdivision has been made into three generic types involving the way in which air is dispersed to give the desired specific surface area. These are bioreactors driven by rotating agitators (stirred tanks), bioreactors driven by gas compression (bubble columns/loop fermenters), and bioreactors driven by circulating liquid (jet loop reactors) (45). The first two are the most important.

3.1. Stirred Tank Bioreactors. Traditionally, stirred tanks have been the most common types of bioreactors for aerobic processes and they remain so even in the face of newer designs. One of the main reasons is their extreme flexibility. Operational designs using controlled air flow rates up to ~1.5 vvm (volume of air/min per unit volume of fermentation fluid) and variable speed motors capable of transmitting powers up to about 5 W/kg with control down to close to zero are suitable for almost any bioreaction. These tanks are also relatively insensitive to fill, ie, to the proportion of liquid added to the bioreactor, and are therefore quite satisfactory for fed batch operations. Control of dissolved oxygen can be carried out by either altering aeration rate and/or agitator power input (via speed control) or for fed batch by adjusting the nutrient feed rate. They are also capable of handling relatively satisfactorily broths that become significantly viscous during the course of a fermentation. In that case, design for good bulk mixing (homogenization) may be the most demanding task that the agitator is required to carry out (46).

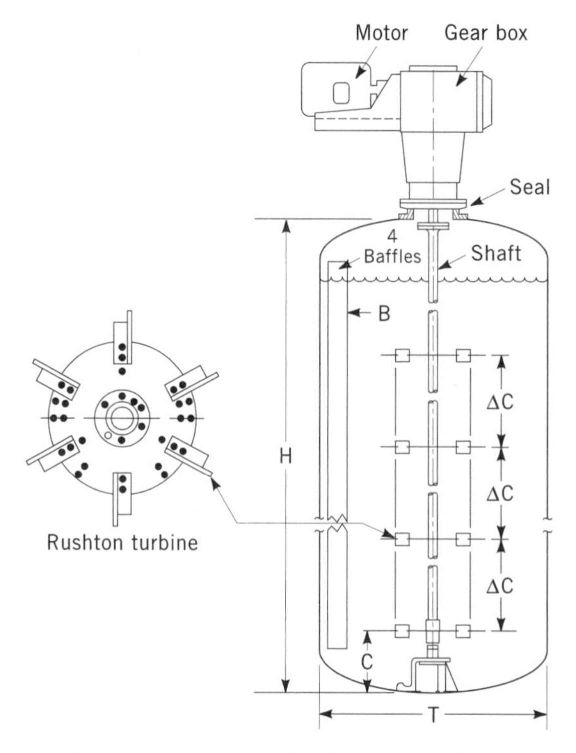

Fig. 3. A large-scale fermenter agitated by Rushton turbines (47) where B/T = 0.1, H/T $\simeq$ 3.3, D/T $\simeq$ 0.35, C/T = 0.25, and ΔC/T = 0.51.

Until recently most industrial scale, and even bench scale, bioreactors of this type were agitated by a set of Rushton turbines having about one-third the diameter of the bioreactor (47) (Fig. 3). In this system, the air enters into the lower agitator and is dispersed from the back of the impeller blades by gas-filled or ventilated cavities (48). The presence of these cavities causes the power drawn by the agitator, ie, the power required to drive it through the broth, to fall and this has important consequences for the performance of the bioreactor with respect to aeration (36). The parameter $k_{\mathrm{L}}a$ has been related to the power per unit volume, P/V, in W/m^3 (or $P/\rho V$ in W/kg) and to the superficial air velocity, v_{s}, in m/s (20), where v_{s} is the air flow rate per cross-sectional area of bioreactor. This relationship in water is

$$k_{\mathrm{L}}a = 2.6 \times 10^{-2}\left(\frac{P}{V}\right)^{0.4} v_{\mathrm{s}}^{0.5} \tag{16}$$

and for electrolyte solutions:

$$k_{\mathrm{L}}a = 2.0 \times 10^{-3}\left(\frac{P}{V}\right)^{0.7} v_{\mathrm{s}}^{0.2} \tag{17}$$

Each equation is independent of impeller type. As pointed out earlier, the absolute $k_{\mathrm{L}}a$ values vary considerably from liquid to liquid. However, similar

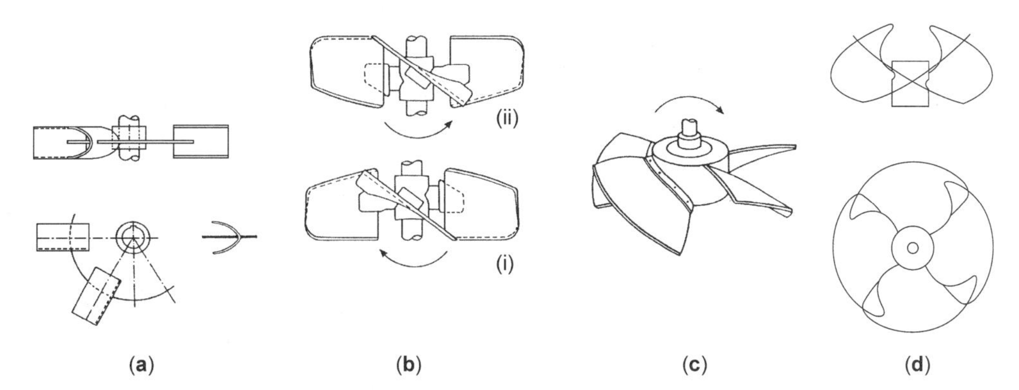

Fig. 4. Examples of agitators (53) progressively replacing the Rushton turbine: (**a**) radial Scaba 6SRGT, (**b**) (i) down-pumping Lightnin' A315[which has also been shown to work effectively in the up-pumping mode (ii) (52)], (**c**) down-pumping Prochem Maxflo T, and (**d**) up-pumping Hayward Tyler (formally APV) B2.

relationships have been found for other fluids, including fermentation broths, and also for hold-up, ε_H. Therefore, loss of power reduces the ability of the Rushton turbines to transfer oxygen from the air to the broth.

There are two other features of Rushton turbines that require consideration (49). First, as these turbines disperse the gas–liquid mixture, they drive it radially outward at each agitator leading to rather poor top-to-bottom mixing. Second, if the air flow rate is too high, they can no longer disperse the air and the impellers are said to become flooded (50). New impellers have been introduced that are better able to handle the large quantities of air needed in large scale fermentations and that do not lose power so significantly. The most common have been the so-called, hollow-blade radial flow impellers, especially the Chemineer CD6 and the Scaba 6SRGT [Fig. 4(**a**)] (49). However, as with Rushton turbines, these other radial impellers in mutiple configurations as found in industrial scale fermenters, also provide relatively poor top-to-bottom bulk blending of nutrients (including air) (49). Therefore there has been another trend which has been aimed at improving this characteristic by using multiple down-pumping, wide blade hydrofoil impellers such as the Lightnin' A315 [Fig. 4(**b**)] and the Prochem Maxflo T [Fig. 4(**c**)] (49). These types do improve vertical mixing but on aeration, they lose power and have very unstable torque, power and flow characteristics (49). A quite counter-intuitive procedure is the utilization of multiple up-pumping, wide blade hydrofoils such as the Hayward Tyler (formerly APV) B2 [Fig. 4(**d**)] (51) or the Lightnin' A315 (52). These give stable and high aerated power characteristics (49), a very high air handling ability without flooding and good vertical blending, even in viscous broths at specific power inputs up to ~2 W/kg (52). The Ekato Intermig (not shown) was proving popular in Europe but it has been shown to give intense vibrational problems because of the large gas-filled cavities that it develops, especially in viscous broths (53) and the manufacturers no longer recommend it for aeration.

These new impellers are progressively taking over from Rushton turbines. However, the reported improvements achieved by them are attributable to better bulk blending compared to Rushton turbines rather than to improved rates of aeration, ie, k_La (46). Indeed, though higher k_La values may be achieved, it is only as a result of increases in either P/V or v_s in equations 16 or 17.

Recently, an equation based on data obtained by the steady-state hydrogen peroxide technique has been obtained (54). The fluids used were water and viscous shear thinning sodium methyl carboxy–cellulose solutions. The impellers used were a Rushton turbine, a Lightnin' A315 , a hollow-blade impeller and a pair of Intermigs. The effect of viscosity was allowed for by the method of Metzner and Otto (53). The equation is

$$k_La = 4.7 \times 10^{-4}(P/V)^{0.34}(v_s)^{0.27}(\mu_a)^{-0.75} \tag{18}$$

where μ_a is in kg/m s. This equation fitted the data to ±20% that is typical for such work (20). However, it is important to recognise that there are many other equations available and it is impossible to determine which is the best.

3.2. Bubble Column and Loop Bioreactors. Air driven bioreactors are said to offer these advantages (44): No opening for a shaft is required and therefore they are less likely to become contaminated; and they are very simple to operate on the very small scale and more economic on the very large scale where huge agitators and motors would otherwise be required (55). Initially, it was also considered that they were less likely to damage bioreactions involving fragile material such as plant (4) or mammalian cells. However, a recent paper on the former, which acknowledges that such systems are often rather viscous, recommends up-pumping hydrofoils (56). Also, for the latter, stresses associated with bursting bubbles rather than rotating agitators have been shown to be the main determinants of cell damage (57).

Examples of air driven bioreactors are given in Figure 5. The bubble column is clearly the simplest of these bioreactors to construct. However, because of its rather ill-defined liquid circulation, air-lift reactors having either internal (draught tube) or external (loop) circulation of broth have been introduced. The major disadvantages of all three types are the poor capability of handling very viscous fermentations, especially those having a yield stress; the inflexibility, especially of the airlift types, which only work well using a fill closely matched to the size of the bioreactor and its internals (this match affects both circulation rates, mixing and mass transfer, and bubble disengagement); and lack of independent control of dO_2 and mixing, since both are closely linked to the aeration rate. Reference 58 presents a very interesting literature review and analysis of bubble columns. In contrast to stirred tank bioreactors, the bubble size may in certain cases be very dependent on the way the air is introduced, ie, on the type of sparger employed. For systems having hindered coalescence, very fine (0.25–1 mm diameter) bubbles can be formed by using porous disks, provided the superficial gas velocity is less than $\sim 10^{-2}$ m/s. These conditions lead to very high k_La values. However, unless coalescence is very repressed, the bubble size grows within ~0.5–1 m to give k_La values similar to those found in coarse bubble systems. There may also be disengagement problems if coalescence does not occur.

Coarse bubble systems are typically found with orifice, perforated disks, or pipe spargers. Under most realistic conditions, bubbles of ~4–6 mm are

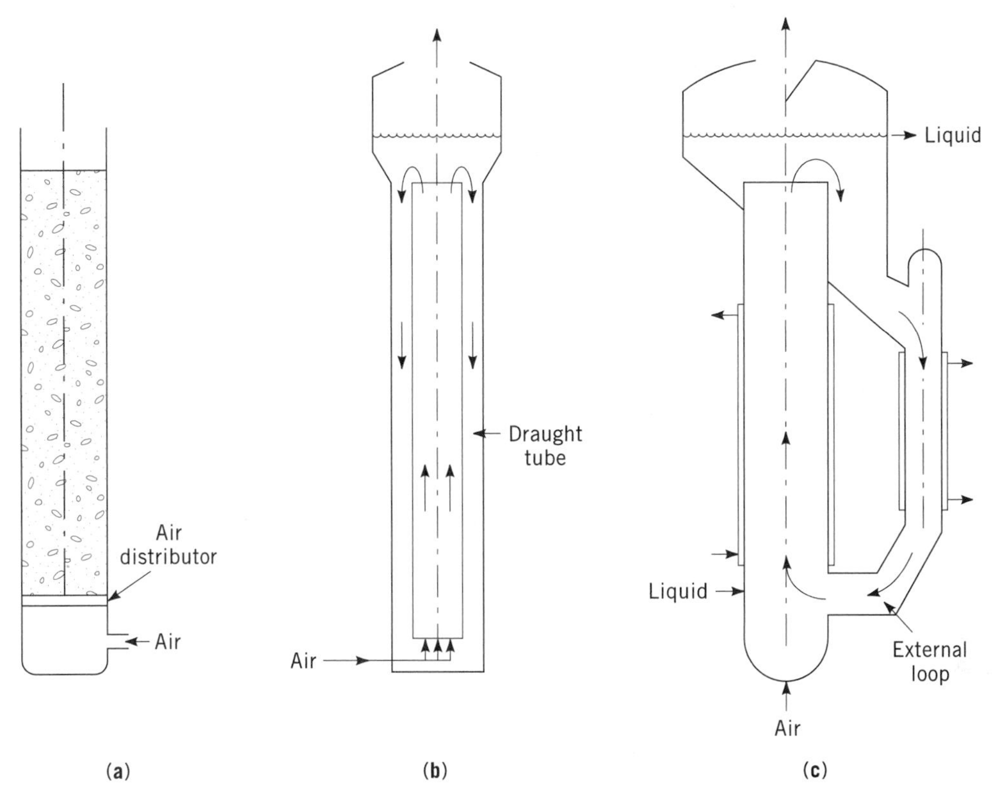

Fig. 5. Examples of air driven bioreactors: (**a**) bubble column, (**b**) draught tube, and (**c**) external loop.

formed. For a wide range of sizes,

$$k_{\mathrm{L}}a = 0.32v_{\mathrm{s}}^{0.7} \tag{19}$$

Figure 6 enables a comparison to be made of $k_{\mathrm{L}}a$ values in stirred bioreactors and bubble columns (58). It can be seen that bubble columns are at least as energy-efficient as stirred bioreactors in coalescing systems and considerably more so when coalescence is repressed at low specific power inputs (gas velocities).

It is also interesting to use Figure 6 to make a comparison of different aeration devices on the basis of energy efficiency. From equation 6 and assuming a constant driving force,

$$N \propto \mathrm{OTR} \propto k_{\mathrm{L}}a \tag{20}$$

and from Figure 6

$$k_{\mathrm{L}}a \propto (P/V)^{0.6} \tag{21}$$

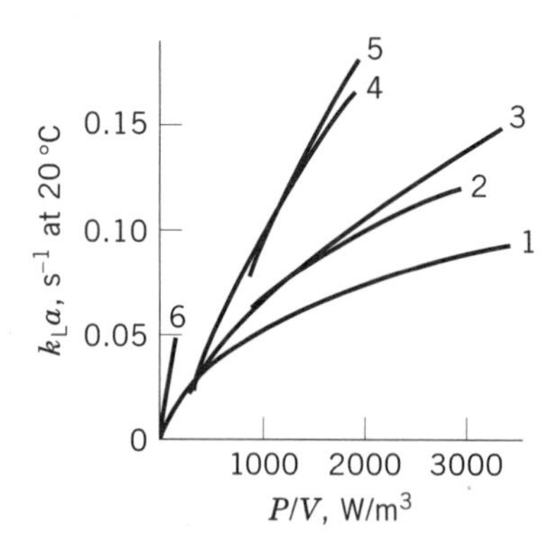

Fig. 6. A comparison of k_La values (58). Represented are 1, stirred bioreactor using water, $v_s = 0.02$ m/s, k_La (eq. 16); 2, stirred bioreactor using water, $v_s = 0.04$ m/s, k_La (eq. 16); 3, bubble column using water, k_La (eq. 19); 4, stirred bioreactor using salt water, $v_s = 0.02$ m/s, k_La (eq. 17); 5, stirred bioreactor using salt water, $v_s = 0.04$ m/s, k_La (eq. 17); and 6, bubble column using salt water (noncoalescing).

Therefore,

$$\frac{\text{OTR}}{P/V} \propto \left(\frac{P}{V}\right)^{-0.4} \tag{22}$$

or

$$\frac{(\text{OTR})V}{P} \propto \left(\frac{P}{V}\right)^{-0.4} \tag{23}$$

so that

$$\left(\frac{\text{kg O}_2}{\text{kW·h}}\right) \propto \left(\frac{P}{V}\right)^{-0.4} \tag{24}$$

From equation 24, it can be seen that the higher the power input per unit volume, the lower the oxygen-transfer efficiency. Therefore, devices should be compared at equal transfer rates. All devices become less energy efficient as rates of transfer increase (3).

External and internal loop airlifts and bubble column reactors containing a range of coalescing and non-Newtonian fluids, have been studied (59,60). It was shown that there are distinct differences in the characteristics of external and internal loop reactors (61). Overall, in this type of equipment

$$k_La \propto \mu_a^{-0.9} \tag{25}$$

showing a greater fall as viscosity increases than in stirred tank bioreactors ($k_La \propto \mu_a^{-0.5} - \mu_a^{-0.7}$) and supporting the contention that such devices are unsuitable for viscous broths. The complete lack of oxygen transfer found in the downcomer in one high viscosity study (59) is also a significant factor and could easily lead to values of $C_L < C_{crit}$ being found in that region.

4. Applications to Different Biological Species

4.1. Mycelial Fermentations. Mycelial fermentations typically become viscous and shear thinning and difficult to mix (53). For such systems, k_La and mixing are inextricably interlinked. Agitators that give better bulk blending per unit of energy can also give higher k_La values by involving more of the fermenter volume in the mass-transfer process (46). In practice, this requirement suggests low power number, large impeller-to-tank diameter ratio impellers such as those shown in Figure 4 should be used. Agitation levels link with mycelial structure and if pelleted growth can be encouraged, for example, using *P. chrysogenum* (12), viscosity does not significantly increase and high levels of k_La can be maintained at satisfactory specific power inputs. Difficulties may arise, however, if the pellets are too large, as a result of diffusion resistance within them (1), leading to oxygen starvation of the potentially active sites at the center of the pellet (9).

Because of the high viscosity, adequate bulk blending, oxygen transfer, and cooling often require high specific power inputs. Recent work has clearly shown that such levels of agitation can cause significant mycelial breakage. In the case of *P. chrysogenum*, such breakdown leads to loss of productivity (62). However, with *Aspergillus oryzae*, though breakdown occurs, productivity is unaffected (63). Models of this breakage have been developed and validated and the reader is referred to the original papers (62,63) for further information as it is outside the scope of this article. Clearly, however, if breakage does not affect productivity, short periods of intense agitation might be used to break hyphae in order to lower viscosity (63). Such an action should then lead to enhanced k_La and allow lower agitation intensity to be used subsequently.

4.2. Xanthan Gum Fermentations. Xanthan gum (and other polysaccharide) fermentations become very viscous toward the end whether batch (64) or fed-batch (65). Therefore, satisfying the oxygen demand at this stage is very demanding of agitation power. Because the broth also possesses a yield stress, bulk blending to maintain dissolved oxygen above the critical concentration throughout the fermenter has a major impact on the productivity and the quality of the gum (66). Stirred bioreactors are therefore preferable using large impeller/tank diameter ratios and more closely placed impellers (66).

4.3. High Oxygen Demanding Fermentations. High oxygen demanding fermentations often require a higher level of oxygen over relatively short periods of time. The use of enriched air, or pure oxygen, as a separate feed stream and/or back pressure for a short period to enhance C_g^* as well as maximizing power and aeration rate to give the highest k_La may be worth considering. In such circumstances, the flexibility of the stirred bioreactor is a distinct advantage.

4.4. Animal Cell Culture. Airlift (67) and bubble column bioreactors have been considered necessary for handling fragile animal cells. However, more recent work has shown that bursting bubbles are much more damaging to these cells than agitation unless protective agents such as Pluronic F68 are included in the media (68,69). On the other hand, sparged aeration in the presence of Pluronic in standard stirred bioreactors have been found not to damage hybridoma cells even at rather high (0.25 kW/m^3) agitator power inputs on a small scale (68,69). Such power inputs even at low aeration rates (say 0.05 vvm) would meet the needs of most industrial animal cell culture processes at the cell densities presently achievable and the use of enriched air and back pressure would further extend the range of cell densities that could be handled. Furthermore, it has recently been shown that the typical very low power inputs initially used industrially ($<\sim$0.01 W/kg) lead to major homogenization problems with respect to pH (70). A trend to higher specific power inputs at the industrial scale in order to overcome these problems is discernable.

4.5. Plant Cell Culture. Airlift bioreactors have been favored for plant cell systems since these cultures were first studied (4). However, they can give rise to problems resulting from flotation of the cells to form a "meringue" on the top and they are often rather viscous (56). It is interesting to note that some reports indicate that stirred bioreactors do not damage such cells (4), and a recent paper (56), after an extensive review of the literature, has recommended the use of up-pumping, wide-blade hydrofoil impellers in agitated bioreactors.

4.6. Seed Priming Bioreactors. Seed priming is a relatively new technique enabling seeds suspended in an osmotica to imbibe moisture and thus be brought to the point of germination (5). However, germination does not occur. Subsequently, on sowing, germination is very rapid and synchronous. While the process is taking place (up to $\sim$14 days), the seeds require oxygen. Both bubble column (71) and stirred bioreactors (5) have been used successfully, although the former requires high air rates to keep seeds in suspension (72). In addition, some seeds such as onions, appear to have a critical oxygen concentration greater than saturation with respect to air that may be due to diffusion limitations within the seed. In that case, enriched air must be used and, in order for the process to be economic, stirred bioreactors are appropriate (71).

4.7. Single-Cell Protein. Systems involving single-cell proteins are often very large throughput, continuous processing operations, such as the Pruteen process developed by ICI. These are ideal for airlift bioreactors of which the pressure cycle fermenter is a special case (55).

4.8. Biological Aerobic Wastewater Treatment. Biological aerobic wastewater treatment is a rather specialized biotechnical application (73). The activated sludge process consists of an aerated bioreactor to which the basic principles of oxygen transfer discussed here apply. Either aerated agitators or air spargers (diffusers) are used. Where the effluent has especially high oxygen demands, however, pure oxygen or oxygen-enriched air is employed. Two examples are the UNOX process (74), which involves agitation and the ICI deep-shaft process (equivalent to a loop fermenter) that is very effective where space is a premium (75). For maintaining the aeration of large quantities of relatively pure water having a low oxygen demand, where space is not a limitation, such as in reservoirs, simple plunging jets having low $k_{L}a$ values but very high energy efficiency are suitable (27). There is also a range of specialized devices in which the area for oxygen transfer is achieved by the use of extended solid surfaces on which a biofilm, irrigated by the water being treated, grows, eg, trickle beds and rotating disk contactors (73). Also, three-phase systems (fluidized beds) have been developed (11). These are bubble columns in which the air flow keeps inert solids such as sand in suspension giving a large surface area on which the biofilm can grow (11).

5. Nomenclature

Symbol	Definition	SI units
a	interfacial area of air per unit volume of liquid	m^2/m^3
A	interfacial area available for mass transfer	m^2
C_{crit}	critical dissolved oxygen concentration	kg/m^3
C_i	concentration of oxygen in the liquid phase at the interface	kg/m^3
C_L	concentration of oxygen in the bulk liquid	kg/m^3
C_g^*	saturation concentration of oxygen in the liquid for an oxygen partial pressure p_g	kg/m^3
d_B	Sauter mean bubble size	m
H	Henry's law constant	$kPa{\cdot}m^3/kg$
J	oxygenation rate	kg/s
k_g	gas film mass-transfer coefficient	$kg/(s{\cdot}m^2{\cdot}kPa)$
k_L	liquid film mass-transfer coefficient	m/s
N	oxygen mass flux	$kg/(m^3{\cdot}s)$
P	power imparted to the liquid	W
P_g	total pressure of the gas phase	kPa
p_g	partial pressure of oxygen in the bulk of the gas phase	kPa
p_i	partial pressure of oxygen in the gas phase at the interface	kPa
T	temperature	°C
v_s	superficial gas velocity	m/s
y	volume (mole) fraction of oxygen in gas phase	dimensionless
ϵ_H	hold-up	dimensionless
σ	surface tension	N/m
θ	temperature coefficient = 1.022	dimensionless
μ_a	apparent dynamic viscosity	$kg/(m \cdot s)$
ρ	broth density	kg/m^3
	Subscripts	
in	air entering the bioreactor	
out	gas leaving the bioreactor	
lm	ln mean	

BIBLIOGRAPHY

"Aeration, Biotechnology" in *ECT* 4th ed., Vol. 1, pp. 645–660, by Alvin W. Nienow, University of Birmingham (U.K.); "Aeration, Biotechnology" in *ECT* (online), posting date: December 4, 2000, by Alvin W. Nienow, University of Birmingham.

CITED PUBLICATIONS

1. J. E. Bailey and D. F. Ollis, *Biochemical Engineering Fundamentals*, 2nd ed., McGraw-Hill Book Co., New York, 1986.

2. K. Kargi and M. Moo-Young, in M. Moo-Young, ed., *Comprehensive Biotechnology*, Vol. 2, Pergamon Press, Oxford, 1985, Chapt. 2.
3. K. van't Riet, *Trends Biotechnol.* **1**, 113 (1983).
4. A. H. Scragg and co-workers, *Proceedings of the International Conference on Bioreactors and Biotransformations*, NEL/Elsevier, London, 1987, p. 12.
5. A. W. Nienow and P. A. Brocklehurst, *Proceedings of the International Conference on Bioreactors and Biotransformations,* NEL/Elsevier, London, 1987, p. 52.
6. M. Lavery and A. W. Nienow, *Biotechnol. Bioeng.* **30**, 368 (1987).
7. A. P. J. Sweere, K. Ch. M. M. Luyben, and N. W. F. Kossen, *Enzyme Microb. Technol.* **9**, 386 (1987).
8. B. Atkinson and F. Mavituna ,*Biochemical Engineering and Biotechnology Handbook*, Macmillan, London, 1983.
9. B. Metz and N. W. F. Kossen, *Biotechnol. Bioeng.* **19**, 781 (1977).
10. B. Atkinson and I. S. Daoud, *Adv. Biochem. Eng.* **4**, 42 (1976).
11. C. Webb, in C. F. Forster and D. A. J. Wase, eds., *Environmental Biotechnology*, Ellis Horwood, Chichester, 1987, Chapt. 9.
12. K. Schugerl and co-workers, *Proceedings of the 2nd International Conference on Bioreactor Fluid Dynamics*, British Hydromechanics Research Association, Cranfield, U.K., 1988, p. 229.
13. J. A. Howell and B. Atkinson,*Water Res.* **10**, 304 (1976).
14. A. W. Nienow, in Ref. 11, Chapt. 10.
15. P. V. Danckwerts, *Gas–Liquid Reactions*, McGraw-Hill Book Co., New York, 1970.
16. N. M. G. Oosterhuis and N. W. F. Kossen, *Biotechnol. Bioeng.* **26**, 546 (1984).
17. R. Manfredini, V. Cavallera, L. Marini, and G. Donati, *Biotechnol. Bioeng.* **25**, 3115 (1983).
18. Y. H. Lee and G. T. Tsao, *Adv. Biochem. Eng.* **13**, 35 (1979).
19. N. P. D. Dang, D. A. Karrer, and I. J. Dunn, *Biotechnol. Bioeng.* **19**, 953 (1977).
20. K. van't Riet, *Ind. Eng. Chem. (Proc. Des. Dev.)* **18**, 357 (1979).
21. D. I. C. Wang and co-workers, *Fermentation and Enzyme Technology*, Wiley-Interscience, New York, 1979.
22. G. André, C. W. Robinson, and M. Moo-Young, *Chem. Eng. Sci.* **38**, 1845 (1983).
23. J. T. Davies, *Turbulence Phenomena*, Academic Press, New York, 1972.
24. V. Machon, A. W. Pacek, and A. W. Nienow, *Trans I. Chem. E., Part A* **75**, 339 (1997).
25. A. Prins and K. van't Riet, *Trends Biotechnol.* **5**, 296 (1987).
26. M. Zlokarnik, *Adv. Biochem. Eng.* **8**, 134 (1978).
27. J. A. C. van de Donk, R. G. J. M. Lans, and J. M. Smith, *Proceedings of the Third European Conference on Mixing*, British Hydromechanics Research Association, Cranfield, 1979, p. 289.
28. B. C. Buckland and co-workers, in Ref. 12, p. 1.
29. S. P. S. Andrews, *Trans Inst. Chem. Eng.* **60**, 3 (1982).
30. T. Martin, C. M. McFarlane, and A. W. Nienow, *Proceedings of the Eighth European Conference on Mixing*, Institution of Chemical Engineers, Rugby, U.K., 1994, p. 57.
31. V. Machon, J. Vlcek, A. W. Nienow, and J. Solomon, *Chem. Eng. J.* **14**, 67 (1980).
32. A. D. Hickman and A. W. Nienow, *Proceedings of First International Conference on Bioreactor Fluid Dynamics*, British Hydromechanics Research Association, Cranfield, U.K., 1986, p. 301.
33. M. L. Jackson and C. C. Shen, *AIChEJ.* **24**, 63 (1978).
34. C. M. Chapman, L. G. Gibilaro, and A. W. Nienow, *Chem. Eng. Sci.* **37**, 891 (1983).
35. S. N. Davies and co-workers, *Proceedings of the Fifth European Conf. on Mixing*, British Hydromechanics Research Association, Cranfield, U.K., 1985, p. 27.
36. A. W. Nienow and co-workers, in Ref. 12, p. 159.
37. V. Linek, P. Benes, and J. Sinkule, *Biotechnol. Bioeng.* **35**, 766 (1990).

38. V. Linek, P. Benes, and J. Sinkule, *Biotechnol. Bioeng.* **33**, 301 (1989).
39. B. Bandyapadhyay, A. E. Humphrey, and H. Taguchi, *Biotechnol. Bioeng.* **9**, 533 (1967).
40. C. Langheinrich and co-workers, *Trans I. Chem. E., Part C* **80**, 39 (2002).
41. M. Cooke and co-workers, *Proceedings of the Seventh European Conf. on Mixing*, KVIV, Antwerp, Belgium, 1991, p. 409.
42. J. M. T. Vasconcelos and co-workers, *Trans I. Chem. E., Part A***75**, 467 (1997).
43. A. G. Pedersen, *Bioreactor and Bioprocess Fluid Dynamics*, (Ed. A.W. Nienow), BHR Group/MEP, London, U.K., 1997, p. 263.
44. K. Schugerl, *Int. Chem. Eng.* **22**, 591 (1982).
45. H. Blenke, *Adv. Biochem. Eng.* **13**, 121 (1979).
46. B. C. Buckland and co-workers, *Biotechnol. Bioeng.* **31**, 737 (1988).
47. A. W. Nienow, *Chem. Eng. Progr.* **86**, 61 (1990).
48. M. M. C. G. Warmoeskerken and J. M. Smith, in Ref. 12, p. 179.
49. A. W. Nienow, *Trans I. Chem. E., Part A* **74**, 417 (1996).
50. A. W. Nienow, M. M. C. G. Warmoeskerken, J. M. Smith, and M. Konno, in Ref. 35, p. 143.
51. D. Hari-Prajitno and co-workers,*Can. J. Chem. Eng.* **76**, 1056 (1998).
52. A. W. Nienow, *Proceedings of the Third International Symposium on Mixingin Industrial Processes*, Osaka University, Japan, 1999, p. 425.
53. A. W. Nienow, *App. Mech. Rev.* **51**, 3 (1998).
54. M. K. Dawson, Ph.D. thesis, University of Birmingham, U.K., 1992.
55. S. R. L. Smith, *Philos. Trans. R. Soc. (London), Ser. B.* **290**, 341 (1980).
56. P. M. Doran, *Biotechnol. Prog.* **15**, 319 (1999).
57. A. W. Nienow, *Trans I. Chem. E. Part C* **78**, 145 (2000).
58. J. J. Heinen and K. van't Riet, *Proceedings of the Fourth European Conf. on Mixing*, British Hydromechanics Research Association, Cranfield, U.K., 1982, p. 195.
59. M. K. Popovic and C. W. Robinson, *AIChEJ.* **35**, 393 (1989).
60. M. Y. Chisti and M. Moo-Young, *J. Chem. Tech. Biotechnol.* **42**, 211 (1988).
61. M. Y. Chisti, *Airlift Bioreactors*, Elsevier Applied Science, New York, 1989.
62. P. Justen, G. C. Paul, A. W. Nienow, and C. R. Thomas, *Biotechnol. Bioeng.* **59**, 762 (1998).
63. A. Amanullah and co-workers, *Biotechnol. Bioeng.* **77**, 815 (2002).
64. A. Amanullah and co-workers, *Biotechnol. Bioeng.* **57**, 95 (1997).
65. A. Amanullah, S. Satti, and A. W. Nienow, *Biotechnol. Prog.* **14**, 265 (1998).
66. A. Amanullah, B. Tuttiet, and A. W. Nienow, *Biotechnol. Bioeng.* **57**, 198 (1998).
67. L. A. Wood and P. W. Thompson, in Ref. 32, p. 157.
68. S. Oh, A. W. Nienow, M. Al-Rubeai, and A. N. Emery, *J. Biotechnol.* **12**, 45 (1989).
69. N. Kioukia, A. W. Nienow, A. N. Emery, and M. Al-Rubeai, *J. Biotechnol.* **38**, 243 (1995).
70. C. Langheinrich and A. W. Nienow, *Biotechnol. Bioeng.* **66**, 171 (1999).
71. W. Bujalski, A. W. Nienow, and D. Gray, *Ann. Appl. Biology* **115**, 171 (1989).
72. W. Bujalski and A. W. Nienow, *Scientia Hortic.* **46**, 13 (1991).
73. C. F. Forster and D. W. M. Johnstone, in Ref. 11, Chapt. 1.
74. A. J. Blatchford, E. M. Tramontini, and A. J. Griffiths, *Water Pollut. Control* **81**, 601 (1982).
75. J. Walker and G. W. Wilkinson, *Ann. N.Y. Acad. Sci.* **326**, 181 (1979).

Alvin W. Nienow
University of Birmingham

AEROGELS

1. Introduction

Aerogels, solid materials that are so porous that they contain mostly air, were first prepared in 1931. Kistler used a technique known as supercritical drying to remove the solvent from a gel (a solid network that encapsulates its solvent) such that, as stated in his paper in *Nature*, "no evaporation of liquid can occur and consequently no contraction of the gel can be brought about by capillary forces at its surface (1)." In the six decades since then, there have been significant advances made both in the use of new precursors to form gels and in the removal of solvent from them. These advances have greatly simplified the preparation of aerogels and, in turn, improved their economic viability for commercial applications. Increasingly an aerogel is defined in terms of its properties and not the way in which it is prepared.

Almost all applications of aerogels are based on the unique properties associated with a highly porous network. Envision an aerogel as a sponge consisting of many interconnecting particles. The particles are so small and so loosely connected that the void space in the sponge, the pores, can make up for over 90% of the sponge volume. As an example, a silica aerogel contains particles that are of the order of 10 nm and each particle is connected to two or three other particles on average. Such a material has a typical density of about 100 kg/m^3 and accessible surface area of about 1000 m^2/g.

The ability to prepare materials of such low density, and perhaps more importantly, to vary the density in a controlled manner, is indeed what make aerogels attractive in many applications such as thermal insulation, detection of high energy particles, and catalysis. Thus, this article begins with a discussion of sol–gel chemistry used to form the wet gel. The intention is not to discuss in detail the fundamental chemistry involved, but to provide the basic principles that explain the formation of a porous network and the effect of preparative variables on its microstructure (often referred to as nanostructure because the relevant length scale is on the order of nanometer), ie, factors that impact on density. The general applicability of these principles are illustrated by examples of inorganic, organic, and inorganic–organic gels.

The section on preparation and manufacturing continues to focus on the microstructure of a gel, in particular its evolution during the removal of solvent. In addition to the original supercritical drying used by Kistler, there are now safer and more cost-effective methods that do not involve supercritical drying. More importantly, these methods ensure that the products possess the defining characteristics of an aerogel (ultrafine pores, small interconnected particles, and high porosity).

This article also aims at establishing the structure–property–application relationships of aerogels. Selected examples are given to show what some desirable properties are and how they can be delivered by design based on an understanding of the preparation and preservation of a gel's microstruc ture.

2. Sol–Gel Chemistry

2.1. Inorganic Materials. Sol–gel chemistry involves first the formation of a sol, which is a suspension of solid particles in a liquid, then of a gel, which is a diphasic material with a solid encapsulating a solvent. A detailed description of the fundamental chemistry is available in the literature (2–4). The chemistry involving the most commonly used precursors, the alkoxides ($M(OR)_m$), can be described in terms of two classes of reactions:

Hydrolysis

$$-M-OR + H_2O \longrightarrow -M-OH + ROH$$

Condensation

$$-M-OH + XO-M- \longrightarrow -M-O-M + XOH$$

where X can either be H or R, an alkyl group.

The important feature is that a three-dimensional gel network comes from the condensation of partially hydrolyzed species. Thus, the microstructure of a gel is governed by the rate of particle (cluster) growth and their extent of crosslinking or, more specifically, by the *relative* rates of hydrolysis and condensation (3).

The gelation of silica from tetraethylorthosilicate (TEOS) serves as an example of the above principle. Under acidic conditions, hydrolysis occurs at a faster rate than condensation and the resulting gel is weakly branched. Under basic conditions, the reverse is true and the resulting gel is highly branched and contains colloidal aggregates (5, 6). Furthermore, acid-catalyzed gels contain higher concentrations of adsorbed water, silanol groups, and unreacted alkoxy groups than base-catalyzed ones (7). These differences in microstructure and surface functionality, shown schematically in Figure 1, result in different responses to heat treatment. Acid- and base-catalyzed gels yield micro- (pore width less than 2 nm) and meso-porous (2–50 nm) materials, respectively, upon heating (8). Clearly an acid-catalyzed gel which is weakly branched and contains surface functionalities that promote further condensation collapses to give micropores. This example highlights a crucial point: *the initial microstructure and surface functionality of a gel dictates the properties of the heat-treated product.*

Besides pH, other preparative variables that can affect the microstructure of a gel, and consequently, the properties of the dried and heat-treated product include water content, solvent, precursor type and concentration, and temperature (9). Of these, water content has been studied most extensively because of its large effect on gelation and its relative ease of use as a preparative variable. In general, too little water (less than one mole per mole of metal alkoxide) prevents gelation and too much (more than the stoichiometric amount) leads to precipitation (3, 9). Other than the amount of water used, the rate at which it is added offers another level of control over gel characteristics.

The principles discussed so far are valid for silicates as well as nonsilicates, although there are more data available for the former. The alkoxides of

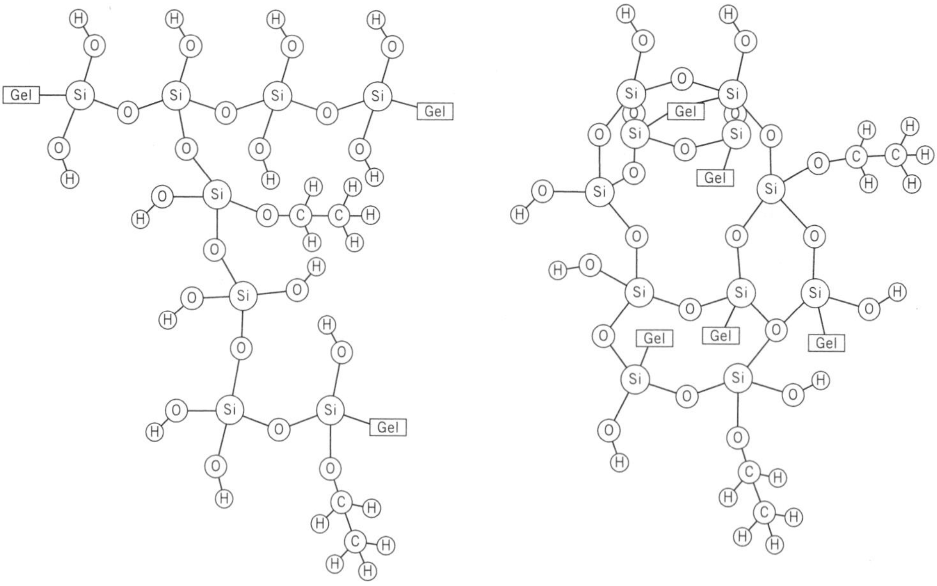

Fig. 1. Schematics of (**a**) acid-catalyzed and (**b**) base-catalyzed silica gels showing the differences in microstructure and surface functional groups. Reproduced from Ref. 7. Courtesy of the American Ceramic Society.

transition metals tend to be more reactive than silicon alkoxides because both hydrolysis and condensation are nucleophilic substitution reactions and their cations have a more positive partial charge than silicon (3). This difference in reactivity presents both a challenge and an opportunity in the preparation of two-component systems. In a two-component system, the minor component can either be a network modifier or a network former. In the latter case, the distribution of the two components, or mixing, at a molecular level is governed by the *relative* precursor reactivity. Qualitatively good mixing is achieved when two precursors have similar reactivities. When two precursors have dissimilar reactivities, the sol–gel technique offers several strategies to prepare well-mixed two-component gels. Two such strategies are prehydrolysis (10), which involves prereacting a less reactive precursor with water to give it a head start, and chemical modification (11), which involves slowing down a more reactive precursor by substituting some of its alkoxy groups with bulkier, less reactive groups such as acetate. The ability to control microstructure *and* component mixing is what sets sol–gel apart from other methods in preparing multicomponent solids.

2.2. Organic Materials. The first organic aerogel was prepared by the aqueous polycondensation of resorcinol with formaldehyde using sodium carbonate as a base catalyst (12). Figure 2**a** shows that the chemistry is similar to the sol–gel chemistry of inorganic materials. Subsequent to the reaction between resorcinol and formaldehyde, the functionalized resorcinol rings (ie, those possessing–CH_2OH groups) condense with each other to form nanometer-sized clusters, which then crosslink to form a gel. The process is influenced by typical

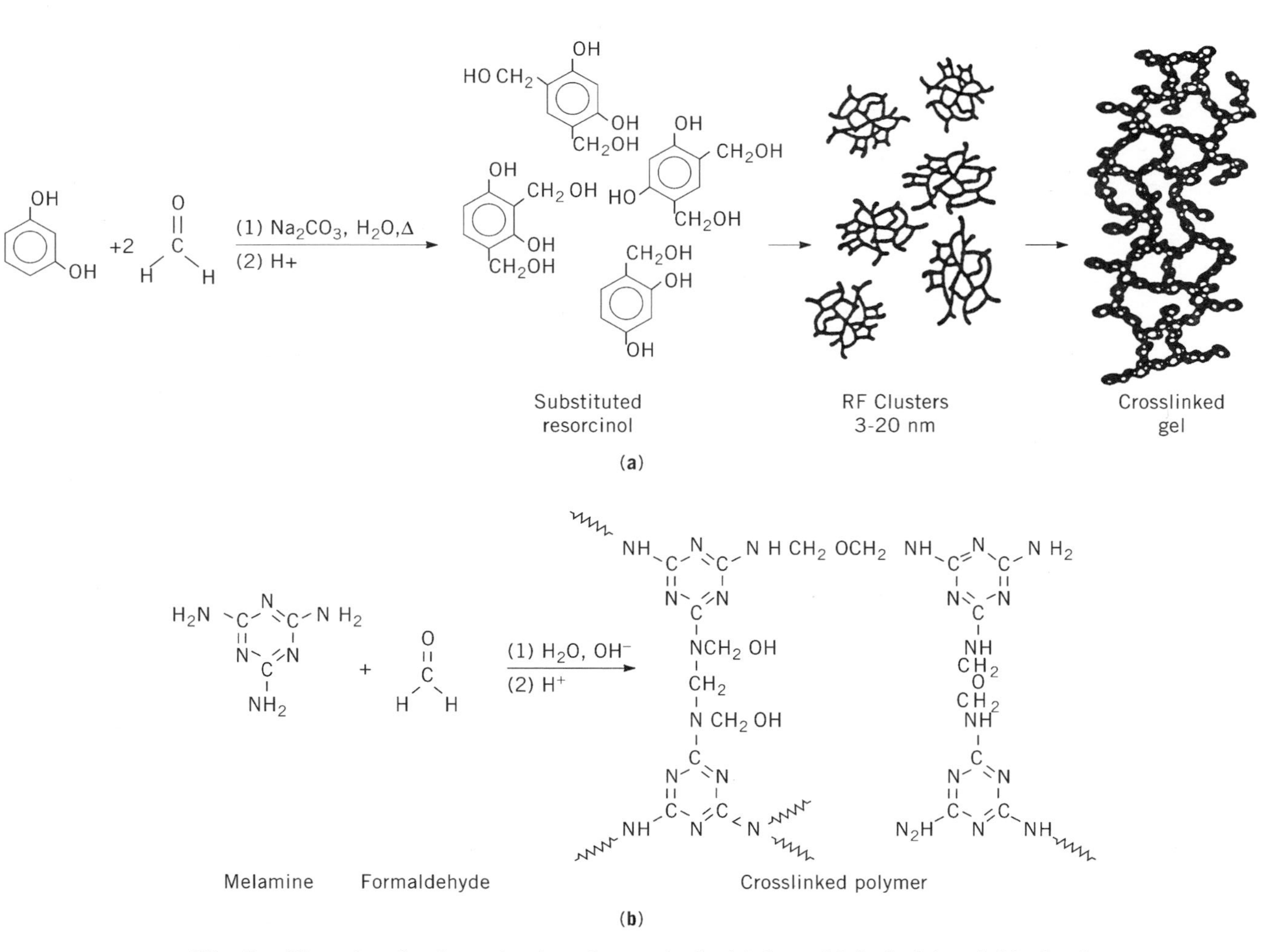

Fig. 2. The sol–gel polymerization of resorcinol with formaldehyde (**a**) and (**b**) of melamine with formaldehyde. Reproduced from Refs. 13 and 14, respectively. Courtesy of the Materials Research Society.

sol–gel parameters such as pH, reactant concentration, and temperature. The resorcinol–catalyst molar ratio turns out to be the dominant factor that affects the microstructure of the gel (13).

Resorcinol–formaldehyde gels are dark red in color and do not transmit light. A colorless and transparent organic gel can be prepared by reacting melamine with formaldehyde using sodium hydroxide as a base catalyst (14). The chemistry, shown in Figure **2b**, involves first the formation of $-CH_2OH$ (hydroxymethyl) groups and then the formation of $-NHCH_2NH-$ (diamino methylene) and $-NHCH_2OHC_2H-$ (diamino methylene ether) bridges through cross-linking reactions. The solution pH is again a key parameter, it is necessary to add hydrochloric acid in the second part of the process to promote condensation and gel formation.

Both the preparations of resorcinol–formaldehyde and melamine–formaldehyde gels are aqueous-based. Since water is deleterious to a gel's structure at high temperatures and immiscible with carbon dioxide (a commonly used supercritical drying agent), these gels cannot be supercritically dried without a tedious solvent-exchange step. In order to circumvent this problem, an alternative synthetic route of organic gels that is based upon a phenolic–furfural reaction using an acid catalyst has been developed (15). The solvent-exchange step is eliminated by using alcohol as a solvent. The phenolic–furfural gels are dark brown in color.

Carbon aerogels can be prepared from the organic gels mentioned above by supercritical drying with carbon dioxide and a subsequent heat-treating step in an inert atmosphere. For example, the following heating schedule has been used in flowing nitrogen (13, 15): 22 to 250°C in 2 h, held at 250°C for 4 h, 250 to 1050°C in 9.5 h, held at 1050°C for 4 h, and cooled to room temperature in 16 h. This treatment results in a volume shrinkage of the sample of 65–75% and a mass of 45–50%. Despite these changes, the carbon aerogels are similar in morphology to their organic precursors, underscoring again the importance of structural control in the gelation step. Furthermore, changing the sol–gel conditions can lead to aerogels that have a wide range of physical properties. As a specific example, surface area for the phenolic–furfural aerogels is about $385 \pm 16\ m^2/g$ over a density range of 100–250 kg/m^3, whereas the corresponding carbon aerogels have surface areas of $512 \pm 40\ m^2/g$ over a density range of 300–450 kg/m^3.

2.3. Inorganic–Organic Hybrids. One of the fastest growing areas in sol–gel processing is the preparation of materials containing both inorganic and organic components. The reason is that many applications demand special properties that pure materials can seldom provide. The combination of inorganic and organic materials is, thus, an attractive way to deliver materials that have desirable physical, chemical, and structural characteristics. In this regard, sol–gel chemistry offers a real advantage because its mild preparation conditions do not degrade organic polymers, as would the high temperatures that are associated with conventional ceramic processing techniques. The voluminous literature on the sol–gel preparation of inorganic–organic hybrids can be found in several recent reviews (16–20) and the references therein; only a qualitative sketch is given in this section.

There are several ways to classify inorganic–organic materials, all of which depend on the strength of interaction between the two components. The interaction can range from nonexistent or weak, such as organic species embedded (or entrapped) in an inorganic network, to very strong, such as systems involving covalent bonds. For weakly interacting systems, the general observation is that most molecules can be entrapped in sol–gel matrices and, once entrapped, these molecules retain most of their characteristic physical and chemical properties (18). The entrapment of bioorganic materials such as enzymes, whole cells, antibodies, and other proteins (21) falls into this category.

For strongly interacting systems, chemical bonding can be induced by using functionalized precursors. There are three basic types of precursors: inorganically functionalized preformed organic polymers, organically functionalized oxides, and precursors containing both inorganic and organic functional groups. Examples of commonly used precursors are organofunctional metal alkoxides $(RO)_n$–E–X–A (19) and bridged polysilsesquioxanes (20). In $(RO)_n$–E–X–A, A is a functional organic group and X is a hydrolytically stable spacer linking A and the metal alkoxide which provides the inorganic function. The use of such precursors has allowed the control of very small domain sizes, often in the nanometer range, in the preparation of inorganic–organic materials (16,19,20). The challenge is to achieve a high degree of mixing of the two phases, thus, enabling the manipulation of interfacial properties at a molecular level. Another promising strategy to provide better homogeneity between the two phases is to form the inorganic and organic phases simultaneously, leading to what is known as a simultaneous interpenetrating network (16).

All the methods described so far involve introducing the organic phase prior to the formation of a solid phase. There is an interesting alternative to prepare a nanocomposite by a sequential approach (22). In this approach a silica aerogel was first prepared, then carbon was deposited in it by the decomposition by hydrocarbons (eg, methane, acetylene) at a temperature range of 500–850°C. This example demonstrates the feasibility of preparing hybrid materials by depositing an inorganic or organic phase onto an organic or inorganic substrate, respectively.

3. Preparation and Manufacturing

3.1. Supercritical Drying. The development of aerogel technology from the original work of Kistler to about late 1980s has been reviewed (23). Over this period, supercritical drying was the dominant method in preparing aerogels and, for this reason, aerogels are synonymous with supercritically dried materials. As noted earlier, supercritical drying could be an insufficient definition of an aerogel because it might not lead to the defining characteristics. Kistler used inorganic salts and a large amount of water in his work (1,24), making the subsequent salt removal and solvent-exchange steps time-consuming (water has to be removed because it would dissolve the gel structure at high temperatures). A significant time savings came when alkoxides were used as precursors in organic solvents, thereby requiring a minimum amount of water and eliminating the tedious solvent-exchange step (25,26). The introduction of carbon dioxide, which has a lower critical temperature and pressure than alcohols, as a supercritical drying agent allowed the drying step to be done under milder conditions and improved

Table 1. **Important Developments in the Preparation of Aerogels**

Decade	Developments
1930	using inorganic salts as precursors, alcohol as the supercritical drying agent, and a batch process; a solvent-exchange step was necessary to remove water from the gel
1960	using alkoxides as precursors, alcohol as the supercritical drying agent, and a batch process; the solvent exchange step was eliminated
1980	using alkoxides as precursor, carbon dioxide as the drying agent, and a semicontinuous process; the drying procedure became safer and faster. Introduction of organic aerogels
1990	producing aerogel-like materials without supercritical drying at all; preparation of inorganic–organic hybrid materials

its safety (27). The development of a semicontinuous drying process further facilitated the preparation of aerogels (28). Together these advances, summarized in Table 1, have made possible the relatively safe supercritical drying of aerogels in a matter of hours. In recent years, the challenge has been to produce aerogel-like materials without using supercritical drying at all in an attempt to deliver economically competitive products. This topic will be discussed in more detail later.

As stated earlier, the main idea behind supercritical drying is to eliminate the liquid–vapor interface inside a pore, thereby removing the accompanying capillary pressure which acts to collapse a gel network. The value of this approach is demonstrated by the fact that aerogels do have higher porosities, higher specific surfaces areas, and lower apparent densities than xerogels, materials that are prepared by evaporative drying. However, it is incorrect to think that a gel remains static during supercritical drying. Rather, supercritical drying should be considered as part of the aging process, during which events such as condensation, dissolution, and reprecipitation can occur. The extent to which a gel undergoes aging during supercritical drying depends on the structure of the initial gel network. For example, it has been shown that a higher drying temperature changes the particle structure of base-catalyzed silica aerogels but not that of acid-catalyzed ones (29). It is also known that gels that have uniform-sized pores can withstand the capillary forces during drying better because of a more uniform stress distribution. Such gels can be prepared by a careful manipulation of sol–gel parameters such as pH and solvent or by the use of so-called drying control chemical additives (DCCA) (30). Clearly, an understanding of the interrelationship between preparative and drying parameters is important in controlling the properties of aerogels.

The most widely studied supercritical drying variable is temperature simply because different solvents have different supercritical temperatures. Specifically, since alcohols have higher supercritical temperatures than carbon dioxide, there have been many recent reports on the effect of drying agent on the textural properties and crystallization behavior of aerogels (31–33). These results demonstrate nicely the accelerated aging that a gel undergoes at high temperatures and pressures. For this reason carbon dioxide is the drying agent of choice if the goal is to stabilize kinetically constrained structure, and materials prepared by this low-temperature route are referred to by some people as *carbogels*. In general, carbogels are also different from aerogels in surface functionality, in particular hydrophilicity, which impacts on the moisture sensitivity of these materials and their subsequent transformations under heat treatment.

It is less well known, but certainly no less important, that even with carbon dioxide as a drying agent, the supercritical drying conditions can also affect the properties of a product. For example, in the preparation of titania aerogels, temperature, pressure, the use of either liquid or supercritical CO_2, and the drying duration have all been shown to affect the surface area, pore volume, and pore size distributions of both the as-dried and calcined materials (34,35). The specific effect of using either liquid or supercritical CO_2 is shown in Figure 3 as an illustration (36).

Other important drying variables include the path to the critical point, composition of the drying medium, and depressurization (37). The rates of heating and depressurization are especially important in the preparation of monoliths because if the pore liquid does not have sufficient time to flow out of the gel network, it could lead to excessive stresses that cause cracking (38,39). The container of a gel is by itself a source of stresses by preventing the radial expansion and flow of liquid. Quantitative models describing these phenomena are available and have been tested against experimental data for silica gels (38–40).

For some applications it is desirable to prepare aerogels as thin films that are either self-supporting or supported on another substrate. All common coating methods such as dip coating, spin coating, and spray coating can be used to prepare gel films. However, for highly porous films (ie, porosity $> 75\%$), special care is necessary to minimize the rate of solvent evaporation both during and after gel formation. One way to do so is to perform the coating processes within an enclosure that is filled with the saturated vapor of the working solvent and a partial pressure of ammonium hydroxide that catalyzes the gelation of the films (41). The subsequent supercritical drying step can be done in either alcohol or carbon dioxide. The choice depends on the desired properties of the aerogels and, in the case of supported films, the thermal stability of the substrate materials.

In all the processes discussed above, the gelation and supercritical drying steps are done sequentially. Recently a process that involves the direct injection of the precursor into a strong mold body followed by rapid heating for gelation and supercritical drying to take place was reported (42). By eliminating the need of forming a gel first, this entire process can be done in less than three hours per cycle. Besides the saving in time, gel containment minimizes some stresses and makes it possible to produce near net-shape aerogels and precision surfaces. The optical and thermal properties of silica aerogels thus prepared are comparable to those prepared with conventional methods (42).

3.2. Ambient Preparations. Supercritical drying with alcohols incurs high capital and operating costs because the process is run at high temperatures and pressures and needs to remove a large amount of pore liquid. Carbon dioxide allows drying to be done under milder conditions but its use is limited to miscible solvents. Thus, economic and safety considerations have provided a strong motivation for the development of techniques that can produce aerogel-like materials at ambient conditions, ie, without supercritical drying. The strategy is to

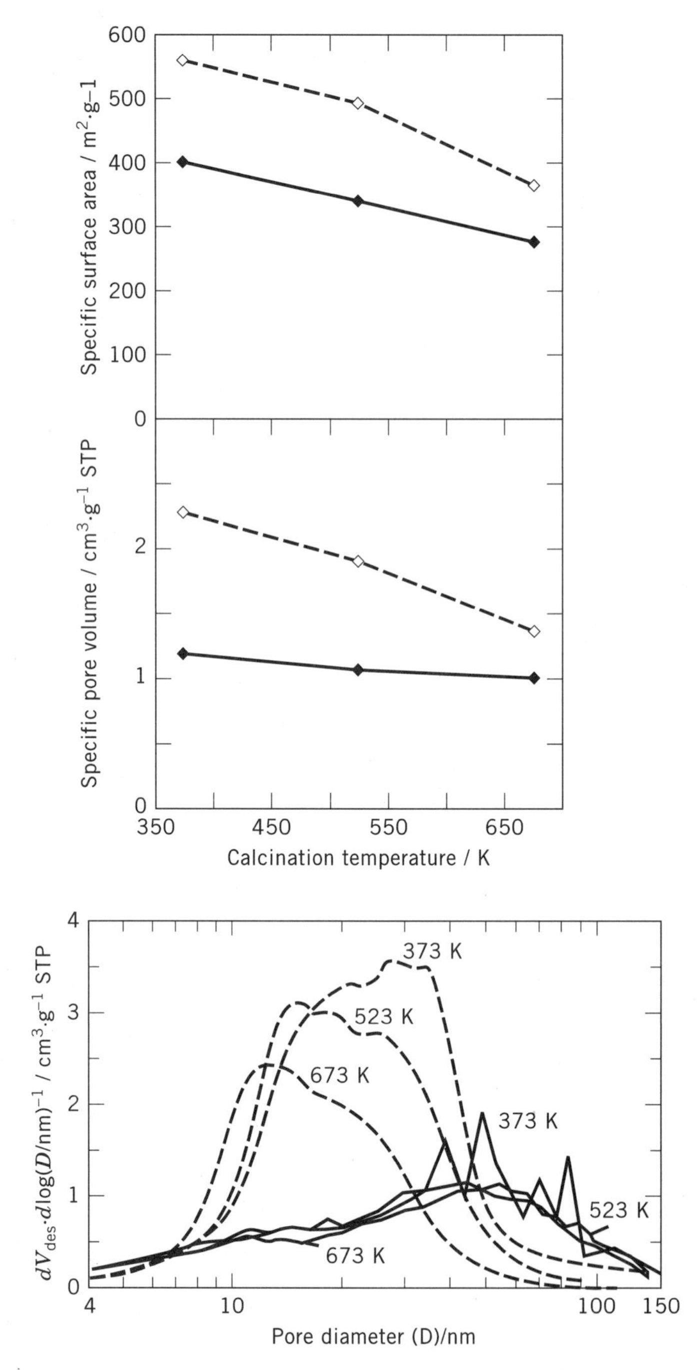

Fig. 3. Effect of using either liquid or supercritical carbon dioxide on the textural properties of titania aerogels calcined at the temperatures shown. (—), dried with liquid carbon dioxide at 6 MPa and 283 K; (- - -), dried with supercritical carbon dioxide at 30 MPa and 323 K. Reproduced from Ref. 36. Courtesy of Marcel Dekker, Inc.

minimize the deleterious effect of capillary pressure which is given by:

$$P = 2\sigma \cos(\theta)/r$$

where P is capillary pressure, σ is surface tension, θ is the contact angle between liquid and solid, and r is pore radius.

The equation above suggests that one approach would be to use a pore liquid that has a low surface tension. Indeed, two-step acid–base or acid–acid catalyzed silica gels have been made, aged in ethanol or water, washed with various aprotic solvents, and finally evaporatively dried at 323 K for 48 hours and then at 383 K for 48 hours (43). The aprotic solvents used and their corresponding surface tension in N/m at room temperature (shown in parentheses) are tetrahydrofuran (23.7×10^{-3}), acetone (23.7×10^{-3}), cyclohexane (25.3×10^{-3}), acetonitrile (29.3×10^{-3}), nitromethane (32.7×10^{-3}), and 1:4 dioxane (33.6×10^{-3}). For both the acid-catalyzed and base-catalyzed gels, it was found that increasing surface tension causes a decrease in surface area and total pore volume. However, for acid-catalyzed gels with an intermediate water wash, the micro pore volume increases with increasing surface tension. It was suggested that the water wash leads to a hydroxylated surface which, upon further condensation, gives a stiffer network and consequently a larger fraction of micropore volume. The important point is that with a pore liquid that has a sufficiently small surface tension, ambient pressure aerogels can have comparable pore volume and bulk density to those prepared with supercritical drying (Fig. 4) (44).

For base-catalyzed silica gels, it has been shown that modifying the surface functionality is an effective way to minimize drying shrinkage (44,46). In particular, surface hydroxyl groups, the condensation of which leads to pore collapse, can be "capped off" via reactions with organic groups such as tetraethoxysilane and trimethylchlorosilane. This surface modification approach (also referred to as surface derivatization), initially developed for bulk specimens, has recently been applied to the preparation of thin films (47,48). The process involves the following steps: (*1*) prepare a base-catalyzed gel, (*2*) wash the gel with hexane, (*3*) replace the surface hydroxyls with organosilicon groups, (*4*) reliquify the surface-modified gel with ultrasound, (*5*) dip-coat the redispersed sols onto a silicon substrate, and (*6*) heat treat the film in air at 450°C. During drying these materials exhibit reversible shrinkage in a gradual dilation, or "spring-back," of the film. The extent of spring-back is a function of processing conditions. Films with porosity in the range of 30–99% can be prepared via a proper control of the washing, surface modification, dip-coating, and heat treatment conditions (48).

In changing surface hydroxyls into organosilicon groups, surface modification has an additional advantage of producing hydrophobic gels. This feature, namely the immiscibility of surface-modified gel with water, has led to the development of a rapid extractive drying process shown in Figure 5 (49). The basic idea is to submerge a wet gel into a pool of hot water that is above the boiling point of the pore fluid. After the pore fluid boils out of the gel, the gel floats to the surface because it is not wetted by water and can be easily recovered. Other working fluids, such as ethylene glycol and glycerol, can be used instead of water as long as they have a high boiling point and do not wet the gel. This

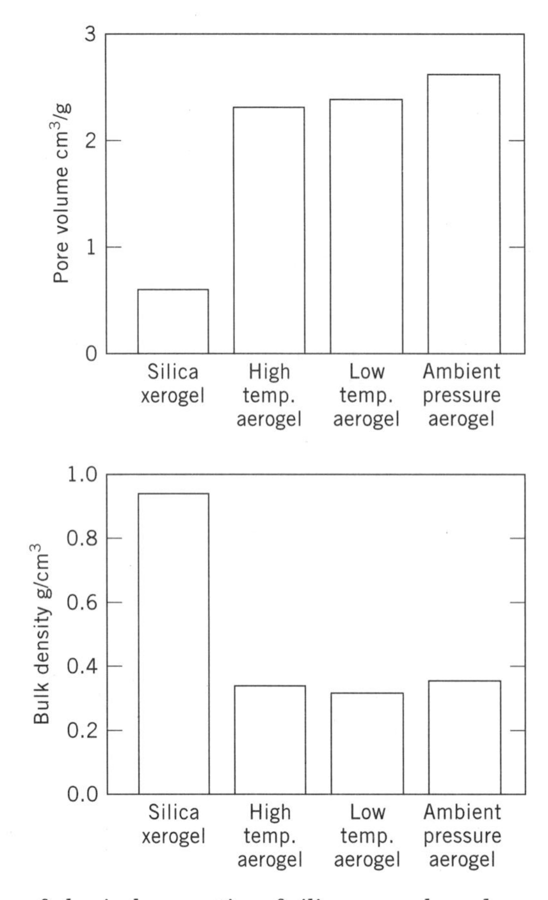

Fig. 4. Comparison of physical properties of silica xerogels and aerogels. Note the similar properties of the aerogels prepared with and without supercritical drying. Reproduced from Ref. 44. Courtesy of the Materials Research Society.

ambient pressure process offers improved heat transfer rates and, in turn, greater energy efficiency without compromising desirable aerogel properties.

Another approach to produce aerogels without supercritical drying is freeze drying, in which the liquid–vapor interface is eliminated by freezing a wet gel into a solid and then subliming the solvent to form what is known as a *cryogel*. Some potential problems are that cracks may develop in the frozen gel and sublimation, if done too fast, can melt the solvent. Cryogels of silica and nickel oxide–alumina have similar, but not identical, properties to aerogels of the same materials (50). In general, there has not been a lot of study on freeze drying and the limited data available suggest that it might not be as attractive as the above ambient approaches in producing aerogels on a commercial scale.

4. Properties

A detailed discussion of the properties of aerogels can be found in several recent review articles (51–55) and the references therein. This section provides a

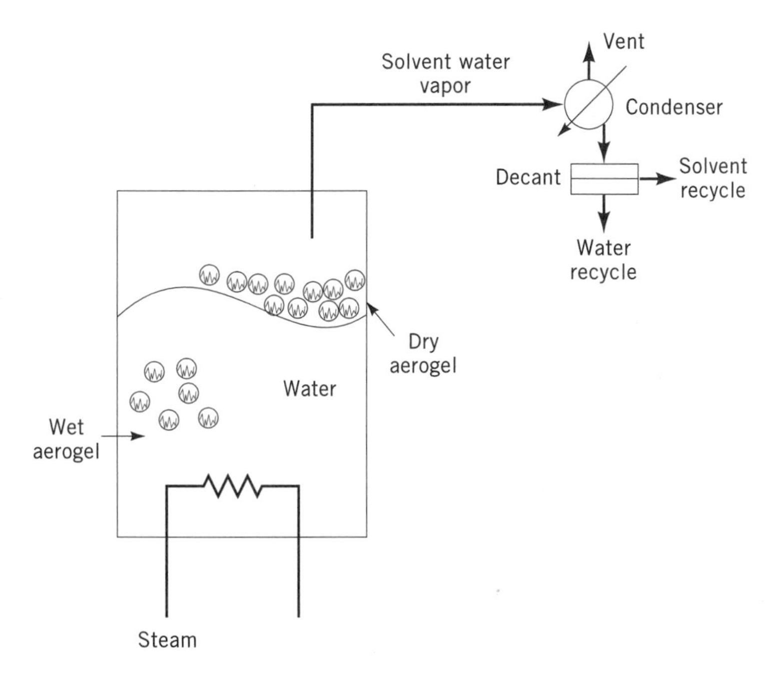

Fig. 5. Schematic diagram of an extractive drying process that produces aerogels at ambient pressure. Reproduced from Ref. 49. Courtesy of the Materials Research Society.

physical basis for these properties by focusing on the microstructure of an aerogel. The intent is to provide a bridge between the two previous sections, which discuss the preparative and drying parameters that affect microstructure, and the next one, which outlines the potential applications made possible by unique structural features. The emphasis is on silica aerogels because they have been the most extensively characterized.

All properties of aerogels are related to the extremely high porosity of these materials. Consider an aerogel as consisting of strings of pearls; each pearl is a structural unit and connected to only two to three other units on average. The size of each structural unit depends on the gelation and drying conditions and varies between 1 to 100 nm in diameter. Correspondingly, pore sizes in an aerogel range from 1 to 100 nm. For silica aerogels skeletal (ie, the solid phase) densities, ρ_s, of 1700–2100 kg/m^3 have been reported, whereas the overall densities, ρ, are in the range of 3–500 kg/m^3 (54). The fact that the overall density is much lower than the skeletal density suggests that an aerogel is full of open space, or pores. Quantitatively the porosity, P, is given by

$$P = 1 - (\rho/\rho_s)$$

Thus, the porosity of an aerogel is in excess of 90% and can be as high as 99.9%. As a consequence of such a high porosity, aerogels have large internal surface area and pore volume.

Since the pores in an aerogel are comparable to, or smaller than, the mean free path of molecules at ambient conditions (about 70 nm), gaseous conduction of heat within them is inefficient. Coupled with the fact that solid conduction is suppressed due to the low density, a silica aerogel has a typical thermal conductivity of 0.015 W/(m·K) without evacuation. This value is at least an order of magnitude lower than that of ordinary glass and considerably lower than that of CFC (chlorofluorocarbon)-blown polyurethane foams (54).

The low density and high porosity of aerogels lead to several other unique properties. Sound travels in silica aerogels at a longitudinal velocity of 100–300 m/s (the corresponding value in ordinary glass is about 500 m/s) (56). The low sound velocity and low density combine to give a low acoustic impedance, which is the product of the two quantities. Silica aerogels also have a refractive index that is very close to unity, meaning light can enter and leave a piece of aerogel without appreciable reflective losses and refractive effects. And for porosity larger than 0.7, silica aerogels have a very low dielectric constant (less than 2) (57). Finally, compared to silica glass, silica aerogels have an extremely small Young's modulus (10^6–10^7 N/m^2, several orders of magnitude less than nonporous materials), and thus, can be compressed easily (51).

From a practical viewpoint it is not only the low density of aerogels but also the variability of density over a wide range that offers interesting possibilities. As discussed in "Inorganic Materials," a silica gel can be formed under either acidic or basic conditions. Densities that can be formed under either acidic or basic conditions. Densities that can be obtained with these one-step processes are in the range of 25–500 kg/m^3. To produce ultralow-density silica aerogels, a two-step process has been developed to extend the lower limit to 3 kg/m^3 (58). In the first step, either tetramethyl- or tetraethyl-orthosilicate is reacted with a substoichiometric amount of water to form a partially hydrolyzed, partially condensed silica precursor. After distilling off the alcohol solvent, this precursor is stored in a nonalcoholic solvent. The second step involves the further hydrolysis of the precursor under basic conditions to form a gel. The ability to vary density over two orders of magnitude is significant because almost all the properties discussed in this section vary with density. Specifically (52,54), the following:

Property	Value
refractive index	$n = 1 + 2.1 \times 10^{-4}\ \rho$; ρ in kg/m^3
sound velocity	$v_s \propto \rho^{\beta}$
Young's modulus	$Y \propto \rho^{\alpha}$
where for silica	$\beta = (\alpha - 1)/2$
	$\alpha = 3.6$; $\beta = 1.3$

Note that scaling exponents depend on preparative conditions.

Table 2 summarizes the key physical properties of silica aerogels. A range of values is given for each property because the exact value is dependent on the preparative conditions and, in particular, on density.

5. Applications

5.1. Thermal Insulation. In addition to their low thermal conductivity, as discussed in the section above, silica aerogels can be prepared to be highly

Table 2. **Typical Values of Physical Properties of Silica Aerogels**

Property	Values
density, kg/m^3	3–500
surface area, m^2/g	800–1000
pore sizes, nm	1–100
pore volume, cm^3/g	3–9
porosity, %	75–99.9
thermal conductivity, $W/(m{\cdot}K)$	0.01–0.02
longitudinal sound velocity, m/s	100–300
acoustic impedance, $kg/(m^2{\cdot}s)$	10^3–10^6
dielectric constant	1–2
Young's modulus, N/m^2	10^6–10^7

transparent in the visible spectrum region. Thus, they are promising materials as superinsulating window-spacer. To take further advantage of its high solar transmission, a silica aerogel layer sandwiched between two glass panes can be used to collect solar energy passively. Figure 6 shows a specific arrangement in which an insulating layer of transparent silica aerogel is placed in front of a

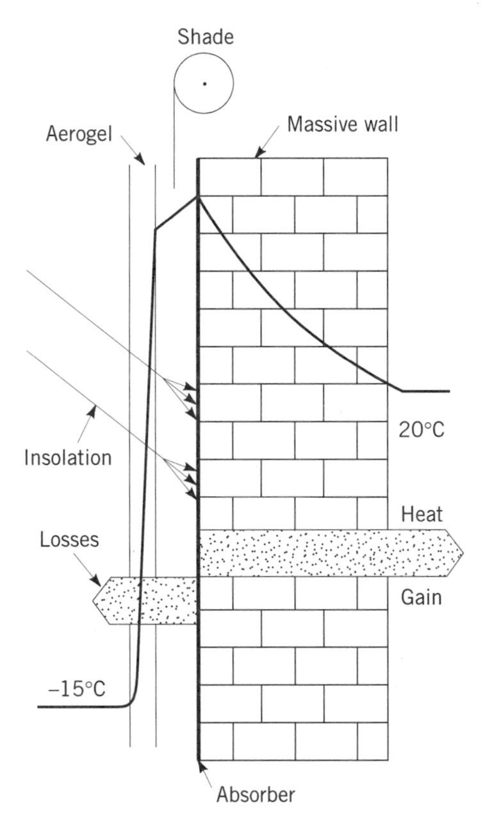

Fig. 6. Schematic diagram showing the use of transparent silica aerogel in passive solar collection. Reproduced from Ref. 53. Courtesy of Elsevier Science-NL.

brick wall, the surface of which is painted black. Solar radiation collected at the black surface is mostly transferred as heat into the house because heat loss through the aerogel layer is minimal. A shade is put into place to prevent overheating if necessary (53). The same principle applies when silica aerogels are used as insulating covers for solar panels.

The thermal conductivity of silica aerogels can be further reduced by minimizing the radiation leakage with an opacifier such as carbon black (54). The introduction of an opacifier makes the material opaque and unsuitable for window insulation. On the other hand, opaque silica aerogels can be used as insulating materials in appliances such as refrigerators and freezers. Compared with CFC-blown insulating foams, which could release chlorine into the atmosphere, silica aerogels pose no such environmental hazard and are nonflammable. Moreover, the thermal conductivity of opaque silica aerogels is a weak function of temperature (52), making them useful as insulating materials for heat-storage systems. Another promising application of silica aerogels is as a filler in vacuum panel because they do not require a high vacuum for good thermal performance. The high surface area of these materials further allow them to act as a "getter" by adsorbing gases in the panel.

The commercial viability of silica aerogels as thermal insulators depends on the ability to produce them at a competitive price. After all, in the 1950s, the production of Monsanto's Santocel stopped after a lower-cost process to manufacture fumed silica was developed (59). Recently an initial economic analysis that consists of six factors in the manufacturing of aerogels: starting material, solvent, energy, wage, equipment, and facility was published (60). The results show that the dominant cost is the cost of the starting material and that aerogels could be competitive with commercial insulating materials on a cost per R value basis. Indeed, BASF has developed a silica aerogel which has the registered trademark Basogel (61). Since supercritical drying, even with carbon dioxide at a lower temperature, is an energy-intensive process, NanoPore, Inc. is developing an ambient approach to make silica aerogels (49,59). Technological progresses in the next several years will be critical in determining whether aerogels can capture a significant share of the commercial insulation market, which is probably their largest potential area of application. At least two U.S. companies are currently developing aerogels as insulating materials. Aspen Systems manufactures silica aerogels in the forms of powders, monoliths, and blankets. Their present (1996) price range is from \$100 to \$2,000 per cubic foot, depending on the size of the order (62). Aerojet Corporation has collaborated with different end-users in evaluating the market potential of organic aerogels (59), which have even lower thermal conductivities than their silica counterparts.

5.2. Catalysis. Kistler explored the catalytic applications of aerogels in the 1930s because of the unique pore characteristics of aerogels (24), but this area of research stayed dormant for about three decades until less tedious procedures to produce the materials were introduced (25,26). Three recent review articles summarize the flurry of research activities since then (63–65). Table 3 is a much abbreviated list of what has been cited in these three articles to demonstrate simply the wide range of catalytic materials and reactions that have been studied.

Table 3. **Example of Materials and Reactions Involving Catalytic Aerogels**

Materials	Reactions	Examples
Type I, simple oxides		silica, alumina, titania, zirconia
Type II, mixed oxides		nickel oxide–alumina, titania–silica, zirconia–silica, chromia–alumina
Type III, ternary oxides		nickel–oxide–silica–alumina, magnesia–alumina–silica, titania–silica–vanadia, alumina–chromia–thoria
Type IV, supported metals		palladium–alumina, ruthenium–silica, platinum–titania, palladium–silica–alumina
Type V, doped oxides (dopant not an oxide)		zirconia–sulfate, zirconia–phosphate, niobia–phosphate, $TiCl_4$–alumina
Types I, II, III, IV	partial oxidation	isobutylene to methyacrolein, acetaldehyde to acetic acid
Type IV	hydrogenation	cyclopendiene to cyclopentene, benzene to cyclohexane
Types I, II, V	isomerization	1-butene to *cis*- and *trans*-2-butene, *n*-butane to isobutane
Type II	epoxidation	1-hexene and cyclohexene to the corresponding epoxides
Type IV	hydrotreating	hydrodenitrogenation and hydrodesulfurization
Type V	polymerization	ethylene to polyethylene
Types II and III	nitric oxide reduction	reduction of NO by ammonia to nitrogen

Most of the studies on catalytic aerogels reported in the open literature involve testing powder samples in experimental reactors. For the potential scale-up to commercial operations, aerogels have several limitations which, ironically, arise from their unique properties. First, even though aerogels have high specific surface area, they also have low densities. The product of the two quantities, which is surface area per unit volume, does not offer a significant advantage over other materials that have been used as adsorbents or catalysts. Second, the low thermal conductivity of aerogels means that it would be difficult to transfer heat in or out of a catalytic packed bed. Third, aerogels are fragile and do not withstand mechanical stress well. Attempts have been made to overcome the last two limitations by supporting or combining aerogels with materials that are either more thermally conducting or more rigid (63). Finally, as in the case of thermal insulation, catalytic aerogels need to be cost-competitive, even though the economic pressure is not as severe because the cost of a catalyst is usually a small fraction of the value of its derived products.

It is commonly believed that catalytic aerogels are interesting because their composition, morphology, and structure can be controlled at a microscopic level. But this feature is generic to any sol–gel-derived materials, so perhaps what set aerogels apart is that during solvent removal, there is another level control over the physical and chemical properties of the products. For example, supercritically dried materials are usually mesoporous and, as such, should be good catalysts for liquid-phase reactions for which there could be diffusional limitations. And for multicomponent materials, the distribution (or mixing state) of the various components could be better preserved without compromising the integrity of the porous network. A nice illustration of these effects is the recent results on epoxidation (66–68).

For the epoxidation of olefins over catalysts containing titania and silica, two factors are considered to be crucial: the accessibility of large pores for the bulky olefins (containing six to ten carbons) and the presence of Ti–O–Si linkages. It has been shown (66–68) that both of these desirable properties can be obtained only by supercritically drying an optimized gel at low temperatures with carbon dioxide. High temperature supercritical drying led to the segregation of titania and destruction of Ti–O–Si linkages. Conventional evaporative drying maintained the density of Ti–O–Si linkages but resulted in microporous xerogels that were inactive. Figure 7 (69) illustrates these observations for the epoxidation of cyclohexene. Besides reporting similar results for other olefins such as cyclododecene, norbonene, and limonene, these researchers were able to establish a semiquantitative correlation between activity and Ti–O–Si connectivity.

The Ti–O–Si linkages in titania–silica also have a large impact on the acidic properties of this mixed oxide. It was recently demonstrated that, over the entire composition range, the extent of mixing as controlled by prehydrolysis changes the acid site density, acid site type (Lewis versus Brønsted), and 1-butene isomerization activity of titania–silica aerogels (70). In fact, these observations appear to be general for other mixed oxide pairs in that sol–gel chemistry affords a higher degree of control over the intimacy of molecular-scale mixing that is not available by other preparative methods (71). Intimate mixing is actually undesirable in some instances, as shown for the selective catalytic reduction of NO with NH_3 (72). For this system, titania crystallites are believed to be good for stabilizing a two-dimensional overlayer of vanadia. Still, for this reaction the role of oxide–oxide interactions in affecting surface acidic properties remains critical (73). The fact that multicomponent materials of specific pore characteristics can be prepared to be either well-mixed or poorly-mixed, depending on the application, represents a unique advantage of catalytic aerogels.

5.3. Scientific Research. There are some applications that require such specific and unique properties that an aerogel is the only alternative. In these situations price is no longer a factor and aerogels have been and will continue to be produced for scientific purposes. One example is the use of silica aerogel monoliths as radiators in Chernekov counters, or detectors, in high energy physics and nuclear astrophysics experiments (23). In order to measure precisely the momentum of elementary particles, which are produced by particle interactions in high energy accelerators, over different momentum ranges, it is necessary to vary the refractive index of radiators. The reason is that a charged particle produces light in passing through a medium only if its velocity is higher than the velocity of light in that medium (52). As pointed out in the section "Properties", the refractive index of silica aerogels is related to density. Thus, by varying the density from about 50 to 300 kg/m^3, which is easily attainable with the

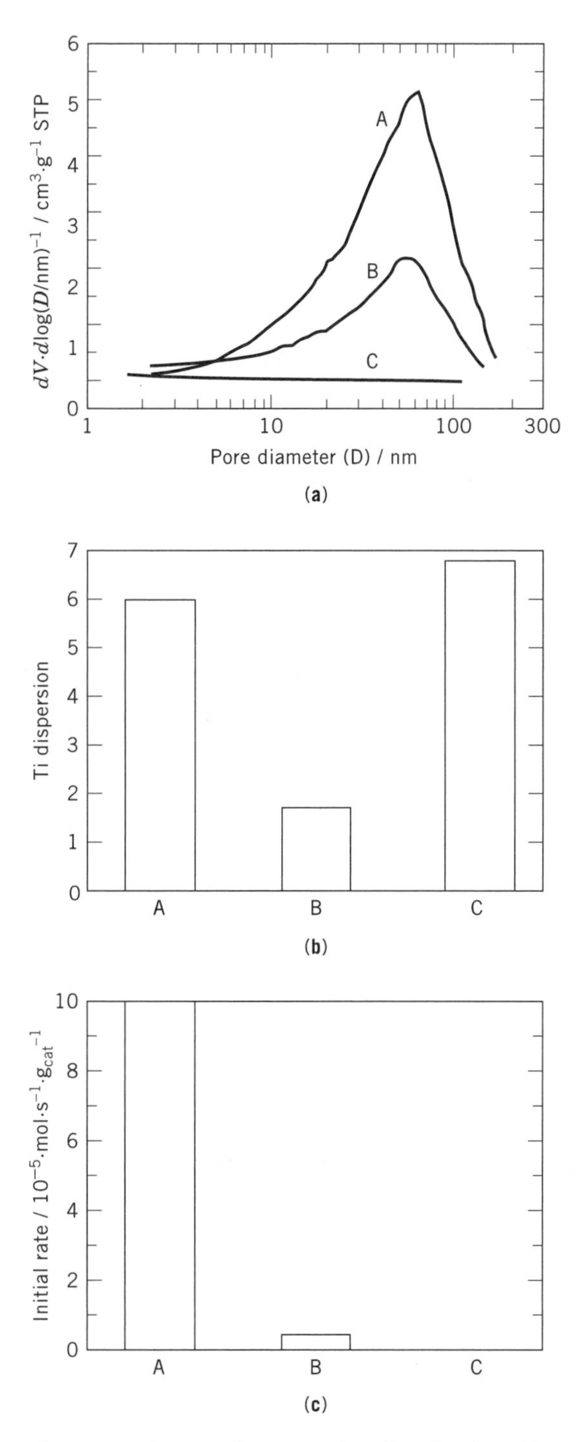

Fig. 7. The effect of preparation on the pore size distribution (**a**), titanium dispersion (**b**), and the activity for epoxidation of cyclohexene (**c**) of titania–silica containing 10 wt % titania and calcined in air at 673 K. Sample A, low-temperature aerogel; Sample B, high-temperature aerogel; Sample C, aerogel. Reproduced from Ref. 69. Courtesy of Marcel Dekker, Inc.

control of sol–gel chemistry, materials of indices of refraction between 1.01 and 1.06 can be delivered to measure low values of momentum. Indeed, two large blocks of silica aerogels have been prepared as particle detectors for use in the CERN Intersecting Storage Rings and the Deutsches Electronen-Synchrotron (DESY) (23).

Aerogels are also among the few materials that can capture cosmic particles intact (74). When a spacecraft approaches a source (eg, the corona of a comet), the cosmic particles have relative speeds of the order of tens of km/s. These hypervelocity particles tend to melt or vaporize upon collision with a solid. Mesoporous materials that are of very low density can capture these particles relatively intact, but a silica aerogel is the only one that allows the captured particles to be located easily because of its transparency. In fact, as a particle enters a silica aerogel at hypervelocity, it creates a carrot-shaped tract that points to where the particle stops. The high surface area of an aerogel provides the added advantage of adsorbing any volatiles that are either in space or generated during capture.

Panels of silica aerogels have already been flown on several Space Shuttle missions (74). Currently a STARDUST mission has been planned by NASA to use aerogels to capture cometary samples (>1000 particles of >15 micron diameter) and interstellar dust particles (>100 particles of >0.1 micron diameter). The mission involves the launch of a spacecraft in February 1999 to encounter the comet Wild 2 in December 2003 and to return the captured samples to earth in January 2006. Silica aerogels of ultralow density, high uniformity, and high purity need to be prepared to meet the primary objectives of the mission. Furthermore, there are challenges to produce nonsilicate aerogels of the same high quality and aerogels that can perform chemical as well as physical capture.

Besides being used as a tool for scientific research, silica aerogels can be the cause for new scientific phenomena. For example, the long-range correlations of the disorder in silica aerogels are believed to be responsible for the intriguing observations of the superfluid transitions in ^{4}He and ^{3}He and on the ordering of ^{3}He–^{4}He mixtures (75).

5.4. Other Applications. There are several other applications that take advantage of the unique properties of silica aerogels shown in Table 2. In a piezoceramic ultrasound transducer, the signal is reduced as sound waves cross the interface of two materials that have very different acoustic impedances (eg, the acoustic impedances of air and piezoelectric ceramics differ by several orders of magnitude). With its tunable acoustic impedance, a layer of silica aerogel sandwiched between a piezoceramic and air minimizes the mismatch and enhances the signal (54). The low dielectric constant of silica aerogels suggests their usefulness as layer materials in integrated circuits. For silica aerogels to replace the currently used silica glasses, several issues such as the control of porosity, mechanical strength, thermal stability, and process integration problems need to be addressed (57). Finally, silica aerogels are effective dehydrating agents because of their high capacity for absorbing moisture and chemical inertness. For household insect control, silica aerogels work by absorbing waxes from the cuticle of insects, which then die from dehydration; their potency can be enhanced with the doping of insecticides (76).

The introduction of organic aerogels has led to interesting areas of applications. Recent data (77,78) showing the thermal properties of these materials and, more importantly, how these properties can be varied with preparative conditions, should help their further development as thermal insulators. Carbon aerogels are also promising electrode materials in electrochemical double-layer capacitors. Large specific capacities and capacitance densities have been demonstrated in a device consisting of two carbon aerogel wafers; this performance is attributed to the high specific surface area and contiguous structure of the carbon electrode (79). These results are encouraging for the use of carbon aerogels in building low cost, high power, and high energy density capacitors (often referred to as "supercapacitors"). Another potential application of carbon–aerogel electrodes is the capacitive deionization of water. In a recently developed process, water with various anions and cations is pumped through an electrochemical cell consisting of a stack of carbon aerogel electrodes (80). The large specific surface area of the electrodes allows the electrostatic capture of ions. After water is purified, the cell can be regenerated by a simple electrical discharge.

6. Summary

As of 1996 the commercial market for aerogels remained very small. However, aerogels have the potential of being marketable both as a commodity chemical (eg, in thermal insulation) and as a specialty chemical (eg, in electronic applications) because of their unique and tailorable properties. The next few years will be critical in assessing whether aerogels can penetrate and grow in either end of the market, as the field is changing rapidly with the development of cost-competitive technologies and novel applications.

BIBLIOGRAPHY

"Aerogels" in *ECT* 4th ed., Suppl. Vol., pp. 1–22, by Edmond I. Ko, Carnegie Mellon University.

CITED PUBLICATIONS

1. S. S. Kistler, *Nature*, **127**, 741 (1931).
2. C. J. Brinker and G. W. Scherer, *Sol-Gel Science: The Physics and Chemistry of Sol-Gel Processing*, Academic Press, New York, 1990.
3. J. Livage, M. Henry, and C. Sanchez, *Prog. Solid State Chem.* **18**, 259 (1988).
4. R. C. Mehrota, *J. Non-Cryst. Solids* **145**, 1 (1992).
5. H. D. Gesser and P. C. Goswami, *Chem. Rev.* **89**, 765 (1989).
6. R. K. Iler, *The Chemistry of Silica*, John Wiley & Sons, New York, 1979, Chapt. 3.
7. J. Y. Ying, J. B. Benziger, and A. Navrotsky, *J. Am. Ceram. Soc.* **76**, 2571 (1993).
8. B. Handy, K. L. Walther, A. Wokaun, and A. Baiker, in P. A. Jacobs, P. Grange, and B. Delmon, eds., *Preparation of Catalysts V*, Elsevier, Netherlands, 1991, 239–246.
9. E. I. Ko, in G. Ertl, H. Knözinger, and J. Weitkamp, eds., *Handbook of Heterogeneous Catalysts*, VCH, Germany, 1997, Vol. 1, Sec. 2.1.4.

10. B. E. Yoldas, *J. Mater. Sci.* **14**, 1843 (1979).
11. J. Livage and C. Sanchez, *J. Non-Cryst. Solids* **145**, 11 (1992).
12. U.S. Pat. 4,873,218 (Oct. 10, 1989); U.S. Pat. 4,997,804 (Mar. 5, 1991), R. W. Pekala.
13. R. W. Pekala and C. T. Alviso, *Mat. Res. Soc. Symp. Proc.* **270**, 3 (1992).
14. R. W. Pekala and C. T. Alviso, *Mat. Res. Soc. Symp. Proc.* **180**, 791 (1990).
15. R. W. Pekala, C. T. Alviso, X. Lu, J. Gross, and J. Fricke, *J. Non-Cryst. Solids* **188**, 34 (1995).
16. B. M. Novak, *Adv. Mater.* **5**, 422 (1993).
17. C. Sanchez and F. Ribot, *New J. Chem.* **18**, 1007 (1994).
18. D. Avnir, *Acc. Chem. Res.* **28**, 328 (1995).
19. U. Schubert, N. Hüsing, and A. Lorenz, *Chem. Mater.* **7**, 2010 (1995).
20. D. A. Loy and K. J. Shea, *Chem. Rev.* **95**, 1431 (1995).
21. D. Avnir, S. Braun, O. Lev, and M. Ottolenghi, *Chem. Mater.* **6**, 1605 (1994).
22. W. Cao, X. Y. Song, and A. J. Hunt, *Mat. Res. Soc. Symp. Proc.* **349**, 87 (1994).
23. A. J. Ayen and P. A. Iacobucci, *Rev. Chem. Eng.* **5**, 157 (1988).
24. S. S. Kistler, S. Swan, Jr., and E. G. Appel, *Ind. Eng. Chem.* **26**, 388 (1934).
25. S. J. Teichner, *CHEMTECH*, **21**, 372 (1991).
26. S. J. Teichner, G. A. Nicolaon, M. A. Vicarini, and G. E. E. Gardes, *Adv. Colloid Interface Sci.* **5**, 245 (1976).
27. U.S. Pat. 4,610,863 (Sept. 9, 1986), P. H. Tewari and A. J. Hunt.
28. U.S. Pat. 4,619,908 (Oct. 28, 1986), C. P. Cheng, P. A. Iacobucci, and E. N. Walsh.
29. P. Wang, A. Emmerling, W. Tappert, O. Spormann, J. Fricke, and H. G. Haubold, *J. Appl. Cryst.* **24**, 777 (1991).
30. D. R. Ulrich, *J. Non-Cryst. Solids* **100**, 174 (1988).
31. D. M. Smith, R. Deshpande, and C. J. Brinker, *Mat. Res. Symp. Proc.* **271**, 553 (1992).
32. G. Cogliati, M. Guglielmi, T. M. Che, and T. J. Clark, *Mat. Res. Symp. Proc.* **180**, 329 (1990).
33. M. Beghi, P. Chiurlo, L. Costa, M. Palladina, and M. F. Pirini, *J. Non-Cryst. Solids* **145**, 175 (1992).
34. C. J. Brodsky and E. I. Ko, *J. Mater. Chem.* **4**, 651 (1994).
35. D. C. M. Dutoit, M. Schneider, and A. Baiker, *J. Porous Mater.* **1**, 165 (1995).
36. M. Schneider and A. Baiker, *Catal. Rev.-Sci. Eng.* **37**(4), 529 (1995).
37. G. W. Scherer, in A. S. Mujumdar, ed., *Drying*, Elservier, the Netherlands, 1992, 92–113.
38. G. W. Scherer, *J. Non-Cryst. Solids*, **145**, 33 (1992).
39. G. W. Scherer, *J. Sol-Gel Sci. Tech.* **3**, 127 (1994).
40. T. Woignier, G. W. Scherer, and A. Alaoue, *J. Sol-Gel Sci. Tech.*, **3**, 141 (1994).
41. L. W. Hrubesh and J. F. Poco, *J. Non-Cryst. Solids*, **188**, 46 (1995).
42. J. F. Poco, P. R. Coronado, R. W. Pekala, and L. W. Hrubesh, *Mat. Res. Soc. Symp. Proc.* **431**, 297 (1996).
43. R. Deshpande, D. M. Smith, and C. J. Brinker, *Mat. Res. Soc. Symp. Proc.* **271**, 553 (1992).
44. C. J. Brinker, R. Sehgal, N. K. Raman, S. S. Prakash, and L. Delattre, *Mat. Res. Soc. Symp. Proc.* **368**, 329 (1995).
45. D. M. Smith, R. Deshpande, and C. J. Brinker, *Mat. Res. Soc. Symp. Proc.* **271**, 567 (1992).
46. D. M. Smith, R. Deshpande, and C. J. Brinker, *Ceram. Trans.* **31**, 71 (1993).
47. S. S. Prakash, C. J. Brinker, A. J. Hurd, and S. M. Rao, *Nature* **374**, 439 (1995).
48. S. S. Prakash, C. J. Brinker, and A. J. Hurd, *J. Non-Cryst. Solids* **190**, 264 (1995).
49. D. M. Smith, W. C. Akerman, R. Roth, A. Zimmerman, and F. Schwertfeger, *Mat. Res. Soc. Symp. Proc.* **431**, 291 (1996).

50. D. Klvana, J. Chaouki, M. Repellin-Lacroix, and G. M. Pajonk, *J. Phys. Colloq.* **C4**, 29 (1989).
51. H. D. Gesser and P. C. Goswami, *Chem. Rev.* **89**, 765 (1989).
52. J. Fricke and A. Emmerling, *J. Am. Ceram. Soc.* **75**(8), 2027 (1992).
53. J. Fricke, *J. Non-Cryst. Solids* **147/148**, 356 (1992).
54. J. Fricke and J. Gross, *Mater. Eng.* **8**, 311 (1994).
55. T. Heinrich, U. Klett, and J. Fricke, *J. Porous Mater.* **1**, 7 (1995).
56. J. Fricke, *Sci. Amer.* **256**(5), 92 (1988).
57. D. M. Smith, J. Anderson, C. C. Cho, G. P. Johnston, and S. P. Jeng, *Mat. Res. Soc. Symp. Proc.* **381**, 261 (1995).
58. T. M. Tillotson and L. W. Hrubesh, *J. Non-Cryst. Solids* **145**, 45 (1992).
59. C. M. Caruana, *Chem. Engr. Prog.* (June 11, 1995).
60. G. Carlson, D. Lewis, K. McKinley, J. Richardson, and T. Tillotson, *J. Non-Cryst. Solids* **186**, 372 (1995).
61. G. Herrmann, R. Iden, M. Mielke, F. Teich, and B. Ziegler, *J. Non-Cryst. Solids* **186**, 380 (1995).
62. *http://www.aspensystems.com/aerogel.html*
63. G. M. Pajonk, *Appl. Catal.* **72**, 217 (1991).
64. G. M. Pajonk, *Catal. Today*, **35**, 319 (1997).
65. M. Schneider and A. Baiker, *Catal. Rev.-Sci. Eng.* **37**(4), 515 (1995).
66. D. Dutoit, M. Schneider, and A. Baiker, *J. Catal.* **153**, 165 (1995).
67. R. Hutter, T. Mallat, and A. Baiker, *J. Catal.* **153**, 177 (1995).
68. R. Hutter, T. Mallat, and A. Baiker, *J. Catal.* **157**, 665 (1995).
69. From Ref. 65, p. 544.
70. J. B. Miller, S. T. Johnston, and E. I. Ko, *J. Catal.* **150**, 311 (1994).
71. J. B. Miller and E. I. Ko, *Catal. Today*, **35**, 269 (1997).
72. B. E. Handy, A. Baiker, M. Schramal-Marth, and A. Wokaun, *J. Catal.* **133**, 1 (1992).
73. R. J. Willey, C. T. Wang, and J. B. Peri, *J. Non-Cryst. Solids* **186**, 408 (1995).
74. P. Tsou, *J. Non-Cryst. Solids* **186**, 415 (1995).
75. M. Chan, N. Mulders, and J. Reppy, *Physics Today* (Aug. 30, 1996).
76. http://www.ent.agri.umn.edu/academics/classes/ipm/chapters/ware.htm
77. X. Lu, R. Caps, J. Fricke, C. T. Alviso, and R. W. Pekala, *J. Non-Cryst. Solids* **188**, 226 (1995).
78. V. Bock, O. Nilsson, J. Blumm, J. Fricke, *J. Non-Cryst. Solids* **185**, 233 (1995).
79. S. T. Mayer, R. W. Pekala, and J. K. Kaschmitter, *J. Electrochem. Soc.* **140**(2), 446 (1993).
80. J. C. Farmer, D. V. Fix, G. V. Mack, R. W. Pekala, and J. F. Poco, *J. Electrochem. Soc.* **143**(1), 159 (1996).

EDMOND I. KO
Carnegie Mellon University

AEROSOLS

1. Introduction

Classically, aerosols are particles or droplets that range from ~0.15 to 5 μm in size and are suspended or dispersed in a gaseous medium such as air. However, the term aerosol, as used in this discussion, identifies a large number of products which are pressure-dispensed as a liquid or semisolid stream, a mist, a fairly dry to wet spray, a powder, or even a foam. This definition of aerosol focuses on the container and the method of dispensing, rather than on the form of the product.

Aerosol products were developed in the early 1940s and were the outgrowth of the so-called "bug spray" developed during World War II. This development was credited with saving the lives of many of World War II GIs who were stationed in some of the malaria infested jungles of the world. The "bug spray" was extremely effective in controlling the spread of malaria by killing the Anopheles mosquito, which was the carrier for this disease. It did not take too long to apply the principles of the "bug spray" to the development of an entire group of consumer products that were advantageous when sprayed. These products were all similar in that they depended on the power of a liquified or compressed gas to dispense the product as a spray. This concept was then applied to products where it was desirable to dispense the product as a foam, or as a semisolid.

The aerosol container has enjoyed commercial success in a wide variety of product categories. Insecticide aerosols were introduced in the late 1940s. Additional commodities, including shave foams, hair sprays, antiperspirants, deodorants, paints, spray starch, colognes, perfumes, whipped cream, and automotive products, followed in the 1950s. Medicinal metered-dose aerosol products have also been developed for use in the treatment of asthma, migraine headaches, and angina. Food aerosols included whipped toppings and creams, cheese spreads, hors d'oeuvres, flavored syrups, and a host of similar products.

Aerosol technology may be defined as involving the development, preparation, manufacture, and testing of products that depend on the power of a liquefied or compressed gas to expel the contents from a container. This definition can be extended to include the physical, chemical, and toxicological properties of both the finished aerosol system and the propellants.

The production of aerosols has increased both in the United States and worldwide, in spite of a substantial decline during the middle to late 1970s when the use of chlorofluorocarbons (CFCs) (see FLUORINE COMPOUNDS, ORGANIC INTRODUCTION) as propellants was seriously restricted. Hydrocarbons

Table 1. **Production of Aerosols in the United States, 1985–2000, Millions of Units**

	Year						
Aerosol product category	1985	1989	1990	1995	1998	1999	2000
personal products[a]	879	1015	1050	1007	941	952	930
household products	630	680	680	725	785	802	802
automotive, industrial	342	475	415	420	433	444	446
paints, finishes	290	350	350	374	468	446	440
insect sprays	190	197	190	184	185	208	201
food products	140	175	175	252	298	323	335
animal products	15	8	8	6	2	3	3
miscellaneous	23	12	15	34	23	25	43
Total	*2509*	*2912*	*2883*	*3002*	*3135*	*3203*	*3200*

[a] Medical Aerosols are included in the personal products category and in 1999 amounted to ~14 million units and in 2000 amounted to ~28 million units.

and compressed gases (qv) have since replaced the CFCs as propellants and aerosols continue to be used by the general public. As seen in Table 1, U.S. aerosol production reached 3200 million units in 2000, which was about the same as in 1999. Although the personal products category continued to grow through 1995, a slight decrease was noted from 1996 to date, most likely due to changes in hair styles that affected the use of hair sprays and mousses.

Personal products were the fastest growing segment of the aerosol industry and still represent the largest of the categories. However, the sales of these products have declined during the past few years and a further decline may be seen in the future. Other areas of growth in this category occurred in underarm deodorants, antiperspirants, pharmaceuticals, and industrial products, which should keep this category as the largest segment of the aerosol industry.

2. Advantages of Aerosol Packaging

Aerosol products are hermetically sealed, ensuring that the contents cannot leak, spill, or be contaminated. The aerosol packages can be considered to be tamperproof. Aerosols deliver the product in an efficient manner generating little waste, often to sites of difficult access. By control of particle size, spray pattern, and volume delivered per second, the product can be applied directly without contact by the user. For example, use of aerosol pesticides can minimize user exposure and aerosol first-aid products can soothe without applying painful pressure to a wound. Spray contact lens solutions can be applied directly and aerosol lubricants (qv) can be used on machinery in operation. Some preparations, such as stable foams, can only be packaged as aerosols. Spray shaving creams and furniture polish are examples of stable foams.

The use of metered-dose valves in aerosol medical applications permits an exact dosage of an active drug to be delivered to the respiratory system where it can act locally or be systemically absorbed. For example, inhalers prescribed for asthmatics produce a fine mist that can penetrate into the bronchial tubes (see ANTIASTHMATIC AGENTS). Recent developments include the administration of insulin and other hormones by oral inhalation; thereby eliminating an injection.

3. Formulation of Aerosols

Aerosols are unique. The various components are all part of the product, and in the aerosol industry, the formulating chemist must be familiar with the entire package assembly and each of its components. All aerosols consist of product concentrate, propellant, container, and valve (including an actuator and dip tube). There are many variations of these components, and only when each component is properly selected and assembled does a suitable aerosol product result. A typical aerosol system is shown in Figure 1.

The aerosol formulater must be knowledgeable about the availability and usage of propellents, various valves and containers, including pressure limitations and construction features, as well as any other components necessary for the product concentrate system. In contrast, the formulation of a nonaerosol

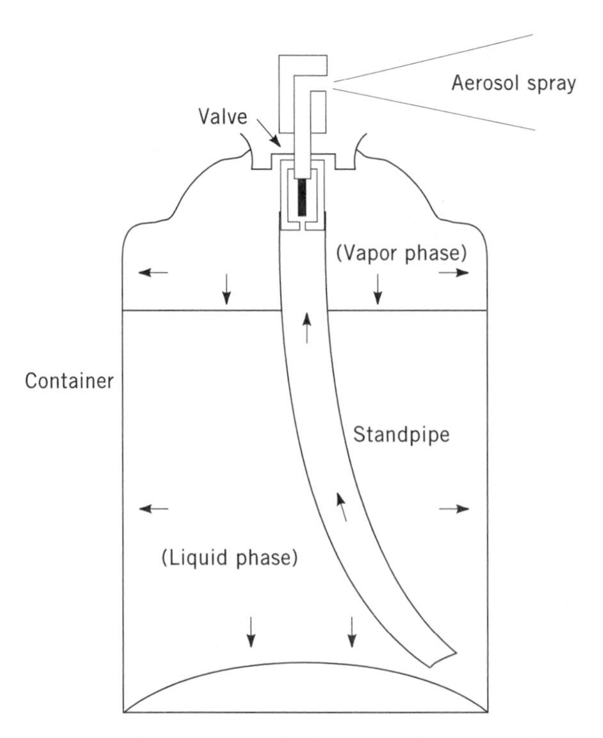

Fig. 1. Solution-type aerosol system in which internal pressure is typically (35–40 psi) at 21°C. To convert kPa to psi, multiply by 0.145.

product is not, to any great extent, affected by either the container or package closure. Nonaerosol products packaged using a pump type system may be an exception, however, as the closure can also play a role in the way a product formulation will be dispensed.

3.1. Product Concentrate. The product concentrate for an aerosol contains the active ingredient and any solvent or filler necessary. Various propellent and valve systems, which must consider the solvency and viscosity of the concentrate–propellent blend, may be used to deliver the product from the aerosol container. Systems can be formulated as solutions, emulsions, dispersions, dry powders, and pastes.

Solutions. To deliver a spray, the formulated aerosol product should be as homogeneous as possible. That is, the active ingredients, the solvent, and the propellant should form a solution. Because the widely used hydrocarbon propellants do not always have the desired solubility characteristics for all the components in the product concentrate, special formulating techniques using solvents such as alcohols (qv), acetone (qv), and glycols (qv), are employed.

The rate of spray is determined by propellant concentration, the solvent used, valve characteristics, and vapor pressure. The pressure must be high enough to deliver the product at the desired rate under the required operating conditions. For example, a windshield ice remover that is likely to be used around 0°C must be formulated to provide an adequate pressure at that temperature. Spray dryness or wetness and droplet size depend upon propellant concentration.

Generally, aerosol packaging consists of many delicately balanced variables. Even hardware design plays an important part. For example, valves that produce considerable breakup are used for the warm sensation desired in some personal products.

Emulsions. Aerosol emulsions (qv) may be oil in water (o/w), such as shaving creams, or water in oil (w/o), such as air fresheners and polishes. These aerosols consist of active ingredients, an aqueous or nonaqueous vehicle, a surfactant, and a propellant, and can be emitted as a foam or as a spray.

Foams. Systems that dispense foams (qv) are generally o/w emulsions, although nonaqueous solvents can also be used as the external phase. When the propellant is a hydrocarbon such as an isobutane–propane blend, as little as 3–4% in a 90–97% emulsion concentrate is sufficient to produce a suitable foam. Although the majority of the propellant is emulsified, some vaporizes and is present in the head space. The resultant pressure is generally on the order of 276 kPa (40 psig). When the valve is depressed, the pressure forces the emulsion up the dip tube and out of the container. Depending on the formulation, either a stable foam, such as would be expected in a shaving cream, or a quick-breaking foam, which collapses in a relatively short period of time, appears. The propellant is an important part of the formulation and is generally considered to be part of the immiscible phase. When the propellant is included in the internal phase, a foam is emitted; when the propellent is in the external phase, the product is dispensed as a spray. Figure 2 illustrates these situations.

When the propellant is in the internal phase (Fig. 2**a**), the propellant vapor, upon discharge, must pass through the emulsion formulation in order to escape into the atmosphere. In traveling through this emulsion, the trapped propellant forms a foam matrix. These systems, are typically oil-in-water emulsions.

An emulsion system in which the propellant is in the external or continuous phase is shown in Figure 2**b**. As the liquefied propellant vaporizes, it escapes directly into the atmosphere, leaving behind droplets of the formulation that are emitted as a wet spray. This system is typical of many water-based aerosols or w/o emulsions.

Extended stability testing is a necessity for emulsion systems in metal containers because of the corrosion potential of water and some of the other ingredients. In most cases where a stable emulsion exists, there is less corrosion potential in a w/o system because the water is the internal phase.

Quick-breaking foams consist of a miscible solvent system such ethanol (qv) [64-17-5] and water, and a surfactant that is soluble in one of the solvents but not in both. These foams are advantageous for topical application of pharmaceuticals because, once the foam hits the affected area, the foam collapses, delivering the product to the wound without further injury from mechanical dispersion. This method is especially useful for treatment of burns. Some personal products such as nail polish remover, after-shave lotion, and foot and body lotions have also been formulated as quick-breaking foams.

Two advantages of foam systems over sprays (qv) are the increased control of the area to which the product is delivered and the decreased incidence of airborne particle release.

Sprays. Aerosol spray emulsions are of the water-in-oil type. The preferred propellant is a hydrocarbon or mixed hydrocarbon–hydrofluorocarbon.

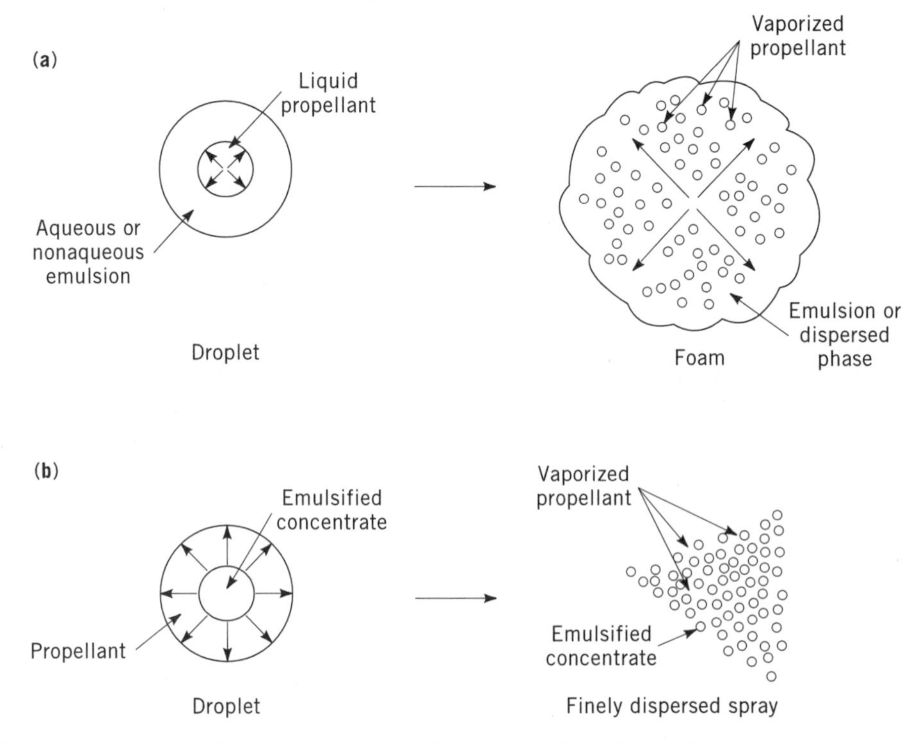

Fig. 2. Aerosol emulsion droplets containing propellant (**a**) in the internal phase with subsequent formation of aerosol foam and (**b**) in the external phase with subsequent formation of a wet spray.

About 25–30% propellent, miscible with the oil, remains in the external phase of the emulsion. When this system is dispensed, the propellant vaporizes, leaving behind droplets of the w/o emulsion (Fig. 2**b**). A vapor tap valve, which tends to produce finely dispersed particles, is employed. Because the propellant and the product concentrate tend to separate on standing, products formulated using this system, such as pesticides and room deodorants, must be shaken before use.

Dispersions. In a powder aerosol, the powder is dispersed or suspended in propellant using dispersants (qv) (oily vehicles) and suspending agents. Moisture content should be below 300 ppm and compacting, agglomeration, and sedimentation need to be minimized so that a fine powder can be uniformly dispensed without clogging of the valve. Powders must have a particle size of <40 μm (325 mesh screen) to pass through the valve orifices. Sedimentation rate can be substantially reduced by adjusting the density of either the propellant or the powder. Techniques include using a mixture of propellants of varying densities and adding an inert powder to the active ingredients. The use of surfactants (qv) as dispersing agents can also serve to lubricate the valve to prevent its sticking.

Pharmaceutical powder aerosols have more stringent requirements placed upon the formulation regarding moisture, particle size, and the valve. For metered-dose inhalers (MDIs), the dispensed product must be delivered as a spray having a relatively small (3–6 μm) particle size so that the particles can be deposited at the proper site in the respiratory system. On the other hand, topical powders must be formulated to minimize the number of particles in the 3–6-μm range because of the adverse effects on the body if these materials are accidently inhaled.

Pastes. Aerosols utilizing a paste as the product concentrate base differ from other formulations in that the product and the propellant do not come in contact with one another. The paste is placed in a bag that is attached to the valve system and fitted into the container. The propellant is then placed between the bag and the inside wall of the container so that the propellant presses against the outside of the bag, dispensing the contents through the valve.

3.2. Propellants. The propellant, said to be the heart of an aerosol system, maintains a suitable pressure within the container and expels the product once the valve is opened. Propellants may be either a liquefied halocarbon, hydrocarbon, or halocarbon–hydrocarbon blend, or a compressed gas such as carbon dioxide (qv), nitrogen (qv), or nitrous oxide.

Liquefied Gas Propellants. One of the advantages in using a liquefied gas propellant is that the pressure in the aerosol container remains constant until the contents are completely expelled. The disadvantages are that the hydrocarbons are flammable.

Chlorofluorocarbons. Prior to 1978 most aerosol products contained CFC propellants. Since that time, the use of chlorinated fluorocarbons for aerosols has been seriously curtailed. These compounds have been implicated in the depeletion of the ozone (qv) layer and are considered to be greenhouse gases (see AIR POLLUTION; ATMOSPHERIC MODELING). Starting in 1978–1979 most household aerosols were reformulated using an environmentally acceptable propellant such as a hydrocarbon.

The 1990 Clean Air Act regulates the production and use of CFCs, hydrochlorocarbons, hydrochlorofluorocarbons (HCFCs), and hydrofluorocarbon (HFC) substitutes. CFC and halon (Class I substances) usage was phased out in steps until total phaseout occured on January 1, 1997.

In the United States, use of CFC propellants, designated as Propellants 11, 12, and 114, is strictly limited, to specialized medicinal aerosol products such as MDIs. for asthma. Most of the countries of the world (except Third World countries) have also banned the use of CFC's for all aerosols except pharmaceutical and medicinal use. The U.S. Food and Drug Administration (FDA) does not intend to prohibit the use of these CFCs in the exempted MDIs until such time as a sufficient number of alternative inhalers become available and their use has been accepted by both patient and physician.

The physical properties and chemical names of these propellents are given in Table 2.

Fluorocarbons are assigned numbers based on their chemical composition and, in general, these numbers are preceded by a manufacturer's trademark. In the numbering system, the digit on the right denotes the number of fluorine atoms in the compound; the second digit from the right indicates the number of hydrogen atoms plus 1; and the third digit from the right indicates the number of carbon atoms less 1. In the case of isomers, each has the same number. The most symmetrical is indicated by the number without any letter following it. As the

Table 2. **Physical Properties of Chlorofluorocarbon and Hydrocarbon Propellants**

	Propellant					
Property	11	12	114	A-108	A-31	A-17
chemical name	trichloromono-fluoromethane	dichlorodifluoro-methane	1,2-dichloro-1,1,2,2-tetra-fluoroethane	propane	isobutane	*n*-butane
CAS Registry Number	[75-69-4]	[75-71-8]	[76-14-2]	[74-98-6]	[75-28-5]	[106-97-8]
formula	CCl_3F	CCl_2F_2	$CClF_2CClF_2$	C_3H_8	$HC(CH_3)_3$	C_4H_{10}
molecular weight	137.4	120.9	170.9	44.1	58.1	58.1
boiling point, °C	23.8	−29.8	3.6	−42.2	−10.2	−0.6
vapor pressure, kPa[a]						
21°C	194	585	190	846	315	214
54°C	269	1349	507	1893	763	556
liquid density, g/mL, 21°C	1.485	1.325	1.468	0.5005[b]	0.5788[b]	0.5571[b]
flammability limit, vol % in air	nonflammable	nonflammable	nonflammable	2.3–7.3	1.8–8.4	1.6–6.5

[a] To convert kPa to psi, multiply by 0.145.
[b] At 68°C.

isomers become more unsymmetrical, the letters a, b, and c are appended. If a molecule is cyclic, the number is preceded by C.

Hydrocarbons. Hydrocarbons such as propane, butane, and isobutane, butane, and isobutane, have been the replacement for CFCs in most household and industrial aerosols. They are assigned numbers based upon their vapor pressure in psia at 21°C. For example, as shown in Table 2, aerosol-grade propane is known as A-108, n-butane as A-17. Blends of hydrocarbons, eg, A-46, and blends of hydrocarbons and hydrochlorocarbons or HCFCs are also used. The chief drawback to the use of hydrocarbon propellants is their flammability.

Hydrocarbons have, for the most part, replaced CFCs as propellants. Most personal products such as hair sprays, deodorants, and antiperspirants, as well as household aerosols, are formulated using hydrocarbons or some form of hydrocarbon–halocarbon blend. Blends provide customized vapor pressures and, if halocarbons are utilized, a decrease in flammability. Some blends form azeotropes that have a constant vapor pressure and do not fractionate as the contents of the container are used.

As with fluorocarbons, a range of pressures can be obtained by mixing various hydrocarbons in varying proportions. As the composition of the hydrocarbons is likely to vary somewhat, depending on their source, blending of hydrocarbons must be based on the final pressure desired and not on the basis of a stated proportion of each component, the pressure of which will depend on its purity. Table 3 lists some commonly used blends that are commercially available.

Hydrofluorocarbons and Hydrochlorofluorocarbons. The properties of HFC and HCFC propellants are given in Table 4. Propellant 22 is nonflammable and can be mixed to form nonflammable blends. Some of these propellants are scheduled for phase-out by 2015–2030. Propellants 142b and 152a, especially the latter is used as the propellant of choice for hair mousse which is available as a foam.

Compressed Gas Propellants. The compressed gas propellants, so named because they are gaseous in conventional aerosol containers, are nontoxic, nonflammable, low in cost, and very inert. When used in aerosols, however, the pressure in the container drops as the contents are depleted. Although the problem is lessened when the contents are materials in which the propellant is

Table 3. **Commonly Used Hydrocarbon Blends**

Designation	Pressure, psig at 70°F (21.1°C)	Composition (mol%)		
		n-Butane	Propane	Isobutane
A-108	108 ± 4	traces	99	1
A-31	31 ± 2	3	1	96
A-17	17 ± 2	98	traces	2
A-24	24 ± 2	49.2	0.6	50
A-40	40 ± 2	2	12	86
A-46	46 ± 2	2	20	78
A-52	52 ± 2	2	28	70
A-70	70 ± 2	1	51	48

Table 4. **Properties of Hydrofluorocarbon and Hydrochlorofluorocarbon Propellants**

	Propellant		
Property	22	142b	152a
chemical name	chlorodifluoro-methane	1-chloro-1,1-difluoroethane	1,1-difluoroethane
CAS Registry Number	[75-68-6]	[75-68-3]	[75-37-6]
formula	$CHClF_2$	$C_2H_3ClF_2$	$C_2H_4F_2$
molecular weight	86.5	100.5	66.1
boiling point, °C	−40.8	−9.44	−23.0
vapor pressure, kPa[a]			
21°C	834	200	434
54°C	2048	669	1220
density, g/mL at 21°C	1.21	1.12	0.91
solubility in water,wt % at 21°C	3.0	0.5	1.7
Kauri-butanol value	25	20	11
flammability limit, vol % in air	nonflammable	6.3–14.8	3.9–16.9
flash point, °C	none	none	<−50°

[a] To convert kPa to psi, multiply by 0.145.

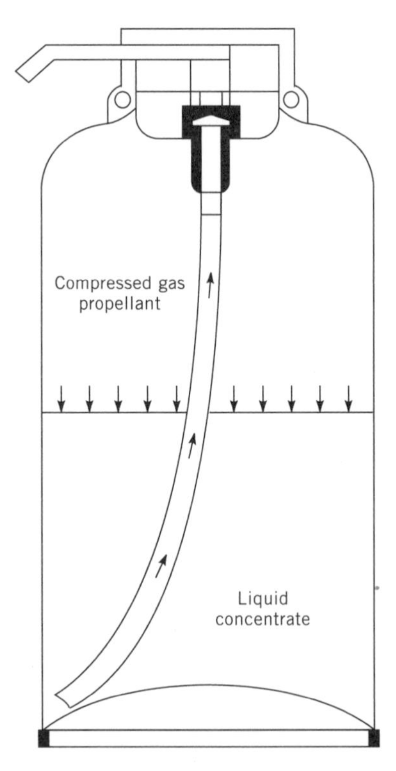

Fig. 3. Compressed gas aerosol using an insoluble gas as propellant.

Table 5. **Properties of the Compressed Gas Propellants**

Property	Carbon dioxide	Nitrous oxide	Nitrogen
CAS Registry Number	[124-38-9]	[10024-97-2]	[7727-37-9]
formula	CO_2	N_2O	N_2
molecular weight	44.0	44.0	28.0
boiling point, °C	−78	−88	−196
critical temperature, °C	31	37	−111
vapor pressure, kPa[a] at 21°C	5772	4966	
solubility in water, vol gas/vol liq at 21°C and 101.3 kPa[a]	0.82	0.6	0.016
flammability limits, vol % in air	nonflammable	nonflammable	nonflammable
flash point, °C	none	none	none

[a] To convert kPa to psi, multiply by 0.145.

somewhat soluble, this pressure drop may cause changes in the rate and characteristics of the aerosol spray. A compressed gas aerosol system is illustrated in Figure 3.

Considerable developmental effort is being devoted to aerosol formulations using the compressed gases given in Table 5. These propellants are used in some food and industrial aerosols. Carbon dioxide and nitrous oxide, which tend to be more soluble, are often preferred. When some of the compressed gas dissolves in the product concentrate, there is partial replenishment of the headspace as the gas is expelled. Hence, the greater the gas solubility, the more gas is available to maintain the initial conditions.

Compressed gas systems were originally developed simply to provide a means of expelling a product from its container when the valve was depressed. Semisolid products such as a cream, ointment, or caulking compound are dispensed as such. A liquid concentrate and a compressed gas propellant (Fig. 3) produce a spray when a mechanical breakup actuator is used. Nitrogen, insoluble in most materials, is generally used as the propellant.

Aerosols using an insoluble gas are not intended to be shaken before use. Shaking causes some of the propellant to be dispersed in the liquid concentrate. Although the product may then be dispersed to a greater extent, greater loss of propellant also results. If enough propellant is lost, the product will become inoperative.

When gases that are somewhat soluble in a liquid concentrate are used, both concentrate and dissolved gas are expelled. The dissolved gas then tends to escape into the atmosphere, dispersing the liquid into fine particles. The pressure within the container decreases as the product is dispersed because the volume occupied by the gas increases. Some of the gas then comes out of solution, partially restoring the original pressure. This type of soluble compressed gas system has been used for whipped creams and toppings and is ideal for use with antistick cooking oil sprays. It is also used for household and cosmetic products either where hydrocarbon propellants cannot be used or where hydrocarbons are undesirable.

Other Propellants. Dimethyl ether (DME) [115-10-6] is also used as an aerosol propellant. DME is soluble in water, as shown in Table 6. Although

Table 6. **Properties of DME Propellant**

Property	Value
CAS Registry Number	[115-10-6]
formula	CH_3OCH_3
molecular weight	46.07
boiling point, °C	−24.8
vapor pressure, kPa[a]	
21°C	434
54°C	1200
density, g/mL at 21°C	0.66
solubility in water, wt % at 21°C and autogenous pressure	34
Kauri-butanol value	60
flammability limits in air, vol %	3.4−1.8
flash point, °C	−41

[a] To convert kPa to psi, multiply by 0.145.

this solubility reduces DMEs vapor pressure in aqueous systems, the total aerosol solvent content may be lowered by using DME as a propellant. The chief disadvantage is that DME is flammable and must be handled with caution.

Alternative Liquified Gas Propellants (HCFCs and HFCs). Many aerosols were developed originally using CFCs 11, 12, and 114. These propellants have found widespread use due to their inertness, nonflammability, and nontoxicity. Unfortunately, the CFCs have been implicated in depleting the ozone layer, and their use as an aerosol propellant has practically been eliminated.

Topical pharmaceutical aerosols have been successfully reformulated with Propellants 152a, 142b, 22, DME, hydrocarbons, and compressed gases. Suitable valves are available which, together with modifications in formulation and propellant blends, produce topical aerosol pharmaceuticals, that are satisfactory and acceptable.

Several new liquified gas materials have been developed to replace the CFCs as refrigerants, foaming agents and in other nonpharmaceutical uses. Propellants 134a and 227 were developed as a substitute for Propellant 12 in MDIs and have survived many of the short- and long term toxicity studies. To date, no suitable replacement has been found for Propellants 11 and 114. Propellant 114 is not essential for use with MDIs, but most of the present suspension formulations require a minimum amount of Propellant 11. Propellant 11 is used to form a slurry with the active ingredient and dispersing agent, which is impossible to accomplish with Propellants 134a and R-227 (unless these propellants are chilled well below their boiling point and handled as a cold fill). The hydrochlorocarbons (HFCs) are extremely poor solvents and will not dissolve a sufficient amount of the currently used FDA approved surfactants (oleic acid, sorbitan trioleate, and soya lecithin).

It also has been noted that many of the currently used dispersing agents are not compatible with these newer materials used in valves. The gaskets and sealing compounds used in metered-dose valves present compatibility problems to the formulator; however, other gaskets have been developed and were found to be satisfactory. Several of the critical properties of these newer propellants are shown in Table 7.

Table 7. **Properties of HFCs**

Property	Tetrafluoroethane	Heptafluoropropane
molecular formula	CF_3CH_2F	CF_3CHFCF_3
numerical designation	134a	227
molecular weight	102	170
boiling point (1 atm)		
°F	−15.0	−3.2
°C	−26.2	−16.5
vapor pressure (psig)		
70°F	71.1	43 (at 20°C)
130°F	198.7	—
liquid density (g/mL) (21.1°C)	1.22	1.41
flammability	nonflammable	nonflammable
solubility in water (% w/w)	0.150	0.058

4. Components

4.1. Containers.

Aerosol containers, made to withstand a certain amount of pressure, vary in both size and materials of construction. They are manufactured from tin-plated steel, aluminum, and glass. The most popular aerosol container is the three-piece tin-plated steel container. Glass containers, which are usually plastic coated, generally have thicker walls than conventional glass bottles. They are limited to a maximum size of ∼120 mL and are generally used for pharmaceutical and cosmetic aerosols.

Steel. The steel container's most usual form is cylindrical with a concave (or flat) bottom and a convex top dome with a circular opening finished to receive a valve with a standard 2.54-cm (1″) opening. The three pieces (body, bottom, and top) are produced separately and joined together by high speed manufacturing. The size of the container is described by its diameter and height to top seam, in that order. Hence, a 202 × 509 container is 54.0 mm ($2\ \frac{2}{16}$ in.) in diameter by 141.3 mm ($5\ \frac{9}{16}$ in.) high. Tables of available sizes and overflow volumes and suggested fill levels can be readily obtained from container manufacturers.

Tin-plated steel was long the mainstay of the U.S. aerosol industry and still represents a very large volume. The tin coating provides both protective internal and external surfaces and the means for soldering the flat body plate into a leak-proof cylinder. Both tin and lead solders have been used in the past but at present has been largely replaced by welded side-seam construction.

The welding process has a slight financial advantage because it eliminates the need for tin. In addition, the welded joint is esthetically more desirable, it is less than one-half the width of the soldered joint, and it does not weaken during prolonged storage at elevated temperatures.

In order to increase resistance of the container to the effect of the product or to protect the product from the tin plating, an inert, internal organic coating can be applied.

Aluminum. The majority of aluminum containers are of monobloc (one-piece) construction, impact extruded from a slug of lubricated aluminum alloy. These containers are widely used for many products and are available in a vast array of heights and diameters. Because these containers lend themselves to additional shaping, many unusual shapes can be found in the marketplace. They may also be coated after the extrusion process.

Aluminum containers are recommended for many applications because of the very hard, corrosion-resistant oxide coating. They are deficient in only one respect: Once the protective skin has been penetrated, aluminum corrosion accelerates.

Two-piece aluminum containers are also available. These consist of impact-extruded upper shells having a seamed-on aluminum bottom. The valve opening is machined from the solid aluminum rather than rolled. All coatings must be applied to the can body after the forming operation.

Glass. Glass containers present completely different design considerations from metal ones. They are totally nonreactive with the product, free of potentially leaking seams (the valve joint may be an exception), transparent or opaque as desired, and can be beautiful in design. The larger sizes attract attention on store shelves. On the other hand, they can break in manufacture, in shipping, and in usage. Although the extent and hazards of breakage can be reduced by means of a thick vinyl coating, glass containers are heavy, making them most costly to ship. Glass containers are processed more slowly and have higher scrap rates, and are generally limited to low pressure product systems.

4.2. Valves. The dispensing valve and actuator serve to close the opening through which the product and frequently the propellant entered the container, to retain the pressure within the container and to dispense the product in the precise form and dosage intended by the manufacturer and expected by the consumer. An aerosol valve, shown in Figure 4, consists of seven components. Many variations exist both for special purposes and to avoid existing patents.

The *mounting cup* (ferrule for bottle valves) mechanically joins the valve to the container. The mounting cup may be made from a variety of materials, but is typically tin-plated steel or aluminum coated on the underside. It contains the gasket that provides the seal. Soft gasketing material is applied wet and bonded in place or, more frequently in larger cans, cut rubber, polyethylene, or polypropylene gaskets are used.

The *housing* physically holds the valve pieces together by means of a mechanical lock (crimp) and fits into the pedestal of the mounting cup. It is made from any of a number of common thermoplastics and contains the metering orifices for both the liquid and vapor phases of the effluent. Many valves do not meter vapor; the flow of the liquid is controlled by other means. A vapor tap may also be present to reduce the flammability of the product when it is emitted or to produce a finer, drier spray.

The plastic *stem* is the movable segment of the valve. It provides the opening mechanism and usually contains another metering orifice as shown.

The *spring* ensures a solid closing action and is usually wound from stainless steel wire. The *dip tube* conducts the product from the container to the valve. It is usually extruded from polyethylene or polypropylene and has an inside diameter of >2.54 mm, although it can be provided in capillary sizes having diameters down to 0.25 mm. These small tubes are used to reduce flow rate and may function in place of the liquid metering orifice in the valve housing.

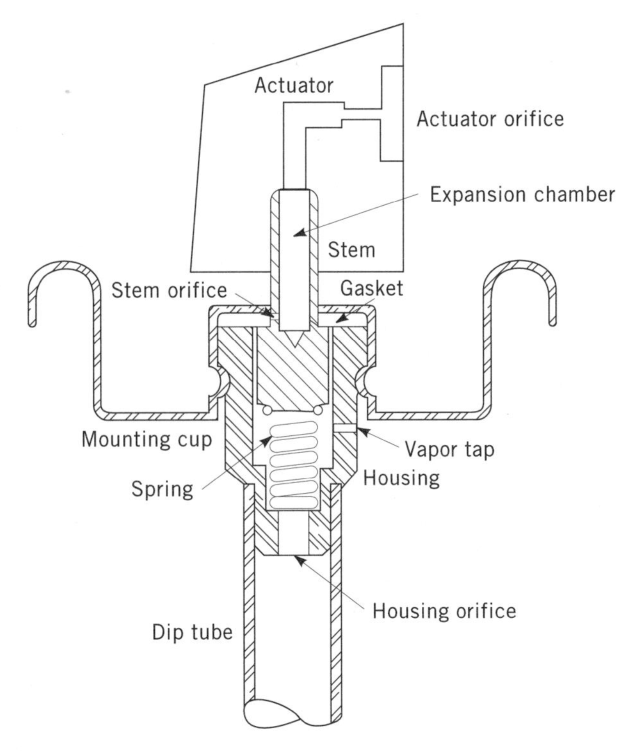

Fig. 4. Aerosol valve components.

The *actuator* contains the final orifice and a finger pad or mechanical linkage for on–off control. The spray pattern is largely affected by the construction of the actuator, particularly by the chamber preceding the orifice. Actuators are often termed mechanical breakup and nonmechanical breakup depending on the complexity of this chamber. Mechanical breakup actuators are of more expensive two-piece construction. Actuators are usually molded from polyethylene or polypropylene; the breakup insert may be almost any material, including metal.

Valves may differ depending on the form of the product. Manufacturers stock a wide array of standard components. In addition, most are willing to produce unique combinations at no additional charge if no major tooling changes are required. The largest numbers of aerosol valves are used to produce sprays. These are actuated by either tilting forward or depressing vertically. Foam valves differ from spray valves primarily in the actuator, which is relatively wide open. Small actuator orifices throw the foam and are used for products such as rug shampoos and decorative snow. Internally, foam valves have no vapor tap and contain relatively large orifices.

Metering valves are used extensively in the pharmaceutical field for inhalers and other products requiring controlled dosage delivery. They typically deliver 50 to 150 mg of product per stroke with good repeatability. The metering is achieved by plugging the body orifice with the downward stroke of the stem, allowing only the product in the housing (metering chamber) to escape.

Codispensing valve (and container) systems are used for products that consist of two reactive ingredients that must be kept separate until dispensed. They can deliver the product components in stoichiometrically correct amounts, thoroughly mixing them in the process. Successful use of these systems requires proper metering throughout the life of the container, safety considerations (should the internal seals between reactants fail), acceptably low permeation across internal membranes, end-use excess of the preferred component, and the usual product–package compatibility. The codispensing valves are the most technically demanding systems in the marketplace, but to date, have not been used commercially to any great extent.

4.3. Barrier–Type Systems. These systems separate the propellant from the product itself. The pressure on the outside of the barrier serves to push the contents from the container. The following types are available.

Piston Type. Since it is difficult to empty the contents of a semisolid from an aerosol container completely, a piston-type aerosol system has been developed. This systems utilizes a polyethylene piston fitted into an aluminum container. The concentrate is placed into the upper portion of the container. The pressure from nitrogen 621–690 kPa ($\sim$90–100 psig), or a liquified gas, pushes against the other side of the piston and, when the valve is opened, the product is dispensed. The piston scrapes against the sides of the container and dispenses most of the product concentrate. The piston-type aerosol system is shown in Figure 5. This sytem has been used successfully to package cheese spreads, cake decorating icings, and ointments. Since the products that use this system are semisolid and viscous, they are dispensed as a lazy stream rather than as a foam or spray. This system is limited to viscous materials since limpid liquids, such as water or alcohol, will pass between the wall of the container and the piston.

Piston Bag Type. This system consists of a collapsible plastic bag fitted into a standard, three-piece, tin plate container. The product is placed within the bag and the propellant is added through the bottom of the container. Since the product is placed into a plastic bag, there is no contact between the product and the container wall except for any product that may escape by permeation through the plastic bag.

Limpid liquids, such as water, can be dispensed either as a stream or fine mist depending on the type of valve used, while semisolid substances are dispensed as a stream. In order to prevent the gas from pinching the bag and preventing the dispensing of product, the inner plastic bag is accordion pleated. This system can be used for a variety of different pharmaceutical and nonpharmaceutical systems, including topical pharmaceutical products as a cream, ointment or gel.

A modification of this system dispenses the product as a gel that will then foam. By dissolving a low boiling liquid such as isopentane or pentane in the product, a foam will result when the product is placed on the hands and the warmth of the hands will cause vaporization of the solvent. This system, as well as the piston system, is used in postfoaming shave gels.

Other variations of these systems include using a laminated pouch that has been sealed onto a 1 in. valve. This unit is placed into an aluminum can, sealed and pressurized to 103–138 kPa ($\sim$15–20 psig). The product is injected into the

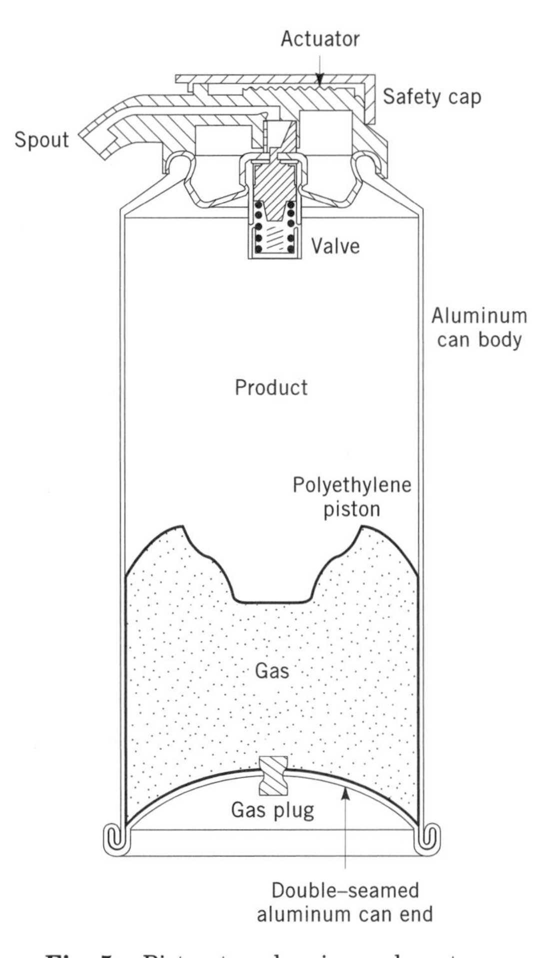

Fig. 5. Piston-type barrier pack system.

pouch, and develops a pressure of 276–345 kPa (~40–50 psig) as the inner bag expands. Another system includes filling the product into a latex bag that then expands. The energy caused by the stressed bag will release the product when the valve is opened. These systems have been used to dispense a variety of personal care products including some pharmaceutical gel products.

5. Filling of Aerosols

All aerosols are produced by either a cold or pressure-filling process. The cold fill process has been used for some aerosols that contain a metered-dose valve, although pressure filling is now the preferred method. Generally, a concentrate is prepared which is filled into the aerosol container and then the valve is added. For the most part, pressure filling is carried out either by an under-the-cup filler or through the valve. If an under-the-cup filler is used, a vacuum is drawn, the propellant is added (under the valve cup), and then the valve is sealed in place.

Where filling is done through the valve stem, the product is first filled into the container, a valve is crimped into place, and, at the same time, a vacuum is drawn in the can. The propellant filler then forms a seal around the head of the can and under high pressure the propellant is forced through the actuator and valve stem into the container. The contents are then checked for leaks and an overcap is added to complete the process.

Nitrogen, carbon dioxide, and nitrous oxide, which are only slightly soluble in the product concentrate are packaged by placing the product into the container and then sealing a valve in place. The compressed gas is added through the valve using a gasser-/shaker-type filler.

The piston-type barrier system is filled by adding the product to the containers. The valve is then sealed into place. The propellant is added through a small hole in the bottom of the can, and the opening is sealed using a rubber plug. The propellant can be either nitrogen at 621 kPa (~90 psig) or with ~5–10 g of a liquid gas propellant. Valves with relatively large openings, such as foam valves, are frequently used.

The plastic bag type system consists of a collapsible plastic bag fitted into a standard three-piece, tin-plated container such as a 202 × 214, 202 × 406, or 202 × 509 can. The product is placed within the bag, the valve is sealed, and the propellant is added through the bottom of the container, which is fitted with a one-way valve. There is no limitation on the viscosity of the product but compatibility with the plastic bag must be considered. A free-flowing liquid can be dispensed either as a stream or a fine spray, depending on the type of valve employed. A viscous material is often dispensed as a stream. This system has been used for caulking compounds, postfoaming gels, and depilatories.

6. Economic Aspects

According to the Clean Air Act of 1990, as well as regulations of the FDA and the EPA, the use of CFCs in nonessential, nonexempted products has been prohibited since 1996, which included all commercial aerosol products except for a few medical aerosols (inhalers used for asthma) as well as refrigerators, freezers, and chillers. Today most of these aerosols contain a logo and the words, "Contains NO CFCs which Deplete the Ozone Layer". (See Fig. 6).

Fig. 6. No CFC Logo.

BIBLIOGRAPHY

"Aerosols" in *ECT* 2nd ed., Vol. 1, pp. 470–480, by M. S. Sage, Sage Laboratories, Inc.; in *ECT* 3rd ed., Vol. 1, pp. 582–597, by Antoine Kawam and John B. Flynn, The Gillette Company; in *ECT* 4th ed., Vol. 1, pp. 670–685, by John Sciarra, Sciarra Aeromed Development Corpation; "Aerosols" in *ECT* (online), posting date: December 4, 2000, by John J. Sciarra, Sciarra Aeromed Development Corporation.

GENERAL REFERENCES

Aerosol Guide, 8th ed., Aerosol Division, Chemical Specialties Manufacturers Association, Washington, D.C., 1995.

Anon, U.S. Aerosol Production 2000, Spray Technology and Marketing, **11**(7), 30 (2001).

J. J. Daly, Jr., Properties and Toxicology of CFC Alternatives, *Aerosol Age* **35**(2), 26.

M. A. Johnson, *The Aerosol Handbook*, Wayne Dorland Company, Mendham, N.J., 1982.

Handbook of Pharmaceutical Excipients, Propellants, American Pharmaceutical Association, Washington, D.C.; Pharmaceutical Press, London, United Kingdom, 2000, pp. 132–137, 184–187, 234–237, 355–352, 560.

P. A. Sanders, *Handbook of Aerosol Technology*, 2nd ed., Van Nostrand Reinhold Co., Inc., New York, 1979.

J. J. Sciarra, in H. Liebermann, H. Rieger, M. Banker, eds., *Aerosol Suspensions and Emulsions in Pharmaceutical Dosage Forms: Disperse Systems*, Vol. 2, 2nd ed., Dekker, New York, 1996.

J. J. Sciarra, C. J. Sciarra, Aerosols, in A. R. Gennaro, *Remington The Science and Practice of Pharmacy*, 20th ed., Lippincott williams and Williams, 2000, pp. 963–979.

J. J. Sciarra and L. Stoller, *The Science and Technology of Aerosol Packaging*, John Wiley & Sons, Inc., New York, 1979.

John J. Sciarra
Christopher J. Sciarra
Sciarra Laboratories, Inc.

AIR POLLUTION

1. Introduction

Air pollution is the presence of any substance in the atmosphere at a concentration high enough to produce an undesirable effect on humans, animals, vegetation, or materials, or to significantly alter the natural balance of any ecosystem. Air pollutants can be solids, liquids, or gases, and can be produced by anthropogenic activities or natural sources. In this article, only nonbiological material is considered and the discussion of airborne radioactive contaminants is limited to radon [10043-92-2], which is discussed in the context of indoor air pollution.

Concern about the effects of atmospheric contaminants on human health and the environment can be traced for centuries. For example, laws in Israel in the first and second centuries AD required that tanneries be located downwind of towns (1). In the fourteenth century, protests over the burning of sea coal resulted in the taxation of coal, and men were tortured for producing a pestilent odor. In the United States, the first air pollution regulations also dealt with coal. Cities such as Chicago, St. Louis, and Cincinnati passed smoke ordinances in the nineteenth century. In the early twentieth century, concern about the impacts of air pollution can be traced to severe episodes, which demonstrated that air pollutants can be hazardous to human health and can even cause death at high enough concentrations. For example, a week-long air stagnation in the Meuse Valley in Belgium in 1930 led to the death of 60 people and respiratory problems for a large number of others. In 1948, similar conditions in Donora, Pennsylvania, resulted in nearly 7000 illnesses and 20 deaths, and in 1952, 4000 deaths were attributed to a 4-day "killer fog" in London, England. The episodes dramatized the acute health effects of high concentrations of air pollutants, but these are rare events.

Over the past several decades concern about air pollutants has evolved; the current focus is on the effects of long term, chronic exposures to nonlethal concentrations of air pollutants, the effects of air pollution on global and regional climate, and the effects of air pollutants on global and regional atmospheric cycles (eg, stratospheric ozone depletion and acid deposition).

Health effects associated with chronic exposure to air pollution is a worldwide problem. The World Health Organization (WHO) has estimated that ~2.7 million deaths are attributable to air pollution throughout the world each year.

Table 1. **Air Quality in Mega-Cities (2)**

City	SO_2	SPM	Pb	CO	NO_2	O_3	Year in which data were collected
Athens	[a]	no data	no data	[a]	[b]	[a]	1995
Bangkok	[a]	[b]	[a]	no data	[a]	no data	1995
Beijing	[b]	[b]	no data	no data	no data	no data	1994
Bucharest	[a]	[c]	[b]	no data	[c]	no data	1995
Calcutta	[a]	[b]	no data	no data	[a]	no data	1995
Caracas	[a]	[c]	[b]	no data	[c]	no data	1995
Delhi	[a]	[b]	no data	no data	[c]	no data	1995
Johannesburg	[a]	[a]	[b]	[a]	[a]	[a]	1994
London	[a]	[a]	[a]	[a]	[c]	[a]	1995
Los Angeles	[a]	[c]	[a]	[a]	[a]	[c]	1995
Mexico City	[a]	[b]	no data	no data	[c]	[b]	1993
Santiago	[a]	[b]	no data	[a]	[c]	[a]	1995
Sofia	[a]	[b]	[a]	no data	[b]	no data	1995
Shanghai	[c]	[b]	no data	no data	no data	no data	1994
Sydney	no data	[a]	[a]	[a]	[a]	[a]	1995
Tokyo	[a]	[c]	[a]	[a]	[c]	[a]	1995
Xian	[c]	[b]	no data	no data	no data	no data	1994

[a] Low pollution—WHO guidelines are normally met.
[b] Moderate to heavy pollution—WHO guidelines exceeded by up to a factor of 2.
[c] Serious problem—WHO guidelines exceeded by more than a factor of 2.

Among the air pollutants of greatest concern are ozone, suspended particulate matter, nitrogen dioxide, sulfur dioxide, carbon monoxide, lead, and other toxins (detailed descriptions of these air pollutants are provided in subsequent sections). Of these pollutants, ozone is one of the most prevalent air pollutants in large cities and has been associated with increased respiratory illness and decreased lung function, particularly in children. Suspended particulate matter is the air pollutant most responsible for mortality worldwide. A review of the distribution and concentration of air pollutants, worldwide, is available from the WHO and is summarized in Table 1.

In addition to concerns about localized health impacts of urban air pollution, there has been a growing recognition of regional and global impacts of air pollutants on the natural balances of the earth's systems and on climate. This article will describe three examples of impacts of air pollutants on natural balances of the earth's systems: stratospheric ozone depletion, global climate change, and acid deposition.

The remainder of this article is organized into three major sections, which describe the sources and impacts of specific air pollutants, the regional and global impacts of air pollutants, and the air quality management and regulatory system in the United States.

2. Air Pollutants

Air pollution is a complex mixture of many chemical species, and any description of air pollution must account for that heterogeneity. The description of air pollutants presented here begins with a discussion of photochemical smog, defined broadly, then focuses on specific pollutants that are the targets of regulations in the United States. These include volatile organic compounds (VOCs), oxides of nitrogen (NO_x, primarily NO and NO_2), sulfur oxides (SO_x, primarily SO_2), carbon monoxide, lead, and air toxics. Finally, a description of indoor air pollutants is provided.

2.1. Photochemical Smog. Photochemical smog is a complex mixture of constituents that are emitted directly to the atmosphere (primary pollutants) and constituents that are formed by chemical and physical transformations that occur in the atmosphere (secondary pollutants). Ozone (O_3), is generally the most abundant species formed in photochemical smog. Ozone is a secondary pollutant formed by the reactions of hydrocarbons and NO_x. Extensive studies have shown that O_3 is both a lung irritant and a phytotoxin. It is responsible for crop damage and is suspected of being a contributor to forest decline in Europe and in parts of the United States. There are, however, a multitude of other photochemical smog species that also have significant environmental consequences. The most important of these pollutants are particles, hydrogen peroxide (H_2O_2), [7722-84-1], peroxyacetyl nitrate (PAN) ($C_2H_3NO_5$), [2278-22-0], aldehydes, and nitric acid (HNO_3), [7697-37-2]. All of these pollutants are secondary, produced in the atmosphere by the reactions of precursor species (although some types of particles, such as diesel soot, are emitted directly).

Since many components of photochemical smog are secondary, government regulatory agencies have attempted to reduce the incidence of smog episodes by controlling the emissions of precursor species. In the case of ozone, these precursor species are reactive hydrocarbons and nitrogen oxides. The mechanism for the production of ozone is generally initiated by the photolysis of nitrogen dioxide. In the presence of sunlight, $h\nu$, NO_2 photolyzes, producing NO and atomic oxygen. The atomic oxygen reacts with O_2 to produce O_3

$$NO_2 + h\nu \rightarrow NO + O\,(^3P) \tag{1}$$

$$O + O_2 + M \rightarrow O_3 + M \tag{2}$$

$$NO + O_3 \rightarrow NO_2 + O_2 \tag{3}$$

where O (^{3}P) is a ground-state oxygen atom and M is any third body molecule (most likely N_2 or O_2 in the atmosphere) that remains unchanged in the reaction. This process produces a steady-state concentration of O_3 that is a function of the concentrations of NO and NO_2, the solar intensity, and the temperature.

$$[O_3] = k\,[NO_2]/[NO]$$

where $[O_3]$, $[NO_2]$, and [NO] are the atmospheric concentrations of ozone, nitrogen dioxide, and nitric oxide and k is a constant dependent on temperature and solar intensity.

Although these reactions are extremely important in the atmosphere, the steady-state O_3 produced by the reactions of nitrogen oxides alone is much lower than the observed concentrations, even in clean air. In order for ozone to accumulate, there must be a mechanism that converts NO to NO_2 without consuming a molecule of O_3, as does reaction 3. Reactions involving hydroxyl radicals and hydrocarbons constitute such a mechanism. In clean air OH may be generated by

$$O_3 + h\nu \rightarrow O_2 + O\,(^1D) \tag{4}$$

$$O(^1D) + H_2O \rightarrow 2\,OH \tag{5}$$

where O(^{1}D) is an excited form of an O atom that is produced from a photon at a wavelength between 280 and 310 nm. This seed OH can then participate in a chain reaction with hydrocarbons. The reactions with methane are shown below.

$$OH + CH_4 \rightarrow H_2O + CH_3 \tag{6}$$

$$CH_3 + O_2 + M \rightarrow CH_3O_2 + M \tag{7}$$

$$CH_3O_2 + NO \rightarrow CH_3O + NO_2 \tag{8}$$

One of the outcomes of reactions 6–8 is the conversion of NO into NO_2. NO_2 can then photolyze producing O_3 (eqs. 1 and 2) and less NO is available to scavenge the ozone (eq. 3), resulting in a higher steady state ozone concentration. In addition, the CH_3O radical continues to react:

$$CH_3O + O_2 \rightarrow HCHO + HO_2 \quad (9)$$

$$HO_2 + NO \rightarrow NO_2 + OH \quad (10)$$

Reactions 9 and 10 result in an additional NO to NO_2 conversion and the regeneration of the hydroxyl radical.

Further, the formaldehyde photodissociates:

$$HCHO + h\nu \rightarrow H_2 + CO \quad (11)$$

$$\rightarrow HCO + H \quad (12)$$

$$HCO + O_2 \rightarrow HO_2 + CO \quad (13)$$

$$H + O_2 \rightarrow HO_2 \quad (14)$$

and the HO_2 from both equations 13 and 14 can form additional NO_2. Moreover, CO can be oxidized:

$$CO + OH \rightarrow CO_2 + H \quad (15)$$

and the H radical can form another NO_2 (eqs. 14 and 10). Thus, the oxidation of one CH_4 molecule is capable of producing three O_3 molecules and two OH radicals. The routes involve both the direct reactions of methane and the reactions of its oxidation products.

Finally, the chain reactions can be terminated by radical–radical recombination, or by radical reactions with more stable species. Two examples are given below.

$$HO_2 + HO_2 \rightarrow H_2O_2 + O_2 \quad (16)$$

$$OH + NO_2 \rightarrow HNO_3 \quad (17)$$

Examination of reactions 6–17 reveals that the chemical sequence initiated by the reaction of hydroxyl radical with a hydrocarbon can lead to the enhancement of ozone concentration, by converting NO to NO_2, and the generation of additional free radicals. The chemistry is complex and depends on the reactions of both the original hydrocarbon species and its reaction products. Because of the complexity of the reactions it is common to characterize the ozone formation potential of hydrocarbons using the parameters shown in Table 2. These parameters are the rate of reaction of a hydrocarbon with hydroxyl radical, and the incremental reactivity of the hydrocarbon.

The rate of reaction of the hydrocarbon with hydroxyl radical characterizes the rate at which the initial reaction (analogous to reaction 6) occurs, and is expressed in Table 2 as the rate constant for the bimolecular reaction with hydroxyl radical, in units of cm^3/molecule/s. In general, internally bonded olefins are the most reactive, followed in decreasing order by terminally bonded olefins,

Table 2. **Reactivities of VOCs**

Compound	CAS Registry Number	Rate constant for reaction with OH × 10^{12} (cm^3/molecule/s/)	Incremental reactivity (grams ozone formed per gram VOC added to a base mixture)
methane	[74-82-8]	0.01	0.0139
isopentane	[78-78-4]	3.7	1.67
n-butane	[106-97-8]	2.44	1.33
toluene	[108-88-3]	3.8	3.97
propane	[74-98-6]	1.12	0.56
ethane	[78-84-0]	0.254	0.31
n-pentane	[109-66-0]	4.0	1.54
ethylene	[74-85-1]	8.5	9.08
m-xylene	[108-38-3]	20	10.61
p-xylene	[106-42-3]	10	4.25
2-methylpentane	[107-83-5]	5.3	1.80
isobutane	[75-28-5]		1.35
propylene	[115-07-1]	26.3	11.58
isoprene	[78-79-5]	101	10.69

multialkyl aromatics, monoalkyl aromatics, C_5 and higher paraffins, C_2–C_4 paraffins, benzene, acetylene, and ethane (3–5).

The incremental reactivity (6, 7) characterizes the ozone formation potential of the hydrocarbon and all of its reaction products. It is expressed as grams of ozone formed per gram of hydrocarbon added to a mixture and is determined by adding an incremental amount of hydrocarbon to a base mixture of hydrocarbons typically found in urban areas, and determining the incremental amount of ozone formed. This incremental reactivity depends on the composition of the base mixture. The values shown in Table 2 are the maximum values of the incremental reactivities for each of the hydrocarbons. Values tend to be highest for species that produce reaction products that are also highly reactive.

The ozone formation reactions summarized in reactions 1–17 are driven by solar radiation. Once the sun sets, O_3 formation ceases and, in an urban area, ozone is rapidly scavenged by freshly emitted NO (eq. 3). On a typical summer night, however, a nocturnal inversion begins to form around sunset, usually below a few hundred meters and consequently, the surface-based NO emissions are trapped below the top of the inversion. Above the inversion to the top of the mixed layer (usually $\sim$1500 m), O_3 is depleted at a much slower rate. The next morning, the inversion dissipates and the O_3-rich air aloft is mixed down into the O_3-depleted air near the surface. This process, in combination with the onset of photochemistry as the sun rises, produces the sharp increase in surface O_3 shown in Figure 1. The overnight O_3 depletion is less in the more rural areas than in a large urban area such as New York City, which is a result of the lower overnight levels of NO in rural areas. Even in the absence of NO or other O_3 scavengers (olefins, eg), O_3 decreases at night near the ground faster than aloft because of its deposition on the ground, buildings, and trees. At the remote mountaintop sites, Whiteface and Utsayantha, there is no overnight decrease in O_3 concentrations.

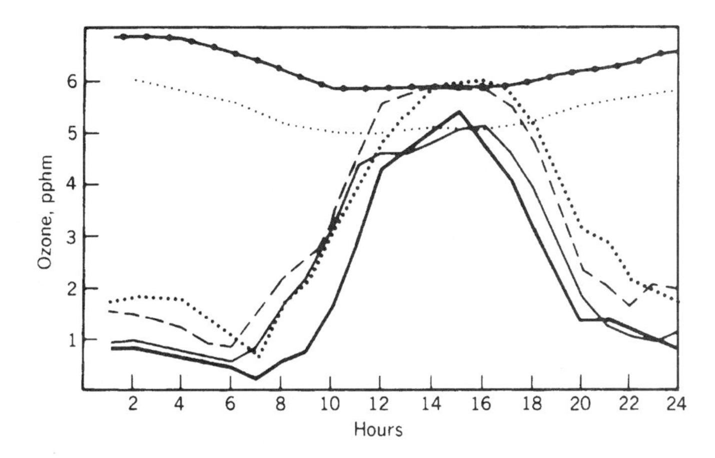

Fig. 1. Ozone formation over the course of a day. Courtesy of Air and Waste Management Association.

Although photochemical smog is a complex mixture of many primary and secondary pollutants and involves a myriad of atmospheric reactions, there are characteristic pollutant concentration versus time profiles that are generally observed within and downwind of an urban area during a photochemical smog episode. In particular, the highest O_3 concentrations are generally found 10–100 km downwind of the major emission sources, unless the air is completely stagnant. This fact, in conjunction with the long lifetime of O_3 in the absence of high concentrations of NO, means that O_3 is a regional air pollution problem. In the Los Angeles basin, high concentrations of O_3 are transported throughout the basin and multiday episodes are exacerbated by the accumulation of O_3 aloft that is then mixed to the surface daily. On the east coast, a typical O_3 episode is associated with a high pressure system anchored offshore producing a southwesterly flow across the region. As a result, emissions from Washington, D.C. travel and mix with emissions from Baltimore and over a period of a few days continue traveling northeastward through Philadelphia, New York City, and Boston. Under these conditions, the highest O_3 concentrations typically occur in central Connecticut (9).

In order to reduce O_3 in a polluted atmosphere, reductions in the VOC and NO_x precursors are required. However, the choice of whether to control VOC, or NO_x, or both depends on the local VOC/NO_x ratio. At low VOC/NO_x ratios, O_3 formation is suppressed through equations 3 and 17. Consequently, in this case reducing NO_x emissions, emitted mainly as NO, reduces the amount of O_3 (eq. 3) and OH (eq. 17) scavenged, increasing the O_3 concentrations.

The dependence of ozone formation on VOC/NO_x ratios is often presented using an ozone isopleth diagram. An example of this type of diagram is illustrated in Figure 2. The diagram plots contours of the maximum ozone concentration achieved as a function of the initial VOC and NO_x concentrations in the mixture. The key features of the ozone isopleth diagram are summarized in Figure 3. The region in the upper left is the NO_x-inhibition region where a decrease

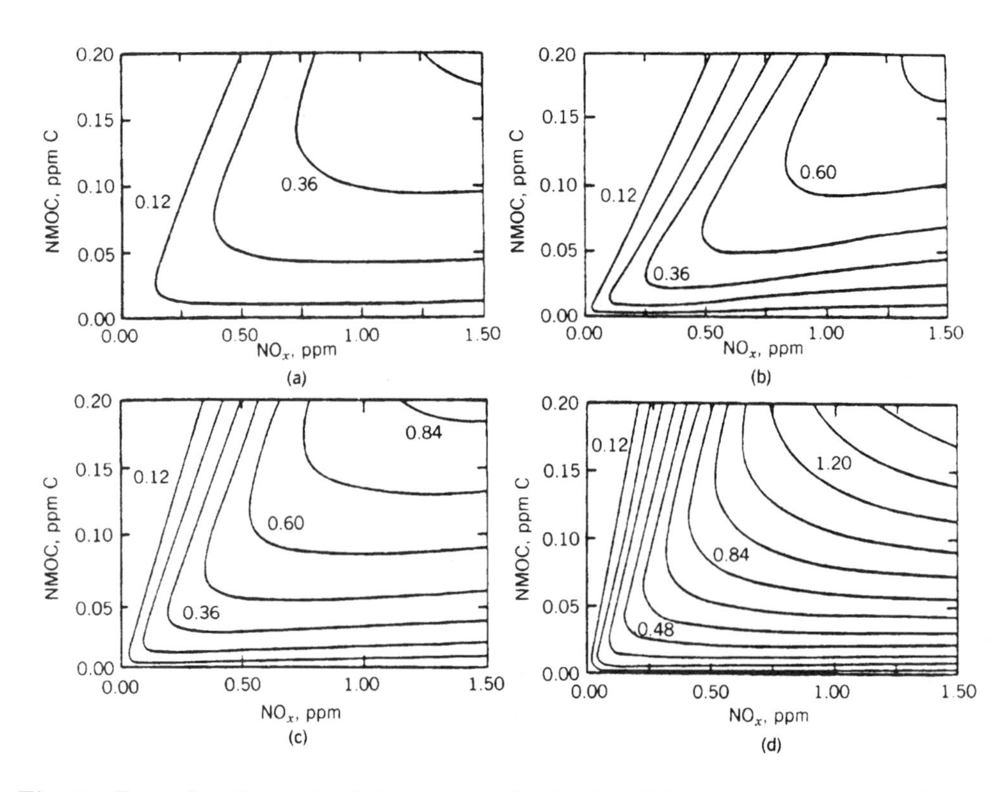

Fig. 2. Examples of ozone isopleths generated using four different chemical mechanisms (**a**) EPA; (**b**) FSM; (**c**) CBII; (**d**) ELSTAR; NMOC = nonmethane organic compounds (10). Courtesy of Pergamon Press.

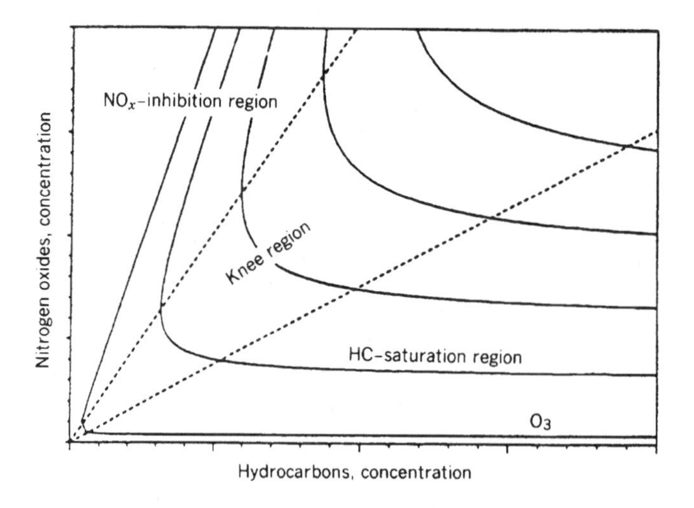

Fig. 3. Typical O_3 isopleth diagram showing the three chemical regimes; HC = hydrocarbons (11). Courtesy of Pergamon Press.

in NO_x alone results in an increase in O_3, but a decrease in VOC decreases O_3. The region at the bottom right is the hydrocarbon (HC) or VOC saturation region where reducing VOCs has no effect on the O_3 level. Here, a reduction in NO_x results in lower O_3. In the middle is the knee region, where reductions in either NO_x or VOC reduce O_3. The upper boundary of this region varies from day to day and from place to place as its location is a function of the reactivity of the VOC mix and the sunlight intensity.

As a guideline, VOC controls are generally the most efficient way to reduce O_3 in areas having a median VOC/NO_x ratio of 10:1 or less (measured between 6 and 9 a.m.); whereas areas with a higher ratio may need to consider NO_x reductions as well (12). The 1990 Clean Air Act Amendments require that O_3 nonattainment areas reduce both VOC and NO_x from major stationary sources unless the air quality benefits are greater in the absence of NO_x reductions. Large cities in the northeast tend to have ratios $<10:1$; cities in the south (Texas and eastward) tend to have ratios $>10:1$. Determining a workable control strategy is further complicated by the transport issue. For example, on high O_3 days in the northeast, the upwind air entering Philadelphia and New York City frequently contains high concentrations of O_3 as a result of emissions from areas to the west and south (13). Consequently, control strategies must be developed on a coordinated, multistate regional basis.

While the focus of this discussion has been ozone, ozone can be considered as a surrogate for the wide variety of chemical species found in photochemical smog. For example, H_2O_2, formed via reaction 16, is important because when dissolved in cloud droplets it is an important oxidant, responsible for oxidizing SO_2 to sulfuric acid (H_2SO_4), [7664-93-9], the primary cause of acid precipitation. The oxidation of many VOCs produces acetyl radicals, CH_3CO, which can react with O_2 to produce peroxyacetyl radicals, $CH_3(CO)O_2$, which react with NO_2

$$CH_3(CO)O_2 + NO_2 \leftrightarrow CH_3(CO)O_2NO_2(PAN) \tag{18}$$

At high enough concentrations, PAN is a potent eye irritant and phytotoxin. On a smoggy day in the Los Angeles area, PAN concentrations are typically 5–10 ppb; in the rest of the United States PAN concentrations are generally a fraction of a ppb. An important formation route for formaldehyde (HCHO), [50-00-0], is the oxidation of hydrocarbons (analogous to reaction 9). However, ozonolysis of olefinic compounds and some other reactions of VOCs can also produce HCHO and other aldehydes. Aldehydes are important because they are temporary reservoirs of free radicals (see eqs. 11 and 12). HCHO is a known carcinogen. Nitric acid is formed by OH attack on NO_2 and by a series of reactions initiated by $O_3 + NO_2$. Nitric acid is important because it is the second most abundant acid in precipitation. In addition, in southern California it is the major cause of acid fog.

Particles are the major cause of the haze that is often associated with smog. The three most important components of particles produced in smog are organics, sulfates, and nitrates. Organic particles are formed when large VOC molecules, especially aromatics and cyclic alkenes, react and form condensable products. Sulfate particles are formed by a series of reactions initiated by the attack of OH on SO_2 in the gas phase or by liquid-phase reactions. Nitrate particles are formed by

$$HNO_{3(g)} + NH_3 \leftrightarrow NH_4NO_{3(s)} \tag{19}$$

or by the reactions of HNO_3 with NaCl or alkaline soil dust.

2.2. Volatile Organic Compounds. VOCs include any organic carbon compound that exists in the gaseous state in the ambient air. In some of the older literature the term VOC is used interchangeably with nonmethane hydrocarbons (NMHC). Reactive organic gases is also a term that is sometimes used to refer to a subset of VOCs that are reactive with hydroxl radical. VOC sources may be any process or activity utilizing organic solvents, coatings, or fuel. Emissions of VOCs are important: some are toxic by themselves, and most are precursors of O_3 and other species associated with photochemical smog. As a result of control measures designed to reduce O_3, VOC emissions are declining in the United States. Figure 4 shows estimates of nationwide emissions of VOCs (14). Emissions peaked ~1970 and have declined by ~40% from that peak. Major sources continue to be industrial processes, solvent use (including solvents used in architectural coatings such as paints and varnishes), non-road sources (such as marine and garden equipment engines) and vehicular sources. This national compilation of VOC emission sources may underestimate the contribution from on-road vehicles, but exhibits the general trends observed in the emissions.

2.3. Nitrogen Oxides (NO_x). Most of the NO_x is emitted as NO, which is then oxidized to NO_2 in the atmosphere (see eqs. 3 and 8). All combustion processes are sources of NO_x. At the high temperatures generated during combustion, some N_2 is converted to NO in the presence of O_2 and, in general, the higher the combustion temperature, the more NO_x produced. Emissions of NO_x have remained relatively constant since the mid-1970s, and the major sources

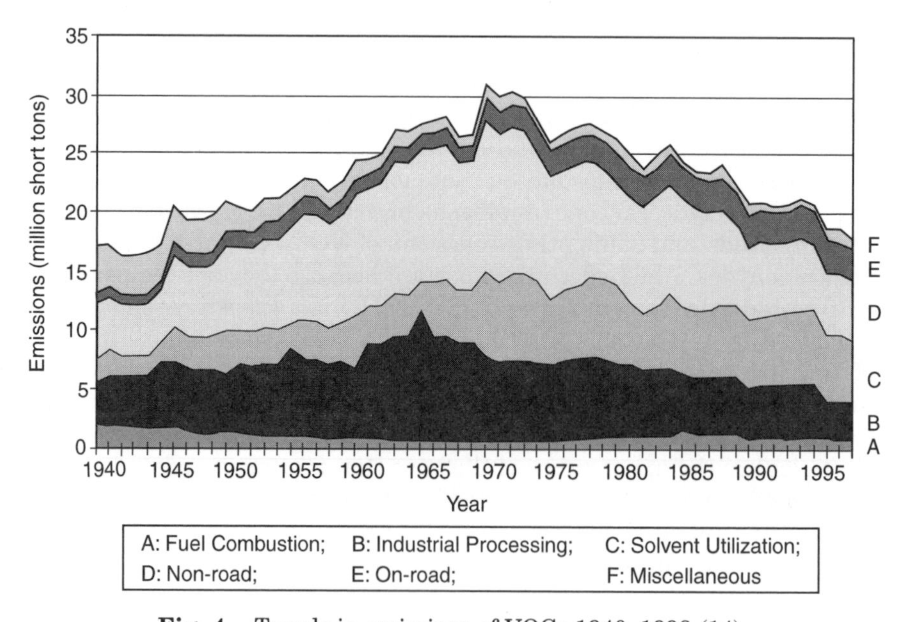

Fig. 4. Trends in emissions of VOCs 1940–1998 (14).

continue to be combustion in industrial facilities, electric power generation, non-road engines (especially construction equipment) and on-road vehicles.

NO_x emissions remain an important issue throughout the United States. As emissions of VOCs have decreased, reducing emissions of NO_x has become a more important component of ozone control strategies in the United States. In addition to being an essential ingredient of photochemical smog and a precursor to HNO_3, itself an ingredient of acid precipitation and fog, NO_2 is the only important gaseous species in the atmosphere that absorbs visible light. In high enough concentrations it can contribute to a brownish discoloration of the atmosphere (see Fig. 5.)

2.4. Sulfur Oxides (SO_x). The combustion of sulfur-containing fossil fuels, especially coal, is the major source of SO_x. Between 97 and 99% of the SO_x emitted from combustion sources is in the form of SO_2. The remainder is mostly sulfur trioxide (SO_3), [7446-11-9], which in the presence of atmospheric water [7732-18-5] vapor is immediately transformed into H_2SO_4, a liquid particulate. Both SO_2 and H_2SO_4 at sufficient concentrations produce deleterious effects on the respiratory system. In addition, SO_2 is a phytotoxin. Control strategies designed to reduce the ambient levels of SO_2 have been highly successful. In the 1960s, most industrialized urban areas in the eastern United States had an SO_2 air quality problem. Now there are no areas that exceed the SO_2 national ambient air quality standards. Over the period 1990–1999, nationwide emissions of SO_2 declined 21% and ambient concentrations decreased 36% (14). However, additional SO_2 reductions are required because of the role that SO_2 plays in acid deposition. In addition, there is some concern over the health effects of H_2SO_4 particles, which are emitted directly from some sources as well as being formed in the atmosphere.

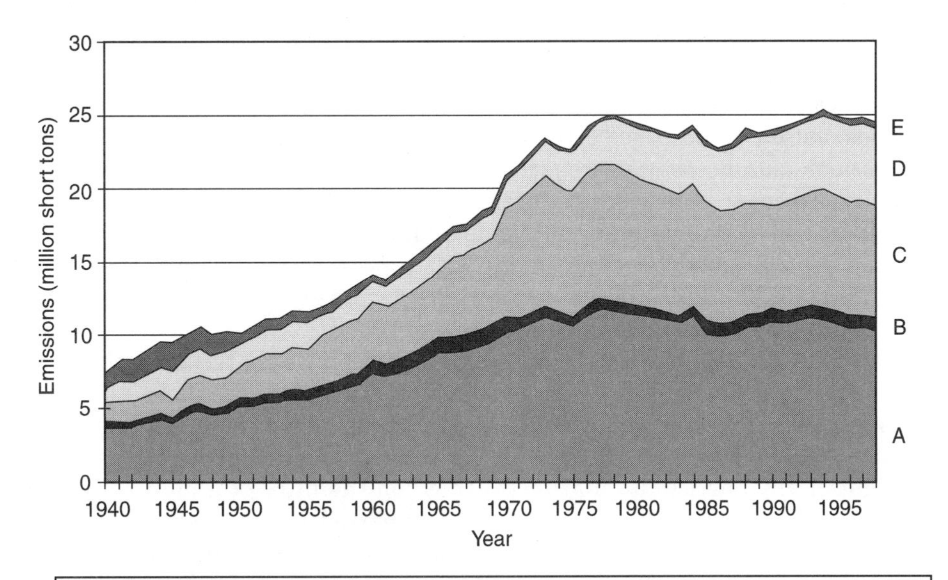

Fig. 5. Trends in emissions of NO_x 1940–1998 (14).

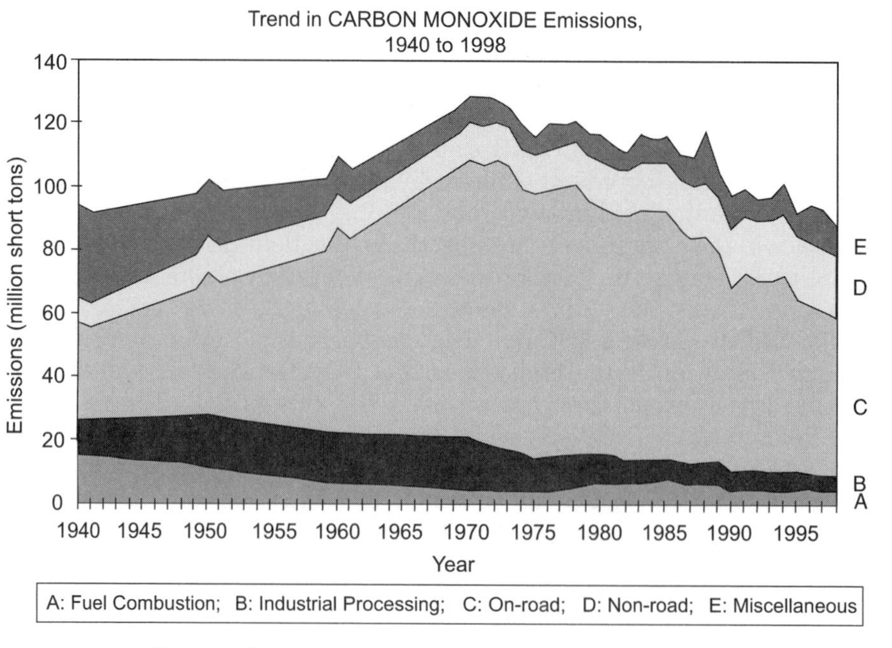

Fig. 6. Trends in emissions of CO, 1940–1998 (14).

2.5. Carbon Monoxide. Carbon monoxide is emitted during any combustion process. Transportation sources account for about two-thirds of the CO emissions nationally, but, in certain areas, significant quantities of CO come from woodburning fireplaces and stoves. CO is absorbed through the lungs into the blood stream and reacts with hemoglobin [9034-51-9] to form carboxyhemoglobin, which reduces the oxygen carrying capacity of the blood.

As shown in Figure 6, emissions of CO in the United States peaked in the late 1960s, but have decreased consistently since that time as transportation sector emissions significantly decreased. Between 1968 and 1983, CO emissions from new passenger cars were reduced by 96%. This has been partially offset by an increase in the number of vehicle-miles traveled annually. Even so, there has been a steady decline in the CO concentrations across the United States. From 1980–1999, CO concentrations decreased by an average of 57% and during the 1990s the decrease was 36% (14).

2.6. Particulate Matter. Solid- and liquid-phase material in the atmosphere is variously referred to as particulate matter, particulates, particles, and aerosols. These terms are often used interchangeably. The original air quality standards in the United States were for total suspended particulates, (TSP), the weight of any particulate matter collected on the filter of a high volume air sampler. On the average, these samplers collect particles that are less than about 30–40 μm in diameter, but collection efficiencies vary according to both wind direction and speed. In 1987, the term PM_{10}, particulate matter having an aerodynamic diameter of 10 μm or less, was introduced. The 10-μm diameter was chosen because 50% of the 10-μm particles deposit in the respiratory tract below the

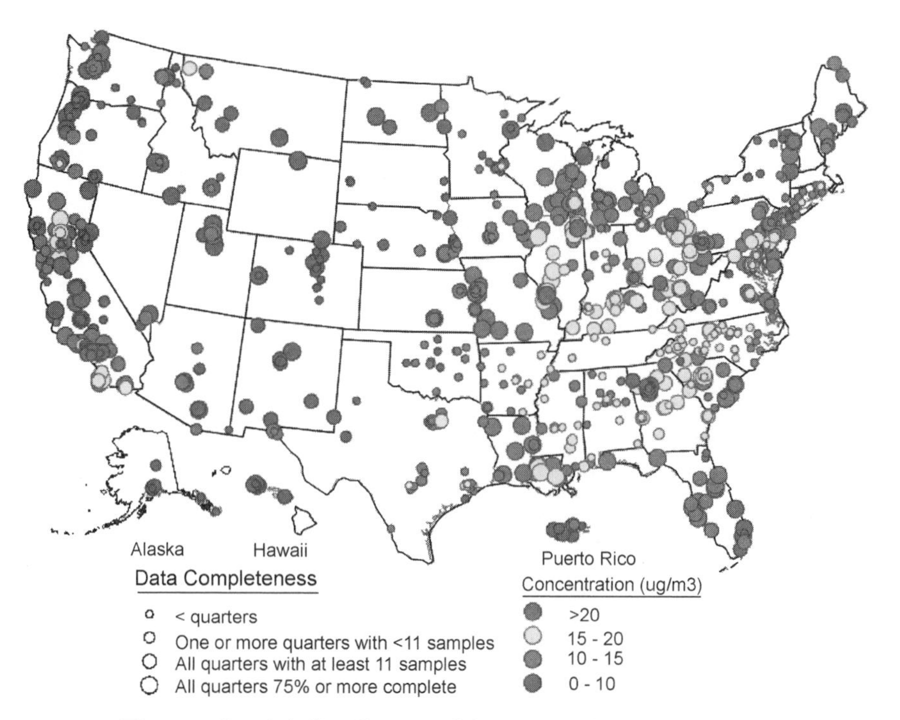

Fig. 7. Spatial distribution of fine particulate matter (14).

larynx during oral breathing. The fraction deposited decreases as particle diameter increases above 10 μm. Beginning in 1997, the term $PM_{2.5}$ (sometimes called fine particulate matter), particulate matter having an aerodynamic diameter of 2.5 μm or less, was introduced. Air quality standards for fine particulate matter were introduced in 1997 in recognition of growing evidence of the health impacts of particles that can penetrate very deeply into the respiratory system. Data on $PM_{2.5}$ concentrations are not yet widely available, but some preliminary measurements are provided in Figure 7. Ambient concentrations of PM_{10} declined by 18% between 1990 and 1999. Data on ambient concentrations of PM_{10} indicate a decline of 18% between 1990 and 1999 (14).

Atmospheric particulate matter can be classified into three size modes: nuclei, accumulation, and large or coarse-particle modes. Characteristics are given in Figure 8. The bulk of the aerosol mass usually occurs in the 0.1–10-μm size range, which encompasses most of the accumulation mode and part of the large-particle mode. The nuclei mode is transient as nuclei, formed by combustion, nucleation, and chemical reactions. The nulceation mode particles coagulate and grow into the accumulation mode. Particles in the accumulation mode are relatively stable because they exceed the size range where coagulation is important, and they are too small to deposit out of the atmosphere quickly. Consequently, particles "accumulate" in this mode. Particles larger than $\sim$2.5 μm begin to have appreciable rates of sedimentation from the atmosphere, so their lifetimes in the atmosphere decrease significantly as particle size increases. The sources of large particles are mostly mechanical processes, such as dust entrainment by wind.

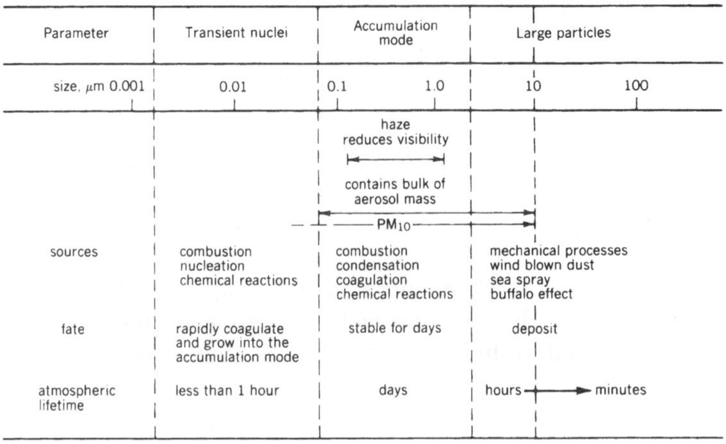

Parameter	Transient nuclei	Accumulation mode	Large particles
size, μm 0.001	0.01	0.1 1.0	10 100
		haze reduces visibility	
		contains bulk of aerosol mass	
		PM_{10}	PM_{10}
sources	combustion nucleation chemical reactions	combustion condensation coagulation chemical reactions	mechanical processes wind blown dust sea spray buffalo effect
fate	rapidly coagulate and grow into the accumulation mode	stable for days	deposit
atmospheric lifetime	less than 1 hour	days	hours → minutes

Fig. 8. Some important aerosol characteristics. Data modified from Ref. 15, which was adopted in part from Ref. 16.

Figure 9 shows the mass size distribution of typical ambient aerosols. Note the mass peaks in the accumulation mode and between 0.5 and 1.0 μm. The minimum in the curves at ~1–2.5 μm results from a lack of sources for these particles. Coagulation is not significant for the accumulation-mode particles, and particles produced by mechanical process are > 2.5 μm. Consequently, particles less than ~ 2.5 μm have different sources from particles > 2.5 μm and it is convenient to classify PM_{10} into a coarse-particulate-mass mode (CPM, diameter range of 2.5–10 μm) and a fine-particulate-mass mode (FPM, diameter < 2.5 μm). By knowing the relative amounts of CPM and FPM as well as the chemical composition of the major species, information on the PM_{10} sources can be deduced. In urban areas the CPM and FPM are usually comparable in mass; in rural areas the FPM is generally lower than in urban areas, but higher than the CPM mass. A significant fraction of the rural FPM is generally transported from upwind sources, whereas most of the CPM is generated locally.

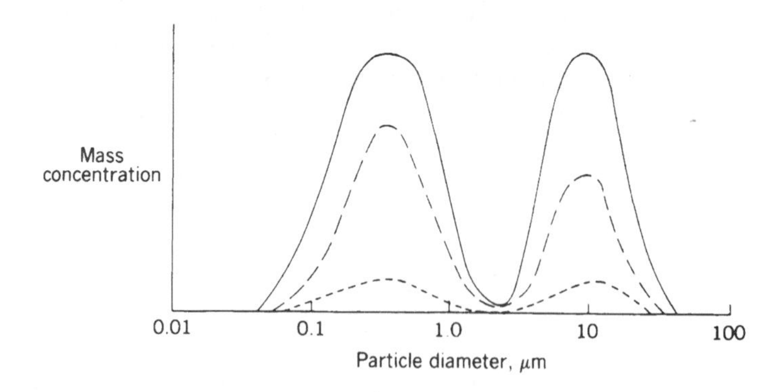

Fig. 9. Size distributions of atmospheric particles in (– –) urban, (— —) rural, and (- -) remote background areas.

Sulfate (SO_4^{2-}), carbon (as organic carbon, OC, and elemental carbon, EC), and nitrate (NO_3^-) compounds generally account for 70–80% of the FPM. In the eastern United States, SO_4^{2-} compounds are the dominant species, although very little SO_4^{2-} is emitted directly into the atmosphere. Thus most of the sulfate is a secondary aerosol formed from the oxidation of SO_2, and in the eastern United States coal-burning emissions are the source of most of the SO_2.

In the atmosphere, the principal SO_2 oxidation routes include homogeneous oxidation by OH, and the heterogeneous oxidation in water droplets by H_2O_2, O_3, or, in the presence of a catalyst, O_2. Atmospheric particles that have been identified as catalysts include many metal oxides and soot. The water droplets include cloud and dew droplets as well as aerosols that contain sufficient water: under high relative-humidity conditions, hygroscopic salts deliquesce, and form liquid aerosols. Sulfuric acid is the initial SO_2 oxidation product. This rapidly reacts with any available ammonia [7664-41-7], NH_3 to form ammonium bisulfate [7803-63-6], NH_4HSO_4. If sufficient NH_3 is present, the final product is ammonium sulfate [7783-20-2], $(NH_4)_2SO_4$. In some urban areas in the western United States, NO_3^- is more abundant than SO_4^{2-}. The NO_3^- in the FPM exists primarily as ammonium nitrate, NH_4NO_3, [6484-52-2] (see eq. 19). However, acidic SO_4^{2-} (H_2SO_4 or NH_4HSO_4) readily reacts with NH_4NO_3 and abstracts the NH_3, leaving behind gaseous HNO_3. Consequently, unless there is sufficient NH_3 to completely convert all of the SO_4^{2-} to $(NH_4)_2SO_4$, NH_4NO_3 does not accumulate in the atmosphere.

Organic compounds are a major constituent of the FPM at all sites. The major sources of OC are combustion and atmospheric reactions involving gaseous VOCs. As is the case with VOCs, there are hundreds of different OC compounds in the atmosphere. A minor but ubiquitous aerosol constituent is elemental carbon (EC), which is the nonorganic, black constituent of soot. Combustion and pyrolysis are the only processes that produce EC, and diesel engines and wood burning are the most significant sources.

Crustal dust and water make up most of the remaining FPM mass. Crustal dust is composed of aerosolized soil and rock from the earth's crust. Although this is natural material, human activities (traffic, which entrains street dust, construction activities, agricultural and land-use practices, etc) affect the rate at which crustal material is aerosolized. Since it is aerosolized by frictional processes, the diameter of most of the crustal dust is > 2.5 μm and typically accounts for most of the CPM and particle mass < 10 μm. In global average crustal material, the major elements contained in decreasing order are O, Si, Al, Fe, Ca, Na, K, and Mg. Consequently, the crustal mass can be estimated from Si measurements alone. However, the relative amounts of the elements do vary spatially.

2.7. Lead. Lead is of concern because of its tendency to be retained by living organisms. When excessive amounts accumulate in humans, lead can inhibit the formation of hemoglobin and produce life-threatening lead poisoning. In smaller doses, lead is also suspected of causing learning disabilities in children. From 1980 to 1999, nationwide Pb emissions decreased 95%. This reduction is a direct result of the removal of lead compounds such as tetraethyllead $(C_2H_5)_4Pb$ [78-00-2] from fuels, primarily gasoline.

2.8. Air Toxics. There are thousands of commercial chemicals used in the United States. Hundreds are emitted into the atmosphere and have some potential to adversely affect human health at certain concentrations; some are known or suspected carcinogens. Identifying all of these substances and promulgating emissions standards is beyond the present capabilities of existing air quality management programs. Consequently, toxic air pollutants (TAPs) need to be prioritized based on risk analysis, so that those posing the greatest threats to health can be regulated.

The 1970 Clean Air Act required that EPA provide an ample margin of safety to protect against hazardous air pollutants (HAPs) by establishing national emissions standards for certain sources. From 1970 to 1990, over 50 chemicals were considered for designation as HAPs, but EPAs review process was completed for only 28 chemicals. However, in the 1990 Clean Air Act Amendments, 189 substances are listed that EPA must regulate by enforcing maximum achievable control technology (MACT). The Amendments mandate that EPA issue MACT standards for all sources of the 189 substances. In addition, EPA must determine the risk remaining after MACT is in place and develop health-based standards that would limit the cancer risk. EPA may add or delete substances from this list.

2.9. Indoor Air Pollution. Indoor air pollution, the presence of air pollutants in indoor air, is of growing concern in offices and residential buildings. Partly in response to the "energy crisis" of the early 1970s, buildings are now constructed more air-tight. Unfortunately, air-tight structures create a setting conducive to the accumulation of indoor air pollutants. Numerous sources and types of pollutants found indoors can be classified into eight categories: tobacco smoke, radon, emissions from building materials, combustion products from inside the building, pollutants which infiltrate from outside the building, emissions from products used within the home, pollutants formed by reactions indoors, and biological pollutants. Concentrations of the pollutants depend on strength of the indoor sources, the ventilation rate of the building, and the outdoor pollutant concentration.

Tobacco smoke contains a variety of air pollutants. In a survey of 80 homes in an area where the outdoor TSP varied between 10 and 30 $\mu g/m^3$, the indoor TSP was the same, or less, in homes having no smokers. In homes having one smoker, the TSP levels were between 30 and 60 $\mu g/m^3$, while in homes having two or more smokers, the levels were between 60 and 120 $\mu g/m^3$ (17). In other studies, indoor TSP levels exceeding 1000 $\mu g/m^3$ have been found in homes with numerous smokers. In addition to TSP, burning tobacco emits CO, NO_x, formaldehyde [50-00-0], benzopyrenes, nicotine [54-11-5], phenols, and some metals such as cadmium [7440-43-9] and arsenic [7440-38-2] (18).

Radon-222 ^{222}Rn, [14859-67-7], is a naturally occurring, inert, radioactive gas formed from the decay of radium-226 ^{226}Ra, [13982-63-3]. Because Ra is a ubiquitous, water-soluble component of the earth's crust, its daughter product, Rn, is found everywhere. A major health concern is radon's radioactive decay products. Radon has a half-life of 4 days, decaying to polonium-218 ^{218}Po, [15422-74-9], with the emission of an α particle. It is ^{218}Po, an α-emitter having a half-life of 3 min, and polonium-214 ^{214}Po, [15735-67-8], an α-emitter having a half-life of 1.6×10^{-4} s, that are of most concern. Polonium-218 decays to lead-214 ^{214}Pb, [15067-28-4], a β-emitter having $t_{1/2} = 27$ min, which decays to bismuth-214 ^{214}Bi, [14733-03-0], a β-emitter having $t_{1/2} = 20$ min, which decays to ^{214}Po. Radon is an inert gas that, when inhaled, is not retained in the lungs. But the Rn daughters, when inhaled, either by themselves or attached to an airborne particle, are retained and the subsequent α-emissions irradiate the surrounding lung tissue.

Radon can enter buildings through emissions from soil, water, or construction materials. The soil route is by far the most common, and construction material the least common, although there have been isolated incidents where construction materials contained high levels of Ra. The emission rate of Rn depends on the concentration of Ra in the soil, the porosity of the soil, and the permeability of the building's foundation. For example, Rn is transported faster through cracks and sumps in the basement floor than through concrete. In the ambient air Rn concentrations are typically 9.25–37 mBq/L, whereas the mean concentration in U.S. residences is about 44 mBq/L (19). However, it is estimated that there are 1 million residences that have concentrations exceeding 0.3 Bq/L or 300 mBq/L, which is the level for remedial action recommended by the National Council on Radiation Protection and Measurements (19). The highest values ever measured in U.S. homes exceeded 37 Bq/L (20). Remedial action consists of (*1*) reducing the transport of Rn into the building by sealing cracks with impervious fillers and installing plastic or other barriers that have proven effective; (*2*) removing the daughters from the air by filtration; and (*3*) increasing the infiltration of outside air using an air-exchanger system.

Of the pollutants emitted from construction materials within the home, asbestos [1332-21-4] has received the most attention. Asbestos is a generic term for a number of naturally occurring fibrous hydrated silicates. By EPA's definition, a fiber is a particle that possesses a 3:1 or greater aspect ratio (length: diameter). The family of asbestos minerals is divided into two types: serpentine [12168-92-2] and amphibole. One type of serpentine, chrysotile, $Mg_6Si_4O_{10}(OH)_8$, [12001-29-5], accounts for 90% of the world's asbestos production. The balance of the world's production is accounted for by two of the amphiboles: amosite $Fe_5Mg_2Si_8O_{22}(OH)_2$, [12172-73-5], and crocidolite, $Na_2(Fe^{3+})_2(Fe^{2+})_3Si_8O_{22}(OH)_2$, [12001-28-4]. Three other amphiboles, anthophyllite $(Mg,Fe)_7Si_8O_{22}(OH)_2$, [77536-67-5], tremolite $Ca_2Mg_5Si_8O_{22}(OH)_2$, [77536-68-6] and actinolite $Ca_2(Mg,Fe)_5Si_8O_{22}(OH)_2$, [77536-66-4], have been only rarely mined. The asbestos minerals differ in morphology, durability, range of fiber diameters, surface properties, and other attributes that determine uses and biological effects. Known by ancients as the magic mineral because of its ability to be woven into cloth, its physical strength, and its resistance to fire, enormous heat, and chemical attack, asbestos was incorporated into many common building products (21).

All forms of asbestos were implicated in early studies linking exposure to airborne fibers and asbestosis (pulmonary interstitial fibrosis), lung cancer, and mesothelioma (a rare form of cancer of the lung or abdomen). However, most of the asbestos-related diseases are now thought to result from exposure to airborne amphiboles rather than chrysotile, the most common asbestos type, and to fibers greater than or equal to 5 μm in length (22). In the 1970s, the spray-on application of asbestos was banned and substitutes were found for many products. Nevertheless, asbestos was used liberally in buildings for several decades, and many of them are still standing. Asbestos in building materials does not spontaneously shed fibers, but when the materials become damaged by normal decay, renovation, or demolition, the fibers can become airborne and contribute to the indoor air pollution problem.

Formaldehyde (HCHO), [50-00-0], another important pollutant emanating from building material, is important because of irritant effects and suspected carcinogenicity. Traces of formaldehyde can be found in the air in virtually every modern home. Mobile homes and houses insulated using urea–formaldehyde [9011-05-6] foam, an efficient insulation material that can be injected into the sidewalls of conventional homes, have the highest concentrations. In 1982, use of the foam was banned in the United States. Higher formaldehyde emissions can occur in mobile homes using particle board held together using an urea–formaldehyde resin, which can also be a problem in a conventional house, but it is usually exacerbated in a mobile home because of the low rate of air exchange. Plywood is also a source of formaldehyde as the layers of wood are held together using a similar urea–formaldehyde resin adhesive. In general, however, particle board contains more adhesive per unit mass, so the emissions are greater. Other sources of indoor formaldehyde are paper products, carpet backing, and some fabrics.

Whenever unvented combustion occurs indoors or when venting systems attached to combustion units malfunction, a variety of combustion products will be released to the indoor environment. Indoor combustion units include: nonelectric stoves and ovens, furnaces, hot water heaters, space heaters, and wood-burning fireplaces or stoves. Products of combustion include CO, NO, NO_2, fine particles, aldehydes, polynuclear aromatics, and other organic compounds. Especially dangerous sources are unvented gas and kerosene [8008-20-6] space heaters that discharge pollutants directly into the living space. The best way to prevent the accumulation of combustion products indoors is to make sure all units are properly vented and properly maintained.

Pollutants from outdoors can also be drawn inside under certain circumstances such as incorrectly locating an air intake vent downwind of a combustion exhaust stack. High outdoor pollutant concentrations can also infiltrate buildings. Unreactive pollutants like CO diffuse through any openings in a building and pass unaltered through any air-intake system. Given sufficient time, the indoor/outdoor ratio for CO approaches 1.0 if outside air is the only CO source. For reactive species such as ozone, which is destroyed on contact with most surfaces, the indoor/outdoor ratio is usually < 0.5, but this ratio varies considerably depending on the ventilation rate and the internal surface area within the building (23).

Air contaminants are emitted to the indoor air from a wide variety of activities and consumer products. Most indoor activities produce some types of pollutants. When using volatile products, care should be exercised to minimize exposure through proper use of the product and by providing adequate ventilation.

Indoor air pollutants can react to form other air pollutants, perhaps most significantly, particles. Olefins found in common cleaning products, such as terpenes, can react with ozone that infiltrates from outdoors to produce particle concentrations that can exceed outdoor air quality standards.

Biological indoor air pollutants include airborne bacteria, viruses, fungi, spores, molds, algae, actinomycetes, and insect and plant parts. Microorganisms, many of which multiply in the presence of high humidity, can produce infections, disease, or allergic reactions; the nonviable biological pollutants can produce allergic reactions. The most notable episode was the 1976 outbreak of Legionella (Legionnaires') disease in Philadelphia where the American Legion convention attendees inhaled Legionella virus from a contaminated central air conditioning system. A similar incident in an industrial environment occurred in 1981: >300 workers came down with "Pontiac fever" as a result of inhalation exposure to a similar virus aerosolized from contaminated machining fluids (24). Preventive maintenance of air management systems and increased ventilation rates reduce the concentrations of all species, and should consequently reduce the incidence of adverse affects.

3. Regional and Global Impacts of Air Pollution

Photochemical smog is most severe in urban areas and has some impact at regional scales. There has been a growing recognition, however, that some air pollutants have impacts at regional to global scales. Three examples of impacts of air pollutants on natural balances of the earth's systems at regional and global scales are acid deposition, stratospheric ozone depletion, and global climate change.

3.1. Acid Deposition. Acid deposition, the deposition of acids from the atmosphere to the surface of the earth, can be dry or wet. Dry deposition involves acid gases or their precursors or acid particles coming in contact with the earth's surface and then being retained. The principal species associated with dry acid deposition are SO_2(g), and acid sulfate particles (H_2SO_4 and NH_4HSO_4), and HNO_3(g). Measurements of dry deposition are quite sparse. On the other hand, there are abundant data on wet acid deposition. Wet acid deposition, acid precipitation, is the process by which acids are deposited by rain or snow. The principal dissolved acids are H_2SO_4 and HNO_3. Other acids, such as HCl and organic acids, usually account for only a minor part of the acidity although organic acids can be significant contributors in remote areas.

The pH of rainwater in equilibrium with atmospheric CO_2 is 5.6, a value frequently cited as the natural background pH. However, in the presence of other naturally occurring species such as SO_2, SO_4^{2-}, NH_3, organic acids, sea salt, and alkaline crustal dust, the natural values of unpolluted rainwater vary between 4.9 and 6.5 depending on time and location. Across the United States, the mean annual average precipitation pH varies from 4.2 in western Pennsylvania to 6.0 in the upper midwest (see Fig. 10). In general, precipitation of the lowest pH occurs in the summer. Precipitation pH is generally lowest in the eastern United States within and downwind of the largest SO_2 and NO_x emissions areas. In the East, SO_4^{2-} concentrations in precipitation are higher during the summer than in winter, but the NO_3^- values are about the same year round. Consequently, the lower pH in summer precipitation results mostly from the higher SO_4^{2-} concentrations. On the average in the eastern United States, ~60% of the wet-deposited acidity can be attributed to SO_4^{2-} and 40% to NO_3^- (25).

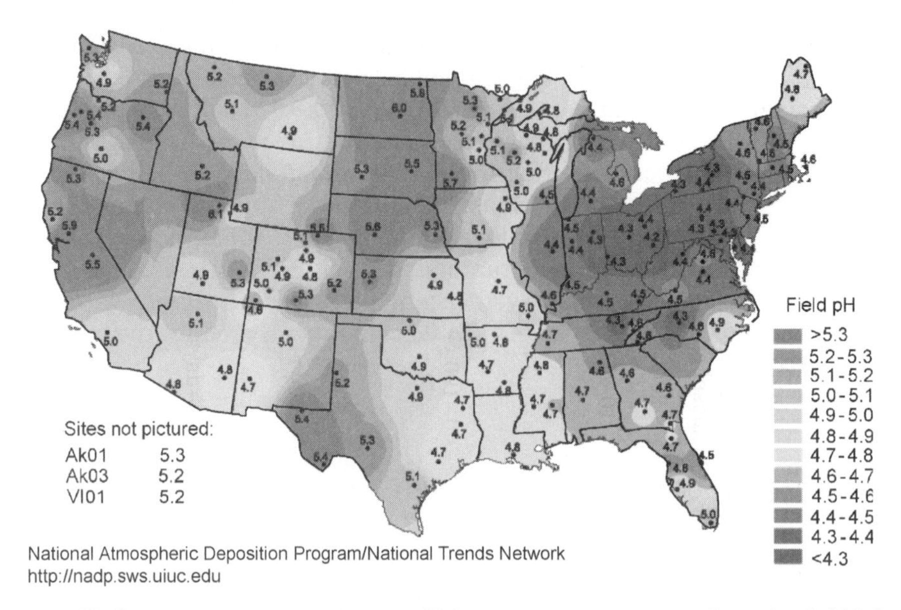

Fig. 10. Hydrogen ion concentration as pH from measurements made at the field laboratories, 1999. Spatial distribution of acid deposition (25).

3.2. Global Warming (The Greenhouse Effect).

The atmosphere allows solar radiation from the sun to pass through without significant absorption of energy. Some of the solar radiation reaching the surface of the earth is absorbed, heating the land and water. Infrared radiation is emitted from the earth's surface, but certain gases in the atmosphere absorb this infrared (ir) radiation, and redirect a portion back to the surface, thus warming the planet and making life, as we know it, possible. This process is often referred to as the *greenhouse effect*. The surface temperature of the earth will rise until a radiative equilibrium is achieved between the rate of solar radiation absorption and the rate of ir radiation emission. Human activities, such as fossil fuel combustion, deforestation, agriculture and large-scale chemical production, have measurably altered the composition of gases in the atmosphere. Recent intergovernmental panels and scientific assessments (see, eg, ref. 26) have concluded that these alterations have caused a warming of the earth-atmosphere system by enhancement of the greenhouse effect.

Table 3 is a list of the most important greenhouse gases along with their anthropogenic sources, emission rates, concentrations, residence times in the atmosphere, relative radiative forcing efficiencies, and estimated contribution to global warming. The primary greenhouse gases are water vapor, carbon dioxide, methane, nitrous oxide, chlorofluorocarbons, and tropospheric ozone. Water vapor is the most abundant greenhouse gas, but is omitted because it is not primarily due to anthropogenic sources. Carbon dioxide contributes significantly to global warming due to its high emission rate and concentration. The major factors contributing to global warming potential of a chemical are ir absorptive capacity and residence time in the atmosphere. Gases with very high absorptive capacities and long residence times can cause significant global warming even

Table 3. **Greenhouse Gases and Global Warming Contribution**[a]

Gas	Source (natural and anthropogenic)	Estimated anthropogenic emission rate	Preindustrial global concentration	Approximate current concentration	Estimated residence time in the atmosphere	Radiative forcing efficiency (absorptivity capacity) ($CO_2 = 1$)	Estimated contribution to global warming
carbon dioxide (CO_2)	fossil fuel combustion; deforestation	6000 *M* t/yr	280 ppm	355 ppm	50–200 yr	1	50%
methane (CH_4)	anaerobic decay (wetlands, landfills, rice patties) ruminants, termites, natural gas, coal mining, biomass burning	300–400 *M* t/yr	0.8 ppm	1.7 ppm	10 yr	58	12–19%
nitrous oxide (N_2O)	estuaries and tropical forests; agricultural practices, deforestation, land clearing, low temperature fuel combustion	4–6 *M* t/yr	0.385 ppm	0.31 ppm	140–190 yr	206	4–6%
chlorofluorocarbons (CFC-11 and CFC-12)	refrigerants, air conditioners, foam blowing agents, aerosol cans, solvents	1 *M* t/yr	0	0.0004–0.001 ppm	65–110 yr	4860	17–21%
tropospheric ozone (O_3)	photochemical reactions between VOCs and NO_x from transportation and industrial sources	not emitted directly	NA[b]	0.022 ppm	hours–days	2000	8%

[a] from ref. 27, M = million.
[b] Not applicable = NA.

though their concentrations are extremely low. A good example of this phenomenon is the chlorofluorocarbons, which are, on a pound for pound basis, >1000 times more effective as greenhouse gases than carbon dioxide.

For the past four decades, measurements of the accumulation of carbon dioxide in the atmosphere have been taken at the Mauna Loa Observatory in Hawaii, a location far removed from most human activity that might generate carbon dioxide. Based on the current level of CO_2 of 360 parts per million (ppm), levels of CO_2 are increasing at the rate of 0.5%/year (from ~320 ppm in 1960). Atmospheric concentrations of other greenhouse gases have also risen. Methane has increased from ~700 ppb in preindustrial times to 1721 ppb in 1994, while N_2O rose from 275 to 311 ppb over the same period.

While it is clear that atmospheric concentrations of carbon dioxide, and other global warming gases are increasing, there is significant uncertainty regarding the magnitude of the effect on climate that these concentration changes might induce. The relationship between atmospheric concentrations of global warming gases and climate can be separated into climate forcing (the change in net radiation due to changes in atmospheric concentrations of global warming gases) and climate response (the change in climate due to changes in heating). There is significantly less uncertainty in the understanding of climate forcing than of climate response; the Intergovernmental Panel on Climate Change (IPPC) regularly issues scientific assessments of both climate forcing and climate response.

3.3. Stratospheric Ozone Depletion. In the stratosphere, O_3 is formed naturally when O_2 is dissociated by (uv) solar radiation in the region 180–240 nm:

$$O_2 + \text{uv} \rightarrow O + O \tag{20}$$

and the atomic oxygen then reacts with molecular oxygen according to equation 2. Ultraviolet radiation in the 200–300 nm region can also dissociate O_3:

$$O_3 + \text{uv} \rightarrow O_2 + O \tag{21}$$

Equation 21 represents the reaction responsible for the removal of uv-B radiation (280–330 nm) that would otherwise reach the earth's surface. There is concern that any process that depletes stratospheric ozone will consequently increase uv-B (in the 293–320-nm region) reaching the surface. Increased uv-B is expected to lead to increased incidence of skin cancer and it could have deleterious effects on certain ecosystems. The first concern over O_3 depletion was from NO_x emissions from a fleet of supersonic transport aircraft that would fly through the stratosphere and cause reactions according to equations 3 and 22 (28):

$$NO_2 + O \rightarrow NO + O_2 \tag{22}$$

The net effect of this sequence is the destruction of two molecules of O_3 as one is lost in NO_2 formation and the O of equation 22 would have combined with O_2 to form another. In addition, the NO acts as a catalyst. It is not consumed, and therefore can participate in the reaction sequence many times.

In the mid-1970s, it was realized that the CFCs in widespread use because of their chemical inertness, would diffuse unaltered through the troposphere and into the mid-stratosphere where they, too, would be photolyzed by uv (< 240 nm) radiation. For example, CFC-12 can photolyze:

$$CF_2Cl_2 + uv \rightarrow CF_2Cl + Cl \tag{23}$$

$$CF_2Cl + O_2 \rightarrow CF_2O + ClO \tag{24}$$

forming Cl and ClO radicals that then react with ozone and O:

$$Cl + O_3 \rightarrow ClO + O_2 \tag{25}$$

$$ClO + O \rightarrow Cl + O_2 \tag{26}$$

In this sequence the Cl also acts as a catalyst and two O_3 molecules are destroyed. It is estimated that before the Cl is finally removed from the atmosphere in 1–2 yr by precipitation, each Cl atom will have destroyed ~100,000 O_3 molecules. The estimated O_3-depletion potential of some common CFCs, hydrofluorocarbons, (HFCs), and hydrochlorofluorocarbons, (HCFCs), are presented in Table 4. The O_3-depletion potential is defined as the ratio of the emission

Table 4. **Ozone Depleting Potentials for Industrially Important Compounds**[a]

Compound	Formula	τ (years)[b]	ODP (relative to CFC-11)
methyl bromide	CH_3Br		0.6
tetrachloromethane	CCl_4	47	1.08
1,1,1-trichloroethane	CH_3CCl_3	6.1	0.12
CFC-11	CCl_3F	60	1.0
CFC-12	CCl_2F_2	120	1.0
CFC-13	$CClF_3$		1.0
CFC 113	CCl_2FCClF_2	90	1.07
CFC 114	$CClF_2CClF_2$	200	0.8
CFC 115	CF_3CClF_2	400	0.5
halon 1201	$CHBrF_2$		1.4
halon 1202	CBr_2F_2		1.25
halon 1211	$CBrClF_2$		4.0
halon 1301	$CBrF_3$		16.0
halon 2311	$CHClBrCF_3$		0.14
halon 2401	$CHBrFCF_3$		0.25
halon 2402	$CBrF_2CBrF_2$		7.0
HCFC-22	$CHClF_2$	15	0.055
HCFC-123	$C_2F_3HCl_2$	1.7	0.02
HCFC-124	C_2F_4HCl	6.9	0.022
HCFC-141b	$C_2FH_3Cl_2$	10.8	0.11
HCFC-142b	$C_2F_2H_3Cl$	19.1	0.065
HCFC-225ca	$C_3HF_5Cl_2$		0.025
HCFC-225cb	$C_3HF_5Cl_2$		0.033

[a] Refs. 29, 30.

[b] τ is the tropospheric reaction lifetime (29,30).

rate of a compound required to produce a steady-state O_3 depletion to the amount of CFC-11 required to produce the depletion. The halons, bromochlorofluorocarbons or bromofluorocarbons that are widely used in fire extinguishers, are also ozone-depleting compounds. Although halon emissions, and thus the atmospheric concentrations, are much lower than the most common CFCs, halons are of concern because they are generally more destructive to O_3 than the CFCs.

The strongest evidence that stratospheric O_3 depletion is occurring comes from the discovery of the Antarctic ozone hole. In recent years during the spring, O_3 depletions of 60% or more integrated over all altitudes and 95–100% in some layers have been observed over Antarctica. During winter in the southern hemisphere, a polar vortex develops that prevents the air from outside of the vortex from mixing with air inside the vortex. The depletion begins in August, as the approaching spring sun penetrates into the polar atmosphere, and extends into October. When the hole was first observed, existing chemical models could not account for the rapid O_3 loss, but attention was soon focused on stable reservoir species for chlorine. These compounds, namely, HCl and $ClNO_3$, are formed in competing reactions involving Cl and ClO that temporarily or permanently remove Cl and ClO from participating in the O_3 destruction reactions. For example,

$$Cl + CH_4 \rightarrow HCl + CH_3 \quad (27)$$

$$ClO + NO_2 + M \rightarrow ClNO_3 + M \quad (28)$$

where M is again any third body molecule, remaining unchanged, in the atmosphere.

Heterogeneous reactions, which break down these reservoir species, can occur on the surfaces of cloud particles and play a central role in explaining the magnitude of ozone depletion. Normally the stratosphere is too dry to allow the formation of clouds, however, within the polar vortex, temperatures as low as $-90°C$ allow the formation of polar stratospheric ice clouds. Two main types of clouds have been recognized. Type I clouds contain crystals of nitric acid trihydrate (Type Ia) or sulfuric acid/nitric acid mixtures (Type Ib) and water. Type II clouds are largely frozen water ice. These clouds catalyze the reactions of chlorine reservoir species, such as HCl, and $ClNO_3$. The reservoir species decompose, forming reactive species such as Cl_2 and nitric acid.

$$HCl(s) + ClNO_3 \rightarrow Cl_2 + HNO_3(s) \quad (29)$$

The Cl_2 will photolyze producing atomic chlorine, which participates in gas phase reactions to deplete ozone. The nitric acid remains in the ice, leading to a net removal of nitrogen oxides from the gas phase, which are a sink for ClO (reaction 28). Additional reactions, such as those involving N_2O_5 also occur, and this decomposition of reservoir species dominates the chemistry of ozone depletion.

Regulations designed to mitigate the destruction of stratospheric ozone began in 1976 when the United States banned the use of CFCs as aerosol propellants. No further steps were taken until 1987 when the United States and some 50 other countries adopted the Montreal Protocol. The impact of these

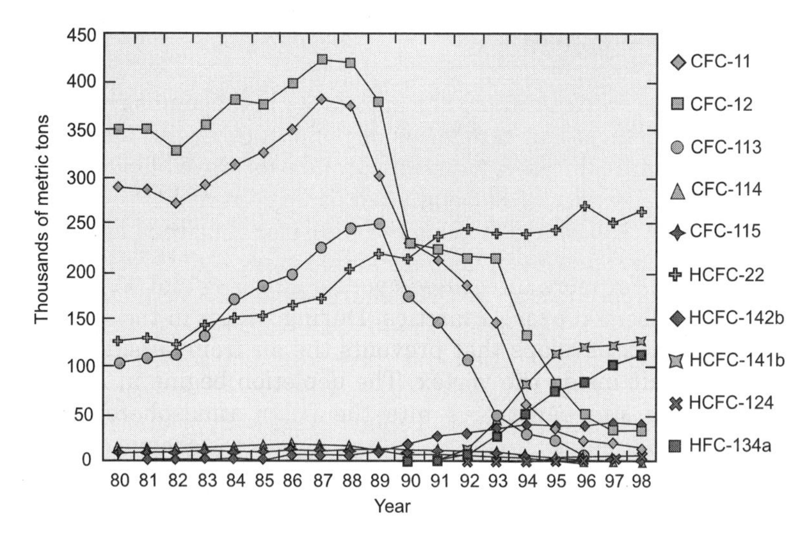

Fig. 11. Annual production rates of ozone depleting compounds 1980–1998 (31).

regulations can be seen in the data presented in Figures 11 and 12. Releases of stratospheric ozone depleting gases reached a peak in the mid-1970s (CFC-11 and CFC 12 combined were ~700 million kg.) Levels have been decreasing since ~1990 (1995 data: 300 million kg., same level as 1966). Yet, atmospheric concentrations are declining relatively slowly, due to the persistence of the ozone depleting gases in the atmosphere.

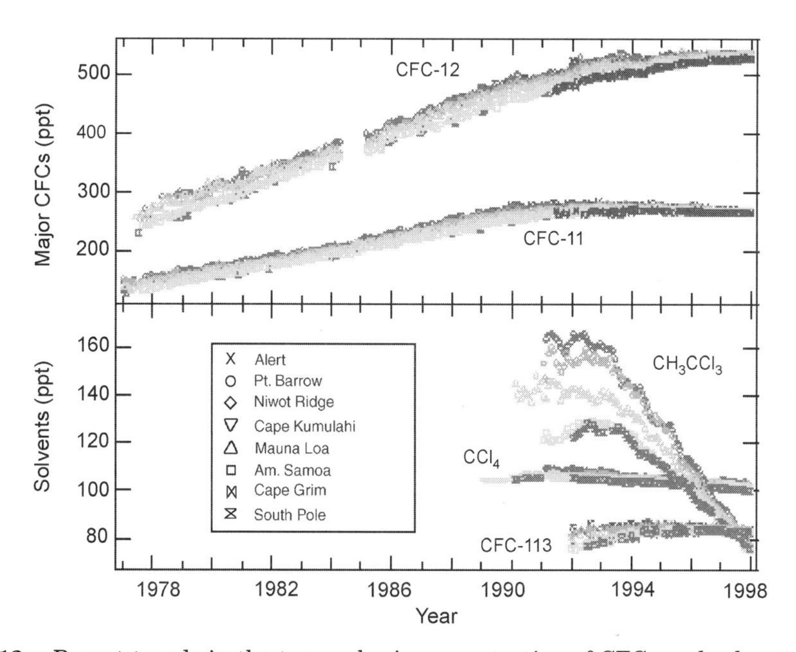

Fig. 12. Recent trends in the tropospheric concentration of CFCs and solvents (32).

4. Air Quality Management

In the United States, the framework for air quality management is the Clean Air Act (CAA), which defines two categories of pollutants: criteria and hazardous. For the hazardous air pollutants, emissions standards are set for selected sources of emissions. Once a substance is designated by EPA as a HAP, EPA has to promulgate a NESHAP (National Emission Standard for Hazardous Air Pollutants), designed to protect public health with an ample margin of safety.

For the criteria pollutants, the CAA charges the EPA with identifying those air pollutants that most affect public health and welfare, and setting maximum allowable ambient air concentrations for these air pollutants. The air pollutants for which national ambient air quality standards are set are referred to as criteria pollutants. Six chemical species (Table 5) have both primary and secondary National Ambient Air Quality Standards (NAAQS). The primary standards are intended to protect the public health with an adequate margin of safety. The secondary standards are meant to protect public welfare, such as damage to crops, vegetation, and ecosystems or reductions in visibility.

The NAAQS apply uniformly across the United States whereas emissions standards for criteria pollutants depend on the severity of the local air pollution problem and whether an affected source already exists or is proposed. In addition, individual states have the right to set their own ambient air quality and emissions standards (which must be at least as stringent as the federal standards) for all pollutants and all sources except motor vehicles. With respect to motor vehicles, the CAA allows the states to choose between two sets of emissions standards: the Federal standards or the more stringent California ones.

To determine if NAAQS are met, states are required to monitor the criteria pollutants' concentrations in areas that are likely to be near or to exceed the standards. If an area exceeds a NAAQS for a given pollutant, it is designated as a nonattainment area for that pollutant, and the state is required to establish an SIP.

The SIP is a strategy designed to achieve emissions reductions sufficient to meet the NAAQS within a deadline that is determined by the severity of the local pollution problem. Areas that receive long (6 years or more) deadlines must show continuous progress by reducing emissions by a specified percentage each year. For SO_2 and NO_2, the initial SIPs were very successful in achieving the NAAQS. For other criteria pollutants, particularly O_3 and to a lesser extent CO, however, many areas have gone through multiple rounds of SIP preparations with little hope of meeting the NAAQS in the near future. If a state misses an attainment deadline, fails to revise an inadequate SIP, or fails to implement SIP requirements, EPA has the authority to enforce sanctions such as banning construction of new stationary sources and withholding federal grants for highways.

In nonattainment areas, the degree of control on small sources is left to the discretion of the state and is largely determined by the degree of required emissions reductions. Large existing sources must be retrofitted with reasonable available control technology (RACT) to minimize emissions. All large new sources and existing sources that undergo major modifications must meet EPAs new source performance standards at a minimum. Additionally, in nonattainment areas, they must be designed using lowest achievable emission rate

Table 5. **Criteria Pollutants and the NAAQS**

	Primary standard (human health related)		Secondary (welfare related)	
Pollutant	type of average	concentration[a]	type of average	concentration
CO [−57%][c] {−22%}[d]	8 h[b]	9 ppm (10 mg/m^3)	no secondary standard	
	1 h[b]	35 ppm (40 mg/m^3)	no secondary standard	
Pb [−94%] {−95%}	maximum quarterly average	1.5 μg/m^3	same as primary standard	
NO_2 [−25%] {+1%}	annual arithmetic mean	0.053 ppm (100 μg/m^3)	same as primary standard	
O_3 [−20%]	1 h[e]	0.012 ppm (235 μg/m^3)	same as primary standard	
	8 h[f]	0.08 ppm (157 μg/m^3)	same as primary standard	
PM_{10} [−18] {−55%}	annual arithmetic mean	50 μg/m^3	same as primary standard	
	24-h[g]	150 μg/m^3	same as primary standard	
$PM_{2.5}$	annual arithmetic mean[h]	15 μg/m^3	same as primary standard	
	24 h[i]	65 μg/m^3	same as primary standard	
SO_2 [−50%] {−28%}	annual arithmetic mean	0.03 ppm (80 μg/m^3)	3 h[b]	0.50 ppm (1300 μg/m^3)
	24 h[b]	0.14 ppm (365 μg/m^3)		

[a] Parenthetical value is an equivalent mass concentration.
[b] Not to be exceeded more than once per year.
[c] Air quality concentration, % change 1980–1999.
[d] Emissions, % change 1980–1999.
[e] Not to be exceeded more than once per year on average.
[f] Three-year average of annual fourth highest concentration.
[g] The preexisting form is exceedance based. The revised form is the ninty-ninth percentile.
[h] Spatially averaged over designated monitors.
[i] The form is the ninty-eight percentile.

Source: 40 Code of Federal Register (CFR) Part 50, revised standards issued July 18, 1997. Adapted from U.S. EPA (14)

(LAER) technology, and emissions offsets must be obtained. Offsets require that emissions from existing sources within the area be reduced below legally allowable levels so that the amount of the reduction is greater than or equal to the emissions expected from the new source. RACT usually is less stringent than LAER: It may not be feasible to retrofit certain sources using the LAER technology.

Large sources of SO_2 and NO_x may also require additional emission reductions because of the 1990 Clean Air Act Amendments. To reduce acid deposition, the amendments require that nationwide emissions of SO_2 and NO_x be reduced.

BIBLIOGRAPHY

"Smokes and Fumes" in *ECT* 1st ed., Vol. 12, pp. 558–573, by G. P. Larson, Air Pollution Control District of Los Angeles County; "Smokes, Fumes, and Smog" in *ECT* 2nd ed., Vol. 18, pp. 400–415, by R. B. Engdahl, Battelle Memorial Institute; "Pollution—Air Pollution" in *ECT* 2nd ed., Supplement Vol., pp. 730–737, by G. P. Sutton, Envirotech Corp., "Air Pollution" in *ECT* 3rd ed., Vol. 1, pp. 624–649 by P. R. Sticksel, R. B. Engdahl, Battelle Memorial Institute; "Air Pollution" in *ECT* 4th ed., Vol. 1, pp. 711–749, by George T. Wolff, General Motors Research Laboratories; "Air Pollution" in *ECT* (online), Posting date: December 4, 2000, by George T. Wolff, General Motors Research Laboratories.

CITED PUBLICATIONS

1. Y. Mamane, *Atm. Environ.* **21**, 1861 (1987).
2. World Health Organization, *www.who.int* (2001).
3. R. Atkinson, *Gas-phase Tropospheric Chemistry of Organic Compounds, Monograph 2, Journal of Physical and Chemical Reference Data*, American Institute of Physics, New York, 1994.
4. R. Atkinson *Kinetics and Mechanisms of the Gas-Phase Reactions of the Hydroxyl Radical with Organic Compounds*, Monograph 1, Journal of Physical and Chemical Reference Data, American Institute of Physics, New York, 1989.
5. R. Atkinson *J. Phys. Chem. Ref. Data.* **26**, 215 (1997).
6. W. P. L. Carter "The SAPRC-99 Chemical Mechanism and Updated Reactivity Scales", Final report to California Air Resources Board on Contracts 92-329 and 95-308, available at *http://pah.cert.ucr.edu/~carter/* (2001).
7. W. P. L. Carter, "Development of Ozone Reactivity Scales for Volatile Organic Compounds", *J. Air Waste Management Asso.* **44**, 881 (1994).
8. W. N. Stasiuk, Jr. and P. E. Coffey, *J. Air Pollut. Control Assoc.* **24**, 564 (1974).
9. G. T. Wolff and co-workers, *Environ. Sci. Technol.* **11**, 506 (1977).
10. A. M. Dunker, S. Kumar, and P. H. Berzins, *Atmos. Environ.* **18**, 311 (1984).
11. N. A. Kelly and R. G. Gunst, *Atmos. Environ.* **24A**, 2991 (1990).
12. *Catching Our Breath. Next Steps for Reducing Urban Ozone*, U.S. Office of Technology Assessment, Washington, D.C., 1989, pp. 101–102.
13. G. T. Wolff, P. J. Lioy, G. D. Wight, and R. E. Pasceri, *J. Air Pollut. Control Assoc.* **27**, 460 (1977).
14. U.S. Environmental Protection Agency *National Air Quality* & Emissions Trends Report, 1999, EPA 454/R-01-004, (*http://www.epa.gov/oar/aqtrnd99/*).
15. G. T. Wolff, *Ann. N.Y. Acad. Sci.* **338**, 379 (1980).

16. K. Willeke and K. T. Whitby, *J. Air Pollut. Control Assoc.* **25**, 529 (1975).
17. J. D. Spengler and co-workers, *Atmos. Environ.* **15**, 23 (1981).
18. California Department of Consumer Affairs, *Clean Your Room, Compendium on Indoor Air Pollution*, Sacramento, Calif., 1982, p. III.EI(III.E.II.).
19. Mueller Associates, Inc., Syscon Corporation, and Brookhaven National Laboratory, *Handbook of Radon in Buildings*, Hemisphere Publishing Corporation, New York, 1988, p. 95.
20. H. W. Alter and R. A. Oswald, *J. Air Pollut. Control Assoc.* **37**, 227 (1987).
21. P. Brodeur, *New Yorker* **44**, 117 (Oct. 12, 1968).
22. B. T. Mossman and co-workers, *Science* **247**, 294 (1990).
23. C. J. Weschler, H. C. Shields, and D. V. Naik, *J. Air Waste Manage. Assoc.* **39**, 1562 (1989).
24. A. Herwaldt and co-workers, *Ann. Intern. Med.* **100**, 333 (1984).
25. *National Atmospheric Deposition Program, 2000 Annual Summary*, NADP, Washington, D.C., 2001
26. National Research Council, *Climate Change Science*, National Academy Press, Washington, D.C., 2001.
27. Intergovernmental Panel on Climate Change, *Climate Change 1995—The Science of Climate Change*, Cambridge University Press, 1996.
28. P. J. Crutzen, *Q. J. Royal Meteorol. Soc.* **96**, 320 (1970).
29. World Meteorological Organization, "*Halocarbon Ozone Depletion and Global Warming Potential, Scientific Assessment of Stratospheric Ozone*", WMO, Report 20 (1989).
30. World Meteorological Organization, "*Ozone Depletion and Chlorine Loading Potential, Scientific Assessment of Ozone Depletion*", WMO, Report 25, (1991).
31. AFEAS, "*Alternative Fluorocarbons Environmental Acceptability Study*", Washington, D.C. *http://www.afeas.org/* Sept. 2000.
32. National Oceanic and Atmospheric Administration, Boulder, CO. *http://www.cmdl.noaa.gov/* Sept. 2000.

GENERAL REFERENCES

References 14, 25, 26, and 27, and the following books and reports constitute an excellent list for additional study.

J. H. Seinfeld and S. N. Pandis, *Atmospheric Chemistry and Physics*, Wiley-Interscience, New York, 1998.

B. J. Finlayson-Pitts and J. N. Pitts, Jr., *Chemistry of the Upper and Lower Atmosphere*, Academy Press, New York, 2000.

T. E. Graedel, D. T. Hawkins, and L. D. Claxton, *Atmospheric Chemical Compounds Sources, Occurrence and Bioassay*, Academic Press, New York, 1986.

National Research Council, *Rethinking the Ozone Problem in Urban and Regional Air Pollution*, National Academic Press, Wahington, D.C., 1992.

J. H. Seinfeld, "Urban Air Pollution: State of the Sciences", *Science* **243**, 745 (1989).

David T. Allen
University of Texas

AIR POLLUTION AND CONTROL, INDOOR

1. Introduction

Indoor air quality (IAQ) has become a growing environmental issue over the past 20 years. An increasing number of health and comfort problems have been reported in office buildings, schools, residences, and similar nonindustrial settings. Many books and scientific articles have dealt with the nature of IAQ problems. The first comprehensive review of indoor air was published in 1981 (1). Compilations of more recent research articles can be found in scientific and technical journals and in proceedings of the recent triennial international conferences on IAQ and climate (see General References section).

Because the human body processes far more air than other environmental media, relatively small concentrations of contaminants can be of concern. (A typical adult breathes 15–20 m^3/day or about 20–25 kg of air; total intake of food and water is about 3–4 kg/day.) Whereas we often think of water and food contamination in parts per million (ppm), many air contaminants are of concern at concentrations in micrograms per cubic meter ($\mu g/m^3$), which is roughly equivalent to parts per billion. The quality of the air we breathe while indoors is especially important, since on average we spend ~90% of our time inside buildings, not including industrial workplaces (2). Unfortunately, studies have shown that indoor concentrations of many contaminants are higher than outdoor concentrations (3,4).

This article deals with air quality inside residential, commercial, and institutional buildings, which are the spaces normally associated with "indoor air" in the literature. Industrial workplaces present quite different exposures (typically, 40 h/week) and populations (typically, healthy workers). Principles of industrial hygiene address quite different sources, exposures, and protective options than those that are dealt with in most residential, commercial, or institutional situations. Therefore, while many aspects of industrial hygiene practice are applicable, requirements for providing good indoor air quality for the general population are generally stricter.

2. Concerns

2.1. Historical Concerns.

Bad odors, smoke, dampness, and infectious diseases have been indicators of poor indoor air quality for centuries (5). Controlling body odors (bioeffluents) has been the basis of ventilation standards since the eighteenth century (6). While these contaminants continue to be problems, the range of concerns broadened considerably in the 1970s—first in Europe, then in North America—as buildings were made tighter and operated with less ventilation to conserve energy. Air quality complaints became more focused on synthetic chemicals and microbial organisms such as molds and mildews (7). Specific contaminants that drew attention to indoor air quality in the 1960s and 1970s included asbestos (from various insulation and building materials), formaldehyde (from faulty foam insulation and pressed-wood products such as particleboard), carbon monoxide (from unvented or leaking vented combustion appliances) and radon (from soil gas infiltrated into homes).

2.2. Current Concerns. As more sophisticated measurements were made in the 1980s and 1990s, it became apparent that there are hundreds of measurable organic compounds in indoor air, present either as gases or associated with particles. See, eg, ref. 3 for a typical listing. Microbial contaminants and their associated gas-phase organic products of respiration and decomposition have also been investigated increasingly in recent years, but sampling and analytical methods are more cumbersome, so data are less complete. See ref. 7 for more detailed discussion of microbial contaminants.

Field investigations of residences, office buildings, and schools show that these contaminants come predominantly from indoor sources such as new materials, cleaning materials, office machines and appliances, and moist areas with favorable conditions for microbial growth. Furthermore, indoor concentrations are highly variable with time and place within a building. Any of hundreds of substances can be the most important with respect to concentration or potential health impact in a given space, at a given time.

The health concerns themselves are numerous. They range from vague dissatisfaction to frank irritation to chronic disease. Common terms found in the literature to classify the health and comfort effects or symptoms are (8):

- **Odor**.
- **Irritation** of eyes, nose, upper airways, throat, and skin.
- **Respiratory function decreases** in nonasthmatics including wheezing, cough, chest tightness, and shortness of breath.
- **Neurological symptoms** including nausea, dizziness, headache, loss of coordination, tiredness, and loss of concentration.
- **Immunological reactions** including inflammatory reactions, delayed hypersensitivity, and immediate hypersensitivity (allergic) reactions.
- Aggravation of **asthma**.
- **Cancer**.
- **Respiratory infections**.
- **Increased susceptibility** to infections or adverse responses to chemical substances.

See references 9–11 for current discussions of the health effects associated with indoor air quality.

Many of the effects or symptoms listed above are included in terms like "sick building syndrome" or "multiple chemical sensitivity (or intolerance)" that are commonly used to describe reactions to or perceptions of the indoor environment. These terms have evolved from complaints of various odors and forms of sensory irritation (especially of mucous membranes in the eyes, nose, throat) by occupants of residences, schools and office buildings. See ref. 12 for a review of sick building syndrome studies and ref. 13 for a discussion of multiple chemical sensitivity/intolerance. Cullen (14) has coined the generally accepted definition of multiple chemical sensitivity.

In addition to frank and perceived health effects, there has recently been increasing concern about monetary impacts of poor indoor air quality: Worker productivity in non-industrial workplaces has been shown in some studies to be measurably affected (15–17). However, more study is needed to separate the effects of air contaminants from the effects of other indoor environmental conditions such as lighting levels, temperature, and relative humidity. The literature (eg, ref. 18) shows a clear improvement in perceived air quality when temperature is reduced from higher temperatures to 20–21°C (68–70°F).

2.3. What Constitutes Poor Indoor Air Quality. Indoor air quality is a public health issue without a widely accepted definition of good air quality or a statement of health goals. Presumably the goal of ambient air quality standards—to protect public health, including the health of sensitive populations such as asthmatics, children and the elderly (19)—is also applicable to indoor air.

Private- and public-sector definitions of indoor air quality have been established for people who own and operate buildings. The American Society of Heating, Refrigerating, and Air-conditioning Engineers (ASHRAE) defines acceptable IAQ as: "Air in which there are no known contaminants at harmful concentrations as determined by cognizant authorities and with which a substantial majority (80% or more) of the people exposed do not express dissatisfaction" (20).

In a guideline document for building owners and managers, the U.S. Environmental Protection Agency has defined good indoor air quality for commercial buildings to include: "introduction and distribution of adequate ventilation air, control of airborne contaminants, maintenance of acceptable temperature and relative humidity... Good indoor air quality enhances occupant health, comfort, and workplace productivity" (21).

What constitutes poor IAQ therefore depends on many factors, including the physiological and psychological health of the breather (ie, the indoor occupant). Certain combinations of contaminant sources, inadequate ventilation, poor maintenance, and susceptible occupants lead to poor indoor air quality. Residential, commercial, and institutional buildings therefore need to be designed, operated and maintained for those who are sensitive or susceptible, but not necessarily for those who are extremely sensitive. The same requirements apply to the materials buildings are made from, and the products buildings contain.

Although their toxicities vary over several orders of magnitude, most of the individual contaminants discussed in the indoor air literature seem to be of concern at concentrations ranging roughly from 1–1000 ppb (1–5000 $\mu g/m^3$) for gases and from 1–100 $\mu g/m^3$ for particle-associated constituents. Given the potentially synergistic activities of contaminants in a mixture, guideline concentrations for total organic vapors from 300 μg (22) to 4000 $\mu g/m^3$ (23) have been suggested.

Poor thermal conditions can also affect the perception of air quality. See, eg, ref. 24 and its cited references for more information on this interrelationship. As a minimum, it seems prudent to maintain temperature and relative humidity within ASHRAE Standard 55 thermal comfort guidelines (25).

2.4. Standards and Guidelines for Indoor Air Quality. Governmental regulatory bodies have set very few indoor air quality standards. The U.S. Food and Drug Administration (FDA) has set an indoor limit of 100 $\mu g/m^3$ of ozone for spaces where ozone is being generated (26), and the EPA guideline of 4 pCi/L for radon has become a *de facto* standard, but there are few others.

Several indoor air quality guidelines have been suggested; the suggestions are often simply a list of outdoor (ambient) air quality standards or workplace standards, such as appended to ASHRAE's ventilation standard 62 (20). Researchers who specialize in sensory irritation effects have proposed Recommended Indoor Levels (RILs) for ~70 nonreactive volatile organic chemicals (27) and ~30 microbial organic chemicals (28). See ref. 29 for a more comprehensive review of guidelines and specific values for vapors of selected organic acids, phenols, and glycol ethers. Guidelines for mold contamination are most often based on observational data; see, eg, ref. 30.

Lacking definitive IAQ standards, outdoor (ambient) air quality standards serve as a starting point. However, fewer than 50 chemicals are covered by USEPA standards (31) or WHO/European guidelines (32). EPA National Ambient Air Quality Standards cover only six species, one of which—particulate matter—is not composition specific. Furthermore, these standards may be insufficiently protective, especially where cumulative exposure to a contaminant is of concern. This is because the typical person in developed societies spends ~90% of his/her time indoors. Since indoor concentrations are often higher than outdoor concentrations, indoor exposure (concentration × time) can be 10 or more times outdoor exposure.

Industrial workplace guidelines such as Threshold Limit Values (TLVs) are generally not applicable to residential, commercial, or institutional settings. They allow much higher concentrations than would be acceptable to the broad range of people who occupy typical "indoor" spaces. They may be useful in evaluating short-term exposures to people who maintain and renovate buildings during periods when cleaning or coating materials are being used and emissions are high. Concentrations should be much lower when normal occupants are in the building.

Minimum ventilation rates required by building codes, many of which are based on ASHRAE Standard 62, Ventilation for Acceptable Indoor Air Quality (20), have historically been based primarily on the amount of outdoor air required to maintain body odor levels acceptable to at least 80% of visitors to the space (5). A section in the Scope of Standard 62 recognizes the complexity of IAQ problems in modern buildings by stating the following: "Acceptable indoor air quality may not be achieved in all buildings meeting the requirements of this standard for one or more of the following reasons: (*a*) because of the diversity of sources and contaminants in indoor air; (*b*) because of the many other factors that may affect occupant perception and acceptance of indoor air quality, such as air temperature, humidity, noise, lighting, and psychological stress; and (*c*) because of the range of susceptibility in the population" (33).

Perhaps the primary way indoor air quality will be regulated in the future is through emission standards—established by regulatory bodies or private sector organizations—for materials and products. Outdoor air is regulated to a large extent by emissions standards, so it will not be surprising to see indoor air take a similar path. The main difference may be the greater involvement of the private sector; see the following section on source management for more discussion.

2.5. Methods of Providing Good Indoor Air Quality.

There are three basic ways to reduce exposures to indoor contaminants:

- Manage **sources** to prevent or reduce emissions.
- Provide **ventilation** to dilute and exhaust contaminants effectively.
- Remove contaminants by **air cleaners**.

Source management includes selecting, using, and maintaining building materials and furnishings, consumable products and equipment, and is generally the most effective option once building ventilation rates meet standards.

Although direct exhaust ventilation of areas with strong sources such as smoking lounges, kitchens, bathrooms, and copy machine rooms can be very beneficial, general ventilation of a building at rates above standards provides relatively little reduction of exposure. General ventilation provides especially little reduction from sources that are physically close to people, because the effect of dilution is so small over short distances.

Highly effective air cleaning can be very expensive, will depend on effective distribution of air in the space being treated, and is not generally feasible for gases and vapors at the relatively low average concentrations in buildings.

3. Contaminants in Modern Buildings

3.1. We Breathe a Complex Mixture.

Table 1 summarizes the types of contaminants found in indoor air. Note that there is a wide range of natural and synthetic substances in both the gaseous and particulate phases. In many respects the list is similar to the spectrum found in outdoor air, but most substances are present at higher concentrations. This is because there are many sources indoors, and dispersion from indoor sources is less in the confinements of a building than in the relatively unrestricted outdoors. In the extreme, when a source is very close to a person, the person can be in the "plume" of emissions, where the concentration is very high. See references 50–53 for example summaries of data in indoor air contaminant concentrations.

The comparison with industrial exposures is less similar. Indoor contaminants are more numerous but generally at lower concentrations than in industrial workplaces. Despite the generally lower concentrations, and the fact that contaminants are usually present below any known sensory irritation or other health effects, reactions to and perceptions of poor indoor air quality are common.

Researchers and regulators have speculated that there are additive or synergistic effects among the many chemicals present. Odor studies of bioeffluents have shown that individual species may be present below their odor threshold concentrations, but the total mixture of them can lead to the perception of poor indoor air quality (54). Some sensory irritation research based on both animal experiments (55,56) and human studies (57,58) also support this theory, but the agonistic effects found to date have been smaller than would explain people's responses in the field.

Table 1. **Types of Contaminants in Indoor Air**

Type	Examples	Typical indoor concentration (BDL = below detection limit)	References
Products of Combustion			
gaseous oxides	nitrogen oxides, sulfur oxides	2–20 μg/m^3. NO_x higher when indoor combustion present; SO_x higher only when S-containing fuels burned	34
	carbon monoxide	BDL-1 ppm. Concentrations above about 2 ppm may indicate an outdoor source or leaking heating system.	
smoke	cigarette; wood, kerosene, coal (emissions from unvented appliances or leakage from vented appliances)	0–100 μg/m^3	
other products of incomplete combustion	gaseous and particulate species ranging from low MW, volatile substances like aldehydes to high MW, semivolatile substances that are mostly constituents of particles (eg, PAHs)	when present due to infiltration from outdoor air, generally <1 μg/m^3. When present due to indoor sources such as cooking, cigarette smoking, or poorly ventilated wood or coal burners, concentrations can be >1 μg/m^3 for extended periods, and up to 100 μg/m^3 for several hours. See references for further details.	35
Synthetic Chemical Contaminants			
vapor-phase	"volatile organic compounds" (VOCs)	100–1000 μg/m^3 (total). Can be higher in new or newly renovated buildings.	36
particle-phase	Pesticides, phthalates, PCBs, latex, synthetic fibers	BDL-1 μg/m^3 (total).	37–40
Natural Contaminants			
human bioeffluents	ammonia, hundreds of gaseous organics	see references.	41
microbial vapor-phase byproducts	hundreds of gaseous organics	See articles on MVOCs in Proceedings of the 9th International Conference on IAQ and Climate (2002).	
biogenic particles	allergens from animals, pollen, insect fragments, molds, fungi, bacteria (eg, TB, Legionella), viruses	data limited; see references.	42–47
soil gases[a]	radon, humic gases	radon: 0.5–5 pCi/L. average in U.S. homes is ~1.5 pCi/L.	
Natural and Synthetic Fibers			
	asbestos, synthetic vitreous fibers, cellulose, carpet and upholstery	see references.	48,49

[a] Soil gases, which are drawn into buildings mostly by pressure differentials between the ground and indoors, can also contain volatile synthetic chemicals from leaking underground storage tanks or pipelines, waste sites, etc.

Table 2. **Categories of Indoor Sources**

Source category	Examples
combustion sources	smoking, unvented appliances (gas, kerosene), leaking vented appliances
materials	building materials, furnishings, consumer products (sources of organic vapors and particles); moist materials (sources of microbes and their products of respiration and decay)
activities	painting, polishing, spraying, cooking, use of machines or solvents
outdoor sources	infiltration of contaminants via outdoor air or soil gas, evaporation of contaminants from domestic water, tracking in pesticides on shoes or clothing
indoor chemical reactions	ozone reactions with unsaturated hydrocarbons emitted from cleaning agents, flooring materials, and furniture[a]

[a]Refs. 59–61.

3.2. Sources. Many different types of materials are used in the construction, furnishing, maintenance, and operation of a building. In addition, there are various activities by occupants. As potential sources of indoor air quality problems, these items can be grouped into five categories: combustion sources, materials, activities, outdoor sources, and indoor chemical reactions. Examples are listed in Table 2. At a given time, emissions from an individual source in any of these categories can dominate the impact on indoor air quality in a building.

In recent years, incidences of buildings with mold contamination have been especially newsworthy. Materials with sufficient organic matter and moisture can be breeding places for various microbials that generate undesirable odors. When drying conditions cause spore formation, activities and ventilation systems can spread airborne spores that cause allergic reactions or disease. A recent review article provides a good overview of this issue and references for further details (62). Many research papers at the 9th International Conference on Indoor Air Quality and Climate, held in Monterey, Calif. in July 2002, reported on this issue. See "General References" section.

There is a growing body of data—some published, much proprietary—on emissions of gaseous and particulate contaminants from a variety of indoor sources. The most common way to generate such data is to put sample of products into chambers through which controlled amounts of clean air are passed. Concentrations of emitted contaminants in the air exiting the chambers are measured. In the most detailed studies, emissions are measured as a function of time, temperature, airflow rate, amount of sample per unit volume of chamber, and relative humidity. ASTM and the European Community have published guideline procedures for emissions testing (63,64). See refs. 65 and 66 for summaries of publicly available data.

Another potential source of indoor contaminants is air-phase chemical reactions. Although the intensity of ultraviolet (uv) light is generally much lower indoors than outdoors, there are apparently indoor conditions in which measurable formation of secondary pollutants occurs. Investigations into this phenomenon have accelerated since the mid-1990's, when Weschler and others published preliminary research results (59–61). Several papers at the 9th International

Conference on Indoor Air Quality and Climate presented updated research results. See "General References" section.

4. Ventilation as a Control Method

4.1. Types of Ventilation. Historically, ventilation has been the primary mode of indoor air quality and climate control (67). Only in the past two decades have the complementary roles of air cleaning (other than for equipment protection) and source management been recognized and practiced.

Ventilation, as defined by ASHRAE (68), is "the process of supplying air to or removing air from a space for the purpose of controlling air contaminant levels, humidity, or temperature within the space." Ventilation can be either natural or mechanical. Natural ventilation is "ventilation provided by thermal, wind, or diffusion effects through doors, windows, or other intentional openings in the building." Mechanical ventilation is defined as "ventilation provided by mechanically powered equipment, such as motor-driven fans and blowers, but not by devices such as wind-driven turbine ventilators and mechanically operated windows." Infiltration and exfiltration, defined as air leakage inward or outward "through cracks and interstices and through ceilings, floors, and walls of a space or building", occurs in addition to intentional ventilation. While infiltration/exfiltration increases the air exchange of a space, it is so irregular and unpredictable that it is not normally considered in assessing the effectiveness of air quality control.

Most single-family residences and small multiple-family units in North America are ventilated by infiltration/exfiltration or naturally ventilated, except for exhaust fans in bathrooms and kitchens. The heating and air-conditioning systems of these buildings are usually designed as closed systems (although the duct systems are notoriously leaky, and often provide substantial, unintentional, mechanical ventilation). Most commercial and institutional buildings in North America are mechanically ventilated by heating, ventilating, and air-conditioning (HVAC) systems; natural ventilation of such buildings is more common in Europe.

Ventilation works in two fundamental ways to control indoor air quality: by dilution, and by direct exhaust. The dilution mode relies on relatively low concentrations of contaminants in the air supplied to the space to dilute contaminants generated in the space. Outdoor air is generally cleaner, hence bringing in outdoor air is an important part of general dilution ventilation. Direct exhaust ventilation is common in areas where there are strong indoor sources; air from those areas is captured and exhausted directly from the building. Good overall ventilation design and operating practice is to exhaust air near strong sources, transfer air from low-concentration to higher concentration areas in a building, locate air intakes to provide good-quality outdoor air, and operate the ventilation system during periods of high emissions such as maintenance activities. For an overview of ventilation strategies, see reference 69.

As with any strategy for indoor air quality control, the fundamental goal of ventilation is, as ASHRAE puts it, to provide "indoor air quality that will be acceptable to human occupants" and to "minimize the potential for adverse health effects"(68). See reference 70 for an excellent overview of ventilation, health, and comfort.

4.2. ASHRAE Ventilation Standard. ASHRAE Standard 62, Ventilation for Acceptable Indoor Air Quality, provides state-of-knowledge guidance from the scientific and technical communities on ventilation system design and operation practices that will help provide good air quality in commercial and residential buildings. Its state-of-knowledge guidance is the best available basis for ventilation system requirements in mechanical sections of building codes for North America. It is also used widely internationally.

Standard 62 has evolved considerably since its first publication in 1973 (71). It is continually reviewed and updated to reflect new information from research and field experience. Recent addenda have created a new section on construction-generated contaminants and ventilation system start-up, a new section on ventilation system operation and maintenance, and new requirements for equipment-related particle filtration.

Table 3 summarizes selected values of minimum ventilation rates for selected spaces out of ~90 different spaces in commercial, institutional, and residential buildings covered by the standard. This table is meant only to illustrate the variety of spaces and rates covered; see the full standard for complete listings and details. Note that the ventilation rates listed are for the outdoor air portion only. With recirculation, the total ventilation rates in buildings are often ~5 times higher.

Currently, the lowest per-person outdoor air rate is 8 L/s; the highest is for smoking lounges, at 30 L/s per person. Spaces with highly irregular or low occupancies or strong sources (such as the example above of malls and arcades) have requirements based on floor area. ASHRAE is considering a revision of these prescriptive rates to separately account for contaminants from people and from other sources; the separate rates would then be added to obtain the total outdoor air rate.

Table 3. **Selected Outdoor Air Requirements for Ventilation**[a]

	Outdoor air requirement[b]			
Space type	cfm per person	L/s per person	cfm per ft^2	L/s per m^2
fast food cafeteria	20	10		
bars, lounges	30	15		
hotel bedroom			30 cfm per room	15 L/s per room
office space	20	10		
malls, arcades			0.20	1.00
classrooms	15	8		
hospital patient rooms	25	13		
residential living areas	≥15 (and 0.35 air changes/h)	≥7.5 (and 0.35 air changes/h)		

[a] From ASHRAE Standard 62-2001

[b] cfm = cubic feet/minute; L/s = liters/second; ft^2 and m^2 refer to floor area.

In addition to the prescriptive rates described above, Standard 62 provides an Indoor Air Quality Procedure for designers who have information on sources, emission rates and compositions, toxicity data for the emissions, and health-based criteria. In principle, this procedure enables greater rigor in calculating ventilation rates for healthful indoor air and enables consideration of the benefits of "clean" materials and air cleaning systems. At present, its application is limited by the lack of health-based criteria for maximum acceptable concentrations of many of the air contaminants found indoors.

4.3. Ventilation and IAQ Models. Various computer-based models for ventilation calculations have been developed. One of the most widely used is CONTAM (72). Characteristics of the building, meteorological conditions, HVAC system design, and indoor sources can be entered, then air flows and contaminant concentrations throughout the building over time are calculated by a series of mass balance calculations.

There are also a number of simpler mass-balance models that are designed more for IAQ calculations. These require the user to enter ventilation flows. See ref. 73 for a review. There are also computational fluid dynamics (CFD) models used by researchers to estimate conditions at individual points in a space. For most practical uses, however, the space-average concentrations provided by mass-balance models are sufficient.

4.4. The Relationship With Energy. Based on the U.S. Department of Energy data summarized by ref. 74, ~36% of total primary energy consumed in the United States is used to heat, cool, light and operate equipment in residential and commercial buildings. Even at current low energy prices, this costs >$230 billion per year. During the short-lived energy cost spikes in the 1970s, energy conservation was increased by tightening buildings to reduce infiltration and by reducing ventilation rates. Those energy conservation efforts tended to increase indoor relative humidity and contaminant levels, and highlighted the important relationship between energy conservation and indoor air quality. As real energy prices increase in the future—as they inevitably will—there will be increased interest in reducing ventilation rates. Given what we now know about indoor air quality, any future reductions in ventilation rates will have to be accompanied by either reduction in emissions from indoor sources, or increased use of air cleaning, or both.

4.5. Effectiveness of Ventilation. The effectiveness of ventilation in controlling indoor air quality has limitations. As noted previously, exhaust ventilation can be quite effective if the capture efficiency is high. Dilution ventilation generally works best if airflow is from low concentration areas toward high concentration areas, but this may not be practical, especially if the locations of major sources change with time. Furthermore, since much of ventilation air is recirculated, contaminants tend to get distributed throughout the building (or at least the portion of the building served by a given air handling system). As also mentioned previously, dilution ventilation has very little effect on emissions from sources close to a person (eg, clothing, cosmetics, office machines, bedding or flooring materials for infants).

Most ventilation systems are designed for complete mixing of air supplied to the space. If a space with poor air quality is being supplied with 50% of the rate specified by code, increasing the rate to code requirements will only reduce contaminant concentrations by ~50%. Increasing the rate to double the code requirement (even if mechanically possible) will only provide another 50% reduction. That level of reduction further assumes that the outdoor air that is being used for the dilution has a zero concentration of the contaminant, is similarly free of additional contaminants, and is not overly humid. The bottom line with respect to ventilation effectiveness is this: Substantial reductions (say of 80–95%) of contaminant concentrations will require reducing emissions from indoor sources or cleaning the air. These approaches are the subject of the following sections.

5. Source Management as a Control Method

5.1. Prevention of Emissions. In principle, preventing IAQ problems by managing indoor sources can be very effective. The objective is to use materials and products with low (or no) emissions of substances that might cause odor, sensory irritation, or other health problems. Over the past two decades, research and development of product testing procedures and exposure prediction models have produced useful tools for evaluating the acceptability of building materials, furnishings, and other products used in buildings. Emission testing and prediction of occupant exposures has become a key step in the design of some buildings (75). This section describes how materials and products can be evaluated to help prevent IAQ problems.

Emissions prevention involves many parties, throughout the whole life cycle of a building. Manufacturers of materials and products need to know how to test and certify that their products are not just low emitting, but have low adverse health impact. Specifiers such as architects, builders, and purchasing agents for building owners need information on low emitting and low impact building materials, furnishings, and consumable products. Public health officials who are increasingly involved with air quality problems in buildings need similar information to pass along to the general public.

5.2. General Selection Criteria for Indoor Materials. Any building materials, furnishings, maintenance materials, or other contents of a building are selected with various physical, aesthetic, economic and environmental criteria in mind. Examples of such criteria are listed in Table 4. Hundreds (sometimes thousands) of material selections need to be made for a typical building, and several of these criteria often conflict.

Historically, environmental criteria have not played a major role in materials selection. In recent years, however, the balance has shifted to give environmental considerations greater weight. For example, the American Institute of Architects (AIA) has developed an Environmental Resource Guide. Intended for architects, it attempts to describe life-cycle environmental impacts of materials used in buildings. The AIA life-cycle concept includes total environmental impacts during production, use, and disposal of the materials (77).

5.3. Emissions Criteria for Indoor Materials. From an indoor environmental point of view, there are several characteristics that could be considered in describing an "ideal" material. Such characteristics are listed in Table 5.

Table 4. **Selection Criteria for Indoor Materials**[a]

Attribute	Examples
physical	strength
	durability
	heat transmission
	light transmission
	maintainability
	effectiveness (eg, as a cleaning agent)
aesthetic	color
	texture
	odor
	noise
economic	initial cost
	maintenance cost
	operating cost (eg, energy)
environmental	emissions to air
	water and waste impacts
	support of microbial growths
	life-cycle impacts

[a] See Ref. 76.

Not many materials will have all of these characteristics, of course. Compromises will usually have to be made, and many that at first seem undesirable may turn out to be quite acceptable. For example, a material with a high initial emission rate but rapid decay may be quite acceptable if used in a manner that does not lead to high exposures, either during use or from reemissions from interior surfaces that have adsorbed the emissions. Examples of material and product selection criteria that have been used in both the public and private sectors are summarized in ref. 78.

Whether the emissions from any source will lead to "acceptable" indoor concentrations and occupant exposures will depend on a number of considerations,

Table 5. **Desirable Characteristics of Indoor Materials**

Factor	Desirable characteristic
emission properties	low emission rates
	low toxicity of emissions
"sink" properties[a]	nonsorbent
	if sorbent, not reemitting
	if not reemitting, nonnutrient
microbial properties	hydrophobic
	nonnutrient
	cleanable
physical properties	as needed for the application
aesthetic properties	as desired for the application
cost	reasonable

[a] Ability of a material to adsorb or absorb vapors from the air. Sinks can be essentially irreversible, or reversible (reemitting). Sinks that reemit sorbed contaminants can be significant sources when the strength of original sources has decreased.

viz: (*a*) emission rates; (*b*) the toxicity or irritation potential of substances emitted; (*c*) ventilation rates; (*d*) physical relationships between the source, the persons present, and the space they occupy (the proximity of the source to people breathing its emissions can greatly affect the amount of dispersion and dilution of emissions, and therefore the concentrations actually breathed); and (*e*) the sensitivity of the occupants.

These factors are highly variable from building to building, and often from area to area within a building. Therefore, product acceptability is essentially a situation-specific issue. The best currently available approach to evaluating indoor materials and products is to test their emission rates, and predict pollutant concentrations in the building where they are to be used. Various indoor air quality prediction models (79–81) are available, but for most purposes the simpler models such as (79) are sufficient.

If emissions data are not available for a particular product of interest, various emissions models are available for estimating emissions rates. See ref. 82 and 83 for a review of these models.

Starting from the best available IAQ criteria, modeling can estimate the maximum advisable emission rate (eg, milligrams per hour of chemical or physical pollutants, or colony forming units per hour of microbial pollutants). The emission rate then usually needs to be divided by the amount of material or product to be used; that will yield an emission factor (emission rate of contaminant per unit of product). The most common units for emission factors are μg/h per m^2, kg, or number of items used.

This maximum advisable emission factor can then become the IAQ basis for selection of indoor materials and products. Since emissions from many materials change greatly with time, the time dependency of the emission factor should be considered. For most situations, emissions at the time when building occupants are first exposed are the most relevant.

In spite of the lack of a solid data base on acceptable concentrations of complex mixtures of substances, many builders, architects and consumers are asking for lists of, or guidelines on, "low emitting" or "clean" materials and products. At present, however, there are not enough published emission rate data to list specific recommended products. Therefore, classification schemes such as shown in Table 6 [adapted from ref. 78] can be used for judging whether products are acceptably low emitting. Acceptability in this case is based primarily on the work of Mølhave that suggested a mucous membrane irritation threshold for total organic vapors in the range of 160–5000 μg/m^3 (84); Seifert's suggested target level of 300 μg/m^3 (85); and Alarie's suggestion of 4000 μg/m^3 for nonreactive organic vapors (86). Assuming 1000 μg/m^3 to represent the range of their recommendations, and accounting for multiple sources with one dominant source, a maximum contribution from any single source type of 500 μg/m^3 is used in Table 6.

Values in Table 6 should be taken as gross guidelines and careful note should be made of the explanatory comments. Many users may want to adopt the conceptual framework of the table, and establish a classification scheme to meet their specific applications.

Manufacturers need to make clean products, but the definition of "clean" is—and will remain—elusive. As a general guideline, it seems prudent to try

Table 6. **A Classification of Low Emitting Materials and Products**[a,b]

Material or product	Maximum emissions[c]
flooring materials	0.6 mg/h per m^2
floor coatings	0.6 mg/h per m^2[d,e]
wall materials	0.4 mg/h per m^2
wall coatings	0.4 mg/h per m^2[d]
movable partitions	0.4 mg/h per m^2
office furniture	2.5 mg/h per workstation
office machines (central)	0.25 mg/h per m^3 of space
ozone emissions	0.01 mg/h per m^3 of space
office machines (personal)	2.5 mg/h per workstation
ozone emissions	0.1 mg/h per workstation

[a] See Ref. 78.
[b] Based on emissions of total organic vapors, except as noted.
[c] This column lists default values for use where predictive modeling of IAQ impacts is not done. For specific indoor situations, modeling is generally preferable to using these defaults, and may yield very different values for maximum emissions. Values for particularly noxious organic compounds will also be lower than those shown.
[d] Many varnishes, paints, waxes and other wet coatings have emission factors substantially higher than this, immediately after application. These coatings might still be considered "low-emitting" if their emission factors drop below this level within several hours. However, the frequent presence of other surfaces that adsorb coating vapors and subsequently re-emit them complicates the classification of coatings.
[e] Basic Assumptions: Indoor air is well mixed; ventilation rate is 0.5 exchange of outdoor air per hour; maximum prudent increment in indoor concentration of organic vapors from any single source type is 500 $\mu g/m^3$; maximum prudent increment of ozone is 20 $\mu g/m^3$ (0.01 ppm); volume of concern for dispersion of emissions from furniture and machines at workstations is 10 m^3.

to keep a product's emissions to rates that—with normal use in occupied spaces—would lead to indoor concentrations less than $\sim$500 $\mu g/m^3$ for low toxicity gases and vapors, and below the ambient air quality standard for sulfur dioxide (the most restrictive standard) for moderate-toxicity gases and vapors. Using the same logic for particles, a product's emissions would lead to indoor concentrations below the particulate matter ambient air quality standards. For practical purposes, this may amount to eliminating volatile, moderate-toxicity substances from indoor products.

Occupant exposures to emissions from periodic building maintenance and other work and personal activities can also be high. These periodic, short-term exposures, in addition to initial emissions from new materials, are thought by some to be the triggering causes of many "sick" buildings that are investigated—without success—several weeks or months after complaints are first expressed.

Degradation of materials leading to particle emissions and biocontamination of materials that have become soiled and wet can also be potentially severe sources of IAQ problems. Therefore, maintainability, durability, and susceptibility to microbial growth are important (and often overlooked) factors in selection of materials for good indoor air quality.

5.4. Emissions Testing. If emission factors are to be used in material selection, data from emissions testing are needed. Proper testing involves the use of environmental chambers with carefully controlled airflows and environmental conditions. Many researchers have been using such chambers in recent years, and limited amounts of data are publicly available. See ref. 87 for a listing of publications through 1997 and ref. 88 for typical values of volatile organic compound emissions derived from the published literature.

A few commercial testing laboratories such as Air Quality Sciences in Atlanta do emission testing for product manufacturers, designers, or building owners. Several manufacturers are also set up to test their own products. They generally follow the ASTM guidelines mentioned previously (89). An important alternative is to use a mouse bioassay to obtain data on sensory irritation effects of emissions (86). This has the benefit of giving a direct biological response to complex mixtures of contaminants. Procedures for this type of testing are described in ASTM Standard 981-84 (90).

Emission rate testing and exposure modeling has become a significant step in the design of some new and renovated office buildings. When coupled with appropriate attention to ventilation system design and operation, aesthetic and ergonomic factors, and building maintenance, occupant complaints that characterize "sick buildings" are almost certain to be reduced. However, this process is currently hampered by the lack of data on emissions and clear understanding of relationships between exposure to emitted substances and health or sensory irritation.

5.5. Product Testing and Labeling. Either type of testing—chemical emission rate or animal bioassay—can be used as the basis for labeling indoor materials and products. Labels can show numerical values for emission factors or intensities of biological responses under standard testing conditions. They are conceptually similar to energy efficiency labels on household appliances or gasoline mileage ratings of automobiles in that they would serve as an indicator of the quality of a material or product for indoor use, from the standpoint of IAQ impacts. A notable example is the "green label" program of the Carpet and Rug Institute (97). A summary of their emissions limits, based on periodic testing of products from production lines, is shown in Table 7.

5.6. Building Design, Operation, and Maintenance. Many indoor air quality (IAQ) problems can be prevented during building design. While the importance of proper design of ventilation systems has long been recognized, selection of materials and products is now receiving increased attention. It is important to avoid sources with emissions that are too great to be diluted and exhausted by ventilation, or removed by affordable air cleaning devices. This is especially true for sources that are close to occupants, such as office furniture, furnishings, office supplies, and personal care products.

Prevention of IAQ problems can often be most cost-effectively accomplished at the design, operation, and maintenance stages of the lifecycle of a building. This is widely recognized in publications by architects (3), ventilation engineers (92), and regulators (93). Selection of materials, as discussed above, is important at the design stage. Levin presents an excellent outline of a step-by-step process for good design in ref. 94.

Building operation and maintenance activities are important in assuring overall cleanliness, moisture control (to prevent microbial problems), and the use of appropriate cleaning materials and methods. Operation and maintenance guidelines have been published by ASHRAE for both the commissioning/start-up

Table 7. **"Green Label" Emissions Criteria Established by the Carpet and Rug Institute**[a]

Product	Emitted contaminant	Maximum emission factor[b]
carpet	4-phenylcyclohexene	0.05
	styrene	0.4
	formaldehyde[c]	0.05
	TVOCs[d]	0.5
carpet adhesives	formaldehyde	0.05
	2-ethyl-1-hexanol	3.0
	TVOCs	10.0
carpet cushions	4-phenylcyclohexene	0.05
	formaldehyde	0.05
	BHT (butylated hydroxytoluene)	0.30
	TVOCs	1.00
vacuum cleaners	dust	emissions that create 100 μg/m^3 of dust in the room[e]

[a] *www.carpet-rug.com*
[b] Emission factor units, except for vacuum cleaners, are mg/h of contaminant per m^2 of product.
[c] The formaldehyde criterion for carpet is intended to prove that none is used in the product.
[d] Total Volatile Organic Compounds. Usually sampled by thermally desorbable solid sorbents and analyzed by gas chromatography–mass spectrometry.
[e] Specifically, this criterion is worded: "The dust containment protocol evaluates the total amount of dust particles released into the surrounding air by the action of the brush rolls, through the filtration bag, and any air leaks from the vacuum cleaner system. This protocol requires that a vacuum cleaner will release into the surrounding environment no more than 100 micrograms of dust particles per cubic meter of air...".

and occupancy phases of a building's life cycle (92,95,96). Others have also addressed the impacts of cleaning and other operations and maintenance activities (97,98).

6. Air Cleaning as a Control Method

6.1. General Comments. The third approach to providing good indoor air quality is air cleaning. Air cleaners can either be stand-alone units for treating air in a single room or a portion of it, or central-system units that are built into the heating, ventilating, and air-conditioning (HVAC) system. For the indoor situations covered by this article, mask-type filters are not considered.

If ventilation conditions meet standards and indoor sources are being managed to the extent practical, air cleaning may provide additional health benefit. This finding is especially true if there are indoor sources that are particularly difficult to manage. In general, however, air cleaning has many of the same types of limitations as ventilation: Large volumes of air need to be processed to remove or destroy very small concentrations of contaminants. To be effective, air cleaner intakes need to be close to sources of key contaminants, the device needs to process a sufficient amount of air relative to the volume of the space being treated, and it must have reasonable collection or destruction efficiency and capacity.

As with the other options, air cleaners also need to be easy to operate and maintain, and be reasonably economical. A particular restriction on central-system air cleaners is pressure drop. Since very large volumes of air are handled in HVAC systems and power requirements for air handling are large, there is very little margin for incremental pressure drops in most systems. A pressure drop of 2.5 cm (1 in.) of water is considered the approximate practical limit for air cleaners. General HVAC system filters, which are installed to remove dust to protect equipment and prevent build-up on heat exchanger surfaces, normally have pressure drops <2.5 cm.

Air cleaners can be designed to remove particles or remove/destroy gaseous contaminants. Devices for general particle removal, infectious particle removal, and gaseous contaminant treatment are discussed in the following sections. Devices that operate by generating ions or ozone are also marketed, but their effectiveness is suspect. Ozone-generating devices should not be used in occupied spaces (99).

6.2. Particle Filtration. There are two general types of particle removal devices used in buildings: fabric filters and electrostatic precipitators. Designs have been adapted from similar devices used in industrial applications. The most common form for filters is the pleated type, although extended-area bag filters are used in some large buildings. Electrostatic precipitators, sometimes referred to as "electronic" air cleaners when marketed for residential applications, are physically similar to, but lighter in construction than industrial units; they also operate at lower voltages. Air cleaners for commercial buildings are often sold in modules designed to handle $\sim$1 m^3/s ($\sim$2000 cfm) of air, and installed in banks when greater airflows are needed.

Good general references to design, operation and maintenance of these devices for commercial and institutional buildings can be found in (101). EPA and Consumers Union have published reports on the performance of residential devices (102,103).

ASHRAE Standard 62, in addressing commercial large residential, and institutional buildings, requires particle filters or air cleaners with a minimum efficiency reporting value (MERV) of 6 or more "...upstream of all cooling coils or other devices with wetted surfaces through which air is supplied to an occupiable space" (104). That standard also allows reduction of ventilation rates when acceptable air cleaners are used (105).

Although testing methods for particle removal efficiency have been available for many years, a widely accepted test method that measures efficiency as a function of particle size has only been in place since 1999. ASHRAE Standard 52.2 (106) is an important improvement, since it measures filter efficiency for particles from 0.3 to 10-μm diameter, in 12 size ranges. This covers the respirable range nicely, and includes the 0.3–1-μm sizes that are difficult to collect by low to medium efficiency filters. Standard 52.2 gives efficiency ratings in MERV values that range from 2 (for low efficiency filters such as furnace filters in residential applications) to 6–9 (for medium efficiency filters such as pleated types) to 16 (hospital grade) to 18 (HEPA) and 20 (ULPA).

ASHRAE Standard 52.2 does not cover electrostatic precipitator-type air cleaners. In principle, "ESPs" can be very effective in removal of the full range of respirable particles, and have much lower pressure drop than filters. Ozone generation from arcing can occur, but most commercial units have adjustable voltage to control that. Some low-cost units for residential and commercial building applications have poor airflow distribution, which allows some particles to bypass the charging zone and compromises performance (107,108). A testing procedure similar to Standard 52.2, with provisions to also measure ozone generation, is needed for these devices.

6.3. Control of Infectious Particles. Person-to-person transmission of bacterial and viral respiratory diseases occurs mainly via airborne particles that are evaporative residuals of droplets created by coughing and sneezing. These "droplet nuclei" are typically 1–3 μm in diameter. Like other particles, their concentrations can be controlled by ventilation, source management (ie, care of infected persons), and particle air cleaners.

Ventilation rates that meet code requirements for body odor and humidity control (typically, ~8–10 L/s per person) have been widely assumed to be sufficient for control of infectious agents, but some studies have shown otherwise (109). Increasing ventilation above code requirements incurs energy and cost penalties, and the increased dilution leads only to modest reductions.

Particle air cleaners with high removal efficiencies and recirculation rates can reduce droplet nuclei concentrations if located and operated so as to properly distribute the clean exhaust air. The effective ventilation rate of the space is increased, perhaps by a factor of 2–4, but this may not be sufficient for spaces occupied by infected persons. Noise and drafts can also be a drawback to these devices (110).

Another type of control system is ultraviolet germicidal irradiation (UVGI). Air is passed over lamps emitting light in the uv-C region (typically, ~254-nm wavelength). These devices are often installed in the upper-room mode, where lamps are mounted near the ceiling. Natural convection in the room circulates the air across the lamps. The theory and application of this technology is thoroughly reviewed in (111,112). A method for evaluating the efficacy of UVGI has been developed (113), and chamber studies have indicated that UVGI, both by itself and in combination with air filtration, can be effective in removing or inactivating airborne bacteria (114). Further experiments and field studies are needed to determine efficacy in practice.

6.4. Gas/Vapor Removal. The possibilities for removing gases or vapors are much more limited than for particles. The most frequently considered (and used) approach is adsorption, particularly by activated carbon. However, the capacity of activated carbon for gases and vapors in the low ppm to high ppb range is too limited to make it a practical option for most applications. There are statements in the indoor air literature that "...activated carbon can absorb large quantities of indoor air pollutants, even at the low concentrations of these substance found in indoor air. As an example, it is not uncommon for activated carbon to adsorb as much as 20 to 30 percent of its weight in contaminants on exposure to indoor air pollutants before losing its ability to adsorb additional contaminant" (115).

However, the relevant performance criterion is capacity before reasonable breakthrough. Careful experiments with hexane at 11, 2, and 0.4 ppm inlet concentrations in air at relative humidity of 50% have been reported for a typical coconut shell carbon. At 30% breakthrough, the carbon had capacities of 10, 6, and 3% contaminant, respectively (116). Therefore, for design purposes it would seem advisable to count on a capacity of no >2–3% contaminants by weight, which translates into bed lives of a few months. At roughly $1500/year for each module serving 1 m^3/s, this is a costly option. See ref. 117 for example calculations, including costs.

Activated carbon has been used successfully to reduce ozone emissions from copy machines. In this application, the contaminant concentration is relatively high, its affinity for carbon is great, and the volume of air treated is small compared to room air recirculation rates. (Catalytic beds are now also used for that application.)

Chemisorption has been considered for many years as a possible technology. In theory, adsorbed contaminants are chemically converted to substances that do not desorb. One of the major barriers to this technology is developing an adsorbent and reactive surface for the wide range of contaminants found in indoor air. Potassium permanganate-impregnated alumina, developed for industrial applications, has been marketed for many years to reduce indoor concentrations of acid gases such as nitric oxide, sulfur dioxide and hydrogen sulfide as well as formaldehyde (118). There is little well-documented performance data from the field, so the effectiveness of this chemisorption option is unknown.

Room temperature catalysis is another, conceptually similar, technology that has been investigated for some time as a way to reduce indoor concentrations of vapor-phase organic compounds. The version of this technique that has gotten most attention is photocatalytic oxidation. Current technology requires far too much uv energy—and therefore far too much cost—to be practical (117).

BIBLIOGRAPHY

1. National Research Council, *Indoor Pollutants*, National Academy Press, Washington, D.C., 1981.
2. Klepeis and co-workers, *Analysis of the National Human Activity Pattern Survey (NHAPS) Respondents from a Standpoint of Exposure Assessment*, EPA/600/R-96/074, U.S. Environmental Protection Agency, Washington, D.C., 1996.
3. S. K. Brown, in T. Salthammer, ed., *Organic Indoor Air Pollutants: Occurrence-Measurement-Evaluation*, Wiley-VCH, Weinheim, 1999, pp. 171–184.
4. W. G. Tucker, in *Inside IAQ: EPA's* Indoor Air Quality Research Update, U.S. Environmental Protection Agency, Research Triangle Park, NC, 1–7 (Spring/Summer 1998).
5. J. E. Janssen, *ASHRAE J.* **41**, 48 (October 1999).
6. P. O. Fanger and B. Berg-Munch, in *Proceedings of an Engineering Foundation Conference on Management of Atmospheres in Tightly Enclosed Spaces*, 45–50, ASHRAE, Atlanta, Ga., 1983.
7. H. A. Burge, in J. D. Spengler, J. M. Samet, and J. F. McCarthy, eds., *Indoor Air Quality Handbook*, McGraw-Hill, New York, Vol. 45, 2001.
8. W. G. Tucker, in W. G. Tucker, B. P. Leaderer, L. Mølhave, and W. S. Cain, eds. "Sources of Indoor Air Contaminants: Characterizing Emissions and Health Impacts," *Annals of The New York Academy of Sciences*, New York, Vol. 641, 1992, p. 1.

9. *Indoor Air*, Supp. 4 (1998).
10. R. B. Gammage and B. A. Berven, eds., *Indoor Air and Human Health*, 2nd ed., CRC/Lewis Publishers, Boca Raton, Fla., 1996.
11. Report of the ALA/ATS Workshop: Achieving Healthy Indoor Air, *Am. J. Respir. Crit. Care Med.* **156**, S33 (1997).
12. H. S. Brightman and N. Moss, in J. D. Spengler, J. M. Samet, and J. F. McCarthy, in Ref. 7, p. 2.
13. C. S. Miller and N. A. Ashford, in J. D. Spengler, J. M. Samet, and J. F. McCarthy, in Ref. 7, p. 27.
14. M. R. Cullen, in M. R. Cullen, ed., *Workers with Multiple Chemical Sensitivity; Occupational Medicine: State of the Art Reviews*, Vol. 2(4), Hanley & Belfus, Philadelphia, 1987, pp. 655–661.
15. W. J. Fisk, in J. D. Spengler, J. M. Samet, and J. F. McCarthy, in Ref. 7, p. 4.
16. W. J. Fisk, *ASHRAE J.*, 56 (May 2002).
17. F. Nunes and co-workers, in J. J. K. Jaakola, R. Ilmarinen, and O. Seppanen, *Proceedings of the 6th International Conference on Indoor Air Quality and Climate*, Helsinki University of Technology, Vol. 1, 1993, pp. 53–58.
18. J. J. K. Jaakola, O. P. Heinonen, and O. Seppanen, *Environ. Int.* **15**, 163 (1989).
19. U.S. Environmental Protection Agency, National Ambient Air Quality Standards, *www.epa.gov/airs/criteria.html.*
20. *ANSI/ASHRAE Standard 62-2001: Ventilation for Acceptable Indoor Air Quality*, American Society of Heating, Refrigerating, and Air-Conditioning Engineers, Inc., Atlanta, Ga., 2001.
21. USEPA (United States Environmental Protection Agency), *Building Air Quality: A Guide for Building Owners and Managers*, Washington, D.C. (1991).
22. B. Seifert, in D. S. Walkinshaw, ed., *Proceedings of the 5th International Conference on Indoor Air Quality and Climate*, Indoor Air Technologies, Inc., Ottawa, Vol. 5, 1990, pp. 43–44.
23. Y. Alarie, G. D. Nielsen, and M. M. Schaper, in J. D. Spengler, J. M. Samet, and J. F. McCarthy, eds., in Ref. 7, pp. 23, 29.
24. L. Fang, G. Clausen, and P. O. Fanger, *Indoor Air* **8**(2), 80 (1998).
25. *ANSI/ASHRAE Standard 55-1992: Thermal Environmental Conditions for Human Occupancy*, American Society of Heating, Refrigerating, and Air-Conditioning Engineers, Inc., Atlanta, 1992.
26. U.S. Food and Drug Administration, *Maximum Acceptable Level of Ozone*, *Code of Federal Regulations*, Title 21, Part 801.415, Washington, D.C., 1988.
27. Y. Alarie, M. Schaper, G. D. Nielsen, and M. H. Abraham, *SAR/QSAR Env. Res.* **5**, 151 (1996).
28. A.-L. Pasanen, A. Korpi, J.-P. Kasanen, and P. Pasanen, *Environ. Int.* **24**, 703 (1998).
29. *Indoor Air*, Supp. 5 (1998).
30. J. H. Park, P. L. Schleiff, M. D. Attfield, J. M. Cox-Ganser, and K. Kreiss, *Proceedings of the 9th International Conference on Indoor Air Quality and Climate*, Vol. 5, pp. 27–32, ISBN 0-9721832-0-5, Indoor Air 2002, Santa Cruz, Calif., 2002.
31. U.S. Environmental Protection Agency, *National Ambient Air Quality Standards*, Washington, D.C. (2002). Available at *www.epa.gov/airs/criteria.html.*
32. World Health Organization, *Update and Revision of WHO Air Quality Guidelines for Europe*, Copenhagen (1999). Available at *www.who/dk/eh/pdf/airqual.pdf.*
33. Ref. 20, p. 3.
34. M. L. Burr, in J. D. Spengler, J. M. Samet, and J. F. McCarthy, eds., *Indoor Air Quality*, Handbook, McGraw Hill, New York, Vol. 29, 2001, p. 28.
35. R. Rudel, in J. D. Spengler, J. M. Samet, and J. F. McCarthy, in Ref. 34, Vol. 34, p. 4.

36. L. A. Wallace, in J. D. Spengler, J. M. Samet, and J. F. McCarthy, in Ref. 34, Vol. 33, p. 5.
37. Ref. 35, pp. 13–14.
38. R. G. Lewis, in J. D. Spengler, J. M. Samet, and J. F. McCarthy **35**(5), 13 (2001).
39. D. J. Vorhees, in J. D. Spengler, J. M. Samet, and J. F. McCarthy, in Ref. 34, Vol. 36, p. 15.
40. M. C. Swanson, C. E. Reed, L. W. Hunt, and J. W. Yunginger, in J. D. Spengler, J. M. Samet, and J. F. McCarthy, in Ref. 34, Vol. 41, p. 7.
41. R. A. Duffee and M. A. O'Brien, in J. D. Spengler, J. M. Samet, and J. F. McCarthy, in Ref. 34, Vol. 21, p. 9.
42. T. A. Myatt and D. K. Milton, in J. D. Spengler, J. M. Samet, and J. F. McCarthy, in Ref. 34, Vol. 42, pp. 4, 8.
43. T. A. E. Platts-Mills, in J. D. Spengler, J. M. Samet, and J. F. McCarthy, in Ref. 34, Vol. 43, p. 9.
44. M. L. Muilenberg, in J. D. Spengler, J. M. Samet, and J. F. McCarthy, in Ref. 34, Vol. 44, p. 6.
45. H. A. Burge, in J. D. Spengler, J. M. Samet, and J. F. McCarthy, in Ref. 34, Vol. 45, pp. 7, 13, 26.
46. E. A. Nardell, in J. D. Spengler, J. M. Samet, and J. F. McCarthy, in Ref. 34, Vol. 47.
47. B. E. Barry, in J. D. Spengler, J. M. Samet, and J. F. McCarthy, in Ref. 34, Vol. 48.
48. Health Effects Institute, *Asbestos in Public and Commercial Buildings: A Literature Review and Synthesis of Current Knowledge*, Health Effects Institute-Asbestos Research, Cambridge, Mass. 1991.
49. T. Schneider, in J. D. Spengler, J. M. Samet, and J. F. McCarthy, in Ref. 34, p. 13.
50. U.S. Environmental Protection Agency, *The Total Exposure Assessment Methodology (TEAM) Study: Summary and Analysis, Vols. I–IV*, EPA 600/6-87/002a,b,c,d, Washington, D.C., 1987.
51. S. K. Brown, M. R. Sim, M. J. Abramson, and C. N. Gray, *Indoor Air* **4**, 123 (1994).
52. U.S. Environmental Protection Agency, *Building Assessment, Survey, and Evaluation (BASE) Study, http://www.epa.gov/iaq/largebldgs/index.html.*
53. W. G. Tucker, in *Inside IAQ: EPA's Indoor Air Quality Research Update,* U.S. Environmental Protection Agency, Research Triangle Park, NC, 1-7 (Spring/Summer 1998).
54. P. O. Fanger, in J. D. Spengler, J. M. Samet, and J. F. McCarthy, eds., *Indoor Air Quality Handbook*, McGraw-Hill, New York, Vol. 22, 2001.
55. G. D. Nielsen, *Critical Rev. Toxicol.* **21**, 183 (1991).
56. Y. Alarie, G. D. Nielsen, and M. M. Schaper, in J. D. Spengler, J. M. Samet, and J. F. McCarthy, in Ref. 34, Vol. 23, p. 28.
57. J. E. Cometto-Muniz, W. S. Cain, and H. K. Hudnell, *Perception Psychophysics* **59** (5), 665 (1997).
58. L. Molhave, *Ann. N.Y. Acad. Sci.* **641**, 46 (1992).
59. C. J. Weschler and H. C. Shields, *Atmospheric Environ.* **31**, 3487 (1997).
60. C. J. Weschler and H. C. Shields, *Env. Sci. Technol.* **31**, 3719 (1997).
61. P. Wolkoff and co-workers, *Atmos. Environ.* **33**, 693 (1999).
62. S. Armstrong and J. Liaw, *ASHRAE J.*, 18–24 (Nov. 2002)
63. American Society for Testing and Materials, *Standard Guide for Small-Scale Environmental Chamber Determinations of Organic Emissions from Indoor Materials/Products*, ASTM D5116-90, Philadelphia, 1990.
64. Commission of the European Communities, *COST Project 613, Report No. 8: Guidelines for the Characterization of Volatile Organic Compounds Emitted from Indoor*

Materials and Products Using Small Test Chambers, EUR 13593 EN, CEC Joint Research Centre, Ispra, Italy, 1991.
65. W. G. Tucker, in J. D. Spengler, J. M. Samet, and J. F. McCarthy **31**(5), 4 (2001).
66. W. G. Tucker, in *Inside IAQ: EPA's Indoor Air Quality Research Update*, U.S. Environmental Protection Agency, Research Triangle Park, NC, 2-5 (Fall/Winter 1998).
67. J. E. Janssen, *ASHRAE J.* **41**, 47 (Sept. 1999).
68. *ANSI/ASHRAE Standard 62-2001: Ventilation for Acceptable Indoor Air Quality*, American Society of Heating, Refrigerating, and Air-Conditioning Engineers, Inc., Atlanta, 3,4, 2001.
69. M. W. Liddament, in J. D. Spengler, J. M. Samet, and J. F. McCarthy, eds., *Indoor Air Quality Handbook*, McGraw-Hill, New York, Vol. 13, 2001.
70. W. S. Cain, J. M. Samet, and M. J. Hodgson, *ASHRAE J.* **37**, 38 (1995).
71. D. Stanke, *ASHRAE J.* **41**, 40 (1999).
72. G. N. Walton, *CONTAM96 User Manual*, NISTR6056, National Institute of Standards and Technology, Gaithersburg, Md., 1997.
73. L. E. Sparks, in J. D. Spengler, J. M. Samet, and J. F. McCarthy, in Ref. 69, Vol. 58.
74. R. C. Diamond, in J. D. Spengler, J. M. Samet, and J. F. McCarthy, in Ref. 69, Vol. 6.
75. H. Levin, in M. Maroni, ed., *Proceedings of Health Buildings '95*, ISBN 88-900086-0-1, pp. 5–24, 1995.
76. W. G. Tucker, in *Report on the NATO / CCMS Meeting on Energy* & Building Sciences in Indoor Air Quality, August 6–8, 1990, Sainte-Adele, Quebec. Institut de Recherche en Sante et en Securite du Travail du Quebec (IRSST), Montreal, ISBN 2-551-12514-6, pp. 57–36, 1990.
77. See *www.aiaonline.org* for *Environmental Resource Guide*.
78. W. G. Tucker, in D. S. Walkinshaw, ed., *Indoor Air '90: Proceedings of the 5th International Conference on Indoor Air Quality and Climate*, Indoor Air Technologies, Inc., Ottawa, Vol. 3, pp. 251–256, 1990.
79. Z. Guo, *Simulation Tool Kit for Indoor Air Quality and Inhalation Exposure (IAQX)*, U.S. Environmental Protection Agency, Research Triangle Park, N.C., 2001.
80. L. E. Sparks, *IAQ Model for Windows RISK Version 1.0 User Manual*, EPA/600/R-96-037 (NTIS PB96-501929), U.S. Environmental Protection Agency, Research Triangle Park, N.C., 1996.
81. G. N. Walton, *Contam96 User Manual*, NISTR6056, National Institute of Standards and Technology, Gaithersburg, Md., 1996.
82. Z. Guo, Review of Indoor Emission Source Models: Part 1, Overview, *Environmental Pollution* (in press, for publication late 2002).
83. Z. Guo, Review of Indoor Emission Source Models: Part 2, Parameter Estimation, *Environmental Pollution* (in press, for publication late 2002).
84. L. Molhave, in R. B. Gammage and S. V. Kaye, eds., *Indoor Air and Human Health*, Lewis Publishers, Chelsea, Mich., 1985, pp. 403–414.
85. B. Seifert, in D. S. Walkinshaw, ed., *Proceedings of the 5th International Conference on Indoor Air Quality and Climate*, Indoor Air Technologies, Inc., Ottawa, Vol. 5, 1990, pp. 43–44.
86. Y. Alarie, G. D. Nielsen, and M. M. Schaper, in J. D. Spengler, J. M. Samet, and J. F. McCarthy, eds., *Indoor Air Quality Handbook*, McGraw-Hill, New York, Vol. 23, 2001, p. 29.
87. W. G. Tucker, in *Inside IAQ: EPA's Indoor Air Quality Research Update*, U.S. Environmental Protection Agency, Research Triangle Park, N.C., 2-5 (Fall/Winter 1998).
88. W. G. Tucker, in J. D. Spengler, J. M. Samet, and J. F. McCarthy **31** (12), 4 (2001).
89. American Society for Testing and Materials, *Standard Guide for Small-Scale Environmental Chamber Determinations of Organic Emissions from Indoor Materials / Products*, ASTM D5116-90, Philadelphia, 1990.

90. American Society for Testing and Materials, *Standard Test Method for Estimating Sensory Irritancy of Airborne Chemicals*, ASTM E 981-84, Philadelphia, 1984.
91. See *www.carpet-rug.com/....* For emissions criteria.
92. *ANSI/ASHRAE Standard 62-2001: Ventilation for Acceptable Indoor Air Quality*, American Society of Heating, Refrigerating, and Air-Conditioning Engineers, Inc., Atlanta, Ga., 2001.
93. USEPA (United States Environmental Protection Agency), *Building Air Quality: A Guide for Building Owners and Managers*, Washington, D.C., 1991.
94. H. Levin, in J. D. Spengler, J. M. Samet, and J. F. McCarthy, in Ref. 86, Vol. 60, p. 4.
95. *ASHRAE Guideline 1: The HVAC Commissioning Process*, American Society of Heating, Refrigerating, and Air-Conditioning Engineers, Inc., Atlanta, 1996.
96. *ASHRAE Guideline 4: Preparation of Operating and Maintenance Documentation for Building Systems*, American Society of Heating, Refrigerating, and Air-Conditioning Engineers, Inc., Atlanta, 1993.
97. J. Kildeso and T. Schneider, in J. D. Spengler, J. M. Samet, and J. F. McCarthy, in Ref. 86, Vol. 64.
98. T. Nathanson, in J. D. Spengler, J. M. Samet, and J. F. McCarthy, in Ref. 86, Vol. 63.
99. American Lung Association: *www.lungusa.org/cleaners/air_clean_toc.html*
100. B. McDonald and M. Ouyang, in J. D. Spengler, J. M. Samet, and J. F. McCarthy, eds., *Indoor Air Quality Handbook*, McGraw-Hill, New York, Vol. 9, 2001.
101. *ASHRAE Handbook: Equipment*, American Society of Heating, Refrigerating, and Air-Conditioning Engineers, Inc., Atlanta, Ga., 2000.
102. U.S. Environmental Protection Agency: *www.epa.gov/iaq/iaqinfo.html*
103. *Consumer Reports*, Consumers Union, Yonkers, N.Y., pp. 42–46 (Jan. 2000).
104. *ANSI/ASHRAE Standard 62-2001: Ventilation for Acceptable Indoor Air Quality*, American Society of Heating, Refrigerating, and Air-Conditioning Engineers, Inc., Atlanta, Ga., Vol. 5.8, 2001, p. 5.
105. Ref. 104, Section 6.2.3, p. 14.
106. *ASHRAE Standard 52.2: Method of Testing General Ventilation Air Cleaning Devices for Removal Efficiency by Particle Size*, American Society of Heating, Refrigerating, and Air-Conditioning Engineers, Inc., Atlanta, Ga., 1999.
107. J. T. Hanley, D. D. Smith, P. A. Lawless, D. S. Ensor, and L. E. Sparks, *Proceedings of the 5th International Conference on Indoor Air Quality and Climate*, Indoor Air Technologies, Inc., Ottawa, Vol. 3, 1990, pp. 145–150.
108. P. A. Lawless, A. S. Viner, D. S. Ensor, and L. E. Sparks, *Proceedings of the 5th International Conference on Indoor Air Quality and Climate*, Indoor Air Technologies, Inc., Ottawa, Vol. 3, 1990, pp. 187–192.
109. E. A. Nardell, in J. D. Spengler, J. M. Samet, and J. F. McCarthy, in Ref. 100, Vol. 11, p. 4.
110. Ref. 109, pp. 7,8.
111. M. W. First, W. Chaisson, E. Nardell, and R. Riley, *ASHRAE Trans.* **105**, 869 (1999).
112. *Ibid.*, 877 (1999).
113. S. L. Miller and J. M. Macher, *Aerosol Sci. Tech.* **33**, 274 (2000).
114. E. Kujundzic, C. Howard, M. Hernandez, and S. Miller, *Proceedings of the 9th International Conference on Indoor Air Quality and Climate*, Vol. 2, 718–723, ISBN 0-9721832-0-5, Indoor Air 2002, Santa Cruz, Calif., 2002.
115. D. M. Underhill, in J. D. Spengler, J. M. Samet, and J. F. McCarthy, in Ref. 100, Vol. 10, p. 10.
116. D. W. VanOsdell, M. K. Owen, L. B. Jaffe, and L. E. Sparks, *J. Air Waste Manage. Assoc.* **46**, 883 (1996).
117. D. B. Henschel, *J. Air Waste Manage. Assoc.* **48**, 985 (1998).
118. Ref. 115, pp. 14–16.

GENERAL REFERENCES

J. D. Spengler, J. M. Samet, and J. F. McCarthy, eds., *Indoor Air Quality Handbook*, McGraw-Hill, New York, 2001. A comprehensive general reference covering all aspects of indoor environmental quality.

Indoor Air: International Journal of Indoor Environment and Health. A quarterly journal devoted to research on indoor air quality. The official journal of the International Society of Indoor Air Quality and Climate.

ASHRAE Journal. Contains frequent articles on indoor air quality and thermal comfort; focuses mostly on ventilation-related aspects.

Other journals such as *Environmental Science and Technology, Journal of the Air and Waste Management Association, Environmental Pollution*, and *Atmospheric Environment* also have occasional articles on indoor air quality. In addition, there are numerous trade periodicals and newsletters.

Proceedings of the International Conferences on Indoor Air Quality and Climate. Contain concise summaries of the current science and technology of indoor air quality, as presented at the triennial Indoor Air conferences. The most recent conference was in July 2002; information on proceedings is available at *www.indoorair2002.org*.

There are many sites on the internet that contain information on indoor air quality, but they should be used with caution. Some of the useful sites are listed below:

American Conference of Governmental Industrial Hygienists, *www.acgih.org* (See especially information on TLVs.)

American Institute of Architects, *www.aiaonline.com* (Use search engine for construction industry news on indoor air quality issues.)

American Lung Association, *www.lungusa.org/air/*

American Society of Heating, Refrigerating, and Air-Conditioning Engineers (ASHRAE), *www.ashrae.org*

Building Owners and Managers Association, *www.boma.org* (Use search engine for indoor air.)

Carpet and Rug Institute *www.carpet-rug.com/* (See especially "green label" testing programs.)

Housing and Urban Development, *www.hud.gov* (Use search engine for indoor air.)

National Institute of Standards and Technology, *www.nist.gov/* (From subject index, go to "indoor air quality".)

National Resources Defense Council, *www.nrdc.org* (Use search engine for indoor air.)

North American Insulation Manufacturers Association, *www.naima.net*

Sheet Metal and Air Conditioning Contractors of North America, *www.smacna.org* (Use search engine for indoor air.)

U.S. Environmental Protection Agency, *www.epa.gov/iaq; www.epa.gov/iaq/iaqinfo.html#iaqinfo.*

W. Gene Tucker
James Madison University